Handbook of Conformal Mappings and Applications

Handbook of Conformal Mappings and Applications

Prem K. Kythe

CRC Press is an imprint of the
Taylor & Francis Group, an informa business
A CHAPMAN & HALL BOOK

CRC Press
Taylor & Francis Group
6000 Broken Sound Parkway NW, Suite 300
Boca Raton, FL 33487-2742

First issued in paperback 2020

ISBN-13: 978-1-138-74847-7 (hbk)
ISBN-13: 978-0-367-73159-5 (pbk)

Library of Congress Cataloging-in-Publication Data

Names: Kythe, Prem K., author.
Title: Handbook of conformal mappings and applications / Prem K. Kythe
(Professor Emeritus of Mathematics, University of New Orleans, New
Orleans, LA).
Description: Boca Raton, Florida : CRC Press, [2019]
Identifiers: LCCN 2018035976| ISBN 9781138748477 (hardback : alk. paper) |
ISBN 9781315180236 (e-book)
Subjects: LCSH: Conformal mapping. | Mappings (Mathematics)
Classification: LCC QA646 .K945 2019 | DDC 515/.9--dc23
LC record available at https://lccn.loc.gov/2018035976

Visit the Taylor & Francis Web site at
http://www.taylorandfrancis.com

and the CRC Press Web site at
http://www.crcpress.com

MATRI FILIORVM MEORVM
QVAE IN CAELO EST CVM DEO
HVNC LIBRVM
D · D · D

Contents

Note: All Map numbers and the related mapping functions, given in this Table of Contents, will not be repeated in the Index at the end of the book.

Part 2: Numerical Conformal Mapping

Preface

After the famous Inaugural Dissertation in 1851 when Riemann announced his theorem, now known as the *Riemann Mapping Theorem*, the milestone for the subject of conformal mapping started in a book form by Bieberbach [1880], which was followed by more than a dozen books on the subject of conformal transformations. The notable among them, in chronological order, are the ones by Lewent [1925], Carathéodory [1932], Kober [1945, 1957], Nehari [1952], Gibbs [1958], Jenkins [1958], Ahlfors [1966], Trefethen [1986], Pommerenke [1992a], Kythe [1998], Papamichael and Stylianopoulos [2010], Mathews and Howell [2008], and Schinzinger and Laura [1991/2003]. The Russian school also produced some remarkable articles and books, notably those by Lavrent'ev [1934], Kantorovich [1937], Krylov [1937], Goluzin [1937], Malentiev [1937], Muratov [1937], followed by Gurevich [1965], Ivanov and Trubestkov [1995], and Ivanov and Papov [2002]. Needless to say, all these books are classic and present great and admirable work on their topics of interest.

This *Handbook on Conformal Mappings and Applications* has many features not available in other books, handbooks, or mathematical encyclopedias. The following section provides the details of the salient features of this handbook.

Salient Features

This book contains twenty-five chapters and ten appendices, and there are over 300 conformal maps, all detailed in the Table of Contents, and over 500 illustrations, some with multiple figures. Although the book does not claim to be exhaustive, all the basic conformal maps compiled and discussed in this book can generate other conformal maps by the process of composition and chain property. A few examples of these kinds of maps are presented in the book.

The book is divided into three parts: (i) theory and conformal maps, defined on their geometric and algebraic properties, throughout the book, but mostly in Chapters 1 through 7; (ii) numerical methods to solve various integral equations that arise as a result of conformal mappings, covering Chapters 8 through 15; and (iii) applications covering Chapters 16 through 25. The details of these parts are given in the following section.

Overview

The main purpose of this book is to provide a self-contained and systematic introduction to the theory and numerical computations involving conformal mappings of simply or multiply connected regions onto the upper half-plane, unit disk, or other canonical regions. It provides a comprehensive and systematic compilation and analysis of the basic theory and related numerical analysis with applications to different areas in mathematical physics and engineering. The prerequisites, besides the theory of univalent functions and conformal mapping, include sound knowledge of methods of numerical analysis, theory of Fredholm and Stieltjes integral equations, and programming languages like MATLAB, Mathematica, or Fortran.

Besides an exposition on the history of the subject in Chapter 1, the basic results from complex analysis and the theory of harmonic and univalent functions are presented in Chapter 2, where the Riemann mapping theorem, conformality and uniqueness, analytic continuation, chain property of conformal maps, the Schwarz reflection principle, Riemann sphere, Bieberbach conjecture, Mercator's projection, and Taylor's series approximations are discussed. The conformal maps are presented based on their geometry, which leads to the linear and bilinear transformations, cross ratio, maps involving Cassini's ovals, cardioid, lemniscates, straight lines and circles, ellipses and hyperbolas, and covers Chapter 3. Conformal maps based on algebraic functions (monomials and polynomials), are treated in Chapter 4, and those belonging to the exponential family of functions are contained in Chapter 5. The Joukowski airfoils and related maps are presented in Chapter 6, and the Schwarz-Christoffel transformations, including polygons, trigons, triangles, rectangles and other regions, and polygons with round corners are discussed in Chapter 7.

The second part of the book consists of numerical conformal mapping involving the following topics: Schwarz-Christoffel integrals and numerical methods of their solution are discussed in Chapter 8. The problems associated with nearly circular regions are treated in Chapter 9. Various integral equation formulations of the conformal mapping problem are discussed in Chapters 8 through 10. These equations, namely, Lichtenstein's, Gershgorin's, Carrier's, Banin's, and Warschawski-Siefel's, are mostly Fredholm equations of the first or second kinds and provide boundary correspondence functions and their computation by iterative methods. Chapter 11 deals with the classical Theodorsen's integral equation, discusses convergence of the iterative method, provides proofs for the convergence theorem due to Warschawski, and investigates the cases of starlike and exterior regions. A trigonometric interpolation method is outlined and the modern iterative and Newton's methods are presented. Chapter 12 deals with Symm's integral equations for the interior and exterior regions, based on a single-layer potential. The orthonormal polynomial method, Lagrange interpolation, and spline approximation method are discussed.

A detailed account of airfoils is presented in Chapter 13. Various methods, like James's method for single-element airfoils, von Karman-Trefftz transformation and Garrick's method of conjugated functions for two-element airfoils are explained with related algorithms and examples. An important aspect of the conformal mapping problem, related to the location and behavior of corner singularities on the boundary and pole-type singularities of the mapping functions near the boundary in simply and doubly connected regions, is discussed. Doubly connected regions are studied in Chapter 14, and the related integral equation based on the dipole distribution is presented with an algorithm for numerical computation. Chapter 15 deals with multiply connected regions. Based on the dipole distribution, Mikhlin's integral equation that determines the density function and related boundary correspondence is solved by a fast Poisson solver. This method is very efficient and solves problems for both simply and multiply connected regions. Although this part has relied on the author's previous book [1998], the chapters have been improved and enlarged with additional recent developments.

Exact solutions of boundary value problems for simple regions, such as circles, squares or annuli, can be determined with relative ease even when the boundary conditions are rather complicated. Green's functions for such simple regions are known. However, for regions with complex structure the solution of a boundary value problem often becomes more difficult, even for a simple case, like the Dirichlet problem. One approach to solve these difficult problems is to conformally transform a given region onto simpler canonical regions. This will, however, result in change not only in the region and the associated boundary conditions but also in the governing differential equation. As compared to the simply connected regions, conformal mapping of multiply connected regions suffers from severe limitations, one of which is the fact that equal connectivity of regions is not a sufficient condition to effect a reciprocally connected map of one region onto another. There are, though, a few methods that carry out such mappings where most of the computational details are done numerically.

The third part of the book, from Chapters 16 through 25, deals with applications. Grid generation and cascade configurations are presented in Chapter 16. Field theories including the definition of initial and boundary conditions and their classifications are presented in Chapter 17, which provides detailed analysis of parabolic, hyperbolic, and elliptic types of equations, and especially Laplace's, Poisson's, and Helmholtz equations. Their general solutions, Green's functions and fundamental solutions are discussed, and Laplace's equation under conformal transformation is explained. The concept of sources and sinks is also included.

Fluid flows are discussed in Chapter 18, including viscous laminar flows, external flows, ideal fluid flows, potential flows, and boundary layer flows; streamlines and circulation and stagnation points. Conformal mapping of flow patterns and Joukowski maps and the geometry of airfoils and the question of lift are explained in this chapter. Heat transfer problems by method of separation of variables and under conformal transformations are discussed in Chapter 19, which includes Poisson's integral formulas, diffusion equation, hypersonic flows, high temperature effects, shock layer, flow through a constriction, and transient problems. Vibration and acoustics, including wave propagation and dispersion, and damped and thermal waves are explained, and vibrations of strings and membranes are discussed in detail.

Chapter 20 deals with acoustics, harmonics and acoustic waveguides. Electromagnetic field, electrostatic field and electric potential are discussed in Chapter 21, including detailed analysis of electromagnetic waves, electric capacitors, capacitance, and AC circuits, and Laplace's and Poisson's equations dealing with electric potential and field between two infinite plates. Chapter 22 deals with transmission lines and waveguides, with conformal mapping techniques, Helmholtz equation, rib-shaped waveguides, coplanar waveguides, and nonuniform waveguides.

Elastic medium is discussed in Chapter 23. Application of finite element, boundary integral and boundary element methods to applicable problems in fluid flows, heat and mass transfer, and electric potential, waveguides, motor design, and other usable applications are discussed in Chapter 24, and Chapter 25, the last chapter, provides information about various computational resources available on the internet and the references used in the book.

There are ten appendices, which are mentioned in the Table of Contents and pertain to certain topics used in the book, to wit, Green's identities, Cauchy's principal-value integral, Riemann mapping theorem including the Riemann-Hilbert problem, Gudermannians, five numerical tables, elliptic functions, Gauss-Jacobi rule, orthogonal polynomials, special finite elements, and the Schwarz formula. The book ends with an extended Bibliography and Index.

Intended Readers

Besides the researchers in applicable and applied mathematics and physics, some readers of this book may also include persons who are engineers mainly interested in the study of acoustics, plane elasticity, electromagnetic theory, fluid flows, inlet configurations, transonic flow problems, cascade of blades in airfoil and wing designs, heat transfer, ion optics, solidification, solid propellant rocket motors, plates, and vibrations, electric transmission and waveguides. The intended readers, in general, cover the following three categories. First, they are students ready for a graduate course in this subject. For them, this handbook can be used as a reference book. The second category is that of graduate students engaged in research. For them, the book should be useful, because it is filled with a vast amount of information on methodology and an extensive bibliography on the subject. The third category consists of scientists and researchers in various areas of applied mathematics, engineering and physics. For them, the book is a vital source of information in classical as well as modern trends in research on the subject of their interest as well as in numerical methods. Research scientists, physicists, engineers, and mathematicians will appreciate this book with comprehensive and up-to-date account of many methods available on the subject.

Acknowledgments

I take this opportunity to thank Mr. Sarfraz Khan, Executive Editor, Taylor & Francis, for his support, and Mr. Callum Fraser for coordinating the book project. I also thank Project Editor Michele A. Dimont for doing a great job of editing the text. Thanks are due to the reviewers and to some of my colleagues who made certain very valuable suggestions toward improvement of the book. Lastly, I thank my friend Michael R. Schäferkotter for help and advice freely given whenever needed.

Prem K. Kythe

Notations, Definitions, and Acronyms

A list of the notations, definitions, abbreviations, and acronyms used in this book is given below.

a, elastic modulus in R^n; thermal conductivity

$(a)_n$, Pochhammer's symbol $= \dfrac{\Gamma(a+n)}{\Gamma(n)} = a(a+1)\dots(a+n-1)$

AC, alternating current

ARROW, anti-resonant reflecting waveguide

$\bar{A}$, closure of a set A

$A \backslash B$, complement of a set B with respect to a set A

$A \times B$, product of sets A and B

$A(\rho_1, \rho_2)$, annulus $\{\rho_1 < |w| < \rho_2\}$

$A^{(e)}$, area of a two-dimensional element

$\mathbf{A}^T$, $(\mathbf{A})^T$, transpose of a matrix $\mathbf{A}$

Area(G), area of a region G

$b(w,u)$, bilinear form

b_i, coefficients in a linear triangular shape functions

$B(a,r)$, open disk $\{|z-a| < r\}$

$\bar{B}(a,r)$, closed disk $\{|z-a| \le r\}$

BE(M), boundary element (method)

BKM, Bergman kernel method

$B(0,1)$, open unit disk $\{|z| < 1\}$

$B(m,n)$, beta function

$\mathbf{b}$, row vector for b_i

$\mathbf{B}$, matrix; also, magnetic field (chapter 21)

$\mathbf{B}^{(e)}$, matrix of the values of b_i, c_i

const, constant

cap (D), capacity of a region D

c_i, coefficients in a linear triangular element

c_v, specific heat at constant volume

CPS, coplanar slot

CB-CPW, circuit-backed coplanar waveguide

CPW, coplanar waveguide
CSK1, Cauchy singular equations of the first kind
CSK2, Cauchy singular equations of the second kind
Ch., Chapter
C, capacitance
C_{ij}, material (elastic) constants
C, connectivity matrix; matrix of the elastic constants C_{ij}
comp(B), complement of a set B
$C^0(D)$, class of functions continuous on a region D
$C^k(D)$, class of continuous functions with k-th continuous derivative, on a region D, $0 \leq k < \infty$
$C^\infty(D)$, class of continuous functions infinitely differentiable on a region D
C-function, same as a C^0-function
$\mathbb{C}$, complex plane
$\mathbb{C}_\infty$, extended complex plane
dof, degree(s) of freedom
dn, a Jacobian elliptic function
D, region in the z-plane; dimension of the geometry in fluid flow
D, displacement density (Chapter 22)
D^*, complement of a set D
diam(E), transfinite diameter of a set E
$\bar{D}$, closure of D
e.g., for example
erf, error function
erfc, complementary error function
EII, effective internal impedance
ECPWFG, embedded coplanar waveguide with finite grounds (embedded in a dielectric)
et al., (et alii) and others
E, modulus of elasticity, Young's modulus
E_k, elastic kinetic energy
E_p, elastic potential energy
EI, flexural rigidity
Eq(s), equation(s) (when followed by an equation number)
Ext(Γ), region exterior to the boundary Γ
E, electric field density (Chapters 21 and 22)
FEM, finite element method
FEM-BI, finite element boundary integral (method)
FFT, fast Fourier transform
FK2, Fredholm integral equation of the second kind
FSPM, flux-switching permanent magnet (machines)
$\langle f, g\rangle = \iint_D f(z)\overline{g(z)}\, dS_z$, inner product (see (2.3.7))
$\mathfrak{f}$, frequency of an AC generator
$f^{(e)}$, local vector element
$\hat{f}$, axial body force
f_x, f_y, body forces
FE, finite element

F, electric force
f_Ω, mapping of a doubly connected region Ω onto the annulus $A(\rho_1, \rho_2)$
$\mathbf{F}$, electromagnetic force
F_0, F_L, shear forces at $x = 0$ and $x = L$
$\mathbf{f}$, force vector
$\mathbf{F}$, force or load vector ($\mathbf{F} = \mathbf{f} + \mathbf{Q}$)
$\mathbf{F}^{(e)}$, force or load vector of an element $\Omega^{(e)}$
g, shear modulus (torsion); gravity; loss coefficient of a conductor
$\mathbf{g}$, vector; body forces
$\bar{g}$, complex conjugate of an analytic function g
gd, Gudermannian
G, region in the w-pane; also, conductance (Chapters 21 and 22)
$G(z, z_0)$, Green's function
HPM, homotopy perturbation method
$h = J_c^2 \rho$, watts per unit volume
H^1, Lipschitz condition
H^α, Hölder condition of order α
$H(t)$, Heaviside function 0 for < 0, and $= 1$ for $t > 0$
$H(z, z_0)$, Cauchy kernel
$\mathbf{H}$, magnetic field (Chapter 22)
i.e., that is
iff, if and only if
IBEM, boundary element modeling (method)
Int(Γ), region interior of Γ
$I(t)$, AC current
$I(\Gamma, z_0)$, index of a contour Γ (winding number)
$\Im\{\cdot\}$, imaginary part of a complex quantity
I, moment of inertia
$I(u)$, functional; total energy of an elastic mechanical system
I_m, modified Bessel function of the first kind and order m
$\mathbf{j}$, unit vector in the y direction; also, current density (Chapter 21)
$\mathbf{J}$, Jacobian matrix
J, Jacobian of the transformation $w = f(z)$
$J(z, t)$, current (Chapter 22)
KCL, Kirchhoff's current law
KVL, Kirchhoff's voltage law
k, thermal conductivity; permeability coefficient (aquifer)
k_x, k_y, thermal conductivity in the x and y direction
$\mathbf{k}$, unit vector in the z direction
$k^{(e)}$, value of k on an element e
K, constant value of a metal property; consistency coefficient
$\mathbf{K}^{(e)}$, stiffness matrix of an element $\Omega^{(e)}$
$\mathbf{K}$, global stiffness matrix
$\mathbf{K}_b$, matrix
$K(k)$, elliptic integral
$K(z, a)$, Bergman kernel

$\mathcal{K}$, conjugation operator
$\mathcal{K}^0(D)$, class of functions $f \in L^2(D)$, $f(a) = 0$, $a \in D$
$\mathcal{K}^1(D)$, class of functions $f \in L^2(D)$, $f(a) = 1$, $a \in D$
$l(w)$, linear functional
$l^{(e)}$, length of the interval $\left[x_1^{(e)}, x_2^{(e)}\right]$
$l_k(z)$, Lagrange's interpolation functions
L, length of an interval; length unit; linear differential operator
L^2, Hilbert space of square-integrable functions
L^∞, Hilbert space of 2π-periodic and bounded functions
$L^2(D)$, class of square-integrable functions defined on a region D
$\mathcal{L}^1(\Gamma)$, class of functions $f \in L^2(\Gamma)$, $f(a) = 1$, $a \in D$
$\mathcal{L}$, Lagrange function; Laplace transform
$\mathbf{L}$, matrix
m , magnification of a linear transformation; $M = 1/m$, $m > 0$
$M = \rho_2/\rho_1$, conformal modulus of a doubly connected region Ω
M, bending moment; polar moment of a cross-sectional area; Mach number
M_0, M_L, bending moment at $x = 0$ and $x = L$, respectively
$\mathbf{M}$, matrix
$\mathbf{M}^{(e)}$, matrix (radially symmetric element)
near-circle, nearly-circular contour
NAH, near-field acoustical holography (algorithm)
n_x, n_y, n_z, direction cosines of $\mathbf{n}$
$N(s,t)$, Neumann kernel
$\mathbf{n}$, outward normal vector
$\mathbf{n} \cdot \nabla, = n_x \dfrac{\partial}{\partial x} + n_y \dfrac{\partial}{\partial y} + n_z \dfrac{\partial}{\partial z}$
$\mathbb{N}$, natural numbers (integers ≥ 1)
o.d.e., ordinary differential equation
ONP, orthonormal polynomial method
p, pressure; perimeter; also $= u_x = \partial u/\partial x$
psi, lbs/in^2
p.v., or p.-v., Cauchy's principal-value (integral)
P, vertical point load; vector
$P_n(x)$, Legendre polynomials of degree n
$P_n(z)$, complex polynomials
q, heat source; point charge; also $= u_y = \partial u/\partial y$
q_n, heat flux
$\dot{q}$, rate of heat generation
$\mathbf{q}$, temperature gradient
$\mathbf{Q}$, heat flux vector (chapter 17)
Q, amount of heat flux, or heat flow (chapter 19)
Q_1, shear force
Q_2, bending moment
$Q_i^{(e)}$, shear force at node i of an element $\Omega^{(e)}$
$Q_n(z)$, orthonormal polynomials
$\mathbf{Q}^{(e)}$, vector of secondary degrees of freedom (boundary terms)

r, radial distance
$r(\tilde{u})$, scalar residual (error) in the Galerkin method
(r, θ, z), cylindrical polar coordinates
RM, Ritz method
R, radius; rectangle
R_j, errors $(j = 1, \dots , n + 1)$
Re, Reynolds number
$\Re\{\cdot\}$, real part of a complex quantity
$\mathbf{R}^{(e)}$, residual or error vector
$\mathbf{R}$, global error vector
$\mathbb{R}^n$, Euclidean n-space
$\mathbb{R}$, real line; set of real numbers
$\mathbb{R}^+$, set of positive real numbers
s, variable of the Laplace transform; arc length; second(s) (time)
sym, symmetric (matrix)
SONAH, statistically optimal near-field acoustical holography (algorithm)
sn, a Jacobian elliptic function
$S(z, a)$, Szegö kernel
TE, transverse electric mode
TM, transverse magnetic mode
TEM, transverse electromagnetic mode
t, time
T, temperature, temperature distribution; tension (in a string)
T_b, base temperature of a fin
T_∞, ambient temperature
$T_n(z)$, Chebyshev polynomials of the first kind
$\mathbf{T}$, temperature vector
$\mathbf{T}^{(e)}$, temperature vector for an element $\Omega^{(e)}$
u, dependent variable; stress function; displacement; mean velocity
$u(x, y)$, potential function
u_∞, free stream velocity
$u_i^{(e)}$, value of u at node i
$u_a^{(e)}$, linear interpolation function for the interval $\left[x_1^{(e)}, x_2^{(e)}\right]$
$\mathbf{u}$, velocity vector, displacement vector
$\dot{\mathbf{u}}$, vector of the first time derivatives of $\mathbf{u}$
$\ddot{\mathbf{u}}$, vector of the second time derivatives of $\mathbf{u}$
$u^{(e)}$, approximation of u on an element $\Omega^{(e)}$
U, unit disk
U^*, region exterior of the unit circle
$U_n(z)$, Chebyshev polynomials of the second kind
U_0, inlet velocity
U_e, nodal value of U at a global node e, $e = 1, \dots , N$
$\mathbf{U}$, vector of global values of displacement u
$u_t, = \dfrac{\partial u}{\partial t}$; $u_{xx}, = \dfrac{\partial^2 u}{\partial x^2}$
VGCPW, V-groove conductor-based coplanar waveguide

VK2, Volterra integral equation of the second kind
VSWR, voltage standing wave ratio
v_r, radial component of velocity $\mathbf{v}$
VM, variational method
V, three-dimensional solid or volume
V_0, free stream velocity
$V(t)$, AC voltage
$\mathbf{v}$, velocity vector
$V(z_1, \dots, z_n)$, Vandermonde determinant
w_i, weights in Gaussian quadrature
$w = f(z) = u + i\,v = R\,e^{i\Theta}$, image of z under f
w_∞, point at infinity in the w-plane
$\{w, z\}$, Schwarzian derivative
W, Sobolev space of 2π-periodic and absolutely continuous functions f with $f' \in L^2[0, 2\pi]$
$(W, \|\cdot\|)$, Banach space
$x_1^{(e)}, x_2^{(e)}$, end points of a line element
x^0, initial guess (Newton's method)
$\|\mathbf{x}\|$, norm of a vector $\mathbf{x}$
$z = x + i\,y = r\,e^{i\theta}$, complex variable
z_∞, point at infinity in the z-plane
z^*, point symmetric to z
$\bar{z}$, point conjugate to z ($\bar{z} = x - iy$)
(z_1, z_2, z_3, z_4), cross-ratio
$\{z_k\}_1^\infty$, a set of distinct points in D
$z_{k-1}\widehat{\,}z_k$, arc joining the points z_i and z_{i+1}
Z^+, set of nonnegative integers
α, thermal capacitance; flow angle
β, film coefficient; convective heat transfer coefficient
γ, specific heat ratio ($= c_p/c_v$)
$\tilde{\gamma}$, reparameterization of γ
$\Gamma = \partial D$, boundary of a region D (a Jordan contour)
Γ_1, Γ_2, disjoint portions of boundary Γ ($\Gamma_1 \cup \Gamma_2 = \partial\Omega$)
δ, tolerance
δS, surface element
δ_{ij}, Kronecker delta ($= 1$ if $i = j$; -1 if $i \neq j$)
$\delta(t - t')$, Dirac delta function
Δt, time step
ϵ, emissivity; permittivity; permeability (a dielectric constant)
ε, strain; penalty parameter
$\boldsymbol{\varepsilon}$, strain vector
ε_{xy}, shear strain
θ, strain $\left(= -du/dx \right)$; angle of twist (torsion)
ϑ, nondimensional temperature
$\vartheta(x|\tau)$, theta function
λ, eigenvalue
Λ, source strength (scaling factor)

μ, dynamic viscosity of a fluid
$\mu(\zeta)$, density function
μPa, micropascal
ν, kinematic viscosity of a fluid ($= \mu/\rho$); Poisson's ratio (elasticity)
ρ, density
σ, stress; Stefan-Boltzmann constant
σ_0, uniform load
$\sigma_n(z)$, Szegö polynomials
$\boldsymbol{\sigma}$, stress vector
τ, shear stress
τ_{xy}, τ_{yz}, average shear stresses
$\phi(x, y)$, potential function of a flow
$\phi_i^{(e)}$, basis function, interpolation shape function
Φ_e, global shape functions for $e = 1, \dots, N+1$
$\boldsymbol{\phi}^{(e)}$, vector of the shape functions for an element $\Omega^{(e)}$
$\psi(x, y)$, stream function
ψ, stream function
ω, radian frequency
$\Omega^{(e)}$, triangular element
$\partial\Omega$, boundary of the domain Ω
$\partial\Omega^{(e)}$, boundary of an element $\Omega^{(e)}$
$\boldsymbol{\omega}$, vorticity vector
$\|\nabla u\|$, norm of gradient vector $\left(= \sqrt{u_x^2 + u_y^2 + u_z^2}\right)$
$\nabla\mathbf{f}$, gradient matrix of $\mathbf{f}$
$\nabla \cdot \mathbf{F}$, divergence of a vector $\mathbf{F}$ (div $\mathbf{F}$)
∇^2, Laplacian $\left(= \frac{\partial^2}{\partial x^2} + \frac{\partial^2}{\partial y^2} + \frac{\partial^2}{\partial z^2}\right)$
∇, grad $= \mathbf{i}\frac{\partial}{\partial x} + \mathbf{j}\frac{\partial}{\partial y} + \mathbf{k}\frac{\partial}{\partial z}$
$\frac{dp}{dx}, \frac{\partial p}{\partial x}$, pressure gradient
$\mathbf{0}$, null vector
$\partial, \bar{\partial}$, partial differential operators $\partial f = \frac{\partial f}{\partial z}, \bar{\partial} f = \frac{\partial f}{\partial \bar{z}}$
$\partial B(a, r)$, circle $|z - a| = r$
$\partial_\infty D$, boundary of D in $\mathbb{C}_\infty$
$-\!\!\!\!\!\int$, Cauchy's principal-vaue (p.-v.) integral
∇, gradient vector $= \mathbf{i}\frac{\partial}{\partial x} + \mathbf{j}\frac{\partial}{\partial y}$
$\langle \nabla u, \nabla v\rangle$, inner (scalar) product
$\mapsto$, reads 'maps to', means 'maps onto'
■ end of a proof, or an example

References are in the following format: Author[yr], or Author [year: #], where # stands for page number(s), or section number(s), or problem number(s). The format Author[year; ... ; year] refers to multiple years of publications.

Part I: Theory and Conformal Maps

1

Introduction

Research in computational conformal mappings has lately taken two major directions. One direction involves the conformal mapping from a standard region, like the unit disk or the upper half-plane, onto the problem region, whereas in the other it is from the problem region onto a standard region. In the former case one solves a nonlinear integral equation involving the conjugate operator (e.g., Theodorsen's integral equation), by fast Fourier transform, polynomial approximation, iteration, or Newton's method. In the latter case the integral equation, derived from the Dirichlet problem, is linear or singular linear if it is derived from potential theory (e.g., Symm's integral equation). Depending on the nature of the problem region, these methods often use the Schwarz-Christoffel transformations.

Various kinds of applications of conformal mappings have been and are being studied. It is an important area and we will provide relevant information from different publications.

The historical development of different methods for computational conformal mapping of simply and multiply connected regions is sketched below.

1.1 Historical Background

The oldest transformation, known as the *stereographic projection* of the sphere, was used by Claudius Ptolemy (ca. 150 A.D.) to represent the celestial sphere. Later, in a totally different mapping of a sphere onto a plane, known as Mercator's projection, the spherical earth is cut along a meridian circle and conformally mapped onto a plane strip. Gerhard Kremer published the first world map in 1569 using this technique, and ever since all sea maps are generated by this method. These two projections, however, do not provide similar maps of the same region, which shows that conformal mapping does not imply similarity of figures. Lambert [1772] was the first mathematician who studied the mathematical projection of maps. These and other similar considerations enabled Joseph-Louis Lagrange in 1779 to obtain all conformal representations of a portion of the earth's surface onto a plane where all circles of longitudes and latitudes are represented by circular arcs.

Gauss [1822] was the first to state and completely solve the general problem of determining all conformal mappings that transform a very small neighborhood of a point on an arbitrary analytic surface onto a plane area. However, Gauss' work did open the harder problem of finding the way whereby a given finite portion of a surface can be mapped onto a portion of the plane. A breakthrough came in 1851 when Riemann gave the fundamental result, known as the *Riemann mapping theorem*, which has since been a turning point for all subsequent developments in the theory of conformal mapping. In the proof of this theorem he assumed that a variational problem, now known as the Dirichlet problem, possesses a solution. It was fifty years later that Hilbert [1901] proved the existence of the solution of the Dirichlet problem. In the mean time the validity of Riemann's result was established rigorously by Schwarz [1890] by using a number of theorems from the theory of logarithmic potential. A detailed proof of the Riemann mapping theorem and the Riemann-Hilbert problem is available in Appendix C.

After the basic theoretical aspects of the theory of functions of a complex variable and conformal mapping were established by Cauchy, Riemann, Schwarz, Christoffel, Bieberbach, Carathéodory, Goursat, Koebe, and others in the nineteenth and early decades of the twentieth century, the first numerical research into developing a method for mapping a region bounded by finitely many Jordan curves Γ_i onto an n-sheeted Riemann surface where the curves Γ_i correspond to rectilinear slits was done by Burnside [1891]. A minimizing principle was established by Bieberbach [1914], namely, that among all suitably normed conformal maps of a given simply connected region the one with the least area is the conformal map onto a circle. In particular, this principle evolved as a result of minimizing the integral $\iint_D \|f'(z)\|^2\,dx\,dy$, where f is regular in D and normalized by $f(0) = 0$, $f'(0) = 1$, and the area theorem. Bieberbach used the Ritz method to find an approximate solution in the form of a polynomial for the above integral and used it to construct the conformal map of a simply connected region onto a circle. An exposition of the Ritz method can be found in Kantorovich and Krylov [1936]. The estimates obtained by this method were later improved by Höhndorf [1926] and Müller [1938] in the problem of a conformal map of a nearly circular region onto a circle. Other improvements on the Bieberbach method were produced by Landau [1926], Julia [1926] and Kantorovich and Krylov [1936] who also developed a graphical method of conformal mapping due to Melent'ev [1937].

The first integral equation method was developed by Lichtenstein [1917] who solved the problem of conformally mapping a simply connected region bounded by a Jordan contour onto a circle by reducing it to the solution of an integral equation. Other attempts in this direction were made by Krylov and Bogolyubov [1929] who reduced the Dirichlet problem to an integral equation which is approximately solved by the Fredholm method with error estimates. Nyström [1930] gave a method for an approximate solution of integral equations, which is useful in conformal mapping. For conformal mappings of nearly circular regions the first numerical work was done by Fock [1929] who determined numerically the mapping function for a circular quadrilateral with angles $\pi/2, \pi/2, \pi/2, \alpha\pi$ $(0 < \alpha < 1)$. An important result, now known as Theodorsen's integral equation, was developed by Theodorsen [1931] and improved upon by Theodorsen and Garrick [1933] for conformally mapping nearly circular regions onto a circle. Another integral equation, known as Gershgorin's equation, was developed by Gershgorin [1933] as a result of conformally mapping a simply connected region bounded by a Jordan contour onto a circle, which is solved by the Nyström method. An exposition on Gershgorin's integral equation and its application to the mapping of a simply connected region onto a circle, of a doubly connected region onto an annulus, and of a multiply connected region, in general, onto slit planes is available in Kantorovich and Krylov [1936]. Various applications of conformal mapping are available in the monograph by Kantorovich and Krylov [1936], where both harmonic and biharmonic functions are studied. The Dirichlet and Neumann problems in $\mathbb{R}^2$ are solved by conformal mapping for different boundary conditions, and the general Hilbert problem is also investigated. Later Banin [1943] developed a method of approximately replacing Gershgorin's integral equation by a system of linear differential equations. Krylov [1938] also reduced the problem of conformally mapping an n-connected region bounded by n Jordan contours onto various canonical regions to the problem of solving a system of simultaneous integral equations.

Successive approximations in the integral equation method for simply and multiply connected regions were used by Kantorovich [1933; 1937]. Following Theodorsen's integral equation method, Warschawski [1945] reduced the problem of conformal mapping of a simply connected region bounded by a Jordan contour onto the unit circle to that of solving a nonlinear integral equation which is then solved by the method of successive approximations where the precise estimates for the convergence are also provided. Goluzin [1934] investigated the region exterior to the circles $C_1, \dots, C_n$ and determined a harmonic function $u(x,y)$ for this region such that it takes preassigned values on C_i, $i = 1, \dots, n$, and $u(\infty) < \infty$. The problem is then reduced to that of solving a finite system of functional equations which are solved by successive approximations. He applied this method to solve the Neumann problem for Laplace's equation for such regions and determined the Green's function for these regions and the conformal maps that carry these regions onto slit planes. The Schwarz method is used to develop an integral equation which is solved by successive approximations in Nevanlinna [1939] and Epstein [1948].

Julia [1927] determined a sequence of polynomials which converges to a suitably normed mapping of a simply connected region. An application of orthogonal polynomials to conformal mapping of simply connected regions was first developed by Szegö [1921]. He has defined a set of polynomials $\sigma_0(z), \sigma_1(z), \dots, \sigma_n(z), \dots$ (subsequently known as the Szegö polynomials) for a closed Jordan contour Γ in the z-plane, satisfying the following two properties: (i) $\sigma_n(z)$ is a polynomial of degree n, i.e., $\sigma_n(z) = a_n z^n + a_{n-1} z^{n-1} + \cdots + a_1 z + a_0$, such that $a_n > 0$, and (ii) $\frac{1}{l} \int_\gamma \sigma_m(z) \overline{\sigma_n(z)}\, ds = \delta_{m,n}$, where l is the length of Γ, s its arc length, and $\delta_{m,n}$ the Kronecker delta. Then the series $S(z,a) = \sum_{n=0}^{\infty} \sigma_n(z) \overline{\sigma_n(a)}$ converges uniformly and absolutely in every closed subregion of the interior of Γ. A formula is then obtained which defines the mapping function of the interior of Γ onto the unit disk in terms of $S(z,a)$. Szegö also gave a mapping function for the exterior of Γ by the polynomials $\sigma_n(z)$. Further developments in the Szegö polynomial method were produced by Smirnov [1928] and later by Kantorovich and Krylov [1936] who investigated the Szegö and Bochner-Bergman type polynomials and applied them to the minimizing problem. Some results about approximate mapping functions for a simply connected region by means of certain polynomials were given by Schaginyan [1944].

The problem of minimizing a functional was first solved by Hadamard [1908] which later became known as the Hadamard variational method. Löwner [1923] also developed an important variational method, whereas Julia [1926] provided another characterization of the mapping function by a minimum principle. The problem of reducing the mapping problem to that of minimizing of a functional was solved by Douglas [1931] who used the Riemann mapping theorem and the Osgood-Carathéodory theorem. Kufarev [1935-1937] investigated a minimal problem for a single-valued analytic function in an annulus and discussed the mapping of the minimizing function. Later, in 1947 he used the Löwner method to study a polygonal problem region which consists of the whole plane cut by a broken polygonal line with finitely many sides, one of which extends to infinity.

The problem of mapping a doubly connected region onto an annulus was reduced to that of minimizing an area integral by Khajalia [1940], who also showed that if the region is accessible from without, then there exists a sequence of minimal rational functions that converges uniformly to the desired mapping function. The problem of mapping a simply connected region bounded by a Jordan contour is reduced by Shiffman [1939] to that of minimizing a functional, almost similar to that of Douglas. This problem deals with the Plateau problem, and the electrostatic characterization of the resulting functional provides an effective method for determining the conformal maps. The Hadamard formula and the variation of domain functions were used by Schiffer [1946] to derive a new variational method.

In 1931, Grötsch solved the problem of conformally mapping a multiply connected region onto some canonical regions based on the assumption that the solution of a similar problem for a simply connected region is known. Iterative methods were established by Goluzin [1939] who used the mapping of a multiply connected region conformally onto some canonical regions, thereby reducing the problem to a sequence of conformal maps of simply connected regions. Heinhold [1954] investigated the problem of conformally mapping the simply connected region lying in the exterior of the unit circle onto the exterior of the unit circle.

Green's function method for an arbitrary region was established by Leja [1934; 1936] where the approximating functions were found closely related to Lagrange polynomials. A set of polynomials was also obtained which were used as mapping function for a region D, containing the point at infinity, onto the exterior of the unit circle. It was determined that the map is univalent if D is simply connected.

The Schwarz-Christoffel formula was used by Bergman [1923-24] to map a half-plane onto a particular polygon where a method for determining the parameters in the formula from the lengths of the sides of the polygon is given. In 1925 he investigated the conformal map of a special polygon onto a rectangle and computed level curves and their orthogonal trajectories which were presented in tabular and graphical forms. Bergman also gave the first punch-card machine method in 1947 to solve the torsion problem where the orthogonal polynomials were applied to solve Laplace's equation numerically. The notion of the kernel functions was developed by Bergman [1922] where the existence of a complete orthonormal system with respect to a region is established and the kernel of the system is related to the conformal map of the region. Based on Bergman's kernel function and complete orthonormal sets of functions, Zarankiewicz [1934] found a method for effectively constructing the conformal map of a doubly connected region onto an annulus; in 1934 he published details of this method. Schiffer [1946] found an expression for Bergman's kernel function $K(z, a)$ of a region in terms of Green's function for the region and gave formulas for the variation of $K(z, a)$. In 1948 he extended the concept of kernel functions and orthonormal sets of functions to a wider class of functions. Further study of the relationship between the kernel function and conformal mappings of regions was done by Bergman and Schiffer [1948-1951].

Hodgkinson and Poole [1924] used elliptic functions to map doubly connected regions of certain types onto the whole plane with two slits on the real axis. Using hyperelliptic integrals, a generalization of the solution by Hodgkinson and Poole was given by Vladimirsky [1941] for the problem of conformal mapping of a doubly connected region bounded by rectangular segments or circular arcs, who also extended the solution to n-tuply connected regions ($n > 2$). Hodgkinson [1930] also established the relationship between the theory of Lamé differential equations and the Schwarz theory of conformal mapping. An up-to-date survey of conformal mapping of multiply connected regions onto canonical domains was published by Keldyš [1939]. Important formulas for various conformal maps of n-tuply connected regions in terms of the kernel function of the region were established by Nehari in 1949. These formulas served as tools for numerical computation of conformal maps since the kernel functions are constructed more easily than the mapping function. Gerabedian and Schiffer [1949] obtained many significant relations between various domain functions of an n-tuply connected region and solved some minimal problems.

The first systematic construction of conformal maps by the method of networks was done by Liebmann [1918]. The Dirichlet problem in $\mathbb{R}^2$ and $\mathbb{R}^3$ was first solved by this method by Phillips and Wiener [1923]. The boundary problems in $\mathbb{R}^2$ with Laplace's equation were solved numerically by this method by Luysternik [1947].

Relaxation methods were first applied to the problems of conformal mapping by Gandy and Southwell [1940] and Southwell [1946]. Several examples of technical interest were given, in which regions of arbitrary shape are mapped onto the interior and exterior of circles and onto rectangles.

The small parameters method developed by Kantorovich [1933] was later used by Rosenblatt and Turski [1936] for conformal mapping of a special type of region. Rosenblatt [1943] constructed the conformal maps of regions onto the unit disk by the Kantorovich method of small parameters and applied it to the dynamic problems of airfoils.

Special conformal mappings were studied by Rothe [1908] who constructed the mapping function for a circular quadrilateral with angles $\pi/2, \pi/2, \pi/2, \rho\pi$ $(\rho = 1/\sqrt{20})$. Hodgkinson's work [1924] on a similar problem has been mentioned above. Wirtinger [1927] derived an explicit formula for computing the conformal map of a triangle with circular arcs and arbitrary angles. The mapping of an ellipse onto a circle and of a region bounded by two symmetrically placed ellipses onto an annulus were investigated by Zmorovich [1935] who also gave an approximate solution in the latter problem. Catalogs of various types of explicit mappings are available in von Koppenfels' papers [1937; 1939]. Other authors on special conformal mappings during this period were Muratov [1937], Krylov [1937], Melent'ev [1937], and Goluzin [1937]. Later Jeffreys and Jeffreys [1946] and Wittich [1947] produced special conformal maps. A very useful dictionary of conformal mappings, which is still a source work in this area, was produced by Kober [1945-1948].

The theory of homogeneous boundary problems for analytic functions was presented by Hilbert [1924] for a boundary that is a single Jordan contour. This problem was then transformed into a Fredholm integral equation, for which Green's functions of the Neumann problem must be determined. Hilbert did not give a complete solution of this integral equation. A complete solution by means of Cauchy integrals for a particular case had already been given by Plemelj [1908]. The solution for the general case, based on the Plemelj method, was later provided by Khvedelidze [1941]. Earlier, Picard [1927] had studied the homogeneous Hilbert problem.

The nonhomogeneous Hilbert problem which is solved to determine a sectionally analytic function $\Phi(z)$ of finite degree at infinity such that $\Phi^+(\zeta) = G(\zeta)\Phi^-(\zeta) + g(\zeta)$ on the boundary Γ, where $G(\zeta)$ and $g(\zeta)$ are defined on the boundary Γ and satisfy the Hölder condition there, was first considered by Carleman [1922]. Later Privalov [1934] investigated it for the case when the boundary Γ is a rectilinear contour, $G(\zeta)$ and $g(\zeta)$ are Lebesgue-integrable, $G(\zeta)$ is bounded, and the limiting value of the solution is taken along a nontangential path. He used Picard's method but did not find a complete solution. A complete solution was first given by Gakhov [1937] for the case when Γ is a single contour. Later Khvedilidze [1941] gave a complete solution of the nonhomogeneous Hilbert problem. Carleman's and Gakhov's methods are essentially alike. In 1851 Riemann considered a very general boundary problem which is now known as the Riemann boundary problem. This problem was also studied and solved by Hilbert in 1904, and therefore it is also called the Riemann-Hilbert problem. It deals with determining an analytic function in a given region with boundary values involving a relation between its real and imaginary parts. Hilbert reduced this problem to a singular equation and then applied it to the solution of two Dirichlet problems. Later, Noether [1921] used this solution to study the subject of singular integral equations.

One of the widely investigated practical aspects of conformal mapping was the development of the airfoil theory which started with the pioneering work by Joukowski [1890]. He studied the flows around a variety of so-called Joukowski airfoils. The developments in this area are very pertinent for computational aerodynamic flows. Extensive research to develop computational methods for solving the direct and inverse problems of airfoil theory has been done by Theodorsen [1931], Glauert [1948], Andersen, Christiansen, Møller and Tornehave [1962], Timman [1951], James [1971], Ives [1976], and Halsey [1979; 1982]. In single and multi-element airfoils, Theodorsen's integral equation has been solved with the von Karman-Trefftz transformation and the FFT by Garrick's method of conjugate functions (Garrick [1949]) by Ives [1976]. James's method (James [1971]) is another approach in the conformal mapping of single and multiple-element airfoils (Halsey 1979; 1982]).

1.2 Modern Developments

During the past decades several methods have evolved for numerical and computational evaluation of mapping functions. The results of some of these methods provide us with an explicit form of a function which approximately evaluates the mapping function for a certain source region. This is possible because of an important result which states that an analytic function defined on a simply connected and bounded region $D \subset \mathbb{C}$ can always be uniformly approximated on every compact subset of D with any preassigned accuracy by means of a polynomial. In most applications the mapping function is continuous in $\bar{D}$, and it can be uniformly approximated in D by a polynomial.

One major approach in developing methods for numerical conformal mapping is based on the following interpretation of the Riemann mapping theorem: there exists a conformal mapping $f : D \mapsto U$ with $f(z_0) = 0$ and $f'(z_0)$ nonzero real, where $z_0 \in D$, and this function has a power series expansion $f(z) = c_1\,(z - z_0) + \sum_{n=2}^{\infty} c_n\,(z - z_0)^n$, with c_1 nonzero real and $z_0 \in D$, which converges uniformly in every closed disk with center z_0 and contained in D. However, a polynomial which is a good approximation of f in D is not the same as a truncated power series. If a polynomial $p(z) = c_1'\,(z - z_0) + \sum_{n=2}^{N} c_n'\,(z - z_0)^n$ approximates f with accuracy $\varepsilon > 0$, then it is necessary that every term of $p(z)$ must approximate the corresponding term of the power series with accuracy $\varepsilon > 0$ on the set $D \cap B(z_0, R)$, where $R = |z - z_0|$ is the radius of convergence of the power series. All this means is that a polynomial p which is a good approximation of the power series starts in the same way as the power series, but the relative error in the coefficients increases with increasing n.

Another direction for developing numerical methods for conformal mapping is based on computing a table of values for the mapping function f at several points of D. The mapping function f always maps the boundary of the source region D in a simple way onto the boundary of the unit disk U. Once the mapping of the boundary is known, the mapping of D itself is determined by the Cauchy formula with positive orientation of the boundary curve. Thus, it is sufficient to determine the mapping of the boundary of D onto the boundary of U, and once we obtain a parametric representation of the boundary of D, we determine only one real-valued function of a real variable t. But in this approach it so happens that this unknown real-valued function satisfies an integral equation. Hence, solving the problem of conformal mapping reduces to that of solving an integral equation.

The widely used current computational techniques are based on the integral equation methods where an integral equation is developed to relate the boundaries of the problem region and the standard region like the unit disk. Once the boundaries are discretized at n points, the integral equation reduces to an algebraic system of equations. The majority of ongoing research in computational conformal mapping is divided basically into two groups: one where the maps are constructed from a standard region such as the unit disk onto the problem region, and the other where the maps are constructed the other way around. In the first group the integral equation is nonlinear and involves the conjugation operator, which can be solved by FFT on a discrete mesh in $O(n \log n)$ operations. This method evolves with the numerical solution of Theodorsen's integral equation, or a related equation, which is solved by using the fixed-point iteration method. The recent development of Newton's method is faster for sufficiently smooth boundaries and the choice of a good initial guess. The basic work in this group has been produced by Wegmann [1978; 1986], Hübner [1986], and Gutknecht [1986]. The first quadratically convergent algorithm for Theodorsen's equation was presented by Wegmann [1978] based on the following two ideas: an induction scheme along tangents to the boundary of the problem region, and a computation scheme similar in formulation to the Riemann-Hilbert problem for the unit disk at each iteration. This method is efficient because the solution of the Riemann-Hilbert problem can be represented by Cauchy integrals. A generalization of this algorithm to the conformal mapping of an annulus onto a doubly connected region was published by Wegmann in 1986. This algorithm is also based on the Riemann-Hilbert problem and is so far the fastest known for this problem. Hübner [1979; 1986] studied Newton's method for the solution of Theodorsen's integral equation. His method is also based on the solution of the Riemann Hilbert problem. He established the quadratic convergence of Newton's method and obtained a quadratically convergent conformal mapping. An extensive survey of almost all known methods for numerical conformal mapping of the unit disk onto a simply connected region is available in Gutknecht's work [1986]. He has derived integral and integro-differential equations involving the conjugation operator for the boundary correspondence function. Then various iterative schemes for solving these equations are presented in this work. The general theory is described by specific methods, especially the successive conjugation methods of Theodorsen [1931], Timman [1951], Freiberg [1951], the projection method of Bergström [1958], Newton's method of Vertgeim [1958], Wegmann [1978], and Hübner [1979].

In the other group where the maps are constructed from the problem region onto the unit disk, the integral equations, mostly derived from the Dirichlet problem, are generally linear, and require $O(n^2 \log n)$ operations. In cases where the geometry is simpler, they may require a smaller number of operations. The methods in this group are based on Symm's equation which is a singular integral equation of the first kind derived by using a single-layer potential as the basis of conformal mapping. Symm [1966; 1969] investigated an integral equation method, like the one he developed for the boundary integral equation method, for computing the conformal mapping of a simply connected region onto the unit disk. Berrut [1976; 1985; 1986] solved Symm's equation numerically by a Fourier method. This equation is a Fredholm integral equation of the second kind for the derivative of the boundary correspondence function for the conformal mapping of a Jordan region with a piecewise twice differentiable boundary onto the unit disk. Kerzman and Trummer [1986] presented a new method to compute the Riemann mapping function numerically. The solution of the integral equation of the second kind is expressed in terms of the Szegö kernel and is based on an earlier work of Kerzman and Stein [1978] on the Cauchy kernel, Szegö kernel and Riemann mapping function, and of Kerzman and Trummer [1986] on a method for the numerical solution of the conformal mapping problem.

A variant of Symm's integral equation which is suitable for conformal mapping of both simply and multiply connected regions was presented by Reichel [1985]. It uses a Fourier-Galerkin technique to produce an extremely fast iterative solution of $O(n^2 \log n)$ which is due to the singularity of the kernel so that the linear algebraic system becomes block diagonally dominant. Another fast method for solving Symm's equation for multiply connected regions was presented by Mayo [1986]. This integral equation formulation is based on a similar formulation in Mikhlin [1957] and has the advantage of reducing the problems to integral equations of the second kind with unique solutions and boundary kernels. Because the solutions are periodic, the trapezoidal rule can be applied effectively. Once the integral equation is solved, a rapid method is available to determine the mapping function in the interior of the region. This method uses a fast Poisson solver for the Laplacian, thus avoiding the time-consuming computation of integrals at singular points near the boundary.

Other significant modern research includes work by Hoidn [1982] which deals with conformal mapping of simply connected regions where the singularity problem near and on the boundary is solved by reparameterization of the boundary curve. Then this method is applied to Symm's equations which are solved by using spline functions. The singularity problem has been solved by Levin, Papamichael and Sideridis [1978], Hough and Papamichael [1981; 1983], Papamichael and Kokkinos [1981; 1982], Papamichael and Warby [1984], and Papamichael, Warby and Hough [1983; 1986] by integral equation methods as well as expansion methods based on the Bergman kernel or on the Ritz approximation. No research is available in this area for multiply connected regions of higher connectivity. The Chebyshev approximation in conjunction with linear programming has been used by Hartman and Opfer [1986] for conformal mapping of simply connected regions onto the unit disk. A simple approximation formula has been derived by Zemach [1986] for the boundary mapping function which gives a remarkably good fit for mappings of regions with highly distorted boundaries. This method is based on reducing the nonlocal integral equation for the mapping function to a local equation depending on the nature of the distorted regions. The Schwarz-Christoffel formula has always been used in mappings of polygons and related regions, but the treatment of the singularity problem in these cases has been only recently investigated by Barnard and Pearce [1986], Elcrat and Trefethen [1986], and Trefethen and Williams [1986]. Boundary problems for analytic functions and integral equations with transformations have been discussed by Lu [1994], and a comprehensive theoretical account on conformal mapping and boundary problems can be found in Wen [1992].

The advantages of integral equation formulation in conformal mapping can be summarized as follows: All integral equations obtained in any conformal mapping problem (except Arbenz's integral equation, see §13.3) are Fredholm integral equations of the second kind with bounded kernel, except for Symm's integral equation which is of the first kind with a kernel that has a logarithmic singularity. The Fredholm integral equations of second kind are never ill-conditioned, and there are reliable error estimates available for them. However, the drawback with the equations of second kind is that the kernel has singularity at points in the neighborhood of the boundary (but not on the boundary itself unless it has a corner singularity) where computational difficulties often arise. This situation, on the other hand, does not occur with the equations of the first kind.

The best strategy for developing a computational method based on an integral equation formulation is to make sure that the solution is periodic and unique. This permits an effective use of the trapezoid rule which is highly accurate on smooth contours. Another feature to look for is that the mapping onto canonical regions (unit disk, annulus, or slit disks) produces systems of linear equations and avoids solving systems of nonlinear equations as in Fornberg's, Guteknecht's, or Wegman's methods.

Some classical applications of conformal mappings to steady-state problems of mathematical physics and especially for the solution of Laplace's equation can be traced to the beginning of the twentieth century. A noteworthy contribution to the theory of elasticity is by Muskhelishvili [1963]. Modern contributions can be found in areas of fluid flow, heat conduction, solidification, electromagnetics, ion optics, acoustics, vibrations, wave guides, and grid generation, to name a few; a detailed review and biography of the applications through 1972 is available in Laura [1975]. The problem of flow and heat transfer in conduits of arbitrary shape in space vehicles was investigated by Sparrow and Haji-Sheikh [1966]. This study was extended to noncircular conduits with uniform wall temperature by Casarella et al. [1971]. Unsteady heat conduction problems in bars of arbitrary cross section were investigated by Laura et al. [1964; 1965; 1968]. Ives [1976] analyzed the incompressible flow between two concentric circles and computed the streamlines by using Garrick's method of conjugate functions. The problem of solidification of steady-state and transient frozen layers in rectangular channels has been solved by Siegel, Goldstein and Savino [1970]. In transient solidification the shape of a frozen region is determined by mapping it onto a potential plane and then computing the time-dependent conformal map between the potential and the physical plane. The thermoelastic problem of uniform heat flow distributed by an isolated hole of ovaloid form was investigated by Florence and Goodier [1960] and extended by Deresiewicz [1961] to holes which are mapped onto the unit circle and approximated by polynomials.

Wilson [1963] and Richardson [1965] used conformal mapping to determine the stresses in solid propellant rocket grains by solving an integral equation of the Fredholm type and a system of coupled integral equations. The method of computing the conformal mapping function and the related eigenvalue problem for plane regions with irregular boundaries was published by Laura [1968]. The ion problem connected with the trajectory of a charged particle in a plane electric field and a normal magnetic field was solved by Naidu and Westphal [1966] by using the Schwarz-Christoffel transformation. Conformal mapping techniques of simply and doubly connected regions have been used, among others, by Kusin and Merkulov [1966], Laura [1967], Laura et al. [1972], and Schinzinger and Laura [2003].

Conformal mapping has been applied to acoustic waveguides of complicated cross section where the Galerkin method is applied to obtain a functional approximation for the solution of the boundary value problem. The grain of a solid propellant rocket motor with a starlike internal propagation is in the form of a circular cylinder bounded by a thin case. To solve any boundary value or eigenvalue problem, the grain cross section is conformally mapped onto a circle or an annulus. Studies on the shear vibrations of such rocket motors were done by Baltrukonis et al. [1965] and Laura and Shahady [1966], and Schinzinger and Laura [2003]. Conformal mapping techniques are used in a study on the Rayleigh-Taylor instability for ideal fluid by Menikoff and Zemach [1980; 1983]. Grid generation for cascades of blades and inlet flows has been investigated by Inoue [1983; 1985]. The Kirchhoff flow problem past a polygonal obstacle was solved by Trefethen [1986]. Other contemporary applications can be found in the book on numerical conformal mapping edited by Trefethen [1986].

Computational conformal mapping in the present decade is still progressing steadily. Fast algorithms are being developed. Computational conformal mapping is being used in engineering problems that require grid generation and related domain simplification. Although the use of conformal mapping in reducing or eliminating singularities in solutions of integral and integro-differential equations is one important issue, the choice of an initial good guess in Newton's method is another. Much has developed since the evolution of computer technology around 1955, and this subject has reached maturity.

1.3 In Retrospect

Finally, the following quote from Bernhard Riemann's Inaugural Dissertation, Göttingen, 1851, is worthy of presentation.

> "Die Ausführung dieser Theorie, welche, wie bemerkt, einfache durch Grössenoperationen bedingte Abhängigkeitsgesetze ins Licht zur setzen bestimmt ist, unterlassen wir indess jetzt, da wir die Betrachtung des Ausdrucken einer Function gegenwärtig ausschliessen.
>
> Aus demselben Grunde befassen wir uns hier auch nicht damit, die Brauchbarkeit unserer Sätze als Grundlagen einer allgemeinen Theorie dieser Abhängigkeits-gesetze darzuthun, wozu der Beweis erfordert wird, dass der hier zu Grunde gelegte Begriff einer Function einer veränderlichen complexen Grösse mit dem einer durch Grssenoperationen ausdrückbaren Abhängigkeit vollig zusammenfällt."

This book starts with the Riemann mapping theorem and its implications developed since its acceptance as "one of the most important theorems of complex analysis" (Ahlfors [1953: 172]).

REFERENCES USED: Ahlfors [1953], Andersen et al. [1962], Barnard and Pearce [1986], Berrut [1986], Elcrat and Trefethen [1986], Freiberg [1951], Gaier [1964; 1983], Goluzin [1969], Gutknecht [1986], Hartman and Opfer [1986], Hoidn [1986], Hübner [1979; 1986], Ives [1982], James [1971], Jawson [1963], Jawson and Symm [1977], Kantorovich and Krylov [1958], Kerzman and Stein [1978], Kerzman and Trummer [1986], Kober [1957], Laura [1975], Lawrentjew and Schabat [1967], Lu [1994], Mayo [1986], Mikhlin [1957], Nehari [1949; 1952], Papamichael et al. [1981; 1983; 1984; 1986], Schinzinger and Laura [2003], Seidel [1952], Symm [1966], Trefethen [1980], Trefethen and Williams [1986], Wegmann [1979; 1986], Wen [1992], Zemach [1986].

2

Conformal Mapping

Some basic concepts and results from complex analysis are presented. They include harmonic functions, Cauchy's theorem, Cauchy kernel, Riemann mapping theorem, analytic continuation, and the Schwarz reflection principle. Proofs for most of the results can be found in textbooks.

2.1 Definitions

Let $\mathbb{R}^n$ denote the Euclidean n-space, and $\mathbb{R}^+$ the set of nonnegative real numbers. The complement of a set B with respect to a set A is denoted by $A \backslash B$ (or compl(B) if the reference to set A is obvious), the product of the sets A and B by $A \times B$, and the closure of a set A by $\bar{A}$.

A complex-valued function f is said to belong to the class $C^k(D)$ if it is continuous together with its kth derivatives, in a domain D, $0 \leq k < \infty$. In this case we often say that f is a $C^k(D)$-function, or that f is a C^k-function on D, where a C^0-function is written simply as a C-function. The function f in the class $C^k(D)$, for which all kth derivatives admit continuous continuations in the closure $\bar{D}$, form the class of functions $C^k(\bar{D})$. The class $C^\infty(D)$ consists of functions f which are infinitely differentiable on D, i.e., continuous partial derivative of all orders exist. These classes are linear sets. Thus, every linear combination $\lambda f + \mu g$, where λ and μ are arbitrary complex numbers, also belongs to the respective class.

Let $\mathbb{C}$ denote the complex plane. If $a \in \mathbb{C}$ and $r > 0$, then

$$
\begin{aligned}
B(a,r) &= \{z \in \mathbb{C} : |z-a| < r\}, \\
\bar{B}(a,r) &= \{z \in \mathbb{C} : |z-a| \leq r\}, \\
\partial B(a,r) &= \{z \in \mathbb{C} : |z-a| = r\},
\end{aligned} \tag{2.1.1}
$$

denotes, respectively, an open disk, a closed disk, and a circle, each of radius r and centered at a. The open unit disk $B(0,1)$ is sometimes denoted by U. A connected open set $A \subseteq \mathbb{C}$ is called a *region* (or domain). The extended complex plane is denoted by $\mathbb{C}_\infty$. Then $\partial_\infty D$

is the boundary of a set D in $\mathbb{C}_\infty$, i.e.,

$$\partial_\infty D = \begin{cases} \partial D & \text{if } D \text{ is bounded,} \\ \partial D \cup \{\infty\} & \text{if } D \text{ is unbounded.} \end{cases}$$

If D is a region in $\mathbb{C}_\infty$, then the following statements are equivalent: (a) D is simply connected, (b) $\mathbb{C}_\infty \backslash D$ is connected, and (c) $\partial_\infty D$ is connected. Regions that have more than one layer over the complex plane are called Riemann surfaces.

Let $z = x + iy$ be the complex variable in the z-plane, where x and y are arbitrary real numbers. Similarly, let $w = u + iv$ be the complex variable in the w-plane, where u and v are arbitrary real numbers. Then a function $w = f(z)$ defined on a domain D in the z-plane and taking values in the w-plane is called a (complex) *function of a complex variable.* Let $z_0 \in D$ be a nonisolated point of D, so that every neighborhood of z_0 contains a point of D other than z_0 itself. Then a complex number A is said to be the *derivative* of the function $w = f(z)$ at a point $z = z_0$ relative to the domain D, denoted by $f'_D(z_0)$, if, given any $\varepsilon > 0$, there exists a $\delta > 0$ such that $0 < |z - z_0| < \delta$, $z \in D$, implies

$$\left| A - \frac{f(z) - f(z_0)}{z - z_0} \right| < \varepsilon. \tag{2.1.2}$$

In this case we say that $w = f(z)$ is *differentiable* at $z = z_0$ with derivative $A = f'_D(z_0)$, where A is the limit of the differential quotient $\dfrac{f(z) - f(z_0)}{z - z_0}$ in the direction $z \to z_0$ in the deleted neighborhood $0 < |z - z_0| < \delta$. Note that above differential quotient is bounded by some constant C, and thus, $|f(z) - f(z_0)| \le C|z - z_0|$ for sufficiently small $|z - z_0|$, $z \in D$. Hence,

$$\lim_{z \to z_0} f(z) = f(z_0).$$

Thus, $f(z)$ is continuous at $z = z_0$ relative to the domain D.

2.1.1 Analytic Functions. A function is said to be *analytic* (synonymously, *holomorphic* or *regular*) on a domain D in the z-plane if $f(z)$ is differentiable on D, i.e., it is differentiable at every point of D. The derivative of $f(z)$ is denoted by $f'(z)$, without explicit reference to the domain D. Thus, a function $f(z)$ is said to be *analytic* on a domain D if $f(z)$ is analytic at a point $z = z_0$, where $z_0 \in D$.

The increment of a complex function $w = f(z)$ differentiable relative to a domain D can be written as

$$\Delta w = f(z_0 + h) - f(z) = f'_D(z_0)h + \varepsilon(h)h, \quad h = z - z_0, \tag{2.1.3}$$

where $\varepsilon(h) \to 0$ as $h \to 0$. Suppose we denote the quantity $f'_D(z_0)h$, called the *principal linear part* of the increment Δw, by $dw(z_0)$, or briefly by dw or df, while writing the increment h as an independent variable $d_D z$ or simply dz. Then the quantities dw, df, dz are called the *differentials*, and the derivative can be written as the ratio of differentials

$$f'(z_0) = \frac{dw}{dz} = \frac{df}{dz}. \tag{2.1.4}$$

Some of the properties of the differential are:
(i) $d(kf) = k df$;
(ii) $d(f \pm g) = df \pm dg$;
(iii) $d(fg) = f\,dg + g\,df$;
(iv) $d\Big(\frac{f}{g}\Big) = \frac{g\,df - f\,dg}{g^2(z_0)}, \quad g(z_0) \neq 0.$

Let $z = x + i\,y$ be a complex number. Then $\bar{z} = x - i\,y$, $x = \frac{z + \bar{z}}{2}$, and $y = \frac{z - \bar{z}}{2i}$. Also

$$\partial f = \frac{\partial f}{\partial z} = \frac{1}{2}\,(f_x - i\,f_y)\,, \quad \bar{\partial} f = \frac{\partial f}{\partial \bar{z}} = \frac{1}{2}\,(f_x + i\,f_y)\,. \tag{2.1.5}$$

A function $f : D \mapsto \mathbb{C}$ is analytic on D iff $\bar{\partial} f = 0$, which is equivalent to the *Cauchy-Riemann equations* for the function $w = f(z) = u(x, y) + iv(x, y)$:

$$u_x = v_y, \quad u_y = -v_x, \tag{2.1.6}$$

or, in polar form ($z = re^{i\theta}$),

$$u_r = \frac{1}{r}\,v_\theta, \quad v_r = -\frac{1}{r}\,u_\theta. \tag{2.1.7}$$

Thus,

$$f'(z) = u_x + i\,v_x = v_y - i\,u_y. \tag{2.1.8}$$

The Cauchy-Riemann equations are necessary conditions for $f(z)$ to be analytic on D. However, merely satisfying the Cauchy-Riemann equations alone is not sufficient to ensure the differentiability of $f(z)$ at a point in D. For example, the function $f(z) = \frac{x^4 - y^4}{x^3 + y^3} + i\,\frac{x^4 + y^4}{x^3 + y^3}$, $f(0) = 0$, $z \in \mathbb{C}$, is not differentiable at the origin, although it satisfies the Cauchy-Riemann equations there: $u_x(0, 0) = 1 = v_x(0, 0) = v_y(0, 0)$, $u_y(0, 0) = -1$, but by limit definition, $f'(0) = (1 + i)/2$ along the line $y = x$.

The following results are obvious: $\partial\,(\log |z|) = \frac{1}{2z}$, $\bar{\partial}\,(\log |z|) = \frac{1}{2\bar{z}}$, $\overline{\partial f} = \bar{\partial} \bar{f}$, and the Laplacian

$$\nabla^2 \equiv \frac{\partial^2}{\partial x^2} + \frac{\partial^2}{\partial y^2} = 4\bar{\partial}\,\partial = 4\partial\,\bar{\partial}. \tag{2.1.9}$$

The Cauchy-Riemann equations (2.1.6) for the function $f(z) = u(x, y) + iv(x, y)$ satisfy the partial differential equations

$$\begin{aligned} u_x\,v_x + u_y\,v_y &= 0, \\ \nabla^2 u = 0, \quad \nabla^2 v &= 0. \end{aligned} \tag{2.1.10}$$

Using the gradient vector $\nabla \equiv \mathbf{i}\frac{\partial}{\partial x} + \mathbf{j}\frac{\partial}{\partial y}$, the first equation in (2.1.10) can be written as the inner (scalar) product:

$$\big\langle \nabla u, \nabla v \big\rangle = 0. \tag{2.1.11}$$

Then the Cauchy-Riemann equations yield $|\nabla u| = |\nabla v| = |f'(z)|$. Eq (2.1.11) also signifies the orthogonality condition for the families of level curves defined by $u(x, y) = \text{const}$ and $v(x, y) = \text{const}$.

If $w = u + iv$, then

$$\Re\{z\} \cdot \Re\{w\} = \Re\left\{\frac{z^2 + |z|^2}{2z} w\right\}, \quad \Im\{z\} \cdot \Im\{w\} = \Im\left\{\frac{z^2 - |z|^2}{2iz} w\right\}. \tag{2.1.12}$$

Example 2.1. Let $f(z) = \bar{z} = x - iy$. Its real and imaginary parts do not satisfy the Cauchy-Riemann equations, and hence $\bar{z}$ does not have a complex derivative. In general, the function $f(z, \bar{z})$ that depends on $\bar{z}$ is not complex-differentiable. ■

Let E denote a closed bounded infinite set of points in the z-plane. For the points $z_1, z_2, \dots, z_n \in E$ the Vandermonde determinant is defined by

$$V(z_1, z_2, \dots, z_n) = \prod_{\substack{i,j=1 \\ i \neq j}}^{n} (z_i - z_j), \quad n \geq 2. \tag{2.1.13}$$

Let us define the numbers

$$d_n = V_n^{2/n(n-1)}, \tag{2.1.14}$$

where $d_{n+1} \leq d_n$. Since d_n does not exceed the diameter[1] of the set E for any n, it follows that the sequence $\{d_n\}$ approaches a finite limit as $n \to \infty$. This limit is called the *transfinite diameter* of the set E and is denoted by $\text{diam}(E)$, i.e., $\text{diam}(E) = \lim_{n\to\infty} d_n$. When the set E has finitely many points, we take $\text{diam}(E) = 0$. ■

Example 2.2. If $f(z) = z\bar{z}$, then $f'(z)$ exists only at $z = 0$. To prove, note that $f(z) = |z|^2 = x^2+y^2$, so that $u = x^2+y^2$ and $v = 0$. Then $\dfrac{\partial u}{\partial x} = 2x$, $\dfrac{\partial u}{\partial y} = 2y$, $\dfrac{\partial v}{\partial x} = 0 = \dfrac{\partial v}{\partial y}$. Hence, the Cauchy-Riemann equations are satisfied only when $x = 0 = y$, i.e., only when $z = 0$. This function not analytic anywhere else. ■

2.1.2 Integration. Let f be a C-function on an open set $D \subset \mathbb{C}$, and let Γ be a piecewise C^1-continuous curve in D. If $|f(z)| \leq M$ for all points $z \in \Gamma$, i.e., for all $z = \gamma(t)$ for $t \in [a, b]$, where $M > 0$ is a constant, then

$$\left|\int_\Gamma f\, dz\right| \leq M\, l(\Gamma), \tag{2.1.15}$$

where

$$l(\Gamma) = \int_a^b |\gamma'(t)|\, dt = \int_a^b \sqrt{x'(t)^2 + y'(t)^2}\, dt \tag{2.1.16}$$

is the arc length of the path Γ. In general, by applying the triangle inequality, we get

$$\left|\int_\Gamma f\, dz\right| \leq \int_\Gamma |f|\, |dz| = \int_a^b |f(\gamma(t))|\, |\gamma'(t)|\, dt. \tag{2.1.17}$$

If $\gamma : [a, b] \mapsto \mathbb{C}$ defines a piecewise smooth contour Γ and F is a function defined and analytic on a region containing Γ, then

$$\int_\Gamma F'(z)\, dz = F(\gamma(b)) - F(\gamma(a)). \tag{2.1.18}$$

[1] The diameter of the set E is defined as $\sup_E\{|z_i - z_j| : z_i, z_j \in E\}$, $i, j = 1, 2, \dots, n$ $(i \neq j)$.

If $\gamma(a) = \gamma(b)$, then

$$\int_\Gamma F'(z)\, dz = 0. \tag{2.1.19}$$

This result is known as the *fundamental theorem for line integrals (contour integration)* in the complex plane. Thus, if a function f is defined and analytic on a region $D \subset \mathbb{C}$ and if $f'(z) = 0$ for all points $z \in D$, then f is a constant on D. If f is a C-function on a region D, then the following three statements are equivalent:

(i) If Γ_1 and Γ_2 are two paths in D from a point $z_1 \in D$ to a point $z_2 \in D$, then $\displaystyle\int_{\Gamma_1} f(z)\, dz = \int_{\Gamma_2} f(z)\, dz$, i.e., the integrals are path-independent.

(ii) If Γ is a Jordan contour lying in D, then $\displaystyle\int_\Gamma f(z)\, dz = 0$, i.e., the integrals on a closed contour are zero.

(iii) There exists a function F defined and analytic on D such that $F'(z) = f(z)$ for all $z \in D$, i.e., there exists a global antiderivative of f on D.

A function $\Phi(z)$, analytic inside a region with boundary $\Gamma = \cup_{k=1}^n \Gamma_k$, is said to be *sectionally analytic* on Γ if $\Phi(z)$ is continuous on each Γ_k from both left and right except at the end points where it satisfies the condition $|\Phi(z)| \le \dfrac{C}{|z-c|^\alpha}$, where c is the corresponding end point of Γ_k, and C and α are real constant with $\alpha < 1$.

Let $w = f(z)$ be an analytic function, regular in a simply connected region D, and α be an interior point of D and a simple zero of $f(z)$ such that $f(\alpha) = 0$ and $f'(\alpha) \neq 0$. Moreover, let $z_0 \in D$ be a first approximation of the zero α, i.e., z_0 is close to α. Then

$$\begin{aligned}\alpha &= \sum_{n=0}^{\infty} (-1)^n \frac{f(z_0)^n}{n!} \left[\frac{d^n f^{-1}(w)}{dw^n}\right]_{w=f(z_0)} \\ &- \exp\left\{ -f(z_0) \left[\frac{df^{-1}(w)}{dw}\right]_{w=f(z_0)} \right\},\end{aligned}$$

where $z = f^{-1}(w)$ denotes the inverse function of $f(z)$ and the exponential function operates symbolically on the differential symbol (Blaskett and Scherdtfeger [1945-1946:266]).

2.1.3 Fatou's Lemma. Let $f_n : I \mapsto \mathbb{R}$ be nonnegative, extended real-valued, measurable functions defined in an interval I and such that the sequence $\{f_n\}$ converges pointwise to the function $f : I \mapsto \mathbb{R}$. If $\displaystyle\underline{\lim}_{n\to\infty} \int_\Gamma f_n < \infty$, then f is integrable and

$$\int_I f \le \underline{\lim}_{n\to\infty} \int_\Gamma f_n. \tag{2.1.20}$$

2.2 Jordan Contour

A simple closed curve Γ in $\mathbb{C}$ is a path $\gamma : [a, b] \mapsto \mathbb{C}$ such that $\gamma(t) = \gamma(s)$ iff $t = s$ or $|t - s| = b - a$. In what follows, a simple closed curve will be called a *Jordan contour*. The *Jordan curve theorem* states that if Γ is a simple contour, then $\mathbb{C}\backslash\Gamma$ has two components, one called the interior of Γ, denoted by $\mathrm{Int}(\Gamma)$, and the other called the exterior of Γ, denoted by Ext(Γ), each of which has Γ as its boundary. Thus, if Γ is a Jordan contour, then Int(Γ) and Ext(Γ) ∪ {∞} are simply connected regions.

Let a continuous curve Γ, defined by $\gamma(t) = \alpha(t) + i\beta(t)$, be divided into n arcs $\sigma_k = z_{k-1}\frown z_k$, $k = 1, \dots, n$, where $z_k = \gamma(t_k)$ for $k = 0, 1, \dots, n$, such that the end point of each arc, except the last one, overlaps the initial point of the next arc. If we join each segment $[z_{k-1}, z_k]$, $k = 1, \dots, n$, by straight line segments (see Figure 2.1, where $n = 5$), we obtain a polygonal line L inscribed in Γ. The segments of L are the chords joining the end points of the arcs σ_k, and

$$l = \text{length of } L = \sum_{k=1}^{n} |z_k - z_{k-1}| \,. \tag{2.2.1}$$

The curve Γ is said to be *rectifiable* if

$$\sup_{\mathcal{P}} \sum_{k=1}^{n} |z_k - z_{k-1}| = l < +\infty,$$

where the least upper bound is taken over all partitions $\mathcal{P} = \{a = t_0, t_1, \cdots, t_n = b\}$ of the interval $[a, b]$, $a \le t \le b$. The nonnegative number l is called the length of the curve Γ. The curve is said to be *nonrectifiable* if the sums (2.1.13) become arbitrarily large for suitably chosen partitions.

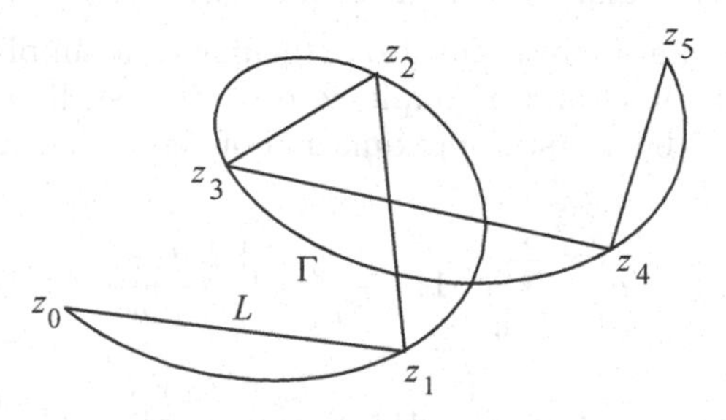

Figure 2.1 Rectifiable curve.

A piecewise smooth curve $\widetilde{\Gamma}$, defined by $\widetilde{\gamma} = [\tilde{a}, \tilde{b}] \mapsto \mathbb{C}$, is called a *reparameterization* of a curve Γ, defined by $\gamma = [a, b] \mapsto \mathbb{C}$ if there exists a function $\alpha \in C^1$, $\alpha : [a, b] \mapsto [\tilde{a}, \tilde{b}]$ with $\alpha'(t) > 0$, $\alpha(a) = \tilde{a}$ and $\alpha(b) = \tilde{b}$, such that $\gamma(t) = \widetilde{\gamma}(\alpha(t))$, $t \in [a, b]$. Then

$$\int_{\Gamma} f = \int_{\widetilde{\Gamma}} f$$

for any C-function f defined on an open set containing the image of γ (which is equal to the image of $\widetilde{\gamma}$).

2.2.1 Hölder Condition. Let $f(t)$ be defined on a Jordan curve Γ (open or closed). If

$$|f(t_1) - f(t_2)| \le A\, |t_1 - t_2|^{\alpha}, \quad 0 < \alpha \le 1, \tag{2.2.2}$$

for arbitrary points $t_1, t_2 \in \Gamma$ $(t_1 \ne t_2)$, where $A > 0$ and α are real constants, then $f(t)$ is said to satisfy the *Hölder condition of order* α, or simply $f(t)$ satisfies the condition H^α, denoted by $f(t) \in H^\alpha$. The condition H^1 is known as the *Lipschitz condition.* If $f(t) \in C(\Gamma)$ and $f(t) \in H^\alpha$, then we say that $f(t)$ is *Hölder-continuous* on Γ. If $f \in C(\Gamma)$ and $f \in H^1$, then $f(t)$ is said to be *Lipschitz-continuous.*

2.3 Metric Spaces

Let S denote the set of all real-valued sequences. Then S is a vector space if addition and scalar multiplication of vectors $s_i,\ t_i \in S$ for $i = 1, 2, \dots$ are defined coordinatewise, i.e., $\{s_i\} + \{t_i\} = \{s_i + t_i\}$, and $\lambda\{s_i\} = \{\lambda s_i\}$ for a scalar λ. The *Fréchet metric* (distance) $d(\{s_i\}, \{t_i\})$ is defined for $\{s_i\}, \{t_i\} \in S$ by

$$d(\{s_i\}, \{t_i\}) = \sum_{i=1}^{\infty} \frac{|s_i - t_i|}{(1 + |s_i - t_i|)\, 2^i}. \tag{2.3.1}$$

The space ℓ^p, $1 \le p < \infty$, is defined by $\ell^p = \Big\{\{s_i\} : \sum\limits_{i=1}^{\infty} |s_i|^p < \infty\Big\}$. Since the sum of any two elements in ℓ^p is also in ℓ^p, then ℓ^p is a vector subspace of S. Define a norm $\|\cdot\|_p$ on ℓ^p by

$$\|\{s_i\}\|_p = \Big(\sum_{i=1}^{\infty} |s_i|^p\Big)^{1/p}. \tag{2.3.2}$$

Then it can be shown that ℓ^p is closed under vector addition and $\|\cdot\|_p$ satisfies the triangle inequality. Let $\{s_i\}$, $\{t_i\} \in \ell^p$ be such that $\sum |s_i|^p = 1$ and $\sum |t_i|^q = 1$, where p and q are called conjugate exponents such that $p > 1$ and $\dfrac{1}{p} + \dfrac{1}{q} = 1$. Then

Hölder inequality:

$$\sum_{i=1}^{\infty} |s_i\, t_i| \le \Big(\sum_{i=1}^{\infty} |s_i|^p\Big)^{1/p} \Big(\sum_{i=1}^{\infty} |t_i|^q\Big)^{1/q}. \tag{2.3.3}$$

Cauchy-Schwarz inequality:

$$\sum_{i=1}^{\infty} |s_i\, t_i| \le \sqrt{\sum_{i=1}^{\infty} |s_i|^2}\sqrt{\sum_{i=1}^{\infty} |t_i|^2}. \tag{2.3.4}$$

Minkowsky's inequality:

$$\Big(\sum_{i=1}^{\infty} |s_i + t_i|^p\Big)^{1/p} \le \Big(\sum_{i=1}^{\infty} |s_i|^p\Big)^{1/p} + \Big(\sum_{i=1}^{\infty} |t_i|^p\Big)^{1/p}, \quad p \ge 1. \tag{2.3.5}$$

Note that $p = 1$ gives the *triangle inequality*.

Let L^2 denote the Hilbert space of all square-integrable analytic functions f in a simply connected region D with boundary Γ. A function $f(z)$ regular in D is said to belong to the class $L^2(D)$ (we say, $f \in L^2(D)$) if the integral $\displaystyle\iint_D |f(z)|^2\, dS_z < +\infty$, where $dS_z = dx\, dy$ denotes an area element in D. The (surface)-norm $\|f\|_{2,D}$ of $f(z)$ is defined by

$$\|f\|_{2,D}^2 = \iint_D |f(z)|^2\, dS_z. \tag{2.3.6}$$

If two functions $f, g \in L^2(D)$, then their inner product is defined by

$$\langle f, g\rangle = \iint_D f(z)\,\overline{g(z)}\, dS_z. \tag{2.3.7}$$

Let Γ be the rectifiable Jordan boundary of the region D, of length l, and let Γ_ρ denote the image of the circle $|w| = \rho$ under the mapping $z = g(w) = f^{-1}(w)$, $0 < \rho < R$. If $f(z)$ is regular in D, then the integral

$$\int_{\Gamma_r} |f(z)|^2 \, ds_z = \rho \int_{\phi=0}^{2\pi} \left| f(g(w)) \sqrt{g'(w)} \right|^2 d\phi, \quad w = \rho \, e^{i\phi}, \tag{2.3.8}$$

where ds_z denotes a line element on Γ, is a monotone increasing function. We say that $f \in L^2(\Gamma)$ if this integral remains bounded as $\rho \to R$, i.e.,

$$\lim_{\rho \to R} \int_{\Gamma_\rho} |f(z)|^2 \, ds_z = \int_\Gamma |f(z)|^2 \, ds_z = \|f\|^2_{2,\Gamma},$$

where $\|f\|^2_{2,\Gamma}$ is the (line)-norm of f. For any two functions $f, g \in L^2(\Gamma)$, we define their inner product as

$$\langle f, g \rangle = \int_\Gamma f(z) \, \overline{g(z)} \, ds_z. \tag{2.3.9}$$

If the disk $B(z_0, r) \subset D$, then

$$|f(z_0)|^2 = \begin{cases} \dfrac{1}{\pi r^2} \iint_D |f(z)|^2 \, dS_z, & \text{if } f \in L^2(D), \\ \dfrac{1}{2\pi r} \int_\Gamma |f(z)|^2 \, ds_z, & \text{if } f \in L^2(\Gamma). \end{cases} \tag{2.3.10}$$

Let a region have a piecewise continuous boundary Γ, and let $f(z), g(z)$ be regular in D, and $f'(z)$ and $g'(z)$ continuous in $\bar{D}$. Then

$$\langle f, g' \rangle = \iint_D f(z) \, \overline{g'(z)} \, dS_z = \frac{1}{2i} \int_\Gamma f(z) \, \overline{g(z)} \, ds_z. \tag{2.3.11}$$

This is known as *Green's formula*. It is useful in converting a surface integral into a line integral. This formula also holds for multiply connected regions. The integral along Γ is taken in the positive sense, i.e., the region D remains to the left as one traverses the contour Γ. The following inequality is also useful: If $f \in L^2(D)$, then $f \in L^2(\Gamma)$, and

$$\iint_D |f(z)|^2 \, dS_z \le \frac{l}{2} \int_\Gamma |f(z)|^2 \, ds_z. \tag{2.3.12}$$

We will denote by L^∞ the Hilbert space of 2π-periodic and bounded functions f with the norm

$$\|f\|_\infty = \max_{[0,2\pi]} |f(x)|, \tag{2.3.13}$$

and by W the Sobolev space of 2π-periodic and absolutely continuous functions f, with $f' \in L^2[0, 2\pi]$. Then

$$\|f\| = \max \left(\|f\|_\infty, \|f'\|_2 \right),$$

and $(W, \|\cdot\|)$ is a Banach space. Thus, if $f, g \in (W, \|\cdot\|)$, then (i) $\|f\| \ge 0$; (ii) $\|f\| = 0$ implies that $f = 0$; (iii) $\|\alpha f\| = |\alpha| \, \|f\|$ for a scalar α; and $\|f + g\| \le \|f\| + \|g\|$ which is known as the triangle inequality.

2.4 Basic Theorems

Some basic theorems from the theory of functions of a complex variable are presented without proofs which can be found in standard textbooks on the subject.

Theorem 2.1. (Cauchy's theorem) *Let f be analytic on a region D, and let Γ be a closed contour which is homotopic to a point in D. Then* $\int_\Gamma f = 0$.

Note that a set is said to be simply connected if every closed contour $\Gamma \subset D$ is homotopic (as a closed curve) to a point in D, i.e., to some constant curve. Some local versions of Cauchy's theorem are as follows:

Theorem 2.2. (Cauchy-Goursat theorem for a disk) *Let $f : B \mapsto \mathbb{C}$ be analytic on a disk $B = B(z_0, r) \subset D$. Then (i) there exists a function $F : B \mapsto \mathbb{C}$ which is analytic on B and is such that $F'(z) = f(z)$ for all $z \in B$ (i.e., f has an antiderivative on B); and (ii) If Γ is a closed contour in B, then* $\int_\Gamma f = 0$.

This theorem also holds if f is continuous on B and analytic on $D \backslash \{z_1\}$ for some fixed $z_1 \in D$.

Theorem 2.3. (Cauchy-Goursat theorem for a rectangle) *Let R denote a rectangle with sides parallel to the coordinate axes, and let f be a function defined and analytic on an open set D containing R. Then* $\int_\Gamma f = 0$.

Even if the function f is analytic on D except at some fixed point $z_1 \in D$ which does not lie on the contour R, and if $\lim_{z \to z_1} (z - z_1) f(z) = 0$, then also $\int_\Gamma f = 0$. The two theorems 1.2.2 and 1.2.3 hold (a) if f is bounded on a deleted neighborhood of z_1, or (b) if f is continuous on D, or (c) if $\lim_{z \to z_1} f(z)$ exists.

The *index* of a curve Γ with respect to a point $z_0 \in \mathbb{C}$ is the integer n that expresses how many times Γ winds around z_0. This index is denoted by $I(\Gamma, z_0)$ and is called the *winding number* of Γ with respect to z_0. Thus,

$$I(\Gamma, z_0) = \frac{1}{2\pi i} \int_\Gamma \frac{dz}{z - z_0}. \tag{2.4.1}$$

In fact,

$$I(\Gamma, z_0) = \begin{cases} \pm n & \text{if } z_0 \in \text{Int}(\Gamma), \\ 0 & \text{if } z_0 \in \text{Ext}(\Gamma). \end{cases}$$

Theorem 2.4. (Cauchy's integral formula) *Let f be analytic on a region D, and let Γ be a simple closed contour in D that is homotopic to a point in D. Let $z_0 \in D$ be a point not on Γ. Then*

$$f(z_0) \cdot I(\Gamma, z_0) = \frac{1}{2\pi i} \int_\Gamma \frac{f(\zeta)}{\zeta - z_0} \, d\zeta. \tag{2.4.2}$$

The integrand in (2.4.2) is known as the *Cauchy kernel* defined by

$$H(z, z_0) = \frac{1}{2\pi i} \frac{f(z)}{z - z_0}. \tag{2.4.3}$$

The formula (2.4.2) is a special case of integrals of the Cauchy type. If we set

$$F(z) = \frac{1}{2\pi i} \int_\Gamma \frac{g(\zeta)}{\zeta - z} \, d\zeta, \tag{2.4.4}$$

where $\Gamma : [a, b] \mapsto \mathbb{C}$ is a simple contour and g a C-function defined on the image $\Gamma([a, b])$, then F is analytic on $\mathbb{C}\backslash\Gamma([a, b])$ and is infinitely differentiable, such that its kth derivative is given by

$$F^{(k)}(z) = \frac{k!}{2\pi i}\int_\Gamma \frac{g(\zeta)}{(\zeta - z)^{k+1}}\, d\zeta, \quad k = 1, 2, \dots . \tag{2.4.5}$$

Then Cauchy's integral formula for the derivatives is

$$f^{(k)}(z) \cdot I(\Gamma, z_0) = \frac{k!}{2\pi i}\int_\Gamma \frac{f(\zeta)}{(\zeta - z)^{k+1}}\, d\zeta, \quad k = 1, 2, \dots . \tag{2.4.6}$$

Let f be analytic on a region D and let $\Gamma = \partial B(z_0, R)$ be a circle lying in D. If $|f| \le M$ for all $z \in \Gamma$, then for $k = 0, 1, \dots$, we have

$$|f^{(k)}(z_0)| \le \frac{k!}{R^k}\, M. \tag{2.4.7}$$

This result is known as *Cauchy's inequality*. A corollary is the *Liouville theorem* which states that the only bounded entire functions are constants. A partial converse of Cauchy's theorem is known as *Morera's theorem* which states that if f is continuous on a region D and if $\displaystyle\int_\Gamma f\, dz = 0$ for every closed contour Γ in D, then f is analytic on D, and $f = F'$ for some analytic function F on D.

Two very useful corollaries of Cauchy's integral formula (2.4.2) are:
(i) The *maximum modulus theorem* which states that if f is a nonconstant analytic function on a region D with a simple boundary Γ, then $|f|$ cannot have a local maximum anywhere in $\text{Int}(\Gamma)$.
(ii) The *mean value property* of an analytic function f defined on the circle $\partial B(z_0, r)$ states that

$$f(z_0) = \frac{1}{2\pi}\int_0^{2\pi} f\left(z_0 + re^{i\theta}\right)\, d\theta. \tag{2.4.8}$$

An application of the maximum modulus theorem is

Lemma 2.1. (Schwarz lemma) *Let f be analytic on the open unit disk U, and suppose that $|f(z) \le 1$ for all $z \in U$ and $f(0) = 0$. Then $|f(z)| \le |z|$ for all $z \in U$ and $|f'(0)| \le 1$. If $|f(z_0)| = |z_0|$ for some $z_0 \in U$, $z_0 \ne 0$, then $f(z) = cz$ for all $z \in U$, where c is some constant such that $|c| = 1$.*

A consequence of the Schwarz lemma is: If $f(z)$ is analytic on the disk $|z| < R$ such that $f(0) = 0$, and $|f(z)| \le M$, where $M > 0$ is a constant, then $|f(z)| \le \dfrac{M}{R}|z|$, $|f'(0)| \le \dfrac{M}{R}$ on this disk, where the equality holds only for $f(z) = e^{i\alpha}\dfrac{M}{R}\, z$, α real (Wen [1992:26-27]).

2.4.1 Singularities. Let f be analytic on a deleted neighborhood of z_0: $\{z : 0 < |z - z_0| < r\}$. In such a case we say that z_0 is an *isolated singularity* of f, and the Laurent series expansion of f at z_0 is

$$f(z) = \cdots + \frac{b_n}{(z - z_0)^n} + \cdots + \frac{b_1}{z - z_0} + a_0 + a_1(z - z_0) + \cdots, \quad 0 < |z - z_0| < r. \tag{2.4.9}$$

If z_0 is an isolated singularity of f and if all but a finite number of b_n in the expansion (2.4.9) are zeros, then z_0 is called a *pole* of f, and if $k \ne 0$ is the highest integer such that

$b_k \neq 0$, then z_0 is called a *pole of order* k. In this case $f(z) = \dfrac{g(z)}{(z-z_0)^n}$, where $g(z)$ is analytic at $z = z_0$ and $g(z_0) \neq 0$. If z_0 is a pole of order 1, it is called a *simple pole*. If infinitely many b_n are nonzero, then z_0 is called an *essential singularity*. A singularity is *essential* if it is not a pole or a branch point; for example, $e^{1/z}$ at $z_0 = 0$.

If f has a removable singularity at z_0, then $f(z) = \sum_{n=0}^{\infty} a_n (z-z_0)^n$ is a convergent power series, where $f(z_0) = a_0$ which is analytic at z_0.

Cauchy's Formula. The deleted neighborhood $0 < |z-a| < r$ is a multiply connected domain, and the function $\dfrac{1}{(z-a)^n}$, $n = 1, 2, \dots$, is analytic on this domain, but not on the whole disk $|z-a| < r$. Then we have

$$\oint_C \frac{dz}{(z-a)^n} = ir^{1-n} \int_0^{2\pi} e^{i(1-n)\theta}\, d\theta = \begin{cases} ir^{1-n} \dfrac{e^{i(1-n)\theta}}{i(1-n)}\Big|_0^{2\pi} = 0 & \text{if } n \neq 1, \\ i\theta\Big|_0^{2\pi} = 2\pi i & \text{if } n = 1. \end{cases} \tag{2.4.10}$$

Branch Points. There are *algebraic branch points* of degree n as in $f(z) = \sqrt[n]{z}$ at $z = z_0$; and there are *logarithmic branch points* of degree ∞ such as $\log z$ at $z = z_0$. In general, the power function $z^a = e^{a \log z}$ is analytic at $z = z_0$ if $a \in \mathbb{Z}$ is an integer; it has an *algebraic branch point* of degree q at the origin if $a = p/q$ is rational, non-integral with $0 \neq p \in \mathbb{Z}$ and $2 \leq q \in \mathbb{Z}$ with no common factor, and is a *logarithmic branch point* of degree ∞ at $z = 0$ for a not rational.

2.4.2 Residues. If f has an isolated singularity at z_0, then the Laurent series expansion of f in a deleted neighborhood of z_0 is given by (2.4.9). The coefficient b_1 is called the *residue* of f at z_0, written as $b_1 = \text{Res}\,(f, z_0)$. If f is analytic in a region $D \in \mathbb{C}$, and if Γ is any circle around z_0 whose interior lies in D, then

$$\int_\Gamma f(z)\, dz = 2\pi i\, b_1. \tag{2.4.11}$$

As an example, using the Laurent series expansion $e^{1/z} = 1 + \dfrac{1}{z} + \dfrac{1}{z^2} + \cdots + \dfrac{1}{n! z^n} + \cdots$, we have $\text{Res}\{e^{1/z}\} = 1$ at $z_0 = 0$.

Theorem 2.5. *If f is analytic in a region $D \in \mathbb{C}$ and has an isolated singularity at $z_0 \in D$, then*

(a) z_0 is an isolated singularity if any one of the following conditions hold: (i) f is bounded in a deleted neighborhood of z_0; (ii) $\lim_{z\to z_0} f(z)$ exists; or (iii) $\lim_{z\to z_0} (z-z_0) f(z) = 0$.

(b) z_0 is a simple pole iff $\lim_{z\to z_0}$ exists and is nonzero; this limit is equal to the residue of f at z_0.

(c) z_0 is a pole of order $\leq k$ (or possibly a removable singularity) iff any one of the following conditions hold: (i) There is a constant $M > 0$ and an integer $k \geq 1$ such that $|f(z)| \leq \dfrac{M}{|z-z_0|^k}$ in a deleted neighborhood of z_0; (ii) $\overset{k+1}{\lim_{z\to z_0}} f(z) = 0$; or (iii) $\overset{k}{\lim_{z\to z_0}} f(z)$ exists.

(d) z_0 is a pole of order $k \geq 1$ iff there is an analytic function g defined on a neighborhood A of z_0 defined by $A \backslash \{z_0\} \subset D$, such that $g(z_0) \neq 0$, and such that $f(z) = \dfrac{g(z)}{(z - z_0)^k}$ for $z \in A$, $z \neq z_0$.

Let f be analytic in a region D and let $z_0 \in D$. Then f has a *zero of order* k at z_0 iff $f(z_0) = 0, \dots, f^{(k-1)}(z_0) = 0, f^{(k)}(z_0) \neq 0$. It is obvious from Taylor's series of f at z_0 that f can be written as $f(z) = (z - z_0)^k g(z)$, where $g(z)$ is analytic at z_0, and $g(z_0) = f^{(k)}(z_0)/k! \neq 0$.

It follows from Theorem 2.5(d) that if $f(z)$ has a pole (of finite order k) at z_0, then $|f(z)| \to \infty$ as $z \to z_0$. However, if f has an essential singularity at z_0, then $|f|$ will not, in general, approach ∞ as $z \to z_0$. In fact, by Picard's theorem (Picard [1879]), if f has an essential singularity at z_0 and if A is any arbitrary small deleted neighborhood of z_0, then for all $\zeta \in \mathbb{C}$, $\zeta \neq z_0$, the equation $f(z) = \zeta$ has infinitely many solutions in A except possibly for one value.

Theorem 2.6. (Casorati-Weierstrass theorem) *Let f have an essential singularity a z_0 and let $w \in \mathbb{C}$. Then there exist $z_1, z_2, \dots, z_n \to z_0$, such that $f(z_n) \to w$.*

Example 2.2. (i) The function $f(z) = \dfrac{\sin z}{z}$ has removable singularity at $z = 0$, because $\lim\limits_{z \to 0} z \cdot (\sin z)/z = \lim\limits_{z \to 0} \sin z = 0$. (ii) The function $f(z) = e^z/z$ has a simple pole at $z = 0$, because $\lim\limits_{z \to 0} z \cdot e^z/z = 1$; this singularity is not removable. (iii) The function $f(z) = (e^z - 1)^2/z$ has a removable singularity at $z = 0$, because $\lim\limits_{z \to 0} z \cdot (e^z - 1)^2/z = 1$. (iv) The function $f(z) = z/(e^z - 1)$ has a removable singularity, because $(e^z - 1)/z = 1 + z/2! + z^2/3! + \cdots \to 1$ as $z \to 0$. (v) If $f(z)$ and $g(z)$ are analytic, both having zeros of order k at z_0, then $f(z)/g(z)$ has a removable singularity at z_0 and $\lim\limits_{z \to z_0} \dfrac{f(z)}{g(z)} = \dfrac{f^{(k)}(z_0)}{g(k)(z_0)}$. (vi) If f is analytic in a region containing a circle Γ and its interior and has a simple zero at z_0 inside or on Γ, then $z_0 = \dfrac{1}{2\pi i} \int_\Gamma \dfrac{z f'(z)}{f(z)}\, dz$.

Also, $f(z) = \dfrac{e^z (z-3)}{(z-1)(z-5)}$ has simple pole at $z = 1$ and $z = 5$.

$f(z) = \dfrac{\cos z}{1 - z}$ has simple pole at $z = 1$. ■

Since it is not convenient to develop the Laurent series expansion of all functions $f(z)$ with all types of singularities, it is desirable to directly develop techniques to determine residues of f.

Case 1. Since f has a removable singularity at z_0 iff $\lim\limits_{z \to z_0} (z - z_0) f(z) = 0$, we have

Theorem 2.7. *If $g(z)$ and $h(z)$ are analytic functions, both having zeros at z_0 of the same order, then $f(z) = \dfrac{g(z)}{h(z}$ has a removable singularity at z_0.*

Some examples are: (i) $e^z/(z-1)$has no singularity at $z_0 = 0$; (ii) $(e^z - 1)/z$ has a removable singularity at 0 because both $e^z - 1$ and z have zeros of order 1 (although each of then vanishes at $z = 0$ but their derivatives do not); and (iii) $z^2/\sin z$ has a removable singularity at $z_0 = 0$ because both numerator and denominator have zeros of order 2.

Case 2. Pole of order k.

Theorem 2.8. *Let f have an isolated singularity at z_0 and let k be the smallest non-negative integer such that $\lim_{z \to z_0} (z - z_0)^k f(z)$ exists. Then $f(z)$ has a pole of order k at z_0. Let $\phi(z) = (z - z_0)^k f(z)$, such that ϕ is defined uniquely at z_0 and analytic at z_0. Then*

$$\mathrm{Res}\,(f, z_0) = \frac{\phi^{(k-1)}(z_0)}{(k-1)!}. \tag{2.4.12}$$

Theorem 2.9. *Let g and h be analytic at z_0, with $g(z_0) \neq 0$, and let $h(z_0) = 0 = \cdots = h^{(k-1)}(z_0)$, $h^{(k)}(z_0) \neq 0$. Then the function $g(z)/h(z)$ has a pole of order k and the residue at z_0 is given by*

$$\mathrm{Res}(g/h, z_0) = \left[\frac{k!}{h^{(k)}(z_0)}\right]^k \times$$
$$\begin{vmatrix} \frac{h^{(k)}(z_0)}{k!} & 0 & 0 & \cdots & 0 & g(z_0) \\ \frac{h^{(k+1)}(z_0)}{(k+1)!} & \frac{h^{(k)}(z_0)}{k!} & 0 & \cdots & 0 & g'(z_0) \\ \frac{h^{(k+2)}(z_0)}{(k+2)!} & \frac{h^{(k+1)}(z_0)}{(k+1)!} & \frac{h^{(k)}(z_0)}{k!} & \cdots & 0 & \frac{g''(z_0)}{2!} \\ \vdots & \vdots & \vdots & \cdots & \vdots & \vdots \\ \frac{h^{(2k-1)}(z_0)}{(2k-1)!} & \frac{h^{(2k-2)}(z_0)}{(2k-2)!} & \frac{h^{(2k-3)}(z_0)}{(2k-3)!} & \cdots & \frac{h^{(k+1)}(z_0)}{(k+1)!} & \frac{g^{(k-1)}(z_0)}{(k-1)!} \end{vmatrix}. \tag{2.4.13}$$

Example 2.3. Let $f(z) = \dfrac{e^z}{\sin^3 z}$. Here $g(z) = e^z, h(z) = \sin^3 z, k = 3$. Then by formula (2.4.13),

$$\mathrm{Res}\left(e^z/\sin^3 z, 0\right) = \frac{3!}{6}\begin{vmatrix} 1 & 0 & 1 \\ 0 & 1 & 1 \\ -\frac{1}{2} & 0 & \frac{1}{2} \end{vmatrix} = \begin{vmatrix} 0 & 0 & 1 \\ -1 & 1 & 1 \\ -1 & 0 & \frac{1}{2} \end{vmatrix} = 1.$$

Case 3. (Essential singularity at z_0) In the case of essential singularity at z_0 there is no simple formula except finding the Laurent series expansion for $f(z)$.

Example 2.4. Since

$$\begin{aligned} f(z) = e^{z+1/z} = e^z \cdot e^{1/z} &= \left(1 + z + \frac{z^2}{2!} + \cdots\right)\left(1 + \frac{1}{z} + \frac{1}{2!z^2} + \cdots\right) \\ &= \frac{1}{z}\left\{1 + \frac{1}{2!} + \frac{1}{2!3!} + \frac{1}{3!4!} + \cdots\right\}, \end{aligned}$$

which gives

$$\mathrm{Res}(f, 0) = 1 + \frac{1}{2!} + \frac{1}{2!3!} + \frac{1}{3!4!} + \cdots,$$

where the resulting series cannot be summed explicitly. ■

A summary of methods to determine residues is given in Table 2.1.

Table 2.1: Summary of methods to determine residues.

Function	Test	Singularity Type	Residue at z_0
1. $f(z)$	$\lim_{z\to z_0} (z-z_0)f(z) = 0$	removable	0
2. $\frac{g(z)}{h(z)}$	g and h have zeros of same order	removable	0
3. $f(z)$	$\lim_{z\to z_0} (z-z_0)f(z)$ exists and is not zero	simple pole	$\lim_{z\to z_0} (z-z_0)f(z)$
4. $\frac{g(z)}{h(z)}$	$g(z_0) \neq 0, h(z_0) = 0,$ $h'(z_0) \neq 0$	simple pole	$\frac{g(z_0}{h'(z_0)}$
5. $\frac{g(z)}{h(z)}$	g has zero of order k, h has zero of order $k+1$	simple pole	$(k+1)\frac{g^{(k)}(z_0)}{h^{(k+1)}(z_0)}$
6. $\frac{g(z)}{h(z)}$	$g(z_0) \neq 0, h(z_0) = 0 =$ $= h'(z_0), h''(z_0) \neq 0$	pole of order 2	$2\frac{g'(z_0)}{h''(z_0)} - \frac{2}{3}\frac{g(z_0)h'''(z_0)}{[h''(z_0)]^2}$
7. $\frac{g(z)}{(z-z_0)^2}$	$g(z_0) \neq 0$	pole of order 2	$g'(z_0)$
8. $\frac{g(z)}{h(z)}$	$g(z_0) = 0, g'(z_0) \neq 0,$ $h(z_0) = 0,\ h'(z_0) =$ $= h''(z_0) = 0, h'''(z_0) \neq 0$	pole of order 2	$3\frac{g''(z_0)}{h'''(z_0)} - \frac{3}{2}\frac{g'(z_0)h^{(iv)}(z_0)}{[h'''(z_0)]^2}$
9. $f(z)$	k is the smallest integer such that $\lim_{z\to z_0} \phi(z_0)$ exists, where $\phi(z) = (z-z_0)^k f(z)$	pole of order k	$\lim_{z\to z_0} \frac{\phi^{(k-1)}(z)}{(k-1)!}$
10. $\frac{g(z)}{h(z)}$	g has zero of order l; and h has zero of order $(k+l)$	pole of order k	$\lim_{z\to z_0} \frac{\phi^{(k-1)}(z)}{(k-1)!}$, where $\phi(z) = (z-z_0)^k\left(f(z)/h(z)\right)$
11. $\frac{g(z)}{h(z)}$	$g(z_0) \neq 0, h(z_0) = \cdots =,$ $h^{(k-1)}(z_0) = 0,\ h^{(k)}(z_0) \neq 0$	pole of order k	residue given by formula (2.4.13) in Theorem 2.9

2.4.3 Boundary Values for Cauchy Integral. If the boundary Γ of a simply connected region D is rectifiable and if $f(z)$ is regular in D and continuous on $\bar{D}$, then Cauchy's integral formula (2.4.2) holds for every $z_0 \in D$. For the boundary values we have the following result: Let the boundary Γ of D be smooth, and let $f(z)$ be regular in D and continuous on $\bar{D}$. Then for a point $z_0 \in \Gamma$ which is not a corner point

$$f(z_0) = \frac{1}{\pi i}\int_\Gamma \frac{f(\zeta)}{\zeta - z_0}\,d\zeta. \tag{2.4.14}$$

In the case when $z_0 \in \Gamma$ is a corner point with inner angle $\alpha\pi$, $0 < \alpha < 2$, the boundary value $f(z_0)$ is given by

$$f(z_0) = \frac{1}{\alpha\pi i} \int_\Gamma \frac{f(\zeta)}{\zeta - z_0}\, d\zeta, \tag{2.4.15}$$

where the integral is taken as a Cauchy p.-v. integral. If D is the region exterior to the Jordan curve Γ which is traversed in the positive sense and if $f(z)$ is regular at $z = \infty$, then the boundary value at a point $z_0 \in \Gamma$ which is not a corner point is given by

$$f(z_0) = -\frac{1}{\pi i} \int_\Gamma \frac{f(\zeta)}{\zeta - z_0}\, d\zeta + 2\, f(\infty), \tag{2.4.16}$$

and at a corner point $z_0 \in \Gamma$ with the inner angle α, $0 < \alpha < 2$, by

$$f(z_0) = \frac{1}{(2-\alpha)\pi i} \int_\Gamma \frac{f(\zeta)}{\zeta - z_0}\, d\zeta + 2\, f(\infty). \tag{2.4.17}$$

The integrals in (2.4.14)-(2.4.17) are taken as Cauchy principal-values (see Appendix B for details on p.-v. integrals)

2.4.4 Argument Principle. (Cauchy's argument principle) This principle relates to the difference between the number of zeros and poles of a meromorphic function to a contour integral of the function's logarithmic derivative, and is defined by

$$\oint_\Gamma \frac{f'(z)}{f(z)}\, dz = 2\pi i(N - P), \tag{2.4.18}$$

where N is the number of zeros and P the number of poles, each counted according to its multiplicity. In other words, it is the number of windings around the point $w = 0$ made by the point $w = f(z)$ as z traverses Γ once in the counterclockwise direction equals $N - P$.

2.4.5 Plemelj Formulas. If Γ is a contour, then

$$\begin{aligned} F^+(\zeta_0) &= \frac{1}{2}\, f(\zeta_0) + \frac{1}{2\pi i} \int_\Gamma \frac{f(\zeta)}{\zeta - \zeta_0}\, d\zeta, \\ F^-(\zeta_0) &= -\frac{1}{2}\, f(\zeta_0) + \frac{1}{2\pi i} \int_\Gamma \frac{f(\zeta)}{\zeta - \zeta_0}\, d\zeta, \end{aligned} \tag{2.4.19}$$

where $F^+(\zeta_0)$ and $F^-(\zeta_0)$ are the limiting values from the right and left of Γ, respectively, $f(\zeta)$ satisfies the Hölder condition on Γ, and ζ_0 does not coincide with those end points where $f(\zeta_0) \neq 0$. If ζ_0 coincides with an end point where $f(\zeta_0) = 0$, then $F^+(\zeta_0) = F^-(\zeta_0) = f(\zeta_0)$. A proof for these formulas can be found in Muskhelishvili [1953/1992].

Another set of useful formulas is as follows:

$$\sin \pi z = \pi z \prod_{n=1}^{\infty} \left(1 - \frac{z^2}{n^2}\right), \tag{2.4.20}$$

$$\pi \cot \pi z = \sum_{n=-\infty}^{\infty} \frac{1}{z-n} = \frac{1}{z} + \sum_{n=1}^{\infty} \left(\frac{1}{z-n} + \frac{1}{z+n}\right), \tag{2.4.21}$$

where the series converges in the Cauchy sense.

2.5 Harmonic Functions

The functions whose Laplacian is zero are known as harmonic functions. Thus, a real-valued function $u(x, y) \in C^2(D)$ is said to be *harmonic* in a region D if $\nabla^2 u = 0$. Some properties of harmonic functions in $\mathbb{R}^2$ are as follows:

(i) The function

$$\frac{1}{r} = \frac{1}{\sqrt{(x - x_0)^2 + (y - y_0)^2}} \tag{2.5.1}$$

is harmonic in a region that does not contain the point (x_0, y_0).

(ii) If $u(x, y)$ is a harmonic function in a simply connected region D, then there exists a conjugate harmonic function $v(x, y)$ in D such that $u(x, y) + i\, v(x, y)$ is an analytic function of $z = x + iy = (x, y)$ in D. In view of the Cauchy-Riemann equations (2.1.6),

$$v(x, y) - v(x_0, y_0) = \int_{x_0, y_0}^{x, y} \left(-u_y\, dx + u_x\, dy\right), \tag{2.5.2}$$

where $(x_0, y_0) = z_0$ is a given point in D. This property is also true if D is multiply connected. However, in that case the conjugate function, $v(x, y)$ can be multiple-valued, as we see by considering $u(x, y) = \log r = \log \sqrt{x^2 + y^2}$ defined on a region D containing the origin which has been indented by a small circle centered at the origin. Then, in view of (2.5.2),

$$v(x, y) - v(x_0, y_0) = \tan^{-1} \frac{y}{x} \pm 2n\pi + \text{const}, \quad n = 1, 2, \dots,$$

which is multiple-valued.

(iii) Since derivatives of all orders of an analytic function exist and are themselves analytic, any harmonic function will have continuous partial derivatives of all orders, i.e., a harmonic function belongs to the class $C^\infty(D)$, and a partial derivative of any order is again harmonic.

(iv) A harmonic function must satisfy the mean-value theorem, where the mean value at a point is evaluated for the circumference or the area of the circle around that point. If u is harmonic on a region containing the closed disk $\bar{B}(z_0, r)$, where $z_0 = x_0 + iy_0$, then

$$u(x_0, y_0) = \frac{1}{2\pi} \int_0^{2\pi} u\left(z_0 + r\, e^{i\theta}\right) d\theta. \tag{2.5.3}$$

(v) In view of the maximum modulus theorem (§1.2), the maximum (and also the minimum) of a harmonic function u in a region D occurs only on the boundary of D. This result is known as

Theorem 2.10. (Maximum Principle) *A nonconstant function which is harmonic inside a bounded region D with boundary Γ and continuous in the closed region $\bar{D} = D \cup \Gamma$ attains its maximum and minimum values only on the boundary of the region.*

Thus, u has a maximum (or minimum) at $z_0 \in D$, i.e., if $u(z) \le u(z_0)$ (or $u(z) \ge u(z_0)$) for z in a neighborhood $B(z_0, \varepsilon)$ of z_0, then $u = $ const in $B(z_0, \varepsilon)$.

(vi) The value of a harmonic function u at an interior point in terms of the boundary values u and $\dfrac{\partial u}{\partial n}$ is given by Green's third identity (A.8).

(vii) If u and U are continuous in $\bar{D}$ and harmonic in D such that $u \le U$ on Γ, then $u \le U$ also at all points inside D. In fact, the function $U - u$ is continuous and harmonic in D.

Hence $U - u \geq 0$ on Γ. Then, in view of the maximum principle (Theorem 2.10), we require that $U - u \geq 0$ at all points inside D.

(viii) If u and U are continuous in $\bar{D}$ and harmonic in D for which $|u| \leq U$ on Γ, then $|u| \leq U$ also at all points inside D. In fact, the three harmonic functions $-U$, u, and U satisfy the relation $-U \leq u \leq U$ on Γ. Then, by (vii), $-U \leq u \leq U$ at all points inside D, or $|u| \leq U$ inside D.

Theorem 2.11. (Harnack theorem) *Suppose that $\{u_n(z)\}$ is a monotone increasing sequence of harmonic functions on a region D, which is convergent at a point $z_0 \in D$. Then $\{u_n(z)\}$ is uniformly convergent on closed sets in D.*

Let $z_0 \neq \infty$ be any point inside D, and let K denote the closed disk $\bar{B}(z_0, R)$ such that $K \subset D$. Then, if $r = |z - z_0| \leq R$, the Harnack inequality

$$u_n(z_0)\,\frac{R-r}{R+r} \leq u_n(z) \leq u_n(z_0)\,\frac{R+r}{R-r} \tag{2.5.4}$$

holds for any annulus $0 < r < R$ with center at z_0, provided that $u_n(z)$ are harmonic and nonnegative on the disk $B(z_0, R)$.

Theorem 2.12. (Identity theorem) *If $f(z)$ is analytic on a region D, $z_0 \in D$, and $\{z_k\}_1^\infty$ is a sequence of distinct points in D such that $z_k \to z_0$, and $f(z_k) = 0$ for $k = 1, 2, \dots$, then $f(z) \equiv 0$ on D.*

This theorem is useful in establishing certain identities, including the concept of analytic continuation.

Example 2.5. (i) The function $f(z) = \dfrac{1}{z} = \dfrac{x}{x^2+y^2} - i\dfrac{y}{x^2+y^2}y = 0$ is harmonic everywhere except at the singularity at $x = 0 = y$.

(ii) The function

$$\begin{aligned} f(z) &= \frac{z-1}{z+1} = \frac{(z-1)(\bar{z}+1)}{(z+1)(\bar{z}+1)} = \frac{|z|^2 + z - \bar{z} - 1}{|z+1|^2} \\ &= \frac{x^2+y^2-1}{(x+1)^2+y^2} = i\,\frac{2y}{(x+1)^2+y^2} \end{aligned}$$

is harmonic everywhere except at the singularity at $x = -1, y = 0$.

(iii) The function $f(z) = \log z = \log\left(r\,e^{i\theta}\right) = \log r + i\theta$ has real and imaginary parts as follows: $\Re\{\log z\} = \log r = \log|z| = \frac{1}{2}\log\left(x^2+y^2\right)$, which is harmonic everywhere except at $x = 0 = y$; and $\Im\{\log z\} = \theta = \arg\{z\}$, which is not defined at $x = 0 = y$ and is a multiple-valued harmonic function everywhere else; it is only specified up to integral multiples of 2π. If $z = x > 0$ is real and positive, then $\log z = \log x$ is the real logarithm provided we choose $\arg\{x\} = 0$. However, if $z = x < 0$, then $\log z = \log|x| + (2k+1)\pi i$, $k \in \mathbb{Z}$, is complex for all $\arg\{x\}$. also, $z = 0$ is called the *logarithmic branch point.* ■

2.5.1 Harmonic Conjugate. Write the Cauchy-Riemann equations (2.1.6) as $\dfrac{\partial v}{\partial x} = -\dfrac{\partial u}{\partial y}, \dfrac{\partial v}{\partial y} = \dfrac{\partial u}{\partial x}$. These equations can be rewritten in vector form as an equation of grad v, known as the *skew gradient of* u, i.e.,

$$\nabla v = \nabla^{\perp} u, \quad \text{where} \quad \nabla^{\perp} = \begin{bmatrix} -u_y \\ u_x \end{bmatrix}. \tag{2.5.5}$$

It is orthogonal to grad u everywhere and is of same length, i.e.,

$$\nabla u \cdot \nabla^{\perp} u = 0, \quad \|\nabla u\| = \|\nabla^{\perp} u\|.$$

This means that the gradient of a harmonic function and that of its harmonic conjugate are mutually orthogonal vector fields having the same Euclidean length:

$$\nabla u \cdot \nabla v = 0, \quad \|\nabla u\| = \|\nabla v\|. \tag{2.5.6}$$

Next, $\nabla^{\perp} u$ has a potential function v iff the corresponding line integral is independent of path, i.e., for every closed curve $C \subset D$, where D is simply connected, and every harmonic function u,

$$\begin{aligned} 0 &= \int_C \nabla v \cdot d\mathbf{x} = \int_C \nabla^{\perp} u \cdot d\mathbf{x} = \int_C \nabla u \cdot \mathbf{n}\, ds \\ &= \iint_D \nabla \cdot \nabla u\, dx\, dy = \iint_D \nabla^2 u\, dx\, dy, \end{aligned}$$

by divergence theorem (Appendix A). This proves the existence of harmonic conjugate functions.

The geometrical significance of Eq (2.5.5) is as follows: The gradient ∇u is normal to the level curves of $u(x, y)$, which are the sets $\{u(x, y) = c\}$ where ∇u assumes a fixed constant value. Also, ∇v is orthogonal to ∇u, which is tangent to the level curves of u. Similarly, ∇v is normal to its level curves. Since their tangent directions ∇u and ∇v are orthogonal, the level curves of u and v form a orthogonal system of plane curves, except at a critical point where $\nabla u = 0$. Then $\nabla u = \nabla^{\perp} u = 0$. Thus, orthogonality of level curves does not necessarily hold at critical points, although critical points of u are the same as those of v and also the same as critical points of $f(z)$ where $f'(z) = 0$.

Note that the harmonic conjugate does not exist only on the punctured plane $\Omega = \mathbb{C}\backslash\{0\}$ because of the logarithmic potential. In fact, if $u(x, y)$ is a harmonic function on $\Omega_R = \{0 < |z| < R\}$, where $0 < R \le \infty$, then there exists a constant c such that $\tilde{u}(x, y) = u(x, y) - c \log \sqrt{x^2 + y^2}$ is also harmonic and possesses a single-valued harmonic conjugate $\tilde{v}(x, y)$. Thus, the function $\tilde{f} = \tilde{u} + \tilde{v}$ is analytic on Ω_R and the original function $u(x, y)$ is the real part of the multiple-valued analytic function $f(z) = \tilde{f}(z) + c \log z$. This fact is useful in the analysis of airfoils.

2.5.2 Capacity. Let D^* denote the complement of the region D that includes the point $z = \infty$. Then the region D^* can be covered by regions $D^{*(n)}$ with boundaries $\Gamma^{*(n)}$, $n = 1, 2, \dots$. The Green's function $G_n(z, \infty)$ of the regions $D^{*(n)}$ is a harmonic function in $D^{*(n)}$ except at the point $z = \infty$ which assumes the value zero on $\Gamma^{*(n)}$ and in a neighborhood of ∞ behaves such that $\lim_{z \to \infty} [G_n(z, \infty) - \log |z|] = \gamma_n$ exists (and is finite). The quantity γ_n is called *Robin's constant* for the regions $D^{*(n)}$. Hence, in a neighborhood of $z = \infty$

$$G_n(z, \infty) = \log |z| + u_n(z) + \gamma_n, \tag{2.5.7}$$

where $u_n(z)$ is a harmonic function in $D^{*(n)}$, including the point $z = \infty$, and $u_n(z) \to 0$ as $z \to \infty$. Since $\overline{D^{*(n)}} \subset D^{*(n+1)}$, the function $G_{n+1}(z, \infty) - G_n(z, \infty)$ is harmonic in $D^{*(n)}$, including the point ∞. In view of the maximum principle (Theorem 2.10), the last statement is true everywhere in the region $D^{*(n)}$, i.e., for all $z \in D^{*(n)}$

$$G_{n+1}(z, \infty) \ge G_n(z, \infty), \quad \gamma_{n+1} \ge \gamma_n. \tag{2.5.8}$$

In view of the Harnack theorem, the sequence of functions $\{u_n + \gamma_n\}$, where $u_n(\infty) = 0$, either converges to ∞ in D^* or converges to a harmonic function $u(z)+\gamma$ such that $u(\infty) = 0$. In the latter case, the quantity γ is called Robin's constant for the region D^*, the quantity $C = e^{-\gamma}$ is called the *capacity* of the region D^* (denoted by cap (D^*)), and the function $G(z, \infty)$ is called Green's function for the region D^*, which assumes nonnegative values everywhere in D^*, but these values need not be zero. For example, at isolated boundary points of D^* the value of $G(z, \infty)$ is positive. In the case when $u_n(z) + \gamma_n \to \infty$ for $z \in D^*$, we have $\gamma_n \to \infty$ because $u_n(\infty) = 0$. In this case the capacity of the region D is taken as zero. The following results are useful:

Theorem 2.13. *For the region D^*, containing the point at infinity, to have a Green's function $G(z, \infty)$, it is necessary and sufficient that the capacity of its boundary Γ be positive.*

Theorem 2.14. *The capacity of an arbitrary closed and bounded region D is equal to its transfinite diameter, i.e.,* cap $(D) = \text{diam}(D)$.

2.6 Univalent Functions

If we place a sphere such that the complex plane is tangent to it and the origin coincides with the south pole, then we can transfer all points of $\mathbb{C}$ to the sphere by projection from the north pole. This is called a *stereographic projection* which is a one-to-one map of $\mathbb{C}$ onto the sphere, such that its image is the whole sphere except the north pole which then corresponds to the point at infinity $z = \infty$.

A mapping f of a region D onto a region G is called *analytic* iff it is differentiable. The mapping f is called *conformal* if it is bijective and analytic. The *conformal mapping theorem* states that if a mapping $f : D \mapsto G$ is analytic and $f'(z_0) \neq 0$ for each $z \in D$, then f is conformal. Thus, f rotates tangent vectors to curves through z_0 by a definite angle θ and magnifies (or contracts) them by a factor r. The mapping f is *conformal* if it is analytic with a nonzero derivative. Two important properties are the following:
(i) If $f : D \mapsto G$ is conformal and bijective (i.e., one-to-one and onto), then $f^{-1} : G \mapsto D$ is also conformal and bijective.
(ii) If $f : D \mapsto G$ and $g : G \mapsto E$ are conformal, then the composition $f \circ g : D \mapsto E$ is conformal and bijective.
Property (i) is useful in solving boundary value problems (e.g., the Dirichlet problem) for a region D. The method involves finding a map $f : D \mapsto G$ such that G is a simpler region on which the problem can be first solved, and then the result for the original problem is provided by f^{-1}. Since the Dirichlet problem involves harmonic functions (see §1.3), the following result on the composition of a harmonic function with a conformal map is useful: If u is harmonic on a region G and if $f : D \mapsto G$ is conformal, then $u \circ f$ is harmonic on D. In fact, let $z \in D$ and $w = f(z) \in G$. Then there is an analytic function g on the open disk $B(w, \rho) \subset G$ such that $u = \Re\{g\}$. Thus, $u \circ g = \Re\{g \circ f\}$, which is harmonic since $g \circ f$ is analytic. A mapping in which both the magnitude and the sense of the angles between the curves and their images are preserved is said to be a conformal mapping of the first kind, but if the sense of the angles is reversed, then it is called a conformal mapping of the second kind.

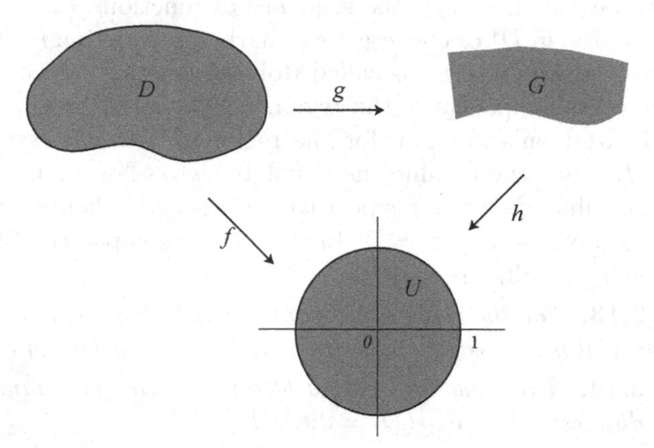

Figure 2.2 Mappings.

Theorem 2.15. (Riemann mapping theorem) *Let $D \subset \mathbb{C}$ be a simply connected region. Then there exists a bijective conformal map $f : D \mapsto U$, where U is the open unit disk. Moreover, the map f is unique provided that $f(z_0) = 0$ and $f'(z_0) > 0$ for $z_0 \in D$.*

A proof of this theorem is available in Appendix C. This theorem implies that if D and G, both contained in $\mathbb{C}$, are any two simply connected regions, then there exists a bijective conformal map $g : D \mapsto G$. If $f : D \mapsto U$ and $h : G \mapsto U$, then $g = h^{-1} \circ f$ is bijective conformal (Figure 2.2). Thus, the two regions D and G are said to be *conformal* if there exists a bijective conformal map between them.

A bijective conformal map is also called a *univalent* map. A function $w = f(z)$ defining a univalent mapping is called a univalent function on a domain D if it is one-to-one and analytic on D. Its inverse image is also a univalent function defined on the image region. In the study of univalent mappings the first question asked is whether a given region can be mapped univalently onto another region. In the case of simply connected regions it is necessary that the two regions have the same connectivity. Once this condition is met, we can inquire about the possibility of univalent conformal mapping of various regions onto a given simply connected region. To determine how a simply connected region is mapped onto a simply connected region, we must know their mappings onto a standard region, such as the unit disk. This enables us to obtain the required mapping first by mapping the given region onto the unit disk and then mapping the unit disk onto the other region. Now the question arises whether an arbitrary simply connected region can be mapped onto the unit disk. It turns out that this is always possible except for two cases, namely, when the region is the entire plane and when it has a single boundary point. Note that the requirement that the boundary orientation be preserved even in the mapping of regions with non-Jordan boundaries is important. Consider, for example, the map $w = \sin z$ which is analytic on the strip $-\pi/4 < x \leq 3\pi/4$ and maps bijectively the two boundaries of this strip onto the two boundaries of the hyperbola $u^2 - v^2 = 1/2$, which make the boundary of a simply connected region (curved strip). But the boundary of this curved strip is not traversed in the positive sense. Other simple examples of functions that are not univalent in the lower or the upper half-plane are $w = z^2$, and $w = \sinh z$, although they are analytic in $\mathbb{C}$ and map bijectively the real axis $\Im\{z\} = 0$ onto the real axis $\Im\{w\} = 0$. Thus, the function $w = f(z)$, analytic on the half-plane $\Im\{z\} > 0$ and continuous on the closed region $\Im\{z\} \geq 0$, grows as $|z| \to \infty$ at most as fast as Cz^2, i.e., the ratio $\dfrac{f(z)}{z^2}$ is bounded for $0 \leq \arg\{z\} \leq \pi$ and a sufficiently large $|z|$.

The Riemann mapping theorem also implies that there exists a unique function $w = F(z)$ that is regular in D, that is normalized at a finite point $z_0 \in D$ by the conditions $F(z_0) = 0$ and $F'(z_0) = 1$, and that maps the region D univalently onto the disk $|w| < 1$. In fact, the function $F(z) = \dfrac{f(z)}{f(z_0)}$ is such a function, where $f(z)$, with $f(z_0) = 0$ and $f'(z_0) > 0$, is the function mentioned in the Riemann mapping theorem, and the radius of the disk onto which the function $w = F(z)$ maps the region D is $R = \dfrac{1}{f'(z_0)}$. If there exists another function $w = F_1(z)$, with $F_1(z_0) = 0$ and $F_1'(z_0) = 1$, that maps D onto a disk $|w| < R_1$, then, by the Riemann mapping theorem, we could have $\dfrac{F_1(z)}{R} = f(z)$, and hence, $\dfrac{1}{R_1} = f'(z_0)$, i.e., $F_1(z) = \dfrac{f(z)}{f'(z_0)} = F(z)$, which proves the uniqueness of the mapping function $w = F(z)$. The quantity $R = \dfrac{1}{f'(z_0)}$ is called the *conformal radius* of the region D at the point $z_0 \in D$.

Since $D^* = \text{Ext}(\Gamma)$ is simply connected, Green's function $G(z, \infty)$ for D^* coincides with $\log|f(z)|$, where the function $w = f(z)$ maps D^* univalently onto $|w| > 1$ such that $f(\infty) = \infty$. Then, Robin's constant γ for the region D^* is equal to $\log|f'(\infty)|$, and $\cap(D^*) = \dfrac{1}{|f'(\infty)|} = R$, where R is the conformal radius of the region D^* (with respect to ∞), i.e., the number R is such that the region D^* is mapped univalently onto $|w| > R$ by a normalized function $w = F(z)$ with $F(\infty) = \infty$ and $F'(\infty) = 1$. Thus, in view of Theorems 2.13 and 2.14, we have the following theorem.

Theorem 2.16. *The capacity, and hence the transfinite diameter of a bounded simply connected region D, is equal to the conformal radius of the region D^* which is the complement of the region D in $\mathbb{C}_\infty$ and contains the point at infinity.*

2.6.1 Conformality and Uniqueness. Conformality is a local phenomenon; it applies to small neighborhoods D of a point z_0 which is mapped from the z-plane onto the w-plane by the function $w = f(z)$ (see Theorem 2.17 below). Suppose that $f(z)$ is analytic at the point z_0 and that $f'(z_0) \neq 0$. Let $f'(z_0) = m\, e^{i\alpha}$, and let $z - z_0 = \Delta r_z\, e^{i\Delta\theta_z}$. Then $f(z) = f(z_0) + f'(z_0)(z - z_0)$, which in the w-plane yields to $w - w_0 = \Delta r_w\, e^{i\Delta\theta_w}$. As z_0 approaches z, where $z, z_0 \in D$, the point w_0 approaches w, which gives $w - w_0 = f'(z_0)(z - z_0)$, or

$$\Delta r_w\, e^{i\Delta\theta_w} = f'(z_0)\,\Delta_z\, e^{i\Delta\theta_z} = m\Delta r_z\, E^{i(\alpha+\Delta\theta_z)}. \tag{2.6.1}$$

Then

$$\begin{aligned} \Delta r_w &= m\Delta r_z, & m &= \bmod f(z_0), \\ \Delta\theta_w &= \alpha + \Delta\theta_z, & \alpha &= \arg\{f'(z_0)\}. \end{aligned} \tag{2.6.2}$$

Note that α depends only on z_0, and not on z. Thus, the angle α that appears in $\Delta\theta_w$ as a rotation remains the same at every $z \in D$. Let us consider a curve C_z in terms of the points z_i, $i = 1, 2, \dots, k$, shown in Figure 2.3. If we map this curve C_z from the z-plane onto the w-plane, it will undergo a small rotation by α and a magnification (compression) by a factor shown in Figure 2.3. Hence, this limiting process with infinitesimally small increments of area elements ΔS_z and ΔS_w leads to

$$\lim_{\Delta z \to 0} \frac{\Delta S_w}{\Delta S_z} = |f'(z)|^2 = \left(\frac{\partial u}{\partial x}\right)^2 + \left(\frac{\partial u}{\partial y}\right)^2$$
$$= \frac{\partial u}{\partial x}\frac{\partial v}{\partial y} - \frac{\partial u}{\partial y}\frac{\partial v}{\partial x} = \begin{vmatrix} \frac{\partial u}{\partial x} & pduy \\ \frac{\partial v}{\partial x} & \frac{\partial v}{\partial y} \end{vmatrix} = J, \tag{2.6.3}$$

where J is the Jacobian of the transformation $w = f(z)$. This transformation is one-to-one when $f'(z) \neq 0$ or $j \neq 0$.

This transformation of incremental length and areas can be summarized, for later use, by the following results:

Transformation of incremental length: $|dw| = |f'(z)|\,|dz|$; (2.6.4)

Transformation of incremental areas $\Delta S_z \mapsto \Delta S_w$: $dw\, d\bar{w} = f'(z)\,\overline{f'(z)}\, dz\, d\bar{z}$,

or $du\, dv = |f'(z)|^2, dx\, dy$; (2.6.5)

Area bounded $= \iint_D |f'(z)|^2\, dx\, dy$; (2.6.6)

Magnification factor : $m = \frac{dw}{dz} = f'(z)$. (2.6.7)

where

$$|f'(z)|^2 = \left(\frac{\partial u}{\partial x}\right)^2 + \left(\frac{\partial v}{\partial x}\right)^2 = \left(\frac{\partial u}{\partial y}\right)^2 + \left(\frac{\partial v}{\partial y}\right)^2. \tag{2.6.8}$$

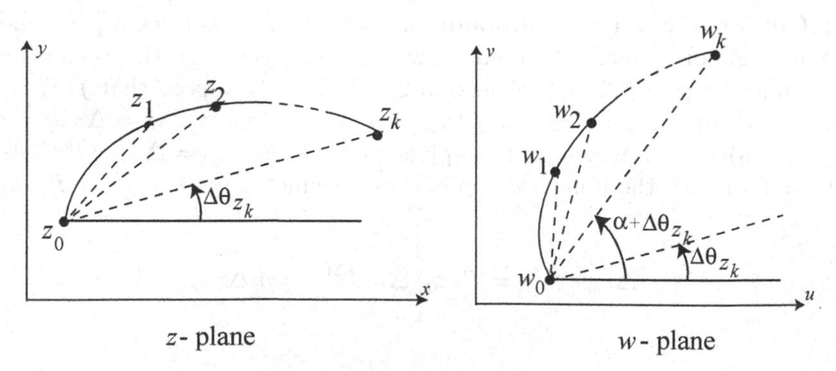

Figure 2.3 Limiting process.

The condition for a well-defined one-to-one mapping is that there must be no critical points, i.e., there must be no point z_0 at which $f'(z_0) = 0$, or $f'(z_0) = \infty$, because otherwise the inverse relationship $\frac{dz}{dw} = \frac{1}{dw/dz}$ as well as $\alpha = \arg\{f'(z_0)$ would become indeterminate.

A domain of transformed points in the w-plane is called *schlicht* or *simple* if none of its points originates from more than one point each in the z-plane.

However, the following example shows that the conformal map guaranteed by the Riemann mapping theorem is not unique, in the sense that in certain cases one can get a choice of more than one conformal map.

Example 2.6. In order to construct a mapping that maps the half-disk $C(0, \frac{1}{2}) = \{|z| < 1, \Im\{z\} > 0\}$ onto the entire unit disk $C(0,1) = \{|z| < 1\}$, the map $w = z^2$ will not work because the image of $B(0, \frac{1}{2})$ omits the positive real axis, thus yielding a disk with a slit $\{|w| < 1, 0 < \arg\{w\} < 2\pi\}$. In order to get the entire unit disk, we may consider the mapping $z = \dfrac{w-1}{w+1}$ (see Map 3.17) which maps the right half-plane $\{\Re\{w\} > 0$ onto the unit disk. It also maps the upper right quadrant $\{0 < \arg\{w\} < \pi/2\}$ onto the half unit disk $C(0, \frac{1}{2})$. Then its inverse $w = \dfrac{z+1}{z-1}$ will then map the half-disk $C(0, \frac{1}{2})$ onto the upper right quadrant.

However, the mapping (3.3.39), $w = \dfrac{iz^2+1}{iz^2-1}$ (see Map 3.44), maps the upper right quadrant onto the unit disk $C(0,1)$. Since

$$w = \frac{iw^2+1}{iw^2-1} = \frac{i\left(\dfrac{z+1}{z-1}\right)+1}{i\left(\dfrac{z+1}{z-1}\right)-1} = \frac{(i+1)(z^2+1)+2(i-1)z}{(i-1)(z^2+1)+2(i+1)z} = -i\frac{z^2+2iz+1}{z^2-2iz+1},$$

we find that, omitting the factor $-i$ which merely rotates the disk by $-\pi/2$, the mapping

$$w = \frac{z^2+2iz+1}{z^2-2iz+1} \tag{2.6.9}$$

is another choice for the solution of the problem. ■

Theorem 2.17. (Local inverse theorem) *If $w = f(z)$ is conformal at z_0, there exists a function f^{-1} such that $z_0 = f^{-1}(w_0)$. Then $w = f(z) = f\left[f^{-1}(w)\right]$, where $z = f^{-1}(w)$, and*

$$\frac{df^{-1}(w)}{dw} = \frac{1}{\dfrac{df(z)}{dz}} = \left(\frac{df(z)}{dz}\right)^{-1}, \quad \textit{provided } \frac{df(z)}{dz} \neq 0 \textit{ or } f \textit{ is conformal.}$$

PROOF. Using $z = x + iy$ and $w = u + iv$, the mapping $w = f(z)$ corresponds to a pair of simultaneous equations $u = u(x,y)$, $v = v(x,y)$ with $du = u_x\, dx + u_y\, dy$ and $= v_x\, dx + v_y\, dy$. Now, if the inverse mapping f^{-1} exists, then we have $x = x(u,y)$, $y = y(u,v)$ with $dx = x_u]\, du + x_v\, dv$ and $dy = y_u\, du + u_v\, dv$. Hence,

$$dx = x_u\left(u_x\, dx + u_y\, dy\right) + x_v\left(v_x\, dx + v_y\, dy\right).$$

Since dx and dy are independent, their coefficients must vanish, i.e.,

$$1 = x_u u_x + x_v v_x, \quad 0 = x_u u_y + x_v v_y, \tag{2.6.10}$$

These equations admit a solution only if

$$\begin{vmatrix} u_x & u_y \\ v_x & v_y \end{vmatrix} \neq 0.$$

The determinant is known as the *Jacobian* defined by

$$J = \frac{\partial(u,v)}{\partial(x,y)} = \begin{vmatrix} u_x & u_y \\ v_x & v_y \end{vmatrix} = u_x v_y - v_x u_y.$$

Thus, f^{-1} exists if $J \neq 0$. Since f is analytic, the Cauchy-Riemann equations (2.1.6) give

$$J = u_x^2 + v_x^2 = \left|\frac{\partial f}{\partial x}\right|^2 = \left|\frac{df}{dz}\right|^2 \neq 0.$$

Returning to the solution of (2.6.10), we have

$$x_u = \frac{1}{J}\begin{vmatrix} 1 & v_x \\ 0 & v_y \end{vmatrix} = \frac{v_y}{J}, \quad x_v = \frac{1}{J}\begin{vmatrix} u_x & 1 \\ u_y & 0 \end{vmatrix} = -\frac{u_y}{J}.$$

Similarly

$$y_u == \frac{v_x}{J}, \quad y_v = \frac{u_x}{J}.$$

Then since f is analytic, we have $x_u = \dfrac{u_x}{J} = y_v$ and $x_v = \dfrac{v_x}{J} = -y_u$. Hence f^{-1} is analytic. ■

Note that this theorem proves only the existence of a local inverse. Nothing global is mentioned. Thus, if f is a many-to-one mapping, then f^{-1} exists only in a neighborhood of one of the branches of the multiple valued function f^{-1}.

2.6.2 Conformal and Isogonal Mappings. Let C_1 and C_2 be two curves in the z-plane that intersect at the point z_0, and let G_1 and G_2 be the images of C_1 and C_2, respectively, under the mapping $w = f(z)$. Then G_1 and G_2 intersect at w_0 which is the image of z_0. Assume that $f(z)$ is analytic at all points of C_1 and C_2 and that z_0 is not a critical point. Let λ_k, $k = 1, 2$, be the parameters that define C_k and G_k. Then the tangents to the two curves at the point of intersection are given by

$$\tau_1(w_0) = f'(z_0)\left.\frac{dz}{d\lambda_1}\right|_{z_0} = f'(z_0)t_1(z_0), \tag{2.6.11}$$

$$\tau_2(w_0) = f'(z_0)\left.\frac{dz}{d\lambda_2}\right|_{z_0} = f'(z_0)t_2(z_0). \tag{2.6.12}$$

Thus, $|\tau_k(w_0)| = |f'(z_0)||t_k(z_0)|$ for $i = 1, 2$ where the angles $\psi_k(w_0) = \theta_k(z_0) + \phi(z_0)$, and thus,

$$\Delta\psi(w_0) = \psi_1(w_0) - \psi_2(w_0) = \theta_1(z_0) - \theta_2(z_0) = \Delta\theta(z_0). \tag{2.6.13}$$

This result states (see Figure 2.4) that the magnitude and the sign of the angle subtended by the tangents to the two curves in the z-plane is the same as the magnitude and sign of the angle subtended by the tangents to the images of these two curves in the w-plane. A mapping under which the magnitude and sign of the angle between the tangents to two curves remains unchanged is called a *conformal mapping*. The sign of the difference between the angles of Eq (2.6.13) is called the *sense* of the angle subtended by the tangents. A transformation for which

$$|\psi_1(w_0) - \psi_2(w_0)| = |\theta_1(z_0) - \theta_2(z_0)|, \tag{2.6.14}$$

but for which the sense of the angle may not be preserved is called an *isogonal mapping*. Hence, a conformal mapping is isogonal but the converse may not be true. An application of the isogonal transformation is given in §21.6.

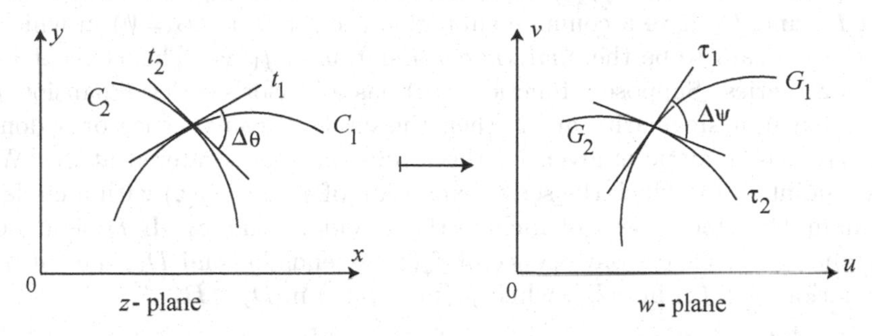

Figure 2.4 Conformal and isogonal mappings.

2.6.3 Conformal Mapping of an Area Element. A differential line element in a region D in the z-plane from the point (x, y) to a point $(x + dx, y + dy)$ is defined by $dz = dx + i\,dy$. Similarly the differential line element in a region G in the w-plane from (u, v) to $(u + du, v + dv)$ is defined by $dw = du + i\,dv$. Consider a rectangular area element A with sides $dz_1 = dx$ and $dz_2 = i\,dy$. Then the images of dz_1 and dz_2 are differential curves in the w-plane defined by $dw_k = du_k + i\,dv_k$, $k = 1, 2$. The area of a rectangle with sides z_1 and z_2 is given by $A = |\Im\{z_1^* z_2)\}|$, and the area element in the z-plane is given by $dA_z = dx\,dy$ and that in the w-plane by $dA_w = du_1\,dv_2 - du_2\,dv_1$, where dz_1 is along a line defined by $y =$ const, and dz_2 is along a line defined by $x =$ const, the images of these two lines in the w-plane are

$$dw_1 = \frac{df}{dz}\Big|_{y=\text{const}} \cdot dz_1 = \frac{\partial f}{\partial x}\,dx, \quad dw_2 = \frac{df}{dz}\Big|_{x=\text{const}} \cdot dz_2 = \frac{1}{i}\frac{\partial f}{\partial y}\,i\,dy = \frac{\partial f}{\partial y}\,dy,$$

where $w = f(z) = u(x, y) + i\,v(x, y)$. Since $\dfrac{\partial f}{\partial x} = \dfrac{\partial u}{\partial x} + i\,\dfrac{\partial v}{\partial x}$ and $\dfrac{\partial f}{\partial y} = \dfrac{\partial u}{\partial y} + i\,\dfrac{\partial v}{\partial y}$, we have

$$dw_1 = \Big(\frac{\partial u}{\partial x} + i\,\frac{\partial v}{\partial x}\Big)dx, \quad dw_2 = \Big(\frac{\partial u}{\partial y} + i\,\frac{\partial v}{\partial y}\Big)dy.$$

Hence, the differential area element in the w plane is given by

$$dA_w = |\Im\{dw_1^* dw_2\}| = \left|\frac{\partial u}{\partial x}\frac{\partial v}{\partial y} - \frac{\partial u}{\partial y}\frac{\partial v}{\partial x}\right| dx\,dy = J(u, v; x, y)\,A_w, \tag{2.6.15}$$

where

$$J(u, v; x, y) \equiv J(w; z) = \left|\frac{\partial u}{\partial x}\frac{\partial v}{\partial y} - \frac{\partial u}{\partial y}\frac{\partial v}{\partial x}\right| = \begin{vmatrix} \dfrac{\partial u}{\partial x} & \dfrac{\partial u}{\partial y} \\ \dfrac{\partial v}{\partial x} & \dfrac{\partial v}{\partial y} \end{vmatrix} = \frac{\partial(u, v)}{\partial(x, y)}$$

is the *Jacobian* of the transformation, which in view of the Cauchy-Riemann equations, reduces to

$$J(w; z) = \Big(\frac{\partial u}{\partial x}\Big)^2 + \Big(\frac{\partial v}{\partial x}\Big)^2 = \Big(\frac{\partial u}{\partial y}\Big)^2 + \Big(\frac{\partial u}{\partial y}\Big)^2. \tag{2.6.16}$$

Note that $J(w; z)$ is positive for all x and y. Since f is analytic in D, we get

$$J(w; z) = |f'(z)|^2 = \left|\frac{dw}{dz}\right|^2. \tag{2.6.17}$$

2.6.4 Analytic Continuation. Analytic continuation is an important aspect of conformal mapping. Let a function $f_1(z)$ be defined on a domain D_1 and another function $f_2(z)$ on D_2. If D_1 and D_2 have a common subregion (i.e., if $D_1 \cap D_2 \neq \emptyset$) in which $f_1(z) = f_2(z)$, then $f_2(z)$ is said to be the *analytic continuation* of $f_1(z)$. This concept is closely related to Taylor's series. Suppose a function $f(z)$ has a Taylor's series expansion $f_1(z)$ at $z = z_1$. If there exists a singularity at z_s, then the circle of convergence or a domain D_1 within which $f_1(z)$ is analytic is given by the radius $|z_s - z_1|$ centered at z_1. Within D_1 there exists a point z_2 for which the series expansion of $f(z)$ is $f_2(z)$ with a circle of convergence or domain D_2 which does not include the singular point z_s. If D_2 is close to the border of D_1 and the circle of convergence of $f_2(z)$ extends beyond D_1, the $f_2(z)$ is the analytic continuation of $f_1(z)$ into D_2, while $f_1(z) = f_2(z)$ in $D_1 \cap D_2$.

The analytic continuation is unique. In fact, suppose that g_1 and g_2 are analytic continuations of f from D_1 into D_2. Then $g_1(z) \equiv g_2(z)$ on $D_1 \cap D_2$, and thus, by identity theorem (§1.3), $g_1 = g_2$ throughout D_2. Thus, in practice we will determine a single function $F(z)$ that is analytic on $D_1 \cup D_2$ and is given by $F(z) = f(z)$ on D_1, and $F(z) = g(z)$ on D_2. For example, let $f(z) = \int_0^\infty e^{-zt}\,dt$. Notice that $f(z)$ is analytic on $\Re\{z\} > 0$ because after evaluating the improper integral we find that $f(z) = 1/z$ is analytic on $\Re\{z\} > 0$. So we take $g(z) = 1/z$. Since $g(z)$ is analytic on $\mathbb{C}\backslash\{0\}$, $g(z)$ becomes the analytic continuation of $f(z)$ from the right half-plane into the whole plane indented at the origin. Similarly, the Laplace transform of $\cos at$ is analytic for $\Re\{z\} > 0$; but the function $\dfrac{z}{z^2 + a^2}$ is its analytic continuation from the right half-plane into the whole plane indented at the points $\pm ia$.

2.6.5 Chain Property. Let $D_0, \dots, D_n$ be regions in $\mathbb{C}$, and let $f_k : D_{k-1} \mapsto D_k$ denote conformal mappings for $k = 1, \dots, n$. Then the mapping $g = f_n \circ \cdots \circ f_1$, defined by

$$g(z) = f_n\left(f_{n-1}\left(\cdots f_2\left(f_1(z)\right)\cdots\right)\right) \tag{2.6.18}$$

is a conformal mapping of D_0 onto D_n. The mapping g is said to be composed of a *chain* of mappings $f_1, \dots, f_n$ and is represented by the scheme

$$D_0 \xrightarrow{f_1} D_1 \xrightarrow{f_2} D_2 \xrightarrow{f_2} \cdots \xrightarrow{f_{n-1}} D_{n-1} \xrightarrow{f_n} D_n. \tag{2.6.19}$$

Thus, the set of regions on $\mathbb{C}$ can be divided into *mapping classes* such that two regions can be mapped conformally onto each other iff they belong to the same mapping class. Figure 2.5, which represents an example of this chain property, is a composite of the following six maps, in that order:

Map 2.1. $w = -i\,z$ (rotation by $-\pi$), which maps the upper-half of the z-plane with a vertical slit from $z = 0$ to $z = i$ onto the right half-plane with a slit from $w = 0$ to $w = 1$.

Map 2.2. $w = z^2$, which maps the right half-plane with a slit from $z = 0$ to $z = 1$ onto the entire w-plane with a slit from $-\infty < u < 1$.

Map 2.3. $w = z^2 - 1$ (translation), which maps the entire z-plane with a slit from $-\infty < x < 1$ onto the entire w-plane with the slit $-\infty < u < 0$ (negative real axis).

Map 2.4. $w = \sqrt{z}$ (see Map 4.26), which maps the entire z-plane with a slit $-\infty < x < 0$ (negative real axis) onto the right-half w-plane.

Map 2.5. $w = iz$ (rotation by π), which maps the right-half z-plane onto the upper half-plane $\Im\{w\} > 0$.

Map 2.6. $w = \dfrac{1-z}{1+z}$ (see Map 3.21), which maps the right-half z-plane onto the unit disk $|w| < 1$.

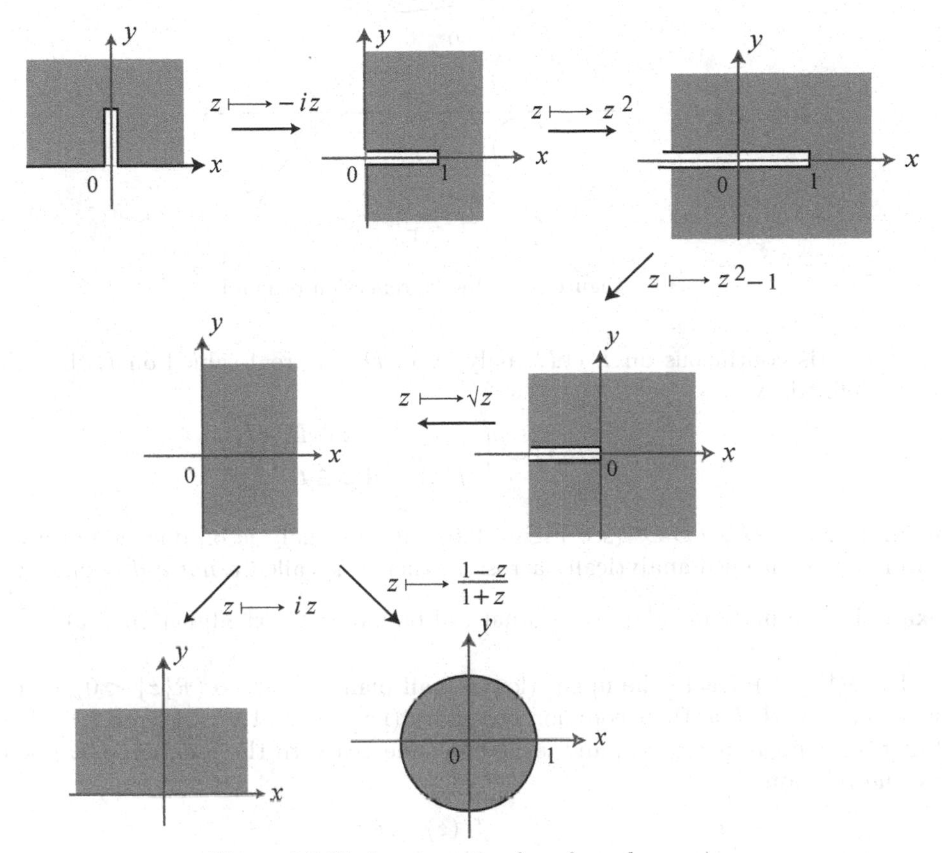

Figure 2.5 Chain property of conformal mappings.

The practical applications of conformal mappings are related to the problem of constructing a function which maps a given region onto a given region. Often we find an explicit expression for the mapping function and determine it by applying the chain property.

Another example of the chain property is that the chain of mappings $f_1 : z \mapsto z - 1$, $f_2 : z \mapsto 2z$, $f_3 : z \mapsto z + 3i/2$ maps the circle $|z - 1| = 1$ onto the circle $|w - 3i/2| = 2$, being composite of the following maps, in that order:

Map 2.7. $w(z) = z - 1$ (translation); **Map 2.8** $w(z) = 2z$ (magnification); and **Map 2.9.** $w(z) = z + \dfrac{3i}{2}$ (translation).

2.6.6 Schwarz Reflection Principle. If $f(z)$ is analytic on a region containing a segment of the real axis and is real-valued on this segment, then $\overline{f(z)} = f(\bar{z})$. This leads to the concept of analytic continuation across the real axis when the function is known only on one side. Let I denote a segment $a < x < b$, and let Γ be a Jordan arc joining a and b and lying in $\Re\{z\} > 0$. Then $C = I \cup \Gamma$ is a contour that encloses a region D lying entirely in $\Re\{z\} > 0$. If we reflect this region into the real axis, we get a region D' that lies in the lower half-plane and is bounded by I and Γ' which is the reflection of Γ into the real axis.

Then the region $I \cup D'$ is symmetric about the real axis.

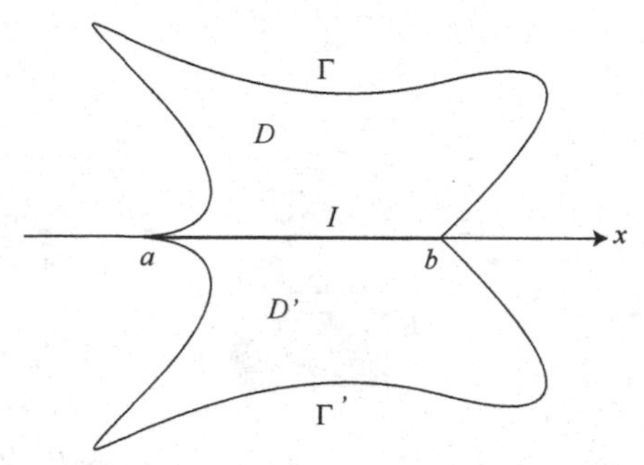

Figure 2.6 Schwarz reflection principle.

If $f(z)$ is continuous on $D \cup I$, analytic on D, and real-valued on I, then the function $F(z)$ defined by

$$F(z) = \begin{cases} f(z), & \text{if } z \in D \cup I, \\ \overline{f(z)}, & \text{if } z \in D' \end{cases}$$

is analytic on $D \cup I \cup D'$ (see Figure 2.6). If f is analytic on one side of a curve Γ and cannot be continued analytically across Γ, then Γ is called a *natural boundary* for f. For example, the unit circle $|z| = 1$ is a natural boundary for the function $f(z) = \sum_{n=0}^{\infty} z^{2^k}$.

Let S^+ (S^-) denote the upper (lower) half-plane $\Re\{z\} > 0$ $(\Re\{z\} < 0)$, respectively, or vice-versa, with L as their common boundary (i.e., the real axis, Figure 2.7). Let $f(z)$ be a function defined at $z \in S^+$, and let it be connected with the function $f^*(z)$ defined in S^- by the relation

$$f^*(z) = \overline{f(\bar{z})}, \tag{2.6.20}$$

i.e., $f(z)$ and $f^*(z)$ take conjugate complex values at points symmetric with respect to the real axis, since the points z and $\bar{z}$ are reflections of each other in L. We can rewrite the relation (2.6.20) as $f^*(z) = \bar{f}(z)$, i.e., if $f(z) = u(x,y) + iv(x,y)$, then $\bar{f}(z) = u(x,-y) - iv(x,-y)$. If $f(z)$ is regular (or meromorphic) in S^+, then $f^*(z) = \bar{f}(z)$ is regular (or meromorphic) in S^-, and the relation (2.6.20) is symmetric as regards f and f^*, i.e.,

$$f(z) = \overline{f^*(\bar{z})}, \quad (f^*(z))^* = f(z).$$

If $f(z)$ is a rational function

$$f(z) = \frac{a_n\, z^n + a_{n-1}\, z^{n-1} + \cdots + a_0}{b_m\, z^m + b_{m-1}\, z^{m-1} + \cdots + b_0}, \tag{2.6.21}$$

then $f^*(z) = \bar{f}(z)$ is obtained by simply replacing the coefficients by their conjugate complex values. Let t be a real number, and assume that $f(z)$ takes a definite limit value $f^+(t)$ as $z \to t$ from S^+. Then $f^{*-}(t)$ exists and

$$f^{*-}(t) = \bar{f}^-(t) = \overline{f^+(t)}, \tag{2.6.22}$$

because $\bar{z} \to t$ from S^- as $z \to t$ from S^+, and hence, $f^*(z) = \overline{f(\bar{z})} \to \overline{f^+(t)}$.

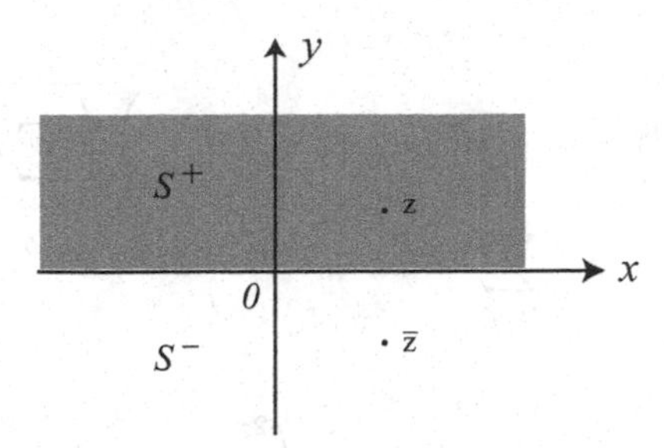

Figure 2.7 Reflections in the real axis.

We will assume that $f(z)$ is regular on S^+, except possibly at infinity, and continuous on L from the left. Define a sectionally regular function $F(z)$ by

$$F(z) = \begin{cases} f(z) & \text{for } z \in S^+, \\ f^*(z) & \text{for } z \in S^-. \end{cases} \tag{2.6.23}$$

Then, in view of (2.6.22)

$$F^-(t) = \overline{F^+(t)}, \quad F^+(t) = \overline{F^-(t)}. \tag{2.6.24}$$

These relations are useful when transforming the boundary conditions in any boundary problem containing $f^+(t)$ and $\overline{f^+(t)}$, or $f^-(t)$ and $\overline{f^-(t)}$ into those involving $F^+(t)$ and $F^-(t)$.

In view of the Schwarz reflection principle, another property is that of extending $f(z)$. If $\Im\{f^*(t)\} = 0$ in any interval I of the real axis, then the function $f^*(z)$ is the analytic continuation of $f(z)$ through the interval I because $f^{*-}(t) = f^+(t)$ on this interval.

2.6.7 Conformal Equivalence. A *conformal equivalence* between two regions D and G in the complex plane is a one-to-one analytic function f with $f(D) = G$. Thus, $f'(z) \neq 0$ for all z in D. Conversely, if $f : D \mapsto \mathbb{C}$ is analytic such that f' never vanishes, then f is not necessarily a conformal equivalence. As an example, consider $f(z) = e^z$. However, if $f'(z) \neq 0$ on D, then f is locally one-to-one and conformal. The Riemann mapping theorem (Theorem 2.15) establishes the conformal equivalence of two regions. If f is a conformal equivalence between the open sets D and G, then

$$\text{Area}(G) = \iint_D |f'|^2 \, dz. \tag{2.6.25}$$

Note that if f is an analytic function, then the Jacobian of f, regarded as a mapping from $\mathbb{R}^2$ into $\mathbb{R}^2$, is $|f'|^2$. Hence, if G is a simply connected region, $g : D \mapsto G$ is a Riemann map, and $g(z) = \sum_n a_n z^n$ in D, then from (2.6.25)

$$\text{Area}(G) = \iint_D |g'|^2 \, dz = \pi \sum_n n|a_n|^2. \tag{2.6.26}$$

In fact, since $g'(z) = \sum_n n a_n z^{n-1}$, then, for $r < 1$,

$$|g'(re^{i\theta})|^2 = \left(\sum_n n a_n r^{n-1} e^{i(n-1)\theta}\right) \overline{\left(\sum_m m a_m r^{m-1} e^{i(m-1)\theta}\right)}$$
$$= \sum_{n,m} n m a_n \bar{a}_m r^{n+m-2} e^{i(n-m)\theta},$$

which converges uniformly in θ. Since $\displaystyle\int_0^{2\pi} e^{i(n-m)\theta}\, d\theta = 0$ for $n \neq m$,

$$\iint_D |g'|^2 = \sum_n n^2 |a_n|^2 2\pi \int_0^1 r^{2n-1}\, dr$$
$$= 2\pi \sum_n n^2 |a_n|^2 \frac{1}{2n}$$
$$= \pi \sum_n n |a_n|^2.$$

This relation is useful in solving the minimum area problem in conformal mapping.

If a function $w = f(z)$ maps conformally a region D onto another region G, and if z_0 is an isolated boundary point of D, then z_0 is a removable singularity or a simple pole of $f(z)$. That is the reason to assume the regions to be without isolated boundary points (Goluzin [1969:205]; Wen [1992:95]).

2.6.8 Riemann Sphere. Riemann, using the stereographic projection from the North Pole, assumed this Pole on the sphere to be mapped onto the point at infinity. If a is a pole of f, then, as $z \to a$, the function $f(z)$ composed with the inverse of stereographic projection tend to the North Pole, so that a meromorphic function in a domain $D \in \mathbb{C}$ is regarded as a continuous image of D onto the sphere. The map at the North Pole is continuous and behaves just like at any other point on the sphere. Thus, at a simple pole where $1/(z-a)$ occurs, the mapping onto the sphere is conformal at a. If higher powers of $1/(z-a)$ occur, the same is true as at any point a where f is analytic but $f' = 0$. The unit sphere in this sense as the extended complex plane via stereographic projection is called the *Riemann sphere.* If a is a pole, we write $f(a) = \infty$. Thus, this mapping technique defined the true mapping significance of the point at infinity.

Theorem 2.18. *Let $f(z)$ be a meromorphic function defined in the entire plane. If $f(z)$ maps the entire plane onto the Riemann sphere, the image of f covers the entire sphere with at most two exceptions. Also, if f is not a rational function, then the same is true for $f(z)$ with $|z| > M$ for $M > 0$.*

Let $f(z)$ be an analytic or meromorphic function in a domain $D \in \mathbb{C}$. A value c of $f(z)$ is said to be *totally ramified*, or *branched*, if, whenever $f(b) = c$, we have $f'(b) = 0$. This means that if f is a mapping from the z-plane onto the w-plane, then f does not define a locally one-to-one conformal map at those points that map onto c. In this case f behaves like the function z^n, $n > 1$ in the neighborhood of the origin, and the image is ramified or branched in a neighborhood of c.

Theorem 2.19. (Nevanlinna [1925]) *An entire function in the z-plane can have at most* *two totally ramified values, whereas a meromorphic function in the z-plane can have at most four totally ramified values.*

Examples that both 2 and 4 numbers are sharp: (i) Complex sine and cosine functions are entire functions defined by $2\cos z = e^{iz} + e^{-iz}$, $2i\sin z = e^{iz} - e^{-iz}$; thus, $\sin^2 z + \cos^2 z = 1$ and $\cos z = \dfrac{d\sin z}{dz}$. Thus, $\sin z = \pm 1$ iff $\cos z = 0$, which shows that the two values ± 1 are totally ramified for $f(z) = \sin z$. Similarly, as Nevanlinna [1925] has shown, Weierstrass's elliptic function $\wp(z)$ has four totally ramified values.

Theorem 2.20. (Bloch [1925]) *Let $w = f(z)$ be analytic in the unit disk and be normalized so that $f'(0) = 0$. Then there exists a constant $B > 0$ with the following property: For every $r < B$ there exists a disk of radius r in the w-plane that is the one-to-one conformal image under f of a domain inside the unit disk in the z-plane.*

The largest value of B is known as *Bloch's constant*; its precise value is unknown.

Corollary 2.1. *Let an arbitrary nonconstant entire function $w = f(z)$ define the mapping of the z-plane onto the w-plane such that for every $R > 0$ there is a disk of radius R in the w-plane that is the one-to-one image under f of some domain in the z-plane.*

2.6.9 Bieberbach Conjecture. It was 99 years ago that Ludwig Bieberbach [1916] mentioned this conjecture in a footnote. This conjecture deals with the bound on the coefficients of the class $\mathcal{S}$ of univalent (schlicht) functions $f : E \mapsto f(E)$, where E is the unit disk, with the series expansion $f(z) = z + \sum_{n=2}^{\infty} a_n z^n$, and normalized by $f(0) = 0$ and $f'(0) = 1$. This normalization is a scaling that also implies that the boundary of $f(E)$ cannot get close to the origin. The Bieberbach conjecture states that $|a_n| \le n$ for all $n = 2, 3, \dots$.

The historical development shows that there was a sustained and broad interest in the Bieberbach conjecture and its impact on creative thinking of mathematicians of the last century. This single phenomenon has been responsible for development of certain beautiful aspects of complex analysis, especially in the geometric-function theory of univalent functions. Although de Branges theorem is deep and meritorious, the work of Fitzgerald [1985], Fitzgerald and Pommerenke [1985], Pommerenke [1985; 1992a, b], and all previous researchers provided significant progress.

Let $\mathcal{S}$ denote the family of univalent functions analytic on the unit disk $|z| < 1$, which are of the form $f(z) = z + \displaystyle\sum_{n=2}^{\infty} a_n z^n$. The coefficients a_n satisfy the inequality $\big|a_n\big| \le n$ for $n = 2, 3, \dots$, and this bound, known as de Branges estimate (see de Branges [1985]), is sharp for the Koebe function $f(z) = \dfrac{z}{(1-z)^2}$. (Wen [1992:54], Kythe [2016]). This result is true only in $\mathbb{C}$, and there is no such conjecture for $\mathbb{C}^n$, $n \ge 2$.

2.6.10 Mercator's Projection. In 1569 Mercator chose to abandon size in favor of shape while constructing his famous map. He used the following definitions for a particular class of maps known in cartography as *cylindrical projections*.

The *equator* is the great circle equidistant from North and South Poles.

The *meridians* are the circular arcs joining the North and South Poles. They are perpendicular to the equator, and are the curves one travels when traveling due north or south from any point.

The *parallels of latitude*, or *parallels* in short, are the circles to the meridians. They are also the circles at fixed distance from the North or South Pole, and they are the curves one traverses when traveling due east or west from any point other than the two Poles.

A *cylindrical projection*[2] is a map constructed as follows: The equator is represented by a horizontal line segment drawn in the center. The length of the segment determines the *scale* of the map along the equator. In other words, if L is the length of the Earth's equator and w the width of the map (which is the length of the horizontal line-segment representing the equator), then all distances along the equator are represented by a fixed factor w/L. The meridians are represented by vertical lines of finite or infinite length,[3] and the parallels are represented by horizontal line segments of the same fixed length w as the equator. Hence, all cylindrical projections have the following two properties which illustrate the above Euler's theorem (Euler [1775]):

(i) The map is in the form of a rectangle or infinite vertical strip representing all the Earth except the Poles, with two vertical sides corresponding to a single meridian, such that every other meridian corresponds to a unique vertical line.

(ii) The quarter circle along a meridian from the equator to the North or South Pole is divided into 90 degrees, and the latitude of any point is the number of degrees along the meridian north or south of the equator. Thus, for example, at latitude ϕ (radian) north or south, the parallel is a circle of radius $R\cos\phi$, where R is the radius of the Earth so that the length of the equator is $L = 2\pi R$. Thus, at latitude ϕ, the parallel is mapped with fixed scale $\dfrac{w}{2\pi R\cos\phi} = \dfrac{w}{L\cos\phi} = s\sec\phi$, where $s = w/L$ is the scale of the map along the equator (see Figure 2.8).

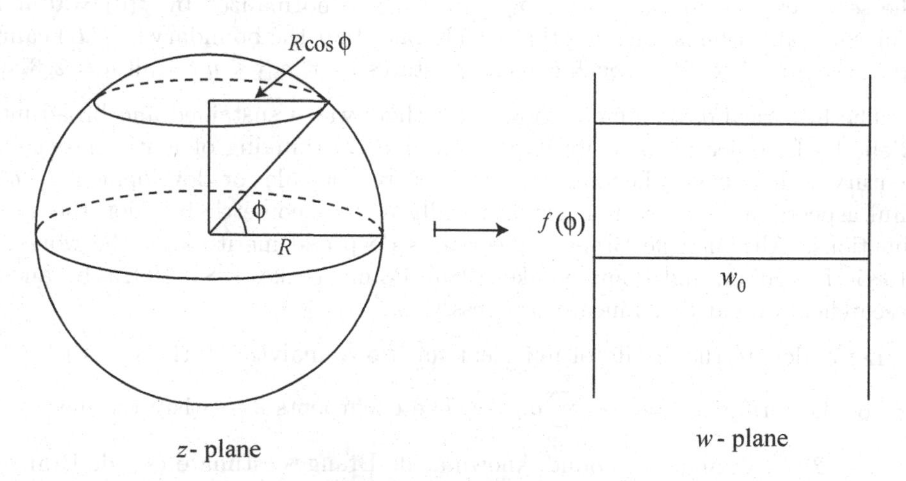

Figure 2.8 Cylindrical projection.

The mathematically true cylindrical projection is described as follows: Let S denote the globe that is a sphere representing the surface of the Earth, and let C denote a cylinder tangent to S along the equator. Now, project S onto C along rays from the center of S (Figure 2.9). Then each meridian on the sphere will map onto a vertical line, and each

---

[2] The word 'projection' used in map-making means any systematic representation in a broad sense; it should not be confused with the narrow mathematical definition, or the colloquial word 'projection' or 'projection' using a slide projector.

[3] The infinite version is a purely theoretical map; the actual finite map is cut off to represent the portion of the Earth between two fixed parallels.

parallel onto a circle on the cylinder parallel to the equator. Next, cut the cylinder along a vertical line and unroll it onto a vertical strip of the plane, thus yielding the so-called true cylindrical projection or central cylindrical projection.

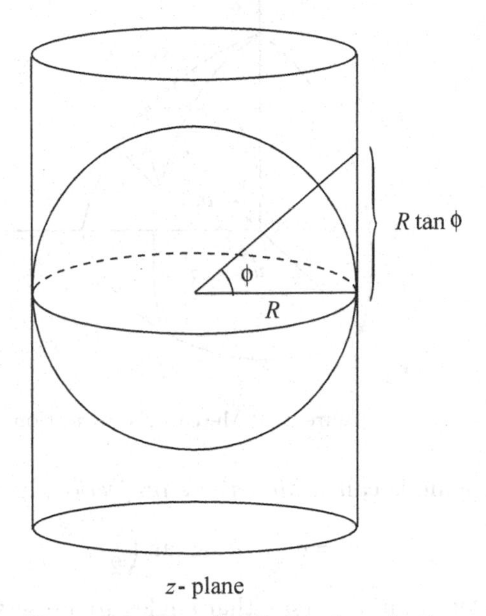

Figure 2.9 Central cylindrical projection.

Much later, Euler [1775] proved the following theorem:

Theorem 2.21. *It is impossible to make an exact scale map of any part of a spherical surface.*

All maps in which the North is the vertical direction on the map use a fixed scale along East-West line; they are the result of the cylindrical projection and cannot have a fixed scale for the entire map. This is because the so-called scale is the factor $\sec\phi$ in the expression $s\sec\phi$ for the scale, which will remain approximately the same over a small portion of the earth's surface and will make very negligible difference. Thus, each meridian on the sphere is mapped onto a vertical line, and each parallel of latitude onto a circle on the cylinder parallel to the equator. Once we cut the cylinder along a suitable vertical line, the resulting flat map becomes a practical central cylindrical projection (Figure 2.9).

If $x = a\,u$ and $y = a\,f(v)$, where $z = x + i\,y$ and $w = u + i\,v$, then the meridians and parallels of the earth will be mapped onto the coordinate lines in the (x, y)-plane. The equator will be mapped onto the x-axis if $f(0) = 0$. To make a rhumb line on the sphere, which cuts all the meridians at the same angle α map onto a straight line in the z-plane cutting the lines x =const at the same angle α, we must have $\cot\alpha = \dfrac{dv}{\cos v\,du} = \dfrac{dy}{dx} =$

$\dfrac{f'(v)\,dv}{du}$, which yields $f(v) = \displaystyle\int_0^v \sec v\, dv = \log \tan\left(\frac{v}{2} + \frac{\pi}{4}\right)$.

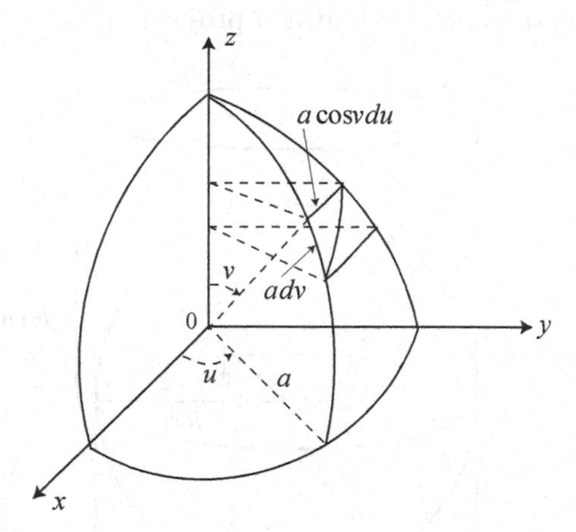

Figure 2.10 Mercator's projection.

The resulting mapping is called *Mercator's projection*, given by

$$z = a\left[u + \log\tan\left(\frac{v}{2} + \frac{\pi}{4}\right)\right]. \tag{2.6.27}$$

This mapping is conformal in the sense that angles are preserved and lengths are multiplied by a factor depending on the latitude. Mercator's projection can also be written as $z = a\left[u + \text{gd}^{-1}(v)\right]$, where $\text{gd}(u)$ denotes the Gudermannian of u (Franklin [1944:140-141]). For definition and other results on Gudermannian, see Appendix D. For interested readers the original work of Mercator is available in Mercator [1569], Lambert [1772], and Osserman [2004].

2.7 Taylor Series Approximations

Let Γ denote the boundary of a domain D in the z-plane, and let $C \equiv B(0,1)$ be the unit circle $|w| = 1$ (see Figure 2.11). To determine a mapping function $z(w)$ that maps the boundary C onto the boundary Γ, we will use the Cauchy integration formula (2.4.2) (with $I(\gamma, z_j) = 1$) to translate any interior point z_j of Γ to the w-plane. Thus, we have

$$w\left(z_j\right) = \frac{1}{2\pi i}\int_\Gamma \frac{w(z)}{z - z_j}\, dw, \tag{2.7.1}$$

where $z = r\, e^{i\theta}$, so that Eq (2.7.1) reduces to

$$w\left(z_j\right) = \frac{1}{2\pi}\int_\Gamma \frac{w(r\, e^{i\theta})}{r\, e^{i\theta} - w_j}\, r\, e^{i\theta}\, d\theta. \tag{2.7.2}$$

Thus, using the Cauchy integral formula to expand $w(z)$ in a Taylor series about any z_j, the nth derivative with respect to z_j is given by

$$w^{(n)}\left(z_j\right) = \frac{n!}{2\pi i}\int_\Gamma \frac{w(z)}{\left(z - z_j\right)^{n+1}}\, dz. \tag{2.7.3}$$

This will give us the Taylor series expansion of $w(z)$ about a point $z = a \in \Gamma$ as

$$w(z) = w(a) + (z-a)w'(a) + \frac{(z-a)^2}{2!}w''(a) + \cdots, \tag{2.7.4}$$

or about $w = 0$ as

$$w(z) = c_0 + c_1 z + c_2 z^2 + \cdots, \tag{2.7.5}$$

where $c_j = z w^{(j)}(0)$, $j = 0, 1, 2, \ldots$.

The inverse function $z = f(w)$ can be approximated as a power series (or a Taylor series) using the expansion (2.7.4) about the point $w = 0$ by reversing the role of w and z. Thus, to map the boundary Γ and its *interior* $\mathrm{Int}(\Gamma)$, which must include the point $z = 0$, onto the unit disk $C \cup \mathrm{Int}(C)$, we can write the approximate series as

$$z \equiv p_n(w) = a_0 + a_1 w + a_2 w^2 + \cdots + a_n w^n. \tag{2.7.6}$$

Similarly, the mapping of the *exterior* region $\mathrm{Ext}(\Gamma)$ that does not include the point $z = 0$ can be approximated by the approximate power series as

$$z \equiv q_m(w) = a_0 + a_{-1} w^{-1} + a_{-2} w^{-2} + \cdots + a_{-m} w^{-m}. \tag{2.7.7}$$

Combining these two power series, we obtain the Laurent series approximation as

$$z = \sum_{j=-m}^{m} a_j w^m. \tag{2.7.8}$$

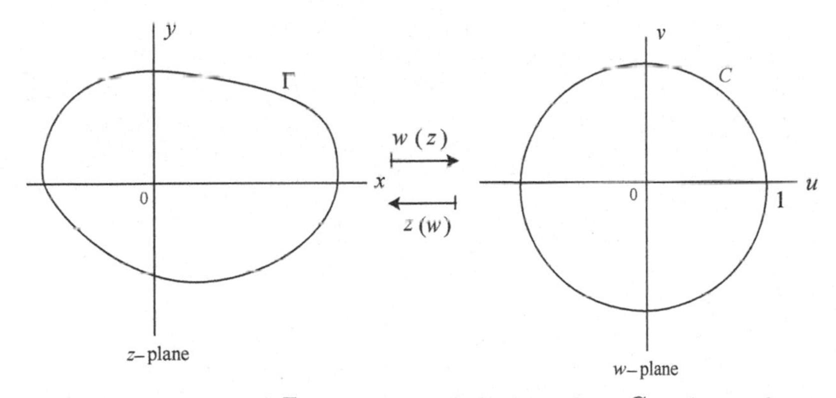

Figure 2.11 Boundary Γ in z-place and the boundary C in the w-plane.

2.7.1 Interior of the Unit Circle. Consider the inverse transformation $z = z(w)$, where $w = e^{i\Theta}$. Then in view of (2.7.3)

$$z = x + iy = c_0 + c_1 w + c_2 w^2 + \cdots = \sum_{j=0}^{\infty} (x_j + i y_j) = \sum_{j=0}^{\infty} c_j\, e^{i\Theta}. \tag{2.7.9}$$

Separating the real and imaginary parts in Eq (2.7.9) we have for the jth term

$$\begin{aligned} c_j &= a_j + i b_j; \\ x_j &= a_j \cos j\Theta - b_j \sin j\Theta; \\ y_j &= b_j \cos j\Theta + a_j \sin j\Theta. \end{aligned} \tag{2.7.10}$$

Map 2.10. The function $w = f(z)$ that maps the unit disk U onto itself such that the point $z_0 \in U$ goes into the origin of the w-plane is given by

$$w = \frac{z - z_0}{1 - 1/\bar{z}_0}. \tag{2.7.11}$$

Accordingly, both $f(z)$ and the associated Bergman kernel function $K(z, z_0)$ have a pole at $z_0 = 1/\bar{z}_0$, and the kernel $K(z, z_0)$ can be represented by the polynomial series

$$K(z, z_0) = \frac{1}{\pi} \sum_{k=1}^{\infty} k \, (z\bar{z}_0)^{k-1}, \tag{2.7.12}$$

which converges rapidly when $|z_0|$ is small; however, the convergence becomes considerably slower the faster $|z_0| \to 1$, i.e., the closer the pole $1/\bar{z}_0$ gets to the boundary of U. For Bergman kernel, see §8.4.1.

REFERENCES USED: Ahlfors [1966], Betz [1964], Bieberbach [1916], Blaskett and Schwerdtfeger [1945-1946], Boas [1987], Carrier, Krook and Pearson [1966], de Branges [1985], Fitzgerald [1985], Franklin [1944], Gaier [1964], Goluzin [1969], Jeffrey [1992], Kythe [1998; 2016], Lawrentjew and Schabat [1967], Marsden and Hoffman [1987], Muskhelishvili [1953/1992], Nehari [1952], Osserman [2004], Schinzinger and Laura [2003], Wen [1992].

3

Linear and Bilinear Transformations

The central problem in the theory of conformal mapping is to determine a function f which maps a given region $D \subset \mathbb{C}$ conformally onto another region $G \subset \mathbb{C}$. The function f does not always exist, and it is not always uniquely determined. The Riemann mapping theorem (§2.6, Theorem 2.15) guarantees the existence and uniqueness of a conformal map of D onto the unit disk U under certain specific conditions. First, we will introduce definitions of certain curves, and some elementary mappings, before we will study linear and bilinear transformations.

3.1 Definition of Certain Curves

Some curves are defined as follows:

3.1.1 Line, or straight line, has the equation $lx + my + p = 0$, where l, m, p are real, and $l^2 + m^2 > 0$. Then $\Re\{\bar{\lambda} z\} = p$, or $\Im\{\mu z\} = p$, where $\lambda = l + i\,m \neq 0$, and $\mu = m + i\,l = i\,\bar{\lambda}$.

The *distance* of a point z_0 from the line $lx + my + p = 0$ is given by $\left|\dfrac{\Re\{\bar{\lambda} z_0\} - p}{\lambda}\right|$.

The angle between the x-axis and the line $lx+my+p = 0$ is given by $\arg\{\lambda + \frac{\pi}{2} + n\pi\}$, $n = 0, \pm 1, \pm 2, \dots$.

3.1.2 Circle *in the extended sense* means a circle or a straight line. There is no distinction between circles and lines in the theory of bilinear transformations. However, a circle in the z-plan is defined by $|z - z_0| = r$ with center at z_0 and radius r; in the w-plane it is defined by $|w - w_0| = R$ with center at w_0 and radius R.

3.1.3 Ellipse has the equation $|z - z_1| + |z - z_2| = k$, $k > |z_1 - z_2|$; its foci are z_1 and z_2; major axis is k; eccentricity is $\dfrac{|z_1 - z_2|}{k}$; exterior of the ellipse is defined as $|z - z_1| + |z - z_2| > k$, and interior as $|z - z_1| + |z - z_2| < k$.

3.1.4 Hyperbola has the equation $|z - z_1| + |z - z_2| = \pm k$, $0 < k < |z_1 - z_2|$, where the plus or minus sign is used according to whether the branch is nearer to the focus z_2 or it is nearer to the focus z_1. The real axis is k; eccentricity $\dfrac{|z_1 - z_2|}{k}$; equations of asymptotes are

$$\left\{ (\bar{z}_1 - \bar{z}_2)\, e^{\pm\theta} \Big(z - \frac{z_1 + z_2}{2} \Big) \right\} = 0, \quad \cos\theta = \frac{k}{|z_1 - z_2|};$$

the exterior, not containing foci is defined by $\big||z - z_1| - |z - z_2|\big| < k$.

3.1.5 Rectangular Hyperbola is the hyperbola of §3.1.4 if $k = \dfrac{|z_1 - z_2|}{\sqrt{2}}$. Moreover, if $z_2 = -z_1$ (i.e., the origin is the center of the curve), then the equation of the rectangular hyperbola is

$$\Re\left\{\frac{z^2}{z_1^2}\right\} = \frac{1}{2},$$

with foci $\pm z_1$. The exterior of the hyperbola is defined by

$$-\infty \le \Re\left\{\frac{z^2}{z_1^2}\right\} < \frac{1}{2}.$$

Equations of asymptotes are (three forms):

$$\Re\left\{\frac{z^2}{z_1^2}\right\} = 0;\ \Re\left\{\frac{z}{z_1}\right\} = \pm\Im\left\{\frac{z}{z_1}\right\} = \frac{1}{2};\ \Re\left\{\frac{z(1 \pm i)}{z_1}\right\} = 0.$$

3.1.6 Parabola has the equation $|z - z_0| = |k - \Re\{\lambda z\}||\lambda|^{-1}$, $\lambda \neq 0$, $k \gtrless \Re\{\lambda z_0\}$. It has focus at $z = z_0$; directrix $\Re\{\lambda z\} = k$; latus rectum $2|k - \Re\{\lambda z_0\}||\lambda|^{-1}$; and vertex $z = z_0 + \dfrac{k - \Re\{\lambda z_0\}}{2\lambda|}$.

3.1.7 Cassini's ovals or Cassinians, presented in Figure 3.1, are defined by $|z-z_1||z-z_2| = \alpha$, $\alpha > 0$; its foci are z_1, z_2. For $\alpha = \frac{1}{4}|z_1 - z_2|^2 = 1$, the curve is a *lemniscate*.

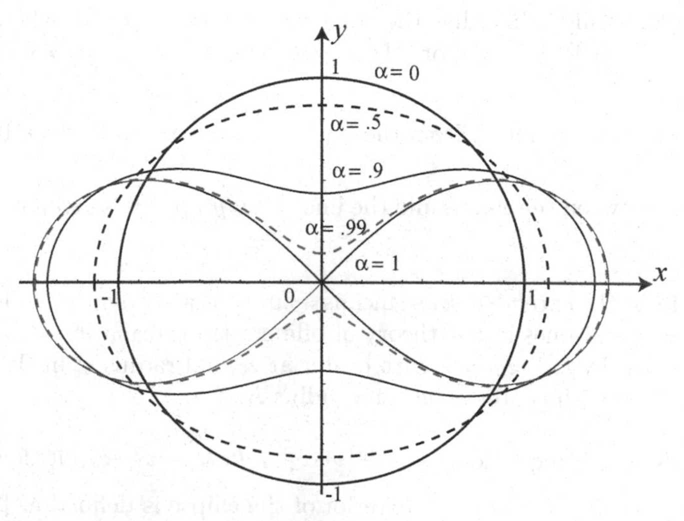

Figure 3.1 Cassini's ovals and lemniscate.

Cassini's ovals are plotted using the definition with rectangular coordinates:

$$F(x, y, \alpha) = \left[(x + \alpha)^2 + y^2\right] \left[(x - \alpha)^2 + y^2\right] = 1$$

for values of $\alpha = 0, 0.5, 0.9, 0.99$, and 1. For $\alpha = 0$ the curve becomes the unit circle. The region does not remain simply connected for $\alpha = 1$; it is a doubly connected region and its boundary curve is called the *lemniscate*.

3.1.8 Cardioid and Limaçons. The cardioid, defined by $r = 2a(1+\cos\theta)$, is the curve in Figure 3.2(a) that is described by a point P of a circle of radius a as it rolls on the outside of a fixed circle of radius a. It is a special case of the limaçon of Pascal, which has the equation $r = b + a\cos\theta$. As described in Figure 3.2(b), the line $0Q$ joining the origin to any point Q on a circle of diameter a passes through 0; then the limaçon is the locus of all points P such that $PQ = b$. Figure 3.2(b) represents the case when $2a > b$, and Figure 3.2(c) the case when $2a < b$. If $b \geq 2a$, the curve is convex. If $b = a$, the limaçon becomes the cardioid.

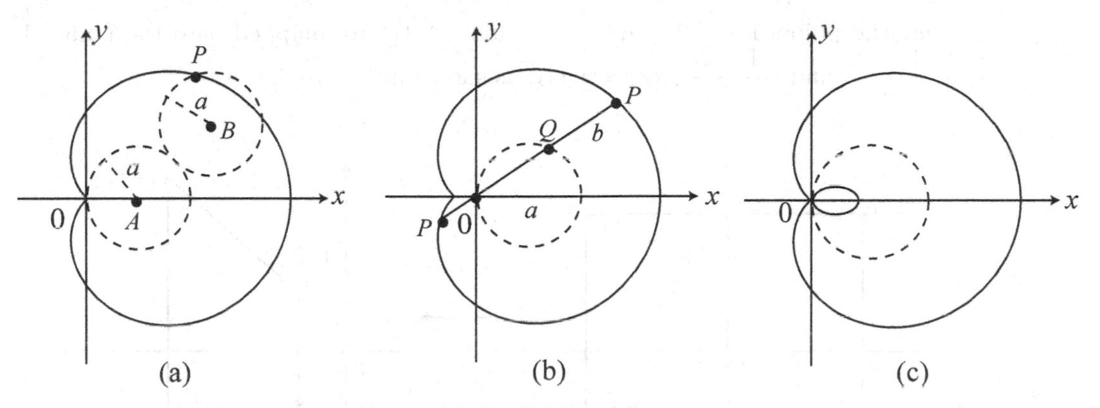

Figure 3.2 Cardioid and limaçons.

3.2 Bilinear Transformations

The *bilinear (linear-fractional, or Möbius) transformation* is of the form

$$w = f(z) = \frac{az+b}{cz+d}, \tag{3.2.1}$$

where a, b, c, d are complex constants such that $ad - bc \neq 0$ (otherwise the function $f(z)$ would be identically constant). If $c = 0$ and $d = 1$, or if $a = 0$, $d = 0$ and $b = c$, then the function (3.2.1) reduces to a linear transformation $w = az + b$, or an inversion $w = \frac{1}{z}$, respectively. The transformation (3.2.1) is also written as

$$w = \frac{a}{c} + \frac{bc - ad}{c(cz+d)}, \tag{3.2.2}$$

that can be viewed as composed of the following three successive functions:

$$z_1 = cz + d, \quad z_2 = \frac{1}{z_1}, \quad w = \frac{a}{c} + \frac{bc - ad}{c} z_2,$$

which shows that the mapping (3.2.1) is a linear transformation, followed by an inversion which is followed by another linear transformation. The bilinear transformation (3.2.1) maps the extended z-plane conformally onto the extended w-plane such that the pole at

$z = -d/c$ is mapped into the point $w = \infty$. The inverse transformation

$$z = f^{-1}(w) = \frac{b - dw}{-a + cw} \tag{3.2.3}$$

is also bilinear defined on the extended w-plane, and maps it conformally onto the extended z-plane such that the pole at $w = a/c$ is mapped into the point $z = \infty$. Note that $f'(z) = \dfrac{ad - bc}{(cz + d)^2} \neq 0$; also $[f^{-1}(w)]' = \dfrac{-ad + bc}{(cw - a)^2} \neq 0$. A bilinear transformation carries circles into circles (in the extended sense), as explained in §3.1.

Example 3.1. The bilinear transformation that maps the square of side 2, centered at the origin of the z-plane onto the parallelogram shown in Figure 3.3 is given by $w = \dfrac{az + b}{cz + d}$ such that the points $1 + i$, $1 - i$, $-1 - i$, and $-1 + i$ are mapped onto the points $1 - 2i$, $1 + 2i$, $\dfrac{1 + 2i}{5}$, and $-\dfrac{1 + 2i}{5}$, respectively, as shown in Figure 3.3. ■

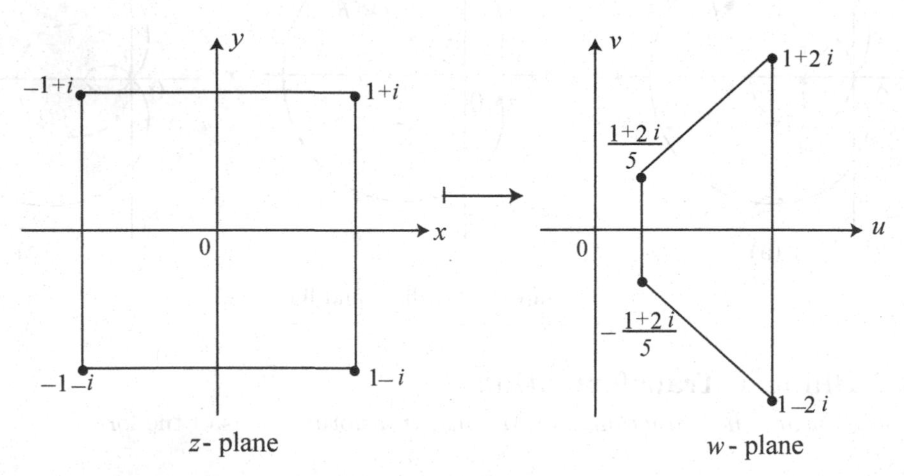

Figure 3.3 Conformal and isogonal mappings.

Example 3.2. The transformation that maps the half-plane $\Im\{z\} > \Re\{z\}$ onto the interior of the circle $|w - 1| = 3$ is

$$w = \frac{(1 + 3i)z + 4i}{z + i}.$$

To prove it, let us regard the given half-plane as the 'interior' of the 'circle' through ∞ defined by the line $\Im\{z\} = \Re\{z\}$. The three points in clockwise order are $z_1 = \infty, z_2 = 0$, and $z_3 = -1 - i$. The three points on the circle $|w - 1| = 3$ in clockwise order are $w_1 = 1 + 3i$, $w_2 = 4$, and $w_3 = -2$. The linear fractional transformation that maps three points in order is defined by Eq (3.2.1), where from the images of ∞ and 0 we have $a = 1 + 3i$ and $b/c = 4$, i.e., $b = 4i$ and $c = i$. Substituting these into Eq (3.2.1), we get the desired mapping.

Since $z_0 = i$ is in the half-plane $\Im\{z\} > \Re\{z\}$, its image under the above map is $w_0 = 2.5 + 1.5\,i$; thus, as $w_0 - 1 = 1.5 + 1.5\,i$ has a modulus $1.5 \times \sqrt{2} \approx 2.12132 < 3$, we find that w_0 is in the interior of $|w - 1| = 3$ (see Figure 3.4). ■

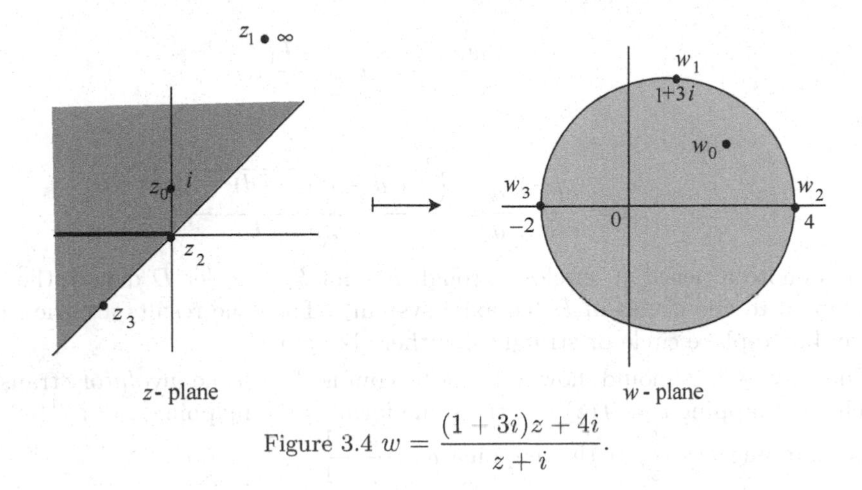

Figure 3.4 $w = \dfrac{(1+3i)z+4i}{z+i}$.

3.2.1 Fixed Points. The fixed points of a mapping are those points at which $w = z$. For example, the mapping $w = z^3$ has a fixed point $z = w = 1$. Let F_1 and F_2 be the *fixed points* of the bilinear transformation (3.2.1), $ad - bc \neq 0$. Then F_1 and F_2 are the roots of the equation

$$cF^2 + (d-a)F - b = 0.$$

We will ignore the *identical transformation* $w = z$, and consider the following cases:

(i) If $F_1 = F_2 = \infty$, we have $w = z + b$, $b \neq 0$, p, q real. Then (i) any line $\Im\{\bar{b}z\} = p$ is mapped onto itself, but onto no other 'circle' (in the extended sense); and (ii) the line $\Re\{\bar{b}z\} = q$ is mapped onto $\Re\{\bar{b}w\} = q - |b|^2$.

(ii) If $F_1 = F_2 \neq \infty$, we have

$$\frac{1}{w-F} = \frac{1}{z-F} + k,$$

where

$$c \neq 0,\ F = \frac{a-d}{2c} = \frac{2b}{d-a},\ k = \frac{2c}{a+d}.$$

The transformation is 'parabolic'. The straight line $\Im\{k(z-r)\} = 0$ and every circle touching this straight line at F is mapped onto itself; no other circle has this property. The set of 'circles' orthogonal to the above set is, as a whole, mapped onto itself.

(iii) If $F_1 \neq F_2 = \infty$, we have $w = F_1 = \alpha(z - F_1)$, where $\alpha = a/d$, $\alpha \neq 0, 1$, $c = 0$, and $F_1 = \dfrac{b}{d-a}$. Let D denote the set of concentric circles centered at F_1, and E denote the set of straight lines through F_1. Then the entire set D is mapped onto itself; so is E. None of the 'circles' other than those stated below is mapped onto itself:

(a) α real, but $\alpha \neq -1$: then every line of E is mapped onto itself;

(b) $\alpha = -1$: then every circle of D, and every line of E, is mapped onto itself;

(c) $|\alpha| = 1$, but $\alpha \neq -1$: then every circle of D is mapped onto itself;

(d) α not real, $|\alpha| \neq 1$: then no circle or line is mapped onto itself (this is known as the *loxodromic* transformation).

(iv) If $F_1 \neq F_2$, $F_1 \neq \infty$, $F_2 \neq \infty$, we have

$$\frac{w - F_1}{w - F_2} = \alpha \frac{z - F_1}{z - F_2},$$

where

$$\alpha = \frac{cF_2 + d}{cF_1 + d} = \frac{\left(a + d - \sqrt{(a-d)^2 + 4bc}\right)^2}{4(ad - bc)}.$$

Let E denote a pencil of 'circles' through F_1 and F_2, and let D denote the set of 'circles' orthogonal to the circles of E (co-axial system). Then the results are the same as in (iii) above, but replace circle or straight line there by 'circle'.

The fixed points should, however, not be confused with the *involutory* transformation for which the mapping $w = f(z)$ has the same form as the mapping $z = f^{-1}(w)$. An example is the mapping $w = \frac{1}{z}$, or the mapping $w = \frac{z+1}{z-1}$.

3.2.2 Linear Transformation. $w = az + b$, $a \neq 0$. The special cases are as follows:

(a) $w = az$: (i) if $a > 0$, then there is *magnification* by ratio a centered at 0; (ii) if $a = e^{i\tau}$, τ real, then there is *rotation* by the angle τ about 0; and (iii) if $a = m\, e^{i\tau}$, $m > 0$, and τ real, then there is *rotation* by the angle τ about 0, followed by a *magnification* of ratio m centered at 0.

(b) $w = z + b$: There is a *translation* by the vector $\overline{0b}$.

(c) $w = az + b$, $a \neq 1$: Then $F_0 = \frac{b}{1-a}$, $a = me^{i\tau}, m > 0, \tau$ real. There is *rotation* about F_0 by the angle τ, followed by *magnification* by ratio m centered at F_0.

The correspondence of the above linear transformations between the z-plane and the w-plane is as follows:

(i) The line $x = p$ is mapped onto the line $\Re\{\frac{w-b}{a}\} = p$.

(ii) The line $y = q$ is mapped onto the line $\Im\{\frac{w-b}{a}\} = q$.

(iii) The line $lx + my = p$ is mapped onto the line $\Re\{(l - im)\frac{w-b}{a}\} = p$.

(iv) The line $\Re\{az + b\} = p$ is mapped onto the line $u = P$.

(v) The line $\Im\{az + b\} = q$ is mapped onto the line $v = Q$.

(vi) The line $\Re\{(l - im)(az + b) = p$ is mapped onto the line $lu + mv = P$.

(vii) The circle $|z - z_0| = r$ is mapped onto the circle $|w - (aw_0 + b)\,| = |a|R$.

(vii) The circle $\left|z - \frac{w_0 - b}{a}\right| = \frac{r}{|a|}$ is mapped onto the circle $|w - w_0| = R$.

3.2.3 Composition of Bilinear Transformations. We will consider the general bilinear transformation (3.2.1) with $ad - bc \neq 0$. Note that if $ad - bc = 0$, then we have $w =$ const. The general bilinear transformation (3.2.1) consists of the following three successive transformations:

(i) translation of z: $z_1 = z + d/c$;

(ii) inversion of z_1: $z_2 = 1/z_1$; and

(iii) translation, rotation, and scale change of z_2: $w = \dfrac{a}{c} - \dfrac{ad-bc}{c^2} z_2$.

Map 3.1. If $a = -d$, the transformation is *involutory*, i.e., w and z may be interchanged. The magnification is $\left|\dfrac{dw}{dz}\right| = \left|\dfrac{ad-bc}{(cz+d)^2}\right|$, or

$$|f'(z)|^2 = \left(\frac{\partial u}{\partial x}\right)^2 + \left(\frac{\partial v}{\partial x}\right)^2 = \left(\frac{\partial u}{\partial y}\right)^2 + \left(\frac{\partial v}{\partial y}\right)^2.$$

The scale or magnification factor m is then defined as $m = |dw/dz| = |f'(z)|$, where $m = |m|$, and $M = 1/m$. The critical points are $z = -d/c$ and $z = \infty$; at these points the transformation is *not* conformal in the usual sense. The details of the transformation are as follows:

(i) The points $z = 0;\ -b/a;\ z_\infty = -d/c$ are mapped onto the points $w = b/d;\ 0;\ \infty$.

(ii) The point at infinity $z = \infty$ is mapped onto the point $w_\infty = a/c$.

Map 3.2. The bilinear transformation $w = f(z)$ that transforms the three points z_k into the points w_k, $k = 1, 2, 3$, can be expressed as (Saff and Snider [1976:329])

$$\begin{vmatrix} 1 & z & w & zw \\ 1 & z_1 & w_1 & z_1 w_1 \\ 1 & z_2 & w_2 & z_2 w_2 \\ 1 & z_3 & w_3 & z_3 w_3 \end{vmatrix} = 0.$$

Map 3.3. Let z be fixed and let $\Re\{z\} \geq 0$. Define a sequence of bilinear transformations

$$T_0(w) = \frac{a_0}{z + a_0 + b_1 + w}, \quad T_k(w) = \frac{a_k}{z + b_{k+1} + w}, \quad k = 1, \dots, n-1,$$

such that each $a_j > 0$ is real, and each b_j is zero or purely imaginary for $j = 0, 1, 2, \dots, n-1$. We use induction to prove that the chain of transformations defined by $\zeta = S(w) = (T_0 \circ \cdots \circ T_{n-2} \circ T_{n-1})(w)$ maps the half-plane $\Re\{w\} > 0$ onto a region contained in the disk $|\zeta - 1/2| < 1/2$. Thus, using the *Wallis criterion*, if $P(z) = z^n + c_1 z^{n-1} + c_2 z^{n-2} + \cdots + c_n$ is a polynomial of degree $n > 0$ with complex coefficients $c_k = p_k + i\,q_k$, $k = 1, 2, \dots, n$, if $Q(z) = p_1 z^{n-1} + i\,q_2 z^{n-2} + p_3 z^{n-3} + i\,q_4 z^{n-4} + \cdots$, and if $Q(z)/P(z)$ can be written as a continued fraction, then we have $Q(z)/P(z) = (T_0 \circ T_{n-2} \cdots \circ T_{n-1})(z)$ (Saff and Snider [1976:330]).

Theorem 3.1. (Schwarzian derivative) (i) *The Schwarzian derivative (also called the Schwarz differential operator), defined by*

$$\{w, z\} = \left(\frac{w''}{w'}\right)' - \frac{1}{2}\left(\frac{w''}{w'}\right)^2 = \frac{w'''}{w'} - \frac{3}{2}\left(\frac{w''}{w'}\right)^2,$$

is invariant under bilinear transformations. (Nehari [1952:199]; Wen [1992:76]).

(ii) *The Schwarzian derivative of the function $w = w(z)$ that maps the upper half-plane $\Im\{z\} > 0$ conformally onto a circular triangle with interior angles $\alpha\pi$, $\beta\pi$ and $\gamma\pi$ is given by*

$$\{w, z\} = \frac{1-\alpha^2}{2z^2} + \frac{1-\beta^2}{2(1-z)^2} + \frac{1-\alpha^2-\beta^2+\gamma^2}{2z(1-z)}.$$

(Nevanlinna and Paatero [1969:342]).

3.3 Cross-Ratio

A cross-ratio between four distinct finite points z_1, z_2, z_3, z_4 is defined by

$$(z_1, z_2, z_3, z_4) = \frac{z_1 - z_2}{z_1 - z_4} \cdot \frac{z_3 - z_4}{z_3 - z_2}. \tag{3.3.1}$$

If z_2, z_3, or z_4 is a point at infinity, then (3.3.1) reduces to

$$\frac{z_3 - z_4}{z_1 - z_4}, \quad \frac{z_1 - z_2}{z_1 - z_4}, \quad \text{or} \quad \frac{z_1 - z_2}{z_3 - z_2},$$

respectively. The cross-ratio (z, z_1, z_2, z_3) is invariant under bilinear transformations.

Theorem 3.2. *A bilinear transformation is uniquely defined by a correspondence of the cross-ratios, i.e.,*

$$(w, w_1, w_2, w_3) = (z, z_1, z_2, z_3), \tag{3.3.2}$$

which maps any three distinct points z_1, z_2, z_3 in the extended z-plane into three prescribed points w_1, w_2, w_3 in the extended w-plane. The cross-ratio (z, z_1, z_2, z_3) is the image of z under a bilinear transformation that maps three distinct points z_1, z_2, z_3 into $0, 1, \infty$.

3.3.1 Symmetric Points. The points z and z^* are said to be *symmetric* with respect to a circle C (in the extended sense) through three distinct points z_1, z_2, z_3 iff

$$(z^*, z_1, z_2, z_3) = \overline{(z, z_1, z_2, z_3)}. \tag{3.3.3}$$

The mapping that carries z into z^* is called a *reflection* with respect to C. Two reflections obviously yield a bilinear transformation.

If C is a straight line, then we choose $z_3 = \infty$, and the condition for symmetry (3.3.3) gives

$$\frac{z^* - z_1}{z_1 - z_2} = \frac{\bar{z} - \bar{z}_1}{\bar{z}_1 - \bar{z}_2}. \tag{3.3.4}$$

Let z_2 be any finite point on the line C. Then, since $|z^* - z_1| = |z - z_1|$, the points z and z^* are equidistant from the line C. Moreover, since $\Im\{\frac{z^* - z_1}{z_1 - z_2}\} = -\Im\{\frac{z - z_1}{z_1 - z_2}\}$, the line C is the perpendicular bisector of the line segment joining z and z^*. If C is the circle $|z - a| = R$, then

$$\begin{aligned}
\overline{(z, z_1, z_2, z_3)} &= \overline{(z - a, z_1 - a, z_2 - a, z_3 - a)} \\
&= \left(\bar{z} - \bar{a}, \frac{R^2}{z_1 - a}, \frac{R^2}{z_2 - a}, \frac{R^2}{z_3 - a}\right) \\
&= \left(\frac{R^2}{\bar{z} - \bar{a}}, z_1 - a, z_2 - a, z_3 - a\right) \\
&= \left(\frac{R^2}{\bar{z} - \bar{a}} + a, z_1, z_2, z_3\right).
\end{aligned}$$

Hence, in view of (3.3.3), we find that the points z and $z^* = \dfrac{R^2}{\bar{z} - \bar{a}} + a$ are symmetric, i.e.,

$$(z^* - a)(\bar{z} - \bar{a}) = R^2. \tag{3.3.5}$$

Note that $|z^* - a||z - a| = R^2$; also, since $\dfrac{z^* - a}{z - a} > 0$, the points z and z^* are on the same ray from the point a (Figure 3.5). Also, the point symmetric to a is ∞.

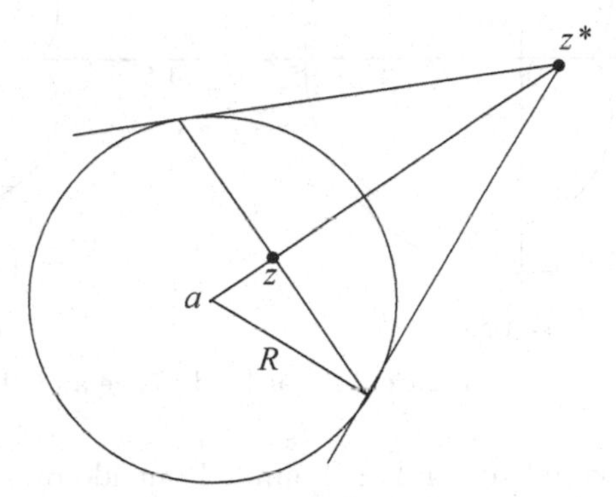

Figure 3.5 Symmetry with respect to a circle.

A GENERALIZATION of this result is as follows: If Γ denotes an analytic Jordan curve with parametric equation $z = \gamma(s)$, $s_1 < s < s_2$, then for any point z *sufficiently close* to Γ, the point

$$z^* = \gamma\,\overline{(\gamma^{-1}(z))} \tag{3.3.6}$$

defines a symmetric point of z with respect to Γ (Sansone and Gerretsen, [1960: 103]; Papamichael, Warby and Hough [1986]). Some examples are as follows:
(i) If Γ is the circle $x^2 + y^2 = a^2/9$, then $z^* = a^2/9\bar{z}$.

(ii) If Γ is the ellipse $\dfrac{(x + a/2)^2}{a^2} + y^2 = 1$, then

$$z^* = -\frac{a}{2} + \frac{(a^2+1)(\bar{z} + a/2) + 2ia\sqrt{a^1 - 1 - (\bar{z} + a/2)}}{a^2 - a}.$$

(iii) If Γ_1 is a cardioid defined by $z = \gamma(s) = \left(\dfrac{1}{2} + \cos\dfrac{s}{2}\right) e^{is}$, $-\pi < s \le \pi$, then from (3.3.6) we cannot write an explicit expression for symmetric points with respect to Γ_1. However, for any real t,

$$\gamma\,(\pm\, it) = \left(\frac{1}{2} + \cosh\frac{t}{2}\right) e^{\mp t}$$

defines two real symmetric points with respect to Γ_1, provided the parameter t satisfies the equation $\gamma(it)\,\gamma(-it) = a^2$, i.e., $\dfrac{1}{2} + \cosh\dfrac{t}{2} = a$ which has the roots

$$t = 2\,\cosh^{-1}(a - 1/2) = \pm 2\,\log\,\rho, \tag{3.3.7}$$

where $\rho = a - \frac{1}{2} + \sqrt{\left(a - \frac{1}{2}\right)^2 - 1}$.

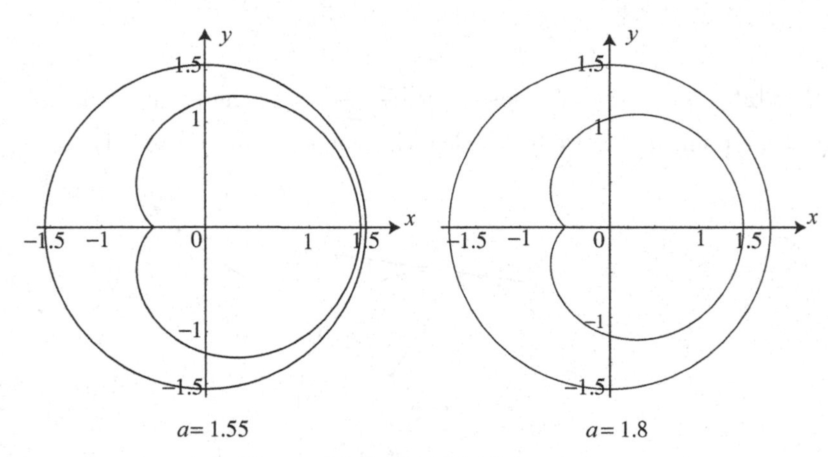

Figure 3.6 Cardioid inside a circle.

(iv) If a doubly connected region Ω is bounded outside by the circle $\Gamma_2 = \{z : |z| = a,\, a > 1.5\}$ and inside by the cardioid Γ_1 defined above in (iii) (see Figure 3.6), then it follows from (3.3.7) that there is one pair of real common symmetric points $\zeta_1 \in \text{Int}(\Gamma_1)$ and $\zeta_2 \in \text{Ext}(\Gamma_2)$ such that $\zeta_1 = a/\rho^2$ and $\zeta_2 = a\rho^2$, where ρ is defined in (3.3.7).

3.3.2 Symmetry Principle. The *symmetry principle* states that if a bilinear transformation maps a circle C_1 onto a circle C_2, then it maps any pair of symmetric points with respect to C_1 into a pair of symmetric points with respect to C_2. This means that bilinear transformations preserve symmetry.

A practical application of the symmetry principle is to find bilinear transformations which map a circle C_1 onto a circle C_2. We already know that the transformation (3.3.2) can always be determined by requiring that the three points $z_1, z_2, z_3 \in C_1$ map onto three points $w_1, w_2, w_3 \in C_2$. But a bilinear transformation is also determined if a point $z_1 \in C_1$ should map into a point $w_1 \in C_2$ and a point $z_2 \notin C_1$ should map into a point $w_2 \notin C_2$. Then, by the symmetry principle, the point z_2^* which is symmetric to z_2 with respect to C_1 is mapped into the point w_2^* which is symmetric to w_2 with respect to C_2, and then the desired bilinear transformation is given by

$$(w, w_1, w_2, w_2^*) = (z, z_1, z_2, z_2^*)\,. \tag{3.3.8}$$

Let U^+ denote the region $|z| < 1$, U^- the region $|z| > 1$, and let $C = \{|z| = 1\}$ be their common boundary (Figure 3.7). Let $f(z)$ be a function defined on U^+. Then this function can be related to the function $f^*(z)$ defined in U^- in the same manner as in (2.6.3) for half-planes, except that now the conjugate complex points are replaced by points inverse with respect to the circle C according to the relation (3.3.5). Thus,

$$f^*(z) = \overline{f\left(\frac{1}{\bar{z}}\right)} = \bar{f}\left(\frac{1}{z}\right). \tag{3.3.9}$$

This relation is symmetrical, i.e.,

$$f(z) = \overline{f^*\left(\frac{1}{\bar{z}}\right)}, \quad (f^*(z))^* = f(z). \tag{3.3.10}$$

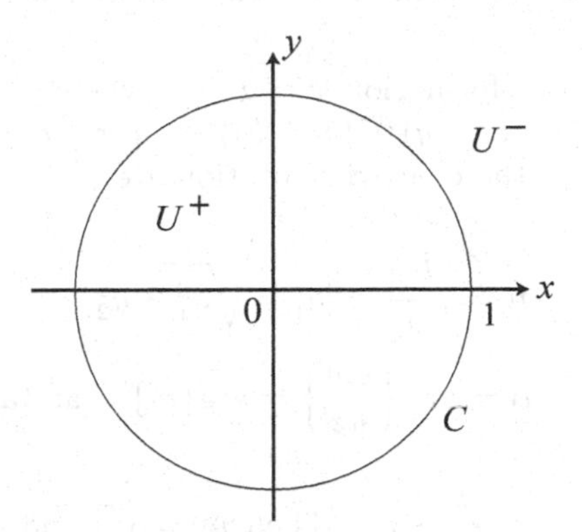

Figure 3.7 Unit disk.

If $f(z)$ is regular or meromorphic in U^+, then $f^*(z)$ is regular or meromorphic in U^-. Also, if $f(z)$ is a rational function defined by (2.6.4), then

$$f^*(z) = \frac{\bar{a}_n\, z^{-n} + \bar{a}_{n-1}\, z^{-n+1} + \cdots + \bar{a}_0}{\bar{b}_m\, z^{-m} + \bar{b}_{-m+1}\, z^{m-1} + \cdots + \bar{b}_0}. \tag{3.3.11}$$

Moreover, if $f(z)$ has the power series expansion

$$f(z) = \sum_{k=-\infty}^{\infty} a_k\, z^k \quad z \in U^+,$$

then

$$f^*(z) = \sum_{k=-\infty}^{\infty} \bar{a}_k\, z^{-k}, \quad z \in U^-. \tag{3.3.12}$$

If $f(z)$ has a zero (pole) of order k at $z = \infty$ $(z = 0)$, so does $f^*(z)$. Let us assume that $f(z)$ approaches a definite limit value $f^+(t)$ as $z \to t \in C$ from U^+. Then $f^{*-}(t)$ exists and

$$f^{*-}(t) = \bar{f}^-\left(\frac{1}{t}\right) = \overline{f^+(t)}, \tag{3.3.13}$$

because $1/\bar{z} \to t$ from U^+ as $z \to t$ from U^-, and hence $f^*(z) = \bar{f}\left(\frac{1}{z}\right) = \overline{f\left(\frac{1}{\bar{z}}\right)} \to \overline{f^+(t)}$. If $f(z)$ is regular in U^+ except possibly at infinity and continuous on C from the left, then let $F(z)$ be sectionally regular and be defined by

$$F(z) = \begin{cases} f(z) & \text{for } z \in U^+, \\ f^*(z) & \text{for } z \in U^-. \end{cases} \tag{3.3.14}$$

Then, $F^*(z) = F(z)$, and, as in (2.6.7),

$$F^-(t) = \overline{F^+(t)}, \quad F^+(t) = \overline{F^-(t)}. \tag{3.3.15}$$

Moreover, in view of the Schwarz reflection principle, if $\Im\{f^+(t)\} = 0$ on some part of the circle C, then $f^*(z)$ is the analytic continuation of $f(z)$ through this part of C.

3.3.3 Special Cases. We will discuss the following special cases of the transformations (3.2.1) and (3.2.3).

Map 3.4. (Linear transformation $w = az + b$, where $a = a_1 + ia_2$ and $b = b_1 + ib_2$. Then $w = u + iv = (\alpha_1 + ia_2)(x + iy) + (b_1 + b_2) = (a_1 x - a_2 y + b_1) + i(a_1 y + a_2 x + b_2)$. The magnification factor and the clockwise rotation are

$$m = \frac{|dw|}{|dx|} = |a| = \sqrt{a_1^2 + a_2^2}, \tag{3.3.16}$$

$$\alpha = \arg\left\{\frac{dw}{dz}\right\} = \arg\{m\} = \arctan\left(\frac{a_2}{a_1}\right). \tag{3.3.17}$$

Let $a = 1 + i$ and $b = -\frac{1}{2} - i\frac{3}{2}$. Then $m = \sqrt{2}$ and $\alpha = \pi/4$. The rectangular lines $u = k_u$ and $v = v_k$ are represented in the z-plane by the rectangular lines $y = x - \left(\frac{1}{2} + k_u\right)$ and $y = -x + \left(\frac{3}{2} + k_v\right)$, respectively. The transformation is presented in Figure 3.8.

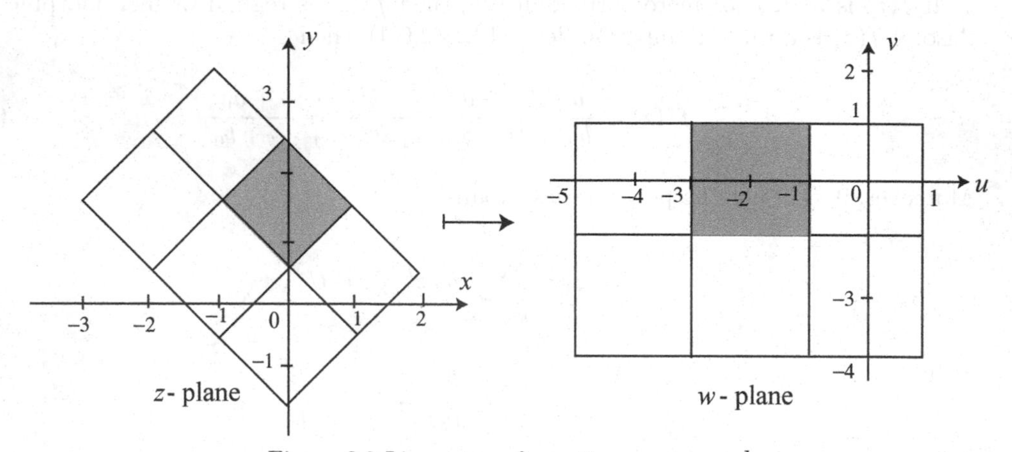

Figure 3.8 Linear transformation $w = az + b$.

In general, the mapping $w = az + b$ translates by b, rotates by $\arg\{a\}$ and magnifies (or contracts) by $|a|$.

Example 3.3. Consider the transformation $w = (1+i)z + (1+2i) = \sqrt{2}\, e^{i\pi/4} z + 1 + 2i$ in the z-plane. It maps the rectangle bounded by the lines AB: $x = 0$; BC: $y = 0$; CD: $x = 2$; and DA: $y = 1$ in the z-plane onto the rectangle bounded by the lines A'B': $u + v = 3$; B'C': $u - v = -1$; C'D': $u + v = 7$; and D'A': $u - v = -3$, respectively, such that the points $z = (0,1), (0,0), (2,0), (2,2)$ are mapped onto the points $w = (0,3), (1,2), (3,4), (2,5)$, respectively. The rectangle ABCD is translated by $(1 + 2i)$, rotated by an angle $\pi/4$ in the counterclockwise direction, and contracted by $\sqrt{2}$. ■

Map 3.5. The transformation $w = 2z - 2i$ maps the circle $|z - i| = 1$ onto the circle $|w| = 2$. The transformation that maps the disk $|z - i| = 1$ onto the exterior of the circle $|w| = 2$ is $w = \dfrac{2}{z - i}$. In fact, let $w = g(z) = \dfrac{az + b}{cz + d}$. Choose $g(i) = \infty$, so $g(z) = \dfrac{az + b}{z - i}$ without loss of generality. Take three points on the circle $|z - i| = 1$ as $0, 1 + i$, and $2i$; then $g(0) = ib, g(1 + i) = a(1 + i)$, and $g(2i) = 2a - ib$. Since all these three points must lie on $|w| = 2$, the simplest choice is $a = 0$ and $b = 2$. Then $g(z) = \dfrac{2}{z - i}$.

Map 3.6. The inversion $w = \dfrac{1}{z}$ is an involutory transformation with fixed points $F_1 = 1$ and $F_2 = -1$. The mapping from z-plane to w-plane, or inversely, is presented in Figure 3.9, where the radius $\overline{0A}$ is 1; the angle $\angle\, z0\bar{z}$ is bisected by the real axis; $a\bar{z} : \bar{z}B = Aw : Bw$; p, q and r are real; and the points $\bar{z}$ and w are inverse with respect to the unit circle.

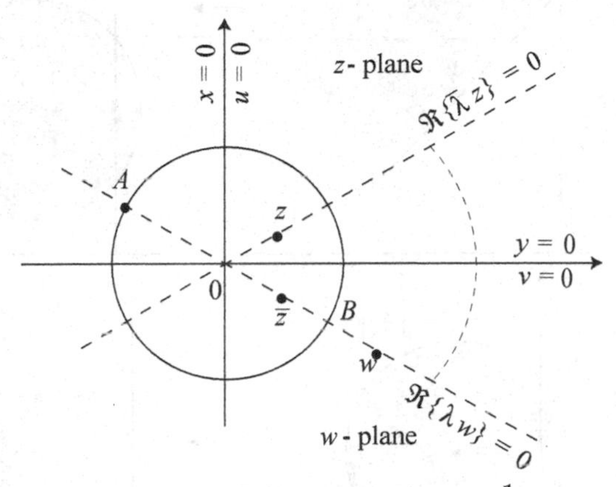

Figure 3.9 Transformation $w = \dfrac{1}{z}$.

There are five special cases under the inversion $w = \dfrac{1}{z}$. They are:

(a) The interior (exterior) of the circle of D in the z-plane is mapped onto the interior (exterior) of the same circle, with points $z_\infty = 0; \infty; i; -i$ in the z-plane mapped into the points $\infty; w_\infty = 0; -i; i$, respectively, in the w-plane.

(b) The line $\Re\{\bar{\lambda} z\} = 0$ in the z-plane is mapped onto the line $\Re\{\lambda w\} = 0$ in the w-plane; and the circle $|z| = r$ in the z-plane is mapped onto the circle $|w| = 1/|z|$ in the w-plane. These two cases are geometrically obvious and no figures are needed for them.

The remaining three cases are presented in Figure 3.10.; details, with $z = p, q, r, r_p, r_q$ real, and $w = P, Q, R, R_p, R_q$ real, are as follows:

(c) The line $\Re\{\bar{\lambda} z\} = 0$ is mapped onto the circle

$$\left| w - \frac{\bar{\lambda}}{2p} \right| = \left| \frac{\lambda}{2p} \right|.$$

See Figure 3.10(c).

(d) The circle $|z - z_0| = |z_0| \neq 0$ is mapped onto the line $\Re\{z_0\} = \frac{1}{2}$.

See Figure 3.10(d).

(e) The circle $|z - z_0| = r \neq |z_0|$ is mapped onto the circle

$$\left| w - \frac{\bar{z}_0}{|z_0|^2 - r^2} \right| = \frac{r}{\left| |z_0|^2 - r^2 \right|}.$$

See Figure 3.10(e).

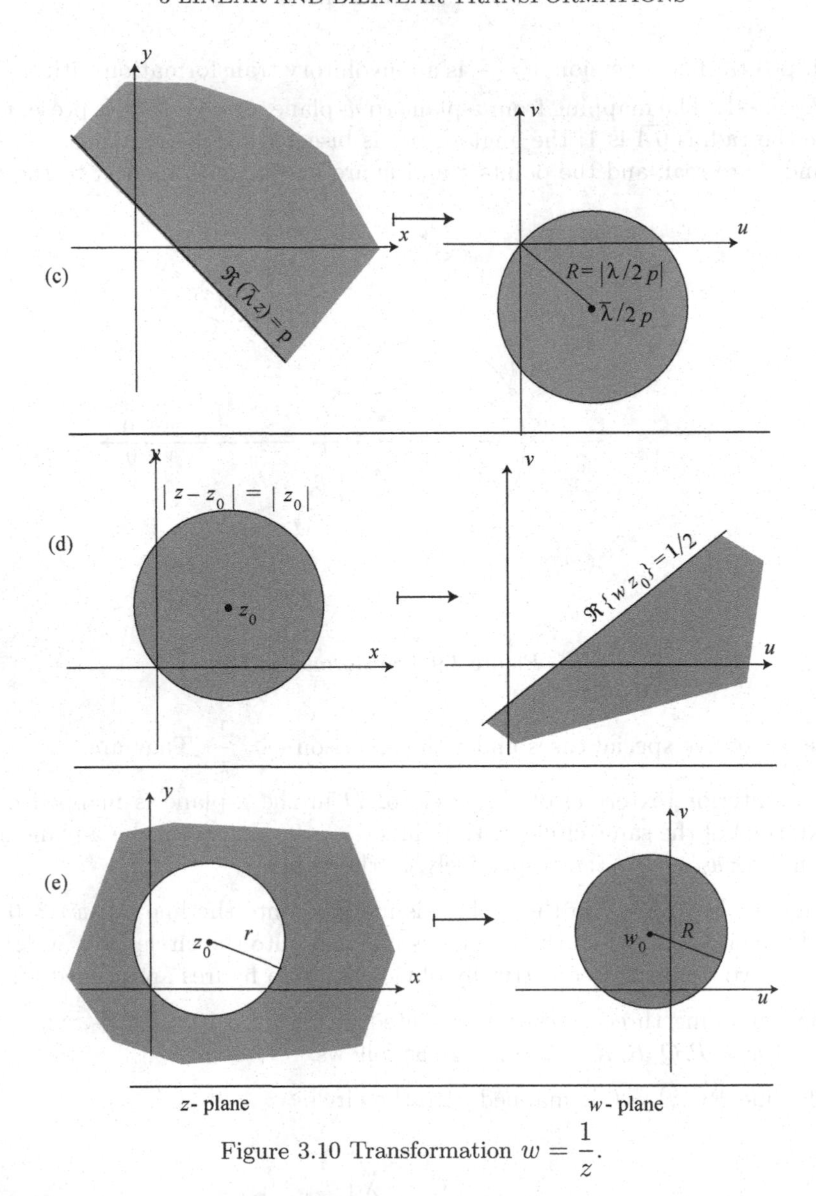

Figure 3.10 Transformation $w = \dfrac{1}{z}$.

Map 3.7. The transformation $w = \dfrac{1}{z}$, or $z = \dfrac{1}{w}$, which maps the interior of a circle in the z-plane onto the exterior of the circle in the w-plane and conversely, is shown in Figure 3.11.

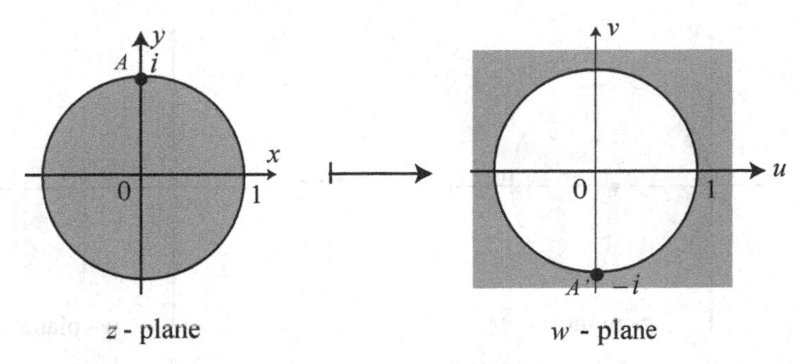

Figure 3.11 Circle in the z-plane onto the exterior of a circle.

Since

$$w = u + iv = \frac{x}{x^2 + y^2} - \frac{y}{x^2 + y^2}\, i, \quad z = x + iy = \frac{u}{u^2 + v^2} - \frac{v}{u^2 + v^2}\, i,$$

we get

$$u = \frac{x}{x^2 + y^2}, \quad v = -\frac{y}{x^2 + y^2}.$$

A line through the origin in the z-plane is mapped onto a line through the origin in the w-plane. However, a line not passing through the origin in the z-plane is mapped onto a circle in the w-plane which passes through the origin, as the following example shows.

Example 3.4. The equation $ax + by + c$, $c \neq 0$ represents a straight line in the z-plane. Under the inversion map $w = 1/z$, this line is mapped onto a circle in the w-plane which passes through the origin. The details are as follows: The mapped curve is

$$\frac{au}{u^2 + v^2} - \frac{bu}{u^2 + v^2} + c = 0.$$

Thus, $au - bv + c\left(u^2 + v^2\right) = 0$. Dividing by c, we obtain

$$u^2 + v^2 + \frac{a}{c}u = \frac{b}{c}v = 0,$$

which is the equation of the required circle. ■

Map 3.8. The transformation $w = \frac{1}{z}$, which maps the interior of the circle $|z - 1| = 1$ in the z-plane onto the right plane $u \geq \frac{1}{2}$, $-\infty < v < \infty$ in the w-plane, is shown in Figure 3.12.

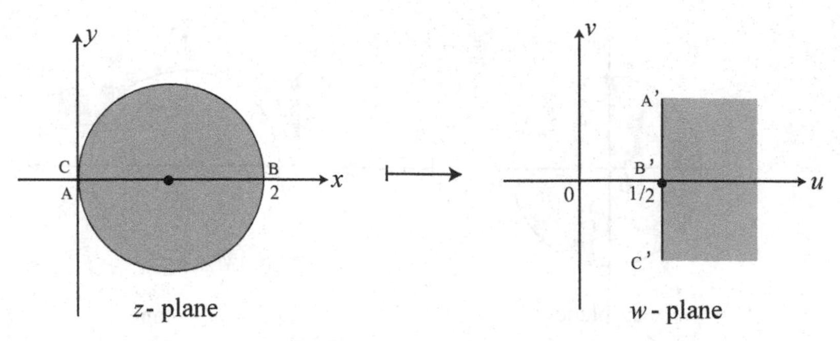

Figure 3.12 Circle in the z-plane onto a rectangle.

Map 3.9. The transformation that maps the exterior of the unit circle in the z-plane onto the interior of the unit circle in the w-plane is defined by

$$w = 1/z,$$

and is presented in Figure 3.13.

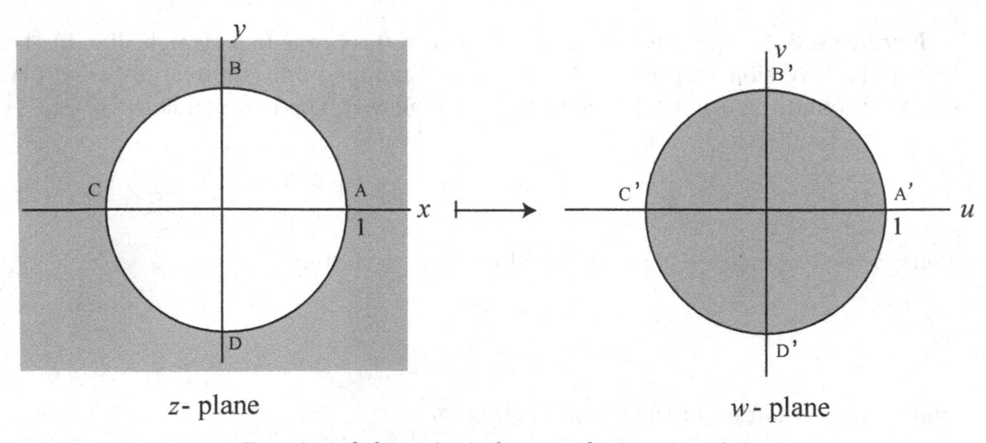

Figure 3.13 Exterior of the unit circle onto the interior of the unit circle.

Map 3.10. The transformation $w = k/z$ is known as the *reciprocal* or the *inverse transformation* for real k. We have $w = R\,e^{i\Theta} = k\left(r\,e^{i\theta}\right)^{-1} = \frac{k}{r}\,e^{-i\theta}$, which gives $R = k/r$, and $\Theta = -\theta$. The fixed points are: $F_{1,2} = \pm\sqrt{k}$, and the critical points are $z = 0, \infty$. It is a useful transformation in electric transmission through a single wire, discussed in Example 22.4 with $k = 3600$. Two particular cases for $k = -\alpha$, $\alpha > 0$, and $k = 1$ are given in Maps 3.6, 3.7, 3.8, and 3.9, respectively.

Map 3.11. The transformation $w = a/z$ maps the exterior of a circle of radius a: $x^2 + y^2 \geq a^2$, in the z-plane onto the unit disk $|w| \leq 1$ in the w-plane.

Map 3.12. The transformation $w = z + \frac{1}{z}$, which maps the exterior of the upper semi-circle in the z-plane onto the upper-half of the w-plane, is presented in Figure 3.14.

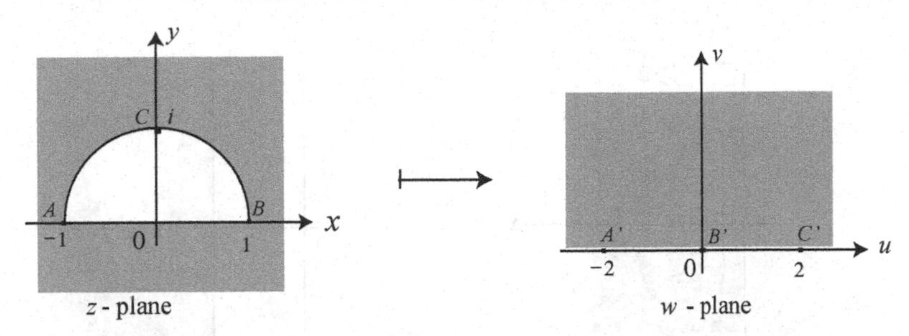

Figure 3.14 Exterior of a circle in the z-plane onto the upper-half of the w-plane.

Map 3.13. The transformation $w = z + \dfrac{1}{z}$, or $z = \dfrac{w + \sqrt{w^2 - 1}}{2}$, which maps the interior of the upper semi-circle in the z-plane onto the lower-half of the w-plane, is presented in Figure 3.15.

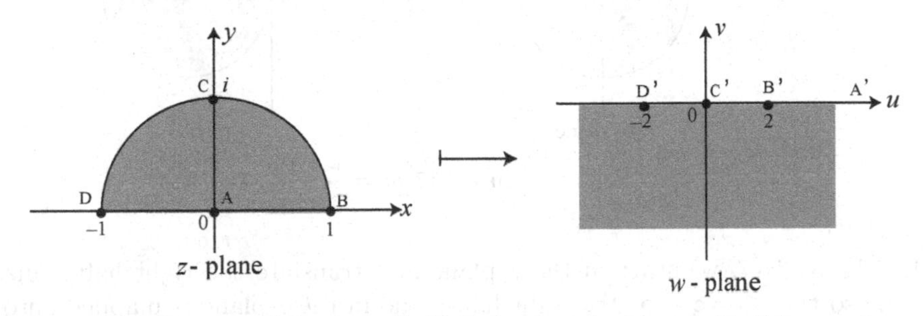

Figure 3.15 Interior of the upper half circle onto the lower-half of the w-plane.

Map 3.14. The transformation $w = z + \dfrac{1}{z}$, which maps the annular region between two upper semi-circles in the z-plane onto the upper-half of the ellipse in the w-plane, where the ellipse (B'C'D') is defined by

$$\left(\frac{ku}{k^2+1}\right)^2 + \left(\frac{kv}{k^2-1}\right)^2 = 1,$$

is presented in Figure 3.16.

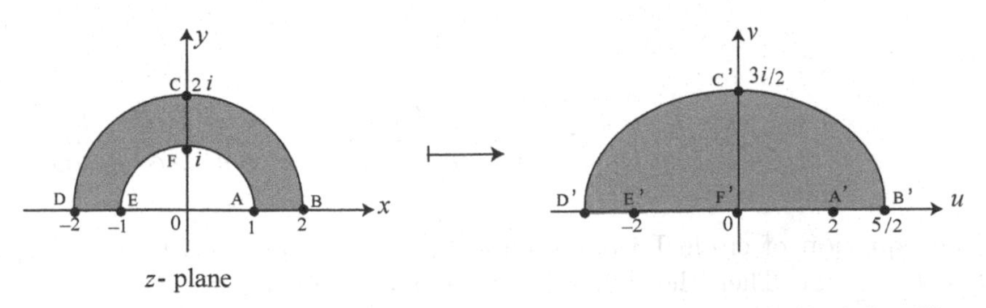

Figure 3.16 Annulus between two semi-circles onto the upper-half of the ellipse.

Map 3.15. If $a = c = 1, d = -b = \alpha$, then $w = \dfrac{z - \alpha}{z + \alpha}$. For this bilinear transformation the above three successive cases are presented in Figure 3.17; these three cases may be treated as separate transformations (see cases 3.11 and 3.44 also).

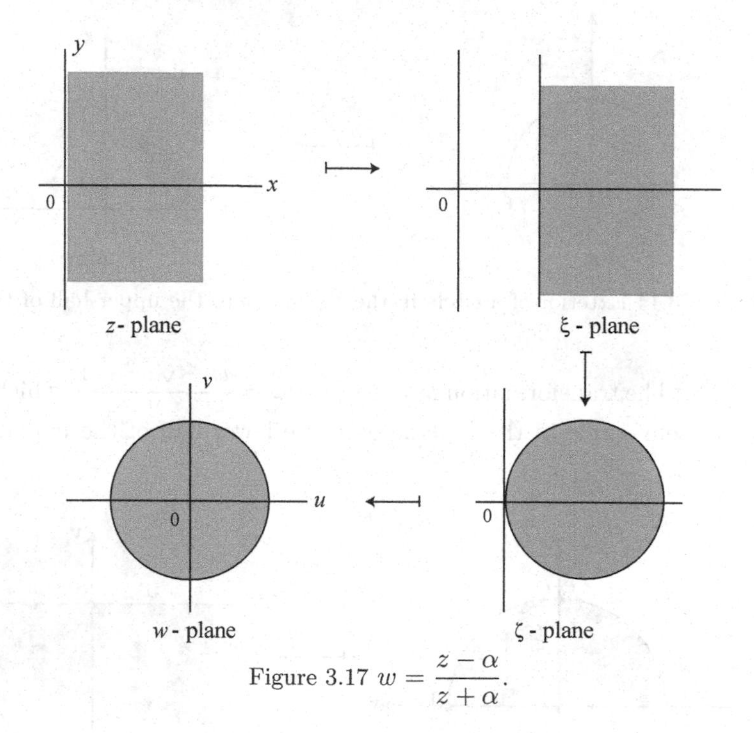

Figure 3.17 $w = \dfrac{z-\alpha}{z+\alpha}$.

In Figure 3.17, we start in the z-plane and translate the right half-plane by α in the ξ-plane so that $\xi = z + \alpha$; the right half-plane in the ξ-plane is mapped onto the circle in the ζ-plane by $\zeta = 1/\xi = 1/(z+\alpha)$; finally, the circle in the ζ-plane is mapped onto the circle in the w-plane so that $w = 1 + 2\alpha\zeta = 1 - \dfrac{2\alpha}{z+\alpha} = \dfrac{z-\alpha}{z\alpha}$.

We can also select suitable parameters for this transformation; for example, given two circles, marked I with radius r' and marked II with radius r'', such that the distance between their centers is h, and the distance of their centers as c_1 and c_2 from the origin of the coordinates (see Figure 3.18), we have the case of translation along the real axis, since

$$w = \frac{z-\alpha}{z+\alpha} = \frac{x+iy-\alpha}{x+iy+\alpha} = Re^{i\Theta}, \tag{3.3.18}$$

where

$$R = \left|\frac{z-\alpha}{z+\alpha}\right|, \quad R^2 = \frac{(x-\alpha)^2+y^2}{(x+\alpha)^2+y^2} = \frac{x^2+y^2+\alpha^2-2\alpha x}{x^2+y^2+\alpha^2+2\alpha x}. \tag{3.3.19}$$

The equation of circle I from Figure 3.18 is $(x-c_1)^2 + (y^2 = (r')^2$, or $x^2 + y^2 = (r')^2 - c_1^2 + 2xc_1$. Then the radius R' of the map of circle I is

$$R'^2 = \frac{r'^2 - c_1^2 + 2xc_1 + \alpha^2 - 2\alpha x}{r'^2 - c_1^2 + 2xc_1 + \alpha^2 + 2\alpha x} = \frac{r'^2 - c_1^2 + \alpha^2 + 2x(c_1-\alpha)}{r'^2 - c_1^2 + \alpha^2 + 2x(c_1+\alpha)}. \tag{3.3.20}$$

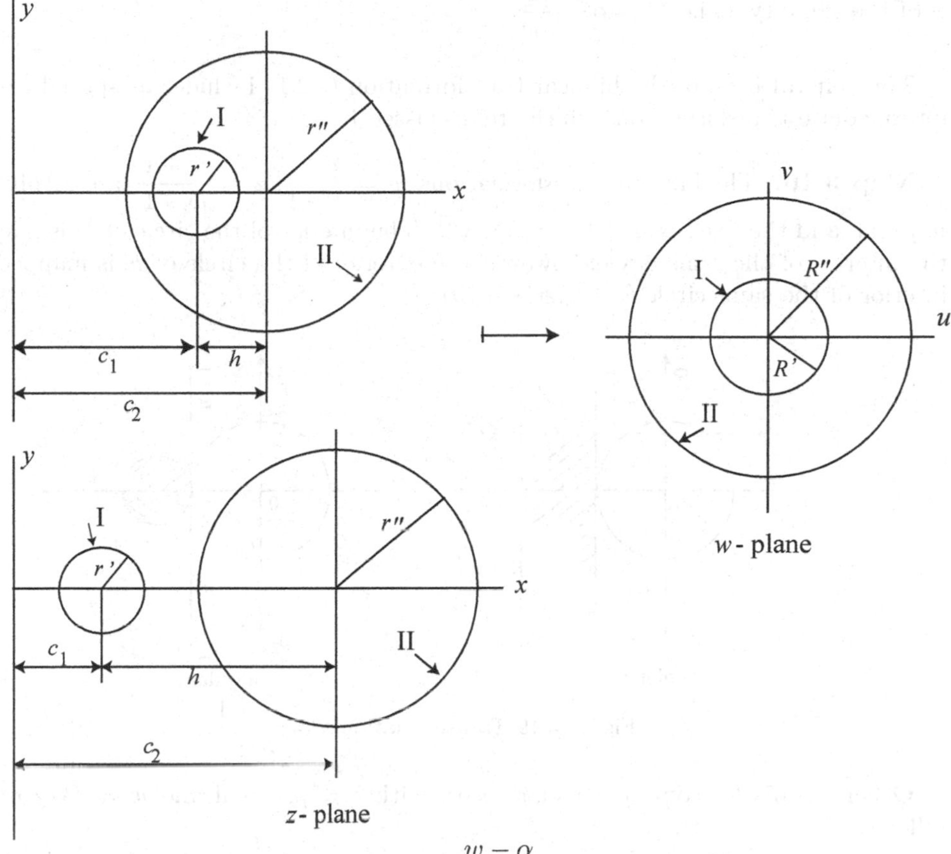

Figure 3.18 $w = \dfrac{w-\alpha}{w+\alpha}$, Two circles.

If we select α and c_1 such that the relation $\alpha^2 = c_1^2 - r'^2$ is satisfied, then R' becomes independent of x, i.e.,

$$R' = \left(\frac{c_1-\alpha}{c_1+\alpha}\right)^{1/2} = \frac{r'}{c_1+\alpha}, \quad r' = \{(c_1-\alpha)(c_1+\alpha)\}^{1/2}. \tag{3.3.21}$$

The circle II can be treated similarly by selecting $\alpha_2^2 = c_2^2 - r''^2$, thus finally giving

$$R'' = \frac{r''}{c_2+\alpha}. \tag{3.3.22}$$

If we substitute $h = c_2 - c_1$, then this transformation can be written as

$$w = \frac{z-\alpha}{z+\alpha} \quad \text{where } \alpha^2 = c_1^2 - r'^2, \tag{3.3.23}$$

and

$$c_1 = \frac{r''^2 - r'^2 - h^2}{2h}, \quad c_2 = \frac{r''^2 - r'^2 + h^2}{2h} = c_1 + h, \quad R' = \frac{r'}{c_1+\alpha}, \quad R'' = \frac{r''}{c_2+\alpha}. \ \blacksquare$$

A critical point for this transformation is $z = -\alpha$, and the inverse of this transformation is of the same type, i.e., $z = \alpha \dfrac{1+w}{1-w}$.

The general form of the bilinear transformation (3.2.1) includes as special cases all the intermediate transformations which are discussed below.

Map 3.16. The bilinear transformations $w = \dfrac{z+1}{z-1}$, $z = \dfrac{w+1}{w-1}$, are involutory, with p, q real, and the fixed-points $F_{1,2} = 1 \pm \sqrt{2}$. The interior of the circle of D is mapped onto the interior of the same circle; however, the exterior of the circle of E is mapped onto the interior of the same circle (see Figure 3.19).

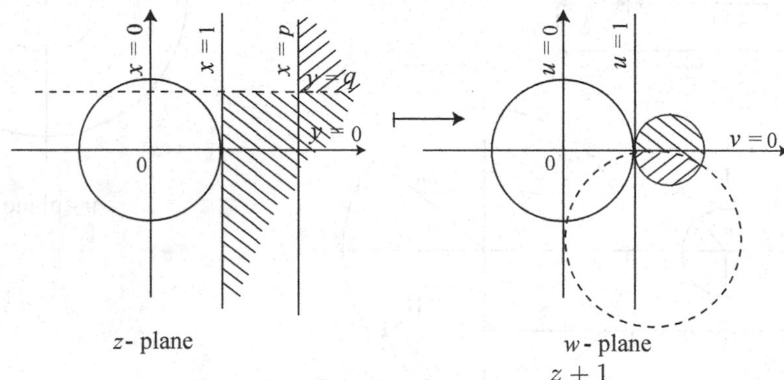

Figure 3.19 Transformation $w = \dfrac{z+1}{z-1}$.

Other details for some particular cases, with $z = p, q$ real and $w = P, Q$ real, are as follows:

(a) The points $z = 0; z_\infty = 1; \infty$ are mapped onto the points $w = -1; \infty; w_\infty = 1$.

(b) The half-plane $x \geq 0$ is mapped onto the region $|w| \geq 1$.

(c) The half-plane $y \geq 0$ is mapped onto the half-plane $v \leq 0$.

(d) The region $|z| \leq 1$ is mapped onto the half-plane $u \leq 0$.

(e) The line $x = 1$ in D is mapped onto the line $u = 1$.

(f) The line $x = p$, $p \neq 1$ is mapped onto the circle $\left| w - \dfrac{p}{p-1} \right| = \dfrac{1}{|p-1|}$.

(g) The line $y = q, q \neq 0$ is mapped onto the circle $\left| w - \dfrac{q-1}{q} \right| = \dfrac{1}{|q|}$.

Map 3.17. The transformation $w = \dfrac{1+z}{1-z} = e^{i\pi} \dfrac{z+1}{z-1}$, is the inverse of the Map 3.16. It maps the upper half-plane $\Im\{z\} > 0$ onto the unit disk $B(0,1)$.

Map 3.18. The transformation $w = \dfrac{1+z}{1-z}$, or $z = -\dfrac{1-w}{1+w}$ maps the unit upper semi-circle in the z-plane onto the first quadrant of the w-plane, such that the points $z = -1, 0, 1$

map into the points $w = 0, 1, -1$, respectively, as shown in Figure 3.20.

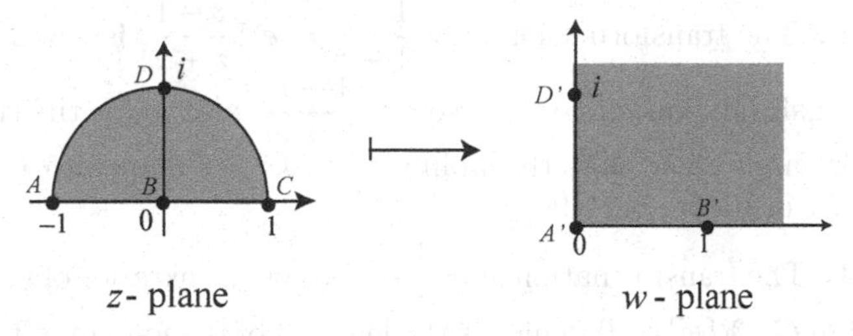

Figure 3.20 Upper semi-circle in the z-plane onto the first quadrant of the w-plane.

Map 3.19. The transformation $w = \dfrac{z-1}{z+1}$, or $z = \dfrac{1+w}{1-w}$ maps the right-half of the z-plane onto the unit circle, such that the points $z = 0, i, -i$ map into the points $w = -1, i, -i$, respectively. This map is shown in Figure 3.21(a).

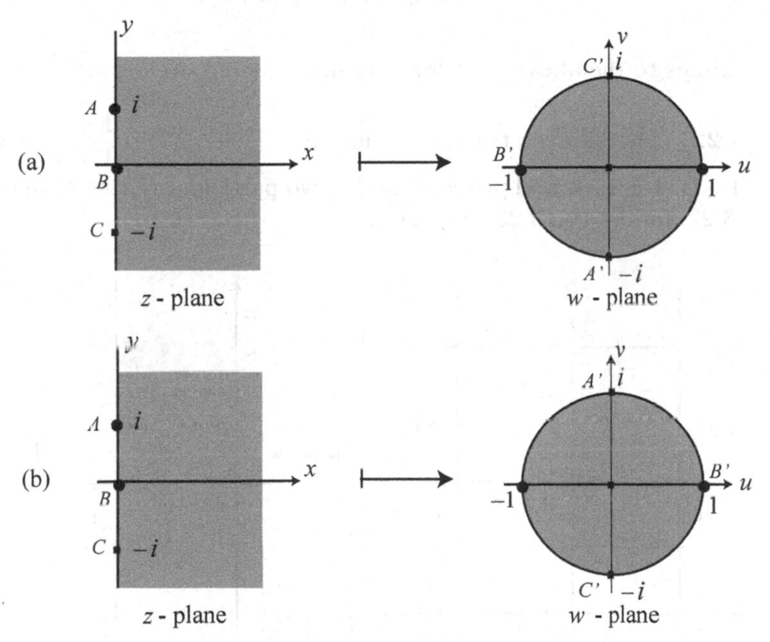

Figure 3.21 Right-half z-plane onto the unit circle.

Example 3.5. To find the linear fractional transformation that maps the interior of the circle $|z - i| = 2$ onto the exterior of the circle $|w - 1| = 3$, we need only three points on the first circle in clockwise order and three on the second circle in counterclockwise order. Let $z_1 = -i, z_2 = -2 + i$, and $z_3 = 3i$ and $w_1 = 4$, $w_2 = 1 + 3i$, and $w_3 = -2$. Then we get

$$\frac{(w-4)(3+3i)}{(w+2)(-3+3i)} = \frac{(z+i)(-2-2i)}{(z-3i)(-2+2i)},$$

or

$$w = \frac{z-7i}{z-i}.$$

The center of the first circle is $z = i$, which, under this transformation, is mapped onto the point $w = \infty$, and thus it lies outside the second circle. ■

Map 3.20. The transformation $w = \dfrac{1-z}{1+z} = e^{i\pi}\dfrac{z-1}{z+1}$, is the inverse of the Map 3.18. Thus, this transformation, or its inverse $z = \dfrac{1-w}{1+w}$, also maps the right-half of the z-plane onto the unit circle such that the points $z = 0, i, -i$ map into the points $w = 1, -i, i$, respectively, as in Figure 3.21(b).

Map 3.21. The transformation $w = \dfrac{1-z}{1+z}$ maps the exterior of the unit circle $D : |z| > 1$ onto the region $G : \Re\{w\} < 0$, which is the left-half of the w-plane. To see this, first consider the left of the regions D and G, and choose any three points on the unit circle that give negative (clockwise) orientation, say, $z_1 = 1, z_2 = -i, z_3 = -1$. Similarly the three points $w_1 = 0, w_2 = i, w_3 = \infty$ in the left region of G. Then the bilinear transformation is given by

$$\frac{w-0}{i-0} = \frac{(z-1)(-i+1)}{(z+1)(-i-1)},$$

which simplifies to the above transformation.

Map 3.22. The bilinear transformation $w = \dfrac{z-i}{z+i}$, $z = i\dfrac{1+w}{1-w}$, has the fixed-points $F_{1,2} = \frac{1}{2}(1-i)\left(1 \pm \sqrt{3}\right)$, and p, q, P, Q real. Two particular cases (A) and (B) are presented in Figure 3.22 and Figure 3.23, respectively.

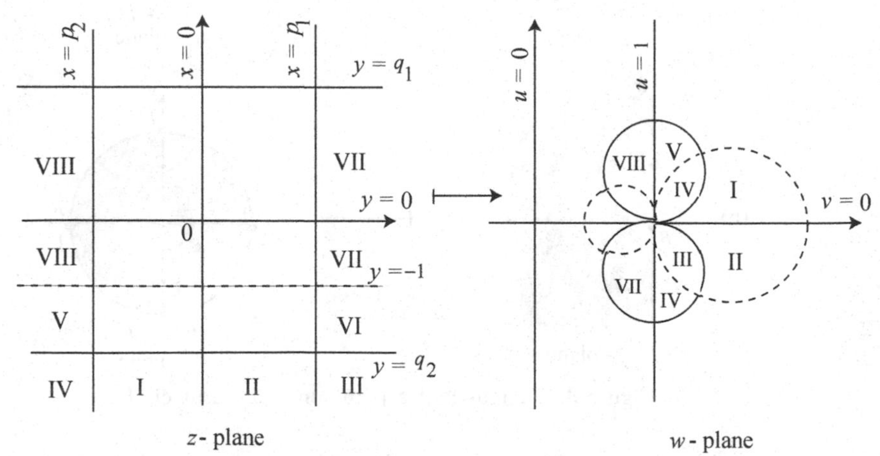

Figure 3.22 Transformation $w = \dfrac{z-1}{z+1}$ (Case A).

The details of the case A, with $z = p, q$ real and $w = P, Q$ real, are as follows:

(a) The points $z = 0, i, -i; z_\infty = \infty$ are mapped onto the points $w = -1, 0, \infty; w_\infty = 1$.

(b) The half-plane $x > 0$ is mapped onto the half-plane $v < 0$.

(c) The half-plane $y > 0$ is mapped onto the region $|w| < 1$.

(d) The region $|z| < 1$ is mapped onto the half-plane $u < 0$.

(e) The line $x - y = 1$ is mapped onto the line $u - v = 1$.

(f) The line $x = p$, $p \neq 0$ in D is mapped onto the circle $\left|w - \left(1 - \frac{1}{p}\right)\right| = \frac{1}{|p|}$.

(g) The line $y = -1$ is mapped onto the line $u = 1$.

(h) The line $y = q, q \neq -1$ is mapped onto the circle $\left|w - \frac{q}{q+1}\right| = \frac{1}{|q+1|}$.

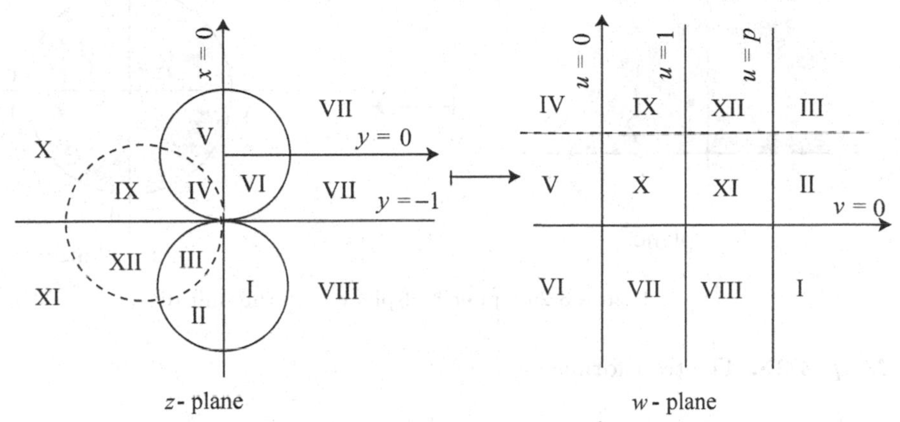

Figure 3.23 Transformation $w = \frac{z-1}{z+1}$ (Case B).

The details of the case B, with $z = p, q$ real and $w = P, Q$ real, are as follows:

(a) The line $y = -1$ is mapped onto the line $u = 1$.

(b) The circle $\left|z - \frac{ip}{1-p}\right| = \frac{1}{|1-p|}$ is mapped onto the line $u = P$, $P \neq 1$.

(c) The circle $\left|z - \left(-i - \frac{1}{q}\right)\right| = \frac{1}{|q|}$ is mapped onto the line $v = Q$, $Q \neq 0$.

Map 3.23. The transformation $w = \frac{z-i}{z+i}$, or $z = i\frac{1+w}{1-w}$, which maps the upper-half z-plane onto the circle, is presented in Figure 3.24.

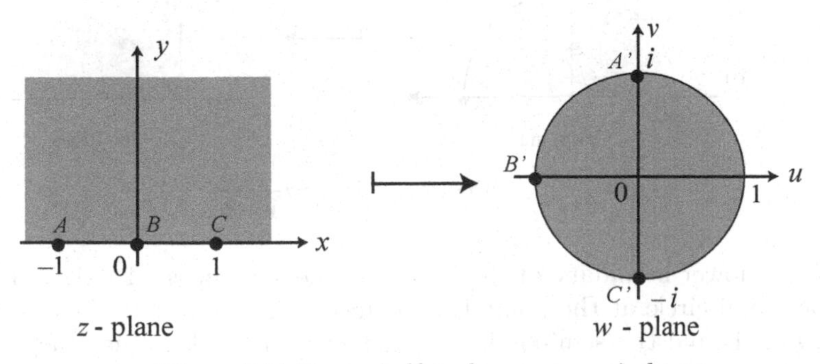

Figure 3.24 Upper-half z-plane onto a circle.

Map 3.24. The bilinear transformation $w = \frac{i+z}{i-z}$ is a composition of the following three transformations: (i) $w = -\frac{1}{s}$ (Map 3.6); (ii) $s = \frac{\zeta - i}{\zeta + i}$ (Map 3.23); and (iii) $\zeta = z$

(translation). It maps the upper half-plane $\Im\{z\} >: 0$ in the z-plane onto the unit disk $|w < 1$ in the w-plane (see Figure 3.25).

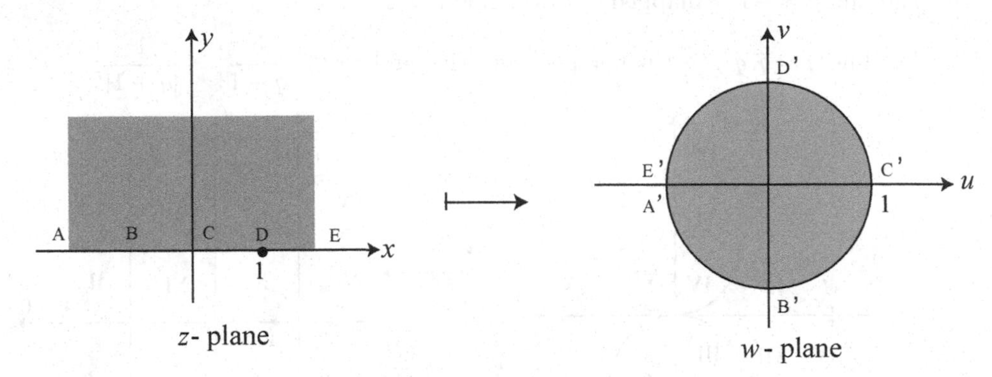

Figure 3.25 Upper half-plane onto the unit disk.

Map 3.25. The transformation

$$w(z) = i\frac{1-z}{1+z}$$

maps the unit disk onto the upper half-plane $\Im\{w\} > 0$.

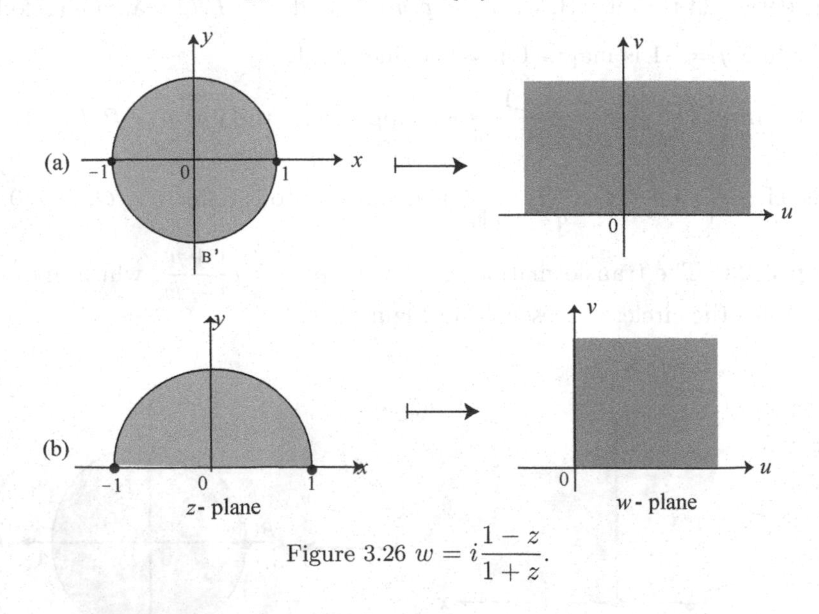

Figure 3.26 $w = i\dfrac{1-z}{1+z}$.

Since the lower boundary of the semi-disk, i.e., the interval $[-1, 1]$, is perpendicular to the upper semi-circle at the point 1, and since w is conformal at $z = 1$, the images of the interval $[-1, 1]$ and the semi-circle intersect at right angle. Since they both pass through the point -1, which is mapped onto ∞, we conclude that these images are perpendicular lines. Thus, using $w(0) = i$ and $w(i) = 1$, the interval $[-1, 1]$ in the z-plane is mapped onto the upper half of the imaginary v-axis, and the semi-circle is mapped on the right half of the real u-axis. Finally, testing the image of an interior point, say $i/2$, we find that $w(i/2) = 4/5 + 3i/5$, which is a point in the first quadrant. Hence, the upper semi-disk in the z-plane is mapped onto the first quadrant $\{u > 0, v > 0\}$ of the w-plane, since the boundary is mapped onto boundary and interior points onto interior points (Figure 3.26).

Map 3.26. The transformation

$$w = -i\frac{z+i}{z-i}, \quad \text{or} \quad z = i\frac{w-i}{w+i},$$

maps the points $z = i, -i, 0$, respectively, onto the points $w = \infty, 0$, and a point with $v > 0$. Thus, this map guarantees that the circle in the z-plane transforms into a straight line, which passes through the origin $w = 0$ and the interior of the circle is mapped onto the upper half of the w-plane.

Map 3.27. The bilinear transformation $w = f(z) = \dfrac{z}{2z-8}$ maps the region Int (Γ), where $\Gamma = \{|z-2| = 2\}$, conformally onto the region $\Re\{w\} < 0$, such that the point $z = 2$ goes into $w = -1/2$ (Saff and Snider [1976:317]).

Map 3.28. The bilinear transformations $w = \dfrac{z}{z-1}$, $z = \dfrac{w}{w-1}$ are involutory, with $z_\infty = 1, w_\infty = 1$, the fixed-points $F_1 = 0, F_2 = 2$, and p, q real. The interior (exterior) of circles of D and E map onto exterior (interior) of circles. The details of different particular cases, with $z = p, q$ real and $w = P, Q$ real, are as follows:

(a) The half-plane $y \geq 0$ is mapped onto the half-plane $v \leq 0$.
(b) The half-plane $x \geq 0$ is mapped onto the region $\left|w - \frac{1}{2}\right| \geq \frac{1}{2}$.
(c) The region $|x| \leq 1$ is mapped onto the half-plane $u \leq \frac{1}{2}$.
(d) The half-plane $x \leq \frac{1}{2}$ is mapped onto the region $|w| \leq 1$.
(e) The line $x = 1$ is mapped onto the line $u = 1$, passing through $w = 1$.
(f) The line $x = p$, $p \neq 1$ is mapped onto the circle $\left|w - \dfrac{2p-1}{2(p-1)}\right| = \dfrac{1}{|2p-2|}$ passing through $w = 1$.
(g) The line $y = q$ is mapped onto the circle $\left|w - (1 - \dfrac{1}{2q})\right| = \dfrac{1}{2|q|}$ passing through $w = 1$.

Map 3.29. The function $w = f(z) = \dfrac{z}{z-(1+i)}$ maps the lens-shaped region bounded by the circles $|z-1| = 1$ and $|z-i| = 1$ onto the region bounded by the rays $\arg\{w\} = 3\pi/4$ and $\arg\{w\} = 5\pi/4$ such that $f(0) = 0$ and $f(1+i) = \infty$.

Map 3.30. The transformation $w = \dfrac{z}{iz+2}$ maps the disk $|z-i| = 1$ onto the real line. To see this, since the points $z = 0, 1+i, 2i$ lie on the given circle in the z-plane and since the real line in the w-plane passes through the points $w = 0, 1, \infty$, so we choose $h(z)$ such that $h(0) = 0, h(1+i) = 1$, and $h(2i) = \infty$. Then $h(z) = \dfrac{z}{iz+2}$. Since $h(i) = i$, so h maps the disk $|z-i| < 1$ onto the upper half-plane.

Map 3.31. CIRCLE ONTO CIRCLE. The transformation is

$$w - w_0 = \Re\left\{e^{i\tau}\,\frac{z - z_0 - r\alpha}{\bar{\alpha}z - \alpha z_0 - r}\right\}, \tag{3.3.24}$$

where τ is real, $|\alpha| \neq 1$, and is presented in Figure 3.27.

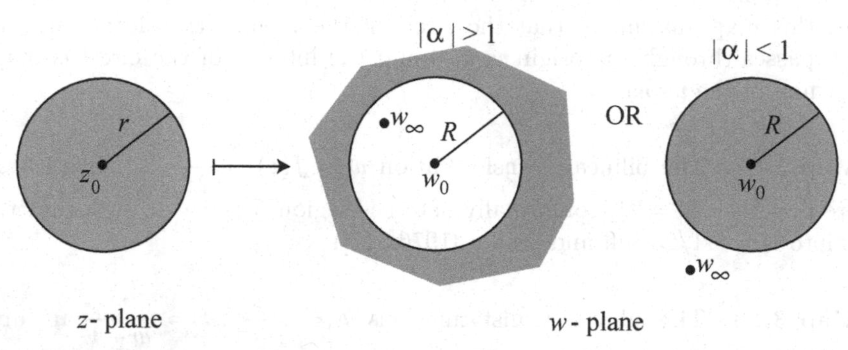

Figure 3.27 Circle onto circle.

Map 3.32. Mapping of the unit circle onto itself:

$$w = e^{i\tau}\,\frac{z - \alpha}{\bar{\alpha}z - 1}, \quad \tau \text{ real and } |\tau| \neq 1. \tag{3.3.25}$$

Map 3.33. The transformation $w = e^{i\theta}\,\dfrac{z - z_0}{1 - \bar{z}_0 z}$, which maps a circle onto the circle in the w-plane such that the point z_0 maps into the point $w = 0$, is shown in Figure 3.28.

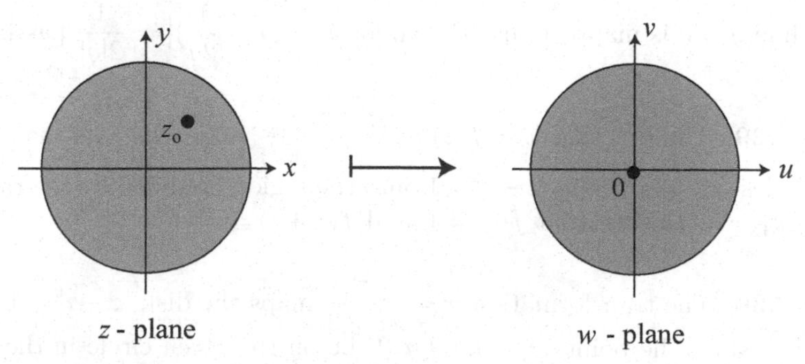

Figure 3.28 Circle onto a circle.

Map 3.34. The function $f(z) = \rho\, e^{i\alpha}\,\dfrac{z - z_1}{z - z_2}$ maps the region enclosed by two arcs onto the region enclosed by the sector shown in Figure 3.29, such that a fixed point $\zeta \neq z_{1,2}$ on Γ_1 goes into a point ω on the u-axis, where $\alpha = \arg\left\{\dfrac{\zeta - z_1}{\zeta - z_2}\right\}$ and $\rho = \omega\left|\dfrac{\zeta - z_1}{\zeta - z_2}\right|$ (Pennisi [1963:321]).

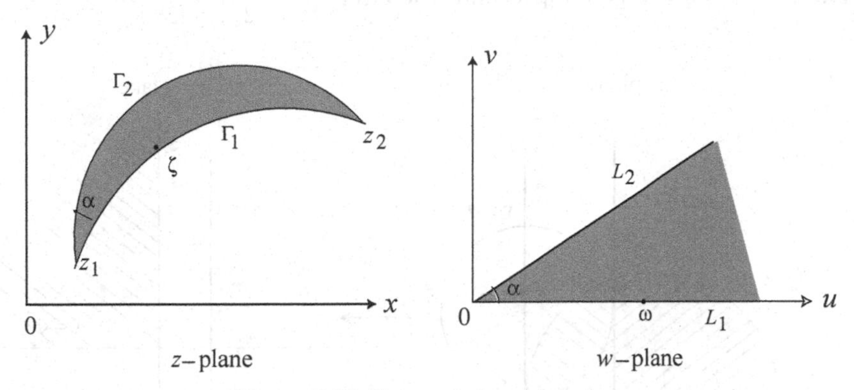

Figure 3.29 Crescent-shaped region.

Map 3.35. The transformation $w = \dfrac{1}{1-z}$, $z = \dfrac{w-1}{w}$, with $z_\infty = 1; w_\infty = 0$; the fixed points $F_1 = \frac{1}{2} + \frac{i}{2}\sqrt{3}$, and $F_2 = \dfrac{1}{F_1}$, and p, q, P, Q are real. The interior of the circle in the z-plane is mapped onto the interior of the circle in the w-plane. The details of the transformation, with $z = p, q$ real and $w = P, Q$ real, are as follows:

(a) The half-plane $y \geq 0$ is mapped onto the half-plane $v \geq 0$.

(b) The half-plane $x \geq 0$ is mapped onto the region $\left|w - \frac{1}{2}\right| \geq \frac{1}{2}$.

(c) The region $|z| \leq 1$ is mapped onto the half-plane $u \geq \frac{1}{2}$.

(d) The region $|z - 1| \leq 1$ is mapped onto the region $|w| \geq 1$.

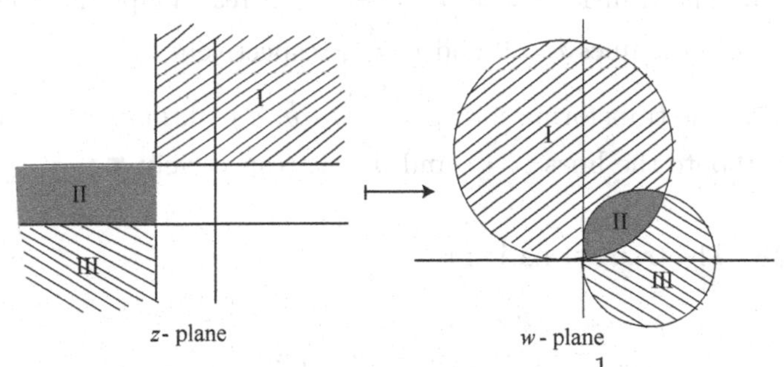

Figure 3.30 Transformation $w = \dfrac{1}{1-z}$.

There are two special cases, one presented in Figure 3.30, and the other in Figure 3.31; their respective details, with $z = p, q$ real and $w = P, Q$ real, are as follows:

For Figure 3.30:

(a) The line $x = 1$ is mapped onto the line $u = 0$.

(b) The line $x = p \neq 1$ is mapped onto the circle $\left|w - \dfrac{1}{2(1-p)}\right| = \dfrac{1}{2|1-p|}$.

(c) The line $y = q \neq 0$ is mapped onto the circle $\left|w - \dfrac{1}{2q}\right| = \dfrac{1}{2|q|}$.

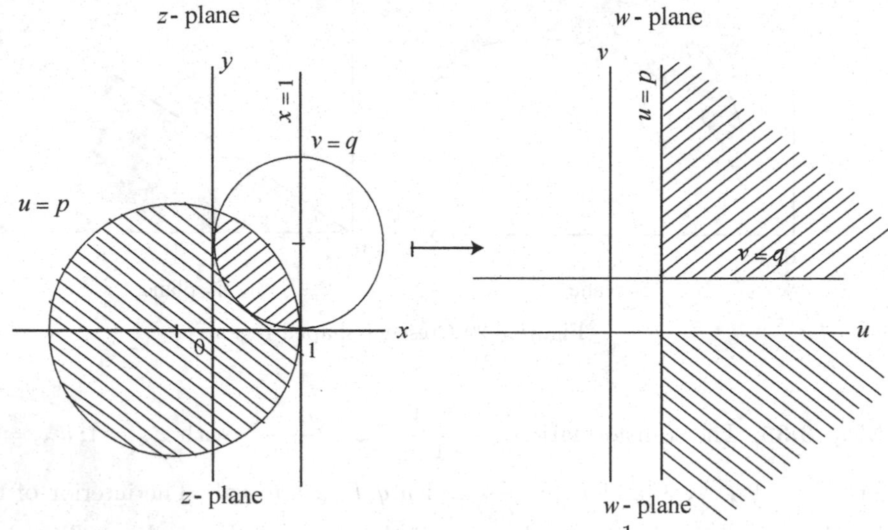

Figure 3.31 Transformation $w = \dfrac{1}{1-z}$.

For Figure 3.31:

(a) The circle $\left|z - \left(1 - \dfrac{1}{2p}\right)\right| = \dfrac{1}{2|p|}$ is mapped onto the line $u = P \neq 0$.

(ib) The circle $\left|z - \left(1 + \dfrac{1}{2q}\right)\right| = \dfrac{1}{2|q|}$ is mapped onto the line $v = Q \neq 0$.

Map 3.36. The transformation $w = \dfrac{1}{z} - a$, a real, maps the line $y = 0$ and the circle $\left|z + \frac{1}{2}i\right| = \frac{1}{2}$ onto the lines $v = 0$ and $v = 1$, respectively.

Map 3.37. The transformation $w = \dfrac{2i\,r}{z} - a$, a real, maps the line $x = 0$ and the circle $|z - r| = r > 0$ onto the lines $v = 1$ and $v = 1$, respectively. ■

Map 3.38. The transformation is

$$w = \frac{r_2}{r_2 - r_1}\left(\frac{2r_1}{z} + i\right). \tag{3.3.26}$$

It is presented in Figure 3.32.

The details of the transformation for this case with $z = p, q$ real and $w = P, Q$ real, are as follows:

(a) The points $z = 0;\ \infty;\ 2ir_1$ are mapped onto the points $w = \infty;\ \dfrac{ir_2}{r_2 - r_1};\ 0$.

(b) The circle $|z - ir_1| = r_1$ is mapped onto the line $v = 0$.

(c) The circle $|z - ir_2| = r_2$ is mapped onto the line $v = 1$.

(d) The line $x = 0$ is mapped onto the line $u = 0$.

(e) The line $y = 0$ is mapped onto the line $v = \dfrac{r_2}{r_2 - r_1}$.

(f) The point $z_3 = ir_1\left(1 - e^{i\theta}\right)$ is mapped onto the point $w_3 = \dfrac{r_2}{r_2 - r_1}\cot\dfrac{\theta}{2}$.

(g) The small circle touching both circles and passing through z_3 is mapped onto the circle $\left|w + w_3 - \frac{1}{2}i\right| = \frac{1}{2}$.

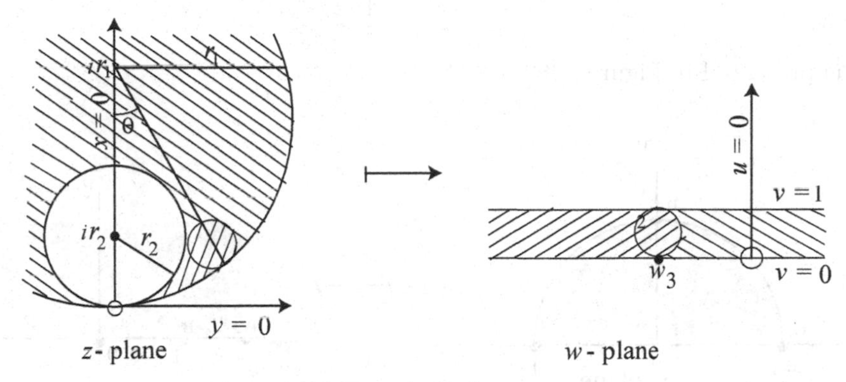

Figure 3.32 Two circles in outer contact.

Map 3.39. The transformation

$$w = \frac{i\,r_2}{r_1 + r_2}\left(1 - \frac{2r_1}{z}\right) \tag{3.3.27}$$

is presented in Figure 3.33.

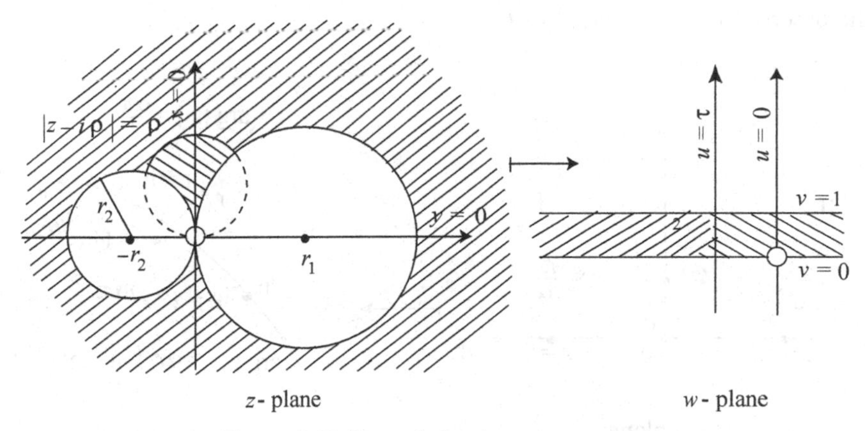

Figure 3.33 Two circles in outer contact.

The details of the transformation for this case with $z = p, q$ real and $w = P, Q$ real, are as follows:

(a) The points $z = 0;\ \infty;\ 2r_1$ are mapped onto the points $w = \infty;\ \dfrac{ir_2}{r_1 + r_2};\ 0$.

(b) The circle $|z - r_1| = r_1$ is mapped onto the line $v = 0$.

(c) The circle $|z + r_2| = r_2$ is mapped onto the line $v = 1$.

(d) The circle $|z - i\rho| = r_2$ is mapped onto the line $u = \tau$, where $\tau = -\dfrac{r_1 r_2}{\rho(r_1 + r_2)}$.

(e) The lines $x = 0$; $y = 0$ are mapped onto the lines $v = \dfrac{r_2}{r_2 - r_1}$; $u = 0$.

Map 3.40. The transformation that maps the upper-half of the unit circle in the z-plane onto the upper half-plane $\Im\{w\} > 0$ is defined by

$$w = \left(\frac{1+z}{1-z}\right)^2,$$

and is presented in Figure 3.34.

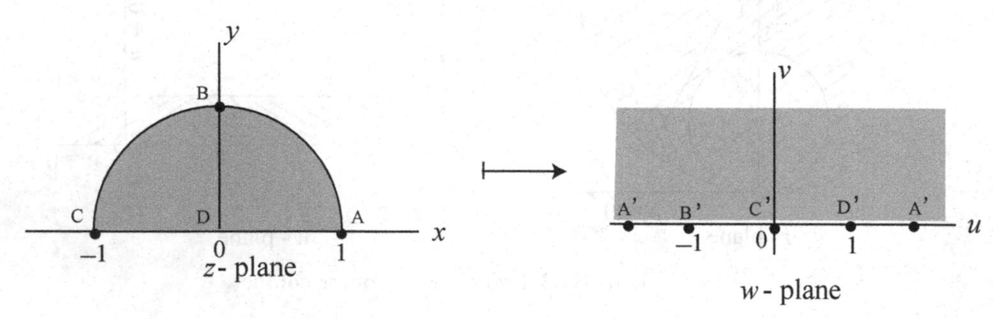

Figure 3.34 Upper-half of the unit circle onto the upper half-plane.

Map 3.41. The transformation that maps the upper half plane $\Im\{z\} > 0$ onto a round corner in the w-plane is $w = (z+1)^\alpha + (z-1)^\alpha$, $1 < \alpha < 2$; it is presented in Figure 3.35. The transformation maps the line segment $y = 0, -1 \le x \le 1$ in the z-plane onto the part $-2^\alpha \cos(\alpha 1)\pi \le u \le 2^\alpha$, $-2^\alpha \sin(\alpha - 1)\pi \le v \le 0$, i.e., the part DCB, of $(v\cos(\alpha\pi) - u\sin(\alpha\pi))^{1/\alpha} + (-v)^{1/\alpha} = 2\,(-\sin(\alpha\pi))^{1/\alpha}$. Note that for $\alpha = 3\pi/2$, the curve is the asteroid $u^{2/3} + (-v)^{2/3} = 2$.

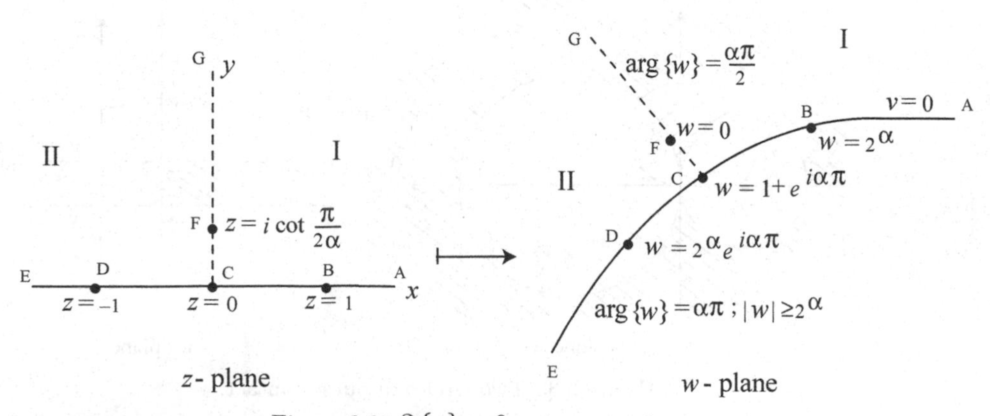

Figure 3.35 $\Im\{z\} > 0$ onto a round corner.

Map 3.42. The transformation that maps the sector of angle π/m of a circle in the z-plane onto the upper half-plane $\Im\{w\} > 0$ is defined by

$$w = \left(\frac{1+z^m}{1-z^m}\right)^2,$$

and is presented in Figure 3.36.

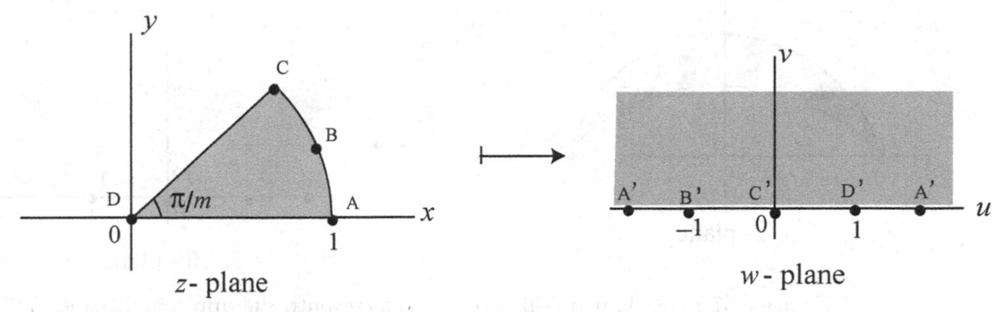

Figure 3.36 Sector of angle π/m of a circle onto the upper half-plane.

Map 3.43. The transformation

$$w = \frac{iz^2 + 1}{iz^2 - 1}, \tag{3.3.28}$$

conformally maps the upper half-plane $U = \Im\{z > 0\}$ onto the unit disk $C(0, 1)$. To prove it, first note that the transformation $z = g(w) = \dfrac{w-1}{w+1}$ (Map 3.17) maps the right half-plane $\Re\{w\} > 0$ onto $B(0, 1)$. Also, multiplication by $i = e^{i\pi/2}$, with $z = h(w) = iw$ rotates the complex plane by $\pi/2$ and thus maps the right half-plane $C(0, \frac{1}{2})$ onto the upper half-plane U, while its inverse $w = h^{-1}(z) = -iz$ will map U to the right half-plane. Hence, to map the upper half plane U onto the unit disk, the composition of these two mappings gives the required map:

$$w = g \circ h^{-1}(z) = \frac{-iz - 1}{-iz + 1} = \frac{iz + 1}{iz - 1}. \tag{3.3.29}$$

Similarly, we already know (see Maps 4.6) that the square map $w = z^2$ maps the upper right quadrant $\{0 < \arg\{z\} < \pi/2\}$ onto the upper half-plane U. Thus, composition of this map with the previously constructed map, where we replace z by w in (3.3.29), we obtain the final required mapping (3.3.28).

Map 3.44. The transformation that maps the lens-shaped region of angle π/m, where this region is bounded by two circular arcs, in the z-plane onto the upper half-plane $\Im\{w\} > 0$ is defined by

$$w = e^{2mi \operatorname{arccot}(p)} \left(\frac{z+1}{z-1}\right)^m, \tag{3.3.30}$$

and is presented in Figure 3.37.

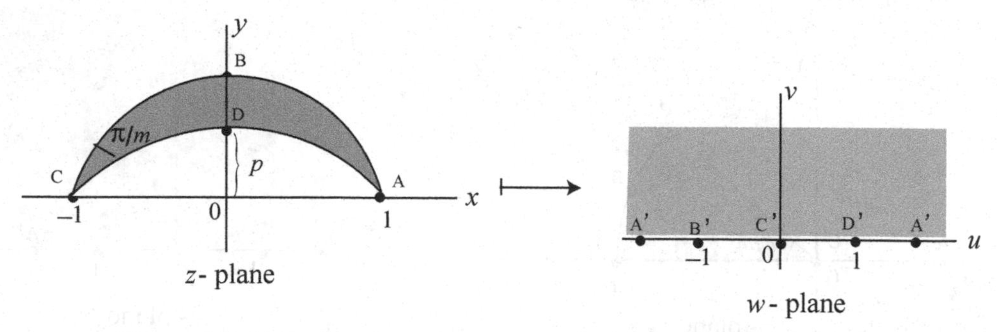

Figure 3.37 Lens-shaped region of angle π/m onto the upper half-plane.

Map 3.45. The transformation $w = \dfrac{z-a}{az-1}$, where $a > 1$, $-1 < x_2 < x_1 < 1$, $R_0 > 1$, and

$$a = \frac{1 + x_1 x_2 + \sqrt{(1-x_1^2)(1-x_2^2)}}{x_1 + x_2},$$
$$R_0 = \frac{1 - x_1 x_2 + \sqrt{(1-x_1^2)(1-x_2^2)}}{x_1 + x_2}, \tag{3.3.31}$$

maps the region between the two circles in the z-plane onto the annular region in the w-plane, as shown in Figure 3.38.

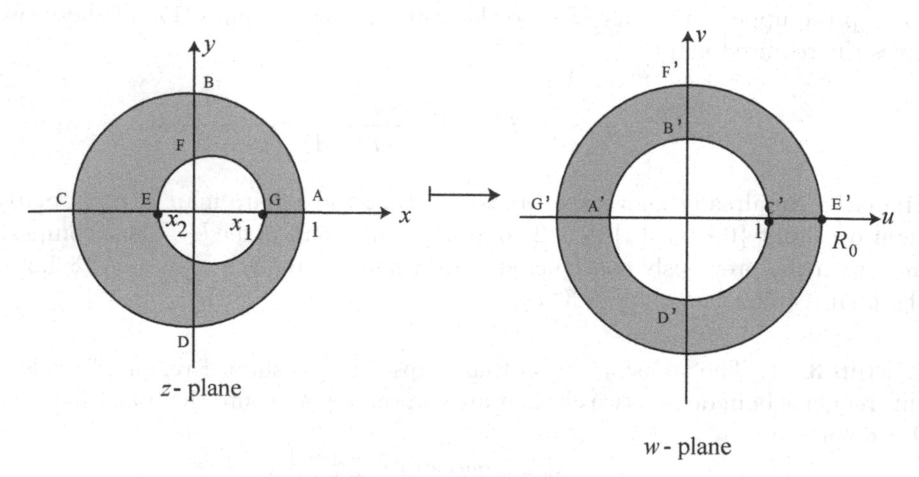

Figure 3.38 Map 3.45.

Map 3.46. The transformation $w = \dfrac{z-a}{az-1}$, where a and R_0 are the same as in (3.3.31), with $x_2 < a < x_1$, and $0 < R_0 < 1$ when $1 < x_2 < x_1$, maps the region exterior to the two circles in the z-plane onto the annular region in the w-plane, as shown in Figure 3.39.

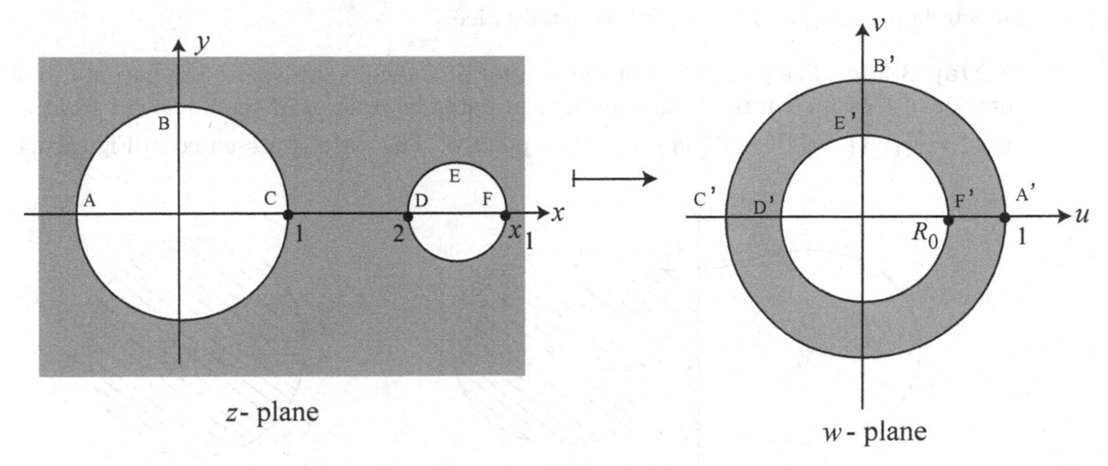

Figure 3.39 Map 3.46.

Map 3.47. The transformation $w = e^{i\lambda}\,\dfrac{z - z_0}{z + \bar{z}_0}$, where λ is a real number, maps the upper half-plane $-\infty < x < \infty, 0 \le y < \infty$, in the z-plane onto the unit disk $|w| \le 1$ in the w-plane.

Map 3.48. The transformation $w = e^{i\lambda}\,\dfrac{z - z_0}{1 - \bar{z}_0 z}$, where λ is a real number, maps the unit disk $x^2 + y^2 \le 1$, in the z-plane onto the unit disk $|w| \le 1$ in the w-plane.

Map 3.49. The transformation $w = e^{i\beta}\,\dfrac{z - \alpha}{\bar{\alpha} z - 1}$ for some $|\alpha| < 1$, $-\pi < \beta < \pi$, maps the unit disk onto itself.

Map 3.50. CASSINI'S OVALS. There are four cases:

Map 3.50a. Either part of the interior of Cassini's ovals $|z - z_1||z - z_2| = C$, with foci $z_1 = \pm\sqrt{w_1}$, is mapped onto the interior of the circle $|w - w_1| = C$, $C \le |w_1|$. This case is presented in Figure 3.40, where $w_1$is taken as a *real* and positive number.

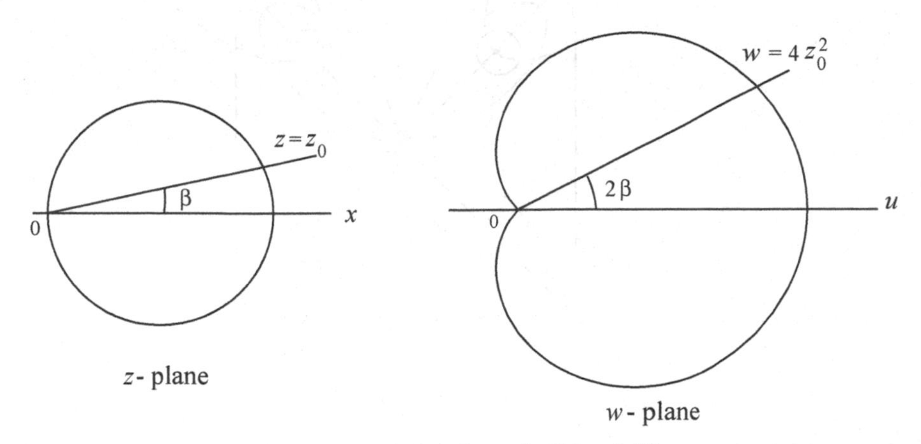

Figure 3.40 Cassini's ovals, Map 3.50a.

Map 3.50b. Cassini's ovals $|z - z_1||z - z_2| = C$, with foci $z_1 = \pm\sqrt{w_1}$, is mapped onto the circle $|w - w_1| = C$, $C > |w_1|$, counted twice.

Map 3.50c. The part $x > 0$ of the interior of Cassini's ovals, and the part $x < 0$ of the interior of Cassini's ovals, both are mapped onto the interior of the circle cut from $w = 0$ to $w' = \Re\{w_1\} = \left(C^2 - \Im\{w_1\}^2\right)^{1/2}$, i.e., point a. This case is presented in Figure 3.41.

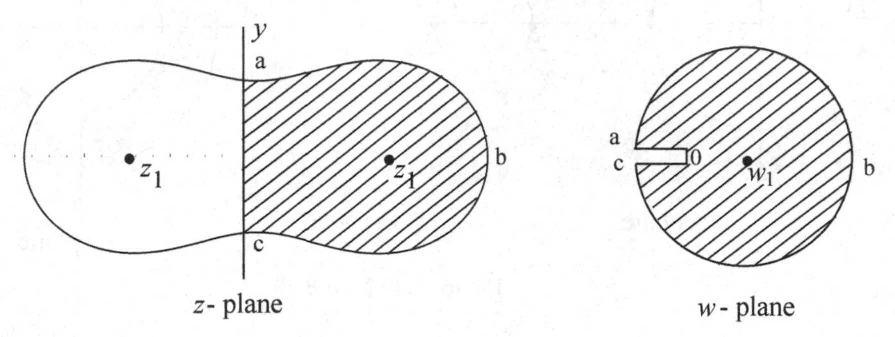

Figure 3.41 Cassini's ovals, Map 3.50c.

Map 3.50d. (Generalized Cassini's Ovals) The details of the transformation are as follows:

(a) Each part of $|z - z_1||z - z_2| \cdots |z - z_n| = C$, with foci $z_j = e^{2i\pi j/n} \sqrt[n]{w_0}$, $j = 1, 2, \ldots, n$, is mapped onto the circle $|w - w_0| = C$, $0 < C \leq |w_0|$.

(b) Closed curve $|z - z_1||z - z_2| \cdots |z - z_n| = C$ (which is approximately circle for large C) is mapped onto the circle $|w - w_0| = C$, counted n times, $0 < |w_0| < C$.

(c) Region bounded by this curve and the lines $z = r\, e^{i\theta}$, $z = r\, e^{i(\theta + 2\theta/n}$, θ fixed, is mapped onto the interior of $|w - w_0| = C$, $C > |w_0|$, cut from 0 to the point at which the line $w = r^n\, e^{in\theta}$ meets the circle.

This transformation for $n = 4$ is presented in Figure 3.42.

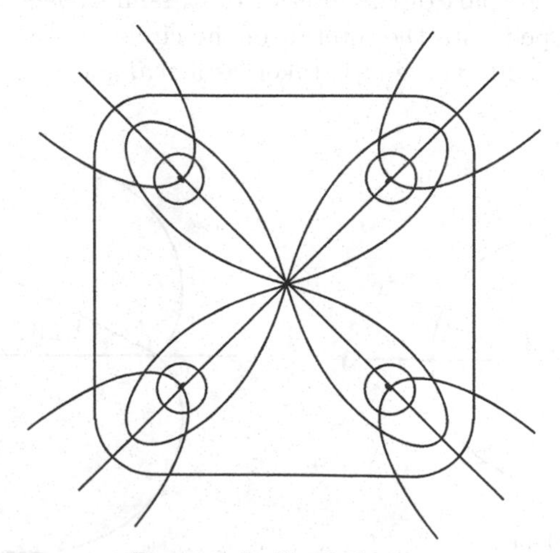

Figure 3.42 Cassini's ovals.

Map 3.51. CARDIOID AND LIMAÇON. The mapping from z-plane onto the w-plane is presented in Figure 3.43.

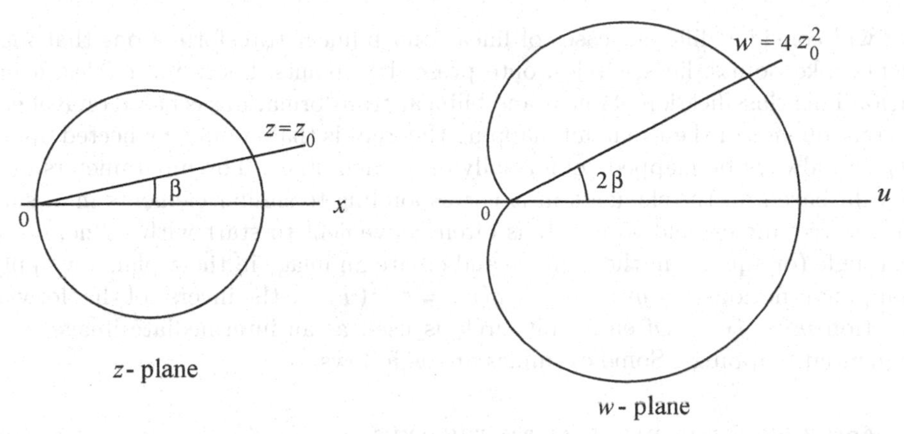

Figure 3.43 Circle onto cardioid.

The details of this transformation are as follows:

(a) The circles $\left\{ \begin{matrix} |z - z_0| = |z_0| \\ |z + z_0| = |z_0| \end{matrix} \right\}$ are mapped onto the cardioid $w = 2z_0^2(1 + \cos\theta)\, e^{i\theta}$, $-\pi \leq \theta < \pi$.

(b) The circles $\left\{ \begin{matrix} |z - z_0| = c \\ |z + z_0| = c \end{matrix} \right\}$, $c > 0, z_0 \neq 0$, are mapped onto the limaçon $w - w_1 = 2c(z_0 + c\cos\theta)\, e^{i\theta}$, $w_1 = z_0^2 - c^2$, $-\pi \leq \theta < \pi$.

Map 3.52. Cardioid and Generalized Cardioids ($\alpha > 1$), and Lemniscates ($0 < \alpha < 1$) (Figure 3.44).

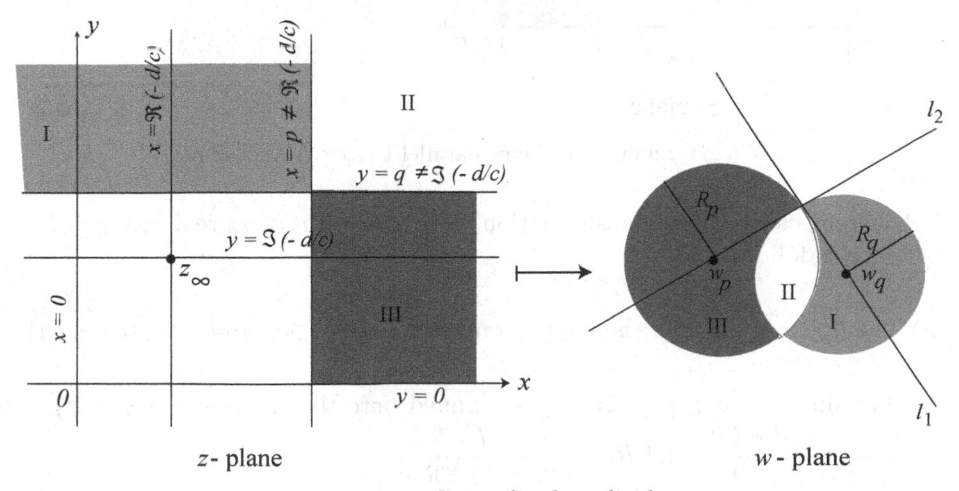

Figure 3.44 Generalized cardioid.

The details of this transformation are as follows:

The line $\Re\{e^{-i\theta}\, z\} = s$, $s > 0$ is mapped onto each part of $R^{\beta} = s^{-1}\cos(\beta\theta + \phi)$, $|\theta - \alpha(2k\pi - \phi| \leq \alpha\pi/2$, $k = 0, \pm1, \pm2, \ldots$; the curves are closed, with a cusp at 0, with tangents $w = R\, e^{i\alpha\left(k\pi + \frac{\pi}{2} - \phi\right)}$, $k = 0, \pm1, \pm2, \ldots$; there are n parts if $\alpha^{-1} = n \in \mathbb{Z}$.

3.4 Straight Lines and Circles

We will consider different cases of linear and bilinear transformations that map given elements, like points, lines, circles, onto prescribed points, lines, and circles, using the cross-ratio. This classification of linear and bilinear transformations is based on Kober [1957:2-32]. A consequence of the Riemann mapping theorem is that simply connected domains D_z and D_w can always be mapped conformally onto each other. Three parameters may be chosen at will, based on the elements in w corresponding to similar elements in z, such as points, circles, rectangles, and so on. It is often convenient to start with a line, or a circle, or a rectangle (or square) in the w plane and create an image in the z-plane by applying various mapping functions $z = g(w)$, where $g(w) = f^{-1}(w)$ are the inverse of the 'forward' mapping function $w = f(z)$. Often a unit circle is used as an intermediate image in constructing sequential mappings. Some examples are as follows.

Map 3.53. LINES PARALLEL TO THE AXES. $z_\infty = -d/c$, $w_\infty = a/c$ (see Figure 3.45).

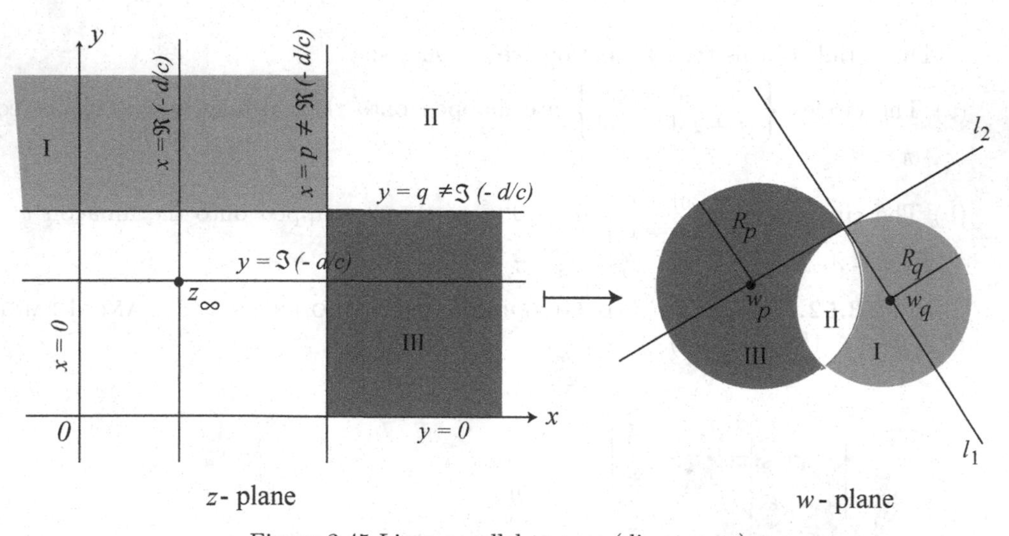

Figure 3.45 Lines parallel to axes (direct map).

Different cases of this transformation, with $z = p, q, r, r_p, r_q$ real and $w = P, Q, R, R_p, R_q$ real, are as follows:

(a) The line $x = -\Re\{d/c\}$ is mapped onto the line $l_1 : \Re\{c\left(\bar{a}\bar{d} - \bar{b}\bar{c}\right)(cw - a)\} = 0$.

(b) The line $x = p$, $p \neq -\Re\{d/c\}$ is mapped onto the circle $|w - w_p| - R_p$, where $w_p = \dfrac{a\bar{c}p + \frac{1}{2}\left(a\bar{d} + b\bar{c}\right)}{p|c|^2 + \Re\{c\bar{d}\}}$, and $R_p = \left|\dfrac{ad - bc}{2p|c|^2 + 2\Re\{\bar{c}d\}}\right|$.

(c) The line $y = -\Im\{d/c\}$ is mapped onto the line l_2: $\Im\{c\left(\bar{a}\bar{d} - \bar{b}\bar{c}\right)(cw - a)\} = 0$.

(d) The line $y = q$, $q \neq -\Im\{d/c\}$ is mapped onto the circle $|w - w_q| = R_q$, where $w_q = \dfrac{a\bar{c}p + \frac{1}{2}i\left(a\bar{d} - b\bar{c}\right)}{q|c|^2 + \Im\{\bar{c}d\}}$, and $R_q = \left|\dfrac{ad - bc}{2q|c|^2 + 2\Im\{\bar{c}d\}}\right|$.

The inverse mapping is presented in Figure 3.46.

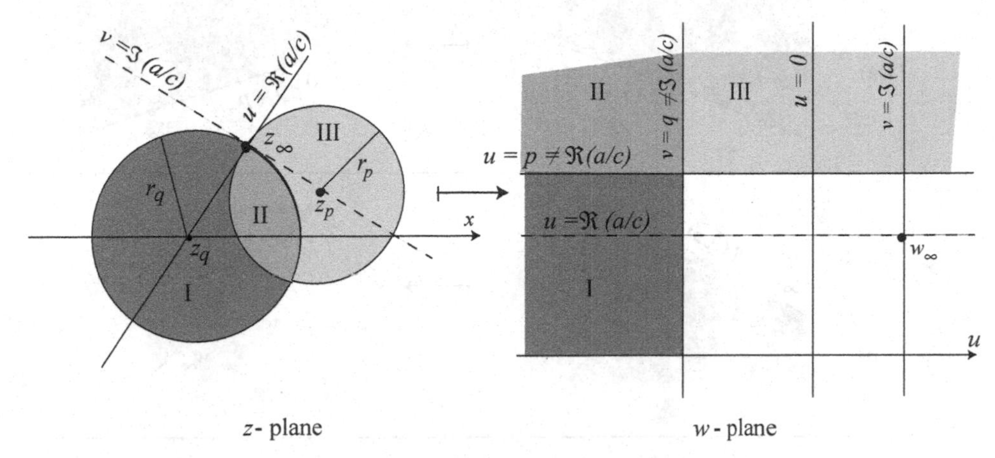

Figure 3.46 Lines parallel to axes (inverse map).

The details of different cases with $z = p, q, r, r_p, r_q$ real and $w = P, Q, R, R_p, R_q$ real, are as follows:

(a) The line $\Re\{c\left(\bar{a}\bar{d} - \bar{b}\bar{c}\right)(cz + d)\} = 0$ is mapped onto the line $u = \Re\{a/c\}$.

(b) The circle with center $z_p = \dfrac{\frac{1}{2}\left(\bar{a}d + b\bar{c}\right) - \bar{c}dp}{p|c|^2 - \Re\{a\bar{c}\}}$ and radius $r_p = \left|\dfrac{ad - bc}{2P|c|^2 - 2\Re\{a\bar{c}\}}\right|$, is mapped onto the line $u = P$, $P \neq \Re\{a/c\}$.

(c) The line $\Im\{c\left(\bar{a}\bar{d} - \bar{b}\bar{c}\right)(cz + d)\} = 0$ is mapped onto the line $v = \Im\{a/c\}$.

(d) The circle with center $z_q = \dfrac{\frac{1}{2}i\left(\bar{a}d - b\bar{c}\right) - \bar{c}dq}{q|c|^2 - 2\Im\{a\bar{c}\}}$ and radius $r_q = \left|\dfrac{ad - bc}{2q|c|^2 - 2\Im\{a\bar{c}\}}\right|$, is mapped onto the line $v = Q$, $Q \neq \Im\{a/c\}$.

Map 3.54. Other lines and circles. $z_\infty = -d/c$, $w_\infty = a/c$ (see Figure 3.47.)

The details of the transformation with $z = p, q, r, r_p, r_q$ real and $w = P, Q, R, R_p, R_q$ real, are as follows:

(1a, b.) The line $\Re\{\bar{\lambda}z\} = p$ passing through z_∞ is mapped onto the line $\Re\{\bar{\Lambda}w\} = P$ passing through w_∞.

(2.) The line $\Re\{\bar{\lambda}z\} = p$ not passing through z_∞ is mapped onto the circle $|w - w_0| = R$ passing through w_∞.

(3.) The circle $|z - z_0| = r$ passing through z_∞ is mapped onto the line $\Re\{\bar{\Lambda}w\} = P$ not passing through w_∞.

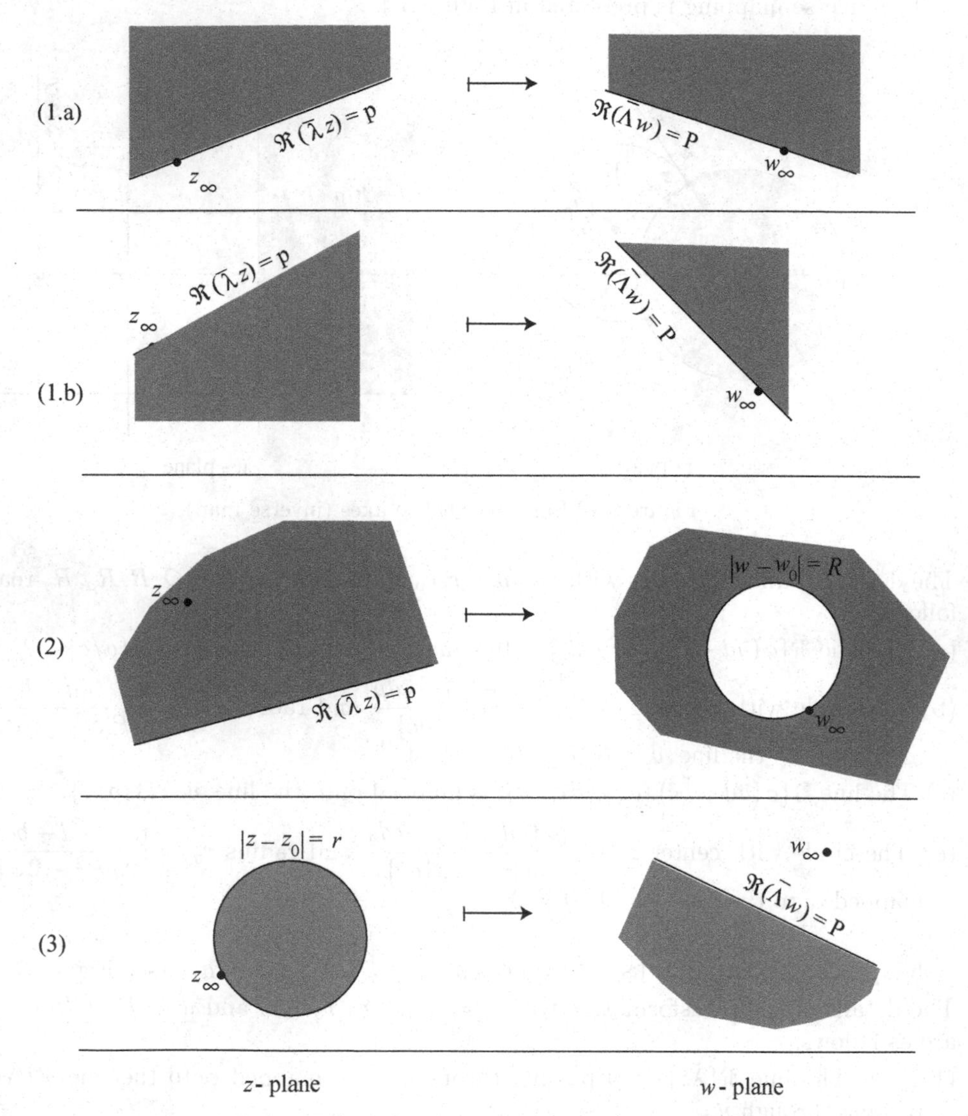

Figure 3.47 Other lines and circles (direct map).

Map 3.55. The circle $|z_0 + d/c| = r$ not passing through z_∞ is mapped onto the circle $|w - w_0| = R$ not passing through w_∞ (see Figure 3.48).

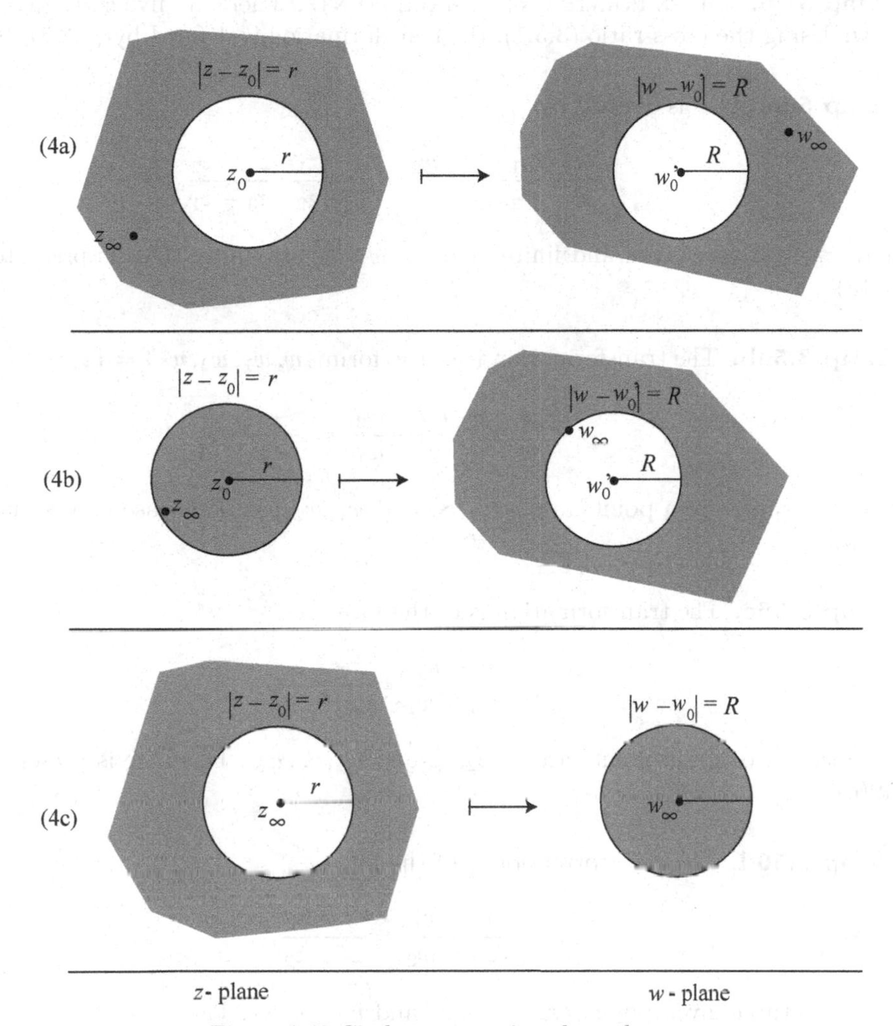

Figure 3.48 Circles not passing through z_∞.

The details of the transformation with $z = p, q, r, r_p, r_q$ real and $w = P, Q, R, R_p, R_q$ real, are as follows, where

$$w_0' = \frac{(az_0 + b)(\bar{a}\bar{z}_0 + \bar{d}) - a\bar{c}r^2}{|cz_0 + d|^2 - |c|^2r^2}, \quad R = \frac{r|ad - bd|}{|cz_0 + d|^2 - |c|^2r^2}.$$

(a) The circle $|z - z_0| = r$ not passing through z_∞ is mapped onto the circle $|w - w_0'| = R$ not passing through w_∞.

(b) The circle $|cz_0 + d| > |c|r$ is mapped onto the circle $|w - w_0| = R$, where $w_0' = \dfrac{az_0 + b - \bar{c}R^2 s}{cz_0 + d}$, $R = r|s|$, and $s = \dfrac{ad - bc}{|cz_0 + d|^2 - |cr|^2}$.

(c) The circle $0 < |cz_0 + d| < |c|r$ is mapped onto the circle $|w - w_0| = R$, where w_0', and R, as in (4a).

(d) The point $z_0 - d/c = z_\infty$ is mapped onto the point $w_0 = a/c = w_\infty$; $R = \left|\dfrac{ad - bc}{c^2 r}\right|$.

Map 3.56. THREE POINTS ONTO THREE POINTS. There are five cases (a)-(e) discussed below. Using the cross-ratio (3.3.1), the transformation is defined by (3.3.2). Then

Map 56a. This is defined by

$$\frac{w - w_1}{w - w_2}\frac{w_3 - w_2}{w_3 - w_1} = \frac{z - z_1}{z - z_2} \cdot \frac{z_3 - z_2}{z_3 - z_1}, \tag{3.4.1}$$

where z_1, z_2, z_3 are given and finite, and w_1, w_2, w_3 are finite. This is presented in Figure 3.49(a).

Map 3.56b. The transformation is of the form $\{w, w_1, w_2, w_3\} = \{z, z_1, z_2, \infty\}$, i.e.,

$$\frac{w - w_1}{w - w_2}\frac{w_3 - w_2}{w_3 - w_1} = \frac{z - z_1}{z - z_2}, \tag{3.4.2}$$

where the three given points are z_1, z_2, ∞ and w_1, w_2, w_3. This case is presented in Figure 3.49(b).

Map 3.56c. The transformation is of the form

$$\frac{w - w_3}{w_3 - w_1} = \frac{z_2 - z_1}{z - z_2}, \tag{3.4.3}$$

where the three given points are z_1, z_2, ∞ and w_1, ∞, w_3. This case is presented in Figure 3.49(c).

Map 3.56d. The transformation is of the form

$$\frac{w - w_1}{w_1 - w_2} = \frac{z - z_1}{z_1 - z_2}, \tag{3.4.4}$$

where the three given points are z_1, z_2, ∞ and w_1, w_2, ∞. This case is presented in Figure 3.49(d).

Map 3.56e. The transformation is of the form

$$\frac{w - w_1}{w - w_2} = \frac{z - z_1}{z - z_2}\frac{z_3 - z_2}{z_3 - z_1}, \tag{3.4.5}$$

where the three given points are z_1, z_2, z_3 and w_1, w_2, ∞. This case is presented in Figure 3.49(e).

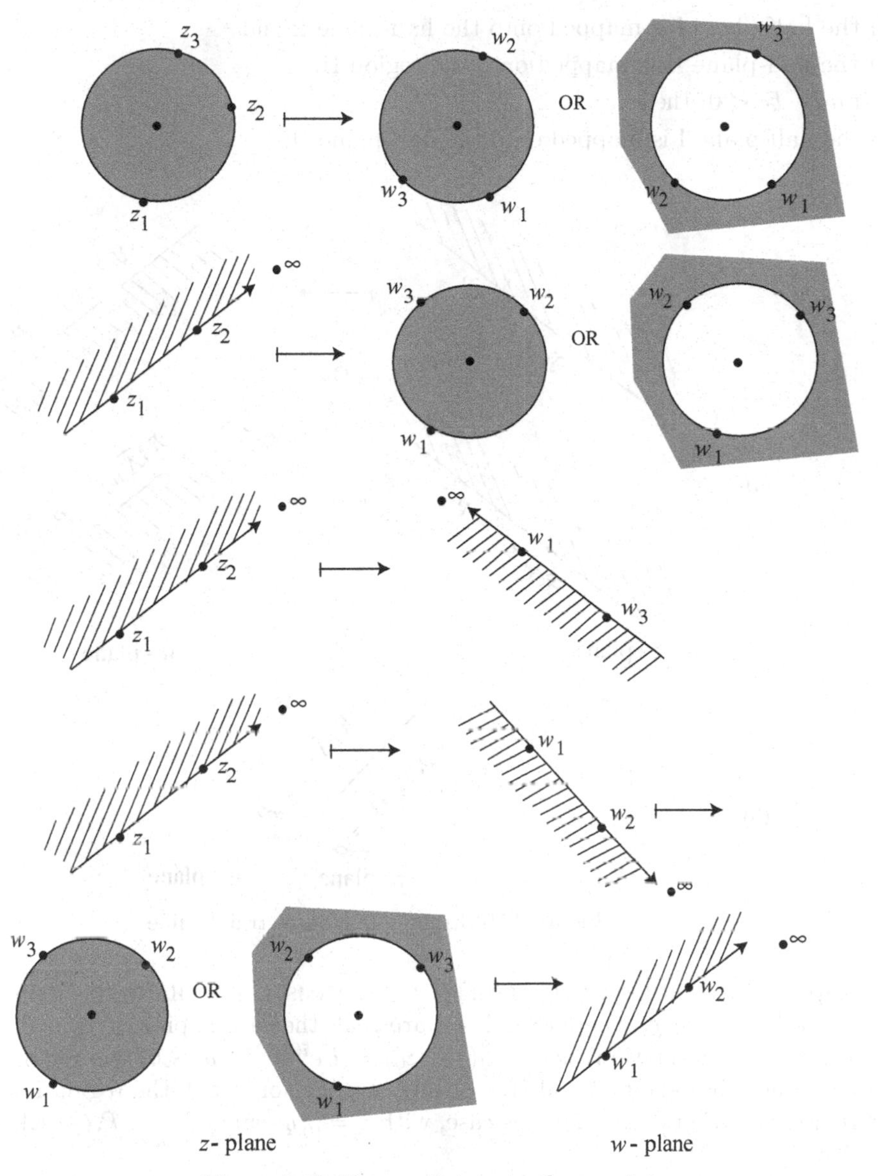

Figure 3.49 Three points onto three points.

Map 3.57. Straight line onto straight line. The transformation is

$$\bar{\lambda} w = P + \frac{a\left(\bar{\lambda} z - p\right) - i\,b}{i\,c\left(\bar{\lambda} z - p\right) + d}, \tag{3.4.6}$$

where λ, p, Λ and P are preassigned. This transformation is presented in Figure 3.50(a). In the case of the mapping of a line onto itself, the above holds except that $\lambda = \Lambda$ and $p = P$, i.e., $\Re\{\bar{\lambda} z\} = p$ is the same as $\Re\{\bar{\Lambda} w\} = P$ (Figure 50(b)).

The details for the case 12a, with $z = p, q$ real and $w = P, Q$ real, are as follows:

If $ad - bd > 0$, then

(a) the half-plane I is mapped onto the half-plane I; and

(b) the half-plane II is mapped onto the region II.

If $ad - bc < 0$, then

(c) the half-plane I is mapped onto the half-plane II.

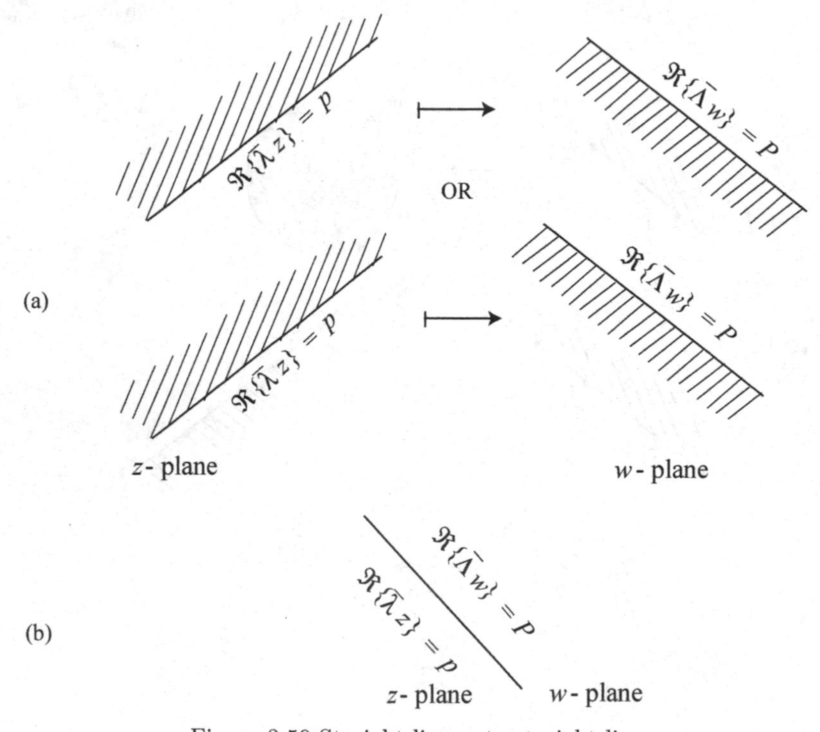

Figure 3.50 Straight line onto straight line.

Map 3.58. ANGLE ONTO ITSELF, WITH ARMS INTERCHANGED. Given, z_0, $\alpha \neq 0$ $(-\pi < \alpha < \pi)$, and τ, where α and *tau* are real, the case is presented in Figure 3.51. For involutory transformation $(w - w_0)(z - z_0) = b\,e^{i(\alpha+2\tau)}$, $b > 0$, the regions I and II are mapped onto the regions I and II, respectively; but for $b < 0$, the region I is mapped onto the region II. Other details for this case, with $z = p, q$ real and $w = P, Q$ real, are as follows:

The points $z_0; z_1 = \infty; z_2 = z_0 + ae^{i\tau}$, where a is real, $a > 0$ in Figure 3.51, are mapped onto the points $w_0 = \infty; w_1 = z_0; w_2 = z_0 + \dfrac{b}{a}\, e^{i(\alpha+\tau)}$.

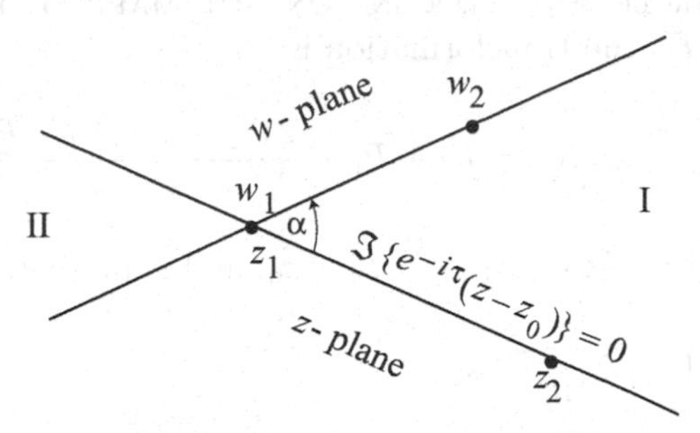

Figure 3.51 Angle onto itself.

Map 3.59. Straight line onto circle. The transformation is

$$w - w_0 = \Re\left\{e^{i\tau}\,\frac{\bar{\lambda}z - p + \beta}{\bar{\lambda}z - p - \beta}\right\}, \quad \tau \text{ real}; \ \Re\{\beta\} \neq 0, \tag{3.4.7}$$

given in Figure 3.52.

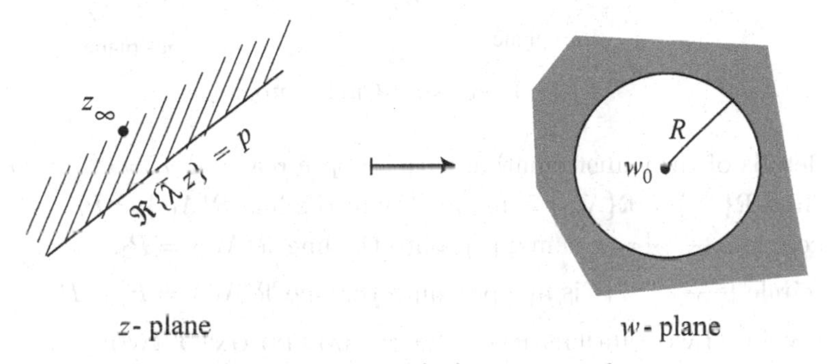

Figure 3.52 Straight line onto circle.

Map 3.60. Circle onto straight line. The transformation is

$$\bar{\Lambda} w = P + i\,a + b\frac{z - z_0 + r\,e^{i\tau}}{z - z_0 - r\,e^{i\tau}}, \tag{3.4.8}$$

where a, b, τ are real, $b \neq 0$, given in Figure 3.53.

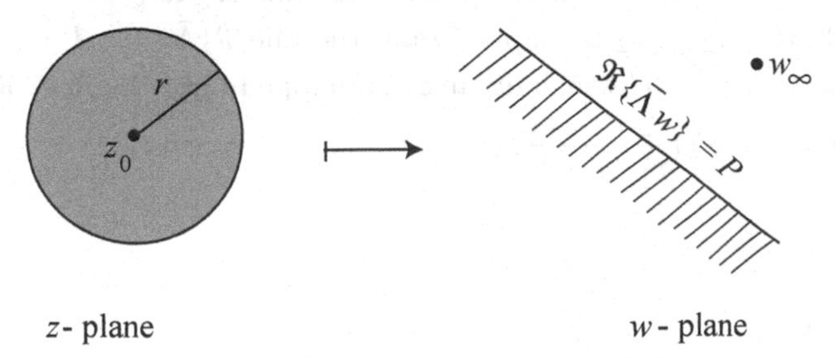

Figure 3.53 Circle onto straight line.

Map 3.61. CIRCLE AND LINE IN-CONTACT MAPPED ONTO TWO PARALLEL LINES. Given $z_0, z_2, \Lambda, P_1, P_2$, the transformation is

$$\bar{\Lambda} w = i\,a + P_1 + \frac{2\,(z_2 - z_0)\,(P_2 - P_1)}{z - z_0}, \tag{3.4.9}$$

where a is real, and $r = |z_2 - z_0|, \lambda = z_2 - z_0$, and is presented in Figure 3.54.

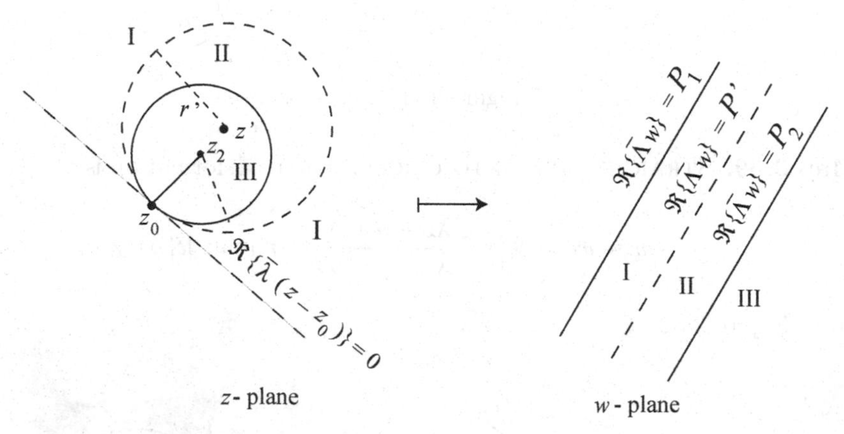

Figure 3.54 Circle onto circle.

The details of the transformation with $z = p, q$ real and $w = P, Q$ real, are as follows:
(a) The line $\Re\{\bar{\lambda} z\} = \Re\{\bar{\lambda} z_0\}$ is mapped onto the line $\Re\{\bar{\Lambda} w\} = P_1$.
(b) The circle $|z - z_2| = r$ is mapped onto the line $\Re\{\bar{\Lambda} w\} = P_2$.
(c) The circle $|z - z'| = r'$ is mapped onto the line $\Re\{\bar{\Lambda} w\} = P' = P_1 + \frac{r}{r'}\,(p_2 - p_1)$.

Map 3.62. TWO CIRCLES IN-CONTACT MAPPED ONTO TWO PARALLEL LINES. (Inner Contact) Given $r_1 > 0$, $r_2 > 0$, z_1, z_2, and a real, the transformation is

$$\bar{\Lambda} w = P_1 + i\,a + \gamma \frac{z + z_0 - 2z_1}{z_0 - z}, \tag{3.4.10}$$

where $\gamma = r_2 \dfrac{P_2 - P_1}{r_1 - r_2}$. The details are provided in Figure 3.55 and where $z_0 = \dfrac{r_1 z_2 - r_2 z_1}{r_1 - r_2}$.

The details of the transformation with $z = p, q$ real and $w = P, Q$ real, are as follows:
(a) The circle $|z - z_1| = r_1$ is mapped onto the line $\Re\{\bar{\Lambda} w\} = P_1$.
(b) The circle $|z - z_2| = r_2$ is mapped onto the line $\Re\{\bar{\Lambda} w\} = P_2$.
(c) The circle $|z - z'| = r'$ touching at z_0 is mapped onto the line $\Re\{\bar{\Lambda} w\} = P'$, where $P' = P_1 + (P_2 - P_1) \dfrac{r_2}{r_1} \dfrac{r' - r_1}{r_2 - r_1}$.

(d) The line $\Re\{\bar{\lambda}\,(z - z_0)\} = 0$ touching at z_0, where $\lambda = z_1 - z_2$, is mapped onto the line $\Re\{\bar{\Lambda}w\} = P_0$, where $P_0 = \dfrac{P_1 r_1 - P_2 r_2}{r_1 - r_2}$.

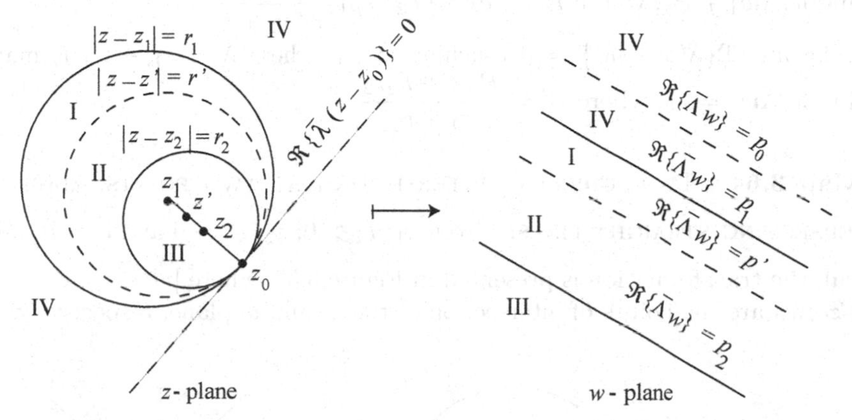

Figure 3.55 Two circles in contact mapped onto two parallel lines (inner contact).

Map 3.63. Two circles in outer contact mapped onto parallel lines. (Outer Contact) Given $|z_1 - z_2| = r_1 + r_2$, transformation is

$$\bar{\Lambda}w = P_1 + i\,a + \gamma\frac{z + z_0 - 2z_1}{z_0 - z}, \tag{3.4.11}$$

where a is real, and $\gamma = r_2\dfrac{P_1 - P_2}{r_1 + r_2}$. This case is presented in Figure 3.56, where $z_0 = \dfrac{r_1 z_2 + r_2 z_1}{r_1 + r_2}$.

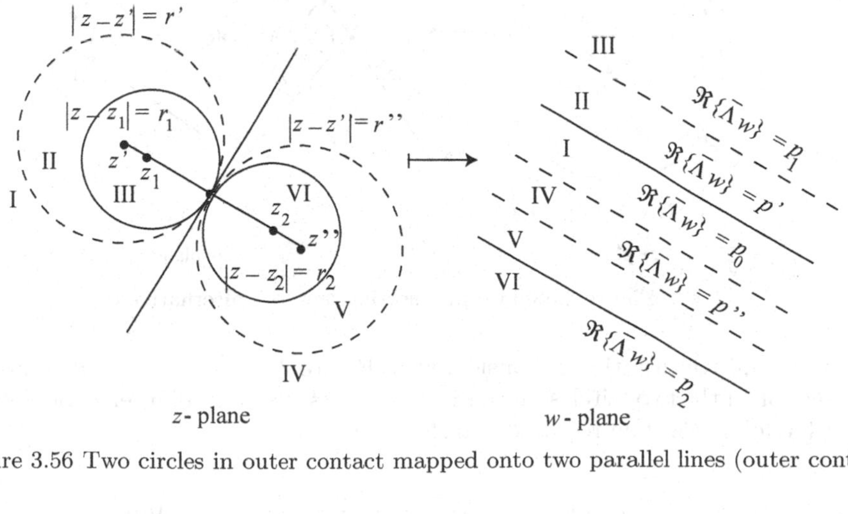

Figure 3.56 Two circles in outer contact mapped onto two parallel lines (outer contact).

The details of the transformation with $z = p, q$ real and $w = P, Q$ real, are as follows:

(a) The circle $|z - z_1| = r_1$ is mapped onto the line $\Re\{\bar{\Lambda}w\} = P_1$.

(b) The circle $|z - z_2| = r_2$ is mapped onto the line $\Re\{\bar{\Lambda}w\} = P_2$.

(c) circle $|z - z'| = r'$ touching at z_0 is mapped onto the line $\Re\{\bar{\Lambda}w\} = P'$, where $P' =$

$P_2 + (P_1 - P_2)\frac{r_1}{r'}\frac{r' + r_2}{r_1 + r_2}$.

(d) The circle $|z - z''| = r''$ touching at z_0, where $\lambda = z_1 - z_2$, is mapped onto the line $\Re\{\bar{\Lambda}w\} + P''$, where $P'' = P_1 + (P_2 - p_1)\frac{r_2}{r''}\frac{r'' + r_1}{r_1 + r_2}$.

(e) The line $\Re\{\bar{\lambda}(z - z_0)\} = 0$ touching at z_0, where $\lambda = z_1 - z_2$, is mapped onto the line $\Re\{\bar{\Lambda}w\} = P_0$, where $P_0' = \frac{P_1 r_1 + P_2 r_2}{r_1 + r_2}$.

Map 3.64. TWO 'CIRCLES', INTERSECTING AT TWO POINTS, MAPPED ONTO TWO INTERSECTING STRAIGHT LINES. Given z_1; $r_1 > 0$; z_2; $r_2 > 0$ and $\alpha = \beta$, i.e., $\frac{\bar{\Lambda}_1}{\bar{\Lambda}_2}\frac{z_0 - z_1}{z_0 - z)2}$ is real, the transformation is presented in Figure 3.57, where $|r_1 - r_2| < |z_1 - z_2| < r_1 + r_2$, and z_0, w_0 are the points of intersection of the z- and w-plane, respectively.

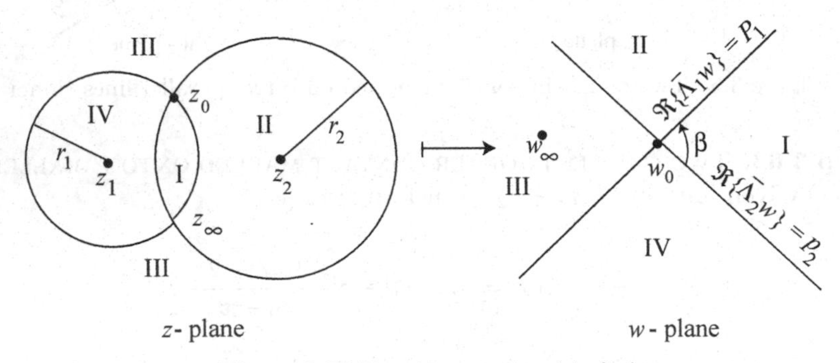

Figure 3.57 Two intersecting 'circles'.

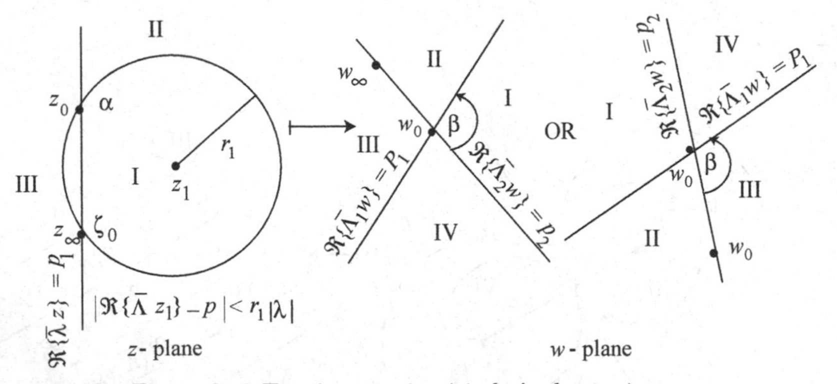

Figure 3.58 Two intersecting 'circles', alternative case.

An alternative situation is considered in Figure 3.58, where z_0 and ζ_0 are the points of intersection of the two 'circles' in the z-plane; w_0 is the point of intersection of $\Re\{\Lambda_1 w\} = P_1$ and $\Re\{\Lambda_1 w\} = P_2$. The required transformation is

$$w = w_0 + K\Lambda_1 \frac{(z - z_0)(z_1 - \zeta_0)}{(z - \zeta_0)(z_0 - \zeta_0)}, \quad K \neq 0 \text{ real.} \tag{3.4.12}$$

The details of the alternative case with $z = p, q$ real and $w = P, Q$ real, are as follows:

(a) The points $z = 0;\ \infty;\ 2r_1$ is mapped onto the points $w = \infty;\ \dfrac{ir_2}{r_1 + r_2};\ 0.$

(b) The circle $|z - z_1| = r_1$ is mapped onto the line $\Re\{\bar{\Lambda}_1 w\} = P_1$.

(c) The circle $|z - z_2| = r_2$ is mapped onto the line $\Re\{\bar{\Lambda}_2 w\} = P_2$.

(d) The circle $|z + r_2| = r_2$ is mapped onto the line $\Re\{\bar{\Lambda}_2 w\} = P_2$, or line $\Re\{\bar{\lambda} z\} = p$.

(e) The points $z = z_0; \zeta_0; z = \infty$ are mapped onto the points $w = w_0;\ \infty;\ w_\infty = w_0 + K\Lambda_1 \dfrac{z_1 - \zeta_0}{z_0 - \zeta_0}$.

(f) The circle passing through z_0, ζ_0 is mapped onto the lines passing through w_0.

(g) The circle orthogonal to $|z - z_0| = r_1$, and to $|z - z_2| = r_2$ or to $\Re\{\bar{\lambda} z\} = p$ is mapped onto the circle with center w_0.

Map 3.65. The transformation presented in Figure 3.59 is

$$w - \frac{K}{\sigma} \frac{z(i\,\sigma - z_1)}{z - i\,\sigma}, \quad \sigma > 0, K > 0. \tag{3.4.13}$$

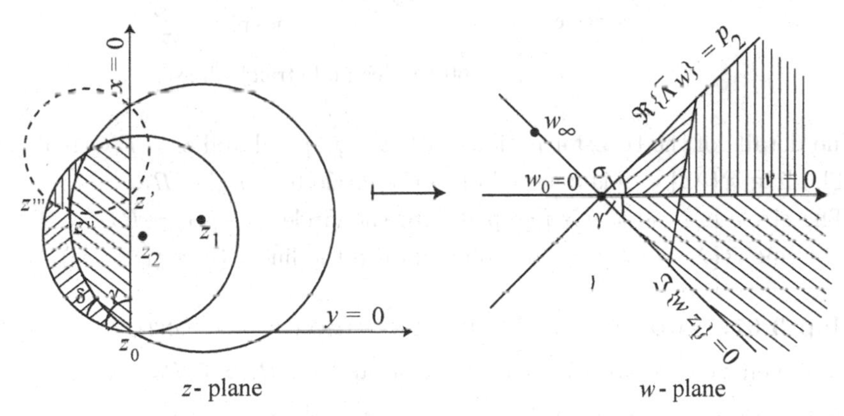

Figure 3.59 Two intersecting 'circles'.

The details of the transformation, with $z = p, q$ real and $w = P, Q$ real, are as follows:

(a) The points $\zeta_0 = i\sigma; z_0 = 0; z'; z''; z'''$ are mapped onto the points $w = \infty; w_0 = 0; w'; w''; w'''$.

(b) The point $z = \infty$ is mapped onto the $w_\infty = \dfrac{K}{\sigma}(i\,\sigma - z_1) = -\dfrac{K}{\sigma}\bar{z}_1$.

(c) The circle $|z - z_1| = r_1$, $\Im\{z_1\} = \frac{1}{2}\sigma$ is mapped onto the line $v = 0$.

(d) The circle $|z - z_2| = r_2$, $\Im\{z_2\} = \frac{1}{2}\sigma$ is mapped onto the line $\Re\{\bar{\Lambda}_2 w\} = P_2$.

(e) The line $x = 0$ is mapped onto the line $\Im\{wz_1\} = 0$.

(f) The circle (z', z'', z''', ζ_0) is mapped onto the line (w', w'', w''').

(g) The curvilinear triangle $\{z_0, z', z''\}$ is mapped onto the right triangle $\{w, w', w''\}$.

(h) The curvilinear triangle $\{z_0, z'', z'''\}$, each with sum of angles, is mapped onto the right triangle $\{w_0, w'', w'''\}$.

Map 3.66. CIRCLES AND STRAIGHT LINE, WITHOUT COMMON POINT, ONTO TWO CONCENTRIC CIRCLES. The transformation, presented in Figure 3.60, is

$$w - w_1 = R_1\, e^{i\tau} \frac{(z - z_0)|\lambda| - \lambda(A + \sigma)}{(z - z_0)|\lambda| - \lambda(A - \sigma)}, \tag{3.4.14}$$

where τ is real, $A = \dfrac{p - \Re\{\bar{\lambda} z_0\}}{|\lambda|}$, and $\sigma = \sqrt{A^2 - \rho^2} > 0$ or $\sigma = -\sqrt{A^2 - \rho^2}$.

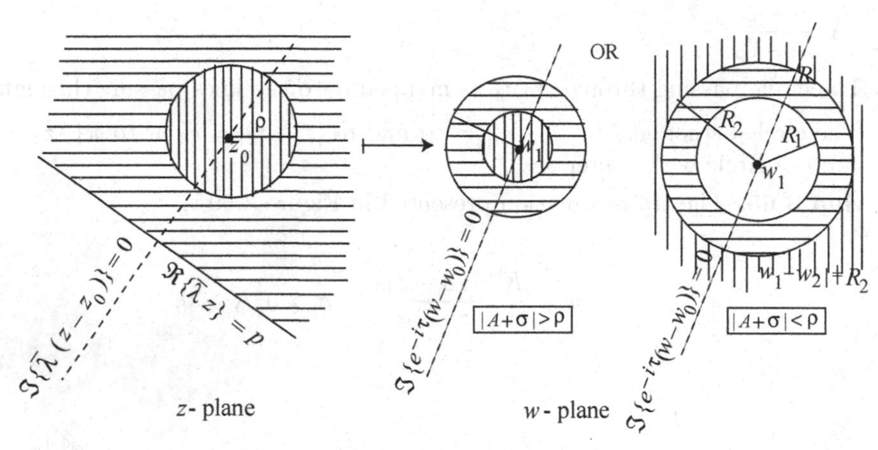

Figure 3.60 Circles and straight lines.

The details of the transformation, with $z = p, q$ real and $w = P, Q$ real, are as follows:
(a) The line $\Re\{\bar{\lambda} z\} = p$ is mapped onto the circle $|w - w_0| = R_1$.
(b) The circle $|z - z_0| = \rho$ is mapped onto the circle $|w - w_1| = R_2$.
(c) The line $\Im\{\bar{\lambda}(z - z_0)\} = 0$ is mapped onto the line $\Im\{e^{-i\tau}(w - w_1)\} = 0$.

Map 3.67. TWO CIRCLES, WITHOUT COMMON POINT, ONTO TWO CONCENTRIC CIRCLES. Given $z_1, z_2, r_1 > 0, r_2 > 0$, $z_1 \neq z_2$, and $w_0, R_1 > 0, R_2 > 0$, and $\dfrac{R_2}{R_1} = \dfrac{r_2}{r_1}\left|\dfrac{t}{s - t}\right|$, where s, t are the real roots of the equations $st = r_1^2$, and $(d-s)(d-t) = r_2^2$, $d = |z_2 - z_1| > 0$, the transformation is

$$w - w_0 = t \frac{R_1}{r_1}\, e^{i\theta}\, \frac{d(z - z_1) - s(z_2 - z_1)}{d(z - z_1) - t(z_2 - z_1)}, \quad \theta \text{ real}. \tag{3.4.15}$$

This transformation is presented Figure 3.61.

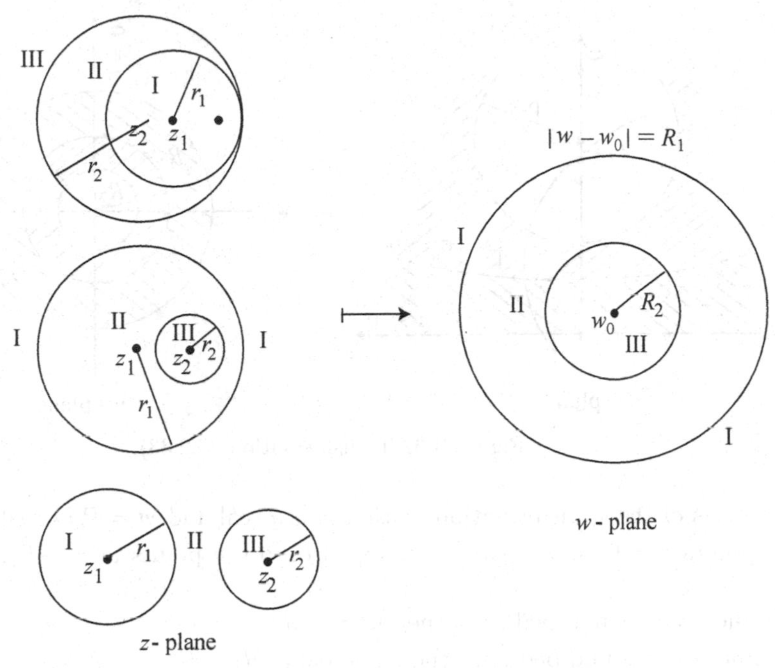

Figure 3.61 Transformation (3.2.35).

The details of this transformation, with $z = p, q$ real and $w = P, Q$ real, are as follows:
(a) The circle $|z - z_1| = r_1$ is mapped onto the circle $|w - w_0| = R_1$.
(b) The circle $|z - z_2| = r_2$ is mapped onto the circle $|w - w_0| = R_2$.
(c) The radical axis of these circles is mapped onto the circle $|w - w_0| = R_1|t|/r_1$.
(d) The point $z_\infty = z_1 + (t/d)(z_2 - z_1)$ is mapped onto the point $w = \infty$.
(e) The point $z_0 = z_1 + (s/d)(z_2 - z_1)$ is mapped onto the point $w = w_0$.
(f) Any 'circle' passing through z_∞ and z_0 is mapped onto the line passing through w_0.

Map 3.68. The transformation, presented in Figure 3.62, is

$$w = R_1 \frac{z - i\sigma}{z + i\sigma}, \tag{3.4.16}$$

where $\sigma = \sqrt{k^2 - \rho^2}$, $0 < \rho < k$.

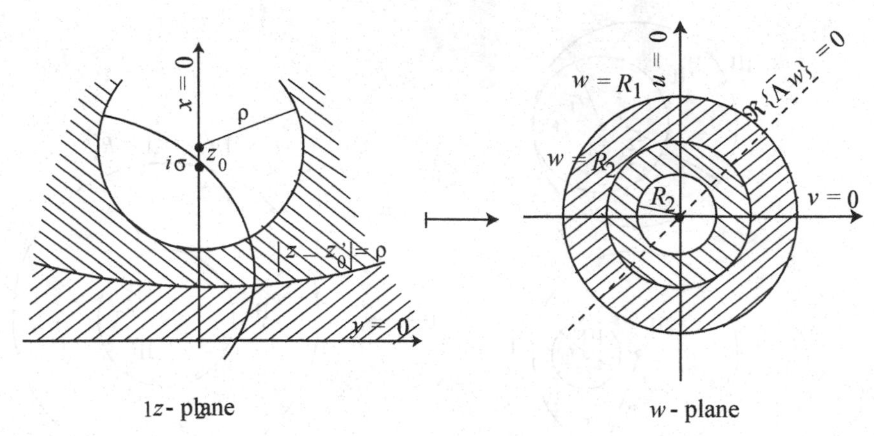

Figure 3.62 Transformation (3.2.33).

The details of the transformation, with $z = p, q$ real and $w = P, Q$ real, are as follows:
(a) The points $z = 0; \infty; i\sigma; -i\sigma$ are mapped onto the points $w = -R_1; R_1; 0; \infty$, respectively.
(b) The line $x = 0$ is mapped onto the line $v = 0$.
(c) The line $y = 0$ is mapped onto the circle $|w| = R_1$.
(d) The circle $|z - ik| = \rho'$, where $z_0' = i\sigma \dfrac{R_1^2 + R^2}{R_1^2 - R^2}$, $\rho' = \dfrac{2\sigma R_1 R}{|R_1^2 - R^2|}$, and $R_1 \neq R$, is mapped onto the circle $|w| = R_2 = \dfrac{R_1 \rho}{|k + \sigma|}$.
(e) The circle $|z| = \sigma$ is mapped onto the line $u = 0$.
(f) The circle $|z - \sigma \dfrac{\Im\{\Lambda\}}{\Re\{\Lambda\}}| = \sigma |\dfrac{\Lambda}{\Re\{\lambda\}}|$, where $\Re\{\Lambda\} \neq 0$, intersecting $x = 0$ at $\pm i\sigma$, is mapped onto the line $\Re\{\bar{\Lambda} w\} = 0$.

Map 3.69. The transformation

$$w = \frac{tR_1}{r_1} \frac{z - s}{z - t}, \tag{3.4.17}$$

where s and t are roots of the equations $st = r_1^2$ and $(z_2 - s)(z_2 - t) = r_2^2$, is presented in Figure 3.63.

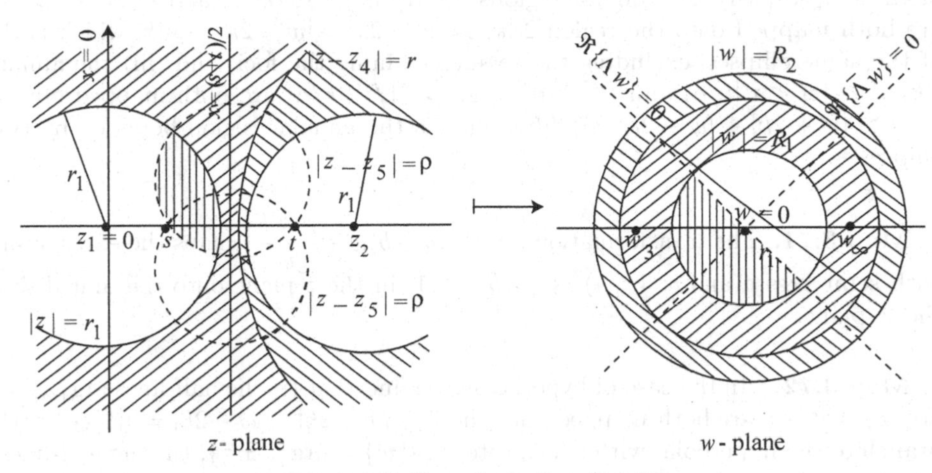

Figure 3.63 Transformation (3.4.17).

The details of this transformation, with $z = p, q$ real and $w = P, Q$ real, are as follows:

(a) The circles $|z - z_{1,2}| = r_{1,2}$, $z_1 = 0, z_2 > 0$ are mapped onto the circles $|w| = R_{1,2}$; $\dfrac{R_2}{R_1} = \dfrac{r_2}{r_1}\dfrac{t}{r_2 - t}$.

(b) The points $z_0 = s; z_\infty = t; z = \infty$; and $z_3 = \dfrac{s+t}{2}$ are mapped onto the points $w_0 = 0, \infty; w_\infty = \dfrac{tR_1}{r_1}$; and $w_3 = -\dfrac{tR_1}{r_1}$.

(c) The circle $|z - z_0| = r$, where $z_4 = r_1^2 \dfrac{R_1^2 - R^2}{tR_1^2 - sR_1^2}$ and $r = \dfrac{r_1 R_1 R|t-s|}{|sR_1^2 - tR_1^2|}$, is mapped onto the circle $|w| = R \neq \infty$.

(d) The line $x = \dfrac{s+t}{2}$ is mapped onto the circle $|w| = \dfrac{R_1 t}{r_1} = w_\infty$.

(e) The circle $|z - z_5| = \rho$, where $z_5 = \dfrac{t\Lambda + s\bar\Lambda}{2\Re\{\Lambda\}}$ and $\rho = \left|\dfrac{\Lambda(s-t)}{2\Re\{\Lambda\}}\right|$, $\Re\{\Lambda\} \neq 0$ is mapped onto the line $\Re\{\bar\Lambda w\} = 0$.

(f) The circle $|z - \bar z_5| = \rho$ is mapped onto the line $\Re\{\Lambda w\} = 0$.

(g) The circle $y = 0$ is mapped onto the line $v = 0$.

(h) The curvilinear rectangular quadrilateral formed by $|z| = r_1, |z - z_2| = r_2$ between $|w| = R_1$, $|w| = R$ and $|z - z_5| \leq \rho$, is mapped onto the half-ring left of $\Re\{\bar\Lambda\} = 0$.

(i) The region $|z - z_2| \leq r_2$ is mapped onto the region $|w| \geq R_2$.

3.5 Ellipses and Hyperbolas

For ellipses and hyperbolas, we will set t, t', and τ fixed, $0 < t < t' < 1$, and $0 < \tau < \pi/2$; also $l = t + 1/t$, $l' = t' + 1/t'$, and $\sigma = \arg\{k/\alpha\}$. The ellipse has been defined in §3.1.3, and the hyperbolas in §3.1.4.

Map 3.70. In the case of ellipses, we find that (i) the regions $|z| < |k/\alpha|t$ and $|z| > |k/\alpha|t^{-1}$ are both mapped onto the region $|w + 2k| + |w - 2k| > 2l|k|$ which is the exterior of the ellipses; (ii) the annular regions $t|k/\alpha| < |z| < |k/\alpha|$ and $t^{-1}|k/\alpha| > |z| > |k/\alpha|$ are both mapped onto the region $2l|k| > |w+2k| + |w-2k| > 4|k|$ which is the exterior of the same ellipses, excluding the segment joining the foci; and (iii) the annular regions $t|k/\alpha| < |z| < t'|k/\alpha|$ and $t^{-1}|k/\alpha| > |z| > (1/t')|k/\alpha|$ are both mapped onto the region $2l|k| > |w+2k| + |w-2k| > 2l'|k|$ which is the annular region bounded by two confocal ellipses.

Map 3.71. The transformation $z = \frac{1}{2}\left[(a-b)w + \dfrac{a+b}{w}\right]$ maps the exterior of an ellipse with semi-axes a and b: $(x/a)^2 + (y/b)^2 \geq 1$, in the z-plane onto the unit disk $|w| \leq 1$ in the w-plane.

Map 3.72. In the case of hyperbolas, we find that (i) the half-lines $\arg\{z\} = \sigma+\tau$ and $\arg\{z\} = \sigma - \tau$ are both mapped onto the line $|w+2k| - |w-2k| = 4|k|\cos\tau$ which is one branch of the hyperbola, with asymptotes $\arg\{w\} = \arg\{k \pm \tau\}$; (ii) the half-lines $\arg\{z\} = \sigma+\tau+\pi$ and $\arg\{z\} = \sigma-\tau+\pi$ are both mapped onto the line $|w+2k|-|w-2k| = -4|k|\cos\tau$ which is the other branch of the hyperbola; (iii) the half-lines $\arg\{z\} = \sigma + \pi/2$ and $\arg\{z\} = \sigma - \pi/2$ are both mapped onto the line $\Re\{\bar{k}w\} = 0$; (iv) the half-line $\arg\{z\} = \sigma$ is mapped onto the half-line $|w| > 2|k|$, $\arg\{w/k\} = 0$, counted twice; and (v) the half-line $\arg\{z\} = \sigma + \pi$ is mapped onto the half-line $|w| > 2|k|$, $\arg\{w/k\} = \pi$, counted twice. These properties are presented in Figure 3.64.

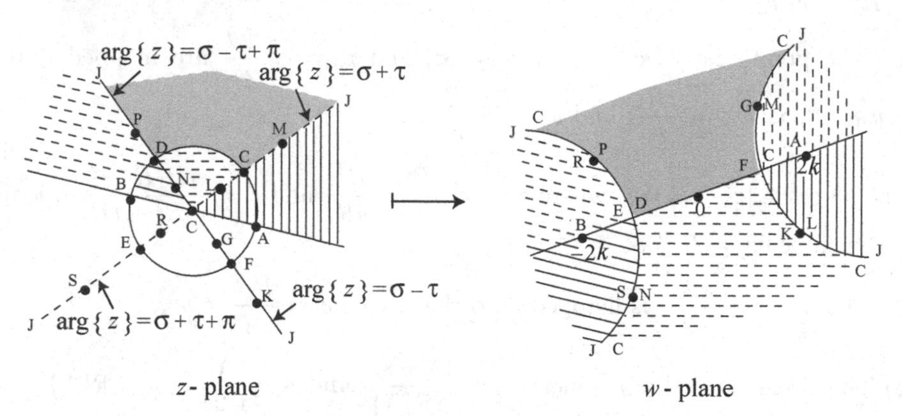

Figure 3.64 Mapping of different regions onto different regions of a hyperbola.

REFERENCES USED: Ahlfors [1966], Boas [1987], Carrier, Krook and Pearson [1966], Ivanov and Trubetskov [1995], Kantorovich and Krylov [1958], Kober [1957], Kythe [1996], Nehari [1952], Nevanlinna and Paatero [1969], Pennisi et al. [1963], Papamichael, Warby and Hough [1986], Saff and Snider [1976], Schinzinger and Laura [2003], Silverman [1967], Wen [1992].

4

Algebraic Functions

The transformations using the polynomials, and particularly the monomials of the form $f(z) = (z-a)^n$, provide another important aspect of conformal mapping of a given domain $D \subset \mathbb{C}$ in the z-plane onto a domain $G \subset \mathbb{C}$ in the w-plane. As mentioned in Chapter 3, the function $w = f(z)$ does not always exist, and it may not be uniquely defined, since the Riemann mapping theorem guarantees the existence and uniqueness only under certain specific conditions. We will also discuss other transformations, like hyperbolas and Cassini's ovals, mappings by exponential and logarithmic functions, trigonometric and hyperbolic functions, and certain cases of composite transformations.

4.1 Polynomials

The general *complex polynomial* $P_n(z)$ of degree n is an entire function for all $z \in \mathbb{C}$ and has derivatives of all orders such that $P_n^{(k)}(z) = 0$ for $k > n$. If $P_n(z)$ has a zero of multiplicity p at z_0, then $P_n(z) = (z-z_0)^p\, g(z)$, where $g(z)$ is a polynomial of degree $n-p$, $g(z_0) \neq 0$. The special cases $P_1(z) = az+b$, where a and b are complex constants, represent a magnification (or dilatation) by $|a|$ and a rotation by $\arg\{a\}$, followed by a translation by b. A *rational function*, defined as the quotient of two complex polynomials $P_n(z)$ and $Q_m(z)$ of degree n and m, respectively, $n \leq m-1$, with no common factors, is analytic for all z which is not a zero of $Q_m(z)$. If the polynomial $Q_m(z)$ has a zero z_0 of multiplicity p, then the partial fraction development of the rational function corresponding to this zero has the form

$$\frac{A_1}{z-z_0} + \frac{A_2}{(z-z_0)^2} + \cdots + \frac{A_p}{(z-z_0)^p}. \tag{4.1.1}$$

Under the mapping $w = P_n(z)$, $n > 1$, there are at most n points w_0 in the extended w-plane with fewer than n distinct inverse images. In fact, the point $w = \infty$ has just one inverse image. If $w_0 \neq \infty$ has fewer than n distinct inverse images, then the equation

$$P_n(z) = w_0 = P_n(z_0) \tag{4.1.2}$$

must have multiple roots which satisfy the equation $P'(z) = 0$. But since the polynomial $P'(z)$ is of degree at most $(n-1)$, it has at most $n-1$ distinct zeros z'_ν, $1 \leq \nu \leq n-1$.

Hence, Eq (4.1.2) can have a multiple root only at the numbers $P_n(z_1'), P_n(z_2'), \dots, P_n(\infty)$, at most n of which are distinct.

Let z_0 be a root of multiplicity $k > 1$ of Eq (4.1.2). Then under the mapping $w = P_n(z)$, every angle with its vertex at z_0 is enlarged k-times, whereas every angle with vertex at $z = \infty$ is enlarged n-times.

A complex polynomial P of degree n defines a mapping of the z-plane onto the w-plane such that for every point in the w-plane, its inverse image consists of exactly n pints with the exception of at most $n - 1$ points where the inverse image consists of fewer than n points.

Map 4.1. Consider the mapping

$$w = (z - a)^n, \quad n > 1. \tag{4.1.3}$$

This function maps the extended z-plane into the extended w-plane such that every point w has n distinct inverse images, except at the two points $w = 0$ and $w = \infty$, for which the n inverse images coalesce into a single point $z = a$ and $z = \infty$, respectively. For $w \neq 0$ and $w \neq \infty$, the n inverse images of w are obtained by solving (4.1.3) for z, which yields

$$z = a + \sqrt[n]{w} = a + \sqrt[n]{|w|}\left(\cos\frac{\arg\{w\}}{n} + i\,\sin\frac{\arg\{w\}}{n}\right). \tag{4.1.4}$$

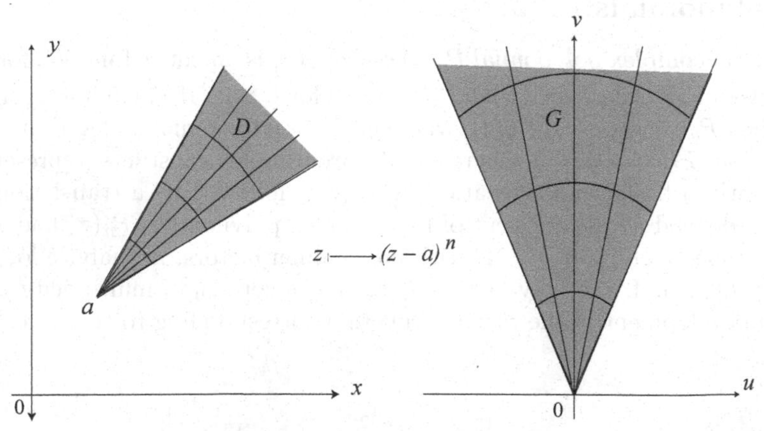

Figure 4.1 Mapping $w = (z - a)^n$.

Thus, the n distinct points z, defined by (4.1.4), are situated at the vertices of a regular n-gon with center at $z = a$. The mapping (4.1.3) is conformal at all points except $z = a$ and $z = \infty$, and every angle with the vertex at one of these two points is enlarged n-times. In fact, since $|w| = |z - a|^n$, $\arg\{w\} = n \arg\{z - a\}$, the circle $|z - a| = r$ is mapped onto the circle $|w| = r^n$. Also, as the point z traverses once around the circle $|z - a| = r$, the image point w traverses n-times the circle $|w| = r^n$ in the same direction, since $\arg\{w\}$ increases continuously by $2n\pi$. Moreover, the ray $\arg\{z - a\} = \theta_0 + 2k\pi$ (k an integer) in the z-plane going from a to ∞ is mapped onto the ray $\arg\{w\} = n\theta_0 + 2m\pi$ (m an integer) in the w-plane going from 0 to ∞.

Let D denote the region $\{z : \theta_0 + 2k\pi < \arg\{z - a\} < \theta_1 + 2k\pi\}$, where k is an integer and $0 < \theta_1 - \theta_0 \leq 2\pi/n$. This region D is called the *interior* of the angle $(\theta_1 - \theta_0)$ which is bounded by the two rays $\arg\{z-a\} = \theta_0+2k\pi$ and $\arg\{z-a\} = \theta_1+2k\pi$, $k = 0, \pm1, \pm2, \dots$. Then the function (4.1.3) maps the region D onto the region $G = \{n\theta_0 + 2m\pi < \arg\{w\} < n\theta_1 + 2m\pi\}$, $m = 0, \pm1, \pm2, \dots$, which is the interior of the angle $n(\theta_1 - \theta_0)$ with vertex at $w = 0$ (Figure 4.1).

The mapping function (4.1.3) produces a conformal and one-to-one map of the interior of an angle onto the interior of another angle which is n-times wider. However, it does not map every circle onto a circle (in the extended sense). For example, for $n = 2$, $a = 0$, the map $w = z^2$ maps every vertical straight line $z = b + it$, where $b \neq 0$ is real and $-\infty < t < \infty$, onto the parabola $v^2 = 4b^2\left(b^2 - u\right)$ which opens to the left, and every horizontal line $z = t + ic$, where $c \neq 0$ is real and $-\infty < t < \infty$, onto the parabola $v^2 = 4c^2\left(c^2 + u\right)$ which opens to the right (Figure 4.2). This mapping is conformal except at $z = 0$, but it is not one-to-one, since every point in the w-plane except $w = 0$ and $w = \infty$ has two inverse images.

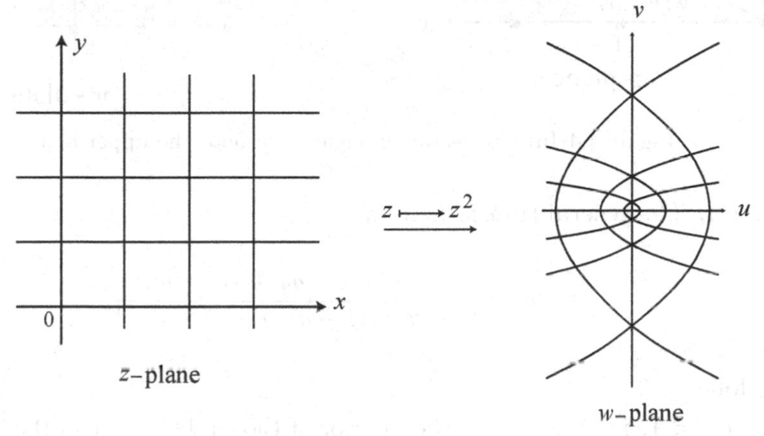

Figure 4.2 Mapping $w = z^2$.

Map 4.2. A particular case of the transformation (4.1.3) is the power transformation $w = z^n$, $n > 0$. Let $z = r\,e^{i\theta}$. Then $w = r^n\,e^{in\theta} = R\,e^{in\Theta}$. Thus, $R = r^n$ and $\Theta = n\theta$; the magnification $m = \left|\dfrac{dw}{dz}\right| = nr^{n-1}$, and the rotation $\alpha = \arg\left\{\dfrac{dw}{dz}\right\} = (n-1)\theta$. For $n = 2$, the polar coordinates are shown in Figure 4.3, in which the region between the rays $\theta = \pi/6$ and $\pi/3$ is mapped onto the region between the rays $\Theta = \pi/3$ and $2\pi/3$, respectively.

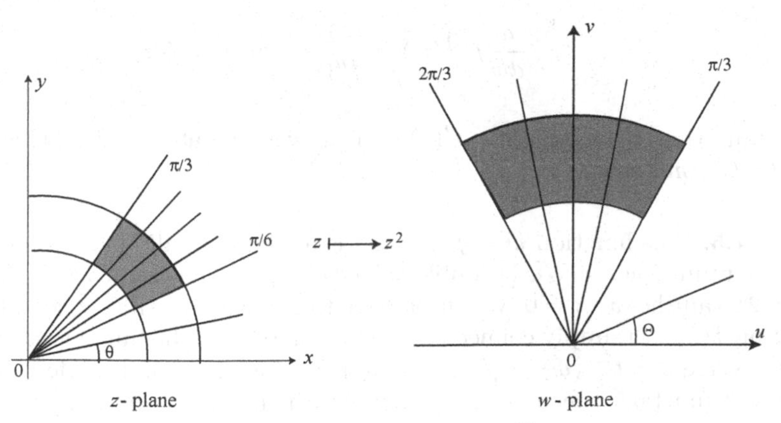

Figure 4.3 Mapping $w = z^2$.

Map 4.3. The transformation that maps an infinite sector of angle π/m, where $m \geq \frac{1}{2}$, in the z-plane onto the upper half-plane $\Im\{w\} > 0$ is defined by

$$w = z^m, \quad m \geq \frac{1}{2}, \tag{4.1.5}$$

and is presented in Figure 4.4.

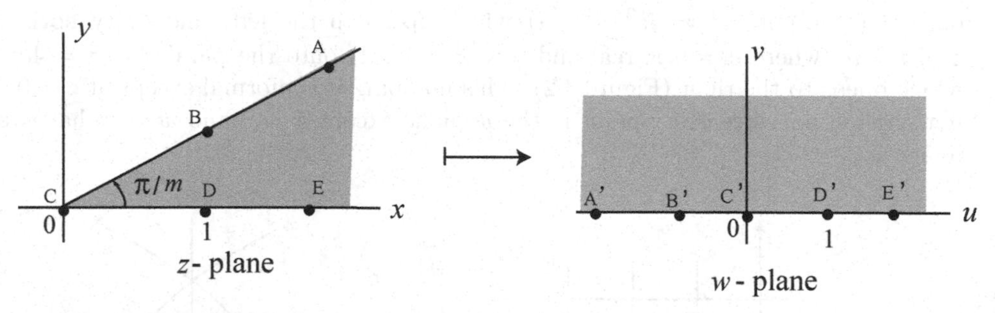

Figure 4.4 Infinite sector of angle π/m onto the upper half-plane.

Map 4.4. The general transformation

$$w = \frac{(z-a_1)(z-a_2)\cdots(z-a_n)}{(1-\bar{a}_1 z)(1-\bar{a}_2 z)\cdots(1-\bar{a}_n z)}, \tag{4.1.6}$$

maps as follows:

(i) for $|a_j| < 1$, $j = 1, 2, \dots, n$, the interior of the circle $|z| = 1$ in the z-plane is mapped onto the unit disk $|w| < 1$, counted n-times.

(ii) for $|a_j| > 1$, $j = 1, 2, \dots, n$, the region $|z| > 1$ in the z-plane is mapped onto the unit disk $|w| < 1$, counted n-times.

Recall that a complex analytic function $w = f(z)$ maps a point $z \in D \subset \mathbb{C}$ to a point $w \in G$ in the w-plane. In order to require that this mapping be univalent, the inverse function $z = f^{-1}(w)$ must be a well-defined map back from G to D so that it is also analytic on all of G. Using the formula for the derivative of an inverse function, we get

$$\frac{d}{dw} f^{-1}(w) = \frac{1}{f''(z)} \quad \text{at } w = f(z).$$

This means that the derivative $f'(z) \neq 0$ at very point $z \in D$. This is known as the *invertibility condition.*

Map 4.5. The function $w = f(z) = z^2$ is analytic on all of $\mathbb{C}$ except at $z = 0$, and its square root function $z = \sqrt{w}$ is double-valued except at the origin $z = 0$. Also, its derivative, $f'(z) = 2z$ vanishes at $z = 0$, which violates the above invertibility condition. However, by restricting $f(z)$ to a simply connected subdomain $G\backslash\{0\}$, the function $w = f(z)$ is univalent and its inverse $z = f^{-1}(w) = \sqrt{z}$ is a well-defined analytic and single-valued branch of the square root function. Since $|w| = |z^2|$ doubles the modules, and $\arg\{w\} = 2\arg\{z\}$, Figure 4.5 illustrates maps of a quarter disk $\{x > 0, y > 0\} = \{0 < \arg\{z\} < \pi/2\}$, upper half-disk $\{y > 0\} = \{0 < \arg\{z\} < \pi\}$, and the square $\{0 < x, y < 1\}$ under the function $w = z^2$.

Note that in the last map, the square is mapped onto a curvilinear triangle such that the edges of the square on the real and imaginary axes map onto the two halves of the straight base of the curvilinear triangle while the other two edges map onto its curved edges.

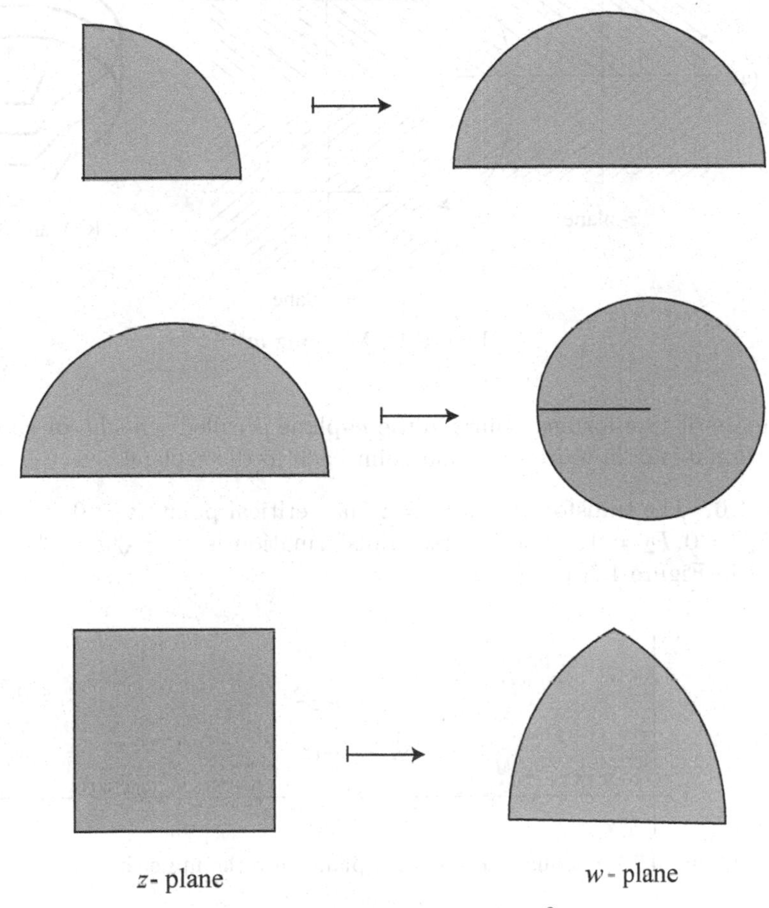

Figure 4.5 Maps under $w = z^2$.

Also note that the case of one-to-one mapping becomes doubtful when the angular rotation exceeds 2π while going from z to w. Again, consider the mapping $w = z^2$. The upper half of the z-plane shown in Figure 4.6(a) is defined by $\Im\{z\} > 0$ or $0 < \theta < \pi$. After the mapping, this half-plane, except for $\theta = 0$ or $\theta = \pi$, becomes the entire w-plane as in Figure 4.6(b), with $w = R^2\, e^{i\Theta}$ or $0 < \Theta(= 2\theta) < 2\pi$, except for $\Theta = 0$ and $\Theta = 2\pi$. But the lower half of the z-plane ($\Im\{z\} < 0$, or $\pi < \theta < 2\pi$) sould similarly be mapped onto the whole w-plane, except $\Theta = 2\pi, 4\pi$, as did the upper part of the z-plane. This is shown in Figure 4.6(c). This situation led Riemann to introduce the connected surfaces, known as the *Riemann surfaces*, shown in Figure 4.6(d).

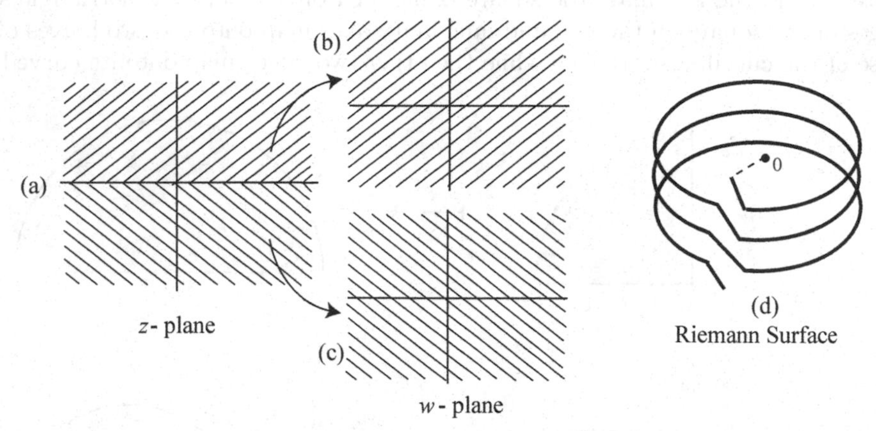

Figure 4.6 Mapping $w = z^2$.

A domain of transformed points in the w-plane is called *schlicht*, or *simple*, if none of its points originates from more than one point each in the z-plane.

Map 4.6. The transformation $w = z^2$ has critical points $z = 0$, $z = \infty$, and the fixed points $F_1 = 0, F_2 = 1$. The inverse transformation is $z = \sqrt{w}$. The transformation is presented in Figure 4.7(a)-(b).

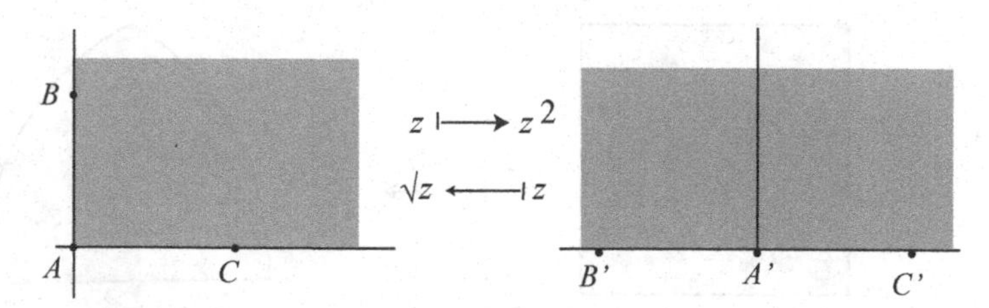

Figure 4.7 First quadrant of the z-plane onto the upper-half of the w-plane.

The details of the transformation when $\theta < \pi$ are as follows:

(a) The half-lines $\left\{\begin{matrix} y = 0,\ 0 \le x < \infty \\ y = 0,\ 0 \ge x > -\infty \end{matrix}\right\}$ are mapped onto the half-line $v = 0,\ 0 \ge u > \infty$.

(b) The half-lines $\left\{\begin{matrix} x = 0,\ 0 \le y < \infty \\ x = 0,\ 0 \ge y > -\infty \end{matrix}\right\}$ are mapped onto the half-line $v = 0,\ 0 \le u < \infty$.

(c) The half-lines $\left\{\begin{matrix} z = r\,e^{i\theta} \\ z = r\,e^{i(\theta+\pi)} \end{matrix}\right\}$, with θ fixed and $0 \le r < \infty$, are mapped onto the half-line $w = R\,e^{i\Theta}$, $0 \le R = r^2 < \infty$, and $\Theta = 2\theta$. .

Moreover, the details of the transformation when $\theta = \pi$ are as follows:

(d) The half-planes $\left\{\begin{matrix} y > 0 \\ y < 0 \end{matrix}\right\}$ are mapped onto the cut w-plane.

(e) The semicircles $\left\{\begin{matrix} |z| = c, y \ge 0 \\ |z| = c, y \le 0 \end{matrix}\right\}$, $c > 0$, are mapped onto the circle $|w| = c^2$, where the point $w = c^2$ is counted twice.

First, we will consider the following cases:

Map 4.7. The transformation $w = z^2$, which maps the first quadrant of the z-plane onto the upper-half of the w-plane, is presented in Figure 4.8.

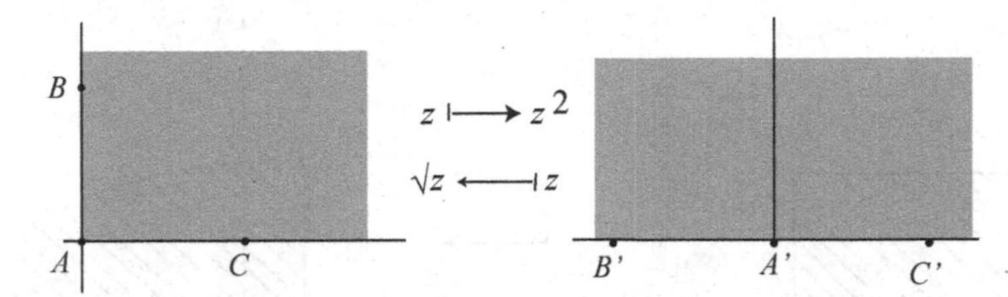

Figure 4.8 First quadrant of the z-plane onto the upper-half of the w-plane.

Map 4.8. The transformation $w = z^2$, which maps the obtuse circle in the upper half z-plane onto the partially complete circle in the w-plane, is presented in Figure 4.9.

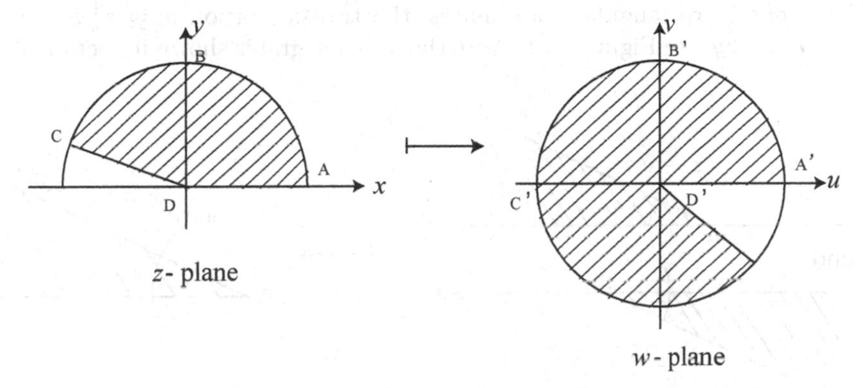

Figure 4.9 Obtuse half-circle onto the partially complete circle.

Map 4.9. The transformation $w = z^2$, which maps the region between two arcs in the first quadrant of the z-plane onto a rectangle in the w-plane, is presented in Figure 4.10.

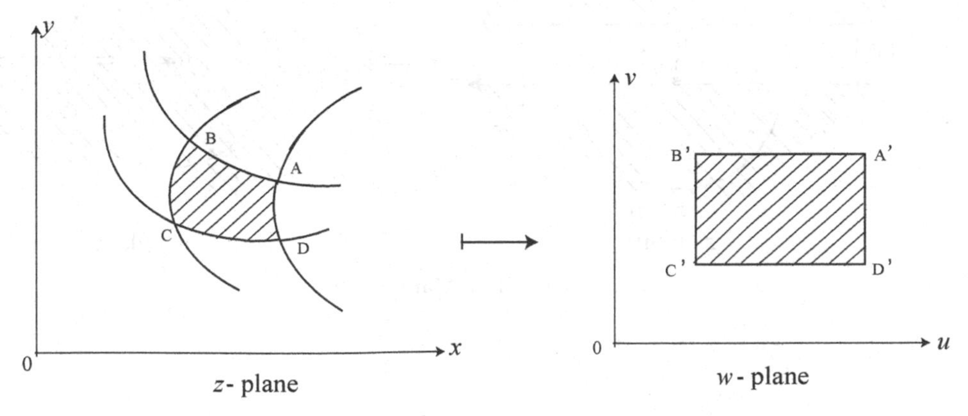

Figure 4.10 Region between two arcs onto a square.

Map 4.10. The transformation $w = z^2$, which maps the rectangle in the first quadrant in the z-plane onto an elliptic region in the upper-half of the w-plane, is presented in Figure 4.11.

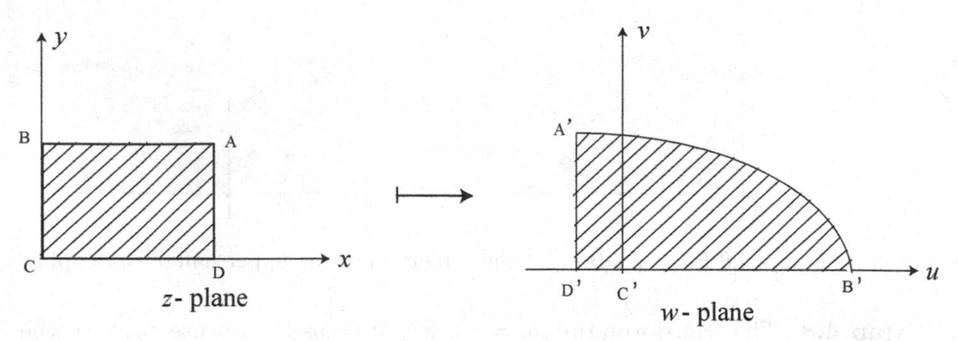

Figure 4.11 Rectangle in the first quadrant onto an elliptic region in the upper half-plane.

Note that in rectangular coordinates, the transformation $w = z^2$ is equivalent to $u = x^2 - y^2$, $v = 2xy$; see Figure 4.12 where the w-plane grid is shown in rectangular coordinates.

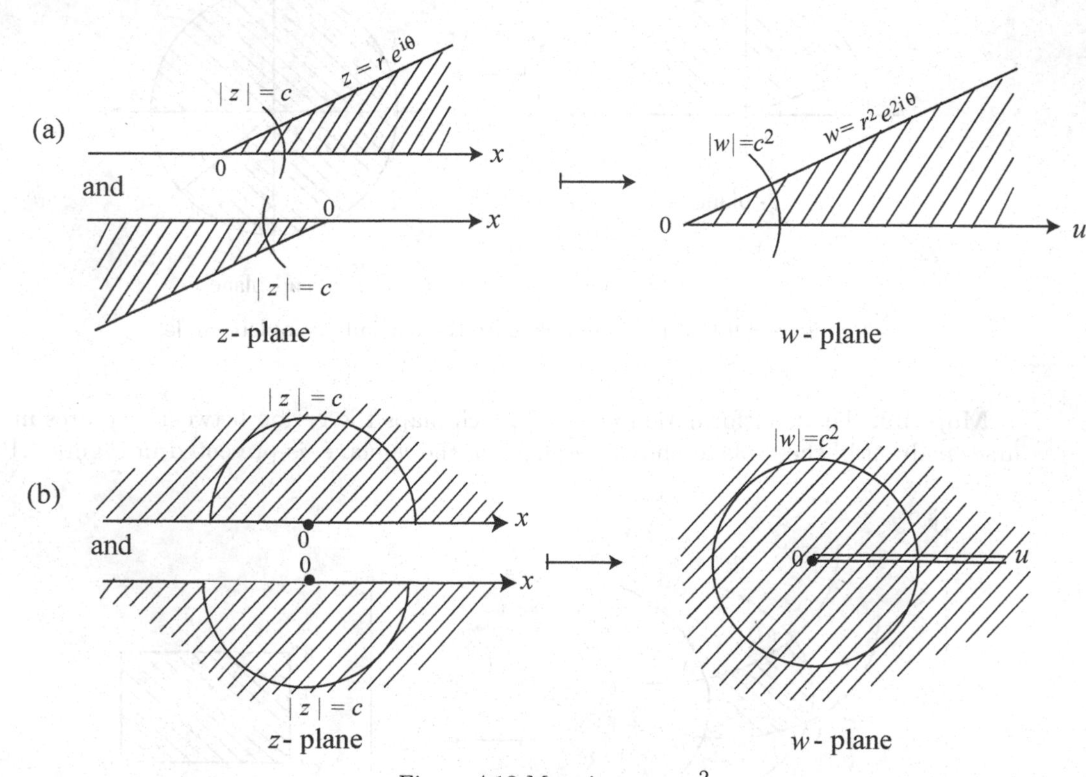

Figure 4.12 Mapping $w = z^2$.

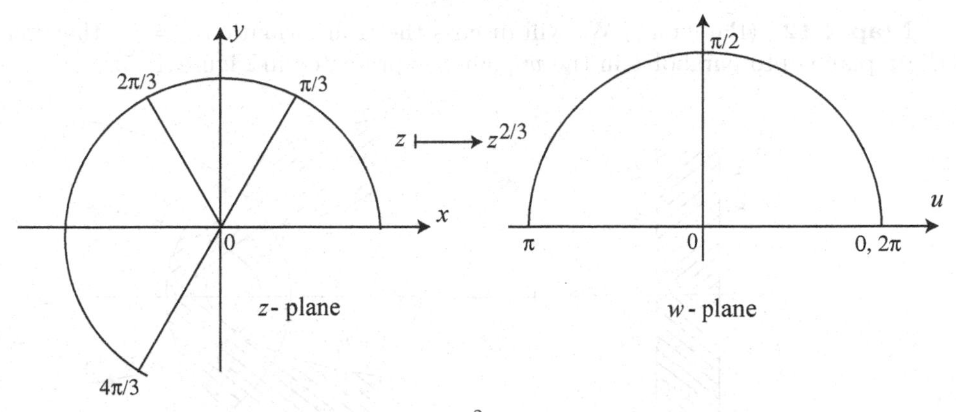

Figure 4.13 Mapping $w = z^2$ in rectangular coordinates.

Map 4.11. Consider the transformation $w = z^2$, and translate the origin of the coordinates in the z-plane from $z = 0$ to $z = 1$ in the transformation $w = z^{1/2}$. Then we find that the images under this transformation become different, as shown in Figure 4.14. The details are as follows:

(a) The circle with radius 1 centered at $z = 1$ has $r = 2\cos\theta$ in the z-plane, which becomes $R = \sqrt{2\,\cos 2\theta}$ (i.e., $\Theta = 2\theta$), which is a lemniscate (see Figure 3.1).

(b) The y-axis $\theta = \pi/2$ is transformed into $\Theta = \pm\pi/4$, and the rays passing through $w = 0$ are inclined at angles of $\pm\pi/4$.

(c) The vertical ray passing through $x = 1$ is transformed into a hyperbola, since $z = 1 + iy$ yieldss $w = \sqrt{1 + iy}$, or $(u + iv)^2 = 1 + iy$, which gives the equation of the rectangular hyperbola $u^2 - v^2 = 1$

(d) A circle of radius $\alpha > 1$ is transformed into a Cassini's oval, since $|z - 1| = \alpha$ becomes $|w - 1||w + 1| = \alpha$.

The translation of the center results in transforming a 'single field' into a double source field if both values of $w = \pm\sqrt{z}$ are considered. ∎

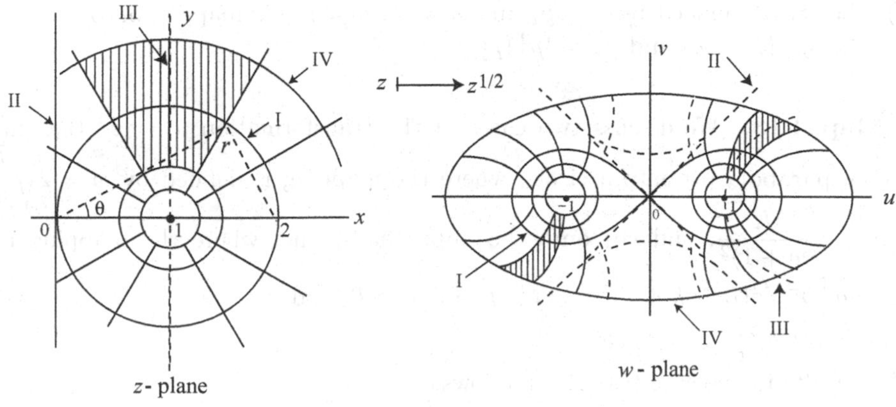

Figure 4.14 Mapping $w = z^{1/2}$.

Map 4.12. (Parabolas) We will discuss the transformation $w = z^2$ that maps lines in the z-plane onto parabolas in the w-plane, as presented in Figure 4.15.

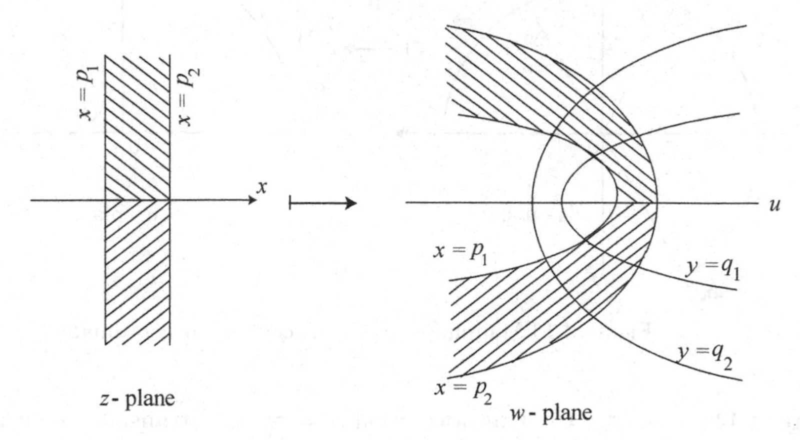

Figure 4.15 Lines onto parabolas.

The details of the transformation are as follows:

(a) The lines $\left\{ \begin{array}{c} \Re\{\bar{\lambda}z\} = s, \\ \Re\{\bar{\lambda}z\} = -s,\ s \gtrless 0 \end{array} \right\}$ are mapped onto parabolas $|w| = |2s^2 - \Re\{\bar{\lambda}^2 w\}||\lambda|^{-2}$, with focus $w = 0$ and directrix $\Re\{\bar{\lambda}^2 w\} = 2s^2$.

(b) The half-plane $\Re\{\bar{\lambda}z\} > 1$ is mapped onto the exterior of this parabola in the w-plane.

(c) The strip $0 < \Re\{\bar{\lambda}z/s\} < 1$ is mapped onto the interior of the parabola, cut along the axis of the parabola from focus to infinity.

(d) The lines $\left\{ \begin{array}{c} x = p \\ x = -p,\ p \gtrless 0 \end{array} \right\}$ are mapped onto the parabola $v^2 = -4p^2\left(u - p^2\right)$.

(e) The lines $\left\{ \begin{array}{c} y = q \\ y = -q,\ q \gtrless 0 \end{array} \right\}$ are mapped onto the parabola $v^2 = 4q^2\left(u + q^2\right)$.

(f) The strip bounded by $x = p_1$ and $x = p_2$, $p_{1,2} > 0$ is mapped onto the region between $v^2 = 4p_1^2\left(p_1^2 - u\right)$ and $v^2 = 4p_2^2\left(p_2^2 - u\right)$.

Map 4.13. We discuss two cases of the transformation $w = z^2$ that maps of exterior of parabola (a) onto a circle, where the mapping is defined by $w = 2\sqrt{\dfrac{K}{z}} - 1$, $K > 0$; $z = \dfrac{4K}{(w+1)^2}$; and (b) onto the upper half-plane, where the mapping is defined by $w = ia\left\{p^{-1/2}(iz + h + p)^{1/2} - 1\right\}$, $p > 0$, $a > 0$, and $z = lw^2 + mw + n$, $l = ipa^{-2}$, $m = -2pa^{-1}$, $n = ih$.

The results in the case (a) are as follows:

(i) The parabola $y^2 = 4K(K - x)$, or $r\cos^2\dfrac{\theta}{2} = K$, is mapped onto the circle $|w| = 1$.

(ii) The region outside the circle in (i) and not containing the focus $z = 0$ is mapped onto the region $|w| < 1$.

(iii) The points $z = K; 4K; \pm 2iK; 2k; \infty$ are mapped onto the points $w = 1; 0; \mp i; \sqrt{2} - 1; -1$.

The results in the case (b) are as follows:

(i) The parabola $x^2 = 4p(y-h)$ is mapped onto the line $v = 0$.

(ii) The region outside the parabola in (i) and not containing the focus $z = i(h+p)$ is mapped onto the half-plane $v > 0$.

(iii) The points $z = i(h+p); ih; i(h-p)$ are mapped onto the points $w = -ia; 0; ia\left(\sqrt{2}-1\right)$.

(iv) The half-line $x = 0$, $-\infty < y < h$ is mapped onto the half-line $u = 0$, $0 < v < \infty$.

Map 4.14. We will consider the transformation $w = z^2$ onto a rectangular hyperbola which is defined by $\Re\{c^{-2}z^2\} = \frac{1}{2}$, with foci at $\pm c$ and the asymptotes $\Re\{c^{-2}z^2\} = 0$. The graph for the hyperbolas mapped onto $u = P$ or $v = Q$ are presented in Figure 4.1, and the details of the transformation are as follows:

(a) Either branch of the hyperbola is mapped onto the line $\Re\{c^{-2}w\} = \frac{1}{2}$.

(b) The interior of either branch is mapped onto the half-plane $\Re\{c^{-2}w\} > \frac{1}{2}$.

(c) The region bounded by a branch of the hyperbola and adjacent parts of the asymptotes is mapped onto the strip $0 < \Re\{c^{-2}w\} < \frac{1}{2}$.

(d) The region bounded by adjacent branches of $\Re\{c_1^{-2}z^2\} = \frac{1}{2}$ and $\Re\{c_1^{-2}z^2\} = \frac{1}{2}$, but real, and $c_2/c_1 \neq \pm 1$, is mapped onto the strip bounded by $\Re\{c_1^{-2}w) = \frac{1}{2}$.

(e) Either branch $x^2 - y^2 = P$, $P \gtrless 0$, is mapped onto the line $u = P$, i.e., $c = \sqrt{2P}$.

(f) Either branch $2xy = Q$, $Q \gtrless 0$, is mapped onto the line $v = Q$, i.e., $c = \sqrt{2iQ}$.

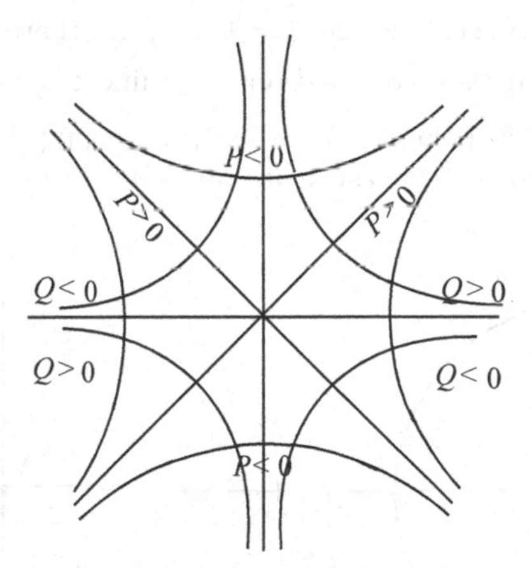

Figure 4.16 Rectangular hyperbolas.

Map 4.15. We will consider the transformation $w = z^2$ onto generalized parabolas and hyperbolas. The graph for the case $w = z^{2/3}$ is presented in Figure 4.17. The details of the transformation are as follows:

(a) The line $\Re\{z\,e^{-i\psi}\} = s$, $s \gtrless 0$ is mapped onto the curve $R^\alpha = s/\cos(\alpha\theta - \psi)$, where $\alpha\left(\pi k - \frac{\pi}{2} + \psi\right) < \theta < \alpha\left(\pi k + \frac{\pi}{2} + \psi\right)$.

(b) The line $x = p$, $p \gtrless 0$ is mapped onto the curve $R^\beta = p/\cos(\alpha\theta)$.

(c) The line $y = q$, $q \gtrless 0$ is mapped onto the curve $R^\beta = q/\sin(\alpha\theta)$.

(d) The curve $r^\alpha = S/\cos(\alpha\theta - \phi)$, where $\alpha\left(\pi k - \frac{\pi}{2} + \phi\right) < \theta < \alpha\left(\pi k + \frac{\pi}{2} + \phi\right)$, is mapped onto the line $\Re\{w\,e^{-i\phi}\} = S,\ S \gtrless 0$.

(e) The curve $r^\alpha = P/\cos(\alpha\theta)$ is mapped onto the line $u = P,\ P \gtrless 0$.

(f) The curve $r^\alpha = Q/\sin(\alpha\theta)$ is mapped onto the line $v = Q,\ Q \gtrless 0$.

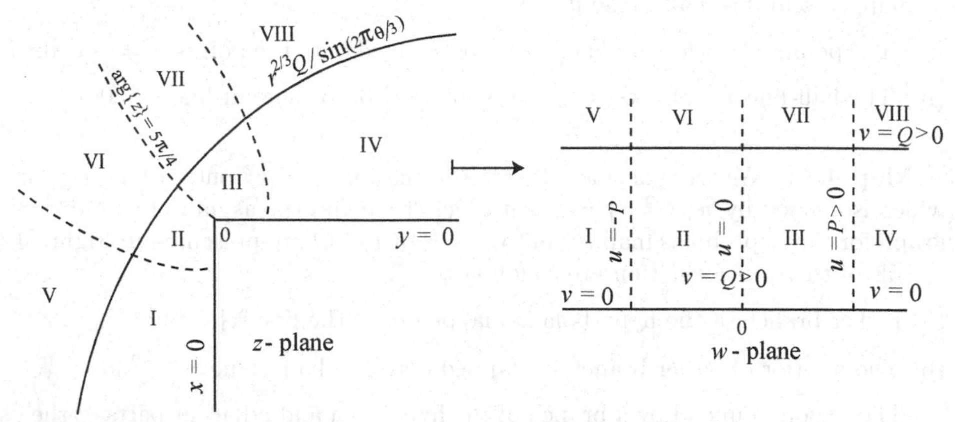

Figure 4.17 Transformation $w = z^{2/3}$.

In the case when $\alpha = n$, where n is an integer, the generalized hyperbola $r^\alpha = S/\cos(\alpha\theta - \phi)$ consists of exactly n branches, each one of which is mapped onto $\Re\{\bar{\Lambda} w\} = S$, where $\Lambda = e^{i\phi}$.

The details of the transformation are as follows:

(i) For the curve $r^\alpha = S/\cos(\alpha\theta - \phi)$, $k = 0, \pm 1, \dots$, there are no asymptotes for $0 < \alpha < 1$, but for $\alpha > 1$, the curve is mapped onto the line $z = r\,e^{i\alpha\left(k\pi + \frac{\pi}{2} + \phi\right)}$.

(ii) The curve $R^\alpha = s/\cos(\alpha\Theta - \theta)$, $k = 0, \pm 1, \dots$, is mapped onto the line $w = R\,e^{i\alpha\left(k\pi + \frac{\pi}{2} + \phi\right)}$ for $0 < \alpha < 1$; but for $\alpha > 1$, there are no asymptotes.

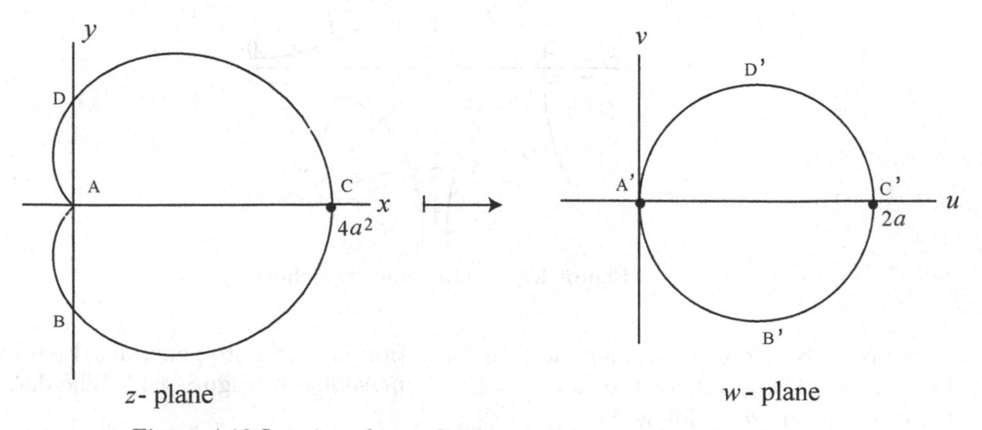

Figure 4.18 Interior of a cardioid onto the interior of the unit circle.

Map 4.16. The transformation that maps the interior of a cardioid in the z-plane onto the interior of a circle of radius a, $|z-a|=a$, in w-plane is defined by

$$w = z^2, \tag{4.1.7}$$

and is presented in Figure 4.18.

Map 4.17. The transformation that maps the triangle bounded by $x=1, y=1$ and $x+y=1$ in the z-plane onto the region bounded on the left by the ellipse $u = \frac{v^2}{4} - 1$, on the right by the ellipse $u = 1 - \frac{v^2}{4}$, and at the bottom by the ellipse $v = \frac{1}{2}\left(1-u^2\right)$, is defined by $w = z^2$. Notice that the angles of the triangle ABC are equal, respectively, to the angles of the curvilinear triangle A'B'C', as shown in Figure 4.19.

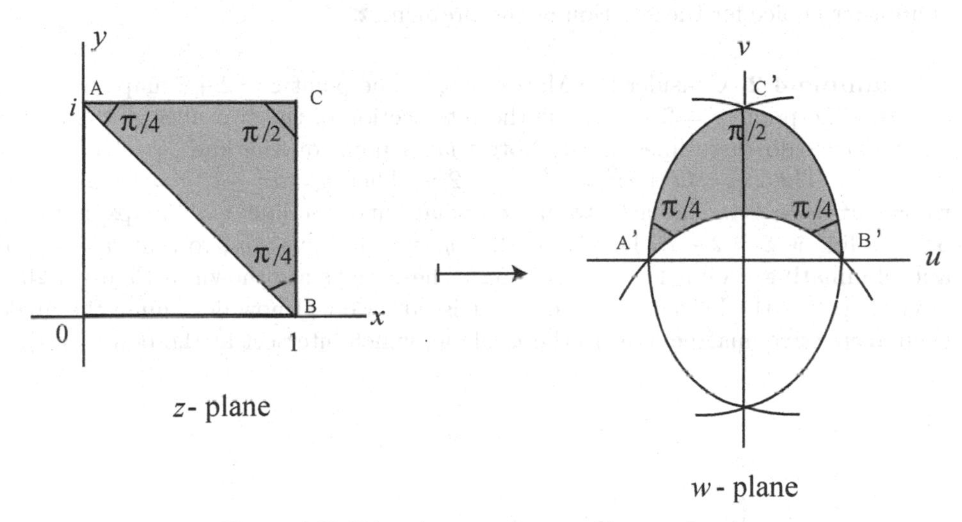

Figure 4.19 Triangle onto the curvilinear triangle.

Map 4.18. The transformation $w = az^2+bz+c$, $a \neq 0$, or the inverse transformation $z = \dfrac{-b \pm \left(4aw+b^2-4ac\right)^{1/2}}{2a}$, is a composite of $\xi = \zeta^2$ and of the two linear transformations $\zeta = z + \dfrac{b}{2a}$, and $w = a\xi + \dfrac{4ac-b^2}{4a}$, with the fixed points $F_{1,2} = \dfrac{1-b \pm \sqrt{(1-b)^2-4ac}}{2a}$.

The following example shows that the conformal map guaranteed by the Riemann mapping theorem is not unique.

Example 4.1. In order to construct a mapping that maps the half-disk $\{|z| < 1/2, \Im\{z\} > 0\}$ onto the entire unit disk $\{|z| < 1\}$, the map $w = z^2$ will not work because the image of the disk $|z| < 1/2$ omits the positive real axis, thus yielding a disk with a slit $\{|w| < 1, 0 < \arg\{w\} < 2\pi\}$. In order to get the entire unit disk, we may consider the mapping $z = \dfrac{w-1}{w+1}$ (see Map 3.16) which maps the right half-plane $\{\Re\{w\} > 0$ onto the unit disk. It also maps the upper right quadrant $\{0 < \arg\{w\} < \pi/2\}$ onto the half unit disk $|z| < 1/2$. Then its inverse $w = \dfrac{z+1}{z-1}$ (see Map 3.15) will map the half-disk $|z| < 1/2$ onto the upper right quadrant.

However, the mapping $w = \dfrac{iz^2+1}{iz^2-1}$ maps the upper right quadrant onto the unit disk $|z| < 1$. Since

$$w = \frac{iw^2+1}{iw^2-1} = \frac{i\left(\frac{z+1}{z-1}\right)+1}{i\left(\frac{z+1}{z-1}\right)-1} = \frac{(i+1)(z^2+1)+2(i-1)z}{(i-1)(z^2+1)+2(i+1)z} = -i\frac{z^2+2iz+1}{z^2-2iz+1},$$

we find that, omitting the factor $-i$ which merely rotates the disk by $-\pi/2$, the mapping

$$w = \frac{z^2+2iz+1}{z^2-2iz+1}$$

is another choice for the solution of the problem. ■

Example 4.2. Consider the Map $w = z^2$. The point $z = 2+i$ maps to $w = (2+i)^2 = 3+4i$. The point $z = 2+i$ lies at the intersection of the two lines $x = 2$ and $y = 1$. To what curves do these lines map? Note that a point on the line $y = 1$ can be written as $z = x+i$. Then $w = (x+i)^2 = x^2-1+2xi$. Then $u = x^2-1$, $v = 2x$, and eliminating x we get $v^2 = 4u+4$. Similarly, to find on what curve the line $x = 2$ maps, note that a point on this line is $z = 2+i$. Then $w = (2+yi)^2 = 4-y^2+4yi$, so that $u = 4-y^2, v = 4y$, and eliminating y we get $v^2 = 64-16u$. These maps are shown in Figure 4.20. Since the angle between the lines $x = 2$ and $y = 1$ is $90°$, we will now determine the angle between their respective image curves in the w-plane, which intersect at the point $(3,4)$.

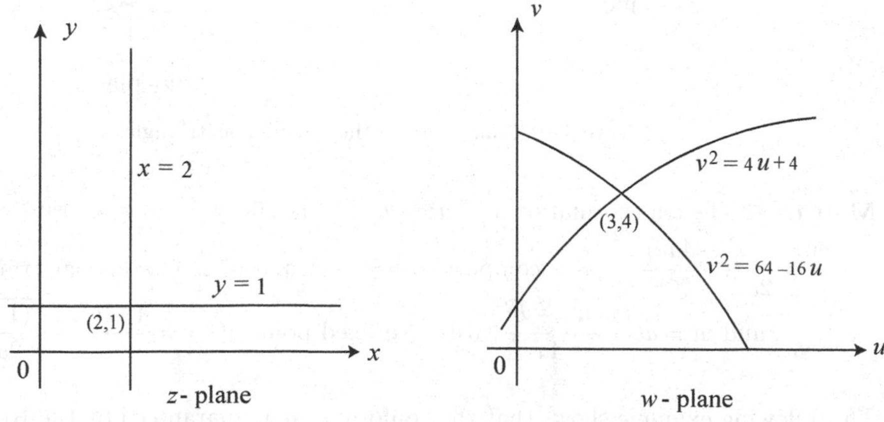

Figure 4.20 $w = z^2$.

The curve $v^2 = 4u+4$ has a gradient $\dfrac{\partial v}{\partial u}$, so that differentiating this equation implicitly we get $\dfrac{\partial v}{\partial u} = \dfrac{2}{v}$, and at the point $(3,4)$ we have $\dfrac{\partial v}{\partial u} = \frac{1}{2}$. Similarly, from the other curve $v^2 = 64-16u$ we get $\dfrac{\partial v}{\partial u} = -\dfrac{8}{v}$, which at the point $(3,4)$ gives $\dfrac{\partial v}{\partial u} = -2$. Hence the angles between image curves is $90°$. ■

Example 4.3. Consider the transformation $w = z^2$. By sketching the curves in the z-plane which map onto lines in the w-plane parallel to the real and imaginary axes, we show that every straight line in the w-plane is the image of a hyperbola (or a pair of straight lines) in the z-plane. Since $z = x + iy, w = u + iv$, so $w = z^2$ gives $u = x^2 - y^2, v = 2xy$. Now $v =$ const $\iff 2xy =$ const. and u const $\iff x^2 - y^2 =$ const, as shown in Figure 4.21. Also, $v = mu + c$ (c const) in the w-plane $\iff 2xy = m(x^2 - y^2) + c$, or $mx^2 - my^2 - 2xy + c = 0$, which has the discriminant $4 + 4m^2 > 0$. Thus, we get a hyperbola or a pair of straight lines. ■

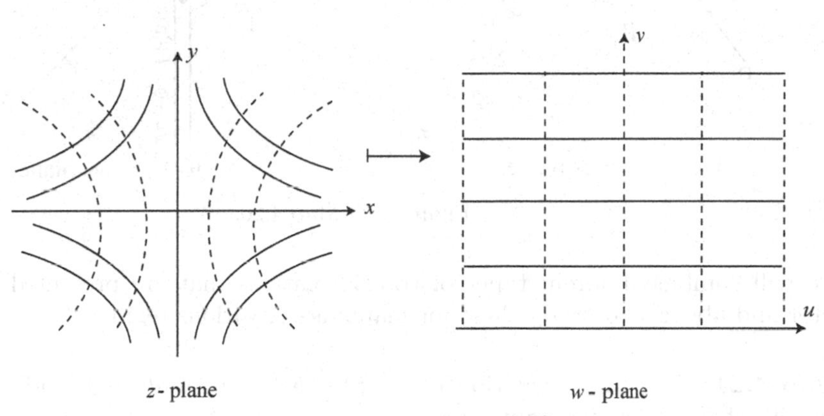

Figure 4.21 Map $w = z^2$

Map 4.19. The transformation $w = a^2 z^2 + b$ maps the first quadrant $0 \le x < \infty, 0 \le y < \infty$ in the z-plane onto the upper half-plane $\Im\{w\} \ge 0$ in the w-plane.

Example 4.4. Consider the map $w = az^N + b$. There are N different sources of the form

$$z = r\,e^{i\theta/N}\,e^{2\pi ik/N}, \quad k = 0, 1, \dots, N-1,$$

that will map to one value of w which can be expressed as $w = ar^N\,e^{i\theta} + b$. For example, let $z_1 = r\,e^{i\theta/N}$ and $z_2 = r\,e^{i\theta/N}e^{2\pi i/N} \ne z_1$. Then $w_1 = ar^N\,e^{i\theta} + b$ and $w_2 = ar^N\,e^{i\theta}e^{2\pi i} + b = ar^N\,e^{i\theta} + b = w_1$. Thus, both z_1 and z_2 map into one point, i.e., $z_1 = z_2 \Rightarrow w_1 = w_2$, but $w_1 = w_2 \not\Rightarrow z_1 = z_2$. Thus, N different values of z have the same image, so the mapping is an N-to-1 mapping. On the other hand, in the case of the mapping $w = az^{1/N} + b$, let $z_1 = r\,e^{i\theta}$ and $z_2 = r\,e^{i\theta}e^{2\pi i} = z_1$. Then $w_1 = r^{1/N}\,e^{i\theta/N}$ and $w_2 = r^{1/N}\,e^{i\theta/N}e^{2\pi i/N} \ne w_1$. Thus, $w_1 = w_2 \Rightarrow z_1 = z_2$ but $z_1 = z_2 \not\Rightarrow w_1 = w_2$. Since there are N different images of each source point z, the mapping is 1-to-N. ■

Map 4.20. We will consider the transformation $w = az^\alpha + bz^\beta$, where $\beta < 0 < \alpha$ and $ab \ne 0$. If $\beta = -\alpha$, we refer to Map 4.22(i), given below. Let $\sigma = \dfrac{\alpha}{\alpha - 1}$, $\phi = (1 - \sigma)\arg\{a\} + \sigma \arg\{-b\}$, $\theta = \dfrac{\sigma}{\alpha}\arg\{-b/a\}$; thus, $\phi = \theta - 0$ if $a > 0 > b$. Also, let $c = \left|\dfrac{b}{a}\,\dfrac{1-\sigma}{\sigma}\right|^{\sigma/\alpha}$, and $d = \left|\left(\dfrac{a}{1-\sigma}\right)^{1-\sigma}\left(\dfrac{b}{a}\right)^{\sigma}\right|$. Then under this transformation, the points $z_0 = \left(-\dfrac{b}{a}\right)^{\sigma/\alpha}$; $z_1 = c\,e^{i(\theta + \sigma\pi/\alpha)}$; and $z_2 = c\,e^{i(\theta - \sigma\pi/\alpha)}$ in the z-plane are mapped onto the points $w_0 = 0$; $w_1 = d\,e^{i(\phi + \sigma\pi)}$; and $w_2 = d\,e^{i(\phi - \sigma\pi)}$, respectively, in the w-plane.

This mapping is presented in Figure 4.22.

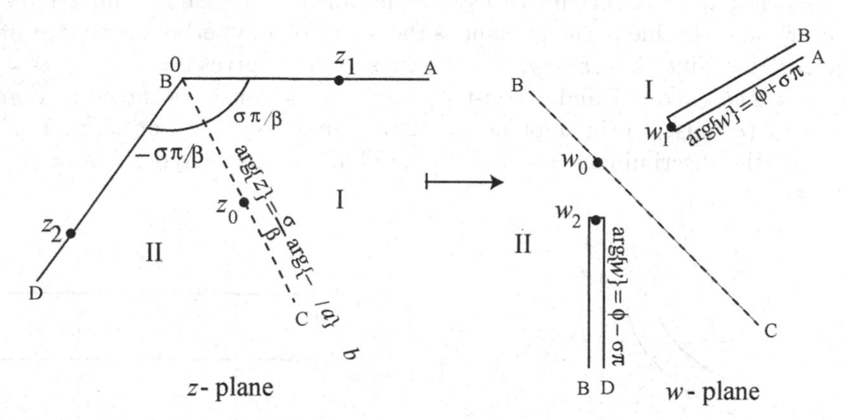

Figure 4.22 Map 4.20.

We will combine different types of transformations that are presented in the previous chapter and above, and study them for mappings of various regions.

Map 4.21. The transformation $w = az^\alpha + bz^\beta$, where $ab \neq 0$, and $\beta > \alpha > 0$, is a composite of the following mappings:

$$w = a\Big(-\frac{b}{a}\Big)^{\alpha/\gamma}\zeta, \quad z = \Big(-\frac{a}{b}\Big)^{1/\gamma}\zeta; \quad \text{and } \xi = \zeta^\alpha - \zeta^\beta, \tag{4.1.8}$$

where $\gamma = \beta - \alpha$. This mapping differs from Map 4.20 in that here $\beta > \alpha > 0$, while $\beta < 0 < \alpha$ in Map 4.20. Compare it with Map 4.22.

Map 4.22. The transformation $w = az^\alpha + bz^{-\alpha}$, where $\alpha > 0$, and $ab \neq 0$, is a sum of the mappings considered in Map 4.24 and Map 4.28 or 4.29. This transformation can also be regarded as a combination of the following mappings:

$$w = 2\sqrt{ab}\,\frac{\xi+1}{\xi-1}, \quad z = \zeta\Big(\frac{b}{a}\Big)^{\alpha/2}, \quad \text{and } \xi = \Big(\frac{\zeta^\alpha+1}{\zeta^\alpha-1}\Big)^2. \tag{4.1.10}$$

The details of different cases under this transformation for $k = 0, \pm1, \pm2, \dots$, are as follows:

(i) The sector area $|z| < \Big|\frac{b}{a}\Big|^{\alpha/2}$ in the z-plane for $\frac{2k\pi}{\alpha} + \frac{\arg\{b/a\}}{2\alpha} < \arg\{z\} < \frac{\arg\{b/a\}}{2\alpha} + \frac{(2k+1)\pi}{\alpha}$, is mapped onto one of the half-planes $\Im\Big\{\frac{w}{\sqrt{ab}}\Big\} \gtrless 0$ in the w-plane.

(ii) The area $|z| > \Big|\frac{b}{a}\Big|^{\alpha/2}$ in the z-plane for $\frac{2k\pi}{\alpha} + \frac{\arg\{b/a\}}{2\alpha} < \arg\{z\} < \frac{\arg\{b/a\}}{2\alpha} + \frac{(2k+1)\pi}{\alpha}$, is mapped onto the other half of the half-planes mentioned in (i) above.

(iii) The area lying between the above two regions (areas), mentioned in (i) and (ii) above, of the circle $|z| = \Big|\frac{b}{a}\Big|^{\alpha/2}$ is mapped onto the line segment joining $-2\sqrt{ab}$ and $2\sqrt{ab}$.

For $\alpha = 1$, see Map 4.37 with $a = b = 1$.

Map 4.23. The multiple-valued function $w = \sqrt[n]{z}$ has the inverse $z = w^n$ which has been discussed above with reverse roles of w and z. Under the mapping $z = w^n$, the interiors $G_1, \dots, G_n$ of the n angles, each equal to $2\pi/n$ radians and formed by the n rays emanating from the point $w = 0$ leads to the n single-valued branches $\left(\sqrt[n]{z}\right)_1, \cdots, \left(\sqrt[n]{z}\right)_n$ of the function $w = \sqrt[n]{z}$, all defined in the region D. These branches, which have nonzero derivatives on D, map D one-to-one and continuously onto G_k, $k = 1, \dots, n$ (Figure 4.23).

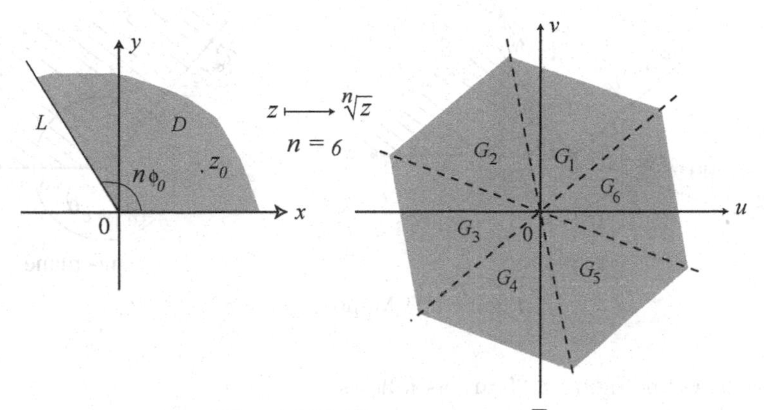

Figure 4.23 The map $w = \sqrt[n]{z}$.

The inverse image in the z-plane of the regions G_k, whose boundaries are marked by solid lines, is the single ray L emanating from $z = 0$ and inclined at an angle $n\phi_0$. The manner in which a branch $\left(\sqrt[n]{z}\right)_k$ changes into the next branch $\left(\sqrt[n]{z}\right)_{k+1}$ can be explained by letting a point $z_0 \neq 0$ in D make a complete circle with center at $z = 0$. We choose the value of $\sqrt[n]{z}$ that is associated with the branch $\left(\sqrt[n]{z}\right)_k$ and represented by the value $w_0 = \sqrt[n]{|z_0|}\, e^{i\theta_0/n} \in G_k$. Then, as the point z_0 moves continuously along the circle $|z| = |z_0|$ in the positive direction, the value of $w = \sqrt[n]{|z_0|}\, e^{i\theta/n}$ varies continuously with θ such that as the point z returns to its original value z_0, the value of w goes to the value $w_1 = \sqrt[n]{|z_0|}\, e^{i(\theta_0+2\pi)/n}$, where w_1 is the value associated with the branch $\left(\sqrt[n]{z}\right)_{k+1}$ on the adjacent region G_{k+1}, where $G_k \cap G_{k+1} = \emptyset$ for $k = 1, \dots, n$. Proceeding in this manner through n windings around $z = 0$, the n branches $\left(\sqrt[n]{z}\right)_k$ undergo the following chain of transformations:

$$
\begin{aligned}
&\left(\sqrt[n]{z}\right)_k \to \left(\sqrt[n]{z}\right)_{k+1}, \\
&\left(\sqrt[n]{z}\right)_{k+1} \to \left(\sqrt[n]{z}\right)_{k+2}, \\
&\cdots\cdots\cdots, \\
&\left(\sqrt[n]{z}\right)_n \to \left(\sqrt[n]{z}\right)_1, \left(\sqrt[n]{z}\right)_1 \to \left(\sqrt[n]{z}\right)_2, \cdots \left(\sqrt[n]{z}\right)_{k-1} \to \left(\sqrt[n]{z}\right)_k .
\end{aligned}
\tag{4.1.11}
$$

$$
\left(\sqrt[n]{z}\right)_k \to \left(\sqrt[n]{z}\right)_{k+1}, \left(\sqrt[n]{z}\right)_{k+1} \to \left(\sqrt[n]{z}\right)_{k+2}, \cdots,
\left(\sqrt[n]{z}\right)_n \to \left(\sqrt[n]{z}\right)_1, \left(\sqrt[n]{z}\right)_1 \to \left(\sqrt[n]{z}\right)_2, \cdots \left(\sqrt[n]{z}\right)_{k-1} \to \left(\sqrt[n]{z}\right)_k .
$$

The points $z = 0$ and $z = \infty$ are the algebraic branch points for the mapping $w = \sqrt[n]{z}$.

Map 4.24. The transformation $w = z^\alpha$ has the inverse transformation $z = w^\beta$, where $\beta = 1/\alpha$, $\alpha > 0$, is known as the angle on the half-plane. The critical points are $z = 0, \infty$. These transformations are presented in Figures 4.24, and 4.25, and the details for Figure 4.23 are as follows:

(a) The half-line $y = 0$, $0 \le x < \infty$ is mapped onto the half-line $v = 0$, $0 \le u < \infty$.

(b) The half-lines $\left\{ \begin{array}{c} z = r\, e^{i\theta} \\ z = r\, e^{i(\theta+2k\pi\beta)} \end{array} \right\}$, $k = \pm 1, \pm 2, \dots$, are mapped onto the half-line $w = Re^{i\Theta}$, $0 < R < \infty$, $\Theta = \alpha\theta$.

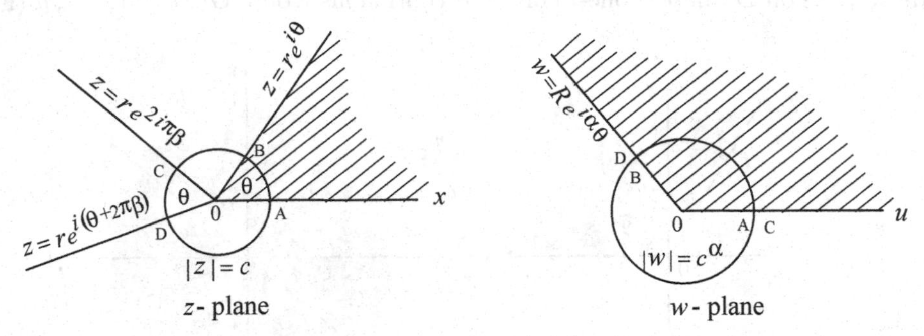

Figure 4.24 Mapping $w = z^\alpha, \alpha > 0$.

The details for Figure 4.25 are as follows:

(a) The angle $0 < \theta < \beta\pi$ is mapped onto the half-plane $u > 0$.

(b) The angle $\beta\pi < \theta < 2\beta\pi$ is mapped onto the half-plane $u < 0$.

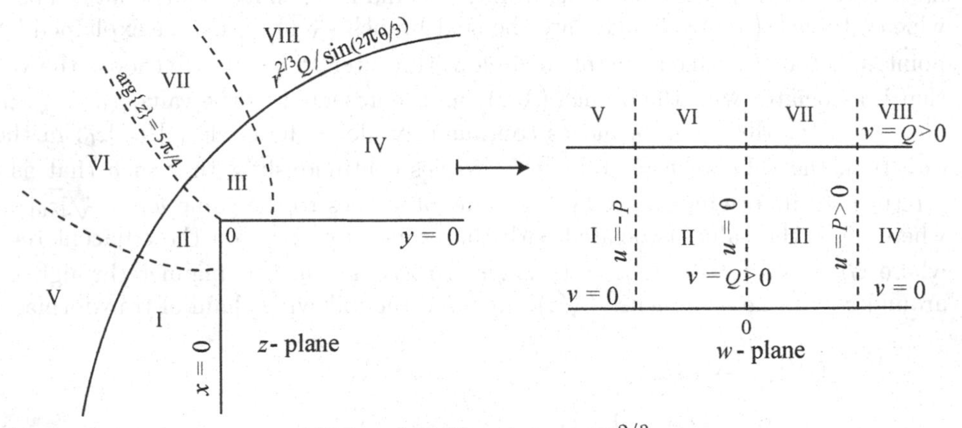

Figure 4.25 Mapping $w = z^{2/3}$.

Map 4.25. The transformation $w = z^{1/2}$ is presented in Figure 4.26, which is, in fact, the inverse of Figure 4.2. The inverse transformation $w^2 = z$ can be written as $(u+iv)^2 = x+iy$, i.e., $x = u^2 - v^2$ and $y = 2uv$, i.e.,

$$\text{For } v = k_v: \quad x = \frac{1}{4k_v^2} y^2 - k_v^2, \tag{4.1.12}$$

which means that the parabolas are symmetric about x-axis open to the right. Similarly,

$$\text{For } u = k_u: \quad -x = \frac{1}{4k_u^2} y^2 - k_u^2, \tag{4.1.13}$$

which means that parabolas are symmetric about the x-axis open to the left (Figure 4.27).

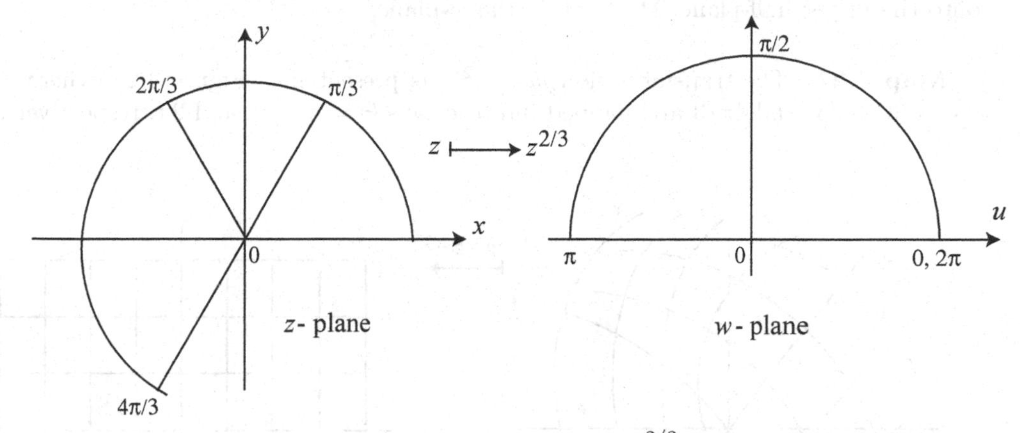

Figure 4.26 Mapping $w = z^{2/3}$.

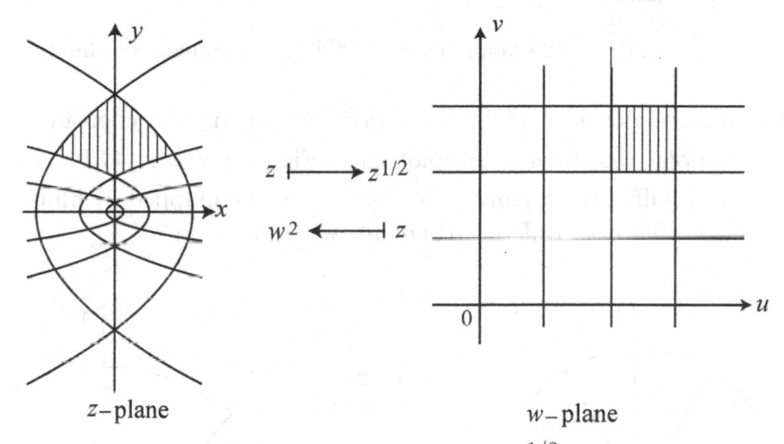

Figure 4.27 Mapping $w = z^{1/2}$.

Definition. The function $z = g(w) = w^{1/2} = |w|^{1/2}\, e^{i \arg g(w)/2}$ is called the *principal square root function.* Fig 4.28 explains this definition by presenting this transformation from the z-pane onto the w-plane.

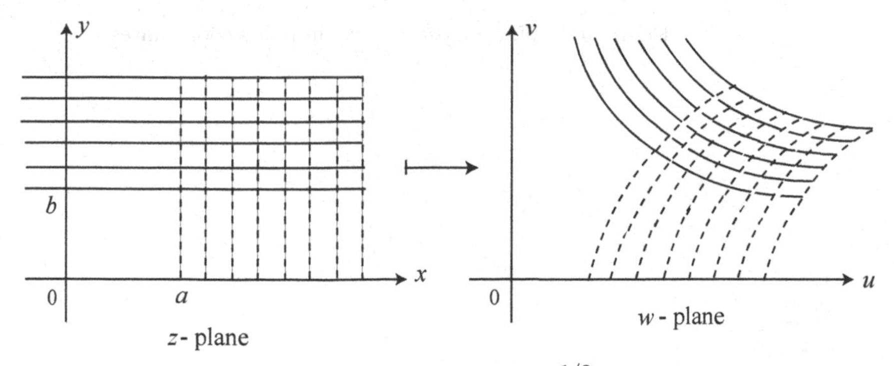

Figure 4.28 Map $z = w^{1/2}$.

Map 4.26. The transformation $w = \sqrt{z}$ maps the z-plane with the cut along the x-axis onto the upper half-plane $\Im\{w\} \geq 0$ in the w-plane.

Map 4.27. The transformation $w = z^{2/3}$ is presented in Figure 4.29, where the rays $\theta = \pi/3, 2\pi/3$, and $4\pi/3$ are mapped into the rays $\Theta = \pi/2, \pi$, and 2π, respectively.

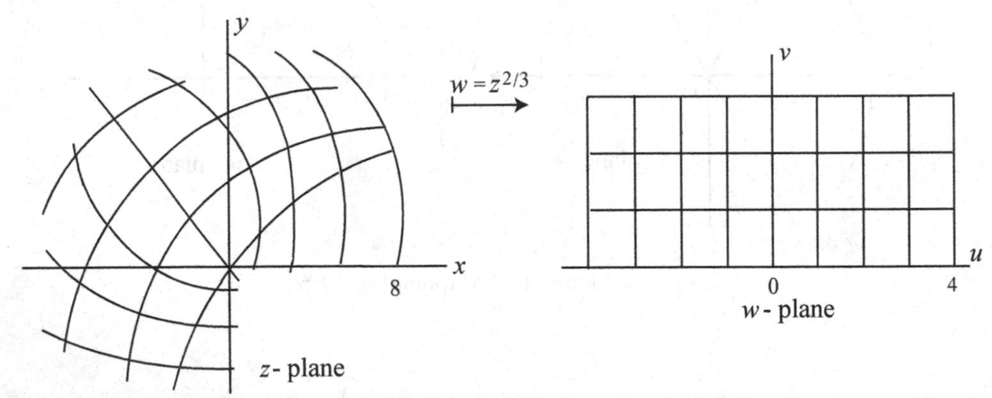

Figure 4.29 Mapping $w = z^{2/3}$ in rectangular coordinates.

The transformation $w = z^{2/3} = (x + iy)^{2/3}$ in rectangular coordinates, and $w = z^{2/3} = \left(r\, e^{i\theta^{2/3}}\right)$ in polar coordinates, are plotted in Figure 4.30 and 4.31, respectively. In Figure 4.31, the lower half of the z-plane ($\pi < \theta < 2\pi$) can be mapped by filling the entire w-plane ($2\pi < \Theta < 4\pi$), but on a different Riemann surface.

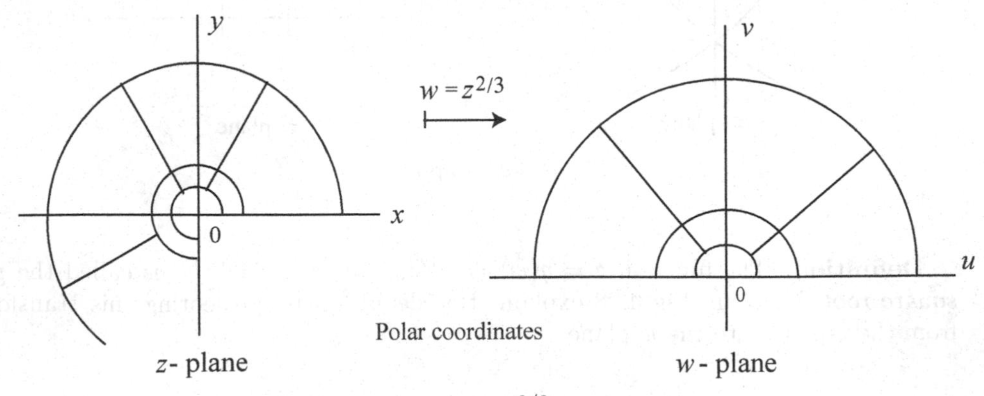

Figure 4.30 Mapping $w = z^{2/3}$ in polar coordinates.

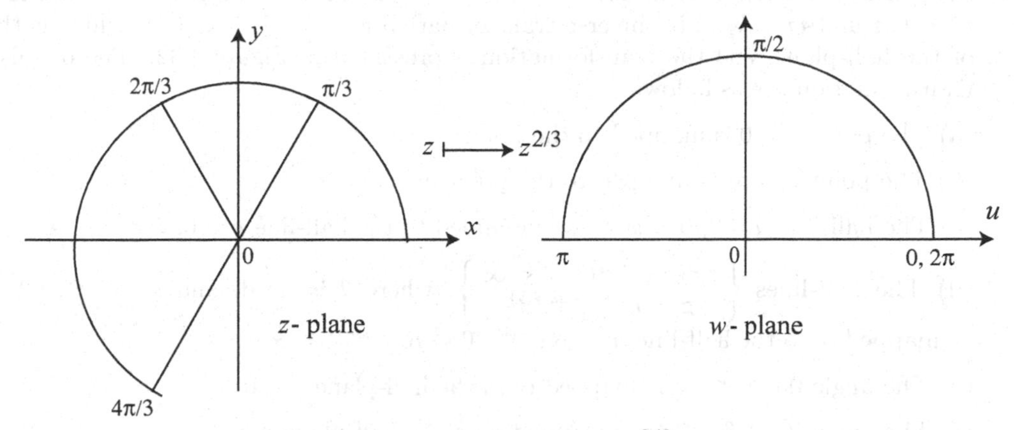

Figure 4.31 Mapping $w = z^{2/3}$.

Map 4.28. The transformation $w = z^{-\alpha}$, $\alpha > 0$, has the inverse transformation $z = w^{\beta}$, $\beta = 1/\alpha$, which is the angle of the half-plane. The transformation is presented in Figure 4.32, and the details are as follows:

(a) The point $z = 0$ is mapped onto the point $w = \infty$.

(b) The point $z = \infty$ is mapped onto the point $w = 0$.

(c) The half-line $y = 0$, $0 < x < \infty$ is mapped onto the half-line $v = 0$, $0 < u < \infty$.

(d) The half-lines $\left\{ \begin{array}{c} z = r\, e^{i\theta},\ 0 < r < \infty \\ z = r\, e^{i(\theta+2k+\beta)}, \end{array} \right\}$, where θ fixed, and $k = \pm 1, \pm 2, \dots$, are mapped onto the half-line $w = R\, e^{-\alpha\theta}$, $0 < R = r^{-\alpha} < \infty$.

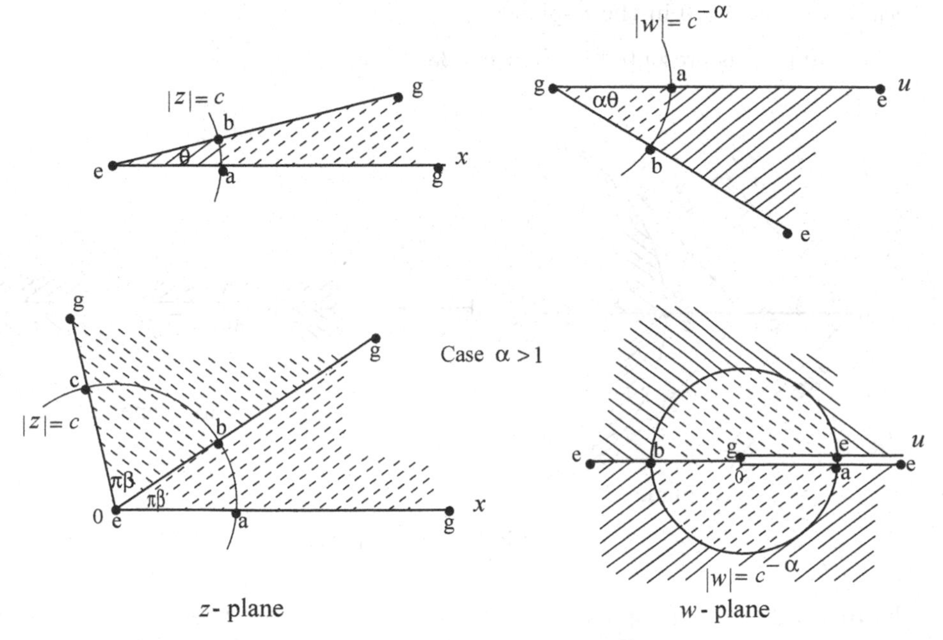

Figure 4.32 Mapping $w = z^{-\alpha}$.

Map 4.29. The transformation $w = z^{-\alpha}$, $\alpha > 1$ is a generalization of the transformation $w = 1/z$ and $w = k/z$. Its inverse transformation is $z = w^{\beta}$, $\beta = 1/\alpha$, which is the angle of the half-plane, and the transformation is presented in Figure 4.32. The details of this transformation are as follows:

(a) The point $z = 0$ is mapped to the point $w = \infty$.

(b) The point $z = \infty$ is mapped to the point $w = 0$.

(c) The half-line $y = 0$, $0 < x < \infty$ is mapped to the half-line $v = 0$, $0 < u < \infty$.

(d) The half-lines $\left\{ \begin{matrix} z = r\,e^{i\theta}, 0 < r < \infty \\ z = r\,e^{i(\theta+2k+\beta)} \end{matrix} \right\}$, where θ is fixed, and $k = \pm1, \pm2, \dots$, are mapped onto the half-line $w = R\,e^{-\alpha\theta}$, $0 < R < r^{-\alpha} < \infty$.

(e) The angle $0 < \theta < \beta\pi$ is mapped onto the half-plane $y < 0$.

(f) The angle $\beta\pi < \theta < 2\beta\pi$ is mapped onto the half-plane $y > 0$.

Map 4.30. The transformation $w = z^{\pi/\alpha}$ maps the interior of an infinite sector with angle α: $0 \le \arg\{z\} \le \alpha, 0 \le |z| < \infty, (0 < \alpha \le 2\pi)$, in the z-plane onto the upper half-plane $\Im\{w\} \ge 0$ in the w-plane.

Map 4.31. We will consider the transformation $w = z^{\alpha} - z^{\beta}$, where $\beta > \alpha > 0$. Let $\gamma = \beta - \alpha$, $z = r\,e^{i\theta}$, $r > 0$ and θ real. The details of this transformation are as follows:

(a) The points $z = 0; 1;$ and $z_0 = (\alpha/\beta)^{1/\gamma}$ in the z-plane are mapped onto the points $w = 0; 0;$ and $w_0 = \dfrac{\gamma}{\beta}\left(\dfrac{\alpha}{\beta}\right)^{\alpha/\gamma}$, respectively, in the w-plane.

(b) The half-line $z_0 \le x < \infty$, $y = 0$ is mapped onto the half-line $w_0 \ge u > -\infty$, $v = 0$ in the w-plane.

(c) Either of the two halves of the curve ABCDE in the z-plane is mapped onto the half-line $w_0 < u < \infty$, $v = 0$ in the w-plane.

The mapping is presented in Figure 4.33.

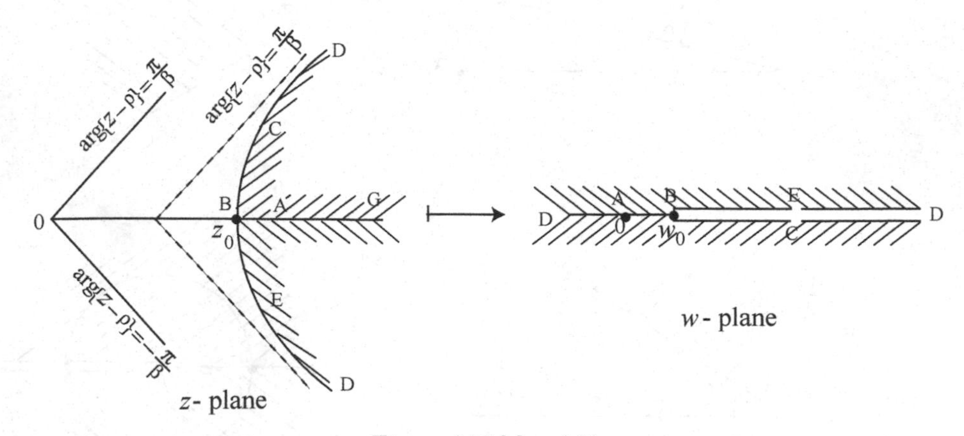

Figure 4.33 Map 4.31.

Map 4.32. The mapping $w = z^{ia}, a > 0$, or $z = w^{-ib}$, where $b = 1/a$, has critical points $z = 0; \infty$. Let $r_0 = e^{2b\pi}, R_0 = e^{-2a\pi}, z = re^{i\theta}, w = Re^{i\Theta}$, and $k, m = 0, \pm1, \pm2, \dots$.

Then the details of this mapping are as follows:

(a) The points $z = z_1 e^{2kb\pi}$ in the z-plane are mapped onto the points $w = z_1^{ia} e^{2ma\pi}$ in the w-plane.

(b) The circle $|z| = c, 0 \le \theta < 2\pi, 1 \le c \le r_0$, in the z-plane is mapped onto the line-segment $\arg\{w\} = a \log c, R_0 \le |w| \le 1$, in the w-plane.

(c) The line-segment $\arg\{z\} = \theta, 1 \le |z| < r_0$, where θ is fixed, $0 \le \theta \le 2\pi$, in the z-plane is mapped onto the circle $|w| = ae^{-a\Theta}, 0 \le \Theta < 2\pi$, in the w-plane.

These details are presented in Figure 4.34.

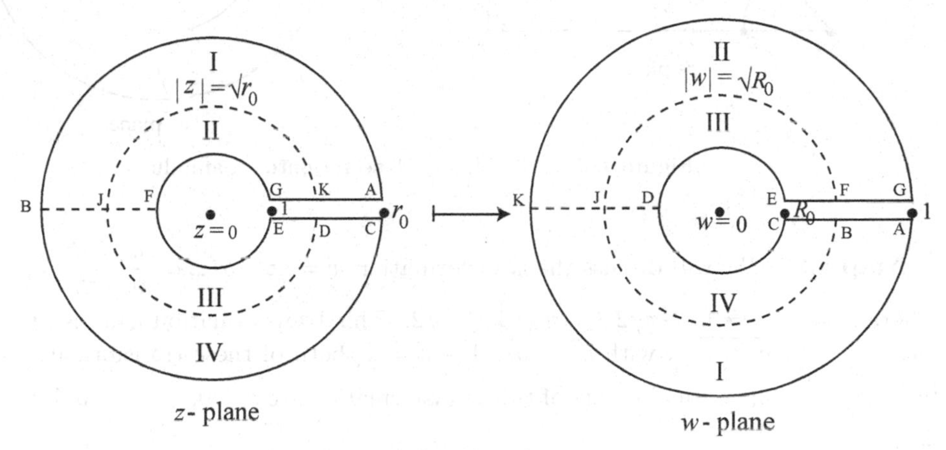

Figure 4.34 $w = z^{ia}, a > 0$.

This mapping also has other properties, when it is used to map a sector of a circle onto an annulus, as shown in Figure 4.35. The details are as follows:

(d) The points $z = e^{i\pi}$ (at F); r_0 (at A); $r_0 e^{i\pi}$ (at B); $e^{i\pi}\sqrt{r_0}$ (at F); $\sqrt{r_0}$ (at K); $e^{2i\pi}\sqrt{r_0}$ (at D) in the z-plane are mapped onto the points $w = \sqrt{R_0}; e^{2i\pi}; e^{2i\pi}\sqrt{R_0}; e^{i\pi}\sqrt{R_0}; e^{i\pi}; R_0 e^{i\pi}$, respectively, in the w-plane.

(e) The sector $0 < \arg\{z\} < \theta$ of the annulus $1 < |z| < r_0, \theta < 2\pi$, in the z-plane is mapped onto the annulus $e^{-a\Theta} < |w| < 1$, cut along the positive real axis, in the w-plane.

(f) The sector $0 < \arg\{z\} < \theta$ of the annulus $c_1 < |z| < c_2, \theta < 2\pi, 1 \le c_1 < c_2 < r_0$, in the z-plane is mapped onto the sector $a \log c_1 < \arg\{w\} < a \log c_2$ of the annulus $e^{-a\Theta} < |w| < 1$ in the w-plane.

These details are shown in Figure 4.35.

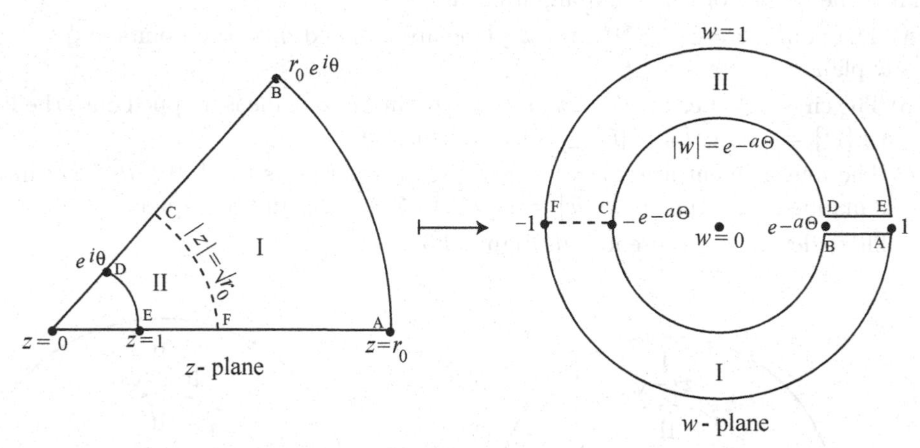

Figure 4.35 $w = z^{ia}, a > 0$, sector onto an annulus.

Map 4.33. We will discuss the transformation $w = az + b/z$, or $\dfrac{w+2k}{w-2k} = \left(\dfrac{z+k/a}{z-k/a}\right)^2$, where $k = \sqrt{ab} \neq 0$, $-\pi/2 \leq \arg\{k\} \leq \pi/2$. This transformation can also be written as $2az = w + \sqrt{w^2 - 4k^2}$, with $k^2 = ab$. For $k = 1$, both of these transformations reduce to $w = az + \dfrac{b}{z}$. The critical points of this transformation are $z = 0, z = \infty$, and $z = \pm\sqrt{b/a} = \pm k/a$.

The details of different cases under this case are as follows:

(a) The points z and $\dfrac{b}{az}$ in the z-plane are mapped onto the point $w = az + \dfrac{b}{z}$.

(b) The points $z = 0; \infty$ are mapped onto the points $w = \infty$.

(c) The points $z = \dfrac{ik}{a}; -\dfrac{ik}{a}$ are mapped onto the point $w = 0$.

(d) The points $z = \dfrac{k}{a}\, e^{i\theta}; \dfrac{k}{a}\, e^{-i\theta}$ are mapped onto the point $w = 2k\cos\theta$.

(e) The points $z = i\dfrac{k}{a}\tan\theta; -i\dfrac{k}{a}\cot\theta$ are mapped onto the point $w = -2ik\cot 2\theta$.

The mapping of the lower or upper half z-plane onto the half w-plane cut plane is shown in Figure 4.36(a); also, the mapping of the interior of the circle in the z-plane onto the cut w-plane is shown in Figure 4.36(b), and that of the exterior of the circle in Figure 4.36(c).

For example, the details of the mapping in the case of Figure 4.36(c) are as follows:

(a) The points $z = z_1 \pm \sqrt{z_1^2 - b/a}$ (marked E, F) in the z-plane are mapped onto the point $w = 2az_1$.

(b) The points $z = z_2 \pm \sqrt{z_2^2 - b/a}$ (marked C, D) in the z-plane are mapped onto the point $w = 2az_2 = 2b/z_1$.

(c) The circular crescent ADBEA, i.e., the region exterior to both circles in the z-plane is mapped onto the exterior of the circle ADBEA in the w-plane.

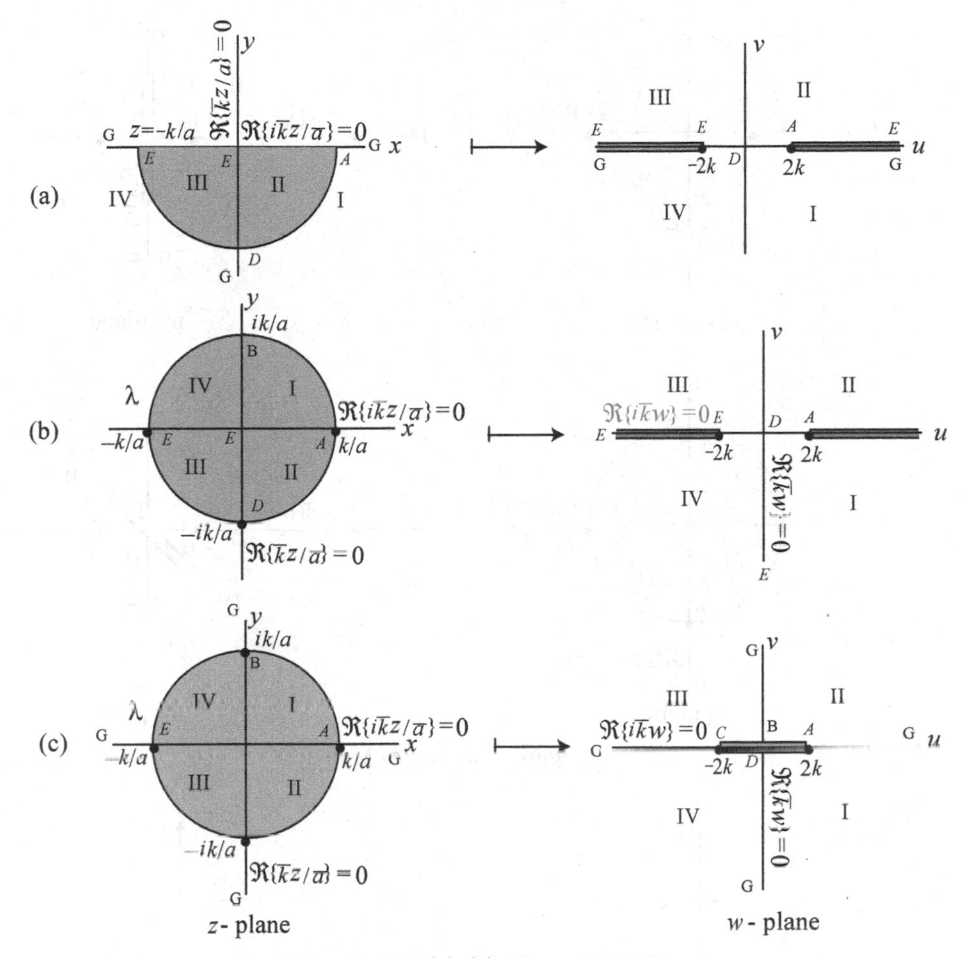

Figure 4.36 (a)-(c) Maps 4.33-4.35.

Map 4.34. The conformal transformation $w = z\,e^{i\alpha} - \dfrac{a^2}{z\,e^{i\alpha}}$ maps the interior of a circle in the z-plane onto the interior of a slanted slit (inclined at an angle α to the x-axis), and exterior of the circle onto the exterior of this slanted slit in the w-plane (Figure 4.36(e)).

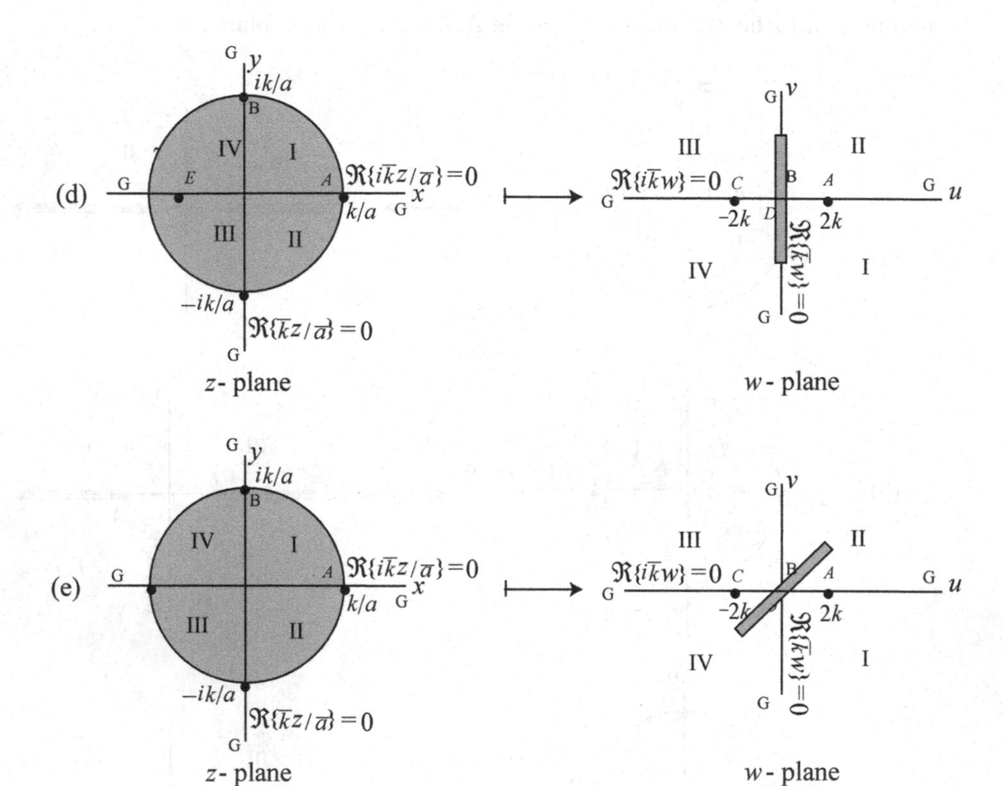

Figure 4.36 (d)-(e) Map 4.35.

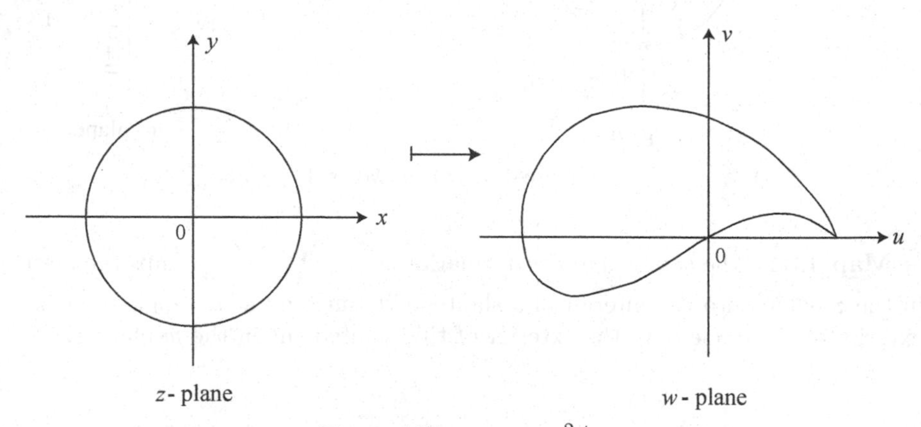

Figure 4.37 $w = z - a^2/z$.

Map 4.35. The conformal transformation $w = z - \dfrac{a^2}{z}$ maps the interior of a circle in the z-plane onto the interior of vertical slit, and exterior of the circle onto the exterior of the vertical slit in the w-plane (Figure 4.36(d)).

The transformation 4.35, with $a \neq -1, 0, 1$, also maps circles which contain $z = 1$ as an interior point and which passes through $z = -1$ onto shapes resembling airfoils, as seen in Figure 4.37. This map creates a cusp at which the associated fluid (air) velocity can be infinite. However, it can be avoided by adjusting the fluid flow in the z-plane, and create lift generated by such an airflow depending on the airfoil shape and air density and speed. See Chapter 6 for more information.

Map 4.36. The transformation $w = \frac{1}{2}z + \dfrac{\beta}{z}, \beta \neq 0$, has fixed points $F_1 = 2k = k/\alpha = \sqrt{2\beta} = -F_2$. This mapping has the following properties:

(a) The exterior of $|z| = \left|\sqrt{2\beta}\right|$ and the interior of $|z| = \left|\sqrt{2\beta}\right|$ in the z-plane are mapped onto the entire plane cut from $-\sqrt{2\beta}$ to $\sqrt{2\beta}$.

(b) The set D of 'circles' orthogonal to the circles through F_1 and F_2 (coaxial system) in the z-plane is mapped entirely onto the similar set D, as a whole, but counted twice in the w-plane.

(c) The set E of circles through F_1 and F_2, as a whole, in the z-plane is mapped onto the similar set E, counted twice, in the w-plane. ∎

Map 4.37. The transformation $w = -\dfrac{z}{a} - \dfrac{a}{z}$ maps the upper half of a circle of radius a: $x^2 + y^2 \leq a^2, x \geq 0, y \geq 0$, in the z-plane onto the upper half-plane $\Im\{w\} \geq 0$ in the w-plane.

Map 4.38. The transformation $w = \dfrac{z}{a} + \dfrac{a}{z}$ maps the upper half-plane with a circular domain of radius a removed: $y \geq 0, x^2 + y^2 \geq a^2$, in the z-plane onto the upper half-plane $\Im\{w\} \geq 0$ in the w-plane.

Map 4.39. The transformation $w = \dfrac{z^2}{a^2} - \dfrac{a^2}{z^2}$ maps the quadrant of a circle of radius a: $x^2 + y^2 \leq a^2, x \geq 0, y \geq 0$, in the z-plane onto the upper half-plane $\Im\{w\} \geq 0$ in the w-plane.

Map 4.40. The transformation $w = z + i/z$ maps the exterior and interior of $|z| = 1$ in the z-plane onto the entire w-plane cut from $-(1+i)\sqrt{2}$ to $(1+i)\sqrt{2}$. ∎

Map 4.41. The transformation that maps the exterior of the parabola $y^2 = 4p(p - x)$ in the z-plane onto the upper half-plane $\Im\{w\} > 0$ is defined by

$$w = i\left(\sqrt{z} - \sqrt{p}\right), \tag{4.1.14}$$

and is presented in Figure 4.38.

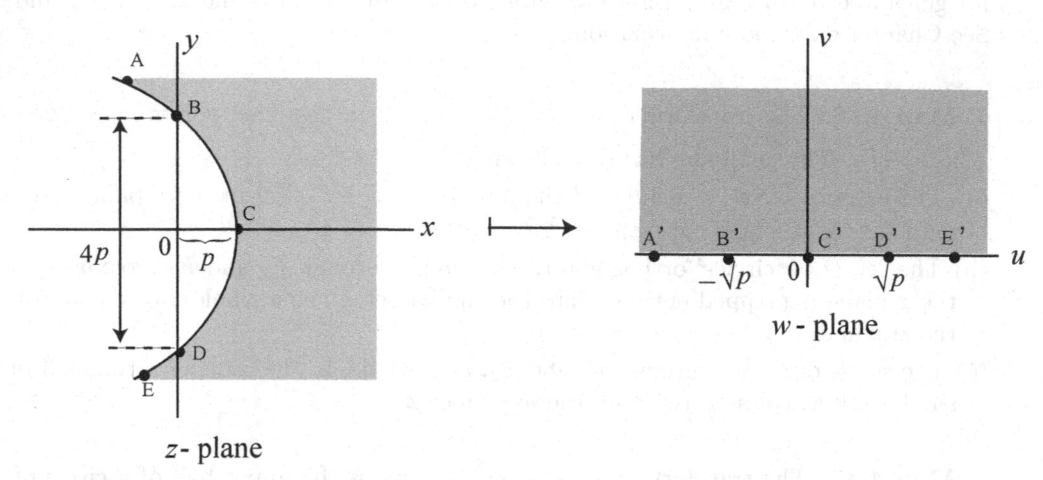

Figure 4.38 Exterior of a parabola onto the upper half-plane.

Map 4.42. The transformation that maps the exterior of the parabola $y^2 = 4p(p - x)$ in the z-plane onto the interior of the unit circle in the w-plane is defined by

$$w = 2\sqrt{\frac{p}{z}} - 1, \tag{4.1.15}$$

and is presented in Figure 4.39.

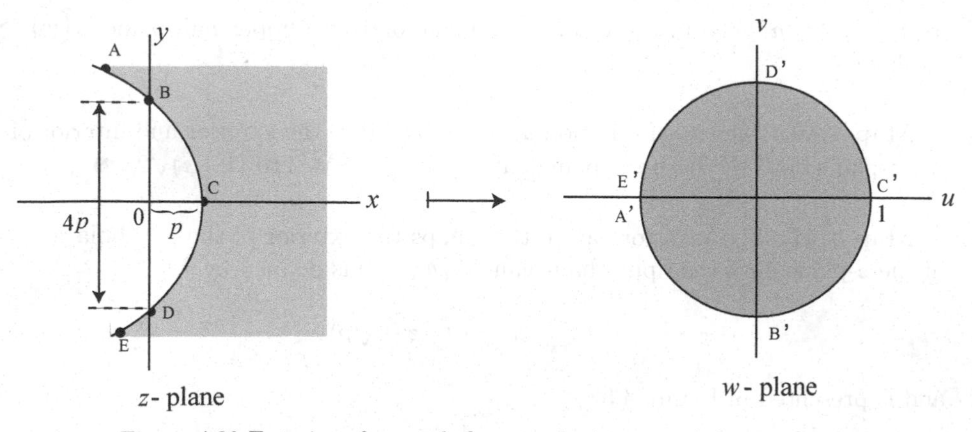

Figure 4.39 Exterior of a parabola onto the interior of the unit circle.

Map 4.43. The transformation $w = \sqrt{z - \frac{1}{2}p} - i\sqrt{\frac{1}{2}p}$ maps the exterior of a parabola: $y^2 - 2px \geq 0$, in the z-plane onto the upper half-plane $\Im\{w\} \geq 0$ in the w-plane.

Next, we consider two cases of the transformation $w = z^\alpha$, $0 < \alpha < 1$. They are:

Map 4.44. The transformation $w = -\left(\frac{z}{a}\right)^{\pi/\beta} - \left(\frac{a}{z}\right)^{\pi/\beta}$ maps the sector of a circle of radius a with angle β: $x^2 + y^2 \leq a^2, 0 \leq \arg\{z\} \leq \beta$, in the z-plane onto the upper half-plane $\Im\{w\} \geq 0$ in the w-plane.

Map 4.45. (a) The transformation $w = \dfrac{az^2+bz+c}{z+d}, a \neq 0, ad^2 - bd + c = \beta \neq 0$, is a composite of $w = \xi + b - 2ad, z = \zeta - d, \xi = a\zeta + \beta/\zeta$.

(b) The transformation $w = \dfrac{az^2+bz+c}{z^2+dz+f}, a \neq 0$, does not reduce to a linear or bilinear transformation; it is equivalent to $\dfrac{w}{a-w} = \dfrac{az^2+bz+c}{(ad-b)z+(af-c)}$.

(c) The transformation $w = \dfrac{i\{b(1+z)^2 + 2(1-z)\}}{2(1+z)}$ is a composite of $w = \xi - i, z = \dfrac{2i}{\zeta} - 1, \xi = \zeta - b/\zeta$. If b is real, $0 < b < 1$, then it maps the line $\zeta = 1$ (CG) in the ζ-plane onto the circle $|z| = 1$ in the z-plane and onto the curve CAGBE which is symmetric with respect to $u = 0$ with a cusp at G for $b = 1$, in the w-plane. It also maps the circle $\zeta - ib/2| = b/2$ in the ζ-plane onto the line $\Re\{z\} = 2/b - 1$ in the z-plane onto the same curve CAGBE in the w-plane. These properties are presented in Figure 4.40 for $b = 3/4$ and $b = -2i$ $(\beta = 1)$. ■

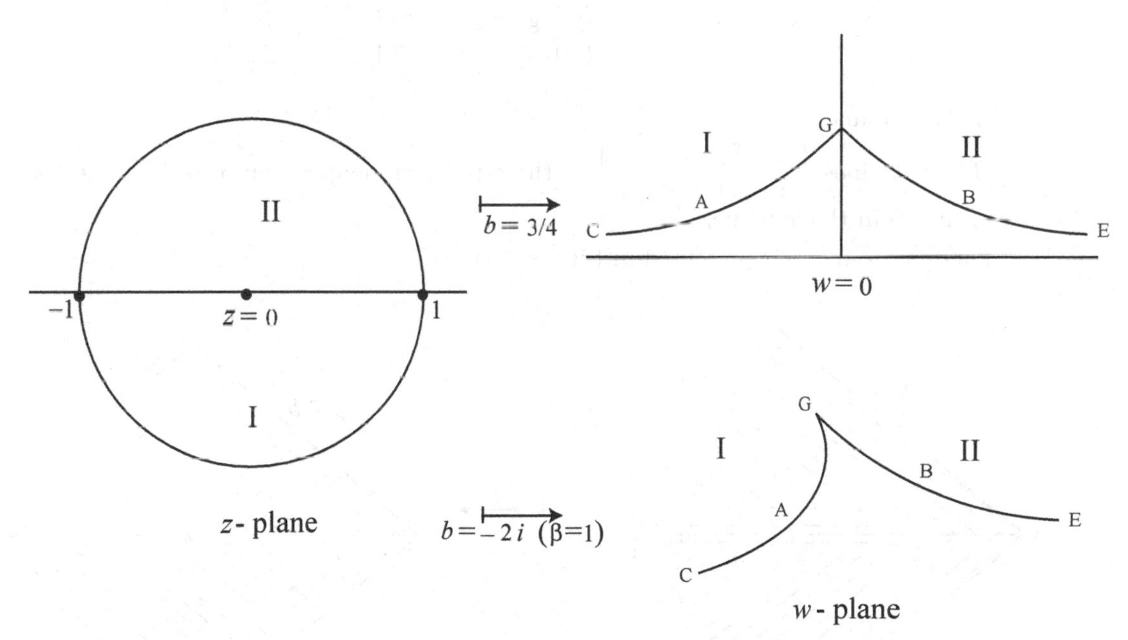

Figure 4.40 Map 4.43 (c).

Map 4.46. The transformation $w = i\,\dfrac{z^2+2az-a^2}{z^2-2az-a^2}$ maps the semi-circle of radius a: $x^2+y^2 \geq a^2, x \geq 0$, in the z-plane onto the unit disk $|w| \leq 1$ in the w-plane.

Map 4.47. The transformation $w = \dfrac{z^2+2cz-c^2}{z^2-2cz-c^2}$, which is a composite of the maps $\zeta = iz, \xi = \left(\dfrac{\zeta+c}{\zeta-c}\right)^2$, and $w = i\dfrac{\xi-i}{\xi+i}$, maps the area of the semicircle $|z| < c, x > 0$, in the z-plane onto the unit disk $|w| < 1$ in the w-plane.

Map 4.48. The transformation $w = \dfrac{\left(1+z^{\pi/\beta}\right)^2 - i\left(1-z^{\pi/\beta}\right)^2}{\left(1+z^{\pi/\beta}\right)^2 + i\left(1-z^{\pi/\beta}\right)^2}$ maps the sector of a unit circle of angle β, $|z| \le 1, 0 \le \arg\{z\} \le \beta$, in the z-plane onto the unit disk $|w| \le 1$ in the w-plane.

Map 4.49. We will consider the transformation $w^b = 1 - z^a$, $a > 0, b > 0, a \ne 1, b \ne 1$, where $|\arg\{z\}| \le \pi/a$, and $|\arg\{w\}| \le \pi/b$. The critical points are $z = 0; 1; \infty$. We will set $\sigma = \dfrac{a\theta}{b} - \dfrac{\pi}{2b}$, where $\theta = \arg\{z\}$ and $\Theta = \arg\{w\}$ are real, and $|w| = R$. The details of different cases under this transformation are as follows:

(a) The points $z = 0; 1; \infty; 2^{1/\sigma}; e^{\pm i\pi a}; e^{i\theta}$, $(0 \le \theta \le \pi/a$; and $e^{i\theta}$, $-\pi/a \le \theta \le 0$ in the z-plane are mapped onto the points $w = 1; 0; \infty; e^{\pm i\pi b}; 2^{1/b}; e^{i\Theta}$, $0 \le \Theta \le \pi/b$; and $e^{i\Theta}$, $-\pi/b \le \Theta \le 0$ in the w-plane.

(b) The segment $0 < x < 1$, $y = 0$ in the z-plane is mapped onto the segment $0 < u, 1$, $v = 0$ in the w-plane.

(c) The half-line $1 < x < \infty$, $y = 0$ in the z-plane is mapped onto the half-lines

$$\left\{ \begin{array}{c} \arg\{w\} = \pi/b \\ \arg\{w\} = -\pi/b \end{array} \right\}$$

in the w-plane.

(d) The half-lines $\left\{ \begin{array}{c} \arg\{z\} = \pi/a \\ \arg\{z\} = -\pi/a \end{array} \right\}$ in the z-plane are mapped onto the half-line $1 < u < \infty$, $v = 0$ in the w-plane.

These mappings are presented in Figure 4.41.

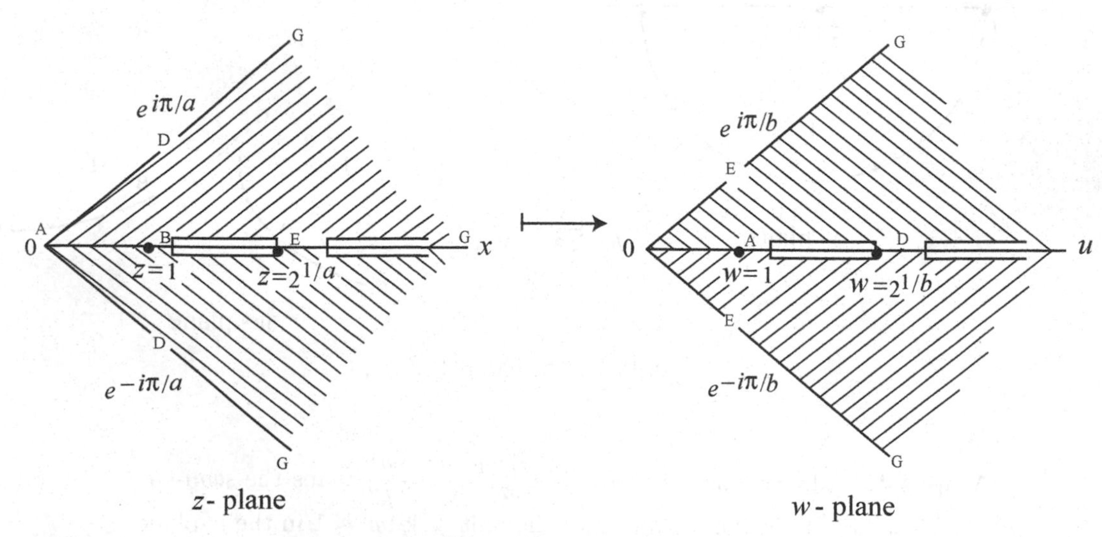

Figure 4.41 Map 4.49.

(e) The half-line $\arg\{z\} = \theta$, θ real and fixed, $0 < \theta < \pi/a$ in the z-plane is mapped on the part $0 > \Theta > \sigma - \pi/(2b)$ of the curve $R^b \cos b(\theta - a) = \sin a\Theta$, with asymptotes $\Theta - \sigma = \pm\dfrac{\pi}{2b}$, and axis of symmetry $\Theta = \sigma$.

(f) The half-line $\arg\{z\} = \theta - \pi/a$ in the z-plane is mapped onto the part $0 < \Theta < \sigma + \pi/(2b)$

of the same curve in the w-plane.

(g) The unit circle $|z| = 1$ in the domain $-\pi/a \le \arg\{z\} \le \pi/a$ in the z-plane is mapped onto the curve $R^b = 2\cos b\Theta$, where $|\Theta| \le \pi/(2b)$.

Note that when $b = 2$, the above curve in the w-plane is one-half of the lemniscate $|w+1||w-1| = 1$, and the curve $R^b \cos b\Theta = \sin a\Theta$ reduces to one branch of the rectangular hyperbola $\Re\Big\{\dfrac{w^2}{1 - e^{2ia\Theta}}\Big\} = \frac{1}{2}$.

Map 4.50. The transformation $w = \left(\dfrac{z+c}{z-c}\right)^2$ or $z = c\,\dfrac{\sqrt{w}+1}{\sqrt{w}-1}$, maps the area of the semicircle onto the upper half-plane. The details are: The point $z = c\,e^{i\theta}$, $0 < \theta < \alpha = \pi$, in the z-plane is mapped onto the point $w = -\cot^2(\theta/2)$ in the w-plane. They are presented in Figure 4.42.

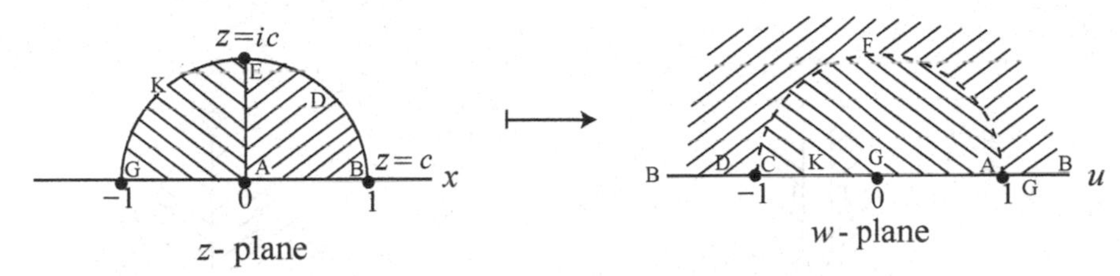

Figure 4.42 Map 4.50 (a).

Map 4.51. Consider the transformation $w = -\left(\dfrac{1+\sqrt{1+z^2}}{z}\right)^2$ or $z = \dfrac{2i\sqrt{w}}{w+1}$. This transformation deals with cut half-planes, cut circles and cut planes on half-planes. The following four cases are analyzed:

Map 4.51a. Upper half-plane, cut from $z = 0$ to $z = 1$, with critical points $z = 0; \infty; 1$, onto the upper half-plane. The details are as follows.

(a) The point $z = 0$ in the z-plane is mapped onto the points $w = 0; \infty$ in the w-plane.

(b) The points $z = \infty; 1$ in the z-plane is mapped onto the points $w = -1; 1$, respectively, in the w-plane.

(c) The half-plane $y = 0$, $0 < x < \infty$, in the z-plane is mapped onto the half-plane $v = 0$, $-\infty < u < -1$ in the w-plane.

(d) The line segment $x = 0+$, $0 < y < 1$ (AKF) in the z-plane is mapped onto the half-line $v = 0$, $1 \le u < \infty$ in the w-plane.

(e) The line segment $x = 0-$, $0 < y < 1$ (FGE) in the z-plane is mapped onto the half-line $v = 0$, $0 \le u \le 1$ in the w-plane.

(f) The half-line $y = 0$, $-\infty < x < 0$, in the z-plane is mapped onto the line segment $v = 0$, $-1 < u < 0$, in the w-plane.

(g) The half-line $x = 0$, $1 < y < \infty$, in the z-plane is mapped onto the semi-circle $|w| = 1$, $v \ge 0$, in the w-plane.

These mappings are presented in Figure 4.43.

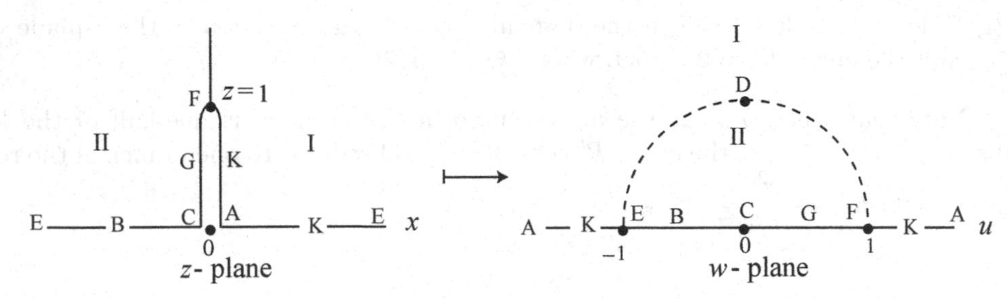

Figure 4.43 Map 4.51a.

Map 4.51b. Upper half-plane, cut from $z = 1$ to $z = \infty$, along the imaginary axis on the upper half-plane. The transformation is $w = \dfrac{z}{\sqrt{1+z^2}}$ or $z = \dfrac{w}{\sqrt{1-w^2}}$, and the critical points are $z = 1; \infty$. The analysis is similar to the previous one and is presented in Figure 4.44.

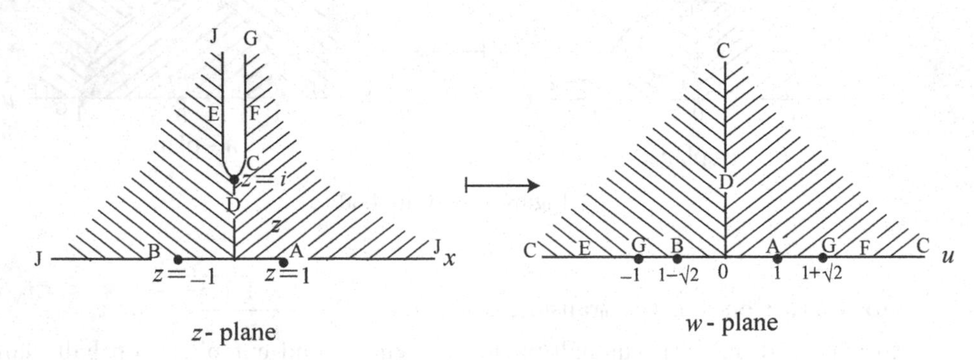

Figure 4.44 Map 4.51b.

Map 4.51c. The interior of the circle $|z| = c$, cut from 0 to c, onto the half-plane is given by $w = \left(\dfrac{\sqrt{z/c}+1}{\sqrt{z/c}-1}\right)^2$ or $\dfrac{z}{c} = \left(\dfrac{\sqrt{w}+1}{\sqrt{w}-1}\right)^2$. This is a particular case of the Map 4.53 for $\alpha = \pi$.

Map 4.51d. Two cases of cut planes onto half-planes are considered. They are:

(a) The transformation is $w^2 = \dfrac{z-1}{z+1}$ or $z = \dfrac{1+w^2}{1-w^2}$, with critical points $z = 1; -1; \infty$. This transformation is a composite of $w = \dfrac{\xi-1}{\xi+1}$ and $z = \frac{1}{2}(\xi + 1/\xi)$, or of $w = i\dfrac{\xi+1}{\xi-1}$ and $-1/z = \frac{1}{2}(\xi + 1/\xi)$. The details of this mapping are as follows.

(i) The points $z = 1; -1; 0; i; -i; \infty$ in the z-plane are mapped onto the points $w = 0; \infty; i; e^{i\pi/4}; e^{3i\pi/4}; 1$, respectively, in the w-plane.

(ii) The segment $-1 < x < 1$, $y = 0$, in the z-plane is mapped onto the half-line $0 < v < \infty$, $u = 0$, in the w-plane.

(iii) The semi-circle $|z| = 1$, $y > 0$, in the z-plane is mapped onto the half-line $\arg w = \pi/4$ in the w-plane.

(iv) The semi-circle $|z| = 1$, $y < 0$, in the z-plane is mapped onto the half-line $\arg w = 3\pi/4$ in the w-plane.

(v) The slit $1 < x < \infty$, $y = 0$, in the z-plane, is mapped onto the segments $\left\{\begin{matrix} 0 < u < 1, v = 0 \\ -1 < u < 0, v = 0 \end{matrix}\right\}$ in the w-plane.

(vi) The slit $-\infty < x < -1$, $y = 0$, in the z-plane, is mapped onto the segments $\left\{\begin{matrix} 1 < u < \infty, v = 0 \\ -\infty < u < -1, v = 0 \end{matrix}\right\}$ in the w-plane.

These details are shown in Figure 4.45.

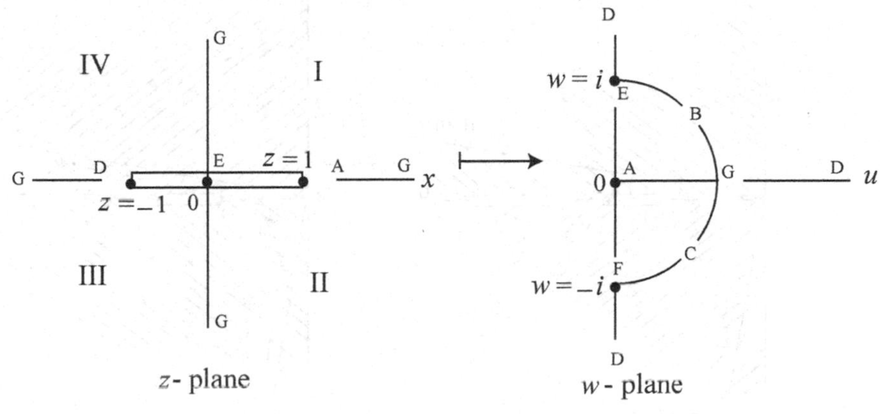

Figure 4.45 Map 4.51d(i).

(b) Plane, cut along the negative part of the real axis, onto the half-plane is given by $w = \sqrt{z}$; see Map 4.25.

Map 4.52. Consider the transformation $w = \sqrt{1 - z^2}$, which is involutory. It is presented in Figure 4.46, and the details of different aspects of this mapping with $\Re\{z\} \geq 0$ and $\Re\{w\} \geq 0$ are as follows.

(i) The points $z = 0; 1; \infty; \pm i$; and $\sqrt{2}$ in the z-plane are mapped onto the points $w = 1; 0; \infty; \sqrt{2}$; and $\pm i$, respectively, in the w-plane.

(ii) The first quadrant $x > 0, y > 0$ in the z-plane is mapped onto the fourth quadrant $u > 0, v < 0$ in the w-plane.

(iii) The fourth quadrant $x > 0, y < 0$ in the z-plane is mapped onto the first quadrant $u > 0, v > 0$ in the w-plane.

(iv) The half-line $\arg\{z\} = \theta$, where θ is fixed, $0 < \theta < \pi/2$, in the z-plane is mapped onto the part in the fourth quadrant of the rectangular hyperbola $\Re\left\{\frac{w}{1 - e^{4i\theta}}\right) = \frac{1}{2}$ through A, with asymptotes $\arg\{w\} = \theta - \pi/2, \theta$.

(v) The half-line $\arg\{z\} = \theta - \pi/2$ in the z-plane is mapped onto the part in the first quadrant of the same hyperbola.

(vi) The half-plane $x > 0$, cut along BEG in the z-plane, is mapped onto the half-plane $u > 0$ cut along ADB.

(vii) Either the second or fourth quadrant $\begin{Bmatrix} x < 0, y > 0 \\ x > 0, y < 0 \end{Bmatrix}$ in the z-plane is mapped onto either the first or third quadrant $\begin{Bmatrix} u > 0, v > 0 \\ u < 0, v < 0 \end{Bmatrix}$ in the w-plane.

(viii) Either the third or first quadrant $\begin{Bmatrix} x < 0, y < 0 \\ x > 0, y > 0 \end{Bmatrix}$ in the z-plane is mapped onto either the fourth or third quadrant $\begin{Bmatrix} u > 0, v < 0 \\ u < 0, v > 0 \end{Bmatrix}$ in the w-plane.

These different aspects are presented in Figure 4.46.

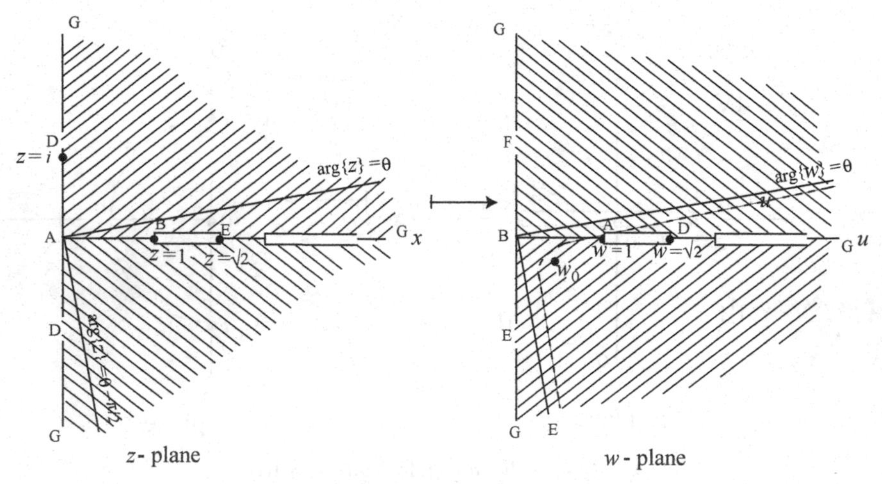

Figure 4.46 Map 4.52.

Note that in the above analysis we can get one-to-one correspondence by considering the above quadrants; otherwise each curve must be counted twice; the details are as follows.

(ix) The curve $\Re\{z^2/z_0^2\} = \frac{1}{2}$, where $z_0 \neq 0$, $z_0^2 \neq 1 + e^{4i\theta_0}$, and $\theta_0 = \arg\{z_0\}$, i.e., a rectangular hyperbola with asymptotes $\arg\{z\} = \theta_0 \pm n\pi/4$, $n = 1, 5$, in the z-plane is mapped onto the curve $\Re\{w^2/w_0^2\} = \frac{1}{2}$, where $w_0^2 + z_0^2 = 1 + e^{4i\Theta_0}$, i.e., a rectangular hyperbola with asymptotes $\arg\{w\} = \Theta_0 + \pm n\pi/4$, $n = 1, 5$, in the w-plane.

(x) The curve $\Re\left\{\dfrac{z^2}{1 + e^{4i\theta_0}}\right\} = \frac{1}{2}$, where $|\theta_0 - k\pi| < \pi/4$, $k = 1, 2$, in the z-plane is mapped onto the lines $\Re\{w\, e^{i(\pm\pi/4 - \Theta_0)}\} = 0$ in the w-plane.

(xi) The hyperbola $|z + 1| - |z - 1| = \pm 2\cos\nu$, where ν is fixed, with asymptotes $\arg\{z\} = \pm(\nu + k\pi)$, $k = 0, 1$, in the z-plane is mapped onto the hyperbola $|w + 1||w - 1| = \pm 2\sin\nu$, with asymptotes $\arg\{w\} = \pm(\pi/2 - \nu + k\pi)$, $k = 0, 1$, in the w-plane (dotted).

(xii) The ellipse $|z + 1||z - 1| = h$, $h > 2$, in the z-plane is mapped onto ellipse $|w + 1||w - 1| = h$ in the w-plane.

(xiii) Cassini's oval $|z - z_0||z + z_0| = c$, $z_0 \neq 0, \pm 1$, and $c > 0$, in the z-plane is mapped onto Cassini's oval $|w + w_0||w - w_0| = c$, where $w_0 = 1 - z_0^2$, in the w-plane.

(xiv) The circle $|z| = \sqrt{c}$, $\sqrt{c} > 0$, in the z-plane is mapped onto Cassini's oval $|w - 1||w + 1| = c$ in the w-plane.

(xv) Cassini's oval $|z - 1||z + 1| = c$, $c > 0$, in the z-plane is mapped onto the circle $|w| = \sqrt{c}$ in the w-plane.

Map 4.53. We will consider the mapping of a sector on the upper half-plane. The transformation is given by $w = \left(\dfrac{(z/c)^{\pi/\alpha}+1}{(z/c)^{\pi/\alpha}-1}\right)^2$, $c > 0$, and $0 < \alpha \le 2\pi$; the inverse transformation is $z = c\left(\dfrac{\sqrt{w}+1}{\sqrt{w}-1}\right)^{\alpha/\pi}$, and the critical points are $z = 0; c; c\,e^{i\alpha}$. The details of the this mapping are as follows:

(a) The points $z = 0$; $c\,e^{i\theta}$ $(0 < \theta < \alpha)$; c; $c\,e^{i\alpha/2}$; $c\,e^{i\alpha}$ in the z-plane is mapped onto the points $w = 1; -\cot^2 \dfrac{\pi\theta}{2\alpha}; \infty; -1; 0$, respectively, in the w-plane.

(b) The line segment $y = 0, 0 \le x < c$, in the z-plane is mapped onto the half-line $v = 0, 1 \le u < \infty$.

(c) The arc $z = c\,e^{i\theta}, 0 < \theta \le \alpha/2$, in the z-plane is mapped onto the half-line $v - 0, -\infty < u \le -1$ in the w-plane.

(d) The arc $z = c\,e^{i\theta}, \alpha/2 < \theta \le \alpha$, in the z-plane is mapped onto the half-line $v - 0, -1 \le u \le 0$ in the w-plane.

(e) The line segment $z = r\,e^{i\alpha}$, $0 \le r \le c$, in the z-plane is mapped onto the line segment $v = 0, 0 \le u \le 1$ in the w-plane.

(f) The line segment $z = r\,e^{i\alpha/2}$, $0 \le r \le c$, in the z-plane is mapped onto the semi-circle $|w| < 1, y \ge 0$ in the w-plane.

These details are presented in Figure 4.47.

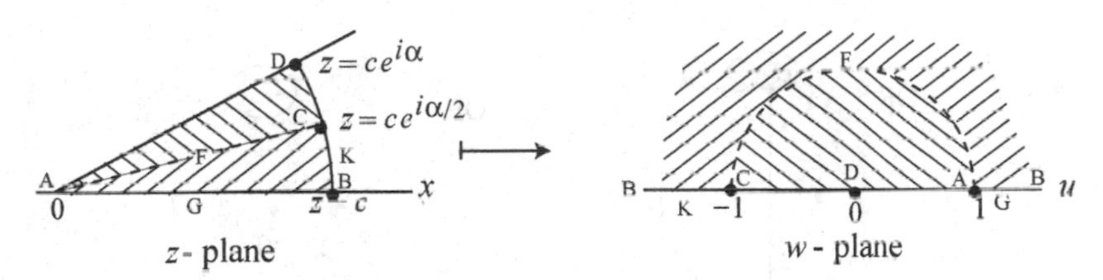

Figure 4.47 Map 4.53.

Map 4.54. The transformation $w = \dfrac{(1+z^3)^2 - i(1-z^3)^2}{(1+z^3)^2 + i(1-z^3)^2}$, which is a composite of the mappings $w = \dfrac{\xi - 1}{\xi + 1}$ and $\xi = \left(\dfrac{z^3+1}{z^3-}\right)^2$, maps (i) the points $z = 0; 1; e^{i\pi/3}; e^{i\pi/6}$ onto the points $w = -1; 1; -1; 1$, respectively, in the w-plane; and (ii) maps the sector area $|z| < 1$, $0 < \arg\{z\} < \pi/3$, in the z-plane onto the unit disk $|w| < 1$ in the w-plane.

4.1.1 Regions Bounded by Two Circular Arcs. The following cases are considered.

Map 4.55. The transformation $w = k\left(\dfrac{z - z_0}{z - \zeta_0}\right)^{\pi/\alpha}$, $k \ne 0$, maps the circular crescent and its vertices at $z = z_0$ and ζ_0, and the angle α in the z-plane onto some half-plane $\Im\{w\} > 0$, as follows:

Map 4.56. The transformation $w = \xi^{\pi/\alpha}$, where $\xi = \varepsilon i K \dfrac{(z - z_0)(z_1 - \zeta_0)}{(z - \zeta_0)(z_0 - \zeta_0)}$, $0 < \alpha < 2\pi$, $K \gtrless 0$, $\varepsilon = \pm 1$, and ε is chosen so that ξ maps the arc with center z_1 on the positive part of the real x-axis. Such mappings are presented in Figure 4.48, where (i) the arc $(z - z_0)$ of the circle $|z - z_1| = |z_0 - z_1|$ is mapped onto the half-plane $v = 0, 0 < u < \infty$; (ii) the arc

$(z-\zeta_0)$ of the circle $|z-z_2| = |z_0 - z_2|$ is mapped onto the half-plane $v = 0, -\infty < u < 0$; and the points z_0, ζ_0 are mapped onto the points $w = 0; \infty$.

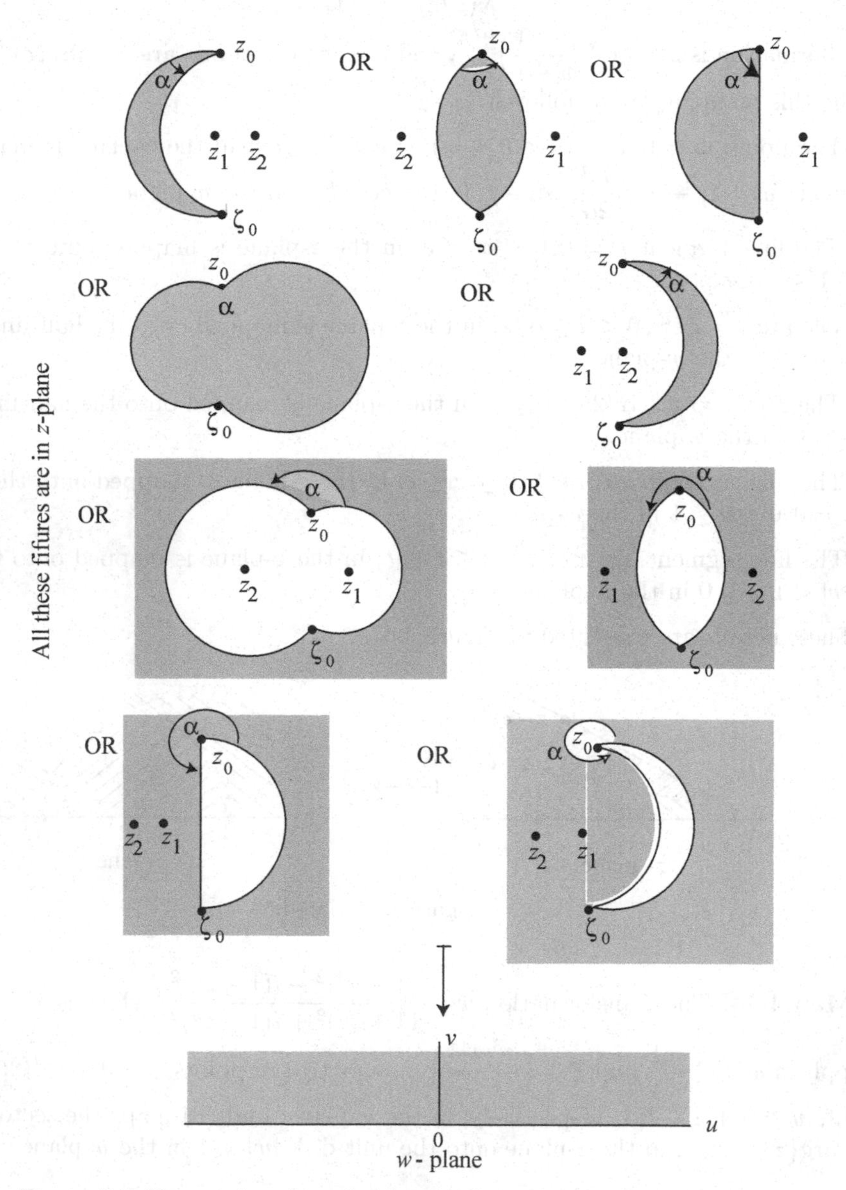

Figure 4.48 Nine regions bounded by circular arcs onto the upper half plane.

Map 4.57. The transformation $w = c\Big(\dfrac{z\bar{z}_1}{i\sigma - z}\Big)^{\pi/\alpha}$, $c > 0, \sigma > 0, \Im\{z_1\} = \frac{1}{2}\sigma$, and $0 < \alpha < \pi - \arg\{z_1\}$, maps the upper and lower halves of the crescent region in the z plane onto the upper half-plane $\Im\{w\} > 0$ as shown in Figure 4.49. This transformation (i) maps the points $z_0; \zeta_0 - i\sigma; z_3 = z_1 - |z_1|; z_4 = z_2 - |z_2|$ in the z-plane onto the points $w = 0; \infty; w_3 = c|z_1|^{\pi/\alpha}; w_4 = -c|z_1|^{\pi/\alpha}$, respectively; (ii) maps the (right) arc $(z_0 z_3 \zeta_0)$ of the circle C_1 with center z_1 in the z-plane onto the half-line $v = 0, 0 \le u \le \infty$; (iii) maps the (left) arc $(z_0 z_4 \zeta_0)$ of the circle C_2 with center z_2 in the z-plane onto the half-line $v = 0, \infty \le u \le 0$; and (iv) maps the arc of an arbitrary circle forming angle β, $0 < \beta < \alpha$, with the circle C_1 at z_0 and with endpoints z_0, ζ_0 onto the half-line $w = \rho\, e^{i\pi\beta}$, $0 \le \rho \le \infty$.

In the case of the set of coaxial circles with limiting points z_0, ζ_0, the above transformation maps the arc, lying in the crescent, of the circle $|\tau z_i| = \sqrt{1+\sigma\tau}$, with center $-i/\tau$, where $\sigma\tau > -$, in the z-plane onto the semi-circle $|w| = c\left|\dfrac{z_1}{\sqrt{1+\sigma\tau}}\right|^{\pi/\alpha}$, $v > 0$.

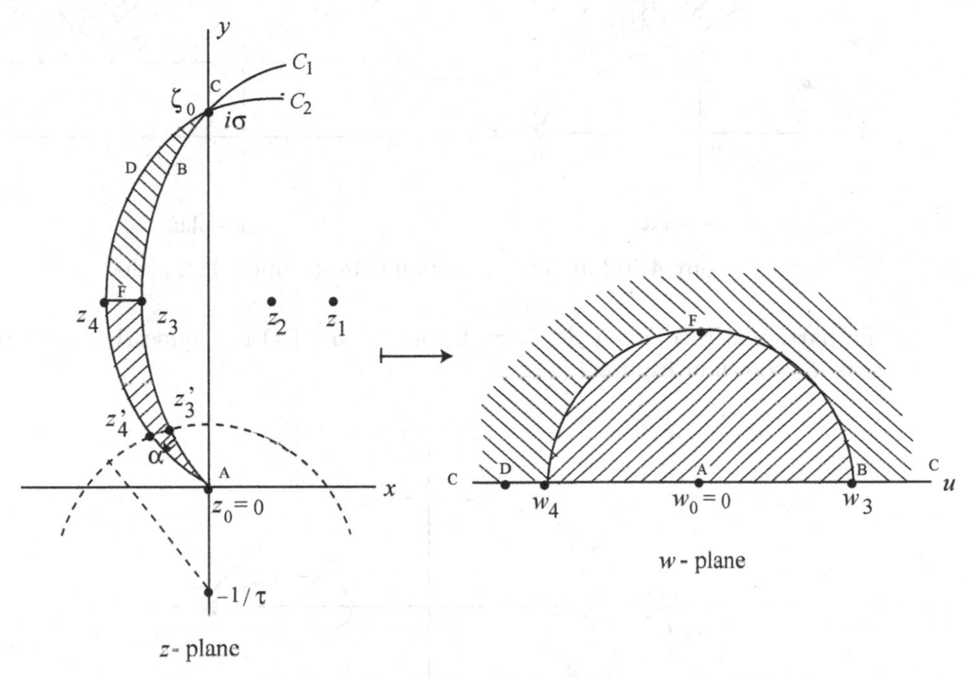

Figure 4.49 Crescent circular arc onto the upper half plane.

Map 4.58. The transformation $w = c\,e^{-i\beta\pi/\alpha}\left(\dfrac{c+z}{c-z}\right)^{\pi/\alpha}$, $c > 0, \beta > 0, 0 < \alpha < \alpha+\beta < \pi$, or $z = c\dfrac{w^{\alpha/\pi} - c^{\alpha/\pi}e^{i\beta}}{w^{\alpha/\pi} + c^{\alpha/\pi}e^{-i\beta}}$, maps the interior of an airfoil in the z-plane onto the upper half-plane $\Im\{w\} > 0$, as shown in Figure 4.50.

This transformation (i) maps the points $z = -c; c$ onto the points $w = 0; \infty$, respectively; (ii) maps the arc ABD in the z-plane onto the half-line $v = 0, 0 \le u \le \infty$; (iii) maps the arc ACD in the z-plane onto the half-line $v = 0, -\infty \le u \le 0$; (iv) maps the region bounded by these two arcs in the z-plane onto the half plane $\Im\{w\} > 0$; (v) maps the airfoil DED, with angle α at D in the z-plane onto the line parallel to $v = 0$ in the upper half plane; and (vi) maps the curve G in the z-plane onto the circle G in the w-plane.

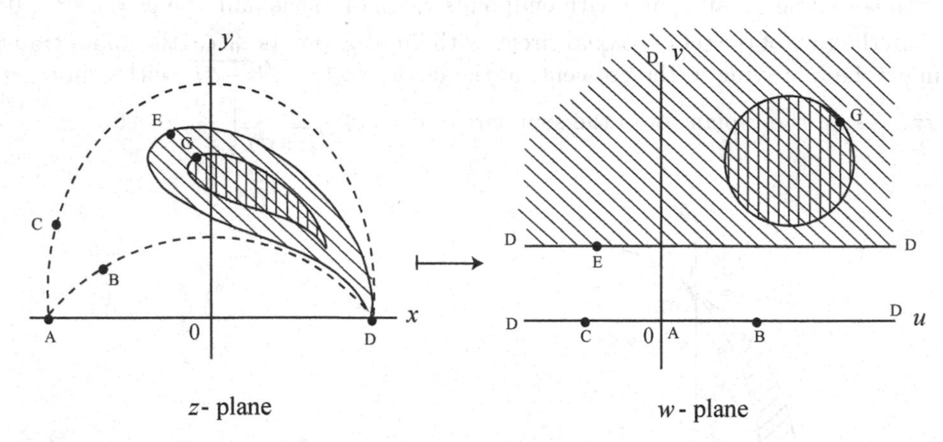

Figure 4.50 Interior of an airfoil onto the upper half plane.

In the particular case when $\beta = -\pi/2$, the airfoil DED is symmetrical with respect to the x-axis, as shown in Figure 4.51.

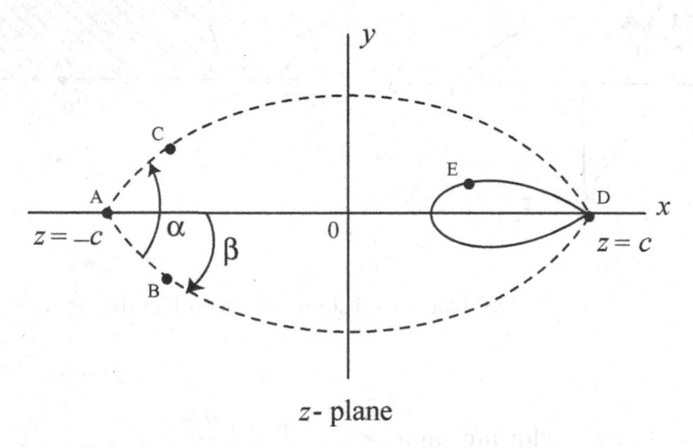

Figure 4.51 Airfoil symmetrical with the x-axis.

REFERENCES USED: Ahlfors [1966], Boas [1987], Carrier, Krook and Pearson [1966], Gaier [1964], Ivanov and Trubetskov [1995], Kantorovich and Krylov [1958], Kober [1957], Kythe [1996], Nehari [1952], Phillips [1943], Polyanin and Nazaikinskiii [2016], Schinzinger and Laura [2003], Silverman [1967], Wen [1992], Wolfram [1996].

5

Exponential Family of Functions

We will introduce the exponential and logarithmic functions, describe their properties, and present different types of conformal mappings that have been developed around these two functions.

5.1 Exponentials

In $\mathbb{R}^2$, the continuous increasing function $\log_a x$ and its inverse function a^x, called the *exponential to the base* a, satisfy the identities $a^{\log_a x} = x$, and $\log_a a^x = x$. In this sense these two functions are sometimes called the *exponentials*. Using the formula

$$\log_a \underbrace{a \cdots a}_{n\text{-times}} = \underbrace{\log_a a + \cdots + \log_a a} - n,$$

the function a^x reduces to $a^n = \underbrace{a \cdots a}_{n\text{-times}}$ for $n = 1, 2, \ldots$. Taking $x = 1/n$ reduces the above second identity to $\log_a a^{1/n} = 1/n$, which implies that $n \log_a a^{1/n} - \log_a \left(a^{1/n}\right) = 1$, so that $\left(a^{1/n}\right)^n = a$, i.e., $a^{1/n}$ is the nth root of a.

The exponential $y = e^x$ is increasing and continuous for all x, and satisfies the identities $a^{x+y} = a^x a^y$, and $(a^x)^y = a^{xy}$ for arbitrary real x and y. By exchanging the roles of a and x in the exponential $y = a^x$, we get the related relation $y = x^a, x > 0$. In terms of exponentials and logarithms we have the formula

$$y = x^a = \left(b^{\log_b x}\right)^a = b^{a \log_b x}, \quad b > 1,$$

and in general, replacing x by pq for $pq > 0$,

$$\begin{aligned}\left(b^{\log_b pq}\right)^a &= \left(b^{\log_b p + \log_b q}\right)^a = b^{a \log_b p + a \log_b q} \\ &= \left(b^{a \log_b p}\right)\left(b^{a \log_b q}\right) = \left(b^{\log_b p}\right)^a \left(b^{\log_b q}\right)^a,\end{aligned}$$

which gives $(pq)^a = p^a q^a$.

The *exponential function* $\exp(z) \equiv e^z$, defined by

$$\exp(z) = 1 + z + \frac{z^2}{2!} + \frac{z^3}{3!} + \cdots + \frac{z^n}{n!} + \cdots, \tag{5.1.1}$$

has the following properties: (i) The series (5.1.1) converges for all z, so that e^z is an *entire function*; (ii) $\exp(1) = e$; (iii) $\exp(x) = e^x$ for x real; (iv) $\exp(iy) = \cos y + i \sin y$ for y real; (v) $\exp(a + b) = \exp(a)\exp(b)$; (vi) $\exp(z + 2n\pi i) = \exp(z)$ for every integer n; and (vii) $|\exp(z)| = e^x \neq 0$ for all z, and, by Euler's formula, $e^z = e^x(\cos y + i\, \sin y)$.

In view of the above property (vii), the equation $e^z = c$, $c \neq 0$ and complex, has infinitely many solutions for every $c \neq 0$ but no solution for $c = 0$.

5.1.1 de Moivre's Theorem. The identity $\left(e^{i\theta}\right)^n = e^{in\theta}$ in trigonometric form yields

$$(\cos\theta + i\,\sin\theta)^n = \cos(n\theta) + i\,\sin(n\theta). \tag{5.1.2}$$

Expanding the left-side of Eq (5.1.2), we obtain *de Moivre's theorem*:

$$\sum_{m=0}^{n} (i)^m \frac{n!}{m!\,(n-m)!} \left(\cos^{n-m}\theta\right)\left(\sin^m\theta\right) = \cos(n\theta) + i\,\sin(n\theta). \tag{5.1.3}$$

Then

$$\cos(n\theta) = \Re\Big\{ \sum_{m=0}^{n} (i)^m \frac{n!}{m!\,(n-m)!} \left(\cos^{n-m}\theta\right)\left(\sin^m\theta\right) \Big\}, \tag{5.1.4a}$$

$$\sin(n\theta) = \Im\Big\{ \sum_{m=0}^{n} (i)^m \frac{n!}{m!\,(n-m)!} \left(\cos^{n-m}\theta\right)\left(\sin^m\theta\right) \Big\}. \tag{5.1.4b}$$

Since $(i)^m = \begin{cases} (-1)^{m/2} & \text{when } m \text{ is even,} \\ i(-1)^{(m-1)/2} & \text{when } m \text{ is odd,} \end{cases}$, we obtain

$$\cos(n\theta) = \sum_{\substack{m=0\\ m\,\text{even}}} (-1)^{m/2} \frac{n!}{m!\,(n-m)!} \left(\cos^{n-m}\theta\right)\left(\sin^m\theta\right), \tag{5.1.5a}$$

$$\sin(n\theta) = \sum_{\substack{m=0\\ m\,\text{odd}}} (-1)^{(m-1)/2} \frac{n!}{m!\,(n-m)!} \left(\cos^{n-m}\theta\right)\left(\sin^m\theta\right), \tag{5.1.5b}$$

For example, for $n = 2$, Eq (5.1.2) becomes

$$(\cos\theta + i\,\sin\theta)^2 = \cos^2\theta - \sin^2\theta + 2i\,\sin\theta\cos\theta.$$

Thus,

$$\cos(2\theta) = \cos^2\theta - \sin^2\theta, \quad \sin(2\theta) = 2\sin\theta\cos\theta.$$

Next, we will discuss conformal maps involving the complex exponential family.

Map 5.1. The map $w = e^z$ is shown in Figure 5.1. Notice that the region $0 \leq y < 2\pi$ is mapped onto the whole w-plane except the origin ($w = 0$); so are mapped similar regions of height 2π, like $2\pi \leq z < 4\pi$ and so on, such that as the horizontal lines in the z-plane sweep out this plane, while their image rays rotate around and around infinite times in the w-plane; the y axis is mapped onto the unit circle $|w| = 1$; and the left-half z-plane maps onto the interior of the unit circle except the origin $w = 0$, while the right-half is mapped onto the exterior of the unit circle $|w| > 1$.

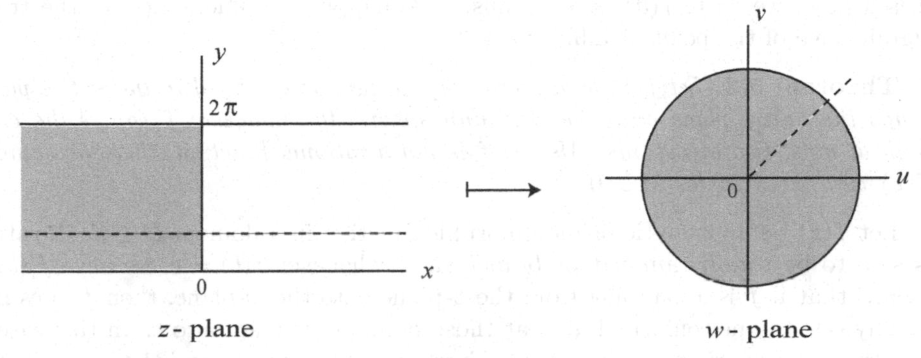

Figure 5.1 $w = e^z$.

Map 5.2. The map $z = \log w$, which is the inverse of the map e^z, defines a mapping of the whole w-plane minus the origin $w = 0$ onto the horizontal strip $0 \le y < 2\pi$ in the z-plane.[1] In polar coordinates the map $z = \log w$, $w = R\,e^{i\Theta}$, is defined by $x = \log R, y = \Theta$. The map $z = \log w$ has the following application to *geographical maps*. If the Earth is mapped by stereographic projection from the North Pole onto the plane, the South Pole is mapped onto the origin, the meridians onto rays emanating from the origin, and the latitudes onto circles centered at the origin. However, if this mapping is in polar coordinates, the meridians map onto horizontal lines and the parallels onto vertical line-segments. Just as both stereographic projection and the complex logarithm preserve angles, so does their composition. Thus, a rotation by 90° in the positive (counterclockwise) direction will result in a mapping of the meridians onto vertical lines, and of parallels onto horizontal line-segments, thus yielding a cylindrical projection and a conformal map, which is the same as Mercator's projection. Hence, the Mercator's projection is obtained from a stereographic projection followed by the complex logarithm mapping and a rotation through 90° in the positive direction. The following three results hold for entire functions:

Theorem 5.1. (Picard's theorem) *Let $f(z)$ be a nonconstant entire function. Then the equation $f(z) = c$ has a solution for every complex number c with at most one exception.*

Theorem 5.2. (Picard's 'big' theorem) *Let $f(z)$ be an entire function which is not a polynomial. Let M be a positive real number. Then for every complex number c, the equation $f(z) = c$ has a solution for $|z| > M$ with at most one exception.*

Corollary 5.1. *For an entire function $f(z)$ which is not a polynomial, the equation $f(z) = c$ has an infinite number of solutions for all values of c with at most one exception.*

Riemann [1851], using the stereographic projection from the North Pole, assumed this Pole on the sphere to be mapped onto the point at infinity. If a is a pole of f, then, as $z \to a$, the function $f(z)$ composed with the inverse of stereographic projection tends to the North Pole, so that a meromorphic function in a domain $D \in \mathbb{C}$ is regarded as a continuous

[1] This is one of the branches of the logarithm function, and the following conclusions are restricted to this branch only.

image of D onto the sphere. The map at the North Pole is continuous and behaves just like at any other point on the sphere. Thus, at a simple pole where $1/(z-a)$ occurs, the mapping onto the sphere is conformal at a. If higher powers of $1/(z-a)$ occur, the same is true as at any point a where f is analytic but $f' = 0$. The unit sphere in this sense as the extended complex plane via stereographic projection is called the *Riemann sphere*. If a is a pole, we write $f(a) = \infty$. Thus, this mapping technique defines the true mapping significance of the point at infinity.

Theorem 5.3. *Let $f(z)$ be a meromorphic function defined in the entire plane. If $f(z)$ maps the entire plane onto the Riemann sphere, the image of f covers the entire sphere with at most two exceptions. Also, if f is not a rational function, then the same is true for $f(z)$ with $|z| > M$ for $M > 0$.*

Let $f(z)$ be an analytic or meromorphic function in a domain $D \in \mathbb{C}$. A value c of $f(z)$ is said to be *totally ramified*, or *branched*, if, whenever $f(b) = c$, we have $f'(b) = 0$. This means that if f is a mapping from the z-plane onto the w-plane, then f does not define a locally one-to-one conformal map at those points that map onto c. In this case f behaves like the function z^n, $n > 1$ in the neighborhood of the origin, and the image is 'ramified' or 'branched' in a neighborhood of c.

Theorem 5.4. (Nevanlinna [1925]) *An entire function in the z-plane can have at most two totally ramified values, whereas a meromorphic function in the z-plane can have at most four totally ramified values.*

Examples that both 2 and 4 numbers are sharp are: (i) Complex sine and cosine functions are entire functions defined by $2\cos z = e^{iz} + e^{-iz}$, $2i \sin z = e^{iz} - e^{-iz}$, respectively; thus, $\sin^2 z + \cos^2 z = 1$ and $\cos z = \dfrac{d \sin z}{dz}$. Thus, $\sin z = \pm 1$ iff $\cos z = 0$, which shows that the two values ± 1 are totally ramified for $f(z) = \sin z$. Similarly, as Nevanlinna [1925] has shown, Weierstrass's elliptic function $\wp(z)$ has four totally ramified values.

Theorem 5.5. (Bloch [1925]) *Let $w = f(z)$ be analytic in the unit disk and be normalized so that $f'(0) = 0$. Then there exists a constant $B > 0$ with the following property: For every $r < B$ there exists a disk of radius r in the w-plane that is the one-to-one conformal image under f of a domain inside the unit disk in the z-plane.*

The largest value of B is known as *Bloch's constant*; its precise value is unknown.

Corollary 5.2. *Let an arbitrary nonconstant entire function $w = f(z)$ define the mapping of the z-plane onto the w-plane such that for every $R > 0$ there is a disk of radius R in the w-plane that is the one-to-one image under f of some domain in the z-plane.*

5.2 Specific Cases of Mappings

Map 5.3. In $\mathbb{C}$, the exponentials are defined by $w = e^z$, or its inverse $z = \log w$. The critical point for $w = e^z$ is at $z = \infty$, and for $z = \log w$ at $w = 0; \infty$. Assuming that p, q, c, R, and θ are real, $R > 0$, and $k = 0, \pm 1, \pm 2, \dots$, we have the following different cases of conformal mappings under the exponentials:

(i) The line segment $x = p$, $c \le y < c + 2\pi$, in the z-plane is mapped onto the circle $|w| = e^p$ in the w-plane.

(ii) The lines $y = 2k\pi$ and $y = (2k+1)\pi$ in the z-plane is mapped onto the half-lines $v = 0$, $0 < u < \infty$ and $v = 0$, $-\infty < u < 0$, respectively, in the w-plane.

(iii) The line $y = q + 2k\pi$, $-\infty < x < \infty$, in the z-plane is mapped onto the half-line

$\arg\{w\} = q$, $0 < |w| < \infty$.

(iv) The infinite strip $2k\pi < y < 2(k+1)\pi$ in the z-plane is mapped onto the plane cut along the positive real axis in the w-plane.

(v) The infinite strip $(2k-1)\pi < y < (2k+1)\pi$ in the z-plane is mapped onto the plane cut along the negative real axis in the w-plane.

(vi) The infinite strip $c < y < c + 2\pi$ in the z-plane is mapped onto the plane cut along the half-line $\arg\{w\} = c$ in the w-plane.

These mappings are shown in Figure 5.2.

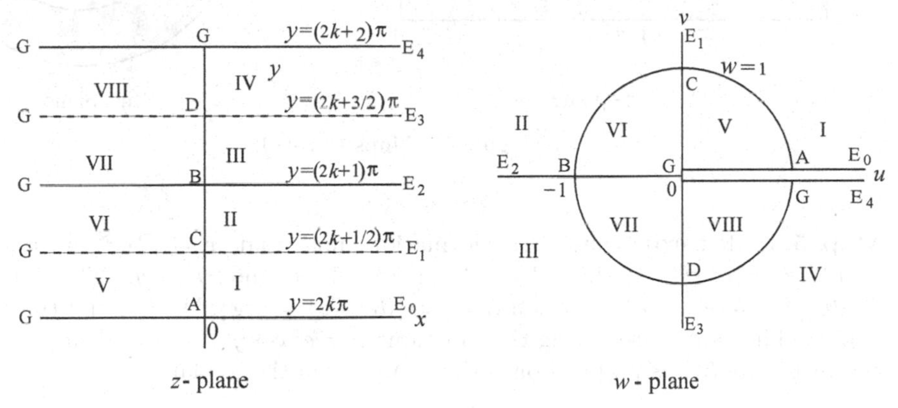

Figure 5.2 Exponential mappings.

Further cases of mappings under this transformation are as follows:

(vii) The curve $\cos(y - c) = p\,e^{-x}$, where $\log|p| \gtrless x < \infty$, and

$$\begin{cases} (2k - \frac{1}{2})\pi < y - c < (2k + \frac{1}{2})\pi, & \text{for } p > 0, \\ (2k + \frac{1}{2})\pi < y - c < (2k + 1 + \frac{1}{2})\pi, & \text{for } p < 0, \end{cases}$$

which are the asymptotes

$$\begin{cases} y = c + (2k \pm \frac{1}{2}) & \text{for } p > 0, \\ y = c + (2k + 1 \pm \frac{1}{2})\pi & \text{for } p < 0, \end{cases}$$

in the z-plane, is mapped onto the line $\Re\{w\,e^{-ic}\} = p$, $p \geq 0$, in the w-plane.

(viii) The line $x + my = p$, $m \gtrless 0$, in the z-plane is mapped onto the spiral $R = e^{-p}\,e^{-m\Theta}$, $w = Re^{i\Theta}$ in the w-plane.

(ix) The rectangle bounded by $y = q_1, q_1$ and $x = p_1, p_2$, $0 < q_2 - q_1 < 2\pi$, in the z-plane is mapped onto the part bounded by $\arg\{w\} = q_1$ and $\arg\{w\} = q_2$, of the annular region bounded by $|w| = e^{p_1}$ and $|w| = e^{p_2}$ in the w-plane.

(x) The rectangle in (ix) but with $q_2 - q_1 = 2\pi$ in the z-plane is mapped onto the annular region bounded by $|w| = e^{p_1}$ and $|w| = e^{p_2}$ cut along $\arg\{w\} = q_1$ in the w-plane.

The above details from (vii-x) are presented in Figure 5.3.

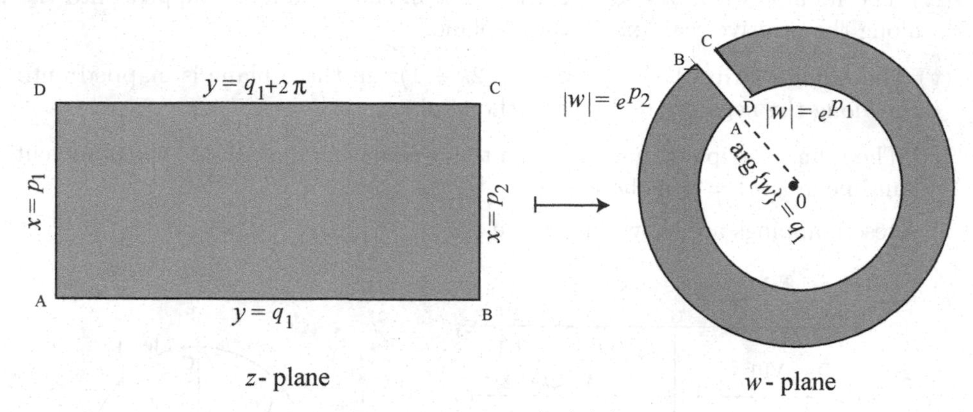

Figure 5.3 Maps (vii)-(x).

Map 5.4. For exponential transformation, note that $w = R\,e^{i\Theta}$; $z = x + iy$; thus, $e^z = e^{x+iy} = e^x\,e^{iy}$, so that $R = e^x$, or $x = \ln R$, and $\Theta = y$. The magnification is $m = |dw/dz| = |e^z| = e^x$; rotation $\alpha = \arg\{dw/dz\} = \arg\{e^z\} = y$; and the critical point is $z = \infty$. The grid representing the functions $u = e^x \cos y$ and $v = e^x \sin y$ in the z-plane, shown in Figure 5.4, is mapped onto the rectangle in the w-plane.

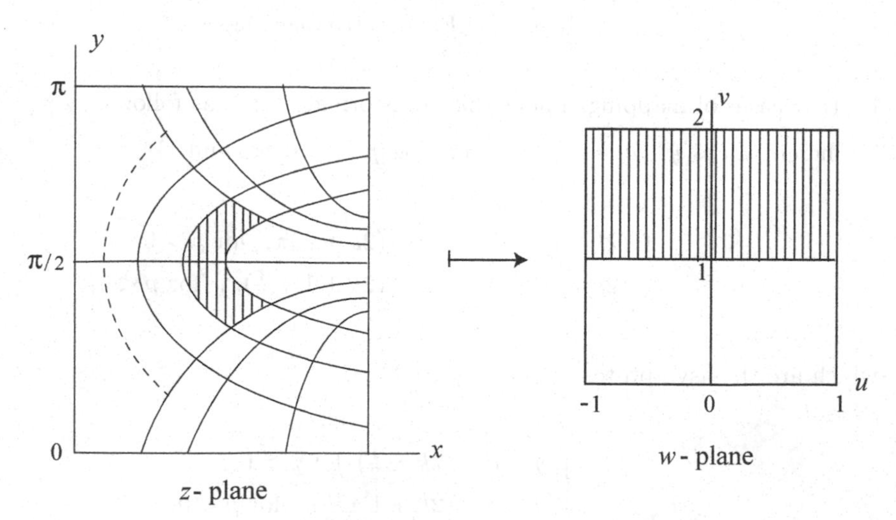

Figure 5.4 $w = e^z$.

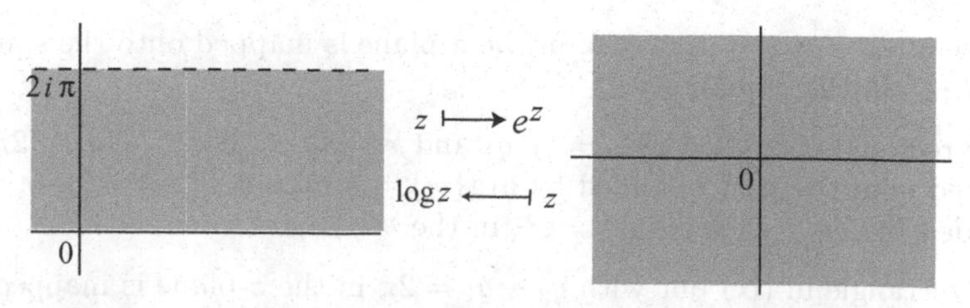

Figure 5.5 $w = e^z$ and $z = \log w$.

The transformation $w = e^z$ and its inverse $z = \log w$ map the upper half-plane $0 \le y < 2\pi$ in the z-plane onto the entire w-plane, as shown in Figure 5.5.

Map 5.5. The transformation $w = e^z$ maps the infinite strip in the z-plane onto the upper half-plane $\Im\{w\} > 0$, as shown in Figure 5.6.

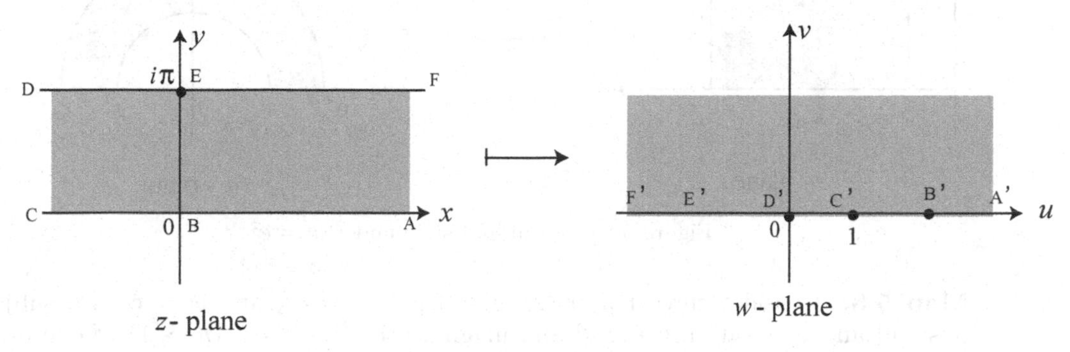

Figure 5.6 Infinite strip under $w = e^z$.

Map 5.6. The transformation $w = e^z$ maps the half-strip $-\infty < x < 0$ in the z-plane onto the unit semi-circle in the w-plane, as shown in Figure 5.7.

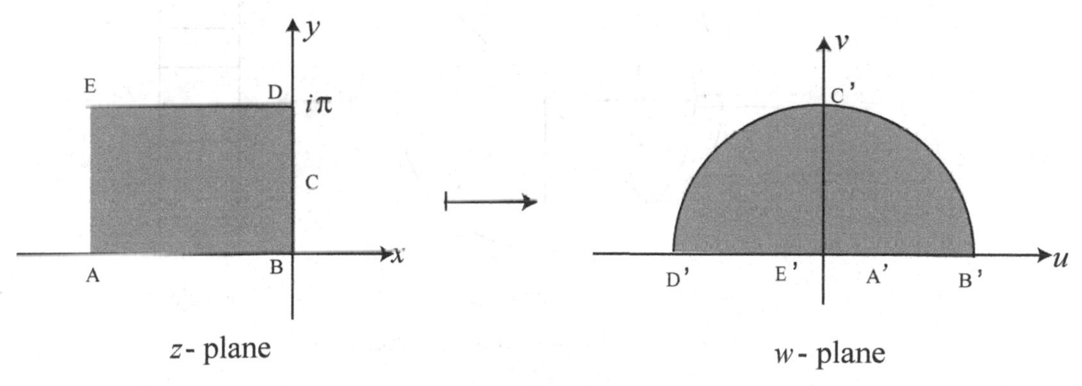

Figure 5.7 Half-strip under $w = e^z$.

Map 5.7. The transformation $w = e^z$ maps the vertical half-strip $0 < y < \infty$ in the z-plane onto the annulus in the upper half of the w-plane, as shown in Figure 5.8.

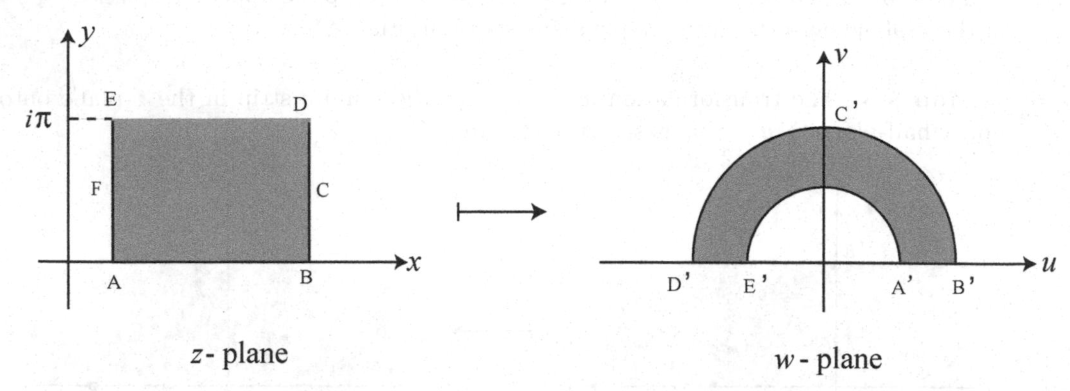

Figure 5.8 Vertical half-strip under $w = e^z$.

Map 5.8. Consider three mappings $w = \log z$, $w = e^z$, and $w = \cosh z$, subjected to a simultaneous rotation $\theta \mapsto n\theta$ and magnification $r \mapsto nr^n$ $(n > 1)$ of one of both boundaries (or compression if $n < 1$), where $\theta = \arg\{z\}$. These three mappings are shown in Figure 5.9(a)-(c), where the curvilinear grids are transformed into rectangular grids, and the arrows indicate the direction of the simultaneous rotation and magnification.

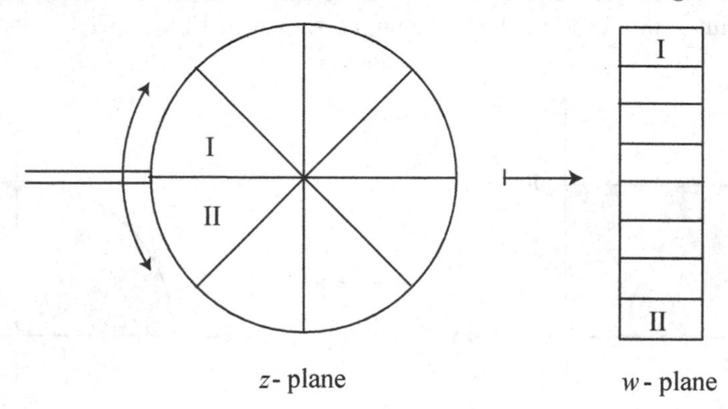

Figure 5.9(a) $w = \log z$.

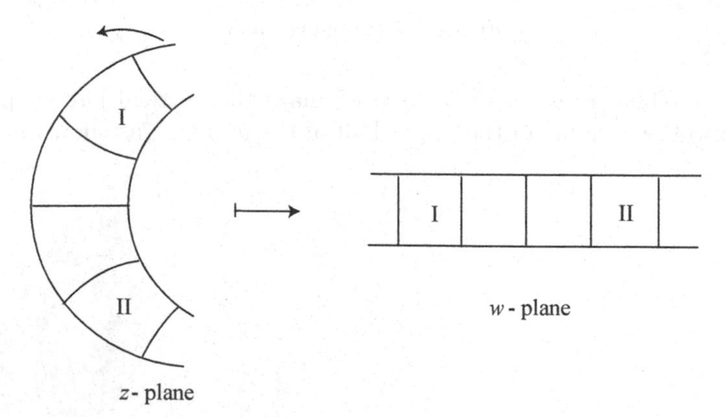

Figure 5.9(b) $w = e^z$.

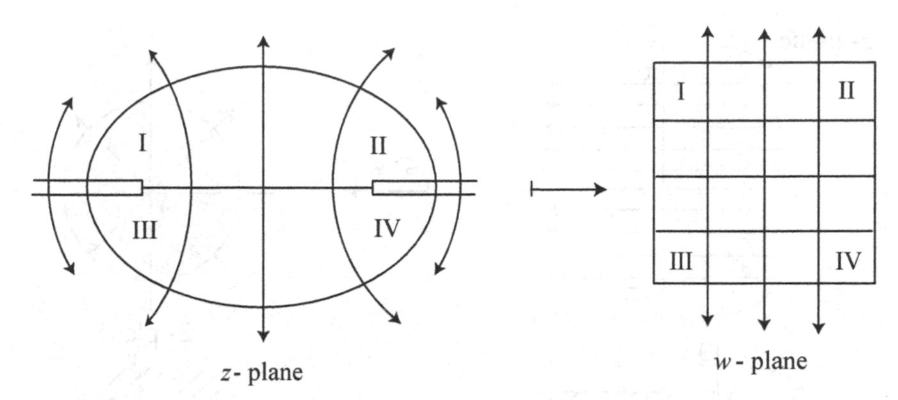

Figure 5.9(c) $w = \cosh z$.

Map 5.9. We will consider functions related to e^z and e^{iz}.

(i) The function $\zeta = e^z$ maps the left-half and the right-half of the z-plane onto the interior and the exterior, respectively, of the unit circle in the ζ-plane (see Figure 5.10(a)).

(ii) The function $w = \dfrac{\zeta + \zeta^{-1}}{2} = \cosh z$ maps the first and the fourth quadrant of the z-plane onto the upper and lower half, respectively, of the w-plane slit from -1 to 1 on the real axis (see Figure 5.10(b)).

(iii) The mapping by the function $w = \dfrac{\zeta - \zeta^{-1}}{2} = \sinh z$ is shown in Figure 5.10(c).

(iv) Let $\zeta = e^{iz}$. Then the mapping by the function $w = \dfrac{\xi + \xi^{-1}}{2} = \cos z$ is shown in Figure 5.10(d).

(v) The mapping of the function $w = \dfrac{\xi - \xi^{-1}}{2i} = \sin z$ is shown in Figure 5.10(e). ■

Map 5.10. The transformation that maps an infinite strip of width a in the z-plane onto the upper half-plane $\Im\{w\} > 0$ is defined by

$$w = e^{\pi z/a}, \tag{5.2.1}$$

and is presented in Figure 5.10.

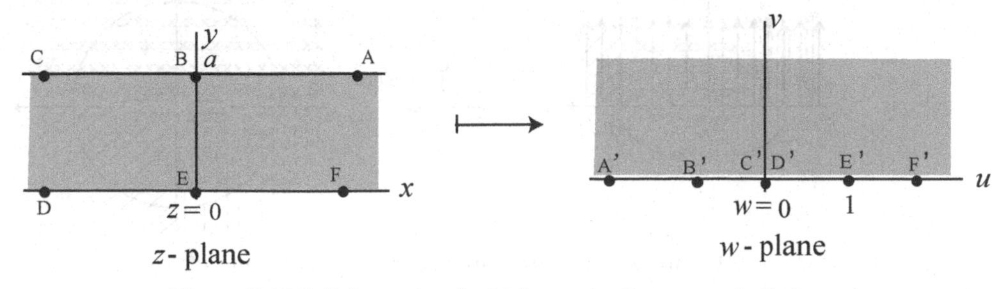

Figure 5.10 Infinite strip of width a onto the upper half-plane.

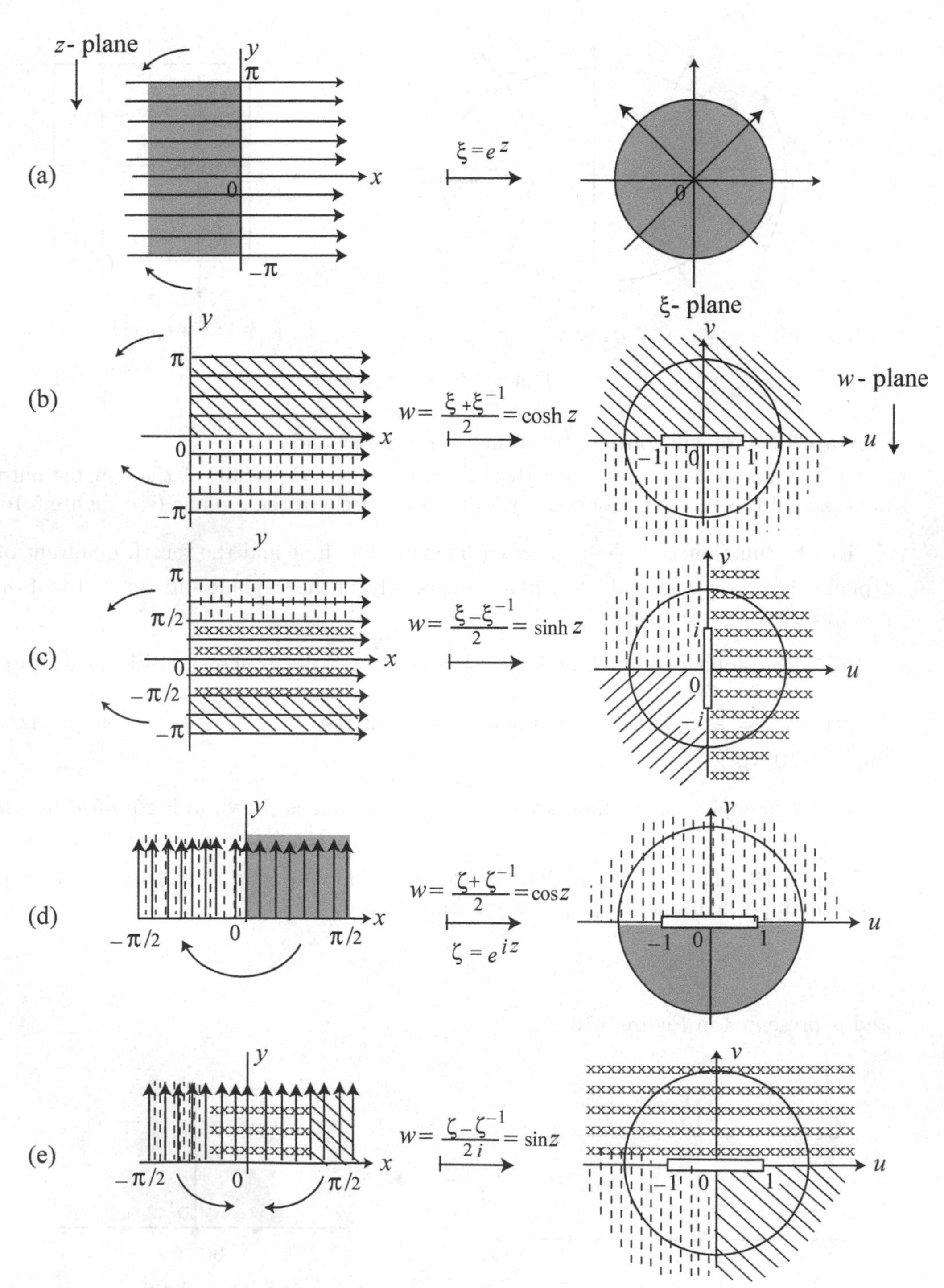

Figure 5.11 (a)-(e) Mappings related to e^z and e^{iz}.

Map 5.11. Mappings composed of the transformation $w = e^{az}$ are presented in Figure 5.11(a)-(e) for different mapping functions: (a) $w = e^z$; (b) $w = \cosh z$; (c) $w = \sinh z$; (d) $w = \cos z$; and (e) $w = \sin z$.

Map 5.12. The transformation $w = e^{\pi z/a}$ maps the infinite strip of width a: $-\infty < x < \infty, 0 \le y \le a$, in the z-plane onto the upper half-plane $\Im\{w\} \ge 0$ in the w-plane.

Maps 5.13. The transformation $w = e^{az} - ce^{bz}$, where a, b, c are real and $a > b > 0, c > 0$, has critical points $z = \infty$; $\dfrac{\log bc - \log a + 2k\pi i}{a-b}$, $k = 0, \pm 1, \pm 2, \dots$.

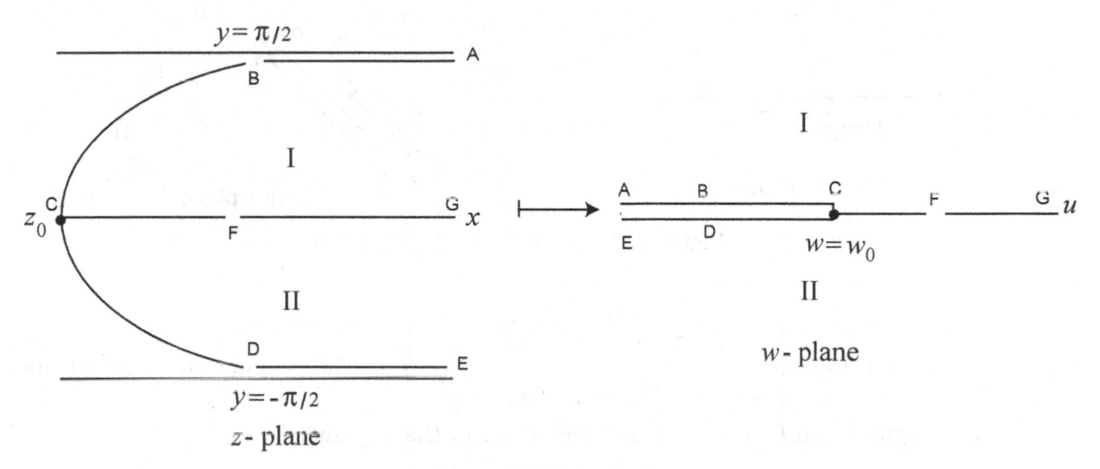

Figure 5.12 Map 5.13.

For $-\pi/a < y < \pi/a$, the details of this transformation are as follows:

(a) The point $z_0 = \dfrac{\log(bc/a)}{a-b}$ in the z-plane is mapped onto the point $w = -\dfrac{a-b}{b}\left(\dfrac{bc}{a}\right)^{a/(a-b)}$ in the w-plane.

(b) The half-line $z_0 \le x < \infty$ in the z-plane is mapped onto the half-line $w_0 \le u < \infty$, $v = 0$, in the w-plane.

(c) The curve $\dfrac{\sin ay}{\sin by} = c\, e^{(b-a)x}$ (ABCDE) for $-\pi/a < y < \pi/a$ with asymptotes $y = \pm\pi/a$ in the z-plane is mapped onto the half-line $-\infty < u \le w_0$ in the w-plane. In particular, if $a = 2b = c = 2$, then the curve becomes $\cos y = e^{-x}$, and in this case it is mapped onto the same half-line $-\infty < u \le w_0$ but counted twice.

These details are presented in Figure 5.12.

Map 5.14. The mapping $w = e^{\alpha z} - e^{(\alpha-1)z}$, $0 < \alpha < 1$, is a special case of the mapping in Map 5.15. It has critical points $z = \infty$; $\log \dfrac{1-\alpha}{\alpha} + (2k+1)i\pi$, where $k = 0, \pm 1, \pm 2, \dots$. Setting $b = e^{\alpha}(1-\alpha)^{\alpha-1}$, the details are as follows:

(a) The points $z = 0; \log \dfrac{1-\alpha}{\alpha} + i\pi; \log \dfrac{1-\alpha}{\alpha} - i\pi$ in the z-plane are mapped onto the points $w = 0; be^{i\pi\alpha}; be^{-i\pi\alpha}$, respectively, in the w-plane.

(b) The half-lines $\left\{\begin{array}{l} y=\pi, \log\dfrac{1-\alpha}{\alpha}\le x<\infty \\ y=\pi, -\infty<x\le \log\dfrac{1-\alpha}{\alpha} \end{array}\right\}$ in the z-plane are mapped onto the line segment $\arg\{w\}=\alpha\pi$, $v\le|w|<\infty$, in the w-plane.

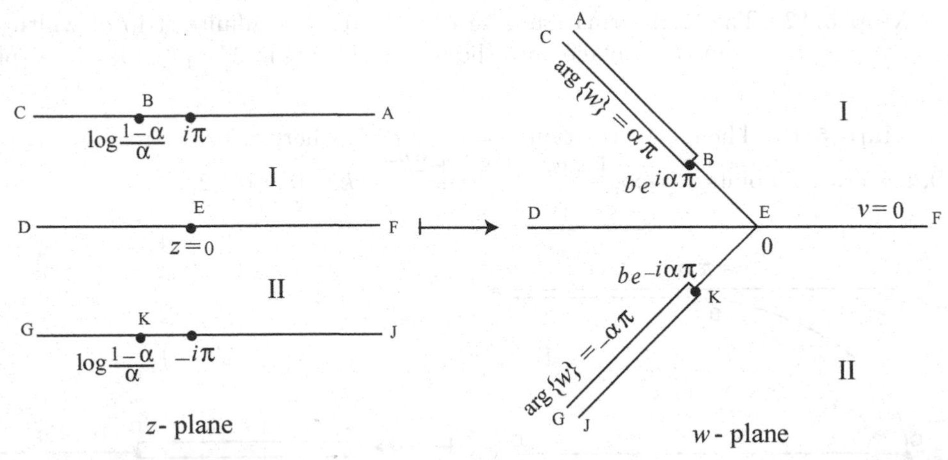

Figure 5.13 $w=e^{\alpha z}-e^{(\alpha-1)z}, 0<\alpha<1$.

(c) The half-lines $\left\{\begin{array}{l} y=-\pi, \log\dfrac{1-\alpha}{\alpha}\le x<\infty \\ y=-\pi, -\infty<x\le \log\dfrac{1-\alpha}{\alpha} \end{array}\right\}$ in the z-plane are mapped onto the line segment $\arg\{w\}=-\alpha\pi$, $v\le|w|<\infty$, in the w-plane.

(d) The line $y=0, -\infty<x<\infty$ in the z-plane is mapped onto the line $v-0, -\infty<x<\infty$, in the w-plane. These details are shown in Figure 5.13.

Map 5.15. The mapping $w=ae^{fz}+be^{gz}$, $f/g<0$ real, and $ab\ne 0$, is a composite of $w=ae^{-fc}\xi, z=\dfrac{\zeta}{f-g}, \xi=e^{\alpha\zeta}-e^{(\alpha-1)\zeta}$, where $c=\dfrac{\log(-b/a)}{g-f}, \alpha=\dfrac{f}{f-g}, 0<\alpha<1$.

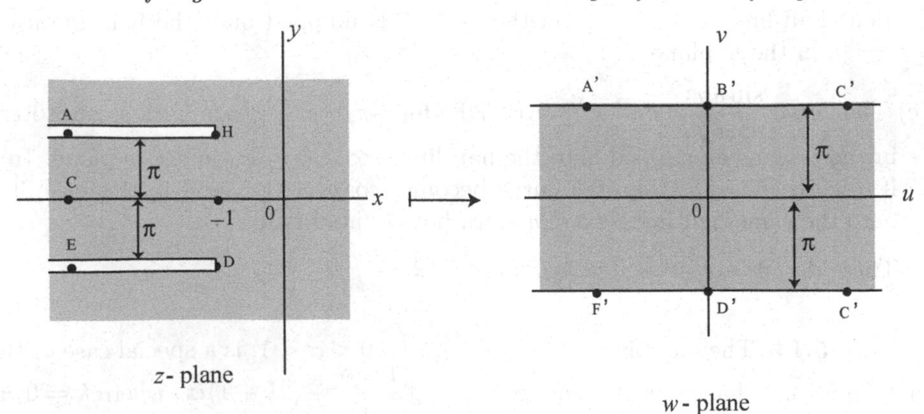

Figure 5.15 Plane with two semi-infinite cuts onto the interior of the infinite strip.

Map 5.16. The transformation that maps the z-plane with two semi-infinite cuts in the

z-plane onto the interior of the infinite strip in w-plane is defined by

$$w = z + e^z, \tag{5.2.2}$$

and is presented in Figure 5.15.

Map 5.17. The transformation $w = \alpha\, e^{az} + \beta\, e^{bz}$, where a/b is real, and $a/b > 1, \alpha\beta \neq 0$, is a composite of the following mappings: $w = \alpha\, e^{a\lambda}\xi$, $z = e^{-i\arg\{\alpha\}}\zeta + \lambda$, and $\xi = e^{|\alpha|\zeta} - e^{|b|\zeta}$, where $\lambda = \dfrac{\log(-\beta/\alpha)}{a-b}$. These different mappings are discussed in previous cases.

Map 5.18. The transformation that maps the interior of the parabola $y^2 = 4p(p-x)$ in the z-plane onto the upper half-plane $\Im\{w\} > 0$ is defined by

$$w = e^{i\pi\sqrt{z/p}}, \tag{5.2.3}$$

and is presented in Figure 5.16.

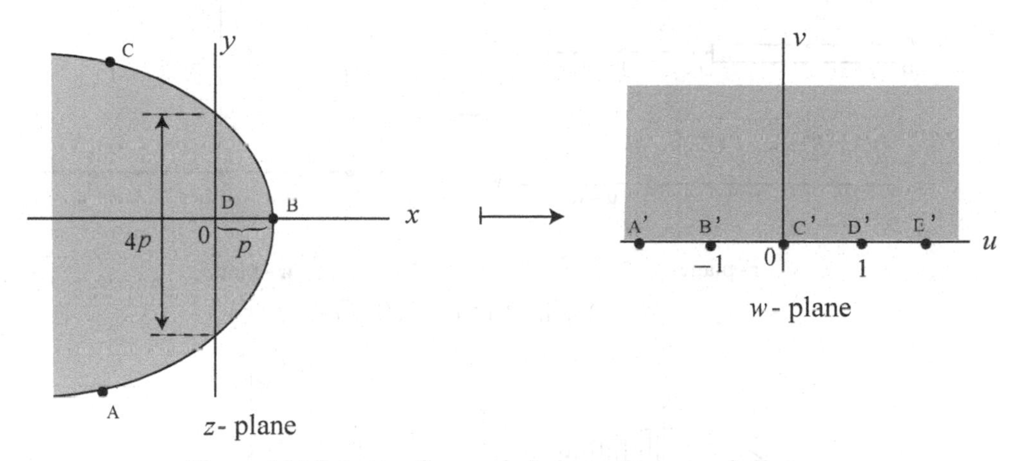

Figure 5.16 Interior of a parabola onto the upper half-plane.

Map 5.19. The transformation $w = \sqrt{1+e^z}$ or $z = \log(w+1)(w-1)$ has the critical points $z = (2k+1)\pi i; \infty$, where $k = 0, \pm 1, \pm 2, \ldots$. The details under this mapping are as follows:

(i) The points $z_0 + 2k\pi i$ in the z-plane are mapped onto the points $w_0 = \pm\sqrt{1+e^{z_0}}$.

If $0 \le y \le 2\pi$, then $v \ge 0$, and the details are as follows:

(ii) The strip $0 < y < 2\pi$ cut along $y = \pi$, $x \le 0$, in the z-plane is mapped onto the half-plane $v > 0$.

(ii) The line $y = 0$ or $y = 2\pi$, respectively, where $-\infty < x < \infty$, in the z-plane is mapped onto half-line $v = 0, 1 < u < \infty$ or $-\infty < u \le -1$, respectively, in the w-plane.

(iii) The half-line $y = x$, $-\infty < x \le 0$, in the z-plane is mapped onto the segments $\begin{Bmatrix} v = 0, 0 \le u < 1 \\ v = 0, 1 < u \le 0 \end{Bmatrix}$ in the w-plane.

(iv) The half-line $y = x, 0 \le x < \infty$ in the z-plane is mapped onto the half-line $u = 0, 0 \le v < \infty$ in the w-plane.

(v) The line segment $x = p, 0 < y < 2\pi$, $p > 0$, in the z-plane is mapped onto the part $v > 0$ of Cassini's oval $|w+1||w-1| = e^p$ in the w-plane.

(vi) The line segment $x = p, 0 < y < \pi$, $p \leq 0$, in the z-plane is mapped onto the part $u > 0, v > 0$ of Cassini's oval $|w+1||w-1| = e^p$ in the w-plane.

(vii) The line segment $x = p, \pi < y < 2\pi$, $p \leq 0$, in the z-plane is mapped onto the part $u < 0, v > 0$ of Cassini's oval $|w+1||w-1| = e^p$ in the w-plane.

(viii) The line $y = q$, $-\infty < z < \infty$, $q \neq \pi$, $0 < q < 2\pi$, in the z-plane is mapped onto the part of the rectangular hyperbola $\Re\{w^2/\left(1 - e^{2iq}\right)\} = \frac{1}{2}$ in the w-plane.

(ix) The line $y = q, 0 < q < \pi$, in the z-plane is mapped onto the part $u > 0, v > 0$ of the hyperbola in the w-plane.

(x) The line $y = q, \pi < q < 2\pi$, in the z-plane is mapped onto the part $u < 0, v > 0$ of the hyperbola in the w-plane.

These details are presented in Figure 5.17. For the general case, see Map 5.26.

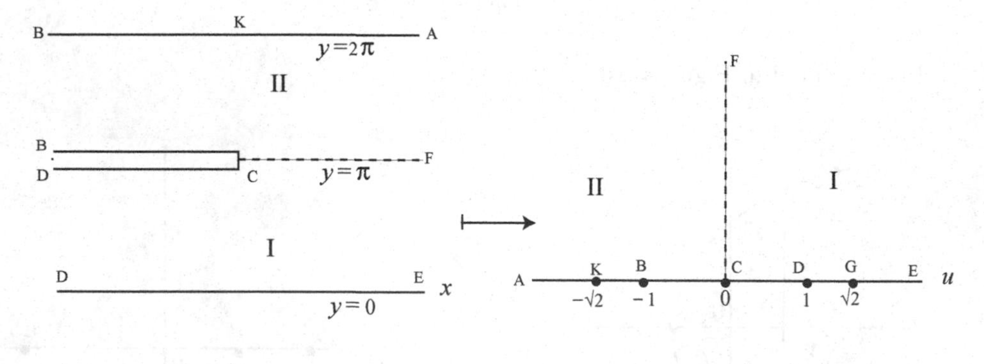

Figure 5.17 $w = \sqrt{1 + e^z}$.

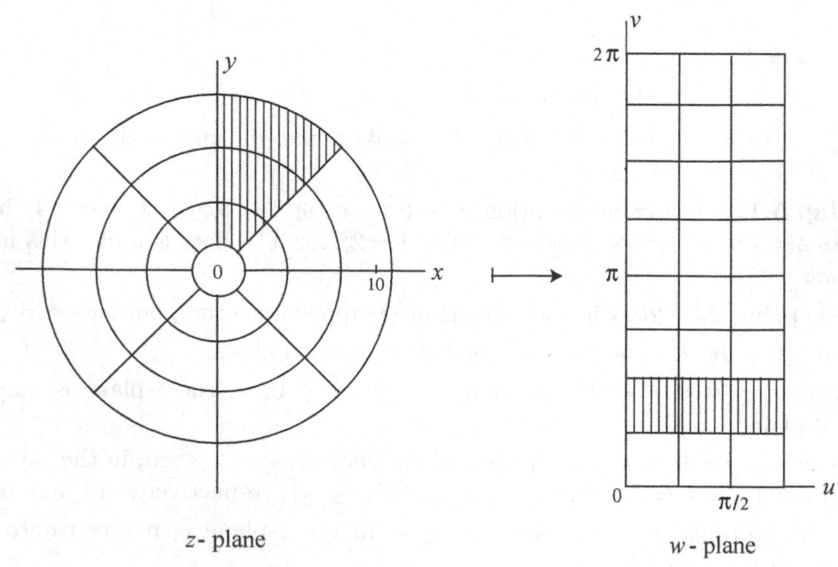

Figure 5.18 $w = \log z$ in polar coordinates.

Map 5.20a. The logarithmic transformation $w = \log(az + b)$, $a > 0, b \geq 0$, represents a single-valued complex velocity function with a nonzero circulation integral, discussed in the case of the Blasius theorem (see §18.10.1). Notice that the derivative $w'(z) = \dfrac{a}{az+b}$ is a single-valued function away from the singularity at $z = -b/a$.

Map 5.20b. The previous logarithmic transformation for $a = 1$ and $b = 0$ gives $w = \log z$, where $w = u + iv$. Since $z = r\, e^{i\theta}$, we have $\log z = \log\left(r\, e^{i\theta}\right)$, so that $\log z = \ln r + i\theta$, $u = \ln r$, and $v = \theta$. The magnification is $m = |dw/dz| = |1/z| = 1/r$; rotation is $\alpha = \arg\{dw/dz\} = -\theta$. The critical points are $z = 0; \infty$. This transformation is shown in Figure 5.18 in polar coordinates.

Map 5.20c. The logarithmic transformation $w = \log z^a = a \log z$, $a \geq 1$, or $a < 0$, produces the mapping properties of the transformation $w = \log z$, namely, the magnification or contraction depending on whether a is positive or negative. For an application, see §22.3.4 (Figure 22.6).

Map 5.21. The mapping $w = \sec^2 z$, or $z = \arccos(1/\sqrt{w})$, or $w = \dfrac{2}{1 + \cos 2z}$ (see Map 5.48 for a general case), has the same critical points as in Map 5.52 (i.e., at $w = \sec z$). The details of this mapping are obvious and are presented in Figure 5.19.

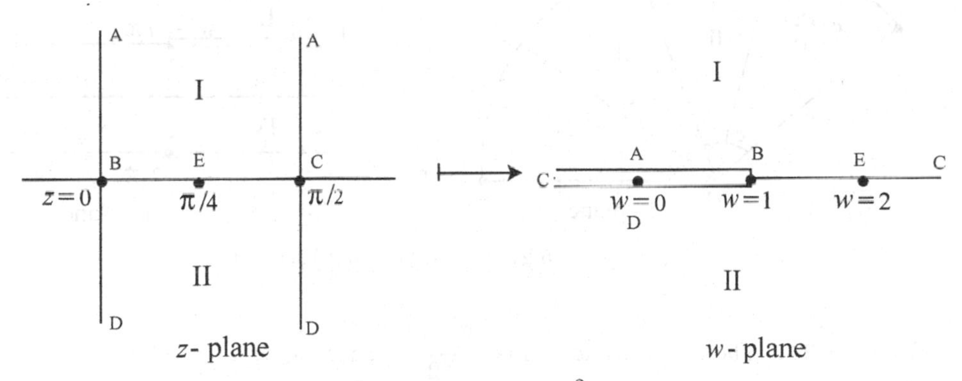

Figure 5.19 $w = \sec^2 z$.

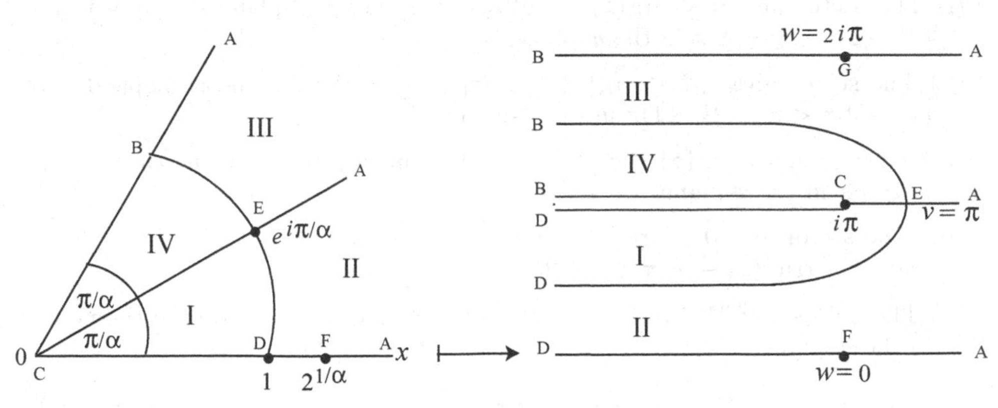

Figure 5.20 $w = \log\left(z^{\alpha} - 1\right), \alpha > 1$.

Map 5.22. The mapping $w = \log(z^\alpha - 1)$, $\alpha > 1$, or $z = (e^w + 1)^{1/\alpha}$, has critical points $z = 0; \infty; e^{2ki\pi/\alpha}$. For $\alpha = 2$, see Map 5.20. Let $k, m = 0, \pm 1, \pm 2, \dots$. Then the details of this mapping are as follows:

(i) The point $z = z_0 e^{2ki\pi/\alpha}$ in the z-plane is mapped onto the point $w = w_0 + 2mi\pi$ in the w-plane.

(ii) If $0 \le \arg\{z\} \le 2\pi/\alpha$, then $0 \le v \le 2\pi$. Then

(a) The points $z = 0; 1; e^{i\pi/\alpha}; e^{1+i\pi/\alpha}; 2^{1/\alpha}; 2^{1/\alpha}e^{2i\pi/\alpha}$ (marked G) in the z-plane are mapped onto the points $w = i\pi; \infty; \log 2 + i\pi; \infty; 0; 2i\pi$, respectively, in the w-plane.

(b) The arc $0 < \theta < 2\pi/\alpha$ of the circle $z = e^{i\theta}$ in the z-plane is mapped onto the curve $2\cos v = -e^{-u}$, $u \le \log 2$, $|v - \pi| < \pi/2$, in the w-plane.

(c) The half-line $\arg\{z\} = \pi/\alpha$ in the z-plane is mapped onto the half-line $v = \pi$, $u \ge 0$.

These details are shown in Figure 5.20.

Map 5.23. The mapping $w = \log(z^\alpha + a)$, $\alpha > 1$, and $a \ne 0$, is a composite of $z = (-a)^{1/\alpha}\zeta$, $w = \xi + \log(-a)$, and $\xi = \log(\zeta^\alpha - 1)$.

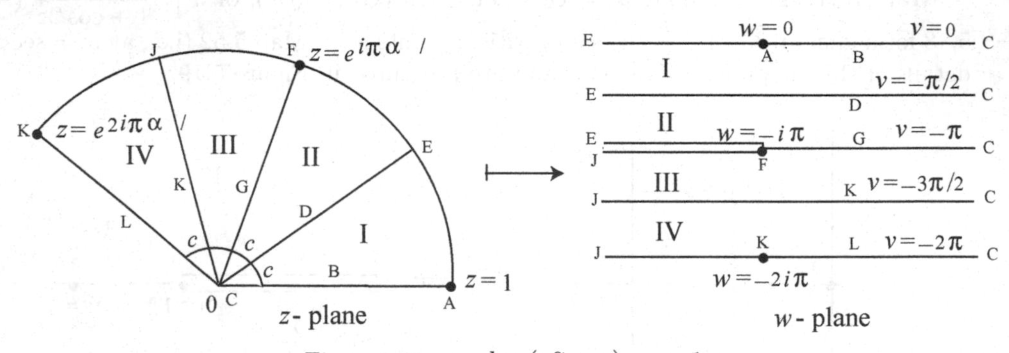

Figure 5.21 $w = \log(z^\alpha + a)$, $\alpha > 1$.

Map 5.24. The mapping $w = \log \dfrac{z^\alpha + z^{-\alpha}}{2}$, $\alpha > 0$, or $z = \left(e^w \pm \sqrt{e^{2w} - 1}\right)^{1/\alpha}$, has critical points $z = 0; \infty; e^{ki\pi/\alpha}$, where $k = 0, \pm 1, \pm 2, \dots$. Let $c = \pi/(2\alpha)$. Then the details of this mapping are as follows:

(i) The sector area $0 < \arg\{z\} < c/2, |z| < 1$, in the z-plane is mapped onto the strip $(2k-1)\pi < v < 2k\pi$ in the w-plane.

(ii) The sector area $c/2 < \arg\{z\} < c, |z| < 1$, in the z-plane is mapped onto the strip $(2k-2)\pi < v < (2k-1)\pi$ in the w-plane.

(iii) The region $0 < \arg\{z\} < c, |z| < 1$, in the z-plane is mapped onto the strip $\left(2k + \frac{1}{2}\right)\pi < v < 2k\pi$ in the w-plane.

(iv) The sector area $0 < \arg\{z\} < mc/2, |z| < 1$, $m = 1, 2, \dots$, in the z-plane is mapped onto the strip $(2k - m)\pi < v < 2k\pi$ in the w-plane.

(v) The points $z_0 e^{2ki\pi/\alpha}$ and $z_0 e^{2ki\pi/\alpha}$ in the z-plane is mapped onto the point $w_0 + 2ki\pi$ in the w-plane.

(vi) Also, the isolated points $e^{i\pi/(3\alpha)}; e^{2i\pi/(3\alpha)}; e^{i\pi/a}; e^{4i\pi/(3\alpha)}; e^{5i\pi/(3\alpha)}; e^{i\pi/(2c)}\left(\dfrac{\sqrt{5}-1}{2}\right)^{1/\alpha}$;

$e^{3i\pi/(2c)}\left(\frac{\sqrt{5}-1}{2}\right)^{1/\alpha}$; $e^{i\pi/(2\alpha)}\left(\sqrt{2}-1\right)^{1/\alpha}$; $e^{3i\pi/(2\alpha)}\left(\sqrt{2}-1\right)^{1/\alpha}$ in the z-plane map onto the points $-\log 2$; $-\log 2 - i\pi$; $i\pi$; $-\log 2 - i\pi$; $-\log 2$; $-\log 2 - i\pi/2$; $-\log 2 - 3i\pi/2$; $-i\pi/2$; $-3i\pi/2$, respectively, in the w-plane.

These details are presented in Figure 5.21.

In the case when 2α is a positive integer m, then, setting $b = e^{i\pi/\alpha}$ in the z-plane so that $-m\pi \le v \le 0$ in the w-plane,

(vii) the cut interior of the circle in Figure 5.22 in the z-plane is mapped onto the strip with $m-1$ slits in the w-plane.

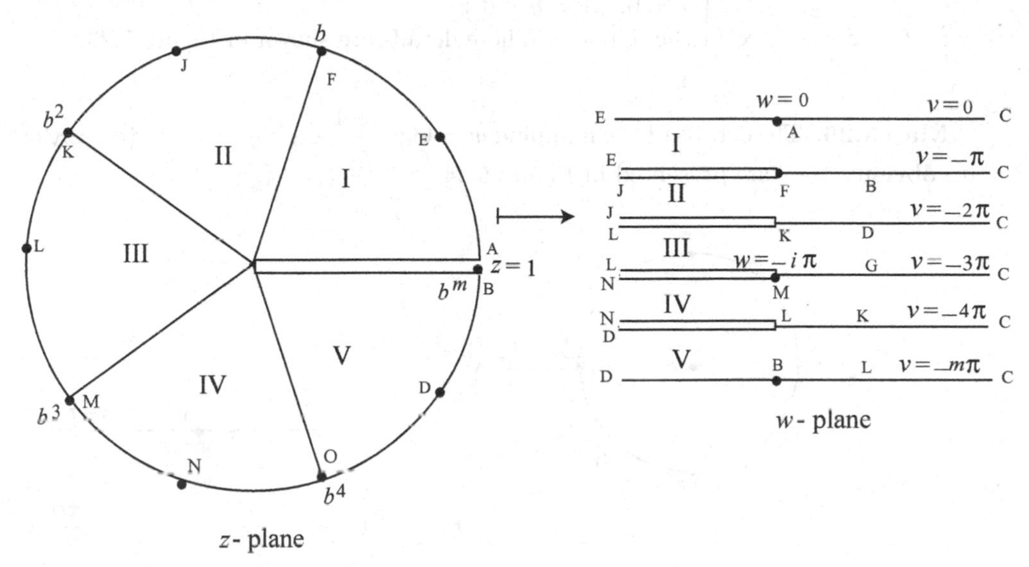

Figure 5.22 Map 5.24, $2\alpha = m$.

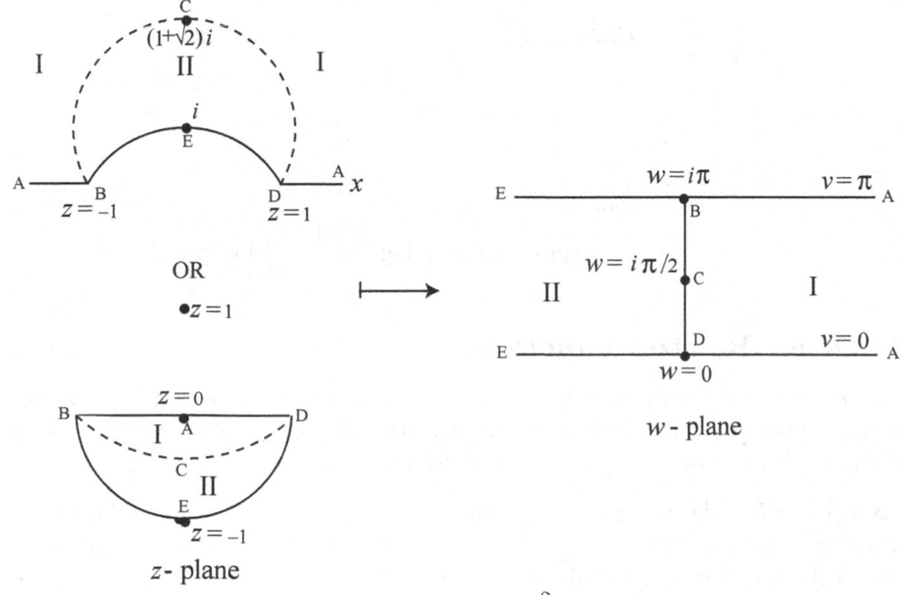

Figure 5.23 $w = \log \dfrac{z^2+1}{2z}$.

Map 5.25. The mapping $w = \log \dfrac{z^2+1}{2z}$, or $z = e^w \pm \sqrt{(e^{2w}-1)}$, is a special case of the mapping 5.22 for $m = 2\alpha = 2$ with one-slit strip (Figure 5.22). An alternative method is as follows: Let $0 \le v \le \pi$. Then,

(a) The points $z = 1; i\left(1 \pm \sqrt{2}\right)$ (marked C); -1 in the z-plane are mapped onto the points $w = 0; i\pi/2; i\pi$, respectively, in the w-plane.

(b) The arc DCB of the circle $|z-1| = \sqrt{2}$ in the z-plane is mapped onto the line-segment $u = 0, 0 \le v \le \pi$ in the w-plane.

(c) The line-segments $\left\{ \begin{matrix} x - 0, 1 < y < \infty \\ x = 0, -1 < y < 0 \end{matrix} \right\}$ in the z-plane are mapped onto the line $v = \pi/2, -\infty < u < \infty$ in the w-plane. These details are shown in Figure 5.23.

Map 5.26. The details of the mapping $w = \log \dfrac{z+1}{z} - \frac{1}{2}\log z$, or $z = \left(e^w \pm \sqrt{e^{2w}-1}\right)^2$, are obvious; they are presented in Figure 5.24.

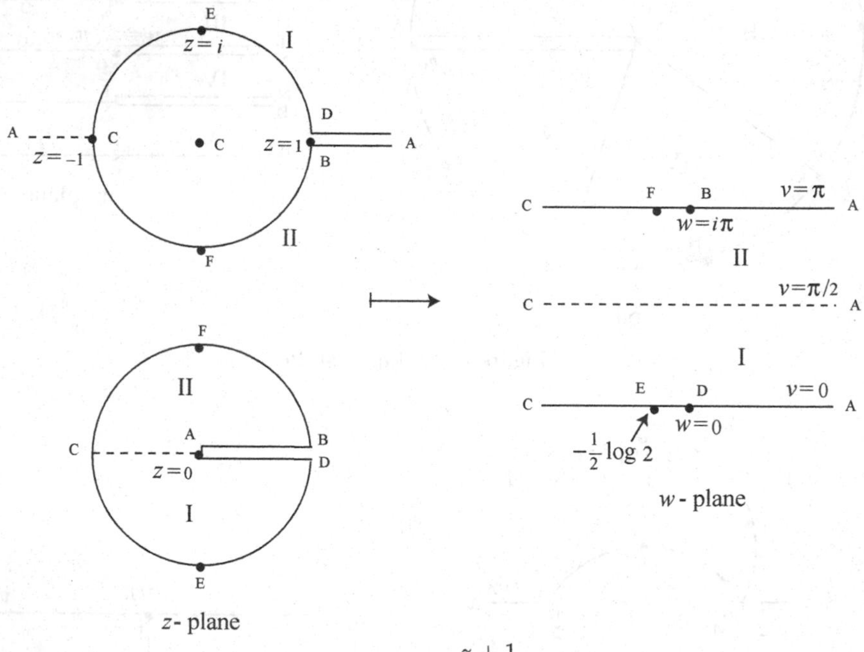

Figure 5.24 $w = \log \dfrac{z+1}{z} - \frac{1}{2}\log z$.

5.3 Other Related Functions

Since $e^z = e^{x+iy} = e^x(\cos y + i \sin y)$, the trigonometric functions are related, and so are the hyperbolic and their inverse functions. We will consider conformal transformations that involve such functions and their compositions.

Map 5.27. The mapping $w = \sin z$, or $z = \arcsin w = -i \log\left\{iw + \sqrt{1-w^2}\right\}$, has critical points $z = \infty; \left(2k + \frac{1}{2}\right)\pi$, where $= 0, \pm 1, \pm 2, \dots$ and p, q are real. The details under this mapping are as follows:

(i) The points $z = k\pi; \left(k + \frac{1}{2}\right)\pi; \left(k + \frac{1}{2} \pm \frac{1}{3}\right)\pi; \left(k + \frac{1}{2} \pm \frac{1}{6}\right)\pi$ in the z-plane are mapped onto the points $w = 0; (-1)^k; (-1)^k/2; (-1)^k\left(\sqrt(3)/2\right)$, respectively, in the w-plane.

(ii) The line $x = k\pi, -\infty < y < \infty$, in the z-plane is mapped onto the line $u = 0$, $-\infty < v < \infty$, for even or odd k.

(iii) The half-lines $\left\{\begin{array}{l} x = \left(k + \frac{1}{2}\right)\pi,\ 0 \le y < \infty \\ x = \left(k + \frac{1}{2}\right)\pi, -\infty < y \le 0 \end{array}\right\}$ in the z-plane are mapped onto the half-line $v = 0$ for $1 \le u < \infty$ if k is even, and for $-\infty < u \le -1$ if k is odd, in the w-plane.

(iv) The strips $\left\{\begin{array}{c} 2k\pi < x < \left(2k + \frac{1}{2}\right)\pi \\ \left(2k + \frac{1}{2}\right)\pi < x < (2k+1)\pi \end{array}\right\}$ in the z-plane are mapped onto the half-plane $u > 0$ cut along DG.

(v) The strips $\left\{\begin{array}{c} (2k+1)\pi < x < \left(2k + \frac{3}{2}\right)\pi \\ \left(2k + \frac{3}{2}\right)\pi < x < (2k+2)\pi \end{array}\right\}$ in the z-plane are mapped onto the half-plane $u < 0$ cut along BG.

These details are presented in Figure 5.25.

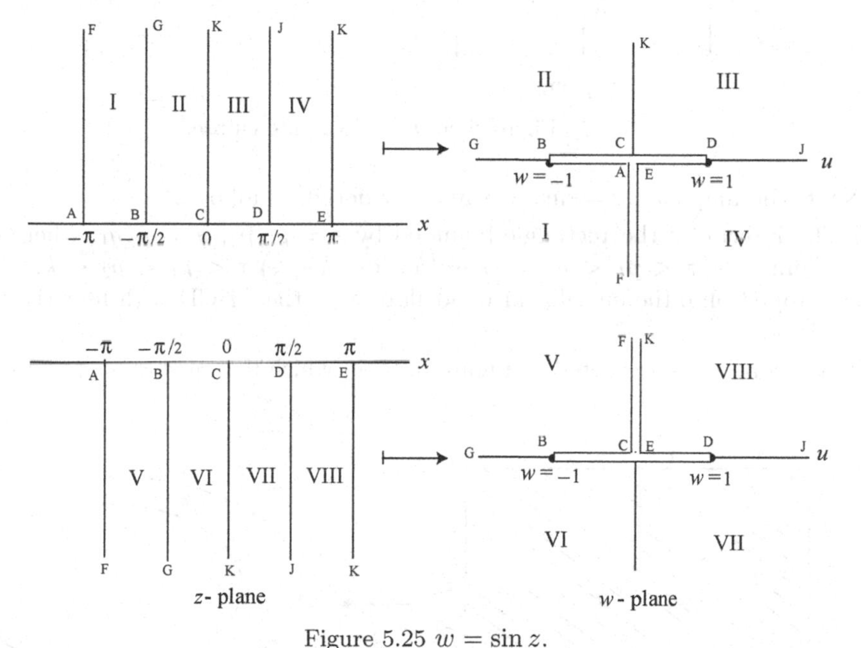

Figure 5.25 $w = \sin z$.

Other details for this mapping function onto an ellipse are as follows:

(vi) The line segment $y = q, q \neq 0$ for $2k\pi \le x < (2k+2)\pi$, in the z-plane is mapped onto the ellipse $|w+1||w-1| = 2\cosh q$ in the w-plane.

(vii) The interiors of the rectangles bounded by $z = \pm\pi, \pm\pi + iq, q > 0$, in the z-plane are mapped onto the interior of the ellipse, except for a slit from -1 to 1 and for a slit along the negative part of $v = 0$ in the w-plane.

(viii) The line $x = p + 2k\pi$, where $2p/\pi$ is not an integer, in the z-plane is mapped onto the branch $|w+1||w-1| = 2\sin p$ of the hyperbola.

These details are presented in Figure 5.26, in which $p' = \pi - p$ is set for $0 < p < \pi/2$, and we have used $p = \pi/4$ in this figure.

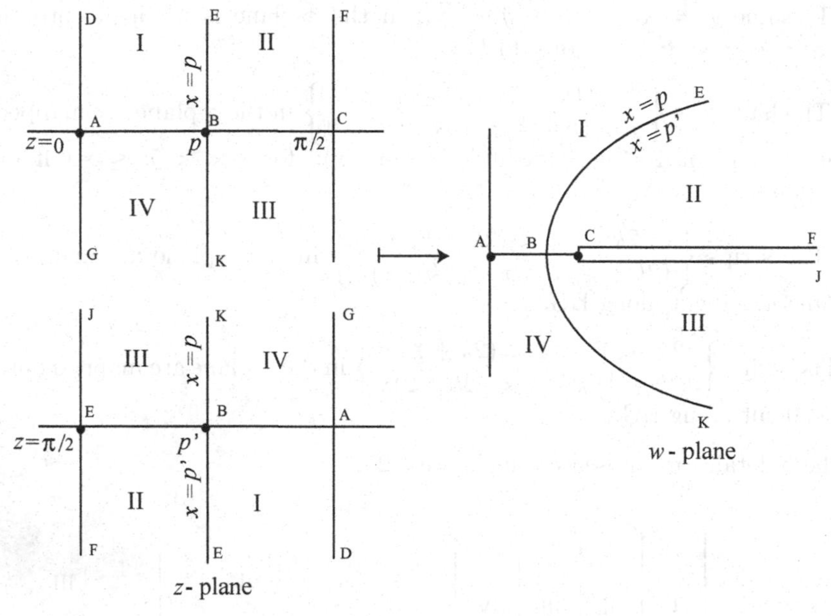

Figure 5.26 $w = \sin z$ onto ellipse.

Next, the mapping $w = \sin z$ has another detail, as follows:

(ix) The interior of the rectangle bounded by $x = p_1, p_2$, $y = q_1, q_1$, where $q_1 q_2 > 0$ and, for some k, $k\pi < p_1 < p_2 < \left(k + \frac{1}{2}\right)\pi$ or $\left(k - \frac{1}{2}\right)\pi < p_1 < p_2 < k\pi$, in the z-plane is mapped onto the curvilinear quadrilateral in the ABCD with four right angles in the w-plane.

This property is presented in Figure 5.27, in which $0 < p_1 < p_2 < \pi/2$ and $q_1 > q_2 > 0$.

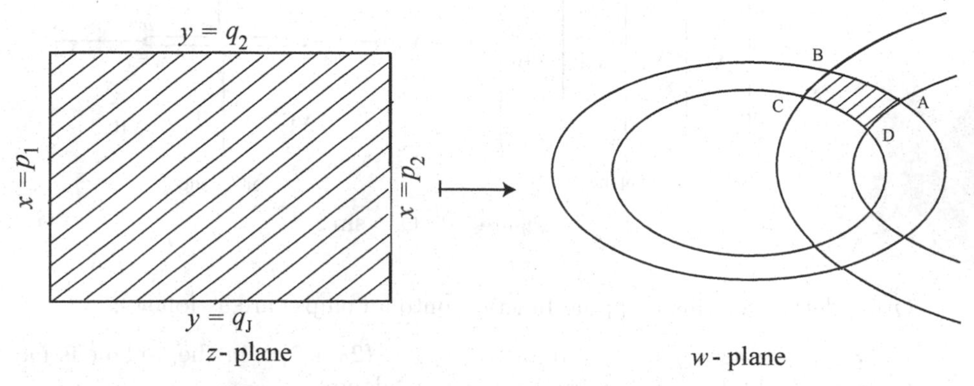

Figure 5.27 Interior of rectangle under $w = \sin z$.

Further, the function $w = \sin z$ maps the trigonal region $-i/2 \le x \le \pi/2, 0 \le y < \infty$ in the z-plane onto the upper-half w-plane, as shown in Figure 5.28.

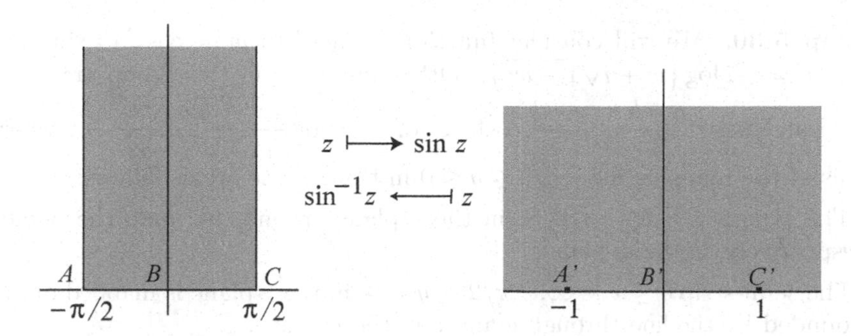

Figure 5.28 Trigonal region under $w = \sin z$.

Note that for the curves in the z-plane, which are mapped onto $R =$ const or $\Theta =$ const (i.e., $w - Re^{i\Theta}$), we have $\log \sin z = \log R + i\Theta$.

Map 5.28. The transformation $w = \sin z$ maps the infinite strip $0 < y < \infty$ (trigonal region) in the first quadrant of the z-plane onto the first quadrant of the w-plane, as shown in Figure 5.29.

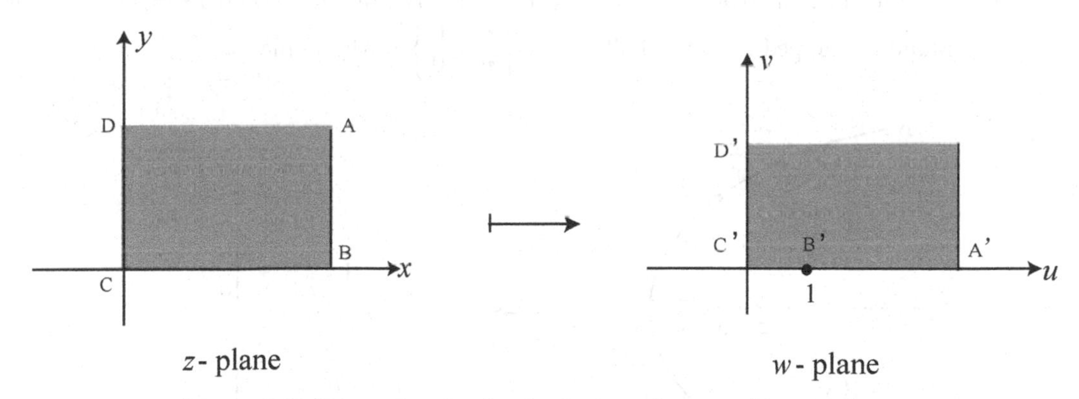

Figure 5.29 Trigonal region in the first quadrant under $w = \sin z$.

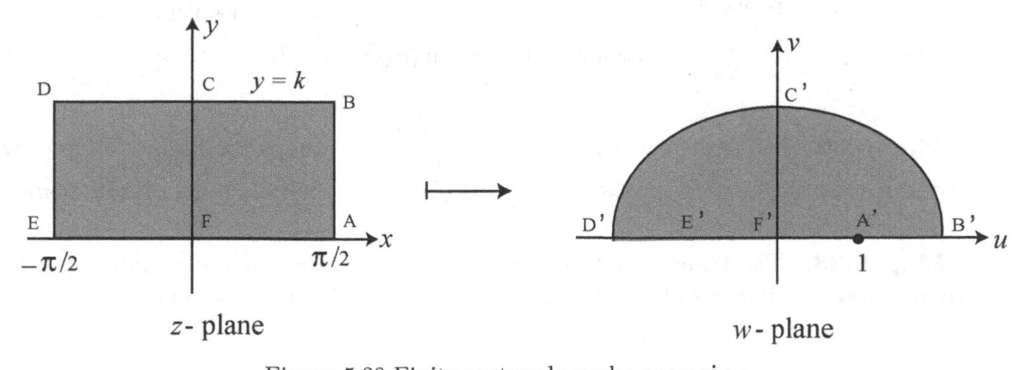

Figure 5.30 Finite rectangle under $w = \sin z$.

Map 5.29. The transformation $w = \sin z$ maps the rectangle $-\pi/2 \le x \le \pi/2$ in the z-plane onto the semi-circle in the w-plane, as shown in Figure 5.30.

Map 5.30. We will consider functions related to $w = \cos z = \sin(z + \pi/2)$, or $z = \arccos w = -i \log\left\{w + i\sqrt{1 - w^2}\right\}$. Other functions in this group are: $w = \sinh z$; $w = \cosh z$; and $w = \frac{a+b}{2} + \frac{a-b}{2} \cosh 2z$, or $z = \log \frac{\sqrt{w-a} + \sqrt{w-b}}{\sqrt{a-b}}$, where $a \ne b$. The details of the mapping for $-\pi/2 \le y \le 0$ in the z-plane are as follows:

(a) The points $z = 0; -i\pi/2; \infty$ in the z-plane are mapped onto the points $w = a; b; \infty$, respectively, in the w-plane.

(b) The semi-strip $0 \le x < \infty$, $-\pi/2 < y < 0$, in the z-plane is mapped onto the half-plane bounded by the line through a and b in the w-plane.

(c) The semi-strip $-\infty < x < 0$, $-\pi/2 < y < 0$, in the z-plane is mapped onto the other half-plane bounded by the line through a and b in the w-plane.

Map 5.31. The mapping $w = \sin\left(a\sqrt{z}\right), a > 0$, has critical points $z = 0; \infty; \left(\frac{2k+1}{2a}\pi\right)^2$. The details of the mapping, which are presented in Figure 5.31, are as follows:

(a) The points $z = 0; c; \left(\frac{2k+1}{2a}\pi\right)^2; 2ic; -2ic$ in the z-plane are mapped onto the points $w = 0; \pm 1; \pm 1; \cosh \pi/2; \pm \cosh \pi/2$, respectively, in the w-plane.

(b) The interior, cut from $z - 0$ to $z = c = (\pi/2a)^2$, of the parabola $^2 = 4c(c - x)$ in the z-plane is mapped onto the half-panes $\left\{\begin{matrix} v > 0 \\ v < 0 \end{matrix}\right\}$ in the w-plane.

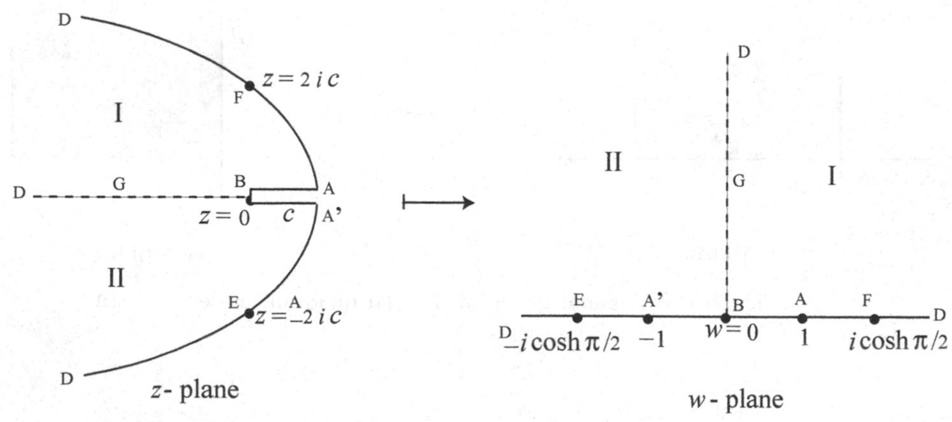

Figure 5.31 $w = \sin\left(a\sqrt{z}\right), a > 0$.

Map 5.32. The transformation $w = \frac{e^{\pi z/a} - e^{\pi z_0/a}}{e^{\pi z/a} + e^{\pi \bar{z}_0/a}}$ maps the infinite strip of width a: $-\infty < x < \infty, 0 \le y < \infty$, in the z-plane onto the unit disk $|w| \le 1$ in the w-plane.

Map 5.33. The transformation that maps the interior of the annulus $A(a, b)$ in the z-plane onto the interior of the rectangle $\ln a \le u \le \ln b$, in the w-plane is defined by

$$w = \ln z, \tag{5.3.1}$$

and is presented in Figure 5.32.

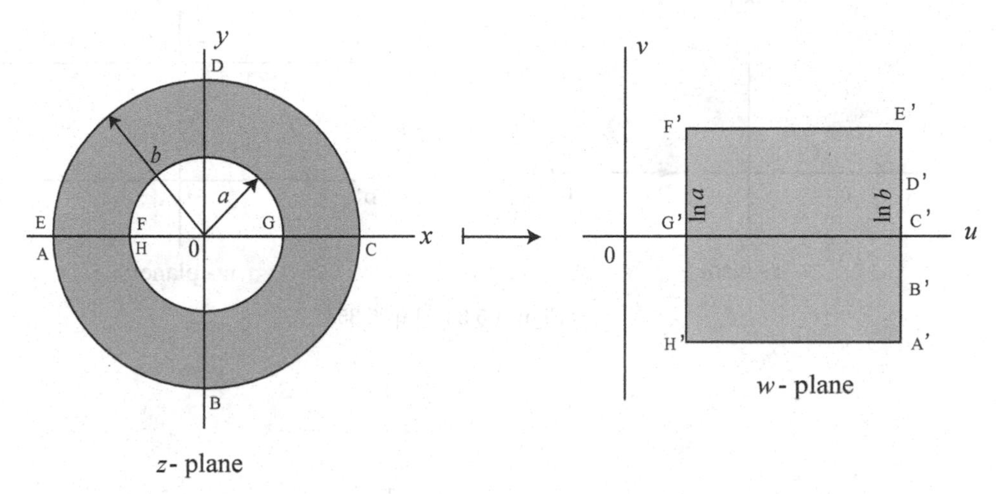

Figure 5.32 Interior of an annulus onto the interior of a rectangle.

Map 5.34. The transformation that maps a semi-infinite strip in the z-plane onto the interior of the infinite strip in w-plane is defined by

$$w = \ln z, \tag{5.3.2}$$

and is presented in Figure 5.33.

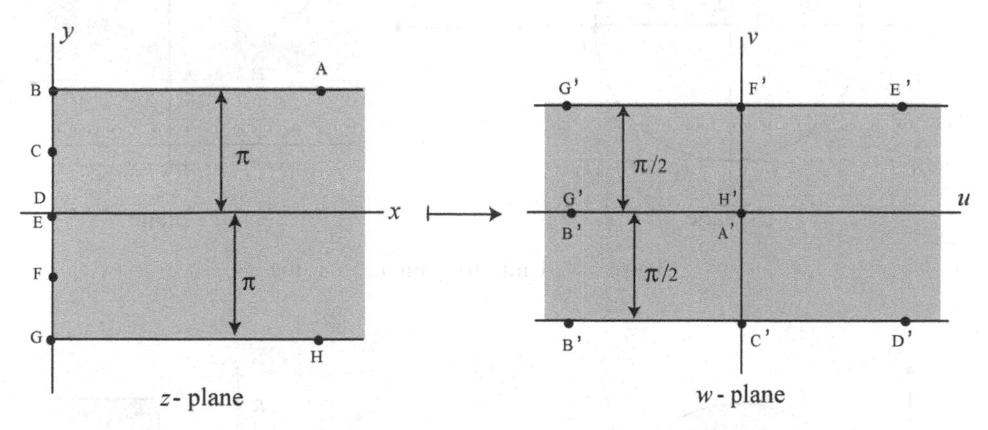

Figure 5.33 Semi-infinite strip onto the interior of the infinite strip.

Map 5.35a. The transformation $w = \log \dfrac{z-1}{z+1}$ maps the upper half-plane $-\infty < x < \infty, 0 < y < \infty$, onto the infinite strip $0 < w < i\pi$, as shown in Figure 5.34.

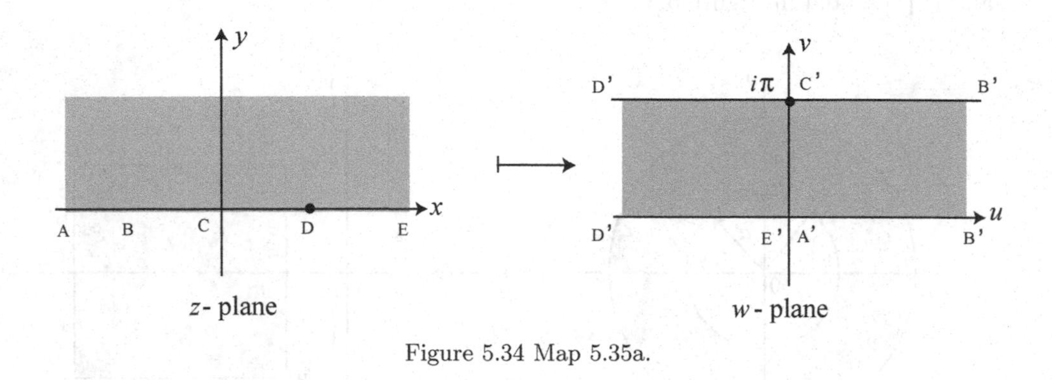

Figure 5.34 Map 5.35a.

Map 5.35b. The transformation $w = \log \dfrac{z-1}{z+1}$ maps the unit circle $x^2+y^2-2y\cot k = 1$ in the z-plane onto the infinite strip $0 < u < \infty$, as shown in Figure 5.35.

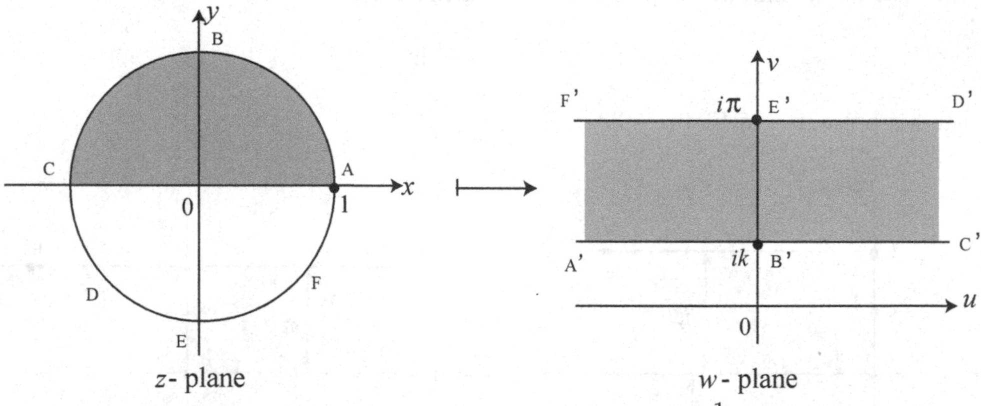

Figure 5.35 Unit circle under $w = \log \dfrac{z-1}{z+1}$.

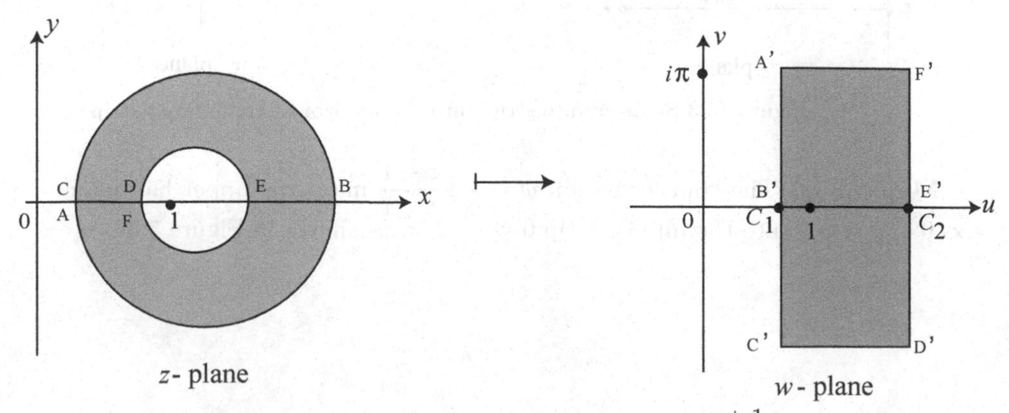

Figure 5.37 Annular region under $w = \log \dfrac{z+1}{z-1}$.

Map 5.36. The transformation $w = \log \dfrac{z+1}{z-1}$ maps the annulus region in the z-plane, with centers of the circles at $z_n = \coth c_n$, and radii $\operatorname{csch} c_n$, $n = 1, 2$, onto the finite rectangle $-i\pi \le v \le i\pi$ in the right-half of the w-plane, as shown in Figure 5.37.

Map 5.37. The transformation that maps the first quadrant in the z-plane onto the horizontal strip bounded by the semi-lines $0 < x < \infty$ and $z = iy$, $0 < x < \infty$, in the z-plane (see Figure 5.36) is

$$w = \log \left[i\, \frac{1-z}{1+z} \right],$$

such that the semi-line $0 < x < \infty$ is mapped onto the line $-\infty < u < \infty$ and the semi-line $z = iy$, $0 < y < \infty$ onto the horizontal line $\log(i\,u) = \ln u + i\,\pi/2$ in the w-plane. The mapping $w = i\, \dfrac{1-z}{1+z}$ is defined as Map 3.25.

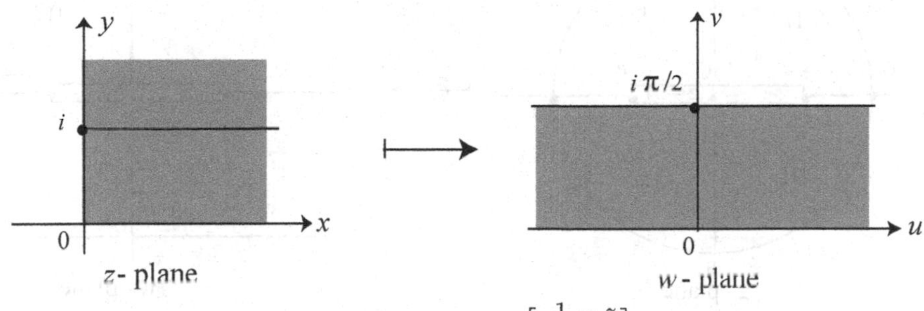

Figure 5.36 $w - \log \left[i\, \dfrac{1-z}{1+z} \right]$.

Map 5.38. The mapping $w = \log \dfrac{ae^z + b}{ce^z + d}$, where $\dfrac{ad}{bc}$ is real and $1 < \dfrac{ad}{bc} < \infty$, is a composite of $w = \beta + \log \dfrac{b}{d} + \xi$, $\zeta = \dfrac{z}{2} - \frac{1}{2} \log \left(-\dfrac{d}{c} \right)$, and $\xi = \log \dfrac{\sinh(\zeta + \beta)}{\sinh \zeta}$, $\beta = \frac{1}{2} \log \dfrac{ad}{bc}$.

Map 5.39. The mapping $w = \log \dfrac{z-m}{z+m} + \log \dfrac{mz-1}{mz+1} = \log \dfrac{z^2+1-2jz}{z^2+1+2jz}$, $0 < m < 1$, $j = \frac{1}{2}(m + 1/m)$, is a composite of $\xi = \frac{1}{2}(z + 1/z)$ and $w = \log \dfrac{\xi - j}{\xi + j}$. It has critical points at $z = m; 1/m; -m; -1/m; \infty$. Let $c = \log \dfrac{1+m}{1-m}$, and $k = 0, \pm 1, \pm 2, \dots$, and this mapping has the following properties:

(i) The points $z; 1/z$ in the z-plane are mapped onto the point $w = w(z) + 2ik\pi$ in the w-plane.

(ii) The points $z = iy$, $-1 < y < 1$ in the z-plane are mapped onto the point $w = 2ik\pi - 2i \arctan \dfrac{2jy}{1-y^2}$ in the w-plane.

(iii) The line-segment $y = 0$, $-m < x < m$, in the z-plane is mapped onto the line $v = 2k\pi$, $-\infty < u < \infty$, in the w-plane.

(iv) The line-segment $y = 0, m < x \leq 1$, in the z-plane is mapped onto the half-line $v = (2k-1)\pi, -\infty < u < -2c$, in the w-plane.

(v) The line-segment $y = 0, -1 \leq x < -m$, in the z-plane is mapped onto the line $v = (2k-1)\pi, 2c \leq u < \infty$, in the w-plane.

(vi) The line-segment $x = 0, 0 \leq y \leq 1$, in the z-plane is mapped onto the line-segment $u = 0, (2k-1)\pi \leq v \leq 2k\pi$, in the w-plane.

(vii) The semi-circle $z = e^{i\theta}, 0 \leq \theta \leq \pi$ and also $\pi \leq \theta \leq 2\pi$, in the z-plane is mapped onto the line-segment $v = (2k-1)\pi, -2c \leq u \leq 2c$, in the w-plane.

These properties are shown in Figure 5.38, in which E is at $-2c + (2k-1)\pi$ and G is at $2c + (2k-1)i\pi$.

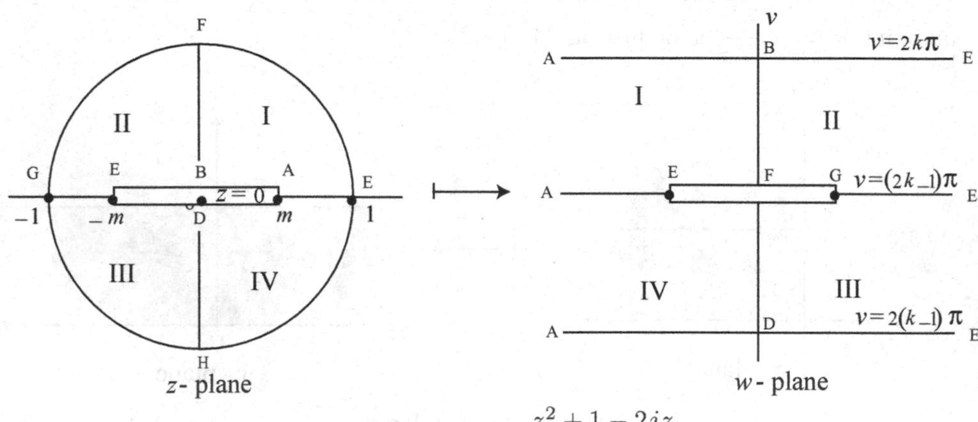

Figure 5.38 $w = \log \dfrac{z^2+1-2jz}{z^2+1+2jz}$.

Other properties are as follows:

(viii) The circle $|z| = 1$ in the z-plane is mapped onto each of the slit EG in Figure 5.39.

(ix) The point $z = e^{i\theta}$ in the z-plane is mapped onto $w = -2\tanh^{-1}\dfrac{2m\cos\theta}{1+m^2} + (2k-1)i\pi$ in the w-plane.

(x) The circle $|z-\rho| = 1-\rho, 0 < \rho < (1-m)/2$, touching the circle $|z| = 1$ at E, counted infinitely many times, in the z-plane is mapped onto the set of airfoils surrounding the slits and exterior to one another, touching EG at E on both sides in the w-plane, as shown in Figure 5.39.

(xi) The region formed by the interiors of this circle, which is counted infinitely many times, where the interiors are connected through the slit ABCD, is mapped onto the set of the above airfoils in the w-plane.

(xii) The region bounded by $|z| = 1$ and $|z-\rho| = 1-\rho$ in the z-plane is mapped onto the interior cut from E to G, of an airfoil in the w-plane.

These properties are presented in Figure 5.39.

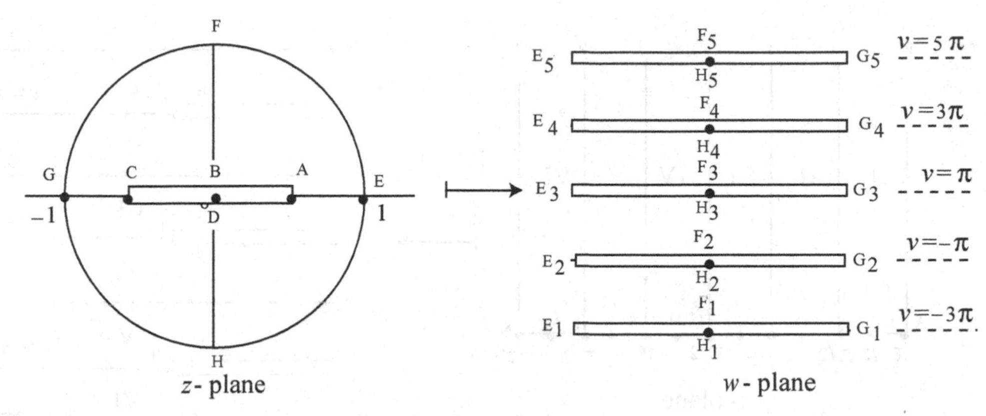

Figure 5.39 Circle onto slits of an airfoil.

Map 5.40. The mapping $w = \log\left\{e^z + \left(e^{2z} - 1\right)^{1/2}\right\}$, or $z = \log\cosh w$, is a composite of $w = \log\zeta$ and $z = \log\dfrac{\zeta^2+1}{2\zeta}$. The details of this mapping are obvious from Figure 5.40.

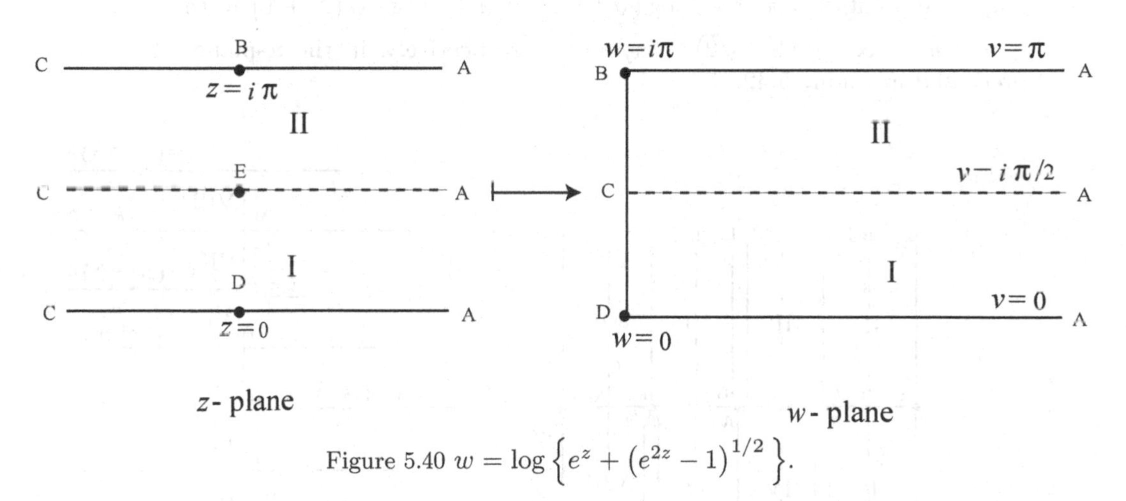

Figure 5.40 $w = \log\left\{e^z + \left(e^{2z} - 1\right)^{1/2}\right\}$.

Map 5.41. The mapping $w = \log\sin z = \log\cosh(iz - i\pi/2)$, or $z = \arcsin e^w$, has critical points $z = \infty; k\pi/2$, where $k = 0, \pm1, \pm2, \ldots$. The details presented in Figure 5.41 are as follows:

The semi-infinite strip $-m\pi < x < n\pi, y > 0$, where $m, n > 0$ are integers, is mapped on the strip $\left(-n + \frac{1}{2}\right)\pi < v < \left(m + \frac{1}{2}\right)\pi$ with $(m + n)$ slits in the w-plane.

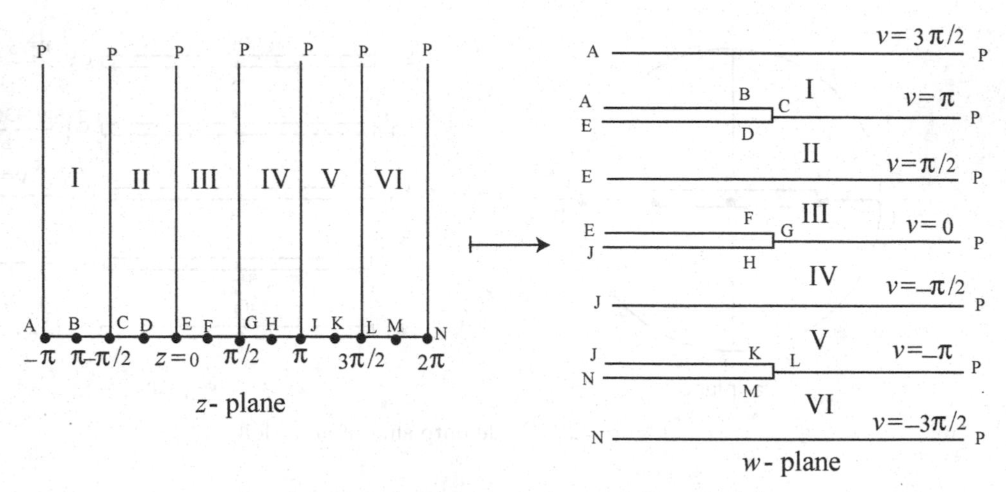

Figure 5.41 $w = \log \sin z = \log \cosh (iz - i\pi/2)$.

Map 5.42. The mapping $w = \log \tan z = \log \coth (iz + i\pi/2) - i\pi/2$, or $z = \arctan e^w$ maps the points $z = \pi/4 - \frac{i}{2} \log \left(\sqrt{2} + 1\right); \pi/4 + \frac{i}{2} \log \left(\sqrt{(2)} + 1\right)$ in the z-plane onto the points $w = \infty; \log \left(1 + \sqrt{2}\right); -i\pi/2; i\pi/2$, respectively, in the w-plane. This mapping is presented in Figure 5.42.

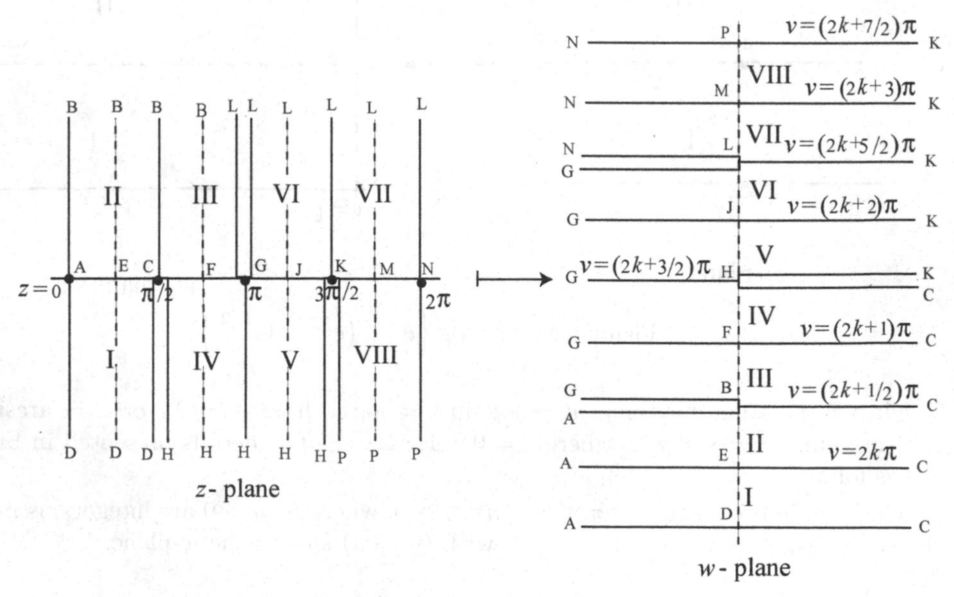

Figure 5.42 $w = \log \tan z$.

Map 5.43. The transformation that maps a semi-infinite strip of width a, $0 \le y < \infty, -a/2 \le x \le a/2$, in the z-plane onto the upper half-plane $\Im\{w\} > 0$ is defined by

$$w = \sin \frac{\pi z}{a}, \tag{5.3.4}$$

and is presented in Figure 5.43.

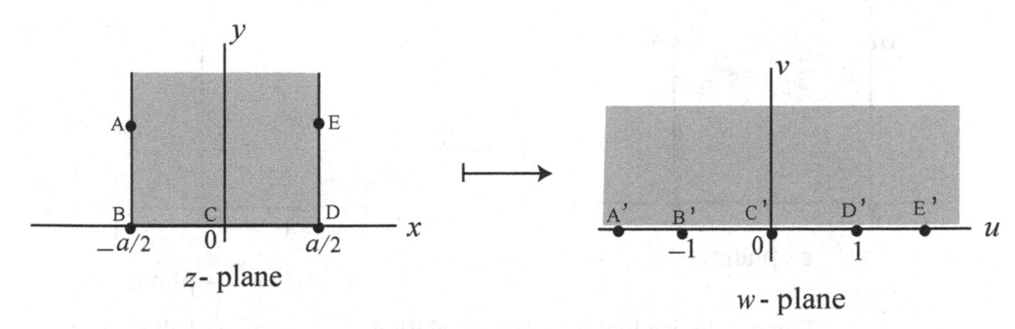

Figure 5.43 Semi-infinite strip of width a onto the upper half-plane.

Map 5.44. The transformation that maps a semi-infinite strip of width a, $0 \leq x \leq a, 0 \leq y < \infty$, in the z-plane onto the first quarter of the w-plane is defined by

$$w = \sin \frac{\pi z}{2a}, \tag{5.3.5}$$

and is presented in Figure 5.44.

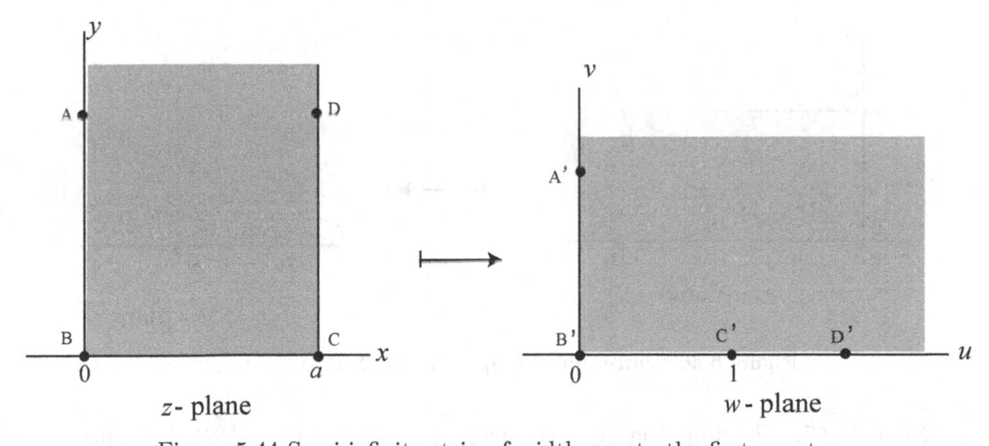

Figure 5.44 Semi-infinite strip of width a onto the first quarter.

Map 5.45. The transformation that maps a semi-infinite strip of width a, $0 \leq y < \infty, 0 \leq x \leq a$, in the z-plane onto the upper half-plane $\Im\{w\} > 0$ is defined by

$$w = \cos \frac{\pi z}{a}, \tag{5.3.6}$$

and is presented in Figure 5.45.

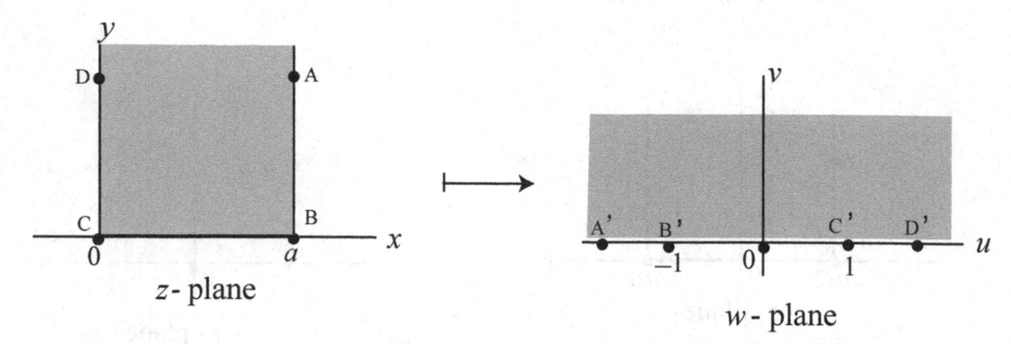

Figure 5.45 Semi-infinite strip of width a onto the upper half-plane.

Map 5.46. The transformation that maps a semi-infinite strip of width a, $0 \le x < \infty, 0 \le y \le a$, in the z-plane onto the upper half-plane $\Im\{w\} > 0$ is defined by

$$w = \cos\frac{\pi z}{a}, \tag{5.3.7}$$

and is presented in Figure 5.46.

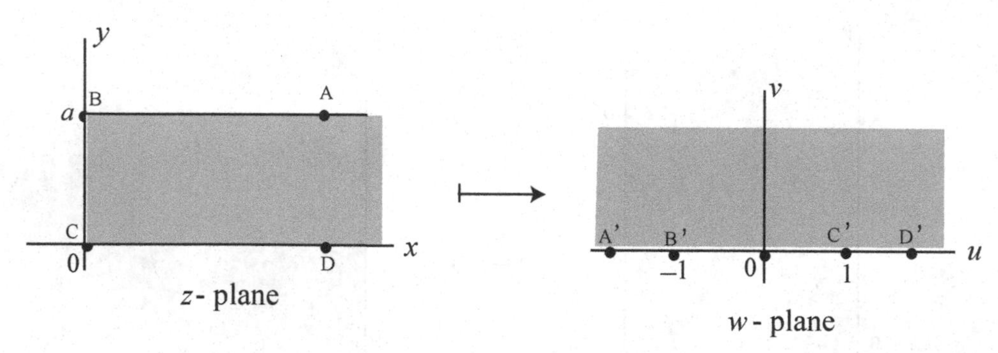

Figure 5.46 Semi-infinite strip of width a onto the upper half-plane.

Map 5.47. The mapping $w = \cos(a\log z), a > 0$, or $z = e^{(\arccos w)/a}$, has critical points $z = 0; \infty; e^{j\pi/a} = r_0^{j/2}$, where j, k are integers, and $r_0 = e^{2\pi/a}$. This mapping has the following properties:

(i) The points $z = \beta r_0^j, \beta \neq 0; \beta^{-1} r_0^j$ in the z-plane are mapped onto the points $w = \cos(a\log\beta + 2iak\pi)$ in the w-plane. (ii) The points $z = 1$ (A); $r_0^{1/4}$ (B); $\sqrt{r_0}$ (C) in the z-plane are mapped onto the points $w = 1; 0; -1$, respectively, in the w-plane.

(iii) The points $z = r_0^{3/4}$ (D); r_0 (E); $r_0 e^{i\pi}$ (F) in the z-plane are mapped onto the points $w = 0; 1; \cosh(a\pi)$, respectively, in the w-plane.

(iv) The points $z = r_0 e^{2i\pi}$ (G); $r_0^{3/4} e^{2i\pi}$ (H); $r_0^{1/2} e^{2i\pi}$ (I) in the z-plane are mapped onto the points $w = \cosh(2a\pi); i\sinh(2a\pi); -\cosh(2a\pi)$, respectively, in the w-plane.

(v) The points $z = r_0^{1/4} e^{2i\pi}$ (J); $e^{2i\pi}$ (K); $e^{i\pi}$ (L) in the z-plane are mapped onto the points $w = -\sinh(2a\pi); \cosh(2a\pi); \coth(a\pi)$, respectively, in the w-plane.

(vi) The line segment GHI in the z-plane is mapped onto the part $v > 0$ of the ellipse $|w-1| + |w+1| = 2\cosh(2a\pi)$ in the w-plane.

(vii) The line segment IJK in the z-plane is mapped onto the part $v < 0$ of the same ellipse $|w-1|+|w+1| = 2\cosh(2a\pi)$ in the w-plane.

(viii) The circle $|z| = r_0^{1/2}, 0 \le \arg\{z\} < 2\pi$, in the z-plane is mapped onto the line-segment $u = 0, -\cosh(2a\pi) < v \le -1$, in the w-plane.

(ix) The circle $|z| = r_0^{3/4}, 0 \le \arg\{z\} < 2\pi$, in the z-plane is mapped onto the line-segment $u = 0, 0 \le v < \sinh(2a\pi)$, in the w-plane.

(x) The circle $|z| = r_0^{\alpha}, 0 \le \arg\{z\} < 2\pi$, where $0 < \alpha < 1, \alpha \ne 1/4, 1/2, 3/4$, in the z-plane is mapped onto the part $-\sin(2\alpha\pi) < v \le 0$ or $0 \le v < -\sin(2\alpha\pi)$ (for $\sin(2\alpha\pi) > 0$ or < 0, respectively) of the hyperbola branch $|w+1| - |w-1| = 2\cos(2\alpha\pi)$ in the w-plane.

These properties are shown in Figure 5.47.

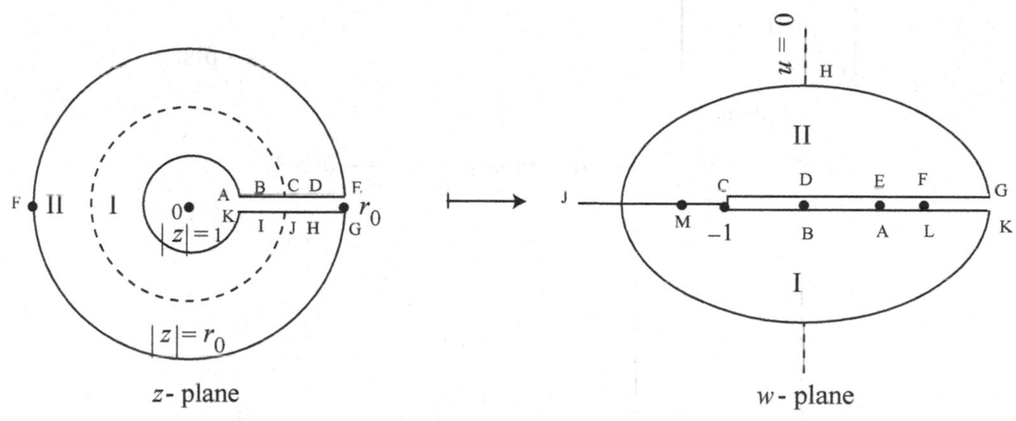

Figure 5.47 $w = \cos(a \log z), a > 0$.

Map 5.48. The mapping $w = \dfrac{1}{a - \cos z}$, a real, $a \ne \pm 1$, or $z = \arccos(a - 1/w)$, has critical points $z = \pm \arccos a + 2k\pi; k\pi; \infty$ for $k = 0, \pm 1, \pm 2, \dots$. The details for different values of a are as follows:

(a) **Case** $a > 1$: Let $z_0 = i\zeta_0$ where $\zeta_0 = a, \zeta_0 > 0$. Then the mapping is shown in Figure 5.48.

(b) **Case** $a < -1$: Then the mapping becomes $w = -\dfrac{1}{|a| - \cos(z+\pi)}$, where $|a| > 1$. It is similar to part (a) above.

(c) **Case** $-1 < a < 1$: The mapping is shown in Figure 5.49.

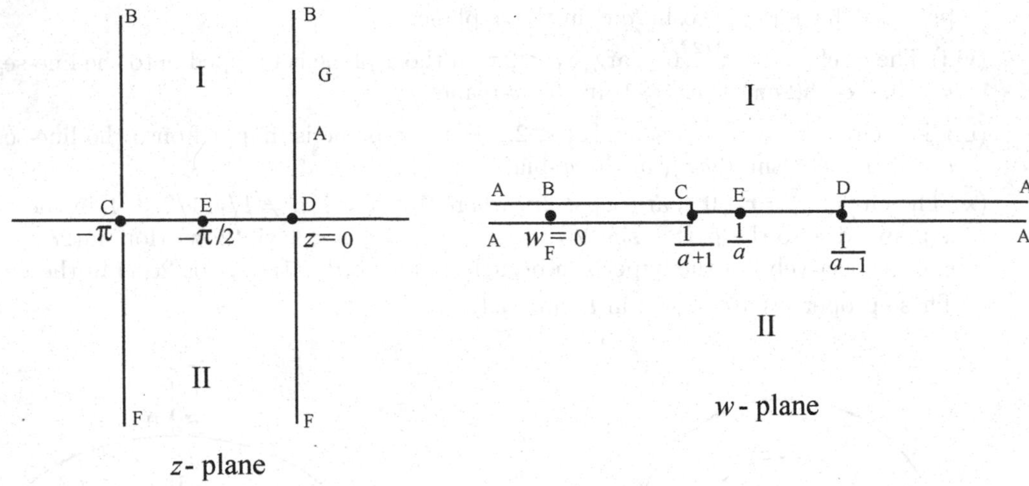

Figure 5.48 $w = \dfrac{1}{a - \cos z}, a > 1.$

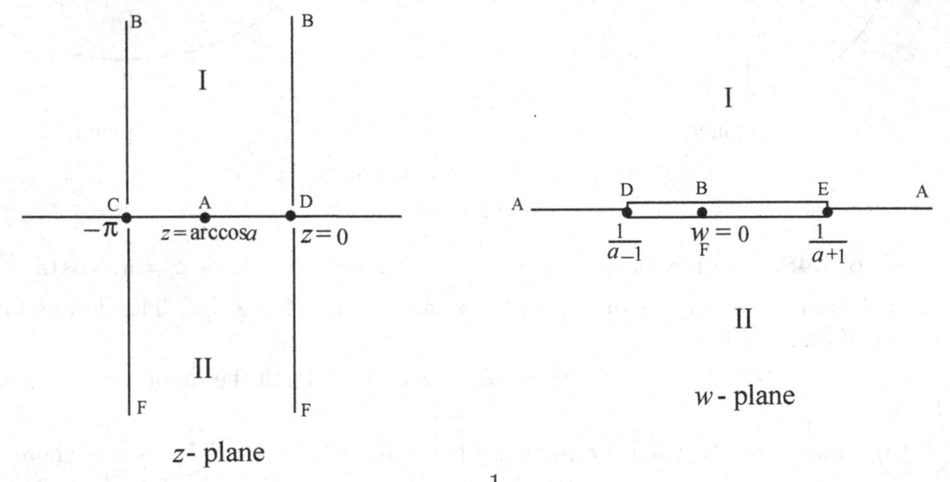

Figure 5.49 $w = \dfrac{1}{a - \cos z}, -1 < a < 1.$

Map 5.49. The transformation $w = \tan\left(\dfrac{\arccos z^2}{4}\right)$, or $z = \dfrac{\left(w^4 - 6w^2 + 1\right)^{1/2}}{1 + w^2}$, maps the region exterior to the slits in the z-plane onto the upper half-plane $\Im\{w\} > 0$. The mapping is shown in Figure 5.50, where $b = \tan \pi/8 = \sqrt{2} - 1$.

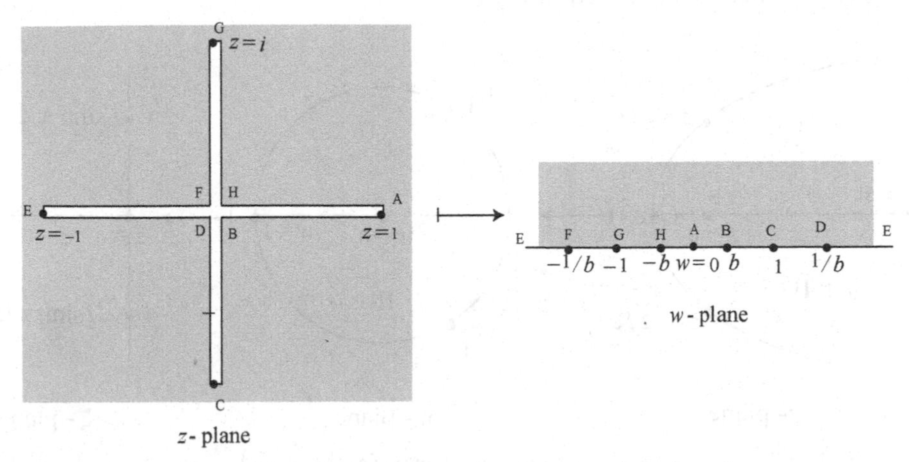

Figure 5.50 Region exterior to slits onto $\Im\{w\} > 0$

Map 5.50. The transformation $w - \tan^2(z/2)$ maps the open strip $0 < y < \pi$ in the second quadrant of the z-plane onto the unit semi-circle in the w-plan, as shown in Figure 5.51.

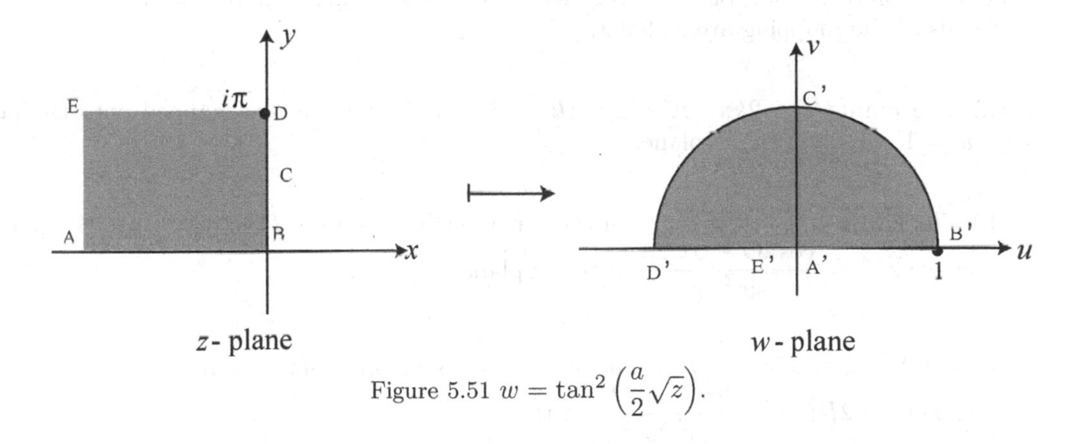

Figure 5.51 $w = \tan^2\left(\frac{a}{2}\sqrt{z}\right)$.

Map 5.51. The mappings $w = \tan^2\left(\frac{a}{2}\sqrt{z}\right)$, and $\xi = \cos(a\sqrt{z})$, where $a \gtrless 0$, has the critical points $z = \infty; (k\pi/a)^2$, $k = 0, \pm 1, \pm 2, \dots$. It maps the interior of parabola onto the interior of a circle or half-plane. We will consider the three planes: z-plane, w-plane, and ξ-plane for the details.

(i) The points $z = 0; c; 2ic; -2ic$ in the z-plane are mapped onto the points $w = 0; 1; w_1 \equiv \frac{1+i\sinh \pi/2}{1-i\sinh \pi/2}; \bar{w}_1$, respectively, in the w-plane, and onto points $\xi = 1; 0; -i\sinh\frac{\pi}{2}; i\sinh\frac{\pi}{2}$, respectively, in the ξ-plane.

(ii) The half-line $y = 0, -\infty < c \leq c$ in the z-plane is mapped onto the line-segment $v = 0, -1 < u \leq 1$ in the w-plane, and onto the half-line $\Im\{\xi\} = 0, 0 \leq \xi < \infty$ in the ξ-plane.

(iii) The interior of the parabola $y^2 = 4c(c - x)$ (focus $z = 0$, vertex $z = c$) in the z-plane is mapped onto the interior of the circle $|w| = 1$ in the w-plane, and onto the half-plane $\Re\{\xi\} > 0$ in the ξ-plane.

These details are presented in Figure 5.52.

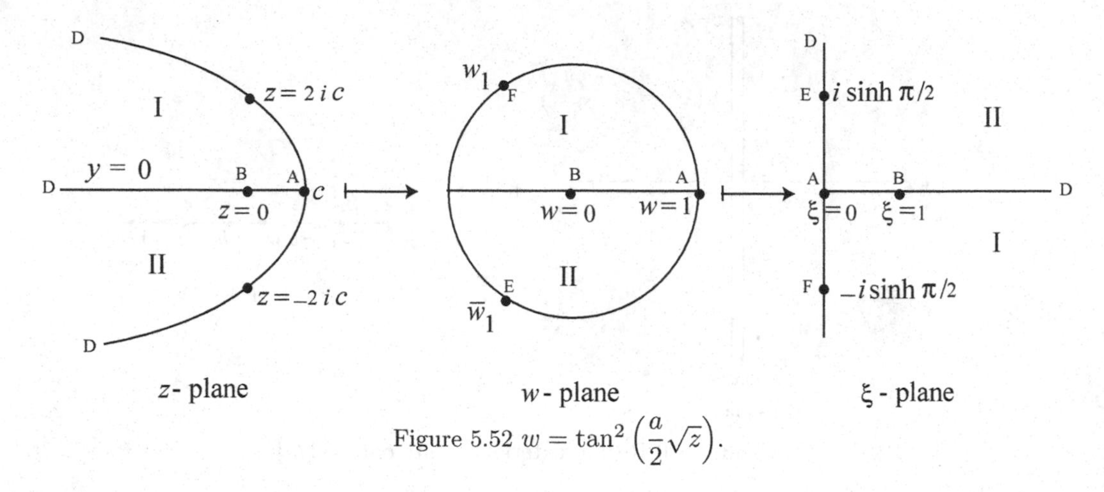

Figure 5.52 $w = \tan^2\left(\frac{a}{2}\sqrt{z}\right)$.

Map 5.52. The mapping $w = \sec z$, or $z = \arccos(1/w) = -i \log \dfrac{1 + \sqrt{1 - w^2}}{w}$, has critical points $z = k\pi; \left(k + \frac{1}{2}\right)\pi; \infty$, where $k = 0, \pm 1, \pm 2, \dots$, and $w = R\,e^{i\Theta}$, $q > 0$. The details of the mapping are as follows:

(a) The points $z = 2k\pi; (2k+1)\pi; \left(k + \frac{1}{2}\right)\pi$ in the z-plane are mapped onto the points $w = 1; -1; \infty$ in the w-plane.

(b) The line $x = p$, $0 < p < \pi/2$ in the z-plane is mapped onto the part $-p < \Theta < p$ of the curve $2R^2 = \dfrac{\cos 2\Theta - \cos 2p}{\sin^2 2p}$ in the w-plane.

(c) The line $x = p$, $\pi/2 < p < \pi$ in the z-plane is mapped onto the part $p < \Theta < 2\pi - p$ of the curve $2R^2 = \dfrac{\cos 2\Theta - \cos 2p}{\sin^2 2p}$ in the w-plane.

(d) The line $x = \pi/4$ in the z-plane is mapped onto the part $u > 0$ of the lemniscate $|w+1||w-1| = 1$ in the w-plane.

(e) The line $x = 3\pi/4$ in the z-plane is mapped onto the part $u < 0$ of the same lemniscate $|w+1||w-1| = 1$ in the w-plane.

(f) The segment $y = q, 0 < x < \pi$, in the z-plane is mapped onto the part $0 < \Theta < \pi$ of the curve $2R^2 = \dfrac{\cosh 2q - \cos 2\Theta}{\sin^2 2q}$ in the w-plane.

(g) The segment $y = -q, 0 < x < \pi$, in the z-plane is mapped onto the part $\pi < \Theta < 2\pi$ of the same curve $2R^2 = \dfrac{\cosh 2q - \cos 2\Theta}{\sin^2 2q}$ in the w-plane.

This mapping is presented in Figure 5.53.

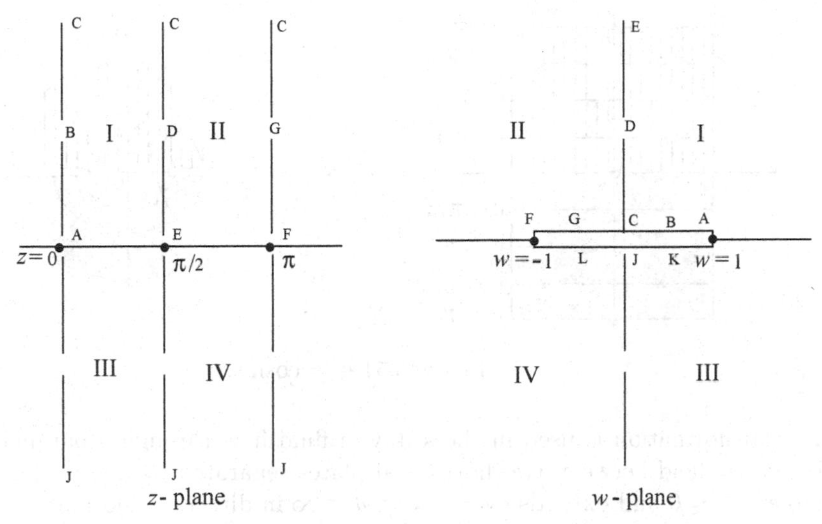

Figure 5.53 $w = \sec z$.

Map 5.53. The mapping function $w = \cosh z$ has magnification $m = |dw/dz| = |\sinh z|$, which for $x = 0$ becomes $m = \frac{1}{2}|\sin y|$. For small x and y, we have $m \approx |y/2|$. The rotation is $\alpha = \arg\{dw/dz\} = \arg\{\sinh z\} = \arg\{iy\}$; for example, $\alpha = \pi/2$ for $y > 0$. The equations for u and v are

$$\frac{u^2}{\cosh^2 x} + \frac{v^2}{\sinh^2 x} = 1, \quad \frac{u^2}{\cos^2 y} + \frac{v^2}{\sin^2 y} = 1.$$

The gridlines $x = k_1$ are mapped onto ellipses $u^2/a^2 + v^2/b^2 = 1$ with half axes $a = \cosh k_1, b = \sinh k_1$, and foci at ± 1. The gridlines $y = k_2$ are mapped onto hyperbolas $u^2/c^2 - v^2/d^2 = 1$ with half axes $c = \cos k_2, d = \sin k_2$, and foci at ± 1. The details of this transformation are as follows:

(a) The horizontal gridlines $y = c$, where $c > 0$ is a constant, in the z-plane are mapped onto the lines which intersect the u-axis at right angles and parallel to the u-axis, except for $y = 0$ and $y = \pi$ in the w-plane.

(b) The horizontal gridlines in the upper-half of the z-plane are mapped onto the horizontal gridlines bent upward in the upper-half of the w-plane.

(c) The horizontal lines in the lower-half of the z-plane are mapped onto the bent gridlines that are mirror images of those in the upper-half of the w-plane.

(d) The inverse transformation can be viewed as splitting the u-axis from $-\infty$ to -1 and then rotating the upper part $y = \pi$ clockwise by $\pi/2$ while a mirrored set of operations continue in the lower half part $y = -\pi$ and the right half $y = 0$ in the z-plane.

The mapping is presented in Figure 5.54.

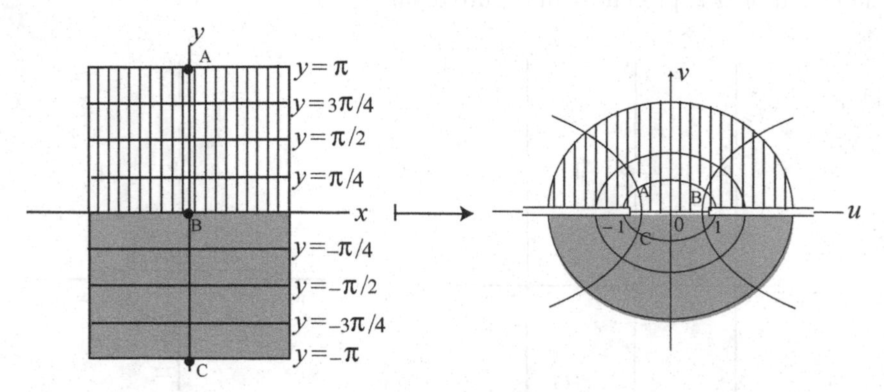

Figure 5.54 $w = \cosh z$.

This transformation is used in the study of fluid flow through a parallel slit in a plane, or the electric field between two horizontal plates separated by a gap. The slit or gap lies in the plane $v = 0$ and extends over $-\infty < u < \infty$ in directions normal to the w-plane.

Map 5.54. The transformation $w = \cosh(\pi z/a)$ maps the semi-infinite strip of width a: $0 \leq x < \infty, 0 \leq y \leq a$, in the z-plane onto the upper half-plane $\Im\{w\} \geq 0$ in the w-plane.

Map 5.55. The transformation $w = i \cosh\left(\pi\sqrt{\frac{1}{2}z/p - \frac{1}{4}}\right)$ maps the interior of the parabola $y^2 - 2px \leq 0$ in the z-plane onto the upper half-plane $\Im\{w\} \geq 0$ in the w-plane.

Map 5.56. The mapping $w = \tanh z = \dfrac{e^{2z} - 1}{e^{2z} + 1}$, or $z = \frac{1}{2}\log\dfrac{1+w}{1-w} - \tanh^{-1} w$, has critical points $z = \pm i\pi/2; \pm 3i\pi/2, \ldots; \infty$. There are six cases:

Case (a). Infinite strip and semi-infinite strip: The details are as follows.

(i) The infinite strip $k\pi < y < \left(k + \frac{1}{2}\right)\pi$ in the z-plane is mapped onto the half-plane $v > 0$ in the w-plane.

(ii) The infinite strip $k\pi < y < (k+1)\pi$ in the z-plane is mapped onto the cut plane in the w-plane.

(iii) The infinite half-strip $k\pi < y < \left(k + \frac{1}{2}\right)\pi, x > 0$ in the z-plane is mapped onto the first quadrant $u > 0, v > 0$.

(iv) The infinite half-strip $k\pi < y < (k+1)\pi, x > 0$ in the z-plane is mapped onto the half-plane $u > 0$ cut from $w = 1$ to $w = 0$ in the w-plane.

These details are shown in Figure 5.56.

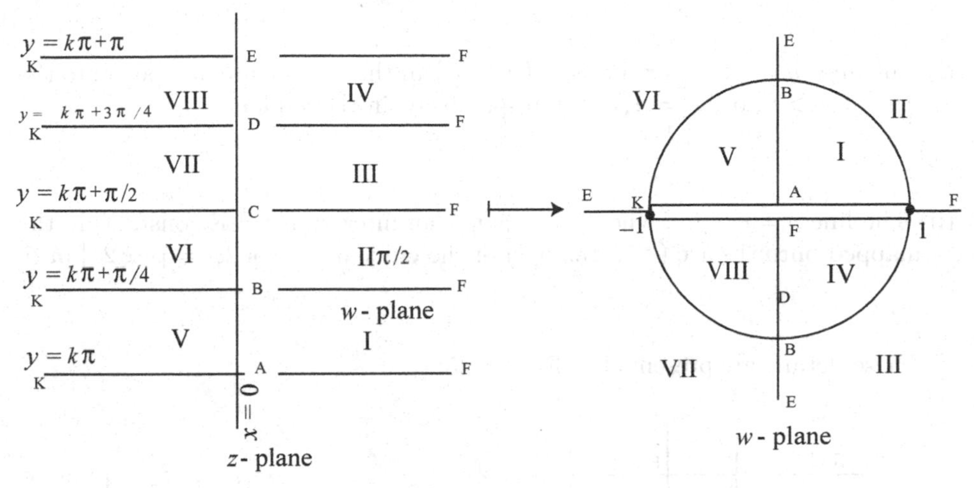

Figure 5.56 $w = \tanh z$, Case (a).

Case (b). Set of coaxial circles passing through $w = -1$ and $w = 1$: The details are as follows.

(i) The points $z = i\phi + ik\pi; ik\pi; \left(k + \frac{1}{2}\right) i\pi$ in the z-plane are mapped onto the points $w = i \tan \phi; 0; \infty$ in the w-plane.

(ii) The line segment $x = 0, \phi \le y < \phi + \pi$, in the z-plane is mapped onto the line $u = 0, -\infty < y < \infty$, where the point $v = \pm\infty$ corresponds to $y_1 = \left(k_1 + \frac{1}{2}\right)\pi$, k_1 being an integer.

(iii) The line $y = k\pi, -\infty < x < \infty$ in the z-plane is mapped onto the line segment $v = 0, -1 < u < 1$ in the w-plane.

(iv) The line $y = \left(k + \frac{1}{2}\right)\pi, -\infty < x < \infty$ in the z-plane is mapped onto the line $v = 0$ in the w-plane, excluding the segment $-1 < u < 1$, where $u = \infty$ corresponds to $x = 0$.

(v) The line $y = \phi + \left(k \pm \frac{1}{2}\right)\pi$ in the z-plane is mapped onto the arc $(-1, -i \cot \phi, 1)$ of the same circle in the w-plane.

These details are presented in Figure 5.57, which is plotted for $0 < \phi < \pi/2c$ and $k = 0$.

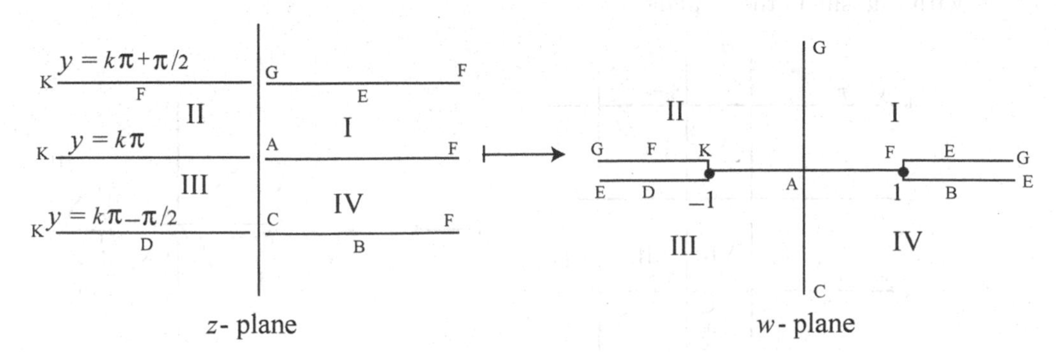

Figure 5.57 $w = \tanh z$, Case (b).

Case (c). Set of coaxial circles $|w - \coth 2p| = |\sinh 2p|^{-1}$ with limit points $-1, 1$. The details are as follows.

(i) The lines $y = \left(k + \frac{1}{2}\right)\pi$ and $y = \left(k + \frac{3\pi}{4}\right)$ in the z-plane are mapped onto the semicircle $|w| + 1, v > 0$, or $|w| = 1, v < 0$, respectively, in the w-plane.

(ii) The line $y = \phi + k\pi$, where $2\phi/\pi$ is not an integer, and ϕ is constant, in the z-plane is mapped onto the arc $(-1, i\tan\phi, 1)$ of the circle $|w + i\cot 2\phi| = |\csc 2\phi|$ in the w-plane.

These details are presented in Figure 5.58.

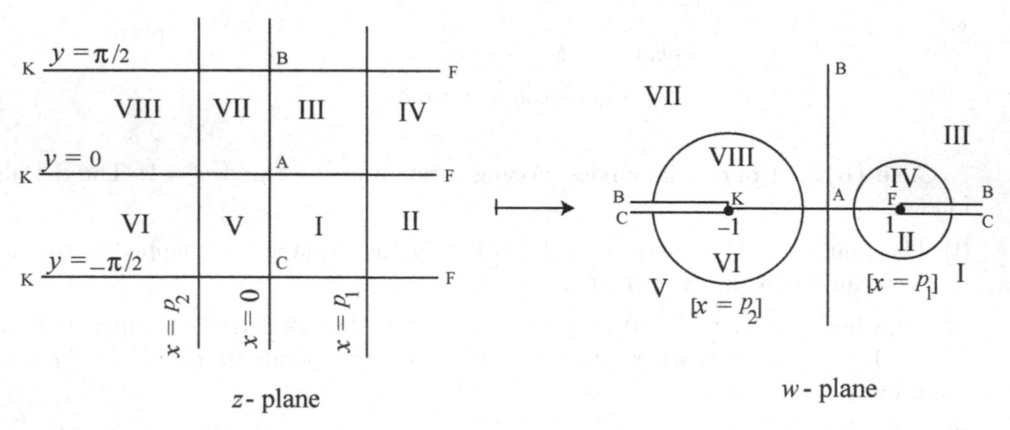

Figure 5.58 $w = \tanh z$, Case (c).

Case (d). $\phi = 0$: The details are (see Figure 5.59):

(i) The rectangle (I+III+V+VII) in the z-plane is mapped onto the exterior of two circles with one slit in the w-plane.

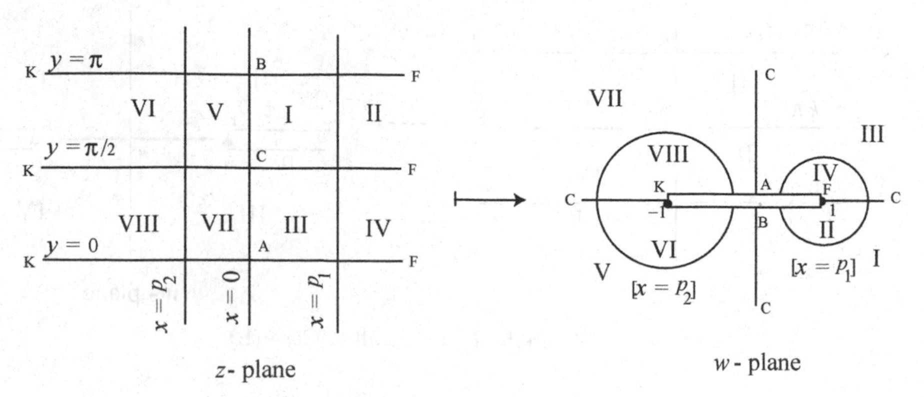

Figure 5.59 $w = \tanh z$, Case (d).

Case (e). $\phi = \pi/2$: See Figure 5.60.

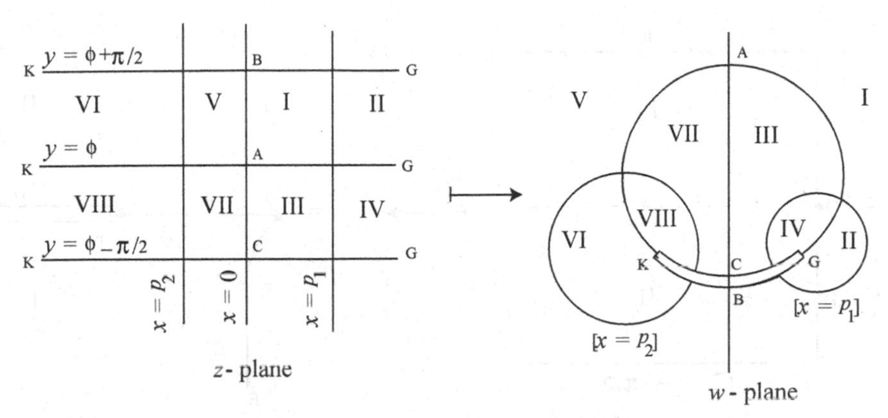

Figure 5.60 $w = \tanh z$, Case (e).

Case (f). $\phi \neq k\pi/2$: The details are as follows:

(i) The rectangle (I+III+V+VII) in the z-plane is mapped onto the exterior of two smaller circles with a slit along the arc of $|w + i\cot 2\phi| = |\csc 2\phi|$ (see Figure 5.60).

Note: Functions related to $w = \tanh z$ are:

(i) $w = \coth z = \tanh(z + i\pi/2)$, or $z = \frac{1}{2}\log\dfrac{w+1}{w-1}$;

(ii) $w = \tan z = -i\tanh z$, or $z = \arctan w = \dfrac{1}{2i}\log\dfrac{1+iw}{1-iw}$;

(iii) $w = \cot z - i\tanh(iz_i\pi/2)$, or $z = \operatorname{arccot} w = \frac{1}{2}\log\dfrac{iw+1}{iw-1}$.

Map 5.57. The mapping $w = \dfrac{\tanh z}{z}$ has critical points $z = \infty$ and at the zeros of the equation $\sinh 2z = 2\pi$, namely at $z = 0; 1.384 \pm 3.7488\,i; 1.6761 \pm 6.95\,i; 1.854 \pm 10.12\,i; 1.9916 \pm 13.277\,i$, and so on. Let $k = 1, 2, \dots$, and a be real. Then the details of this mapping are as follows.

(a) The points $z = 0; \pm 1; a\,i \pm \infty; \pm i\pi/2$, in the z-plane are mapped onto the points $w = 1; \tanh 1; 0; \infty$, respectively, in the w-plane.

(b) The line segment $x = 0, k\pi \leq y < \left(k + \frac{1}{2}\right)\pi$, in the z-plane is mapped onto the half-line $0 \leq u < \infty, v = 0$ in the w-plane.

(c) The line segment $x = 0, \left(k + \frac{1}{2}\right)\pi < y \leq (k+1)\pi$ in the z-plane is mapped onto the half-line $-\infty < u \leq 0, v = 0$ in the w-plane.

(d) The line segments $\left\{\begin{array}{c} x = 0, 0 \leq y < \pi/2 \\ x = 0, -\pi/2 < y \leq 0 \end{array}\right\}$ in the z-plane are mapped onto the half-line $1 \leq u < \infty, v = 0$ in the w-plane.

(e) The strip $0 < y < \pi/, -\infty < x < \infty$ in the z-plane is mapped onto the same region as above but cut from $w = 0$ to $w =$, where the asymptote to the above curve is $u = 4/\pi^2$.

This mapping is shown in Figure 5.61.

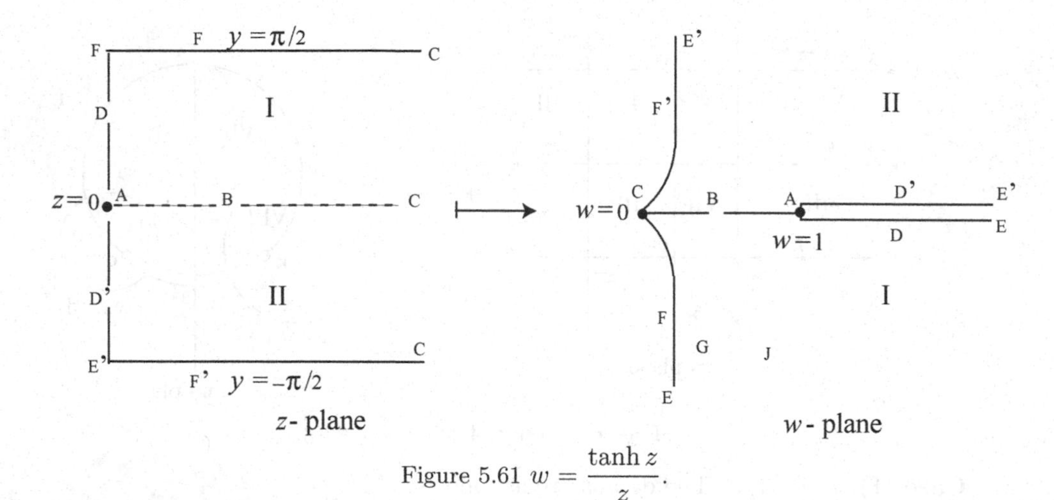

Figure 5.61 $w = \dfrac{\tanh z}{z}$.

Map 5.58. The transformation that maps the interior of the parabola $y^2 = 4p(p - x)$ in the z-plane onto the interior of the unit circle in the w-plane is defined by

$$w = \tanh^2 \frac{\pi}{4}\sqrt{\frac{z}{p}}, \tag{5.3.8}$$

and is presented in Figure 5.62.

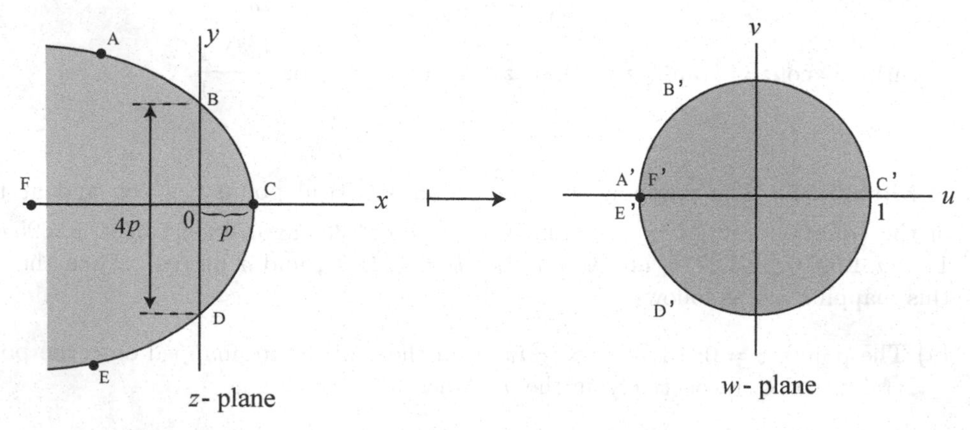

Figure 5.62 Interior of a parabola onto the interior of the unit circle.

Map 5.59. The transformation $w = \coth(z/2)$ maps the infinite rectangle in the right-half of the z-plane, open to the right, onto the right-half of the w-plane, and is presented in Figure 5.63.

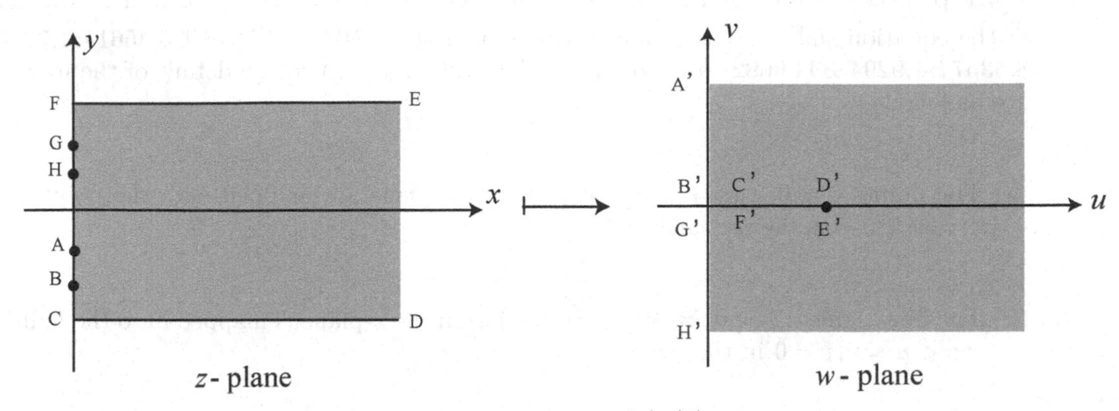

Figure 5.63 $w = \coth(z/2)$.

Map 5.60. The transformation that maps the upper half-plane $\Im\{z\} > 0$ with the unit circle removed, in the z-plane, onto the upper half-plane $\Im\{w\} > 0$ is defined by

$$w = \coth(\pi/z), \tag{5.3.9}$$

and is presented in Figure 5.64.

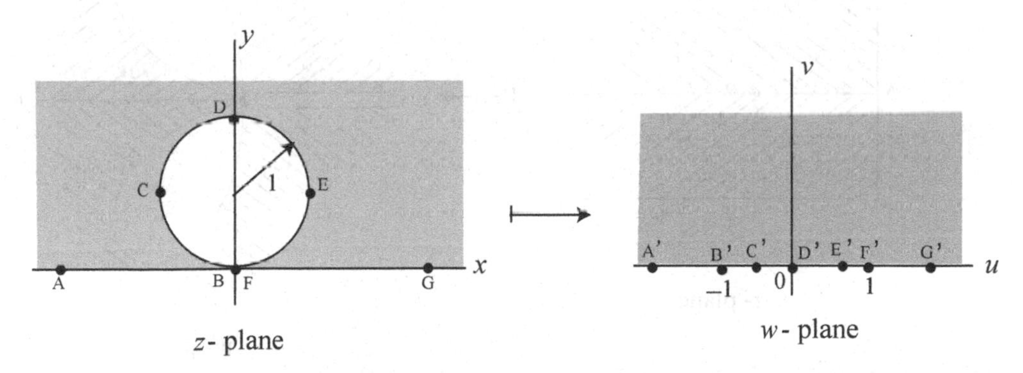

Figure 5.64 Upper half-plane with circle removed onto the upper half-plane.

Map 5.61. The mapping $w = z + c\coth z, c > 0$, has critical points $z = \infty; ik\pi; ik\pi \pm \sinh^{-1}\sqrt{c}$, where $k = 0, \pm 1, \pm 2, \dots$. It maps the points $z = \sinh^{-1}\sqrt{c} \equiv z_0; -z_0; -\frac{1}{2}i\pi$ in the z-plane onto the points $w = \sinh^{-1}\sqrt{c} + (c+1)\sqrt{c} \equiv w_0; -w_0; -\frac{1}{2}i\pi$ in the w-plane. This mapping is shown in Figure 5.55.

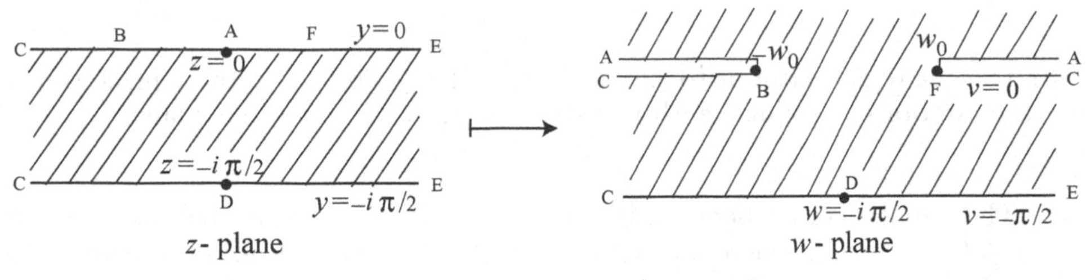

Figure 5.55 $w = z + c\coth z, c > 0$.

Map 5.62. The mapping $w = \dfrac{\coth z}{z}$ has critical points at $z = \infty$ and at the zeros of the equation $\sinh z = -2\pi$, namely, $z \approx 0; 1.1254 \pm 2.1062\, u; 1.5516 \pm 5.3561\, i; 1.7775 \pm 8.5367\, i; 1.9294 \pm 11.6692\, i$ and so on. Let $k = 0, 1, 2, \dots$. Then the details of the mapping are as follows:

(a) The points $z = 0; \pm 1; a\, i \pm \infty; \pm i\pi/2$ in the z-plane are mapped onto the points $w = \infty; \coth 1; 0; 0$ in the w-plane.

(b) The line segment $x = 0, k\pi < y \le \left(k + \frac{1}{2}\right)\pi$, in the z-plane is mapped onto the half-line $-\infty < u \le 0, v = 0$ in the w-plane.

(c) The line segment $x = 0, \left(k + \frac{1}{2}\right)\pi < y \le (k+1)\pi$, in the z-plane is mapped onto the half-line $0 \le u < \infty, v = 0$ in the w-plane.

The mapping is presented in Figure 5.65.

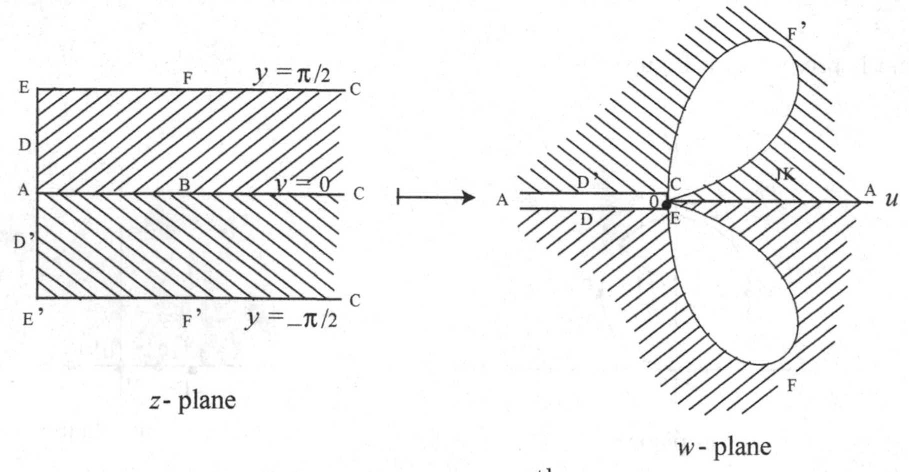

Figure 5.65 $w = \dfrac{\coth z}{z}$.

Map 5.63. To map the region exterior to two disjoint circles, with a cut annular region, we will use the mapping $w = -2\,\mathrm{arccot}(z/c)$, where $0 < c < a, c < b$. Set $r_1^2 = a^2 - c^2$ and $r_2^2 = b^2 - c^2$. The circles have the x-axis ($y = 0$) as the radical axis. Then the details of this mapping are as follows.

(a) The cut region exterior to $|z - ia| = r_1$ and $|z + ib| = r_2$ in the z-plane is mapped onto the interior of rectangle with vertices $i\alpha \pm \pi, -i\beta \pm \pi$ in the w-plane.

(b) The annular region between $|z - ia| = r_1$ and $|z - ib| = r_2$, cut along $x = 0, a - r_1 \le y \le b - r_2$ (AG), where $a \ne b$, in the z-plane is mapped onto the interior of rectangle with vertices $i\alpha \pm \pi, i\beta \pm \pi$.

These details are presented in Figure 5.66.

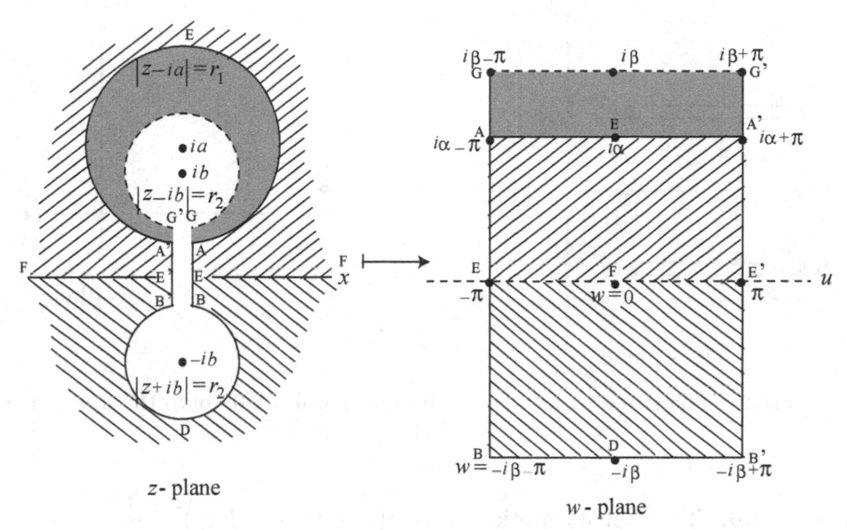

Figure 5.66 $w = -2\,\mathrm{arccot}(z/c)$.

Map 5.64. The transformation $w = i\pi + z - \log z$ maps the upper half-plane $\Im\{z\} > 0$ onto the upper half-plane of finite width $0 < v < 1 + i\pi$ with a slit in the w-plane, as shown in Figure 5.67.

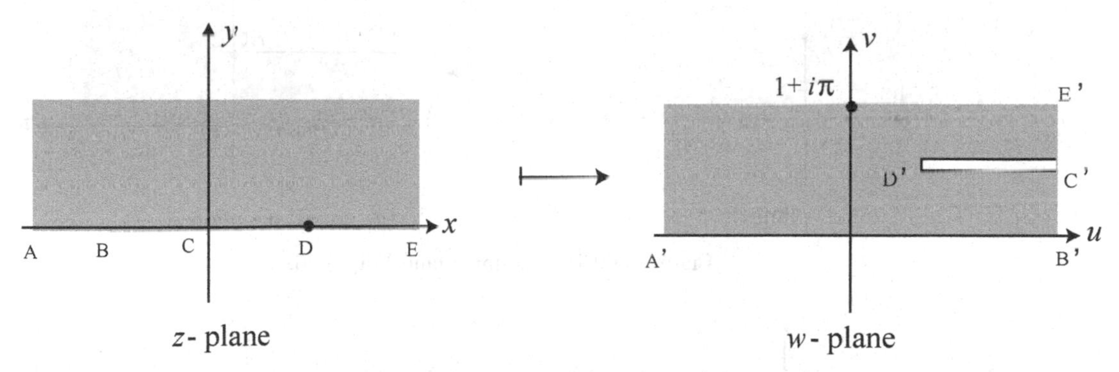

Figure 5.67 Upper half-plane under $w = i\pi + z - \log z$.

Map 5.65. The transformation that maps the z-plane with two semi-infinite parallel cuts onto the upper half-plane $\Im\{w\} > 0$ is defined by

$$w = -i\pi + 2\log z - z^2, \tag{5.3.10}$$

and is presented in Figure 5.68.

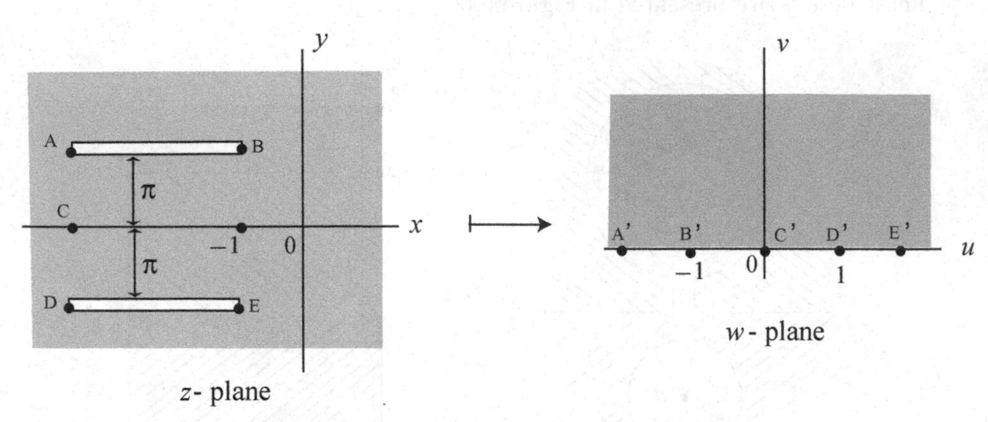

Figure 5.68 Plane with two semi-infinite parallel cuts onto the upper half-plane.

Map 5.66. The transformation $w = z(z+1)^{1/2} + \log \dfrac{(z+1)^{1/2}-1}{(z+1)^{1/2}+1}$ maps the upper half-plane $\Im\{z\} > 0$ onto the upper half first quadrant and finite strip $0 < v \le i\pi$ in the w-plane, as shown in Figure 5.69.

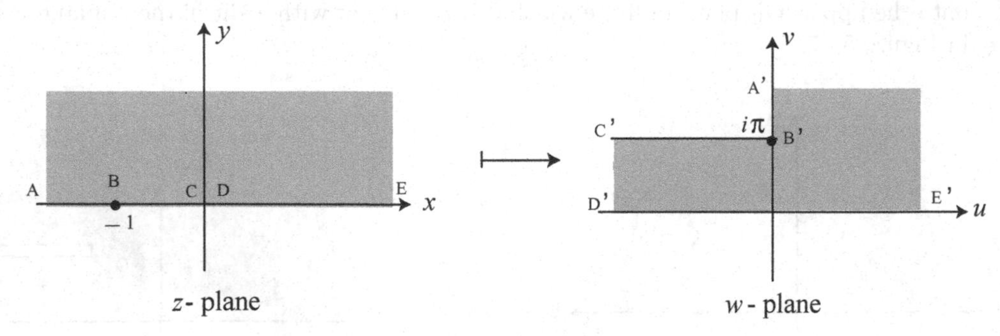

Figure 5.69 Upper half-plane Map 5.66.

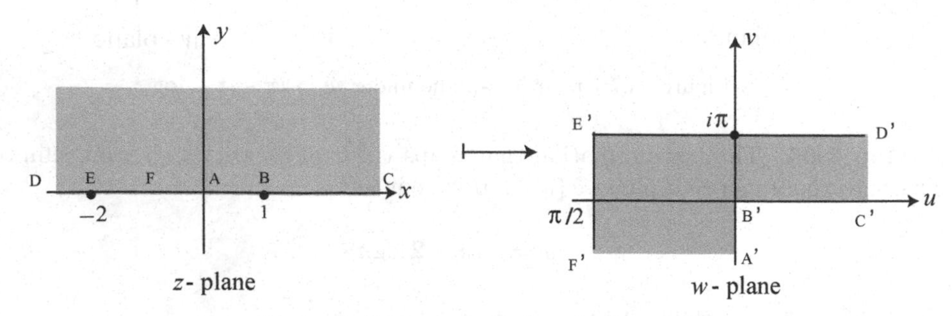

Figure 5.70 Upper half-plane in Map 5.67.

Map 5.67. The transformation $w = \dfrac{i}{k} \log \dfrac{1+ikt}{1-ikt} + \log \dfrac{1+t}{1-t}$, where $t = \left(\dfrac{z-1}{z+k^2}\right)^{1/2}$, maps the upper half-plane $\Im\{z\} > 0$ onto the strip $-\infty < v \le i\pi$ in the left-half and the strip $0 < v \le i\pi$ in the right-half of the w-plane, as shown in Figure 5.70.

Map 5.68. The mapping $w = z - \dfrac{1}{z} + 2c\log z, c > 0$, has critical points $z = 0; \infty; -c + \sqrt{c^2-1}; -c - \sqrt{c^2-1}$. Set $f = i(2 + c\pi)$. This mapping has the following properties:

(i) The points $z = i; -i; 1$ in the z-plane are mapped onto the points $w = f; -f; 0$ in the w-plane.

(ii) The half-line $y = 0, 0 < x < \infty$, in the z-plane is mapped onto the line $v = 0, -\infty < u < \infty$ in the w-plane.

(iii) The half-line $x = 0, 0 < y < \infty$, in the z-plane is mapped onto the curve $v = c\pi + 2\cosh\dfrac{u}{2c}, -\infty < u < \infty$ in the w-plane.

(iv) The half-line $x = 0, -\infty < y < 0$, in the z-plane is mapped onto the curve $v = -c\pi - 2\cosh\dfrac{u}{2c}, -\infty < u < \infty$ in the w-plane.

These properties are presented in Figure 5.71.

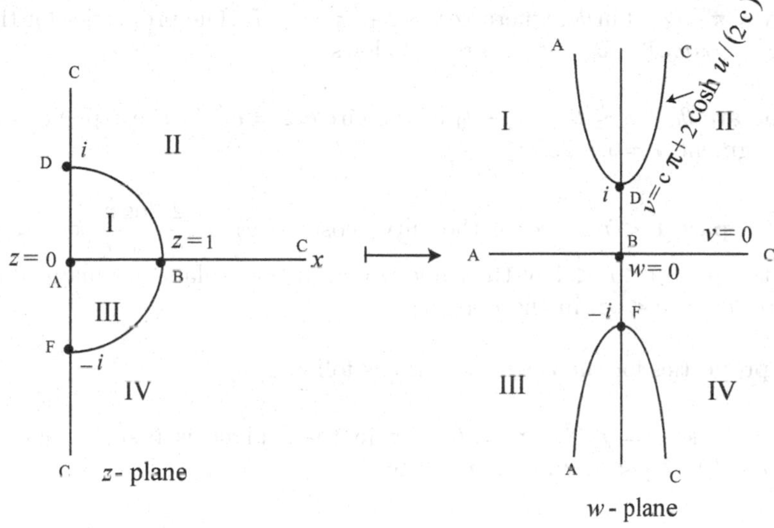

Figure 5.71 $w = z - \dfrac{1}{z} + 2c\log z, c > 0$.

Other properties for the case $c = 1$ are as follows:

(v) The circle $z = e^{i\theta}, -\pi < \theta < \pi$, in the z-plane is mapped onto the line-segment $u = 0, -2\pi < v < 2\pi$ in the w-plane.

(vi) The part $1 \le r < \infty$ of the curve CKGLA (Figure 5.72), defined by $\cos(\pi - \theta) = \dfrac{2\log r}{r - 1/r}, \pi/2 < \theta \le \pi$, and also the part $0 < r \le 1$ of the same curve, in the z-plane are mapped onto the half-line $u = 0, 2\pi \le v < \infty$ in the w-plane.

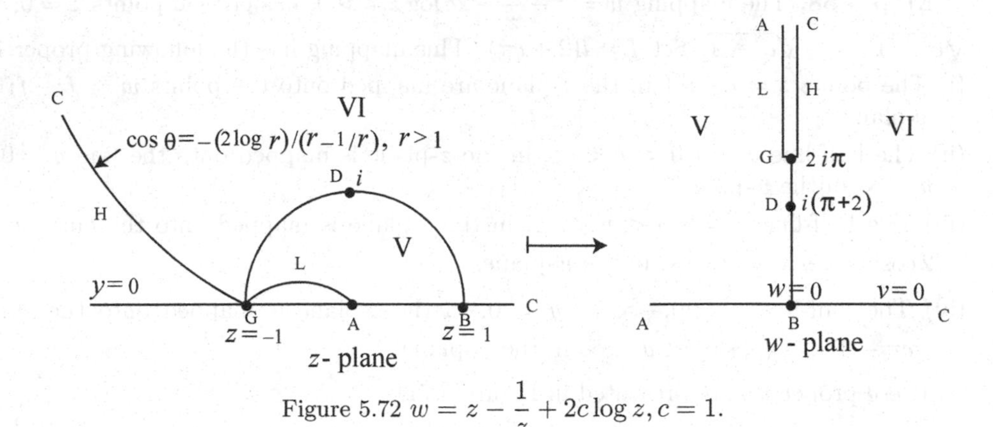

Figure 5.72 $w = z - \dfrac{1}{z} + 2c \log z, c = 1$.

Let $\lambda = \pi - \theta_0 + \tan\theta_0$, where $2c\lambda > 2 + c\pi = -if$. The properties for the case $0 < c < 1$, where $c = \cos\theta_0, 0 < \theta_0 < \pi/2$, are as follows:

(vii) The arc $\theta_0 - \pi \leq \theta \leq \pi - \theta_0$ of the circle $z = e^{i\theta}$ in the z-plane is mapped onto the line-segment $u = 0, -2c\lambda \leq v \leq 2c\lambda$.

(viii) The part $1 < r < \infty$ of the curve $\cos(\pi - \theta) = \dfrac{2c \log r}{r - 1/r}, \pi/2 < \theta < \pi - \theta_0$, and also the part $0 < r < 1$ of the same curve, in the z-plane are mapped onto the half-line $u = 0, 2c\lambda < v < \infty$, in the w-plane.

The properties for the case $c > 1$ are as follows:

(ix) The circle $z = re^{t\theta}, -\pi < \theta < \pi$ in the z-plane is mapped onto the line-segment $u - 0, -2c\pi < v < 2c\pi$ in the w-plane.

(x) The part $r_0 < r < \infty$ of the curve $\cos(\pi - \theta) = \dfrac{2c \log r}{r - 1/r}, \pi/2 < \theta < \pi$, as well as the part $0 < r < 1/r_0$ of the same curve, in the z-plane are mapped onto the half-line $u = 0, 2c\pi < v < \infty$ in the w-plane.

(xi) The segment $-1 < x < -e^{-\alpha}$ of $y = 0$, as well as the segment $-e^{-\alpha} < x < -1/r_0$ of $y = 0$, in the z-plane are mapped onto the segment $-2(\alpha \cosh\alpha - \sinh\alpha) < u < 0$ of $v = 2c\pi$ in the w-plane.

(xii) The segment $-e^{-\alpha} < x < -1$ of $y = 0$, as well as the segment $-e^{-\alpha} < x < -r_0$ of $y = 0$, in the z-plane are mapped onto the segment $0 < u < 2(\alpha \cosh\alpha - \sinh\alpha)$ of $v = 2c\pi$ in the w-plane.

(xiii) The points $z = -e^{\alpha}; -e^{-\alpha}$ in the z-plane are mapped onto the points $w = 2(\alpha \cosh\alpha - \sinh\alpha) \pm 2ic\pi; 2(\sinh\alpha - \alpha\cosh\alpha) \pm 2ci\pi$.

These results are presented in Figure 5.73. Note that in the case of $0 < c < 1$, the point $z_0 = -e^{-i\theta} = -c + i\left(1 - c^2\right)^{1/2}$ and $\bar{z}_0 - = -e^{-i\theta}$; in the case $c > 1$, the circle $|z| = 1$ (i.e., $u = 0$) is counted infinitely many times; so is circle $u = p_1$, where $0 < p_1 < p_2 < \cdots < \cdots$, and $z = -1/z + 2c \log z = w$.

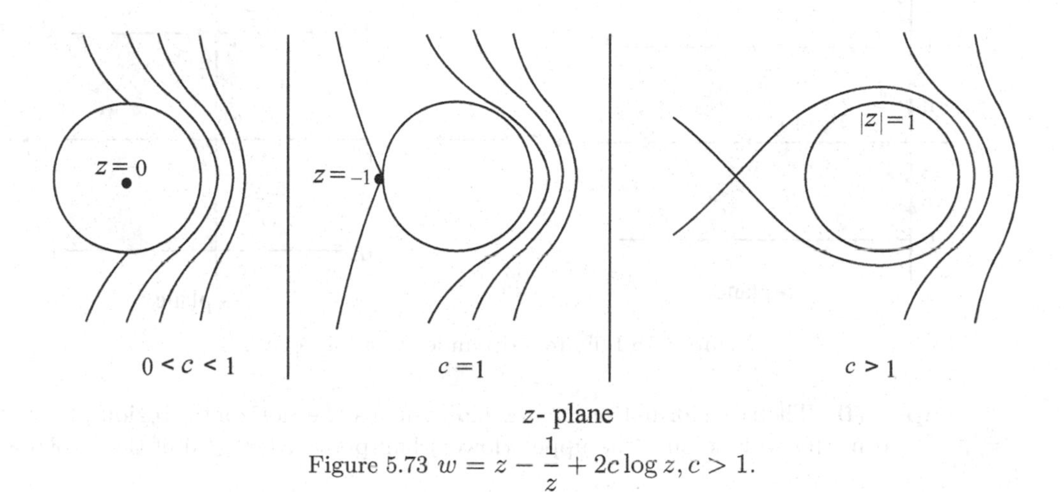

Figure 5.73 $w = z - \dfrac{1}{z} + 2c \log z, c > 1$.

Map 5.69. The mapping $w = \log \coth \dfrac{z}{2}$, or $z = \log \coth \dfrac{w}{2}$, has critical points $z = \infty; ki\pi$, where $k = 0, \pm 1, \pm 2, \dots$. This mapping is involutory. Under this mapping, the points $z = ki\pi; \log\left(1 + \sqrt{2}\right); i\pi/2; -i\pi/2$ in the z-plane are mapped onto the points $w = \infty; \log\left(1 + \sqrt{2}\right); -i\pi/2; i\pi/2$, respectively, in the w-plane. The mapping is presented in Figure 5.74.

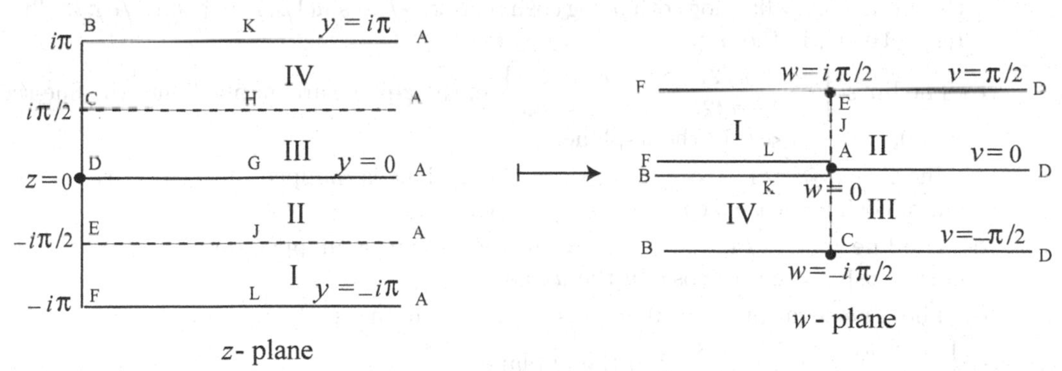

Figure 5.74 $w = \log \coth \dfrac{z}{2}$.

The transformation $w = \log \coth(z/2)$ maps the infinite strip in the right half z-plane onto the infinite strip $-\infty < u < \infty, i\pi/2 \leq v \leq i\pi/2$ in the w-plane, as shown in Figure 5.75.

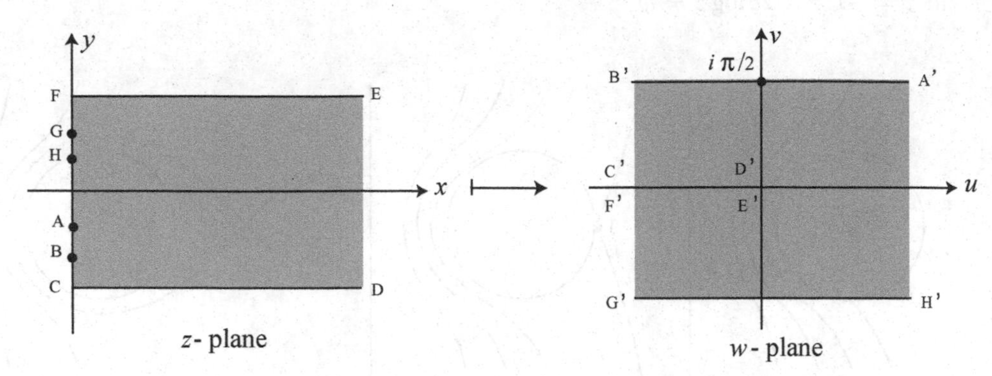

Figure 5.75 Infinite strip under $w = \log\coth(z/2)$.

Map 5.70. The transformation $w = \pm\sinh z$ maps the horizontal region $-\pi/2 < y < \pi/2$, $x \geq 0$ in the z-plane onto the upper (lower) half-plane $\Im\{w\} \gtrless 0$ of the w-plane.

Map 5.71. The mapping $w = \log\dfrac{\sinh(z+\beta)}{\sinh z} = \log(a + \coth z) - \log\sqrt{(a^2-1)}$, $\beta > 0, a = \coth\beta > 1$, or $z = \frac{1}{2}\log\dfrac{\sinh\frac{1}{2}(w+\beta)}{\sinh\frac{1}{2}(w-\beta)} - \frac{1}{2}\beta$, has critical points $z = ik\pi; -\beta + ik\pi; \infty$, where $k = 0, \pm1, \pm2, \dots$. The details of this transformation are as follows:

(a) The points $z = -\frac{\beta}{\pm}i\pi/2; -\beta \pm i\pi/2; \pm i\pi/2; -\frac{1}{2}\beta; -\beta/2 + i\pi/4; -\frac{1}{2}\beta - i\pi/4; -\frac{1}{2}\beta - \frac{i}{2}\arcsin 1/a; -\frac{1}{2}\beta + \frac{i}{2}\arcsin(1/a)$ in the z-plane, $-\pi/2 \leq y \leq \pi/2$, are mapped onto the points $w = 0; -\log\coth\beta; \log\cosh\beta; \pm i\pi; -i\arcsin(1/a); i\arcsin(1/a); i\pi/2; -i\pi/2$, respectively, in the w-plane, $-\pi \leq v \leq \pi$.

(b) The lines $\left\{\begin{array}{c} y = \pi/2, -\infty < x < \infty \\ y = -\pi/2. -\infty < x < \infty \end{array}\right\}$ in the z-plane are mapped onto the line-segment $v = 0, = \beta < u < \beta$ in the w-plane.

(c) The line $y = \pi/4, -\infty < x < \infty$, in the z-plane is mapped onto the part $v < 0$ of the curve $\cosh u = \cosh\beta\cos v$ in the w-plane.

(d) The line $y = -\pi/4, -\infty < x < \infty$, in the z-plane is mapped onto the part $v > 0$ of the curve $\cosh u = \cosh\beta\cos v$ in the w-plane.

(e) The line-segment $y = 0, = \beta < x < \beta$ in the z-plane is mapped onto the lines $\left\{\begin{array}{c} v = \pi, -\infty < u < \infty \\ v = -\pi, -\infty < u < \infty \end{array}\right\}$ in the w-plane.

(f) The line-segment $x = -\beta/2, 0 < y < \pi/2$, in the z-plane is mapped onto the half-line $u = 0, -\pi < v < 0$ in the w-plane.

(g) The line-segment $x = -\beta/2, -\pi/2 < y < 0$, in the z-plane is mapped onto the half-line $u = 0, 0 < v < \pi$ in the w-plane.

This mapping is presented in Figure 5.76.

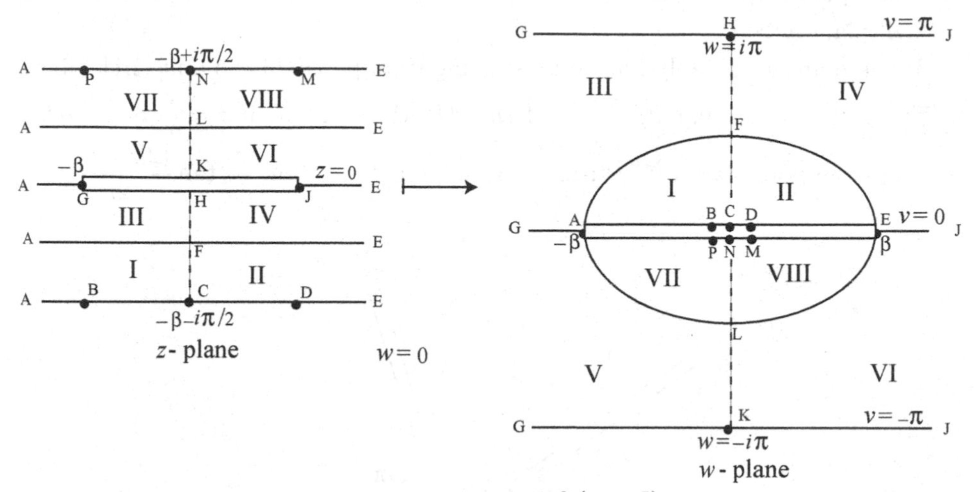

Figure 5.76 $w = \log \dfrac{\sinh(z+\beta)}{\sinh z}$.

Map 5.72. The transformation $w = k \log \dfrac{k}{1-k} + \log 2(1-k) + i\pi - k \log(z+1)^{-(1-k)} \times \log(z-1)$, $x_1 = 2k-1$, maps the open rectangle in the upper half-plane $-\infty < x < \infty$, $y > 0$, onto the infinite strip $0 < u < \infty, 0 < v \leq i\pi$, with a slit at $z = ki\pi$ in the first quadrant in the w-plane, as shown in Figure 5.77.

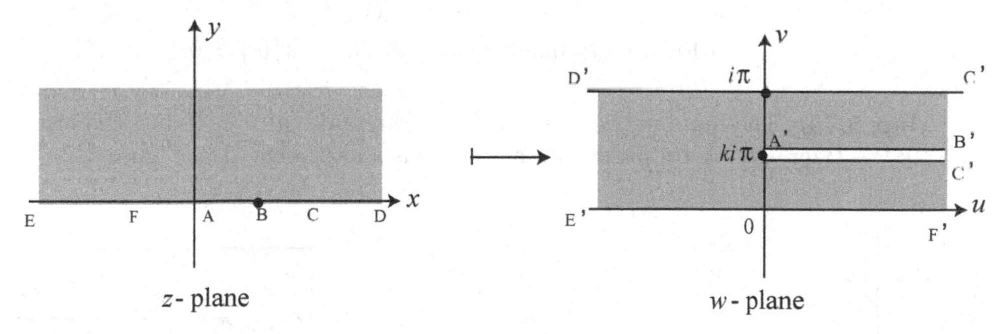

Figure 5.77 Map 5.72.

Map 5.73. The transformation $w = \tan\left(\dfrac{1}{n} \arccos z^{n/2}\right)$, $z = \{\cos(n \arctan w)\}^{-2/n}$, $n \in \mathbb{N}$, maps the z-plane with n equally spaced slits onto the upper half plane $\Im\{w\} > 0$. The mapping is shown in Figure 5.78.

This mapping has the following properties:

(a) The critical points are $z = 0; \infty; e^{2\pi ki/n}$, $k = 1, 2, \dots, n$.

(b) The point $z = 0$ is mapped onto the points $w = \tan \dfrac{(2k-1)\pi}{2n}$, $k = 1, 2, \dots, n$.

(c) The points $z = e^{-2\pi ki/n}$, $k = 1, 2, \dots, n$, are mapped onto the points $w = \tan \dfrac{k\pi}{n}$.

(d) The plane with n slits defined by $\arg\{z\} = \dfrac{2\pi ki}{n}$, $0 < |z| \le 1$, is mapped onto the half-plane $v > 0$.

If there are n infinitely long slits, starting from points of the unit circle, along $\arg\{z\} = e^{2\pi ki/n}$, $|z| \ge 1$, $k = 0, 1, 2, \dots, n-1$ $(n \in \mathbb{N})$, the mapping is $z = \left\{\cos(n \arctan w)\right\}^{-2/n}$. It maps the points $z = e^{2\pi ki/n}; \infty$ onto the points $w = \tan\dfrac{k\pi}{n}; \tan\dfrac{(2k+1)\pi}{2n}$.

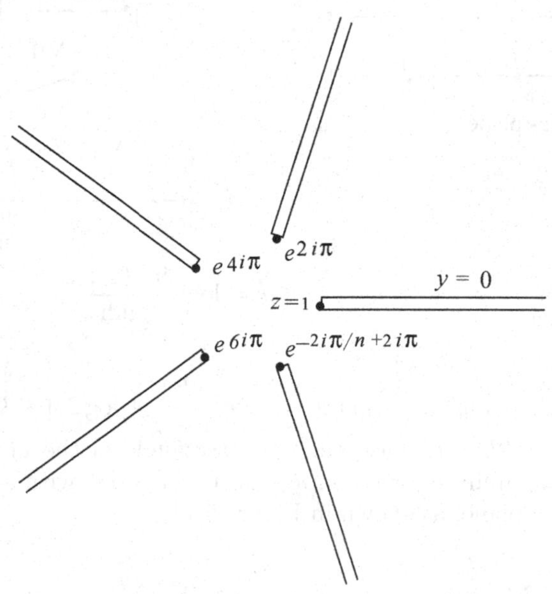

Figure 5.78 Infinitely long slits onto $\Im\{w\} > 0$.

Map 5.74. The mapping $w = \tanh^{-1} z - b\arctan(z/b), b > 0$, has the critical points $z = 0; 1; -1; ib; -ib; \infty$. Its properties are obvious and presented in Figure 5.79.

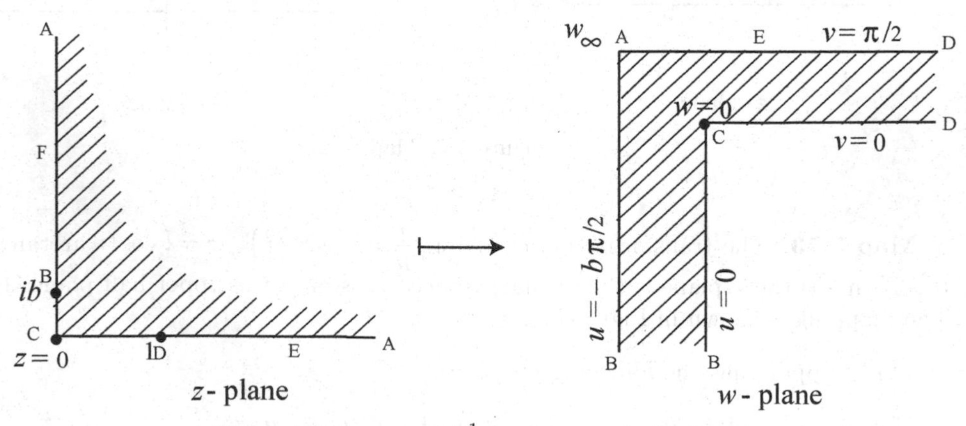

Figure 5.79 $w = \tanh^{-1} z - b\arctan(z/b), b > 0$.

Map 5.75. The transformation that maps a right-angled channel with right angle bend in the z-plane onto the upper half-plane $\Im\{w\} > 0$ is defined by

$$w = \frac{2}{\pi}\{\tanh^{-1} p\sqrt{z} - p\arctan\sqrt{z}\}, \tag{5.3.11}$$

and is presented in Figure 5.80.

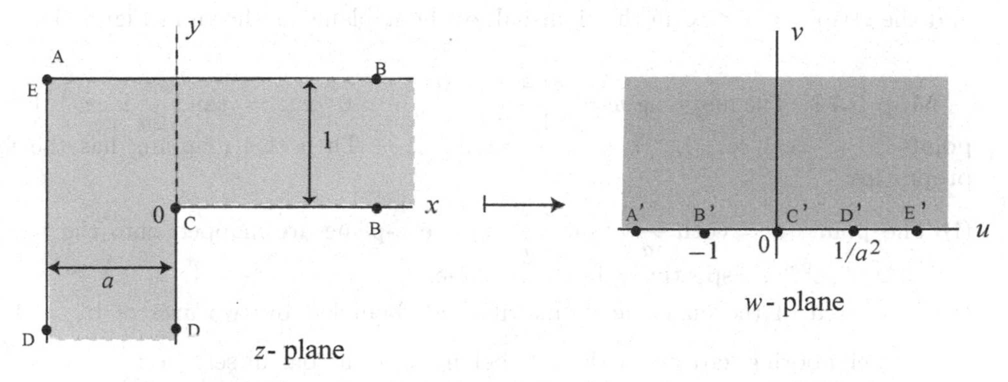

Figure 5.80 Channel with right angle bend onto the upper half-plane.

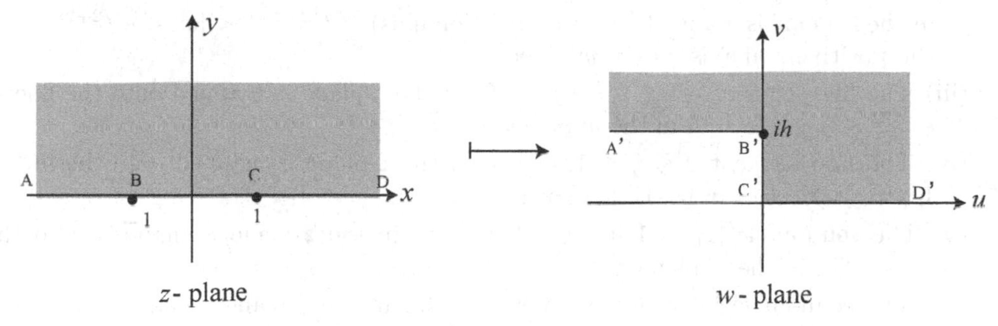

Figure 5.81 Upper half-plane in Map 5.76.

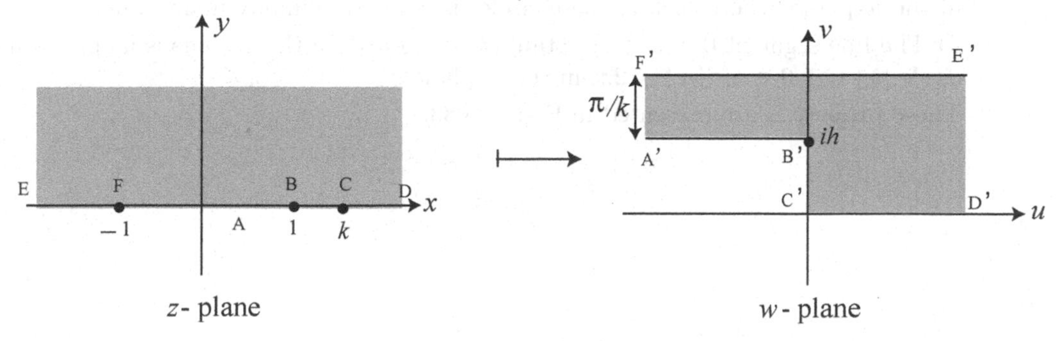

Figure 5.82 Upper half-plane in Map 5.77.

Map 5.76. The transformation $w = \frac{h}{\pi}\left[\left(z^2-1\right)^{1/2} + \cosh^{-1} z\right]$ maps the upper half-plane $\Im\{z\} > 0$ onto the strip $h \le v < \infty$ in the left-half and the strip $0 < v < \infty$ in the right-half of the w-plane, as shown in Figure 5.81.

Map 5.77. The transformation $w = \cosh^{-1}\left(\frac{2z-k-1}{k-1}\right) - \frac{1}{k}\cosh^{-1}\left(\frac{(k+1)z-2k}{(k-1)z}\right)$ maps the upper half-plane $\Im\{z\} > 0$ onto the strip AF of finite width π/k in the left-half and the strip $0 < u < \infty$ in the right-half of the w-plane, as shown in Figure 5.82.

Map 5.78. The mapping $w = \left(\frac{1+z}{1-z}\right)^{a/(i\pi)}$, $a > 0$, or $z = i\tan\left(\frac{\pi}{2a}\log w\right)$, has critical points $z = -1; 1; \infty$. Let $k, m = 0, \pm 1, \pm 2, \dots$. Then this mapping has the following properties:

(i) The points $z = \coth\frac{k\pi^2}{a}; \tanh\frac{k\pi^2}{a}$ in the z-plane are mapped onto the points $w = e^{(2m+1)a}; e^{2ma}$, respectively, in the w-plane.

(ii) The area of the curvilinear quadrilateral, bounded by two arcs of $|z| = 1$ and by two neighboring 'circles' of the set, belonging to a coaxial set, $\Gamma_k : \left|z - \coth\frac{2k\pi^2}{a}\right| = \left(\sinh\frac{2|k|\pi^2}{a}\right)^{-1}$, (where for $k = 0$ we take the line $x = 0$), with limiting points $z = \pm i$, in the z-plane is mapped onto the ring (annulus) $e^{-a/2+2ma} < |w| < e^{a/2+2ma}$, cut along the positive real axis, in the w-plane.

(iii) The line-segment $-1 \le y \le 0, x = 0$, in the z-plane is mapped onto the line-segment $e^{-a/2} \le u \le 1, v = 0$, in the w-plane.

(iv) The line-segment $0 \le y \le 1, x = 0$, in the z-plane is mapped onto the line-segment $1 \le u \le e^{a/2}, v = 0$, in the w-plane.

(v) The semi-circle $|z| = 1, 0 \le \arg\{z\} < \pi$, in the z-plane is mapped onto the circle $|w| = e^{a/2}$ in the w-plane.

(vi) The semi-circle $|z| = 1, \pi \le \arg\{z\} < 2\pi$, in the z-plane is mapped onto the circle $|w| = e^{-a/2}$ in the w-plane.

(vii) The line segment $-1 < x < 1, y = 0$, in the z-plane is mapped onto the circle $|w| = 1$ in the w-plane, where each of these circles is covered infinitely many times.

(viii) The line segment $0 \le x < z_1 = \tanh(\pi^2/2), y = 0$, in the z-plane is mapped onto the circle $|w| = 1, 0 < \arg\{w\} \le 2\pi$, in the w-plane.

These properties are presented in Figure 5.83.

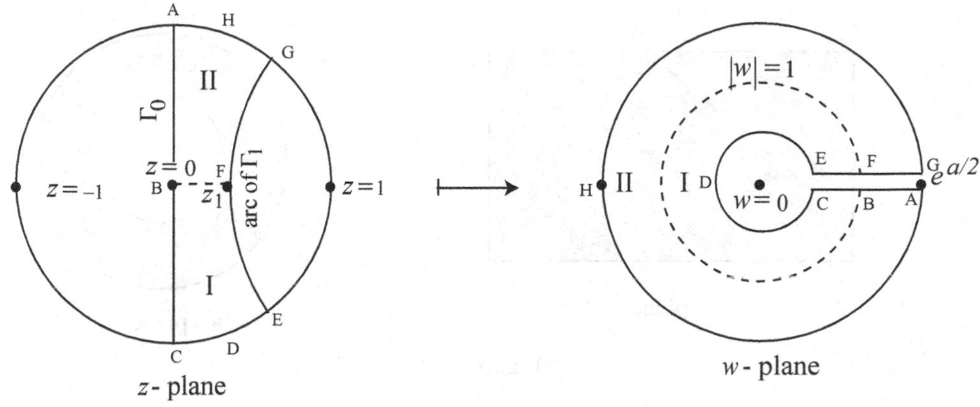

Figure 5.83 $w = \left(\frac{1+z}{1-z}\right)^{a/(i\pi)}, a > 0.$

5.4 Complex Exponential Function

A final comment about the map $w = e^z$ (Map 5.1) is appended here. The complex exponential $w = f(z) = e^z$, or $u = e^x \cos y$, $v = e^x \sin y$, satisfies the condition $f'(z) = e^z \neq 0$ everywhere. However, it is not one-to-one because $e^{z+2\pi i} = e^z$, and thus, the points which differ by an integer multiple of $2\pi i$ are all mapped into the same point. Under the exponential map, the horizontal line $\Im\{z\} = b$ is mapped onto the curve $w = e^{x+ib} = e^x(\cos b + i \sin b)$, which, for $-\infty < x < \infty$, traces out the ray emanating from the origin making an angle $\arg\{w\} = b$ with the real axis (see Figure 5.1). Hence, the exponential map will map a horizontal strip

$$S_{a,b} = \{a < \Im\{z\} < b\}$$

onto a wedge-shaped region $G_{a,b} = \{a < \arg\{w\} < b\}$ and is one-to-one provided $|b - a| < 2\pi$. In particular, the horizontal strip

$$S_{-\pi/2,\pi/2} = \{-\pi/2 < \Im\{z\} < \pi/2\}$$

of width π and centered around the real axis is mapped one-to-one to the right half-plane $G_{-\pi/2,\pi/2} = \{-\pi/2 < \arg\{w\} < \pi/2\} = \Im\{w\} > 0$, while the horizontal strip

$$S_{-\pi,\pi} = \{\pi < \Im\{z\} < \pi\}$$

of width 2π is mapped onto the region

$$G_{-\pi,\pi} = \{-\pi < \arg\{w\} < \pi\} = \mathbb{C}\backslash\{\Im\{z\} = 0, \Re\{z\} \leq 0\},$$

with a slit in the complex plane $\mathbb{C}$ along the negative real axis.

The vertical lines $\Re\{z\} = a$ are mapped to the circles $|w| = e^a$. Thus, a vertical strip $a < \Re\{z\} < b$ is mapped onto the annulus $e^a < |w| < e^b$, though many-to-one, since the strip is continuously wrapped around the annulus. The rectangle $R = \{a < x < b, \pi < y < \pi\}$ of height 2π is mapped one-to-one on a annulus cut along the negative real axis (see Figure 5.84). Finally, no domain is mapped onto the unit disk $U = \{|w| < 1\}$, or any other region containing the origin, since the exponential function is never zero.

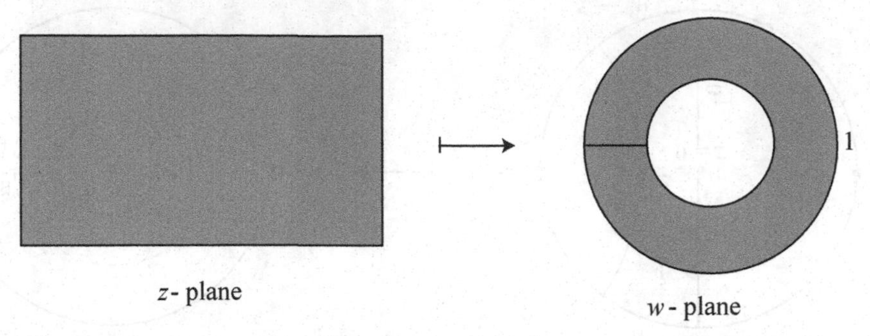

Figure 5.84 $w = e^z$.

Also note that the curve $z(t) = e^{it} = \cos t + i \sin t$, $0 \le t \le 2\pi$, parameterizes the unit circle $|z| = 1$ in the complex plane. Its complex tangent $\dot{z}(t) = i\, e^{it} = i\, z(t)$ is obtained by rotating $z(t)$ through 90°. The physical interpretation is as follows: if the curve is regarded as the trajectory of a particle in the complex plane, the function $z(t)$ defines the position of the particle at time t, and the tangent $\dot{z}(t)$ represents its instantaneous velocity, whereas its modulus $|\dot{z}(t)| = \sqrt{\dot{x}^2 + \dot{y}^2}$ gives the particle's speed, and its phase $\arg\{\dot{z}\}$ measures the direction of motion.

REFERENCES USED: Ahlfors [1966], Boas [1987], Carrier, Krook and Pearson [1966], Kantorovich and Krylov [1958], Kober [1957], Kythe [1998], Nehari [1952], Nevanlinna [1925], Schinzinger and Laura [2003].

6

Joukowski Airfoils

The research developed by the Russian hydro- and aero-dynamics scientist Nikolai Joukowski (Николаи Жыковцкий) to determine the exact force exerted by a flow on a body around which it is flowing eventually led to the theoretical foundation for practical aircraft construction, and the methods of conformal mapping played an important role in modern aviation.

In practical applications of airfoils, nearly circular approximations are used. This topic is discussed in §13.1. The single-, two-, and multi-element airfoils are discussed in §13.2.1, §13.2.3, and §13.2.4.

6.1 Joukowski Maps

The Joukowski transformation is discussed in the following cases.

Map 6.1. The mapping $w = \frac{1}{2}(z + \dfrac{1}{z})$, known as the *Joukowski transformation*, leads to the mapping $w = \cosh\xi$, where $e^{\xi} = z$ or $\xi = \log z$. Since $\xi = \frac{1}{2}\left(e^{\xi} + e^{-\xi}\right)$, we get $w = \frac{1}{2}\left(z + 1/z\right) = \cosh\xi$. For $w = \cosh z$, see Maps 5.10(b) and 5.53. The properties of this mapping are shown in Figure 6.1.

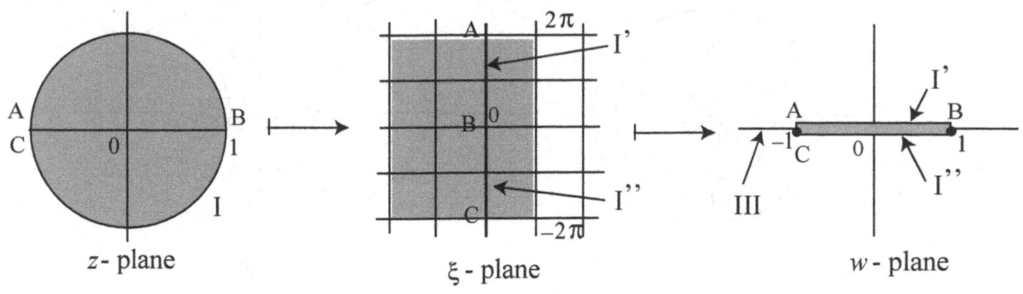

Figure 6.1 $w = \cos(\log z) = \frac{1}{2}\left(z + 1/z\right)$.

The inverse mapping is $z = w \pm \left(w^2 - 1\right)^{1/2}$; the two roots of this inverse mapping suggest two Riemann surfaces for w with transitions at $w = \pm 1$. The unit circle $|z| = 1$ (marked by I in Figure 6.1) is mapped onto the line segments I' and I" in the ξ-plane, and onto the line segment $v = 0, -1 \leq u \leq 1$ in the w-plane (the slit AB, CD); the region exterior to this unit circle is mapped onto the w-plane minus the slit AB, CD.

Map 6.2. The Joukowski function

$$w = f(z) = \frac{1}{2}\left(z + \frac{1}{z}\right) \tag{6.1.1}$$

is a second order rational function which satisfies the condition $f(z) = f(1/z)$. It means that every point of the w-plane except $w = \pm 1$ has only two distinct inverse images z_1 and z_2 such that

$$z_1\, z_2 = 1, \tag{6.1.2}$$

since the two points $z + 1 \neq z_2$ are transformed by (6.1.1) into one and the same point in the w-plane, i.e., $z_1 + \dfrac{1}{z_1} = z_2 + \dfrac{1}{z_2}$, and then $z_1 - z_2 = \dfrac{z_1 - z_2}{z_1\, z_2}$ only if $z_1\, z_2 = 1$. The function (6.1.1) is analytic in $\mathbb{C}_\infty$ except at $z = 0$ which is a simple pole for this function. The derivative

$$\frac{dw}{dz} = \frac{1}{2}\left(1 - \frac{1}{z^2}\right)$$

is nonzero at all points except $z = \pm 1$. Thus, the mapping by this function is conformal everywhere except at $z = \pm 1$. The regions of univalence for the Joukowski function are $U = \{|z| < 1\}$ and $U^* = \{|z| > 1\}$, both of which are mapped conformally by this function onto one and the same region in the w-plane which is determined as follows: Consider the mapping of the circle $|z| = r$ by the function (6.1.1). With $z = r\, e^{i\theta}$ $0 \leq \theta < 2\pi$, we find that

$$u(r,\theta) = \frac{1}{2}\left(r + r^{-1}\right)\cos\theta, \quad v(r,\theta) = \frac{1}{2}\left(r - r^{-1}\right)\sin\theta. \tag{6.1.3}$$

If we eliminate θ in (6.1.3), we get

$$\frac{4u^2}{\left(r + r^{-1}\right)^2} + \frac{4v^2}{\left(r - r^{-1}\right)^2} = 1. \tag{6.1.4}$$

Hence, the function (6.1.1) maps concentric circles $|z| = r$ conformally onto confocal ellipses with semi-axes $\left(r \pm r^{-1}\right)/2$ and foci at ± 1 on the u-axis (Figure 6.2).

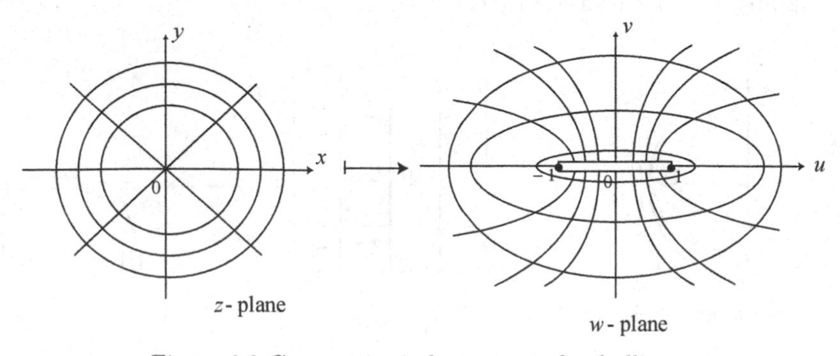

Figure 6.2 Concentric circles onto confocal ellipses.

Moreover, as $r \to 1$, the ellipse (6.1.4) reduces to the segment $[-1, 1]$ of the u-axis traversed twice. As $r \to 0$, the ellipse is transformed into a circle of infinitely large radius. Hence, the Joukowski function (6.1.1) maps the region U in the z-plane conformally onto the w-plane slit along the segment $[-1, 1]$ of the u-axis. The boundary $|z| = 1$ is mapped onto this segment such that the upper semicircle is mapped onto the lower edge of the slit and the lower semicircle onto the upper edge of the slit. The region U^* in the z-plane is mapped onto the second sheet of the w-plane slit along the segment $[-1, 1]$ of the u-axis, the upper semicircle $|z| = 1$, $\Im\{z\} > 0$ onto the upper edge, and the lower semicircle $|z| = 1$, $\Im\{z\} < 0$ onto the lower edge of the slit. Thus, the Joukowski function (6.1.1) maps the extended z-plane conformally onto the Riemann surface of the inverse function $z = g(w) = w + \sqrt{w^2 - 1}$, which is a two-sheeted surface made up of two sheets of the w-plane slit along the segment $[-1, 1]$ of the real axis.

To determine the image of a rays $\arg\{z\} = \theta_0$, we eliminate r from Eqs (6.1.3) and replace θ by θ_0. This gives

$$\frac{u^2}{\cos^2 \theta_0} - \frac{v^2}{\sin^2 \theta_0} = 1, \tag{6.1.5}$$

which shows that the rays $\arg\{z\} = \theta_0$ are transformed into branches of the hyperbola (6.1.5) with foci at ± 1 (Figure 6.2). The Joukowski function defines the orthogonal system of polar coordinates in the z-plane in terms of an orthogonal system in the w-plane such that the confocal families of ellipses and hyperbolas in the w-plane are orthogonal. For $\theta = 0$, we find from (6.1.3) that $u = \left(r + r^{-1}\right)/2$, $v = 0$, $0 \le r < 1$, which represents the interval $1 < u \le +\infty$. The infinite interval $-\infty \le u < 1$ is the image of the ray $\theta = \pi$. For $\theta = \pi/2$, we have $u = 0$, $v = -\left(r - r^{-1}\right)/2$, $0 \le r < 1$, which represents the negative imaginary axis $-\infty \le v < 0$. The positive imaginary axis $0 < v \le +\infty$ is the image of the ray $\theta = -\pi/2$. Thus, the horizontal diameter of the unit disk U is mapped onto the real axis going from -1 to $+1$ through the point at infinity and excluding the points ± 1. The vertical diameter of U is mapped onto the entire imaginary axis including the point at infinity but excluding the origin.

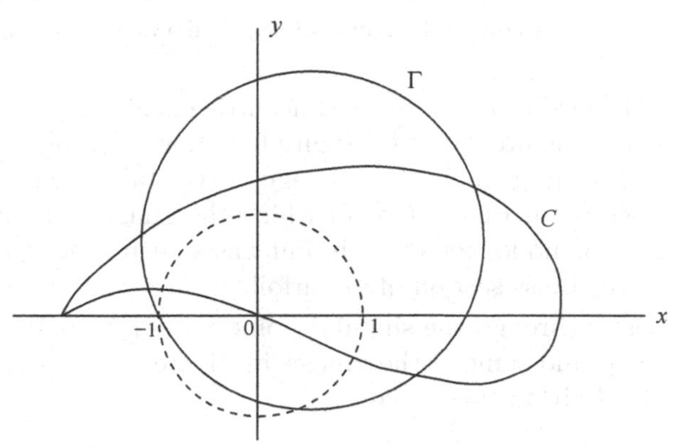

Figure 6.3 Joukowski profile.

The geometrical interpretation of the mapping

$$w = z + \frac{1}{z} \tag{6.1.6}$$

is as follows: If Γ is a circle in the z-plane passing through the point $z = -1$, such that the point $z = 1$ lies inside Γ, then the function (6.1.6) conformally maps the region exterior to

Γ onto the region exterior to the Joukowski profile C (Figure 6.3). The shape of the curve C is obtained from the circle Γ by making the point z trace out the circle Γ and adding the vectors z and $1/z$.

Map 6.3. We reconsider the Joukowski Map 6.2. Since

$$\frac{dw}{dz} = \frac{1}{2}\Big(1 - \frac{1}{z^2}\Big) = 0 \quad \text{iff } z = \pm 1,$$

the Map 6.2 is conformal except at the points $z = \pm 1$ and also at the singularity $z = 0$ where it is not defined. If $z = e^{i\theta}$ lies on the unit circle, then $w = \frac{1}{2}\left(e^{i\theta} + e^{-i\theta}\right) = \cos\theta$ lies on the real axis with $-1 \le w \le 1$. Thus, this map reduces the unit circle down to the real line segment $[-1, 1]$. The exterior of the unit circle is mapped onto the rest of the w-plane, as do the nonzero points inside the unit circle. In the case of the inverse transformation $z = w \pm \sqrt{w^2 - 1}$, we find that every w except $w = \pm 1$ is an image of two different points z, while for w not on the line segment $[-1, 1]$, one point (with the minus sign) lies inside and one (with the plus sign) lies outside the unit circle, but for the case $-1 < w < 1$, both points lie on the unit circle and a common vertical line. Thus, this inverse transformation maps the exterior of the unit circle $|z| > 1$ conformally onto the exterior of the line segment $[-1, 1]$, i.e., onto the domain $\mathbb{C} \backslash [-1, 1]$ (see Figure 6.4).

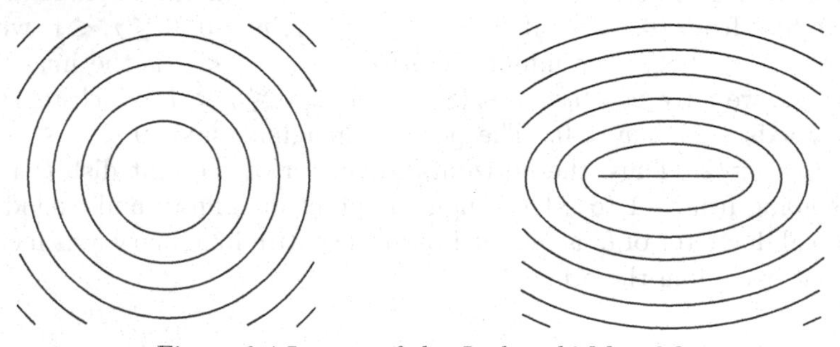

Figure 6.4 Inverse of the Joukowski Map 6.2.

This Joukowski transformation maps concentric circles $|z| = r \neq 1$ onto ellipses with foci at ± 1 in the w-plane, as presented in Figure 6.2. However, the case of circles not centered at the origin is interesting in that the image curves take a wide variety of shapes, some of which are presented in Figure 6.5, in which the image of a circle passing through the singular point $z = 1$ is no longer smooth, but has a cusp at $w = 1$, and some image curves take the shape of the cross-section of an airfoil.

If the circle passes through the singular point $z = -1$, then its image is not smooth and has a cusp at $w = 1$, and some of the images in Figure 6.5 are shaped like the cross-section through an idealized airfoil (plane wing).

A few specific details for the Joukowski transformation (6.1.1), $w = \frac{1}{2}\left(z + \frac{1}{z}\right)$ or $z = w + \sqrt{w^2 - 1}$, for a circle in the z-plane are presented in Figures 6.6 (a)–(e):

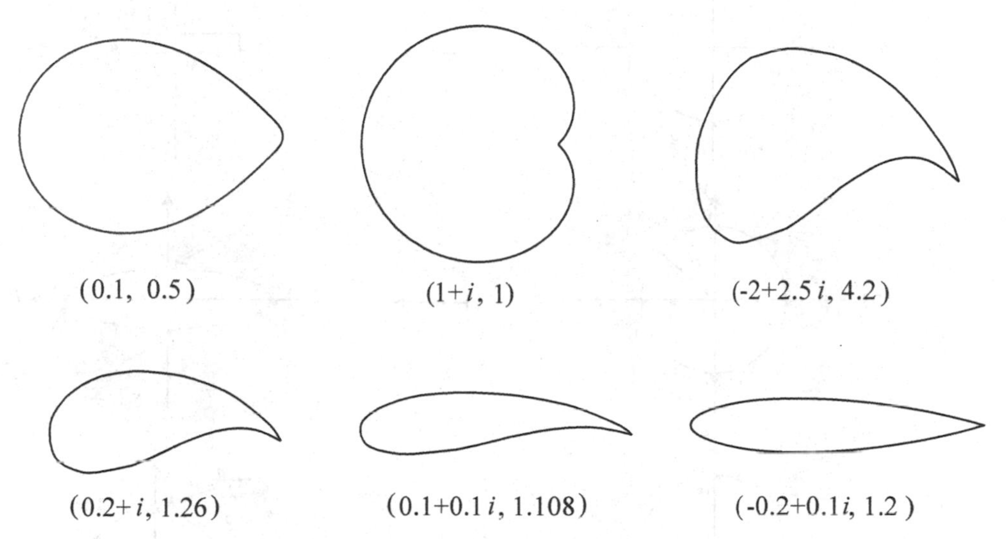

Figure 6.5 Airfoils obtained by Joukowski Map from circles (center, radius).

Map 6.4a. Figure 6.6 (a) shows the mapping from the exterior of the unit disk in the z-plane onto the entire w-plane with a slit from $w = -1$ to $w = 1$. The mapping function is the inverse function $z = w + \sqrt{w^2 - 1}$.

Map 6.4b. Figure 6.6 (b) shows the mapping from the exterior of the circle of radius c in the z-plane onto the exterior of an ellipse w-plane with major and minor axis $(c + 1/c)$ and $(c - 1/c)$ respectively. The mapping function is the inverse function $z = w + \sqrt{w^2 - 1}$.

Map 6.4c. Figure 6.6 (c) shows the mapping from the exterior of the circle in the z-plane passing through the points $(0, a)$ and $(0, b)$, $a > b$, and the real axis through the points $x = \pm 1$, onto the circular arc through the points$-1, (a - b)/2, 1$ in the w-plane. The mapping function is the inverse function $z = w + \sqrt{w^2 - 1}$.

Map 6.4d. Figure 6.6 (d) shows the mapping from the exterior of the circle with radius 0.1 in the z-plane onto the airfoil with center 0.1 and radius 0.5 in the w-plane. The mapping function is the inverse function $z = w + \sqrt{w^2 - 1}$.

Map 6.4e. Figure 6.6 (e) shows the mapping from the exterior of the circle with radius 0.1 in the z-plane onto the airfoil with center 0.2 and radius 1.2 in the w-plane. The mapping function is the inverse function $z = w + \sqrt{w^2 - 1}$.

Notice that if in Maps 6.4a through 6.4e the mapping is from the interior of the circle onto their respective image domains in the w-plane, the mapping function will be the inverse function $z = w - \sqrt{w^2 - 1}$.

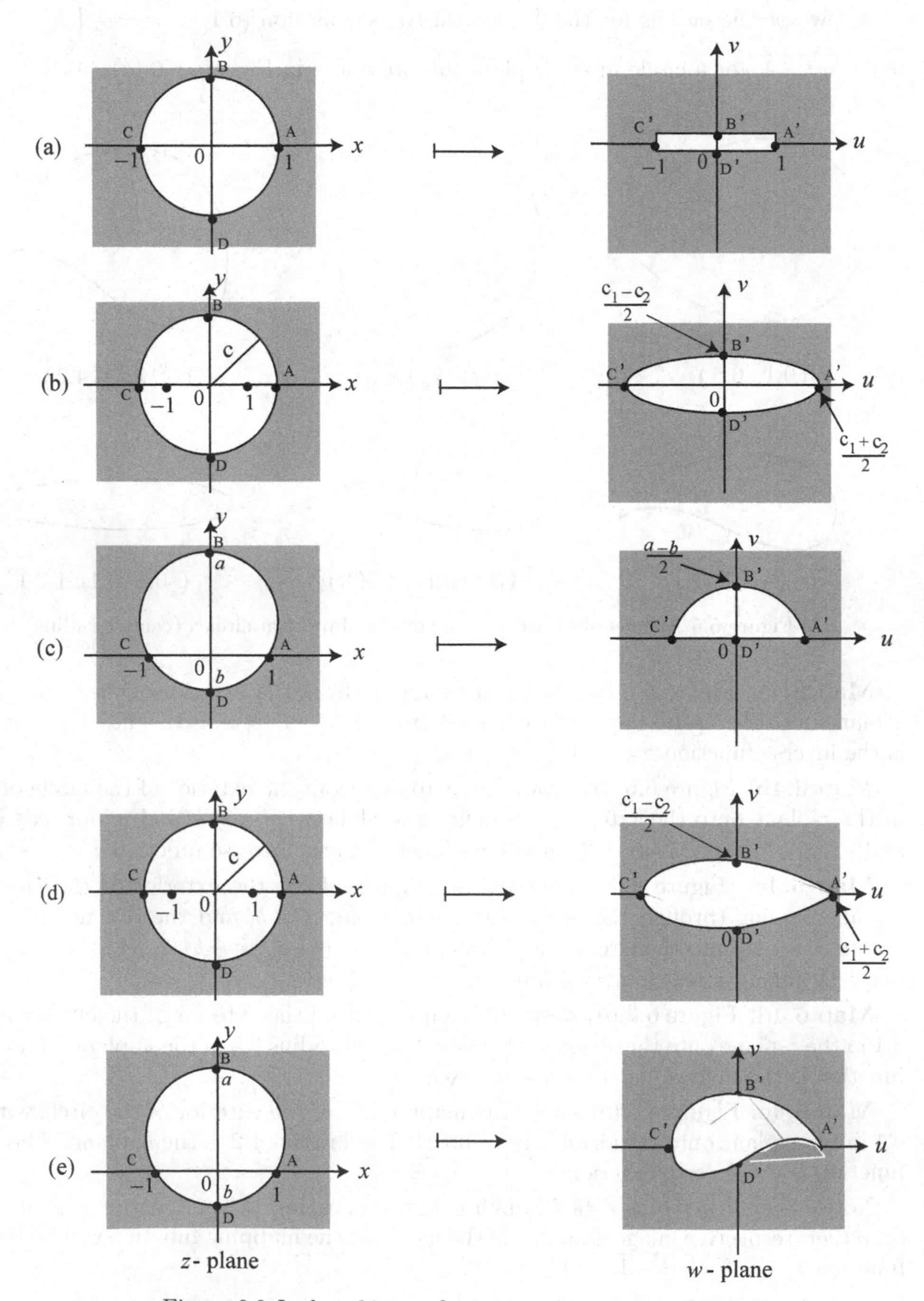

Figure 6.6 Joukowski transformations for exterior of a circle.

Map 6.5. The transformation that maps the half-plane with semi-circle removed in the z-plane onto the upper half-plane $\Im\{w\} > 0$ is defined by

$$w = \frac{a}{2}\Big(z + \frac{1}{z}\Big), \tag{6.1.7}$$

and is presented in Figure 6.7.

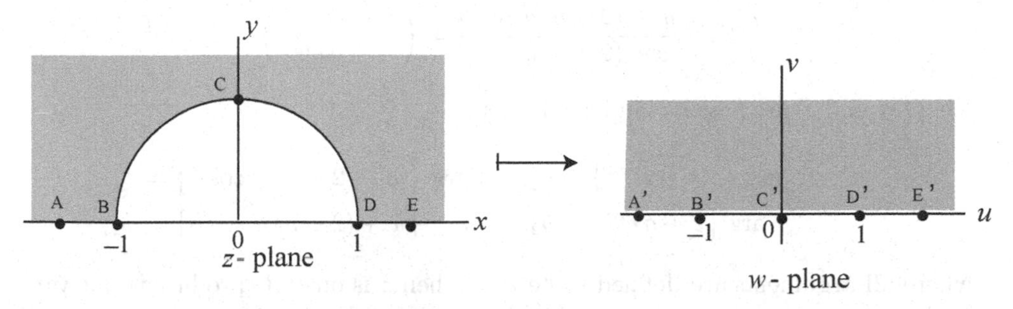

Figure 6.7 Half-plane with semi-circle removed onto the upper half-plane.

Map 6.6. Consider the transformation

$$w = \frac{1}{2}\left(\frac{a+b}{a}\,z + \frac{a(a-b)}{z}\right). \tag{6.1.8}$$

Then for $z = a\,e^{i\theta}$ on the circle $\Gamma : \{|z| = a\}$ we have $u + i\,v = a\,\cos\theta + i\,b\,\sin\theta$, i.e., the contour C is an ellipse with semi-axes a and b and eccentric angle θ. If the mapping is taken as

$$w = z + \frac{a^2}{z}, \tag{6.1.9}$$

which is obtained by setting $b = 0$ in (6.1.8), then $u = a\,\cos\theta$, $v = 0$, i.e., the circle Γ is mapped onto the two sides of the straight line from $(2a, 0)$ to $(-2a, 0)$ and back. This mapping can be written as

$$\frac{w+2a}{w-2a} = \left(\frac{z+a}{z-a}\right)^2, \tag{6.1.10}$$

which yields $\dfrac{dw}{dz} = 1 - \dfrac{z^2}{a^2}$. Now consider

$$\frac{dw}{dz} = \prod_{k=1}^{n}\left(1 - \frac{z_k}{z}\right), \tag{6.1.11}$$

where $\sum z_k = 0$, $|z_1| = a$, $|z_k| < a$ if $k \neq 1$. The mapping (6.1.10) is a special case of (6.1.11) for $n = 2$, $z_1 = a$, and $z_2 = -a$. Then the contour C has a cusp at w_1, and in the neighborhood of this point we have $\dfrac{dw}{dz} = (z - z_1)\,g(z)$, where $g(z)$ is regular, $g(z_1) \neq 0$. Thus, $w - w_1 = \dfrac{1}{2}\,g(z_1)\,(z - z_1)^2 + \cdots$, which means that, as z traverses the circle Γ and passes through z_1, w approaches w_1 and then recedes along a curve with the same tangent. If instead of Γ we take a larger circle that passes through $z = a$ but slightly beyond $z = -a$, and transform this circle, then we have the mapping in Figure 6.8 (in which a is at $x = -1$), with a cusp at $w = 2a$ and a rounded end at w a little less than $-2a$.

The Joukowski airfoils are mappings of the type $w = z \sum\limits_{j=0}^{\infty} \dfrac{c_j}{z_j}$, where c_j may be complex and $c_0 \neq 0$ (c_0 is generally taken as 1), they all have cusps. But an airplane wing is not a cusp. Glauert [1929/1948] removed this problem as follows: Consider the mapping

$$\frac{w-(2-n)\,a\,\cos\beta}{w+(2-n)\,a\,\cos\beta}=\left(\frac{z-a\,e^{-i\beta}}{z+a\,e^{i\beta}}\right)^{2-n}, \tag{6.1.12}$$

where n and β are positive and small. Set

$$\arg\left\{z-a\,e^{-i\beta}\right\}=\theta_1,\qquad \arg\left\{w-(2-n)\,a\,\cos\beta\right\}=\phi_1,$$
$$\arg\left\{z+a\,e^{i\beta}\right\}=\theta_2,\qquad \arg\left\{w+(2-n)\,a\,\cos\beta\right\}=\phi_2,$$

where all arguments are defined to be zero when z is on AB (produced) but vary continuously as z traverses a contour outside the circle (see Figure 6.8).

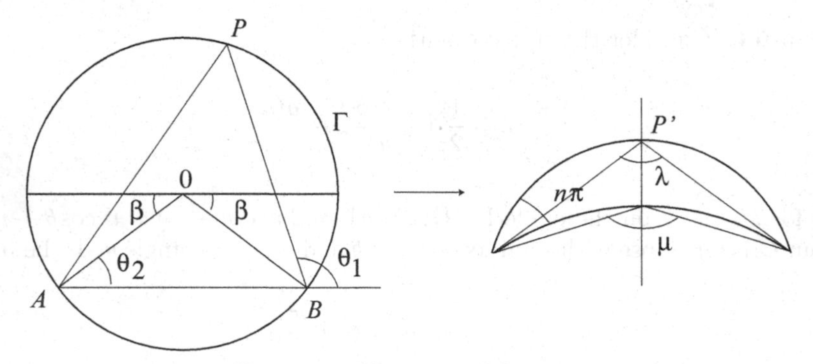

Figure 6.8 Glauert modification.

Then at P, $\theta_1-\theta_2=\dfrac{\pi}{2}-\beta$, and $\phi_1-\phi_2=(2-n)(\theta_1-\theta_2)$ for points outside Γ. The image point P' in the w-plane is on a circular arc through the points $\pm(2-n)\,a\,\cos\beta$, subtending an angle $\lambda=(2-n)\left(\dfrac{\pi}{2}-\beta\right)$. In the case when P moves near B and then travels around a small semicircle about B, then θ_2 increases by π, and in this case $\phi_1-\phi_2=(2-n)\left[\left(\dfrac{\pi}{2}-\beta\right)-\pi\right]=-(2-n)\left(\dfrac{\pi}{2}+\beta\right)<0$. We add 2π to this angle to make it positive. Then the lower arc of Γ is mapped onto the lower circular arc in the w-plane that subtends an angle $\mu=2\pi+(\phi_1-\phi_2)=2\pi-(2-n)\left(\dfrac{\pi}{2}+\beta\right)$. If $\mu<\pi$, the lower arc in the w-plane is concave downward. The two circular arcs in the w-plane intersect at an angle $n\pi$ (Figure 6.9). If instead of Γ, we take a circle passing through A but a little beyond B, then we obtain a rounded leading edge. For large z the Glauert mapping (6.1.12) can be approximated by using the series

$$w=z+i\,a\sin\beta+\frac{(1-n)(3-n)}{3}\,\cos^2\beta\,\frac{a^2}{z}+\cdots. \tag{6.1.13}$$

The symmetric Joukowski airfoil is defined as follows: Let $a>0$. Consider the circles

$$\Gamma:\{|z|=a\},\quad \Gamma_1:\left\{|z+a|=a-c,\ -\infty<c<a,\ c\neq 0,\frac{a}{2}\right\},$$
$$\Gamma_2:\left\{\left|z+\frac{a}{2}\right|=\frac{a}{2}\right\}.$$

(See Map 6.7). Then the mapping (6.1.9) for $a>0$,
(i) maps the circle Γ_1 onto the symmetric airfoil Γ_1' : $A'G'F'D'A'$ with cusp at A';

(ii) maps the circle Γ_2 and the line $x = -c$ onto the circle $\Gamma_2' : O'B'A'C'O'$ with a cusp at A' and the line $u = -a$ as its asymptote, respectively; and
(iii) maps the region Int (Γ_1) bijectively, with $0 < c < a/2$, onto the region Ext (Γ_1').

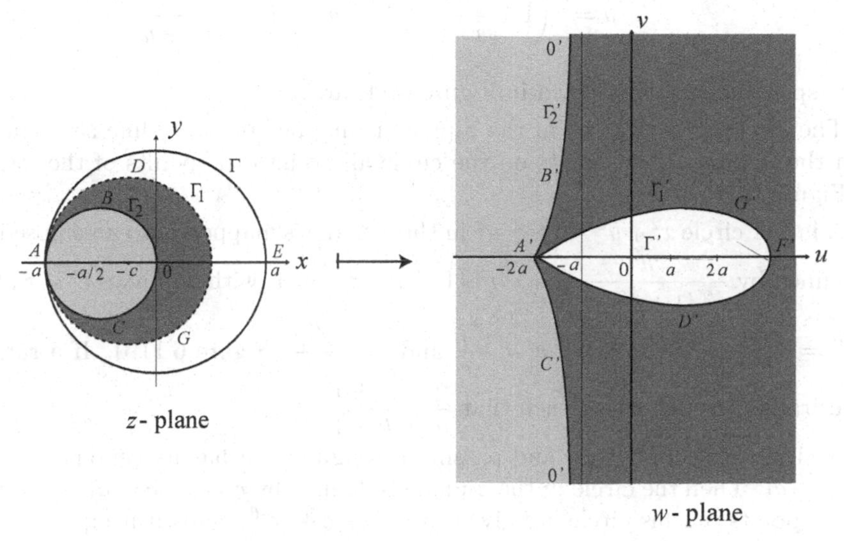

Figure 6.9 Symmetric Joukowski airfoil.

This transformation continues in Map 6.8.

Map 6.7. The transformation that maps the exterior of an ellipse in the z-plane onto the interior of the unit circle in the w-plane is defined by

$$w = \frac{1}{2}\left(z\, e^{-\alpha} + z^{-1}\, e^{\alpha}\right), \tag{6.1.14}$$

and is presented in Figure 6.10.

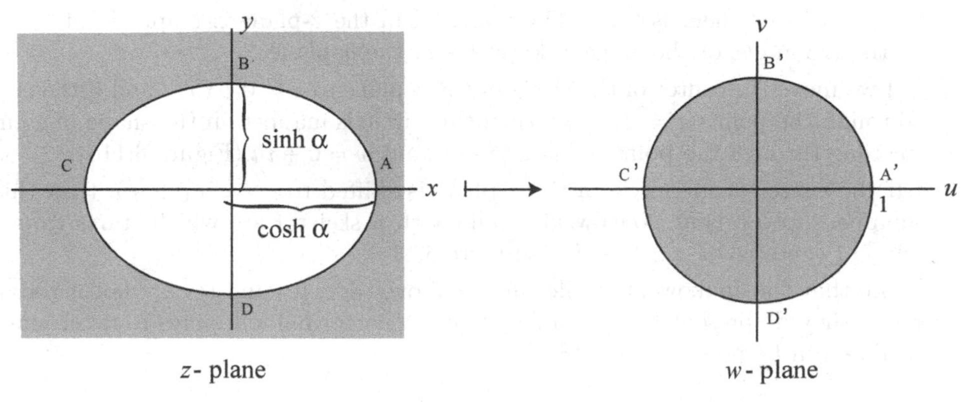

Figure 6.10 Exterior of an ellipse onto the interior of the unit circle.

Map 6.8. The mapping $w = z + \dfrac{a^2}{z}$, which is a generalization of the Maps 6.5, is resolved as follows. Since $w = u + iv$, we have

$$u = x\Big(1 + \frac{a^2}{x^2+y^2}\Big), \quad v = y\Big(1 - \frac{a^2}{x^2+y^2}\Big). \tag{6.1.15}$$

Some special cases, illustrated in Figure 6.11, are:

(a) The circle $x^2 + y^2 = a^2$ in the z-plane is mapped onto the line segment $u = 2x, v = 0$ in the w-plane. The points on the circle all go into the points of the strip of length $4a$ (Figure 6.11a).

(b) A larger circle $x^2 + y^2 = r^2 > a^2$ in the z-plane is mapped onto an ellipse in the w-plane, defined by $\dfrac{u^2}{r^2\left(1 + a^2/r^2\right)^2} + v^2 r^2\left(1 - a^2/r^2\right)^2 = 1$, with half-axes $c' = r\left(1 + a^2/r^2\right)$ and $c'' = r\left(1 - a^2/r^2\right)$: thus, $u = \dfrac{c'x}{r}$ and $v = \dfrac{c''y}{r}$ (Figure 6.11b). If a ratio $c'/c'' = k$ is desirable, then choose r such that $\dfrac{r}{a} = \dfrac{k+1}{k-1}$.

(c) A circle of radius $r > a$ and passing through $x = a$ has its center at $y = 0, x = -c = -|r - a|$. Then the circle in the z-plane is defined by $r = a + c = a(1+d)$, where $d = c/a$. The points on this circle are given by $z = -c + re^{i\theta}$. Substituting $x = -c + r\cos\theta$ and $y = r\sin\theta$ into Eqs (6.1.15) we get

$$\begin{aligned} u &= a\big[-d + (1+d)\cos\theta\big]\Big(1 + \frac{1}{1 + 2d(1+d)(1+\cos\theta)}\Big) = 2a\cos\theta, \\ v &= a\big[-d + (1+d)\sin\theta\big]\Big(1 + \frac{1}{1 - 2d(1+d)(1-\cos\theta)}\Big) = 2ad\sin\theta(1-\cos\theta). \end{aligned} \tag{6.1.16}$$

The approximations for the equations (6.1.16) are valid only for a small ratio of thickness/length $\le (3/4)\sqrt{2}d = 1.299d$ (see Truckenbrodt [1980:163]). As in Figure 6.11c, the basic circle II in the z-plane is mapped onto the straight line segment as in (a) above. The circle I and the basic circle II are tangent to each other at $z = a$ (point III). The angle between them is zero. The point III in the z-plane is mapped onto the tail end, with zero angle, of the Joukowski profile in the w-plane.

(d) If we move the center of the circle in the z-plane to $z = 0 + ih/2$ and if the circle passes through the points $z = \pm a + i0$, then this circle is mapped in the shape of a circular arc passing through the points $w = \pm 2a + i0$ and $w = 0 + ih$ (Figure 6.11d).

(e) If the center of the circle in the z-plane is shifted to $z = -c_1 + ic_2$, then this circle is mapped onto a bent Joukowski profile with a skeletal arc which starts from the basic circle I centered at $z = 0 + ih/2$ (Figure 6.11e).

Note that the Joukowski profiles in the above cases (c) and (e) are not aerodynamically perfect, they do present the possibility of using conformal mappings to determine the wing lift. This will be presented in §18.10.4.

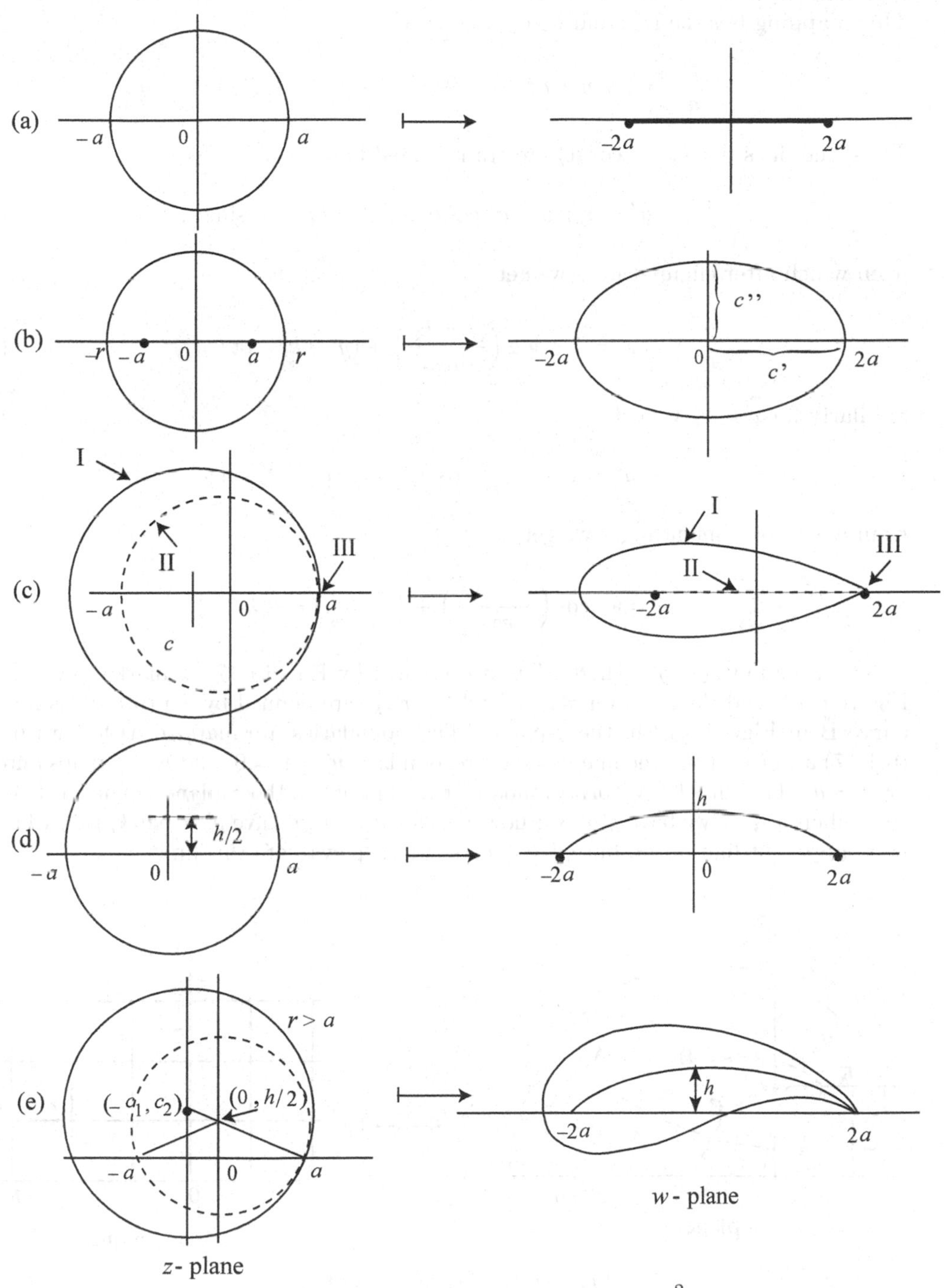

Figure 6.11 Joukowski profiles $w = z + \dfrac{a^2}{z}$.

Map 6.9. The mapping $z = \dfrac{a}{\pi}\left(1 + w + w^2\right)$, known as *Maxwell's transformation*, is a composite of rectangular-to-polar transformations related to the exponential function e^w, which does not contribute much when real parts of w are negative, and $z = w$ which is more dominant in that region. The critical points are $w = \infty; (2n+1)i\pi$ for $n = 0, \pm 1, \pm 2, \dots$. This mapping has the real and imaginary parts

$$\frac{\pi}{a}x = 1 + u + e^u \cos v \equiv x', \quad \frac{\pi}{a}y = v + e^u \sin v \equiv y'.$$

Thus, the lines $v = k_v = \text{(const)}$ are transformed into

$$x' = 1 + u + e^u \cos k_v, \quad y' = k_v + e^u \sin v_k, \tag{6.1.17}$$

from which after eliminating u we get

$$x' = 1 + \log\left(\frac{y' - k_v}{\sin k_v}\right) + (y' - k_v)\cot k_v. \tag{6.1.18}$$

Similarly for $u = k_u$ we get

$$x' = 1 + k_u e^{k_u} \cos v, \quad y' = v + e^{k_u} \sin v, \tag{6.1.19}$$

from which by eliminating v we get

$$y' = \arccos\left(\frac{x' - k_u}{e^{k_u}}\right) + \left\{e^{2k_u} - (x' - k_u)^2\right\}^{1/2}. \tag{6.1.20}$$

The function $z(x', y')$ where x', y' are defined by Eq (6.1.17) is marked by curve A in Figure 6.12, and the function $z(x', y')$ where x', y' are defined by Eq (6.1.19) is marked by curve B in Figure 6.12 in the z-plane. The boundaries are mapped as follows from Eqs (6.1.17) and (6.1.19): The line $v = 0$ corresponds to $y' = y = 0$, and $v = \pi$ maps onto $y = \pi$ or $y = a$. The line $u = 0$ corresponds to the curve C in the z-plane (Figure 6.12). In the case when $u \ll 0$, we have $y' = k_v$; however, for $u \gg 0$ we have $y' = (\sin k_v)\, e^u$, where there is no apparent limit on u, but edge effects can help evaluate this limit.

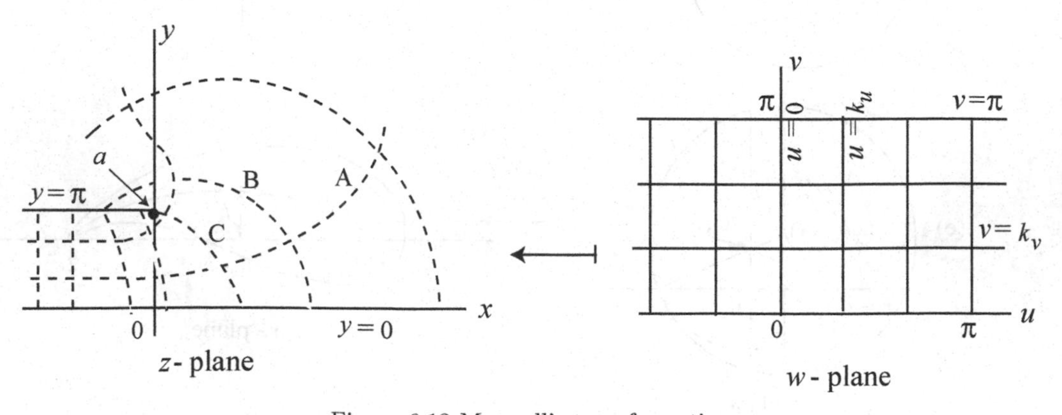

Figure 6.12 Maxwell's transformation.

Map 6.10. For the Joukowski map $w(z) = \frac{1}{2}\left(z + \frac{1}{z}\right)$, let $S_r = \{z : |z| = r, 0 \le \arg\{z\}\}$ for fixed $r > 1$. Let the image of a point $z = r\, e^{i\theta}$ on S_r be $w = u + iv$. Then

$$w = \frac{1}{2}\left(r\, e^{i\theta} + \frac{1}{r}\, e^{-i\theta}\right) = \frac{1}{2}\left(r + \frac{1}{r}\right)\cos\theta + \frac{i}{2}\left(r - \frac{1}{r}\right)\sin\theta,$$

which gives

$$u = \frac{1}{2}\Big(r + \frac{1}{r}\Big)\cos\theta, \quad v = \frac{1}{2}\Big(r - \frac{1}{r}\Big)\sin\theta,$$

so that

$$\cos\theta = \frac{u}{\frac{1}{2}(r + 1/r)}, \quad \sin\theta = \frac{v}{\frac{1}{2}(r - 1/r)}.$$

As θ traces from 0 to π, the function w traces the upper part of the ellipse

$$\frac{u^2}{\left[\frac{1}{2}(r+1/r)\right]^2} + \frac{v^2}{\left[\frac{1}{2}(r+1/r)\right]^2} = \cos^2\theta + \sin^2\theta = 1.$$

Map 6.11. Consider the general Joukowski transformation

$$w = f(z) = z + \frac{c^2}{z}, \tag{6.1.23}$$

where $f'(\pm c) = 0$ but $f''(\pm c) \neq 0$. Thus, the angle between the two short line elements which intersect at either $z = c$ or $z = -c$ are doubled by this mapping. The inverse transformation

$$z = \frac{1}{2}w + \Big(\frac{1}{4}w^2 - c^2\Big)^{1/2} \tag{6.1.24}$$

is multiple valued and has branch points at $w = \pm 2c$. This means that we can cut the w-plane along the real axis between $w = 2c$ and $w = -2c$, and interpret $\left(\frac{1}{4}w^2 - c^2\right)^{1/2}$ as that branch of the function which behaves like $\frac{1}{2}w$ (as opposed to $-\frac{1}{2}w$) as $|w| \to \infty$. This will ensure that $z \sim w$ for large $|w|$.

Example 6.1. (circular disk) Recall that the Joukowski Map 6.1 reduces the unit circle $|z| = 1$ to the real line segment $[-1, 1]$ in the w-plane. It suggests that it will map a fluid flow outside the unit disk onto the flow past the line segment, which has the complex potential $\Theta(w) = w$. Let $g(z)$ denote the complex potential function. Then the resulting complex potential is $g(z) = \Theta \circ g(z) = \frac{1}{2}(z + 1/z)$, where the factor $\frac{1}{2}$ indicates that the corresponding flow is half as fast. ■

Example 6.2. (airfoils) An affine map $w = az + b$ (Map 3.4) maps the unit disk $|z| \leq 1$ onto the disk $|w - b| \leq |a|$ with center at b and radius $|a|$. The boundary of this circle will continue to pass through the point $w = 1$ provided $|a| = |1 - b|$. Also, the angular component of a produces rotation, resulting in the streamlines around the new disk at an angle $\phi = \arg\{a\}$ asymptotically. If we apply the Joukowski transformation

$$\zeta = \frac{1}{2}\Big(w + \frac{1}{w}\Big) = \frac{1}{2}\Big(az + b + \frac{1}{az + b}\Big) \tag{6.1.25}$$

will map the disk $|w - b| \leq |a|$ onto an airfoil. The resulting complex potential for the flow past the airfoil is obtained by substituting the inverse map

$$z = \frac{w - b}{a} = \frac{\zeta - b + \sqrt{\zeta^2 - 1}}{a},$$

onto the disk potential $\Theta(\zeta) \circ g(z)$, where

$$\Theta(\zeta) = \frac{\zeta - b + \sqrt{\zeta^2 - 1}}{a} + \frac{a(\zeta - b - \sqrt{\zeta^2 - 1})}{1 - 2b\zeta + b^2}. \tag{6.1.26}$$

Then, if we replace ζ by $e^{i\phi}\,\zeta$ in Eq (6.1.26), the airfoil is tilted by the (attack) angle $\phi = \arg\{a\}$ to the horizontal. The major flaw with such airfoils is that they do not produce lift.

Note that the rotation $w = e^{i\phi} z$ through an angle ϕ maps the disk potential $\Theta \circ g(z)$ to the complex potential $e^{-i\phi}\, w^+ + e^{-i\phi}\, w^-$, with streamlines no longer asymptotically horizontal, but rather tilted at an angle ϕ. For more, see Example 18.21, §18.9. ■

REFERENCES USED: Ahlfors [1966], Boas [1987], Carathéodory [1932], Kantorovich and Krylov [1958], Kober [1957], Kythe [1998], Nehari [1952], Schinzinger and Laura [2003], Truckenbrodt [1980].

7

Schwarz-Christoffel Transformations

As mentioned earlier, in conformal mapping we always try to determine a function f which maps a given region $D \subset \mathbb{C}$ conformally onto another region $G \subset \mathbb{C}$. However such a function f does not always exist, and it is not always uniquely determined. Recall that the Riemann mapping theorem guarantees the existence and uniqueness of a conformal map of D onto the unit disk U under certain specific conditions. In this chapter the Schwarz-Christoffel transformations are presented. There are three types of equivalent definitions for this transformation, and they all lead to integral equations which are then solved as boundary value problems.

7.1 Schwarz-Christoffel Transformations

The *Schwarz-Christoffel formula* helps solve an important class of boundary value problems that involve regions with polygonal boundaries. This transformation has an integral representation and is used to conformally map the interior of a given polygon of n sides ($n \geq 2$) in the w-plane onto the upper half-plane $\Im\{z\} > 0$ or the unit disk U in the z-plane.

Map 7.1. Let $\Gamma = \bigcup_{j=1}^{n} \Gamma'_j$ denote the sum of the sides Γ_j of the polygon $G = \text{Int}(\Gamma)$ in the w-plane with vertices w_j and exterior angles $\pi\alpha_j$, $|\alpha_j| < 1$, $\sum_{j=1}^{n} \alpha_j = 2$ for $j = 1, \dots, n$ (Figure 7.1). This transformation is defined by

$$\frac{dw}{dz} = C_1 (z - z_1)^{-\alpha_1} (z - z_2)^{-\alpha_2} \cdots (z - z_n)^{-\alpha_n}, \tag{7.1.1}$$

or

$$\begin{aligned} w(z) &= C_1 \int_{z_0}^{z} (z - z_1)^{-\alpha_1} (z - z_2)^{-\alpha_2} \cdots (z - z_n)^{-\alpha_n} \, dz + C_2 \\ &= C_1 \int_{z_0}^{z} \prod_{j=1}^{n} (t - z_j)^{-\alpha_j} \, dt + C_2, \end{aligned} \tag{7.1.2}$$

where the complex constant C_1 provides the scaling, and the complex constant C_2 gives the position of the orientation of the polygon, the integration is carried out along any path in D that joins $z_0 \in D$ to z, and the principal branch is used for the multiple-valued function $(t - z_j)^{-\alpha_j}$ in the integrand such that $0 < \arg\{t - x_j\} < \pi$, $j = 1, \dots, n$, for $\Im\{z\} > 0$. These branches are a direct analytic continuation into the upper half-plane of the real-valued functions $(x - x_j)^{-\alpha_j}$, where $x > x_j$. Then the integral (7.1.2) is a single-valued analytic function in the upper half-plane $\Im\{z\} > 0$, and the points x_j lying on the x-axis are the singularities of the Schwarz-Christoffel integral (7.1.2). This function $w = f(z)$ defines a conformal mapping of G onto D provided the points x_i are suitably chosen.

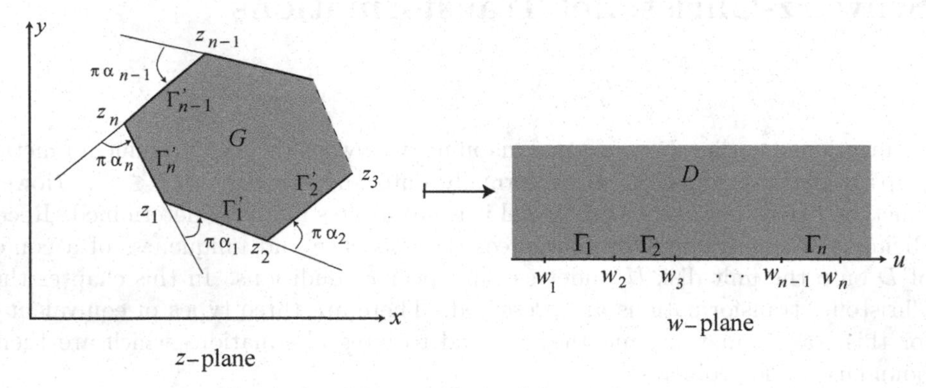

Figure 7.1 Schwarz-Christoffel transformation.

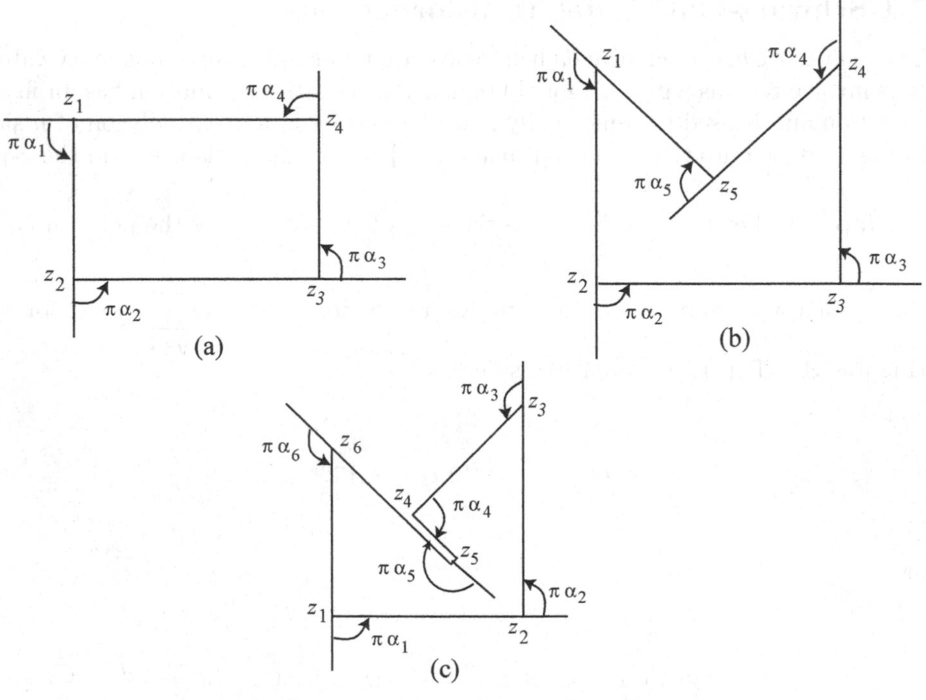

Figure 7.2 Signs of external angles.

The following facts about this transformation must be noted.

(i) The sides of the polygon G must not intersect one another; self-contacts are permitted and the vertices may lie at infinity. Thus, the sides must form the boundary of a simply-connected region G which is the interior of the polygon. If this region lies to the left when we traverse the sides from one vertex to another in order, then it is mapped onto the upper half-plane $y > 0$. The angles $\pi\alpha_j$ denote the change of direction when we pass through the vertex z_j to z_{j+1} . If $\alpha_{n+1} = 0$, then the polygon has n vertices $z_1, z_2, \dots, z_n$. Some cases are presented in Figure 7.2.

(ii) The angles $\pi\alpha_j$, which represent the change of direction at z_j (taken counterclockwise positive), are measured by traversing the sides in order so that the interior of the polygon lies to the right.

(iii) For change of direction when passing through a vertex at infinity, draw a sufficiently large circle that contains all the finite vertices in its interior. Instead of passing through a vertex, we pass from the 'side' p to the 'side' q along that arc which, joining p to q, lies in the interior of the polygon, and find the total change of direction.

(iv) Certain transformations are equivalent from the topological point of view. Thus, when $w = f(z)$ maps the interior G of the polygon conformally onto the upper half of the z-plane, then $\tilde{w} = \tilde{f}(z)$ has the same property iff it can be represented in the form

$$\tilde{w} = f\left(\frac{az+b}{cz+d}\right),$$

where a, b, c, d are real such that $ad - bc \neq 0$. For example, the transformation $w = e^z$ maps the strip $0 < v < \pi$ (i.e., a polygon with two vertices both at infinity, with $\alpha_1 = \alpha_2 = 1$, also called the *trigonal region*, see Map 7.3 below) onto $y > 0$.

(v) Not every differential equation of the form (7.1.1) defines a Schwarz-Christoffel transformation. For example, $\dfrac{dz}{dw} = w^{-1}(w-2)^{-\alpha_2}(w-3)^{-\alpha_3}$ does not, whenever $-2 \leq \alpha_2 + \alpha_3 < -1$. However, for every polygon defined as above, a Schwarz-Christoffel transformation can always be constructed.

In general, there are five major steps that can be used to determine the Schwarz-Christoffel transformation for a given mapping problem.

STEP 1. Label the vertices of the polygon in the z-plane as A, B, ..., or $1, 2, \dots$. To each vertex j, assign the corresponding exterior angle $\pi\alpha_j$. Note that we have to turn at vertex j on the boundary (or to get back on the boundary if the polygon opens at infinity) while traversing in the counterclockwise direction. Check that the sum of all α_j equals 2.

STEP 2. Select a suitable location on the x-axis to correspond to any three or less vertices in the z-plane. The sets of three points, viz.,$-1, 0, 1$; $0, 1, \infty$; or $-\infty, 0, 1$, are often chosen. Assign the vertices with exterior angles 2π or π to $-\infty$, and 0.

STEP 3. Formulate and solve the Schwarz-Christoffel integral equation (7.1.2) or (7.1.5), leaving the term corresponding to $\pm\infty$. Use boundary conditions on the polygon in the z-plane to determine the integration constants A and B as well as any x_j which were not specified in Step 2. The remaining points x_j (which total $n-3$) can be determined from $n-3$ ratios between the sides of the polygon which determine its shape. The length of a side is given by the Schwarz-Christoffel integral evaluated between the endpoints of the corresponding section of the x-axis.

STEP 4. The specific rules for the parallel lines are as follows: (i) If the transition between two parallel lines in the z-plane occurs at a location corresponding to $x_k = 0$ on the x-axis, then the constant B is determined from

$$d = -i\pi B \prod_{j \neq k} (0 - x_j)^{-\alpha_j}, \tag{7.1.3}$$

where d is the complex separation between the lines, excluding $x_j = \infty$. But if the transition is made to occur at $x = \infty$, then $d = i\pi B$.

STEP 5. Transform the x-axis configuration into the w-plane.

7.1.1 Triangles. Before we establish formulas (7.1.1) and (7.1.2), we will consider the case of a triangular region G. The mapping of the interior of a polygon onto the upper half-plane is defined by the Schwarz-Christoffel transformation which uses the angular change encountered at corners when one traces the boundary of the polygon counterclockwise. The domain D in the z-plane is mapped onto the domain G in the w-plane. The polygonal boundary of G is restricted to straight line segments Γ_i, $i = 1, 2, \dots, n$, in the w-plane.

Map 7.2. Let us choose an arbitrary point $A_1 : x = x_1$ on the x-axis in the segment marked Γ_1 (Figure 7.3(a), where the region G is a three-sided polygon). For $x < x_1$, we have $(x - x_1) = |x - x_0|\, e^{i\pi}$ since x is always smaller than x_1 and $\arg\{x - x_1\} = \pi$. As we move in the direction of increasing x, we find that $(x - x_1) = |x - x_0|\, e^{i0}$ as soon as we cross $x = x_1$ and enter the segment Γ_2. Note that $\arg\{x - x_1\}$ has undergone a step change of $-\pi$ at the point A_2, since we are tracing the polygonal contour in anticlockwise direction. Then the function $w - w_1 = (z - x_1)^{1/2}$ will map the upper half of the z-plane onto the interior of the right angle sector in the w-plane, and the mapping function $w = (z - x_1)^{1/2}$ is analytic at all values of z. Similarly, a function $w = (z - x_1)^{1/\rho}$ would produce a sector of angle π/ρ. The function $w = (z - x_1)^{1/\rho}$ produces a one-to-one mapping except at the values of w for which the derivative $\dfrac{dw}{dz} = \dfrac{1}{\rho}\,(z - x_1)^{(1/\rho)-1} = 0$. This can occur at $z = x_1$ or $x = \infty$ depending on the value of ρ. Problems arising from $x = \infty$ are handled by a procedure that will be explained later (see Eq (7.1.6)).

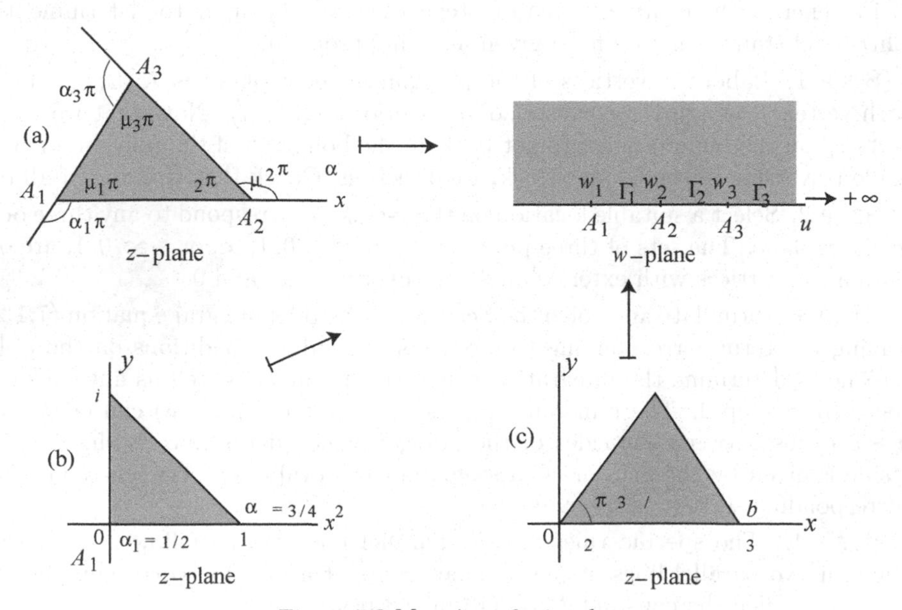

Figure 7.3 Mapping of triangles.

Since the Riemann mapping theorem permits us to choose three points, we take the points A_1, A_2, A_3 along with segments $\Gamma_1, \Gamma_2, \Gamma_3$ on the x-axis. Thus, in general, we have the exterior angles $\pi\alpha_1, \pi\alpha_2, \pi\alpha_3$ with the corresponding interior angles $\pi - \pi\alpha_1 = \pi\mu_1, \pi - \pi\alpha_2 = \pi\mu_2, \pi - \pi\alpha_3 = \pi\mu_3$, respectively, at the points A_1, A_2, A_3. As we trace the contour and pass into the segment Γ_2, approaching the point A_2, any increment dw has an argument equal to π/ρ. As we pass around A_2 into the segment Γ_3, the dw-increments change their slopes to $\rho + \alpha_2$. Thus, the change in the argument of dw at A_2 is equal to $\pi\alpha_2$ and the interior angle π/ρ is equal to $\pi - \pi\alpha_2$. Hence $1/\rho - 1 = -\alpha_2$, which gives

$$\frac{dw}{dz} = \frac{1}{\rho}\left(z - z_2\right)^{(1/\rho)-1} = C_0\left(z - z_2\right)^{-\alpha_2},$$

where $C_0(=1/\rho)$ is a complex constant that gives scaling. Following the above process for the third vertex, we find that the mapping is given by

$$\frac{dw}{dz} = C_0\left(z - z_1\right)^{-\alpha_1}\left(z - z_2\right)^{-\alpha_2}\left(z - z_3\right)^{-\alpha_3}. \tag{7.1.4}$$

This is the required mapping of a regular triangular region onto the upper half-plane. The above argument can be continued to the n vertices of a regular polygon and establish the transformation given by (7.1.1).

Note that the sum of the exterior angles is 2π, while the sum of all interior angles is $(n-2)\pi$.

If all the numbers x_j are finite, then the function $f(z)$ defined by (7.1.2) remains bounded in the neighborhood of the singularities x_i. To see that the Schwarz-Christoffel integral (7.1.2) remains bounded as $z \to \infty$, we rewrite this integral as

$$\begin{aligned} w = f(z) &= C_1 \int_{z_0}^{z} t^{-(\alpha_1+\cdots+\alpha_n)}\left(1 - \frac{x_1}{t}\right)^{-\alpha_1}\cdots\left(1 - \frac{x_n}{t}\right)^{-\alpha_n} dt + C_2 \\ &= C_1 \int_{z_0}^{z} \frac{1}{t^2}\prod_{j=1}^{n}\left(1 - \frac{x_j}{t}\right)^{-\alpha_j} dt + C_2, \end{aligned} \tag{7.1.5}$$

which shows that the integral is convergent as $z \to \infty$. Hence, the Schwarz-Christoffel integral (7.1.2) defines a univalent function of z in $D = \Im\{z\} > 0$ that maps D conformally onto the bounded polygon G in the w-plane. If G is convex, then $\alpha_j > 0$. Thus, there are $(n-1)$ independent numbers α_j, and $(n-3)$ independent numbers x_j.

To see the structure of the boundary $\Gamma = \bigcup_{j=1}^{n}\Gamma_j$ of the polygon G, consider the derivative of the Schwarz-Christoffel integral (7.1.2), which is (7.1.1), i.e.,

$$f'(z) = C_1(z - x_1)^{-\alpha_1}\cdots(z - x_n)^{-\alpha_n}.$$

Notice that $f' \neq 0$ everywhere in $\Im\{z\} \geq 0$ except at the singularities x_i where it vanishes or becomes unbounded. As z varies along every interval $x_k < x < x_{k+1}$, $k = 1, \ldots, n-1$, the value of $\arg\{f'(z)\}$ does not change, since

$$\arg\left\{(x - x_k)^{-\alpha_k}\right\} = \begin{cases} -\pi\alpha_k, & \text{if } x < x_k, \\ 0, & \text{if } x > x_k. \end{cases}$$

Geometrically, $\arg\{f'(z)\}$ determines the size of the angle through which the tangent to a Jordan curve passing through x_k must be rotated in order to obtain the tangent of the image of this curve at the point $w_k = f(x_k)$. Hence the segments $x_k < x < x_{k+1}$, $k = 1, \dots, n-1$, of the real axis are mapped by $f(z)$ into rectilinear segments in the w-plane. The points x_k of the real axis are transformed into the vertices w_k in the w-plane, where the polygon Γ is made up of the polygonal lines through $w_1, w_2, \dots, w_n$ with straight line segments as sides. As the point z traverses the entire real x-axis in the positive sense, the corresponding point w travels completely counterclockwise through the polygonal lines of Γ.

7.1.2 Size of the Angles. The size of the angle between adjacent segments of the polygon Γ can be determined as follows: Consider the variation of $\arg\{f'(z)\}$ as z passes through the point x_i in the positive direction (Figure 7.4). Then the angle between the vectors $\overrightarrow{w_{i-1}\, w_i}$ and $\overrightarrow{w_i\, w_{i+1}}$ is equal to $\pi\alpha_i$. For $\mu_i < 1$, where $\mu_i = 1 - \alpha_i$, the transition from the vector $\overrightarrow{w_{i-1}\, w_i}$ to the direction of the vector $\overrightarrow{w_i\, w_{i+1}}$ occurs in the positive sense (Figure 7.4(a)), whereas for $\mu_i > 1$ it occurs in the negative sense (Figure 7.4(b)), although the angle of transition in the positive sense in both cases from the direction of the vector $\overrightarrow{w_{i-1}\, w_i}$ to the direction of the vector $\overrightarrow{w_i\, w_{i+1}}$ is $\pi\alpha_i$. If a polygonal line does not intersect itself, it becomes the boundary of a closed polygon. Also the sum of all interior angles of this closed polygon is equal to $\sum_{i=1}^{n} \pi\mu_i = \sum_{i=1}^{n} \pi\,(1 - \alpha_i) = (n-2)\pi$.

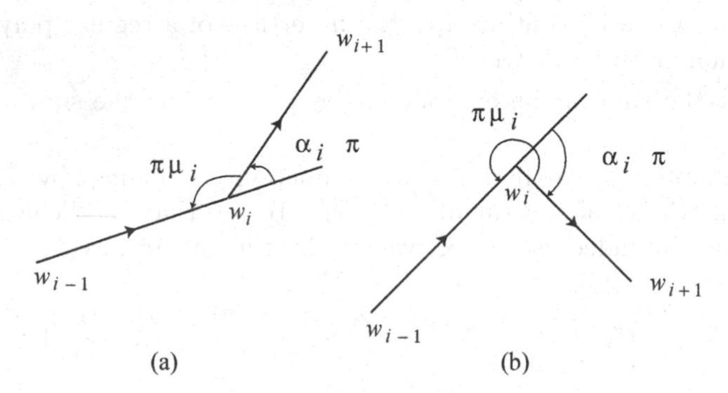

Figure 7.4 Interior and exterior angles.

For practical purposes, while constructing a conformal map of D onto G one can specify any three points x_i, x_j, x_k of the real axis that go into the three prescribed vertices w_i, w_j, w_k of the polygon Γ. When one of the points x_i, say x_n, coincides with the point at infinity, the vertices of the polygon Γ correspond to the points $x'_1, \dots, x'_{n-1}, \infty$, and the formula (7.1.2) becomes

$$w = f(z) = C'_1 \int_{z_0}^{z} \prod_{i=1}^{n-1} (t - x'_i)^{-\alpha_i}\, dt + C_2. \tag{7.1.6}$$

Note that formula (7.1.6) is similar to (7.1.2) except that the term corresponding to the point $x_n = \infty$ has been dropped. The inverse Schwarz-Christoffel transformation is given by

$$z(w) = C \int_{w_0}^{w} \prod_{i=1}^{n} (t - w_i)^{-\mu_i}\, dt + C_0. \tag{7.1.7}$$

Note that the mapping of the upper half z-plane onto the *exterior* of the polygon G is given by

$$z(w) = C_1 \int_{z_0}^{z} \prod_{j=1}^{n} (t - w_j)^{\alpha_j} \frac{dt}{(1+t^2)} + C_2. \tag{7.1.8}$$

The *exterior* of the unit circle is mapped onto the exterior of the polygon G by

$$z(w) = C_1 \int_{w_0}^{w} \prod_{j=1}^{n} (t - z_j)^{\alpha_j} \frac{dt}{t^2} + C_2. \tag{7.1.9}$$

So far we have presented the mapping of the interior of a polygon onto the upper half z-plane. But if we want to map the interior of a polygon onto the interior of the unit circle, we will use the additional transformation that maps the upper half-plane onto the unit disk $|w| < 1$. This transformation is given by

$$w(z) = \frac{z - i}{z + i}, \quad \text{or} \quad z(w) = i\,\frac{1 + w}{1 - w} \quad \text{(Map 3.22)}. \tag{7.1.10}$$

Similarly, for mapping onto a regular strip in the w-plane we have the transformation

$$w(z) = \log z, \quad \text{or} \quad z(w) = e^{w} \quad \text{(Map 5.8)}. \tag{7.1.11}$$

If the n-sided polygon is regular, then the mapping of $|w| < 1$ onto the interior of such a regular polygon is

$$z(w) = C_1 \int_{w_0}^{w} \frac{dt}{(1 - t^n)^{2/n}} + C_2, \quad n \geq 3, \tag{7.1.12}$$

where in case $w = z = 0$, we get $C_2 = 0$, and C_1 is determined from

$$\frac{z(w)}{C_1} = \int_0^{w} \frac{dt}{(1 - t^n)^{2/n}} = w + \sum_{j=1}^{\infty} \frac{2(2+n)(2+2n)\cdots(2+(j-1)n)}{w^{jn+1}}. \tag{7.1.13}$$

For the series expression, see Kober [1957:183]. The length L of each side of the regular polygon is given by (Sansone and Gerretson [1960, vol. II:153])

$$L = \frac{1}{n} 2^{1-4/n} \frac{\Gamma^2\left(\frac{1}{2} - \frac{1}{n}\right)}{\Gamma\left(1 - \frac{2}{n}\right)}, \tag{7.1.14}$$

which for $n = 3$ (triangle) is $L = \frac{1}{2\pi}\Gamma^3\left(\frac{1}{3}\right) \approx 3.059908$, and for $n = 4$b (square) $L = \frac{1}{4\sqrt{\pi}}\Gamma^2\left(\frac{1}{4}\right) \approx 1.854075$; the values of C_1 are as follows: $C_1 = 1.0788$ for $n = 4$; $C_1 = 1.0515$ for $n = 5$; $C_1 = 1.0376$ for $n = 6$; $C_1 = 1.0279$ for $n = 7$; and $C_1 = 1.0220$ for $n = 8$.

Map 7.3. The inverse Schwarz-Christoffel transformation of the polygonal region G in the z-plane onto the upper half-plane $\Im\{w\} > 0$ is given by

$$\frac{dz}{dw} = C_1 \left(w - w_1\right)^{-\alpha_1} \left(w - w_2\right)^{-\alpha_2} \cdots \left(w - w_n\right)^{-\alpha_n}, \tag{7.1.15}$$

or

$$z(w) = C_1 \int_{w_0}^{w} \prod_{j=1}^{n} \left(t - w_j\right)^{-\alpha_j} dt + C_2, \tag{7.1.16}$$

which are obtained from (7.1.2) and (7.1.3) by interchanging z and w.

Map 7.4. The Schwarz-Christoffel transformation of the unit disk in the z-plane onto a polygonal region G with exterior angles $\alpha_k \pi$ at vertices $w_k = f(z_k)$ for each $k = 1, 2, \ldots, n$, is defined by

$$\frac{dw}{dz} = C_1 \left(1 - \frac{z}{z_1}\right)^{-\alpha_1} \left(1 - \frac{z}{z_2}\right)^{-\alpha_2} \cdots \left(1 - \frac{z}{z_n}\right)^{-\alpha_n}, \tag{7.1.17}$$

or

$$w(z) = C_1 \int_{z_0}^{z} \prod_{k=1}^{n} \left(1 - \frac{z'}{z_k}\right)^{-\alpha_k} dz' + C_2. \tag{7.1.18}$$

Some authors, e.g., Papamichael and Stylianopoulos [2010: 25], take the interior angles at each vertex w_k as $\alpha_k \pi$, in which case the exterior angles are $(1-\alpha_k)\pi$ for each $k = 1, 2, \dots, n$. In this situation the quantity $-\alpha_k$ in the above transformations must be replaced by $\alpha_k - 1$. Then, for example, the map (7.1.8) becomes

$$w(z) = C_1 \int_{z_0}^{z} \prod_{k=1}^{n} \left(1 - \frac{z'}{z_k}\right)^{\alpha_k - 1} dz' + C_2. \tag{7.1.19}$$

Note that the Maps 7.3 and 7.4 are distinct cases of Map 7.1; they are provided here separately for convenience.

Note that formulas (7.1.2), (7.1.15), and (7.1.18) apply to polygons with vertices at infinity and polygons with slits, i.e., vertices with angle 2π. They are also used in mapping of the exterior of the unit circle onto exterior of the polygon. The images $w_j, j = 1, 2, \dots, n$, of the vertices z_j of the polygon, where $w_j = f(z_k)$, are not known *a priori*, and their determination is known as the *parameter problem* which is discussed in §8.1.

7.1.3 Transition between Two Parallel Lines in the z-plane. Such a transition is always chosen to occur at a location corresponding to $\xi_k = 0$ in the ζ-plane. Then the constant C_1 can be found from formula

$$d = -i\pi \prod_{j \neq k} (\xi_k - \xi_j)^{\mu_j} = -i\pi \prod_{j \neq k} (0 - \xi_j)^{\mu_j},$$

where d is the complex separation of the two parallel lines. But if this transition is made at $\xi = \infty$, then

$$d = i\pi C_1. \tag{7.1.20}$$

7.2 Specific Transformations

Some specific cases of Schwarz-Christoffel Transformations are as follows.

Map 7.5. To map the upper half-plane $\Im\{z\} > 0$ onto a triangle $A_1A_2A_3$ with interior angles $\mu_1\pi$, $\mu_2\pi$ and $\mu_3\pi$, respectively (Figure 7.3(a)), such that the vertex A_1 corresponds to $x_1 = 0$, A_2 to $x_2 = 1$ and A_3 to $x_3 = \infty$, we get, by using (7.1.2), the mapping function

$$w = C_1 \int_0^z t^{-\alpha_1} (1 - t)^{-\alpha_2} \, dt + C_2. \tag{7.2.1}$$

If the triangle is *right-angled isosceles* (Figure 7.3(b)) such that the vertices A_1, A_2, A_3 are at $w = 0$, $w = 1$, and $w = i$, respectively, then from (7.2.1) the transformation becomes

$$w = C_1 \int_0^z t^{-1/2} (1 - t)^{-3/4} \, dt + C_2. \tag{7.2.2}$$

Since $w = 0$ corresponds to $z = 0$, we get $C_2 = 0$, and since $w = 1$ corresponds to $z = 1$, we find from (7.2.2) that

$$C_1 = \frac{1}{\int_0^1 t^{-1/2}\,(1-t)^{-3/4}\,dt}. \tag{7.2.3}$$

For an *equilateral* triangle of side b (Figure 7.3(c)), let $x_1 = -1$ correspond to the vertex w_1, $x_2 = 1$ to $w = 0$, and $x_3 = \infty$ to $w_3 = b$. Since $\alpha_1 = \alpha_2 = \alpha_3 = 2/3$, the transformation is given by

$$w = \int_1^z (t+1)^{-2/3}(t-1)^{-2/3}\,dt. \tag{7.2.4}$$

When $z = -1$, we set $t = x$. Then for $-1 < x < 1$, we have $x + 1 > 0$ and $\arg\{x+1\} = 0$, but $|x-1| = 1-x$, and $\arg\{x-1\} = \pi$. Thus,

$$\begin{aligned} w_1 &= \int_{-1}^1 (x+1)^{-2/3}\,e^{-2i\pi/3}\,(1-x)^{-2/3}\,dx = e^{i\pi/3}\int_{-1}^1 \frac{dx}{(1-x^2)^{2/3}} \\ &= 2e^{i\pi/3}\,C_1\,B\left(\frac{1}{2},\frac{1}{3}\right) = b\,e^{i\pi/3}, \end{aligned} \tag{7.2.5}$$

where $b = 2B\left(\frac{1}{2},\frac{1}{3}\right)$, and B denotes the beta function of its arguments. For the vertex w_3, note that it is on the positive u-axis, i.e.,

$$w_3 = \int_1^\infty (x+1)^{-2/3}(x-1)^{-2/3}\,dx = \int_1^\infty \frac{dx}{(x^2-1)^{2/3}}.$$

But w_3 is also defined by (7.2.4) when z goes to ∞ along the negative u-axis, i.e.,

$$\begin{aligned} w_3 &= \int_1^\infty (x+1)^{-2/3}(x-1)^{-2/3}\,dx \\ &= w_1 + e^{-4i\pi/3}\int_{-1}^{-\infty} [|x+1|\,|x-1|]^{-2/3}\,dx \quad \text{using (7.2.5)} \\ &= b\,e^{i\pi/3} + e^{-i\pi/3}\int_1^\infty \frac{dx}{(x^2-1)^{2/3}} \\ &= b\,e^{i\pi/3} + w_3\,e^{-i\pi/3}, \end{aligned}$$

which yields $w_3 = b$.

Map 7.6. (Trigonal region) A trigon (*Zweieck* in German) is an open polygon of three sides with three vertices. Unlike a regular triangle, its two sides are parallel and extend to infinity, so it looks like an infinitely deep slot. We solve two problems: (a) We will determine the Schwarz-Christoffel transformation that maps a trigon in the z-plane onto the upper-half ζ-plane; and (b) we will determine the transformation that maps the region between two parallel lines $v = 0$ and $v = pi$ in the w-plane onto the upper half-plane $\Im\{\zeta\} > 0$, as presented in Figure 7.5(a)-(b).

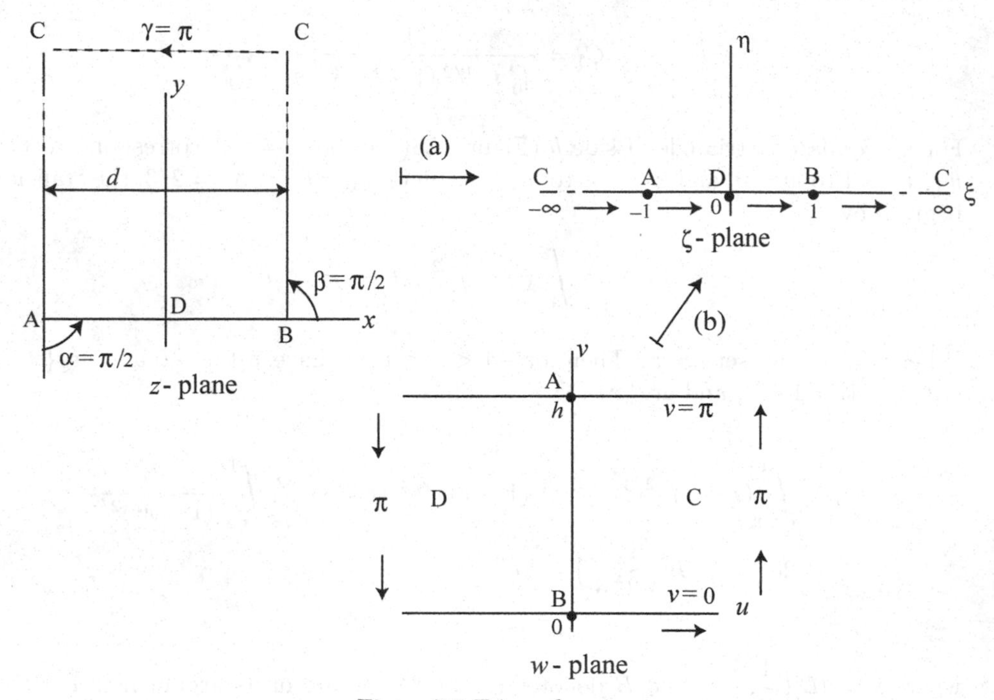

Figure 7.5 Trigonal region.

(a) Let the exterior angle at the vertex A be $\pi\alpha$ and at the vertex B be $\pi\beta$; however, the exterior angle at the vertex C is π, since this vertex is at infinity. Then using formula (7.2.1) we have

$$w(\zeta) = C_1 \int_{\zeta_0}^{\zeta} (t-\zeta_1)^{-\alpha} (t-\zeta_2)^{-\beta}\, dt + C_2. \tag{7.2.6}$$

Let us choose $\zeta_1 = 0, \zeta_2 = 1$. Then the above integral becomes

$$w(\zeta) = C_1 \int_{\zeta_0}^{\zeta} \frac{dt}{t^{\alpha}(t-1)^{\beta}} + C_2.$$

Since $w = 0$ at $\zeta = 0$ and $w = 1$ at $\zeta = 1$, we get $C_2 = 0$, and then we have

$$1 = C_1 \int_0^1 \frac{dt}{t^{\alpha}(t-1)^{\beta}} = C_1 \frac{\Gamma(\alpha)\Gamma(\beta)}{\Gamma(\alpha+\beta)},$$

or

$$C_1 = \frac{\Gamma(\alpha+\beta)}{\Gamma(\alpha)\Gamma(\beta)} = \frac{1}{B(\alpha,\beta)},$$

where $B(\alpha,\beta)$ is the beta function of the arguments. However, this does not complete the integration because C_1 is unknown. However, for the trigonal region in Figure 7.5 with $\alpha = \beta = \pi/2$, we find that

$$w(\zeta) = C_1 \int_0^1 \frac{dt}{\sqrt{t^2-1}} = C \int_0^1 \frac{dt}{\sqrt{1-t^2}} = C \arcsin(\zeta), \tag{7.2.7}$$

where $C = -C_1$. Using the boundary conditions $z_1 = -d/2$ at $\zeta_1 = -1$, and $z_2 = d/2$ at $\zeta_2 = 1$, we find that $-\dfrac{d}{2} = C\left(-\frac{1}{2}\pi\right)$ and $\dfrac{d}{2} = C\left(\frac{1}{2}\pi\right)$ yield $C = d/\pi$, which gives

$$w(\zeta) = \frac{d}{\pi}\arcsin(\zeta), \quad \text{or} \quad \zeta(z) = \sin\frac{\pi z}{d}. \tag{7.2.8}$$

(b) For the mapping from the ζ-plane to the w-plane, let $\zeta_1 = -1$ and $\zeta_2 = 1$ correspond to $w = 0$ and $w_2 = ih$, respectively. Since all the exterior angles at w_1 and w_2 are zero and at C at both ends it is π, the Schwarz-Christoffel transformation is

$$w(\zeta) = B_1\int_{\zeta_0}^{\zeta}\frac{dt}{(t+1)^0(t-1)^0(t-0)^1} + B_2 = B_1\log(\zeta) + B_2,$$

where $\zeta_0 = 1$ (point B). We use the following conditions: at B, $\zeta = 1$ and $w = 0$, which gives $B_2 = 0$; at A, $\zeta = -1$ and $w = ih$, which gives $B_1 = h/\pi$. Hence, the transformation is

$$w(\zeta) = \frac{h}{\pi}\log(\zeta). \tag{7.2.9}$$

Map 7.7. To map the upper half-plane $\Im\{z\} > 0$ onto a rectangle $A_1\,A_2\,A_3\,A_4$ with vertices at the points $w = \pm a$, $\pm a + ib$, where $2a$ and b are the width and the height of the rectangle (Figure 7.6), note that in the formula (7.1.2) with $n = 4$, only three of the four points x_1, x_2, x_3, x_4 may be chosen arbitrarily. Since the rectangle is symmetric about the v axis, we can choose the x's symmetrically. Thus, for the right-half rectangle OA_1A_2B let $w = 0, a, ib$ correspond to $z = 0, 1, \infty$, respectively, and let the preimage of A_2 be $z = 1/k$, $0 < k < 1$. Similarly, for the left-half rectangle OBA_3A_4 let $w = 0, -a, ib$ correspond to $z = 0, -1, -\infty$, respectively, with the preimage of A_3 as $-1/k$. Then the formula (7.1.2) yields

$$\begin{aligned} w &= C_1\int^{z}\prod_{i=1}^{4}(t-x_i)^{-\alpha_i}\,dt + C_2 \\ &= C_1\int_0^{z}\frac{dt}{\sqrt{t-1}\,\sqrt{t+1}\,\sqrt{t-1/k}\,\sqrt{t+1/k}} \\ &= C_1\int_0^{z}\frac{dt}{\sqrt{t^2-1}\,\sqrt{t^2-1/k^2}} = C_1\int_0^{z}\frac{dt}{\sqrt{t^2-1}\,\sqrt{1-k^2t^2}}, \end{aligned} \tag{7.2.10}$$

where $C_2 = 0$ since $z = 0$ goes into $w = 0$, and C_1 is an arbitrary constant. Now, to evaluate the constants C_1 and k, note that since $z = 1$ goes into $w = a$, we get from (7.2.10)

$$a = C_1\int_0^1\frac{dt}{\sqrt{1-t^2}\,\sqrt{1-k^2t^2}}. \tag{7.2.11}$$

Also, since $z = 1/k$ goes into $w = a + ib$, using (7.2.11) we find that

$$\begin{aligned} a + ib &= C_1\int_0^{1/k}\frac{dt}{\sqrt{1-t^2}\,\sqrt{1-k^2t^2}} \\ &= C_1\left(\int_0^1 + \int_1^{1/k}\right)\frac{dt}{\sqrt{1-t^2}\,\sqrt{1-k^2t^2}} \\ &= a + i\,C_1\int_1^{1/k}\frac{dt}{\sqrt{1-t^2}\,\sqrt{1-k^2t^2}}, \end{aligned}$$

where C_1 is an arbitrary complex constant. Thus, with $C_1 = 1$

$$b = \int_1^{1/k} \frac{dt}{\sqrt{1-t^2}\,\sqrt{1-k^2t^2}}. \tag{7.2.12}$$

Hence, we can determine C_1 and k from (7.2.11) and (7.2.12) if a and b are prescribed. But if k is preassigned and we take $C_1 = 1$, then the values of a and b are determined from (7.2.11) and (7.2.12). Then the mapping function becomes

$$w = \int_0^{z} \frac{dt}{\sqrt{1-t^2}\,\sqrt{1-k^2t^2}}, \tag{7.2.13}$$

which is known as the elliptic integral of the first kind. The inverse function is called the elliptic sine function $z = \operatorname{sn} w = \operatorname{sn}(w; k)$ which is a Jacobian elliptic function. The function sn is a $2a$- and $2ib$-periodic function. For more material on the function sn and other Jacobian elliptic functions, see Appendix F. If we denote the value of a, given by (7.2.11), by $K(k)$, where $0 < k < 1$, then the ratio a/b of the sides of the rectangle is given by

$$\frac{a}{b} = 2\,\frac{K(k)}{K\left(\sqrt{1-k^2}\right)}. \tag{7.2.14}$$

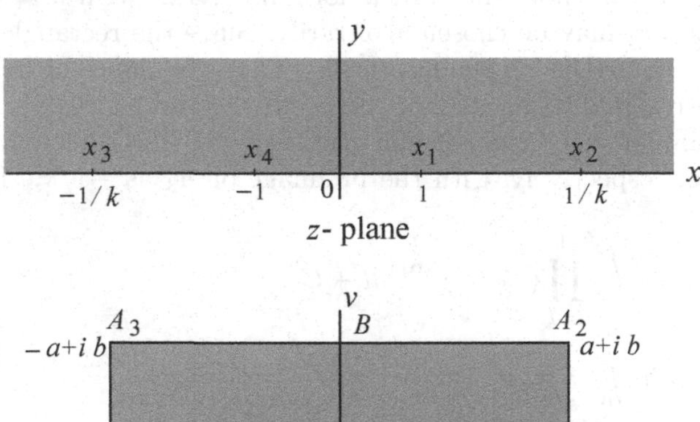

Figure 7.6 $\Im\{z\} > 0$ onto a rectangle.

An evaluation of $K(k)$ is accomplished by the Landen transformation: Set $k_1 = \sqrt{1-k^2}$, and $k' = \dfrac{1-k_1}{1+k_1}$. Then

$$K(k) = \frac{2}{1+k_1}\,K(k'). \tag{7.2.15}$$

In fact, by separating the integral (7.2.11) into two parts, we get

$$K(k) = \left(\int_0^{1/\sqrt{1+k_1}} + \int_{1/\sqrt{1+k_1}}^{1}\right) \frac{dt}{\sqrt{(1-t^2)(1-k^2t^2)}} \equiv I_1 + I_2.$$

If we set $t = \sqrt{\dfrac{1-t^2}{1-k^2t^2}}$, where $1/\sqrt{1+k_1} \le t \le 1$, and $t = \sqrt{\dfrac{1-t^2}{1-k^2t^2}}$, $0 \le t \le 1/\sqrt{1+k_1}$, then

$$I_2 = \int_{1/\sqrt{1+k_1}}^{1} \frac{dt}{\sqrt{(1-t^2)(1-k^2 t^2)}} = \int_0^{1/\sqrt{1+k_1}} \frac{dt}{\sqrt{(1-t^2)\,(1-k^2t^2)}}.$$

Moreover, if we set $\tau = (1+k_1)\,\dfrac{t\sqrt{1-t^2}}{\sqrt{1-k^2t^2}}$, $0 \le t \le 1/\sqrt{1+k_1}$, then $0 \le \tau \le 1$, and

$$I_1 = \frac{1}{1+k_1}\, K\left(\frac{1-k_1}{1+k_1}\right).$$

Hence,

$$I_1 + I_2 = 2\,I_1 = \frac{2}{1+k_1}\, K\left(\frac{1-k_1}{1+k_1}\right),$$

which proves (7.2.15). Since

$$k' = \frac{1-k_1}{1+k_1} = \frac{1-k_1^2}{(1+k_1)^2} = \frac{k^2}{(1+k^2)^2} < k^2 < 1,$$

after n applications of the Landen transformation we find that the quantities $k'^{(n)} \to 0$ very rapidly as $n \to \infty$. This means that we can use the Landen transformation to evaluate the integral in (7.2.11) step-by-step with smaller and smaller values of k, so that after a finite number of steps the value of the integral (7.2.11) becomes equal to $\arcsin(1) = \pi/2$. However, for values of k closer to 1, the convergence becomes very slow. In that case we can use the inverse Landen transformation

$$K(k') = \frac{1}{2}\,(1+k_1)\,K(k), \tag{7.2.16}$$

where $k_1 = \dfrac{1-k'}{1+k'}$, $k = \sqrt{1-k_1^2} = \dfrac{2\sqrt{k'}}{1+k'}$, and then we can evaluate $K(k)$ approximately with sufficient accuracy from the asymptotic formula

$$K(k) \approx \ln\frac{4}{k_1} = \frac{1}{2}\ln\frac{16}{1-k^2} \quad \text{for } k \text{ close to 1.} \tag{7.2.17}$$

Further, if we set $t = \sin\phi$ in (7.2.11), then

$$K(k) = \int_0^{\pi/2} \frac{d\phi}{\sqrt{1-k^2\sin^2\phi}} = \frac{2}{1+k_1}\int_0^{\pi/2} \frac{d\phi}{\sqrt{1-k'^2\sin^2\phi}},$$

or

$$\begin{aligned} K(k) &= \int_0^{\pi/2} \frac{d\phi}{\sqrt{\cos^2\phi + k_1^2\sin^2\phi}} \\ &= \frac{2}{1+k_1}\int_0^{\pi/2} \frac{d\phi}{\sqrt{\cos^2\phi + (1+k'^2)\sin^2\phi}}. \end{aligned}$$

Let $k_1 = \dfrac{d_0}{c_0}$, where $c_0 = 1$ and $d_0 = k_1$. Then

$$1 - k'^2 = 1 - \left(\frac{1-k_1}{1+k_1}\right)^2 = \left(\frac{2\sqrt{k_1}}{1+k_1}\right)^2 = \left(\frac{2\sqrt{c_0\, d_0}}{c_0+d_0}\right)^2 = \left(\frac{d_1}{c_1}\right)^2,$$

where $d_1 = \sqrt{c_0\, d_0}$, and $c_1 = (c_0 + d_0)/2$, which yields

$$\begin{aligned} K(k) &= \int_0^{\pi/2} \frac{d\phi}{\sqrt{c_0^2 \cos^2\phi + d_0^2 \sin^2\phi}} \\ &= \frac{2}{1 + d_0/c_0} \int_0^{\pi/2} \frac{d\phi}{\sqrt{\cos^2\phi + (d_1/c_1)^2 \sin^2\phi}} \\ &= \int_0^{\pi/2} \frac{d\phi}{\sqrt{c_1^2 \cos^2\phi + d_1^2 \sin^2\phi}}. \end{aligned}$$

Repeating this process n-times, we get

$$K(k) = \int_0^{\pi/2} \frac{d\phi}{\sqrt{c_n^2 \cos^2\phi + d_n^2 \sin^2\phi}}, \tag{7.2.18}$$

where c_n and d_n are determined recursively from

$$c_n = \frac{1}{2}\left(c_{n-1} + d_{n-1}\right), \quad d_n = \sqrt{c_{n-1}\, d_{n-1}}.$$

Hence, as $n \to \infty$, the sequences $\{c_j\}$ and $\{d_j\}$ for $j = 0, 1, 2, \dots$, and $c_0 > d_0$, converge to the same limit, i.e.,

$$\lim_{n\to\infty} c_n = \lim_{n\to\infty} = M(k), \tag{7.2.19}$$

which yields

$$K(k) = \frac{\pi}{2\, M(k)}.$$

Proof of this result is given in Andersen et al. [1962:163]. Hence, from (7.2.14) the ratio between the sides of the rectangle is given by

$$\frac{a}{b} = 2\,\frac{K(k)}{K(k_1)} = 2\,\frac{M(k_1)}{M(k)}, \tag{7.2.20}$$

where $M(k_1)$ is obtained in the same manner as (7.2.19) by applying the above recursion to $k_1 = \sqrt{1-k^2}$. For more on elliptic integrals and Jacobian elliptic functions, see Phillips [1943, Ch. 1 and 2]. ■

Map 7.8. To map the trapezoid $A_1\, A_2\, A_3\, A_4$ in the z-plane onto the upper half-plane (Figure 7.7), the mapping function from (7.2.14) is given by

$$w = C_2 + C_1 \int_0^z (\zeta+1)^{-1/6}(\zeta-1)^{-1/3}(\zeta+k)^{-2/3}(\zeta-3)^{-5/6}\, d\zeta. \tag{7.2.21}$$

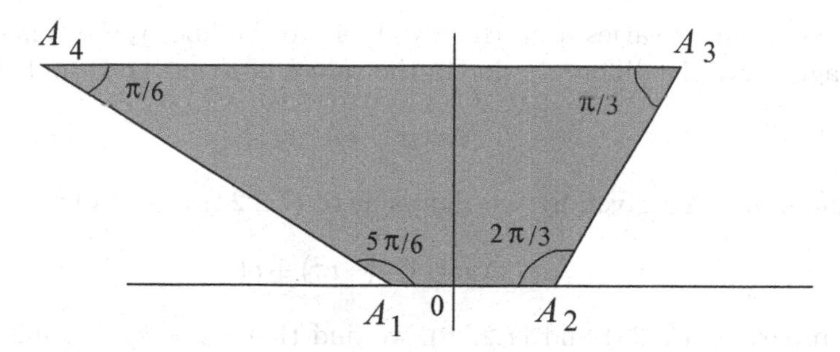

Figure 7.7 Trapezoid onto the upper half-plane.

Map 7.9. A special case of the above transformation is the mapping of the upper half-plane $\Im\{z\} > 0$ onto the interior of a square of side a. We start with the unit disk $|w| < 1$ in the w-plane, and use the transformation (7.1.11) which gives

$$z(w) = \int_0^w \frac{dt}{\sqrt{1-t^4}}, \tag{7.2.22}$$

with the length of the square's side equal to $\sqrt{2}$ in the z-plane.

Map 7.10. To map the upper half-plane $\Im\{z\} > 0$ onto an arbitrary quadrilateral $A_1A_2A_3A_4$ with interior angles $\mu_1\pi$, $\mu_2\pi$, $\mu_3\pi$, and $\mu_4\pi$, respectively, such that the angle $\mu_1\pi$ at A_1 is the smallest and the ratio of the side A_4A_1 to the side A_1A_2 is λ. Without loss of generality, let the vertices A_1, A_2, A_3, A_4 correspond to the points $x_1 = -1$, $x_2 = 1$, $x_3 = k$, $x_4 = 3$. Then, by (7.1.2), the transformation is given by

$$w = f(z) = C_1 \int_1^z (t+1)^{-\alpha_1}(t-1)^{-\alpha_2}(t-k)^{-\alpha_3}(t-3)^{-\alpha_4}\,dt + C_2. \tag{7.2.23}$$

Newton's method for numerical evaluation of improper integrals in (7.2.23) and an approximate value of k is given in §8.2. ■

Map 7.11. To map a horizontal parallel strip with a rectilinear horizontal cut onto the upper half-plane (Figure 7.8), note that the given strip is equivalent to a quadrilateral with vertices at A_1, A_2 and A_3, all at infinity, and A_4 at the origin. For the vertices at A_1, A_2, A_3 we have $\alpha_1 = \alpha_2 = \alpha_3 = 1$, and for A_4 we have $\alpha_4 = -1$. We choose $x_1 = 1$, $x_2 = \infty$, $x_4 = 0$, and let $w = 0$ correspond to $z = 0$. Then C_2 in (7.1.2) is zero. With this choice the point A_3 would correspond to some point $x_3 = -k$ on the negative x-axis ($k > 0$). Then from (7.1.2) the required transformation with proper choice of principal values for the integrand is given by

$$\begin{aligned} w &= C_1 \int_0^z \frac{t\,dt}{(1-t)(k+t)} \\ &= C_1 \int_0^z \frac{1}{1+k}\left(\frac{1}{1-t} - \frac{k}{k+t}\right) dt \\ &= C_1 \left[\log(1-z) + k\log\left(1+\frac{z}{k}\right)\right], \end{aligned} \tag{7.2.24}$$

where C_1 is an arbitrary constant. To determine C_1 and k if a and b are given, let $z = 1 - \varepsilon\, e^{i\theta}$, $-\pi \le \theta < 0$, $\varepsilon > 0$ and small, and define a half-circle C_ε. Then, as z moves along

C_ε, its image point w varies from the ray $A_4 A_1$ to the line $A_1 A_2$. Thus, for the end point of the image curve the difference Δw in the values of w on $A_4 A_1$ and $A_1 A_2$ is

$$\Delta w = -ia + o(1). \tag{7.2.25}$$

But the difference Δw given by the right side of (7.2.24) is given by

$$\Delta w = C_1 (-i\pi) + o(1). \tag{7.2.26}$$

Hence, comparing (7.2.25) and (7.2.26), we find that $C_1 = a/\pi$. Similarly, as z moves on the other half-circle C'_ε with center at $z = -k$, i.e., $z + k = \varepsilon\, e^{i\theta}$, the value of w varies from its value on $A_2 A_3$ to $A_3 A_4$, and in this case the difference is $\Delta w = -ib + o(1)$, which from (7.2.24) is equal to $-C_1 k i\pi + o(1)$. This yields $b = C_1 k\pi = ka$, or $k = b/a$. Thus, the required mapping is given by

$$w = \frac{a}{\pi} \log(1 - z) + \frac{b}{\pi} \log\left(1 + \frac{a}{b} z\right). \blacksquare \tag{7.2.27}$$

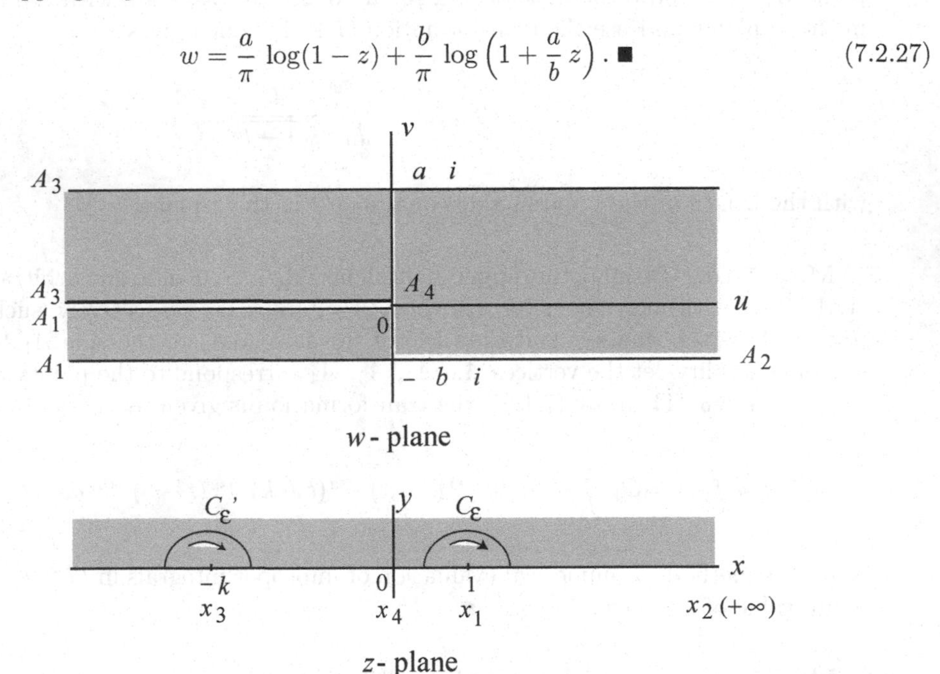

Figure 7.8 Map 7.11.

Map 7.12. To map the upper half-plane $\Im\{z\} > 0$ onto a set of two parallel lines, we use Map 7.11 by treating the w-plane as an intermediate ζ-plane ($\zeta = \xi + i\eta$), and the set of two parallel lines in the w-plane which are h distance apart (Figure 7.9). Let the images of the two parallel lines be at $\xi_1 = -1$ and $\xi_2 = 1$, which correspond to $w = 0$ and $w = ih$. The exterior angles in the w-plane are $\theta_1 = 0 = \theta_2$ at A and B, and $\theta_3 = \theta_4 = \pi$ at C and D. The infinitely distant point C is equally distant in the w-plane, while D translates into $\Re\{\xi\} = \infty$. Thus, the Schwarz-Christoffel transformation is

$$\begin{aligned} w(\zeta) &= B_1 \int_{\zeta_0}^{\zeta} \frac{dt}{(t-\zeta_1)^{\theta_1/\pi}\,(t-\zeta_2)^{\theta_2/\pi}\,(t-\zeta_4)^{\theta_4/\pi}} + B_2 \\ &= B_1 \int_{\zeta_0}^{\zeta} \frac{dt}{(t+1)^0\,(t-1)^0\,(t-0)^1} + B_2 \\ &= B_1 \log\zeta + B_2, \end{aligned} \tag{7.2.28}$$

where $\xi_0 = \zeta_0 = 1$ (point B). To determine the integration constants, $\xi = \zeta = 1$ and $w = 0$ at B, which gives $B_2 = 0$; and $\xi = \zeta = -1$ and $w = ih$ at A, which gives $ih = B_1 \log(-1) = B_1 \log\left(e^{i\pi}\right)$, thus $B_1 = h/\pi$. Hence, $w(\zeta) = \dfrac{h}{\pi} \log \zeta$.

Map 7.13. A simple reorganizing of the above case yields the following result: The Schwarz-Christoffel transformation $w = f(z) = \dfrac{2}{\pi} \arcsin z$ maps the upper half-plane $\Im\{z\} > 0$ onto the semi-infinite strip $|\Re\{w\}| < 1, \Im\{w\} > 0$, such that $f(0) = 0$ and $f(1) = 1$.

Map 7.14. Map the semi-infinite strip $u > 0$, $0 < v < a$, as the limit of the triangle OCA as $\theta \to \pi/2$ onto the upper half-plane $\Im\{z\} > 0$ (Figure 7.9), such that $z = -1, 1$ correspond to $w = ia, 0$, respectively.

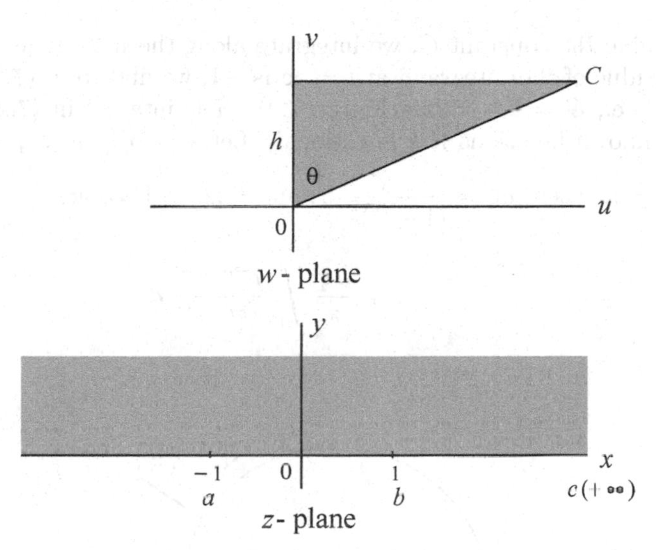

Figure 7.9 Triangle as limit of the upper half-plane.

The transformation (7.1.2) yields

$$
\begin{aligned}
w = f(z) &= C_2 + \lim_{\theta \to \pi/2} C_1 \int^z \frac{dt}{(t+1)^{1/2}(t-1)^{1-\theta/\pi}} \\
&= C_1 \int_1^z \frac{dt}{\sqrt{t^2 - 1}} = B \cosh^{-1} z,
\end{aligned}
\tag{7.2.29}
$$

where $C_2 = 0$ since $w = 0$ corresponds to $z = 1$. From (7.2.29) we have $w/C_1 = \cosh^{-1} z$, which gives $z = \cosh(w/C_1)$, and since $w = ia$ corresponds to $z = -1 = \cosh i\pi$, we get $C_1 = a/\pi$. Hence, from (7.2.29)

$$
w = \frac{a}{\pi} \cosh^{-1} z,
$$

and the required transformation is given by

$$
z = \cosh \frac{\pi w}{a}. \blacksquare \tag{7.2.30}
$$

Map 7.15. To map the trapezoidal region in the z-plane, shown in Figure 7.10, onto the upper half-plane $\Im\{w\} > 0$, formula (7.2.1) gives

$$z = C \int_1^w \frac{dt}{t^{1-k}\,(t-1)^k}, \quad 0 < k < 1. \tag{7.2.31}$$

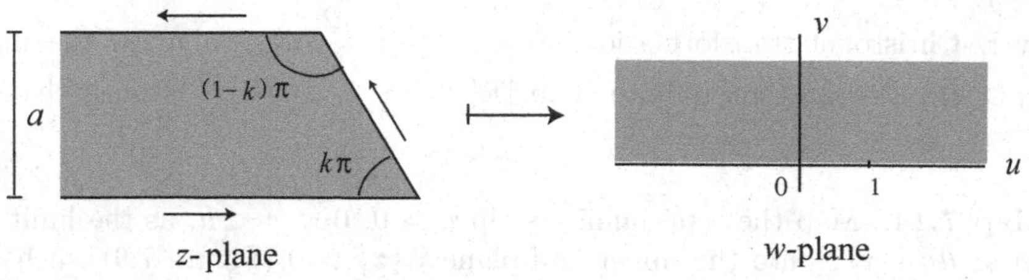

Figure 7.10 Semi-infinite strip onto $\Im\{w\} > 0$.

To determine the constant C, we integrate along the half-circle $w = R\,e^{i\theta}$, $0 < \theta < \pi$. Since the residue of the integrand at $t = \infty$ is -1, we find from (7.2.31), as $R \to \infty$, that $ai = -Ci\pi$, i.e., $C = -a/\pi$ (see Figure 7.9). The integral in (7.2.31) can be evaluated in terms of known functions if k is rational. Let $k = p/q$, $p < q$, where $p, q \in \mathbb{R}^+$. Set $\left(\dfrac{\zeta - 1}{\zeta}\right)^{1/q} = t$. Then, $d\zeta = \dfrac{qt^{q-1}}{(1-t^q)^2}\,dt$, and (7.2.31) becomes

$$z = \frac{a\,q}{\pi} \int_0^t \frac{t^{-p+q-1}}{t^q - 1}\,dt. \tag{7.2.32}$$

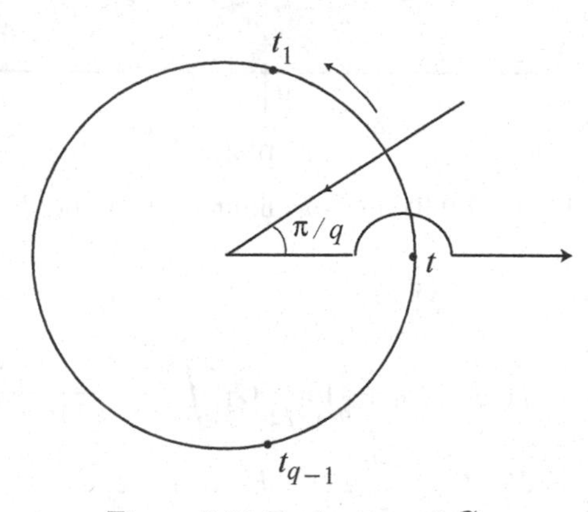

Figure 7.11 Evaluation of C .

Now, the q poles of the integrand in (7.2.35) are the q-th zeros of $(t^q - 1)$, i.e., they are at $t_n = e^{2ni\pi/q}$, $n = 0, 1, \dots, q-1$. Thus,

$$\frac{t^{-p+q-1}}{t^q - 1} = \sum_{n=0}^{q-1} \frac{t_n^{-p+q-1}}{q\,t_n^q - 1}\,\frac{1}{t - t_n} = \frac{1}{q} \sum_{n=0}^{q-1} \frac{1}{t_n^p}\,\frac{1}{t - t_n},$$

and

$$\int_0^t \frac{1}{t - t_n}\,dt = \ln\left|t - t_n\right| + \text{const} = \ln\left(1 - \frac{t}{t_n}\right),$$

where the constant is zero because the integrand is zero at $t = 0$ and the principal value of the logarithm is taken. Thus, the required transformation becomes

$$z = \frac{a}{\pi} \sum_{n=0}^{q-1} \frac{1}{t_n^q} \ln\left(1 - \frac{t}{t_n}\right). \tag{7.2.33}$$

As a special case, if $\mu = 1/2$, i.e., if $p = 1$, $q = 2$, the transformation (7.2.33) reduces to

$$z = \frac{a}{\pi} \ln \frac{1-t}{1+t}, \quad t = \left(\frac{t-1}{t}\right)^{1/2}. \ \blacksquare \tag{7.2.34}$$

In the next chapter we will discuss the problem of approximately computing the values of the $(2n+2)$ parameters involved in the Schwarz-Christoffel formula (7.1.2). This discussion involves a numerical solution of improper integrals known as Schwarz-Christoffel integrals.

Map 7.16. The Schwarz-Christoffel transformation

$$w = f(z) = \frac{1}{\pi} \sqrt{2(1-z)} + \frac{1}{\pi} \log \frac{\sqrt{1-z} - \sqrt{2}}{\sqrt{1-z} + \sqrt{2}} - i \tag{7.2.35}$$

maps the upper half-plane $\Im\{z\} > 0$ onto the region consisting of the fourth quadrant plus the strip $0 < v < 1$ in the w-plane, $w = u + i\,v$, such that $f(1) = 0$.

Map 7.17. The mapping of the exterior of a thin straight line slit in the z-plane onto the exterior of the unit circle $|w| > 1$ can be carried out using the Schwarz-Christoffel transformation. Let the segment AB of length $4a$ be symmetrical with respect to the y-axis. The exterior angles at its endpoints A and B are each equal to π. Let $w_1 = -1$ and $w_2 = 1$. Then, using the transformation (7.1.9) of the required mapping is

$$\begin{aligned} z(w) &= C_1 \int_{w_0}^{w} (t+1)(t-1) \frac{dt}{t^2} \\ &= C_1 \int_{w_0}^{w} \left(1 - \frac{1}{t^2}\right) dt + C_2' = C_1\left(w + \frac{1}{w}\right) + C_2'. \end{aligned}$$

Since $z = \pm 2a$ correspond to $w = \pm 1$, respectively, we have $C_1 = a$ and $C_2' = 0$. Hence, the required transformation is

$$z(w) = a\left(w + \frac{1}{w}\right), \quad \text{or} \quad w(z) = \frac{1}{2a}\left[z \pm \sqrt{z^2 - 4a^2}\right], \tag{7.2.36}$$

where the positive square-root is used so that $z = \infty$ corresponds to $w = \infty$. Note that the *negative* square-root in (7.2.36) will map the exterior of the slit onto the interior of the unit circle $|w| < 1$.

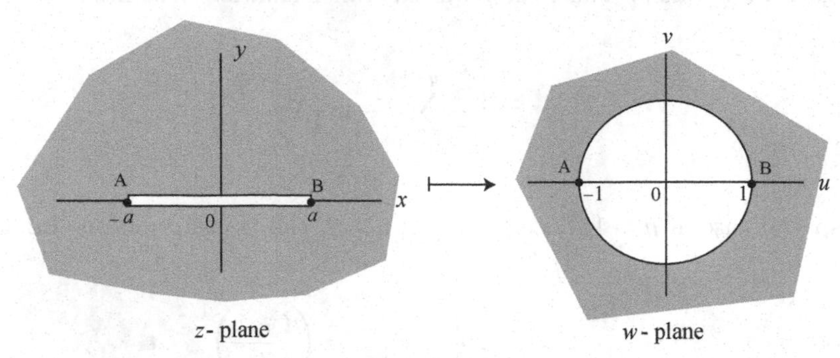

Figure 7.12 Exterior of a slit onto the exterior of unit circle.

7.3 Regions Exterior to and between Rectangles

We will consider cases of mapping the regions exterior to rectangles and between rectangles.

Map 7.18. Consider the exterior of a rectangle as in Figure 7.13, where the mapping from the z-plane onto the ζ-plane is defined by

$$z(\zeta) = \int_0^\zeta t^{1/2}(t^2-1)^{-1/2}\,dt, \tag{7.3.1}$$

where, according to Bieberbach [1914] and von Koppenfels and Stallman [1959], the semi-infinite lines BA and DE in Figure 7.13 can be taken as axes of symmetry.

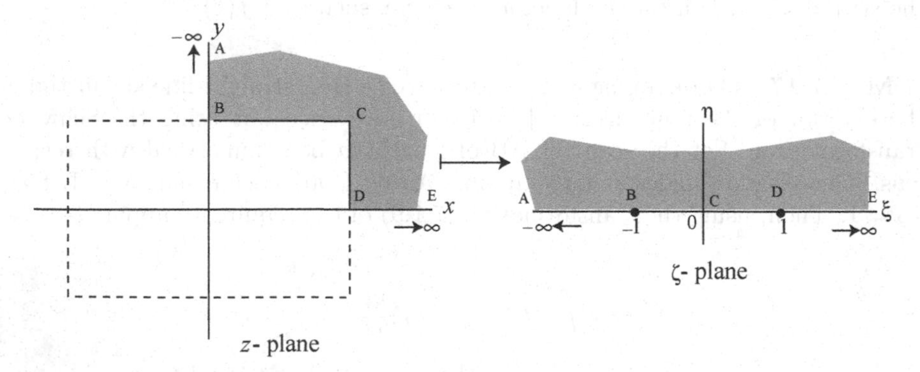

Figure 7.13 Exterior of a rectangle onto the upper half-plane.

Map 7.19. The symmetry mentioned in the above case can also be used to determine the transformation that maps the region between two rectangles (one an inner and the other an outer rectangle) with a common center at $z = 0$ onto the upper half-plane $\Im\{\zeta\} > 0$, as shown in Figure 7.14. The Schwarz-Christoffel transformation is

$$z(\zeta) = C_1 \int_0^\zeta \left(\frac{t^2-\xi_2^2}{(t^2-1)\,(t^2-k^2)\,(t^2-\xi_1^2)}\right)^{1/2} dt, \tag{7.3.2}$$

where the values of the constant C_1 and the modulus k can be found from the given dimensions of the rectangles.

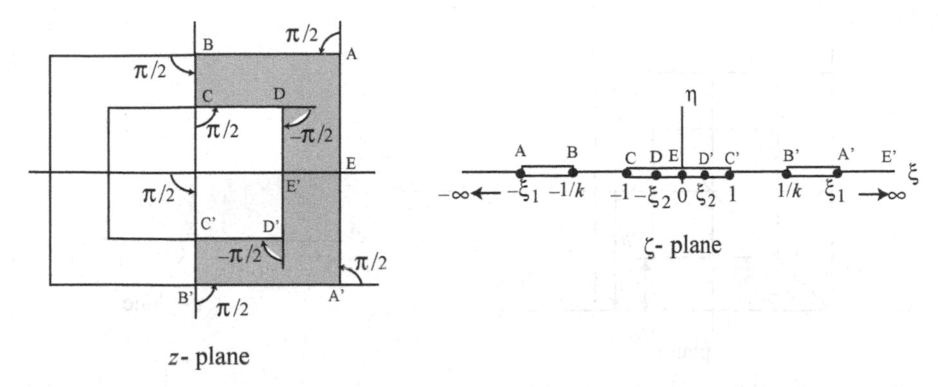

Figure 7.14 Region between two rectangles onto the upper half-plane.

Let the value of the integral (7.3.2) be taken as

$$I\left(\zeta_m,\zeta_n\right)=\int_{\zeta_m}^{\zeta_n}\left(\frac{|t^2-\xi_2^2|}{\left|\left(t^2-1\right)\left(t^2-k^2\right)\left(t^2-\xi_1^2\right)\right|}\right)^{1/2}dt. \tag{7.3.3}$$

Then, using Figure 7.14, we have: $a/2=C_1I\left(\xi_2,1\right);b/2=C_1I\left(0,\xi_2\right);d/2=$ segment (B-C)+ segment (D-E); $C_1=I\left(0,\xi_2\right)+I\left(1,1/k\right)$; and $c/2=$ segment (B-A) $=C_1I\left(1/k,\xi_2\right)$. These values can be now used to determine the three unknowns ξ_1,ξ_2 and k from the known ratios of $a/d, b/d$ and c/d. Moreover, we can determine C_1 using the relation $I(0,1)=a/2+b/2=K(k)$, a complete elliptic integral of the first kind. Finally, the transformation $w=\log(\zeta)$ will map the upper half-plane $\Im\{\zeta\}>0$ onto a parallel strip of length $K(k)$ and height $K'(k)$.

As an application, this method has been widely used to determine the characteristic impedance of various thin sheet registers and transmission line cross-sections (see parameter problem, and a numerical method in §22.3).

Map 7.20. To map two concentric squares symmetric about a diagonal axis as onto a parallel strip in the w-plane shown in Figure 7.15, we consider two cases:

Map 7.20a. For the mapping shown in Figure 7.15, the Schwarz-Christoffel transformation is

$$\begin{aligned}z(\zeta)&=C_1\int_0^{\zeta}\frac{dt}{(t-0)^{3/4}(t-1)^{1/2}(t-1/k)^{1/2}}\\&=C\int_0^{\zeta}\frac{dt}{t^{3/4}(1-t)^{1/2}(1-kt)^{1/2}},\end{aligned} \tag{7.3.4}$$

where $C=C_1/k$. Using Bowman's method (Bowman [1933]), we carry out one more intermediate mapping $\zeta=(w^*)^2$, and then set $w^*=\sin w$. This will provide a simple

parallel strip of length of the sides K and K' in the w-plane.

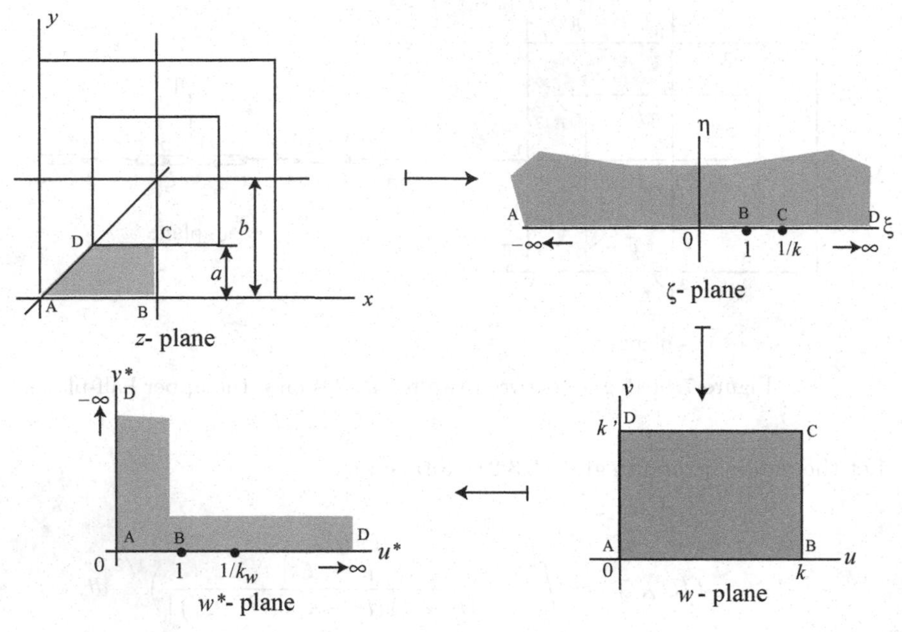

Figure 7.15 Two concentric squares, Map 7.20a.

Map 7.20b. For the mapping of the inner square set like a diamond inside a corner square as shown in Figure 7.16, the Schwarz-Christoffel transformation is

$$z(\zeta) = C_1 \int_0^\zeta \frac{dt}{(t-0)^{1/2}(t-1)^{1/4}(t-i/k)^{1/2}}. \tag{7.3.5}$$

Using the same Bowman's method, the transformation to map it onto the parallel strip of sides K and K' in the w-plane is given by

$$\zeta = \text{sn}^2(w). \tag{7.3.6}$$

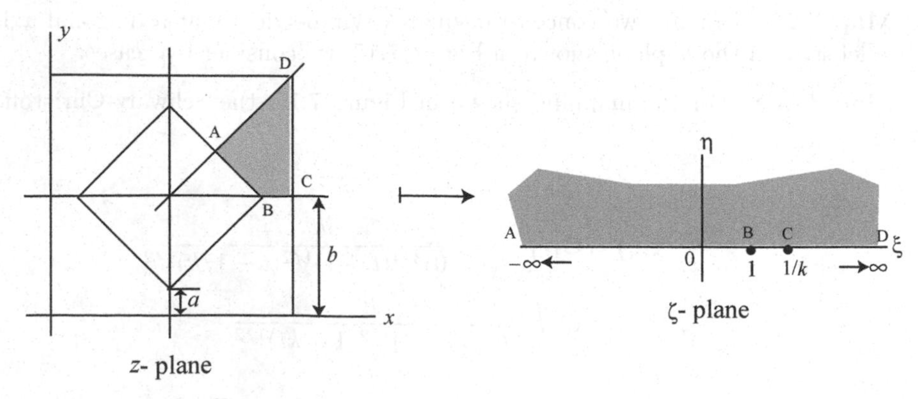

Figure 7.16 Two concentric squares, Map 7.20b.

7.4 Polygons with Round Corners

Map 7.21. In the real world problems, one usually finds round corners in manufactured items. Sharp corners are idealized or mathematically simplified cases. Some of the causes of round corners are: (i) wear and tear by constant use of the physical object; and (ii) production of round cornered objects on purpose to reduce the high potential gradients which can result in physical failures especially in electrically operated objects. We will consider an example of polygons with round corners as presented in Figure 7.17, and use Schwarz-Christoffel transformation to determine the field between the right angle (unrounded corner at C) and the y-axis. First, for the right angle corner at C we have the transformation

$$z(w) = C_1 \int \frac{dt}{(t-0)^1 (t-1)^{-1/2}} + C_2 = C_1 \int \frac{\sqrt{t-1}}{t}\, dt + C_2, \tag{7.4.1}$$

which corresponds to the points A, B, C as marked in the z-plane and w-plane in Figure 7.17. Note that $z = 0$ does *not* correspond to $u = 0$, and, thus, C_2 need not be zero. Also, the angle $3\pi/2$ at the point A in the z-plane is avoided by moving this point to infinity in the w-plane. Thus, the solution for the unrounded or right-angle corner at C is

$$z(w) = 2C_1 \Big(\sqrt{t-1} + \frac{i}{2} \log \frac{i - \sqrt{t-1}}{i + \sqrt{t-1}} \Big) + C_2. \tag{7.4.2}$$

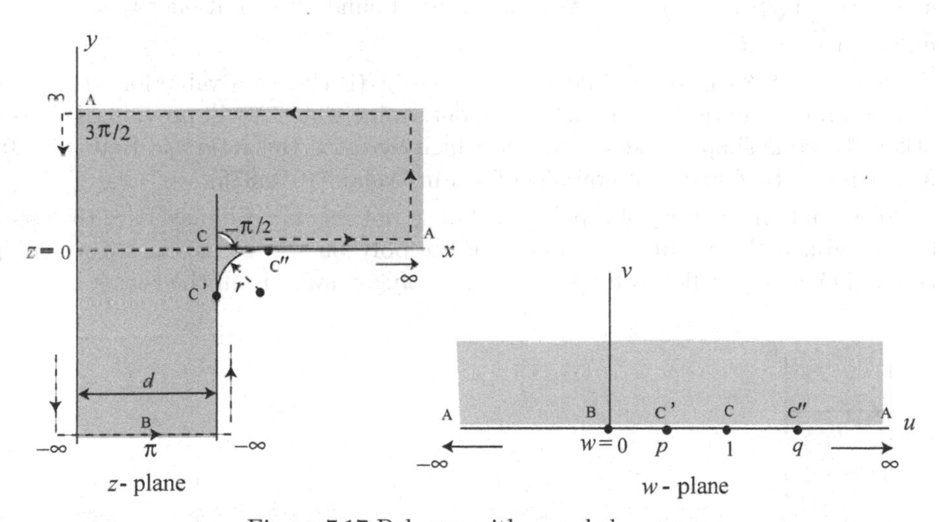

Figure 7.17 Polygon with rounded corner.

The jump across the gap of width d between $z = 0$ and $z = d$ (corner C) occurs at B, which corresponds to $w = 0$. Using formula (7.1.18) we get

$$d = -i\pi C_1 (0-1)^{1/2}, \quad \text{or} \quad C_1 = d/k. \tag{7.4.3}$$

Since the point $z = d$ (point C) in the z-plane is mapped onto the point $w = 1$ (C in the w-plane), we have the boundary condition $z(1) = d$, using which we find that $C_2 = 2d$, thus, giving the solution for the right-angle corner at C as

$$z(w) = \frac{2\pi}{d} \Big(\sqrt{e^w - 1} + \frac{i}{2} \log \frac{i - \sqrt{e^w - 1}}{i + \sqrt{e^w - 1}} \Big) + 2d. \tag{7.4.4}$$

Next we resolve the problem for the round corner at C in two steps: Firstly, we solve the problem with the sharp (unrounded corner); then we find an equipotential line near this corner which has the shape nearly equal to the rounding. It then becomes the new boundary.

Secondly, in a general setting we may replace the right corner by a circular arc. In the case of Figure 7.17 the point C is replaced by a quarter arc of a circle of radius r such that this arc is tangent to the polygon at C' and C" with coordinates $z = d - ir$ and $z = d + r - i0$, respectively. In the w-plane, the point C' corresponds to $u = p$ and the point C'' to $u = q$.

Thus, the factor of the type $(w - u_j)^{-\alpha_j}$ is replaced by $(u - p)^{-\alpha_j} + \lambda(u - q)^{\alpha_j}$. Hence, the transformation is

$$z(w) = C_1 \int \frac{\sqrt{t-p} + \lambda\sqrt{t-q}}{t}\, dt + C_2, \tag{7.4.5}$$

which after integrating yields

$$z(w) = 2C_1\left[\sqrt{u-p} - \sqrt{p}\arctan\left(\sqrt{(u-p)/p}\right)\right] + \lambda\left[\sqrt{u-q} - \sqrt{q}\arctan\left(\sqrt{(u-q)/q}\right)\right] + C_2. \tag{7.4.6}$$

Although the weight factor λ is not known, the constant C_1 can be evaluated using (7.1.18) as $C_1 = \dfrac{d}{\pi}\left(\sqrt{p} + \lambda\sqrt{q}\right)^{-1/2}$. Also, using the boundary condition $z = d - ir$ at $u = p$, we find that $C_2 = d$.

The weight λ can be evaluated in two ways: (i) choose a value for either p or q on the xi-axis and use optimization method to determine which λ will provide the best fit; (ii) the other choice is simply take $q = 1 - p$, which by using the ratio r/d leads to a function for λ which can be found in a graphical form in Weber [1950:375].

Note that in the case of a polygon that is not regular, we may have to rescale the field to reestablish the original dimensions of the polygon. Note that such a rounding does not run sufficiently parallel to the sides of the polygon away from the corner.

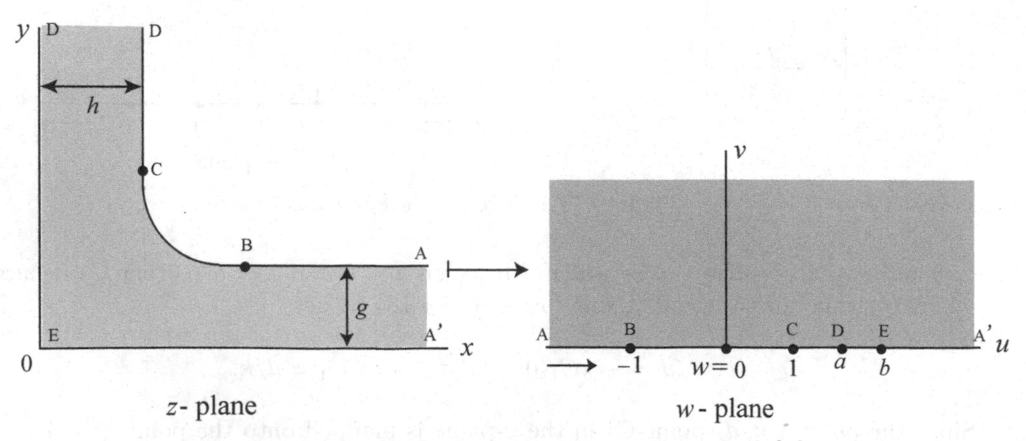

Figure 7.18 Region with a curved boundary-line onto $\Im\{z\} > 0$.

Map 7.22. The transformation $\dfrac{dz}{dw} = \dfrac{\sqrt{w+1} + c\sqrt{w-1}}{(w-a)\sqrt{w-b}}$, where $b > a > 1$, and $c = \sqrt{\dfrac{b+1}{b-1}} > 0$, maps the region with a curved boundary-line in the z-plane onto the upper half-plane $\Im\{w\} > 0$. This mapping is shown in Figure 7.18, where the curve is almost a quarter of a circle, $g = (1+c)\pi$, and $h = \left(\sqrt{\dfrac{a+1}{b-a}} + c\sqrt{\dfrac{a-1}{b-a}}\right)\pi$.

Map 7.23. The above transformation (Map 7.22), when combined with $w = b + e^{-i\pi}\,\zeta^2$, maps the region bounded by $x = 0$ and the two curved boundary-lines onto the upper half-plane $\Im\{\zeta\} > 0$ (i.e., $\eta > 0$), where $\zeta = \xi + i\eta$, and g and h are defined as above. This mapping is presented in Figure 7.19.

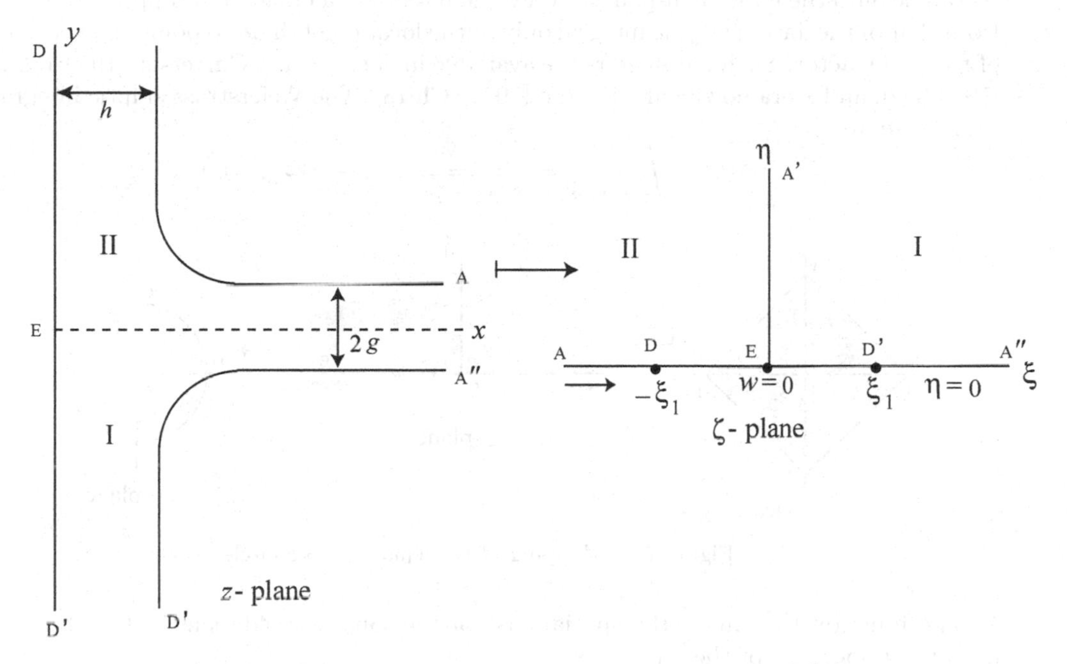

Figure 7.19 Region with two curved boundary-lines onto $\Im\{\zeta\} > 0$.

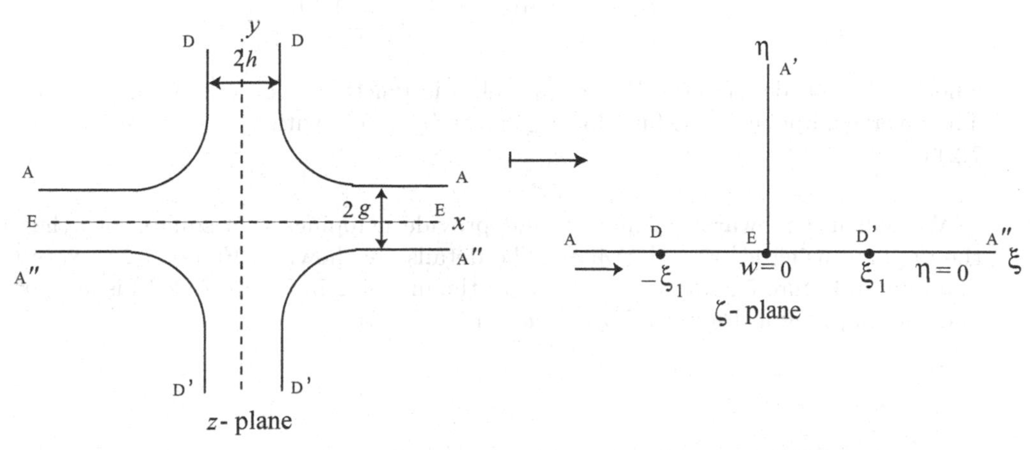

Figure 7.20 Region with four curved boundary-lines onto $\Im\{\zeta\} > 0$.

Map 7.24. The above transformation (Map 7.23), when combined with $w = b + \dfrac{a-b}{4}(\zeta + 1/\zeta)^2$, maps the region bounded by the four curved boundary-lines onto the upper half-plane $\Im\{\zeta\} > 0$ (i.e., $\eta > 0$), where $\zeta = \xi + i\eta$, and g and h are defined as above. This mapping is presented in Figure 7.20.

7.5 Weierstrass Integral Equation

Map 7.25. The Weierstrass elliptic integral equation is another method to map a rectangle onto the upper half-plane or onto the interior of a circle. This approach is different from that of the Jacobi elliptic integral only in the location of the zero point and specification of the scale factor. Such parameters are available in Sansone and Garretson [1969:132], Betz [1964:352], and Abramowitz and Stegun [1972: Ch 18]. The Weierstrass elliptic integral and its inverse are

$$z(\xi) = \int_{\xi}^{\infty} \frac{d\xi}{\sqrt{4\xi^3 - g_2\xi - g_3}}, \qquad \xi = \wp\,(z; g_2, g_3)\,. \tag{7.5.1}$$

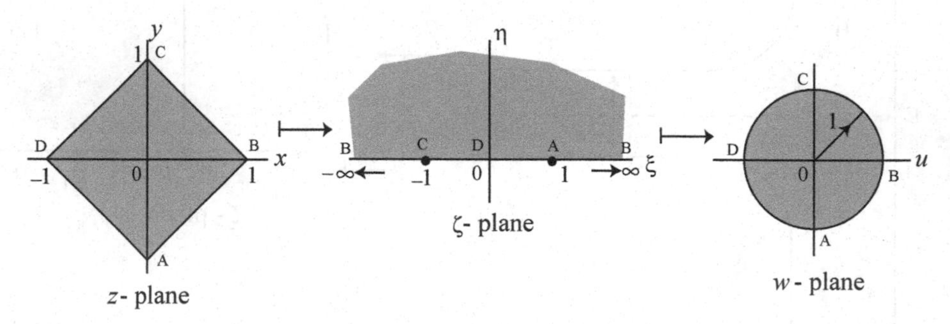

Figure 7.21 Mapping of the square onto a circle.

The function that maps the special case of the square of side length L in the z-plane onto the upper half of the ζ-plane is

$$\zeta = \wp\,(z; 4, 0)\,, \quad L = \frac{1}{4\sqrt{2\pi}}\,\Gamma^2\left(\frac{1}{4}\right), \tag{7.5.2}$$

where $\wp(z; 4, 0)$ denotes the Weierstrass elliptic function of order 2 with periods 4 and 0. The inverse mapping is the function $z(\xi)$ given by (7.5.1) with $g_2 = 4$ and $g_3 = 0$ (see Figure 7.21).

We will use Schwarz's symmetry and provide mappings of a square into the circle in the w-plane using $w^2 = \wp^{-1}(z; 4, 0)$. The details are shown in Figure 7.22. Note that the mapping in Figure 7.22(a) is $\xi = \wp(z; 4, 0)$; the mapping in Figure 7.22(b) is $w = \wp^2(z; 4.0)$; and the mapping in Figure 7.22(c) is $w = 1/\wp(z; 4, 0)$.

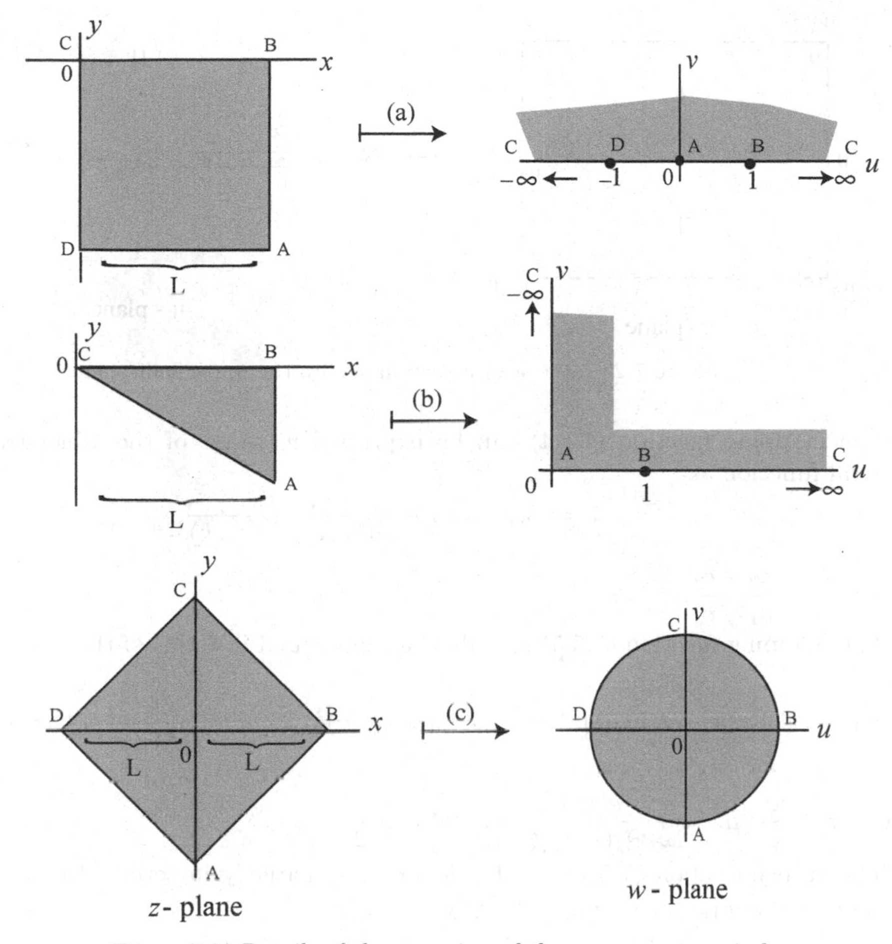

Figure 7.22 Details of the mapping of the square onto a circle.

An application of these mappings is

Map 7.26. Consider the transformation of the interior of a rectangle onto the half-plane, as presented in Figure 7.23. The Schwarz-Christoffel transformation is

$$z = \int_{-\infty}^{w} \frac{dt}{\sqrt{4\,(t-e_1)\,(t-e_2)\,(t-e_3)}} = \int_{-\infty}^{w} \frac{dt}{\sqrt{4t^3 - g_2 t - g_3}}, \tag{7.5.3}$$

where

$$\begin{aligned} e_1 &= \frac{1}{3}\Big(\frac{\pi}{2\omega_1}\Big)^2 \big(e_3^4(\tau) + e_0^4(\tau)\big), \quad e_2 = \frac{1}{3}\Big(\frac{\pi}{2\omega_1}\Big)^2 \big(e_2^4(\tau) + e_0^4(\tau)\big), \\ e_3 &= -\frac{1}{3}\Big(\frac{\pi}{2\omega_1}\Big)^2 \big(\Theta_2^4(\tau) + \Theta_3^4(\tau)\big), \quad \tau = \frac{\omega_3}{\omega_1}, \quad i\tau < 0; -\infty < e_< e_2 < e_1 < \infty \\ g_2 &= \frac{2}{3}\Big(\frac{\pi}{2\omega_1}\Big)^2 \big(\Theta_2^8(\tau) + \Theta_3^8(\tau) + \Theta_0^8(\tau)\big), \quad g_3 = 4e_1e_2e_3. \end{aligned} \tag{7.5.4}$$

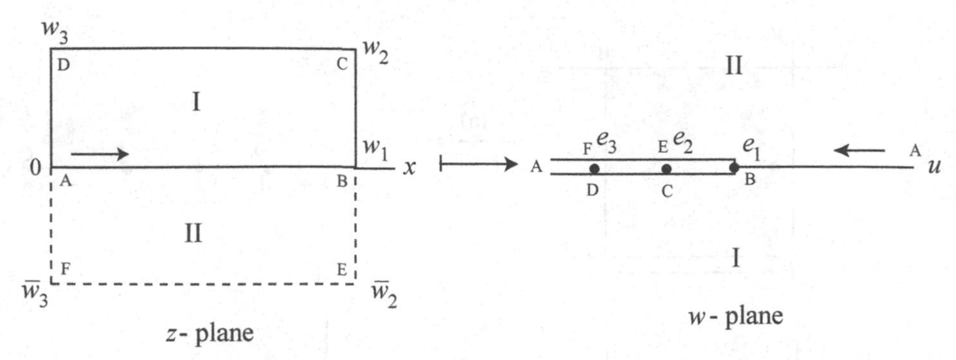

Figure 7.23 Interior of a rectangle onto the upper half-plane.

The mapping function (7.5.4) can be expressed in terms of the Weierstrass elliptical integral function as

$$w = \wp(z) = e_3 + \frac{e_1 - e_3}{\operatorname{sn}^2\left(z\sqrt{e_1 - e_3}, k\right)}, \tag{7.5.5}$$

where $k^2 = \dfrac{e_2 - e_3}{e_1 - e_3}$.

The mapping function (7.5.5) can also be represented in terms of the θ-series as

$$w = \wp(z) = e_1 + \left(\mu\vartheta_2(\xi|\tau)^2\right) = e_2 + \left(\mu\vartheta_3(\xi|\tau)^2\right) = e_3 + \left(\mu\frac{\Theta_2(\tau)}{\Theta_0(\tau)}\vartheta_0(\xi|\tau)\right)^2, \tag{7.5.6}$$

where $\tau = \dfrac{\omega_3}{\omega_1}, \mu = \dfrac{\vartheta_1'(\tau)}{2\omega_1\Theta_2(\tau)\vartheta_1(\xi|\tau)}$, and $\nu = \dfrac{z}{2\omega_1}$.

The transformation (7.5.4) can be defined alternatively in terms of e_1, e_2, e_3, all real, with $e_3 < e_2 < e_1$, $e_1 + e_2 + e_3 = 0$, and

$$k^2 = \frac{e_2 - e_3}{e_1 - e_3}, \quad k'^2 = \frac{e_1 - e_2}{e_1 = e_3},$$
$$\omega_1 = \frac{K}{\sqrt{e_1 - e_3}} \text{ for } \omega_1 > 0; \quad \omega_3 = \frac{iK}{\sqrt{e_1 - e_3}} \text{ for } \omega_3/i > 0,$$

where

$$\begin{aligned} K &= \int_0^1 \frac{dt}{\sqrt{1-t^2}\sqrt{1-k^2t^2}} = \int_0^{\pi/2} \frac{d\phi}{\sqrt{1-k^2\sin^2\phi}} \\ &= \frac{\pi}{2}\left\{1 + \sum_{n=1}^{\infty}\left(\frac{1\cdot 3\cdot 5\cdots(2n-1)}{2\cdot 4\cdot 6\cdots 2n}\right)^2 k^{2n}\right\} = F\left(k, \frac{\pi}{2}\right); \\ K' &= \int_0^{1/k} \frac{dt}{\sqrt{-1+t^2}\sqrt{1-k^2t^2}} = \int_0^{\pi/2} \frac{d\phi}{\sqrt{1-k'^2\sin^2\phi}} \\ &= \int_0^1 \frac{dt}{\sqrt{-1+t^2}\sqrt{1-k^2t^2}} = F\left(k', \frac{\pi}{2}\right). \end{aligned} \tag{7.5.7}$$

7.6 Rectangles and Other Regions

Map 7.27. The transformation that maps a square onto the lower half-plane, such that the four right triangles of angles $\pi/2, \pi/4, \pi/4$ within the square marked I, II, III, and IV, are mapped onto the four regions of the w-plane separated by the arcs of a semicircle in the w-plane (Figure 7.24), is given by

$$w = \wp(z; 4, 0), \tag{7.6.1}$$

where $g_2 = 4, g_3 = 0, e_1 = -e_3 = 1, e_2 = 0$, and

$$\omega_1 = \int_0^1 \frac{dt}{\sqrt{1-t^4}} = \frac{\pi}{2\sqrt{2}}\Big\{1 + \sum_{n=1}^{\infty}\Big(\frac{1\cdot 3\cdot 5\cdots(2n-1)}{2\cdot 4\cdot 6\cdots 2n}\Big)^2 2^{-n}\Big\},$$

and $\omega_3 = i\omega_1$.

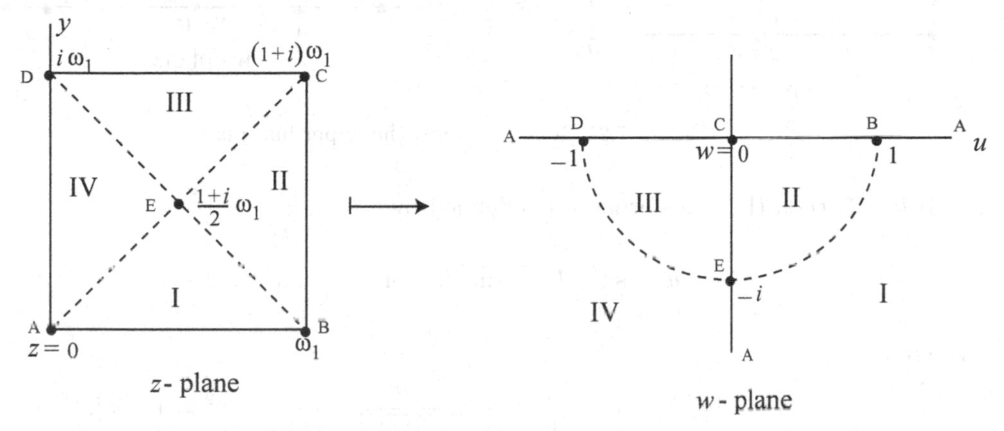

Figure 7.24 Square onto the lower half-plane.

Note that triangles with angles $\pi/2, \pi/4, \pi/4$; $\pi/3, \pi/3, \pi/3$; $2\pi/3, \pi/6, \pi/6$; $\pi/2, \pi/3, \pi/6$ are the only four cases of finite triangles in the z-plane that are mapped onto a half-plane under a conformal mapping function $w = f(z)$; see Map 7.5.

Map 7.28. To map a rectangle onto a half-plane or onto a quarter of a plane, as shown in Figure 7.25, is given by

$$z = \alpha \int_0^w \frac{dt}{\sqrt{(1-t^2)(1-k^2t^2)}}, \tag{7.6.2}$$

or

$$w = \operatorname{sn}\Big(\frac{z}{\alpha}, k\Big) = \operatorname{sn}\frac{z}{\alpha}, \tag{7.6.3}$$

where

$$\alpha = \frac{a}{K}, \quad K = \int_0^1 \int_0^w \frac{dt}{\sqrt{(1-t^2)(1-k^2t^2)}}. \tag{7.6.4}$$

In this case both a, b are positive, so that

$$\tau = \frac{ib}{a}, \; k = \frac{\Theta_2(\tau)}{\Theta_3(\tau)} = \sqrt{\lambda(\tau)}, \;\; 0 < k < 1.$$

As seen from Figure 7.25, the interior of the rectangle ABDE (marked I, II) is mapped onto the quadrant $u > 0, v > 0$ (first quadrant) in the w-plane; the interior of the rectangle ABB'A' (marked II, II') is mapped onto the upper half-plane $v > 0$ in the w-plane; and the line-segment $y = b/2$, $-a \le x \le a$ in the z-plane is mapped onto the semi-circle $|w| = \sqrt{e}, v \ge 0$ in the w-plane.

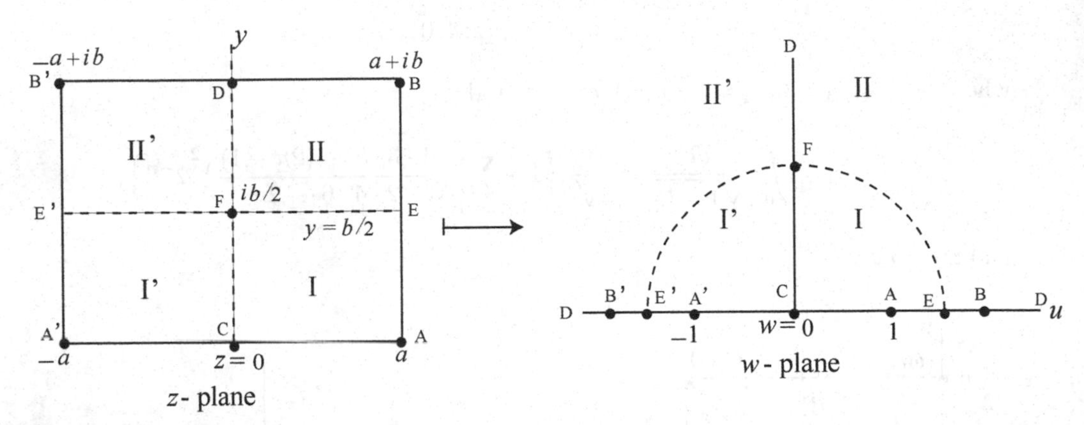

Figure 7.25 Rectangle onto the upper half-plane.

If $k < 1$, then the transformation is defined by

$$w = \operatorname{sn}(z, k) = \operatorname{sn}(z), \quad \text{or} \quad w = \operatorname{sn}^2(z, k), \tag{7.6.5}$$

where

$$a = K, \quad b = K' = \int_0^1 \frac{dt}{\sqrt{(1-t^2)\,(1-k'^2t^2)}}, \quad k'^2 = 1 - k^2,$$

and K is the same as defined in (7.6.1). In this case, the interior of the rectangle with vertices at $z = 0, K, K + iK', iK'$ is mapped onto the upper half-plane $v > 0$.

Map 7.29. To map a square onto a quarter of a plane, the transformation is

$$w = \operatorname{sn}\left(z, \frac{1}{\sqrt{2}}\right), \tag{7.6.6}$$

with $K = K' = 1$. This mapping is described in Figure 7.26 as follows:

(i) The points $z = 0$ (marked A); K (marked B); $(1 + i)K$ (marked C); iK (marked D); $\frac{1+i}{2}K$ (marked E); and $\frac{iK}{2}$ (marked F) in the z-plane are mapped onto the points $w = 0$ (A); 1 (B); $\sqrt{2}$ (C); ∞ (D); $\sqrt[4]{2}\,e^{i\pi/8}$ (E); and $i\sqrt[4]{2}$ (F), respectively, in the w-plane.

(ii) The line-segment $y + \frac{1}{2}K$, $-K \le x \le K$, in the z-plane is mapped onto the semi-circle $|w| = \sqrt[4]{2}v \ge 0$ in the w-plane.

(iii) The segment $x > 0, y > 0$ of the line $x + y = 1$ (BED) in the z-plane is mapped onto part $u > 0, v > 0$ of the hyperbola $u^2 - v^2 = 1$.

(iv) The segment $x < 0, y > 0$ of the line $y - x = 1$ (B'E'D) in the z-plane is mapped onto part $u < 0, v > 0$ of the same hyperbola.

(v) The segment $0 < x < K$ of the line $y = x$ (AEC) in the z-plane is mapped onto part $u > 0, v > 0$ of the lemniscate $|w-1||w+1| = 1$.

(vi) The segment $-K < x < 0$ of the line $y = -x$ (AE'C') in the z-plane is mapped onto part $u < 0, v > 0$ of the above lemniscate.

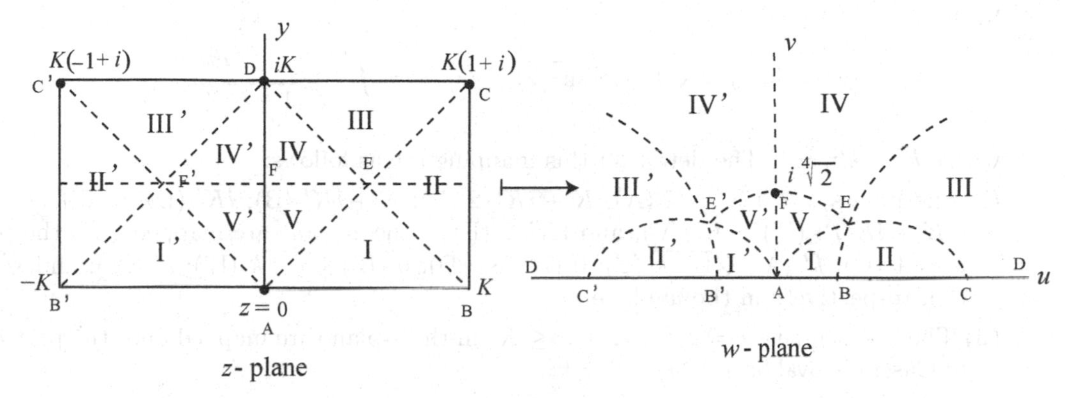

Figure 7.26 Square onto a quarter plane.

Map 7.30. The transformation of a rectangle, with four sub-rectangles marked I, II, III, and IV, in the z-plane onto the right-half of a Cassini's oval with a slit from $0 \leq w \leq 1$ is given by

$$w = \operatorname{cn} z = \sqrt{1 - \operatorname{sn}^2 z}, \quad z = \int_w^1 \frac{dt}{\sqrt{(1-t^2)(1-k^2t^2)}}, \quad k^2 + k'^2 = 1. \tag{7.6.7}$$

This transformation is presented in Figure 7.27.

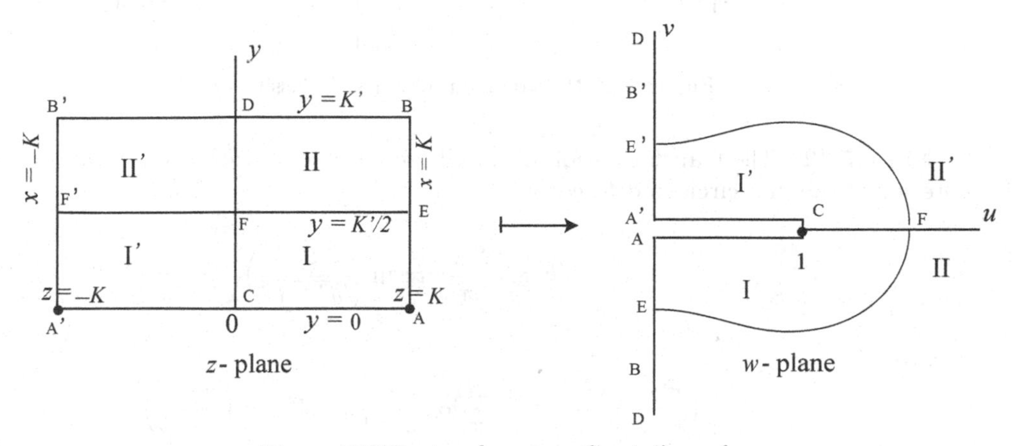

Figure 7.27 Rectangle onto a Cassini's oval.

The details of this mapping are as follows:

(i) The points $z = 0$ (C); K (A); $-K$ (A'); iK' (D); $K+iK'$ (B); $-K+iK'$ (B'); $iK'/2$ (F); $K+iK'/2$ (E); and $-K+iK'/2$ (E'), in the z-plane are mapped onto the points $w = 1$ (C); 0 (A); 0 (A') ∞ (D); $-ik'/k$ (B); ik'/k (B'); $(k+1)^{1/2}k^{-1/2}$ (F); $-i(1-k)^{1/2}k^{-1/2}$ (E); and $i(1-k)^{1/2}k^{-1/2}$ (E'), respectively, in the w-plane.

(ii) The line-segment $y = K'/2, -K \le x \le K$ in the z-plane is mapped onto the part $u \ge 0$ of Cassini's oval $|w+1||w-1| = k^{-1}$.

Map 7.31. The transformation of the same rectangle as in Map 7.29 onto the right-half of a closed Cassini's oval with a slit from $w = 0$ to $w = 1$ (B to C in Figure 7.28) is given by

$$w = \mathrm{dn}(z) = \sqrt{1 - k^2 \mathrm{sn}^2(z)}, \quad \text{or} \quad z = \int_w^1 \frac{dt}{\sqrt{(1-t^2)(t^2-k'^2)}}, \tag{7.6.8}$$

where $k^2 + k'^2 = 1$. The details of this mapping are as follows:

(i) The points $z = 0$ (C); K (A); $K + iK'/2$ (E); $K + iK'$ (B); iK' (D); $-K + iK'$ (B'); $-K + iK'/2$ (E'); $-K$ (A'); and $iK'/2$ (F) in the z-plane are mapped onto the points $w = 1$ (C); k' (A); $\sqrt{1-k}$ (E); 0 (B); ∞ (D); 0 (B') $\sqrt{1-k}$ (E'); k' (A'); and $\sqrt{1+k}$ (F), respectively, in the w-plane.

(ii) The line-segment $y = K', -K \le x \le K$, in the z-plane are mapped onto the part $u \ge 0$ of Cassini's oval $|w+1||w-1| = k$.

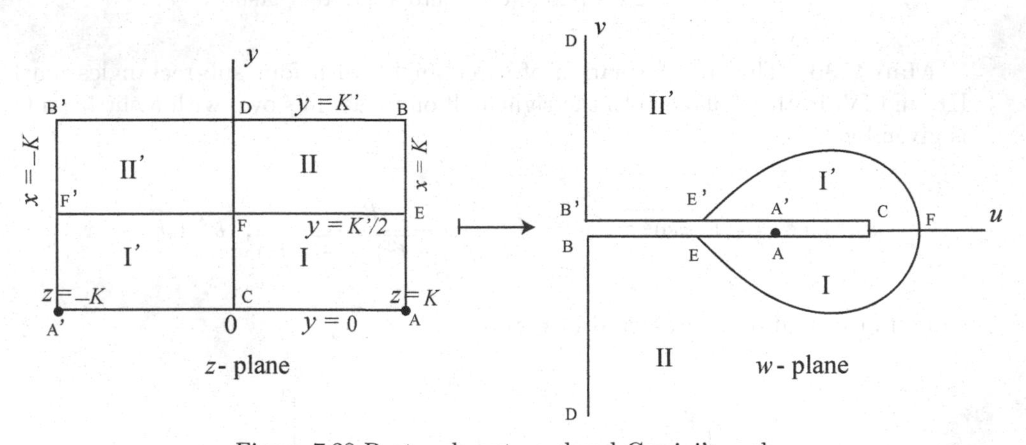

Figure 7.28 Rectangle onto a closed Cassini's oval.

Map 7.32. The transformation of an ellipse with semi-axes a and b, $a > b$, onto the interior of the unit circle is defined by

$$w = \sqrt{k}\,\mathrm{sn}\left(\frac{2K}{\pi} \arcsin \frac{z}{\sqrt{a^2 - b^2}}\right), \tag{7.6.9}$$

where

$$k = \left(\frac{\Theta_2(\tau)}{\Theta_3(\tau)}\right)^2, \quad \tau = \frac{2i}{\pi} \log \frac{a+b}{a-b}, \quad e^{i\pi\tau} = \left(\frac{a-b}{a+b}\right)^2,$$

$$K = \int_0^{\pi/2} \frac{d\phi}{\sqrt{1 - k^2 \sin^2 \phi}} = F\left(k, \frac{\pi}{2}\right).$$

The transformation is presented in Figure 7.29, and the details are as follows:

(i) The points $z = 0; \sqrt{a^2-b^2}; -\sqrt{a^2-b^2}; a;$ and ib in the z-plane are mapped onto the points $w = 0; \sqrt{k}; -\sqrt{k}; 1;$ and i, respectively, in the w-plane.

(ii) The ellipse $\frac{x^2}{a^2} + \frac{y^2}{b^2} = 1$ in the z-plane is mapped onto the unit circle $|w| = 1$.

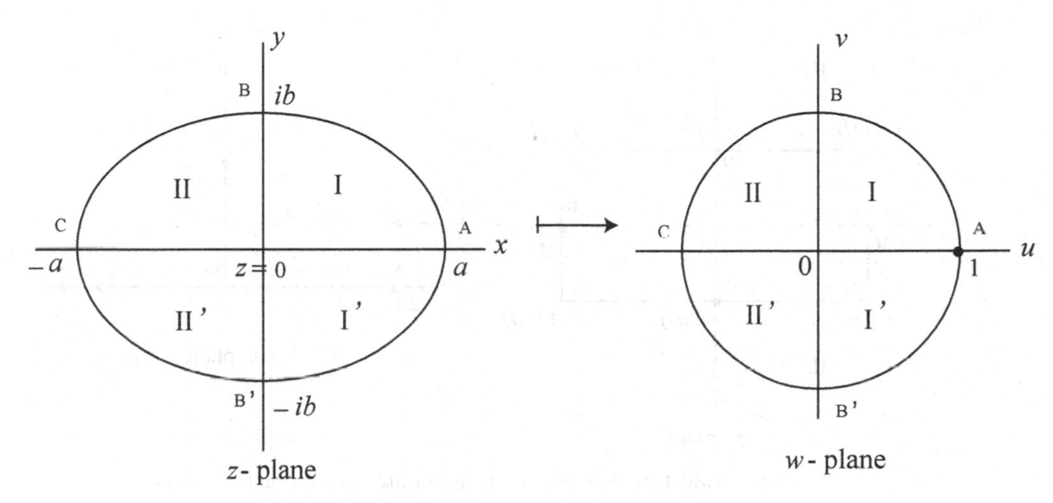

Figure 7.29 Ellipse onto the unit circle.

Map 7.33. The Schwarz-Christoffel transformation that maps the exterior of a rectangle of sides a, b, $a > b$, and the vertices at $\pm a \pm ib$ onto the upper half of the w-plane such that the boundary of the rectangle is mapped onto the boundary of the upper half of the unit circle is given by

$$\frac{dz}{dw} = C\left(w^2 - c^2\right)^{1/2}\left(w^2 - c^{-2}\right)^{1/2}\left(w^2 + 1\right)^{-2}, \tag{7.6.10}$$

or

$$z = \frac{c}{k}\left\{\operatorname{sn}\xi + \left(\frac{E}{K} - k'^2\right)\xi\right\}, \quad w = \frac{1 - \operatorname{dn}(\xi, k)}{k\operatorname{sn}(\xi, k)} = \frac{1 - \operatorname{dn}\xi}{k\operatorname{sn}\xi}, \tag{7.6.11}$$

where

$$E = \int_0^{\pi/2} \sqrt{1 - k^2\sin^2\phi}\, d\phi, \quad k'^2 = 1 - k^2,$$

$$E' = \int_0^{\pi/2} \sqrt{1 - k'^2\sin^2\phi}\, d\phi, \quad C = -\frac{ibK}{E - k'^2K};$$

ξ is an auxiliary variable, and $\operatorname{sn}(\xi)$ is the Jacobian zeta function

$$\operatorname{sn}(\xi) = \frac{\vartheta_0'\left(\frac{\xi}{2K}\middle|\tau\right)}{2K\vartheta_0\left(\frac{\xi}{2K}\middle|\tau\right)}, \quad \tau = \frac{iK'}{K},$$

K and K' are defined by (7.6.1); k depends on b/a and is the root of $\left(E - k'^2K\right)a = \left(E' - k^2K;\right)b$; and $c = \frac{1 - \sqrt{1 - k^2}}{k}$, $0 < c < 1$. If $a = b$, then the rectangle reduces to a square of side a; then $k = 1/\sqrt{2}$, and $c = \tan\pi/8 = \sqrt{2} - 1$. This transformation from the

z-plane is presented in Figure 7.30.

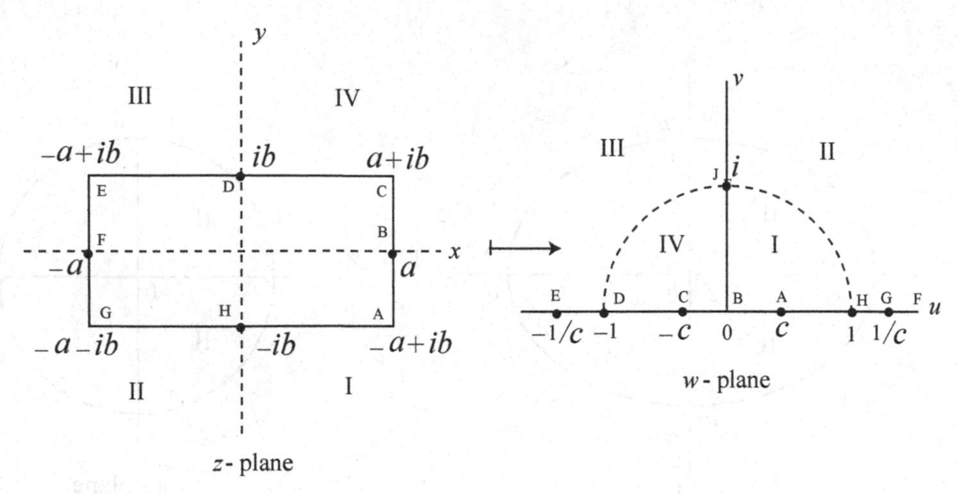

Figure 7.30 Exterior of a rectangle onto the upper half-plane.

This mapping from the ξ-plane is shown in Figure 7.31.

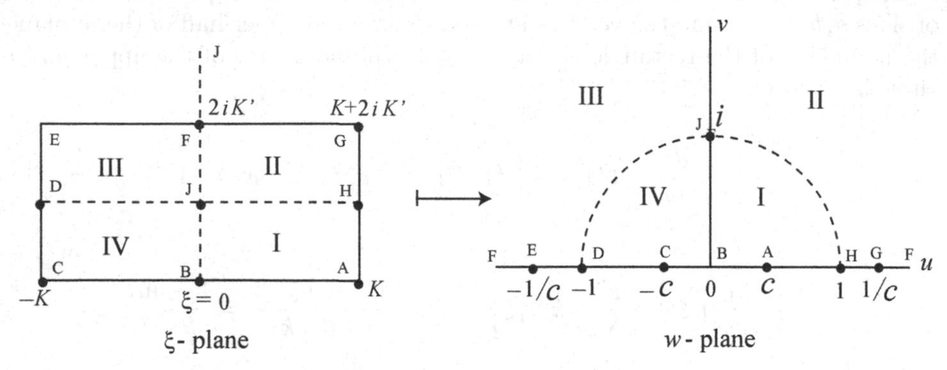

Figure 7.31 Rectangle of Figure 7.30 in the ξ-plane.

Map 7.34. The Schwarz-Christoffel transformation that maps the region exterior to two semi-infinite strips (trigonal regions) of width $2a$ and separated by $2b$, $a > 0, b > 0$, in the z-plane onto the lower half of the w-plane is given by

$$\frac{dz}{dw} = C\left(w^2 - k^2\right)^{1/2}\left(w^2 - 1\right)^{1/2} w^{-2}, \tag{7.6.12}$$

or

$$z = -C\Big\{\xi\big(k'^2 - \frac{2E}{K}\big) - 2\,\mathrm{sn}(\xi) - \frac{\mathrm{cn}(\xi)\,\mathrm{dn}(\xi)}{\mathrm{sn}(\xi)}\Big\} - b, \tag{7.6.13}$$

where $w = 1/\mathrm{sn}(\xi)$ and $C = -\dfrac{ia}{2E - k'^2 K}$; and depending on b/a only, k is a root of $\dfrac{b}{a} = \dfrac{K'k'^2 - 2K' + 2E'}{2(Kk'^2 - 2E)}$. This transformation is presented in Figure 7.32.

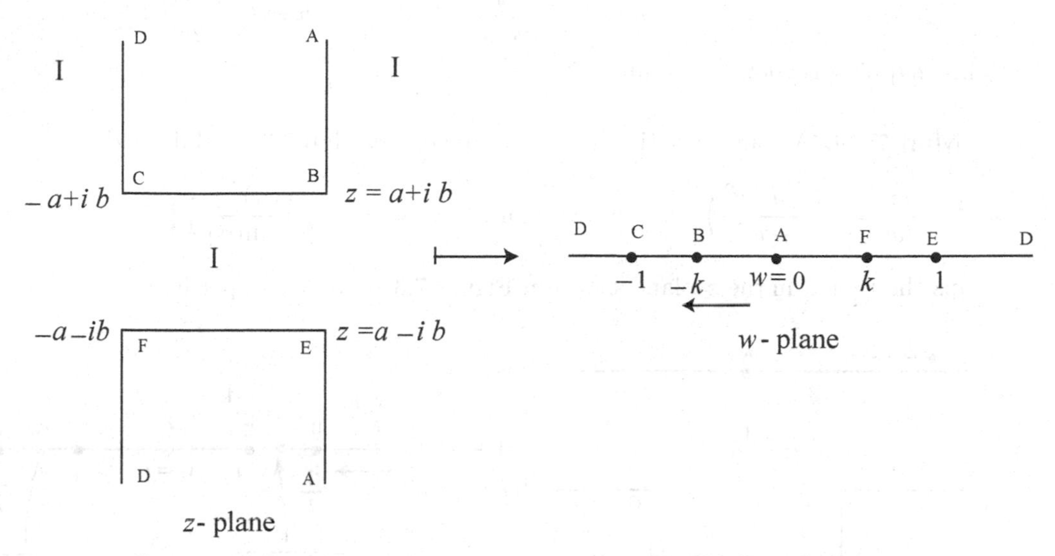

Figure 7.32 Region exterior to two semi-strips onto the lower half-plane.

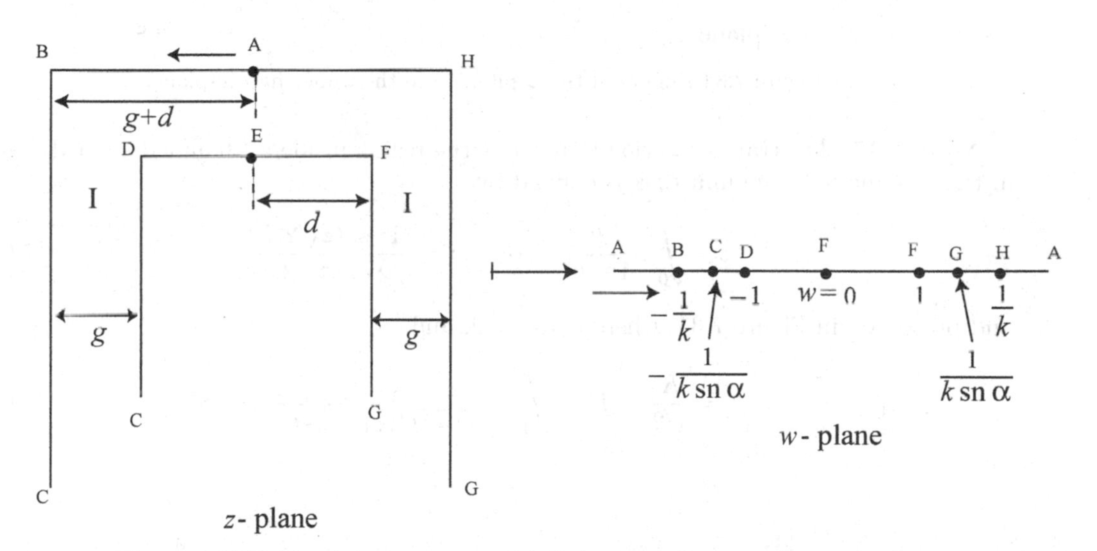

Figure 7.33 Region interior to a semi-infinite strip, but exterior to another one.

Map 7.35. The Schwarz-Christoffel transformation that maps the region interior to a semi-infinite strip, but exterior to another one in the z-plane onto the upper half of the w-plane is defined by

$$\frac{dz}{dw} = \left(\frac{1-w^2}{1-k^2w^2}\right)^{1/2} \left(1 - k^2w^2\,\mathrm{sn}^2(\alpha)\right)^{-1}, \tag{7.6.14}$$

or

$$w = \mathrm{sn}(\xi), \quad z = \xi - \frac{\mathrm{dn}(\alpha)}{k^2\,\mathrm{sn}(\alpha)\,\mathrm{cn}(\alpha)} \prod(\xi,\alpha), \quad 0 < k < 1, \tag{7.6.15}$$

where

$$\alpha = K - K'/2, \quad \prod(\xi,\alpha) = \frac{\xi}{2K}\frac{\vartheta_0'\left(\frac{\alpha}{2K}\middle|\tau\right)}{\vartheta_0\left(\frac{\alpha}{2K}\middle|\tau\right)} + \frac{1}{2}\log\frac{\vartheta_0\left(\frac{\xi-\alpha}{2K}\right)}{\vartheta_0\left(\frac{\xi+\alpha}{2K}\right)},$$

$$\tau = \frac{iK'}{K}, \quad d = K\Big(\frac{1+k}{2k}\Big), \quad g = \frac{\pi}{2k}, \quad h = K'\frac{1+K}{2k}.$$

This mapping is shown in Figure 7.33.

Map 7.36. A transformation similar to the one in Map 7.34 and defined by

$$\frac{dz}{dw} = \Big(\frac{1-k^2w^2}{1-w^2}\Big)^{1/2}, \quad \text{or} \quad w = \mathrm{sn}(\xi), \ z = \xi - \frac{\mathrm{cn}(\alpha)}{\mathrm{sn}(\alpha)\,\mathrm{dn}(\alpha)} \prod(\xi, \alpha), \tag{7.6.16}$$

maps the region in the z-plane shown in Figure 7.34 onto the upper half of the w-plane.

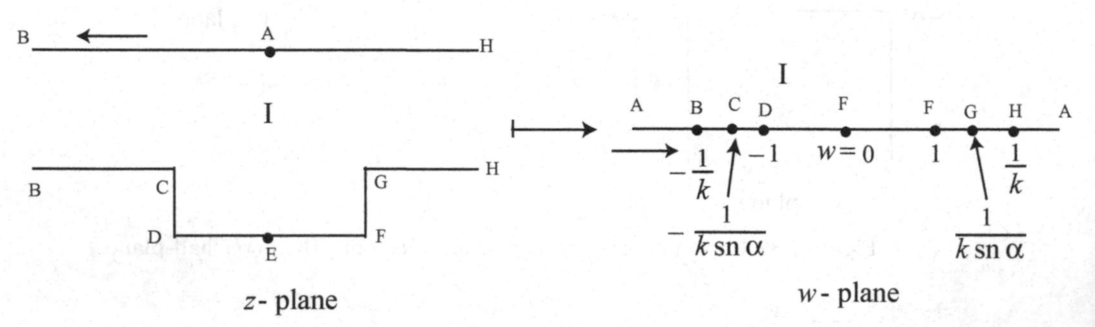

Figure 7.34 Region in the z-plane onto the upper half w-plane.

Map 7.37. The transformation that maps the region inside a (diamond-shaped) square in the z-plane onto the unit disk is defined by

$$z = \int_0^w \frac{dt}{1-t^4}, \quad \text{or} \quad w = \frac{1}{\sqrt{2}} \frac{\mathrm{sn}(z\sqrt{2}, k)}{\mathrm{dn}(z\sqrt{2}, k)}, \tag{7.6.17}$$

and presented in Figure 7.35, where $k = 1/\sqrt{2}$, and

$$\omega_1 = \int_0^1 \frac{dt}{1-t^4} = \frac{K}{\sqrt{2}}, \quad K = \int_0^1 \frac{dt}{\sqrt{(1-t^2)(1-k^2t^2)}} \approx 1.8541. \tag{7.6.18}$$

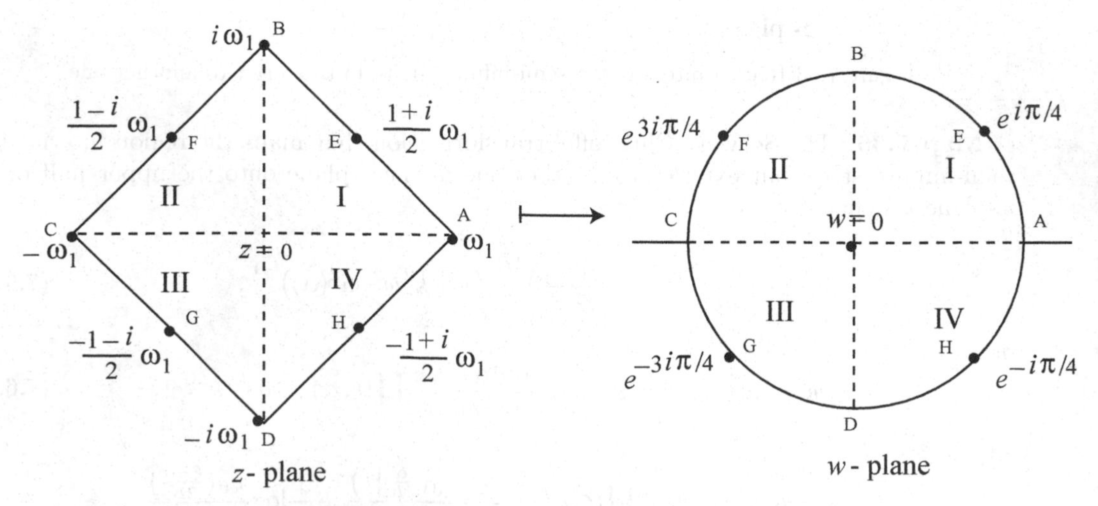

Figure 7.35 Region inside a square onto the unit disk in the w-plane.

Another form of the transformation (7.6.17) is

$$w^2 = \frac{1}{\wp(z;4,0)}, \tag{7.6.19}$$

which means that $g_2 = 4, g_3 = 0, e_1 = -e_3 = 1, e_2 = 0$; and $\omega_3 = i\omega_1$, where ω_1 is defined in (7.6.18).

Map 7.38. The transformation that maps a triangle with angles $\pi/2, \pi/4, \pi/4$ onto the half-plane is defined by

$$w = \sqrt{2}\,\mathrm{sn}(z\sqrt{2})\,\mathrm{dn}(z\sqrt{2}). \tag{7.6.20}$$

The details of this transformation are: (i) The points $z = -\omega_1; \omega_1; i\omega_1$ in the z-plane are mapped onto the points $w = -1; 1; \infty$, respectively; and (ii) the triangle (CAB in Figure 7.35) is mapped onto the half-plane $v > 0$, where ω_1 and k are defined in Map 7.36. Compare this transformation with (7.2.2).

Map 7.39. The transformation that maps a regular polygon with n vertices $(n \geq 3)$ in the z-plane onto the unit circle in the w-plane is defined by

$$z = \int_0^w \frac{dt}{(1-t^n)^{2/n}} = w + \sum_{j=1}^{\infty} \frac{2\cdot(2+n)\cdot(2+2n)\cdots(2+(j-1)n)}{j!\,n^j(jn+1)}. \tag{7.6.21}$$

The details of this transformations are: (i) The points $z = \omega\, e^{2mi\pi/n}; \dfrac{\omega}{\cos(\pi/n)}\, e^{(2m+1)i\pi/n}$, $m = 0, 1, \dots, n-1$, in the z-plane, where

$$\omega = \int_0^1 (1-t^n)^{-2/n}\, dt, \tag{7.6.22}$$

are mapped onto the points $w = e^{2mi\pi/n}; e^{(2m+1)i\pi/n}$, respectively, in the w-plane; (ii) the line-segment $\arg\{z\} = 2m\pi/n, 0 < |z| < \omega$, in the z-plane is mapped onto the segment $\arg\{w\} = 2m\pi/n, 0 < |w| < 1$; and (iii) the line-segment $\arg\{z\} = (2m+1)\pi/n$, $0 < |z| < \omega/(\cos\pi/n)$ in the z-plane is mapped onto the segment $\arg\{w\} = (2m+1)\pi/n, 0 < |w| < 1$. Compare it with (7.1.3).

Map 7.40. The transformation that maps an quilateral triangle in the z-plane onto the half-plane is defined by

$$\begin{aligned} w &= \int_0^z \frac{dt}{(t-t^2)^{2/3}} = f(z), \\ \text{or}\quad z - z^2 &= \wp^3\Big(i\frac{w+w_2}{3}; 0, 1\Big), \\ \text{or}\quad z &= \frac{1}{2} + \frac{A\,\mathrm{sn}(\zeta,k)\,\mathrm{dn}(\zeta,k)}{(1+\mathrm{cn}(\zeta,k))^2}, \end{aligned} \tag{7.6.23}$$

where

$$\begin{aligned} &k = \frac{\sqrt{3}}{2\sqrt{2}}; \quad \zeta = \frac{2w - w_2}{\sqrt[4]{27}\sqrt[3]{2}}; \quad A = \sqrt[4]{27}; \\ &w_1 = 0; \quad w_2 = \int_0^1 \frac{dt}{(t-t^2)^{2/3}} \approx 5.298, \quad w_3 = e^{i\pi/3} w_2. \end{aligned} \tag{7.6.24}$$

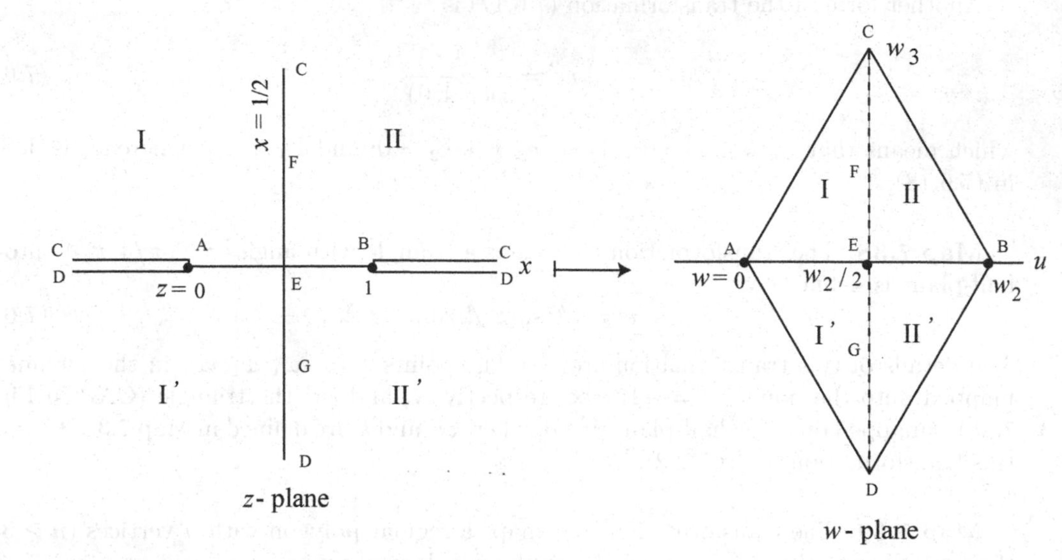

Figure 7.36 Equilateral triangles onto half-planes.

This transformation is presented in Figure 7.36, where the two equilateral triangles are each in the upper and lower half-planes joined at their common base at the u-axis. Compare this transformation with (7.2.4).

When z is real and $0 \le z \le 1$,

$$f(z) = \frac{w_2}{2} \mp \frac{\sqrt[4]{27}}{\sqrt[3]{4}} \int_0^{\gamma} \frac{d\phi}{\sqrt{1 - k^2 \sin^2 \phi}}, \tag{7.6.25}$$

where the minus or plus sign is taken according as $0 \le z \le \frac{1}{2}$ or $\frac{1}{2} \le z \le 1$; and

$$k = \sin \frac{5\pi}{12} = \frac{\sqrt{6} + \sqrt{2}}{4}; \quad \tan \gamma = 3^{-1/4} \{1 - \left(4z - 4z^2\right)^{1/3}\}, \quad 0 \le \gamma \le 1.842.$$

Map 7.41. Another case of Schwarz-Christoffel transformation of two triangles, one each in the upper and lower half-planes and joined at their common base on the u-axis onto the upper and lower half-planes with a slit $-\infty < x \le 1$ is shown in Figure 7.37. The transformation in this case is given by

$$w = \frac{2}{\sqrt[4]{27}} \int_z^{\infty} \frac{dt}{(t^3 - t^2)^{2/3}}, \quad \text{or} \quad z = c\, \frac{1 + \text{cn}(w^2)}{\text{sn}(w)\,\text{dn}(w)}, \tag{7.6.26}$$

where $k = \dfrac{1 + \sqrt{3}}{2\sqrt{2}}$, and $c = \dfrac{\sqrt[4]{3}}{6}$.

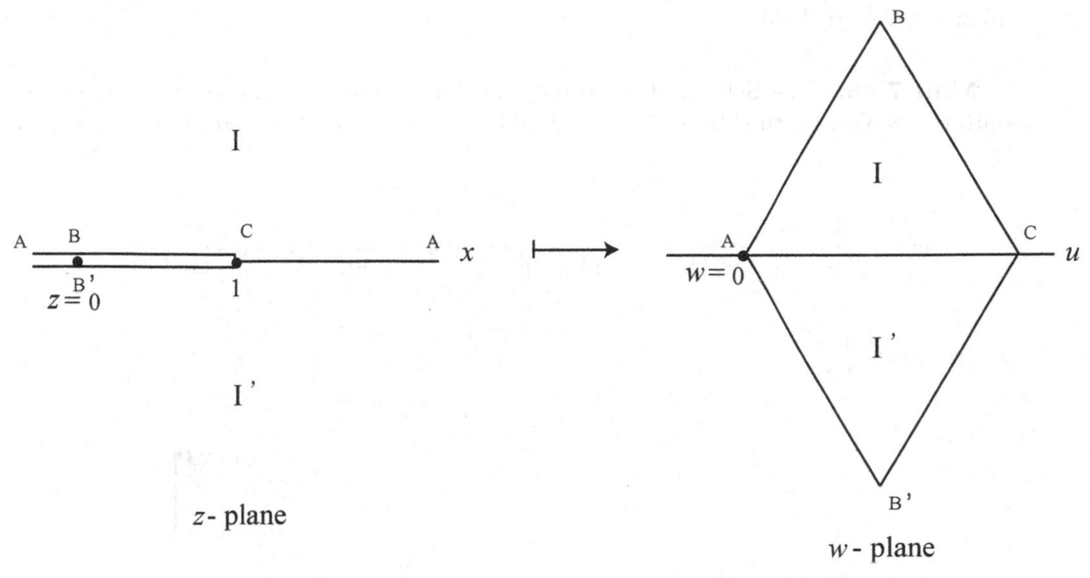

Figure 7.37 Equilateral triangles onto half-planes.

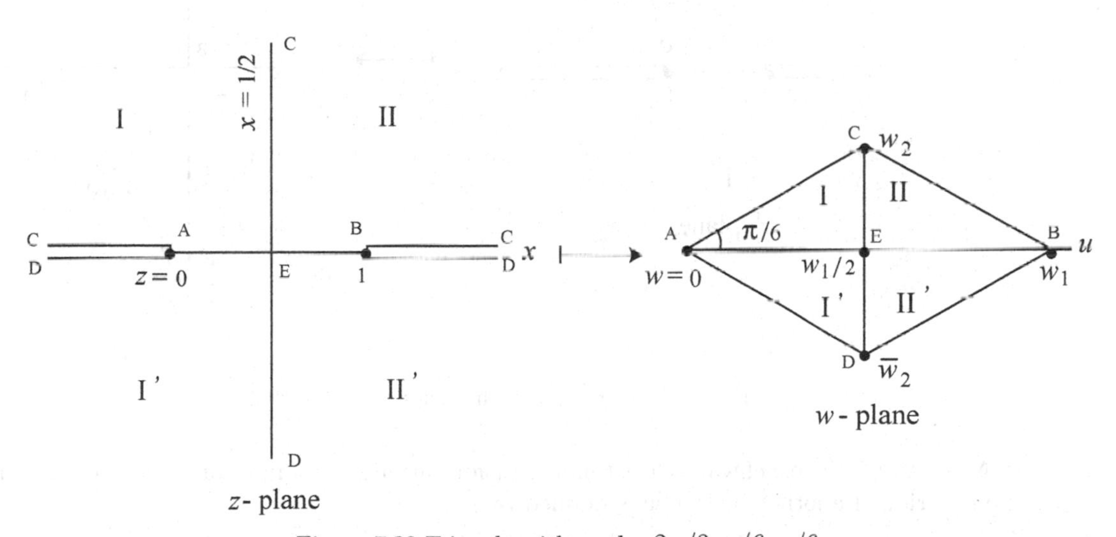

Figure 7.38 Triangle with angles $2\pi/3, \pi/6, \pi/6$.

Map 7.42. The Schwarz-Christoffel transformation of a triangle with angles $\frac{2\pi}{3}, \frac{\pi}{6}, \frac{\pi}{6}$ onto the z-plane with slits $-\infty < x \leq 0$ and $1 \leq x < \infty$, quarter-by-quarter in both planes as presented in Figure 7.38, is given by

$$w = \int_0^z \frac{dt}{(t-t^2)^{5/6}}, \quad \text{or} \quad z - z^2 = \frac{1}{16}\,\wp^{-3}\Big(\frac{w}{3\sqrt[3]{2}}; 0, 1\Big), \tag{7.6.27}$$

$$w_1 = \int_0^1 \frac{dt}{(t-t^2)^{5/6}}, \quad w_2 = \frac{w_1\, e^{i\pi/6}}{\sqrt{3}}. \tag{7.6.28}$$

This transformation maps the quarter of the z-plane onto the interior of a triangle with angles $\pi/2, \pi/3, \pi/6$; and the half-plane $y > 0$ onto the interior of the triangle ABC with angles $\pi/6, \pi/6, 2\pi/3$.

Map 7.43. The Schwarz-Christoffel transformation of a triangle with angles $\frac{\pi}{2}, \frac{\pi}{3}, \frac{\pi}{6}$ onto the z-plane with slits $-\infty < x \le 0$ and $1 \le x < \infty$, as shown in Figure 7.39, is defined by

$$w = \frac{1}{\sqrt[4]{108}} \int_0^z \frac{dt}{t^{1/2}(1-t)^{2/3}(1+t)^{5/6}}, \quad \text{or} \quad \frac{1-z}{1+z} = 1 - \sqrt{3}\,\frac{\text{sn}^2(w)}{\text{cn}^2(w)}, \tag{7.6.29}$$

where $k = \dfrac{1+\sqrt{3}}{2\sqrt{2}}$.

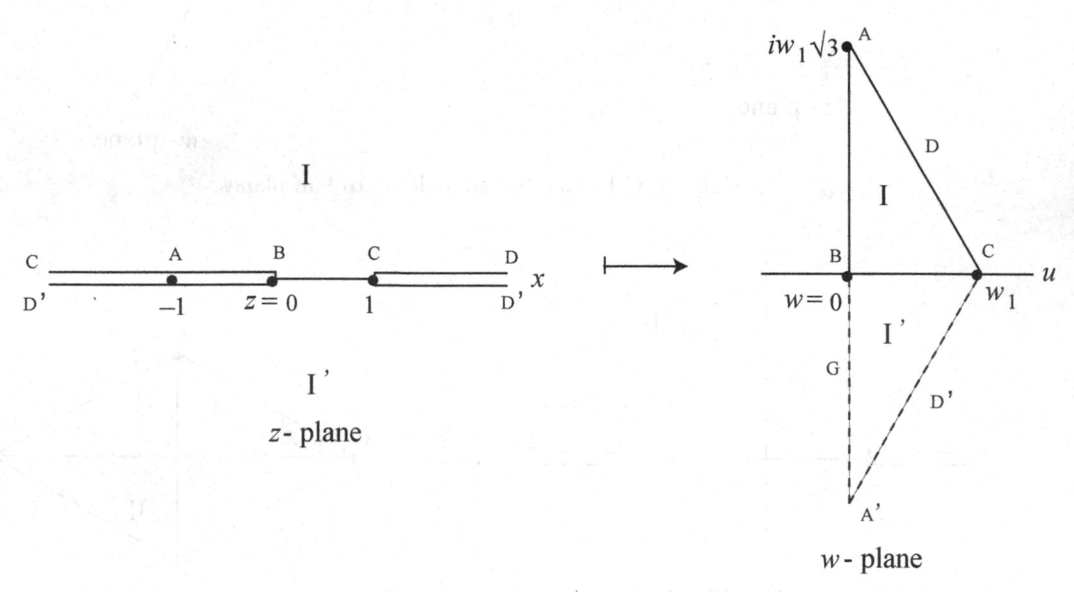

Figure 7.39 Triangle with angles $\pi/2, \pi/3, \pi/6$.

Map 7.44. The Schwarz-Christoffel transformation of the interior of a rectangle onto the exterior of another rectangle is defined by

$$w = \zeta(z) + e_2 z, \quad \text{or} \quad \frac{dw}{ds} = -\frac{1}{2}\,(s-e_1)^{-1/2}\,(s-e_2)^{1/2}\,(s-e_3)^{-1/2}, \quad s = \wp(z), \tag{7.6.30}$$

where

$$\begin{aligned} &\omega_1 > 0, \quad \frac{\omega_3}{i} > 0; \quad e_1 > e_2 > e_3; \\ &\zeta(z) = \frac{1}{z} - \int_0^z \left(\wp(t) - t^{-2}\right) dt, \quad \zeta'(z) = -\wp(z); \\ &\tau = \omega_3/\omega_1; \quad \zeta(\omega_1) = \eta, \ \zeta(\omega_3) = \eta'; \ \eta\omega_3) - \eta'\omega_1 = i\pi/2. \end{aligned} \tag{7.6.31}$$

This transformation maps the points $z = \omega_1$ (A); $-\omega_1$ (E); ω_3 (C); $-\omega_3$ (G); $\omega_1 + \omega_3$ (B); $\omega_3 - \omega_1$ (D); $-\omega_1 - \omega_3$ (F); $\omega_1 - \omega_3$ (H); and 0 (K) in the z-plane onto $w = \eta + e_2\omega_1$ (A); $-\eta - e_2\omega_1$ (E); $\eta' + e_2\omega_3$ (C); $-\eta' - e_2\omega_3$ (G); $eta + \eta' + e_2\,(\omega_1 + \omega_3)$ (B); $\eta' - \eta + e_2\,(\omega_3 = \omega_1)$

(D); $-\eta - \eta' - e_2(\omega_1 + \omega_3)$ (F); $\eta - \eta' + e_2(\omega_1 - \omega_3)$ (H); and ∞ (C), respectively, in the w-plane.

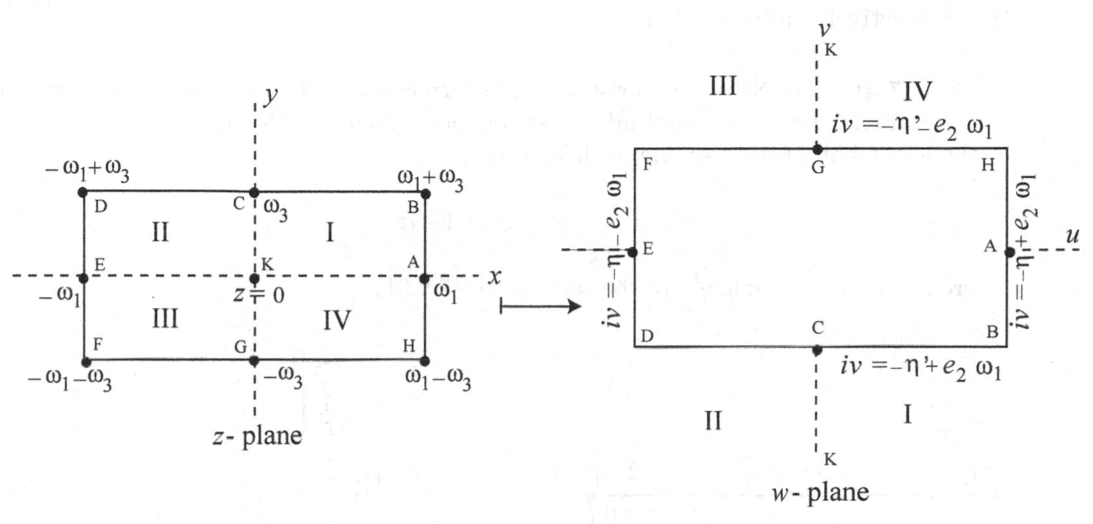

Figure 7.40 Interior of a rectangle onto exterior of another rectangle.

Note that if $\omega_3 = i\omega_1$, then $e_2 = e_1 + e_3 = 0, w = \zeta(z)$, and the rectangle becomes a square.

Map 7.45. The Schwarz-Christoffel transformation of the interior of a rectangle onto the region exterior to two semi-infinite strips, one opening in the right-half and the other in the left-half of the w-plane, is defined by

$$w = \zeta(z) + e_1 z, \quad \text{or} \quad \frac{dw}{ds} = -\frac{1}{2}(s-e_1)^{1/2}(s-e_2)^{-1/2}(s-e_3)^{-1/2}, \quad s = \wp(z), \tag{7.6.32}$$

where $\omega_1, \omega_3; e_1, e_2, e_3; \eta, \eta'$ are the same as in (7.6.31).

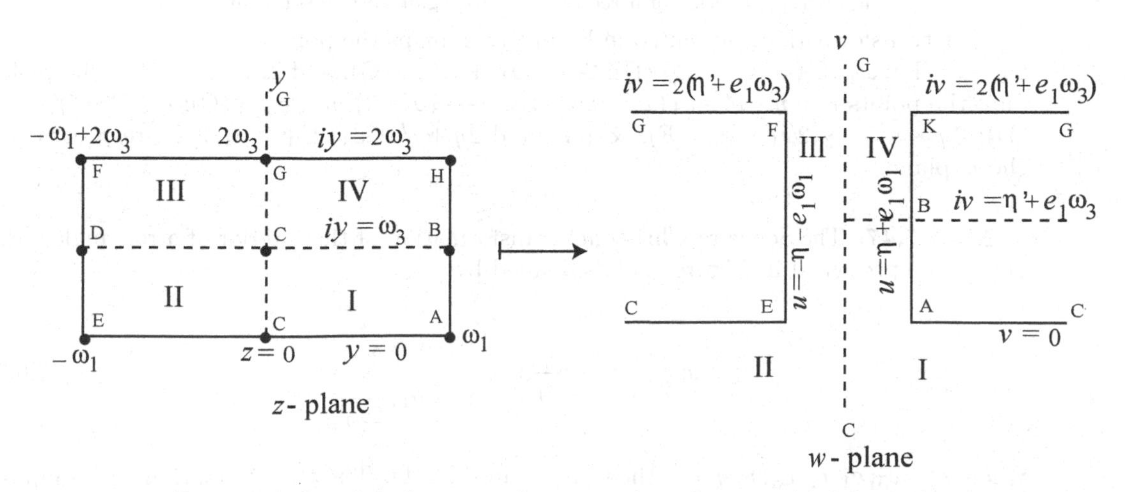

Figure 7.41 Interior of a rectangle onto region of two semi-infinite strips.

This transformation, presented in Figure 7.41, maps the points $z = \omega_1$ (A); $\omega_1 + \omega_3$ (B);

$\omega_1+2\omega_3$ (K); ω_3 (C); $2\omega_3$ (G); 0 (C); $-\omega_1$ (E); $-\omega_1+\omega_3$ (D); and $-\omega_1+\omega_3$ (F) in the z-plane onto the points $w = \eta+e_1\omega_1$ (A); $\eta+\eta'+e_1(\omega_1+\omega_3)$ (B); $\eta+2\eta'+e_1(\omega_1+2\omega_3)$ (K); $\eta'+e_1\omega_3$ (C); ∞ (G); ∞ (C); $-\eta-e_1\omega_1$ (E); $-\eta+\eta'+e_1(\omega_3-\omega_1)$ (D); and $-\eta+2\eta'+e_1(2\omega_3-\omega_1)$ (F), respectively, in the w-plane.

Map 7.46. The Schwarz-Christoffel transformation of the interior of a rectangle onto the region exterior to two semi-infinite strips, one opening in the upper-half and the other in the lower-half of the w-plane, is defined by

$$w = \zeta(z) + e_3 z, \tag{7.6.33}$$

where $\omega_1, \omega_3; e_1, e_2, e_3; \eta, \eta'$ are the same as in (7.6.31).

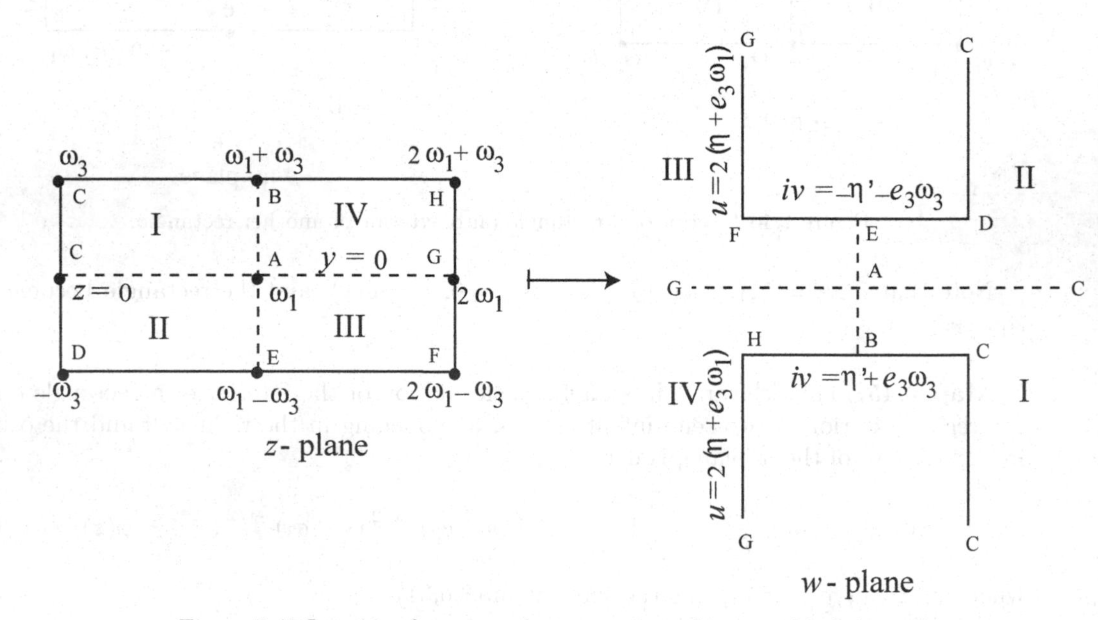

Figure 7.42 Interior of a rectangle onto region of two semi-infinite strips.

This transformation, presented in Figure 7.42, maps the points $z = \omega_1 - \omega_3$ (E); ω_1 (A); $\omega_1+\omega_3$ (B); ω_3 (C); 0 (C); $-\omega_3$ (D); $2\omega_1-\omega_3$ (F); $2\omega_1$ (G); and $2\omega_1+\omega_3$ (K) in the z-plane onto the points $w = \eta-\eta'+e_3(\omega_1-\omega_3)$ (E); $\eta+e_3\omega_1$ (A); $\eta'+e_3\omega_3$ (C);∞ (C); $-\eta'-e_3\omega_3$ (D); $2\eta-\eta'+e_3(2\omega_1-\omega_3)$ (F); ∞ (G); and $2\eta+\eta'+e_3(2\omega_1+\omega_3)$ (K), respectively, in the w-plane.

Map 7.47. The Schwarz-Christoffel transformation of the interior of a rectangle onto a cut plane, presented in Figure 7.43, is defined by

$$w = \zeta(z) - \frac{\eta}{\omega_1} z = \frac{1}{2\omega_1} \frac{\vartheta_1'\left(\dfrac{z}{2\omega_1}|\tau\right)}{\vartheta_1\left(\dfrac{z}{2\omega_1}|\tau\right)}, \tag{7.6.34}$$

where $\omega_1, \omega_3; e_1, e_2, e_3; \eta, \eta'$ are the same as in (7.6.31). The transformation is a composite of

$$\frac{dw}{ds} = -\frac{s+\eta/\omega_1}{\sqrt{4(s-e_1)(s-e_2)(s-e_3)}}, \quad \text{and} \quad s = \wp(z). \tag{7.6.35}$$

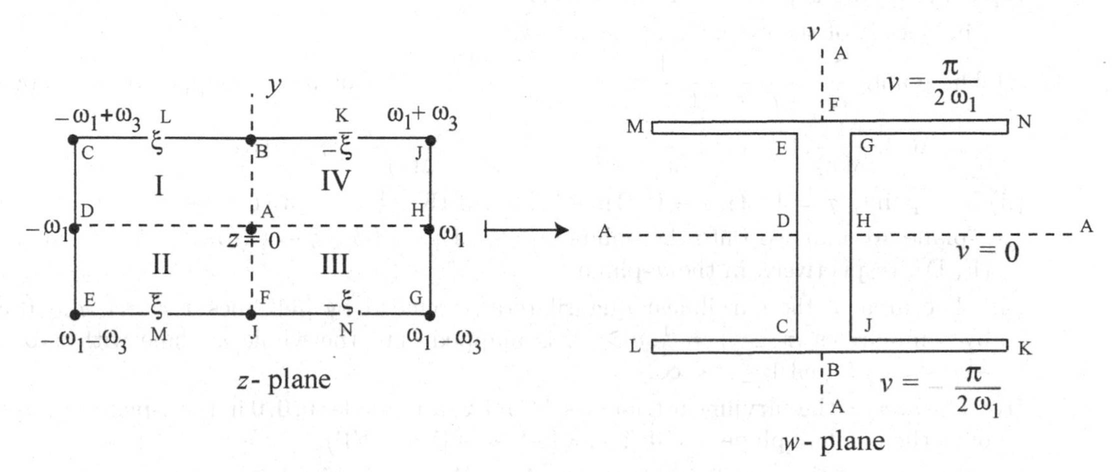

Figure 7.43 Interior of a rectangle onto a cut plane.

This transformation maps (i) the points $z = \omega_3 - \omega_1$ (C); $\omega_1 + \omega_3$ (J); ω_3 (B) onto $w = -\dfrac{i\pi}{2\omega_1}$ (C, J, B); the points $z = -\omega_3 - \omega_1$ (E); $\omega_1 - \omega_3$ (G); $-\omega_3$ (F) onto the points $w = \dfrac{i\pi}{2\omega_1}$ (E, G, F); the points $z = -\omega_1$ (D); ω_1 (K); 0 (A) onto the points $w = 0$ (D); 0 (K); ∞ (A); and the points $x = \xi$ (L); $\bar{\xi}$ (M); $-\bar{\xi}$ (K); and $-\xi$ (N) onto the points $w_P = \zeta(\xi)$; $\bar{w}_P$ (M); $-\bar{w}_P$ (K); and $-w_P$ (N) in the w-plane; and (ii) the line segments $y = 0$, $\omega_1 < x < 0$ or $0 < x < \omega_1$, are mapped onto the half-lines $v = 0$, $-\infty < u < 0$ or $0 < u < \infty$ in the w-plane.

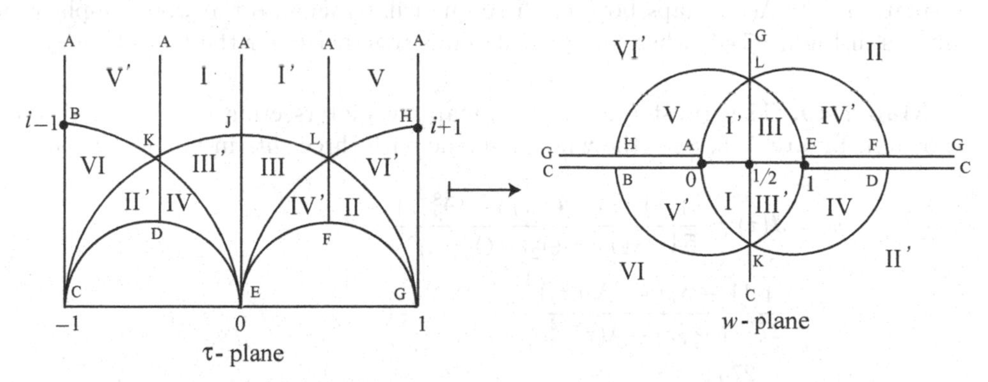

Figure 7.44 Curvilinear triangles onto two intersecting unit circles.

Map 7.48. We will consider curvilinear triangles with angles (i) $0, 0, 0$; (ii) $\pi/3, \pi/3, 0$; (iii) $\pi/2, \pi/3, 0$; (iv) $\pi/2, 0, 0$; and (v) $2\pi/3, 0, 0$ in the τ-plane, and their mapping onto two intersecting unit circles $|w| = 1$ and $w - 1| = 1$. The transformation for this mapping is defined by

$$w = \lambda(\tau) = \frac{\Theta_2^4(\tau)}{\Theta_3^4(\tau)} = k^2, \tag{7.6.36}$$

where $\lambda(\tau)$ is the modular function (see Appendix F for definition), and $\Im\{\tau\} > 0$ ($\tau = x + iy$). Let $\alpha, \beta, \gamma, \delta$ be integers, where a and δ are odd, and β and γ are even, such that

$\alpha\delta - \gamma\beta = 1$. The critical points of the mapping are $\tau = \infty$ and any point on $y = 0$. A typical mapping is presented in Figure 7.44.

The details of the mapping are as follows:

(i) The points $\dfrac{\gamma + \delta\tau}{\alpha + \beta\tau}; \dfrac{\tau}{\tau+1}; \dfrac{1}{\tau}; \tau + 1; \dfrac{1}{\tau_1}; \dfrac{\tau+1}{\tau}$ in the z-plane are mapped onto the points $w = \lambda(\tau); \dfrac{1}{\lambda(\tau)}; 1 - \lambda(\tau); \dfrac{\lambda(\tau)}{\lambda(\tau) - 1}; \dfrac{1}{1 - \lambda(\tau)}; \dfrac{\lambda(\tau) - 1}{\lambda(\tau)}$, respectively.

(ii) The points $\tau = 1$ (J); $i - 1$ (B); $-\frac{1}{2} + \frac{i}{2}\sqrt{3}$ (K); $\frac{1}{2} + \frac{i}{2}\sqrt{3}$ (L); $\pm\frac{1}{2} + \frac{i}{2}$ (F, D) in the z-plane are mapped onto the points $w = \frac{1}{2}$ (J); -1 (B); $\frac{1}{2} - \frac{i}{2}\sqrt{3}$ (K); $\frac{1}{2} + \frac{i}{2}\sqrt{3}$ (L); 2 (F, D), respectively, in the w-plane.

(iii) The area of the curvilinear quadrilateral, bounded by half-lines $x = \pm 1, y \geq 0$ and by semi-circles $\left|\tau \pm \frac{1}{2}\right| = \frac{1}{2}, y \geq 0$, is mapped onto the whole w-plane with two slits $-\infty < x \leq 0$ and $1 \leq x < \infty$.

(iv) The area of the curvilinear triangles ACJGA, with angles $0, 0, 0$ in the z-plane is mapped onto the cut half-plane $u < 0$ (i.e., VI+V'+I+I'+V+VI').

(v) The area of the curvilinear triangles AKJLA, with angles $\pi/3, \pi/3, 0$ in the z-plane is mapped onto the region (I+I').

(vi) The area of the curvilinear triangles AJLA, with angles $\pi/2, \pi/3, 0$ in the z-plane is mapped onto the region I'.

(vii) The area of the curvilinear triangles AJLGA, with angles $\pi/2, 0, 0$ in the z-plane is mapped onto the quadrant $u < 0, v > 0$ (i.e., I'+V+VI').

(vii) The area of the curvilinear triangles AJELA, with angles $0, 2\pi/3, 0$ in the z-plane is mapped onto the region (I'+III).

Note that the transformation $\tau' = \dfrac{\gamma + \delta\tau}{\alpha + \beta\tau}$ maps the area of the quadrilateral, mentioned in (iii) above, onto another curvilinear quadrilateral with angles $0, 0, 0, 0$. Next, the transformation $w = \lambda(\tau')$ maps both of these quadrilaterals onto the whole w-plane, with two slits as in Figure 7.44, where all quadrilaterals together cover the half-plane $y > 0$.

Map 7.49. The transformation that maps two intersecting unit circles in the ζ-plane, shown in Figure 7.45, onto the whole w-plane with three slits in line is defined by

$$\begin{aligned} w = J(\tau) &= \frac{1}{54} \frac{\Theta_0^8(\tau) + \Theta_2^8(\tau) + \Theta_3^8(\tau)}{\left[\pi\Theta_0(\tau)\Theta_2(\tau)\Theta_3(\tau)\right]^8} \\ &= \frac{4}{27} \frac{\left(1 - \lambda(\tau) + \lambda^2(\tau)\right)^3}{\left(\lambda^2(\tau) - \lambda(\tau)\right)^2} \\ &= \frac{27g_2^3}{27g_2^3 - g_3^2}, \quad y > 0, \ \tau = x + iy; \quad \pi\Theta_0(\tau)\Theta_2(\tau)\Theta_3(\tau) = \theta_1'. \end{aligned} \tag{7.6.37}$$

The critical points of this transformation are: $\tau = \infty$, and any point on the axis $y = 0$. The transformation (7.6.37) is a composition of $\zeta = \lambda(\tau)$ and $w = \dfrac{4}{27} \dfrac{\left(1 - \zeta + \zeta^2\right)^3}{\left(\zeta^2 - \zeta\right)^2}$.

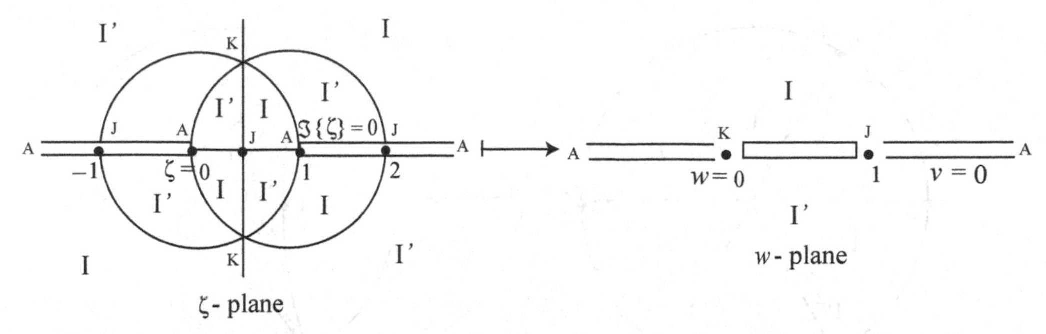

Figure 7.45 Two intersecting unit circles onto the whole plane with three slits in line.

Map 7.50. The transformation that maps the regions over the upper half of the unit circle $|\tau| = 1$ in the τ-plane onto the whole $w = J(\tau)$-plane with one slit $-\infty < u \leq 0$ is presented in Figure 7.46; it has the following properties: (i) The points $\tau = i$ (J); $\frac{1}{2} + \frac{i}{2}\sqrt{3}$ (L); $-\frac{1}{2} + \frac{i}{2}\sqrt{3}$ (K) are mapped onto the points $w = 1$ (J); 0 (K); 0 (L), respectively, in the w-plane; and (ii) the point $\tau' = \dfrac{\gamma + \delta\tau}{\alpha + \beta\tau}$ is mapped onto the point $w = J(\tau) = J(\tau')$.

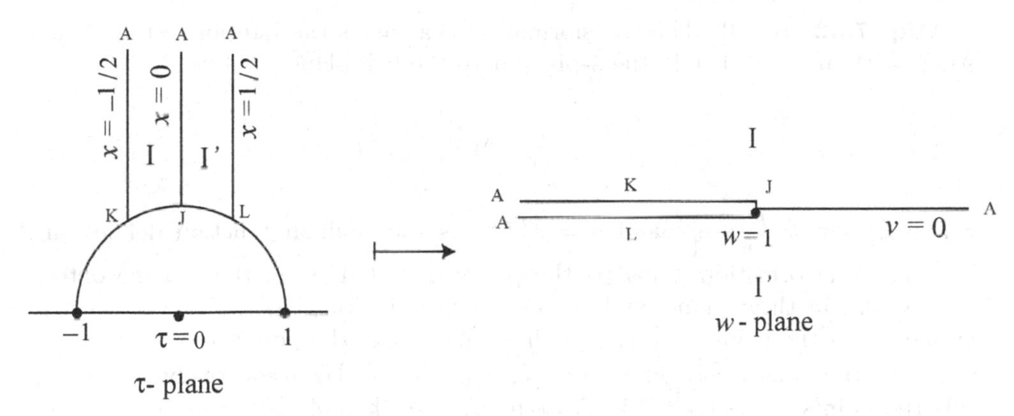

Figure 7.46 Upper half of the unit circle onto the whole plane with one slit.

Map 7.51. The transformation that maps the equilateral equiangular circular triangle, with angle $\alpha, 0 \leq \alpha < \pi$, in the z-plane onto a circle in the w-plane, as shown in Figure 7.47, is defined by

$$w = z\,\frac{\Gamma\left(\frac{2}{3}\right)\Gamma\left(\frac{5}{6}+\beta\right)F\left(\frac{5}{6}-\beta, \frac{1}{2}-\beta, \frac{4}{3}; z^3\right)}{\Gamma\left(\frac{4}{3}\right)\Gamma\left(\frac{1}{6}+\beta\right)F\left(\frac{1}{6}-\beta, \frac{1}{2}-\beta, \frac{2}{3}; z^3\right)}, \tag{7.6.38}$$

where $\beta = \alpha/(2\pi)$, and

$$F(a,b,c;z) = 1 + \frac{ab}{1!c}z + \frac{a(a+1)b(b+1)}{2!c(c+1)}z^2 + \frac{a(a+1)(a+2)b(b+1)(b+2)}{3!c(c+1)(c+2)}z^3 + \cdots, \quad |z| < 1, \tag{7.6.39}$$

is the Gaussian hypergeometric series.

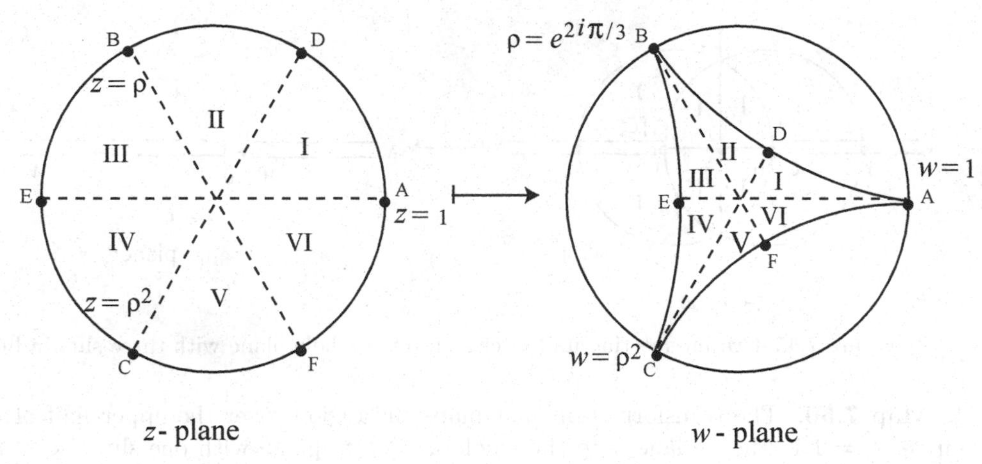

Figure 7.47 Equilateral equiangular circular triangle onto a circle.

Notice that in Figure 7.47, the radius ADB is equal to $\frac{\sqrt{3}}{2}\left|\cos\left(\frac{\alpha}{2}+\frac{\pi}{3}\right)\right|^{-1}$. For $\alpha = \pi/3$, see Map 7.39.

Map 7.52. $\alpha = 0$: The transformation that maps the interior of the circular triangle ABC, with angles $0, 0, 0$, in the z-plane onto the half-plane $y > 0$ is

$$z = \lambda\Big(\frac{\rho^2 - w\rho}{w-1}\Big), \tag{7.6.40}$$

where $\lambda(\tau), \tau = \dfrac{\rho^2 - w\rho}{w-1}$ and $\rho = e^{2i\pi/3}$, is the modular function defined in Appendix F. This transformation maps (i) the points $z = 0; 1; \infty$ in the z-plane onto the points $\tau = \infty; 0; 1$ in the τ-plane, and the points $w = 1$ (A); ρ (B); ρ^2 (C), respectively, in the w-plane; (ii) the points $z = 2; \frac{1}{2}$ in the z-plane onto the points $\tau = \frac{1+i}{2}; i$ in the τ-plane, and onto the points $w = \sqrt{3} - 2$ (E); $(2 - \sqrt{3})\, e^{i\pi/3}$ (D), respectively, in the w-plane; and (iii) the points $z = -1; e^{i\pi/3}$ in the z-plane onto the points $\tau = i + 1; -\rho^2 = e^{i\pi/3}$ in the τ-plane, and onto the points $w = (2 - \sqrt{3})\, e^{-i\pi/3}$ (F); 0 (C), respectively, in the w-plane.

Map 7.53. The transformation $w = \left(\dfrac{\zeta^{\pi/\alpha} + 1}{\zeta^{\pi/\alpha} - 1}\right)^2$, where $\zeta = \dfrac{z_1 - \zeta_0}{z_1 - z_0}\dfrac{z - z_0}{z - \zeta_0}$, with given z_0, z_1, ζ_0 and the angle α; z_0, ζ_0 finite, and $0 < \alpha < 2\pi$, maps the curvilinear triangle with two right angles, and the third angle $\alpha \neq \pi$, shown in Figure 7.48, in the z-plane, onto the upper half-plane $\Im\{w\} > 0$. For other cases, see Map 4.50 and 4.53.

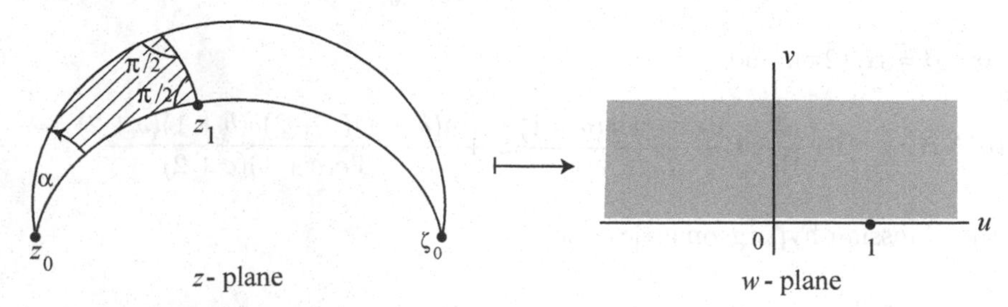

Figure 7.48 Curvilinear triangle with two right angles onto the upper half-plane.

Map 7.54. The transformation that maps the interior of a triangle in the z-plane onto the upper half-plane $\Im\{w\} > 0$ is defined by

$$w = \int_0^z t^{\alpha/\pi-1}\,(1-t)^{\beta/\pi-1}\,dt, \tag{7.7.1}$$

and is presented in Figure 7.49.

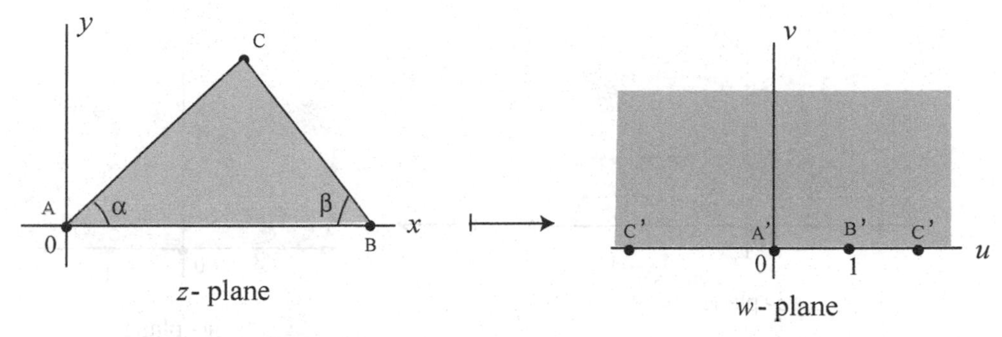

Figure 7.49 Interior of a triangle onto the upper half-plane.

Map 7.55. The transformation that maps the interior of a rectangle in the z-plane onto the upper half-plane $\Im\{w\} > 0$ is defined by

$$w = \int_0^z \frac{dt}{\sqrt{(1-t^2)\,(1-k^2t^2)}}, \tag{7.7.2}$$

and is presented in Figure 7.50.

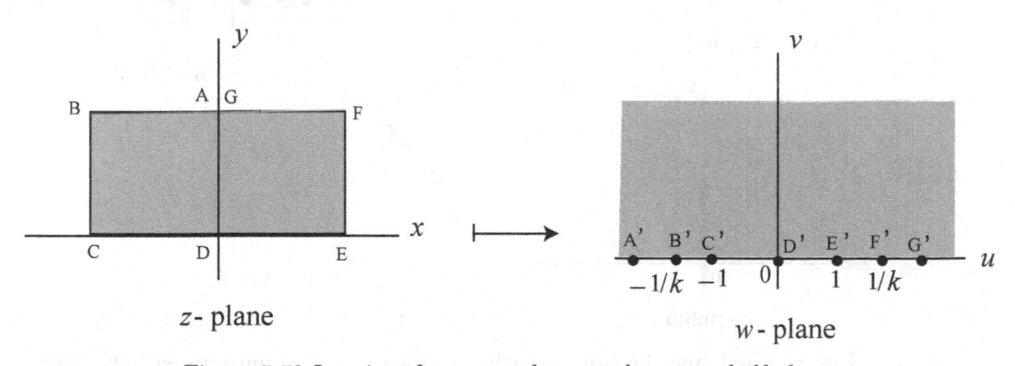

Figure 7.50 Interior of a rectangle onto the upper half-plane.

Map 7.56. The Schwarz-Christoffel transformation that maps the upper half of the z-plane with the path AOCB, where C is at $w = ia$, onto the upper half-plane $\Im\{w\} > 0$, as shown in Figure 7.51, is defined by

$$\frac{dw}{dz} = C_1(z-0)^{-1/2}(z-1)^{-3\pi/2}C_1\sqrt{\frac{1-z}{z}} = K\sqrt{\frac{1-z}{z}},$$

which gives

$$w = K\int\sqrt{\frac{1-z}{z}}\,dz.$$

To integrate, take $z = \sin^2\theta$, which leads to $w = K\left(\arcsin\sqrt{z} + \sqrt{z(1-z)}\right) + C_2$. The constants are evaluated using the boundary conditions: $w(0) = 0$ and $w(1) = ib$, which yields

$$w = \frac{2ib}{\pi}\left(\arcsin\sqrt{z} + \sqrt{z(1-z)}\right). \tag{7.7.3}$$

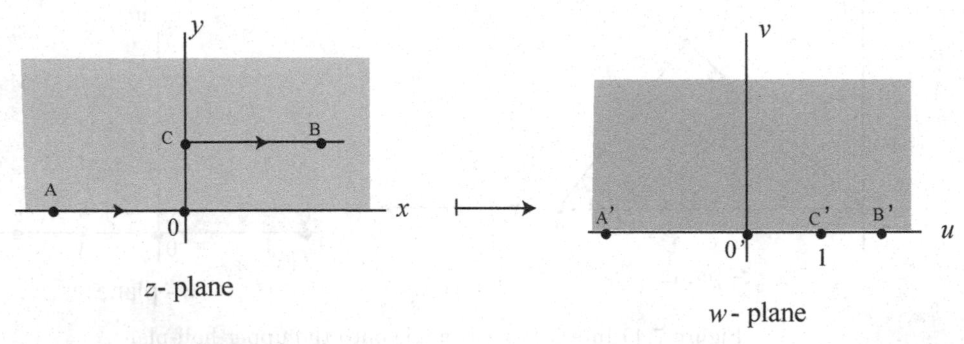

Figure 7.51 Upper half-plane onto upper half-plane.

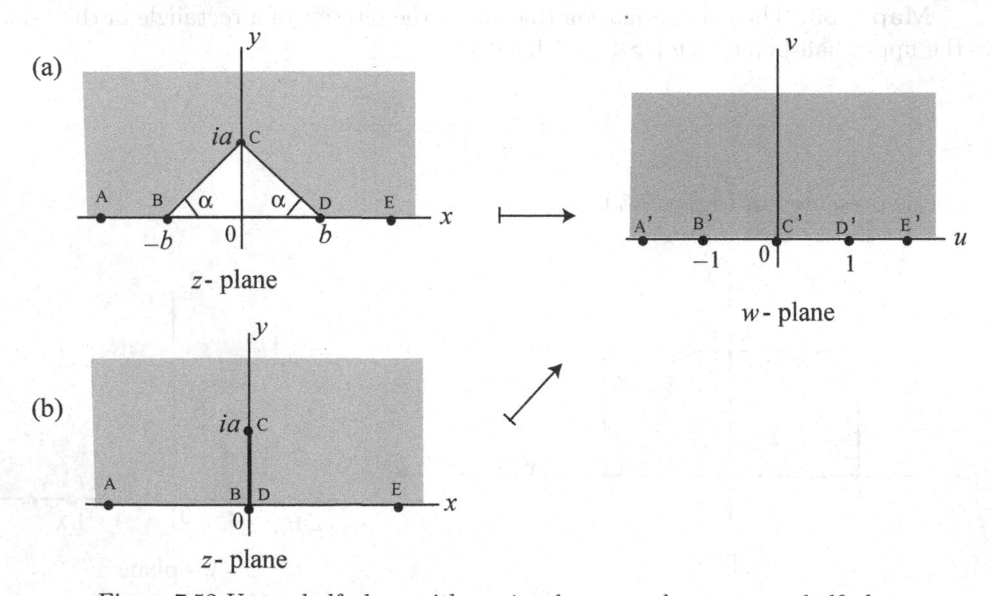

Figure 7.52 Upper half-plane with a triangle removed onto upper half-plane.

Map 7.57. The Schwarz-Christoffel transformation that maps the upper half of the z-plane with an isosceles triangle of angle α removed onto the upper half-plane $\Im\{w\} > 0$, as presented in Figure 7.52, is defined by

$$w = K\int_0^z \frac{t^{2\alpha/\pi}}{(1-t^2)^{\alpha/\pi}}\,dt + C_2,$$

where using the boundary condition $w(0) = ia$ we obtain $C_2 = ia$, and using the boundary condition $w(1) = b$ the value of K can be expressed in terms of the gamma function as

$$K = \frac{(b - ia)\sqrt{\pi}}{\Gamma\left(\frac{\alpha}{\pi} + \frac{1}{2}\right)\gamma\left(1 - \frac{\alpha}{\pi}\right)}.$$

The case $b = 0$ is shown in Figure 7.52(b). In this case $\alpha \to \pi/2$, and the transformation from part (a) reduces to

$$w = ia - ia \int_0^z \frac{t}{\sqrt{1 - t^2}}, dt = ia\sqrt{1 - z^2} = a\sqrt{z^2 - 1}.$$

Map 7.58. The function

$$w = f(z) = \sqrt{k(q)}\,\mathrm{sn}\left(\frac{2K}{\pi}\arcsin z, q\right), \quad q = \left(\frac{a-b}{a+b}\right)^2, \tag{7.7.4}$$

maps the interior of the ellipse $\dfrac{x^2}{a^2} + \dfrac{y^2}{b^2} = 1$ onto the unit disk $|w| < 1$, such that the foci of the ellipse go into the points $w = \pm\sqrt{k(q)}$, and

$$f'(0) = \frac{2}{a+b}\left(\sum_{n=0}^{\infty} q^{n(n+1)}\right)\cdot\left(1 + 2\sum_{n=1}^{\infty} q^n\right)$$
$$= \begin{cases} 1.0165984 & \text{for } a/b = 1.2, \\ 1.2376223 & \text{for } a/b = 2, \\ 2.372368 & \text{for } a/b = 5. \end{cases}$$

Map 7.59. Let the field of an infinite two-dimensional capacitor be as shown in the Figure 7.53, with the values of the potential on the curves Γ_1 and Γ_2 as 0 and 1. Then the function $w = f(z)$ that maps the upper half-plane onto the triangle $A_1A_2A_3$ is given by

$$w = C_0 \int_{-1}^{z} \zeta^{-1}(1 + \zeta)^{\alpha}\, d\zeta + C_1,$$

where $C_1 = ih$ and $C_0 = h/\pi$. (Sveshnikov and Tikhonov [1978:217-218].)

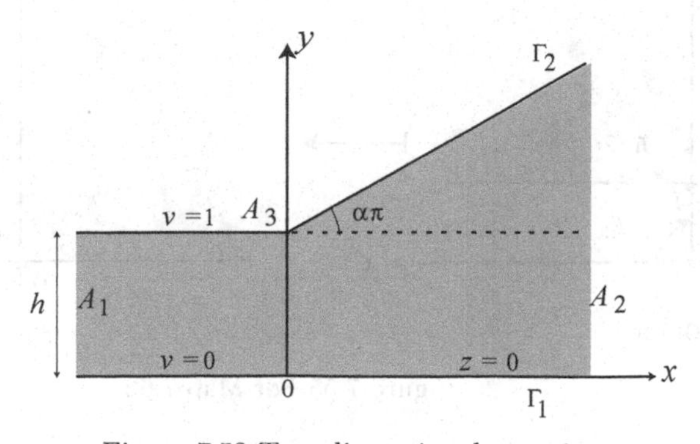

Figure 7.53 Two-dimensional capacitor.

Map 7.60. The function $w = \int_0^z \frac{dt}{\sqrt{t(1-t^2)}}$ maps the upper half-plane $\Im\{z\} > 0$ onto the interior of the square of side $\frac{\Gamma^2(1/4)}{2\sqrt{2\pi}}$. [(Phillips 1966:65].)

Map 7.61. The Schwarz-Christoffel transformation that maps the region (shaded) in Figure 7.54 onto the upper half-plane $\Im\{w\} > 0$ such that $z = 0$ goes into $w = 0$ is given by

$$z = \frac{a}{\mu\pi} \int_1^w \left(\frac{\zeta - 1}{\zeta}\right)^\mu d\zeta.$$

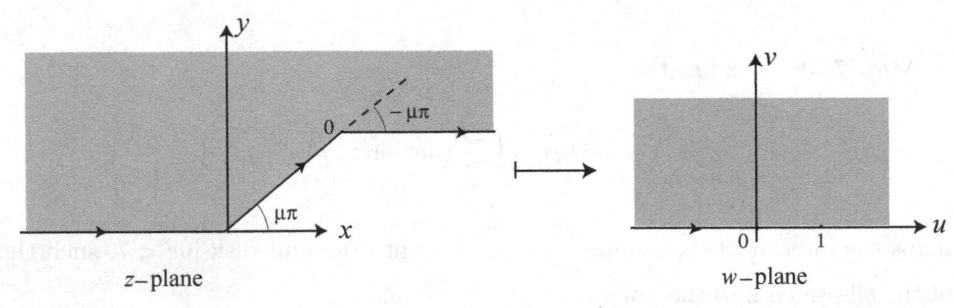

Figure 7.54 The shaded region.

Let $\mu = p/q$, $p < q$ $(p, q \in \mathbb{R}^+)$, and set $t = \left(\frac{\zeta - 1}{\zeta}\right)^{1/q}$, as in Map 7.15. Show that the mapping function becomes

$$z = \frac{a}{\pi}\left[-\frac{1}{\mu}\frac{t^p}{t^q - 1} + \sum_{n=0}^{q-1} t_n^p \log\left(1 - \frac{t}{t_n}\right)\right],$$

where $t_n = e^{2i\pi n/q}$, $n = 0, 1, \dots, q-1$, are the q zeros of $t^q - 1$. [Use $\frac{t^{p-1}}{t^q - 1} = \frac{1}{q}\sum_{n=0}^{q-1} \frac{t_n^p}{t - t_n}$.] In particular, for $\mu = 1/2$, $p = 1$, $q = 2$, show that the mapping function is given by $z = \frac{a}{\pi}\left[\frac{2t}{1-t^2} + \log\frac{1-t}{1+t}\right]$ (von Koppenfels [1939: 210-211]).

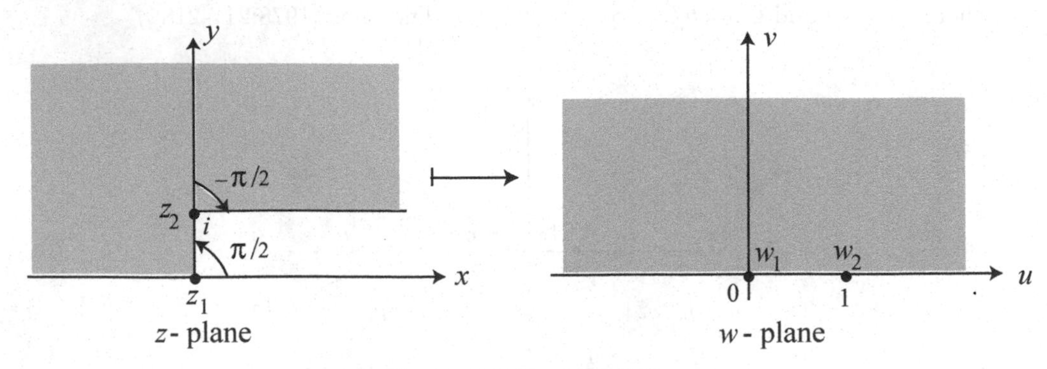

Figure 7.55 For Map 7.62.

Map 7.62. This is a special case of Map 7.61 when $\mu = \frac{1}{2}$. The region shown in Figure 7.55 is mapped onto the upper half-plane $\Im\{w\} > 0$, such that the three vertices z_1, z_2, z_3 are mapped onto the points $w_1 = 0$, $w_2 = 1$, and $w_3 = \infty$, respectively. Then the Schwarz-Christoffel transformation is given by

$$w = A' \int_{z_0}^{z} t^{-1/2}(t-1)^{1/2}\, dt + B. \tag{7.7.5}$$

If we write $A'(z-1)^{1/2} = A(1-z)^{1/2}$, then the mapping (7.7.5) becomes

$$w = A \int_{z_0}^{z} \sqrt{\frac{1-z}{z}}\, dz + B. \tag{7.7.6}$$

If we set $z = \sin^2\theta$, then $\dfrac{1-z}{z} = \dfrac{1-\sin^2\theta}{\sin^2\theta}$, so that the mapping (7.7.6) becomes

$$w = A \int_{z_0}^{z} (1+\cos 2\theta)\, d\theta + B = \theta + \sin\theta\cos\theta.$$

Again, since $\sin\theta = \sqrt{z}$, the map (7.7.6) becomes

$$w = A \int_{z_0}^{z} \sqrt{\frac{1-z}{z}}\, dz + B = A\left(\arcsin\sqrt{z} + \sqrt{z(1-z)}\right) + B,$$

which after applying the conditions that $z_1 = 0$ corresponds to $w_1 = 0$ and $z_2 = i$ corresponds to $w_2 = 1$, gives $B = 0$, and $A = 2i/\pi$. Hence, the Schwarz-Christoffel transformation for this problem is

$$w = \frac{2i}{\pi}\left(\arcsin\sqrt{z} + \sqrt{z(1-z)}\right). \tag{7.7.7}$$

Map 7.63. The Schwarz-Christoffel transformation that maps the rectangular region in the z-plane onto the upper half-plane $\Im\{w\} > 0$ (see Figure 7.56) is given by

$$w(z) = A \int_{z_0}^{z} \frac{dt}{\sqrt{(t-z_1)(t-z_2)(t-z_3)}} + B, \tag{7.7.8}$$

where the points z_1, z_2, z_3 are mapped onto the points w_1, w_2, w_3 respectively, and the point z_4 is mapped onto the point at infinity.

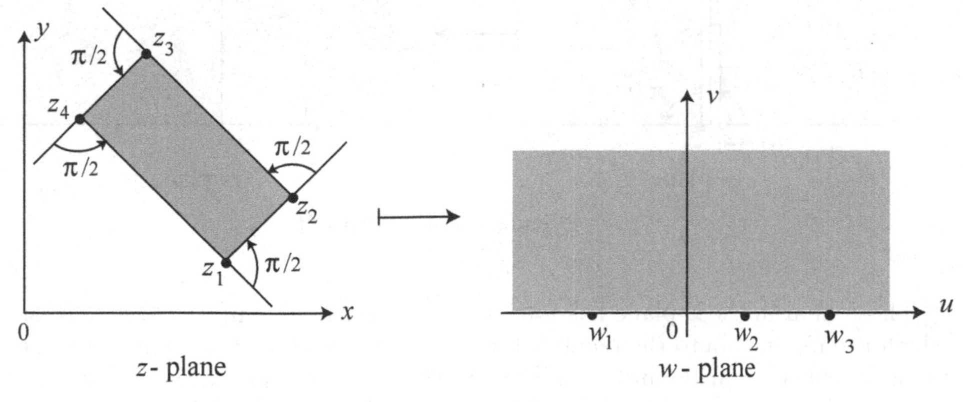

Figure 7.56 For Map 7.63.

Map 7.64. Consider the region in the z-plane as shown in Figure 7.57. There are three vertices at infinity and two are finite at the points z_2 and z_4, with exterior angles as marked in the figure. Notice that the sum of the exterior angles is $\pi - 3\pi/4 + \pi - \pi/4 + \pi = 2\pi$. The Schwarz-Christoffel transformation that maps the region in the z-plane onto the upper half-plane $\Im\{w\} > 0$ is given by

$$w = A \int_{z_0}^{z} \frac{(t - z_2)^{3/4}(t - z_4)^{1/4}}{(t - z_1)(t - z_3)}\, dt + B. \tag{7.7.9}$$

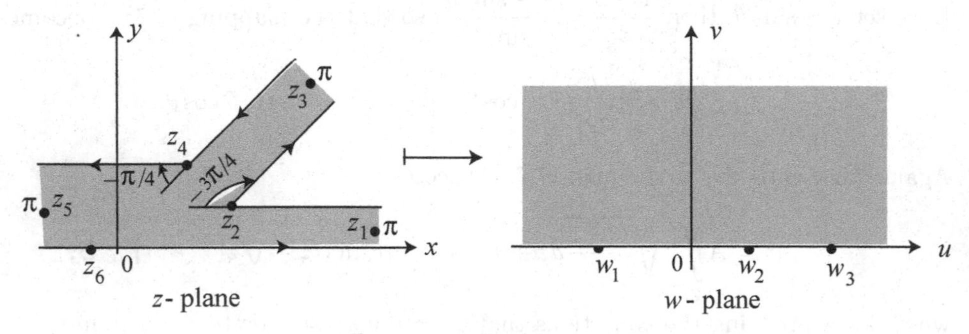

Figure 7.57 For Map 7.64.

Map 7.65. The function $\pi w = \cosh^{-1} z - \arcsin(1/z) + \pi/2$ maps the region in the positive quadrant of the w-plane bounded by the lines $u = 0$, $v = 0$, $u = 1$ $(v > 1)$, and $v = 1$ $(u > 1)$, onto the upper half-plane $\Im\{z\} > 0$. (Phillips [1966:66].)

Map 7.66. We will determine the Schwarz-Christoffel transformation that maps the shaded region with a slit from $z - 0$ to $z = i$ in the z-plane onto the upper half-plane $\Im\{w\} > 0$ (Figure 58).

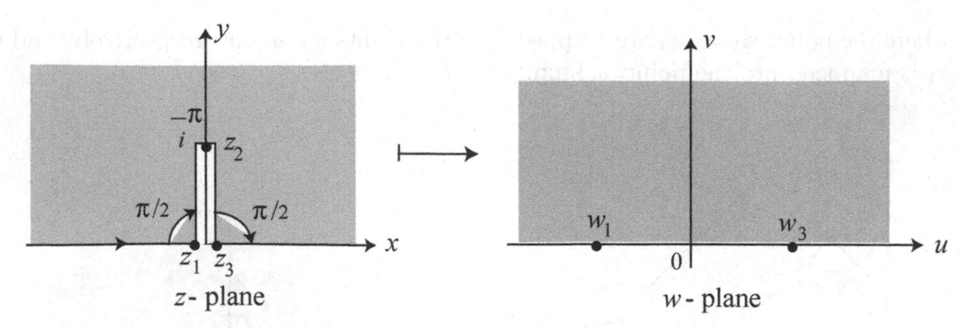

Figure 7.58 For Map 7.66.

The region in the z-plane has four vertices, namely, $z = 0, z_2 = i, z_3 = 0$, and $z_4 = \infty$, which are mapped onto the points $w_1 = -1, w_3 = 1$, and $w_4 = \infty$, but the point w_2 cannot be specified arbitrarily since it is the fourth point. All we know is that $-1 < w_2 < 1$. The exterior angles at z_1, z_2, z_3, z_4 are, respectively, $\pi/2, -\pi, \pi/2, 2\pi$, and the sum of these angles is 2π. Then the mapping function is

$$\begin{aligned} w &= A' \int_{z_0}^{z} \frac{t - x_2}{\sqrt{t^2 - 1}}\, dt + B = A \int_{z_0}^{z} \frac{t - x_2}{\sqrt{1 - -t^2}}\, dt + B \\ &= A \int_{z_0}^{z} \frac{t}{\sqrt{1 - t^2}}\, dt - \int_{z_0}^{z} \frac{x_2}{\sqrt{1 - t^2}}\, dt + B \\ &= -A\sqrt{1 - z^2} - x_2 A \arcsin z + B, \end{aligned} \tag{7.7.10}$$

which, after using the conditions that $w = -1$ when $z = 0$, yields $B = \dfrac{\pi x_2 A}{2}$, and then using (7.7.10), we get $B = 0$, which gives $x_2 A = 0$. Since $A \neq 0$ (otherwise the mapping becomes constant), we use the condition that $w = i$ when $z = x_2 = 0$ in (7.7.10) and find that that $A = -i$, so that the mapping is

$$w = i\sqrt{1 - z^2}.$$

Map 7.67. The Schwarz-Christoffel transformation that maps the region (shaded) in the z-plane onto the upper half-plane $\Im\{w\} > 0$ is given by

$$w = \frac{1 - b}{2} \log(z - 1) + \frac{1 + b}{2} \log(z + 1),$$

where b satisfies the equation

$$\frac{i\pi(1 - b)}{2} + \frac{1 - b}{2} \log(1 - b) + \frac{1 + b}{2} \log(1 + b) = a\, i.$$

Note that Map 7.11 is a particular case of this mapping.

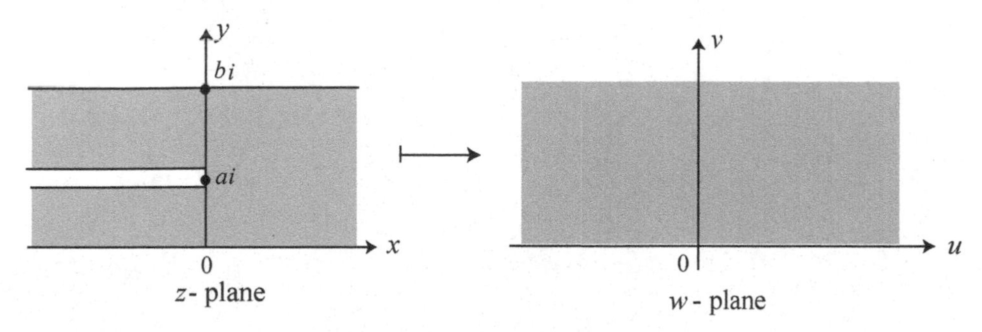

Figure 7.59 For Map 7.67.

7.7 Inverse Schwarz-Christoffel Mapping

The inverse of the Schwarz-Christoffel mapping is defined by Eq (7.1.7). However, if this inverse is defined by $z = g(w)$, it can be found, in principle, by solving the first order, nonlinear differential equation

$$g'(w) = \frac{1}{f'(z)}\bigg|_{z=g(w)} = \frac{1}{C_1}\Big(g(w) - z_1\Big)^{\alpha_1}\Big(g(w) - z_2\Big)^{\alpha_2} \cdots \Big(g(w) - z_n\Big)^{\alpha_n}. \tag{7.7.1}$$

In case $g'(w)$ contains a single factor, Eq (7.7.1) becomes

$$g'(w) = \frac{1}{C_1}\Big(g(w) - z_1\Big)^{\alpha_1},$$

which, when divided by $(g - z_1)_1^{\alpha}$ and integrated, yields

$$g(w) = \begin{cases} z_1 + \Big(\dfrac{1-\alpha_1}{C_1}\Big)^{1/(1-\alpha_1)} w^{1/(1-\alpha_1)} & \text{if } \alpha_1 \neq 1, \\ z_1 + e^{w/C_1} & \text{if } \alpha_1 = 1. \end{cases} \tag{7.7.2}$$

If $g'(w)$ contains two or more factors, Eq (7.7.1) becomes a difficult, and more likely an impossible, nonlinear differential equation to solve in closed form.

REFERENCES USED: Ahlfors [1966], Andersen et al. [1962], Betz [1964], Boas [1987], Carrier, Krook and Pearson [1966], Ivanov and Trubetskov [1995], Kantorovich and Krylov [1958], Kober [1957], Kythe [1996], Nehari [1952], Phillips [1943; 1966], Polyanin and Nazaikinskiii [2016], Sansone and Gerretsen [1969], Schinzinger and Laura [2003], Silverman [1967], von Koppenfels [1959], Wen [1992], Whittaker and Watson [1962].

Part 2: Numerical Conformal Mapping

8

Schwarz-Christoffel Integrals

In practical applications of conformal mapping of a standard region (the half-plane or the unit disk) onto a problem region which is in the form of a polygon, it becomes necessary to determine approximately the $(2n+2)$ parameters $\alpha_1, \dots, \alpha_n$, $x_1, \dots, x_n$, and the constants C_1 and C_2 that appear in the Schwarz-Christoffel formula (7.1.1). Evaluation of these quantities is known as the *parameter problem*. We have seen that the mapping functions obtained by using the Schwarz-Christoffel formula involve certain improper integrals which are known as Schwarz-Christoffel integrals. We will discuss methods for numerical solution of these integrals and present Newton's method for the general case of mapping the upper half-plane onto a quadrilateral.

8.1. Parameter Problem

If the values of all $(2n+2)$ parameters in the Schwarz-Christoffel formula (7.1.1) or (7.1.6) are known, then the polygon is uniquely defined, and the coordinates w_k of the vertices are given by

$$w_k = C_1 \int_0^{x_k} \prod_{i=1}^{n} (\zeta - x_i)^{-\alpha_i}\, d\zeta + C_2, \quad k = 1, \dots, n. \tag{8.1.1}$$

The length of the side joining the vertices w_k and w_{k+1} of the polygon is determined by

$$|w_k, w_{k+1}| = |w_{k+1} - w_k| = |C_1| \int_{x_k}^{x_{k+1}} \prod_{i=1}^{n} (\zeta - x_i)^{-\alpha_i}\, d\zeta. \tag{8.1.2}$$

Thus, it is very easy to figure out the behavior of the $(2n+2)$ parameters in the formula (7.1.1) or (7.1.6). In fact, the parameters $\alpha_1, \dots, \alpha_n$ are related to the quantities $\mu_1, \dots, \mu_n$ which are the ratios of the interior angles of the polygon to π. The parameter C_2 affects the location of the vertices of the polygon. Any change in C_2 changes the coordinates of every vertex by the same amount through homothety and translation and displaces the polygon as a unit. The parameter C_1 which as a factor in the formula (8.1.2) affects the lengths of the sides of the polygon. Any change in C_1 changes all sides of the polygon by the same amount and rotates the polygon as a unit. Any change in the parameters $x_1, \dots, x_n$ also produces relative change in the lengths of the sides of the polygon.

Let the ratio of the second, third, ... , $(n-2)$th side of the polygon to the first side be denoted by $\lambda_2, \lambda_3, \dots, \lambda_{n-2}$, respectively, i.e.,

$$\lambda_j = \frac{|w_{j+1} - w_j|}{|w_2 - w_1|}, \quad j = 2, 3, \dots, n-2. \tag{8.1.3}$$

Consider the function

$$\tilde{w} = \int_0^z \prod_{i=1}^{n} (\zeta - x_i)^{-\alpha_i} \, d\zeta. \tag{8.1.4}$$

If we choose the numbers $x_1, \dots, x_n$ in this function such that the relations (8.1.3) are satisfied, where $\alpha_i = 1 - \mu_i$, the function $\tilde{w}$ will define a conformal mapping of the upper half-plane onto a polygon $\widetilde{G}$. In order to pass from the polygon $\widetilde{G}$ to G, we use the linear transformation

$$w = C_1 \, \tilde{w} + C_2, \tag{8.1.5}$$

which does the following: It transfers any vertex, say $\tilde{w}_k$, of the polygon $\widetilde{G}$ to the corresponding vertex w_k of the polygon G, then rotates the polygon $\widetilde{G}$ about the vertex $\tilde{w}_k$ so that its sides become parallel to the sides of the polygon G, and finally, without changing the position of the vertex $\tilde{w}_k$, it changes the lengths of all sides of the polygon $\widetilde{G}$ such that the polygon $\widetilde{G}$ coincides with G. The constants C_1 and C_2 in (8.1.3) can be determined by comparing the location and sides of the polygons $\widetilde{G}$ and G.

In order to determine the parameters $x_1, \dots, x_n$, there are only $(n-3)$ equations, since three of these points can be chosen arbitrarily. We can use a Möbius transformation to carry the upper half-plane onto itself such that the three points of the x-axis go into three preassigned points of the u-axis. Let these three preassigned points be denoted by p_1, p_2, p_3. Let us denote the improper integrals thus obtained from formula (7.1.1) by I_m, $m = 1, 2, \dots, n-2$, and define them as

$$\begin{aligned}
I_1 &= \int_{p_1}^{p_2} (\zeta - p_1)^{-\alpha_1} (p_2 - \zeta)^{-\alpha_2} (x_3 - \zeta)^{-\alpha_3} \cdots (x_{n-1} - \zeta)^{-\alpha_{n-1}} \\
&\qquad \times (p_3 - \zeta)^{-\alpha_n} \, d\zeta, \\
I_2 &= \int_{p_2}^{x_3} (\zeta - p_1)^{-\alpha_1} (\zeta - p_2)^{-\alpha_2} (x_3 - \zeta)^{-\alpha_3} \cdots (x_{n-1} - \zeta)^{-\alpha_{n-1}} \\
&\qquad \times (p_3 - \zeta)^{-\alpha_n} \, d\zeta, \\
&\ \ \vdots \\
I_{n-2} &= \int_{x_{n-2}}^{x_{n-1}} (\zeta - p_1)^{-\alpha_1} (\zeta - p_2)^{-\alpha_2} (\zeta - x_3)^{-\alpha_3} \cdots (\zeta - x_{n-2})^{-\alpha_{n-1}} \\
&\qquad \times (p_3 - \zeta)^{-\alpha_n} \, d\zeta.
\end{aligned} \tag{8.1.6}$$

These integrals are known as *Schwarz-Christoffel integrals*. Then, in view of (8.1.3),

$$I_j\,(x_3, x_4, \dots, x_{n-1}) = \lambda_j \, I_1\,(x_3, x_4, \dots, x_{n-1}), \quad j = 2, 3, \dots, n-2. \tag{8.1.7}$$

In view of the Riemann mapping theorem (§2.6), once the three points x_1, x_2, x_3 are chosen arbitrarily, the system of equations (8.1.7) has a unique solution.

In order to solve the system (8.1.7) numerically, we use Newton's method which is as follows: Let $\tilde{x}_3, \tilde{x}_4, \dots, \tilde{x}_{n-1}$ denote the solution. Let us take the initial guess for these values as $x_3^{(0)}, x_4^{(0)}, \dots, x_{n-1}^{(0)}$ which are assumed to be sufficiently close to $\tilde{x}_3, \tilde{x}_4, \dots, \tilde{x}_{n-1}$. If we expand each equation of the system (8.1.7) into a Taylor series in powers of the difference $\tilde{x}_\nu - x_\nu^{(0)}$, $\nu = 3, 4, \dots, n-1$, and, as the first approximation, truncate these Taylor series after the first power of the differences, then we obtain a system of equations

$$I_j^{(0)} + \sum_{\nu=3}^{n-1} h_\nu^{(1)} \frac{\partial I_j^{(0)}}{\partial x_\nu} = \lambda_j \Big[I_1^{(0)} + \sum_{\nu=3}^{n-1} h_\nu^{(1)} \frac{\partial I_1^{(0)}}{\partial x_\nu} \Big], \quad j = 2, \dots, n-2, \tag{8.1.8}$$

with a nonzero determinant, where $h_\nu^{(1)}$, known as the corrections of the first order, are the perturbed values of the differences $\tilde{x}_\nu - x_\nu^{(0)}$. Then the system (8.1.8) is reduced and solved for $x_\nu^{(1)}$, $\nu = 3, 4, \dots, n-1$. Thus, using the initial values of $h_\nu^{(0)}$, the first approximations are given by

$$x_\nu^{(1)} = x_\nu^{(0)} + h_\nu^{(1)}, \quad \nu = 3, 4, \dots, n-1. \tag{8.1.9}$$

Next, the system (8.1.8) is expanded in Taylor series in powers of the differences

$$\tilde{x}_\nu - x_\nu^{(1)}, \quad \nu = 3, 4, \dots, n-1, \tag{8.1.10}$$

and these series are truncated after the first powers of these differences. Thus, a new system, analogous to (8.1.8), is constructed with unknowns $x_\nu^{(2)}$, $\nu = 3, 4, \dots, n-1$, with the perturbed values of the differences (8.1.10), except that now the new system is computed using the known values of $x_\nu^{(1)}$. It yields the values of $x_\nu^{(2)}$, and the second approximations are given by

$$x_\nu^{(2)} = x_\nu^{(1)} + h_\nu^{(2)}, \quad \nu = 3, 4, \dots, n-1. \tag{8.1.11}$$

This process is continued until we reach arbitrarily close to the solutions $\tilde{x}_\nu$, such that the difference between two consecutive approximate values is within a prescribed tolerance.

Once the system (8.1.8) is solved, we are able to approximate the coordinates of the vertices of the polygon and the lengths of its sides. But in doing so, we must compute the Schwarz-Christoffel integrals of the form

$$\begin{aligned} E = \int_{x_k}^{x_{k+1}} & (\zeta - p_1)^{-\alpha_1} (\zeta - p_2)^{-\alpha_2} (\zeta - x_3)^{-\alpha_3} \cdots (\zeta - x_k)^{-\alpha_k} \\ & \times (x_{k+1} - \zeta)^{-\alpha_{k+1}} \cdots (p_3 - \zeta)^{-\alpha_n} \, d\zeta, \end{aligned} \tag{8.1.12}$$

which are improper because the integrand of each integral becomes unbounded at two points where $\zeta = x_k, x_{k+1}$ which are the limits of integration. These integrals exist because each $\alpha_k > 0$.

8.1.1 Kantorovich's Method. This method is used to solve the above integral E, as follows: Let the integrand in (8.1.12) be denoted by $F(\zeta)$. Then

$$\begin{aligned} F(\zeta) &= (\zeta - p_1)^{-\alpha_1}(\zeta - p_2)^{-\alpha_2}(\zeta - x_3)^{-\alpha_3} \cdots (\zeta - x_k)^{-\alpha_k} \\ &\quad (x_{k+1} - \zeta)^{-\alpha_{k+1}} \cdots (p_3 - \zeta)^{-\alpha_n} \\ &= (\zeta - x_k)^{-\alpha_k} \left[\phi(x_k) + \phi'(x_k)(\zeta - k_k)\right] \\ &\quad + (x_{k+1} - \zeta)^{-\alpha_{k+1}} \left[\psi(x_{k+1}) - \psi'(x_{k+1})(x_{k+1} - \zeta)\right] \\ &\quad + \Big\{F(\zeta) - (\zeta - x_k)^{-\alpha_k} \left[\phi(x_k) + \phi'(x_k)(\zeta - k_k)\right] \\ &\quad - (x_{k+1} - \zeta)\left[\psi(x_{k+1}) - \psi'(x_{k+1})(x_{k+1} - \zeta)\right]\Big\}, \end{aligned} \tag{8.1.13}$$

where

$$\begin{aligned} \phi(\zeta) &= (\zeta - p_1)^{-\alpha_1}(\zeta - p_2)^{-\alpha_2}(\zeta - x_3)^{-\alpha_3} \cdots (\zeta - x_{k-1})^{-\alpha_{k-1}} \\ &\quad (x_{k+1} - \zeta)^{-\alpha_{k+1}} \cdots (p_3 - \zeta)^{-\alpha_n}, \end{aligned} \tag{8.1.14}$$

$$\begin{aligned} \psi(\zeta) &= (\zeta - p_1)^{-\alpha_1}(\zeta - p_2)^{-\alpha_2}(\zeta - x_3)^{-\alpha_3} \cdots (\zeta - x_k)^{-\alpha_k} \\ &\quad (x_{k+2} - \zeta)^{-\alpha_{k+2}} \cdots (p_3 - \zeta)^{-\alpha_n}. \end{aligned} \tag{8.1.15}$$

Let $E = E_1 + E_2$, where

$$\begin{aligned} E_1 = \int_{x_k}^{x_{k+1}} &\Big\{(\zeta - x_k)^{-\alpha_k}\left[\phi(x_k) + \phi'(x_k)(\zeta - x_k)\right] \\ &+ (x_{k+1} - \zeta)^{-\alpha_{k+1}}\left[\psi(x_{k+1}) - \psi'(x_{k+1})(x_{k+1} - \zeta)\right]\Big\}\, d\zeta \end{aligned} \tag{8.1.16a}$$

$$\begin{aligned} E_2 = \int_{x_k}^{x_{k+1}} &\Big\{F(\zeta) - (\zeta - x_k)^{-\alpha_k}\left[\phi(x_k) + \phi'(x_k)(\zeta - x_k)\right] \\ &- (x_{k+1} - \zeta)\left[\psi(x_{k+1}) - \psi'(x_{k+1})(x_{k+1} - \zeta)\right]\Big\}\, d\zeta. \end{aligned} \tag{8.1.16b}$$

The integral E_1 can be evaluated directly in a finite form, and since E_2 has no singularities it can be approximated by any formula for numerical integration of definite integrals, like Simpson's rule.

For numerical parameter problems and related methods, see §8.7.

8.2 Newton's Method

We will discuss Newton's method for mapping the upper half-plane onto an arbitrary quadrilateral (see Map 7.10). In fact, in order to determine k in (7.2.14), we have only one equation from (8.1.8)

$$I_4(k) = \lambda\, I_1(k), \tag{8.2.1}$$

where

$$\begin{aligned} I_4(k) = &\int_3^\infty (\zeta + 1)^{-\alpha_1}(\zeta - 1)^{-\alpha_2}(\zeta - k)^{-\alpha_3}(\zeta - 3)^{-\alpha_4}\, d\zeta \\ &+ \int_{-\infty}^{-1} (-1 - \zeta)^{-\alpha_1}(1 - \zeta)^{-\alpha_2}(k - \zeta)^{-\alpha_3}(3 - \zeta)^{-\alpha_4}\, d\zeta, \end{aligned} \tag{8.2.2}$$

and

$$I_1(k) = \int_{-1}^{1} (\zeta+1)^{-\alpha_1}(1-\zeta)^{-\alpha_2}(k-\zeta)^{-\alpha_3}(3-\zeta)^{-\alpha_4}\,d\zeta. \tag{8.2.3}$$

In $I_4(k)$, set $\zeta = 2 + 1/t$ in the first integral and $\zeta = -1/t$ in the second. Then

$$\begin{aligned} I_4(k) &= \int_0^1 (1+3t)^{-\alpha_1}(1+t)^{-\alpha_2}\,(1+(2-k)t)^{-\alpha_3}\,(1-t)^{-\alpha_4}\,dt \\ &\quad + \int_0^1 (1-t)^{-\alpha_1}(1+t)^{-\alpha_2}(1+kt)^{-\alpha_3}(1+3t)^{-\alpha_4}\,d\zeta. \end{aligned} \tag{8.2.4}$$

Let $F(k) = I_4(k) - \lambda\, I_1(k)$. Then $F'(k) > 0$. To see this, we have from (8.1.13)

$$\begin{aligned} F'(k) &= \alpha_3 \int_0^1 t(1+3t)^{-\alpha_1}(1+t)^{-\alpha_2}\,(1+(2-k)t)^{-\alpha_3-1}\,(1-t)^{-\alpha_4}\,dt \\ &\quad - \alpha_3 \int_0^1 t(1-t)^{-\alpha_1}(1+t)^{-\alpha_2}\,(1+kt)^{-\alpha_3-1}\,(1+3t)^{-\alpha_4}\,dt \\ &\quad + \lambda\,\alpha_3 \int_{-1}^1 (\zeta+1)^{-\alpha_1}(1-\zeta)^{-\alpha_2}(k-\zeta)^{-\alpha_3-1}(3-\zeta)^{-\alpha_4}\,d\zeta. \end{aligned} \tag{8.2.5}$$

Note that $F'(1)$ and $F'(3)$ do not exist because they are not finite. Since

$$\begin{aligned} F''(k) &= \alpha_3\,(1+\alpha_3)\,\Big\{ \int_0^1 t^2(1+3t)^{-\alpha_1}(1+t)^{-\alpha_2} \\ &\quad (1+(2-k)\,t)^{-\alpha_3-2}\,(1-t)^{-\alpha_4}\,dt \\ &\quad + \int_0^1 t^2(1-t)^{-\alpha_1}(1+t)^{-\alpha_2}\,(1+kt)^{-\alpha_3-2}\,(1+3t)^{-\alpha_4}\,dt \\ &\quad - \lambda \int_{-1}^1 (\zeta+1)^{-\alpha_1}(1-\zeta)^{-\alpha_2}(k-\zeta)^{-\alpha_3-2}(3-\zeta)^{-\alpha_4}\,d\zeta \Big\}, \end{aligned} \tag{8.2.6}$$

we find that $F''(k) > 0$ for $1 < k < 3$, and $F''(1) = -\infty$ and $F''(3) = +\infty$. Hence, as k varies from 1 to 3, both $F(k)$ and $F''(k)$ increase continuously from $-\infty$ to $+\infty$ (for details, see Kantorovich and Krylov [1958]). Hence, the function $F(k)$ has a zero in the interval $(1, 3)$. Let the zero of $F(k)$ be denoted by k^*, and that of $F''(k)$ by k^{**}. Now we will solve the equation $F(k) = 0$ by Newton's method. Although the initial guess for the value of k is important in this method, we will show that it can be chosen arbitrarily from below or from above in the interval $(1, 3)$. Let k_0 denote an arbitrary initial guess for the value of k. Then

Theorem 8.1 *In order to solve the equation $F(k) = 0$ by Newton's method, the value of $k_0 \in (1,3)$ can be chosen arbitrarily, independent of the values of k^* and k^{**}.*

PROOF. There are three cases to analyze:

Case 1: If $K^{**} = k^*$, then the initial value k_0 is any number in the interval $(1, 3)$ (see Figure 8.1). In fact, let k_0 be an approximation of k from below. Then $F(k) < 0$, and $F''(k) < 0$ for all $k \in (k_0, k^*)$. Hence, there is an M such that $F'(k) < M$ for all $k_0 \le k \le k^*$. The first correction is given by

$$\delta^{(1)} = -\frac{F(k_0)}{F'(k_0)} > 0. \tag{8.2.7}$$

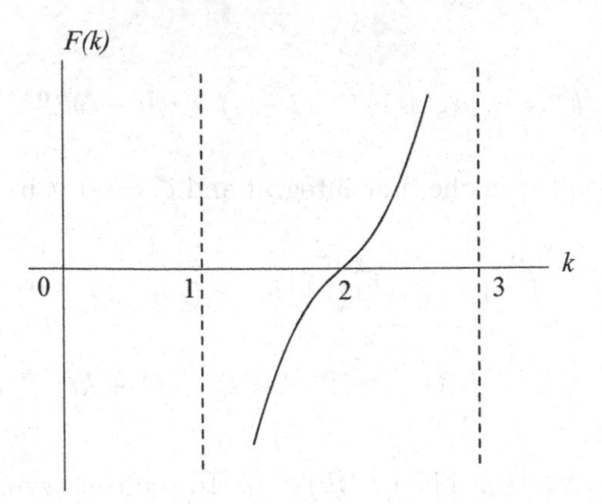

Figure 8.1 Newton's method: Case 1.

Thus, the exact first correction h_1 is positive and satisfies the equation

$$F(k_0) + h_1\, F'(k_0) + \frac{h_1^2}{2}\, F''(\tilde{k}) = 0, \quad k_0 < \tilde{k} < k^*. \tag{8.2.8}$$

Hence, from (8.2.7) and (8.2.8), we have

$$\left[h_1 - \delta^{(1)}\right]\, F'(k_0) = -\frac{h_1^2}{2}\, F''(\tilde{k}),$$

which yields

$$h_2 = h_1 - \delta^{(1)} = -\frac{h_1^2}{2}\, \frac{F''(\tilde{k})}{F'(k_0)}, \tag{8.2.9}$$

which is positive in view of (8.2.8). Thus, the first approximation k_1, like k_0, is an approximation from below, and, therefore, all subsequent approximations will be very small. It is the basic property of Newton's method that the exact corrections h are always positive and decreasing. In fact, we can show that if the difference between h_n and h_{n+1} is sufficiently small, then h_n is itself very small. Assume that

$$h_n - h_{n+1} < \frac{K}{M}\, \varepsilon, \tag{8.2.10}$$

where $\varepsilon > 0$ is arbitrarily small. Since $h_n - h_{n+1} = \delta^{(n)}$, we have $\delta^{(n)} < \dfrac{K}{M}\, \varepsilon$. Also,

$$\delta^{(n)} = -\frac{F\,(k_{n-1})}{F'\,(k_{n-1})} = \frac{|F\,(k_{n-1})\,|}{F'\,(k_{n-1})},$$

or

$$|F\,(k_{n-1})\,| = \delta^{(n)}\, F'\,(k_{n-1})\,. \tag{8.2.11}$$

But h_n can be evaluated from the equation

$$F\,(k_{n-1}) + h_n\, F'(\tilde{k}), \quad k_{n-1} < \tilde{k} < k^*. \tag{8.2.12}$$

Hence, from (8.2.11) and (8.2.12)

$$h_n = \frac{|F\,(k_{n-1})\,|}{F'\left(\tilde{k}\right)} = \frac{\delta^{(n)}\, F'\,(k_{n-1})}{F'(\tilde{k})}, \quad h_n < \frac{K}{M}\,\varepsilon\,\frac{M}{K} = \varepsilon.$$

This shows that the sequence $\{h_1, h_2, \dots, h_n, \dots\} \searrow 0$. This analysis leads to the same conclusion if k_0 is an approximation from above.

Case 2: If $k^{**} > k^*$, then for any initial guess $k_0 < k^*$, the convergence is the same as in case 1. Also, $F(k) < 0$, and $F'(k) < 0$ for $k_0 < k < k^*$ (as in case 1). Now, let k_0 be such that $k^* < k_0 < k^{**}$. Then $F(k_0) > 0$, and $F''(k) < 0$ for all $k \le k_0$. Moreover, $\delta^{(1)} < 0$, and $h_1 < 0$. Now,

$$h_2 = -\frac{1}{2}\, h_1^2\, \frac{F''(\tilde{k})}{F'(k_0)}\, F''(\tilde{k}) < 0,$$

and thus, $h_2 > 0$. This means that the first approximation is an approximation from below, and all subsequent approximations will converge from below to k^* (as in case 1). In the case when $k_0 > k^{**}$, we have $F(k) > 0$ for all $k \in (k^*, k_0)$; $F'(k)\begin{cases} < 0 & \text{for all } k \in (k^*, k^{**}) \\ > 0 & \text{for all } k \in (k^{**}, k_0) \end{cases}$; $\delta^{(1)} < 0$, and $h_1 < 0$. Thus, k_1 can be in any one of the three intervals $(1, k^*)$, (k^*, k^{**}), and (k^{**}, k_0). If k_1 is in the first two intervals, then we have convergence from below to k^*. But if k_1 is in the third interval, then all approximations, although decreasing continuously, still remain greater than k^{**} and approach some limiting value $k_1 \ge k^{**}$. The difference between h_n and h_{n+1} would be sufficiently small for sufficiently large n, and $h_n - h_{n+1} < \dfrac{K}{M}\, \dfrac{k^{**} - k^*}{2}$, where $M = \max\limits_{k \in (k^*, k_0)} F'(k)$. As in case 1, it can be shown that $|h_n| < \dfrac{k^{**} - k^*}{2}$. But, by assumption, we have $|h_n| \ge k^{**} - k^*$. This contradiction shows that if Newton's method starts with $k_0 > k^{**}$, then the approximation will cross k^{**} and fall in the interval where the convergence is established as in case 1.

Case 3: If $k^{**} < k^*$, then this case can be analyzed by taking k_0 in any one of the intervals discussed in cases 1 and 2. We will apply this method in Case 8.3 in the next section.

8.3 Numerical Computations

Now we will solve some parameter problems.

Case 8.1. Consider the integral in the denominator in (7.2.3). Denoting it by E, we have $C_1 = 1/E$, where

$$E = \left(\int_0^{1/2} + \int_{1/2}^1 \right) \zeta^{-1/2} (1 - \zeta)^{-3/4}\, d\zeta = E_1 + E_2.$$

Then, by (8.1.16),

$$\begin{aligned} E_1 &= \lim_{t \to 0} \int_t^{1/2} \zeta^{-1/2} (1 - \zeta)^{-3/4}\, d\zeta \\ &= \lim_{t \to 0} \int_t^{1/2} \left\{ \zeta^{-1/2} \left(1 + \frac{3}{4}\zeta\right) + \zeta^{-1/2} \left[(1 - \zeta)^{-3/4} - 1 - \frac{3}{4}\zeta \right] \right\} d\zeta \\ &= E_{11} + E_{12} \approx 1.59099 + 0.0708022 = 1.66179; \\ E_2 &= \lim_{t \to 1} \int_{1/2}^{t} \zeta^{-1/2} (1 - \zeta)^{-3/4}\, d\zeta = \lim_{t \to 1} \int_{1/2}^{t} \Big\{ (1 - \zeta)^{-3/4} [1 + \frac{1}{2}\,(1 - \zeta)] \\ &\qquad + (1 - \zeta)^{-3/4} [\zeta^{-1/2} - 1 - \frac{1}{2}(1 - \zeta)] \Big\}\, d\zeta \end{aligned}$$

$$= E_{21} + E_{22} \approx 3.53176 + 0.0505577 = 3.58232.$$

Hence $E = E_1 + E_2 \approx 5.24412$, and the constant $C_1 = 1/E \approx 0.19069$. Note that the exact value of $E = \dfrac{\sqrt{\pi}\,\Gamma(1/4)}{\Gamma(3/4)} \approx 0.19068994$. The hypotenuse of the triangle is given by

$$|w_3 - w_2| = C_1 \int_1^\infty \zeta^{-1/2}(1-\zeta)^{-3/4}\, d\zeta \approx 1.41421356 \approx \sqrt{2}$$

with an error of $O\left(10^{-10}\right)$. ■

Case 8.2. In order to determine k in (7.2.10) such that the ratio $\dfrac{w_2 - w_1}{w_3 - w_2} = 2$, note that the system (8.1.7) reduces to only one equation $I_2 = 2\, I_1$, i.e.,

$$\int_{-1}^{1} (1-\zeta^2)^{-1/2}(1-k^2\zeta^2)^{-1/2}\, d\zeta = 2 \int_1^{1/k} (\zeta^2 - 1)^{-1/2}(1-k^2\zeta^2)^{-1/2}\, d\zeta, \tag{8.3.1}$$

which is solved by Newton's method as follows: We have only one correction which we will denote by h with subscript to denote the appropriate number of approximation. Let us take the initial guess $k = 1/2$, and determine h_1 from (8.1.8) which is

$$I_2\left(\frac{1}{2}\right) + h_1 \frac{dI_2\left(\frac{1}{2}\right)}{dk} = I_1\left(\frac{1}{2}\right) + h_1 \frac{dI_1\left(\frac{1}{2}\right)}{dk}. \tag{8.3.2}$$

The free term $I_2(1/2) = \displaystyle\int_0^1 (1-\zeta)^{-1/2}\,\psi(\zeta)\, d\zeta$, where $\psi(\zeta) = (1+\zeta)^{-1/2} + \left(1-\zeta^2/4\right)^{-1/2}$, which yields $I_2(1/2) \approx 2.15652$ on integration. Also,

$$\frac{dI_2\left(\frac{1}{2}\right)}{dk} = \frac{1}{2}\int_0^1 \frac{\zeta^2}{(1-\zeta^2)^{1/2}}\left(1 - \frac{\zeta^2}{4}\right)^{3/2} d\zeta \approx 0.541732,$$

$$\frac{dI_1(1/2)}{dk} \approx -1.79181.$$

Substituting these values in (8.3.2) we find that $h_1 \approx 0.201739$. Hence, the first approximation for k is $k \approx 0.5 + 0.201739 \approx 0.7$. Now, for the second approximation, first we compute h_2 from (8.1.8), i.e.,

$$I_2(0.7) + h_2 \frac{dI_2(0.7)}{dk} = I_1(0.7) + h_2 \frac{dI_1(0.7)}{dk}, \tag{8.3.3}$$

which gives $h_2 \approx 0.00668985$, and the second approximation for k is $k \approx 0.7 + 0.00668985 \approx 0.70669$. However, the exact value of k can be determined in this case by setting $\zeta = \dfrac{1}{\sqrt{1-(1-k^2)x^2}}$ in $I_2(k)$. Then

$$I_2(k) = \int_0^1 (1-x^2)^{-1/2}(1-k'^2x^2)^{-1/2}\, dx,$$

where $k'^2 = 1 - k^2$. Hence, Eq (8.3.1) gives

$$\int_0^1 (1-\zeta^2)^{-1/2}(1-k^2\zeta^2)^{-1/2}\,d\zeta = \int_0^1 (1-x^2)^{-1/2}(1-k'^2x^2)^{-1/2}\,dx.$$

Thus, $k = k'$, which gives $k = 1/\sqrt{2} \approx 0.7071$. A comparison of this value of k with the second approximation for k shows that the error is about 0.04%. ■

Case 8.3. The Schwarz-Christoffel transformation that maps the trapezoid $A_1\,A_2\,A_3\,A_4$ in the z-plane onto the upper half-plane is given by the Map 7.8 (Figure 7.7). The mapping function is (7.2.21), which we rewrite here as

$$w = C_1 \int_0^z (\zeta+1)^{-1/6}(\zeta-1)^{-1/3}(\zeta+k)^{-2/3}(\zeta-3)^{-5/6}\,d\zeta + C_2. \tag{8.3.4}$$

In order to determine the value k, we have from Eq (8.1.8)

$$I_4(k) = 3\,I_1(k), \tag{8.3.5}$$

where from (8.2.4)

$$\begin{aligned} I_4(k) &= \int_0^1 (1+3t)^{-1/6}(1+t)^{-1/3}\,(1+(2-k)t)^{-2/3}\,(1-t)^{-5/6}\,dt \\ &\quad + \int_0^1 (1-t)^{-1/6}(1+t)^{-1/3}(1+kt)^{-2/3}(1+3t)^{-5/6}\,dt \\ &= I_{41}(k) + I_{42}(k), \end{aligned} \tag{8.3.6}$$

and from (8.2.3)

$$\begin{aligned} I_1(k) &= \int_{-1}^1 (\zeta+1)^{-1/6}(1-\zeta)^{-1/3}(k-\zeta)^{-2/3}(3-\zeta)^{-5/6}\,d\zeta \\ &= \left(\int_{-1}^0 + \int_0^1\right)(\zeta+1)^{-1/6}(1-\zeta)^{-1/3}(k-\zeta)^{-2/3}(3-\zeta)^{-5/6}\,d\zeta \\ &= I_{11}(k) + I_{12}(k). \end{aligned}$$

Let $k = 2$ be the initial guess. Then we find that $\delta^{(1)} \approx -0.650694$, $k_1 \approx 1.34931$, $\delta^{(2)} \approx 0.0150829$, $k_2 \approx 1.36439$, $\delta^{(3)} \approx 0.0005$, and $k_3 \approx 1.36554$. Now, in order to compute C_2 and C_1, note that the function

$$w = \int_0^z (\zeta+1)^{-1/6}(\zeta-1)^{-1/3}(\zeta+k)^{-2/3}(\zeta-3)^{-5/6}\,d\zeta \tag{8.3.7}$$

maps the upper half-plane onto a trapezoid $A_1^*\,A_2^*\,A_3^*\,A_4^*$ similar to the given $A_1A_2A_3A_4$. To determine the complex coordinate w_1^* of the vertex A_1^* which corresponds to $z = -1$, we have

$$\begin{aligned} w_1^* &= \int_0^{-1} (\zeta+1)^{-1/6}(\zeta-1)^{-1/3}(\zeta+k)^{-2/3}(\zeta-3)^{-5/6}\,d\zeta \\ &\approx -(-1)^{1/6}(0.24631) = -0.24631\,i. \end{aligned}$$

Similarly, $w_2^* \approx 0.90311\,i$. Hence $w = C_2 + C_1 w^*$ yields

$$-\frac{1}{2} = C_2 - 0.24631\,i\,C_1, \quad \frac{1}{2} = C_2 + 0.90311\,i\,C_1,$$

which gives $C_2 = -0.2851$, and $C_1 = -0.87\,i$, and the required transformation is given by

$$w \approx -0.2851 - 0.87\,i \int_0^z (\zeta+1)^{-1/6}(\zeta-1)^{-1/3}(\zeta+k)^{-2/3}(\zeta-3)^{-5/6}\,d\zeta. \blacksquare \tag{8.3.8}$$

Map 8.1. (Schwarz-Christoffel integral for the unit disk) In view of (7.1.4), the integral in the Schwarz-Christoffel formula (7.1.1) is approximately equal to ζ^{-2} when ζ is close to infinity, whereas the integrand in the formula (7.1.5) is approximately $\zeta^{-\alpha_n}$. These quantities are significant when the region is infinite. However, we can avoid an infinite region by mapping the upper half-plane $\Im\{z\} > 0$ onto the unit disk by the chain of mappings $z_1 = \dfrac{z-i}{z+i}$ and $z = i\,\dfrac{1+z_1}{1-z_1}$, where $dz = \dfrac{2i}{(1-z_1)^2}\,dz_1$. Then the integrand in formula (7.1.1) becomes

$$\begin{aligned}\prod_{j=1}^n (z-x_j)^{-\alpha_j}\,dz &= \prod_{j=1}^n \left(i\,\frac{1+z_1}{1-z_1} - x_j\right)^{-\alpha_j} \frac{2i}{(1-z_1)^2}\,dz_1\\ &= 2i \prod_{j=1}^n \left(\frac{z_1(x_j+i)-(x_j-i)}{1-z_1}\right)^{-\alpha_j} \frac{1}{(1-z_1)^2}\,dz_1\\ &= 2i \prod_{j=1}^n (x_j+i)^{-\alpha_j}\left(z_1 - \frac{x_j-i}{x_j+i}\right)^{-\alpha_j} \frac{dz_1}{(1-z_1)^{2-\alpha_j}}\\ &= C_1 \prod (z_1-b_j)^{-\alpha_j}\,dz_1,\end{aligned} \tag{8.3.9}$$

where $b_j = (a_j - i)/(a_j + i)$, and the exponent $2-\alpha_j$ is zero in the product. Thus, formula (7.1.1) becomes

$$w = C_1 \int_{z_{10}}^{z_1} \prod_{j=1}^n (z_1 - b_j)\,dz_1 + z_0, \tag{8.3.10}$$

where the points b_j lie on the unit circle $|z_1| = 1$. The lower limit z_{10} may be chosen as the center of this circle or a point on its circumference. Then there are two cases to consider:

Case (i) If $z_{10} = 0$, then the integration is carried out along the ray $z_1 = r\,e^{i\theta}$, $0 \le r \le R$, $\theta =$const, and $b_j = e^{i\phi_j}$, $j = 1, \dots, n$. Then the mapping function (8.3.10) reduces to

$$\begin{aligned}w &= C_1\,e^{i\theta} \int_0^R \prod_{j=1}^n \left(re^{i\theta} - e^{i\phi_j}\right)^{-\alpha_j} r\,dr + z_0\\ &= C_1\,e^{i\theta} \int_0^R \prod_{j=1}^n e^{-i\theta\phi_j}\left(r - e^{i(\phi_j-\theta)}\right)^{-\alpha_j} r\,dr + z_0\\ &= C_1\,e^{-i\theta} \int_0^R \prod_{j=1}^n \left(r - e^{i(\phi_j-\theta)}\right)^{-\alpha_j} r\,dr + z_0.\end{aligned} \tag{8.3.11}$$

Case (ii). If $z_{10} = 1$, we choose the path of integration along the circumference. Thus, for a point $z_1 = e^{i\theta}$, the mapping function (8.3.10) becomes

$$\begin{aligned} w &= C_1 \int_0^\theta \prod_{j=1}^n \left(e^{i\theta} - e^{i\phi_j}\right)^{-\alpha_j} ie^{i\theta}\, d\theta + z_0 \\ &= iC_1 \int_0^\theta \prod_{j=1}^n e^{-(\sum \phi_j\alpha_j)/2} \left(e^{i(\theta-\phi_j)/2} - e^{-i(\theta-\phi_j)/2}\right)^{-\alpha_j} d\theta + z_0 \\ &= i(2i)^{-2}\, C_1 \int_0^\theta \prod_{j=1}^n \left(\sin\frac{\theta-\phi_j}{2}\right)^{-\alpha_j} d\theta + z_0 \\ &= K \int_0^\theta \prod_{j=1}^n \left(\sin\frac{\theta-\phi_j}{2}\right)^{-\alpha_j} d\theta + z_0, \quad K = i(2i)^2 C_1. \end{aligned} \tag{8.3.12}$$

If $\sin\dfrac{\theta-\phi_j}{2} < 0$, then we choose the branch

$$\sin\left(\frac{\theta-\phi_j}{2}\right)^{-\alpha_j} = \exp\left\{-\alpha_j \log\left|\sin\frac{\theta-\phi_j}{2}\right| - i\pi\alpha_j\right\}.$$

There is no problem for $\sin\dfrac{\theta-\phi_j}{2} \geq 0$. Thus, there exists a constant argument in (8.3.12) in each interval $\phi_{j-1} < \theta < \phi_j$, $j = 1, \dots, n$, and the length $l_j = |z_{j-1}, z_j| = |z_j - z_{j-1}|$ is given by

$$l_j = |K| \int_{\phi_{j-1}}^{\phi_j} \prod_{j=1}^n \left|\sin\frac{\theta-\phi_j}{2}\right|^{-\alpha_j} d\theta, \quad j = 1, \dots, n. \tag{8.3.13}$$

Therefore, the parameter problem for the unit circle is solved by carrying out the integration in (8.3.13) over a finite interval. ■

8.3.1 Removal of Singularities. The computation of the improper integrals in the formulas (7.1.1) or (7.1.5) can be easily carried out by removing the singularities in the integrand. There are two analytical methods to do this.

Method 1. We will consider the exponent $-\alpha_1$; others can be handled similarly. If $\alpha_1 > 0$ in (7.1.1), we set

$$K_1 = (x_1 - x_2)^{-\alpha_1} \cdots (x_1 - x_n)^{-\alpha_n}$$

and rewrite the integrand in (7.1.1) as

$$\begin{aligned} &\int_{z_0}^{z} \left[(\zeta - x_1)^{1-\alpha_1} (\zeta - x_2)^{-\alpha_2} \cdots (\zeta - x_n)^{-\alpha_n} - (\zeta - x_1)^{-\alpha_1} K_1\right] d\zeta \\ &\qquad + \frac{K_1}{1-\alpha_1} \left[(\zeta - x_1)^{1-\alpha_1} - (z_0 - x_1)^{1-\alpha_1}\right]. \end{aligned} \tag{8.3.14}$$

The integrand in (8.3.14) in the neighborhood of $z = z_1$ is approximately equal to $(\zeta - x_1)^{-\alpha_1}$. In fact, it consists of two factors, the first of which goes to ∞ and the second to 0 as $\zeta \to x_1$. Hence, the product is bounded or even goes to zero.

If we use (8.3.14) for part (ii) of Case 8.4, then we get integrals of the form

$$I = \int_{\phi_1}^{\theta} \left(\sin \frac{\theta - \phi_1}{2}\right)^{-\alpha_1} d\theta = 2 \int_0^{(\theta-\phi_1)/2} (\sin t)^{-a_1}\, dt,$$

where we have set $\theta = \phi_1 + 2t$. Now, let $\phi_1 < \theta < \phi_1 + \pi$. Then, by setting $t = \sin^{-1} u$, $u = \sqrt{x}$, we obtain

$$I = 2 \int_0^{u_0} u^{-\alpha_1} \frac{1}{\sqrt{1-u^2}}\, du = \int_0^{x_0} x^{-\alpha_1/2} (1-x)^{-1/2}\, dx, \quad x_0 < 1, \tag{8.3.15}$$

which is the incomplete beta function.

Method 2. If $\alpha_1 > 0$ in (7.1.1), then we remove the singularity at $\zeta = x_1$ by using the transformation $z_1 = (z - x_1)^{1-\alpha_1}$, i.e.,

$$z = x_1 + z_1^{1/(1-\alpha_1)}. \tag{8.3.16}$$

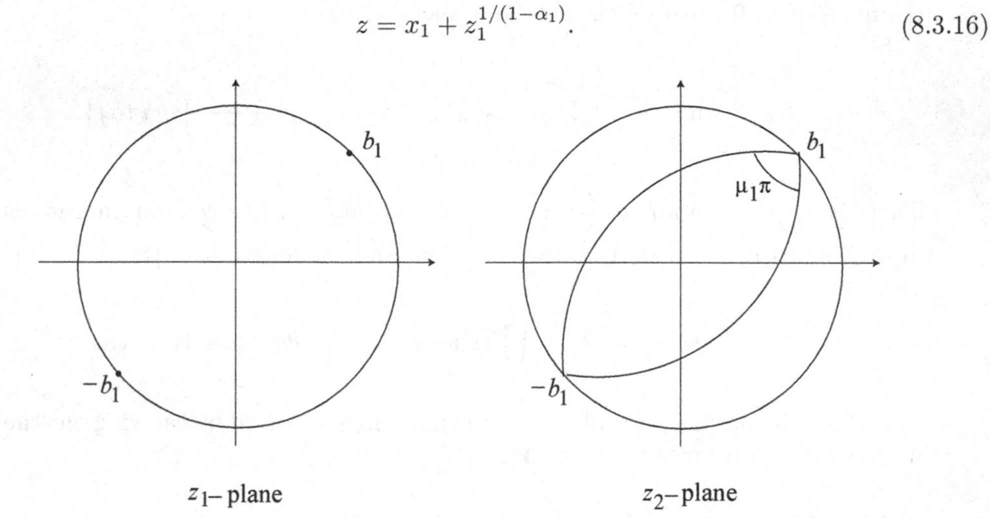

Figure 8.2 A circle onto the region bounded by two circles.

Then the integral (7.1.1) becomes

$$w = z_0 + \frac{C}{\alpha_1} \int_{z_{10}}^{z_1} \left(z_1^{\alpha_1/(1-\alpha_1)} + x_1 - x_2\right)^{-\alpha_2} \cdots \left(z_1^{-\alpha_1/(1-\alpha_1)} x_1 - x_n\right)^{-\alpha_n} dz_1, \tag{8.3.17}$$

which does not have infinity for α_1. The transformation (8.3.16) transforms the half-plane $\Im\{z\} > 0$ onto an angular sector of argument $(1-\alpha_j)\pi = \mu_j \pi$. Note that the mapping (8.3.16) is not suitable for the case of the circle. However, the transformation that maps the circle onto a region bounded by two circles is given by

$$z_1 = b_1 \frac{(b_1 + z_2)^{1/\mu_1} - (b_1 - z_2)^{1/\mu_1}}{(b_1 + z_2)^{1/\mu_1} + (b_1 - z_2)^{1/\mu_1}}, \tag{8.3.18}$$

which is represented in Figure 8.2. ■

8.3.2 Trefethen's Method. Let the N points z_k (called *prevertices*) be taken in the counterclockwise order around the unit circle and two complex constants C and w_c. The Schwarz-Christoffel formula, defined by (7.1.18), is written as

$$w = f(z) = w_c + C \int_0^z \prod_{k=1}^{N} \left(1 - \frac{z'}{z_k}\right)^{-\alpha_k} dz', \tag{8.3.19}$$

where the variables $z_1, \dots, z_n, C$, and w_c are the *accessory* parameters of the Schwarz-Christoffel mapping problem. We will first resolve the parameter problem by determining the values of these accessory parameters so that the lengths of the sides of the image polygon comes out right.

Theorem 8.2. (Schwarz-Christoffel theorem) *Let D be a simply connected region in the z-plane bounded by a polygon P with vertices $z_1, \dots, z_n$ and exterior angles $\alpha_k \pi$, where $-1 \le \alpha_k \le 1$ if z_k is finite, and $1 \le \alpha_k \le 3$ if $z_k = \infty$. Then there exists an analytic function which maps the unit disk in the complex plane conformally onto D, and every such function is written in the form (8.3.19).*

For proof, see Henrici [1974, Theorem 5.12e].

Given any polygon, there are not one, but infinitely many such conformal mappings. To determine a unique map, we may fix exactly any three points z_k, or fix one point z_k and also fix the complex value w_c, or fix w_c and the argument of the derivative $f'(0)$. The numerical computation of the Schwarz-Christoffel integral (8.3.19) is carried out as follows.

Gaier [1964] put forth a comprehensive research in numerical conformal mapping. For Schwarz-Christoffel transformation, he worked on determining the accessory parameters z_k by setting up a constrained nonlinear system of $N - 3$ equations related to the mapping (8.3.19) and solving it iteratively by Newton's method (Gaier [1964: 171]). This method has been tried by at least three researchers, namely Meyer [1979], Howe [1973], and Vecheslavov and Kokolin [1974].

Following Gaier [1964], three innovations are made to cut computing processing time and cost: (i) use of the Gauss-Jacobi quadrature to evaluate the integral in (8.3.19). In this case it was found that this procedure provided very low accuracy in realistic problems. To overcome this problem, a compound form of Gauss-Jacobi quadrature was developed; the details are given in §8.3.2(c). (ii) The computations may be performed not only for bounded polygons but also for polygons with any number of vertices at infinity. This was made possible by taking the unit disk as the model domain rather than the upper half-plane, and evaluating complex contour integrals within the disk rather than along the boundary. Note that the ability to handle unbounded polygonal regions is important for applications, because the use of conformal mapping lies in reducing an unbounded region into a bounded one. (iii) The treatment of constraints in the nonlinear system is accomplished by a simple change of variables to eliminate these constraints directly. This procedure is efficient and eliminates the need for initial guess of the accessory parameters. The details of these procedures are discussed bellow.

(a) Formulation as a nonlinear system (subroutine SCFUN)). The parameters in the map (8.3.19) that need to be fixed at the outset so that the Schwarz-Christoffel transformation may be determined uniquely have the following choices: (i) Fix three of the boundary points z_k, say, $z_1 = 1, z_2 = i, z_3 = -i$. This has the advantage that the resulting nonlinear system has size only $(N - 3) \times (N - 3)$, which for a typical problem with $N = 8$ may lead to a solution in less than one-half the time that a method involving an $(N-1) \times (N-1)$ system would require. Nevertheless, by normalizing the conditions

$$z_N = 1, \quad w_c \text{ an arbitrary point within } G, \tag{8.3.20}$$

we obtain an $(N-1) \times (N-1)$ system. This choice arises from the idea of numerical scaling, so as to allow the vertices to distribute themselves more evenly around the unit circle than they would otherwise do. Recall that a Möbius transformation will transform a map obtained by any normalization to a different region or a different transformation.

The formulation of the nonlinear system is as follows: Since the final map must satisfy N complex conditions, we have

$$w_k - w_c = C \int_0^{z_k} \prod_{j=1}^{N} \left(1 - \frac{z'}{z_j}\right)^{-\alpha_j} dz', \quad 1 \le k \le N. \tag{8.3.21}$$

Thus, $2N$ real conditions must be satisfied; but they are heavily over-determined, because the Schwarz-Christoffel formula (8.3.19) guarantees that the angles will be correct no matter what accessory parameters are chosen. We must reduce the number of operative equations to $N-1$. However, this becomes tricky in the case of unbounded polygons; care is needed in such a situation to have enough information about the polygon P so that no degree of freedom remains in the computed solution. The procedure is as follows:

STEP 1. We require that every connected component of P contains at least one vertex w_k. Even an infinite straight boundary must contain a (degenerate) vertex.

STEP 2. At least one component of P must actually contain two finite vertices, and w_N and w_1 will be taken to be two such vertices, thereby eliminating rotational degrees of freedom.

STEP 3. Define

$$C = \frac{w_N - w_c}{\displaystyle\int_0^{z_N} \prod_{j=1}^{N} \left(1 - \frac{z'}{z_j}\right)^{-\alpha_j} dz'}, \tag{8.3.22}$$

where $z_N = 1$ is fixed permanently by (8.3.20).

STEP 4. Impose the complex condition (we have two real equations $1, 2$)

$$w_1 - w_c = C \int_0^{z_1} \prod_{j=1}^{N} \left(1 - \frac{z'}{z_j}\right)^{-\alpha_j} dz'. \tag{8.3.23}$$

This amounts to two real equations to be satisfied.

STEP 5. Denote by $\Gamma_1, \dots, \Gamma_m$ the distinct components of P, numbered counterclockwise. For each $l \ge 2$, impose one more complex condition: If z_{k_l} is the last vertex of Γ_l in the counterclockwise direction, then we have real equations $3, 4, \dots, 2m$:

$$w_{k_l} - w_c = C \int_0^{z_{k_l}} \prod_{j=1}^{N} \left(1 - \frac{z'}{z_j}\right)^{-\alpha_j} dz'. \tag{8.3.24}$$

STEP 6. Impose $n - 2m - 1$ conditions of side length. For each pair (z_k, z_{k+1}) beginning at $k = 1$ and moving counterclockwise, where both vertices are finite, we require that the following $2m+1, \dots, N-1$ real equations be satisfied:

$$|w_{k+1} - w_k| = \left| C \int_0^{z_{k+1}} \prod_{j=1}^{N} \left(1 - \frac{z'}{z_j}\right)^{-\alpha_j} dz', \tag{8.3.25}$$

until a total of $N-1$ conditions have been imposed.

If P contains at least one vertex at infinity, then every bounded side has already been represented in a condition of the form (8.3.25) except for the side (w_N, w_1), which has already been included in (8.3.20) and (8.3.23). If P is bounded, then at least two sides in counterclockwise order, namely (w_{N-2}, w_{N-1}) and (w_{N-1}, w_N), will not be so represented.

In the above presentation we have not defined the contours for the integrals of Eq (8.3.23), (8.3.24) and (8.3.25). Mathematically it does not matter since the integrands are analytic functions. However, numerically it is an important aspect. The procedure discussed here computes these integrals over straight line segments between two endpoints, and it is justified since the unit disk is strictly convex. The contours involved in computing the integrals (8.3.22), (8.3.23), (8.3.24) and (8.3.25) are presented in Figure 8.3 for the simple case with $N = 10, m = 3$.

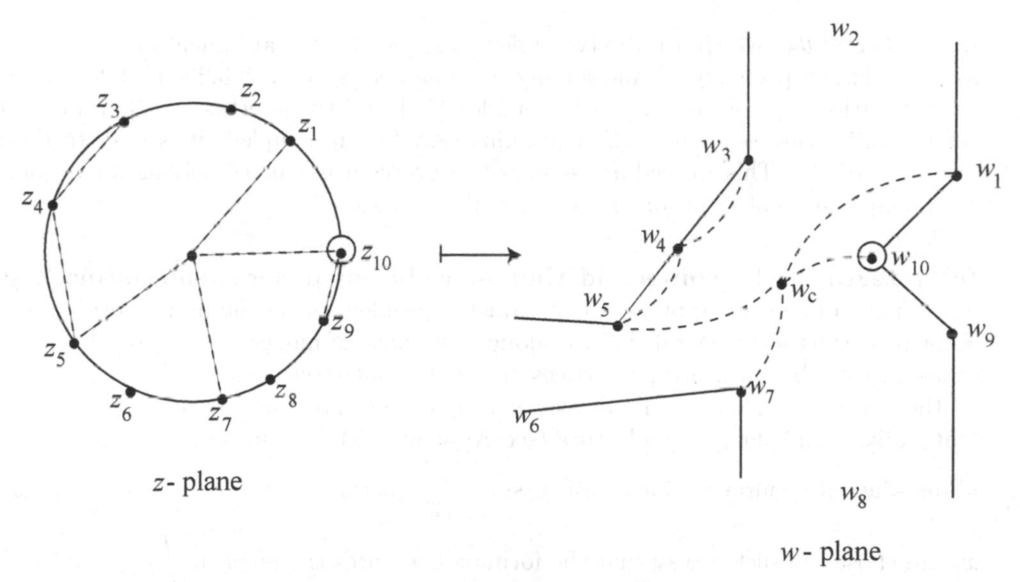

Figure 8.3 Contours of integration within the unit disk.

This figure illustrates, for example, the computation of the following integrations:
(i) one radical integral along (0 to z_{10}) defined by Eq (8.3.22);
(ii) one radical integral along (0 to z_1) determines two real equations to fix w_1 (Eq (8.3.23);
(iii) two radical integrals along (0 to z_5) and (0 to z_7) determine four real equations to fix w_5 and w_7 (Eq (8.3.24);
(iii) three radical integrals along (z_3 to z_4), (z_4 to z_5), and (z_9 to z_{10}) determine three real equations to fix $|w_4 - w_3|$, $|w_5 - w_4|$, and $|w_{10} - w_9|$ (Eq (8.3.25);
thus, the total being $N - 1 = 9$ equations.

The Fortran subroutine SCFUN, available in Trefethen [1979], was used for the above computations.

(b) Transformation of an unconstrained system (subroutine YZTRAN). The nonlinear system of equations (8.3.23)-(8.3.25) apparently involves $N-1$ complex unknown points $z_1, \dots, z_{N-1}$ on the unit circle. In practice we deal not with the points themselves, but their arguments θ_k are given by

$$z_k = e^{i\theta_k}, \quad 0 \le \theta_k \le 2\pi. \tag{8.3.26}$$

The system now depends on $N-1$ real unknowns, and the solution in terms of θ_k is fully determined. However, this system of nonlinear equations must be subject to a set of strict inequality constraints

$$0 < \theta_k < \theta_{k+1}, \quad 1 \le k \le N-1. \tag{8.3.27}$$

These constraints arise from the fact that the vertices z_k must lie in ascending order counterclockwise around the unit circle. To solve this system numerically, we must eliminate these constraints somehow. This is done by transforming Eqs (8.3.23)-(8.3.25) to a system of $N-1$ variables $y_1, \dots, y_{N-1}$, defined by

$$y_k = \log \frac{\theta_k - \theta_{k-1}}{\theta_{k+1} - \theta_k}, \quad 1 \le k \le N-1, \tag{8.3.28}$$

where θ_0 and θ_N, which are the two different names for the argument of $z_N = 1$, are taken as 0 and 2π, respectively. While solving the nonlinear system (8.3.23)-(8.3.25), we begin, at each iteration, by computing a set of angles $\{\theta_k\}$ and then vertices $\{z_k\}$ from the current set of $\{y_k\}$. However, since the equations (8.3.28) are coupled, it is easy to do, though not immediate. This procedure reduces the problem to one of solving an unconstrained nonlinear system of equations in $N-1$ real variables.

(c) Integration by compound Gauss-Jacobi quadrature (subroutine ZQUAD). The main aim for computing the parameter problem is the numerical evaluation of the Schwarz-Christoffel integral (8.3.19) along some path of integration. Typically, one or both endpoints of this path are prevertices z_k on the unit circle, and in this case a singularity of the form $(1 - z/z_k)^{-\alpha_k}$ is always present in the integrand at one or both endpoints. Naturally, Gauss-Jacobi quadrature (see Appendix G) computes such integrals quickly. A Gauss-Jacobi quadrature formula is a sum $\sum_{i+1}^{\text{NPTS}} w_i f(x_i)$, where the weights w_i and nodes x_i are chosen in such a way that the formula computes the integral $\int_{-1}^{+1} f(x)(1-x)|a(1+x)^{\beta}\,dx$ exactly for a polynomial $f(x)$ of as high a degree as possible, and NTPS denotes the number of quadrature nodes per half-interval. The required weights and nodes are computed numerically, using a Fortran program GAUSSQ by Golub and Welsch [1969].

Using Gauss-Jacobi quadrature, good results were obtained for many polygons with a small number of vertices. However, in general, this method of integration became inaccurate for $N = 12$ and NPTS $= 16$, where it produced inaccuracy of about 10^{-2}. The results became much worse in cases of troublesome polygons. To illustrate this situation, consider a problem of Figure 8.4, where we want to compute the integral (8.3.19) along the segment from z_k to some point p. In the parameter problem p might be 0 or z_{k-1}; or it might be any point in the unit disk. A direct application will then involve sampling the integrand at only NPTS nodes between z_k and p. If the singularity z_{k+1} is so close to the path of integration that the distance $\varepsilon = |z_{k+1} - z_k|$ is comparable to the distance between nodes, then Gauss-Jacobi formula will give an inaccurate result. Also, it is found that in Schwarz-Christoffel problems the correct spacing of prevertices z_k around the unit circle is typically very irregular; examples are given in Chapter 7.

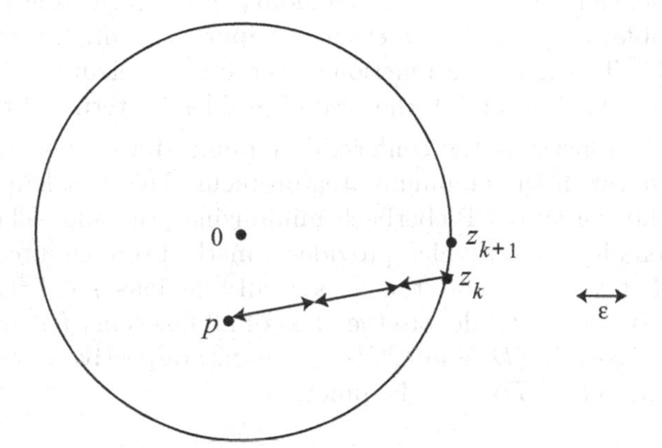

Figure 8.4 Compound Gauss-Jacobi quadrature.

This figure shows the division of an interval of integration into subintervals to maintain the desired resolution. To maintain high accuracy with much speed, a compound Gauss-Jacobi quadrature, provided in Davis and Rabinowitz [1975: 56], is adopted, with the quadrature principle

No singularity z_k shall lie closer to an interval of integration than half the length of that interval.

To achieve this principle, the quadrature subroutine ZQUAD must be able to divide an interval of integration into shorter subintervals as needed, working from the endpoints inward. On the short subinterval adjacent to the endpoint, Gauss-Jacobi quadrature is applied; on longer interval(s) away from the endpoint, pure Gaussian quadrature is applied. The effect of this procedure is that the number of integrand evaluations required to achieve a given accuracy is reduced from $O(1/\varepsilon)$ to $O\left(\log_2(1/\varepsilon)\right)$. For example, in the 12-vertex example of Figure 8.3, the switch to compound Gauss-Jacobi integration decreases the error from 10^{-2} to $2 \cdot 10^{-7}$.

There is only one case where the above-described integration by compound Gauss-Jacobi quadrature fails. This case involves an integration interval with one point very close to some prevertex z_k corresponding to a vertex $w_=\infty$. This integral cannot be computed, considering an interval that begins at z_k, because the integral then becomes infinite. However, a proper procedure in this case is integration by parts, which can reduce the singular integrand to one that is not infinite. Depending on the angle α_k, one would need one to three applications of integration by parts. However, this procedure has not been implemented in the subroutine ZQUAD.

(d) Solution of system by packaged solver (subroutine SCSOLV). After the above discussions, the unconstrained nonlinear system is ready to be computed. A library subroutine NSOIA (by Powell [1968]) is employed; it uses a steepest descent search in early iterations if necessary followed by a variant of Newton's method later on.

Fortran programs of all subroutines are available in Trefethen [1979: 48-55].

8.4 Minimum Area Problem

Gaier's variational method is used to solve two extremal problems in the theory of conformal mapping. The first deals with the conformal mapping of a simply connected region onto a disk, and the second with that of the boundary of the region onto the circumference of the disk. Both problems use the Ritz method for approximating the minimal mapping function by polynomials. This mapping function in the first problem is represented in terms of the Bergman kernel function, and in the second problem in terms of the Szegö kernel.

An extremal property in the conformal mapping of a simply connected region D onto a disk is connected with the minimum area problem. This problem, which we will denote as Problem I, is known as the Bieberbach minimizing principle. The mapping function possesses an extremal property which provides a method to compute an approximate solution for the map. Let $\mathcal{K}^1(D)$ denote the class of all functions $f \in L^2(D)$ with $f(a) = 1$, where $a \in D$. Similarly, let $\mathcal{K}^0(D)$ denote the class of all functions $f \in L^2(D)$ such that $f(a) = 0$. Note that the classes $\mathcal{K}^1(D)$ and $\mathcal{K}^0(D)$ represent, respectively, a closed convex subset and a closed subspace of $L^2(D)$. Let the function

$$w = f(z) = \sum_{n=0}^{\infty} a_n (z-a)^n, \quad |z-a| < R, \tag{8.4.1}$$

which is regular in D, map D onto the disk $B(0, R)$ in the w-plane. Without loss of generality, we will sometimes take the point a as the origin.

We will designate the *minimum area problem* as Problem I, defined as follows:

Problem I: In the class $\mathcal{K}^1$, minimize the integral

$$I = \iint_D |f'(z)|^2 \, dS_z. \tag{8.4.2}$$

The Riemann mapping theorem (Theorem 2.15) guarantees the existence and uniqueness of the solution of this extremal problem.

Theorem 8.3. *Problem I (minimum area problem) has a unique solution $f_0(z) = f'(z)$. The minimum is πR^2.*

PROOF. If $f_0(a) = 0$ and $f_0'(a) = 1$, then in view of (2.6.25)

$$\begin{aligned} \text{area}(D) &= \iint_D |f'(z)|^2 \, dS_z = \int_0^R \int_0^{2\pi} \left| f'\left(r\, e^{i\theta}\right)\right|^2 r \, dr \, d\theta \\ &= \int_0^R \sum_{n=0}^{\infty} |a_n|^2 n^2 2\pi r^{2n-1} \, dr \\ &= \pi R^2 |a_1|^2 + \pi \sum_{n=2}^{\infty} n \, |a_n|^2 \, R^{2n}. \end{aligned} \tag{8.4.3}$$

The above result implies that in the problem of mapping by the function (8.4.1), which is regular in $B(0, R)$ and is such that $f'(a) = a_1$, the area of the mapped region D is always greater than $\pi R^2 \, |a_1|^2$. It is exactly equal to this value if the map $f(z)$ is linear, i.e., if $w = a_0 + a_1 z$. A particular case is when $a_1 = 1$. Then the mapping function is $w = a_0 + z$. If this linear transformation is excluded, then the mapping function can be normalized by the conditions $f(a) = 0$ and $f'(a) = 1$, by considering the function $f(z)/a_1$. In either case the minimum area of D is πR^2. ∎

Before we solve Problem I, we will examine the minimal function $f_0(z)$ closely.

Theorem 8.4. *The function $f_0(z)$ is orthogonal to every function $g \in L^2(D)$ with $g(a) = 0$, i.e.,*

$$\langle f_0, g\rangle = \iint_D f_0(z)\,\overline{g(z)}\,dS_z = 0. \tag{8.4.4}$$

PROOF. For every $\varepsilon > 0$ and $0 \le \theta \le 2\pi$, the function $f_0(z) + \varepsilon g(z)$ belongs to the class $\mathcal{K}^1$. Then

$$\begin{aligned}\iint_D |f_0(z)|^2\,dS_z &\le \iint |f_0(z) + \varepsilon g(z)|^2\,dS_z \\ &= \iint_D |f_0(z)|^2\,dS_z + 2\varepsilon\,\Re\left\{\iint_D f_0(z)\,\overline{g(z)}\,dS_z\right\} + \varepsilon^2 \iint_D |g(z)|^2\,dS_z,\end{aligned}$$

which implies that

$$\Re\left\{\iint_D f_0(z)\,\overline{g(z)}\,dS_z\right\} + \frac{\varepsilon}{2}\iint_D |g(z)|^2\,dS_z \ge 0.$$

If (8.4.4) were false, then the above expression would be negative for sufficiently small $\varepsilon > 0$. ■

8.4.1 Bergman Kernel. If we take $g(z) = f_0(z) - f_0(a)$, then from (8.4.4) for $f \in \mathcal{K}^1(D)$ we have

$$\iint_D f_0(z)\,\overline{f(z)}\,dS_z = \overline{f(a)} \iint_D f_0(z)\,dS_z,$$

and if $f = f_0$, then

$$\|f_0\|^2 = \iint_D |f_0(z)|^2\,dS_z = \overline{f_0(a)} \iint_D f_0(z)\,dS_z.$$

If we introduce the Bergman kernel

$$K(z,a) = \frac{f_0(z)}{\|f_0\|^2}, \tag{8.4.5}$$

where $f_0(z)$ minimizes Problem I, then for every $f \in \mathcal{K}^1(D)$

$$\iint_D \overline{K(z,a)}\,f(z)\,dS_z = f(a). \tag{8.4.6}$$

Hence, every function $f \in L^2(D)$ is the eigenfunction of the integral equation (8.4.6) with eigenvalue $\lambda = f_0(a) = 1$. Then the minimal function $f_0(z)$ is, in view of (8.4.5), given by

$$f_0(z) = \frac{K(z,a)}{K(a,a)}, \tag{8.4.7}$$

where

$$K(a,a) = \iint_D |K(z,a)|^2\,dS_z = \frac{1}{\|f_0\|^2}. \tag{8.4.8}$$

Note that we cannot find $f_0(z)$ directly. We can find $f'(z)$ since it appears in the integrand in (8.4.2). Then the mapping function f is related to the Bergman kernel of D by

$$f(z) = \frac{\iint_D K(z,a)\, dS_z}{K(a,a)}. \tag{8.4.9}$$

8.5 Numerical Methods for Minimum Area Problem

We will study the Ritz method (RM) and the Bergman kernel method (BKM) for the minimum area problem (Problem I).

8.5.1 Ritz Method. The Ritz method is used to find the solution of the above extremal problem approximately in the form of a polynomial. Consider an arbitrary system of linearly independent functions $u_0(z), u_1(z), \dots$, which are regular in D and are such that $\iint_D |u_k(z)|^2\, dx\, dy < +\infty$ for $k = 0, 1, \dots$. We assume that one of these functions, say $u_0(z)$, is such that $u_0(a) \neq 0$. Without loss of generality, we take $a = 0$. Let $\{\phi_n(z)\}$ be a complete set of $L^2(D)$, and denote by $\mathcal{K}_n^0$ and $\mathcal{K}_n^1$ the n-dimensional counterparts of $\mathcal{K}^0$ and $\mathcal{K}^1$, respectively, i.e., if

$$\phi_n(z) = \sum_{k=0}^{n} c_k\, u_k(z), \tag{8.5.1}$$

then $\phi_n \in \mathcal{K}_n^0$ if $\phi(0) = 0$, and $\phi_n \in \mathcal{K}_n^1$ if $\phi_n(0) = 1$, where $c_k = \alpha_k + i\beta_k$, $c_0 \neq 0$, are complex numbers. Now we will consider the integral

$$I(\phi_n) = \iint_D |\phi_n(z)|^2\, dS_z, \tag{8.5.2}$$

which is the same as (8.4.1) except that the integrand in (8.5.2) is $\phi_n(z)$ instead of $f'(z)$.

Problem $\mathbf{I}_n$: In the class $\mathcal{K}_n^1$, minimize the integral $I(\phi_n)$ defined by (8.5.2).

Now we will discuss the existence and uniqueness of the minimal polynomial $\phi_n(z)$, determine $\phi_n(z)$, and approximate $f_0(z)$ by the minimal polynomial $\phi_n(z)$. The numerical value of the integral (8.5.2) is equal to the area of the image of the region D. Then the problem reduces to a choice of the coefficients c_k so that this value is a minimum among the values of the same integral for any other linear combination ψ_n of the functions $u_k(z)$, $k = 0, 1, \dots, n$, subject to the condition $\psi_n(0) = 1$. Suppose that

$$\psi_n(z) = \phi_n(z) + \varepsilon\, \gamma_n(z), \tag{8.5.3}$$

where ε is a complex number and $\gamma_n(z)$ is a linear combination of $u_k(z)$, $k = 0, 1, \dots, n$. The requirement that $\psi_n(0) = 1$ for all ε implies that $\gamma_n(0) = 0$. Now

$$\begin{aligned} I(\psi_n) = \iint_D |\phi_n|^2\, dx\, dy + \bar{\varepsilon} \iint_D \phi_n\, \bar{\gamma}_n\, dx\, dy \\ + \varepsilon \iint_D \bar{\phi}_n\, \phi_n\, dx\, dy + |\varepsilon|^2 \iint_D |\gamma_n|^2\, dx\, dy. \end{aligned} \tag{8.5.4}$$

The sign of the difference $I(\psi_n) - I(\phi_n)$ for small ε will depend on the linear terms in ε and $\bar{\varepsilon}$ because the last term in (8.5.4) is of order $O(\varepsilon^2)$. Thus, $I(\psi_n) - I(\phi_n) \geq 0$ iff the following orthogonality relations hold:

$$\iint_D \phi_n \, \bar{\gamma}_n \, dx \, dy = 0 \quad \text{and} \quad \iint_D \bar{\phi}_n \, \gamma_n \, dx \, dy = 0 \tag{8.5.5}$$

for any linear combination γ_n that satisfies the condition $\gamma_n(0) = 0$. Otherwise, we can always choose an ε which will make $I(\psi_n) < I(\phi_n)$, and this will contradict the minimal properties of $\phi_n(z)$. Note that the integrands in (8.5.5) are complex conjugates of each other, so we can use either as needed.

Conversely, if $\phi_n(z)$ satisfies the orthogonality relations (8.5.5), then $\phi_n(z)$ imparts the integral $I(\phi_n)$ its minimum value among the values imparted by all linear combinations of $\psi_n(z)$ with $\psi_n(0) = 1$. Thus, from (8.5.4) and (8.5.5) we get

$$I(\psi_n) - I(\phi_n) = |\varepsilon|^2 \iint_D |\gamma_n|^2 \, dx \, dy. \tag{8.5.6}$$

Hence, the polynomial $\phi_n(z)$ will be unique if it exists, since the integral on the right side of (8.5.6) is equal to zero only when $\gamma_n = 0$, i.e., when $\psi_n = \phi_n$. Thus, the orthogonality relations (8.5.5) constitute necessary and sufficient conditions for $\phi_n(z)$ to be the minimal polynomial.

We will rewrite the conditions (8.5.5) in a different but equivalent form. Let

$$v_k(z) = u_k(z) - \frac{u_k(0)}{u_0(0)} \, u_0(z), \quad k = 1, \dots, n. \tag{8.5.7}$$

Then each of the functions $v_k(z)$ satisfies the requirements imposed on $\gamma_k(z)$, and the conditions (8.5.5) become

$$\sum_{j=0}^{n} A_{kj} \, c_j = 0, \tag{8.5.8}$$

where

$$A_{kj} = \iint_D \bar{u}_j \, v_k \, dx \, dy = \iint_D u_j \, \bar{v}_k \, dx \, dy, \quad k = 1, \dots, n. \tag{8.5.9}$$

Also, since

$$\sum_{k=0}^{n} u_k(0) \, c_k = 1, \tag{8.5.10}$$

Eqs (8.5.8) and (8.5.10) provide us with a system of $(n+1)$ equations to determine the numbers $c_0, c_1, \dots, c_n$ uniquely.

Note that the functions $u_k(z)$, though linearly independent, are still undetermined and are not subject to limitation. However, it becomes very easy to determine the integrals in (8.5.9) if all $u_k(z)$ are suitably chosen beforehand such that $u_0(0) = 1$, and $u_k(0) = 0$ for $k = 1, \dots, n$. Then $v_k(z) = u_k(z)$, and the integral

$$\iint_D u_j \, \bar{v}_k \, dx \, dy = \iint_D u_j \, \bar{u}_k \, dx \, dy.$$

Also, Eq (8.5.10) degenerates to $c_0 = 1$, which reduces the number of unknowns by one. Moreover, if the system of functions $u_0(z), u_1(z), \dots$ is orthogonal in D, i.e., if

$$\iint_D u_j \, \bar{u}_k \, dx \, dy = 0 \quad \text{for } k \neq j,$$

then

$$\iint_D u_j \, \bar{v}_k \, dx \, dy = 0 \quad \text{for } j \neq 0 \text{ and } k \neq j,$$

and

$$\iint_D u_0 \, \bar{v}_k \, dx \, dy = -\frac{\overline{u_k(0)}}{\overline{u_0(0)}} \iint_D u_0 \, \bar{u}_0 \, dx \, dy,$$

$$\iint_D u_k \, \bar{v}_k \, dx \, dy = \iint_D u_k \, \bar{u}_k \, dx \, dy.$$

Then Eqs (8.5.8)-(8.5.10) simplify to

$$c_k \iint_D |u_k|^2 \, dx \, dy - c_0 \frac{\overline{u_k(0)}}{\overline{u_0(0)}} \iint_D |u_0|^2 \, dx \, dy = 0. \tag{8.5.11}$$

Thus, we have proved

Theorem 8.5. *The problem I_n has a unique solution. The coefficients of the minimal polynomial $\phi_n(z)$, defined by (8.5.1), are determined by the system of linear equations (8.5.8)-(8.5.9). The coefficients A_{kj}, defined by (8.5.9), are hermitian.*

Let the system $\{u_k\}_{k=0}^n$ be taken as the complete set $\{1, z, z^2, \dots, z^n\}$. Then for the minimal polynomial

$$\phi_n(z) = 1 + c_1 \, z + c_2 \, z^2 + \dots + c_n \, z^n, \tag{8.5.12}$$

the system of equations (8.5.8)-(8.5.9), which determine the n coefficients c_k, becomes

$$\begin{bmatrix} A_{10} & A_{11} & A_{12} & \cdots & A_{1n} \\ A_{20} & A_{21} & A_{22} & \cdots & A_{2n} \\ \cdots & \cdots & \cdots & \cdots & \cdots \\ A_{n0} & A_{n1} & A_{n2} & \cdots & A_{nn} \end{bmatrix} \begin{Bmatrix} 1 \\ c_1 \\ c_2 \\ \vdots \\ c_n \end{Bmatrix} = \{0\}, \tag{8.5.13}$$

where the coefficients A_{kj} are given by

$$A_{kj} = \iint_D z^k \, \overline{z^j} \, dS_z, \quad A_{kj} = \bar{A}_{jk}. \tag{8.5.14}$$

Note that if $a \neq 0$, then the coefficients A_{kj} are determined from

$$A_{kj} = \iint_D (z-a)^k \, \overline{(z-a)^j} \, dS_z, \quad A_{kj} = \bar{A}_{jk}. \tag{8.5.15}$$

The computation of the coefficients A_{kj} depends on the region D, although it may sometimes present difficulties.

Example 8.1. Let the region D be starlike with respect to a point $a \neq 0 \in D$, i.e., every ray emanating from the point a intersects the boundary in only one point. Let the equation of the boundary be $r = r(\theta)$. Using the polar coordinates $z - a = r\, e^{i\theta}$, we get

$$\begin{aligned} A_{kj} &= \int_0^{2\pi} \int_0^{r(\theta)} r^{j+k}\, e^{i(j-k)\theta}\, r\, dr\, d\theta = \frac{1}{j+k+2} \int_0^{2\pi} r^{j+k+2}(\theta)\, e^{i(j-k)\theta}\, d\theta \\ &= \frac{1}{j+k+2} \int_0^{2\pi} r^{j+k+2}(\theta)\, \cos(j-k)\theta\, d\theta \\ &\quad + \frac{i}{j+k+2} \int_0^{2\pi} r^{j+k+2}(\theta)\, \sin(j-k)\theta\, d\theta. \end{aligned} \tag{8.5.16}$$

Note that $\Re\{A_{kj}\}$ and $\Im\{A_{kj}\}$ differ from the coefficients of the Fourier series for $r^{j+k+2}(\theta)$ by a factor $\dfrac{\pi}{j+k+2}$, and hence, they can be easily computed. ■

DEFINITION 8.5.1. The system of polynomials $\{\phi_n\}$ is said to be *complete* in the Hilbert space $L^2(D)$ if for every function $f \in L^2(D)$ and every $\varepsilon > 0$ there exists a polynomial ϕ_n such that $\|f - \phi_n\| < \varepsilon$.

Now the question arises, under what additional assumptions on D the polynomials $\phi_n(z)$ form a complete system in $L^2(D)$. Naturally, $\|f_0 - \phi_n\| \searrow 0$ must hold, and thus also $\phi_n(z) \to f_0(z)$ as $n \to \infty$ in any closed subset $\bar{G} \subset D$.

Theorem 8.6. *Let the polynomial $p(z)$ belong to the class $\mathcal{K}_n^1$. Then*

$$\|f_0 - p\|^2 = \iint_D |f_0(z) - p(z)|^2\, dS, \tag{8.5.17}$$

is minimal only if $p(z) = \phi_n(z)$.

PROOF. In view of (8.4.4)

$$\langle f_0, p \rangle = \langle f_0, p - f_0 \rangle + \langle f_0, f_0 \rangle = \langle f_0, f_0 \rangle,$$

thus,

$$\langle f_0 - p, f_0 - p \rangle = \langle f_0, f_0 \rangle - 2\langle f_0, f_0 \rangle + \langle p, p \rangle = \|p\|^2 - \|f_0\|^2,$$

and the result follows since $\|p\|^2$ is minimal if $p = \phi_n$. ■

Hence, among all $p \in \mathcal{K}_n^1$ the polynomial ϕ_n yields the minimum norm $\|f_0 - p\|$, and since $\mathcal{K}_{n+1}^1 \supset \mathcal{K}_n^1$, then $\|f_0 - p\| \searrow 0$ only if the system of polynomial $\{\phi_n\}$ is complete in the space $L^2(D)$. Moreover, in Problem I_n there exists a polynomial $p = \phi_n$ which satisfies the additional condition $p(a) = f(a)$. Also note that the Bieberbach polynomial $\pi_n(z)$ is defined by

$$\pi_n(z) = \int_a^z \phi_{n-1}(\zeta)\, d\zeta \to f(z) \quad \text{as } n \to \infty \text{ in } \bar{G} \subset D. \tag{8.5.18}$$

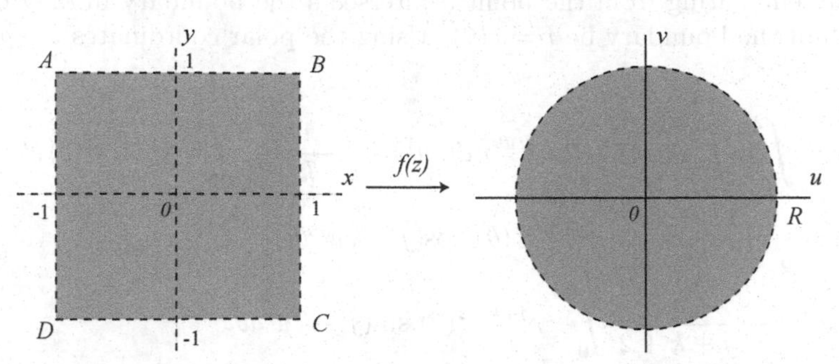

Figure 8.5 Square onto a circle.

Example 8.2. Determine the minimal polynomial $\phi_n(z)$ and the approximate mapping function $f(z)$ that maps the square region $D = \{x, y : -1 < x,\, y < 1\}$ conformally onto $|w| < 1$ (Figure 8.5). From `cs422.nb` (see Notes at the end of this chapter), the minimum polynomial is given by

$$\phi_8(z) = 1 + \frac{97402305}{266254834}\, z^4 + \frac{68765697}{2130038672}\, z^8,$$

which yields the approximate mapping function

$$\begin{aligned} f(z) &\approx \int_0^z \phi_8(t)\, dt \\ &= z + \frac{19480461}{266254834}\, z^5 + \frac{7640633}{2130038672}\, z^9 \\ &\approx z + 0.0731647\, z^5 + 0.00358709\, z^9. \blacksquare \end{aligned}$$

8.5.2 Bergman Kernel Method. In this method the mapping function $f(z)$ is determined approximately from (8.4.19) by first approximating the kernel $K(z, 0)$ by a finite Fourier sum. Let $\{\phi_j^*(z)\}$ denote a complete orthonormal set of $L^2(D)$. Consider the Fourier series expansion of $K(z, 0)$. Then, in view of (8.4.6),

$$\langle K, \phi_j^* \rangle = \iint_D K(z, 0)\, \overline{\phi_j^*(z)}\, dS_z = \overline{\phi_j^*(0)}. \tag{8.5.19}$$

Thus, the kernel has a series expansion

$$K(z, 0) = \sum_{j=1}^{\infty} \overline{\phi_j^*(0)}\, \phi_j^*(z), \tag{8.5.20}$$

which converges in the mean of $L^2(D)$, i.e., the series (8.5.20) converges almost uniformly in D.

Hence, if we have a complete set $\{\phi_j(z)\}$ of $L^2(D)$, then by using (8.4.19) and (8.5.20) we obtain an approximate mapping function $f(z)$ as follows:

(i) Orthonormalize the set $\{\phi_j(z)\}_{j=1}^{n}$ by using the Gram-Schmidt process which yields the set of orthonormal functions $\{\phi_j^*(z)\}_{j=1}^n$. Note that the Gram-Schmidt process requires evaluating $\langle \phi_i, \phi_j \rangle$, which, in view of Green's formula (2.3.11), is given by

$$\langle \phi_i, \phi_j \rangle = \iint_D \phi_i(z)\,\overline{\phi_j(z)}\,dS_z = \frac{1}{2i}\int_\Gamma \phi_i(z)\,\overline{\psi_j(z)}\,dz, \tag{8.5.21}$$

where $\psi_j'(z) = \phi_j(z)$. Then the integrals in (8.5.21) are computed by Gaussian quadrature.
(ii) Truncate the series (8.5.20) after n terms to obtain the approximation $K_n(z,0)$ of $K(z,0)$ as

$$K_n(z,0) = \sum_{j=1}^n \overline{\phi_j^*(0)}\,\phi_j^*(z). \tag{8.5.22}$$

(iii) Use (8.4.19) to obtain the approximate mapping function $f_n(z)$ as

$$f_n(z) = \frac{\int_0^z K_n(z,0)\,dz}{K_n(0,0)}. \tag{8.5.23}$$

(iv) The approximate radius R_n of the disk $|w| < R$ is given by

$$R_n = \frac{1}{\sqrt{\pi\,K_n(0,0)}}, \tag{8.5.24}$$

since $\|f_0\|^2 = \pi\,R^2$ (because $f_0 \in \mathcal{K}^1$).
(v) Thus, from (8.5.23) and (8.5.24) the approximation of the mapping function $F(z)$ that maps D conformally onto the unit disk $|w| < 1$ is given by

$$F_n(z) = \sqrt{\frac{\pi}{K_n(0,0)}}\int_0^z K_n(z,0)\,dz. \tag{8.5.25}$$

The maximum error estimate for $|F_n(z)|$ is given by

$$E_n = \max_j |e_n(z_j)|,$$

where $z_j \in \Gamma$ are the test points on the boundary and $e_n(z) = 1 - |F_n(z)|$. During the computation process the number n of the basis function is increased by one each time and this process is terminated when the inequality $E_{n+1} < E_n$ no longer holds. Then such a number n is taken as the 'optimum number' for the basis functions.

Note that in both RM and BKM we have obtained approximations of the form

$$f_n(z) = \sum_{j=1}^n a_j\,u_j(z),$$

where $u_j'(z) = \phi_j(z)$. In both methods the set of monomials z^{j-1}, $j = 1, 2, \dots$, which is a complete set in $L^2(D \cup \Gamma)$ is the best choice of basis functions in computation. Then this basis gives the polynomials $\phi_j(z)$ defined in (8.5.13).

Map 8.2. The function $w = f(z)$ that maps the unit disk U onto itself such that the point $z_0 \in U$ goes into the origin of the w-plane is given by

$$w = f(z) = \frac{z - z_0}{z - 1/\bar{z}_0}.$$

Thus, both $f(z)$ and the associated Bergman kernel function $K(z, z_0)$ have a pole at $z = 1/\bar{z}_0$. Since the polynomials $\phi_j^*(z) = \sqrt{j/\pi}\, z^{j-1}$, $j = 1, 2, \dots$, form a complete orthonormal basis set of U, the kernel $K(z, z_0)$ can, in view of (8.5.20), be represented by the polynomial series

$$K(z, z_0) = \frac{1}{\pi} \sum_{j-1}^{\infty} j \, (\bar{z}_0 \, z)^{j-1},$$

which converges rapidly when $|z_0|$ is small, but the convergence becomes considerably slower the faster $|z_0| \to 1$, i.e., the closer the pole $1/\bar{z}_0$ gets to the boundary of U. ∎

8.6 Minimum Boundary Problem

An analogous minimum problem in the conformal mapping of a region D onto $|w| < R$ leads to another characterization of the mapping function $f(z)$ by considering the line integral

$$I = \int_\Gamma |f(z)|^2 \, ds, \tag{8.6.1}$$

where Γ is the boundary of D. This problem, studied by Julia [1931], is known as the minimum boundary problem which we will call Problem II. Let Γ be a rectifiable Jordan curve, and let $\mathcal{L}^1(\Gamma)$ denote the class of all functions $f \in L^2(\Gamma)$ with $f(a) = 1$, where $a \in D$ can be taken as the origin.

Problem II: In the class $\mathcal{L}^1(\Gamma)$ minimize the integral (8.6.1).

Theorem 8.7. *Problem II has a unique solution $f_0(z)$, and it is $f_0(z) = \sqrt{f'(z)}$. The minimum is $2\pi R$.*

PROOF. For every function $F \in L^2(\Gamma)$

$$\begin{aligned}\int_\Gamma |F(z)|^2 \, ds &= \lim_{r \to R} \int_{\Gamma_r} |F(z)|^2 \, ds = \lim_{r \to R} \int_{|w|=r} |F(g(w)) \sqrt{g'(w)}|^2 \, |dw| \\ &= \lim_{r \to R} r \int_0^{2\pi} |h\left(r \, e^{i\theta}\right)|^2 \, d\theta,\end{aligned}$$

where $h(w) = F(g(w)) \sqrt{g'(w)}$, $h(0) = 1$, and $z = g(w)$ (see (2.3.8)). If $h(w) = \sum_{n=0}^{\infty} a_n \, w^n$, $a_0 = 1$, then

$$\int_\Gamma |F(z)|^2 \, ds = 2\pi \sum_{n=0}^{\infty} |a_n|^2 \, R^{2n+1} \geq 2\pi R,$$

where the equality holds only for $a_n = 0$, $n > 0$, which yields $F(z) = \sqrt{f'(z)}$ for $h(w) = 1$. ∎

This theorem implies that in the class of all conformal mappings $w = \phi(z)$ of the region D with $\phi(a) = 0$, $\phi'(a) = 1$, the integral $\int_\Gamma |\phi'(z)| \, ds$ is minimum only when $\phi(z) = f(z)$. This is known as the *principle of minimizing the image boundary*.

The conformal map f of D onto $|w| < R$, normalized by $f(a) = 0$, $f'(a) = 1$, is given by

$$f(z) = \int_a^z [f_0(\zeta)]^2 \, d\zeta.$$

The theory for Problem II is developed exactly on the same lines as in §8.4 and §8.5. Thus, as in (8.4.4), we have

Theorem 8.8. *The function* $f_0(z)$ *is orthogonal to every function* $g \in L^2(\Gamma)$ *with* $g(a) = 0$, *i.e.*,

$$\langle f_0, g\rangle = \int_\Gamma f_0(z)\,\overline{g(z)}\,ds = 0. \tag{8.6.2}$$

For the minimal function $f_0(z)$ of Problem II we introduce the Szegö kernel function

$$S(z,a) = \frac{f_0(z)}{\int_\Gamma |f_0(z)|^2\,ds}, \tag{8.6.3}$$

with the properties

$$S(a,a) = \int_\Gamma |S(z,a)|^2\,ds, \quad \text{and} \quad f_0(z) = \frac{S(z,a)}{S(a,a)}. \tag{8.6.4}$$

For any $f \in L^2(\Gamma)$, we apply (8.6.2) to $f(z) - f(a)$ and get

$$\int_\Gamma f_0(z)\,\overline{f(z)}\,ds = \overline{f(a)} \int_\Gamma f_0(z)\,ds,$$

and if $f = f_0$, then, since $h(w) = 1$ implies $f(a) = 1 = \overline{f(a)}$, we have

$$\int_\Gamma |f_0(z)|^2\,ds - \int_\Gamma f_0(z)\,ds.$$

Thus, analogous to (8.4.6) we have: For every function $f \in L^2(\Gamma)$

$$\int_\Gamma \overline{S(z,a)}\,f(z)\,ds = f(a). \tag{8.6.5}$$

Theorems 8.3 and 8.7 together with the definitions (8.4.5) and (8.6.3) yield

$$K(z,a) = 4\pi\,[S(z,a)]^2\,. \tag{8.6.6}$$

Thus, $S(z,a)$ can be evaluated by this method.

8.6.1 Ritz Method for Problem II. Let $\mathcal{L}_n^1$ denote the class of all polynomials $p(z)$ of degree $\le n$ with $p(a) = 1$.

Problem II$_n$: In the class $\mathcal{L}_n^1$ minimize the line integral $\displaystyle\int_\Gamma |p(z)|^2\,ds$.

We will discuss the existence and uniqueness of the minimal polynomial $\phi_n(z)$, determine $\phi_n(z)$, and approximate $f_0(z)$ by the minimal polynomial $\phi_n(z)$ and $f(z)$ by the integral $\displaystyle\int_a^z [\phi_n(\zeta)]^2\,d\zeta$, respectively. As in §8.5, it can be shown that a minimal polynomial $\phi_n(z)$ exists for Problem II, that it is unique, and that it is characterized by

$$\langle \phi_n, g\rangle = 0 \tag{8.6.7}$$

for every polynomial $g(z)$ of degree $\leq n$ with $g(a) = 0$. The proof is analogous to that of (8.4.4).

In view of (8.6.7), the coefficients of the minimal polynomial

$$\phi_n(z) = 1 + a_1 (z - a) + \cdots + a_n (z - a)^n \tag{8.6.8}$$

are determined by

$$\int_\Gamma \left(\sum_{k=0}^{n} a_k (z - a)^k \right) \overline{(z - a)^j}\, ds = 0, \quad j = 1, 2, \dots, n, \tag{8.6.9}$$

and $a_0 = 1$. If we set

$$B_{kj} = \int_\Gamma (z - a)^k \, \overline{(z - a)^j}\, ds, \quad k, j = 0, 1, \dots, \tag{8.6.10}$$

then the coefficients of the minimal polynomial $\phi_n(z)$ defined by (8.6.8) are determined from the (consistent) system of equations

$$\sum_{k=0}^{n} B_{kj}\, a_k = 0, \quad a_0 = 1, \quad j = 1, \dots, n. \tag{8.6.11}$$

For any arbitrary polynomial $p \in \mathcal{L}_n^1$, we have, in view of (8.5.2),

$$\langle f_0 - p \rangle = \langle f_0, p - f_0 \rangle + \langle f_0, f_0 \rangle = \langle f_0, f_0 \rangle,$$

and hence

$$\| f_0 - p \|^2 = \|p\|^2 - \|f_0\|^2, \tag{8.6.12}$$

which implies that $p(z)$ has the minimal property:

$$\int_\Gamma |f_0(z) - p(z)|^2\, ds \quad \text{is minimum in } \mathcal{L}_n^1 \text{ for } p(z) = \phi_n(z). \tag{8.6.13}$$

Again, as in §8.5, $\|f_0 - \phi_n\| \searrow 0$ as $n \to \infty$, since $\mathcal{L}_{n+1}^1 \supset \mathcal{L}_n^1$. Then we ask, under what assumptions on D is the system of polynomials $\{\phi_n\}$ complete in the Hilbert space $L^2(\Gamma)$? An answer was given by Smirnov [1928] as

Theorem 8.9. *The system of polynomials in $L^2(\Gamma)$ is complete iff the boundary Γ of the region D satisfies the condition*

$$\log |g'(w)| = \frac{1}{2\pi} \int_0^{2\pi} \log \left| g'\left(r\, e^{i\theta}\right) \right| \frac{R^2 - r^2}{R^2 - 2Rr \cos(\alpha - \theta) + r^2}\, d\alpha, \tag{8.6.14}$$

for all $r < R$, where $z = g(w)$ is the inverse mapping function which maps the region $|w| < R$ onto the region D in the z-plane.

A proof of this theorem can be found in Goluzin [1957:396] or [1969:449]. The condition (8.6.14) is known as the S-condition. This condition depends only on the region D and not on the normalization of $g(w)$ or on the choice of $a \in D$. All such regions D whose mapping satisfies the S-condition (8.6.14) are said to belong to the *Smirnov class* $\mathcal{S}$. Not all regions with a rectifiable boundary belong to the class $\mathcal{S}$. Besides the rectifiability of Γ, however, it is sufficient for $D \in \mathcal{S}$ if one of the following conditions is met: (i) D is convex or starlike with respect to a point in D; (ii) Γ is piecewise smooth, and its smooth arcs γ_k, $k = 1, \dots, n$, join one another with a nonzero interior angle; (iii) the ratio of the length of any arc γ_k of $\Gamma = \cup \gamma_k$ to the length of its chord does not exceed a fixed limit; (iv) if $D \notin \mathcal{S}$, then the behavior of $\|\phi_n\|^2$ and of $\|f_0 - \phi_n\|^2 = \|\phi_n\|^2 - \|f_0\|^2$ is known; and (v) for any D with a rectifiable boundary Γ, the integral $\int_\Gamma |\phi_n(z)|^2\, ds \searrow 2\pi R\delta$, where

$$\delta = \exp\left\{\frac{1}{2\pi}\int_0^{2\pi} \log\left|g'\left(R\, e^{i\theta}\right)\right|\, d\theta\right\}. \tag{8.6.15}$$

Thus, under the conditions (i), (ii), or (iii) we have $\|f_0 - \phi_n\| \searrow 0$, and, in view of Theorem 8.4, $\phi_n(z) \to f_0(z)$ as $n \to \infty$ for every region $G \subset D$. Hence, for $z \in G \subset D$ the polynomial

$$\pi_{2n+1} = \int_a^z \left[\phi_n(\zeta)\right]^2 d\zeta \quad \to f(z) \quad \text{as } n \to \infty. \tag{8.6.16}$$

Note that the polynomial $\pi_{2n+1}(z)$ is different from the Bieberbach polynomial (8.5.18).

Theorem 8.10. *For the system of polynomials to be complete in a region D, it is necessary and sufficient that an arbitrary function $F(z) \in L^2(\Gamma)$ satisfies the condition*

$$\int_\Gamma |F(z)|^2\, ds = \sum_{k=1}^{n} |a_k|^2, \tag{8.6.17}$$

where a_k are the Fourier coefficients of $F(z)$.

PROOF. Let $p(z)$ be an arbitrary polynomial of degree n:

$$p(z) = \sum_{k=0}^{n} c_k\, u_k(z). \tag{8.6.18}$$

Then for $F(z) \in L^2(\Gamma)$

$$\begin{aligned} \int_\Gamma |F(z) - p(z)|^2\, ds &= \int_\Gamma |F(z)|^2\, ds + \int_\Gamma |p(z)|^2\, ds - 2\,\Re\left\{\int_\Gamma F(z)\,\overline{p(z)}\, ds\right\} \\ &= \int_\Gamma |F(z)|^2\, ds + \sum_{k=0}^{n} |c_k|^2 - 2\,\Re\left\{\sum_{k=0}^{\infty} a_k\, c_k\right\} \\ &= \int_\Gamma |F(z)|^2\, ds - \sum_{k=0}^{n} |c_k|^2 + \sum_{k=0}^{n} |a_k - c_k|^2. \end{aligned} \tag{8.6.19}$$

Thus, the polynomial $p(z)$ defined by (8.6.18) with $c_k = a_k$ attains the minimum of the integral

$$\int_\Gamma |F(z) - p(z)|^2\, ds, \tag{8.6.20}$$

and, in view of (8.6.18), this minimum yields

$$\int_\Gamma |F(z)|^2\, ds - \sum_{k=0}^{n} |c_k|^2 = 0. \tag{8.6.21}$$

Hence, a system of polynomials is complete iff the difference (8.6.21) approaches zero as $n \to \infty$ for an arbitrary function $F \in L^2(\Gamma)$. ∎

Theorem 8.11. *If the S-condition (8.6.14) is satisfied, then an arbitrary function $F(z) \in L^2(\Gamma)$ can be represented in D by a Fourier series*

$$F(z) = \sum_{k=0}^{\infty} a_k\, u_k(z), \tag{8.6.22}$$

which converges uniformly in D, where

$$a_k = \int_\Gamma f(w)\, \overline{u_k(w)}\, ds. \tag{8.6.23}$$

PROOF. Since the minimum of the integral (8.6.20) is attained out of all polynomials $p_n(z)$ by a polynomial defined by (8.6.18) with $c_k = a_k$, $k = 0, 1, \dots, n$, then, if the S-condition is satisfied, we have

$$\lim_{n\to\infty} \int_\Gamma |F(\zeta) - p_n(\zeta)|^2\, ds = 0. \tag{8.6.24}$$

But by Cauchy's formula in D

$$F(z) - p_n(z) = \frac{1}{2i\pi} \int_\Gamma \frac{F(\zeta) - p_n(\zeta)}{\zeta - z}\, d\zeta.$$

If $\bar{G}$ is a closed subset of D and δ is the distance from G to Γ, then for $z \in \bar{G}$

$$\begin{aligned} |F(z) - p_n(z)| &\le \frac{1}{2\pi\delta} \int_\Gamma |F(\zeta) - p_n(\zeta)|\, ds \\ &\le \frac{1}{2\pi\delta} \sqrt{l \int_\Gamma |F(\zeta) - p_n(\zeta)|^2\, ds}, \end{aligned} \tag{8.6.25}$$

where l is the length of Γ, and this, in view of (8.6.24), implies that $p_n(z) \to F(z)$ uniformly on $\bar{G}$ as $n \to \infty$. But $p_n(z)$ is a finite part of the Fourier series (8.6.22), which proves the theorem. ∎

Example 8.3. Let E denote the ellipse $\dfrac{x^2}{a^2} + \dfrac{y^2}{b^2} = 1$. The Bergman kernel for E has the form

$$K(z, a) = \frac{4}{\pi} \sum_{n=0}^{\infty} \frac{(n+1) U_n(z)\, \overline{U_n(a)}}{\rho^{n+1} - \rho^{-n-1}}, \quad \rho = (a+b)^2,$$

where $U_n(z) = \left(1 - z^2\right)^{-1/2} \sin\left((n+1)\cos^{-1} z\right)$ are the Chebyshev polynomials of the second kind and degree n (Nehari [1952: 258-259]). *bull*

Example 8.4. Let the arc Γ be defined by the ellipse $E : \dfrac{(x - x_c)^2}{a^2} + \dfrac{(y - y_c)^2}{b^2} = 1$, $a > b$, and let the parametric equation of Γ be $z = \gamma(s) = z_c + a\, e\, \cos(s - i\, q)$, $0 \le s_1 < s < s_2 < 2\pi$, where $z_c = x_c + i\, y_c$ is the center C , $e = \sqrt{1 - b^2/a^2}$ the eccentricity of the ellipse, $\cosh q = 1/e$, and $s_2 - s_1 < 2\pi$. Then the function $z = \gamma(\zeta)$, $\zeta = s + i\, t$, is univalent in the strip $\{\zeta : \zeta + s + i\,t,\ s_1 < s < s_2,\ -\infty < t < q\}$, and the region G^* is a symmetric subregion of the rectangle $\{\zeta : \zeta = s + i\,t,\ s_1 < s < s_2,\ -q < t < q\}$ (Papamichael, Warby and Hough [1983: 157]). *bull*

8.7 Numerical Parameter Problem

The Schwarz-Christoffel transformation defined by (7.1.7) or (7.1.15) is

$$z(w) = C_1 \int_{w_0}^{w} \prod_{j=1}^{n} (t - w_j)^{-\alpha_j}\, dt + C_2. \tag{8.7.1}$$

The problem is to determine all w_j and the constants C_1 and C_2. There are different methods to evaluate these parameters, e.g., those described in von Koppenfels and Stallman [1959], Kantorovich and Krylov [1964], Gaier [1964], and Henrici [1986: vol. 3, ch. 16], and Bjørstad and Grusse [1987]. We will use the following four methods:

8.7.1 Parameter Method. The parameters w_j, $j = 1, 2, \dots, n$ and C_1, C_2 are determined by using the boundary conditions, i.e., by equating values of $z(w)$ in the z-plane to the values of the integral evaluated at the corresponding points w_j on the v-axis. Although certain cases have been determined exactly in Chapter 6, numeral methods are needed when n is large. First, we assume that we have reduced the problem such that $z(0) = 0$, which yields $C_2 = 0$. Next, let the distance between two adjacent vertices z_k and z_l of the polygon be denoted by Δz_{kl} and the corresponding values determined from integration by ΔF_{kl}, i.e.,

$$\Delta z_{kl} = |z_l - z_k|; \quad \Delta F_{kl} = C_1 \int_{z_k}^{z_l} \prod_{j=1}^{n} (t - w_j)^{-\alpha_j}\, dt. \tag{8.7.2}$$

Now, let the next vertex be z_m. Then equating the ratios of the lengths Δz_{kl} and Δz_{lm} we get

$$\frac{\Delta z_{kl}}{\Delta z_{mk}} = \frac{\Delta F_{kl}}{\Delta F_{mk}}. \tag{8.7.3}$$

This equation results in a set of simultaneous, nonlinear equations which can be solved using the least-error-squared condition. For example, if the error is denoted as $e = \Delta z - \Delta F$, we can minimize the sum of all the e^2. Powell [1964] provided the method of conjugate direction, which was used by Lawrenson and Gupta [1968] to develop a minimization algorithm, who found that displacement of limits in the order of 0.001 and 0.005 were sufficient to get accurate results in small intervals.

8.7.2 van Dyke's Method. (van Dyke [1975: 96]) This method deals with the following situation: Since a vertex, say z_j, of the polygon is mapped onto the point w_j on the u-axis, each finite w_j contributes a term $(t - w_j)^{-\alpha_j}$ in the Schwarz-Christoffel transformation (8.7.1). If an approximate, yet almost accurate, solution is required in the interval $w_j^- < w < w_j^+$, van Dyke's method works with accuracy of the order of 0.001 to 0.005 as determined by Lawrenson and Gupta [1968] and Binns [1971]. This method uses the fact that the factor $(t - w_j)^{-\alpha_j}$ in (8.7.1) predominates as the singular point w_j is approached by w, while the other factors $(t - w_m)^{-\alpha_m}$, $m \neq j$, remain almost constant in their magnitude and direction as integration is taken from w_m^- to w_m^+. Thus, the difference

$$\Delta z_m = z_m^+ - z_m^- = C_1 \int_{w_m^-}^{w_m^+} \prod_{j=1}^{n} (t - w_j)^{-\alpha_j}\, dt \tag{8.7.4}$$

changes to

$$\Delta z_m = C_1 \Big\{ \prod_{m \neq j} (w_m - w_j)^{-\alpha_j} \Big\} \frac{(w - w_m)^{1-\alpha_m}}{1 - \alpha_m} \Bigg|_{w_m^-}^{w_m^+}. \tag{8.7.5}$$

However, this method will fail in the following cases and should be avoided:

(i) The factor $1 - \alpha_m$ (in the denominator in (8.7.5)) will remain nonnegative since any vertex with $|\alpha_m| > 1$ occurring at infinity in the z-plane will be dropped out from the integral. The difficulty would arise only when this factor itself becomes infinitely large when $\alpha_m = 1$. Such situation normally arises in idealized representations, e.g., of capacitor plates or air foils of zero thickness where the exterior angles of π as the boundaries are traversed in the counter-clockwise direction. However, extremely narrow slits are acceptable since they have $\alpha = -\pi$.

(ii) This method will give inaccurate results in the case when the singular points are dense on the u-axis. This situation may require appropriate scaling and placements of such points.

(iii) If the images of the vertices in the z-plane occur on the u-axis at infinity, the integration in (8.7.4) will be taken between infinitely large intervals of w. Imagine the case when a vertex in the z-plane is at infinity. Then the open polygon can be transformed into a symmetric, closed polygon, as seen in Figure 8.6, due to Davis [1979].

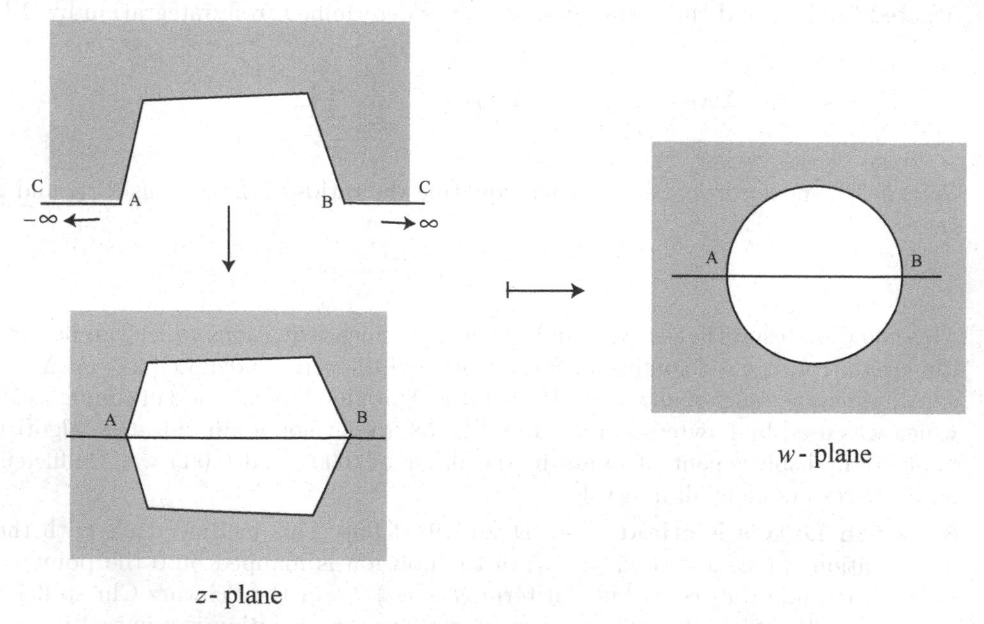

Figure 8.6 Use of symmetry to avoid vertices at infinity.

8.7.3 Trefethen's Method. (Trefethen [1980]) Instead of mapping the z-plane onto the upper-half w-plane, this method is based on the mapping of the polygon in the z-plane onto the unit disk $|w| \leq 1$ (see §7.1.3), and this is the advantage over other methods. Thus, all vertices z_j of the polygon are mapped onto the unit circle. For example, a vertex z_j will correspond to $w_j = e^{i\Theta_j}$, thereby Θ_j will be the only parameter to be determined. For an efficient integration of Eq (8.7.1) the endpoints w_j should not be too close to singular points which are images of the other vertices. To achieve it, a careful spacing of the angles

$\Theta_{j-1}, \Theta_j, \Theta_{j+1}$ should be maintained while retaining their power sequence. In the words of Trefethen, "no singularity shall lie closer to an interval of integration than half the length of that interval." A computer program SCPACK, prepared by Trefethen, evaluates the integrals using Gauss-Jacobi quadrature, and the nonlinear equations (8.7.5) are solved using an iterative procedure that is based on an initial guess and subsequent corrections using Powell's minimization method (Fletcher and Powell [1963], Powell [1964]).

Trefethen has solved some practical problems, like flat polygonal registers (Trefethen [1984]), and Hall generator elements which involves oblique derivative boundary conditions (Trefethen and Williams [1986]). Later, Elcrat and Trefethen [1986] extended this method to open polygons in free-streamline flow problems without using the log-hodograph method (see §8.4 for minimum area problem).

Dias [1986] has applied this method to dividing-flows and point sources (singularities) such as wells. However, instead of using the Gauss-Jacobi quadrature, he has used change of variable to remove singularities, and thus, deals efficiently with vertices at infinity.

8.7.4 Foster-Anderson's Method. (Foster and Anderson [1974]) This method uses the elliptic functions to numerically evaluate the Schwarz-Christoffel integral.

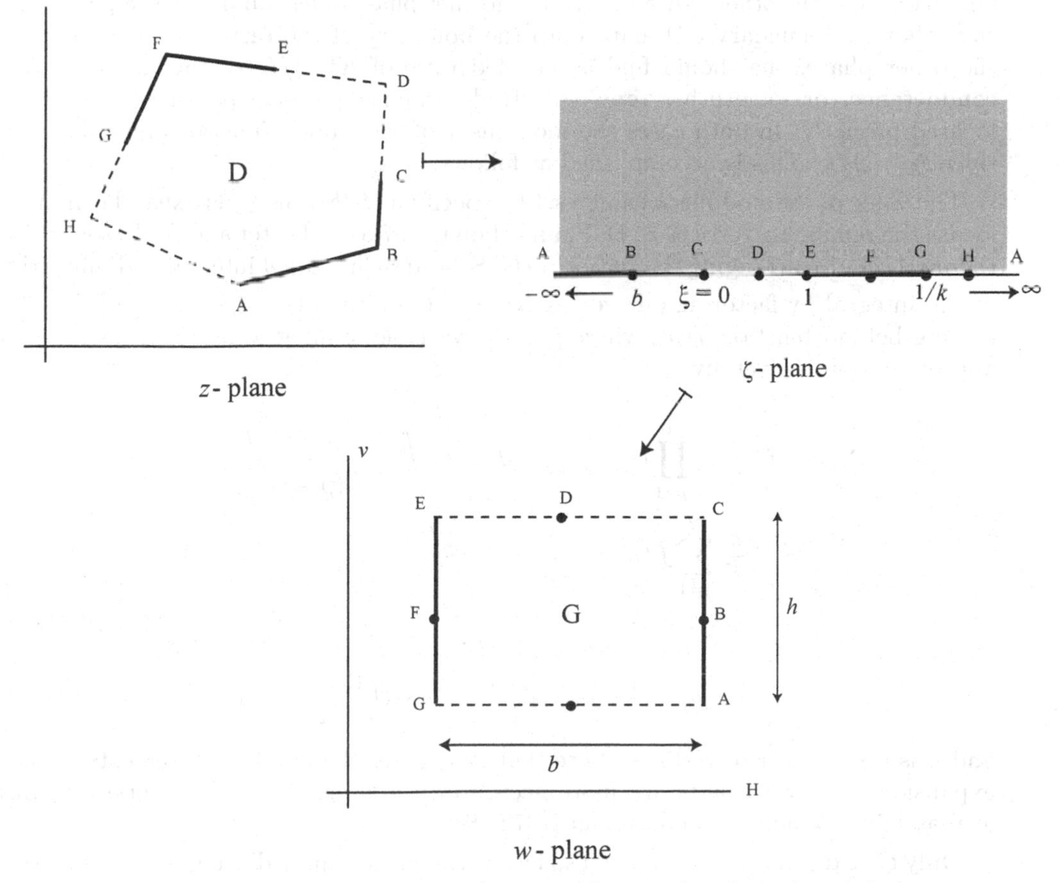

Figure 8.7 Synthesis of a polygon to conform with a desired modulus.

The algorithm for this method has been published by Warner and Anderson [1981], and its details are as follows: Consider a closed polygon with eight vertices in the z-plane, marked A through H in Figure 8.7, enclosing the region D. The boundary ∂D together with the eight vertices is mapped onto the real axis ξ in the ζ-plane ($\zeta = \xi + i\eta$), such that the vertices C, E, and A are mapped onto $\xi_C = 0$, $\xi_E = 1$, and $\xi_A = \infty$, respectively. However, we do not have any information about the mapping of the vertices B, D, F, G and H, except that G is determined for the case when the ζ-plane is mapped onto the rectangle ABCD in the w-plane, where the segments ABC and EFG correspond to opposite sides. This situation can be compared to the case of a conducting plane D in which the segments with heavy lines represent electrodes attached to the plate, and with the dashed lines represent the border, or lines of symmetry, through which no current passes. Then the ratio of the height and base of the rectangle can be defined as a nondimensional modulus by

$$\text{mod} = \frac{h}{b} = \frac{\text{conductance}}{\sigma\tau} = \frac{K'(k)}{K(k)}, \tag{8.7.6}$$

where σ denotes the conductivity and τ the thickness of the plate, and K' and K are elliptic functions (see Appendix F). In the case when a square is required in the w-plane, just set $h = b$, and the modulus will be 1.

According to Foster and Anderson [1974], there are two phases of the problem: one analytical and the other synthetic. In the former phase one should find the point $\xi_G = 1/k$ such that the boundary ∂D maps onto the boundary of the rectangle in the w-plane. In the other phase, one should find a specified ratio of $K'(k)/K(k)$ that is proportional to conductance, or some other topologically determined physical parameter, so that ξ_G is located properly. In both cases the modulus k of the elliptic functions must be evaluated since $\xi_G = 1/k$. This is accomplished as follows:

The value of the modulus k can be set by specifying h/b, thus ξ_G is fixed. Then proceed to locate the remaining vertices B, D, F, and H on the xi-axis. Foster and Anderson [1974] used the method of rearranging the terms of the Schwarz-Christoffel integral and approximating of the integral by factoring out vertex pairs of the form $\left\{ (p - \xi_l)\,(\xi_m - p) \right\}^{-1/2}$, thereby leaving behind function $f(p)$, where p is the variable of integration (same as ξ). Next, an approximation is given by

$$\begin{aligned} \Delta z_m &= C \int_{\xi_l}^{\xi_m} \prod_{j=1}^{n} (t - \xi_j)^{-\alpha_j}\, dt = C \int_{\xi_l}^{\xi_m} \frac{f(p)}{\left\{ (p - \xi_l)\,(\xi_m - p) \right\}^{1/2}}\, dp \\ &= C\,\frac{\pi}{N} \sum_{j=1}^{N} f\left(p_j\right), \end{aligned} \tag{8.7.7}$$

where

$$p_j = \frac{\xi_m + \xi_j}{2} + \frac{\xi_m - \xi_l}{2} \cos\left(\frac{(2j-1)\pi}{2N}\right),$$

and n is the number of vertices. Note that N denotes the number of elements in the series expansion, and the results are more accurate, the larger N is. The series expansion is available in Abramowitz and Stegun [1972: 889].

Only C and $\xi_1, \dots, \xi_n$ are known so far, as can be determined using (8.7.1)-(8.7.3). Thus, we get

$$|z_m - z_j| = C\,\frac{\pi}{N} \sum_{j=1}^{N} f\left(p_j^{lm}\right),$$

or

$$\frac{|z_m - z_l|}{|z_l - z_k|} = \frac{\sum_j f\left(p_j^{lm}\right)}{\sum_j f\left(p_j^{kl}\right)}, \tag{8.7.8}$$

where p^{lm} is the variable of integration in the interval $\xi_l - \xi_m$.

Example 8.5. (Warner and Anderson [1981]) Consider a square-section coaxial line in the z-plane which is mapped onto the ξ-plane (Figure 8.8) considered in Map 7.15. Then applying (8.7.2) to this case, we get

$$\Delta z_{lm} = C_1 \int_{\xi_l}^{\xi_m} \frac{1}{(t-0)^{1/2}(t-1)^{1/2}\,(t-\xi_A)}\,dt, \tag{8.7.9}$$

where ξ_A is at infinity. Thus, for segment CD we have $\Delta z = 1 - a$, or

$$C \int_0^1 \frac{1}{(t-0)(1-t)^{1/2}} \left\{ \frac{1}{(\xi_A - t)^{1/4}} \right\} dt = 1 - a, \tag{8.7.10}$$

giving $CV_1 = 1 - a$, where V_1 is the value of the integral in (8.7.10), and $C = C_1 i^{-3/2}$. Similarly, for the segment DA we have $\Delta z = 1 - a$, or

$$C \int_1^{\xi_A} \frac{1}{(t-1)\,(\xi_a - t)^{1/2}} \left\{ \frac{(\xi_A - 1)^{1/4}}{(t-0)^{1/2}} \right\} dt = a, \tag{8.7.11}$$

which gives $CV_2 = a$, where V_2 is the value of the integral in (8.7.11).

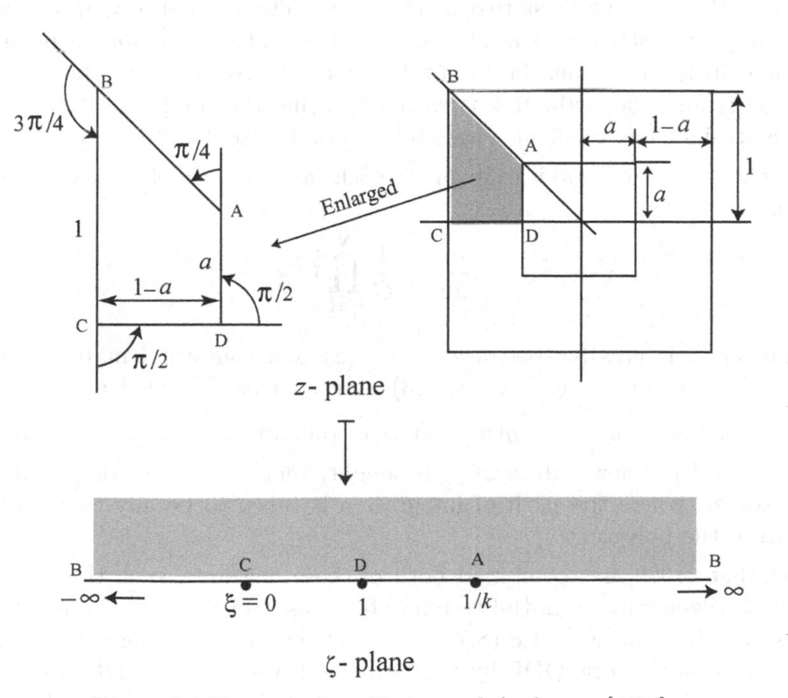

Figure 8.8 Example from Foster and Anderson [1974].

For example, consider the case for which $K'/K = 1.28$ holds for the trapezoidal segment which represents $1/8$ of the region in the z-plane. This fixed $k = 0.5$, and thus, $\xi_A = 1/k^2 = 4$. With this value of ξ_A, integrals (8.7.10) and (8.7.11) give $V_1 = 2.3 = V_3$, and $a = 0.5$. ■

For other examples with different cases with more vertices are available in Warner and Anderson [1981], and Schinzinger and Laura [2003: 293].

8.7.5 Unit Disk onto Polygon. (Subroutine WSC) To evaluate the inverse map 7.4 for a given point z in the unit disk or on the circle, we compute the integral

$$w = w_0 + C \int_{z_0}^{z} \prod_{j=1}^{N} \left(1 - \frac{z'}{z_j}\right)^{-\alpha_j} dz', \tag{8.7.12}$$

where $w_0 = w(z_0)$, and the endpoint z_0 may be a point in the closed disk at which the image $w(z_0)$ is known and finite. There are three possible choices for the point z_0: (i) $z_0 = 0$, which gives $w_0 = w_c$; (ii) $z_0 = z_k$ for some k, which gives $w_0 = w_k$ which is vertex of the polygon P; and (iii) z_0 is some other point in the disk at which w has already been computed. Note that in case (i) and (iii), neither endpoint has a singularity, and the evaluation of (8.7.12) can be done using compound Gauss quadrature. However, in case (ii) a singularity of the form $(1 - z/z_k)^{-\alpha_k}$ is present at one of the endpoints and the other point has no singularity. Hence, the best way to compute $w(z)$ is: If z is close to a singular point z_k (but not the one with $w_k = \infty$), use subroutine ZQUAD; otherwise use SCFUN. In either case use compound Gauss-Jacobi quadrature, taking the same number of nodes as was used in solving the parameter problem. This procedure will evaluate $w(z)$ with complete accuracy by computing all accessory parameters, where quadrature nodes and weights need only be computed once.

8.7.6 Polygon onto Unit Disk. (Subroutine ZSC). To compute the inverse map $z = z(w)$ there are at least two methods: (i) The most direct approach is to regard the direct map $w = w(z)$ as a nonlinear equation and solve it for z, given some fixed value w. The solution may then be found iteratively by Newton's method or a related method, where $w(z)$ must be evaluated at each step using the compound Gauss-Jacobi quadrature along a straight line segment whose initial point is kept fixed throughout the iteration.

(ii) The second method is to invert the Schwarz-Christoffel transformation formula (7.1.4) to yield

$$\frac{dz}{dw} = \frac{1}{C} \prod_{k=1}^{N} \left(1 - \frac{w}{w_k}\right)^{\alpha_k}. \tag{8.7.13}$$

The inversion is possible because $w = w(z)$ is a conformal mapping, which means that $|dw/dz| > 0$ everywhere. Eq (8.7.13) may also be regarded as an ordinary differential equation of the form $\dfrac{dz}{dw} = g(w, z)$ in one complex variable w. If a pair of values (z_0, w_0) is known and the new values $z(w)$ is sought, then z may be computed using a numerical o.d.e. solver, where the path of integration is taken to be any curve from w_0 to w which lies within the polygon P.

Trefethen [1979] has combined both of these methods, such that the second method is first used to generate an initial estimate to be used in the first. Thus, the o.d.e formulation begins, first by expressing Eq (8.7.13) as a system of first-order o.d.e. in two real variables, and then using the code ODE by Shampine and Allen [1973], and integrating along straight line segment from w_c to w, whenever possible. Since the polygon P is not convex, more than one line segment may be required to connect w_0 and w. Note that it will not do to take $w_0 = w_k$ for some vertex w_k without special care, because Eq (8.7.13) is singular at w_k. Thus, the following steps are needed to compute $z = z(w)$ rapidly to full accuracy:

STEP 1. Solve Eq (8.7.13) using the ODE package whenever possible along a line segment from w_c to w; call the result $\tilde{z}$; and

STEP 2. Solve the equation $w(z) = w$ for z by Newton's method, using $\tilde{z}$ as an initial guess.

Fortran programs of all subroutines are available in Trefethen [1979: 48-55].

8.8 Generalized Schwarz-Christoffel Parameter Problem

Trefethen and Williams [1986] have used Schwarz-Christoffel transformations to numerically solve Laplace's equation in a polygonal domain with boundary conditions that the derivative in the direction at a specified oblique angle from the normal is zero along each side of the polygon. The solution is constructed by taking the real part of the mapping function that maps the physical domain onto another (model) domain with straight sides according to the angles given in the boundary conditions.

Given a set of real numbers θ_k, $k = 1, 2, \dots, n$, let $u_n(z) = \dfrac{\partial u}{\partial n}(z)$ denote the inward normal derivative of a function u, and let $u_\sigma = \dfrac{\partial u}{\partial \sigma}$ be its tangential derivative along the side Γ_k in the positive sense, of a polygon D with vertices z_k for each $k = 1, 2, \dots, n$. Then the problem considered in the above work, known as the *oblique derivative problem*, is as follows:

Problem O. Find the functions $u \in C^2(D) \cap C^1(D \cup \bigcup_k \Gamma_k)$ which satisfy Laplace's equation $\nabla^2 u(z) = 0$, $z \in D$, subject to the homogeneous oblique boundary conditions

$$\cos\theta_k u_n(z) + \sin\theta_k u_\sigma(z) = 0. \tag{8.8.1}$$

This problem is illustrated in Figure 8.7(a). If $\theta_k = 0 \pmod{\pi}$, then we have a Neumann condition on the side Γ_k, while $\theta_k = \pi/2 \pmod{\pi}$ gives the tangential condition.

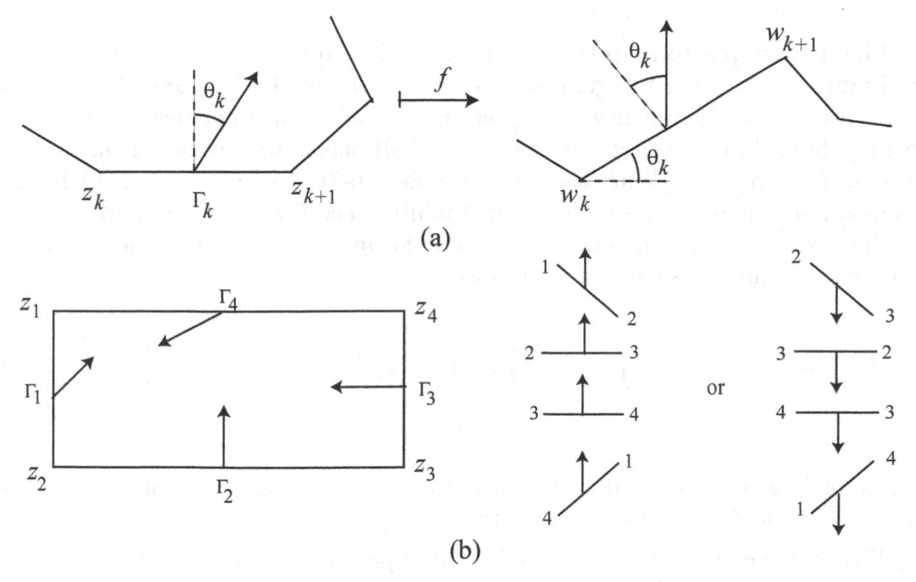

Figure 8.9 Problem O.

This problem for a rectangle bounded by four sides $\Gamma_1, \dots, \Gamma_4$, is presented in Figure 8.9(b); the Neumann condition holds on Γ_1 and Γ_4, oblique derivative conditions at 45° from the normal in the indicated directions. To apply the conformal mapping method, we must find a polygonal image domain $f(D)$ whose sides are oriented as in Figure 8.9 (right-side figure).

The conformal mappings begin in the following order: First, as shown in Figure 8.10, the problem from D onto the upper half t-plane by a conformal mapping ϕ with $\phi(z_k) = \infty$ for some vertex z_k. It is known that ϕ must be the Schwarz-Christoffel mapping

$$\phi^{-1}(t) = C_1 + C_2 \int_{T_\phi}^{t} \prod_{\substack{k=1 \\ k \neq k_\infty}}^{n} (t' - t_k)^{-\alpha_k} \, dt', \tag{8.8.2}$$

where $C_1, C_2 \in \mathbb{C}$, and $T_\infty \in \mathbb{R}$ are constants; $t_k = \phi(z_k)$ are the prevertices on the real axis and $\alpha_k \pi$ is the external angle at the vertex z_k, $-1 \leq \alpha_k \leq 3$. However, the difficulty is that the values of $\{t_k\}$ are not known in advance. Thus, determining ϕ numerically requires an iterative method to determine these values. The Fortran package SCPACK (§24.10) is used for this purpose.

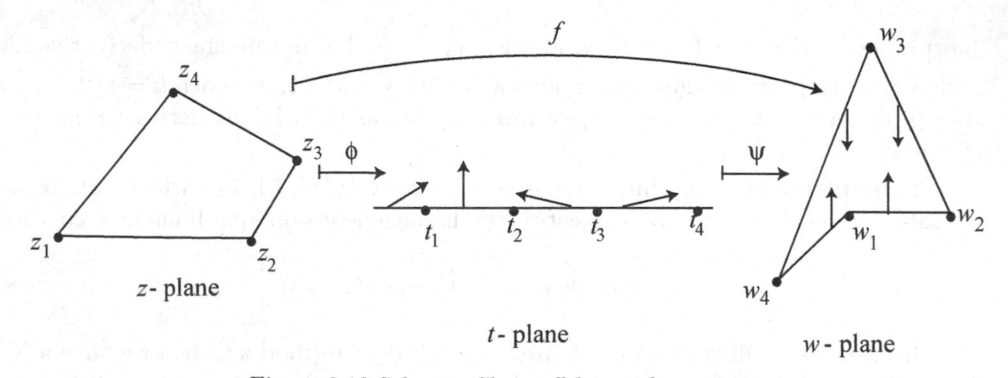

Figure 8.10 Schwarz-Christoffel transformations.

The next step is to map the upper half t-plane onto $f(D)$ by a function $\psi(t)$, as shown in Figure 8.10. This is again another Schwarz-Christoffel transformation involving the same prevertices t_k and new angle parameters α'_k, and other new features. The function ψ may have branch points in the upper half-plane; to handle them, additional factors $(t' - s_k)(t' - \bar{s}_k)$, $s_k \in \mathbb{R}$ are used; also the factors $(t' - s_k)$ are included whenever there are points along some Γ_k where $f'(z) = 0$. Finally, since $f(D)$ is not required to be embeddable in the plane, the parameters α'_k need not lie in $[-1, 3]$. Hence, the general map ψ with algebraic singularities can be written as

$$\psi(t) = C_3 + e^{i\theta_{k_m} - 1} \int_{T_\psi}^{t} p(t') \prod_{\substack{k=1 \\ k \neq k_\infty}}^{n} (t' - t_k)^{m_k - \alpha'_k} \, dt', \quad \alpha'_k = \frac{1}{\pi} (\theta_k - \theta_{k-1}), \tag{8.8.3}$$

where $p(t')$ is a polynomial with real coefficients, each m_k is an integer, T_ψ is any point in (t_{k-1}, ∞), and C_3 is a complex constant.

This is a *generalized Schwarz-Christoffel parameter problem*, where p is linear, which makes it easily solvable. The idea of generalized parameter problems was introduced by

Trefethen. Boundary value problems for elliptic equations in non-smooth domains, especially Poisson's equation on polygonal domains with oblique derivative, have been studied, e.g., by Grisvard [1980], and Kondrat'ev and Olcinin [1983].

8.9 Proofs

To prove the Schwarz-Christoffel formula (7.1.2), which leads to Eq (8.1.1), we will follow the argument from Driscoll and Trefethen [2002], and prove

Theorem 8.12. *Let Γ be a polygon with vertices $w_1, \dots, w_n$ and exterior angles $\alpha_1\pi, \dots, \alpha_n\pi$ in counterclockwise order. Let $w = f(z)$ be any conformal bijective map from the upper half-plane $\Im\{z\} > 0$ onto $G = \text{Int}(\Gamma)$ in the w-plane such that $w(\infty) = w_n$. Then the mapping w can be written in the form (7.1.2), where $z_1 < z_2 < \cdots < z_n$ are real numbers satisfying $w(z_k) = w_k$ for $k = 1, 2, \dots, n$.*

Note that Eq (7.1.2) involves improper contour integrals.

Lemma 8.1. *Let Γ be a polygon and let f map the upper half-plane $H \equiv \Im\{z\} > 0$ conformally onto G. Then f extends continuously to the closure $\bar{H}$. More precisely, there exists a homeomorphism F from $\bar{H}$ onto $\Gamma \cup G$ satisfying $F\big|_H = f$.*

PROOF. Consider a vertex $a = w_k$ of Γ. Then the function $z \mapsto z^{1/(1-\alpha_k)}$ makes the angle straight between the two edges on which a lies. Thus, $g = f^{-1}$ extends to a continuous function ϕ on all of $G \cup \Gamma$. Since ϕ' does not vanish on any open edge of Γ, then, by mean-value theorem, ϕ is injective on each closed edge. In particular, it cannot reverse direction on adjacent edges and thus, $\phi\big|_\Gamma$ is injective. Since $g' \neq 0$ on G, by the open mapping theorem we have $\phi(\Gamma) \cap \phi(G) = \emptyset$, which means that ϕ is injective on all $G \cup \Gamma$. Since Γ is a closed curve, $\phi(\Gamma)$ is defined on the extended real axis. The proof is completed by taking $F = \phi^{-1}$. ■

Lemma 8.2. *Let f, Γ, and G be fixed, and let $a \in \Gamma$ be not a vertex. Then the function g extends to an analytic function ϕ in a disk centered at a such that $g(a) \neq 0$.*

PROOF OF THEOREM 8.12. In view of Lemma 8.1, the map w can be continuously extended to the closed upper half-plane. For $k = 1, 2, \dots, n$, denote the prevertices by $z_k = w^{-1}(w - k)$. Using the Schwarz reflection principle (§2.6.6) we can extend w and w' analytically across the real axis everywhere except the prevertices. According to Lemma 8.1, the function $z \mapsto (z - z_k)^{-\alpha_k} w'(z)$ has an analytic extension in the neighborhood of z_k. Thus, we have

$$w'(z) = (z - z_k)^{-\alpha_k} \psi(z),$$

where $\psi(z)$ is some function analytic in the neighborhood of z_k. Hence, we get

$$\begin{aligned} w'(z) &= (z - z_k)^{-\alpha_k} \psi(z), \\ w''(z) &= -\alpha_k (z - z_k)^{-\alpha_k - 1} \psi(z) + \psi'(z)(z - z_k)^{a_k}, \end{aligned}$$

so that

$$\frac{w''(z)}{w'(z)} = \frac{-\alpha_k (z - z_k)^{-\alpha_k - 1}}{\psi(z)} (z - z_k)^{\alpha_k} \psi(z) + \frac{\psi'(z)(z - z_k)^{-\alpha_k}}{(z - z_k)^{-\alpha_k} \psi(z)} = -\frac{\alpha_k}{z - z_k} + \frac{\psi'(z)}{\psi(z)},$$

which implies that $\dfrac{w''(z)}{w'(z)}$ has a simple pole at $z = z_k$ with residue $-\alpha_k$. Hence, the function $\dfrac{w''(z)}{w'(z)} + \sum\limits_{k=1}^{n} \dfrac{\alpha_k}{z - z_k}$ is an entire function. Also, since all the prevertices are finite,

w is analytic at $z = \infty$. Using the Laurant series expansion we find that $\dfrac{w''(z)}{w'(z)} \to 0$ as $z \to \infty$, such that each term of the summand goes to zero as $z \to \infty$, which means that it is a bounded entire function. Then by Liouville's theorem (§2.4) the expression $\dfrac{w''(z)}{w'(z)}$ is constant and is identically zero. Next, since $(\log(w'))' = \dfrac{w''(z)}{w'(z)}$, we get

$$
\begin{aligned}
(\log(w'))' &= \sum_{k=1}^{n} \frac{-\alpha_k}{z - z_k}, \\
\log(w') &= \int \sum_{k=1}^{n} \frac{-\alpha_k}{z - z_k}, \\
&= \sum_{k=1}^{n} \int \frac{-\alpha_k}{z - z_k} = -\sum_{k=1}^{n} \alpha_k \ln(z - z_k) + C_1 = \sum_{k=1}^{n} \ln(z - z_k)^{-\alpha_k} + C_1 \\
w' &= \exp\Big\{\sum_{k=1}^{n} \ln(z - z_k)^{-\alpha_k} + C_1\Big\} = \exp\Big\{\sum_{k=1}^{n} \ln(z - z_k)^{-\alpha_k} \exp(C_1)\\
&= C \prod_{k=1}^{n} \exp\big\{\ln(z - z_k)^{-\alpha_k}\big\}, \quad \text{where } C = \exp\{C_1\} = C \prod_{k=1}^{n} (z - z_k)^{-\alpha_k}
\end{aligned}
$$

which after integration yields

$$
w = C_1 \int \prod_{k=1}^{n} (z - z_k)^{-\alpha_k} + C_2. \blacksquare
$$

NOTES. cs442.nb:

```
A[j_, k_] := Integrate[ Integrate[ (x+ I*y)^j * (x-I*y)^k,
{x, -1,1}], {y, -1,1}];
MatA = Table[A[j,k], {j,1,8}, {k,1,8}];
MatrixForm[MatA];
B=Table[A[j,0], {j,1,8}];
c=LinearSolve[MatA,- B];
(* These are the coefficients of phi_8[z] *)
phi8[z_] := 1 + c . Table[z^i, {i, 8}];
phi8[z];
(* The mapping function is given by f'[z]=phi8[z] *)
f[z_] := Integrate[phi8[t], {t, 0,z}];
f[z]
```

REFERENCES USED: Ahlfors [1966], Birkoff, Young and Zarantonello [1951], Boas [1987], Carrier, Krook and Pearson [1966], Elcrat [1982], Driscoll and Trefethen [2002], Elcrat and Trefethen [1986], Foster and Anderson [1974], Gaier [1964], Goluzin [1957; 1969], Kantorovich and Krylov [1958], Kythe [1998], Nehari [1952], Papamichael, Warby and Hough [1983]. Robertson [1965], Schinzinger and Laura [2003], Trefethen [1980], Trefethen and Williams [1986], von Koppenfels and Stallmann [1959], Warner and Anderson [1981]

9

Nearly Circular Regions

We will investigate methods for constructing a mapping function for conformal mapping of a simply connected nearly circular region onto a disk. A classical method that involves series expansion in powers of a small parameter for the interior and the exterior regions, known as the *method of infinite systems*, is presented with case studies (designated as Maps), in which successive approximations are used to compute the approximate mapping function.

9.1 Small Parameter Expansions

Let a one-parameter family of Jordan curves Γ_λ in the z-plane be defined by

$$\Gamma_\lambda : z = z(t, \lambda), \tag{9.1.1}$$

where t and λ are real parameters. Let the origin $z = 0$ lie inside all of these curves. Further, let the function

$$w = f(z, \lambda) \tag{9.1.2}$$

map the region D_λ bounded by the curve Γ_λ conformally onto a disk $|w| < R$ in the w-plane, which implies that the function (9.1.2) must satisfy the conditions

$$f(0, \lambda) = 0, \quad f'(0, \lambda) = 1. \tag{9.1.3}$$

Note that if the function $z = z(t, \lambda)$ which defines the boundary curve Γ_λ, where the parameter t defines the position of the point z on Γ_λ, is an analytic function of λ in the neighborhood of some value of λ, say $\lambda = 0$, then the mapping function (9.1.2) can also be regarded as an analytic function in that neighborhood. Thus, the function $f(z, \lambda)$ for any $z \in D_\lambda$ can be expanded in a power series in λ as

$$f(z, \lambda) = f_0(z) + \sum_{n=1}^{\infty} \lambda^n f_n(z), \tag{9.1.4}$$

which converges for sufficiently small $|\lambda|$. Now, to compute the coefficients $f_n(z)$, we know from (9.1.3) that $f_n(0) = 0$ for $n = 0, 1, 2, \dots$, $f_0'(0) = 1$, and $f_n'(0) = 0$ for $n = 1, 2, \dots$.

Thus, $f_0(z) = f(z,0)$, which implies that the function $w = f_0(z)$ maps the region D_0 bounded by Γ_0 exactly onto the disk $|w| < R$.

To compute $f_n(z)$ for $n \geq 1$, let us consider a system of functions $\{u_n(z)\}_{n=1}^{\infty}$, which are analytic in a region D containing all D_λ for sufficiently small $|\lambda|$, such that $u_n(0) = 0$ for $n = 1, 2, \dots$, $u_1'(0) = 1$, and $u_n'(0) = 0$ for $n = 2, 3, \dots$. Then any function $f(z, \lambda)$ analytic on D can be expanded in a series involving $u_n(z)$. Thus, let

$$f(z,\lambda) = u_1(z) + \sum_{n=2}^{\infty} \alpha_n(\lambda)\, u_n(z), \tag{9.1.4}$$

where the coefficients $\alpha_n(\lambda)$ depend only on λ. Hence, the problem of determining the function $f(z,\lambda)$ reduces to that of computing the coefficients $\alpha_n(\lambda)$, which are, in fact, solutions of an infinite system of equations.

However, in practical problems, the function $f(z,\lambda)$ is represented approximately by taking a finite sum in (9.1.4). Then the coefficients $\alpha_n(\lambda)$ are determined by solving a finite system of equations. The details of computation, known as the *Kantorovich-Krylov's method* (Kantorovich and Krylov [1964]) are as follows: Let

$$U_n(z) = u_1(z) + \sum_{j=2}^{n} \alpha_j(\lambda)\, u_j(z) \tag{9.1.5}$$

denote a partial sum of the series (9.1.4). Then $|U_n(z)|^2$ can be expanded in a trigonometric series in $t \in [0, 2\pi)$ as

$$|U_n(z)|^2 = a_0 + \sum_{n=1}^{\infty} \left(a_n \cos nt + b_n \sin nt\right). \tag{9.1.6}$$

This means that the coefficients a_n and b_n in this expansion are quadratic functions of $\alpha_j(\lambda)$. We can choose that all coefficients $a_n(\alpha_j)$ and $b_n(\alpha_j)$ are zero for $n = 1, 2, \dots$, or that only the first $(n-1)$ coefficients $a_n(\alpha_j)$ and $b_n(\alpha_j)$ are zero. In the former case we get an exact determination of the function $f(z,\lambda)$. But in the second case we obtain a system of $(2n-2)$ equations

$$a_k(\alpha_j) = 0, \quad b_k(\alpha_j) = 0 \quad k = 1, 2, \dots, n-1, \tag{9.1.7}$$

which determine the $(n-1)$ unknown complex coefficients $\alpha_j(\lambda)$.

The form of the partial sum $U_n(z)$, defined by (9.1.5), depends on the choice of the system of the functions $u_n(z)$. If we take the system $\{u_n(z)\} = \{z^n\}$, then $U_n(z)$ becomes a polynomial of degree n:

$$U_n(z) \equiv p_n(z) = z + \alpha_2(\lambda) z^2 + \cdots + \alpha_n(\lambda) z^n. \tag{9.1.8}$$

Let us assume that the boundary Γ of the region D is nearly circular and is defined by

$$\Gamma: \, z(t) = e^{it}\left\{1 + \lambda F\left(e^{it}, \lambda\right)\right\}, \tag{9.1.9}$$

where $F(\tau, \lambda)$, where $\tau = e^{it}$, is an analytic function of its arguments for $|\tau|$ close to 1 and λ close to 0. Then $F(\tau,\lambda)$ can be expanded in a Laurent series in τ as

$$F(\tau,\lambda) = \sum_{\nu=-\infty}^{\infty} \beta_\nu(\lambda)\, \tau^\nu = \sum_{\nu=-\infty}^{\infty} \beta_\nu(\lambda)\, e^{i\nu t},$$

where the coefficients $\beta_\nu(\lambda)$ are analytic functions of λ. Then from (9.1.9) the boundary Γ is defined by

$$z(t) = e^{it}\left\{1 + \lambda \sum_{\nu=-\infty}^{\infty} \beta_\nu(\lambda)\, e^{i\nu t}\right\} = \sum_{\nu=-\infty}^{\infty} \beta_\nu^{(1)}(\lambda)\, e^{i\nu t}, \tag{9.1.10}$$

where

$$\beta_\nu^{(1)}(\lambda) = \begin{cases} \lambda\beta_{\nu-1}(\lambda) & \text{for } \nu \neq 1, \\ 1 + \lambda\beta_0(\lambda) & \text{for } \nu = 1. \end{cases} \tag{9.1.11}$$

The kth power of $z(t)$, defined by (9.1.10), is given by

$$z^k(t) = \sum_{\nu=-\infty}^{\infty} \beta_\nu^{(k)}(\lambda)\, e^{i\nu t}, \tag{9.1.12}$$

where $\beta_\nu^{(k)}(0) = \delta_{\nu k}$. Thus, from (9.1.8) and (9.1.12),

$$\begin{aligned} |p_n(z)|^2 &= p_n(z), \overline{p_n(z)} \\ &= \sum_{k,j=1}^{n} \alpha_k \bar{\alpha}_j\, z^k \bar{z}^j, \quad (\text{where } \alpha_1 = 1) \\ &= \sum_{\nu=-\infty}^{\infty} \sum_{k,j=1}^{n} \alpha_k \bar{\alpha}_j \sum_{\substack{p,q=-\infty \\ p-q=\nu}}^{\infty} \beta_p^{(k)}(\lambda)\, \bar{\beta}_q^{(j)}(\lambda)\, e^{i\nu t}. \end{aligned} \tag{9.1.13}$$

Note that the right side of (9.1.13) represents a trigonometric series, whose coefficients depend on α_k. If we denote the free term (corresponding to $\nu = 0$) on the right side of (9.1.13) by R^2, we get

$$R^2 = \sum_{k,j=1}^{n} \alpha_k \bar{\alpha}_j \sum_{p=-\infty}^{\infty} \beta_p^{(k)}(\lambda)\, \bar{\beta}_p^{(j)}. \tag{9.1.14}$$

We will choose the coefficients $\alpha_2, \dots, \alpha_n$ such that the coefficients of $e^{it}, e^{2it}, \dots, e^{(n-1)it}$ are zero, i.e.,

$$\sum_{k,j=1}^{n} \alpha_k \bar{\alpha}_j \sum_{n=-\infty}^{\infty} \beta_n^{(k)}(\lambda)\, \bar{\beta}_{n-m}^{(j)} = 0 \quad \text{for } m = 1, 2, \dots, n-1. \tag{9.1.15}$$

Note that the coefficients of $e^{-it}, e^{-2it}, \dots, e^{-(n-1)it}$ are also zero. Hence, the system (9.1.15) should determine the coefficients $\alpha_2, \dots, \alpha_n$. Moreover, the difference between $f(z)$ and $U_n(z)$ decreases as n increases. The proof for the convergence of $U_n(z)$ to $f(z)$ through the method of successive approximations is given in Kantorovich and Krylov [1964:435]. We will look into some particular regions as cases for which the system (9.1.15) provides simpler solutions.

Map 9.1. To determine the function $w = f(z, \lambda)$ that maps the interior of the ellipse $x = (1 + \lambda^2)\cos t, y = (1 - \lambda^2)\sin t$ conformally onto the disk $|w| < R$, first note that the equation of the ellipse can be written as

$$z(t) = e^{it}\left(1 + \lambda^2 e^{-2it}\right). \tag{9.1.16}$$

We will find the approximate mapping function $U_n(z) = p_n(z)$ accurate to λ^{10}. Then the last coefficient in (9.1.8) is α_{11}. Moreover, since the ellipse has two axes of symmetry, all α_k are real and those with even indices are zero, thus

$$p_n(z) = z + \alpha_3 z^3 + \alpha_5 z^5 + \alpha_7 z^7 + \alpha_9 z^9 + \alpha_{11} z^{11},$$

and

$$\begin{aligned} |p_n(z)|^2 &= p_n(z)\,\overline{p_n(z)} \\ &= z\,\bar z + \alpha_3 \left(z^3\,\bar z + \bar z^3 z\right) + \left[\alpha_5 \left(z^5 \bar z + \bar z^5 z\right) + \alpha_3^2 z^3 \bar z^3\right] \\ &+ \left[\alpha_7 \left(z^7 \bar z + \bar z^7 z\right) + \alpha_5\alpha_3 \left(z^5 \bar z^3 + \bar z^5 z^3\right)\right] \\ &+ \left[\alpha_9 \left(z^9 \bar z + \bar z^9 z\right) + \alpha_7\alpha_3 \left(z^7 \bar z^3 + \bar z^7 z^3\right) + \alpha_5^2 z^5 \bar z^5\right] \\ &+ \left[\alpha_{11} \left(z^{11} \bar z + \bar z^{11} z\right) + \alpha_9\alpha_3 \left(z^9 \bar z^3 + \bar z^9 z^3\right)\right. \\ &\left.+ \alpha_7\alpha_5 \left(z^7 \bar z^5 + \bar z^7 z^5\right)\right] + \left[\alpha_{13} \left(z^{13} \bar z + \bar z^{13} z\right)\right. \\ &\left.+ \alpha_{11}\alpha_3 \left(z^{11}\,\bar z^3 + \bar z^{11}\,z^3\right) + \alpha_9\alpha_5 \left(z^9 \bar z^5 + \bar z^9 z^5\right) + \alpha_7^2 z^7 \bar z^7\right]. \end{aligned} \tag{9.1.17}$$

The combinations $z^k \bar z^j + \bar z^k z^j$ that appear in (9.1.17) have been determined by Mathematica. Thus, substituting them into (9.1.17), equating the free term to R^2, and equating the coefficients of different cosines to zero, we obtain the following system of equations:

$$\begin{aligned} &1 + \lambda^4 + 6\alpha_3\,\lambda^2 \left(1 + \lambda^4\right) + 20\alpha_5\lambda^4 + \alpha_3^2 \left(1 + 9\lambda^4\right) + 10\alpha_3\alpha_5\lambda^2 + \alpha_5^2 = R^2, \\ &\lambda^2 + \alpha_3 \left(1 + 6\lambda^4 + \lambda^8\right) + 5\alpha_3\lambda^2 \left(1 + 4\lambda^4\right) + 3\alpha_3^2\lambda^2 \left(1 + 3\lambda^4\right) \\ &\qquad + 21\alpha_7\lambda^4 + \alpha_3\alpha_5 \left(1 + 25\lambda^4\right) + 7\alpha_3\alpha_7\lambda^2 + 5\alpha_5^2\lambda^2 + \alpha_3\alpha_7 = 0, \\ &\alpha_3\lambda^2 \left(1 + \lambda^4\right) + \alpha_5 \left(1 + 5\lambda^4\right) + 3\alpha_3^2\lambda^4 + 7\alpha_7\lambda^2 + 3\alpha_3\alpha_5\lambda^2 + \alpha_3\alpha_7 = 0, \\ &\alpha_5\lambda^2 + \alpha_3^2\lambda^6 + \alpha_7 \left(1 + 7\lambda^4\right) + 3\alpha_3\alpha_5\lambda^4 + 9\alpha_9\lambda^2 + 3\alpha_3\alpha_7\lambda^2 + \alpha_3\alpha_9 = 0, \\ &\alpha_7\lambda^2 + \alpha_9 = 0, \\ &\alpha_9\lambda^2 + \alpha_{11} = 0. \end{aligned} \tag{9.1.18}$$

These equations, except for the first one, will be solved by the *method of successive approximations*. Thus, transposing $\alpha_3, \alpha_5, \alpha_7, \alpha_9$ and α_{11} we get

$$\begin{aligned} \alpha_3 &= -\left[\lambda^2 + \alpha_3 \left(6\lambda^4 + \lambda^8\right) + 5\alpha_3\lambda^2 \left(1 + 4\lambda^4\right) + 3\alpha_3^2\lambda^2 \left(1 + 3\lambda^4\right)\right. \\ &\qquad \left.+ 21\alpha_7\lambda^4 + \alpha_3\alpha_5 \left(1 + 25\lambda^4\right) + 7\alpha_3\alpha_7\lambda^2 + 5\alpha_5^2\lambda^2 + \alpha_3\alpha_7\right], \\ \alpha_5 &= -\left[\alpha_3\lambda^2 \left(1 + \lambda^4\right) + 5\alpha_5\lambda^4 + 3\alpha_3^2\lambda^4 + 7\alpha_7\lambda^2 + 3\alpha_3\alpha_5\lambda^2 + \alpha_3\alpha_7\right], \\ \alpha_7 &= -\left[\alpha_5\lambda^2 + \alpha_3^2\lambda^6 + 7\alpha_7\lambda^4 + 3\alpha_3\alpha_5\lambda^4 + 9\alpha_9\lambda^2 + 3\alpha_3\alpha_7\lambda^2 + \alpha_3\alpha_9\right], \\ \alpha_9 &= -\alpha_7\lambda^2, \\ \alpha_{11} &= -\alpha_9\lambda^2. \end{aligned}$$

The values of the coefficients $\alpha_3, \alpha_5, \alpha_7, \alpha_9$ and α_{11} starting with initial values zero are computed up to the fifth successive approximation (see Table E.1, Appendix E). Hence,

$$p(z) = z - \left(\lambda^2 + \lambda^6 + 4\lambda^{10}\right) z^3 + \left(\lambda^4 + 3\lambda^8\right) z^5 - \left(\lambda^6 + 5\lambda^{10}\right) z^7 + \lambda^8\, z^9 + \lambda^{10}\, z^{11},$$

which is accurate to λ^{10}. The same fifth successive approximations for α_3, α_5, α_7, α_9, and α_{11} when substituted in the first equation in (9.1.18) yield

$$R^2 = 1 - 4\lambda^4 + 10\lambda^8.$$

To check this result, note that $z = 1 + \lambda^2$ for $t = 0$, thus $p(z) = p\left(1 + \lambda^2\right) = 1 - 2\lambda^4 + 3\lambda^8$, which coincides with $R = \left(1 - 4\lambda^4 + 10\lambda^8\right)^{1/2} = 1 - 2\lambda^4 + 3\lambda^8$.

Note that the ellipse, defined by (9.1.16), can be written as $z = a\,\cos t + i\,b\,\sin t$, where $a = 1 + \lambda^2$ and $b = 1 - \lambda^2$. The function $f(z, \lambda)$ (i) in the case $\lambda^2 = 1/11$, i.e., $a/b = 1.2$, is

$$\begin{aligned} f(z,\lambda) = 1.0165984\,z - 0.0932071\,z^3 + 0.008615\,z^5 - 0.0007963\,z^7 + 0.0000734\,z^9 \\ - 0.0000068\,z^{11} + 0.0000006\,z^{13} + 0.00000005\,z^{15}; \end{aligned}$$

(ii) in the case $\lambda^2 = 1/5$, i.e., $a/b = 1.5$, is

$$\begin{aligned} f(z,\lambda) = 1.081728\,z - 0.226394\,z^3 + 0.049024\,z^5 - 0.010611\,z^7 + 0.002176\,z^9 - 0.000461\,z^{11} \\ + 0.000064\,z^{13} + 0.000013\,z^{15}; \end{aligned}$$

and (iii) in the case $\lambda^2 = 1/3$, i.e., $a/b = 2$, is

$$\begin{aligned} f(z,\lambda) = 1.2373\,z - 0.4787\,z^3 + 0.1948\,z^5 - 0.0791\,z^7 + 0.0247\,z^9 - 0.0091\,z^{11} + 0.0014\,z^{13} \\ + 0.00046\,z^{15}. \end{aligned}$$

(See Andersen et al. [1962:254-255].)

9.2 Method of Infinite Systems

We will present a general approach for the method of §9.1. This is known as the *method of infinite systems*, initially developed by Kantorovich and Krylov [1964] and later summarized in a systematic form with the first computer program (in ALGOL) by Andersen et al. [1962].

Let the boundary Γ_λ of a nearly circular region D_λ in the z-plane, be denoted by $z = G\left(e^{it}, \lambda\right)$, where λ is a small parameter, and have a Laurent series expansion in e^{it} with suitable finite n and m as

$$\begin{aligned} z = G\left(e^{it}, \lambda\right) &= e^{it} \sum_{p=-n}^{n} C_p(\lambda)\, e^{ipt} = e^{it} \sum_{p=-n}^{n} \sum_{q=|p|}^{m} \left(k_{q,p}\, \lambda^q\right) e^{ipt} \\ &= e^{it} \Big\{ \cdots + e^{-2it}\, \left(0 + 0 + k_{2,-2}\,\lambda^2 + k_{3,-2}\,\lambda^3 + \cdots\right) \\ &\quad + e^{-it}\, \left(0 + k_{1,-1}\,\lambda + k_{2,-1}\,\lambda^2 + k_{3,-1}\,\lambda^3 + \cdots\right) \\ &\quad + e^{0}\, \left(1 + k_{1,0}\,\lambda + k_{2,0}\,\lambda^2 + k_{3,-0}\,\lambda^3 + \cdots\right) \\ &\quad + e^{it}\, \left(0 + k_{1,1}\,\lambda + k_{2,1}\,\lambda^2 + k_{3,1}\,\lambda^3 + \cdots\right) \\ &\quad + e^{it}\, \left(0 + 0 + k_{2,2}\,\lambda^2 + k_{3,2}\,\lambda^3 + \cdots\right) + \cdots \Big\}, \end{aligned} \tag{9.2.1}$$

where the nonzero coefficients $k_{p,q}$ are known complex constants except for $k_{0,0}$ which takes the value 1, because $\lambda = 0$ must reduce the boundary Γ_λ to the circle $z = e^{it}$, and small values of λ produce nearly circular boundaries. Note that the expansion (9.2.1) is similar to (9.1.10).

Map 9.2. Let the function $w = f(z,\lambda)$ that maps the region D_λ onto the disk $|w| < R$ (or 1) have a series representation

$$\begin{aligned} w = f(z,\lambda) &= \sum_{p=0}^{\infty} \lambda^p \left(\sum_{q=1}^{p+1} a_q^{(p)} z^q \right) \\ &= \sum_{p=0}^{\infty} \lambda^p \left[a_1^{(p)} z + a_2^{(p)} z^2 + a_3^{(p)} z^3 + \cdots + a_{p+1}^{(p)} z^{p+1} \right] \\ &= a_1^{(0)} z + \lambda \left[a_1^{(1)} z + a_2^{(1)} z^2 \right] + \lambda^2 \left[a_1^{(2)} z + a_2^{(2)} z^2 + a_3^{(2)} z^3 \right] + \cdots . \end{aligned} \tag{9.2.2}$$

Since $\lambda = 0$ gives the identity mapping, we have $a_1^{(0)} = 1$. Also, all coefficients $a_q^{(p)}$ are real since $f'(0,\lambda) = a_1^{(0)} + a_1^{(1)} \lambda + a_1^{(2)} \lambda^2 + \cdots$ is real for all λ. The problem of approximating $f(z,\lambda)$ reduces to that of determining the unknown coefficients $a_q^{(p)}$ from the fact that after $z = G\left(e^{it},\lambda\right)$ from (9.2.1) is substituted, we should have $|w|^2 = w \cdot \bar{w} = R^2$ for all t and every λ up to the desired accuracy in powers of λ.

First, we determine z^q. Thus, from (9.2.1) we get

$$\begin{aligned} z^q = e^{iqt} \cdot \Big\{ \cdots &+ e^{-2it} \left(0 + 0 + k_{2,-2}^{(q)} \lambda^2 + k_{3,-2}^{(q)} \lambda^3 + \cdots \right) \\ &+ e^{-it} \left(0 + k_{1,-1}^{(q)} \lambda + k_{2,-1}^{(q)} \lambda^2 + k_{3,-1}^{(q)} \lambda^3 + \cdots \right) \\ &+ e^{0} \left(1 + k_{1,0}^{(q)} \lambda + k_{2,0}^{(q)} \lambda^2 + k_{3,-0}^{(q)} \lambda^3 + \cdots \right) \\ &+ e^{it} \left(0 + k_{1,1}^{(q)} \lambda + k_{2,1}^{(q)} \lambda^2 + k_{3,1}^{(q)} \lambda^3 + \cdots \right) \\ &+ e^{it} \left(0 + 0 + k_{2,2}^{(q)} \lambda^2 + k_{3,2}^{(q)} \lambda^3 + \cdots \right) + \cdots \Big\}. \end{aligned} \tag{9.2.3}$$

Note that the coefficients $k_{q,p}^{(q)}$ in (9.2.3), although known, are different from $k_{p,q}$ of (9.2.1).

Next, while computing $|w|^2$, we will group together terms that have same powers of λ. Thus, the coefficients of λ^p include only $a_q^{(s)}$, $s = 0,1,2,\dots,p$ ($s > p$ does not occur). This feature provides an application of the *method of successive approximations* to determine $a_q^{(p)}$ under the condition that the coefficient of λ^p in $|w|^2$ must vanish for $p = 0,1,2,\dots$ (which yields a system of equations) and the free term (all terms without λ) must be equal to R^2 (or 1 if the region D_λ is mapped onto the unit disk).

Since $a_p^{(q)}$ does not depend on $a_q^{(s)}$ for $s > p$, any subsequent revision of (9.2.2) to include higher powers of λ than previously taken will only entail determination of new terms that should be added to the free term and to each equation of the above system to be solved by successive approximations.

An *algorithm* to compute (9.2.2) up to λ^N is as follows:
Use all terms in (9.2.1) corresponding to λ^0, λ^1, λ^2, ..., λ^N.
Use all terms in z^2 corresponding to λ^0, λ^1, λ^2, ..., λ^{N-1}.
Continue for $z^3,\dots,z^p$, i.e., use all terms in z^p corresponding to λ^0, λ^1, λ^2, ..., λ^{N-p+1}.
Continue until z^{N+1}, i.e., use all terms in z^N corresponding to λ^0, λ^1.
Use all terms in z^{N+1} corresponding to λ^0.
Note that the sum of all coefficients of λ^0 yields the free term, and the coefficients of $\lambda^1,\lambda^2,\dots,\lambda^N$ equated to zero yield a system of N equations to determine $a_q^{(s)}$, $s = 1,\dots,N$,

by successive approximations. This method will produce an approximate function that maps every single boundary (for a fixed λ) by a power series in z, whose accuracy will depend on the manner in which $a_q^{(p)}$ are computed.

Details for computation of $a_1^{(0)}, a_1^{(1)}, a_2^{(1)}$ and $a_1^{(2)}, a_2^{(2)}, a_3^{(2)}$ are given below. Without using any computational tools, we will determine $f(z,\lambda)$ up to λ^2.

$p = 0$ yields $a_1^{(0)} = 1$. Then

$$z = e^{it}, \quad w = a_1^{(0)}\, z = z, \quad |w|^2 = a_1^{(0)}\, e^{it} \cdot \overline{a_1^{(0)}}\, e^{-it} = R^2.$$

$p = 1$ yields

$$\begin{aligned}
z &= e^{it}\left\{1 + \lambda\left[k_{1,-1}^{(1)}\, e^{-it} + k_{1,0}^{(1)} + k_{1,1}^{(1)}\, e^{it}\right]\right\},\\
z^2 &= e^{2it},\\
w &= z + \lambda\left[a_1^{(1)}\, z + a_2^{(1)}\, z^2\right]\\
&= e^{it}\left\{1 + \lambda\left[k_{1,-1}^{(1)}\, e^{-it} + k_{1,0}^{(1)} + k_{1,1}^{(1)}\, e^{it}\right]\right\}\\
&\quad + \lambda\left[a_1^{(1)}\, e^{it} + a_2^{(1)}\, e^{2it}\right]\\
&= e^{it}\left\{1 + \lambda\left[k_{1,-1}^{(1)}\, e^{-it} + \left(k_{1,0}^{(1)} a_1^{(1)}\right) + \left(k_{1,1}^{(1)} + a_2^{(1)}\right) e^{it}\right]\right\},\\
|w|^2 &= 1 + \lambda\Big[\left(k_{1,-1}^{(1)} + \overline{k_{1,1}^{(1)}} + \overline{a_2^{(1)}}\right) e^{-it} + \Big(k_{1,0}^{(1)} + a_1^{(1)} + \overline{k_{1,0}^{(1)}}\\
&\quad + \overline{a_1^{(1)}}\Big) + \left(k_{1,1}^{(1)} + a_2^{(1)} + \overline{k_{1,-1}^{(1)}}\right) e^{it}\Big]\\
&= 1 + \lambda\Big[\left(\overline{M_1^{(1)}} + \overline{a_2^{(1)}}\right) e^{-it} + \left(M_0^{(1)} + \overline{M_0^{(1)}} + a_1^{(1)} + \overline{a_1^{(1)}}\right)\\
&\quad + \left(M_1^{(1)} + a_2^{(1)}\right) e^{it}\Big] = R^2.
\end{aligned}$$

Hence $a_1 = -M_0^{(1)}$, $a_2 = -M_1^{(1)}$, and

$$w = f(z,\lambda) = z + \lambda\left[L_{-1}^{(1)} + L_0^{(1)}\, z + L_1^{(1)}\, z^2\right].$$

$p = 2$ yields

$$\begin{aligned}
z &= e^{it}\Big\{1 + \lambda\left[k_{1,-1}^{(1)}\, e^{-it} + k_{1,0} + k_{1,1}\, e^{it}\right]\\
&\quad + \lambda^2\left[k_{2,-2}^{(1)}\, e^{-2it} + k_{2,-1}^{(1)}\, e^{-it} + k_{2,0}^{(1)} + k_{2,1}^{(1)}\, e^{it} + k_{2,2}^{(1)}\, e^{2it}\right]\Big\},\\
z^2 &= e^{2it}\left\{1 + \lambda\left[k_{1,-1}^{(2)}\, e^{-it} + k_{1,0}^{(2)} + k_{2,1}^{(2)}\, e^{it}\right]\right\},\\
z^3 &= e^{3it},\\
w &= z + \lambda\left[a_1^{(1)}\, z + a_2^{(1)}\, z^2\right] + \lambda^2\left[a_1^{(2)}\, z + a_2^{(2)}\, z^2 + a_3^{(2)}\, z^3\right],
\end{aligned}$$

which gives

$$
\begin{aligned}
w = e^{it} \Big\{ 1 &+ \lambda \left[L_{-1}^{(1)} e^{-it} + L_0^{(1)} + L_1^{(1)} e^{it} \right] \\
&+ \lambda^2 \big[k_{2,-2}^{(1)} e^{-2it} + k_{2,-1}^{(1)} e^{-it} + k_{2,0}^{(1)} + k_{2,1}^{(1)} e^{it} + k_{2,2}^{(1)} e^{2it} \\
&+ a_1^{(1)} k_{1,-1}^{(1)} e^{-it} + a_1^{(1)} k_{1,0}^{(1)} + a_1^{(1)} k_{2,1}^{(1)} e^{it} \\
&+ a_2^{(1)} k_{1,-1}^{(2)} + a_2^{(1)} k_{1,0}^{(2)} e^{it} + a_2^{(1)} k_{2,1}^{(2)} e^{2it} \\
&+ a_1^{(2)} + a_2^{(2)} e^{it} + a_3^{(2)} e^{it} \big] \Big\} \\
= e^{it}\, 1 &+ \lambda \left[L_{-1}^{(1)} e^{-it} + L_0^{(1)} + L_1^{(1)} e^{it} \right] \\
&+ \lambda^2 \big[K_{-2}^{(2)} e^{-2it} + K_{-1}^{(2)} e^{-it} + \left(K_0^{(2)} + a_1^{(2)} \right) \\
&+ \left(K_1^{(2)} + a_2^{(1)} \right) e^{it} + \left(K_2^{(2)} + a_3^{(2)} \right) e^{2it} \big] \Big\},
\end{aligned}
$$

where

$$
\begin{aligned}
K_{-2}^{(2)} &= k_{2,-2}^{(1)}, \\
K_{-1}^{(2)} &= k_{2,-1}^{(1)} + a_1^{(1)} k_{1,-1}^{(1)}, \\
K_0^{(2)} &= k_{2,0}^{(1)} + a_1^{(1)} k_{1,0}^{(1)} + a_2^{(1)} k_{1,-1}^{(2)} + a_1^{(2)}, \\
K_1^{(2)} &= k_{2,1}^{(1)} + a_1^{(1)} k_{1,1}^{(1)} + a_2^{(1)} k_{1,0}^{(2)} + a_2^{(2)}, \\
K_2^{(2)} &= k_{2,2}^{(1)} + a_2^{(1)} k_{1,1}^{(2)} + a_3^{(2)},
\end{aligned}
$$

and $K_{-j}^{(q)} = \overline{K_j^{(q)}} = K_j^{(q)}$. Hence,

$$
\begin{aligned}
|w|^2 = 1 + \lambda^2 \Big[&\left(M_2^{(2)} + a_3^{(2)} \right) e^{-2it} + \left(M_1^{(2)} + a_2^{(2)} \right) e^{it} \\
&+ 2 \left(M_0^{(2)} + a_1^{(2)} \right) + \left(M_1^{(2)} + a_2^{(2)} \right) e^{it} \\
&+ \left(M_2^{(2)} + a_3^{(2)} \right) e^{2it} \Big] = R^2,
\end{aligned}
$$

where $M_{-j}^{(q)} = \overline{M_j^{(q)}} = M_j^{(q)}$, and

$$
\begin{aligned}
M_2^{(2)} &= K_{-2}^{(2)} + \overline{K_2^{(2)}} + L_{-1}^{(1)} L_1^{(1)} = 2 \left(K_1^{(2)} + L_1^{(1)} \right), \\
M_1^{(2)} &= K_1^{(2)} + \overline{K_{-1}^{(2)}} + L_0^{(1)} \overline{L_{-1}^{(1)}} + \overline{L_0^{(1)}} L_1^{(1)} = 2 \left(K_1^{(2)} + L_0^{(1)} L_1^{(1)} \right), \\
M_0^{(2)} &= K_0^{(2)} + L_{-1}^{(1)} \overline{L_{-1}^{(1)}} + L_0^{(1)} \overline{L_0^{(1)}} = K_0^{(2)} + 2 \left(L_0^{(1)} + L_1^{(1)} \right).
\end{aligned}
$$

Hence, $a_1^{(2)} = -M_0^{(2)}, a_2^{(2)} = -M_1^{(2)}, a_3^{(2)} = -M_2^{(2)}$, and

$$
w = f(z, \lambda) = z - \lambda \left[M_0^{(2)} + M_1^{(2)} \right] - \lambda^2 \left[M_0^{(2)} z + M_1^{(2)} z^2 + M_2^{(2)} z^3 \right].
$$

Map 9.3. We will consider the same problem as in Map 9.1, where the contour Γ_λ is given by (9.1.16), i.e.,

$$
z(t, \lambda) = e^{it} \left(1 + \lambda [0 \cdot e^{-it} + 0 + 0 \cdot e^{it}] + \lambda^2 [1 \cdot e^{-2it} + 0 \cdot e^{-it} + 0 + 0 \cdot e^{it} + 0 \cdot e^{2it}] \right).
$$

The coefficients $a_q^{(p)}$ for $N = 8$ are given by

$p \setminus q$	1	2	3	4	5	6	7	8	9
0	1								
1	0	0							
2	0	0	−1						
3	0	0	0	0					
4	2	0	0	0	1				
5	0	0	0	0	0	0			
6	0	0	−3	0	0	0	−1		
7	0	0	0	0	0	0	0	0	
8	1	0	0	0	5	0	0	0	1

Hence, the mapping function is given by

$$w = f(z,\lambda) = z - \lambda^2 z^3 + \lambda^4 \left(2z + z^5\right) - \lambda^6 \left(3z^3 + z^7\right) + \lambda^8 \left(z + 5z^5 + z^9\right). \tag{9.2.4}$$

This method can be improved as follows: Since

$$[z(t,\lambda)]^2 = e^{2it} \left[1 + 2\lambda^2 e^{-2it} + \lambda^4 e^{-4it}\right], \tag{9.2.5}$$

we can redefine the contour Γ_λ by

$$\Gamma_\mu : Z(\tau,\mu) = e^{i\tau} \left[1 + 2\mu e^{-i\tau} + \mu^2 e^{-2i\tau}\right], \tag{9.2.6}$$

where $\mu = \lambda^2$ and $\tau = 2t$. The new contour Γ_μ yields the mapping function $w = \phi(Z,\mu)$ such that $\phi(z^2,\lambda^2) \equiv [f(z,\lambda)]^2$, since $f(z,\lambda)$ has the form $z \cdot g(z^2,\lambda^2)$, i.e.,

$$[f(z,\lambda)]^2 = z^2 \left[g(z^2,\lambda^2)\right]^2 = Z\,[g(Z,\mu)]^2.$$

Also, since $f\left(z(t,\lambda),\lambda\right) = e^{i\theta}$ implies that $\left[f\left(z(t,\lambda),\lambda\right)\right]^2 = e^{2i\theta}$, we find that

$$Z(\tau,\mu) \cdot \left[g\left(Z(\tau,\lambda),\lambda\right)\right]^2 = e^{2i\theta},$$

and for $z^2 = Z = 0$ we get $[f(z,\lambda)]^2 = 0$, where $\dfrac{\partial}{\partial z}[f(z,\lambda)]^2$ is real. Hence, the functions $\phi(Z,\mu)$ and $Z \cdot [g(Z,\mu)]^2$ yield the same mapping function of the contour Γ_μ onto the unit circle for $0 \le \tau < 2\pi$ for every fixed $\mu = \lambda^2$ as the function $[z(t,\lambda)] \cdot \left[g(z^2(t,\lambda),\lambda^2)\right]^2$ for $0 \le t < \pi$. An advantage of this technique is that the computation of $\phi(z^2,\lambda^2)$ up to the power μ^N provides the function $f(z^2,\lambda^2)$ up to the power λ^{2N}, which while computing $\sqrt{\phi(z^2,\lambda^2)}$, will yield a mapping function with power up to λ^{16} instead of λ^8 as in (9.2.4).

Map 9.4. We will again consider the contour (9.1.16) for the ellipse, which we rewrite as

$$Z = e^{i\tau} \left[1 + \mu \left[2 \cdot e^{-i\tau} + 0 + 0 \cdot e^{i\tau}\right] + \mu^2 \left[1 \cdot e^{-2i\tau} + 0 \cdot e^{it\tau} + 0 + 0 \cdot e^{i\tau} + 0 \cdot e^{2i\tau}\right]\right].$$

Taking $N = 8$ we obtain the coefficients $a_q^{(p)}$ as follows:

$p \setminus q$	1	2	3	4	5	6	7	8
0	1							
1	0	−2						
2	4	0	3					
3	0	−10	0	−4				
4	6	0	20	0	5			
5	0	−28	0	−34	0	−6		
6	8	0	77	0	52	0	7	
7	0	−62	0	−164	0	−74	0	−8

Thus, after replacing μ by λ^2 and τ by $2t$, we have

$$
\begin{aligned}
[f(z,\lambda)]^2 = z^2 &+ \lambda^2\left(-2z^4\right) + \lambda^4\left(4z^2+3z^6\right) + \lambda^6\left(-10z^4-4z^8\right) \\
&+ \lambda^8\left(6z^2+20z^6+5z^{10}\right) + \lambda^{10}\left(-28z^4-34z^8-6z^{12}\right) \\
&+ \lambda^{12}\left(8z^2+77z^6+52z^{10}+7z^{14}\right) \\
&+ \lambda^{14}\left(-62z^4-164z^8-74z^{12}-8z^{16}\right).
\end{aligned}
$$

An application of the binomial expansion to the above expression yields the mapping function as

$$
\begin{aligned}
f(z,\lambda) = z &+ \lambda^2\left(-z^3\right) + \lambda^4\left(2z+z^5\right) + \lambda^6\left(-3z^3-z^7\right) \\
&+ \lambda^8\left(z+5z^5+z^9\right) + \lambda^{10}\left(-7z^3-7z^7-z^{11}\right) \\
&+ \lambda^{12}\left(2z+16z^5+9z^9+z^{13}\right) \\
&+ \lambda^{14}\left(-12z^3-29z^7-11z^{11}-z^{15}\right). \qquad (9.2.7)
\end{aligned}
$$

9.3 Special Cases

Map 9.5. In the case of conformal mapping of the unit disk $|z| < 1$ onto a simply connected region D in the w-plane, the mapping function $z = F(w)$ can be represented by a Taylor series

$$
z = x + i\,y = \sum_{n=0}^{\infty} c_n w^n, \quad c_n = a_n + i\,b_n. \qquad (9.3.1)
$$

In particular when the region D is also the unit disk, bounded by $|w| = 1$, we set $w = e^{i\,\phi}$ in (9.3.1) and obtain

$$
z = \sum_{n=0}^{\infty} c_n\, e^{i\,n\phi}, \qquad (9.3.2)
$$

which, when separated into real and imaginary parts, gives

$$
\begin{aligned}
x = f(\phi) &= \sum_{n=0}^{\infty} \left(a_n \cos n\phi - b_n \sin n\phi\right), \\
y = g(\phi) &= \sum_{n=0}^{\infty} \left(b_n \sin n\phi + a_n \cos n\phi\right).
\end{aligned} \qquad (9.3.3)
$$

Thus, the mapping function (9.3.1) can be represented in the form of a Cauchy integral as

$$z = \frac{1}{2\pi} \int_0^{2\pi} \frac{f(\phi) + i\, g(\phi)}{e^{i\phi}}\, e^{i\phi} d\phi. \tag{9.3.4}$$

The method of infinite systems is useful when the boundary Γ is defined by an implicit function.

Case 9.1. Let Γ be defined implicitly by

$$\gamma(x, y) = 0, \tag{9.3.5}$$

where $\gamma(x, y) = \gamma(z)$ is an analytic function for $z \in \Gamma = \partial B(0, 1)$. Then, by substituting x and y from (9.3.4) in (9.3.5) and expanding the result in a Fourier series, we obtain

$$\begin{aligned} \gamma(x, y) &= \gamma\left(f(\phi), g(\phi)\right) \\ &= \gamma_0(a_j, b_j) + \sum_{n=1}^{\infty} \left\{\gamma_n(a_j, b_j) \cos n\phi + \gamma_n^*(a_j, b_j) \sin n\phi\right\}, \end{aligned} \tag{9.3.6}$$

where $\gamma_0, \gamma_n, \gamma_n^*$ are the Fourier coefficients, and a_j, b_j denote the dependence of γ and γ^* on $a_0, b_0, a_1, b_1, \dots$. Now, by equating the coefficients of cos $n\phi$ and sin $n\phi$ to zero, we obtain an infinite system of equations for (a_j, b_j):

$$\gamma_0(a_j, b_j) = 0, \; \gamma_n(a_j, b_j) = 0, \; \gamma_n^*(a_j, b_j) = 0, \quad n = 1, 2, \dots. \tag{9.3.7}$$

Case 9.2. Let Γ be defined by an implicit complex function

$$\psi(z, \bar{z}) = 0. \tag{9.3.8}$$

Then, by substituting the series (9.3.2) for z in (9.3.8), we find that

$$\psi\left(\sum_{n=0}^{\infty} c_n\, e^{i n\phi}, \sum_{n=0}^{\infty} \bar{c}_n\, e^{-i n\phi}\right) = \sum_{n=-\infty}^{\infty} \psi_n(a_j, b_j)\, e^{i n\theta}, \tag{9.3.9}$$

which leads to an infinite system

$$\psi_n(a_j, b_j) = 0, \quad n = \dots, -1, 0, 1, \dots. \tag{9.3.10}$$

In order to obtain a solution for the unknowns $a_0, b_0, a_1, b_1, \dots$ from the system (9.3.7) and (9.3.10) it is necessary that all related series converge and the derivative $\left.\frac{dz}{dw}\right|_{|w|=1} > 0$.

METHOD. Let a nearly circular boundary Γ_λ be represented by

$$x^2 + y^2 + \lambda\, P(x, y) = 1 \quad \text{in Case 9.1,} \tag{9.3.11}$$

or by

$$z\bar{z} + \lambda\, \Pi(z, \bar{z}) = 1 \quad \text{in Case 9.2,} \tag{9.3.12}$$

where λ is a small real parameter and $P(x,y)$ and $\Pi(x,y)$ satisfy the same conditions as the functions $\gamma(x,y)$ and $\psi(x,y)$, respectively. Note that both equations (9.3.11) and (9.3.12) are equivalent since $2x = z+\bar{z}$ and $2i\,y = z-\bar{z}$. Hence, we can use either equation. Suppose we consider Eq (9.3.12). Then

$$z\bar{z} = \cdots + (c_0\bar{c}_1 + c_1\bar{c}_2 + \cdots)\,e^{-i\theta} + (c_0\bar{c}_0 + c_1\bar{c}_1 + \cdots) + (c_1\bar{c}_0 + c_2\bar{c}_1 + \cdots)\,e^{i\theta} + \cdots. \tag{9.3.13}$$

In $\Pi(z,\bar{z})$ the coefficients of conjugate quantities $e^{in\theta}$ and $e^{-in\theta}$ must also be conjugate. Also, $\Pi(z,\bar{z})$ must be real since in Eq (9.3.12) both $z\bar{z}$ and λ are real. Thus, after substituting the values of z and $\bar{z}$ from (9.3.2), we get

$$\Pi(z,\bar{z}) = \sum_{n=0}^{\infty} \tau_n(a_j,b_j)\,e^{in\theta} + \sum_{n=0}^{\infty} \overline{\tau_n(a_j,b_j)}\,e^{-in\theta}, \tag{9.3.14}$$

where

$$\tau_n(a_j,b_j) = t_n(a_j,b_j) + i\,t_n^*(a_j,b_j), \quad t_0^*(a_j,b_j) = 0. \tag{9.3.15}$$

Substituting the quantities (9.3.13) and (9.3.14) in (9.3.12) and comparing the coefficients of positive powers of $e^{i\theta}$, we obtain the infinite system of equations

$$\begin{aligned} c_0\bar{c}_0 + c_1\bar{c}_1 + c_2\bar{c}_2 + \cdots + \lambda\,t_0(a_j,b_j) &= R^2, \\ c_1\bar{c}_0 + c_2\bar{c}_1 + c_3\bar{c}_2 + \cdots + \lambda\tau_1(a_j,b_j) &= 0, \\ c_2\bar{c}_0 + c_3\bar{c}_1 + c_4\bar{c}_2 + \cdots + \lambda\tau_2(z_j,b_j) &= 0, \\ \cdots \quad \cdots \quad \cdots \quad \cdots \quad &\cdots. \end{aligned} \tag{9.3.16}$$

Note that we will obtain an infinite system conjugate to (9.3.16) if we compare the coefficients of negative powers of $e^{i\theta}$ in Eq (9.3.12). The system (9.3.16), except for the first equation, can be rewritten as

$$\begin{aligned} c_1\bar{c}_1 &= 1 - c_0\bar{c}_0 - c_2\bar{c}_2 - \cdots - \lambda\,t_0(a_j,b_j), \\ c_2 &= -\frac{c_1\bar{c}_0}{\bar{c}_1} - \frac{c_3\bar{c}_2}{\bar{c}_1} - \cdots - \frac{\lambda}{\bar{c}_1}\tau_1(a_j,b_j), \\ c_3 &= -\frac{c_2\bar{c}_0}{\bar{c}_1} - \frac{c_4\bar{c}_2}{\bar{c}_1} - \cdots - \frac{\lambda}{\bar{c}_1}\tau_2(z_j,b_j), \\ \cdots \quad &\cdots \quad \cdots \quad \cdots \quad \cdots. \end{aligned} \tag{9.3.17}$$

Map 9.6. We will use the above method to obtain the function that maps the unit disk onto the interior of the ellipse

$$(1+\lambda)\,x^2 + (1-\lambda)y^2 = 1, \tag{9.3.18}$$

with semi-major and semi-minor axes as $(1+\lambda)^{-1/2}$ and $(1-\lambda)^{-1/2}$, respectively, such that the center $w=0$ goes into the center $z=0$ and the real axis into the real axis. Then $c_0 = 0$, and c_1 will be real. Because of the symmetry of the ellipse about the x-axis, all b_j, $j = 1,2,\ldots$ will be zero. The symmetry of the y-axis also implies that all a_j with even j will be zero. The equation of the ellipse (9.3.18) in the complex form is

$$z\bar{z} + \lambda\frac{z^2+\bar{z}^2}{2} = 1.$$

Hence,

$$\Pi(z, \bar{z}) = \frac{z^2 + \bar{z}^2}{2}.$$

Now, on Γ_λ $(\tau = e^{i\theta})$,

$$\begin{aligned} z^2 &= \left(a_1 e^{i\theta} + a_3 e^{3i\theta} + \cdots\right)^2, \\ &= a_1^2 e^{2i\theta} + 2a_1 a_3 e^{4i\theta} + \left(2a_1 a_5 + a_3^2\right) e^{6i\theta} \\ &\quad + 2\left(a_1 a_7 + a_3 a_5\right) e^{8i\theta} + \left(2a_1 a_9 + 2a_3 a_7 + a_5^2\right) e^{10i\theta} + \cdots, \end{aligned}$$

which yields $(\tau_n = e^{in\theta})$

$$\begin{aligned} &\tau_0 = \tau_1 = \tau_3 = \tau_5 = \ldots = 0, \\ &\tau_2 = -\frac{1}{2}a_1^2, \ \tau_4 = -a_1 a_3, \ \tau_6 = -(a_1 a_5 + \frac{1}{2}a_3^2), \\ &\tau_8 = -(a_1 a_7 + a_3 a_5), \quad \tau_{10} = a_1 a_9 + a_3 a_7 + \frac{1}{2}a_5^2, \end{aligned}$$

and so on. Thus, the system (9.3.17) becomes

$$\begin{aligned} a_1^2 &= 1 - a_3^2 - a_5^2 - a_7^2 - a_9^2 - \cdots, \\ a_3 &= -\frac{1}{a_1}\left(a_5 a_3 + a_7 a_5 + a_9 a_7 + \cdots + \frac{\lambda}{2}a_1^2\right), \\ a_5 &= -\frac{1}{a_1}\left(a_7 a_3 + a_9 a_5 + a_{11} a_7 + \cdots + \lambda a_1 a_3\right), \\ a_7 &= -\frac{1}{a_1}\left(a_9 a_3 + a_{11} a_5 + \cdots + \lambda(a_1 a_5 + \frac{1}{2}a_3^2)\right), \\ a_9 &= -\frac{1}{a_1}\left(a_{11} a_3 + \cdots + \lambda(a_1 a_7 + a_3 a_5)\right), \\ a_{11} &= -\frac{1}{a_1}\left(a_{13} a_3 + \cdots + \lambda(a_1 a_9 + a_3 a_7)\frac{1}{2}a_5^2\right), \quad \cdots. \end{aligned} \tag{9.3.19}$$

If we introduce the notation $A_0 = a_1$, $A_j = \dfrac{a_{2j} + 1}{a_1}$ for $j = 1, 2, \ldots$, then the system (9.3.19) reduces to

$$\begin{aligned} A_0 &= \left(1 + \sum_{j=1}^{\infty} A_j^2\right)^{-1/2}, \\ A_1 &= -\frac{\lambda}{2} - \sum_{j=1}^{\infty} A_j\, A_{j+1}, \\ A_2 &= -\lambda A_1 - \sum_{j=1}^{\infty} A_j\, A_{j+2}, \\ A_3 &= -\lambda\left(A_2 + A_1^2\right) - \sum_{j=1}^{\infty} A_j\, A_{j+3}, \end{aligned}$$

$$A_4 = -\lambda\,(A_3 + A_1\,A_2) - \sum_{j=1}^{\infty} A_j\,A_{j+4},$$

$$A_5 = -\lambda\left(A_4 + A_1\,A_3 + \frac{1}{2}A_2^2\right) - \sum_{j=1}^{\infty} A_j\,A_{j+5}, \quad \cdots .$$

Holding the first equation of the above system, thus treating A_0 as undetermined, we will use the method of §9.2 and solve the remaining equations in this system by successive approximations. Let the initial values be taken as $A_j = 0$ for $j = 1, 2, \ldots$. The fifth approximations for A_1, A_2, A_3, A_4, and A_5 are available in Table E.2 (Appendix E). Then the first equation of the above system yields

$$\begin{aligned} A_0 &= 1 - \frac{1}{2}\left(A_1^2 + A_2^2 + \cdots\right) + \frac{\frac{1}{2}\cdot\frac{3}{2}}{2!}\left(A_1^2 + A_2^2 + \cdots\right)^2 + \cdots \\ &= 1 - \frac{1}{8}\lambda^2 + \frac{3}{128}\lambda^4, \end{aligned} \tag{9.3.20}$$

which is accurate to λ^5. Hence, the mapping function is given by

$$\begin{aligned} z &= A_0\,w\Big[1 + \sum_{n=1}^{\infty} A_n\,w^{2n}\Big] \\ &= \left(1 - \frac{1}{8}\lambda^2 + \frac{3}{128}\lambda^4\right)\left\{w - \left(\frac{\lambda}{2} - \frac{\lambda^3}{4} + \frac{3\lambda^5}{32}\right)w^3 \right. \\ &\quad \left. + \left(\frac{\lambda^2}{2} - \frac{9\lambda^4}{16}\right)w^5 - \left(\frac{5\lambda^3}{8} - \frac{9\lambda^5}{8}\right)w^7 + \frac{7\lambda^4}{8}w^9 + \frac{21\lambda^5}{16}w^{11}\right\}, \end{aligned} \tag{9.3.21}$$

which is accurate to λ^5.

Map 9.7. We will compute the function that maps the family of squares

$$z\,\bar{z} + k\left(\frac{z^2 - \bar{z}^2}{4}\right)^2 = 1 \tag{9.3.22}$$

in the z-plane onto the unit circle $|w| = 1$ such that the point $w = 0$ goes into the point $z = 0$. Note that for $k = 1$, Eq (9.3.22) reduces to $\left(x^2 - 1\right)\left(y^2 - 1\right) = 0$ which represents the sides of the square of Figure 9.1.

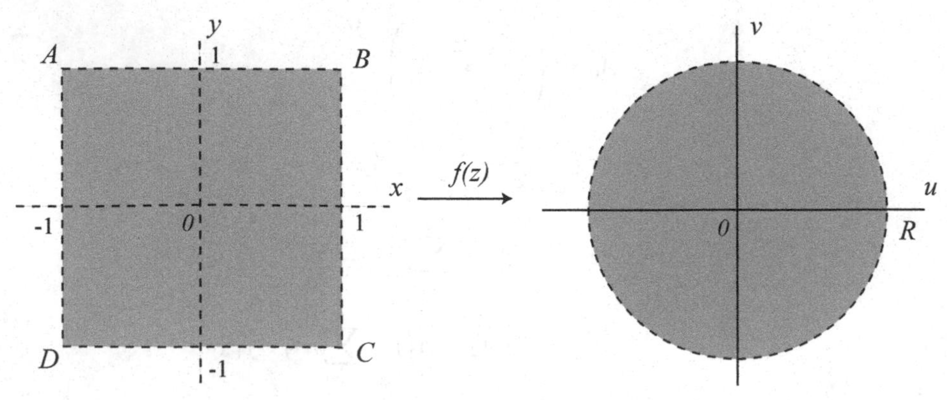

Figure 9.1 Map 9.7.

Obviously, $c_0 = 0$, and since the real axes are preserved, $\arg\{c_1\} = 0$. The square (9.3.22) is symmetric about the x and y axes and also about the lines $y = \pm x$. Hence, $b_j = 0$, $a_{2j} = 0$, and $a_{4j-1} = 0$ for $j = 1, 2, \dots$. Then, from (9.3.2)

$$z = \sum_{n=1}^{\infty} a_{4n-3}\, e^{i(4n-3)\theta}, \tag{9.3.23}$$

which gives

$$\begin{aligned}\Pi(z,\bar z) &= \left(\frac{z^2-\bar z^2}{4}\right)^2\\ &= \frac{1}{16}\left[a_1^2\, e^{2i\theta} + 2a_1\, a_5\, e^{6i\theta} + \left(2a_1\, a_9 + a_5^2\right)\, e^{10i\theta} + \cdots\right.\\ &\quad \left. - a_1^2\, e^{-2i\theta} - 2a_1\, a_5\, e^{-6i\theta} + \left(2a_1\, a_9 + a_5^2\right)\, e^{-10i\theta} + \cdots\right]^2.\end{aligned}$$

Thus,

$$\begin{aligned}\tau_0(a_j,b_j) &= -\frac{1}{2}\left[\left(\frac{a_1^2}{2}\right)^2 + (a_1a_5)^2 + \left(a_1a_9 + \frac{1}{2}\,a_5^2\right)^2 (a_1a_{13}+a_5a_9)^2 + \cdots\right],\\ \tau_4(a_j,b_j) &= -\frac{1}{2}\left[\frac{a_1^4}{8} + \frac{a_1^3a_5}{2} + \left(a_1a_5a_9 + \frac{1}{2}\,a_1a_5^3\right) + \cdots\right],\\ \tau_8(a_j,b_j) &= -\frac{1}{2}\left[-\frac{a_1^3a_5}{2} + \left(\frac{a_1^3a_9}{2} + \frac{a_1^2a_5^2}{2}\right) + \left(a_1^2a_5a_9 + \frac{a_1a_5^3}{2}\right) + \cdots\right],\\ \tau_{12}(a_j,b_j) &= -\frac{1}{2}\left[-\left(\frac{a_1^3a_9}{2} + \frac{a_1^2a_5^2}{4}\right) - \frac{a_1^2a_5^2}{2} + \left(\frac{a_1^3a_{13}+a_1^2a_5a_9}{2}\right) + \cdots\right],\end{aligned}$$

which yields the system (9.3.19) as

$$\begin{aligned}a_1^2 &= 1 - a_5^2 - a_9^2 - a_{13}^2 - \cdots,\\ a_3 &= -\frac{a_3a_7}{a_1} - \frac{a_7a_{13}}{a_1} - \cdots + \frac{k}{2a_1}\left[\frac{a_1^4}{4} + a_1^2a_5^2 + \left(a_1a_9 + \frac{a_5^2}{2}\right)^2 + (a_1a_{13}+a_5a_9)^2 + \cdots\right],\\ a_5 &= -\frac{a_5a_9}{a_1} - \frac{a_9a_{13}}{a_1} - \cdots + \frac{k}{2a_1}\left[-\frac{a_1^4}{8} + \frac{a_1^3a_5}{2} + \left(a_1^2a_9 + \frac{1}{2}\,a_1a_5^3\right)\right.\\ &\qquad \left. + (a_1a_{13}+a_5a_9)\left(a_1a_9 + \frac{1}{2}\,a_1a_5^2\right) + \cdots\right],\\ a_9 &= -\frac{a_5a_{13}}{2} - \cdots + \frac{k}{2a_1}\left[-\frac{a_1^3a_5}{2} + \frac{a_1^3a_9}{2} + \frac{a_1^2a_5^2}{4} + \left(a_1^2a_5a_{13} + a_1a_5^2a_9\right) + \cdots\right],\\ a_{13} &= -\frac{a_5a_{17}}{2} - \cdots + \frac{k}{2a_1}\left[-\frac{a_1^3a_9}{2} - \frac{a_1^2a_5^2}{2} - \frac{a_1^2a_5^2}{2} + \frac{1}{2}\left(a_1^3a_{13} + a_1^2a_5a_9\right) + \cdots\right].\end{aligned}$$

If we take the initial values as $a_1 = 1$, $a_5 = a_9 = \cdots = 0$, then computing up to the third successive approximations (see Table E.3, Appendix E), we get the approximate mapping function as

$$\begin{aligned}z &= \left(1 + \frac{k}{16} + \frac{3k^2}{256} + \frac{3k^2}{1024}\right) w - \left(\frac{k}{16} + \frac{7k^2}{256} + \frac{11k^2}{1024}\right) w^5\\ &\quad + \left(\frac{k^2}{64} + \frac{27k^3}{2048}\right) w^9 - \frac{11k^3}{2048}\, w^{13}.\end{aligned} \tag{9.3.24}$$

For $k = 1$, this becomes

$$z = 1.077\, w - 0.1006\, w^5 + 0.0288\, w^9 - 0.0054\, w^{13},$$

which compares with the exact solution (H.30) with a maximum error of the order of 10^{-3}.

If the boundary Γ_λ is defined by the parametric equations

$$x = g(t, \lambda), \quad y = h(t, \lambda), \tag{9.3.25}$$

then we can expand the functions g and h as trigonometric series

$$\begin{aligned} g(x, \lambda) &= \alpha_0 + \sum_{n=1}^{\infty} \left(\alpha_n \cos nt + \beta_n \sin nt\right), \\ h(t, \lambda) &= \gamma_0 + \sum_{n=1}^{\infty} \left(\gamma_n \cos nt - \delta_n \sin nt\right). \end{aligned} \tag{9.3.26}$$

Since the series in (9.3.26) are conjugate, we obtain the complex form for the equation of Γ_λ as

$$\begin{aligned} z = x + iy &= \pi_0 + \sum_{n=1}^{\infty} \left(\pi_n \cos nt + \rho_n \sin nt\right) \\ &= \pi_0 + \sum_{n=1}^{\infty} \left\{ \frac{\pi_n - i\rho_n}{2}\, e^{int} + \frac{\pi_n + i\rho_n}{2}\, e^{-int} \right\}, \end{aligned} \tag{9.3.27}$$

where $\pi_n = \alpha_n + i\gamma_n$, and $\rho_n = \beta_n - i\delta_n$. We will assume that the curve (9.3.27) has the same form as (9.2.1), where the coefficients π_n and ρ_n depend on λ. If we take $\lambda = 0$ in (9.3.27), then this equation for Γ_λ reduces to

$$z = G\left(e^{it}, \lambda\right) = \pi_0 + \sum_{n=1}^{\infty} \frac{\pi_n - i\rho_n}{2}\, e^{int}. \tag{9.3.28}$$

The function $w = G\left(e^{it}, \lambda\right)$, where e^{it} is a point on the unit circle is assumed to be analytic in w and λ near the values $\lambda = 0$. The parameter t represents the polar angle of the point in the w-plane such that $w = |w|\, e^{it}$ for any $w \in U$. However, the parameter t, in general, does not coincide with the argument θ taken for the values of x and y in (9.3.3). Therefore, we will substitute in (9.3.28)

$$t = \theta + \lambda\, \psi_1(\theta) + \lambda^2\, \psi_2(\theta) + \cdots, \tag{9.3.29}$$

where $\psi_j(\theta)$, $j = 1, 2, \dots$, are real, periodic functions, yet to be determined. Note that for $\lambda = 0$, the series (9.3.29) reduces to $t = \theta$. Now, the functions $\psi_j(\theta)$ must be determined for $j = 1, 2, \dots$ such that the coefficient of λ^j in the series (9.3.27) does not contain any term in negative powers of $e^{i\theta}$. This process is explained in the next case.

Let the boundary Γ_λ be defined in the parametric form by $x = \cos t + \dfrac{\lambda}{2} \cos 2t$, $y = \sin t + \dfrac{\lambda}{2} \sin 2t$ (Figure 9.2, where the dotted curve is the unit circle); this boundary in the complex form is defined by

$$z = e^{it} + \frac{\lambda}{2} \left[e^{i(n+1)t} + e^{-i(n-1)t} \right]. \tag{9.3.30}$$

Substituting for t from (9.3.29), we obtain

$$
\begin{aligned}
z = e^{i\theta} &+ \lambda\Big[i\psi_1\, e^{i\theta} + \frac{1}{2}\Big\{e^{i(n+1)\theta} + e^{-i(n+1)\theta}\Big\}\Big] \\
&+ \lambda^2\Big[\Big(i\psi_2 - \frac{1}{2}\psi_1^2\Big)\, e^{i\theta} + \frac{1}{2}\, i(n+1)\psi_1\, e^{i(n+1)\theta} \\
&- \frac{1}{2}\, i(n-1)\psi_1\, e^{i(n-1)\theta}\Big] + \cdots .
\end{aligned} \tag{9.3.31}
$$

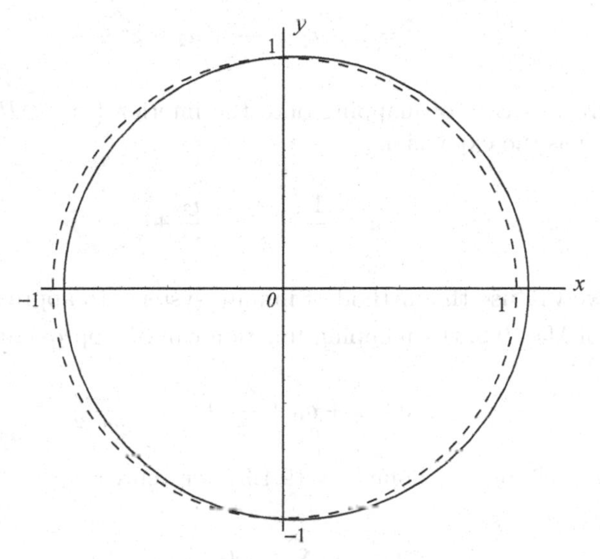

Figure 9.2 Boundary Γ_λ (solid curve).

It is obvious from this expression that z will contain only positive powers of $e^{i\theta}$ if

$$
\begin{gathered}
i\psi_1\, e^{i\theta} + \frac{1}{2}\Big\{e^{i(n+1)\theta} + e^{-i(n+1)\theta}\Big\} = e^{i(n+1)\theta}, \\
\Big(i\psi_2 - \frac{1}{2}\psi_1^2\Big)\, e^{i\theta} + \frac{i}{2}\psi_1\Big[(n+1)\, e^{i(n+1)\theta} - (n-1)\, e^{i(n-1)\theta}\Big] = e^{i(2n+1)\theta},
\end{gathered}
$$

which yields

$$
\begin{aligned}
\psi_1 &= \frac{i}{2}\left(e^{-in\theta} - e^{in\theta}\right) = \sin n\theta, \\
\psi_2 &= i\,\frac{2n-1}{8}\left(e^{-2in\theta} - e^{2in\theta}\right) = \frac{2n-1}{4}\sin 2n\theta.
\end{aligned}
$$

Substituting these values in (9.3.31), we get

$$
z = e^{i\theta} + \lambda\, e^{i(n+1)\theta} + \frac{2n+1}{4}\lambda^2\left[e^{i(2n+1)\theta} - e^{i\theta}\right], \tag{9.3.32}
$$

which is accurate up to $O(\lambda^2)$. Hence, the approximate mapping function is given by

$$
z = \left(1 - \frac{2n+1}{4}\lambda^2\right) w + \lambda\, w^{n+1} + \frac{2n+1}{4}\lambda^2\, w^{2n+1}. \tag{9.3.33}
$$

9.4 Exterior Regions

We will approximate the function that maps the region exterior to the boundary (9.1.9) or (9.2.1) onto the exterior or interior of the circle $|w| = R$, assuming that the point at infinity $z = \infty$ goes into the point $w = \infty$ or into the origin $w = 0$, respectively. There are two cases to consider:

Map 9.8. In the case of mapping onto the exterior $\{|w| > R\}$, the mapping function $w = f(z)$ with $f(\infty) = 1$ has an expansion

$$w = z + a_0 + \frac{a_1}{z} + a_2 + z^2 + \cdots \tag{9.4.1}$$

Map 9.9. In this case of mapping onto the interior $\{|w| < R\}$, the mapping function with $f(\infty) = 0$ has the expansion

$$w = \frac{1}{z} + \frac{a_2}{z^2} + \frac{a_3}{z^3} + \cdots . \tag{9.4.2}$$

In both cases we will use the method of infinite systems to approximate $w = f(z)$.

In the case of Map 9.8, the mapping function can be approximated by taking the first n terms in (9.4.1) :

$$w = z + a_0 + \frac{a_1}{z} + \cdots + \frac{a_{n-2}}{z^{n-2}}. \tag{9.4.3}$$

In order to find $|w|^2$ on the boundary (9.1.9), we represent $1/z$ in the form of the series (9.1.10). Thus,

$$\frac{1}{z} = \sum_{\nu=-\infty}^{\infty} \beta_\nu^{(-1)}(\lambda)\, e^{i\nu t}, \tag{9.4.4}$$

where the coefficients $\beta_\nu^{(-1)}(\lambda)$ are regular functions of the parameter λ, such that

$$\beta_\nu^{(-1)}(0) = \begin{cases} 0 & \text{for } \nu \neq -1 \\ 1 & \text{for } \nu = -1. \end{cases}$$

First, we compute $|w|^2$ from (9.4.1) and substitute in it the value of z from (9.1.10) and the value of $1/z$ from (9.4.4). This will yield a trigonometric series, in which we equate the free terms to R^2 (i.e., those terms which are independent of trigonometric functions), and set the coefficients of the first n terms of this series to zero. This will yield a system of equations exactly as in §9.1, which can be solved by the method of successive approximations to determine approximate values of $a_0, a_1, \dots, a_n$, and R.

Map 9.10. We will consider the mapping of the region exterior to the ellipse

$$z(t) = e^{it}\left(1 + \lambda e^{-2it}\right)$$

onto the exterior $|w| > R$. Since the region is symmetric about the coordinate axes, the mapping function has an expansion about the point at infinity

$$w = z + \frac{a_1}{z} + \frac{a_3}{z^3} + \cdots , \tag{9.4.5}$$

where all a_j are real and $w(\infty) = \infty$. Let us approximate w by a polynomial of the form

$$w = z + \frac{a_1}{z} + \frac{a_3}{z^3} + \frac{a_5}{z^5} + \frac{a_7}{z^7} + \frac{a_9}{z^9}.$$

Then

$$\begin{aligned} |w|^2 &= z\bar{z} + a_1\left(\frac{z}{\bar{z}} + \frac{\bar{z}}{z}\right) + \left[a_3\left(\frac{z}{\bar{z}^3} + \frac{\bar{z}}{z^3}\right) + a_1^2\frac{1}{z\bar{z}}\right] \\ &+ a_5\left[\left(\frac{z}{\bar{z}^5} + \frac{\bar{z}}{z^5}\right) + a_1a_3\left(\frac{1}{z^3\bar{z}} + \frac{1}{z\bar{z}^3}\right)\right] \\ &+ a_7\left[\left(\frac{z}{\bar{z}^7} + \frac{\bar{z}}{z^7}\right) + a_1a_5\left(\frac{1}{z^5\bar{z}} + \frac{1}{z\bar{z}^5}\right) + a_3^2\frac{1}{z^3\bar{z}^3}\right] \\ &+ a_9\left[\left(\frac{z}{\bar{z}^9} + \frac{\bar{z}}{z^9}\right) + a_1a_7\left(\frac{1}{z^7\bar{z}} + \frac{1}{z\bar{z}^7}\right) \right. \\ &\left. + a_3\,a_5\left(\frac{1}{z^5\bar{z}} + \frac{1}{z\bar{z}^5}\right)\right]. \end{aligned} \tag{9.4.6}$$

$$\begin{aligned} z\bar{z} &= 1 + \lambda^2 + 2\lambda\cos 2t, \\ \frac{z}{\bar{z}} + \frac{\bar{z}}{z} &= -2\left[\left(\lambda - \lambda^3\right) - \cos 2t - \left(\lambda - \lambda^3\right)\cos 4t\right], \\ \frac{z}{\bar{z}^3} + \frac{\bar{z}}{z^3} &= 2\left[6\lambda^2 - 3\lambda\cos 2t + \left(1 - 3\lambda^2\right)\cos 4t + \lambda\cos 6t\right], \\ \frac{1}{z\bar{z}} &= \left(1 + \lambda^2\right) - 2\lambda\cos 2t + 2\lambda^2\cos 4t, \\ \frac{z}{\bar{z}^5} + \frac{\bar{z}}{z^5} &= -2\left(5\lambda\cos 4t - \cos 6t - \lambda\cos 8t\right), \\ \frac{1}{z^3\bar{z}} + \frac{1}{z\bar{z}^3} &= -2\left(3\lambda - \cos 2t + \lambda\cos 4t\right), \\ \frac{z}{\bar{z}^7} + \frac{\bar{z}}{z^7} &= 2\cos 8t, \\ \frac{1}{z^5\bar{z}} + \frac{1}{z\bar{z}^5} &= 2\cos 4t, \\ \frac{1}{z^3\bar{z}^3} &= 1. \end{aligned}$$

Hence, substituting these values in (9.4.6), equating the terms independent of e^{it} to R^2, and equating the coefficient of cosines to zero, we obtain

$$\begin{gathered} 1 + \lambda^2 - 2\left(\lambda - \lambda^3\right)a_1 + 12\lambda^2 a_3 + \left(1 + \lambda^2\right)a_1^2 - 6\lambda a_1 a_3 + a_3^2 = R^2, \\ \lambda + a_1 - 3\lambda a_3 - \lambda a_1^2 + a_1 a_3 = 0, \\ \left(\lambda - \lambda^3\right)a_1 + \left(1 - 3\lambda^2\right)a_3 + \lambda^2 a_1^2 - 5\lambda a_5 - \lambda a_1 a_3 + a_1 a_5 = 0, \\ \lambda a_3 + a_5 = 0, \quad \lambda a_5 + a_7 = 0, \end{gathered} \tag{9.4.7}$$

which except for the first equation is rewritten as

$$\begin{aligned} a_1 &= -\lambda + 3\lambda a_3 + \lambda a_1^2 - a_1 a_3, \\ a_3 &= 3\lambda^2 a_3 - \left(\lambda - \lambda^3\right)a_1 - \lambda^2 a_1^2 + 5\lambda a_5 + \lambda a_1 a_3 - a_1 a_5, \\ a_5 &= -\lambda a_3, \quad a_7 = -\lambda a_5. \end{aligned}$$

Choosing the initial values for a_1, a_3, a_5, a_7 as zero, the successive approximations for these coefficients are available in Table E.4 (Appendix E), where we have retained the values up to the fourth approximation. Hence, the mapping function accurate up to λ^4 is given by

$$w = z - \frac{\lambda - 5\lambda^3}{z} + \frac{\lambda^2 - 11\lambda^4}{z^3} - \frac{\lambda^3}{z^5} + \frac{\lambda^4}{z^7},$$

and the approximate value of R^2 from the first equation in (9.4.7) is $R^2 = 1 - 4\lambda^2 - 2\lambda^8$, which yields the radius $R = 1 + 2\lambda^2 - 3\lambda^4$ (compare this value of R with that obtained in Map 9.1).

Map 9.9 can be analyzed analogously by taking the approximate function as

$$w = \frac{a_1}{z} + \frac{a_2}{z^2} + \cdots + \frac{a_n}{z^n}, \tag{9.4.8}$$

and following the above method step-by-step, where $a_1 = 1$ for a nearly circular boundary of the type (9.1.9) or (9.2.1).

Map 9.11. We will map the exterior of the square $\{-1 \leq x, y \leq 1\}$ onto the disk $|w| < R$. The equation of the square in complex form is

$$z\bar{z} + \frac{z^2 - \bar{z}^2}{4} = 1.$$

We will, however, analyze the family of curves

$$z\bar{z} + \lambda\frac{z^2 - \bar{z}^2}{4} = 1, \tag{9.4.9}$$

where $\lambda = 1$ gives the above square. Since the squares are symmetric about the coordinate axes and about the diagonals $y = \pm x$, the function w has, from (9.4.8), the form

$$z = a_1 e^{-i\theta} + a_5 e^{3i\theta} + a_9 e^{7i\theta} + \dots, \tag{9.4.10}$$

where a_j are real. Now,

$$\begin{aligned}\frac{z^2 - \bar{z}^2}{4} = \frac{1}{2}\Big\{&-\left(\frac{1}{2}a_9^2 + \dots\right)e^{-14i\theta} - (a_5 a_9 + \dots)\, e^{-10i\theta} \\ &- \left(a_1 a_9 + \frac{1}{2}a_5^2\right)e^{-6i\theta} - \left(a_1 a_5 - \frac{1}{2}a_1^2\right)e^{-2i\theta} \\ &+ \left(a_1 a_5 - \frac{1}{2}a_1^2\right)e^{2i\theta} - \left(a_1 a_9 - \frac{1}{2}a_5^2\right)e^{6i\theta} \\ &+ (a_5 a_9 + \dots)\, e^{10i\theta} - \left(\frac{1}{2}a_9^2 + \dots\right)e^{14i\theta}\Big\}.\end{aligned} \tag{9.4.11}$$

After substituting (9.4.10) and (9.4.11) in Eq (9.4.9) and comparing the coefficients of different exponential powers, we get

$$a_1^2 + a_5^2 + a_9^2 + \cdots = R^2 + \frac{\lambda}{2}\Big[\Big(a_1 a_5 - \frac{1}{2} a_1^2\Big)^2 + \Big(a_1 a_9 + \frac{1}{2} a_5^2\Big)^2$$
$$+ a_5^2 a_9^2 + \frac{1}{4} a_9^4 + \dots \Big],$$
$$a_1 a_5 + a_5 a_9 + \cdots = -\frac{\lambda}{2}\Big[\frac{1}{2}\Big(a_1 a_5 - \frac{1}{2} a_1^2\Big)^2 - \Big(a_1 a_9 + \frac{1}{2} a_5^2\Big)$$
$$\times \Big(a_1 a_5 + \frac{1}{2} a_1^2\Big)^2 - a_5 a_9 \Big(a_1 a_9 + \frac{1}{2} a_5^2\Big) - \frac{1}{2} a_9^2 a_5 a_9 - \cdots \Big]$$
$$a_1 a_9 + \cdots = -\frac{\lambda}{2}\Big[\Big(a_1 a_5 - \frac{1}{2} a_1^2\Big)\Big(a_1 a_9 + \frac{1}{2} a_5^2\Big) - a_5 a_9 \Big(a_1 a_5 + \frac{1}{2} a_1\Big)$$
$$- \frac{1}{2} a_9^2 \Big(a_1 a_9 + \frac{1}{2} a_5^2\Big) - \dots \Big].$$

Taking the initial value of $a_1 = 1$, and $a_3 = a_5 = \cdots = 0$ and using the method of successive approximations up to the third approximation (see Table E.5, Appendix E), we find that the approximate mapping function is given by

$$w = \Big(1 + \frac{\lambda}{16} + \frac{7\lambda^2}{256} + \frac{9\lambda^3}{1024}\Big)\frac{1}{z} - \Big(\frac{\lambda}{16} + \frac{7\lambda^2}{256} + \frac{9\lambda^3}{1024}\Big) z^3 + \frac{\lambda^3}{2048} z^7.$$

If we set $\lambda = 1$, we obtain

$$w = \frac{1125}{1024}\frac{1}{z} - \frac{203}{2048} z^3 + \frac{1}{2048} z^7.$$

Map 9.12. Let E denote the nearly circular ellipse $b^2u^2 + a^2v^2 = a^2b^2$, where $b = 1$ and $a = 1 + \varepsilon$. The function $f(z) = z + \frac{\varepsilon}{2} z\,(1 + z^2) + o(\varepsilon)$ maps the unit disk $|z| < 1$ onto the region Int (E) (Nehari [1952: 265]).

Map 9.13. The function

$$w = f(z) = z + \frac{\varepsilon z}{2\pi}\int_0^{2\pi} \frac{e^{it} + z}{e^{it} - z}\, p(t)\, dt + o(\varepsilon)$$

maps the unit disk $|z| < 1$ onto a nearly circular region whose boundary has the polar equation $r = 1 + \varepsilon\, p(\theta)$, where $p(\theta)$ is bounded and piecewise continuous and $\varepsilon > 0$ is a small parameter (Nehari [1952: 263]).

Map 9.14. Let the boundary Γ of a simply connected region be defined in polar coordinates by $r = 1 + \varepsilon\, g(\theta)$, $\varepsilon > 0$, where $g(\theta)$ has a finite Fourier series expansion of the form

$$g(\theta) = a_0 + \sum_{j=1}^{n} (a_j \cos j\theta + b_j \sin j\theta).$$

The function

$$f(z) = z + \varepsilon\, z + \Big[a_0 + \sum_{j=1}^{n} (a_j - i\, b_j)\, z^j\Big] + o(\varepsilon)$$

maps the unit disk $|z| < 1$ onto the nearly circular region $\mathrm{Int}\,(\Gamma)$ (Nehari [1952: 265]).

REFERENCES USED: Andersen et al. [1962], Goluzin [1937], Kantorovich and Krylov [1964], Nehari [1952].

10

Integral Equation Methods

We will discuss certain integral equations which arise in the problem of computing the function $w = f(z)$ that maps a simply connected region D, with boundary Γ and containing the origin, conformally onto the interior or exterior of the unit circle $|w| = 1$. In the case when Γ is a Jordan contour, we obtain Fredholm integral equations of the second kind $\phi(s) = \pm \int_\Gamma N(s,t)\,\phi(t)\,dt + g(s)$, where $\phi(s)$, known as the *boundary correspondence function*, is to be determined and $N(s,t)$ is the Neumann kernel. We will discuss an iterative method for numerical computation of the Lichtenstein-Gershgorin equation and present the case of a degenerate kernel and also of the Szegö kernel. The case when Γ has a corner yields Stieltjes integral equations and is presented in Chapter 13.

10.1 Neumann Kernel

Let $\Gamma : z = \gamma(s)$, $0 \le s \le L$, be a Jordan contour with continuously turning tangent and positive orientation with respect to a simply connected region $D = \text{Int}\,(\Gamma)$. We say that Γ belongs to the class Γ'_α (and write $\Gamma \in \Gamma'_\alpha$) if $z'(s)$ satisfies a Hölder condition of order α, $0 < \alpha \le 1$. Similarly, $\Gamma \in \Gamma''_\alpha$ if $z''(s)$ satisfies a Hölder condition of order α. We assume that $w = f(z)$ maps the region D univalently onto the unit disk $|w| < 1$, such that a point $z_0 \in D$ goes into $w = 0$ and a boundary point $z = e^{i\theta}$ goes into a point $w = e^{i\phi}$. The Neumann kernel $N(s,t)$ is defined for $t \ne s$ by

$$N(s,t) = \frac{\sin(\tau - \theta_s)}{\pi\, r_{st}} = \frac{1}{\pi}\frac{\partial}{\partial t}\theta_s(t) = -\frac{1}{\pi}\frac{\partial r_{st}}{\partial n_t}, \tag{10.1.1}$$

where $\tau = \tau(t)$ is the tangent angle, $\theta_s = \theta_s(t) = \arg\{\gamma(t) - \gamma(s)\}$, $r_{st} = |\gamma(t) - \gamma(s)|$, and n_t is the interior normal at $\gamma(t)$ (Figure 10.1).

This kernel first appeared in the solution of the Dirichlet problem by Carl Neumann [1877]. Some of its properties are as follows:

1. $N(s,t)$ is continuous for $t \ne s$, but, in general, it is not bounded as $t \to s$.
2. If $\Gamma \in \Gamma'_\alpha$, then $N(s,t)\,|s-t|^{1-\alpha}$ and $\dfrac{\partial}{\partial s}N(s,t)\,|s-t|^{2-\alpha}$ are bounded for $0 \le s, t \le L$.

3. If $\Gamma \in \Gamma''_\alpha$, then $\dfrac{\partial}{\partial s} N(s,t)\,|s-t|^{1-\alpha}$ is bounded for $0 \le s,\, t \le L$.
4. If Γ has a continuous curvature, then for every s_0

$$\lim_{s,t \to s_0, s_0} N(s,t) = \frac{1}{2}\kappa(s_0), \tag{10.1.2}$$

where $\kappa(s_0)$ is the curvature of Γ at the point $\gamma(s_0)$.
5. The kernel is normalized by

$$\int_\Gamma N(s,t)\,dt = 1, \quad \text{or} \quad \int_\Gamma N(t,s)\,dt = -1, \tag{10.1.3}$$

since an application of the formula (2.4.14) to $f(z) \equiv 1$ gives

$$1 = \frac{1}{\pi}\int_\Gamma \frac{\sin(\tau - \theta_s)}{r_{st}}\,dt - \frac{i}{\pi}\int_\Gamma \frac{\cos(\tau - \theta_s)}{r_{st}}\,dt. \tag{10.1.4}$$

The integral in (10.1.3) takes a Cauchy p.v.
6. From (10.1.1) we have

$$\begin{aligned}\frac{\partial}{\partial s} N(s,t) &= \frac{1}{\pi}\frac{\partial}{\partial s}\frac{\partial}{\partial t}\arg\{\gamma(t)-\gamma(s)\} = \frac{1}{\pi}\frac{\partial}{\partial t}\frac{\partial}{\partial s}\arg\{\gamma(t)-\gamma(s)\} \\ &= \frac{1}{\pi}\frac{\partial}{\partial t}\frac{\partial}{\partial s}\arg\{\gamma(s)-\gamma(t)\} = \frac{\partial}{\partial t} N(t,s),\end{aligned} \tag{10.1.5}$$

where $\dfrac{\partial}{\partial s}\dfrac{\partial}{\partial t} = \dfrac{\partial}{\partial t}\dfrac{\partial}{\partial s}$ is permitted because these mixed derivatives exist and are continuous for $t \ne s$.

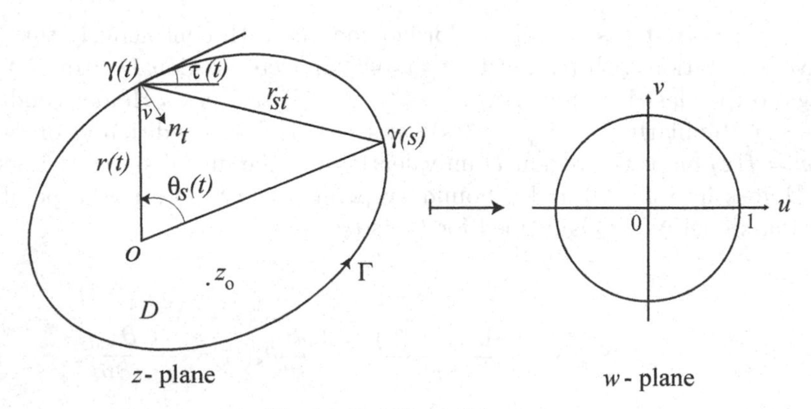

Figure 10.1 Normal kernel.

The proofs of these properties can be constructed as in Gaier [1964: 4]. This kernel plays an important role in certain integral equations that arise in conformal mapping.

Let δ denote the length of Γ. Then for all functions $f(s) \in C[0,L]$, the quadratic functional

$$\langle f, f\rangle = \int_\Gamma\int_\Gamma f(s)\,f(t)\,\log\frac{\delta}{r_{st}}\,ds\,dt, \tag{10.1.6}$$

is positive-definite. For a fixed δ we introduce the Hilbert space H which is obtained by completing the set of all functions $f(s) \in C[0,L]$ with the norm $\|f(s)\|^2$ defined by (10.1.6).

Let T denote an operator on H such that for a continuous function $f(t) \in H$

$$T\,f = \int_\Gamma N(s,t)\,f(t)\,dt. \tag{10.1.7}$$

The kernel $N(s,t)$ can be made symmetric by $\log \dfrac{\delta}{r_{st}}$. Thus,

$$M(s,t) = \int_\Gamma N(s,x)\,\log \frac{\delta}{r_{st}}\,dx \tag{10.1.8}$$

is symmetric, i.e., $M(s,t) = M(t,s)$. This implies that the operator T is hermitian: $\langle T\,f, g\rangle = \langle f, T\,g\rangle$. It can also be shown that T is a completely continuous operator on H. This means that if λ_i, $i = 1, 2, \dots$, denote all eigenvalues of an equation $\phi = \lambda T\,\phi$, where each λ_i is counted according to its multiplicity, and if h_i is a set of associated eigenfunctions such that $\langle \phi_i, \phi_j\rangle = \delta_{ij}$, where δ_{ij} is the Kronecker delta, then

$$\lambda_1 = 1 < |\lambda_2| \le |\lambda_3| \le \cdots, \tag{10.1.9}$$

and

$$\frac{1}{|\lambda_2|} \le \sup \frac{\langle T\,\phi, \phi\rangle}{\|\phi\|} \tag{10.1.10}$$

for all $\phi \in H$ with $\langle \phi, \phi_1\rangle = 0$, which implies that $\displaystyle\int_\Gamma \phi(s)\,ds = 0$ since $\phi_1(s) = \text{const.}$

10.2 Interior Regions

First, we will derive the three following integral equations when the function $w = f(z)$ maps $\text{Int}\,(\Gamma)$ conformally onto the unit disk $|w| < 1$.

10.2.1 Lichtenstein's Integral Equation. Let $w = f(z)$, $f(0) = 0$, map a simply connected region D conformally onto the unit disk $|w| < 1$ such that a boundary point $z = e^{i\theta}$ goes into a boundary point $w = e^{i\phi}$. We will consider the function

$$F(z) = \log \frac{f(z)}{z}, \tag{10.2.1}$$

which is analytic in D and continuous on $\bar{D}$, where

$$\arg\{F(z)\} = \arg\left\{f\left(e^{i\theta}\right)e^{-i\theta}\right\} = \phi(s) - \theta(s).$$

An application of Cauchy's formula (2.4.14) on $F(z)$ yields

$$\log \frac{f(z)}{z} = \frac{1}{i\pi}\int_\Gamma \log \frac{f(\zeta)}{\zeta}\,\frac{d\zeta}{\zeta - z}, \qquad z \in \Gamma. \tag{10.2.2}$$

If we set $\zeta = \gamma(t)$, $z = \gamma(s)$, then $\dfrac{d\zeta}{\zeta - z} = \dfrac{e^{i(\phi(t)-\theta_s(t))}}{r_{st}}\,dt$. Thus, equating the imaginary parts on both sides of (10.2.2), we obtain

$$\begin{aligned}\phi(s) - \theta(s) &= \frac{1}{\pi}\int_\Gamma [\phi(s) - \theta(s)]\,\frac{\sin\left[\phi(t) - \theta_s(t)\right]}{r_{st}}\,dt\\ &\quad + \frac{1}{\pi}\int_\Gamma \log r(t)\,\frac{\cos\left[\phi(t) - \theta_s(t)\right]}{r_{st}}\,dt,\end{aligned}$$

which gives *Lichtenstein's integral equation*

$$\phi(s) - \theta(s) = \frac{1}{\pi} \int_\Gamma N(s,t) \; [\phi(t) - \theta(t)] \; dt + g(s), \tag{10.2.3}$$

where

$$g(s) = \frac{1}{\pi} \int_\Gamma \log \, r(t) \, \frac{\cos [\phi(t) - \theta_s(t)]}{r_{st}} \, dt, \tag{10.2.4}$$

and $r(t) = |\gamma(t)|$. The integrals in (10.2.3) and (10.2.4) take Cauchy p.v.'s. The integral equation (10.2.3), derived by Lichtenstein [1917], is periodic in the angular deformation $\phi(s) - \theta(s)$.

10.2.2 Gershgorin's Integral Equation. Let D' denote the region obtained from D by indenting a disk $B(0, \varepsilon)$ whose boundary is denoted by Γ' (Figure 10.2). The function $F(z)$, with $F(\gamma(0+)) = i \, \phi(0)$, is single-valued on D'. Then, in view of (12.4.14), for $z = \gamma(s)$, $s \neq 0$, we have

$$\log \, f(z) = \frac{1}{i\pi} \int_{\Gamma + \Gamma'} \frac{\log \, f(\zeta)}{\zeta - z} \, d\zeta. \tag{10.2.5}$$

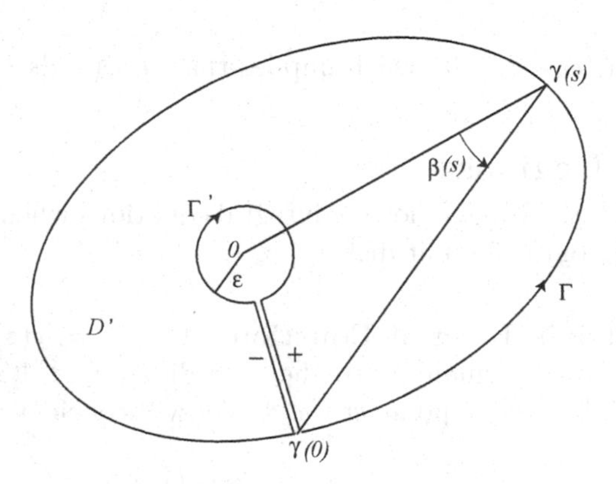

Figure 10.2 Gershgorin's contour.

Since $|f(\zeta)| \leq A \, |\zeta|$, where $A > 0$ is a constant, the contribution of the integral over Γ' is of order $O(\varepsilon \, \log \, \varepsilon) = o(1)$. Also, along the cut the integral has the value $\log \, f(z^+) = \log \, f(z^-) - 2i\pi$. Thus, as $\varepsilon \to 0$, the integral along the cut approaches $-2i\pi \, \frac{1}{i\pi} \int_{\zeta=0}^{\zeta=\gamma(0)} \frac{d\zeta}{\zeta - z}$, whose imaginary part is equal to $-2 \, \arg \left\{ \frac{\gamma(0) - \gamma(s)}{0 - \gamma(s)} \right\} = -2 \, \beta(s)$ (see Figure 10.2). Hence, equating the imaginary parts on both sides of (10.2.5) we get

$$\phi(s) = \Im \left\{ \frac{1}{i\pi} \int_\Gamma \frac{i \, \phi(t) \cdot e^{i(\phi(t) - \theta_s(t))}}{r_{st}} \, dt \right\} - 2 \, \beta(s)$$

or

$$\phi(s) = \int_\Gamma N(s,t) \, \phi(t) \, dt - 2 \, \beta(s). \tag{10.2.6}$$

This is known as *Gershgorin's integral equation* (Gershgorin [1933]).

10.2.3 Carrier's Integral Equation. Carrier [1947] considered the problem when $w = f(z)$ maps the region D conformally onto $|w| < 1$ such that two interior points P and Q in D go into two points $w = \pm a$, $0 < a < 1$, respectively, i.e., $f(P) = a$ and $f(Q) = -a$. The function $f(z)$ and the quantity a are uniquely determined. In fact, if we consider the function

$$F(z) = \log \left\{ \frac{f(z) - a}{f(z) + a} \cdot \frac{f(z) - a^{-1}}{f(z) + a^{-1}} \right\} - i\pi \tag{10.2.7}$$

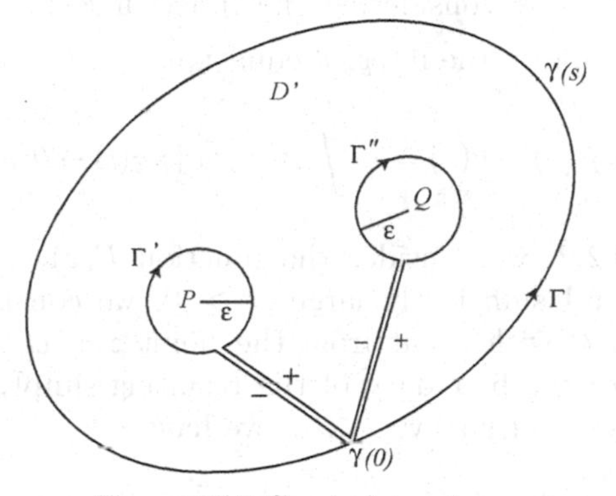

Figure 10.3 Carrier's contour.

in the region D' bounded by Γ, Γ' and Γ'' (Figure 10.3), then, in the case when the boundary of D' is a Jordan contour (i.e., it has no corners), we find by (2.4.14) that

$$\begin{aligned} F(z) &= \frac{1}{i\pi} \int_{\Gamma+\Gamma'+\Gamma''} \frac{F(\zeta)}{\zeta - z}\, d\zeta, \quad z - \gamma(s), \quad s \neq 0, \\ &= \frac{1}{i\pi} \int_\Gamma \frac{F(\zeta)}{\zeta - z}\, d\zeta - 2 \int_{\zeta=P}^{\zeta=\gamma(0)} \frac{d\zeta}{\zeta - z} + 2 \int_{\zeta=Q}^{\zeta=\gamma(0)} \frac{d\zeta}{\zeta - z} \\ &\equiv I_1 + I_2 + I_3, \end{aligned} \tag{10.2.8}$$

because the value of $F(z)$ is given by $F(z^+) = F(z^-) - 2\,i\pi$ along the cut from P to $\gamma(0)$, and by $F(z^+) = F(z^-) + 2\,i\pi$ along the cut from Q to $\gamma(0)$. Since $I_1 + I_2 = 2 \log \left| \dfrac{P - \gamma(s)}{Q - \gamma(s)} \right|$ which is real, we find that $\Im\{F(z)\} = 0$ for $z \in \Gamma$. In fact, on $|w| = 1$, we have from (10.2.7)

$$\arg \left\{ \frac{w - a}{w + a} \cdot \frac{w - a^{-1}}{w + a^{-1}} \right\} = \pi.$$

Hence, taking $\Phi(s) = \Re\{F(z)\}$, we find from (10.2.8) that

$$\Phi(s) = \int_\Gamma N(s,t)\, \Phi(t)\, dt + 2 \log \left| \frac{P - \gamma(s)}{Q - \gamma(s)} \right|, \quad s \neq 0, \tag{10.2.9}$$

which is known as *Carrier's integral equation.* This equation describes the problem of the potential flow of an inviscid fluid past a periodic array of airfoils of arbitrary shape (more on this problem in Chapter 13).

10.3 Exterior Regions

Although the problem of conformally mapping the region Ext (Γ) onto $|w| > 1$ can be reduced to that of the interior regions of §10.2 by applying the Schwarz reflection principle

(§2.6.6), the following direct method produces faster converging results in numerical computations. As before, we assume that Γ is a Jordan contour, the arc length s is measured in the positive sense, and at $z = \infty$ the mapping function has the series representation

$$f_E(z) = A\,z + a_0 + \frac{a_1}{z} + \frac{a_2}{z^2} + \cdots, \qquad A > 0. \tag{10.3.1}$$

First, as in §10.2.1, by considering the function $F(z) = \log \dfrac{f_E(z)}{z}$ and applying the formula (2.4.16), we obtain the integral equation

$$\phi_E(s) - \theta(s) = -\int_\Gamma N(s,t)\,[\phi_E(t) - \theta(t)]\,dt - g(s). \tag{10.3.2}$$

Secondly, as in §10.2.2, we consider the function $F(z) = \log f_E(z)$. Then, for a fixed $z = \gamma(s)$, $s \neq 0$, and sufficiently large $R > 0$, we consider the region between Γ and the circle $|\zeta - z| = R$ with a cut from the point $z = a$ to $z = \gamma(s) + R = \zeta_R$ (Figure 10.4). Let Γ^* denote the boundary of the resulting simply connected region. Obviously, $F(z^-) = F(z^+) + 2i\pi$, so that by (2.4.16) we have

$$\begin{aligned} \log f_E(z) &= -\frac{1}{i\pi}\int_{\Gamma^*} \frac{\log f_E(\zeta)}{\zeta - z}\,d\zeta \\ &= -\frac{1}{i\pi}\int_\Gamma \frac{\log f_E(\zeta)}{\zeta - z}\,d\zeta - 2\int_{\zeta=a}^{\zeta=\zeta_R} \frac{d\zeta}{\zeta - z} + \frac{1}{i\pi}\int_{|\zeta - z|=R} \frac{\log f_E(\zeta)}{\zeta - z}\,d\zeta \\ &\equiv I_1 + I_2 + I_3. \end{aligned} \tag{10.3.3}$$

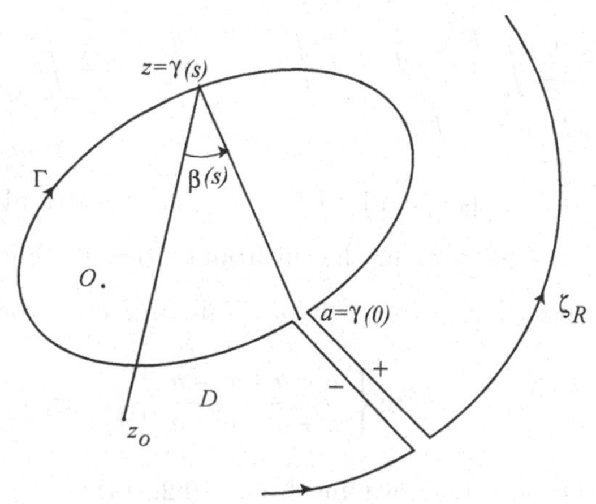

Figure 10.4 Integration contour.

Note that $\Im\{I_2\} = 2\,[\arg\{(z - a) - \pi\}]$. Since, in view of (10.3.1), with $\zeta - z = R\,e^{i\phi_E}$,

$$\frac{1}{i\pi}\int_{|\zeta - z|=R} \log\frac{f_E(\zeta)}{\zeta - z}\,\frac{d\zeta}{\zeta - z} = 2\,\log A,$$

which is real, we get

$$\Im\{I_3\} = \Im\left\{\frac{1}{i\pi}\int_{|\zeta - z|=R} \frac{\log(\zeta - z)}{\zeta - z}\,d\zeta\right\} = 2\,\pi.$$

Hence, equating the imaginary parts on both sides of (10.3.3), we obtain the integral equation

$$\phi_E(s) = -\int_\Gamma N(s,t)\,\phi_E(t)\,dt + 2\,\arg\{\gamma(s) - a\}. \tag{10.3.4}$$

If $f_E(z_0) = \infty$ for a finite point $z_0 \in D$ and $f_E(\infty) = \rho_\infty\, e^{i\varphi_\infty}$, then the mapping of D onto $|w| > 1$ is univalent only if $f_E(a) = 1$. In this case we obtain the integral equation

$$\phi_E(s) = -\int_\Gamma N(s,t)\,\phi_E(t)\,dt + 2\,\arg\{\beta(s) - \varphi_\infty\}. \tag{10.3.5}$$

For the external regions under consideration, Eqs (10.3.2) and (10.3.5) are analogues of the integral equations (10.2.3) and (10.2.6).

Now we will derive two integral equations for $\phi'_E(s)$, one from (10.3.4) and the other from (10.2.6). These integral equations will involve the kernel $N(t,s)$ which is conjugate to the Neumann kernel $N(s,t)$. These equations are interesting from a numerical standpoint. As opposed to $\phi_E(s)$, the function $\phi'_E(s)$ is periodic with period L, and hence, an application of the quadrature formula to $\phi'_E(s)$ increases computational precision.

10.3.1 Banin's Integral Equation. Let $\Gamma \in \Gamma'_\alpha$ be a Jordan contour. If the function $f_E(z)$, with the series expansion (10.3.1) at $z = \infty$, maps $\mathrm{Ext}\,(\Gamma)$ conformally onto $|w| > 1$, then we can rewrite (10.3.4) as

$$\phi_E(s) = -\int_\Gamma N(s,t)\,[\phi_E(t) - \phi_E(s)]\,dt - \phi_E(s) + 2\,\arg\{\gamma(s) - a\}. \tag{10.3.6}$$

Since $|\phi_E(t) - \phi_E(s)| = O(|t-s|)$, and $|t-s|^{2-\alpha}\left|\dfrac{\partial}{\partial s} N(s,t)\right|$ is bounded by property 2 of §10.1, we find after differentiating (10.3.6) with respect to s that

$$\phi'_E(s) = -\int_\Gamma \frac{\partial}{\partial s} N(s,t)\,[\phi_E(t) - \phi_E(s)]\,dt + 2\,\arg\{\gamma(s) - a\}, \tag{10.3.7}$$

where the differentiation under the integral sign is justified in view of the Lebesgue convergence theorem. Then, using (10.1.5) and integrating (10.3.7), we get

$$\begin{aligned}
\phi'_E(s) &= -\int_\Gamma \frac{\partial}{\partial t} N(t,s)\,[\phi_E(t) - \phi_E(s)]\,dt + 2\frac{\partial}{\partial s}\,\arg\{\gamma(s) - a\} \\
&= -N(0,s)\,[\phi_E(L) - \phi_E(0)] + \int_\Gamma N(t,s)\,\phi'_E(t)\,dt + 2\frac{\partial}{\partial s}\,\arg\{\gamma(s) - a\} \\
&= -\frac{1}{\pi}\frac{\partial}{\partial s}\theta_0(s)\cdot 2\pi + \int_\Gamma N(t,s)\,\phi'_E(t)\,dt + 2\frac{\partial}{\partial s}\,\arg\{\gamma(s) - a\},
\end{aligned}$$

which yields *Banin's integral equation*

$$\phi'_E(s) = \int_\Gamma N(t,s)\,\phi'_E(t)\,dt, \tag{10.3.8}$$

since $\dfrac{\partial}{\partial s}\,\theta_0(s) = \arg\{\gamma(s) - \gamma(0)\}$.

10.3.2 Warschawski-Stiefel's Integral Equation. If we apply the method outlined above in §10.3.1 to Gershgorin's integral equation (10.2.6), we obtain the following integral equation:

$$\phi'_E(s) = -\int_\Gamma N(t,s)\,\phi'_E(t)\,dt + 2\,\frac{d\theta(s)}{ds}, \tag{10.3.9}$$

which was established independently by Warschawski [1955] and Stiefel [1956]. If we set $\tau(s) = \phi_E(s) - \theta(s)$ in (10.3.9), then the integral equation for $\tau'(s)$ is

$$\tau'(s) = -\int_\Gamma N(t,s)\,\tau'(t)\,dt + k(s), \tag{10.3.10}$$

where

$$k(s) = -\int_\Gamma N(t,s)\,\theta'(t)\,dt + \theta'(s). \tag{10.3.11}$$

10.3.3 Interior and Exterior Maps. In §10.2 we have considered the problem of determining the mapping function $w = f(z)$ which maps the region $D = \text{Int}\,(\Gamma)$ univalently onto the unit disk $U = \{|w| < 1\}$ such that $f(0) = 0$ and $f'(0) > 0$. In §10.3.1 and §10.3.2 we have considered the problem of finding the mapping function $w = f_E(z)$ which maps the region $D^* = \text{Ext}\,(\Gamma)$ univalently onto the region $U^* = \{|w| > 1\}$ such that $f_E(\infty) = \infty$ and $\lim_{z\to\infty} f'_E(z) > 0$. These two problems are related to each other by the inversion transformation $z \mapsto z^{-1}$, which transforms the boundary Γ into a Jordan contour $\hat\Gamma$ and maps D onto $\hat D^* = \text{Ext}\,(\hat\Gamma)$ and D^* onto $\hat D = \text{Int}\,(\hat\Gamma)$. Let $\hat f$ and $\hat f_E$ be the interior and exterior univalent maps associated with $\hat\Gamma$. Then

$$\begin{aligned} f_E(z) &= \left\{\hat f\left(z^{-1}\right)\right\}^{-1}, \\ f(z) &= \left\{\hat f_E\left(z^{-1}\right)\right\}^{-1}. \end{aligned} \tag{10.3.12}$$

Hence, there is no need to consider the interior and exterior mappings as separate problems. From the computational point of view, it is convenient first to determine $f(z)$ and then use the relations (10.3.12) to compute $f_E(z)$. But in integral equation methods it is advantageous to determine $f(z)$ and $f_E(z)$ separately.

In each case the conformal maps are determined from the respective boundary correspondence functions $\phi(s)$ and $\phi_E(s)$.

10.4 Iterative Method

As seen from §10.2 and §10.3, the function $\phi(s) = \arg\{f(\gamma(s))\}$, $0 \le s \le L$, in general, satisfies the integral equation

$$\phi(s) = \lambda \int_0^L N(s,t)\,\phi(t)\,dt + g(s), \tag{10.4.1}$$

where $\lambda = 1$ corresponds formally to the integral equations (10.2.3), (10.2.6) and (10.2.9) for the interior regions, whereas $\lambda = -1$ corresponds to equations (10.3.8) and (10.3.10) for the exterior regions. We will present an iterative scheme for the numerical solution of Eq (10.4.1) for $\lambda = 1$; the case $\lambda = -1$ can be handled by similar iterations. Note that

$\lambda = 1$ is the smallest eigenvalue of the kernel $N(s,t)$. The associated eigenfunction for the homogeneous equation (10.4.1) is a constant. The eigenfunction for the conjugate kernel $N(t,s)$ is the equilibrium distribution $\mu(t)$ with $\int_\Gamma \mu(t)\,dt = 1$. Since $\lambda = 1$ is the only simple pole of $N(s,t)$ on $|\lambda| = 1$ and its principal part at this pole is $\dfrac{\mu(t)}{1-\lambda}$, the function

$$\gamma(s,t;\lambda) = \sum_{i=1}^{\infty} \lambda^i \left[N_{i+1}(s,t) - \mu(t)\right], \tag{10.4.2}$$

where $N_{i+1}(s,t)$ denote the iterated kernels with $N_1(s,t) = N(s,t)$, is analytic for $|\lambda| < |\lambda_2|$, where λ_2 is the next eigenvalue close to 1 ($|\lambda_2| > 1$). Then the series

$$\sum_{i=1}^{\infty} \lambda^i \left[N_{i+1}(s,t) - \mu(t)\right]$$

converges for $|\lambda| < |\lambda_2|$. Since $\int_0^L g(t)\,\mu(t)\,dt = 0$, by the Fredholm theory, it follows that the Neumann series

$$\sum_{i=1}^{\infty} \lambda^i \left[N_{i+1}(s,t) - \mu(t)\right] + g(s) \tag{10.4.3}$$

converges for $|\lambda| < |\lambda_2|$, and for $\lambda = 1$ it represents a solution $\phi(s)$ of Eq (10.4.1). The main result due to Warschawski [1956] is the following:

Theorem 10.1. *Let $\Gamma \in \Gamma'_\alpha$, $0 < \alpha \le 1$. Suppose that $\phi_0(t) \in C[0,L]$ and that $\phi_0(L) - \phi_0(0) = 2\pi$. Then the iterations $\phi_n(s)$ defined by*

$$\phi_{n+1}(s) = \int_\Gamma N(s,t)\,\phi_n(t)\,dt + g(s), \quad n = 0,1,2,\dots, \tag{10.4.4}$$

converge uniformly to the solution $\phi(s)$ of Eq (10.4.1) with $\lambda = 1$, such that $\int_\Gamma \phi(s)\,\mu(s)\,ds = \int_\Gamma \phi_0(s)\,\mu(s)\,ds$. More precisely,

$$|\phi_{n+1}(s) - \phi_n(s)| \le \frac{1}{\pi\,|\lambda_2|^n}\,\|N(s,t)\|\,\|\phi_0' - \phi'\|\,\sqrt{\frac{\lambda_2^2}{\lambda_2^2 - 1}}. \tag{10.4.5}$$

In fact, by the Schwarz inequality,

$$\|\phi_0' - \phi'\| \le \sqrt{\int_\Gamma \int_\Gamma \left(\log \frac{\delta}{r_{st}}\right)^2 ds\,dt} \cdot \sqrt{\int_\Gamma \left(\phi_0' - \phi'\right)^2 dt}.$$

The factor $\|\phi_0' - \phi'\|$ in the error estimate (10.4.5) must be small which happens when $\max\limits_{0\le s\le L} |\phi_0' - \phi'|$ is small or when $\int_\Gamma (\phi_0'(t) - \phi'(t))\,dt$ is small. Thus, the factor $\|\phi_0' - \phi'\|$ in the error estimate (10.4.5) plays a useful role.

To prove the uniform convergence of the iterations $\phi_n(s)$ to the solution $\phi(s)$, it suffices to assume that $\phi(s)$ exists and satisfies the condition $\int_\Gamma g(s)\,\phi(s)\,ds = 0$, and that $\phi_n(s) \in C[0, L]$ for all $n = 0, 1, 2, \dots$, where $\phi_0(L) - \phi_0(0) = 2\pi$. Then

$$|\phi_{n+1}(s) - \phi_n(s)| \le \frac{1}{\pi\,|\lambda_2|^{n-1}}\,\|N(s,t)\|\,\|\phi_1' - \phi_0'\|\sqrt{\frac{\lambda^2}{\lambda^2 - 1}}, \tag{10.4.6}$$

which leads to the solution

$$\phi(s) = \phi_0(s) + \sum_{n=0}^{\infty}\left[\phi_{n+1}(s) - \phi_n(s)\right]. \tag{10.4.7}$$

ESTIMATES FOR $|\lambda_2|$: The inequality (10.4.6) gives an estimate for the rate of convergence. A result on the convergence of the derivatives $\phi_n'(s)$, which is due to Warschawski [1956], is as follows:

Theorem 10.2. *If $\Gamma \in \Gamma_\alpha''$, $0 < \alpha \le 1$, then the derivatives $\phi_n'(s)$ converge uniformly to $\phi'(s)$, $0 \le s \le L$. More precisely, the following estimate holds:*

$$|\phi_{n+1}(s) - \phi_n(s)| \le \frac{1}{\pi\,|\lambda_2|^{n-1}}\left\|\frac{\partial N(s,t)}{\partial s}\right\|\sqrt{\frac{\lambda_2^2}{\lambda_2^2 - 1}}. \tag{10.4.8}$$

Let Γ_0 be a Jordan contour and $N_0(s,t)$ denote the associated Neumann kernel. Suppose that the second eigenvalue Λ_2 of $N_0(s,t)$ is known. For example, for an ellipse $b^2x^2 + a^2y^2 = a^2b^2$, $\Lambda_2 = \dfrac{a+b}{a-b}$; for a circle $x^2 + y^2 = a^2$, $\Lambda_2 = \infty$. Then estimates for λ_2 can be given in terms of Λ_2 in the following cases:

(a) Let Γ_0 be close to $\Gamma \in \Gamma_\alpha''$, $0 < \alpha \le 1$, in the sense that either $\Gamma_0 \subset \operatorname{Int}\Gamma$ or $\Gamma \subset \operatorname{Int}\Gamma$. The former situation corresponds to the case of the interior regions (§10.2) when $w = f(z)$ maps $\operatorname{Int}(\Gamma)$ onto the unit disk $|w| < 1$, whereas the latter corresponds to the case of the exterior regions (§10.3) when $w = f(z)$ maps $\operatorname{Ext}(\Gamma)$ onto the unit disk $|w| > 1$ such that $z = \infty$ goes into $w = \infty$. Let

$$q = \frac{\max\limits_{|w|=1} |f'(z)|}{\min\limits_{|w|=1} |f'(z)|}, \quad \text{and} \quad M = \int_\Gamma \left\|\frac{\partial N(s,t)}{\partial t}\right\|^2 dt.$$

If d is the Fréchet distance (see §2.3 for definition) of Γ and Γ_0, then

$$c_1 \le \frac{1}{|\lambda_2|} \le \frac{1}{\Lambda_2} + a\,d\,\lambda_2^2, \tag{10.4.9}$$

where $a = 2\,qM/\pi$, and c_1 is the real root of the cubic equation $d\,a\,x^3 + x/\Lambda_2 = 1$.

(b) Let $\Gamma \in \Gamma_\alpha''$ and $\Gamma_0 \in \Gamma_\alpha'$, and let contours Γ and Γ_0 have the same length δ. Suppose that for some choice of the points corresponding to $s = 0$ on each contour

$$\int_\Gamma\int_\Gamma \left(N(s,t) - N_0(s,t)\right)^2 ds\,dt = \varepsilon^2,$$

$$\int_\Gamma\int_\Gamma \left(\log\frac{1}{r_{st}} - \log\frac{1}{\rho_{st}}\right)^2 ds\,dt = \nu^2,$$

where $\rho_{st} = |z_0(s) - z_0(t)|$, $N_0(s,t)$ is the Neumann kernel associated with Γ_0, and $z_0(s)$ is the parametric representation of Γ_0 in terms of the arc length parameter s, $0 \le s \le L$. Then

$$c_2 \le \frac{1}{|\lambda_2|} \le \frac{1}{\Lambda_2} + \frac{a\,\lambda^2}{2\pi}\left[\nu^2\left(\frac{1}{\lambda_2} + \frac{1}{\Lambda_2}\right) + \varepsilon\, B\right], \tag{10.4.10}$$

where $B = \sqrt{\int_\Gamma\int_\Gamma \left(\log \frac{1}{\rho_{st}}\right)^2 ds\,dt}$, c_2 is the real root of the cubic equation

$$\left(\frac{a\,\delta}{2\pi\,\Lambda_2} + \varepsilon\,B\right) x^3 + \frac{a\,\delta}{2\pi}\,x^2 + \frac{x}{\Lambda_2} = 1,$$

and a is the same as in (10.4.9).

(c) If $\Gamma \in \Gamma'_\alpha$ for $1/2 < \alpha < 1$, i.e., Γ is the boundary of a nearly circular region, then

$$\frac{1}{\lambda_2} \le \frac{1}{2}\left[\int_\Gamma N_2(s,t)\,ds - 1\right], \tag{10.4.11}$$

where, by the Schwarz inequality,

$$N_2(s,t) = \int_\Gamma\int_\Gamma N(s,x)N(x,s)\,dx\,ds \le \int_\Gamma\int_\Gamma N^2(s,x)\,dx\,ds,$$

and

$$\frac{1}{\lambda_2^2} \le \frac{1}{2}\left[\int_\Gamma\int_\Gamma N^2(s,x)\,dx\,ds - 1\right] = \frac{1}{2}\left[\int_\Gamma\int_\Gamma \left(N(s,x) - \frac{1}{\delta}\right)^2 dx\,ds\right]. \tag{10.4.12}$$

Since the kernel $N_0(s,t) - \frac{1}{\delta}$ for a circle of circumference δ, the condition that the last integral in (10.4.12) be less than unity implies that Γ is a near circle.

(d) Neumann's lemma states that if Γ is a convex Jordan contour, then $N(s,t) \ge 0$ and there exists a constant κ, $0 < \kappa < 1$, known as the *Neumann constant*, which depends only on Γ and has the following property: Let $g^*(s) = \int_\Gamma N(s,t)\,g(t)\,dt$, where the function $g(t)$ is bounded and integrable on Γ, $0 \le t \le L$. Let $m \le g(t) \le M$ on $[0,L]$ and $m^* \le g^*(s) \le M^*$ on $[0,L]$. Then

$$M^* - m^* \le (M - m)\,(1 - \kappa). \tag{10.4.13}$$

A Jordan contour is said to be *nearly convex* if it satisfies the following criterion: There exists a convex Jordan contour Γ_0 such that (i) Γ_0 has the same length as Γ, and (ii) if $N(s,t)$ and $N_0(s,t)$ are the kernels associated with Γ and Γ_0, respectively, then for all $s \in [0,L]$

$$\int_\Gamma |N(s,t) - N_0(S,t)|\,dt \le \varepsilon < \kappa.$$

If Γ is nearly convex, and if $\phi_0(s) \in C[0,L]$ is an arbitrary function, then the iterations (10.4.4) satisfy the inequality

$$|\phi_{n+1}(s) - \phi_n(s)| \le (1+\varepsilon)\,V\,(1+\varepsilon-\kappa)^n, \tag{10.4.14}$$

where V is a constant; $V \le \omega_0 + 2\pi$, and ω_0 is the oscillation of $\phi_0(s)$ in $[0, L]$. If $\phi_0(s)$ is nondecreasing and $\omega_0 = 2\pi$, then $V = 2\pi$. Finally, if $\phi_0(s)$ is an approximation of the solution of Eq (10.4.1), i.e., if it is known a priori that $|\phi_0(s) - \phi(s)| \le \eta$, $0 \le s \le L$, for some solution of (10.4.1), then $V \le 2\eta$. If Γ_0 is a circle, then $\kappa = 1$. The Neumann constant κ characterizes a nearly circular region in a manner different from that presented in Chapter 9.

For computational purposes, the best method for numerically solving the integral equation (10.4.1) is to discretize the integral and replace the equation by a matrix equation. Thus, the problem becomes one of matrix inversion. To do this, we partition the boundary Γ into n parts at the points

$$t_j = j\,\frac{L}{n}, \quad j = 0, 1, \dots, n, \quad (t_0 = t_n),$$

and obtain for the values $\phi(t_j)$ of $\phi(s)$ at the n points $t_k = kL/n$, $k = 1, 2, \dots, n$, the following system of n linear equations where the integral in (10.4.1) is replaced by a sum:

$$\phi(t_k) = \sum_{j=1}^{n} N(t_k, t_j)\, \phi(t_j)\, \frac{L}{n} + g(t_k),$$

or

$$\sum_{j=1}^{n} \left[\delta_{jk} - N(t_k, t_j)\right] \phi(t_j) = \frac{n}{L}\, g(t_k). \tag{10.4.15}$$

In practical applications, since the boundary Γ cannot be divided into partitions of equal length, it is useful to take more partition points on those portions of Γ where the curvature is positive and larger. This is accomplished by transforming the arc length parameter t into an integration variable τ such that $t = \psi(\tau)$, $s = \psi(\sigma)$, $\psi'(\tau) > 0$, $0 \le s, t \le L$, $0 \le \tau \le l$, and $\psi(\tau)$ is small (large) according as the curvature is large (small). This substitution transforms Eq (10.4.1) into

$$\phi(\psi(\tau)) = \int_0^l N^*(\sigma, \tau)\, \phi^*(\tau)\, d\tau + g^*(\sigma), \tag{10.4.16}$$

which, after discretization with partitions of equal length in τ, yields the matrix equation

$$\sum_{j=1}^{n} \left[\delta_{jk} - N^*(\sigma_k, \tau_j)\right] \phi^*(\tau_j) = \frac{n}{L}\, g^*(\tau_k),$$

which is similar to (10.4.15).

In this method the matrix inversion is of $O(n^3)$, which becomes considerably large if n is large. To overcome this difficulty, an iterative method is used where the computations are of order $O(mn^2)$, m being the number of iterations. This iteration starts with a function $\phi(s) = \phi_0(s)$, called the *initial guess*, that is taken close to the correct solution. Then, with this function the right side of Eq (10.4.1) is computed, which yields

$$\phi_1(s) = \int_0^L N(s, t)\, \phi_0(t)\, dt + g(s),$$

and the iterative process is repeated n times. It leads to Eq (10.4.4) for $n = 1, 2, \dots$, which, by Theorem 10.1, converges uniformly to the solution $\phi(s)$.

In numerical computation, the rate of convergence is fast only if the region D is nearly circular, i.e., if it can be approximated in polar coordinates by the function $r = r(\theta)$, $0 \le \theta \le 2\pi$, $r(0) = r(2\pi)$, which belongs to the class $C^2[0, 2\pi]$, is almost constant for all θ, i.e., $r(\theta) \approx$ const, and has a small first derivative $r'(\theta) \ll 1$. Thus, the algorithm for solving Eq (10.4.1) is as follows:
1. Check if Γ is a Jordan contour with no corners. In case Γ has corners, they should be first analyzed by the methods of §13.4.
2. Use elementary conformal mapping (like, log, exp, sin, cos functions) to make the region D "circular" (see Figure 10.5).
3. Carry out the iterations (10.4.4) using the discretized formula (10.1.15) or (10.1.16).
4. Stop the iterations when the difference $|\phi_m(s) - \phi_{m-1}(s)| < \varepsilon$, where $\varepsilon > 0$ is a preassigned quantity (called the *tolerance*).

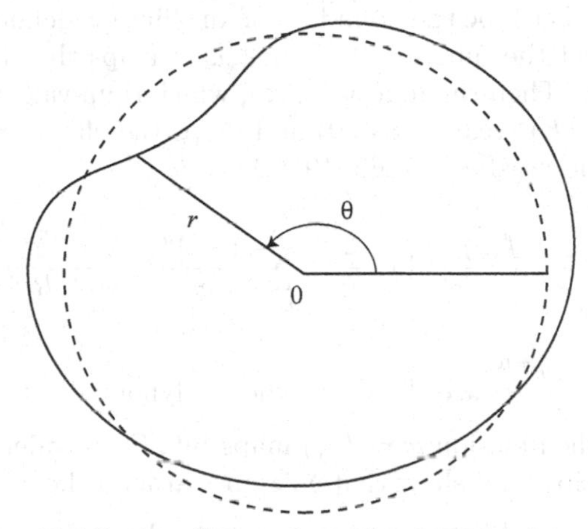

Figure 10.5 Region D transformed into a circular region.

If c denotes the rate of convergence, i.e., if it is the largest number greater than unity such that

$$|\phi_{n+1}(s) - \phi_n(s)| \le \frac{1}{c}\, |\phi_n(s) - \phi_{n-1}(s)|$$

for all s and n, then an upper bound for the error made by taking $\phi_m(s) \approx \phi(s)$ is given by

$$|\phi(s) - \phi_m(s)| \le \frac{\varepsilon}{c} + \frac{\varepsilon}{c^2} + \cdots = \frac{\varepsilon}{c-1}. \tag{10.4.17}$$

The value of c may be approximately estimated during the iteration process. Since $c > 1$, we find from (10.4.17) that the error is smaller than ε. However, in the entire computation, besides this error, we have the discretization as well as round-off errors.

The eigenvalue λ_1 is important in numerical computations. Ahlfors [1952] has given a simple estimate for $1/\lambda_1$ which is called the *convergence factor*. If the boundary Γ is defined in polar coordinates, as above, by $r = r(\theta)$ and if $v_0 = \max v$, where v is the angle between the radius vector and the normal (Figure 10.1), then the Ahlfors estimate is given by

$$\lambda_1 \ge \frac{1}{\sin v_0}. \tag{10.4.18}$$

This estimates for an ellipse E, defined in the z-plane by

$$x = a\cos t, \quad y = b\sin t, \quad 0 \le t \le 2\pi, \tag{10.4.19}$$

with foci ± 1, semi-axes a and b, $a > b$, $a^2 - b^2 = 1$, and axes-ratio $k = a/b$, and for Cassini's oval $|z^2 - 1| = k$ are given below (Andersen et al. [1962:190]):

FOR THE ELLIPSE:

$k = 1.2$: $\csc v_0 = 5545$, and $C = -11.0$,
$k = 1.6$: $\csc v_0 = 2282$, and $C = -4.4$,
$k = 2.0$: $\csc v_0 = 1667$, and $C = -3.0$.

FOR CASSINI'S OVAL:

$k = 1.2$: $\csc v_0 = 1.2$, and $C = -2.0$,
$k = 2.0$: $\csc v_0 = 2.0$, and $C = -3.8$,
$k = 5.0$: $\csc v_0 = 5.0$, and $C = -10.0$,

where $|C|$ denotes the rate of convergence which is sufficiently large for nearly circular regions. The negative sign for C indicates that the iterations "oscillate."

Example 10.1. Let Γ be the boundary of an ellipse E defined in the z-plane by (10.4.19). It is well known that the function $2\,z = w + w^{-1}$ maps the circle $|w| = R = a + b$, $R > 1$, conformally onto E. Then the function $f(z)$, which is univalent in Int (E) and $|f(z)| = 1$ on E, is regular in Int (E) (real at $z = 0$) and maps the ellipse E conformally onto the circle $|w| = R$ such that it satisfies (Szegö [1950])

$$\log \frac{f(z)}{z} = \log \frac{2}{R} + \sum_{n=1}^{\infty} \frac{(-1)^n}{n} \frac{2\,R^{-2n}}{R^{2n} + R^{-2n}}\, T_{2n}(z), \tag{10.4.20}$$

where $T_n(z) = \dfrac{w^n + w^{-n}}{2}$ are the Chebyshev polynomials of the first kind.

(a) To prove that the function $w = f(z)$ maps Int (E) univalently onto $|w| < R$, we will use the argument principle and show that as z goes around the ellipse E, the point w describes the circle $|w| = R$ exactly once and in the same direction, i.e., $\dfrac{d}{d\phi} \Im \{\log f(z)\} > 0$. We find from (10.4.20), with $w = R\,e^{i\phi}$, that

$$\begin{aligned}
\Im \{\log f(z)\} &= \Im \{\log z\} + \sum_{n=1}^{\infty} \frac{(-1)^n}{n} \frac{2\,R^{-2n}}{R^{2n} + R^{-2n}} \Im \{T_{2n}(z)\} \\
&= \Im \left\{\log \frac{R\,e^{i\phi} + R^{-1}\,e^{-i\phi}}{2}\right\} \\
&\quad + \sum_{n=1}^{\infty} \frac{(-1)^n}{n} \frac{2\,R^{-2n}}{R^{2n} + R^{-2n}} \frac{R^{2n} - R^{-2n}}{2} \sin 2n\phi.
\end{aligned}$$

Thus,

$$\begin{aligned}
\frac{d}{d\phi} \Im \{\log f(z)\} &= \Im \left\{ \frac{i\,\left(R\,e^{i\phi} - R^{-1}\,e^{-i\phi}\right)}{R\,e^{i\phi} + R^{-1}\,e^{-i\phi}} \right\} \\
&\quad + 2 \sum_{n=1}^{\infty} (-1)^n \frac{R^{2n} - R^{-2n}}{R^{2n} + R^{-2n}} R^{-2n} \cos 2n\phi \\
&\equiv A_1 + A_2.
\end{aligned}$$

Since $A_1 = \Re \left\{ \dfrac{1 - R^{-2}\,e^{-2i\phi}}{1 + R^{-2}\,e^{-2i\phi}} \right\} = 1 + 2 \displaystyle\sum_{n=1}^{\infty} (-1)^n R^{-2n} \cos 2n\phi$, we get

$$\frac{d}{d\phi}\Im\{\log f(z)\} = A_1 + A_2 = 1 + 2\sum_{n=1}^{\infty}(-1)^n R^{-2n}\cos 2n\phi$$
$$+ 2\sum_{n=1}^{\infty}(-1)^n \frac{R^{2n}-R^{-2n}}{R^{2n}+R^{-2n}} R^{-2n}\cos 2n\phi$$
$$= 1 + 2\sum_{n=1}^{\infty}(-1)^n \frac{2}{R^{2n}+R^{-2n}}\cos 2n\phi$$
$$= \lim_{\rho\to 1-}\Big\{1 + 2\sum_{n=1}^{\infty}(-1)^n \frac{2\,\rho^n}{R^{2n}+R^{-2n}}\cos 2n\phi\Big\}.$$

If we define $R^2 = e^{\alpha\pi}$, then since $\displaystyle\int_0^\infty \frac{\cos\alpha x}{\cosh(x/2)}\,dx = \frac{\pi}{\cosh\alpha\pi}$, we have for $n = 0, 1, \dots$

$$\frac{2}{R^{2n}+R^{-2n}} = \frac{2}{e^{n\alpha\pi}+e^{-n\alpha\pi}} = \frac{1}{\cosh n\alpha\pi} = \frac{1}{\pi}\int_0^\infty \frac{\cos n\alpha x}{\cosh(x/2)}\,dx.$$

Hence,

$$1 + 2\sum_{n=1}^{\infty}(-1)^n \frac{2\,\rho^n}{R^{2n}+R^{-2n}}\cos 2n\phi$$
$$= \frac{1}{\pi}\int_0^\infty \frac{1}{\cosh(x/2)}\left(1 + 2\sum_{n=1}^{\infty}(-\rho)^n \cos n\alpha x\,\cos 2n\phi\right)dx$$
$$= \frac{1}{2\pi}\int_0^\infty \frac{1}{\cosh(x/2)}\Big(\frac{1-\rho^2}{1+2\rho\cos(\alpha x+2\phi)+\rho^2}$$
$$+ \frac{1-\rho^2}{1+2\rho\cos(\alpha x-2\phi)+\rho^2}\Big)\,dx > 0,$$

by (6.4.9) and (6.4.10).

(b) We will determine the exact solution for the boundary map $f(z)$. Let $w = e^{i(\pi/2-x)} = i\,e^{-ix}$, and $z = \cos(\pi/2-x) = \sin x$. Then

$$\log\frac{f(z)}{z} = \log\frac{2}{R} + \sum_{n=1}^{\infty}\frac{1}{n}\,\frac{2\,R^{-2n}}{R^{2n}+R^{-2n}}\cos 2nx.$$

Since

$$\sum_{n=1}^{\infty}\frac{1}{n}\,\frac{2\,R^{-2n}}{R^{2n}+R^{-2n}}\cos 2nx = \log\operatorname{sn}\left(\frac{2Kx}{\pi}, k\right) - \log\left(2\,q^{1/4}\right)$$
$$+ \frac{1}{2}\log k - \log\sin x,$$

(see Whittaker and Watson [1962:509, Ex. 3]), where $q = R^{-4}$, sn is a Jacobian elliptic function of modulus k, $0 < k < 1$ (see §7.1), we find that

$$f(z) = \sqrt{k}\,\operatorname{sn}\left(\frac{2\,K\,x}{\pi}, k\right) = \sqrt{k}\,\operatorname{sn}\left(\frac{2K}{\pi}\sin^{-1} z, k\right), \tag{10.4.21}$$

which is the Schwarz formula.

(c) For the interior regions (§10.2) the boundary correspondence function is given by

$$\begin{aligned}\phi(t) &= \arg\{f(z)\} = \arg\{z\} + \sum_{n=1}^{\infty} \frac{(-1)^n}{n}\, R^{-2n}\, \frac{R^{4n}-1}{R^{4n}+1}\, \sin 2nt \\ &= \tan^{-1}\left\{ \frac{\left(R^2-1\right)^2 - 2\,\cos^2 t}{\left(R^2+1\right)^2 - 2\,\sin^2 t}\, \tan t \right\} - 2 \sum_{n=1}^{\infty} \frac{(-1)^n}{n}\, R^{-2n}\, \frac{\sin 2nt}{R^{4n}+1}.\end{aligned} \tag{10.4.22}$$

If we set $L_j = 2 \sum_{n=j}^{\infty} \frac{(-1)^n}{n}\, R^{-2n}\, \frac{\sin 2nt}{R^{4n}+1}$, then

$$|L_j| \le \frac{2\,R^6}{R^6-1}\, \frac{1}{j\,R^{6j}}, \tag{10.4.23}$$

and the function $\phi(t)$, defined by (10.4.22), can be approximated by

$$\begin{aligned}\phi(t) &\approx \phi_j(t) \\ &= \tan^{-1}\left\{ \frac{\left(R^2-1\right)^2 - 2\,\cos^2 t}{\left(R^2+1\right)^2 - 2\,\sin^2 t}\, \tan t \right\} - 2 \sum_{n=1}^{j-1} \frac{(-1)^n}{n}\, R^{-2n}\, \frac{\sin 2nt}{R^{4n}+1},\end{aligned} \tag{10.4.24}$$

with an error (uniform in t, $0 \le t \le 2\pi$) given by (10.4.23). In computations, j in (10.4.24) is usually estimated so that

$$\frac{2\,R^6}{R^6-1}\, \frac{1}{j\,R^{6j}} < 10^{-12}.$$

Thus, j becomes larger, the closer R is taken to 1. Todd and Warschawski [1955] have taken the minimum value of R in their computations as $R = \sqrt{1.5}$; then $j = 22$. The function actually computed is $\phi_{22} - \pi$ for the values of $t = 0°\,(3°)\,90°$. Since these values are symmetric about the y-axis, they are extended in other quadrants. Then the results obtained by computation from (10.4.24) are compared with those from (10.4.21).

(d) In the case of exterior regions (§10.3), $\phi(t) = t$. Note that $\lambda_2 = \dfrac{a+b}{a-b}$ for the ellipse E. The function

$$G(z,0) = \Re\left\{ \log \frac{1}{f(z)} \right\} = \log \frac{1}{|f(z)|}$$

is Green's function of Int (E) with respect to the origin. ■

Example 10.2. We will determine the correspondence between the boundaries in the case of mapping the ellipse E, defined by (10.4.19), onto the unit circle. The boundary correspondence function ϕ is normalized so that the origin is preserved and $\phi(0) = -\pi$. Using the parameter t instead of the arc length s, we find from (10.4.19) that

$$\frac{ds}{dt} = \sqrt{a^2\,\sin^2 t + b^2\,\cos^2 t} = b\,\sqrt{k^2\,\sin^2 t + \cos^2 t}.$$

Then, Gershgorin's integral equation (10.2.6) can be written as

$$\phi(\tau) = \frac{1}{\pi} \int_0^{2\pi} N(\tau,t)\,\phi(t)\,dt + g(\tau), \tag{10.4.25}$$

where

$$g(\tau) = -2\,\beta\,(s(\tau)) = 2\,\tan^{-1}\left\{\frac{k\,\sin\tau}{k^2\,(\cos\tau - \cos^2\tau) - \sin^2\tau}\right\}, \tag{10.4.26}$$

and

$$\begin{aligned} N(\tau,t) &= \frac{k/2}{\sqrt{k^2\,\sin^2\dfrac{t-\tau}{2} + \left[k^2\,\sin t\,\sin\dfrac{t-\tau}{2} + \cos t\,\cos\dfrac{t-\tau}{2}\right]^2}} \\ &\quad \times \sqrt{\frac{k^2\,\sin^2 t + \cos^2 t}{k^2\,\sin^2\dfrac{t-\tau}{2} + \cos^2\dfrac{t-\tau}{2}}} \\ &= \frac{k}{(k^2+1) - (k^2-1)\,\cos(t+\tau)}. \end{aligned} \tag{10.4.27}$$

The branch of arctan in (10.4.26) is chosen such that $g(0) = \lim_{\tau\to 0^+} g(\tau) = -\pi$ and $g(2\pi) = \lim_{\tau\to 2\pi^-} g(\tau) = \pi$. Since $\phi(2\pi-\tau) = -\phi(\tau)$, we can write Eq (10.4.25) as

$$\begin{aligned} \phi(\tau) &= \frac{k}{\pi}\int_0^\pi \left[\frac{k_1\,\phi(t)}{1-k_2\,\cos(t+\tau)} - \frac{k_1\,\phi(t-\tau)}{1+k_2\,\cos(t+\tau)}\right] dt \\ &\quad + 2\,\tan^{-1}\left\{\frac{\sin\tau}{k(1-\cos\tau)\,(k_3\,\cos\tau - k^{-2})}\right\}, \end{aligned} \tag{10.4.28}$$

where

$$k_1 = \frac{1}{k^2+1}, \quad k_2 = \frac{k^2-1}{k^2+1}, \quad k_3 = 1 - \frac{1}{k^2}.$$

Eq (10.4.28) can be represented in the operator form as

$$\phi(\tau) = T\,\phi + g(\tau), \tag{10.4.29}$$

where T is the integral operator in (10.4.28) and $g(\tau)$ is the arctan term. If the initial guess is taken as $\phi_0 = g(\tau)$, then the iterations

$$\phi_{n+1}(\tau) = T\,\phi_n + g(\tau), \tag{10.4.30}$$

lead to the equation

$$\phi_{n+1} = g_0 + g_1 + \cdots + g_n, \quad g_0 = g, \quad T\,g_n = g_{n+1}. \tag{10.4.31}$$

Thus, first we must compute $\phi_0 = g(\tau)$, and then compute the integrals $T\,g_n$ which are replaced by an appropriate approximate quadrature. For $k = 1.2$, Todd and Warschawski [1955] found that at a fixed t, the functions $g(\tau)$ and $N(\tau,t)$ behave approximately like $g(\tau) \approx 1.5\,(1+\cos\tau)$ and $N(\tau,t) \approx 0.5 + 0.1\,\cos\tau$. Weddle's rule was used for quadrature (see Birkoff et al. [1950; 1951]. The results for ϕ_n for $k = 1.2$ are computed against t at a step-size of 3° from $t = 0°$ to $t = 180°$ and presented in Figure 10.6 (t along the horizontal axis). Similar computations can also be carried out for $k = 2$ and 5. The results match the exact value given by (10.4.22). These computations, though carried out for an ellipse, are well suited for other types of nearly circular regions. ■

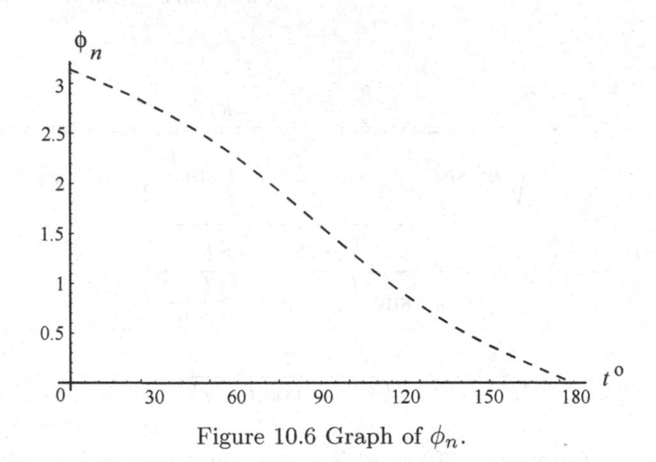

Figure 10.6 Graph of ϕ_n.

Example 10.3. The integral equation for the dipole distribution is given by

$$\mu(s) = \int_0^L N(s,t)\,\mu(t)\,dt + g(s), \tag{10.4.32}$$

where the dipole strength on the boundary Γ is $2\pi\,\mu(s)$, $g(s)$ denotes the boundary value of the potential function, and t, $0 \le t \le L$, is the arc length parameter along the positive direction of Γ. The integral equation (10.4.32) has the same form as (10.4.1). Let us assume that the distribution $\mu(s)$ has already been determined by solving Eq (10.4.25). Then the Dirichlet problem and, hence, the problem of conformal mapping is reduced to that of quadratures. To see this, note that the potential $u(P)$ at a point $P = (x, y) \in D$ is given by

$$u(P) = \int_0^L N(P,t)\,\mu(t)\,dt, \tag{10.4.33}$$

where $N(P,t) = -\dfrac{\partial}{\partial n_t} r_{Pt}$ (Figure 10.7). The kernel $N(P,t)$ becomes unbounded as P approaches the boundary Γ. To avoid this difficulty, we proceed as follows: Consider a point P' near Γ (Figure 10.7). Let t' be a point on Γ such that the normal $n_{t'}$ passes through the point P'. Then, we can rewrite (10.4.33) as

$$\begin{aligned} u(P) &= \mu(t') + \int_0^L N(P,t)\;[\mu(t) - \mu(t')]\;dt \\ &= \mu(t') + \int_{-L/2}^{L/2} N(P,t+t')\;[\mu(t+t') - \mu(t')]\;dt, \end{aligned} \tag{10.4.34}$$

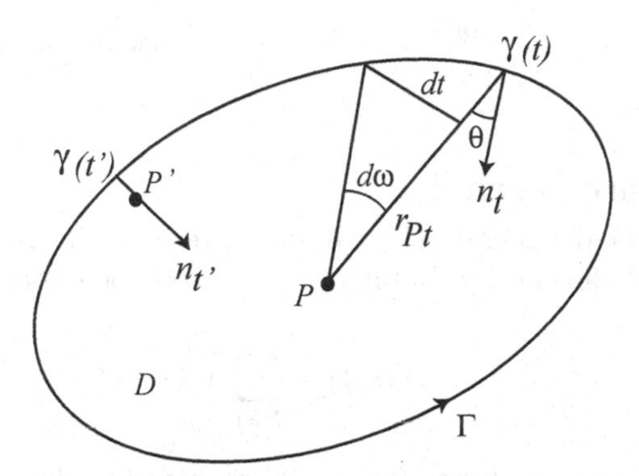

Figure 10.7 Integration of $u(P)$.

where t is replaced by $t'+t$, and $N(P,t'+t) = -\dfrac{\partial}{\partial n_{t'+t}} r_{Pt}$. The integral in (10.4.34) is finite for all t, and $\mu(t'+t)-\mu(t') \approx t\,\mu'(t')$ for numerically small values of t. Then (10.4.34) can be written as

$$u(P) = \mu(t') - \int_{-L/2}^{L/2} \Big\{ [\mu(t'+t) - \mu(t')] \frac{\partial}{\partial n_{t'+t}} r_{Pt} + [\mu(t'-t) - \mu(t')] \frac{\partial}{\partial n_{t'-t}} r_{Pt} \Big\}\, dt, \tag{10.4.35}$$

since $\displaystyle\int_0^L \frac{\partial}{\partial n_t} r_{Pt} - \int_0^{2\pi} d\omega = 2\pi$, and $\displaystyle\int_0^L \frac{\partial}{\partial n_t} \log r_{Pt} = \int_0^{\pi} d\omega = \pi$ for $\gamma(s) \in \Gamma$ (Figure 10.7). ■

Example 10.4. If U and U^* denote the region $|z| < 1$ and $|z| > 1$, respectively, $z = e^{i\theta}$, and if $D = \text{Int}\,(\Gamma)$, as before, then for certain curves the following data is useful (Gaier [1964:264]):

1. Eccentric circle in the w-plane: Let $D: |w-a| < b$, $b > a > 0$. Then
$\Gamma: \rho = \rho(\phi) = a\cos\phi + \sqrt{b^2 - a^2\sin^2\phi}$;
Mapping $D \mapsto U: z = \dfrac{b\,w}{a\,w + b^2 - a^2}$;
Boundary correspondence function: $\tan\theta = \dfrac{\sin\phi}{c\,\rho(\phi) + \cos\phi}$, $c = \dfrac{a}{b^2 - a^2}$.

2. Inverted ellipses: The region D is formed by inverting the exterior of the ellipse with half-axes $1/p$ and 1 into the unit circle.
Mapping $U \mapsto D: w = \dfrac{2\,p\,z}{(1+p) + (1-p)\,z^2}$;
$\Gamma: \rho = \rho(\phi) = \sqrt{1 - (1-p)^2\cos^2\phi}$, $0 < p < 1$;
Boundary correspondence function: $\tan\phi = p\tan\theta$.

3. Ellipses, with boundary $\Gamma = \{z = a\cos t + i\,b\sin t,\ a^2 - b^2 = 1\}$;
For the mapping $\text{Int}\,(\Gamma) \mapsto U$, the boundary correspondence function is defined by (10.4.22);
For the mapping $\text{Ext}\,(\Gamma) \mapsto U^*$, the boundary correspondence function is defined by $\phi(t) = t$.

The boundary correspondence function determines the correspondence between the boundaries of the two regions where one is the problem region and the other the image region. It is denoted by ϕ when a given simply connected region in the z-plane is mapped onto a region in the w-plane ($z = r\,e^{i\theta} \mapsto w = \rho\,e^{i\phi}$, which is the notation used here), but by θ if $w = \rho\, e^{i\phi} \mapsto z = r\, e^{i\theta}$. This function is usually normalized so that the origin is preserved. ■

10.5 Degenerate Kernel

A kernel is said to be degenerate if it can be represented in the semi-discrete form of a finite sum of products. A degenerate Neumann kernel can be expressed as

$$N(s,t) = \sum_{k=1}^{n} \alpha_k(s)\, \beta_k(t), \tag{10.5.1}$$

where it is assumed that the functions $\alpha_k(s)$ are linearly independent. Otherwise the number of terms in the expression for the kernel in (10.5.1) would reduce. For such a kernel we can determine a complete solution of the Fredholm integral equation of the form

$$\phi(s) = \lambda \int_0^L N(s,t)\, \phi(t)\, dt + g(s). \tag{10.5.2}$$

Let us assume the required solution of Eq (10.5.2) with $\lambda = 1$ in the form

$$\phi(s) = g(s) + \sum_{i=1}^{n} A_i\, \alpha_i(s), \tag{10.5.3}$$

where A_i are constants yet to be determined. We will use the notation:

$$\int_0^L \beta_i(t)\, g(t)\, dt = f_i, \quad \int_0^L \alpha_i(t)\, \beta_j(t)\, dt = b_{j,i}. \tag{10.5.4}$$

Then substituting (10.5.3) for $\phi(s)$ in Eq (10.5.2), we get

$$\begin{aligned} \sum_{i=1}^{n} A_i\, \alpha_i(s) - \lambda \int_0^L \left(\sum_{i=1}^{n} \alpha_i(s)\, \beta_i(t) \right) g(t)\, dt \\ - \lambda \int_0^L \sum_{i=1}^{n} \sum_{j=1}^{n} A_j\, \alpha_i(s)\, \beta_i(t)\, \alpha_j(t)\, dt = 0. \end{aligned} \tag{10.5.5}$$

Equating the coefficients of $\alpha_i(s)$ and using the notation (10.5.4), we find that

$$A_i - \lambda \sum_{j=1}^{n} A_j\, b_{i,j} = \lambda\, f_i, \quad i = 1, 2, \cdots, n. \tag{10.5.6}$$

The determinant of the system (10.5.6) in the unknowns A_i is

$$D(\lambda) = \begin{vmatrix} 1 - \lambda\, b_{1,1} & -\lambda\, b_{1,2} & \cdots & -\lambda\, b_{1,n} \\ -\lambda\, b_{2,1} & 1 - \lambda\, b_{2,2} & \cdots & -\lambda\, b_{2,n} \\ \cdots & \cdots & \cdots & \cdots \\ -\lambda\, b_{n,1} & -\lambda\, b_{n,2} & \cdots & 1 - \lambda\, b_{n,n} \end{vmatrix}. \tag{10.5.7}$$

Then, by Cramer's rule,

$$A_i = \lambda \frac{\sum\limits_{k=1}^{n} D_{i,k}\, f_k}{D(\lambda)}, \quad i = 1, 2, \cdots, n, \tag{10.5.8}$$

where $D_{i,k}$ is the algebraic complement of the kth row and ith column. Hence, from (10.5.3) the approximate solution $\widetilde{\phi}(s)$ of Eq (10.5.2) is given by

$$\begin{aligned}\widetilde{\phi}(s) &= g(s) + \lambda \sum_{i=1}^{n} \frac{\sum_{k=1}^{n} D_{i,k}\, f_k}{D(\lambda)}\, \alpha_i(s) \\ &= g(s) + \lambda \frac{\sum_{k=1}^{n} \left(\sum_{i=1}^{n} D_{i,k}\, \alpha_i(s) \right) \int_0^L \beta_k(t)\, g(t)\, dt}{D(\lambda)} \\ &= g(s) + \lambda \int_0^L \frac{\sum_{k=1}^{n} \sum_{i=1}^{n} D_{i,k}\, \alpha_i(s)\, \beta_k(t)}{D(\lambda)}\, g(t)\, dt \\ &= g(s) + \lambda \int_0^L \frac{D(s,t,\lambda)}{D(\lambda)}\, g(t)\, dt,\end{aligned} \tag{10.5.9}$$

where

$$D(s,t,\lambda) = \sum_{i=1}^{n} \sum_{k=1}^{n} D_{i,k}\, \alpha_i(s)\, \beta_k(t). \tag{10.5.10}$$

Thus, the resolvent $\gamma(s,t,\lambda)$ of Eq (10.5.2) is given by

$$D(s,t,\lambda) = \frac{D(s,t,\lambda)}{D(\lambda)}. \tag{10.5.11}$$

In practical applications, an arbitrary kernel $N(s,t)$ can be replaced approximately by a degenerate kernel $\widetilde{N}(S,t)$, which will help solve the resulting approximate equation

$$\widetilde{\phi}(s) - \lambda \int_0^L \widetilde{N}(s,t)\, \widetilde{\phi}(t)\, dt = g(s), \tag{10.5.12}$$

rather than the original equation (10.5.2). A suitable degenerate kernel close to the one given in an integral equation can always be found by taking it as a finite part of its Taylor series of its Fourier series, or by a trigonometric interpolation scheme (see §11.7). An error estimate in replacing the given kernel by an approximate degenerate kernel is contained in the following result:

Theorem 10.3. *Given two kernels $N(s,t)$ and $\widetilde{N}(s,t)$ such that*

$$\int_0^L \left| N(s,t) - \widetilde{N}(s,t) \right| dt < h,$$

let the resolvent $\gamma(s,t,\lambda)$ of Eq (10.5.2) satisfy the inequality

$$\int_0^L |\gamma(s,t,\lambda)|\, dt < B.$$

If the conditions $|g(s) - \tilde{g}(s)| < \varepsilon$ and $1 - |\lambda|\, h\, (1 + \lambda B) > 0$ are satisfied where $\tilde{g}(s)$ is an approximation of $g(s)$, then Eq (10.5.2) has a unique solution $\phi(s)$, and

$$\left| \phi(s) - \widetilde{\phi}(s) \right| < \frac{M\, |\lambda|\, (1 + |\lambda|\, B)^2}{1 - |\lambda|\, h\, (1 + \lambda B)} + \varepsilon\, (1 + |\lambda|\, B), \tag{10.5.13}$$

where M is the maximum modulus of $g(s)$.

A proof of this result can be found in Kantorovich and Krylov [1958: 143], or Berezin and Zhidkov [1965: 647].

Example 10.5. (a) To solve the integral equation

$$\phi(s) = \int_0^{1/2} \sin st\, \phi(t)\, dt + g(s), \tag{10.5.14}$$

with no assumptions on $g(s)$ at this time, expand $\sin st$ in its Taylor series and get

$$\sin st = st - \frac{s^3t^3}{3!} + \frac{s^5t^5}{5!} - \cdots .$$

If we replace $\sin st$ in (10.5.14) by the first two terms of this series expansion, then Eq (10.5.14) reduces to

$$\widetilde{\phi}(s) = \int_0^{1/2} \left(st - \frac{s^3t^3}{6} \right) \widetilde{\phi}(t)\, dt + g(s), \tag{10.5.15}$$

which has an algebraic kernel. We will assume a solution of the form

$$\widetilde{\phi}(s) = as + bs^3 + g(s).$$

Then substituting it in (10.5.12) we get

$$as + bs^3 - sf_1 - s^3 f_2 - \frac{as}{24} - \frac{bs}{160} + \frac{as^3}{768} + \frac{bs^3}{5376} = 0,$$

where

$$f_1 = \int_0^{1/2} t\, g(t)\, dt, \quad f_2 = -\frac{1}{6} \int_0^{1/2} t^3\, g(t)\, dt.$$

Equating the coefficients of a and b to zero, we obtain

$$\begin{gathered} \frac{23\, a}{24} - \frac{b}{160} - f_1 = 0, \\ \frac{a}{160} + \frac{5377\, b}{5376} - f_2 = 0, \end{gathered}$$

and

$$\widetilde{\phi}(s) = g(s) + as + bs^3 = g(s) + \int_0^{1,2} \gamma(s,t,1)\, dt, \tag{10.5.16}$$

where the resolvent $\gamma(s,t,1)$ is given by

$$\gamma(s,t,1) = 1.043277\, (1.000186\, s\,t - 0.0010416\, s^3\, t - 0.0010416\, s\, t^3 - 0.1597222\, s^3\, t^3).$$

Since $\displaystyle\int_0^{1/2} |\gamma(s,t,1)|\, dt < \frac{1}{12}$, we can take $B = 1/12$ in the above estimate (10.5.16). Also, since

$$\int_0^{1/2} \left| N(s,t) - \widetilde{N}(s,t) \right| dt \le \int_0^{1/2} \frac{s^5\, t^5}{120}\, dt \le \frac{1}{46080} \left(\frac{1}{2}\right)^5 < \frac{1}{1474560},$$

we can take $h = \dfrac{1}{1474560} \approx \dfrac{3}{4 \cdot 10^6}$. Then

$$\left|\phi(s) - \widetilde{\phi}(s)\right| < M \frac{\dfrac{3}{4\cdot 10^6}\,(1+1/12)}{1 - \dfrac{3}{4\cdot 10^6}\,(1+1/12)} < \frac{M}{10^6}.$$

In particular, if $g(s) = 1 + \dfrac{1}{s}\left(\cos\dfrac{s}{2} - 1\right) = 1 - \dfrac{s}{8} + \dfrac{s^3}{384} - \cdots$, then $M = 1$, and the approximate solution is

$$\phi(s) \approx 1 + 0.0000009\, s - 0.0000002\, s^3,$$

which has an error of $O\left(10^6\right)$.

(b) To solve the equation

$$\phi(s) = \int_0^1 \sin st\, \phi(t)\, dt + g(s), \tag{10.5.17}$$

we take $\widetilde{\phi}(s) = a\,s + b\,s^3 + g(s)$. Then

$$a\,s + b\,s^3 = \int_0^1 \left(s\,t - \frac{s^3\,t^3}{6}\right)\left(a\,s + b\,s^3 + g(s)\right)\, dt$$

yields the system of equations $\dfrac{2\,a}{3} - \dfrac{b}{5} - f_1 = 0,\ \dfrac{a}{30} + \dfrac{43\,b}{42} - f_2 = 0$. The resolvent is

$$\gamma(s,t,1) = \frac{3225}{2171}\,s\,t - \frac{105}{2171}\,s\,t^3 - \frac{105}{2171}\,s^3\,t - \frac{350}{2171}\,s^3\,t^3.$$

Since $\displaystyle\int_0^1 |\gamma(s,t,1)|\, dt \le \frac{445}{668} < 1$, we take $B = 1$. Also, since

$$\int_0^1 \left|N(s,t) - \widetilde{N}(s,t)\right|\, dt \le \frac{1}{720},$$

we take $h = 1/720$. Then

$$\left|\phi(s) - \widetilde{\phi}(s)\right| < M \frac{\dfrac{1}{720}\,(2)}{1 - \dfrac{1}{720}\,(2)} < \frac{M}{10^3}.$$

In particular, if $g(s) = 1 + \dfrac{1}{s}\,(1 - \cos s) = 1 - \dfrac{s}{2} + \dfrac{s^3}{24} - \cdots$, then $M = 1$, and the approximate solution is

$$\phi(s) \approx 1 + 0.001545\, s,$$

which has an error of $O\left(10^3\right)$. ■

Example 10.6. To solve

$$\phi(s) + \int_0^1 s\,\left(e^{st} - 1\right)\,\phi(t)\, dt = e^s - s,$$

we take

$$\widetilde{N}(s,t) = s\,\left(e^{st}-1\right) = s^2 t + \frac{1}{2}\,s^3t^2 + \frac{1}{6}\,s^4t^3.$$

Then we will solve the equation

$$\widetilde{\phi}(s) + \int_0^1 \widetilde{N}(s,t)\,\widetilde{\phi}(t)\,dt = e^s - s,$$

where we have

$$\widetilde{\phi}(s) = e^s - s + as^2 + bs^3 + cs^4.$$

Substituting this expression for $\widetilde{\phi}(s)$ into the above equation, we obtain the system

$$\begin{aligned} \frac{5}{4}\,a - \frac{1}{5}\,b + \frac{1}{6}\,c &= -\frac{2}{3}, \\ \frac{1}{5}\,a - \frac{13}{6}\,b + \frac{1}{7}\,c &= \frac{9}{4} - e, \\ \frac{1}{6}\,a - \frac{1}{7}\,b + \frac{49}{8}\,c &= 2e - \frac{29}{5}, \end{aligned}$$

whose solution is $a = -0.501019$, $b = -0.167126$, $c = 0.0418054$. Thus

$$\widetilde{\phi}(s) = e^s - s - 0.501019\,s^2 - 0.167126\,s^3 - 0.0418054\,s^4.$$

The exact solution of the given equation is $\phi(s) \equiv 1$. Note that $\widetilde{\phi}(0) = 1$, $\widetilde{\phi}(1/2) = 0.999963$, and $\widetilde{\phi}(1) = 1.00833$. ∎

10.6. Szegö Kernel

Let D be a bounded simply connected region with a Jordan boundary Γ. The Cauchy kernel

$$H(z,a) = \frac{1}{2i\pi}\,\frac{1}{z-a} \qquad a \in D, \tag{10.6.1}$$

defined by (2.4.3), represents an analytic function $f(z)$ on D with

$$f(a) = \frac{1}{2i\pi}\int_\Gamma \frac{f(z)}{z-a}\,dz = \frac{1}{2i\pi}\int_0^L \frac{f(z)}{z-a}\,\gamma'(s)\,ds, \quad a \in D, \tag{10.6.2}$$

where $z = \gamma(s)$, $dz = \gamma'(s)\,ds$, $0 \le s \le L$, is the parameterization of Γ in terms of the boundary arc s. The Szegö kernel, defined by (4.3.3), also represents an analytic function f on D with

$$f(a) = \int_\Gamma S(z,a)\,f(z)\,dz, \quad a \in D. \tag{10.6.3}$$

Let $\mathcal{H}^2(\Gamma)$ denote a closed subspace of $L^2(\Gamma)$, containing boundary values of analytic functions on D, and let $\mathcal{S} : L^2(\Gamma, ds) \mapsto \mathcal{H}^2(\Gamma)$ denote the orthogonal projection. Then for any $f \in L^2(\Gamma)$

$$\mathcal{S}\,f(a) = \int_\Gamma S(z,a)\,f(z)\,dz, \quad a \in D. \tag{10.6.4}$$

This relation implies (10.6.3) because $f = \mathcal{S} f$ in (10.6.3). The Szegö kernel coincides with the Cauchy kernel

$$H(z,a) = \frac{1}{2i\pi}\frac{1}{z-a}\gamma'(s), \tag{10.6.5}$$

iff D is a disk. In fact, the Szegö kernel for the unit disk U, denoted by $S_U(z,a)$, is given by

$$S_U(z,a) = \frac{1}{2\pi}\frac{1}{1-a\bar{z}}, \quad |a|<1, |z|=1. \tag{10.6.6}$$

The following result holds for an analytic function $f : D \mapsto U$ such that $f(a) = 0$ and $f'(a) > 0$ real, where $a \in D$. Such a function is called the Riemann mapping function (Theorem 2.15).

Theorem 10.4. *For a given $a \in D$ let an analytic function $w = f(z)$ be the Riemann mapping function. Then*

$$f'(z) = \frac{2\pi}{S(a,a)}\, S^2(z,a), \quad z \in D. \tag{10.6.7}$$

PROOF. The following transformation connects S to S_U:

$$S(z,\zeta) = \sqrt{f'(\zeta)}\, S_U\left(f(\zeta), f(z)\right) \overline{\sqrt{f'(z)}} \quad \text{for } z,\zeta \in D. \tag{10.6.8}$$

Set $\zeta = a$. Then $f(a) = 0$, and (10.6.6) yields

$$S(z,a) = \frac{1}{2\pi}\sqrt{f'(z)}\,\overline{\sqrt{f'(a)}},$$

which implies (10.6.7). ∎

We use (10.6.7) to compute the Riemann mapping function by the formula

$$f(z) = \frac{1}{i}\frac{f'(z)}{|f'(z)|}\gamma'(s). \tag{10.6.9}$$

This formula is used by Kerzman and Trummer [1986] for numerical computation of $f(z)$. Now we will define the Kerzman-Stein kernel $A(z,a)$ in terms of the Cauchy kernel by

$$\begin{aligned} A(z,a) &= \bar{H}(z,a) - H(z,a) \in C^\infty(\Gamma \times \Gamma) \\ &= -\frac{1}{2i\pi}\,\Re\left\{\frac{\gamma''(s)}{\gamma'(s)}\right\}. \end{aligned} \tag{10.6.10}$$

Note that $A(z,z) = 0$ since $\gamma''(s)$ is orthogonal to $\gamma'(s)$. Let $\mathcal{A} : L^2(\Gamma) \mapsto L^2(\Gamma)$ define the integral operator

$$\mathcal{A} f(a) = \int_\Gamma A(z,a)\, f(z)\, ds, \quad a, z \in \Gamma,\ z = \gamma(s). \tag{10.6.11}$$

This operator is compact, and $\mathcal{A}$ is self-adjoint. The operator $(1 - \mathcal{A})$ is bijective on $L^2(\Gamma) \mapsto L^2(\Gamma)$, and hence it has a bounded inverse $(1-\mathcal{A})^{-1} : L^2(\Gamma) \mapsto L^2(\Gamma)$. The geometric interpretation of the kernel $A(z,a)$ is as follows: Let both z and a lie on a closed contour in Γ, with tangent vectors (complex numbers) $\gamma'(s)$ and $\gamma'(t)$, $z = \gamma(s)$ and $a = \gamma(t)$, respectively. Then the vector (complex number) $\overline{\gamma'(s)}$ is the reflection of $\gamma'(a)$ in the chord joining z and a (Figure 10.8). Then

$$A(z,a) = \frac{1}{2i\pi}\,\frac{1}{z-a}\,\left[\gamma'(s) - \gamma'(t)\right].$$

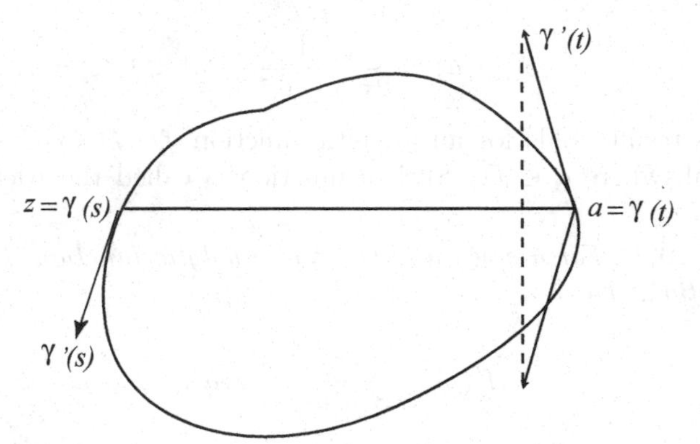

Figure 10.8 Vector $\gamma'(s)$.

This relation is obvious if $\arg\{z-a\} = 0$ or π, i.e., if the chord joining z and a is horizontal. Otherwise, it is proved by rotation. Since a circle is the only closed contour where a chord joining any two boundary points meets the circumference at the same angle at both points, we have $A(z,a) \equiv 0$ for all z, a on the boundary iff the boundary is a circle. If we expand $z = \gamma(s)$ in a Taylor series at $a = \gamma(t)$, we obtain

$$z = \gamma(s) = \gamma(t) + \gamma'(t)\,(s-t) + \frac{1}{2}\,\gamma''(t)\,(s-t)^2 + O(h^3), \quad h = s-t, \tag{10.6.12}$$

which yields

$$z - a = \gamma'(t)\,(s-t)\,\left[1 + \frac{1}{2}\,\frac{\gamma''(t)}{\gamma'(t)}\,(s-t) + O(h^2)\right],$$

and

$$\frac{1}{z-a} = \frac{1}{\gamma'(t)\,(s-t)}\,\left[1 - \frac{1}{2}\,\frac{\gamma''(t)}{\gamma'(t)}\,(s-t) + O(h^2)\right].$$

By differentiating (10.6.12), we get

$$\gamma'(s) = \gamma'(t) + \gamma''(t)\,(s-t) + O(h^2),$$

which together with (10.6.13) gives

$$\frac{\gamma'(s)}{z-a} = \frac{1}{s-t} + \frac{1}{2}\,\frac{\gamma''(t)}{\gamma'(t)} + O(h). \tag{10.6.13}$$

Similarly,

$$\frac{\gamma'(t)}{z-a} = \frac{1}{s-t} - \frac{1}{2}\,\frac{\gamma''(t)}{\gamma'(t)} + O(h). \tag{10.6.14}$$

Subtracting (10.6.13) and (10.6.14) we find that the singularities (which are real) of $1/(s-t)$ cancel, and then using (10.6.5), we obtain the Kerzman-Stein kernel $A(z,a)$ defined by

(10.6.10). This kernel is continuous and skew-symmetric, i.e., $A(z, a) = -\overline{A(z, a)}$. The following result (Kerzman and Stein [1986]) is useful:

Theorem 10.5. *The Szegö kernel $S(z, a)$, as a function of z, is the unique solution of the integral equation*

$$S(z, a) + \int_\Gamma A(z, a)\, S(z, a)\, d\gamma_\zeta = \overline{H(a, z)}, \quad z, \zeta \in \Gamma, \tag{10.6.15}$$

where γ denotes the arc length on Γ.

If we use a parameterization $z = \gamma(s)$ on Γ, $0 \le s \le L$, the integral equation (10.6.15) becomes

$$\theta(s) + \int_0^L k(s, t)\, \theta(t)\, dt = g(s), \quad 0 \le s \le L, \tag{10.6.16}$$

where

$$\begin{aligned} \theta(s) &= \left|\gamma'(s)\right|^{1/2} S\left(\gamma(s), a\right), \\ g(s) &= \left|\gamma'(s)\right|^{1/2} \overline{H\left(a, \gamma(s)\right)}, \\ k(s, t) &= \left|\gamma'(s)\right|^{1/2} A\left(\gamma(s), \gamma(t)\right) \left|\gamma'(t)\right|^{1/2}. \end{aligned} \tag{10.6.17}$$

This integral equation is solved by the Nyström method (see Atkinson [1976]; Delves and Mohamed [1985]), which is as follows: Since all functions in this equation are periodic, we take n equi-spaced collocation points $s_j = (j-1)L/n$, $j = 1, \dots, n$, and use the trapezoidal rule. This gives

$$\theta\left(s_j\right) + \frac{L}{n} \sum_{m=1}^{n} k\left(s_j, s_m\right) = g\left(s_j\right). \tag{10.6.18}$$

Let $B_{jm} = \dfrac{L}{n} k\left(s_j, s_m\right)$ define the skew-hermitian matrix $\mathbf{B}$, and let $x_j = \theta\left(s_j\right)$, be written in matrix form as

$$(I + \mathbf{B})\, \mathbf{x} = \mathbf{y}, \tag{10.6.19}$$

which is solved by an iterative method based on the generalized conjugate gradient method (GCM), the details of which can be found in Trummer [1986].

The discretized form (10.6.18) of Eq (10.6.16) gives the interpolation formula

$$\theta(s) = g(s) - \frac{L}{n} \sum_{m=1}^{n} k\left(s, s_m\right) x_m. \tag{10.6.20}$$

Once the solution θ of Eq (10.6.16) is computed, the boundary correspondence function $\phi(s)$, defined by $f(z) = f\left(\gamma(s)\right) = e^{i\phi(s)}$, $z = r\, e^{i\theta}$, can be computed from the formula

$$\phi(s) = \arg\left\{ -\, i\, \theta^2(s)\, \gamma'(s) \right\}. \tag{10.6.21}$$

Trummer [1986] has applied this method to the following six conformal mapping problems:
1. Γ is the inverted ellipse, defined by $z(s) = e^{is} \sqrt{1 - (1 - p^2)\cos^2 s}$, $0 < p \le 1$, where $\tan s = p \tan \phi(s)$ (see §10.4).
2. Γ is the ellipse $z(s) = e^{is} - \varepsilon\, e^{-is}$, $0 \le \varepsilon < 1$, with eccentricity$= (1 - \varepsilon)/(1 + \varepsilon)$, where

$\phi(s) = s + 2 \sum_{n=1}^{\infty} \frac{1}{n} \frac{\varepsilon^n}{1+\varepsilon^{2n}} \sin 2ns.$

3. Γ is the epitrochoid ('apple') $z(s) = e^{is} + \frac{\alpha}{2} e^{2is}$, $0 \le \alpha < 1$, where $\phi(s) = s$.
4. Γ is Cassini's oval $|z - \alpha||z + \alpha| = 1$, $0 \le \alpha < 1$, or

$$z(s) = e^{is} \sqrt{\alpha^2 \cos 2s + \sqrt{1 - \alpha^4 \sin^2 2s}},$$

where $\phi(s) = s - 0.5 \arg\{h(s)\}$, $h(s) = \sqrt{1 - \alpha^4 \sin^2 2s}$.
5. Γ is the unit square, where $\cos \phi(s) = \mathrm{dn}(Ky)$.
6. Γ is the stadium with the boundary composed of two semicircles joined by two line segments, all of the same length.

This method requires some programming. For details see Trummer [1986].

Example 10.7. The results for the ellipse with eccentricity 0.5 are presented in Figure 10.9. ■

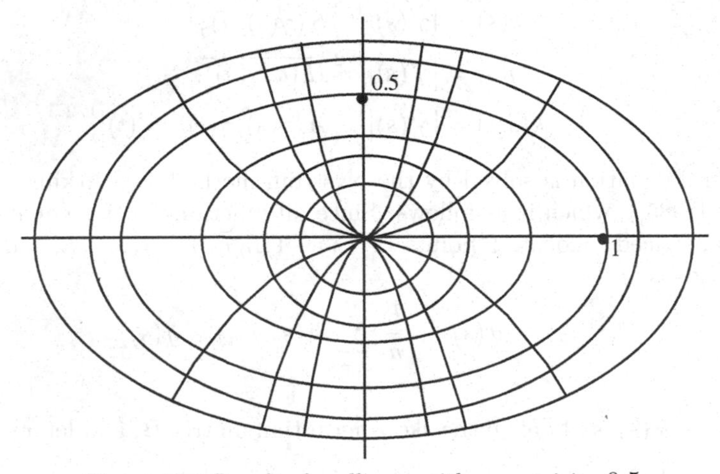

Figure 10.9 Results for ellipses with eccentricity 0.5.

Example 10.8. The approximate solution of the integral equation

$$\phi(x) - \int_0^{1/2} \sin xy \, \phi(y) \, dy = f(x),$$

is $\phi(x) \approx 1 + 0.0000009\, x - 0.0000002\, x^3$. Use $\sin xy \approx xy - \frac{x^3 y^3}{6} + \frac{x^5 y^5}{120}$, and $f(x) = 1 + \frac{1}{x}\left(\cos \frac{x}{2} - 1\right)$. Note that the exact solution is $\phi(x) = 1$.] (Kantorovich and Krylov [1958:145].) ■

Example 10.9. The approximate solution of the integral equation

$$y(x) + \int_0^1 x\,(e^{xs} - 1)\; y(s)\, ds = e^x - x$$

is $y(x) \approx e^x - x - 0.501\, x^2 - 0.1671\, x^3 - 0.0422\, x^4$, and the exact solution $y(x) \equiv 1$. (Berezin and Zhidkov [1965:653].) ■

Example 10.10. The function $u = \log 1/r$, $|z| = r$, which is a potential function, regular for $r \neq 0$, has the following properties:
(i) It yields the force flux of 2π at $r = 0$. [Hint: Evaluate $\lim_{\varepsilon\to 0} \frac{\partial u}{\partial n}\, ds$, where n is the outward normal.]
(ii) Define a source of strength q at a point $\zeta = (\xi, \eta)$ on the boundary Γ of a simply connected region D by

$$u(\xi,\eta) = \frac{q}{2\pi} \log \frac{1}{\sqrt{(x-\xi)^2 + (y-\eta)^2}}.$$

A source of strength $-q$ is called a *sink* of strength q. By combining a source and a sink of the same strength q and letting the distance between them tend to zero while keeping the moment constant, the potential is given by

$$u(\zeta) = \frac{qx}{2\pi\,(x^2+y^2)} = \frac{q\,\cos r}{2\pi r}.$$

Note that this potential is known as a dipole of strength q which is also the moment along the x-axis. (Andersen et al. [1962:173].) ■

Example 10.11. If s denotes the arc length on the boundary Γ of a simply connected region D, $0 \le s \le L$, and if the dipole density of Γ is $2\pi\nu(s)$, then the potential $u(z)$ is defined in D and in $D^* = \operatorname{Ext}(\Gamma)$ by

$$u(z) = \int_\Gamma \nu(s) \frac{\partial}{\partial n_s}\left(\log \frac{1}{r_{sz}}\right) ds, \quad z \in D, \tag{10.6.22}$$

where n_s is the inward normal at a point $\zeta \in \zeta(s)$ and $r_{sz} = |z-\zeta|$. The function $u(z)$ is regular in D and D^* but is discontinuous on Γ. If the unknown potential function $u(z)$ is determined by the boundary values $u_i(s) = g(s)$ from (10.6.22), then the dipole distribution $\nu(s)$ satisfies the integral equation

$$\nu(s) = \frac{1}{\pi}\left[g(s) - \int_0^L \nu(t) \frac{\partial}{\partial n_t}\left(\log \frac{1}{r_{tz}}\right) dt\right],$$

where n_t is the inward normal at a point $\zeta = \zeta(t)$, $0 \le t \le L$, $t \neq s$. (Andersen et al. [1962:173].) ■

10.6.1 Generalized Szegö and Bergman Kernels. These kernels have been generalized to domains in $\mathbb{C}^n$, $n \ge 2$, as follows: For very simple domains, like the unit ball, the Bergman and Szegö kernels can be computed explicitly. If $D \subset \mathbb{C}^n$ is the unit ball, then the Bergman kernel is defined by

$$B(z,\xi) = \frac{n!}{\pi^n} \frac{1}{(1 - z\cdot\bar{\xi})^{n+1}}, \tag{10.6.23}$$

and the Szegö kernel is given by

$$S(z,\xi) = \frac{(2n-1)!}{2\pi^n} \frac{1}{(1 - z\cdot\bar{\xi})^{n}}. \tag{10.6.24}$$

For the derivation of these formulas, see, e.g., Stein [1972]. For certain more complex domains Greiner and Stein [1978] have found an explicit formula for the Szegö kernel defined in regions of the type $D_k = \{(z, z_1) \in \mathbb{C}^2 : \Im\{z_1\} > |z|^{2k}\}$ for any positive integer k. Thus, for $\xi = \big(z, t + i(|z|^{2k} + \mu\big)$ and $\omega = \big(w, s + i(|w|^{2k} + \nu\big)$, where $\mu, \nu > 0$, the Szegö kernel is given by

$$\begin{aligned} S(\xi, \omega) = \frac{1}{4\pi^2}\Big[\Big(\frac{i}{2}(s-t) + \frac{|z|^{2k} + |w|^{2k}}{2} + \frac{\mu+\nu}{2}\Big) - z\bar{w}\Big]^{-2} \\ \times \frac{1}{4\pi^2}\Big[\frac{i}{2}(s-t) + \frac{|z|^{2k} + |w|^{2k}}{2} + \frac{\mu+\nu}{2}\Big]^{(1-k)/k}. \end{aligned} \tag{10.6.25}$$

Diaz [1987] has shown that for these domains, the Szegö kernel is bounded in L^p, $1 < p < \infty$.

This result has been generalized to domains in $\mathbb{C}^n$ by Frances and Hanges [1995], as follows: The Szegö kernel in domains of the type $D = \{z, \xi, w) \in \mathbb{C}^{n+m+1} : \Im\{w\} > \|z\|^{2p} + \|\xi\|^{2p}\}$ is given by

$$S(z, \xi, t; z', \xi', t') = \sum_{k=1}^{n+1} c_k \frac{(A - z \cdot \bar{z}')^{(k/p)-n-1}}{\big[(A - z \cdot \bar{z}')^{1/p} - \xi \cdot \bar{\xi}'\big]^{m+k}}, \tag{10.6.26}$$

where $A = \frac{1}{2}\left[\|z\|^2 + \|z'\|^2 + \|\xi\|^2 + \|\xi'\|^2 - i(t - t')\right]$.

REFERENCES USED: Ahlfors [1952], Andersen et al. [1962], Atkinson [1976], Berezin and Zhidkov [1965], Birkoff et al. [1950; 1951], Carrier [1947], Carrier, Krook and Pearson [1966], Gaier [1964], Gershgorin [1933], Kerzman and Stein [1978; 1986], Kythe [1998], Lichtenstein [1917], Kantorovich and Krylov [1958], Trummer [1986], Todd and Warschawski [1955], Warschawski [1955; 1956], Whittaker and Watson [1962].

11

Theodorsen's Integral Equation

We will present Theodorsen's integral equation and establish the convergence of the related iterative method for the standard case of mapping the unit circle onto the interior (or exterior) of almost circular and starlike regions, both containing the origin. A trigonometric interpolation scheme is presented, and Wegmann's iterative and Newton's method for numerically solving this equation are discussed. The last two methods are based on a certain Riemann-Hilbert problem, which turns out to be a linearized form of a singular integral equation of the second kind. Unlike the classical iterative method, the solution of the linearized problem in Wegmann's method for the conformal map of the unit circle can be represented explicitly in terms of integral transforms, which leads to a quadratic convergent Newton-like method that avoids the numerical solution of a system of linear equations and thus becomes more economical. Theodorsen's integral equation has specific significance in the theory of airfoils.

11.1 Classical Iterative Method

Let Γ denote a Jordan curve defined in the polar coordinate system by $\rho = \rho(\phi)$, $0 \le \phi \le 2\pi$, where $\rho(\phi) \in C^1$, such that for some ε $(0 < \varepsilon < 1)$ and a constant $a > 0$

$$\frac{a}{1+\varepsilon} \le \rho(\phi) \le a(1+\varepsilon), \tag{11.1.1}$$

and

$$\left|\frac{\rho'(\phi)}{\rho(\phi)}\right| \le \varepsilon. \tag{11.1.2}$$

Any Jordan curve Γ that satisfies these two conditions is called a nearly circular contour or a near circle. Let us assume that a function $w = f(z) = \rho\, e^{i\phi}$, $z = e^{i\theta}$, with $f(0) = 0$ and $f'(0) > 0$, maps the unit disk $U = \{|z| < 1\}$ onto Int (Γ) which is starlike with respect to the origin. Then $f\left(e^{i\theta}\right) = \rho\left(\phi(\theta)\right)\, e^{i\phi(\theta)}$ defines the boundary correspondence function $\phi : [0, 2\pi] \mapsto \mathbb{R}$. If ϕ is known, then f is known. Consider the function

$$F(z) = \log \frac{f(z)}{z} = \log \left|\frac{f(z)}{z}\right| + i \arg \left\{\frac{f(z)}{z}\right\},$$

which, defined as real-valued $\log f'(0)$ for $z = 0$, is single-valued and analytic in U, and continuous on $U \cup \partial U$. If we set $z = e^{i\theta}$, then $\arg\left\{\frac{f(z)}{z}\right\} = \arg\left\{f\left(e^{i\theta}\right) e^{-i\theta}\right\} = \phi(\theta) - \theta$. Thus,

$$F\left(e^{i\theta}\right) = \log \rho[\phi(\theta)] + i\,[\phi(\theta) - \theta]. \tag{11.1.3}$$

Then, in view of the Schwarz formula which states that

$$w\left(\rho e^{i\theta}\right) = iv(0) + \frac{1}{2\pi}\int_0^{2\pi} u\left(R e^{i\phi}\right) \frac{R e^{i\theta} + \rho e^{i\phi}}{R e^{i\theta} - \rho e^{i\phi}}\, d\theta, \tag{11.1.4}$$

we obtain with $\rho = R = 1$ and $v(0) = 0$

$$\phi(\theta) - \theta = \frac{1}{2\pi}\int_0^{2\pi} \log \rho\left(\phi(\theta)\right) \cot \frac{\phi - \theta}{2}\, d\theta. \tag{11.1.5}$$

Note that not only the mapping function $f(z)$ is defined in the form $\rho = \rho(\phi)$ on Γ, but the relation (11.1.5) represents an integral equation for the unknown function $\phi(\theta)$, i.e.,

$$\phi(\theta) - \theta = -\frac{1}{2\pi}\int_0^{\pi} \left[\log \rho\left(\phi(\theta + t)\right) - \log \rho\left(\phi(\theta - t)\right)\right] \cot \frac{t}{2}\, dt. \tag{11.1.6}$$

This is known as *Theodorsen's integral equation.* Once the function $\phi(\theta)$ is determined, the function $F(z)$ and then the mapping function $f(z)$ can be computed. The term $\arg\left\{\frac{f(z)}{z}\right\}_{z=0} = \arg\left\{f'(0)\right\}$ is not added to the right side of (11.1.6) because it is zero. Theodorsen [1931] showed that ϕ is a solution of Eq (11.1.6). Gaier [1964: 66] proved that this equation has exactly one solution which is continuous and strongly monotone. The Riemann mapping theorem (§2.6) guarantees the existence of a continuous solution of this integral equation. We will also show that this solution is unique.

Theodorsen's method for solving the integral equation (11.1.6) for nearly circular regions is based on the iterations

$$\begin{aligned} &\phi_0(\theta) = 0, \\ &\phi_n(\theta) - \theta = -\frac{1}{2\pi}\int_0^{\pi} \left\{\log \rho\left[\phi_{n-1}(\theta + t)\right] - \log \rho\left[\phi_{n-1}(\theta - t)\right]\right\} \cot \frac{t}{2}\, dt, \\ &\qquad n = 1, 2, \dots . \end{aligned} \tag{11.1.7}$$

The functions $\phi_n(\theta)$ are absolutely continuous, and $\phi_n'(\theta) \in L^2[0, 2\pi]$. In fact, this is obviously true for $n = 0$. Suppose that this statement is true for some $n \geq 0$. Since, in view of (11.1.2), the function $\log \rho(\phi)$ has a bounded difference quotient and $\phi_n(\theta)$ is absolutely continuous, it follows that $\log \rho\left(\phi_n(\theta)\right)$ is also absolutely continuous. Also, since

$$\left(\frac{\rho'\left(\phi_n(\theta)\right)}{\rho\left(\phi_n(\theta)\right)}\, \phi_n'(\theta)\right)^2 \leq \varepsilon^2\, [\phi_n'(\theta)]^2, \tag{11.1.8}$$

the integral $\displaystyle\int_0^{2\pi} \left[\frac{\rho'\left(\phi_n(\theta)\right)}{\rho\left(\phi_n(\theta)\right)}\, \phi_n'(\theta)\right]^2 d\theta$ exists. Hence, the function $\phi_n(\theta) - \theta$ which is the conjugate of $\log \rho\left(\phi_n(\theta)\right)$ exists and is absolutely continuous. The integrands in (11.1.7) are singular at $t = 0$ where the integrals take the Cauchy principal values. In what follows we will use the notation

$$\sigma\left(\phi(\theta)\right) = \frac{\rho'\left(\phi(\theta)\right)}{\rho\left(\phi(\theta)\right)}, \quad \text{and} \quad p(\phi) = \frac{d}{d\phi}\left[\frac{\rho'(\phi)}{\rho(\phi)}\right]. \tag{11.1.9}$$

Both σ and p are Hölder-continuous.

11.2 Convergence

The following result holds for the convergence of Theodorsen's iterative method (11.1.7).

Theorem 11.1. *The sequences* $\{\phi_n(\theta)\}$ *and* $\{\phi_n'(\theta)\}$ *converge uniformly to* $\phi(\theta)$ *and* $\phi'(\theta)$, *respectively, as* $n \to \infty$.

This result will in turn establish that $\log \rho[\phi_n(\theta)]$ converges uniformly to $\log \rho[\phi(\theta)]$ as $n \to \infty$, so that the functions

$$F_0\left(e^{i\theta}\right) = \log a, \quad F_n\left(e^{i\theta}\right) = \log \rho[\phi_{n-1}(\theta)] + i\left(\phi_n(\theta) - \theta\right), \quad n \geq 1, \tag{11.2.1}$$

will compute $f\left(e^{i\theta)}\right)$ to any desired accuracy. Let the functions $F_n(z)$ be analytic on U and assume boundary values $F_n\left(e^{i\theta}\right)$ on $|z| = 1$. In view of the maximum modulus principle, the uniform convergence of $F_n\left(e^{i\theta}\right)$ to $F\left(e^{i\theta}\right)$ implies the uniform convergence of $F_n(z)$ to $F(z)$ on $|z| \leq 1$. Hence, the functions $f_n(z) = z\, e^{F_n(z)}$ converge uniformly to the mapping function $f(z) = z\, e^{F(z)}$.

To prove the above theorem, we will first derive the estimates for the differences $|\phi_n(\theta) - \phi(\theta)|$ and $|\phi_n'(\theta) - \phi'(\theta)|$ in terms of ε and n, and show that these differences approach zero as $n \to \infty$.

Theorem 11.2. *If* Γ *is a near circle and if* $\phi_n(\theta)$ *and* $\phi(\theta) = \arg\left\{f\left(r\, e^{i\theta}\right)\right\}$ *are defined by (11.1.7) and (11.1.3), then*

$$|\phi_n(\theta) - \phi(\theta)| \leq 2\left(\frac{\pi}{1-\varepsilon^2}\right)^{1/4} \varepsilon^{(n+2)/2}. \tag{11.2.2}$$

Note that the bound in (11.2.2) goes to zero as $n \to \infty$ since $0 < \varepsilon < 1$. This will establish the convergence of $\phi_n(\theta)$ to $\phi(\theta)$.

Theorem 11.3. *If* Γ *is a near circle and if* $\sigma \in H^1$, *then*

$$|\phi_n(\theta) - \phi(\theta)| \leq \sqrt{2\pi A(n+1)}\, \varepsilon^{n+1}, \tag{11.2.3}$$

where $A = 4^\varepsilon\, e^{\varepsilon^2}$.

This result provides a bound that converges to zero more rapidly as $n \to \infty$.

Theorem 11.4. *If* Γ *is a near circle and if* $\sigma \in H^1$ *and* $p(\phi) \in H^1$, *then*

$$|\phi_n'(\theta) - \phi'(\theta)| \leq \sqrt{2\pi c_n}\, [A(n+1)]^{3/2}\, \varepsilon^{n+1}, \tag{11.2.4}$$

where

$$c_1 = 1 + \varepsilon, \quad c_n = (1+\varepsilon) \prod_{k=2}^{n} \left(1 + \varepsilon^k \sqrt{2\pi A k}\right), \tag{11.2.5}$$

and for all n

$$c_n \leq (1+\varepsilon)\, e^{2\varepsilon^2 \sqrt{\pi A}\,(1-\varepsilon)^{-3/2}}. \tag{11.2.6}$$

The last inequality shows that c_n is bounded if $0 < \varepsilon < 1$. The proofs for these theorems are given in Warschawski [1945]. We will outline these proofs in the next section. It should be noted that the estimates for the difference $|F_n(z) - F(z)|$ for $|z| \leq 1$ are obtained from those for $|\phi_n(\theta) - \phi(\theta)|$ given above. Thus, in view of (11.1.2), we have

$$\left|F_n\left(e^{i\theta}\right) - F\left(e^{i\theta}\right)\right| \leq \sqrt{\varepsilon^2\, [\phi_{n-1}(\theta) - \phi(\theta)]^2 + [\phi_n(\theta) - \phi(\theta)]^2},$$

and

$$|F_n(z) - F(z)| \le \max_\theta \left|F_n\left(e^{i\theta}\right) - f\left(e^{i\theta}\right)\right| \quad \text{for } |z| \le 1.$$

In the case of Theorem 11.2 this yields

$$|F_n(z) - f(z)| \le 2\left(\frac{\pi^2}{1-\varepsilon}\right)^{1/4} \varepsilon^{(n+2)/2}\sqrt{1+\varepsilon},$$

and in the case of Theorem 11.3

$$|F_n(z) - f(z)| \le 2\,\varepsilon^{n+1}\sqrt{\pi A\left(n+\frac{1}{2}\right)}.$$

Hence, for $0 < \varepsilon < 1$ the iterations $F_n(z)$ converge uniformly to $F(z) = \log\dfrac{f(z)}{z}$ for $|z| \le 1$. The above theorems constitute the classical theory for the convergence of the numerical method for solving Theodorsen's integral equation by iterations in terms of the boundary correspondence function ϕ, which provides the required boundary map f.

11.3 Theodorsen's Method

Theodorsen's method converges strongly like a geometric series. First, we obtain the bounds for the square means

$$M_n^2 = \frac{1}{2\pi}\int_0^{2\pi} [\phi_n(\theta) - \phi(\theta)]^2\, d\theta, \quad {M'_n}^2 = \frac{1}{2\pi}\int_0^{2\pi} [\phi'_n(\theta) - \phi'(\theta)]^2\, d\theta,$$
$$M''^2_n = \frac{1}{2\pi}\int_0^{2\pi} [\phi''_n(\theta) - \phi''(\theta)]^2\, d\theta. \tag{11.3.1}$$

Then the results of these theorems are obtained by using the inequalities

$$|\phi_n(\theta) - \phi(\theta)| \le \sqrt{2\pi M_n M'_n}, \quad |\phi'_n(\theta) - \phi'(\theta)| \le \sqrt{2\pi M'_n M''_n}. \tag{11.3.2}$$

The following results are also needed:

Lemma 11.1. *If the function $g(\theta) \in L^2[0, 2\pi]$ is real-valued, 2π-periodic and square-integrable (in Lebesgue's sense) on $0 \le \theta \le 2\pi$, and if $\bar{g}(\theta)$ is a conjugate function of $g(\theta)$, then*

$$\frac{1}{2\pi}\int_0^{2\pi} [\bar{g}(\theta)]^2\, d\theta + \alpha^2 = \frac{1}{2\pi}\int_0^{2\pi} [g(\theta)]^2\, d\theta + \beta^2, \tag{11.3.3}$$

where

$$\alpha^2 = \frac{1}{2\pi}\int_0^{2\pi} g(\theta)\, d\theta, \quad \beta^2 = \frac{1}{2\pi}\int_0^{2\pi} \bar{g}(\theta)\, d\theta.$$

Lemma 11.2. *If $g(\theta) \in L^2[0, 2\pi]$ is a real-valued, absolutely continuous, 2π-periodic and square-integrable function, then for any θ_0*

$$[g(\theta)]^2 - [g(\theta_0)]^2 \le 2\pi M M', \tag{11.3.4}$$

where

$$M^2 = \frac{1}{2\pi}\int_0^{2\pi} [g(\theta)]^2\, d\theta, \quad M'^2 = \frac{1}{2\pi}\int_0^{2\pi} [g'(\theta)]^2\, d\theta.$$

The factor 2π *in (11.3.4) is the best possible.*

Lemma 11.3. *If* Γ *is nearly circular, then the function* $\phi(\theta) = \arg\{f\,(r\,e^{i\theta})\}$*, defined by (11.1.3), is absolutely continuous and* $[\phi'(\theta)]^2 \in L^2[0, 2\pi]$ *in Lebesgue's sense such that*

$$\frac{1}{2\pi}\int_0^{2\pi} [\phi'(\theta)]^2\,d\theta \le \frac{1}{1-\varepsilon^2}. \tag{11.3.5}$$

Lemma 11.4. *If* $g(\theta) \in H^\alpha$, $0 < \alpha \le 1$, *is a* 2π*-periodic function, then any conjugate function of* $g(\theta)$ *also satisfies a Hölder condition.*

Lemma 11.5. *If* Γ *is a near circle and if* $\sigma(\phi) \in H^1$,† *then*

$$\frac{1}{A\sqrt{1+\varepsilon^2}} \le \phi'(\theta) \le A, \tag{11.3.6}$$

$$\sqrt{\frac{1}{2\pi}\int_0^{2\pi} [\phi''(\theta)]^2\,d\theta} < A^{3/2}\,\varepsilon\,\min\left(1+\varepsilon; \sqrt{2}\right). \tag{11.3.7}$$

Lemma 11.6. *If* $u(t) \in C^1[0, 2\pi]$ *is a* 2π*-periodic function and if* $v(t) \in C^1[0, 2\pi]$ *is a function conjugate to* $u(t)$*, then for every* θ

$$\int_0^{2\pi} \left(\frac{u(t)-u(\theta)}{\sin\dfrac{t-\theta}{2}}\right)^2 dt = \int_0^{2\pi} \left(\frac{v(t)-v(\theta)}{\sin\dfrac{t-\theta}{2}}\right)^2 dt.$$

Proof of Lemma 11.1 is available in Zygmund [1935: Eq (4)]; of Lemma 11.4 in Privaloff [1916] or Zygmund [1935: 156]; of Lemma 11.5 in Warschawski [1950]; and of Lemma 11.6 (on conjugate functions) in Warschawski [1945].

PROOF OF LEMMA 11.2. Note that for $0 \le \theta \le 2\pi$, $0 \le \theta_0 \le 2\pi$,

$$g^2(\theta) - g^2(\theta_0) = 2\int_{\theta_0}^{\theta} g(t)\,g'(t)\,dt = 2\int_{\theta_0}^{\theta-2\pi} g(t)\,g'(t)\,dt. \tag{11.3.8}$$

Since

$$\left|\int_{\theta_0}^{\theta} |g\,g'|\,dt\right| + \left|\int_{\theta_0}^{\theta-2\pi} |g\,g'|\,dt\right| = \int_{\theta-2\pi}^{\theta} |g\,g'|\,dt = \int_0^{2\pi} |g\,g'|\,dt,$$

one of the two integrals in (11.3.8) does not exceed $\dfrac{1}{2}\displaystyle\int_0^{2\pi} |g\,g'|\,dt$. Hence, by the Schwarz inequality,

$$[g(\theta)]^2 - [g(\theta_0)]^2 \le \int_0^{2\pi} |g\,g'|\,dt \le 2\pi M M'.$$

Also, applying (11.3.4) with $g(\theta) = \cos^n\theta$ $(\theta_0 = \pi/2)$ and letting $n \to \infty$, we find that the constant 2π cannot be replaced by any smaller one. ■

† The space H^1 consists of 2π-periodic functions $w \in H$ such that the derivative w' is an element of the Hilbert space $H = L_2(0, 2\pi)\backslash\mathbb{R}$.

PROOF OF LEMMA 11.3. Since Γ is rectifiable, the function $F\left(e^{i\theta}\right)$ is absolutely continuous (this follows from a theorem of F. and M. Riesz [1923]). Hence,

$$\frac{d}{d\theta} F\left(e^{i\theta}\right) - i = \sigma\left(\phi(\theta)\right)\, \phi'(\theta) + i\, \phi'(\theta)$$

exists a.e. for $0 \le \theta \le 2\pi$ and is integrable. Moreover, the function $\dfrac{\partial}{\partial\theta} F(z) - i = u(z) + i\, v(z)$, $z = r\, e^{i\theta}$, has the Poisson integral representation in the unit disk as

$$u(z) + i\, v(z) = \frac{1}{2\pi} \int_0^{2\pi} \left[u\left(r\, e^{it}\right) + i\, v\left(r\, e^{it}\right)\right] \frac{1 - r^2}{1 + r^2 - 2r\, \cos(t - \theta)}\, dt. \tag{11.3.9}$$

For almost all $\theta \in [0, 2\pi]$ we have

$$\lim_{r \to 1} u\left(r\, e^{i\theta}\right) = \sigma\left(\phi(\theta)\right)\, \phi'(\theta) = u\left(e^{i\theta}\right),$$
$$\lim_{r \to 1} v\left(r\, e^{i\theta}\right) = \phi'(\theta) = v\left(e^{i\theta}\right),$$

and since Int (Γ) is starlike (see §11.5), $\phi'(\theta) \ge 0$. Then by (11.1.2)

$$v\left(e^{i\theta}\right) \pm u\left(e^{i\theta}\right) \ge \phi'(\theta)\,(1 - \varepsilon) \ge 0.$$

Thus, in view of (11.3.9) we have $v(z) + u(z) \ge 0$ and $v(z) - u(z) \ge 0$ for $|z| < 1$. Hence, $v^2(z) - u^2(z) \ge 0$ for $|z| < 1$. Also,

$$\frac{1}{2\pi} \int_0^{2\pi} \left[v^2\left(r\, e^{i\theta}\right) - u^2\left(r\, e^{i\theta}\right)\right] d\theta = 1.$$

Then, taking the limit as $r \to 1$ and using Fatou's lemma (§2.1.3) , we get

$$\frac{1}{2\pi} \int_0^{2\pi} [\phi'(\theta)]^2 \left\{[1 - [\sigma\left(\phi(\theta)\right)]^2\right\} d\theta \le 1,$$

which, in view of (11.1.2), yields (11.3.5). ■

PROOF OF THEOREM 11.2. (a) First we determine an estimate for M_n. In view of (11.1.3) and (11.1.7), we have

$$\int_0^{2\pi} [\phi(\theta) - \theta]\, d\theta = 0, \qquad \int_0^{2\pi} [\phi_n(\theta) - \theta]\, d\theta = 0. \tag{11.3.10}$$

Now, applying Lemma 11.1 with $g(\theta) + i\, \bar{g}(\theta) = F_n\left(e^{i\theta}\right) - F\left(e^{i\theta}\right)$ and noting that $\beta = 0$ because of (11.3.10), we get

$$\begin{aligned} M_n^2 &= \frac{1}{2\pi} \int_0^{2\pi} [\phi_n(\theta) - \phi(\theta)]^2\, d\theta \\ &\le \frac{1}{2\pi} \int_0^{2\pi} \left[\log\, \rho\left(\phi_{n-1}(\theta)\right) - \log\, \rho\left(\phi(\theta)\right)\right]^2 d\theta. \end{aligned} \tag{11.3.11}$$

Since by (11.1.2)

$$\left|\log\, \rho\left(\phi_{n-1}(\theta)\right) - \log\, \rho\left(\phi(\theta)\right)\right| \le \varepsilon\, \left|\phi_{n-1}(\theta) - \phi(\theta)\right|,$$

from (11.3.11) we obtain

$$M_n^2 \le \varepsilon^2 \frac{1}{2\pi} \int_0^{2\pi} [\phi_{n-1}(\theta) - \phi(\theta)]^2 \, d\theta \le \varepsilon^2 M_{n-1}^2,$$

or

$$M_n \le \varepsilon M_{n-1}, \quad M_n \le \varepsilon^n M_0.$$

For $n = 0$ we find from (11.3.3) by using (11.1.1) that

$$M_0^2 = \frac{1}{2\pi} \int_0^{2\pi} [\phi(\theta) - \theta]^2 \, d\theta \le \frac{1}{2\pi} \int_0^{2\pi} \left[\log \frac{\rho(\phi(\theta))}{a} \right]^2 d\theta \le \varepsilon^2,$$

which yields the estimate

$$M_n \le \varepsilon^{n+1}. \tag{11.3.12}$$

(b) Now we determine an estimate for M_n'. Since $F_n\left(e^{i\theta}\right)$ and $F\left(e^{i\theta}\right)$ are absolutely continuous and $\frac{d}{d\theta} F_n\left(e^{i\theta}\right)$ and $\frac{d}{d\theta} F\left(e^{i\theta}\right)$ belong to the class $L^2[0, 2\pi]$, the imaginary part of $\frac{d}{d\theta}\left[F_n\left(e^{i\theta}\right) - F\left(e^{i\theta}\right)\right]$ is a conjugate function of the real part. Then

$$\int_0^{2\pi} \frac{d}{d\theta} F\left(e^{i\theta}\right) d\theta = F\left(e^{i\theta}\right)\Big|_{\theta=0}^{2\pi} = 0, \quad \int_0^{2\pi} \frac{d}{d\theta} F_n\left(e^{i\theta}\right) d\theta = 0. \tag{11.3.13}$$

Hence, using Lemma 11.1 with $g(\theta) + i\,\bar{g}(\theta) = \frac{d}{d\theta}\left[F_n\left(e^{i\theta}\right) - F\left(e^{i\theta}\right)\right]$, we obtain

$$\begin{aligned} {M_n'}^2 &= \frac{1}{2\pi} \int_0^{2\pi} [\phi_n'(\theta) - \phi'(\theta)]^2 \, d\theta \\ &= \frac{1}{2\pi} \int_0^{2\pi} \left\{\sigma\left(\phi_{n-1}(\theta)\right) \phi_{n-1}'(\theta) - \sigma\left(\phi(\theta)\right) \phi'(\theta)\right\}^2 d\theta. \end{aligned} \tag{11.3.14}$$

In view of (11.1.2), then

$${M_n'}^2 \le 2\varepsilon^2 \frac{1}{2\pi} \int_0^{2\pi} \left[\phi'^2_{n-1}(\theta) - \phi'^2(\theta)\right]^2 d\theta. \tag{11.3.15}$$

By Lemma 11.3, we have $\frac{1}{2\pi} \int_0^{2\pi} \phi'^2(\theta)\, d\theta \le \frac{1}{1 - \varepsilon^2}$. Moreover, again using Lemma 11.1 with $g(\theta) + i\,\bar{g}(\theta) = \frac{d}{d\theta} F_n\left(e^{i\theta}\right)$, and (11.1.2), we get

$$\begin{aligned} \frac{1}{2\pi} \int_0^{2\pi} [\phi'_n(\theta) - 1]^2 \, d\theta &= \frac{1}{2\pi} \int_0^{2\pi} \left[\sigma\left(\phi_{n-1}(\theta)\right) \phi_{n-1}'(\theta)\right]^2 d\theta \\ &\le \varepsilon^2 \frac{1}{2\pi} \int_0^{2\pi} \phi'^2_{n-1}(\theta)\, d\theta, \end{aligned}$$

or, after suppressing the argument θ in the integrands,

$$\frac{1}{2\pi} \left\{ \int_0^{2\pi} [\phi'_n - 1]^2 \, d\theta - 2 \int_0^{2\pi} \phi_n' \, d\theta + 2\pi \right\} \le \varepsilon^2 \frac{1}{2\pi} \int_0^{2\pi} \phi'^2_{n-1} \, d\theta.$$

Since $\int_0^{2\pi} {\phi'}_n \, d\theta = 2\pi$, we have

$$\frac{1}{2\pi} \int_0^{2\pi} {\phi'}_n^2 \, d\theta - 1 \leq \varepsilon^2 \frac{1}{2\pi} \int_0^{2\pi} {\phi'}_{n-1}^2 \, d\theta.$$

If we set $m_n^2 = \frac{1}{2\pi} \int_0^{2\pi} {\phi'}_n^2 \, d\theta$, then $m_n^2 \leq 1 + m_{n-1}^2$, and hence, $m_n^2 \leq 1 + \varepsilon^2 + \varepsilon^4 + \cdots + \varepsilon^{2n} m_0^2$. Since $m_0^2 = \frac{1}{2\pi} \int_0^{2\pi} d\theta = 1$, we get

$$m_n^2 \leq (1 - \varepsilon^2)^{-1}. \tag{11.3.16}$$

Thus, from (11.3.15), (11.3.5) and (11.3.16) we find that

$${M'_n}^2 \leq \frac{4\varepsilon^2}{1 - \varepsilon^2}. \tag{11.3.17}$$

(c) Finally, we determine an estimate for $|\phi_n(\theta) - \phi(\theta)|$. We set $g(\theta) = \phi_n(\theta) - \phi(\theta)$ in Lemma 11.2. Then, since $\int_0^{2\pi} g(\theta) \, d\theta = 0$, there exists a value θ_0 such that $g(\theta_0) = 0$, which yields

$$|\phi_n(\theta) - \phi(\theta)| \leq \sqrt{2\pi M_n {M'}_n}.$$

This gives (11.2.2) after using (11.3.12) and (11.3.17), which completes the proof of Theorem 11.2. ■

Proof of Theorem 11.3. (a) First, we determine an estimate for M'_n. Using (11.3.14), we have by Minkowsky's inequality for $n \geq 1$

$$\begin{aligned} M'_n &= \sqrt{\frac{1}{2\pi} \int_0^{2\pi} \left[\left(\sigma\left(\phi_{n-1}\right) - \sigma(\phi) \right) \phi' + \sigma\left(\phi_{n-1}\right) \left(\phi'_{n-1} - \phi' \right) \right]^2 d\theta} \\ &\leq \sqrt{\frac{1}{2\pi} \int_0^{2\pi} \left[\sigma\left(\phi_{n-1}\right) - \sigma(\phi) \right]^2 \phi'^2 \, d\theta} + \sqrt{\frac{1}{2\pi} \int_0^{2\pi} \left(\phi'_{n-1} - \phi' \right)^2 \left[\sigma\left(\phi_{n-1}\right) \right]^2 d\theta}. \end{aligned}$$

Since $0 < \phi'(\theta) \leq A$ (Lemma 11.5) and $\sigma(\phi) \in H^1$, from (11.1.2) we obtain

$$M'_n \leq \varepsilon A \sqrt{\frac{1}{2\pi} \int_0^{2\pi} \left(\phi_{n-1} - \phi \right)^2 d\theta} + \varepsilon M'_{n-1} = \varepsilon \left(A M_{n-1} + M'_{n-1} \right),$$

and, then by (11.3.12)

$$M'_n \leq \varepsilon \left(A \varepsilon^n + M'_{n-1} \right) \quad \text{for } n \geq 1. \tag{11.3.18}$$

For $n = 0$ from (11.1.2) and Lemma 11.1 we get

$${M'}_0^2 = \frac{1}{2\pi} \int_0^{2\pi} \left(\phi' - 1 \right)^2 d\theta = \frac{1}{2\pi} \int_0^{2\pi} \left[\sigma(\phi) \, \phi' \right]^2 d\theta \leq \varepsilon^2 \frac{A}{2\pi} \int_0^{2\pi} \phi' \, d\theta = \varepsilon^2 A. \tag{11.3.19}$$

Now we will prove by induction that

$$M'_n \le A(n+1)\,\varepsilon^{n+1}. \tag{11.3.20}$$

In fact, since (11.3.19) holds for $n = 0$, we assume that (11.3.20) is true for some $n > 0$. Then from (11.3.18)

$$M'_{n+1} \le \varepsilon\,\left[A\,\varepsilon^{n+1} + A(n+1)\,\varepsilon^{n+1}\right] = A(n+2)\,\varepsilon^{n+2},$$

which proves (11.3.20) for $(n+1)$.
(b) Next, we determine an estimate for $|\phi_n(\theta) - \phi(\theta)|$. In fact, by using Lemma 11.2, (11.3.12), and (11.3.20)

$$|\phi_n(\theta) - \phi(\theta)| \le \sqrt{2\pi M_n M'_n} \le \varepsilon^{n+1}\,\sqrt{2\pi A(n+1)}. \;\blacksquare \tag{11.3.21}$$

Proof of Theorem 11.4. Since $p \in H^1$, the function $F\left(e^{i\theta}\right) \in C^2[0,2\pi]$ (see Warschawski [1935: Theorem III]). Similarly, all functions $F_n\left(e^{i\theta}\right) \in C^2[0,2\pi]$, where $F_n\left(e^{i\theta}\right)$ are defined by (11.2.1). Also,

$$\begin{aligned}
\frac{dF}{d\theta} &= \sigma(\phi)\,\phi' + i\,(\phi' - 1),\\
\frac{d^2F}{d\theta^2} &= p(\phi)\,\phi'^2 + \sigma(\phi)\,\phi'' + i\,\phi'',\\
\frac{dF_n}{d\theta} &= \sigma\left(\phi_{n-1}\right)\,\phi'_{n-1} + i\,\left(\phi'_n - 1\right),\\
\frac{d^2F_n}{d\theta^2} &= p\left(\phi_{n-1}\right)\,{\phi'_{n-1}}^2 + \sigma\left(\phi_{n-1}\right)\,\phi''_{n-1} + i\,\phi''_n.
\end{aligned} \tag{11.3.22}$$

(a) Now we will estimate

$$M''_n = \sqrt{\frac{1}{2\pi}\int_0^{2\pi} \left(\phi''_n - \phi''\right)\,d\theta},$$

and prove that

$$M''_n \le A^2(n+1)^2\,c_n\,\varepsilon^{n+1} \quad \text{for } n \ge 1. \tag{11.3.23}$$

Proof of (11.3.23) is as follows: Since, by (11.3.13),

$$\frac{1}{2\pi}\int_0^{2\pi} \frac{d^2}{d\theta^2}\left[F_n\left(e^{i\theta}\right) - F\left(e^{i\theta}\right)\right]\,d\theta = 0,$$

we find by applying Lemma 11.1 that

$$\begin{aligned}
{M''_{n+1}}^2 = \frac{1}{2\pi}\int_0^{2\pi} &\Big\{\left(p(\phi_n) - p(\phi)\right)\,\phi'^2 + p(\phi_n)\,\left({\phi'_n}^2 - \phi'^2\right)\\
&+ \left(\sigma(\phi_n) - \sigma(\phi)\right)\,\phi'' + \sigma(\phi_n)\,\left(\phi''_n - \phi''\right)\Big\}^2\,d\theta.
\end{aligned} \tag{11.3.24}$$

Since $\sigma(\phi) \in H^1$, we have $|p(\phi)| \le \varepsilon$, and since $\phi'(\theta) \le A$ (Lemma 11.5) and $p \in H^1$, we find from (11.1.2) that

$$M''_{n+1}{}^2 \le \frac{\varepsilon^2}{2\pi} \int_0^{2\pi} \left[A^2|\phi_n - \phi| + |\phi'_n{}^2 - \phi'^2| + |\phi''|\,|\phi_n - \phi| + |\phi''_n - \phi''|\right]^2 d\theta,$$

which, by Minkowsky's inequality, yields

$$\begin{aligned} M''_{n+1} \le \varepsilon \Big[A^2 M_n &+ \sqrt{\frac{1}{2\pi}\int_0^{2\pi} \left(\phi'_n{}^2 - \phi'^2\right)^2 d\theta} \\ &+ \sqrt{\frac{1}{2\pi}\int_0^{2\pi} (\phi''_n)^2 (\phi_n - \phi)^2\, d\theta} + M''_n\Big], \end{aligned} \tag{11.3.25}$$

where M_n and M'_n are defined in (11.3.1). Since $\sqrt{\frac{1}{2\pi}\int_0^{2\pi} (\phi''(\theta))^2\, d\theta} \le \varepsilon\, A^{3/2} \min(1 + \varepsilon; \sqrt{2})$ (Lemma 11.5), we have

$$M''_0 = \sqrt{\frac{1}{2\pi}\int_0^{2\pi} (\phi'')^2\, d\theta} \le \sqrt{2}\,\varepsilon\, A^{3/2},$$

and since $|\phi_n(\theta) - \phi(\theta)| \le \varepsilon^{n+1}\sqrt{2\pi\, A(n+1)}$ by (11.2.3), we get

$$\begin{aligned} \sqrt{\frac{1}{2\pi}\int_0^{2\pi} (\phi_n - \phi)^2\, (\phi'')^2\, d\theta} &\le \sqrt{2}\,\varepsilon^{n+2}\, A^{3/2}\sqrt{2\pi A(n+1)} \\ &= 2\,\varepsilon^{n+2}\, A^2 \sqrt{\pi(n+1)}. \end{aligned} \tag{11.3.26}$$

Thus, using Lemma 11.2 with $g(\theta) = \phi'_n(\theta) - \phi'(\theta)$, we find that

$$(\phi'_n - \phi')^2 \le 2\pi M'_n\, M''_n,$$

which, after taking the square root and using $\phi'(\theta) \le A$, yields $|\phi'_n + \phi| \le |\phi'_n - \phi'| + 2|\phi'| \le \sqrt{2\pi M'_n\, M''_n} + 2A$. Hence,

$$\sqrt{\frac{1}{2\pi}\int_0^{2\pi} \left(\phi'_n{}^2 - \phi'^2\right)^2 d\theta} \le M'_n \left(\sqrt{2\pi M'_n\, M''_n} + 2A\right). \tag{11.3.27}$$

Now, integration by parts gives

$$\begin{aligned} \frac{1}{2\pi}\int_0^{2\pi} (g'(\theta))^2\, d\theta &= \left| g(\theta)\, g'(\theta)\Big|_0^{2\pi} - \int_0^{2\pi} g(\theta)\, g''(\theta)\, d\theta \right| \\ &\le \int_0^{2\pi} |g(\theta)\, g''(\theta)|\, d\theta \\ &\le \sqrt{\int_0^{2\pi} (g(\theta))^2\, d\theta \cdot \int_0^{2\pi} (g''(\theta))^2\, d\theta}, \end{aligned}$$

which, after setting $g(\theta) = \phi_n(\theta) - \phi(\theta)$, yields the inequality $M'_n \le \sqrt{M'_n M''_n}$. Applying this inequality to (11.3.27), we get

$$\sqrt{\frac{1}{2\pi}\int_0^{2\pi} \left(\phi_n'^2 - \phi'^2\right)^2 d\theta} \le 2AM'_n + M''_n \sqrt{2\pi M_n M'_n}. \tag{11.3.28}$$

Hence, from (11.3.25) we find, after using (11.3.12), (11.3.28), (11.3.26), and (11.3.20), that for $n \ge 2$

$$\begin{aligned}
M''_{n+1} &\le \varepsilon \Big\{A^2 \varepsilon^{n+1} + 2A^2(n+1)\varepsilon^{n+1} + 2A^2\varepsilon^{n+2}\sqrt{\pi(n+1)} \\
&\quad + \left(1 + \varepsilon^{n+1}\sqrt{2\pi A(n+1)}\right) M''_n\Big\} \\
&= A^2\varepsilon^{n+2}\left\{1 + 2(n+1) + 2\varepsilon\sqrt{\pi(n+1)} + \left(1 + \varepsilon^{n+1}\sqrt{2\pi A(n+1)}\right)\frac{M''_n}{A^2\varepsilon^{n+1}}\right\} \\
&\le A^2\varepsilon^{n+2}\Big\{1 + 2(n+1) + 2\varepsilon\sqrt{\pi(n+1)} \\
&\quad + \left(1 + \varepsilon^{n+1}\sqrt{2\pi A(n+1)}\right)(n+1)^2 c_n\Big\} \quad \text{by (11.3.23)} \\
&\le A^2\varepsilon^{n+2} c_{n+1}\left\{1 + 2(n+1) + (n+1)^2\right\} \\
&= A^2(n+2)^2 c_{n+1}\varepsilon^{n+2},
\end{aligned} \tag{11.3.29}$$

since

$$1 + 2(n+1) + 2\varepsilon\sqrt{\pi(n+1)} < [1 + 2(n+1)](1+\varepsilon) < [1 + 2(n+1)]\, c_{n+1}$$

for $n \ge 2$, where we have assumed that (11.3.20) is true for $n \ge 2$. To show that (11.3.23) is true for $n = 1$, note that from (11.3.24) for $n = 0$

$$\begin{aligned}
M''_1 &= \sqrt{\frac{1}{2\pi}\int_0^{2\pi} \left[p(\theta) - p(\phi(\theta))]\phi'^2 + p(\theta)\left(1 - \phi'^2\right) + \sigma(\phi(\theta))\,\phi''\right]^2 d\theta} \\
&\le \varepsilon\left[A^2 M_0 + \sqrt{\frac{1}{2\pi}\int_0^{2\pi}\left(1 - \phi'^2\right)^2 d\theta} + M''_0\right],
\end{aligned}$$

by Minkowsky's inequality and (11.1.2), where $\phi_0 = \theta$, $\sigma(\phi) \in H^1$ and $p \in H^1$. Since $M''_0 \le A^2\varepsilon(1+\varepsilon)$, and $\phi'(\theta) \le A$ by Lemma 11.5, we get, in view of (11.3.12) and (11.3.20),

$$M''_1 \le \varepsilon^2\left[A^2 + (1+A)A + A^2(1+\varepsilon)\right] = A^2\varepsilon^2\left(2 + \frac{1+A}{A} + \varepsilon\right).$$

Since $A \ge 1$ and $\dfrac{1+A}{A} \le 2$, we find that

$$M''_1 \le A^2\varepsilon^2(4+\varepsilon) < 4A^2\varepsilon^2(1+\varepsilon). \tag{11.3.30}$$

We will establish that (11.3.23) holds for $n = 2$. In fact, by applying (11.3.29) with $n = 1$ and replacing M''_1 by $A^2\varepsilon^2(4+\varepsilon)$ from (11.3.30), we get

$$M''_2 \le A^2\varepsilon^3\left[5 + 2\varepsilon\sqrt{2\pi} + \left(1 + 2\varepsilon^2\sqrt{\pi A}\right)(4+\varepsilon)\right]. \tag{11.3.31}$$

Since $2\sqrt{2\pi} < 6$ and $1 + 2\varepsilon^2\sqrt{\pi A} > 1$, we find from (11.3.31) that

$$\begin{aligned} M_2'' &\le A^2\,\varepsilon^3\,(5 + 6\varepsilon + 4 + \varepsilon)\,\left(1 + 2\varepsilon^2\sqrt{\pi A}\right) \\ &< 9(1+\varepsilon)\,\left(1 + 2\varepsilon^2\sqrt{\pi A}\right)\,A^2\,\varepsilon^3 = 9A^2\,c_2\,\varepsilon^3. \end{aligned} \tag{11.3.32}$$

(b) Next, we determine an estimate for $|\phi_n'(\theta) - \phi'(\theta)|$. By applying Lemma 11.1 with $g(\theta) = \phi_n'(\theta) - \phi'(\theta)$, we find from (11.3.20) and (11.3.23) that

$$|\phi_n'(\theta) - \phi'(\theta)| \le \sqrt{2\pi}\,c_n\,[A(n+1)]^{3/2}\,\varepsilon^{n+1}. \tag{11.3.33}$$

(c) Finally, we estimate c_n. Note that

$$\prod_{k=2}^{n}\left(1 + \varepsilon^k\sqrt{2\pi A k}\right) \le e^{\sqrt{2\pi A}\,\sum_{k=2}^{n}\varepsilon^k\sqrt{k}}.$$

Then, by the Schwarz inequality

$$\sum_{k=2}^{n}\varepsilon^k\sqrt{k} = \varepsilon\sum_{k=2}^{n}\varepsilon^{(k-1)/2}\left(\sqrt{k}\,\varepsilon^{(k-1)/2}\right) \le \varepsilon\sqrt{\sum_{k=2}^{n}\varepsilon^{k-1}\sum_{k=2}^{n}k\,\varepsilon^{k-1}}.$$

Hence,

$$\sum_{k=2}^{n}\varepsilon^k\sqrt{k} \le \varepsilon\sqrt{\frac{\varepsilon}{1-\varepsilon}\left(\frac{1}{1-\varepsilon^2} - 1\right)} < \frac{\varepsilon^2\sqrt{2}}{(1-\varepsilon)^{3/2}},$$

which yields (11.2.6). This completes the proof of Theorem 11.4. ■

PROOF OF THE UNIQUENESS OF THE SOLUTION OF THEODORSEN'S INTEGRAL EQUATION: We will show that if Γ is a near circle, then the integral equation (11.1.6) has at most one continuous solution. Let us assume that there exist two such solutions $\phi_1(\theta)$ and $\phi_2(\theta)$. Since

$$\int_0^{2\pi}[\phi_1(\theta) - \theta]\,d\theta = 0, \qquad \int_0^{2\pi}[\phi_2(\theta) - \theta]\,d\theta = 0,$$

we have by Lemma 11.1

$$M^2 = \frac{1}{2\pi}\int_0^{2\pi}[\phi_1(\theta) - \phi_2(\theta)]^2\,d\theta \le \frac{1}{2\pi}\int_0^{2\pi}[\log\rho(\phi_1(\theta)) - \log\rho(\phi_2(\theta))]^2\,d\theta.$$

Since, in view of (11.1.2),

$$|\log\rho(\phi_1(\theta)) - \log\rho(\phi_2(\theta))| \le \varepsilon\,|\phi_1(\theta) - \phi_2(\theta)|,$$

we find that $M^2 \le \varepsilon^2 M^2$, which yields $M = 0$ because $0 < \varepsilon < 1$. Hence, $\phi_1(\theta) = \phi_2(\theta)$. ■

11.4 Integral Representation

An integral representation for the function $\phi_n'(\theta)$ is given by the following result.

Theorem 11.5. *If* Γ *is a near circle and if the function* $\rho(\phi)$ *which defines* Γ *satisfies the condition* H^1, *then* $\phi'(\theta)$ *is continuous and*

$$\phi_1'(\theta) - 1 = -\frac{1}{2\pi}\int_{\theta-\pi}^{\theta+\pi} [\sigma(t) - \sigma(\theta)] \cot\frac{t-\theta}{2}\, dt, \tag{11.4.1}$$

and for $n \geq 1$

$$\begin{aligned}\phi_{n+1}'(\theta) - 1 = -\frac{1}{2\pi}\int_{\theta-\pi}^{\theta+\pi} &[\sigma(\phi_{n-1}(t)) - \sigma(\phi_{n-1}(\theta))]\, \phi_{n-1}'(t) \\ &\times \cot\frac{t-\theta}{2}\, dt - \sigma(\phi_{n-1}(\theta))\, \sigma(\phi_{n-2}(\theta))\, \phi_{n-2}'(\theta).\end{aligned} \tag{11.4.2}$$

PROOF. The integrand in (11.4.1) is continuous in both t and θ, except at $t = \theta$, and is bounded because $\sigma \in H^1$. Hence, the integral (11.4.1) is a continuous function of θ, represents a conjugate function of $\sigma(\theta)$, and is equal to $(\phi_1'(\theta) - 1)$ at least for almost all θ and, because of continuity, for all θ. This proves (11.4.1).

We will prove (11.4.2) by induction. Let us assume that $\phi_k'(\theta)$ is a continuous function for $k = 1, 2, \ldots, n$. We will prove that (11.4.2) holds for $(n+1)$ and that $\phi_{n+1}'(\theta)$ is continuous.

The absolute continuity of $F_{n+1}(e^{i\theta}) = \log \rho(\phi_n(\theta)) + i(\phi_{n+1}(\theta) - \theta)$ implies that $(\phi_{n+1}'(\theta) - 1)$ is conjugate of $\sigma(\phi_n(\theta))\, \phi_n'(\theta)$, and for almost all θ we have

$$\begin{aligned}\phi_{n+1}'(\theta) - 1 &= -\frac{1}{2\pi}\int_0^{\pi} \sigma(\phi_n(\tau))\Big|_{\tau=\theta-t}^{\theta+t} \cot\frac{t}{2}\, dt \\ &= -\frac{1}{2\pi}\int_0^{\pi} [\sigma(\phi_n(\theta+t)) - \sigma(\phi_n(\theta))]\, \phi_n'(\theta+t) \cot\frac{t}{2}\, dt \\ &\quad + \frac{1}{2\pi}\int_0^{\pi} [\sigma(\phi_n(\theta-t)) - \sigma(\phi_n(\theta))]\, \phi_n'(\theta-t) \cot\frac{t}{2}\, dt \\ &\quad - \sigma(\phi_n(\theta))\, \frac{1}{2\pi}\int_0^{\pi} [\phi_n'(\theta+t) - \phi_n'(\theta-t)] \cot\frac{t}{2}\, dt \\ &\equiv I_1 + I_2 + I_3.\end{aligned}$$

Note that, since $\sigma(\phi) \in H^1$ and $\phi'(\theta)$ is continuous, the integrals I_1 and I_2 represent continuous functions of θ. The integral I_3, without the factor $\sigma(\phi_n(\theta))$, is equal to $\sigma(\phi_{n-1}'(\theta))\, \phi_{n-1}'(t)$, since $(\phi_n'(\theta) - 1)$ is conjugate to this function. If we set $\tau = \theta + t$ in I_1 and $\tau = \theta - t$ in I_2, we get

$$\begin{aligned}\phi_{n+1}'(\theta) - 1 = &-\frac{1}{2\pi}\int_{\theta-\pi}^{\theta+\pi} [\sigma(\phi_n(\tau)) - \sigma(\phi_n(\theta))]\, \phi_n'(\tau) \cot\frac{t-\theta}{2}\, d\tau \\ &- \sigma(\phi_n(\theta))\, \sigma(\phi_{n-1}(\theta))\, \phi_{n-1}'(\theta),\end{aligned}$$

which defines a continuous function of θ. ■

Theorem 11.5 establishes conditions under which the images Γ_n of the unit circle under the mapping function $w = f_n(z) = z\, e^{F_n(z)}$ are starlike with respect to the origin. We will discuss starlike regions in the next section. The advantage in assuming starlike contours is that $\phi_n(\theta)$ becomes a monotone increasing function and, therefore, possesses a unique inverse function $\theta = \theta(\phi)$. This helps us compute the approximate inverse mapping function $z = e^{i\theta_n(\phi)}$ that maps the unit circle onto Γ_n.

11.5 Starlike Regions

A region D is said to be *starlike* with respect to a point $z_0 \in D$ if every ray from the point z_0 intersects the boundary ∂D in exactly one point. The functions $f_n(z) = z\, e^{F_n(z)}$ map the unit circle $|z| = 1$ onto Jordan curves Γ_n. Since the functions $f_n(z)$ are approximations of the mapping function $f(z)$, it is important to assume that the contours Γ_n enclose regions that are starlike with respect to the origin, as we have done in §11.3 in the proof of Lemma 11.3 and remarked about the advantage of such an assumption at the end of §11.4. The functions $f_n(z)$ are starlike with respect to the origin if $\Re \left\{ \dfrac{z\, f_n'(z)}{f_n(z)} \right\} > 0$, which, in view of (11.3.22), leads to the condition $(\phi_n' - 1) < 1$.

Example 11.1. We will determine the conditions on ε under which the regions bounded by Γ_n are starlike. In fact, we will show that if Γ is a near circle and if $\sigma(\phi) \in H^1$, then the region bounded by the contour Γ_1 is starlike with respect to the origin if $\varepsilon \le (2 \log 2)^{-1} \approx 0.72$, by Γ_2 if $\varepsilon \le 0.34$, by Γ_3 if $\varepsilon \le 0.31$, and by Γ_n for $n \ge 4$ if $\varepsilon \le 0.295$.

These results are obtained by evaluating the values of ε for which $|\phi_n'(\theta) - 1| \le 1$, so that $\phi'(\theta) \ge 0$, which implies that $\phi_n(\theta)$ is a monotone increasing function. Thus, since $\sigma(\phi) \in H^1$, we have, in view of (11.4.1),

$$|\phi_1(\theta) - 1| \le \frac{1}{2\pi} \int_{\theta-\pi}^{\theta+\pi} |\sigma(t) - \sigma(\theta)| \cot \left| \frac{t-\theta}{2} \right| dt \le \frac{\varepsilon}{2\pi} \int_{\theta-\pi}^{\theta+\pi} (t-\theta) \cot \frac{t-\theta}{2}\, dt = 2\,\varepsilon \log 2,$$

where σ is defined by (11.1.9). Then the region bounded by Γ_1 is starlike if $\phi_1'(\theta) \ge 0$, i.e., if $2\,\varepsilon \log 2 \le 1$, which gives $\varepsilon \le (2 \log 2)^{-1}$.

Assuming that $\phi_n'(\theta) \ge 0$ for some $n \ge 1$, provided that $\varepsilon \le \varepsilon_0 < 1$, we will evaluate $\phi_{n+1}'(\theta)$. Thus, since $\sigma(\phi) \in H^1$, we have by (11.4.2) and (11.1.2),

$$\begin{aligned} |\phi_{n+1}'(\theta) - 1| &\le \frac{\varepsilon}{2\pi} \int_{\theta-\pi}^{\theta+\pi} [\phi_n(t) - \phi_n(\theta)]\, \phi_n'(t) \cot \frac{t-\theta}{2}\, dt \\ &\quad + \varepsilon^2 \left|\phi_{n-1}'(\theta)\right| \equiv \varepsilon\, m_n^2 + \varepsilon^2 \left|\phi_{n-1}'(\theta)\right|. \end{aligned} \tag{11.5.1}$$

Note that $\phi_n(t) - \phi_n(\theta)$ has the same sign as $(t - \theta)$ since $\phi_n'(t) \ge 0$. Now, integrating by parts, we find that

$$m_n^2 = \frac{1}{2\pi}\int_{\theta-\pi}^{\theta+\pi} [\phi_n(t) - \phi_n(\theta)]\, \phi_n'(t) \cot\frac{t-\theta}{2}\, dt = \frac{1}{2\pi}\int_{\theta-\pi}^{\theta+\pi} \left(\frac{\phi_n(t)-\phi_n(\theta)}{2\sin\frac{t-\theta}{2}}\right)^2 dt$$
$$= \frac{1}{2\pi}\int_{\theta-\pi}^{\theta+\pi} \left(\frac{\phi_n(t)-t-[\phi_n(\theta)-\theta]+t-\theta}{2\sin\frac{t-\theta}{2}}\right)^2 dt,$$

which, by Minkowsky's inequality, yields

$$\begin{aligned} m_n &\le \sqrt{\frac{1}{2\pi}\int_{\theta-\pi}^{\theta+\pi} \left(\frac{\phi_n(t)-t-[\phi_n(\theta)-\theta]}{2\sin\frac{t-\theta}{2}}\right)^2 dt} \\ &\quad + \sqrt{\frac{1}{2\pi}\int_{\theta-\pi}^{\theta+\pi} \left(\frac{t-\theta}{2\sin\frac{t-\theta}{2}}\right)^2 dt} \equiv \sqrt{J_1}+\sqrt{J_2}. \end{aligned} \tag{11.5.2}$$

Now, by Lemma 11.6,

$$\begin{aligned} J_1 &= \frac{1}{2\pi}\int_{\theta-\pi}^{\theta+\pi} \left(\frac{\log\rho\left(\phi_{n-1}(t)\right)-\log\rho\left(\phi_{n-1}(\theta)\right)}{2\sin\frac{t-\theta}{2}}\right)^2 dt \\ &\le \frac{\varepsilon}{2\pi}\int_{\theta-\pi}^{\theta+\pi} \left(\frac{\phi_{n-1}(t)-\phi_{n-1}(\theta)}{2\sin\frac{t-\theta}{2}}\right)^2 dt \\ &= \varepsilon^2\, m_{n-1}^2. \end{aligned}$$

Also, integrating by parts, we get $J_2 = 2\log 2 \equiv c^2$. Hence, from (11.5.2)

$$m_n \le \varepsilon\, m_{n-1} + c. \tag{11.5.3}$$

Since $c = m_0$, we find that

$$m_n \le c\left(1+\varepsilon+\varepsilon^2+\cdots+\varepsilon^n\right) = c\,\frac{1-\varepsilon^{n+1}}{1-\varepsilon}.$$

Thus, by (11.5.1),

$$\begin{aligned} \left|\phi_{n+1}'(\theta)-1\right| &\le \varepsilon\, m_n^2 + \varepsilon^2\, \left|\phi_{n-1}'(\theta)\right| \\ &\le 2\varepsilon\left(\frac{1-\varepsilon^{n+1}}{1-\varepsilon}\right)^2 \log 2 + \varepsilon^2\, \left|\phi_{n-1}'(\theta)\right|. \end{aligned} \tag{11.5.4}$$

Set $n = 1$ in (11.5.4). Then, since $\phi_0'(t) = 1$, we find that

$$\left|\phi_2'(t)-1\right| \le 2\varepsilon\,(1+\varepsilon)^2 \log 2 + \varepsilon^2, \tag{11.5.5}$$

which is less than 1 if $\varepsilon \le 0.34$. If we set $n = 2$ in (11.5.4), then, since $\phi_1'(\theta) \le 1 + 2\varepsilon \log 2$, and $\phi_1'(\theta) > 0$ for $\varepsilon < (2\log 2)^{-1}$, we find that

$$\left|\phi_3'(\theta)-1\right| \le 2\varepsilon\left(1+\varepsilon+\varepsilon^2\right)^2 \log 2 + \varepsilon\,(1+2\varepsilon\log 2),$$

which is less than 1 if $\varepsilon \le 0.31$. Note that, by (11.5.5), $|\phi_2'(\theta| \le 1.7927$ if $\varepsilon = 0.3$. Hence, setting $n = 3$ in (11.5.4) and using this estimate for $\phi_2'(\theta)$, we find that $|\phi_4'(\theta) \le 1$ if $\varepsilon \le 0.3$. Let us assume that $0 < \phi_{n-1}'(\theta) \le 2$ for some $n \ge 1$. Then, from (11.5.4),

$$\left|\phi_{n+1}'(\theta) - 1\right| \le \varepsilon \frac{2 \log 2}{(1-\varepsilon)^2} + 2\varepsilon^2 < 1,$$

if $\varepsilon \le 0.295$. Since this assumption is obviously valid for $n = 1$ and $n = 2$ if $\varepsilon \le 0.295$, it follows that $\left|\phi_{n+1}'(\theta) - 1\right| < 1$ for all $n \ge 4$ if $\varepsilon \le 0.295$. ∎

11.6 Exterior Regions

Theodorsen [1931] considered the case where the exterior of the circle $|\zeta| = R$ is mapped onto the exterior of a nearly circular region D bounded by Γ. In this case the mapping function $\omega = g(\zeta)$ is normalized such that $\lim_{\zeta\to\infty} \frac{\omega}{\zeta} = 1$. This case reduces to that analyzed in §11.1 if we use the transformations $w = 1/\omega$ and $z = R/\zeta$. Let the boundary Γ be defined by $r = r(\Phi)$, $0 \le \Phi \le 2\pi$, where, analogous to (11.1.1) and (11.1.2), for some $b > 0$ and $0 < \varepsilon < 1$,

$$\frac{b}{1+\varepsilon} \le r(\Phi) \le b\,(1+\varepsilon), \tag{11.6.1}$$

and

$$\left|\frac{r'(\Phi)}{r(\Phi)}\right| \le \varepsilon. \tag{11.6.2}$$

Thus, the function $w = f(z) = \dfrac{1}{g(\zeta)}$, $\zeta = R/z$, maps the unit disk $|z| < 1$ onto $\operatorname{Int}(\Gamma)$, where Γ is represented by the equation

$$\rho = \rho(\phi) = \frac{1}{r(\Phi)},$$

such that $\phi = -\Phi$, $\rho(\phi)$ satisfies the conditions (11.1.1) and (11.1.2), and $a = 1/b$. Then, for $\zeta = R\,e^{i\psi}$, we have

$$\arg\{g(\zeta)/\zeta\} = \Phi(\psi) - \psi,$$

where we take $\arg\{g(\zeta)/\zeta\}_{\zeta=\infty} = 0$. Thus, for $\zeta = R\,e^{i\psi}$ and $z = e^{i\theta}$, where $\theta = -\psi$, we have

$$\begin{aligned} \log \frac{g(\zeta)}{\zeta} &= \log r\,[\Phi(\psi)] + i\,[\Phi(\psi) - \psi] - \log R \\ &= -\log \frac{f(z)}{z} - \log R \\ &= -\log \rho\,[\phi(\theta)] - i\,[\phi(\theta) - \theta] - \log R. \end{aligned} \tag{11.6.3}$$

Thus, $\Phi_n(\psi) = -\phi(\theta)$, $\psi = -\theta$, and we can form the iterations in this case, analogous to (11.1.7), as

$$\begin{aligned} &\Phi_0(\psi) = 0, \\ &\Phi_n(\psi) - \psi = \frac{1}{2\pi} \int_0^\pi \{\log r\,(\phi_{n-1}(\psi + t)) - \log r\,(\phi_{n-1}(\psi - t))\} \\ &\qquad\qquad \times \cot\frac{t}{2}\,dt, \quad n = 1, 2, \dots. \end{aligned} \tag{11.6.4}$$

Hence, the estimates for $|\phi_n(\theta) - \phi(\theta)|$ and for derivatives of these differences obtained in §11.3 also hold for $|\Phi_n(\psi) - \Phi(\psi)|$ and derivatives of these differences. See §11.4 for more on Theodorsen's method and its convergence problems.

11.7 Trigonometric Interpolation

In practice, computations are specially simple if functions are expanded in Fourier series and then the method of trigonometric interpolation is applied. We fix $2N$ points on the unit circle in the z-plane by

$$\theta_k = \frac{k\pi}{n}, \quad k = 0, 1, \dots, 2N-1,$$

and denote

$$\phi_k^{(\nu)} = \phi^{(\nu)}(\theta_k), \quad w_k^{(\nu)} = \log \rho(\phi^{(\nu)}(\theta_k)). \tag{11.7.1}$$

Using the Fourier polynomial

$$w_k^{(\nu)} = \frac{\alpha_0^{(\nu)}}{2} + \sum_{n=1}^{N} \left(\alpha_n^{(\nu)} \cos n\theta_k + \beta_n^{(\nu)} \sin n\theta_k \right), \tag{11.7.2}$$

where

$$\begin{aligned} \alpha_n^{(\nu)} &= \frac{1}{N} \sum_{k=0}^{2N-1} w_k^{(\nu)} \cos n\theta_k, \quad n = 0, 1, \dots, N, \\ \beta_n^{(\nu)} &= \frac{1}{N} \sum_{k=0}^{2N-1} w_k^{(\nu)} \sin n\theta_k, \quad n = 0, 1, \dots, N-1. \end{aligned} \tag{11.7.3}$$

Thus, the harmonic function

$$\phi_k^{(\nu+1)} - \theta_k = \sum_{n=1}^{N} \left(\alpha_n^{(\nu)} \sin n\theta_k - \beta_n^{(\nu)} \cos n\theta_k \right) \tag{11.7.4}$$

is conjugate to $w_k^{(\nu)}$. If we substitute the values of $\alpha_n^{(\nu)}$ and $\beta_n^{(\nu)}$ from (11.7.3), then we obtain the formula

$$\phi_k^{(\nu+1)} - \theta_k = -\frac{1}{N} \sum_{n=1}^{N} \left(w_{k+n}^{(\nu)} - w_{k-n}^{(\nu)} \right) \cot \frac{\theta_n}{2}, \tag{11.7.5}$$

where we set $w_{1 \pm 2N} = w_1$. Taking the initial value $\phi_k^{(0)} = \theta_k$, formula (11.7.5) together with (11.7.1) provides an iterative technique to compute the quantities $\phi_k^{(\nu)}$ and $w_k^{(\nu)}$. Note that the finite sum (11.7.5) has a form that corresponds to the integral in (11.1.6). Since the method converges, we can determine the Fourier coefficients α_n and β_n by taking the limiting processes $\phi_k = \lim_{\nu \to \infty} \phi_k^{(\nu)}$ and $w_k = \lim_{\nu \to \infty} w_k^{(\nu)}$, which yields an approximation $\tilde{F}(z)$ for the function $F(z)$ as

$$\tilde{F}(z) = \frac{\alpha_0}{2} + \sum_{n=1}^{N-1} (\alpha_n - i\beta_n) z^n + \alpha_N z^N, \tag{11.7.6}$$

and an approximation $\tilde{f}(z)$ for the mapping function as

$$\tilde{f}(z) = z\, e^{\tilde{F}(z)}. \tag{11.7.7}$$

The function $\tilde{f}(z)$ maps the unit disk onto a region $\tilde{D}$ with a boundary $\tilde{\Gamma}$ which cuts the curve Γ in $2N$ points determined by $z = e^{i\theta_k}$. The quality of the approximation depends on the closeness of the two curves Γ and $\tilde{\Gamma}$ between these $2N$ points. For better closeness, even $4N$ points can be chosen where the values of ϕ_k and w_k already computed can be used as the first approximation. This approximation method with Fourier series is suitable for regions with smooth boundaries.

11.8 Wegmann's Method

This quadratic convergence method deals with the general problem of conformal mapping between any two simply connected regions D and G bounded by Jordan contours Γ and Δ, respectively. Let this conformal map be denoted by $f : D \mapsto G$ such that all such maps f can be extended to $\bar{D}$ together with the homeomorphism $f : \Gamma \mapsto \Delta$. The function f is uniquely defined if $f(z_0) = a_0$ for $z_0 \in D$ and $f(\zeta) = a_1$ for $\zeta \in \Gamma$. Since the mapping function is fully determined by its boundary values, the problem of computing f can be reduced to that of finding the boundary map $f : \Gamma \mapsto \Delta$. Thus, if a parameterization $\gamma(s)$ of Δ is prescribed, then there exists a real function $S(\zeta)$ which satisfies

$$f(\zeta) = \gamma\,(S(\zeta))\,, \tag{11.8.1}$$

where $\zeta = \zeta(t)$ is a parametric representation of the boundary Γ, $\gamma(s)$ is assumed to be a 2π-periodic function, and S is a continuous, multiple-valued function that changes by 2π while winding around Γ once, i.e.,

$$[S(\gamma)]_\Gamma = 2\,\pi. \tag{11.8.2}$$

This mapping problem can be generalized to the case when the boundary Δ is not necessarily a Jordan contour. In that case, let $\gamma(s) \in C^1[0, 2\pi]$ with $\gamma'(s) \neq 0$, where prime denotes differentiation with respect to s, and let $\kappa\,(\geq 1)$ be the winding number of γ' with respect to the origin. Then a function f analytic in the region D and continuous on $\bar{D}$ and satisfying (11.8.1) and (11.8.2) maps D onto a κ-sheeted Riemann surface with $\kappa - 1$ branch points. Since these branch points cannot be determined by the boundary Γ alone, we can fix $\kappa - 1$ additional parameters through $\kappa - 1$ interpolation functions. Then f can be determined if the following additional conditions are satisfied:

$$f(z_j) = a_j, \quad j = 0, 1, \dots, \kappa, \tag{11.8.3}$$

where $z_0 \in \Gamma$, $a_0 \in \Delta$, and $z_1, \dots, z_\kappa \in D$.

The iterative method to determine the approximate function $\tilde{f}$ that maps the boundary Γ conformally onto the boundary Δ requires that $\tilde{f}_\nu(\zeta) = \gamma\,(S_\nu(\zeta))$ prior to the νth iteration. Then at the νth step this map is updated by shifting the function values along the tangent to the boundary curve, i.e., by determining a real-valued function $u_\nu(\zeta)$ such that

$$\gamma\,(S_\nu(\zeta)) - u_\nu(\zeta)\,\gamma'\,(S_\nu(\zeta)) = h_{\nu+1}(\zeta), \tag{11.8.4}$$

where the function $h_{\nu+1}$ is analytic in D and continuous on $\bar{D}$ with $h_{\nu+1}(z_\nu) = a_\nu$, $\nu = 1, 2, \dots, n$. For the boundary point $z_0 \in \Gamma$ (the case $\nu = 0$) the prescribed value a_0 lies on

Δ, i.e., $a_0 = \gamma(s_0)$. But since $h_{\nu+1}(z_0)$ lies on the tangent through the point $\gamma\left(S_\nu(z_0)\right)$, the condition $h_{\nu+1}(z_0) = a_0$, in general, is not satisfied. Therefore, we replace this condition by

$$u_\nu(z_0) = s_0 - S_\nu(z_0). \tag{11.8.5}$$

Note that the function $h_{\nu+1}$ is an approximation of the boundary map f, although the values that it takes on Γ, in general, do not lie on Δ. The function $h_{\nu+1}$, however, yields a new approximation for the parameter mapping function S as

$$S_{\nu+1}(\zeta) = S_\nu(\zeta) + u_\nu(\zeta). \tag{11.8.6}$$

Thus, after starting with an arbitrary function $S_1(\zeta)$, the conditions $S_\nu(z_0) = s_0$ and $u_\nu(z_0) = 0$ are satisfied for all $\nu \geq 2$, and hence, the condition $h_{\nu+1}(z_0) = a_0$ is satisfied.

If we multiply (11.8.4) by $\dfrac{1}{2i\pi}\dfrac{d\zeta}{\zeta - z}$ and integrate over Γ, we get

$$\begin{aligned} &\frac{1}{2i\pi}\int_\Gamma \frac{\gamma\left(S_\nu(\zeta)\right) + u_\nu(\zeta)\,\gamma'\left(S_\nu(\zeta)\right)}{\zeta - z}\,d\zeta \\ &\qquad = \begin{cases} 0 & \text{for } z \notin \bar{D}, \\ \frac{1}{2}\left[\gamma\left(S_\nu(z)\right) + u_\nu(z)\,\gamma'\left(S_\nu(z)\right)\right] & \text{for } z \in \Gamma, \\ a_\nu & \text{for } z = z_\nu,\ \nu = 1, \dots, \kappa. \end{cases} \end{aligned} \tag{11.8.7}$$

Note that the formula (11.8.7) is a linearization of the equation

$$\frac{1}{2i\pi}\int_\Gamma \frac{\gamma\left(S(\zeta)\right)}{\zeta - z}\,d\zeta = \begin{cases} 0 & \text{for } z \notin \bar{D}, \\ \frac{1}{2}\gamma\left(S(z)\right) & \text{for } z \in \Gamma, \\ a_\nu & \text{for } z = z_\nu,\ \nu = 1, \dots, \kappa. \end{cases} \tag{11.8.8}$$

The middle part of Eq (11.8.8) is a nonlinear singular integral equation of the second kind for the parameter function S associated with the boundary map f. The method of solution for this equation is linearization and the use of Newton's method. This fact is used to prove convergence which is locally quadratic. A proof is available in Wegmann [1984].

The numerical computation of the iterative method begins with the discretization (11.8.7) of Eq (11.8.8) for the boundary Γ which is used to compute the updates u_ν. However, in the particular case when Γ is the unit circle, explicit representations of the solution in terms of integrals can be obtained. Details for this case are as follows: All functions are discretized at $n = 2N$ equidistant points $\zeta_i = e^{i\phi_i} \in \Gamma$, $\phi_i = \phi_0 + i\pi/N$, $i = 1, \dots, n$. The integrals are evaluated by trigonometric interpolation. Thus, for example, if

$$F(z) = \frac{1}{2i\pi}\int_\Gamma \frac{\phi(\zeta)}{\zeta - z}\,d\zeta, \tag{11.8.9}$$

then the function ϕ is represented as a polynomial in ζ and ζ^{-1}, i.e.,

$$\phi(\zeta_i) = \sum_{i=-N}^{N} \alpha_j\,\zeta_i^i = F^+(\zeta_i) + F^-(\zeta_i), \tag{11.8.10}$$

where

$$F^+(\zeta_i) = \sum_{i=0}^{N} \alpha_j\, \zeta_i^i, \quad F^-(\zeta_i) = \sum_{i=-N}^{N} \alpha_j\, \zeta_i^i. \tag{11.8.11}$$

If we define

$$G(z) = \begin{cases} e^{F(z)-F(0)/2} & \text{for } |z| < 1, \\ z^{2\kappa}\, e^{F(z)-F(0)/2} & \text{for } |z| > 1, \end{cases} \tag{11.8.12}$$

$$\psi(z) = \frac{G(z)}{2i\pi} \int_\Gamma \frac{\hat{g}(\zeta)\, \Im \left\{ \dfrac{\hat{f}(\zeta)}{\hat{g}(\zeta)} \right\}}{G^+(\zeta)} \frac{d\zeta}{\zeta - z}, \tag{11.8.13}$$

where the function $\hat{f}(\zeta) = \gamma\,(S(\zeta))$ and $\hat{g}(\zeta) = \gamma'\,(S(\zeta))$ are Hölder-continuous, and

$$h_0(z) = \frac{1}{2} \left[\psi^+(\zeta) + \overline{\psi^-(\zeta)} \right], \tag{11.8.14}$$

then Wegmann [1984] has proved the following:

Theorem 11.6. *Let ζ and γ be Hölder-continuous functions with $\zeta(t) \neq 0$ and $\gamma'(s) \neq 0$ for all s and t, where $\zeta = \zeta(t)$ and $\gamma = \gamma(s)$ are parametric representations of the boundaries Γ and Δ, respectively, of simply connected regions D and G. Let $S(\zeta)$ be a Hölder-continuous function satisfying (11.8.1). Let the winding number κ of γ' be positive, and let the points $z_1, \dots, z_\kappa \in D$ and $z_0 \in \Gamma$ and the complex numbers $a_1, \dots, a_\kappa$ and a real number u_0 be given. Then there exists a unique Hölder-continuous real function $u(\zeta)$ defined on Γ and a unique complex function h analytic in D and continuous on $\bar{D}$ such that*

$$\gamma\,(S(\zeta)) + u(\zeta)\, \gamma'\,(S(\zeta)) = h(\zeta), \tag{11.8.15}$$

satisfying the conditions

$$h(z_\nu) = a_\nu \quad \text{for } \nu = 1, 2, \dots, \kappa, \tag{11.8.16}$$

and

$$u(z_0) = a_0. \tag{11.8.17}$$

In the case when Γ is the unit circle, the solution of the linearized integral equation in (11.8.7) can be represented explicitly in terms of integral transforms.

Thus, the integrals are computed by using the discrete Fourier transform. Note that if FFT is used instead, then the transformation of ϕ to $F^\pm$ takes $O(n \log n)$ operations. The foregoing outline of the method and this theorem show how this iterative method differs from that discussed in §11.1 and §11.2.

The iterative process is carried out in the following two steps:
STEP 1: Assuming that the functions $\gamma(s)$, $\gamma'(s)$ and $\phi_0(s) = \arg\{\gamma'(s)\}$ are defined explicitly and the initial value of $S_1(\zeta_0)$ is available (initial guess), once S_ν has been computed for some $\nu \geq 1$, then compute

$$\begin{aligned} \hat{f}_\nu\,(\zeta_i) &= \gamma\,(S_\nu(\zeta_i)), \quad \hat{g}_\nu(\zeta_i) = \gamma'\,(S_\nu(\zeta_i)), \\ \phi(\zeta_i) &= 2\,\phi_0\,(S_\nu(\zeta_i)) - 2\,\kappa\,\theta_i. \end{aligned} \tag{11.8.18}$$

STEP 2: Use one of the methods described below.

METHOD 1: Determine ψ and h_0 from (11.8.13) and (11.8.14), and set

$$h_{\nu+1}(z) = h_0(z) + P(z)\,G(z), \tag{11.8.19}$$

where the polynomial P is chosen such that

$$\begin{aligned} P(z_0)\,G^+(z_0) &= (s_0 - S_\nu(z_0))\,\hat{g}_\nu(z_0) + \hat{f}_\nu(z_0) - h_0(z_0), \\ P(z_i)\,G(z_i) &= a_i - h_0(z_i) \quad \text{for } i = 1, \dots, \kappa. \end{aligned} \tag{11.8.20}$$

Note that in this discretization z_0 is equal to one of the ζ_i. Then use Cauchy's formula (2.4.2) and integral representation (11.8.9) to compute $h_0(z_i)$ and $F(z_i)$ for $i \geq 1$. Thus, $G(z_i) = e^{F(z_i)-F(0)/2} \neq 0$, and P is taken as an interpolation polynomial. Finally, set

$$S_{\nu+1}(\zeta_i) = S_\nu(\zeta_i) + \Re\left\{\frac{h_{\nu+1}(\zeta_i) - \hat{f}_\nu(\zeta_i)}{\hat{g}_\nu(\zeta_i)}\right\}. \tag{11.8.21}$$

METHOD 2: Instead of (11.8.13), compute ψ for $\nu \geq 2$ by the formula

$$\psi(z) = \frac{G(z)}{\pi} \int_\Gamma \frac{\hat{g}_\nu(\zeta)\,\Im\left\{\left(\hat{f}_\nu(\zeta) - h_\nu(\zeta)\right)\,\hat{g}_\nu(\zeta)\right\}}{G^+(\zeta)}\,\frac{d\zeta}{\zeta - z}. \tag{11.8.22}$$

Then h_0 is computed from (11.8.14), which leads to

$$h_{\nu+1}(z) = h_\nu(z) + h_0(z) + P(z)\,G(z), \tag{11.8.23}$$

where the polynomial P is computed from

$$P(z_i) = -\frac{h_0(z_i)}{G(z_i)} \quad \text{for } i = 0, 1, \dots, \kappa. \tag{11.8.24}$$

As noted by Wegmann [1986], the first method gives more accurate results than the second.

Example 11.2. The result of the conformal mapping of the unit circle $\zeta(t) = e^{it}$ onto the inverted ellipses (see Example 9.4), defined by $\rho(s) = \sqrt{1 - (1 - p^2)\cos^2 s}$, $0 < p < 1$, are shown in Wegmann [1986] for $p = 0.4, 0.6$, and 0.8, where the error after the 10th iteration is $\|S_{11} - S\| \approx 1.8 \times 10^{-2}$, $5.7 \times 10{-}5$, and 10^{-8}, respectively, for the above values of p with $n = 40$. The exact boundary correspondence function is given by $\tan s = p\,\tan\theta$, where $\theta = \arg\{z\}$, and the boundary map is defined by $w = \dfrac{2\,p\,z}{(1+p) + (1-p)\,z^2}$. ■

11.9 Newton's Method

As explained in §11.1 and §11.8, the function $f\left(r\,e^{i\theta}\right) = \rho\left(\phi(\theta)\right)\,e^{i\phi(\theta)}$ defines the boundary correspondence function $\phi : [0, 2\pi] \mapsto \mathbb{R}$ for the boundary mapping by f. Let $\mathcal{K}$ denote the conjugation operator

$$\mathcal{K}\,[h]\,(\theta) = \frac{1}{2\pi} - \int {}_0^{2\pi} h(\theta)\,\cot\frac{\theta - t}{2}\,dt, \tag{11.9.1}$$

where the integral takes a Cauchy p.v. Then Theodorsen's integral equation (11.1.5) can be written as

$$\phi(\theta) = \theta + \mathcal{K}\left[\log \rho\left(\phi(\theta)\right)\right](\theta), \tag{11.9.2}$$

or, if we set $\Psi(\theta) = \phi(\theta) - \theta$, then

$$\Psi(\theta) = \mathcal{K}\left[\log \rho\left(\Psi(\theta) + \theta\right)\right](\theta), \tag{11.9.3}$$

which, according to Gaier [1964] and Wegmann [1984], has a unique 2π-periodic solution $\Psi^*(\theta)$. Wegmann [1984] has also proved that for $f \in L^2$ and $\mathcal{K} f \in L^\infty$

$$\|\mathcal{K} f\|_\infty \le \sqrt{\frac{\pi}{6} \int_0^\pi f'(x)\, dx} = \sqrt{\frac{\pi}{3}}\, \|f\|_2 < 2\, \|f\|_2. \tag{11.9.4}$$

We will assume that the 2π-periodic function ρ is absolutely continuous on $\mathbb{R}$ and $|\sigma| =< \varepsilon$ almost everywhere (we say that ρ satisfies the ε-condition), where σ is defined in (11.1.9). The solution $\Psi^*(\theta)$ of Eq (11.9.3) is given explicitly as a consequence of the Riemann-Hilbert problem (see Appendix C). As we have seen in §11.1 and §11.3, a numerical solution of the integral equation (11.9.3) can be obtained by first discretizing this equation and then applying an iterative method. The convergence of this method depends on the ε-condition ($\varepsilon < 1$), except in the case of certain symmetric curves with corners and pole singularities. Thus, it can be shown, as in Gaier [1964], Warschawski [1955], Hübner [1979] and Gutknecht [1981; 1983], that $\|\Psi_n - \Psi^*\|_2 \to 0$ and $\|\Psi_n - \Psi^*\|_\infty \to 0$ as $n \to \infty$.

Let $\mathcal{F} : W \mapsto W$, where W is the Sobolev space of 2π-periodic and absolutely continuous functions f with $f' \in L^2$. Define the operator

$$\mathcal{F}\left(\Psi(\theta)\right) = \Psi(\theta) - \left(\mathcal{K}\left[\log \rho\left(\Psi(\theta) + \theta\right)\right]\right)(\theta). \tag{11.9.5}$$

Then Newton's method for solving the equation (11.9.3) is

$$\begin{aligned} &\Psi_0 \in W, \\ &\mathcal{F}'\left(\Psi_n\right)\left[\Psi_{n+1} - \Psi_n\right] = -\mathcal{F}\left(\Psi_n\right), \quad n = 0, 1, \dots, \end{aligned} \tag{11.9.6}$$

where $\mathcal{F}'\left(\Psi_n\right)$ is the F-derivative* of $\mathcal{F}$ in the Banach space $(W, \|\cdot\|)$ at Ψ_n. The operator $\mathcal{F}(\Psi)$ has an inverse for any $\Psi \in W$. Using the solution of a Riemann-Hilbert problem (see §C.3), Hübner [1986] has shown that

$$\Psi_{n+1} - \psi_n = -\left(\mathcal{F}'\left(\Psi_n\right)\right)^{-1} \mathcal{F}\left(\Psi_n\right). \tag{11.9.7}$$

Thus, to determine Ψ_{n+1} from Ψ_n in formula (11.9.7), we require two applications of the conjugation operator $\mathcal{K}$. For numerical computation, we discretize (11.9.7) instead of (11.9.3). Then by using FFT we first approximate Ψ_n by a vector in $\mathbb{R}^{2n}$ and then compute an approximate value of Ψ_{n+1} in the same space. This method, however, is different from Wegmann's (§11.8), as shown in Example 11.3.

Hübner [1986] has proved the following two results on the convergence of Newton's method.

---

*If p is Hölder-continuous and $\Psi \in W$, then the F-derivative of $\mathcal{F}$ at Ψ in $(W, \|\cdot\|)$ exists and is given by $\left(\mathcal{F}'(\Psi)\Omega\right)(\theta) = \Omega(\theta) - \left(\mathcal{K}\left[\sigma(\Psi(\theta) + \theta) \cdot \Omega(\theta)\right]\right)(\theta)$ for some $\Omega \in W$.

Theorem 11.7. *If p is 2π-periodic and Lipschitz-continuous, then Newton's method for Theodorsen's integral equation (11.9.3) converges locally and quadratically in $(W, \|\cdot\|)$.*

Theorem 11.8. *If $|\sigma| < 1/3$ and $p \in L^\infty$, then Newton's method for Theodorsen's equation (11.9.3) converges globally in $(W, \|\cdot\|_2)$.* Proofs of these theorems are given in Hübner [1986].

Example 11.3. Wegmann's and Newton's methods are not identical for boundary curves defined in polar coordinates. As an example, take $\rho(\phi) \equiv 1$ and $\Psi_0 = \sin\theta$. Then $\mathcal{F}(\Psi_0(\theta)) = \Psi_0 - \mathcal{K}[1] = \sin\theta$. Hence, $\Psi_1 = 1$, which is the exact solution of Eq (11.9.3). ■

An algorithm for Newton's method is available in Hübner [1986:19-30].

11.10 Kantorovich's Method

Let a finite region D (the point $z = \infty$ is not in D) be mapped onto a circle by a function w that can be expressed as a Laurent series of the type (2.7.6). To select suitable coefficients in this Laurent series, Kantorovich (Kantorovich and Krylov [1964:414]) have used the inverse transformation $z = z(w)$, where $w = e^{i\theta}$, and derived the coefficients as in (2.7.8). Assuming that the boundary ∂D is defined by a real, implicit function $f(x, y) = 0$ which can be expanded as a Fourier series

$$f(x, y) = f_0 + \sum_{j=1}^{\infty} \{f_j \cos j\theta + g_j \sin j\theta\}, \qquad (11.11.1)$$

where f_j and g_j are functions of a_j and b_j as in (2.7.8). The function $f(x, y)$ can also be expressed as $F(z, \bar{z}) = 0$, where $x = (z + \bar{z})/2, y = (z - \bar{z})/2i$, i.e.,

$$z = \sum_{j=0}^{\infty} c_k e^{ij\theta}, \quad \bar{z} = \sum_{j=0}^{\infty} \bar{c}_j \, e^{-ij\theta},$$

thus giving

$$F(z, \bar{z}) = \sum_{j=-\infty}^{\infty} d_j \, e^{ij\theta}, \qquad (11.10.2)$$

where the coefficients d_j, which are functions of the coefficients a_j, b_j of (2.7.8), must be all zero so that the function $F(z, \bar{z}) = 0$ for all θ, i.e., $d_j(a_j, b_j) = 0$ for $j = \dots, -2, -1, 0, 1, 2, \dots$.

In the case when the boundary ∂D is the unit circle, we have $f(x, y) = x^2 + y^2 - 1 = 0$; $F(z, \bar{z}) = |z|^2 - 1$. In the case of nearly circular boundary ∂D, we have

$$f(x, y) = x^2 + y^2 + \lambda p(x, y) - 1 = 0, \quad F(z, \bar{z}) = |z|^2 + \lambda p(z, \bar{z}) - 1 = 0. \qquad (11.10.3)$$

Note that the term $|z|^2$ has the Laurent series expansion

$$|z|^2 = z\bar{z} = \cdots + (c_0\bar{c}_1 + c_1\bar{c}_2 + \cdots)\, e^{-i\theta} + (c_0\bar{c}_0 + c_1\bar{c}_1 + \cdots) + (c_1\bar{c}_0 + c_2\bar{c}_1 + \cdots)\, e^{i\theta} + \cdots. \qquad (11.10.4)$$

Since $e^{ij\theta}$ and $e^{-ij\theta}$ are complex conjugates of each other, the function $p(x, y)$ can be expressed as

$$p(z, \bar{z}) = q_0 + \sum_{j=1}^{\infty} \{q_j \, e^{ij\theta} + \bar{q}_j \, e^{-ij\theta}\}, \qquad (11.10.5)$$

where $q_j = r_j(a_j, b_j) + i s_j(a_j, b_j)$, $j = 0, 1, 2, \ldots$, and $s_0(a_j, b_j) = 0$. If the boundary ∂D has m axes of symmetry, then all coefficients except q_{mj+1} are zero.

Next, substituting (11.10.4) and (11.10.5) into Eq (11.10.3), and equating terms containing similar powers of $e^{i\theta}$ and $e^{-i\theta}$, we obtain the following set of simultaneous equations:

$$\begin{aligned} c_0\bar{c}_0 + c_1\bar{c}_1 + c_2\bar{c}_2 + \cdots + \lambda r_0(a_j, b_j) &= 1, \\ c_1\bar{c}_0 + c_2\bar{c}_1 + c_3\bar{c}_2 + \cdots + \lambda r_1(a_j, b_j) &= 0, \\ c_2\bar{c}_0 + c_3\bar{c}_1 + c_4\bar{c}_2 + \cdots + \lambda r_2(a_j, b_j) &= 0, \\ \cdots\cdots \quad \cdots\cdots \quad \cdots\cdots \quad \cdots\cdots \quad &\cdots\cdots \end{aligned} \tag{11.10.6}$$

Note that if the x-axis is the axis of symmetry, all b_j are zero; then the $a_j, c_j, \bar{c}_j$ determine $|z|2$ and $p(z, \bar{z})$.

The next simplification can be obtained by using $z(0) = 0$ and $z'(0) = c_1 > 0$; and we obtain $c_0 = 0$ and $c_1 = a_1 > 0$. Then Eq (11.10.6) can be solved by an *iterative scheme*, such as the Gauss-Seidel method, in the sequence $c_1, c_2, c_3, \ldots$ where their conjugates are treated as 'previous best values'. As initial values, we use those values that correspond to $\lambda = 1$ or $z = w$, so that the zeroth iteration is $c_1^0 = 1, c_2^0 = c_3^0 = \cdots = 0$. Kantorovich and Krylov [1964] have concluded that this method converges for small values of λ. Other significant facts are as follows: (i) Since the initial values of $c_j, j = 1, 2, \ldots$ are taken as zeros, it takes m number of iterations to evaluate m terms in the series approximation; and (ii) the $(m+1)$th approximation follows from the mth approximation (this is iteration).

Example 11.4. Consider the ellipse with two axes of symmetry ($m = 2$):

$$x^2 + y^2 - \lambda(x^2 - y^2) = 1. \tag{11.10.7}$$

Kantorovich and Krylov [1964:420] have derived the mapping function as

$$\begin{aligned} z = \Big(1 - \frac{\lambda^2}{8} + 3\frac{\lambda^4}{128}\Big)\Big\{w + \Big(\frac{\lambda}{2} - \frac{\lambda^3}{4} + 3\frac{\lambda^5}{32}\Big)w^3 + \Big(\frac{\lambda^2}{2} - 9\frac{\lambda^4}{16}\Big)w^5 \\ + \Big(5\frac{\lambda^3}{8} - 9\frac{\lambda^5}{8}\Big)w^7 + 7\frac{\lambda^4}{8}w^9 + 21\frac{\lambda^5}{16}w^{11}\Big\}, \end{aligned} \tag{11.10.8}$$

which is correct up to λ^5. In particular, for $\lambda = 0.25$, this function reduces to

$$z = 0.99\{w + 0.12w^3 + 0.03w^5 + 0.01w^7\}. \tag{11.10.9}$$

Notice that with $m = 2$, the powers of w are $1, 3, 5, 7$. ■

Example 11.5. Let the region D be a 2×2 square centered at the origin and oriented with its sides parallel to the coordinates axes. The boundary ∂D has $m = 4$ axes of symmetry and is defined by

$$|\bar{z}|^2 + \Big(\frac{z^2 - \bar{z}^2}{4}\Big)^2 = 1. \tag{11.10.10}$$

With $\lambda = 1$, it is expressed as

$$z = 1.0807w - 0.1081w^5 + 0.045w^9 - 0.0242w^{13} + 0.017w^{17} - 0.0126w^{21}. \tag{11.10.11}$$

Notice that with $m = 4$, the powers of w in (11.10.11) are $1, 5, 9, 13, 17, 21$. ■

Example 11.6. Kantorovich's method can also be used to map the region $C \cup \text{Int}(C)$ onto the region $\Gamma \cup \text{Ext}(\Gamma)$. Let $w = 0$ be the center of the unit circle C and let this center be transformed to $z = \infty$. Then using a power series of the form

$$z = c_{-1}w^{-1} + c_0 + c_1 w + c_2 w^2 + \cdots , \tag{11.10.12}$$

where $c_1 = a_1 + b_1 = a_1$ so that $b_1 = 0$, and applying the same procedure of the simultaneous equations (11.10.6) where the powers of w will, of course, be different, Kantorovich and Krylov [1964] have given the mapping function in the case of the 2×2 square of Example 11.5 as

$$z = \frac{1125}{1024} w - \frac{203}{2048} w^3 + \frac{1}{2048} w^7 . \blacksquare \tag{11.10.13}$$

11.11 Fornberg's Method of Successive Approximation

Fornberg [1980] used the Fourier transform method for mapping the interior of the boundary Γ in the z-plane, which is noncircular. The function $z(w)$ is approximated by a Taylor series expansion $f(w)$ as defined by Eq (2.7.7), i.e.,

$$f(w) = c_0 + c_1 w + c_2 w^2 + \cdots = \sum_{j=0}^{\infty} c_j \, e^{ij\theta} \quad \text{for } |w| = 1. \tag{11.11.1}$$

Let θ take the values $\theta_j = j/N$, $j = 0, 1, 2, \dots , N-1$. Then $f(w)$ can be rewritten in a form that is a discrete Fourier transform. All coefficients of this transform can be determined by applying the fast Fourier transform. The function $f(w)$ of (11.11.1) resembles the function $z(w)$ expanded directly in as a power series

$$z(w) = \sum_{j=-N/2+1}^{N/2} d_j w^j , \tag{11.11.2}$$

where

$$d_j = \frac{1}{N} \sum_{n=0}^{N-1} z_n \, e^{-2\pi i n/N} , \tag{11.11.3}$$

where d_j must satisfy the condition $d_j = 0$ is for $n \le 0$, to ensure that there are no singularities in the unit disk $|w| \le 1$. Next, the points z_j are relocated and the process is repeated. Fornberg [1980], while applying this method to analyze the ocean waves of nearly circular contour, has shown that the computation is of the order $N log N$.

11.11.1 Taylor's Series Method. Since the integral equations are Fredholm equations of the second kind, of the form

$$\phi(x) - \lambda \int_a^b k(x,s) \phi(s) \, ds = f(x), \quad a \le x \le b, \tag{11.11.4}$$

where the kernel $k(x,s)$ and the free term $f(x)$ are continuous on the interval $[a,b]$, and $\lambda \ne 0$ is a regular value of the kernel, a practical difficulty arises when λ is an eigenvalue of the kernel or when the regular value of λ is close to an eigenvalue. Therefore, we should exclude cases when the kernel is badly behaved, e.g., when $k(x,s) \notin C^1[a,b]$. However,

$\lambda = 1$ for the three important integral equations in numerical conformal mapping, which are named after (i) Lichtenstein and Gershgorin, (ii) Theodorsen, and (iii) Symm.

Equations of the form (11.11.4) arise in various physical problems, such as potential theory, Dirichlet problems, electrostatics, radiation heat transfer, particle transport problems in astrophysics, and reactor theory. Numerical solutions of Eq (11.11.4) by different methods are available in Kythe and Puri [2002].

For an approximate solution of Eq (11.11.4) we develop a method based on Taylor's series expansion. First, we take $a = 0$ and $b = 1$, which can be achieved by a simple transformation, assume that a unique solution of Eq (11.11.4) exists, and the kernel is either (i) $k(x,s) = k(x-s) \in C[a,b]$ and decreases rapidly as $(x-s)$ increases from zero; or (ii) of the form $u(x-s)\,\kappa(x,s)$, where κ is continuous for x and s, and $u(x-s) = O\left(|x-s|\right)^{-\alpha}$, $0 < \alpha < 1$, is weakly singular. In case (i), which is often encountered in conformal mapping, a Taylor polynomial can approximate the function $\phi(s)$, and we have

$$\phi(s) \approx \tilde{\phi}(s) = \tilde{\phi}(x) + \tilde{\phi}'(x)\,(x-s) + \cdots + \frac{\tilde{\phi}^{(n)}(x)}{n!}\,(x-s)^n, \tag{11.11.5}$$

where $\tilde{\phi}(x)$ is the approximate value of $\phi(x)$. This representation is valid because if $E(s)$ denotes the error between $\phi(s)$ and the expansion (11.11.5), then the contribution of the integral $\int_a^b k(x,s)\,E(s)\,ds$ is negligible since $k(x,s)$ decreases rapidly as $(x-s)$ increases. If we substitute (11.11.5) into Eq (11.11.4), we get

$$\begin{aligned} &\Big\{1 - \int_a^b k(x,s)\,ds\Big\}\,\tilde{\phi}(x) - \Big\{\int_a^b k(x,s)(x-s)\,ds\Big\}\,\tilde{\phi}'(x) - \cdots \\ &\cdots - \Big\{\frac{1}{n!}\int_a^b k(x,s)(x-s)^n\,ds\Big\}\,\tilde{\phi}^{(n)}(x) \approx f(x), \quad a < x < b. \end{aligned} \tag{11.11.6}$$

Notice that the expressions within the braces in (11.11.6) are functions of x only, and Eq (11.11.6) is a linear ordinary differential equation of order n with dependent variable $\phi(x)$, which can be solved either analytically or numerically provided we have the appropriate number of boundary (and/or initial) conditions. The problem of determining these boundary conditions is difficult; sometimes they can be determined from the physical constraints such as symmetry or thermal balance, as shown by Perlmutter and Siegel [1963] in the thermal radiation problem from gray surfaces. But, in general, it is not possible to determine them.

This method can, however, be modified in such a way that the boundary (initial) conditions are not required. This modification, presented by Ren, Zhang, and Qiao [1999], is "simple yet effective" and leads to an accurate approximate solution of a Fredholm equation of the second kind (and also of a Volterra equation of the second kind). Assume that $f \in C^n[a,b]$ and that $k(x,s) = k(x-s) \in C^n[a,b]$, which implies that $\phi \in C^n[a,b]$. Let the solution $\phi(x)$ of Eq (11.11.4) be approximated by the Taylor polynomial (11.11.5). Then, differentiating both sides of Eq (11.11.4), we find that

$$\begin{aligned} \phi'(x) - \int_a^b k'_x(x,s)\,\phi(s)\,ds &= f'(x), \\ &\vdots \\ \phi^{(n)}(x) - \int_a^b k_x^{(n)}(x,s)\,\phi(s)\,ds &= f^{(n)}(x), \quad a < x < b, \end{aligned} \tag{11.11.7}$$

where $k_x^{(n)}(x,s) = \dfrac{\partial^n k(x,s)}{\partial x^n}$. After substituting $\phi(x)$ for $\phi(s)$ in the integrals in (11.11.7) and replacing $\phi(x)$ by $\tilde{\phi}(x)$, we get

$$\begin{aligned} &\tilde{\phi}'(x) - \Big\{ \int_a^b k_x'(x,s)\,ds \Big\}\, \tilde{\phi}(x) = f'(x), \\ &\quad\vdots \\ &\tilde{\phi}^{(n)}(x) - \Big\{ \int_a^b k_x^{(n)}(x,s)\,ds \Big\}\, \tilde{\phi}(x) = f^{(n)}(x), \quad a < x < b. \end{aligned} \tag{11.11.8}$$

Then substituting for $\tilde{\phi}'(x), \dots, \tilde{\phi}^{(n)}(x)$ from (11.11.8) into Eq (11.11.6), we find that

$$\begin{aligned} &\Big\{ \Big(1 - \int_a^b k(x,s)\,ds\Big) - \Big(\int_a^b k(x,s)(x-s)\,ds\Big)\Big(\int_a^b k_x'(x,s)\,ds\Big) - \cdots \\ &\qquad - \Big(\frac{1}{n!}\int_a^b k(x,s)(x-s)^n\,ds\Big)\Big(\int_a^b k_x^{(n)}(x,s)\,ds\Big)\Big\} \\ &= f(x) + \Big(\int_a^b k(x,s)(x-s)\,ds\Big) f'(x) + \cdots \\ &\qquad\qquad + \Big(\frac{1}{n!}\int_a^b k(x,s)(x-s)^n\,ds\Big) f^{(n)}(x), \end{aligned}$$

which gives

$$A_n(x)\,\tilde{\phi}(x) = F_n(x), \tag{11.11.9}$$

where

$$\begin{aligned} &A_n(x) = 1 - G_0^{(0)}(x) - \sum_{j=1}^n G_j^{(0)}(x)\, G_0^{(j)}(x), \\ &G_p^{(m)}(x) = \frac{1}{p!}\int_a^b k_x^{(m)}(x,s)\,(x-s)^p\,ds, \\ &G_0(x) = f(x), \quad A_0(x) = 1 - G_0^{(0)}(x), \\ &F_n(x) = f(x) + \sum_{j=1}^n G_j^{(0)}(x)\, f^{(j)}(x), \quad F_0(x) = f(x). \end{aligned} \tag{11.11.10}$$

Hence, $\tilde{\phi}(x)$ can be computed directly from formula (11.11.9).

There are three integral equations, with their respective methods: (i) Lichtenstein-Gershgorin's; (ii) Theodorsen's; and (iii) Symm's. These and their methods are discussed in the sequel. Gutknecht [1986] has mentioned that all these three integral equations have one common feature: they require an auxiliary function to develop the conjugate pairs $u(x,y)$ and $v(x,y)$ such that $w(z) = u + iv$ will satisfy the boundary conditions. This auxiliary function is $\log\{f(z)/z\}$, which can be defined, using $\dfrac{f(z)}{z} = \dfrac{w}{z} = \dfrac{R\,e^{i\Theta}}{r\,e^{i\theta}}$, as

$$\log\frac{f(z)}{z} = \big\{\log|w| - \log|z|\big\} = i\,\{\Theta - \theta\}. \tag{11.11.11}$$

11.12 Lichtenstein-Gershgorin Method

To solve the problem of conformally mapping a domain D in the z-plane onto the unit disk $|w| < 1$ such that the boundary Γ of D maps onto the circle $|w| = 1$ and $z = 0$ is mapped into $w = 0$, we use the Cauchy integral formula (2.4.2) to determine the mapping function $w = f(z)$. Let the boundary Γ be traversed in the counter-clockwise direction starting and ending at the point $z(0)$ on the boundary in increments $|ds|$, where $0 \le s \le L$ (Figure 11.1). Then at any point $z = z(s)$, the tangent to the boundary relative to the real axis is defined by an angle $\phi(s) = \arg\{z'(s)\}$, where $z'(s) = |dz/ds| = 1 = e^{i\phi(s)}$, and $|dz| = ds$ along the boundary. The corresponding function in the w-plane is $\Theta(s) = \arg\{f(z(s)\}$, which leads to the required mapping function $w = e^{i\Theta(s)}$. Thus, using the Cauchy integral formula we get

$$w = f(z) = \frac{1}{2\pi i}\int_\Gamma \frac{f(\zeta)}{\zeta - z}\,d\zeta = \frac{1}{2\pi i}\int_0^L \frac{f(z(s))}{z(s) - z}\,z'(s)\,ds = \frac{1}{2\pi i}\int_0^L e^{i\Theta(s)}\,\frac{z'(s)}{z(s) - z}\,ds. \tag{11.12.1}$$

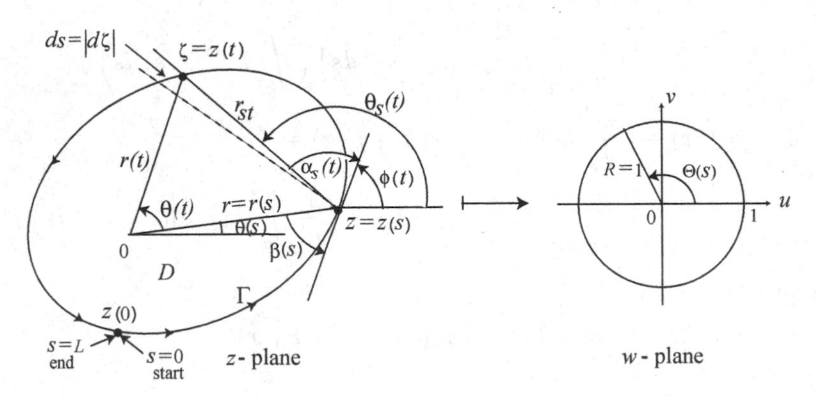

Figure 11.1 Parameters used in domains D and G.

In order to avoid singularities due to $\zeta = z$, we will confine to boundary points z, we use the boundary value for Cauchy's integral (2.4.9) which gives, for any point $z \in \Gamma$ which is not a corner point,

$$f(z) = \text{p.v. } \frac{1}{\pi i}\int_\Gamma \frac{f(\zeta)}{\zeta - z}\,d\zeta, \tag{11.12.2}$$

where p.v. signifies that the principal value for the integral is used. We will use the auxiliary functions of the type (11.11.11) which are analytic inside and on the unit circle. Let $F(z)$ denote this auxiliary function; thus, for $|w| = 1$ (see Figure 11.1),

$$\begin{aligned} F(z) = \log\frac{f(z)}{z} &= \{\log|w| - \log|z|\} + i\{\Theta(s) - \theta(s)\} \\ &= i\{\Theta(s) - \theta(s)\} - \log r. \end{aligned} \tag{11.12.3}$$

Next, applying (11.12.2) to $F(z)$, we get

$$F(z) = \log\frac{f(z)}{z} = \frac{i}{\pi i}\int_\Gamma \log\frac{f(\zeta)}{\zeta}\,\frac{d\zeta}{\zeta - z}, \tag{11.12.4}$$

where $z = z(s)$ and $\zeta = z(t), 0 \le t \le L$, and (from Figure 11.1)

$$\frac{d\zeta}{\zeta - z} = \frac{z'(t)}{z(t) - z(s)} = \frac{1}{r_{st}}\,e^{\phi(t) - \theta_s(t)}\,dt.$$

Substituting these values into (11.12.4) and equating imaginary parts, we find that

$$\begin{aligned}\Theta(s)-\theta(s) &= \frac{1}{\pi}\int_0^L \{\Theta(t)-\theta(t)\}\frac{\sin\{\phi(t)-\theta_s(t)\}}{r_{st}}\,dt \\ &\quad+\frac{1}{\pi}\int_0^L \log r(t)\frac{\cos\{\phi(t)-\theta_s(t)\}}{}dt, \qquad (11.12.5)\end{aligned}$$

and equating the real parts, with $f(z)=|w|=1$,

$$\begin{aligned}\log|z| = \log r(t) &= \frac{1}{\pi}\int_0^L \log r(t)\frac{\sin\{\phi(t)-\theta_s(t)\}}{r_{st}}\,dt \\ &\quad-\frac{1}{\pi}\int_0^L \{\Theta(t)-\theta(t)\}\frac{\cos\{\phi(t)-\theta_s(t)\}}{}dt \\ &\equiv \int_0^L N(s,t)\log r(t)\,dt - g(s), \qquad (11.12.6)\end{aligned}$$

where principal values of these two integrals are used in their evaluation, $N(s,t)$ is the Neumann kernel of the first part of the integral equation (see §10.4), and $g(s)$ denotes the second part. Then Eq (11.12.5) becomes

$$\Theta(s)-\theta(s) = \int_0^L N(s,t)\{\Theta(t)-\theta(t)\}\,dt + g(s), \qquad (11.12.7)$$

which gives Θ implicitly. However, this equation is a linear integral equation of the second kind, which can be solved by fast matrix inversions. A pertinent historical note is that Eq (11.12.7) was developed by Lichtenstein [1917], and independently by Gershgorin [1933] who expressed this equation in the following form:

$$\Theta(s)-\theta(s) = \int_0^L N(s,t)\{\Theta(t)-\theta(t)\}\,dt - 2\beta(s), \qquad (11.12.8)$$

where $\beta(s)$ is defined in Figure 11.1. To solve Eq (11.12.8) numerically, divide the boundary length L into n equal intervals of size $h = L/n$, thereby replacing s by $s_k = kh$, $k = 1,2,\dots,n$; thus, $dt = h = ds$. Then using Euler's method of numerical integration, we get

$$\begin{aligned}\Theta(s)+2\beta(s) &= \int_0^L N(s,t)\Theta(t)\,dt \\ &= h\sum_{k=1}^{n-1} N(s,t_k)\,\Theta(t_k) + \frac{h}{2}\{N(s,0)\Theta(0)+N(s,L)\Theta(L)\}. \qquad (11.12.9)\end{aligned}$$

Since the start point $s=0$ has not been included, let $\Theta(0)=0$. Then $\Theta(L)=2\pi$. Also, $N(s,0)=N(s,L)$. Thus, we have for s_j, $j=1,2,\dots,n-1$:

$$\Theta(s_j)+2\beta(s_j) - h\pi N(s_j,0) \approx h\sum_{k=1}^{n-1} N(s_j,t_k)\,\Theta(t_k). \qquad (19.12.10)$$

Repeating this equation for each j, we obtain a system of simultaneous equations in $\theta_k \equiv \Theta(t_k)$:

$$\sum_{k=1}^{n-1}\{hN(s_j,t_k)-\delta_{jk}\}\Theta_k = 2\beta(s_j) - h\pi N(s_j,0), \qquad (11.12.11)$$

where δ_{jk} is the Kronecker delta.

Now that the values Θ_k have been found in the w-plane, the related values of $r_k = |z_k|$ can be evaluated using Eq (11.12.6). For such an inverse mapping we will use a truncated trigonometric series to establish the relation between $w = e^{i\Theta}$ and $z = x + iy = r\, e^{i\theta}$ as in the Kantorovitch's method (§8.1.1). Let the conjugate functions $x(\Theta)$ and $y(\Theta)$ be expressed as

$$\begin{aligned} x &= \sum_{k=0}^{n} \left(a_k \cos k\Theta + b_k \sin \Theta\right), \\ y &= \sum_{k=0}^{n} \left(a_k \sin k\Theta - b_k \cos \Theta\right), \end{aligned} \tag{11.12. 12}$$

where all coefficients except b_0 are known. The coefficient b_0 depends on the choice of the correspondence between the boundary points z and w. Thus,

$$z = \sum_{k=0}^{n} (a_k - ib_k)\, w^k. \tag{11.12.13}$$

In case we want to map the exterior of the region Γ onto the exterior of the unit circle $|w| = 1$, we have

$$\Theta(s) = -\int_0^L N(s,t)\{\Theta(t) - \theta_s(t)\}\, dt + g(s), \tag{11.12.14}$$

or

$$\Theta(s) = -\frac{1}{\pi}\int_0^L \frac{\Theta(t)\cos\alpha_s(t)}{r_s(t)}\, dt + 2\beta(s), \tag{11.12.15}$$

which can be numerically solved by the above Kantorovich's method (Kantorovich and Krylov [1964: 508]). The above analysis is adapted from Schinzinger and Laura [2003:122].

Note that Eq (11.12.7) has been solved by an iterative method in §10.4.

The (numerical) mapping of a doubly connected region between two closed boundaries Γ_1 and Γ_2 onto an annulus $A(R_1, R_2)$ in the w-plane can be accomplished as follows: Map the interior of Γ_1 by Eq (11.12.9), and the exterior of Γ_2 by Eq (11.12.15). The details can be found in Chapter 14.

For the numerical mapping of a square onto a circle, see Gaier [1964: 57], where the step size $h = L/n$ is taken as 0.1. Banin [1943] was the first to use the Lichtenstein method in a fluid flow problem.

11.13 Theodorsen-Garrick Method

The Theodorsen-Garrick method was developed (Theodorsen [1931], Theodorsen and Garrick [1933], Garrick [1936; 1952]) for mapping of airfoils (see Chapter 6). This method, known as the *method of conjugates*, is very effective in cases of conformal mapping of circular boundaries onto the unit circle. An extensive survey of this method is available in Gaier [1964; 1983] and Laura and Shinzinger [2003:128].

Consider an airfoil given in the z-plane, which under the inverse Joukowski transformation is mapped onto a near-circle in the ζ-plane ($\zeta = \xi + i\eta$), which in turn is mapped onto the unit circle in the w-plane (Figure 11.2). However, we will discuss this mapping in the reverse order. The mapping from the w-plane onto the ζ-plane is defined by

$$\zeta = w\, e^{g(w)}, \quad \text{or} \quad g(w) = \log \frac{\zeta}{w}, \tag{11.13.1.}$$

where $g(z)$ is an analytic function, yet unknown, but which can be represented by a series expansion of the form

$$g(w) = \sum_{j=1}^{\infty} \frac{c_j}{w^j}, \tag{11.13.2}$$

where c_j is a complex quantity.

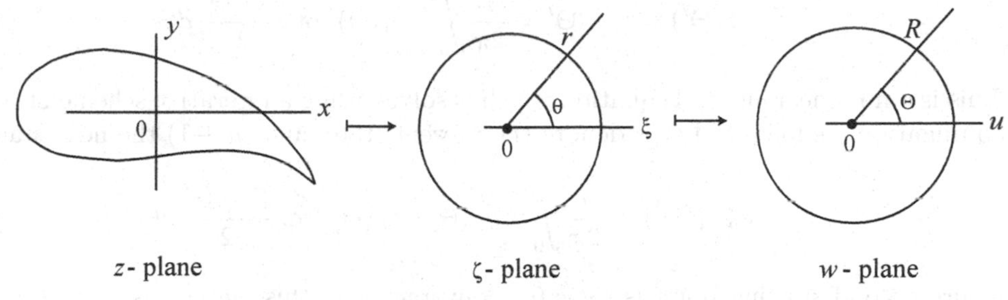

Figure 11.2 Joukowski profile onto the unit circle.

Since we are dealing with the potential fields exterior of the airfoil, we can exclude $w = 0$. Thus, let the circle in the w-plane be given by $w = R\, e^{i\Theta}$, (R constant), and the near-circle by $\zeta = \rho\, e^{i\psi}$, $\rho = a\, e^{\Theta}$, so that $\zeta = a\, e^{\Theta + i\psi}$. Notice that the radius ρ of the near-circle varies because of the term e^{Θ}. Thus, we have

$$g(w) = \log \zeta - \log w = \log \left(a\, e^{\Theta+i\psi}\right) - \log \left(R\, e^{i\Theta}\right) = \left(\Theta - \log \frac{R}{a}\right) + i(\psi - \Theta). \tag{11.13.3}$$

Then c_j/w^j becomes

$$\frac{c_j}{w^j} = \frac{c_j}{R^j} e^{-ij\Theta} = (a_j + ib_j)\,(\cos j\Theta - i \sin j\Theta),$$

and thus,

$$g(w) = \left(\Theta - \frac{R}{a}\right) + i\,(\psi - \Theta) = \sum_{j=1}^{\infty} (a_j + ib_j)\,(\cos j\Theta - i \sin j\Theta). \tag{11.13.4}$$

Equating real and imaginary parts on both sides, we get

$$\begin{aligned} \Theta - \Theta_0 &= \sum_{j=1}^{\infty} (a_j \cos j\Theta + b_j \sin j\Theta), \\ \Theta - \psi &= \sum_{j=1}^{\infty} (a_j \sin j\Theta - b_j \cos j\Theta), \end{aligned} \tag{11.13.5}$$

where Θ_0 is a constant, like a bias of the Fourier series $2\pi\Theta_0 = \int_0^{2\pi} t\, dt$. The expressions on the left sides of these equations justify the distortion of the circle. Also, at the point at infinity, we have $\zeta = w$ and $d\zeta/dw = 1$.

Alternatively, we can represent $g(w) = g_1(\Theta) + i g_2(\Theta)$, where the two functions g_1 and g_2 are complex conjugates related by

$$\begin{aligned} g_1(\Theta') &= \frac{1}{2\pi}\int_0^{2\pi} g_1(\Theta)\, d\Theta - \frac{1}{2\pi}\int_0^{2\pi} g_2(\Theta) \cot\frac{\Theta-\Theta'}{2}\, d\Theta; \\ g_2(\Theta') &= \frac{1}{2\pi}\int_0^{2\pi} g_2(\Theta)\, d\Theta - \frac{1}{2\pi}\int_0^{2\pi} g_1(\Theta) \cot\frac{\Theta-\Theta'}{2}\, d\Theta. \end{aligned} \tag{11.13.6}$$

Next, we determine $g_2(w) = \psi - \Theta \equiv e(\Theta)$ from Eq (11.13.5), which leads to *Theodorsen's integral equation*:

$$e(\Theta') = \psi - \Theta' = \frac{1}{2\pi}\int_0^{2\pi} g_1(\Theta) \cot\frac{\Theta-\Theta'}{2}\, d\Theta. \tag{11. 13.7}$$

This is a nonlinear integral equation which is solved using an iterative scheme of consecutive conjugation, as follows: Let n denote the present state, and $(n+1)$ the next stage. Then

$$e_{n+1}(\Theta') = \frac{1}{2\pi}\int_0^{2\pi} g_1\left(\Theta - e_n(\Theta)\right) \cot\frac{\Theta-\Theta'}{2}\, d\Theta, \tag{11. 13.8}$$

where a good starting point is $e_0 = 0$. Convergence of this method is very restricted, but in practical problems it works fine. Another restriction is that the boundary in the ζ-plane must be starlike (see §11.1 and §11.2), which means that any ray starting from the origin $\zeta = 0$ must intersect the boundary only once.

The Fast Fourier transform (FFT) method has greatly simplified the numerical computation. This method is available in Henrici [1979], Gutknecht [1981; 1983; 1986], Hübner [1982], Wegman [1986], and Trefethen [1986].

11.14 Variational Methods

The variational method uses the area or the boundary of a region. Let l be the length of the boundary of a region D in the z-plane, and let L be the length of the boundary of a region G in the w-plane. We will determine the transformation $w = f(z)$ that maps D onto the unit disk $C : |w| < 1$. Then the length of the circumference of the unit circle $|w| = 1$ will be $L = 2\pi i$, such that, in view of (2.6.4),

$$L = \int_{\partial G} |dw| = \int_{\partial D} |f'(z)||dz| = 2\pi. \tag{11.14.1}$$

Also, in view of the Cauchy integral formula (2.4.2), any function $g(w)$ can be evaluated at a point $w_j \in C$ by

$$g(w_j) = \frac{1}{2\pi i}\int_{\partial G} \frac{g(w)}{w - w_j}\, dw. \tag{11.14.2}$$

First, for all $w_j \in \partial G$, the value of $g(0)$ is given by

$$g(0) = \frac{1}{2\pi i}\int_{\partial G} \frac{|g(w)|}{e^{i\theta}}\, dw = \frac{1}{2\pi}\int_0^{2\pi} g\left(e^{i\theta}\right) d\theta. \tag{11.14.3}$$

Thus,

$$2\pi g(0) \le \int_{\partial G} |g(w)|\, |dw| = \int_0^{2\pi} |g(w)|\, d\theta, \tag{11.14.4}$$

where equality holds only when $|g(w)| = |g(0)| > 0 =$ (constant).

Next, let us assume that $f'(z)$ is not known, and let consider some $f(z)$ whose derivative is, say, $h(z)$. Then from Eq (11.14.1) we get

$$\int_{\partial D} |g(z)||dz| = \int_{\partial G} |g(z)||f^{-1}(z)|^{-1}\,|dw|. \tag{11.14.5}$$

Combining Eqs (11.14.4) and (11.14.5), and assuming that $z = 0$ is mapped onto $w = 0$, we get

$$\int_0^{2\pi} \left|\frac{g(z)}{f'(z)}\right| d\theta \geq 2\pi \left|\frac{g(0)}{f'(0)}\right|, \tag{11.14.6}$$

where equality holds if $g(z)/f'(z)| = |g(0)/f'(0)$, i.e., if $g(z)/|g(0)| = f'(z)/|f'(0)|$, which is known as the *optimality condition*. Hence, the problem reduces to finding the function $g(z)$ that minimizes the integral (11.14.6). The optimality condition can be stated in terms of minimizing the circumference by either of the following criteria:

$$\begin{aligned} I_1 &: \int_{\partial D} |g(z)|\,|dz| \geq 2\pi |g(0)/f'(0)|, \\ I_2 &: \int_{\partial D} |f'(z)|\,|dz| \geq 2\pi. \end{aligned} \tag{11.14.7}$$

Thus, the second criterion leads to the *minimum circumference principle*, stated by Julia [1931], as "Of all the functions $f(z)$, analytic in the region D with $f(0) = 0$, the one which minimizes the integral I_2 maps D onto the unit disk."

Since the area element $du\,dv = |f'(z)|^2\,dx\,dy$, we can compare the above principle with the *minimum area principle*, stated by Bieberbach [1914], as "Of all the functions $f(z)$, analytic in D with $f(0) = 0$ and $f'(0) = 1$, the one which minimizes the integral $I = \iint_D |f'(z)|^2\,dx\,dy$ maps D onto the unit disk."

11.14.1 Minimization Process. Let $g^*(z)$ denote the optimum minimum function. Then

$$\min\{I\} = \int_D |g^*(z)|\,ds = \int_D |g^*(z)|\,|dz|, \tag{11.14.8}$$

where

$$g(z) = \frac{f(z)}{|f'(z_0)|} = f(z) \quad \text{for } z_0 = 0, |f'(0)| = 1.$$

Then we obtain the desired function $f(z)$ by integrating $g^*(z)$, such that the optimum integral $I_1^* = 2\pi$. Next, we use the Ritz method for numerical computation, by setting $g(z) = \{p(z)\}^2$, or $|g(z)| = p(z)\overline{p(z)}$. Let the nth order approximation $p_n(z)$ of $\sqrt{g(z)}$ be given by

$$p_n(z) = P_0(z) + \sum_{j=1}^{n} \alpha_j P_j(z), \tag{11.14.9}$$

where all P_j are linearly independent functions on D. Since $f'(0) = g(0) = 1$, we have $P_0(0) = 1$ and $P_j(0) = 0$ for $j = 1, 2, \dots$. Let s_{jk} denote the scalar product $s_{jk} = \langle P_j, P_k \rangle = \int_D P_j(z)\overline{P_k(z)}\,|dz|$. Then the minimum of I, given by

$$\min\{I\} = I^* = \int_D p_n^* \bar{p}_n^*\,|dz| = \langle p_n^*, \bar{p}_n^* \rangle, \tag{11.14.10}$$

is attained by the function p_n^* which is orthogonal to all other polynomials P_k, i.e., $\langle p_n^*, P_k\rangle = 0$ for $k = 1, 2, \dots, n$ (von Koppenfels and Stallman [1959:185]). Moreover, the polynomials P_k are of the form $(z - z_0)^k = z^k$. Hence, we obtain the following set of simultaneous linear equations in the unknown coefficients α_j defined in Eq (11.14.9), and the solution p_n^* is as follows:

$$p_n^* = \frac{\begin{vmatrix} p_0 & s_{01} & s_{02} & \cdots & s_{0n} \\ p_1 & s_{11} & s_{12} & \cdots & s_{1n} \\ \cdots & \cdots & \cdots & \cdots & \cdots \\ p_n & s_{n1} & s_{n2} & \cdots & s_{nn} \end{vmatrix}}{\begin{vmatrix} s_{01} & s_{02} & \cdots & s_{0n} \\ s_{11} & s_{12} & \cdots & s_{1n} \\ \cdots & \cdots & \cdots & \cdots \\ s_{n1} & s_{n2} & \cdots & s_{nn} \end{vmatrix}}. \tag{11.14.11}$$

This method converges if the boundary of the region D is smooth and D has no slits. The number of computational operations is proportional to n^3. The operations are simple and easily computable, but the method is nonrecursive. The computation can be reduced if orthonormal polynomials are used, since these polynomials have the inner product

$$\langle p_j, p_k\rangle = \int_D g_j(z)\overline{g_k(z)}\,|dz| = \iint_A g_j(z)\overline{g_k(z)}\,dx\,dy, \tag{11.14.12}$$

which satisfies

$$\langle p_j, p_k\rangle = \begin{cases} 0 & \text{if } j \neq k, \\ 1 & \text{if } j = k \text{ for orthonormality.} \end{cases} \tag{11.14.13}$$

The Szegö's method, discussed below, is based on the definition (11.14.12), and Bergman's method, discussed in §8.5.2, is based on (11.14.13).

11.14.2 Orthogonalization. We will discuss two methods, as follows.

METHOD 1. This procedure, defined by (11.14.12), is recursive by solving the set of simultaneous equations (11.14.11). We will discuss the Schmidt method which is well suited for recursion. First, we generate orthonormal functions $p_j, j = 1, 2, \dots, n$, which are linear and defined as

$$p_j(z) = \sum_{k=1}^{n} \alpha_{jk} P_k(z). \tag{11.14.14}$$

We will determine the coefficients a_{jk}. Let $p_1 = P_1 \neq 0$, and determine a_{11} that makes p_1 orthogonal to $p_2 = P_2 + a_{11}P_1$, i.e.,

$$\langle p_2, p_1\rangle = \langle P_2, P_1\rangle + a_{11}\langle P_1, P_1\rangle = 0,$$

where $\langle P_2, P_1\rangle$ is defined in the same way as the definition (11.14.10) thus giving

$$a_{11} = -\frac{\langle P_2, P_1\rangle}{\langle P_1, P_1\rangle}.$$

Proceeding using this recursion we finally obtain the conditions as

$$\langle p_j, p_{j-1}\rangle = \langle P_j, p_{j-1}\rangle + a_{j,j-1}\langle p_{j-1}, p_{j-1}\rangle, \quad j, = 1, 2, \dots, n. \tag{11.14.15}$$

However, if the polynomials p_j are chosen to be orthonormal, then these polynomials are normalized by $p_j^0 = \dfrac{p_j}{\sqrt{p_j, p_j}}$, and thus, the recursive relation becomes

$$p_j = P_j - \left[\langle P_j, p_1^0\rangle p_1^0 + \cdots + \langle P_j, p_{j-1}^0\rangle p_{k-1}^0\right], \tag{11.14.16}$$

where the scalar (inner) product is defined by (11.14.12).

METHOD 2. This method, based on solving simultaneous equations, is discussed in Kantorovich and Krylov [1964:382] , Gaier [1964:132], and Davis and Rabinowitz [1961]. It defines z as the scalar product

$$s_{jk} = \langle z^j, z^k\rangle = \frac{1}{l}\int_D z^j z^k\,|dz|, \tag{11.14.17}$$

where l the length of D, and $s_{jk} = s_{kj}$. Define the determinants

$$D_0 = 1, \quad D_n = \begin{vmatrix} s_{00} & s_{10} & \cdots & s_{n0} \\ s_{01} & s_{11} & \cdots & s_{n1} \\ \cdots & \cdots & \cdots & \cdots \\ s_{0n} & s_{1n} & \cdots & s_{nn} \end{vmatrix}, \quad D_n > 0.$$

Then the Szegö polynomials $p_n(z)$ are defined as

$$p_n(z) = \frac{1}{\sqrt{D_{n-1}D_n}} = \begin{vmatrix} s_{00} & s_{10} & \cdots & s_{n0} \\ s_{01} & s_{11} & \cdots & s_{n1} \\ \cdots & \cdots & \cdots & \cdots \\ s_{0,n-1} & s_{1,n-1} & \cdots & s_{n,n-1} \\ 1 & z & \cdots & z^n \end{vmatrix}. \tag{11.14.18}$$

11.14.3 Minimization of the Perimeter. To minimize the integral

$$I = \int_D |f(z)|\,|dz|, \tag{11.14.19}$$

we proceed as follows: Replace $f(z)$ by $g(z) = |p(z)|^2 = p(z)\overline{p(z)}$, as in §11.14.1, and expand $p(z)$, as in Eq (11.14.14), by $p_n(z) = \sum\limits_{j=0}^{n} C_j P_j(z)$. Then

$$p_n(z)|^2 = \sum_{j=0}^{n} |C_j|^2 = \sum_{j=0}^{n} C_j \bar{C}_j. \tag{11.14.20}$$

Since $f'(0) = g(0) = 1$ by assumption, we get $p_n(0) = \sum\limits_{j=0}^{n} C_j P_j(0) = 1$; however, from (11.14.20) we get $|p_n(0)|^2 = \sum\limits_{j=0}^{n} |C_j|^2 = 1$. Then, by Cauchy-Schwarz inequality, we get

$$|p_n(0)|^2\{|P_0(0)^2| + |P_1(0)|^2 + \cdots + |P_n(0)|^2\} \geq 1,$$

which gives

$$|p_n|^2 \geq \{|P_0(0)^2| + |P_1(0)|^2 + \cdots + |P_n(0)|^2\}, \tag{11.14.21}$$

where $|P_0(0)^2| + |P_1(0)|^2 + \cdots + |P_n(0)|^2$ is the best possible estimate.

Next, let the coefficients of the required polynomial be defined as

$$a_j = \overline{P_j(0)}\{|P_0(0)^2| + |P_1(0)|^2 + \cdots + |P_n(0)|^2\}^{-1}.$$

Then the required polynomial approximation is

$$p_n(z) = \sum_{j=0}^{n} a_j P_j(z) = \frac{\sum_{j=0}^{n} \overline{P_j(0)} P_j(z)}{\sum_{j=0}^{n} P_j(0)\overline{P_j(0)}} = \frac{K_n(0,z)}{K_n(0,0)}, \tag{11.14.22}$$

where $K_n(a,b) = \sum_{j=0}^{n} \overline{P_j(a)} P_j(b)$. The function K_n with $n = \infty$ is known as the Szegö kernel (§10.6). Kantorovich and Krylov [1964:385] have shown that if $w = 0$ corresponds to $z = a$ with $f'(a) = 1$, then the required polynomial is

$$p(z) = \frac{K_n(a,z)}{K_n(a,a)}. \tag{11.14.23}$$

Hence, the solution of the problem is obtained by integrating

$$f(z) = \int_0^z f'(z)\, dz = \int_0^z (g * z)\, dz,$$

where

$$g(z) = p(z)\overline{p(z)} = \frac{K_n^2(0,z)}{K_n^2(0,0)},$$

or

$$f(z) = \frac{1}{K_n^2(0,0)} \int_0^z K_n^2(0,z)\, dz. \tag{11.14.24}$$

If the length of the boundary of the domain D is l, then the radius of the circle $|w| = |f(z)| = R$ is obtained from $2\pi R = I$, which is

$$\frac{R}{2\pi K_n^2(0,0)} \int_D K_n(0,z)\overline{K_n(0,z)}\, |dz| = \frac{l}{2\pi K_n(0,0)},$$

giving the radius R of the circle as

$$R = \frac{l}{2\pi K_n(0,0,)}. \tag{11.14.25}$$

Finally, note that the orthonormalization process is time consuming, but worth pursuing, as it provides a method of computing the Szegö kernel as the 'solution of a new, numerically tractable, integral equation of the second kind' (Kerzmann and Trummer [1986]).

Example 11.7. To determine the minimum polynomial that maps a square of side 2, with vertices at $(\pm 1, \pm i)$, onto a circle C, the inner product s_{kj} defined by (11.14.17) is

$$\begin{aligned} s_{kj} &= \frac{1}{l}\int_D z^j z^k\,|dz| \\ &= \frac{1}{8}\Big\{\int_{-1}^{1}(x-i)^j(x+i)^k\,|dx| + \int_{-1}^{1}(x+i)^j(x-i)^k\,|dx| \\ &\quad + \int_{-1}^{1}(1+iy)^j(1-iy)^k\,|dy| + \int_{-1}^{1}(-1+iy)^j(-1-iy)^k\,|dy|\Big\} \\ &= \begin{cases} \displaystyle\int_0^1 (x^2+1)\,\Re\{(x_i)^{j-k}\}\,|dx|, & \text{if } j-k \text{ is a multiple of 4,} \\ 0, & \text{otherwise.}\end{cases} \end{aligned} \tag{11.14.26}$$

The values of s_{jk} are evaluated as follows:

$$\begin{aligned} &1,\ \ 0,\ \ 0,\ \ 0,\ \ -4/5,\ \ \dots;\ \ 0;\ \ 4/3, 0, 0, 0, \dots; \\ &0,\ \ 0,\ \ 28/15,\ \ 0,\ \ 0,\ \ \dots;\ \ 0,\ \ 0,\ \ ,0,\ \ 96/35,\ \ 0, \dots; \\ &-4/5,\ \ 0,\ \ 0,\ \ 0,\ \ 1328/315, \dots. \end{aligned}$$

Then using Eq 11.14.18, the Szegö polynomials are

$$p_0 = 1, \quad p_1 = \frac{\sqrt{3}}{2}z, \quad p_3 = \sqrt{\frac{35}{96}}z^3, \quad p_4 = \frac{15}{16}\sqrt{\frac{7}{22}}\left(z^4 + \tfrac{4}{5}\right).$$

Also, evaluating the terms $K_n(0, z)$ for $n = 5$, we have

$$K(0, z) = 1 + \frac{315}{1408}\left(z^4 + \tfrac{4}{5}\right), \quad K(0, 0) = \frac{415}{352},$$

thus,

$$\frac{K^2(0, z)}{K^2(0, 0)} = 1 + 0.379518z^4 + 0.189759z^8,$$

and hence, the required polynomial is

$$f(z) = z + \frac{63}{830}z^5 + \frac{441}{110224}z^9 \approx z + 0.075904z^5 + 0.004001z^9. \tag{11.14.27}$$

This result, due to Schinzinger and Laura [2003], can be compared with Example 7.2

The radius of the circle C is obtained from (11.14.26). Since $l = 8$ for the square in the z-plane, we get

$$R = |w| = \frac{8}{2\pi K(0, 0)} = \frac{8}{2\pi(415/352)} \approx 1.08.$$

If the circle C is the unit circle, then multiplying (11.14.27) by the factor $c = 0.92704$, we obtain

$$\begin{aligned} w(z) &= \frac{1}{1.08}\Big\{z + \frac{c^4}{10}z^5 + \frac{c^8}{120}z^9 + \cdots\Big\}, \\ z(w) &= 1.08\Big\{w - \frac{c^4}{10}w^5 + \frac{c^8}{24}w^9 \pm \cdots\Big\}. \blacksquare \end{aligned} \tag{11.14.28}$$

In these approximate methods, Chebyshev's min-max principle, which deals with the minimization of the maximum absolute deviation, is also important as it leads to linear programming, as in Arafeh and Schinzinger [1978]. The mapping of a region in the z-plane onto a rectangle or parallelogram by linear programming procedures is studied in Opfer [1979; 1980; 1982] and Hartman and Opfer [1986].

The following references are useful for solution and numerical treatment of the Fredholm and Volterra integralo equations of the second kind: Atkinson [1997], Baker [1978], Brunner and van der Houwen [1986], Delves and Mohamed [1985], Hackbusch [1995], Kythe and Wei [2004], Reichel [1987; 1989], and Yan [1994].

REFERENCES USED: Carrier, Krook and Pearson [1966], Fornberg [1980], Gaier [1964], Kantorovich and Krylov [1964], Kerzmann and Trummer [1986], Kythe [1998], Privaloff [1916], F. and M. Riesz [1923], Schinzinger and Laura [2003], Theodorsen [1931], Warschawski [1935; 1945; 1950, 1955], Wegmann [1984; 1986], Zygmund [1935].

12

Symm's Integral Equation

A potential-theoretic formulation of the problem of conformally mapping a simply connected region (or its complement) onto the unit disk leads to a Fredholm integral equation of the first kind, known as Symm's integral equation, which has a kernel with a logarithmic singularity. Unlike Fredholm integral equations of the second kind, e.g., Theodorsen's equation, in which the singularity of the kernel at points near but not on the boundary creates computational difficulties, Symm's integral equation is found to be easily solvable by numerical methods, such as the orthonormal polynomials method or its modified form, Lagrange's interpolation method, and spline approximations which are discussed in this chapter. Numerical evaluation of Green's functions is another viable alternative to obtain the approximate mapping function.

12.1. Symm's Integral Equation

We will derive Symm's integral equation for both interior and exterior regions described below.

12.1.1 Interior Regions. The function $w = f(z)$ that maps a given simply connected region D with a Jordan boundary Γ onto the unit disk U such that $f(z_0) = 0$ is given by

$$w = f(z) = e^{\log(z-z_0)+g(z-z_0)+i\,h(z-z_0)}, \tag{12.1.1}$$

(see (6.2.10); also Gram [1962]), such that $\nabla^2 g = 0$, $z \in D$, which yields $g(z) = -\log\left|z - z_0\right|$, $z \in \Gamma$, and h is the conjugate of g. Without loss of generality, we will take z_0 as the origin. Then $f(z) = e^{\log z + g(x,y) + i\,h(x,y)}$, $z = x + i\,y$, where

$$\begin{aligned} &\nabla^2 g = 0, \quad z \in D, \\ &g(x,y) = -\frac{1}{2}\log\left(x^2+y^2\right), \quad z \in \Gamma. \end{aligned} \tag{12.1.2}$$

If $w = u + i\,v$, then from (12.1.1), we get

$$\begin{aligned} u(x,y) &= e^{\log|z|+g}\cos\left(\arg\{z\}+h\right), \\ v(x,y) &= e^{\log|z|+g}\sin\left(\arg\{z\}+h\right). \end{aligned} \tag{12.1.3}$$

Thus, the conformal mapping problem reduces to that of determining the harmonic functions g and h. We will represent the harmonic function $g(x,y)$ as a single-layer logarithmic potential

$$g(x,y) = \int_\Gamma \log|x-\zeta|\,\mu(\zeta)\,d\zeta, \tag{12.1.4}$$

where $\mu(\zeta)$ is a suitable source density function on the boundary Γ (see Kythe [1996:21], and Maiti [1968]). The harmonic conjugate of the representation (12.1.4) is given by

$$h(x,y) = \int_\Gamma \theta(z-\zeta)\,\mu(\zeta)\,d\zeta, \tag{12.1.5}$$

(see Jawson [1963]), where $\theta(z-\zeta) = \arg\{z-\zeta\}$. Hence, the problem further reduces to that of finding the density function $\mu(\zeta)$ such that g satisfies (12.1.2) on Γ. Once $\mu(\zeta)$ is known, the functions g and h can be determined by quadrature at any point in D. Since the function $\mu(\zeta)$ is continuous in $\bar{D}$, it satisfies the integral equation

$$\int_\Gamma \log|z-\zeta|\,\mu(\zeta)\,d\zeta = -\log|z|, \quad z \in D,\ \zeta \in \Gamma, \tag{12.1.6}$$

which is known as *Symm's integral equation* for interior regions. This equation has a unique solution provided $\operatorname{cap}(\Gamma) \neq 1$ (for the existence of the solution, see Jawson [1963]).

Note that although the function $g(x,y)$ is single-valued, but the function $h(x,y)$, in general, is multiple-valued. In fact, suppose that some Jordan contour Γ^* is contained in D such that $0 \in \operatorname{Int}(\Gamma^*)$ and $\displaystyle\int_{\Gamma^*} \frac{\partial g}{\partial n}\,ds = A$, where n denotes the inward normal to Γ^*, and $A \neq 0$ is a constant. Then, in view of the Cauchy-Riemann equations, the value of h will increase by an amount A whenever z traverses Γ^* in a clockwise direction. Hence, h will be single-valued on the contour Γ^* only if $A = 0$, i.e., for any such Γ^* we require that

$$\int_{\Gamma^*} \frac{\partial g}{\partial n}\,ds = 0, \tag{12.1.7}$$

as a condition for the function h to be single-valued.

12.1.2 Exterior Regions. For conformal mapping of the region $D^* = \operatorname{Ext}(\Gamma)$ onto the region $U^* = \{w : |w| > 1\}$ such that $f(0) = 1$ and $f(\infty) = \infty$, the mapping function $w = f_E(z)$ is unique up to a rotation. Let $C = \operatorname{diam}(D) = \lim\limits_{z\to\infty} |f'(z)|^{-1} > 0$ denote the transfinite diameter of D (see §2.1). Then the required mapping function $f_E(z)$ must satisfy the condition

$$f_E'(z) \to \frac{1}{C} \quad \text{as } z \to \infty. \tag{12.1.8}$$

Assuming that $0 \in D$, the function $\dfrac{f_E(z)}{z}$ is regular in D^*, including the point $z = \infty$. Hence, we take

$$f_E(z) = e^{\log z + \psi(z)}, \tag{12.1.9}$$

where $\psi(z)$ is regular in D^*, and in view of (12.1.8), $\psi(z) \to -\log C = \gamma$ as $z \to \infty$, where γ is Robin's constant (§2.5.2). Let $\psi(z) = \beta(z) + \gamma$. Then $\beta(z)$ is regular in D^*, and $\beta(z) \to 0$ as $z \to \infty$. Hence, $\beta(z)$ can be represented as

$$\beta(z) = \hat{g}(x,y) + i\,\hat{h}(x,y), \quad z = x + iy,$$

where $\hat{g}$ and $\hat{h}$ are conjugate harmonic functions in D^* such that $\hat{g} \to 0$ and $\hat{h} \to 0$ as $z \to \infty$. Thus,

$$f_E(z) = e^{\log z + \gamma + \hat{g}(x,y) + i\,\hat{h}(x,y)}, \tag{12.1.10}$$

where the boundary condition $\left|f_E(z)\right| = 1$ for $z \in \Gamma$ becomes

$$\gamma + \hat{g}(x,y) = -\log |z|, \quad z \in \Gamma. \tag{12.1.11}$$

As in §12.1.1, the harmonic functions $\hat{g}(x,y)$ and $\hat{h}(x,y)$ can be represented in the form (12.1.4) and (12.1.5), respectively. Then the boundary condition (12.1.11) for the density function $\mu(\zeta)$ becomes

$$\int_\Gamma \log (z-\zeta)\,\mu(\zeta)\,d\zeta + \gamma = -\log |z|, \quad z \in \Gamma, \tag{12.1.12}$$

and the condition $g(\infty) = 0$ reduces to

$$\int_\Gamma \mu(\zeta)\,d\zeta = 0. \tag{12.1.13}$$

Eqs (12.1.12) and (12.1.13) are coupled integral equations for $\mu(\zeta)$ and γ, and they have a unique solution (see Jawson [1963], and Symm [1967]). Once μ and γ are computed from (12.1.12)-(12.1.13), the mapping function $f_E(z)$ can be determined from (12.1.10). Also note that the region D^* may be mapped onto U by using the inverse transformation $z \mapsto z^{-1}$ such that the point $z = \infty$ goes into $w = 0$.

12.2. Orthonormal Polynomial Method

The orthonormal polynomial (ONP) method, developed by Rabinowitz [1966], is used to compute the density function $\mu(\zeta)$ numerically. The basic idea is to approximate $\mu(\zeta)$ by a step-function. To do this, we partition the boundary Γ into N sections $\Gamma_1, \dots, \Gamma_N$, and assume that $\mu(\zeta) \equiv \mu_j (= \text{const})$ for any point $\zeta \in \Gamma_j$, $j = 1, \dots, N$.

(a) Interior Regions. Then Eq (12.1.6) reduces to

$$\sum_{j=1}^N \left\{ \int_{\Gamma_j} \log |z-\zeta|\,|d\zeta| \right\} \mu_j = -\log |z|, \quad z \in \Gamma. \tag{12.2.1}$$

If we take $z \in \Gamma$ with each of the N nodes $z_k = x_k + i\,y_k$, $k = 1, \dots, N$, then we obtain a system of N linear equations in N unknowns μ_j:

$$\sum_{j=1}^N \left\{ \int_{\Gamma_j} \log |z_k-\zeta|\,|d\zeta| \right\} \mu_j = -\frac{1}{2} \log \left(x_k^2 + y_k^2\right), \quad k = 1, \dots, N. \tag{12.2.2}$$

The solution μ_j, $j = 1, \dots, N$, of this system gives the approximate values of g and h from (12.1.4) and (12.1.5), respectively, as

$$G(x,y) = \sum_{j=1}^N \left\{ \int_{\Gamma_j} \log |z-\zeta|\,|d\zeta| \right\} \mu_j, \quad z \in \bar{D}, \tag{12.2.3}$$

$$H(x,y) = \sum_{j=1}^N \left\{ \int_{\Gamma_j} \theta(z-\zeta)\,|d\zeta| \right\} \mu_j, \quad z \in \bar{D}, \tag{12.2.4}$$

and then the approximate mapping function $f(z)$ can be determined from (12.1.3).

The details for selecting the nodes z_j and evaluating the integrals in (12.2.3) and (12.2.4) are as follows: A convenient way to partition Γ is to take the sections Γ_j with end points $z_{j-1/2}$ and $z_{j+1/2}$ and to take the nodes z_k as any point in each Γ_j. Then, for any $z \in \bar{D}\backslash\{\Gamma_j\}$, we take

$$\int_{\Gamma_j} \log |z-\zeta|\,|d\zeta| = \frac{l_j}{6}\left[\log\left|z_k - z_{j-1/2}\right| + 4\log\left|z_k - z_j\right| \right.$$
$$\left. + \log\left|z_k - z_{j+1/2}\right|\right], \tag{12.2.5}$$

$$\int_{\Gamma_j} \theta(z-\zeta)\,|d\zeta| = \frac{l_j}{6}\left[\theta\left(z_k - z_{j-1/2}\right) + 4\,\theta\left(z_k - z_j\right) \right.$$
$$\left. + \,\theta\left(z_k - z_{j+1/2}\right)\right], \tag{12.2.6}$$

where l_j is the length of Γ_j. If Γ is a simple Jordan contour, the length l_j for each Γ_j is the arc length which can be easily evaluated analytically, and if z_j is the mid-point of Γ_j, the formulas (12.2.5) and (12.2.6) correspond to Simpson's rule. However, if the boundary Γ, in general, is analytic, the length l_j can be approximated by

$$l_j = \left|z_j - z_{j-1/2}\right| + \left|z_j - z_{j+1/2}\right|, \tag{12.2.7}$$

where z_j need not be a midpoint of Γ_j.

When $z \in \Gamma_j$, formulas (12.2.5)-(12.2.6) cannot be used because $\log|z-\zeta|$ has a singularity at $z=\zeta$ and $\theta(z-\zeta)$ is undefined at $z=\zeta$. However, in each case the integrals exist and can be computed approximately. In particular, for $z = z_j$ we take

$$\int_{\Gamma_j} \log|z_j - \zeta|\,|d\zeta| = \left|z_j - z_{j-1/2}\right|\left(\log\left|z_j - z_{j-1/2}\right| - 1\right)$$
$$+ \left|z_j - z_{j+1/2}\right|\left(\log\left|z_j - z_{j+1/2}\right| - 1\right), \tag{12.2.8}$$

and for $z = z_{j\pm 1/2}$ we take

$$\int_{\Gamma_j} \log|z_{j\pm 1/2} - \zeta|\,|d\zeta| = \left|z_{j+1/2} + z_{j-1/2}\right|\left(\log\left|z_{j+1/2} - z_{j-1/2}\right| - 1\right), \tag{12.2.9}$$

and

$$\int_{\Gamma_j} \theta\left(z_{j\pm 1/2} - \zeta\right)\,|d\zeta| = l_j\,\theta\left(z_{j\pm 1/2} - z_j\right). \tag{12.2.10}$$

Thus, formula (12.2.5) with $z = z_k$ for $j \neq k$, and formula (12.2.8) for $j = k$ give the coefficients of μ_j in (12.2.2), whereas the coefficients of μ_j in (12.2.3) and (12.2.4) are given by (12.2.5) and (12.2.6), respectively.

To estimate the error involved in this method, let $F(z)$ denote the approximate mapping function. Then, by hypothesis, $|F(z)| = 1$ at each node $z_j \in \Gamma_j$. The maximum error E_M can be estimated by computing the values of $\big||F(z)| - 1\big|$, $z \in \Gamma$, such that

$$E_M = \sup_{z\in\Gamma} \big||F(z)| - 1\big|. \tag{12.2.11}$$

In view of the maximum modulus theorem (§2.4), $|F(z) - f(z)|$ assumes its maximum value somewhere on the boundary Γ. Also for any z,

$$\begin{aligned} |F(z) - f(z)| &\le \big| |F(z)| - |f(z)| \big| + |f(z)| \, \big|\arg\{F(z)\} - \arg\{f(z)\}\big| \\ &= \big| |F(z)| - |f(z)| \big| + |f(z)| \, |H(x,y) - h(x,y)| \\ &\le \max_{z\in\Gamma} \Big\{ \big| |F(z)| - 1 \big| + |H(x,y) - h(x,y)| \Big\} \\ &\le E_M + \max_{z\in\Gamma} |H(x,y) - h(x,y)|, \end{aligned} \tag{12.2.12}$$

where we have used (12.1.5) and (12.2.4). This inequality implies that if h is known on Γ, then the absolute error in the approximate mapping function $F(z)$ can be determined at any point. But h, in general, is not known for any z, except when the region D has symmetry. In that case h is known at some points of Γ. In fact, if $\theta(z - \zeta)$ is so defined that the axes of symmetry are mapped onto themselves, then $h = 0$ on such axes. Also, we may expect maximum error at some of these points on Γ, which are, e.g., the end points of the major and minor axes of an ellipse or corners of a rectangle or square. Therefore, we take the largest value attained by $|H|$ at such points as an estimate of maximum error in $|H(z) - h(z)|$ for $z \in \Gamma$, which, in view of (12.2.12), accounts for maximum error in $\arg\{F(z)\}$, $z \in \Gamma$. We denote this estimate by E_A. Then, from (12.2.12), the sum $E_M + E_A$ gives an estimate for the upper bound on absolute error in the ONP method.

Example 12.1. Consider Cassini's oval

$$\Gamma : \left[(x+1)^2 + y^2\right] \left[(x-1)^2 + y^2\right] = a^4,$$

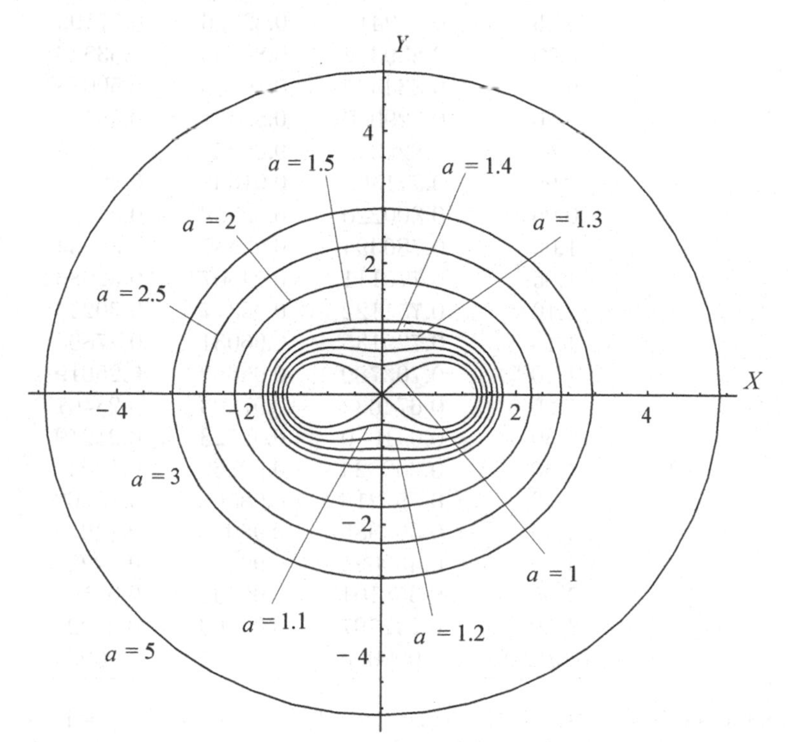

Figure 12.1 Cassini's ovals for $a = 1, 1.1, 1.2, 1.3, 1.4, 1.5, 2, 2.5, 3, 5$.

which is represented in Figure 12.1 for different values of a. The region D is not univalent for $a = 1$. Each contour Γ is symmetric about both coordinate axes, so we will consider the first quadrant. The exact mapping function is given by (Rabinowitz [1966])

$$f(z) = \frac{a\, z}{\sqrt{a^4 - 1 + z^2}}.$$

Tabular data for the approximate mapping function, using the ONP method, are given in Symm [1966] for $a = 1.2$. The functions $U(x, y)$ and $V(x, y)$ computed for $a = 1.3$ are given below in Table 12.1. ■

Table 12.1 Cassini's oval, $\alpha = 1.3$.

x	y	$U(x,y)$	$V(x,y)$
0.00	0.830662	0.00000	0.99999
0.05	0.830937	0.09541	0.99491
0.10	0.831741	0.18887	0.98201
0.15	0.833019	0.27838	0.96047
0.20	0.834681	0.36248	0.93199
0.25	0.836608	0.44009	0.89795
0.30	0.838659	0.51061	0.85981
0.35	0.840675	0.57384	0.81897
0.40	0.842488	0.62991	0.77667
0.45	0.843923	0.67921	0.73395
0.50	0.844805	0.72227	0.69161
0.55	0.844960	0.75971	0.65026
0.60	0.844218	0.79217	0.61029
0.65	0.842412	0.82026	0.57199
0.70	0.839382	0.84456	0.53546
0.75	0.834966	0.86559	0.50075
0.80	0.829006	0.88381	0.46783
0.85	0.821342	0.89963	0.43665
0.90	0.811805	0.91338	0.40709
0.95	0.800220	0.92538	0.37903
1.00	0.786394	0.93587	0.35234
1.05	0.770111	0.94507	0.32686
1.10	0.751122	0.95317	0.30245
1.15	0.729135	0.96031	0.27895
1.20	0.703789	0.96662	0.25619
1.25	0.674634	0.97223	0.23401
1.30	0.641080	0.97723	0.21219
1.35	0.602324	0.98169	0.19049
1.40	0.557216	0.98568	0.16862
1.45	0.503985	0.98973	0.14612
1.50	0.439623	0.99251	0.12225
1.55	0.358101	0.99542	0.09562
1.60	0.242597	0.99806	0.06228
1.6401219	0.000000	1.00001	0.00000

(b) EXTERIOR REGIONS. In this case we solve the coupled equations (12.1.12)-(12.1.13) by the ONP method as follows: With the partitions Γ_j, $j = 1, \dots, N$, Eq (12.1.12) reduces to the system of linear equations

$$\sum_{j=1}^{N} \Big\{ \int_{\Gamma_j} \log |z_k - \zeta| \, |d\zeta| \Big\} \mu_j + \hat{\gamma} = -\frac{1}{2} \log \left(x^2 + y^2\right), \quad k = 1, \dots, N, \tag{12.2.13}$$

where $\hat{\gamma}$ approximates γ and Eq (12.1.13) becomes

$$\sum_{j=1}^{N} \Big\{ \int_{\Gamma_j} |d\zeta| \Big\} \mu_j = 0. \tag{12.2.14}$$

Thus, Eqs (12.2.13)-(12.2.14) form a system of $(N+1)$ linear equations which are solved to determine the $(N+1)$ unknowns $\mu_1, \dots, \mu_N$ and $\hat{\gamma}$. The solution for $\hat{\gamma}$ determines the approximate transfinite diameter $\hat{C}$ of the region D, where $\hat{C} = e^{-\hat{\gamma}}$. The approximations $\hat{G}$ and $\hat{H}$ for the functions g and h are given by

$$\hat{G}(x, y) = \sum_{j=1}^{N} \Big\{ \int_{\Gamma_j} \log |z - \zeta| \, |d\zeta| \Big\} \mu_j,$$

$$\hat{H}(x, y) = \sum_{j=1}^{N} \Big\{ \int_{\Gamma_j} \arg \{z - \zeta\} \, |d\zeta| \Big\} \mu_j,$$

where $z = x + i\,y \in D^* \cup \Gamma$. Then the approximate mapping function $F_E(z)$ is computed from (12.1.10) as

$$f_E(z) \approx F_E(z) = U + i\,V = \frac{z}{\hat{C}}\, e^{\hat{G}(x,y) + i\,\hat{H}(x,y)}. \tag{12.2.15}$$

The choice of the nodes and evaluation of integrals is performed in the same manner as in part (a) above, except for the integral in (12.2.14) which is evaluated by

$$\int_{\Gamma_j} |d\zeta| \approx \left|z_j - z_{j-1/2}\right| + \left|z_j - z_{j+1/2}\right|, \tag{12.2.16}$$

where $z_{j\pm 1/2}$ denote the end points of Γ_j.

ERROR ESTIMATE. Since $f_E(\infty) = \infty$ and $f_E(z) \neq 0$ for $z \in D$, the relative error is given by

$$\frac{\left|F_E(z) - f_E(z)\right|}{\left|f_E(z)\right|} = \left|\frac{F_E(z)}{f_E(z)} - 1\right|.$$

By the maximum modulus theorem, the maximum value of this error is attained on the boundary Γ or at $z = \infty$. In view of Symm [1967], the error at $z = \infty$ is considerably less than the maximum error on Γ. Thus, for $z \in \Gamma$

$$\left|\frac{F_E(z)}{f_E(z)} - 1\right| = \left|\left|F_E(z)\right| e^{i(\hat{H} - h)} - 1\right| \le \left|\left|\frac{F_E(z)}{f_E(z)}\right| - 1\right| + \left|\hat{H} - h\right|.$$

Since $\left|\hat{H} - h\right|$ is generally not known on Γ, although it is of the same order as $\left|\hat{G} - g\right|$, and since, by hypothesis, $\left|\left|F_E(z_j)\right| - 1\right| = 0$, the maximum error E in $F_E(z)$ can be measured by

$$E = \max_j \left|\left|F_E\left(z_{j-1/2}\right)\right| - 1\right|. \tag{12.2.17}$$

Example 12.2. Consider the ellipse $x^2/a^2 + y^2 = 1$, $a = 1.5$, and $N = 100$ in the first quadrant because of the symmetry about both axes. The values of $U(x, y)$ and $V(x, y)$ are given in Table 12.2 below. The transfinite diameter $C = (a+1)/2$ (Pólya and Szegö [1951]), and the exact mapping function is given by (Phillips [1966])

$$f_E(z) = \frac{z + \sqrt{z^2 - a^2 + 1}}{a + 1}. \blacksquare$$

Table 12.2 Ellipse, $a = 1.5$.

x	y	$U(x,y)$	$V(x,y)$
0.00	1.000000	0.00000	0.88989
0.05	0.999444	0.03633	0.88936
0.10	0.997775	0.07267	0.88774
0.15	0.994987	0.10903	0.88503
0.20	0.991071	0.14542	0.88123
0.25	0.986013	0.18184	0.87633
0.30	0.979796	0.21831	0.87032
0.35	0.972397	0.25483	0.86317
0.40	0.963789	0.29142	0.85487
0.45	0.953939	0.32808	0.84540
0.50	0.942809	0.36483	0.83472
0.55	0.930352	0.40167	0.82278
0.60	0.916515	0.43862	0.80961
0.65	0.901234	0.47567	0.79509
0.70	0.884433	0.51285	0.77918
0.75	0.866025	0.55019	0.76184
0.80	0.845905	0.58761	0.74296
0.85	0.823947	0.62521	0.72248
0.90	0.800000	0.66296	0.70025
0.95	0.773879	0.70088	0.67614
1.00	0.745356	0.73896	0.64997
1.05	0.714143	0.77722	0.62152
1.10	0.679869	0.81565	0.59048
1.15	0.642045	0.85425	0.55647
1.20	0.600000	0.89303	0.51892
1.25	0.552771	0.93198	0.47703
1.30	0.498888	0.97110	0.42959
1.35	0.435890	1.01039	0.37452
1.40	0.359011	1.04983	0.30778
1.45	0.256038	1.08942	0.21902
1.50	0.000000	1.12915	0.00000

12.3 Modified ONP Method

The ONP method, discussed in §12.2, for numerically solving Eqs (12.1.4), (12.1.5), and (12.1.6) has been modified by Hayes, Kahaner and Kellner [1972] as follows: Let the parametric representation of the boundary Γ be $\zeta = \zeta(t) = x(t) + i\,y(t)$, $0 \le t \le L$. Then Eq

(12.1.4) and (12.1.5) can be written as

$$g(z) = \int_0^L \mu(t) \log |z - \zeta(t)| \, dt, \tag{12.3.1}$$

$$h(z) = \int_0^L \mu(t) \arg \{z - \zeta(t)\} \, dt, \tag{12.3.3}$$

where $\mu(t) = \mu(\zeta(t))$ and $z \in \bar{D}$. We will assume that $\mu(t) \in C^3(-\infty, +\infty)$ and the boundary Γ has no corners (for corner singularities, see §12.5 and §12.2). We partition Γ into an even number n of uniform sections Γ_j, $j = 1, \dots, n$, of length $\alpha = L/n$ each. Then the nodes on the section Γ_j are uniform as regards the arc length of each Γ_j. First, we define a set of piecewise polynomials $p_j(t)$, $j = 1, \dots, n$, by

$$\begin{aligned} p_1(t) &= \begin{cases} \dfrac{1}{2\alpha^2}(t+\alpha)(t+2\alpha), & -2\alpha \le t \le 0, \\ \dfrac{1}{2\alpha^2}(t-\alpha)(t-2\alpha), & 0 \le t \le 2\alpha, \\ 0, & \text{otherwise}, \end{cases} \\ p_2(t) &= \begin{cases} -\dfrac{1}{2\alpha^2} t(t-2\alpha), & 0 \le t \le 2\alpha, \\ 0, & \text{otherwise}, \end{cases} \\ p_{2k+1}(t) &= p_1(t - 2k\alpha), \quad k = 1, 2, \dots, \frac{n}{2} - 1, \\ p_{2m}(t) &= p_2\left(t - 2(m-1)\alpha\right), \quad m = 1, 2, \dots, \frac{n}{2}. \end{aligned} \tag{12.3.3}$$

The graphs of some of these polynomials are presented in Figure 12.2 for $\alpha = 0.1$.

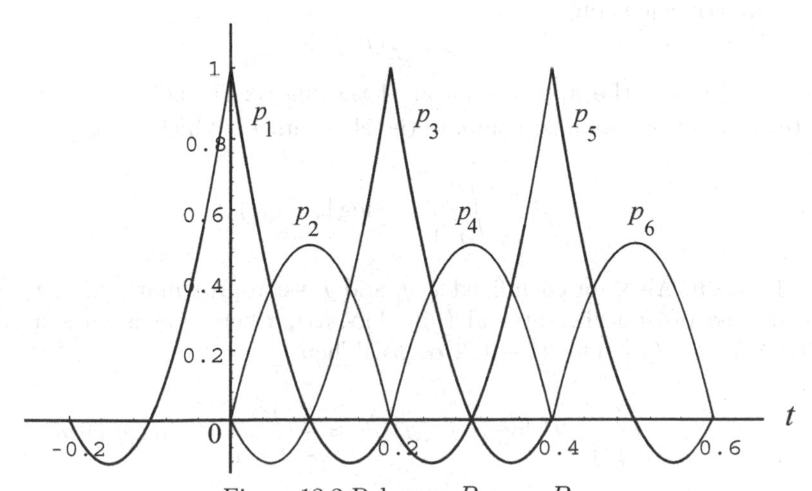

Figure 12.2 Polygons $P_1, \dots, P_6$.

Next, we define the function

$$\tilde{\mu}(t) = \sum_{j=1}^{n} \mu(j\alpha) \, p_j(t). \tag{12.3.4}$$

The following results hold:

(i) $\tilde{\mu}(t)$ is a polynomial of degree two on $[j\alpha, (j+2)\alpha]$, $j = 0, 2, 4, \dots, n-2$.

(ii) $\tilde{\mu}(t) = \mu(t)$ at $t = j\alpha, j = 0, 1, 2, \dots, n$.

$$(iii)\ \mu(t) = \tilde{\mu}(t) + O\left(\alpha^3\right) = \sum_{j=1}^{n} \mu_j\, p_j(t) + O\left(\alpha^3\right), \tag{12.3.5}$$

where $\mu_j = \mu(j\alpha)$ for $j = 1, \dots, n$. Using the approximation (12.3.5) for $\mu(t)$ in (12.3.1), we obtain an approximation for g as

$$\sum_{j=1}^{n} \mu_j \int_0^L p_j(t) \log|z - \zeta(t)|\, dt = g(z) + O\left(\alpha^3\right). \tag{12.3.6}$$

The function $g(z) = -\log|z - z_0|$ for $z \in \Gamma$. Thus, we can evaluate Eq (12.3.6) at the points $z = j\alpha$ for $j = 1, \dots, n$, which yields a system of n linear equations with constant coefficients for the unknowns $\mu_1, \mu_2, \dots, \mu_n$. Set $A = (a_{jk})$ and $B = (b_j)$, where

$$\begin{aligned} a_{jk} &= \int_0^L p_j(t) \log|\zeta(j\alpha) - \zeta(t)|\, dt \quad \text{for } j, k = 1, \dots, n\,, \\ b_j &= -\log|\zeta(j\alpha) - z_0|, \quad \text{for } j = 1, \dots, n. \end{aligned} \tag{12.3.7}$$

Using this matrix notation, Eq (12.3.6) becomes the linear system

$$A\,\mu = B + O\left(\alpha^3\right), \tag{12.3.8}$$

where $O\left(\alpha^3\right)$ is a vector (column matrix). Thus, each component of Eq (12.3.8) is bounded by $O\left(\alpha^3\right)$, and $\mu = \left(\mu_1, \dots, \mu_n\right)^T$, where T denotes the transpose of a matrix. Hence, we solve the matrix equation

$$\tilde{A}\,\tilde{\mu} = B, \tag{12.3.9}$$

where $\tilde{A} = (\tilde{a}_{jk})$ is the approximation of the matrix A and $\tilde{\mu}$ that of the vector μ. To evaluate $\tilde{A}$, however, we must compute the elements $\tilde{a}_{jk}$ which contain integrals of the form

$$\int_{(j-1)\alpha}^{j\alpha} t^j \log|z - \zeta(t)|\, dt \tag{12.3.10}$$

for $j = 1, \dots, n$. Also, for each fixed x, y, and j, we approximate $|z - \zeta(t)|$ by a polynomial $q(t)$ of degree two on the interval $((j-1)\alpha, j\alpha)$, where $q(t)$ is chosen such that $q(t) = |z - \zeta(t)|^2$ for $t = (j-1)\alpha,\ (j-1/2)\alpha,\ j\alpha$. Then

$$\int_{(j-1)\alpha}^{j\alpha} t^j \log|z - \zeta(t)|\, dt \approx \frac{1}{2} \int_{(j-1)\alpha}^{j\alpha} t^j \log\left[q(t)\right] dt, \tag{12.3.11}$$

which can be evaluated explicitly. In some particular cases, e.g., when $|z - \zeta(t)| = 0$ on $[(j-1)\alpha, j\alpha]$, the integral (12.3.10) is computed with a polynomial of higher order.

Then the matrix equation (12.3.9) is solved for the vector $\tilde{\mu} = \left(\tilde{\mu}_1, \dots, \tilde{\mu}_n\right)^T$ which is the approximation for the density function μ, with an error estimate

$$\|\tilde{\mu} - \mu\| \le \|\left(\tilde{A}^{-1} - A^{-1}\right) B\| + \|A^{-1}\|\, O\left(\alpha^3\right), \tag{12.3.12}$$

where the error due to the term $\tilde{A}^{-1} - A^{-1}$ seldom dominates the $\|A^{-1}\|\, O(\alpha^3)$ term, an observation made by Hayes et al. [1972] from certain examples. Once $\tilde{\mu}$ is computed, the functions $g(z)$ and $h(z)$ are obtained from

$$g(z) = \int_0^L \mu(t) \log|z - \zeta(t)|\, dt = \sum_{k=1}^{n} \tilde{\mu}_k \int_0^L p_k(t) \log|z - \zeta(t)|\, dt, \tag{12.3.13}$$

and

$$\begin{aligned} h(z) &= \int_0^L \mu(t) \arg\{z - \zeta(t)\}\, dt \\ &\approx \sum_{j=1}^{n} \mu(j\alpha) \int_0^L p_j(t) \arg\{z - \zeta(t)\}\, dt \\ &= \sum_{j+1}^{n} \mu(j\alpha)\, \left[p_j(t)\, \eta_1(t) - p_j'(t)\, \eta_2(t) + p_j''(t)\, \eta_3(t)\right], \end{aligned} \tag{12.3.14}$$

by integration by parts, where

$$\eta_1(t) = \int_0^L \arg\{z - \zeta(t)\}\, dt, \quad \eta_2(t) = \int_0^L \arg\{z - \zeta(t)\}\, dt,$$
$$\eta_3(t) = \int_0^L \arg\{z - \zeta(t)\}\, dt.$$

Note that $p_j''(t)$ is constant. Thus, (12.3.14) involves integrals of the form (12.3.10).

Hayes et al. [1972] wrote a Fortran IV computer program for this method. This program can easily be adapted to any modern operating system. As examples, they investigated cases of Cassini's oval (Example 12.1; Example 12.11), an ellipse (Example 12.13), a rectangle (Example 12.12), a limaçon (Example 12.9), and an isosceles triangle (Example 12.3 below).

Example 12.3. Consider an isosceles triangle with corners at the points $(0, 1)$, $(2, -1)$ and $(-2, 1)$ such that the point $(0, 0)$ goes into the point $w = 0$ under the conformal mapping $w = f(z)$. The partition is taken with an equal number of nodes on each side. The error E_M, defined by (12.2.11) is found as follows:

$$E_m = \begin{cases} 2 \times 10^{-4} & \text{for } n = 17, \\ 2 \times 10^{-5} & \text{for } n = 33, \\ 10^{-6} & \text{for } n = 65. \ \blacksquare \end{cases}$$

12.4. Lagrange Interpolation

The choice of a suitable basis set used in the RM, BKM or ONP method is an important aspect of any numerical technique for polynomial approximation of the mapping function. Gautschi [1977; 1978; 1979] investigated numerical conditions of various bases for polynomial approximation on the real axis. Reichel [1985] observed that a well-conditioned basis depends on the shape of the simply connected region D bounded by a Jordan contour Γ. For example, the monomials $\phi_j(z) = (z/r)^j$, $j = 0, 1, \ldots$, are well-conditioned for disks $B(0, r)$, but this basis becomes ill-conditioned for ellipses (see Example 12.4 below where a criterion for well-conditioned bases is developed and applied to these cases).

Example 12.4. (*Gautschi criterion*). Let $p_j(z)$, $j = 0, 1, \dots, n$, denote polynomials such that span $\{p_j\}_{j=0}^{n} =$ span $\{z^j\}_{j=0}^{n}$. We will determine the sensitivity of the functions

$$P_n(z) = \sum_{j=0}^{n} a_j\, p_j(z), \quad z \in \bar{D}, \tag{12.4.1}$$

subject to the perturbations in the coefficients a_j. Let $M_n : \mathcal{C}^{n+1} \to \Pi_n$ denote the mapping of the coefficient space $\mathcal{C}^{n+1}$ onto the space Π_n of polynomials of degree $\le n$. Let $\mathbf{a} = (a_0, a_1, \dots, a_n) \in \mathcal{C}^{n+1}$ denote a vector. Then we define

$$(M_n \mathbf{a})(z) = \sum_{k=0}^{n} a_k\, p_k(z), \quad z \in \bar{D}. \tag{12.4.2}$$

Note that $M_n^{-1} P_n(z) = \mathbf{a}$. The maximum norm in $\mathcal{C}^{n+1}$ and Π_n is defined, respectively, by $\|\mathbf{a}\|_\infty = \max\limits_{0 \le k \le n} |a_n|$, and $\|\Pi_n\|_\Gamma = \sup\limits_{z \in \Gamma} |P_n(z)|$. Let $\|M_n\|$ and $\|M_n^{-1}\|$ be the induced operator norms. We are interested in determining how the condition of the map M_n defined by

$$\text{cond } (M_n) = \|M_n\|\, \|M_n^{-1}\| \tag{12.4.3}$$

grows with n for different choices of the polynomial $p_j(z)$. We will examine two choices:
(a) Let D be the unit disk U and $p_j(z) = z^j$, $j = 0, 1, \dots$. Then

$$\|M_n\| = \max_{\|\mathbf{a}\|_\infty = 1} \Big\| \sum_{j=0}^{n} a_j\, z^j \Big\|_\Gamma = n + 1,$$

and

$$\|M_n^{-1}\| = \left(\min_{\|\mathbf{a}\|_\infty = 1} \Big\| \sum_{j=0}^{n} a_j\, z^j \Big\|_\Gamma \right)^{-1} \ge 1.$$

Since

$$\|P_n\|_\Gamma \ge \sqrt{\frac{1}{2\pi} \int_0^{2\pi} \left| P_n\left(e^{i\theta}\right) \right|^2 d\theta} = \sqrt{\sum_{j=0}^{n} |a_n|^2},$$

we find that

$$\|M_n^{-1}]\| \le \left(\min_{\|\mathbf{a}\|_\infty = 1} \sqrt{\sum_{j=0}^{n} |a_j|^2} \right)^{-1} = 1.$$

Hence, cond $(M_n) = n+1$, which shows that the monomial basis $\{z^j\}_{j=0}^{n}$ is well-conditioned for the unit disk. Thus, this basis is well-conditioned for the disk $B(0, r)$.
(b) Let Γ denote the ellipse $E(a, b) = \{(x, y) : x^2/a^2 + y^2/b^2 = 1\}$, $z = x + i\, y$, and $a \ge b$. Using the scaling factor $1/a$, this ellipse is transformed into the ellipse $E(1, b/a) = \{Z : Z = z/a\}$ which has foci at $\pm\xi$, where $\xi = \sqrt{1 - (b/a)^2}$. We use the Chebyshev polynomials T_n of the first kind

$$\begin{aligned} T_n(Z, \xi) = \frac{\xi^n}{2^{n-1}} T_n(Z/\xi) = \frac{1}{2^n} \Big[\left(Z + \sqrt{Z^2 - \xi^2} \right)^n \\ + \left(Z - \sqrt{Z^2 - \xi^2} \right)^n \Big], \quad n = 0, 1, \dots. \end{aligned} \tag{12.4.4}$$

Let the points $Z \in E(1, b/a)$ be taken as

$$Z = \frac{1}{2}\left[\left(1+\frac{b}{a}\right) e^{i\theta} + \left(1-\frac{b}{a}\right) e^{-i\theta}\right], \quad 0 \le \theta \le 2\pi.$$

Substituting them in (12.4.4), we obtain

$$\begin{aligned} T_n(Z,\xi) &= \left[\frac{1}{2}\left(1+\frac{b}{a}\right)\right]^n e^{in\theta} + \left[\frac{1}{2}\left(1-\frac{b}{a}\right)\right]^n e^{-in\theta} \\ &\sim \left[\frac{1}{2}\left(1+\frac{b}{a}\right)\right]^n e^{in\theta} \quad \text{as } n \to \infty. \end{aligned} \tag{12.4.5}$$

Let $\beta^{(n)} = \left(\beta_0^{(n)}, \beta_1^{(n)}, \dots, \beta_{[n/2]}^{(n)}\right)^T$ denote the coefficient vector of $T_n(Z,\xi)$, i.e.,

$$T_n(Z,\xi) = \sum_{k=0}^{[n/2]} \beta_k^{(n)} Z^{n-2k}, \tag{12.4.6}$$

where $[n/2]$ is the greatest integer $\le n/2$. If $\xi = 0$ (the case of the circle $|z| = a$), then $\beta_0^{(n)} = 1$ and $\beta_k^{(n)} = 0$ for all $k \ge 1$. We will not discuss this case because it has been examined in part (a) above. Let $\xi > 0$. Then

$$\|M_n\|^{-1} = \left[\min_{\|\mathbf{a}\|_\infty = 1} \|\sum a_k Z^k\|_{E(1,b/a)}\right]^{-1} \ge \frac{\|\beta^{(n)}\|_\infty}{\|T_n\|_{E(1,b/a)}}. \tag{12.4.7}$$

From Gautschi [1979] we have

$$\|\beta^{(n)}\| \sim \sqrt{\frac{2}{n\pi}}\, \frac{\left(1+\xi^2\right)^{3/4}}{\xi} \left(\frac{1+\sqrt{1+\xi^2}}{2}\right)^n \quad \text{as } n \to \infty. \tag{12.4.8}$$

Hence, from (12.4.5), (12.4.7) and (12.4.8) we find that

$$\|M_n^{-1}\| \ge \sqrt{\frac{2}{n\pi}}\, \frac{\left(1+\xi^2\right)^{3/4}}{\xi} \left(\frac{1+\sqrt{1+\xi^2}}{1+\dfrac{b}{a}}\right)^n \quad \text{as } n \to \infty.$$

Since $\|M_n\| = n+1$, we have

$$\text{cond}\,(M_n) \ge \sqrt{\frac{2n}{\pi}}\, \frac{\left(1+\xi^2\right)^{3/4}}{\xi} \left(\frac{1+\sqrt{1+\xi^2}}{1+\dfrac{b}{a}}\right)^n \quad \text{as } n \to \infty.$$

If we set $\rho = b/a$ and

$$F(\rho) = \frac{1+\sqrt{1+\xi^2}}{1+\dfrac{b}{a}} = \frac{1+\sqrt{2-\rho^2}}{1+\rho},$$

then

$$F(0) = 1 + \sqrt{2}, \quad \text{which corresponds to } \Gamma = [-a, a],$$
$$F(1) = 1,$$
$$F'(\rho) < 0, \quad 0 \le \rho \le 1,$$

which imply that in the case of a nondegenerate ellipse ($\xi > 0$) the condition number cond (M_n) increases exponentially with n for the monomial basis and the growth rate increases with ρ. Hence, if the region is not circular or nearly circular, the choice of the monomial basis is ill-conditioned for the ellipse and may produce computational inaccuracy. ■

Reichel [1985] has shown that Lagrange's interpolation functions

$$l_k(z) = \prod_{\substack{j=0 \\ j \ne k}}^{n} \frac{z - z_j}{z_k - z_j}, \quad k = 0, 1, \dots, n, \tag{12.4.9}$$

where z_k are Fejér points on the boundary Γ, provide a well-conditioned basis which is simple to compute for boundaries that are not circular or nearly circular. In fact, he has proved the following result:

Theorem 12.1. *For Lagrange's functions $l_k(z)$, defined by (12.4.9), the condition number of M_n is*

$$cond\ (M_n) \le \frac{2}{\pi} \log n + \alpha, \tag{12.4.10}$$

where the constant α depends on the shape of the analytic boundary Γ.

Let $w = f_E(z)$ map the region Ext (Γ) conformally onto $U^* = \{w : |w| > 1\}$ such that $f_E(\infty) = \infty$ and $f_E(z_1) = 1$, where z_1 is an arbitrary point on Γ. By analytic continuation the function $f_E(z)$ can be continued to a bijective map from $D^* \cup \Gamma$ onto $|w| \ge 1$. The points z_k, $k = 1, \dots, n$, are called the *Fejér points* if

$$f_E(z_k) = e^{2i(k-1)\pi/n}, \quad k = 1, \dots, n. \tag{12.4.11}$$

To determine a set of Fejér points $\{z_k\}_{k=1}^{n}$ to be used in (12.4.9), we must restrict f_E to Γ. This restriction, however, leads to a *modified Symm's integral equation* which has a unique solution, is solvable for all scalings of the boundary Γ, and differs only in their right-hand sides from Eqs (12.1.12)-(12.1.13) in the case of exterior regions (and from Eq (12.1.6) in the case of interior regions). For exterior regions we have the following result (Reichel [1985]):

Theorem 12.2. *Let $\phi(z) = \arg\{f_E(z)\}$, and let $\gamma = cap\,(\Gamma)$. Then the unique solution (C^*, μ^*) of the modified Symm's integral equations*

$$\begin{aligned} &\int_\Gamma \log|z - \zeta|\,\mu(\zeta)\,|d\zeta| + q = 0, \quad z \in \Gamma, \\ &\int_\Gamma \mu(\zeta)\,|d\zeta| = 1, \end{aligned} \tag{12.4.12}$$

where $\mu^* = \mu^*(s) \in L^2(\Gamma)$ and C^* is a constant, satisfies

$$\phi(z) = 2\pi \int_{z_1}^{z} \mu^*(\zeta)\,|d\zeta|, \quad C^* = -\log\gamma, \tag{12.4.13}$$

where integration in (12.4.13) is carried out in the positive direction.

PROOF. Let the boundary Γ have the parametric representation $s \mapsto \zeta(s)$, $0 \leq s < L$, $\zeta(0) = z_1$, where s is the arc length of Γ. Then the boundary correspondence function $\phi(\zeta)$ may be regarded as a function of s, i.e., we write $\phi(s) = \phi(\zeta(s))$. It is shown in Gaier [1976] that for a rectifiable curve Γ and $\gamma \neq 1$, the integral equation

$$\int_\Gamma \log |z - \zeta| \, \mu(\zeta) \, |d\zeta| = 1 \tag{12.4.14}$$

has a.e. a unique, integrable solution

$$\mu(\zeta) = \frac{1}{2\pi} \frac{1}{\log \gamma} \phi'(s), \quad \zeta = \zeta(s), \tag{12.4.15}$$

where the prime denotes differentiation with respect to s. Thus, Eq (12.4.12) has a unique solution

$$\begin{aligned} \mu^*(\zeta) &= \frac{1}{2\pi} \phi'(s), \quad \zeta = \zeta(s), \\ C^* &= -\log \gamma. \end{aligned} \tag{12.4.16}$$

It is also known (Reichel [1985]) that Eq (12.4.12) has a unique solution for any scaling of Γ, and that μ^* is invariant under scaling and C^* varies continuously with scaling. Hence, (12.4.16) is also a unique solution for $\gamma = 1$. ∎

Let $w = f(z)$ map the region Int (Γ) conformally onto the unit disk U. Then for interior regions we have the following result (Reichel [1985]):

Theorem 12.3. *Let $\phi_j(z) = \arg\{f(z)\}$ for $z \in \Gamma$ with $\phi_j(z_1) = 0$. Then the unique solution (C_j^*, μ_j^*) of the system of modified Symm's equations*

$$\begin{aligned} &\int_\Gamma \log |z - \zeta| \, \mu_j(\zeta) \, |d\zeta| + C_j^* = \log |z|, \quad z \in \Gamma, \\ &\int_\Gamma \mu_j(\zeta) \, |d\zeta| = 1, \end{aligned} \tag{12.4.17}$$

where $\mu_j^* = \mu_j^*(z) \in L^2(\Gamma)$ and C_j^* are constants, satisfies

$$\phi_j(z) = 2\pi \int_{z_1}^{z} \mu_j^*(\zeta) \, |d\zeta|, \quad C_j^* = 0, \tag{12.4.18}$$

where the integration in (12.4.18) is taken in the positive direction of Γ.

PROOF. As in the above proof, we will use Gaier's result. Let $\gamma = \text{cap}(\Gamma)$, and assume that $\gamma \neq 1$. Then

$$\int_\Gamma \log |z - \zeta| \, \mu(\zeta) \, |d\zeta| = \log |z|, \quad z \in \Gamma, \tag{12.4.19}$$

has a unique solution $\mu^*(\zeta)$ such that $\int_\Gamma \mu^*(\zeta) \, |d\zeta| = 1$. Thus, $\mu_j^* = \mu^*$, and $C^* = 0$ is a solution of (12.4.17), and, as in the proof of the previous theorem, this solution is unique and invariant under scaling. Hence, (12.4.18) holds for boundaries for which $\gamma = 1$. ∎

To approximate the mappings $f(z)$ and $f_E(z)$ numerically, the following algorithm is used:

1. Compute $\phi(z)$ and $\phi_j(z)$ as defined in Theorems 12.2 and 12.3.
2. Determine the $(n+1)$ Fejér points z_k by solving the system of equations

$$\phi(z_k) = \frac{2k\pi}{n+1}, \quad k = 0, 1, \dots, n.$$

3. Compute the images w_k of z_k under f by evaluating

$$w_k = e^{i\phi_j(z_k)}, \quad k = 0, 1, \dots, n.$$

4. Determine a polynomial approximation $p_k(z)$ of $\dfrac{f(z)}{z}$ of degree $\le n$ by interpolating $\dfrac{f(z)}{z}$ at the Fejér points z_k, $k = 0, 1, \dots, n$.

Then an approximation of $f(z)$ is given by

$$f_{n+1}(z) = z\, p_n(z). \tag{12.4.20}$$

The accuracy of f_{n+1} depends on that of $\phi_j(z)$, $\phi(z)$ and the interpolation error. Reichel [1985] has found no computational problems with polynomials $p_j(z)$ of degree 80-100, since the basis with Lagrange's interpolation functions is well-conditioned.

Example 12.5. Consider the region D bounded by the contour $\Gamma = \{z = 2\cos t + i\left(\sin t + 2\cos^3 t\right),\ 0 \le t < 2\pi\}$ (see Figure 12.3), where the Fejér points are marked with dots. The following data for the error $E = \||f_n(z)| - 1\|_\Gamma$ is from Reichel [1985]:

n	Basis	E
16	Monomial	3×10^{-2}
32	Monomial	3×10^{-3}
64	Lagrange	9×10^{-6}

and cond $(M_{32}) = 5 \times 10^9$. ■

Example 12.6. Consider the ellipse $\Gamma = \{z = \cos t + i\, b \sin t,\ 0 \le t < 2\pi\}$. The mapping function $f(z)$ onto U with $f(0) = 0$ is known in terms of the elliptic sine function (see Kober [1957:177])

$$f(z) = \sqrt{k}\,\mathrm{sn}\left(\frac{2K}{\pi}\,\sin^{1}\frac{z}{\sqrt{1-b^2}}\right),$$

which shows that the singularities of $f(z)$ close to the boundary are the poles at the points $\zeta_{1,2} = \pm\dfrac{2ib}{\sqrt{1-b^2}}$. We use a Möbius transformation f_M (§3.2), with $f_M(0) = 0$, such that $f_M(z)$ maps the circle of curvature through ib onto the unit circle. It has a pole at $\zeta_1^* = 2ib\,\dfrac{1-b^2/2}{1-b^2}$. Note that $\zeta_1^* = \zeta_1 + O\left(b^4\right)$ as $b \to 0$. Hence, we approximate the function $f(z)\left(z - \zeta_1^*\right)\left(z - \bar{\zeta}_1^*\right)$ by the polynomials defined with Lagrange's interpolation functions as the basis. The nature and location of singularities adjacent to the boundary are studied in detail in Chapter 13. The use of the Möbius transformation in such cases generally provides good approximations of the singularities by locally approximating f by f_M. ■

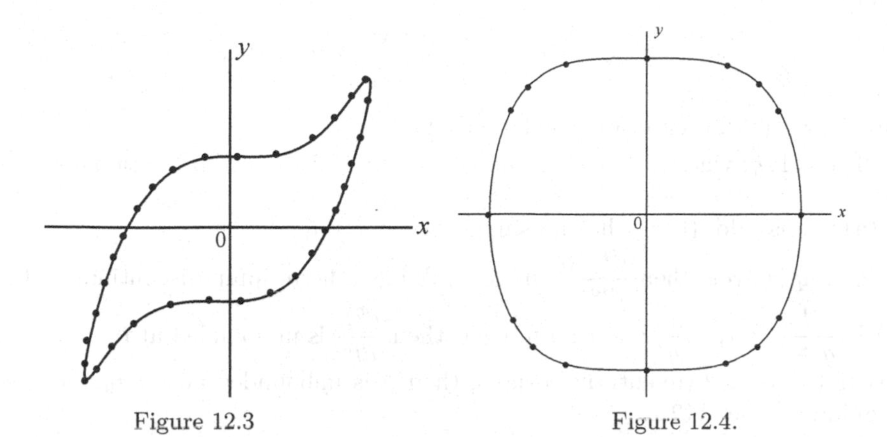

Figure 12.3 Figure 12.4.

Example 12.7. Consider the region (square with round corners) bounded by $\Gamma = \{z = x + i\,y,\ x^4 + y^4 = 1\}$ (Figure 12.4, where 16 Fejér points are marked). In this case the error $E = \| |f_{16}(z)| - 1 \|_\Gamma = 6 \times 10^{-4}$ (see Reichel [1985]). ■

12.5. Spline Approximations

Spline functions of various degrees can be used effectively to approximate the source density $\mu(\zeta)$ at corner singularities of the boundary Γ. In fact, the singular functions are combined with splines and together they approximate μ over the entire boundary. In previous sections we have evaluated μ and the functions g and h to approximate the mapping function $w = f(z)$ for any point $z \in \bar{D}$. If $f(z)$ is required only at the boundary points, then the function h can be determined by simple integration of μ as follows: Let ϕ denote the boundary correspondence function defined by

$$\phi(t) = \arg\{f(\zeta(t)\} = \frac{1}{2i}\,(\log w - \log \bar{w}), \quad z \in \Gamma. \tag{12.5.1}$$

Then, since $|w| = 1$ for $z \in \Gamma$, we have (see Gautschi [1976])

$$\mu(t) = -\frac{1}{2\pi}\frac{d\phi}{dt} = \frac{1}{4i\pi}\left(w\,\frac{\partial \bar{w}}{\partial t} - \bar{w}\,\frac{\partial w}{\partial t}\right). \tag{12.5.2}$$

Let z_0 be a corner of the polygonal boundary Γ with interior angle $\alpha\pi$, $0 < \alpha < 2$. Then the Schwarz-Christoffel formula (7.1.2) implies that in a neighborhood of z_0

$$f(z) = f(z_0) + \sum_{j=1}^{\infty} A_j\,(z - z_0)^{j/\alpha}. \tag{12.5.3}$$

For any point $z = \zeta(t)$ on adjacent sides of the polygon at a corner point $z_0 = \zeta(t_0)$, we may take

$$z - z_0 = \begin{cases} t - t_0, & \text{if } t \ge t_0, \\ (t - t_0)\,e^{i\alpha\pi}, & \text{if } t < t_0. \end{cases} \tag{12.5.4}$$

Then (12.5.2)-(12.5.4) give

$$\mu(t) = \begin{cases} \sum_{j=1}^{\infty} a_j\,(t - t_0)^{-1+j/\alpha}, & t > t_0, \\ \sum_{j=1}^{\infty} (-1)^{j+1}\,a_j\,(t_0 - t)^{-1+j/\alpha}, & t < t_0, \end{cases} \tag{12.5.5}$$

where

$$a_j = \frac{1}{2\alpha\pi} \Im\Big\{ \sum_{k=1}^{j} k\, A_{j-k}\, \bar{A}_k \Big\}, \tag{12.5.6}$$

and $A_0 = f(z_0)$. Hence, we conclude that

(i) If $\alpha = 1/q$, where $q \ge 1$ is an integer, then (12.5.5) has no fractional powers of $(t - t_0)$, and

(a) if q is odd, then μ has no singularity at $t = t_0$;

(b) if q is even, then $\dfrac{d^{q-1}\mu}{dt^{q-1}}$, in general, has a finite jump discontinuity at $t = t_0$;

(ii) If $\dfrac{1}{q+1} < \alpha < \dfrac{1}{q}$, $q \ge 1$ an integer, then $\dfrac{d^q \mu}{dt^q}$ is unbounded at $t = t_0$;

(iii) If $1 < \alpha < 2$ (re-entrant corner), then μ is unbounded at $t = t_0$. For more on corner singularities, see §12.2.

The spline approximation of μ is generally carried out by using spline functions of degree n for most boundaries. However, in a neighborhood of a corner point z_0 with interior angle $\alpha\pi$ where μ exhibits the behavior mentioned in (i)-(iii) above, we use either of the two schemes: (a) continue the spline approximation through z_0 or (b) use a special function which reflects the known asymptotic behavior of μ near z_0 to approximate μ in the neighborhood of this point. In the latter scheme the special functions are combined with the spline approximation subject to appropriate continuity conditions through the point z_0. This yields an augmented spline of degree n. The choice between scheme (a) or (b) depends on the nature and location of the singularity of μ and on the degree of the splines used to approximate μ over the rest of the boundary. Thus, for example, scheme (b) should always be used if $1 < \alpha < 2$ (case (iii) above). However, if $\alpha = 1/2$ (case (i)(b) above) where there is a jump discontinuity in $\dfrac{d\mu}{dt}$, scheme (a) for splines of degree 0 or 1 and scheme (b) for splines of degree $n \ge 2$ should be used.

The details of the numerical method for augmented spline approximations are as follows (Hough and Papamichael [1981]): Assume that μ is an L-periodic function of the real parameter t. Let the corner point z_0, where scheme (b) is to be used, be designated as a singular corner point, and assume that there are $N \ge 1$ such points with arc lengths $t = \tau_m$, $m = 1, \dots, N$, where $0 < \tau_1 < \dots < t_n < L$. Let each $[\tau_m, \tau_{m+1}]$ for $m = 1, \dots, N$ be divided into $(k_m + 3)$ intervals, $m = 1, \dots, N$, where $k_m \ge 0$ and $\tau_{N+1} = \tau_1 + L$. The end points of these intervals are denoted by t_{mj}, where

$$\tau_m = t_{m0} < t_{m1} < \dots < t_{m,k_m+3} = \tau_{m+1}, \quad m = 1, \dots, N. \tag{12.5.7}$$

We will also use the notation

$$t_{1,-1} = t_{N,k_N+2} - L, \qquad t_{m,-1} = t_{m-1,k_{m-1}+2}, \quad m = 1, \dots, N. \tag{12.5.8}$$

We will assume that every corner point on the boundary Γ is an end point in the partition (12.5.7) which may not be uniform (i.e., may have unequal arc lengths). Then the source density μ is approximated by

$$\tilde{\mu}(t) = \begin{cases} r_m(t), & t_{m,-1} < t < t_{m1}, \\ s_m(t), & t_{m1} < t < t_{m,k_m+2}, \end{cases} \tag{12.5.9}$$

where $r_m(t)$ is an appropriate singular function which depends on the nature of the boundary singularity, and $s_m(t)$ is a spline of degree n with knots t_{mj}, $j = 1, \dots, k_m + 2$. Thus, for example, if the singular corner point at $\tau_m = t_{m0}$ has an interior angle $\alpha_m \pi$, $0 < \alpha_m < 2$, then the series (12.5.5) for μ is truncated, and we use

$$r_m(t) = \begin{cases} \sum_{j=1}^{n_m} a_{mj}\,(t - t_{m0})^{-1+j/\alpha_m}, & t_{m0} \le t < t_{m1}, \\ \sum_{j=1}^{n_m} (-1)^{j+1} a_{mj}\,(t_{m0} - t)^{-1+j/\alpha_m}, & t_{m,-1} < t \le t_{m0}, \end{cases} \tag{12.5.10}$$

and

$$s_m(t) = \sum_{j=0}^{n} b_{mj}\,(t - t_{m1})^j + \sum_{j=2}^{k_m+1} c_{mj}\,(t - t_{mj})^n\,\chi\,(t - t_{mj}), \tag{12.5.11}$$

where

$$\chi\,(t - t_{mj}) = \begin{cases} 0, & \text{if } t \le t_{mj}, \\ 1, & \text{if } t > t_{mj}. \end{cases} \tag{12.5.12}$$

The total number of unknown parameters needed to compute $\tilde{\mu}$ is determined from (12.5.9)-(12.4.10) to be equal to $M_0 = (n+1)N + \sum_{m=1}^{N} (k_m + n_m)$. These parameters are determined by the collocation method at a number of points on Γ and also by subjecting $\tilde{\mu}$ to certain continuity conditions.

The collocation equations are formulated as follows: For splines $s_m(t)$ of odd degree n we collocate at the end points (12.5.7) of each interval. For splines of even degree n we collocate at the midpoints of each of these intervals. We can always reduce the number collocation equations in the case of any symmetry of the boundary. Thus, let z_i, $i = 1, \dots, M_1$, denote the chosen collocation points. Then from (12.5.7) we find that $M_1 = 3N + \sum_{m=1}^{N} k_m$, and Symm's integral equation (12.1.6) yields the following collocation equations:

$$\int_0^L \log\left|z_i - \zeta(t)\right| \tilde{\mu}(t)\, dt = -\log|z_i|, \quad i = 1, \dots, M_1, \tag{12.5.13}$$

which in view of (12.5.9)-(12.4.11) reduce to

$$\sum_{m=1}^{N} \Big[\sum_{j=1}^{n_m} A_{mij}\, a_{mj} + \sum_{j=0}^{n} B_{mij}\, b_{mj} + \sum_{j=2}^{k_m+1} C_{mij}\, c_{mj} \Big] = -\log|z_i|, \qquad i = 1, \dots, M_1, \tag{12.5.14}$$

where

$$\begin{aligned} A_{mij} &= (-)^{j+1} \int_{t_{m,-1}}^{t_{m0}} (t_{m0} - t)^{-1+j/\alpha_m} \log\left|z_i - \zeta(t)\right| dt \\ &\quad + \int_{t_{m0}}^{t_{m1}} (t - t_{m0})^{-1+j/\alpha_m} \log\left|z_i - \zeta(t)\right| dt, \\ B_{mij} &= \int_{t_{m1}}^{t_{m,k_m+2}} (t - t_{m1})^j \log\left|z_i - \zeta(t)\right| dt \\ C_{mij} &= \int_{t_{m,j}}^{t_{m,k_m+2}} (t - t_{mj})^n \log\left|z_i - \zeta(t)\right| dt. \end{aligned} \tag{12.5.15}$$

The continuity conditions to be imposed on $\tilde{\mu}$ are based on the assumption that the first l_m derivatives of $\tilde{\mu}$ must be continuous at the points $t_{m,-1}$ and t_{m1} where the type of approximation changes. This leads to

$$r_m^{(k)} = \begin{cases} s_{m-1}^{(k)} & \text{at } t = t_{m,-1}, \text{ where } m = 2, \dots, N, \\ s_m^{(k)} & \text{at } t = t_{m1}, \text{ where } m = 1, \dots, N, \end{cases} \tag{12.5.16}$$
$$r_1^{(k)}(t_{1,-1}) = s_N^{(k)}(t_{N,k_N+2}), \quad k = 1, \dots, l_m,$$

which, in view of (12.5.10) and (12.5.11), yield the continuity equations

$$\begin{aligned}
&\sum_{j=1}^{n_m} a_{mj} \left(-1 + \frac{j}{\alpha_m}\right)\left(-2 + \frac{j}{\alpha_m}\right) \cdots \left(-k + \frac{j}{\alpha_m}\right) \\
&\qquad \times (t_{m1} - t_{m0})^{-1-k+j/\alpha_m} = k!\, b_{mk}, \\
&\sum_{j=1}^{n_m} a_{mj} (-1)^{j+k-1} \left(-1 + \frac{j}{\alpha_m}\right)\left(-2 + \frac{j}{\alpha_m}\right) \cdots \left(-k + \frac{j}{\alpha_m}\right) \\
&\qquad \times (t_{m0} - t_{m,-1})^{-1-k+j/\alpha_m} \\
&\qquad = \sum_{j=0}^{n} b_{m-1,j}\, j(j-1) \cdots (j-k+1)\, (t_{m,-1} - t_{m-1,j})^{j-k} \\
&\qquad + \sum_{j=2}^{k_{m-1}+1} c_{m-1,j}\, n(n-1) \cdots (n-k+1)\, (t_{m,-1} - t_{m-1,j})^{n-k}, \\
&b_{0j} = b_{Nj}, \quad c_{0j} = c_{Nj}, \quad t_{0j} = t_{Nj} - L,
\end{aligned} \tag{12.5.17}$$

where $m = 1, \dots, N$, and $k = 0, 1, \dots, l_m$. The total number of these continuity equations is $M_2 = 2 \sum_{m=1}^{N} (1 + l_m)$. Combining the collocation equations (12.5.14) and the continuity equations (12.5.17), we obtain a linear system of $(M_1 + M_2)$ equations which is solved for the unknown parameters a_{mj}, b_{mj}, and c_{mj}, which in turn determines

$$\tilde{g}(z) + i\,\tilde{h}(z) = \int_0^L \log\,(z - \zeta(t))\,\tilde{\mu}(t)\,dt, \tag{12.5.18}$$

so that the approximate mapping function $f(z)$ can be evaluated from (12.1.1) (with $z_0 = 0$). In fact, substituting (12.5.10)-(12.5.12) in (12.5.18) we get

$$\tilde{g}(z) + i\,\tilde{h}(z) = \sum_{m=1}^{N} \left\{ \sum_{j=1}^{n_m} a_{mj}\, A_{mj}(z) + \sum_{j=0}^{n} b_{mj}\, B_{mj}(z) + \sum_{j=2}^{k_m+1} c_{mj}\, C_{mj}(z) \right\}, \tag{12.5.19}$$

where

$$\begin{aligned}
A_{mj}(z) &= (-1)^{m+j} \int_{t_{m,-1}}^{t_{m0}} (t_{m0} - t)^{-1+j/\alpha_m} \log\,(z - \zeta(t))\,dt \\
&\quad + \int_{t_{m0}}^{t_{m1}} (t - t_{m0})^{-1+j/\alpha_m} \log\,(z - \zeta(t))\,dt, \\
B_{mij}(z) &= \int_{t_{m1}}^{t_{m,k_m+2}} (t - t_{m1})^{j} \log\,(z - \zeta(t))\,dt, \\
C_{mij}(z) &= \int_{t_{m,j}}^{t_{m,k_m+2}} (t - t_{mj})^{n} \log\,(z - \zeta(t))\,dt.
\end{aligned} \tag{12.5.20}$$

Note that the coefficients in (12.1.14) are related to (12.1.19) by $A_{mij} = \Re\{A_{mj}(z)\}$, $B_{mij} = \Re\{B_{mj}(z)\}$ and $C_{mij} = \Re\{C_{mj}(z)\}$. The error E is defined by

$$E = \max_{j=1,\dots,M_1} \left| \left| \tilde{f}\left(z_{j+1/2}\right) \right| - 1 \right|, \tag{12.5.21}$$

where M_1 is the number of collocation points used, and $z_{j+1/2} \in \Gamma$ denotes the end point or midpoint of the interval depending on whether the degree n of splines is even or odd.

Certain reduction in the number of splines occurs in some particular cases. For example, if a corner has interior angle π/q, where q is an odd positive integer, then the corner is treated as a singular corner with (i) $n_m = 2$, when $n = 0$; (ii) $n_m = 2$, when $n = 2, 4$ and $q \geq 2$; (iii) $n_m = 4$, when $n = 2, 4$ and $q < 2$; and (iv) $n_m = 3$, when $n = 1, 3, 5$. Symmetry of any kind always reduces the size of the linear system.

Example 12.8. Consider the rectangles $\Gamma_a = \{(x, y) : |x| < 1, |y| < a\}$ for $a = 1, 0.5, 0.2, 0.1$. The exact mapping function is known (see Map 7.7). Take $M_1 = 128$. The boundary is partitioned into sections of uniform length on each side with 32, 20, 10, and 6 intervals on the side $x = \pm 1$. Because of the symmetry about both coordinate axes, the total number of equations in the linear system reduces to $(17 + n)$ (n even) and $(18 + n)$ (n odd) when $a = 1$, and $(32 + 2n)$ (n even) and $(33 + 2n)$ (n odd) when $a \neq 1$. Hough and Papamichael [1981] have computed the error E as follows:

n	$a = 1$	$a = 0.5$	$a = 0.2$	$a = 0.1$
0	2×10^{-4}	1×10^{-4}	4×10^{-4}	3×10^{-3}
1	9×10^{-6}	2×10^{-5}	2×10^{-4}	1×10^{-3}
2	4×10^{-8}	3×10^{-7}	1×10^{-5}	3×10^{-4}
3	9×10^{-9}	6×10^{-8}	3×10^{-6}	7×10^{-5}
4	3×10^{-9}	3×10^{-9}	6×10^{-7}	8×10^{-5}
5	3×10^{-11}	5×10^{-10}	1×10^{-7}	2×10^{-2}
Symm [1966]	4×10^{-4}	2×10^{-4}	4×10^{-4}	3×10^{-3}
Hayes et al. [1972]	1×10^{-6}	2×10^{-5}	5×10^{-4}	5×10^{-3}
Global cubic spline	9×10^{-5}	6×10^{-5}	3×10^{-6}	7×10^{-5}

Their conclusions on the effectiveness of this method are as follows:
(i) For splines and augmented splines of degree $n \leq 3$ there is no ill-conditioning effect, as is found in other polynomial basis methods, in solving the linear system for parameters used in the approximation.
(ii) If $n \geq 4$, there is a possible loss of significance in computations, which indicates that for higher-degree splines the use of B-splines instead of (12.5.11) may be necessary if the knot spacing is not uniform.
(iii) Augmented cubic splines with three singular terms for each singular corner provide the best results in most problems.
(iv) Every corner of a polygonal boundary must be treated like a singular corner. This makes

the method easy to implement because it introduces appropriate terms from (12.5.10) at every corner.
(v) This method can be generalized to regions with curved boundaries. In such cases, the integrals in (12.5.20) must be evaluated by numerical quadrature, like Gauss quadrature, which gives good approximations for nonsingular integrals, but any logarithmic and fractional power singularities need special treatment.
(vi) Besides the corner singularities, any poles of $f(z)$ that lie very close to the boundary may also introduce inaccuracies in numerical computations (see §12.3). In such cases a suitable choice of knot spacing which may not be equi-spaced is recommended.
(vii) The corner singularity in narrow rectangles ($a \ll 1$) does not produce any computational difficulty. The difficulty arises from the location of the poles of the mapping function $f(z)$. ■

Example 12.9. Consider the limaçon $r = a - \cos\theta$, where $r = 0$ corresponds to $z = 1/2$ for $a = 1$, and to $z = 0$ for $a > 1$ (Figure 12.5).

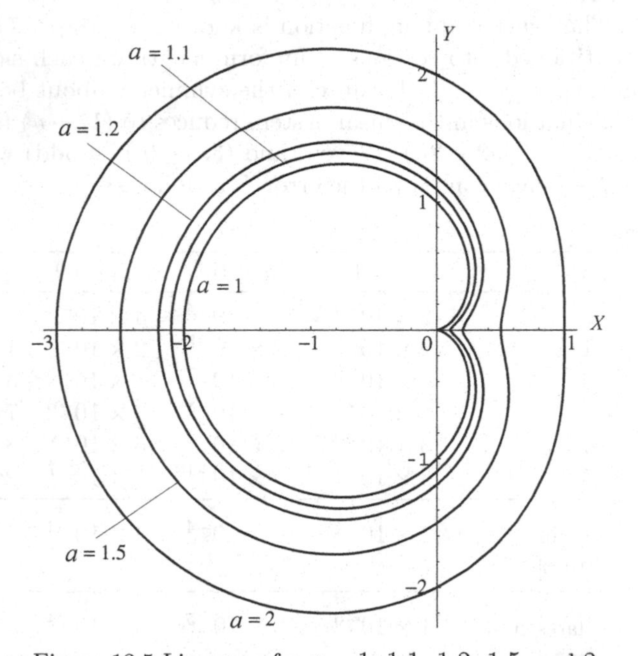

Figure 12.5 Limaçons for $a = 1$, 1.1, 1.2, 1.5, and 2.

The ray $\theta = 0$ is the direction of the polar axis which is the axis of symmetry. The partitions Γ_j, $j = 1, \dots, N$, are taken for $\theta = 0\ (\pi/N)\ 2\pi$, $N = 2n$. Then the errors in the OPN method are presented in Table 12.3.

Table 12.3 Errors in the OPN method for the limaçon.

	For $n = 16$		For $n = 32$	
a	E_M	E_A	E_M	E_A
1.0	0.0723	0.0029	0.0408	0.0014
1.1	0.0800	0.0002	0.0215	0.0003
1.2	0.0261	0.0004	0.0048	0.0004
1.3	0.0116	0.0006	0.0019	0.0004
1.4	0.0064	0.0006	0.0011	0.0004

Note that $E_A \ll E_M$ and the maximum error in each case occurs at the interval point on the polar axis when $r = a-1$ and the boundary is concave. For $a = 1$, there is a singularity at the point, but the concavity of the boundary decreases with an increase in a, and so does the error. An increase in N results in a decrease in error (Symm [1966]). ■

Example 12.10. Consider the ellipse $x^2/a^2 + y^2 = 1$. Take the partitions Γ_j, $j = 1, \dots, N$ for $\theta = 0\ (\pi/N)\ 2\pi$, $N = 4n$. Then the errors in the OPN method are presented in Table 12.4.

Table 12.4 Errors in the OPN method for the ellipse.

	For $n = 16$		For $n = 32$	
a	E_M	E_A	E_M	E_A
1.25	0.00002	0.0024	0.0001	0.0012
2.5	0.0010	0.0005	0.0002	0.0003
5	0.0079	0.0003	0.0012	0.0001
10	0.0525	0.0003	0.0083	0.0001
20	0.2135	0.0018	0.0530	0.0001

Note that $E_A > E_M$ for small a, but as a increases, the error on the whole increases, and E_M becomes prominent. The maximum error occurs at the end points of the minor axis where the nodes are widely spread. The error decreases as N increases (Symm [1966]). ■

Example 12.11. Consider Cassini's oval

$$\left[(x+\alpha)^2 + y^2\right]\left[(x-\alpha)^2 - y^2\right] = 1$$

of Figure 12.1. Take $\alpha = 5/6$ (the near circle case), and determine the errors in the OPN method for $N = 16$ and $N = 32$ for both interior and exterior regions. ■

Example 12.12. Consider the rectangle $-1 \le x \le 1$, $-a \le y \le a$ (see Map 7.7 and Case 8.2 in §8.3, and compare it with Example H.1). To determine the error in mapping problems for the interior and exterior regions, use $C = \frac{a}{2}\left(E' - k^2 K'\right) = \frac{a}{2}\left(E - k'^2 K\right)$, where E and E' are complete elliptic integrals and $k' = \sqrt{1-k^2}$, k being the modulus of elliptic functions, such that $E(k,\phi) = \int_0^{\pi/2} \sqrt{1 - k^2 \sin^2\phi\, d\phi} = \int_0^K \operatorname{dn} u\, du$. ■

Example 12.13. Consider the ellipse $\Gamma : z(t) = e^{it}\left(1 + \lambda^2 e^{-2it}\right)$, where $|\lambda|$ is taken sufficiently small for the region D to be nearly circular. The approximate mapping function obtained by the method of infinite systems is given by (5.2.7). Take $\lambda = 0.1$, obtain the approximate solution by the ONP method, and compare it with the solution (5.2.7). ■

REFERENCES USED: Gaier [1976], Gram [1962], Gautschi [1977; 1978; 1979], Gutknecht [1986], Hayes, Kahaner and Kellner [1972], Jawson [1963], Hough and Papamichael [1981], Kober [1957], Kythe [1998], Muskhelishvili [1953], Phillips [1966], Pólya and Szegö [1951], Rabinowitz [1966], Reichel [1985], Symm [1963; 1966; 1967].

13

Airfoils and Singularities

The Joukowski airfoils and the generalized Joukowski mappings are discussed in Chapter 6. We will study the potential flow analysis of airfoils using James's method (James [1971]), which was very successful for all types of contours that do not have corner singularity. We will develop Joukowski mapping functions, compare numerical solutions of single-element airfoils by both Theodorsen's and James's iterative methods, and look into the mechanism of divergence of Theodorsen's method in those cases where the image boundary is not almost circular, and finally analyze multiple-element airfoils by using von Karman-Trefftz transformations and FFT with Garrick's method of conjugate functions.

13.1 Nearly Circular Approximations

In practical applications, airfoils are not exactly defined by the Joukowski function (6.1.1). They are often approximated by nearly circular regions. This idea can be explained as follows: Let D denote the region bounded by two circular arcs γ_1 and γ_2 with end points a and b. Let θ $(0 < \theta < 2\pi)$ denote the angle between the arcs γ_1 and γ_2. The region D has a central arc γ_c connecting points a and b and bisecting the angle θ (see Figure 13.1(a) if D is the interior region and Figure 13.1(b) if D is the exterior region). The region D is mapped onto the unit disk U by a chain of mappings $f^{-1} \circ g \circ f$ (Figure 13.2), where

$$f(z) = e^{i\alpha}\,\frac{z-a}{z-b}, \tag{13.1.1}$$

α is a real number chosen such that the central arc γ_c is mapped onto the positive real axis in D_1, and

$$g(z) = z^{\pi/\theta}. \tag{13.1.2}$$

Note that the function f^{-1} maps the right half-plane D_2 onto a region similar to D except that the angle between the arcs now becomes π. Thus, f^{-1} maps D_2 onto U which is bounded by the unit circle through a and b and orthogonal to the central arc. Similarly, the chain of mappings $f^{-1} \circ (-g) \circ f$ will map the region D onto the region U^* exterior to the unit circle. These mappings have points a and b as fixed points. In the case of Figure 13.1(b), where D is the exterior region, a circle slightly larger than the boundary of U and touching U at point b is the map of the curve that looks like an airfoil (the curve on the left in Figure 13.3).

Although airfoils are shaped slightly differently rather than regions like D considered above, they can be approximated by such regions. The idea is to map an airfoil approximately onto a nearly circular region such as D, which can then be mapped onto the unit disk.

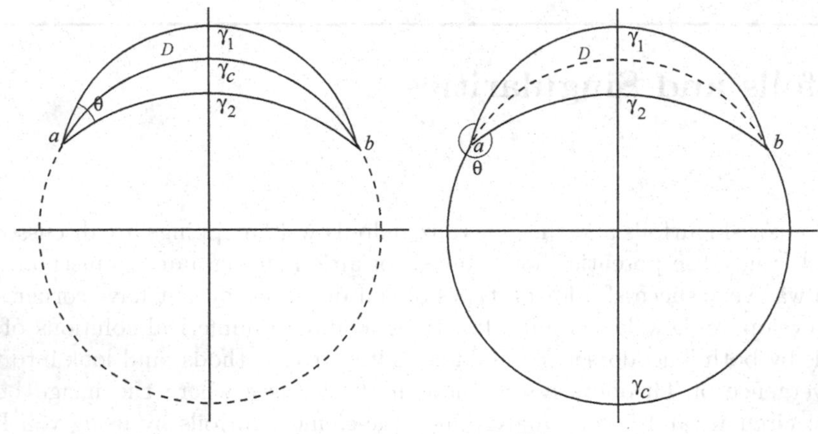

Figure 13.1 Region D.

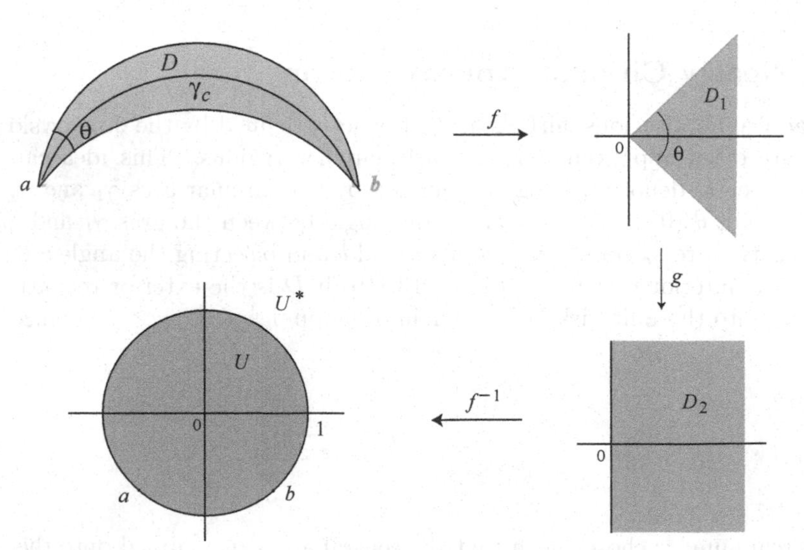

Figure 13.2 Region D onto the unit circle.

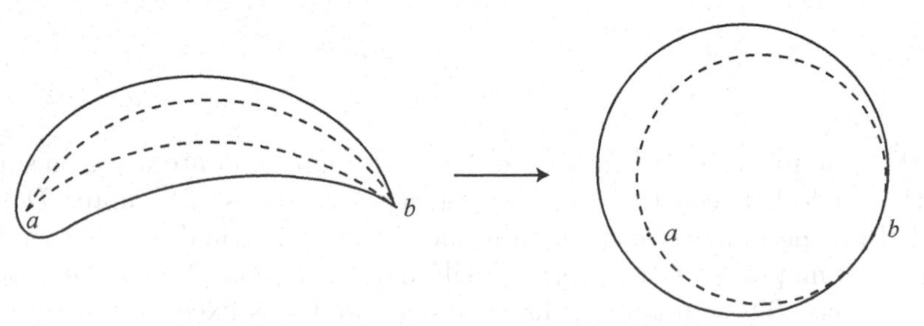

Figure 13.3 Exterior to the region D onto exterior of the unit circle.

13.2 James's Method

As we have seen in §11.6, Theodorsen's method starts with the conformal map of the exterior of the unit circle $|z| > 1$ onto the exterior of a physical boundary in the ζ-plane. The iterations in this method are convergent only if the image boundary is a near-circle (i.e., a nearly-circular contour). An analysis of two-element airfoils using Theodorsen's method is given in §13.2.1 below. However, the limitations of Theodorsen's method do not apply in James's method (James [1971]) which will be presented below in §13.2.1. Note that every multi-element potential flow analysis makes use of the single-element mapping methods at some stage during the computations.

The single-element mapping deals with the problem of conformally mapping the exterior of the unit circle onto the region outside (or inside) a Jordan contour Γ. Contours Γ with corners or boundary curves that do not close can be considered by using preliminary transformations or by including singular functions in the series for the mapping function with augmented bases, as in the RM, BKM or ONP method. A further assumption often made during the approximation process is that the image curve maintains its proper scale and orientation whereas the distant regions are left undisturbed. Mappings are generally made, however, without any prior knowledge as to the scale factor or rotation of the boundary.

13.2.1 Single-Element Airfoils. Let the mapping be represented by a function of the form

$$z = \zeta + a_0 + \frac{c_1}{\zeta} + \frac{a_2}{\zeta^2} + \cdots , \tag{13.2.1}$$

where $z = e^{i\theta}$ and $\zeta = \rho\, e^{i\phi}$ denote the complex coordinates on the unit circle and the physical plane, respectively, and a_j, $j = 0, 1, \dots$, are complex constants. Both Theodorsen [1931] and James [1971] use series closely related to (13.2.1). Their methods can be called the *method of successive conjugates* (Garrick [1949]).

(a) Theodorsen's Method. This method uses the truncated series of the form

$$\log\left(\frac{z}{\zeta}\right) = b_0 + \frac{b_1}{\zeta} + \frac{b_2}{\zeta^2} + \cdots + \frac{b_{N-1}}{\zeta^{N-1}}, \tag{13.2.2}$$

which is applied at equi-spaced points j on the unit circle (clockwise along the perimeter). This yields

$$\log\left(\frac{z}{\zeta}\right)_j = \sum_{k=0}^{N-1} b_k\, e^{i2\pi jk/N}. \tag{13.2.3}$$

The terms on the left side of (13.2.3) are related to the geometric variables by the relation

$$\log\left(\frac{z}{\zeta}\right)_j = \log\rho_j + i\,(\phi_j - \theta_j), \tag{13.2.4}$$

where ρ_j is the radial coordinate of the point j in the ζ-plane and ϕ_j and θ_j are the arguments (positive clockwise) of the points in the ζ- and z-plane, respectively. The real and imaginary parts of $\log(z/\zeta)$ are conjugate harmonic functions, so that if one is known the other can be computed efficiently by Fourier transforms. Thus, the iterations are carried out step-by-step as follows:

STEP 1. Compute the values of θ_j at the defining points in the z-plane.

STEP 2. Approximate the values of ϕ_j at the points in the ζ-plane corresponding to the equi-spaced points in the z-plane ($\phi_j = \theta_j$ is often assumed).
STEP 3. Use the curve-fit coefficients to determine the values of $\log \rho_j$ corresponding to the estimated values of ϕ_j.
STEP 4. Compute the conjugate harmonic function corresponding to the latest values of $\log \rho_j$, and use them to update the estimated values of ϕ_j.
STEP 5. Repeat steps 3 and 4 until the values of ϕ_j converge.
STEP 6. Determine the coefficients of the mapping function from the converged data.

(b) James's Method. This method uses the truncated series of the form

$$\log\left(\frac{dz}{d\zeta}\right) = c_0 + \frac{c_1}{\zeta} + \frac{c_2}{\zeta^2} + \cdots + \frac{c_{N-1}}{\zeta^{N-1}}, \tag{13.2.5}$$

which is applied at equi-spaced points on the circle. This yields

$$\log\left(\frac{dz}{d\zeta}\right)_j = \sum_{k=0}^{N-1} c_k\, e^{i2\pi jk/N}. \tag{13.2.6}$$

The term on the left side is related to the geometric variables by the relation

$$\log\left(\frac{dz}{d\zeta}\right)_j = \log\left|\frac{dz}{d\zeta}\right|_j + i \arg\left\{\frac{dz}{d\zeta}\right\}_j, \tag{13.2.7}$$

$$s_j = \int_0^{\theta_j} \left|\frac{dz}{d\zeta}\right| d\theta, \tag{13.2.8}$$

$$\arg\left\{\frac{dz}{d\zeta}\right\}_j = \tau_j + \theta_j - \frac{3\pi}{2}, \tag{13.2.9}$$

where s_j is the arc length on the contour Γ in the ζ-plane and τ_j is the angle on that contour (for the convention, see Figure 13.4). Then the iterations are performed as follows:
STEP 1. Compute the values of s and τ at the defining points in the z-plane, and determine the curve-fit coefficients of τ versus s.
STEP 2. Approximate the values of $\left|\frac{dz}{d\zeta}\right|_j$ at equi-spaced points on the circle ($\left|\frac{dz}{d\zeta}\right|_j = 1.0$ is usually assumed).
STEP 3. Integrate (13.2.8), and obtain approximate values of s_j.
STEP 4. Use the curve-fit coefficients to determine the values of τ_j corresponding to the approximate values of s_j, and compute $\arg\left\{\frac{dz}{d\zeta}\right\}_j$ from (13.2.9).
STEP 5. Compute the conjugate function to determine the values of $\log\left|\frac{dz}{d\zeta}\right|_j$ corresponding to the latest values of $\arg\left\{\frac{dz}{d\zeta}\right\}_j$, and take the exponential to update the values of $\left|\frac{dz}{d\zeta}\right|_j$.
STEP 6. Repeat steps 3 through 5 until the values of $\left|\frac{dz}{d\zeta}\right|_j$ converge.

STEP 7. Compute the coefficients of the mapping function from the converged data.

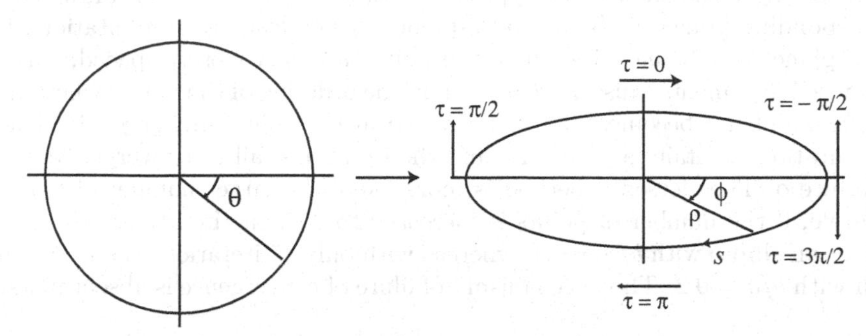

Figure 13.4 James's method.

Halsey [1982] found, by comparing the numerical data and convergence properties of both of these methods, that James's method is more suitable than Theodorsen's method for a larger class of single- and multi-element boundaries. For example, an application of Theodorsen's method to a slightly complicated boundary, such as a cambered ellipse, where polar coordinates are used ($\log \rho$ and ϕ), leads to functions with multiple values. This makes numerical interpolation inaccurate, if not impossible. On the contrary, in James's method the use of intrinsic coordinates (τ and s) always yields a family of interpolated functions so long as the boundary has no corner singularity. Thus, James's method is more suited for mapping complicated boundaries. The three-element looped boundary shown in Figure 13.5 failed to be mapped by Theodorsen's method, but James's method presented no computational difficulty. Even when the use of polar coordinates is not appropriate, James's method still works smoothly in many cases where Theodorsen's method fails. These include elliptic boundaries of different ratios b/a of the minor and major axes. For example, for $b/a = 0.9$ both methods require 10 iterations to reduce the residuals to less than 10^{-5}; for $b/a = 0.4$ Theodorsen's method requires 25 iterations, but James's 15, to reduce the residuals to 10^{-5}; and for $b/a = 0.3$ Theodorsen's method fails to converge, whereas James's method succeeds even down to $b/a = 0.005$ with only a maximum of 25 iterations.

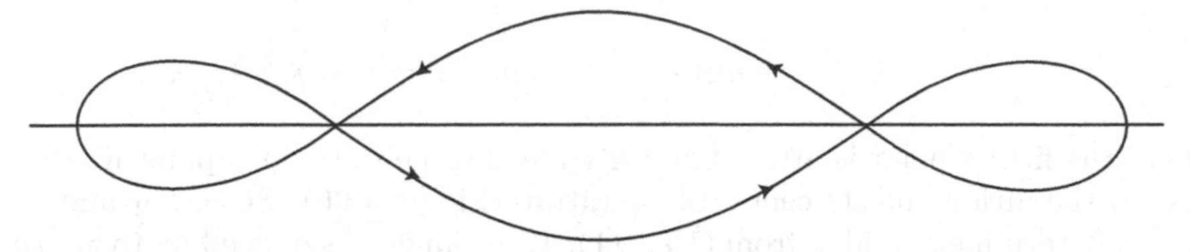

Figure 13.5 Three-element loop.

Halsey [1982] has found that both methods fail to satisfy Warschawski's sufficient conditions (9.2.5)-(9.2.6) for convergence established in §11.2.

A detailed examination of all computed data has revealed the mechanism for the failure of iterations to converge in Theodorsen's method. In general, equi-spaced points on the circle transform into more closely spaced points on the boundary Γ for values of $\left|\frac{dz}{d\zeta}\right| < 1$ and into more sparsely spaced points on the boundary for values of $\left|\frac{dz}{d\zeta}\right| > 1$. Thus, the thinner a boundary gets, the smaller some the values of $\left|\frac{dz}{d\zeta}\right|$ become and the more closely spaced some of the points become. As such, computation of the conjugate harmonic function in Theodorsen's method is an approximation to the difference in arguments between the corresponding points in the z- and ζ-planes, which leads to computation of arguments in the ζ-plane. For boundaries where the points are very closely spaced, small errors in the computed arguments cause a breakdown in the ordering of the points where the argument at a point number j becomes smaller than that at the point number $j-1$. Hence, subsequent computations contain large errors, and the iterations fail to converge. Moreover, this kind of failure of Theodorsen's method is more likely if a larger number of points is used. For example, if the number of points is increased to 257, the iterations fail to converge in the case of an ellipse with $b/a = 0.4$, whereas with only 17 iterations the iterations do converge even with $b/a = 0.2$. This mechanism of failure of convergence is absent in James's method.

13.2.2. von Karman-Trefftz Transformations. These transformations are useful in analyzing potential flow over multi-element airfoils. First, we will consider a single-element airfoil and modify Theodorsen's method by using a von Karman transformation and FFT. There are four basic procedural steps in the problem of conformally mapping any airfoil onto a circle: (i) remove the effects of slope discontinuities in the airfoil contour, and expand the regions of rapid flow changes (e.g., the nose region) by conformally mapping the airfoil point-by-point onto a nearly circular contour; (ii) translate the coordinate system so as to place the centroid of the nearly circular region on or near the origin; (iii) use interpolation and obtain a continuous representation of the nearly circular contour and thus of the airfoil; and (iv) map the nearly circular contour onto a circle.

The analytical and computational tools to perform these four steps are as follows. For step (i) a von Karman-Trefftz transformation is used. Let z denote a complex coordinate in the airfoil plane (z-plane), and ζ a complex coordinate in the nearly circular contour plane (ζ-plane). Then the von Karman-Trefftz transformation is given by

$$\frac{z-\zeta}{z+\zeta} = \left(\frac{z - z_s - \kappa\, z_l}{z + z_s - \kappa\, z_l}\right)^{1/\kappa}, \tag{13.2.10}$$

where z_s and z_l are complex constants, $\kappa = 1 - \tau/\pi$, and τ is the trailing edge included angle. This transformation is singular at

$$z = z_s + \kappa\, z_l \equiv z_{T1}, \quad \text{and} \quad z = z_s - \kappa\, z_l \equiv z_{N1}, \tag{13.2.11}$$

where the firmer point is at the trailing edge and the latter at a point midway between the nose of the airfoil and its center of curvature (Figure 13.6). Since z_{T1} and z_{N1} are known, we can determine z_s and z_l from (13.2.11). If the angle τ is opened up to π (see Figure 13.2), then the airfoil is mapped into a near-circle in the ζ-plane where the circle (dotted) is drawn for comparison, 0 is the origin of the coordinates and C is the approximate centroid which is the origin of the ζ'-plane. Since the transformation (13.2.10) contains a rational exponent, the proper branch should be chosen by 'tracking' the transformation from a point where it is known to the point of interest. A proper choice of the branch in (13.2.10) implies continuity of the argument of the base (the expression within parentheses) along a path that does not cross the boundary of the airfoil. Note that this argument approaches zero as $z \to \infty$.

In step (ii) the origin of the coordinate system is translated to the centroid C of the nearly circular region. First an approximate centroid C is determined by connecting adjacent points of the line segment. This translation is given by

$$\zeta' = \zeta - C. \tag{13.2.12}$$

It will be used to improve the convergence of a series expression used in step (iv).

In step (iii) a continuous representation of the airfoil image, which has so far been defined pointwise, is obtained. To do this, a polar coordinate system is defined in the ζ-plane (Figure 13.6), where $\log \rho$ as a function of ψ is fitted with a periodic cubic spline (see §12.5 for more on splines). This curve fitting technique leads to a smooth definition of the airfoil image in the ζ-plane with a high degree of accuracy.

In step (iv) the Theodorsen-Garrick transformation is used to map the near-circle in the ζ-plane onto a circle in the ζ'-plane, where instead of Theodorsen's integral equation (9.1.6) with a cotangent kernel, we will use the Fourier series analysis to eventually use FFT. That is how Theodorsen's method is modified.

Thus, the Theodorsen-Garrick transformation can be written as

$$\zeta_j' = \zeta \exp\left\{ \sum_{j=0}^{N} (A_j + i\, B_j)\, \zeta^j \right\}. \tag{13.2.13}$$

Note that the near-circle in the ζ-plane is represented by

$$\zeta = \rho(\psi)\, e^{i\psi}. \tag{13.2.14}$$

We will finally map the circle in the ζ'-plane onto the unit circle $w = e^{i\phi}$. Thus, substituting these polar representations in (13.2.14), taking logarithms on both sides, and equating real and imaginary parts, we obtain the following set of equations:

$$\log \rho = A_0 + \sum_{j=1}^{N} (A_j \cos j\phi + B_j \sin j\phi)\,, \tag{13.2.15}$$

$$\psi = \phi + B_0 + \sum_{j=1}^{N} (B_j \cos j\phi - A_j \sin j\phi)\,. \tag{13.2.16}$$

Since $\log \rho$ on the near-circle is known as a function of ψ from the periodic cubic spline fit, the problem reduces to determining the coefficients A_j and B_j by FFT and an iterative scheme, as follows: Choose $2N$ equi-spaced points on the unit circle in the w-plane, starting at the image of the trailing edge. Thus,

$$\phi_k = \frac{(k-1)\pi}{N}, \quad k = 1, \dots, 2N. \tag{13.2.17}$$

To place the trailing edge at $\phi = 0$ in the w-plane, set

$$B_0 + \sum_{j=1}^{N} B_j = \psi_T, \tag{13.2.18}$$

where ψ_T is the value of ψ at the trailing edge in the ζ'-plane. Take $B_N = 0$ to make a closed system. Then, ψ_k is given by

$$\psi_k - \phi_k = \psi_T - \sum_{j=1}^{N-1} (B_j + A_j \sin j\phi_k - B_j \cos j\phi_k)\,, \tag{13.2.19}$$

which can be evaluated by the Fourier technique as follows (assuming A_j and B_j are given for $j = 1, \ldots, N-1$):

$$\begin{aligned} y_1 &= \frac{1}{2}\Big(\psi_T - \sum_{j=1}^{N-1} B_j\Big), \\ y_{j+1} &= \frac{1}{2}\left(B_j + i\,A_j\right), \quad j = 1, \ldots, N-1, \\ y_{2N-j+1} &= \bar{y}_j, \quad j = 1, \ldots, N, \end{aligned} \tag{13.2.20}$$

where the bar denotes the complex conjugate. Then

$$\psi_k - \phi_k = \sum_{j=1}^{2N} y_j \exp\Big\{\frac{i\pi(j-1)(k-1)}{N}\Big\}, \tag{13.2.21}$$

which is known as a discrete Fourier transform (Cooley and Tukey [1965]; Cooley, Lewis and Welch [1970]) and is evaluated by FFT technique in $O\left(N \log_2 N\right)$ operations for $k = 1, \ldots, N$. Note that a direct evaluation of Eq (13.2.21) takes $O\left(N^2\right)$ operations.

We can apply a similar FFT technique to solve Eq (13.2.15) and obtain A_j and B_j by using a trigonometric series fit through the points $(\log \rho_k)$ for $k = 1, \ldots, 2N$ as follows: Define

$$y_j = \frac{1}{2N} \sum_{k=1}^{2N} (\log \rho)_k \exp\Big\{-\frac{i\pi(j-1)(k-1)}{N}\Big\}, \tag{13.2.22}$$

where $(\log \rho)_k$ is the value of $\log \rho$ at the kth point given by Eq (13.2.17). Then

$$\begin{aligned} A_j &= 2\,\Re\{y_{k+1}\}, \quad j = 0, 1, \ldots, N, \\ B_j &= -2\,\Re\{y_{j+1}\}, \quad j = 1, \ldots, N-1. \end{aligned} \tag{13.2.23}$$

This evaluation takes $O\left(N \log_2 N\right)$ operations.

Now, we can summarize the iterative method by the following five-step algorithm:

1. Set $A_j = B_j = 0$ for $j = 1, \ldots, N-1$.
2. Evaluate ψ_j from (13.2.21) by using (13.2.20).
3. Evaluate $(\log \rho)_k$ for $k = 1, \ldots, 2N$, using ψ_k and the cubic spline fit coefficients.
4. Solve for A_0, A_N, and A_j, B_j for $j = 1, \ldots, N-1$, using (13.2.22)-(13.2.23).
5. Using the latest computed values of A_j and B_j, repeat steps 2 through 4 until the values converge.

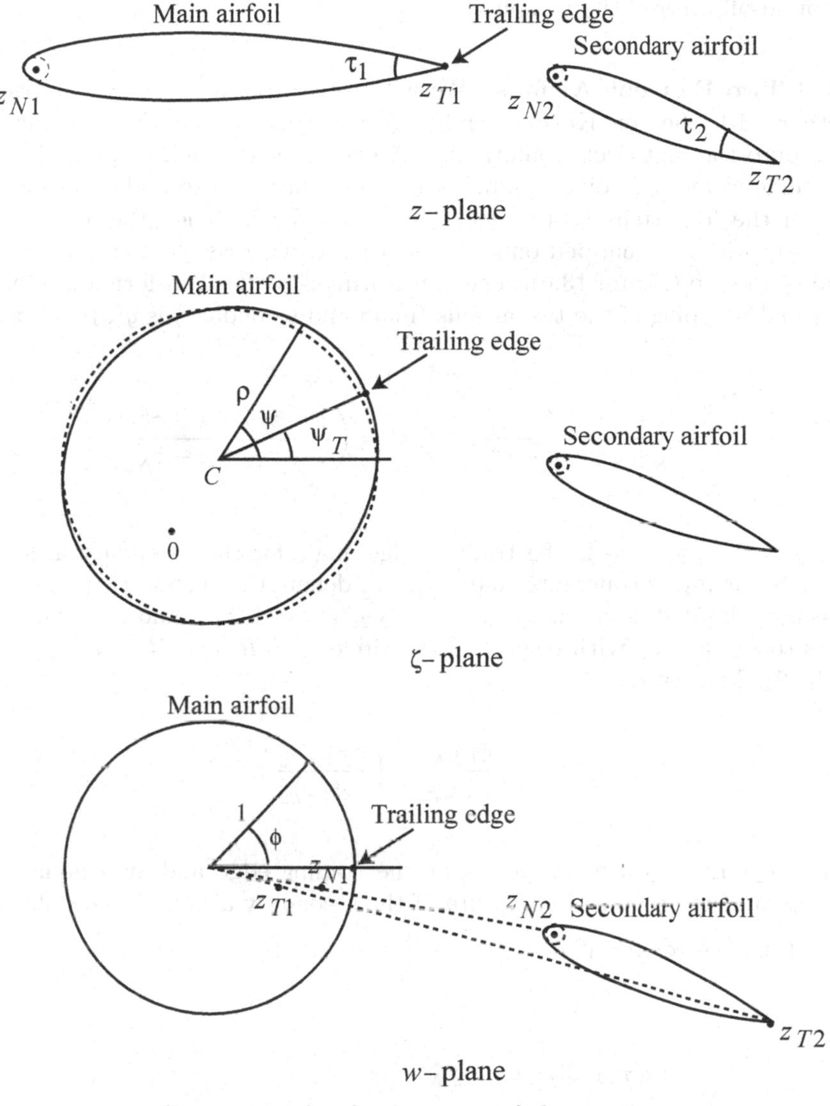

Figure 13.6 Single-element airfoil mappings.

Warschawski's sufficient conditions for convergence (§11.2) for the above algorithm imply that

$$\sqrt{\frac{\rho_{\max}}{\rho_{\min}}} - 1 < \varepsilon, \quad \left| \left(\frac{\partial \log \rho}{\partial \psi} \right)_{\max} \right| < \varepsilon,$$

where $\varepsilon = 0.2954976$, and the maximum and minimum values are taken on the nearly circular contour. Note that the composite accuracy of this method depends on the accuracy at each step. Since (13.2.20) involves trigonometric functions, the computation of A_j, B_j described in step 4 may not produce accurate results. However, an error analysis in step 4 must be carried out to determine the terms that should be taken to approximate the trigonometric functions involved in this step. It is known (Abramovici [1973]) that about 100 terms are enough to approximate $\log \rho$ in terms of the trigonometric series in ψ. Another source of errors lies in step 3 where a periodic cubic spline is used to interpolate $\log \rho$ as a function of ψ. These errors, though small, are due to the definition of the airfoil only at finitely many points j which are mapped onto the circle with an accuracy limited only by the round-off error of the computer.

13.2.3 Two-Element Airfoils. We will use Garrick's approach (Garrick [1936, 1949]), together with the von Karman-Trefftz transformation and FFT, to map a two-element airfoil onto the unit circle conformally. A two-element airfoil mapping is initially identical to a single-element airfoil mapping except that there is a secondary airfoil which is carried through the four steps (i)-(iv) mentioned in §13.2.2. Then the airfoil-like shape of the secondary airfoil is mapped onto the near-circle whereas the mapping of the main airfoil remains a circle (Figure 13.6). The von Karman-Trefftz transformation for a simultaneous conformal mapping of the two airfoils (main and secondary) is $g(\zeta) = f(z)$, where

$$g(\zeta) = \frac{\zeta - \zeta_T}{\zeta - \zeta_N} \cdot \frac{\zeta - \zeta_T^*}{\zeta - \zeta_N^*}, \quad f(z) = \left(\frac{z - z_{T2}}{z - z_{N2}} \cdot \frac{z - z_{T2}^*}{z - z_{N2}^*} \right)^{1/\kappa_2}, \tag{13.2.24}$$

and $\kappa_2 = 2 - \tau_2/\pi$, τ_2 is the trailing edge angle for the secondary airfoil; z_{N2}, z_{T2}, ζ_N, and ζ_T are complex constants, and z_{N2}^*, z_{T2}^* denote the symmetric points to z_{N2}, z_{T2} with respect to the unit circle, i.e., $z_{N2}^* = 1/\bar{z}_{N2}$, $z_{T2}^* = 1/\bar{z}_{T2}$, and ζ_N^*, and ζ_T^* are symmetric points to ζ_N, and ζ_T with respect to the circle $|\zeta| = R$, i.e., $\zeta_N^* = R^2/\bar{\zeta}_N$, and $\zeta_T^* = R^2/\bar{\zeta}_T$ (see §8.2). Moreover,

$$\frac{\zeta_T\, \zeta_N^*}{\zeta_N\, \zeta_T^*} = \left(\frac{z_{T2}\, z_{N2}^*}{z_{N2}\, z_{T2}^*} \right)^{1/\kappa_2}, \tag{13.2.25}$$

where z_{T2} and z_{N2} are the points at the trailing edge and at a point midway between the nose and the center of curvature of the secondary airfoil. From (13.2.24) we find that $\dfrac{d\zeta}{dz} \to 1$ as $\zeta \to \infty$ only if

$$\zeta_T + \zeta_T^* - \zeta_N - \zeta_N^* = \frac{1}{\kappa_2} \left(z_{T2} + z_{T2}^* - z_{N2} - z_{N2}^* \right). \tag{13.2.26}$$

Also, $\dfrac{dz}{df} \to \infty$ when

$$\frac{1}{z - z_{N2}} + \frac{1}{z - z_{T2}} - \frac{1}{z - z_{N2}^*} - \frac{1}{z - z_{N2}^*} = 0, \tag{13.2.27}$$

and $\dfrac{d\zeta}{dg} \to \infty$ when

$$\frac{1}{\zeta - \zeta_N} + \frac{1}{\zeta - \zeta_T} - \frac{1}{\zeta - \zeta_N^*} - \frac{1}{\zeta - \zeta_N^*} = 0. \tag{13.2.28}$$

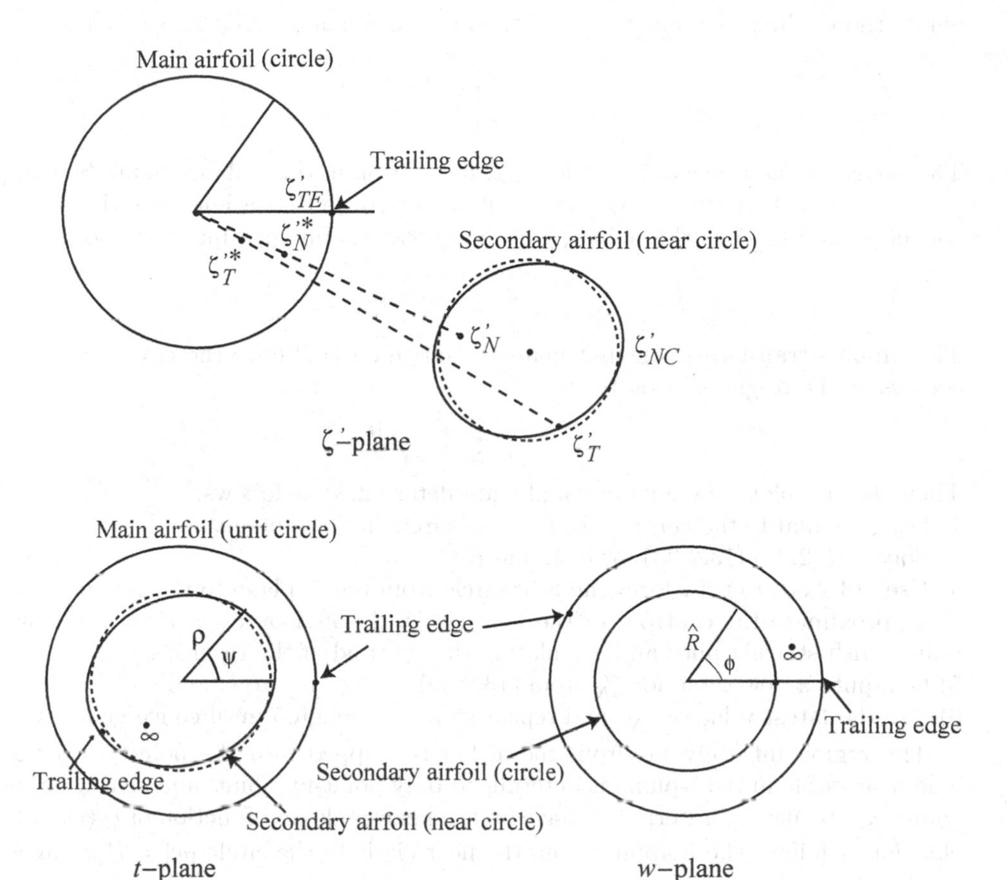

Figure 13.7 Two-element airfoil mappings.

Let the (simple) roots of Eqs (13.2.27) and (13.2.28) be denoted by z_{01}, z_{02} and ζ_{01}, ζ_{02}, respectively. A further analysis shows that $\arg\{f\} = \arg\{g\}$ at these roots which are singular points, although the magnitudes of these roots, in general, are different. Thus, we have

$$\left| \frac{z_{0m} - z_{T2}}{z_{0m} - z_{N2}} \cdot \frac{z_{0m} - z^*_{T2}}{z_{0m} - z^*_{N2}} \right|^{1/\kappa_2} = \left| \frac{\zeta_{0m} - \zeta_T}{\zeta_{0m} - \zeta_{N2}} \cdot \frac{\zeta_{0m} - \zeta^*_T}{\zeta_{0m} - \zeta^*_N} \right|, \quad m = 1, 2, \tag{13.2.29}$$

in order that $|dz/d\zeta|$ or its inverse is regular at these singular points. In this analysis all quantities except a real R and the complex variable ζ_N can be computed from (13.2.24) and (13.2.25). The two unknowns R and ζ_N can be computed from (13.2.25) by using $\bar{\zeta}_T = R^2/\zeta^*_T$.

The above mapping transforms the secondary airfoil into a near-circle and at the same time transforms the main airfoil into a circle of radius $\hat{R}$ (w-plane, Figure 13.7). This mapping is nonsingular at all points except z_N, z_T, z^*_N, z^*_T, ζ_N, ζ_T, ζ^*_N and ζ^*_T. The near-circle is mapped onto the unit circle by the Möbius transformation

$$t = a\,\frac{\zeta' + b}{\zeta' + c}, \tag{13.2.30}$$

where the trailing edge image ζ'_{TE} of the main airfoil is mapped into $w = 1$ if

$$a\,\frac{\zeta'_{TE} + b}{\zeta'_{TE} + c} = 1. \tag{13.2.31}$$

The image of the near-circle of the secondary airfoil in the t-plane should be mapped onto $|w| = r$, $r < 1$, such that its center lies at the origin so that a Fourier series used later will converge rapidly. Also the point ζ'_N in the ζ'-plane is mapped into $t = 0$ so that

$$\zeta'_N + b = 0. \tag{13.2.32}$$

The Möbius transformation that maps the circle $\zeta'| = R$ onto the circle $|t| = 1$ with their centers at the origin is given by

$$\frac{1}{\zeta'_N} + c = 0. \tag{13.2.33}$$

Then the complex constants a, b and c are determined as follows:
1) Set ζ_N equal to the centroid of the near-circle in the ζ-plane.
2) Solve (13.2.31)-(13.2.33) for a, b, and c.
3) Use (13.2.30) to transform the near-circle from the ζ'-plane to the t-plane.
4) Approximate the centroid of the near-circle, denoted by ζ'_{NC}, by connecting adjacent points with straight lines and calculating the centroid of the resulting region.
5) Compute a new value for ζ'_N from (13.2.30), using ζ'_{NC}, a, b, and c.
6) Use the latest value of ζ'_N, and repeat steps 2 through 5 until convergence is achieved.

The region infinitely far from the airfoil is mapped into the point ∞ in the t-plane. The near-circle in the t-plane is determined only point-by-point, and thus a periodic cubic spline can be used to interpolate $\log\rho$ on the near-circle as a function of ψ (as in the single-element airfoils). The mapping from the near-circle to the circle $|w| = \hat{R}$ is carried out by the function

$$w = \sum_{j=1}^{N}\left[(-A_{2j} + i\,B_{2j})\left(\hat{R}t\right)^{j} + (A_{2j} + i\,B_{2j})\left(\frac{\hat{R}}{t}\right)^{j}\right], \tag{13.2.34}$$

where $\hat{R} < 1$ is the image of the nearly circular contour of the secondary airfoil in the w-plane. Note that since w is purely imaginary for $t = e^{i\beta}$ for any real β, the mapping (13.2.34) maps the ζ'-plane onto itself in the t-plane.

13.2.4 Multi-Element Airfoils. The von Karman-Trefftz transformation for simultaneous conformal mapping of n airfoils in the z-plane is given by $g(\zeta) = f(z)$, where

$$g(\zeta) = \prod_{j=1}^{n}\frac{\zeta - \zeta_{Tj}}{\zeta - \zeta_{Nj}}, \quad f(z) = \prod_{j=1}^{n}\left(\frac{z - z_{Tj}}{z - z_{Nj}}\right)^{1/\kappa_j}, \tag{13.2.35}$$

where $\kappa_j = 2 - \tau_j/\pi$, τ_j is the trailing edge included angle of the jth airfoil; z_{Tj} and z_{Nj} denote the complex coordinates of the trailing edge and the point midway between the nose and the center of curvature of the jth airfoil, and ζ_{Tj} and ζ_{Nj} are suitably chosen complex coefficients. The condition $\lim\limits_{\zeta\to\infty}|dz/d\zeta| = 1$ is satisfied only if

$$\sum_{j=1}^{n}\left[\zeta_{Tj} - \zeta_{Nj} - \frac{1}{\kappa_j}\,(z_{Tj} - z_{Nj})\right] = 0. \tag{13.2.36}$$

The coordinate system in the ζ-plane is defined by $\zeta_{T1}+\zeta_{N1}=0$. Note that dz/df becomes unbounded when

$$\sum_{j=1}^{n} \frac{1}{\kappa_j}\left(\frac{1}{z-z_{Tj}}-\frac{1}{z-z_{Nj}}\right)=0, \tag{13.2.37}$$

and $d\zeta/dg$ becomes unbounded when

$$\sum_{j=1}^{n}\left(\frac{1}{\zeta-\zeta_{Tj}}-\frac{1}{\zeta-\zeta_{Nj}}\right)=0. \tag{13.2.38}$$

In general, there exist $(2n-2)$ finite roots of Eq (13.2.37), all different from z_{Tj} and z_{Nj}. There are also $(2n-2)$ finite roots of Eq (13.2.38), all different from ζ_{Tj} and ζ_{Nj}. Let these roots (which are also singularities) be denoted by z_{0j} and ζ_{0j}, $j=1,\dots,2n-2$. Then $dz/d\zeta$ and its inverse will be finite at all these singular points only if

$$g\left(\zeta_{0j}\right)=f\left(z_{0j}\right), \quad j=1,\dots,2n-2. \tag{13.2.39}$$

The $2n$ complex quantities ζ_{Tj} and ζ_{Nj}, $j=1,\dots,n$, are uniquely determined from (13.2.35)-(13.2.39) by using an n-dimensional complex Newton-Raphson iteration scheme. For $n=1$ this mapping reduces to that of the single-element airfoils defined by (13.2.10).

Map 13.1. The conformal mapping

$$w=\frac{lz^2+2mz+n}{pz^2+2qz+r} \tag{13.2.40}$$

degenerates to a bilinear transformation if $(lr-np)^2=4(mr-nq)(lq-mp)$. If this condition is not satisfied, then the above mapping can be reduced to the following forms:

$$\begin{aligned}
&\frac{w-\beta}{w-\alpha}=k\left(\frac{z-\mu}{z-\lambda}\right)^2 \quad \text{when } q^2\neq pr,\ lq\neq mp;\\
&\frac{w-\beta}{w-\alpha}=k\,(z-\mu)^2 \quad \text{when } q^2\neq pr,\ lq\neq mp,\ p\neq 0;\\
&w-\beta=k\left(\frac{z-\mu}{z-\lambda}\right)^2 \quad \text{when } q^2=pr,\ lq\neq mp,\ p\neq 0;\\
&w-\beta=\frac{k}{(z-\lambda)^2} \quad \text{when } q^2=pr,\ lq=mp,\ p\neq 0;\\
&w-\beta=k\,(z-\mu)^2 \quad \text{when } p=0,\ q=0,
\end{aligned} \tag{13.2.41}$$

where α and β are two unequal roots of $lz^2+2mz+n-w\left(pz^2+2qz+r\right)=0$, $z=\lambda$ and $z=\mu$ are the two real values that correspond to $w=\alpha$ and $w=\beta$, respectively, and $\lambda\neq\mu$ since $\alpha\neq\beta$. (Piaggio and Strain [1947].)

Map 13.2. Let the circle Γ pass through the point $z=ia$, $a>0$, and let the point $z=-i\in \operatorname{Int}(\Gamma)$. Then the transformation $w=z+a^2/z$ maps the circle Γ onto a Joukowski airfoil. (Pennisi et al. [1963:335].)

Map 13.3. The transformation $w=z+a^2/z$ maps the circle $|z-i|=\sqrt{2}$ onto one half of the unit circle $|w|<1$, and the circle $|z+i|=\sqrt{2}$ onto the other half of $|w|<1$. (Pennisi et al. [1963:335].)

Map 13.4. The mapping $w = \left(\dfrac{z^n+1}{z^n-1}\right)^2$, where $n > 0$ an integer, maps the sector $|z| < 1$, $z = |z|\,e^{i\theta}$, $0 < \theta < \pi/n$, onto the upper half-plane $\Im\{w\} > 0$. [Hint: Use the chain of mappings $\zeta = z^n$ and $w = \left(\dfrac{\zeta+1}{\zeta-1}\right)^2$, in that order, to transform the sector onto the semi-circle $|\zeta| \le 1$ and $\Im\{\zeta\} \ge 0$ onto $\Im\{w\} \ge 0$.] (Pennisi et al. [1963:336-337].)

Example 13.1. Using the formula for density distribution $\mu(s)$ of a charge on an ideally conducting circular cylinder, given by $\mu(s) = \dfrac{e}{2\pi}\left|\dfrac{dz}{d\zeta}\right|^{-1}_{|\zeta=1}$, where e denotes the charge per unit length of the cylinder, show that the charge density in a strip of width $2a$ in the (x, y)-plane along the segment $-a < x < a$ is given by $\mu(x) = \dfrac{e}{2\pi}\dfrac{1}{\sqrt{a^2-x^2}}$. [Hint: Use the mapping (11.1.9).] (Sveshnikov and Tikhonov [1978:216].) ■

Example 13.2. (a) Teardrop wing profile of NACA0010 is shown in Figure 13.8(a); and (b) a composite elliptic wing profile composed of upper and lower curves from two ellipses with the same major axis but a different minor axis is shown in Figure 13.8(b)). ■

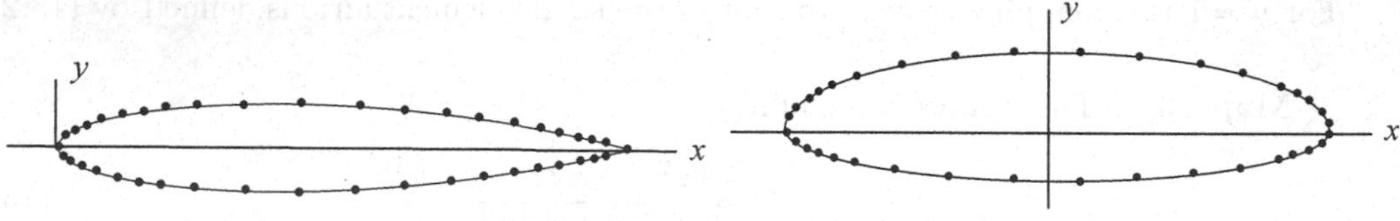

Figure 13.8(a), (b) Airfoils.

13.3 Arbenz's Integral Equation

We will derive the analogues of Gershgorin's integral equation and then obtain Arbenz's integral equation (Arbenz [1958]) which uses Radon's method to determine conformal maps for boundaries with corners and has a unique solution. The cases of interior and exterior mapping functions $f(z)$ and $f_E(z)$ are related to each other through inversion by the relations (10.3.12). We are interested in the behavior of these univalent maps and those of doubly connected regions at singularities on and near the boundary, which are corner-type or pole-type. The nature and location of such singularities are determined.

Let us assume that the boundary Γ of a simply connected region D consists of two Jordan curves Γ_1 and Γ_2 ($\Gamma_1 \cup \Gamma_2 = \Gamma$), and suppose that Γ_1 and Γ_2 meet at a point $z_0 = \gamma(s_0)$ and form a corner with interior angle $\alpha\pi$, $0 < \alpha < 2$. Suppose that z_0 is a regular point of both curves. Let $f(z) \in \mathcal{K}^0(D)$ denote the function that maps D univalently onto the unit disk such that $f(0) = 0$. Let the parametric equation of the boundary Γ be $z = \gamma(s)$, $0 \le s \le L$, which is positively oriented with respect to the region D. Then $f(\gamma(s)) = e^{i\phi(s)}$ and $f_E(\gamma(s)) = e^{i\phi_E(s)}$, where the boundary correspondence functions $\phi(s) = \arg\{f(\gamma(s))\}$ and $\phi_E(s) = \arg\{f_E(\gamma(s))\}$ are continuous principal arguments which play a significant role in integral equation methods. Let the function $\theta(s) = \arg\{\gamma(s)\}$ be defined for $0 \le s \le L$, such that it has at most finitely many jump discontinuities of magnitude less than π in the interval $[0, L]$. This yields finitely many subintervals of $[0, L]$, in each of which $\theta(s)$ is continuous and has bounded variations. Thus, at a corner point on Γ we have $|\theta(s^+) - \theta(s^-)| < \pi$, and the boundary Γ is called a contour with bounded variation. The following result is due to Radon [1919].

Theorem 13.1. *If Γ is a contour with bounded variation, then $\theta_s(t) = \arg\{\gamma(s) - \gamma(t)\}$, defined in (10.1.1), is of bounded variation for every fixed $s \in [0, L]$ and is uniformly bounded for all $s \in [0, L]$.*

The Stieltjes integral equations that arise in Radon's method have the form $\int_0^L \phi(t)\, d_t\theta_s(t) = g(s)$, where $\phi(t)$ is continuous in $[0, L]$, and the subscript t denotes the variable of the Stieltjes integration. In fact,

$$\begin{aligned} \lim_{\varepsilon\to 0} \int_{\tau-\varepsilon}^{\tau+\varepsilon} \phi(t)\, d_t\theta_s(t) &= \phi(t)\,\left[\theta_s(\tau^+) - \theta_s(\tau^-)\right] \\ &= \begin{cases} 0, & \text{if } \tau \neq s, \\ \alpha\pi\, \phi(s), & \text{if } \tau = s, \end{cases} \end{aligned} \tag{13.3.1}$$

where $\alpha\pi = \alpha(s)\pi$, $s \in [0, L]$, denotes the interior angle at the corner point $\gamma(s)$ on the boundary Γ.

To derive the Stieltjes integral equation associated with Gershgorin's integral equation (10.2.6), let $z = \gamma(s)$ be a corner point on Γ. Then, in view of (2.4.10), the left side of Eq (10.2.2) becomes $\alpha \log f(z)$, and then instead of Eq (10.2.6) we obtain

$$\alpha\, \phi(s) = \frac{1}{\pi} \int_0^L \phi(t)\, d_t\theta_s(t) - 2\,\beta(s), \tag{13.3.2}$$

where the integral takes the Cauchy p.v. at $t = s$. Since, by (13.3.1),

$$\begin{aligned} \int_0^L \phi(t)\, d_t\theta_s(t) &= \frac{1}{\pi} \lim_{\varepsilon\to 0} \left(\int_0^{s-\varepsilon} + \int_{s-\varepsilon}^{s+\varepsilon} + \int_{s+\varepsilon}^{L} \right) \phi(t)\, d_t\theta_s(t) \\ &= \int_0^L \phi(t)\, d_t\theta_s(t) + \alpha\pi\, \phi(s), \end{aligned}$$

from (13.3.2) we find that

$$\frac{1}{\pi} \int_0^L \phi(t)\, d_t\theta_s(t) = 2\,\beta(s), \quad s \neq 0. \tag{13.3.3}$$

Since $\displaystyle\int_0^L d_t\theta_s(t) = 0$, the solution of Eq (13.3.3) is determined up to an additive constant, and hence, it is not unique. This situation is avoided in Arbenz's integral equation which can be derived from (13.3.3) as follows: For $s \neq 0$ set

$$\hat{\theta}_s(t) = \begin{cases} \theta_s(t), & t < s, \\ \theta_s(s^-), & t = s, \\ \theta_t(s), & t > s, \end{cases} = \theta_s(t) + \begin{cases} 0, & t < s, \\ \pi, & t > s. \end{cases} \tag{13.3.4}$$

Also, $\theta_s(t) = \hat{\theta}_s(t) - \theta_0(t)$, where

$$\theta_0(t) = \begin{cases} \theta_0(0^+) & \text{for } t = 0, \\ \theta_0(L) = \theta_0(L^-) & \text{for } t = L. \end{cases} \tag{13.3.5}$$

The angle $\theta_s(t)$ is shown in Figure 13.9. Then for $s \neq 0$ we find from (13.3.3) that

$$\begin{aligned}\frac{1}{\pi}\int_0^L \phi(t)\, d_t\theta_s(t) &= \frac{1}{\pi}\int_0^L \phi(t)\, d_t\theta_s(t) + \phi(s) - \frac{1}{\pi}\int_0^L \phi(t)\, d_t\theta_0(t)\\ &= 2\,\beta(s) + \phi(s) - \frac{1}{\pi}\int_0^L \phi(t)\, d_t\theta_0(t).\end{aligned} \tag{13.3.6}$$

To determine the integral in (13.3.6), let $s \to 0^+$ in (13.3.3), and use Hally's theorem which states that $\lim_{s\to 0^+} \theta_s(0) = \theta_0(0) + \pi$. Then the limit value of this integral is given by

$$\frac{1}{\pi}\int_0^L \phi(t)\, d_t\theta_0(t) = \phi(0^+) + 2\,\beta(0^+),$$

which yields

$$\frac{1}{\pi}\int_0^L \phi(t)\, d_t\theta_s(t) = 2\,\beta(s) + \phi(s) - \left[\phi(0^+) + 2\,\beta(0^+)\right].$$

If we require that the boundary correspondence function be $\phi(0^+) = -2\,\beta(0^+)$, then $\phi(s)$ is uniquely determined from the integral equation

$$\frac{1}{\pi}\int_0^L \phi(t)\, d_t\theta_s(t) = \phi(s) + 2\,\beta(s), \quad s \neq 0, \tag{13.3.7}$$

which is known as *Arbenz's integral equation.* Note that the integral in (13.3.7) is not evaluated as a Cauchy p.v. as in (13.3.3).

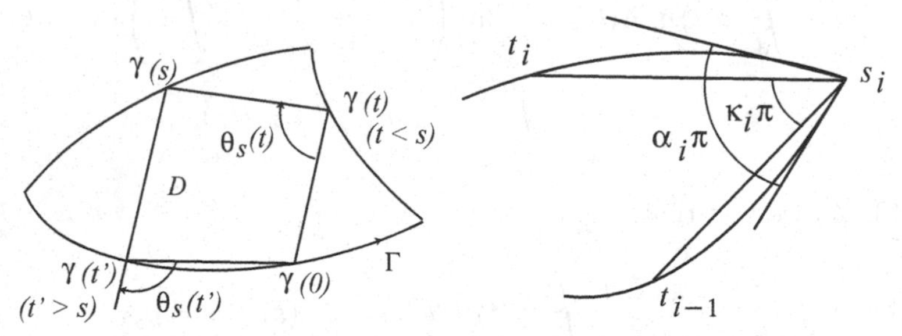

Figure 13.9 Angle $\theta_s(t)$, Figure 13.10 Angle $\kappa + i\pi$.

The discretization method should be used when the boundary Γ is represented geometrically rather than analytically. In order to discretize Arbenz's equation (13.3.7), we partition the interval $[0, L]$ into N subintervals $[t_{k-1}, t_k]$ of equal or unequal length with s_k as an interior point, $t_{k-1} < s_k < t_k$, $k = 1, 2, \dots, N$. In practice, it is useful to take a corner point coincident with a partition point and have more subintervals in its neighborhood. Then for $s = s_i$ Eq (13.3.7) becomes

$$\frac{1}{\pi}\sum_{k\neq i}\int_{t_{k-1}}^{t_k} \left[\phi(t) - \theta(t)\right]\, d_t\theta_{s_i}(t) = \phi\left(s_i\right) + 2\,\beta\left(s_i\right). \tag{13.3.8}$$

The solution $\phi(s)$ is continuous and can be taken as

$$\phi_i = \phi\left(s_i\right) = \sum_{k=1}^{N} a_{ik}\,\phi_k, \tag{13.3.9}$$

where

$$a_{ik} = \frac{1}{\pi} \int_{t_{k-1}}^{t_k} d_t \theta_{s_i}(t). \tag{13.3.10}$$

For $i = k$ the integral in (13.3.10) takes the principal value of Stieltjes integration. In fact, if s_i is a corner point, then the arcs at s_i subtend the interior angle $\alpha_i \pi$. Let $\kappa_i \pi$ denote the angle between the chords (s_i, t_{i-1}) and (s_i, t_i) (Figure 13.10). Then

$$\begin{aligned} a_{ik} &= \frac{1}{\pi} \theta_{s_i}(t_{k-1}, t_k), \quad i \neq k, \\ a_{ii} &= \alpha_i \left(1 - \frac{\kappa_i}{\alpha_i}\right), \end{aligned} \tag{13.3.11}$$

and

$$\sum_{k=1}^{N} a_i k = \frac{1}{\pi} \int_0^L d_t \theta_{s_i}(t) = \alpha_i, \tag{13.3.12}$$

where the integral takes the p.v. of Stieltjes integration.

Map 13.5. (Gaier [1964:57]). For the mapping of the square $\{-1 < x,\ y < 1\}$ onto the unit disk $|w| < 1$, we have discretized the boundary of the square with $N = 40$ subintervals, where $t = 0$ and $t = L$ at the point t_0, although in Figure 13.11 only quarter regions of each boundary are presented because of the symmetry of the square with respect to the x and y axes and the symmetry of $\phi(s)$, i.e., $\phi_{20+j} = \pi + \phi_j$ and $\phi_{10+j} = \pi/2 + \phi_j$.

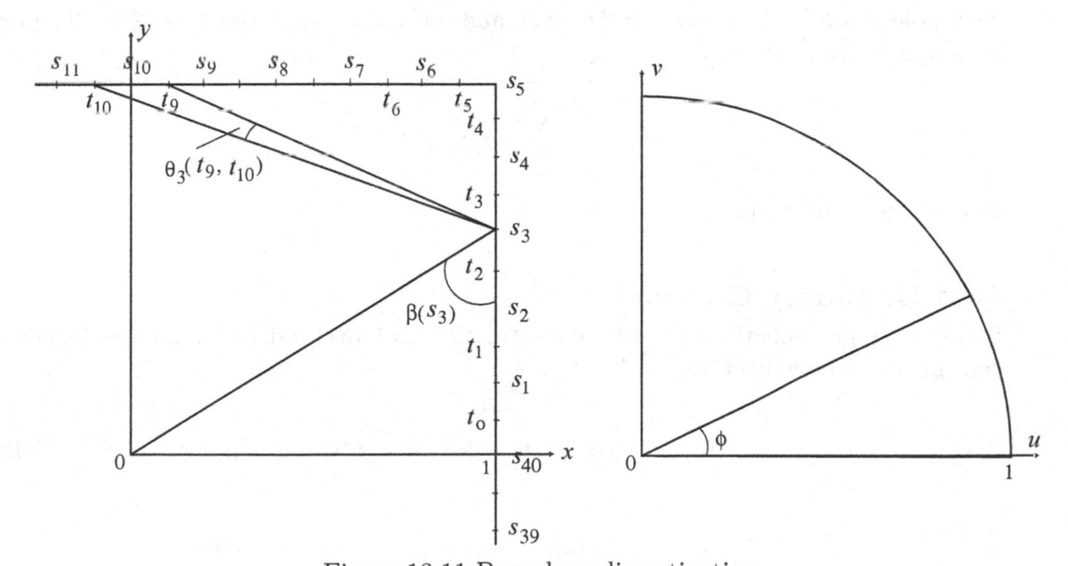

Figure 13.11 Boundary discretization.

Then

$$\frac{1}{\pi} \sum_{k=1}^{40} \left[\arg\{\gamma(t_k) - \gamma(s_i)\} - \arg\{\gamma(t_{k-1}) - \gamma(s_i)\}\right] \phi_k = 2\beta(s_i). \tag{13.3.13}$$

Thus,

$$\frac{1}{\pi} \sum_{k=1}^{40} \theta_i \left(t_{k-1}, t_k\right) \phi_k = 2\,\beta\left(s_i\right). \tag{13.3.14}$$

The coefficient for ϕ_k is 0.5 for $k = i$ when $i = 5, 15, 25, 35$, and is equal to 1 otherwise. The following results, obtained after computing Eqs (13.3.3) and (13.3.6), are compared with the exact solutions in the following table:

i	Eq (13.3.3)	Eq (13.3.6)	Exact
1	0.256503	0.256738	0.256319
2	0.480278	0.481440	0.479890
3	0.648856	0.654261	0.648240
4	0.751901	0.789932	0.751028
5	0.785394	0.785406	0.785398
6	0.818887	0.780878	0.819768
7	0.921933	0.916549	0.922556
8	1.090511	1.089369	1.090906
9	1.314286	1.314071	1.314477

The exact boundary correspondence function is given by

$$\cos\phi = \operatorname{dn}(K\,y), \quad K = K\left(\frac{1}{\sqrt{2}}\right), \tag{13.3.15}$$

where dn is one of the twelve Jacobian elliptic functions which is a meromorphic function with pole at $i\,K'(m)$, $K(m) = K'(1-m)$, and $\operatorname{dn}^2 u = 1 - \operatorname{sn}^2 u$ (see Case 7.2). The inverse mapping is given by

$$z = A \sum_{n=0}^{\infty} \frac{\binom{1/2}{n}}{4n-1}\, w^{-4n+1}, \tag{13.3.16}$$

where A is a constant.

13.4 Boundary Corner

Symm's integral equations (10.1.6) and (10.1.12)-(10.1.13) for determining the density function $\mu(s)$ can be written as

$$\int_0^L \mu(s)\,\log|\gamma(t) - \gamma(s)|\,ds = k(t), \quad t \neq s, \tag{13.4.1}$$

where

$$k(t) = \begin{cases} -\log|\gamma(t)| & \text{for } z = \gamma(s) \in D = \operatorname{Int}(\Gamma), \\ 1 \quad \text{for } z = \gamma(s) \in D^* = \operatorname{Ext}(\Gamma). \end{cases} \tag{13.4.2}$$

This equation has a unique solution provided that the capacity $\operatorname{cap}(\Gamma) \neq 1$ (see §2.5, 2.6). The density function $\mu(s)$ is related to the boundary correspondence function $\phi(s)$ by

$$\phi'(s) = \begin{cases} -2\,\pi\,\mu(s) & \text{for } z \in \operatorname{Int}(\Gamma), \\ 2\,\pi\,\hat{\gamma}\,\mu(s) & \text{for } z \in \operatorname{Ext}(\Gamma), \end{cases} \tag{13.4.3}$$

where $\hat{\gamma}$ is Robin's constant ($\hat{\gamma} = -\log\{\text{cap}(\Gamma)\}$, see §2.5).

If z_0 is a corner point on a portion of the boundary Γ with interior angle $\alpha\pi$, where $\alpha = p/q$ (p and q are relatively prime), then we should analyze the asymptotic behavior of the mapping function in the neighborhood of the point $z_0 \in \Gamma$. If D is a polygonal region, we know from the Schwarz-Christoffel transformation that in the neighborhood of z_0

$$f(z) - f(z_0) = \sum_{k=1}^{\infty} a_k (z - z_0)^{k/\alpha}, \tag{13.4.4}$$

or, if $f \in \mathcal{K}^0(D)$ (i.e., $f(0) = 0$), then

$$f(z) = \sum_{k=1}^{\infty} a_k \left[(z - z_0)^{k/\alpha} - (-z_0)^{k/\alpha}\right], \quad a_1 \neq 0, \tag{13.4.5}$$

(see Copson [1975: 170]). Thus, if $1/\alpha$ is not an integer, then $f(z)$ has a branch point singularity at z_0 (see §8.5). This corresponds to Case 3 in Theorem 13.2 given below. Lichtenstein [1911] was the first to show that if a corner point is located at the origin, then $\dfrac{df(z)}{dz} = z^{1/\alpha-1}\, h(z)$, where $h(z)$ is a continuous function such that $h(0) \neq 0$, and α is irrational. Warschawski [1932; 1955] proved this result for all α. In the case when the two arcs at z_0 are straight line segments with $\alpha = 1$, Lewy [1950] proved the stronger result that $f(z)$ has an asymptotic expansion in powers of z and $\log z$. This result was generalized by Lehman [1957] in the development of the mapping function at an analytic corner as $z \to z_0$, as follows:

Theorem 13.2. *Let $z_0 \in \Gamma$ denote a corner point. Then there are three cases of asymptotic expansions for the mapping function $f(z)$ as $z \to z_0$:*
CASE 1. *If α is rational, $\alpha = p/q$ where p and q are relatively prime, then as $z \to z_0$*

$$f(z) = f(z_0) + \sum_{k,l,m} B_{k,l,m}\, (z - z_0)^{k+l/\alpha} \left(\log (z - z_0)\right)^m, \quad B_{0,1,0} \neq 0, \tag{13.4.6}$$

where k, l and m are integers, $k \geq 0$, $1 \leq l \leq p$ and $0 \leq m \leq k/q$. The terms in (13.4.6) are ordered such that the term containing $B_{k',l',m'}$ always follows the term containing $B_{k,l,m}$ if $k' + l'/\alpha \geq k + l/\alpha$ and $m' < m$.
CASE 2. *If α is irrational, then as $z \to z_0$*

$$f(z) = f(z_0) + \sum_{k,l} B_{k,l}\, (z - z_0)^{k+l/\alpha}, \quad B_{0,1} \neq 0, \tag{13.4.7}$$

where k and l are integers, $k \geq 0$ and $l \geq 1$.
CASE 3. *In the case when the two boundary arcs at z_0 are straight line segments, then the asymptotic expansions (13.4.6) and (13.4.7) simplify to*

$$f(z) = f(z_0) + \sum_{l=1}^{\infty} B_l\, (z - z_0)^{l/\alpha}, \quad B_1 \neq 0. \tag{13.4.8}$$

Cases 1 and 2 correspond to the situation when D is not a polygonal region, whereas Case 3 applies when D is a polygonal region. In this case, to eliminate the effect of a branch point singularity at z_0, if such a singularity exists, we augment the basis $\{\phi_j(z)\}_{j=1}^n$ by the functions (Papamichael and Kokkinos [1981])

$$\phi(z) = \frac{d}{dz}\left\{(z-z_0)^{k/\alpha} - (-z_0)^{k/\alpha}\right\} - d = \frac{k}{\alpha}\,(z-z_0)^{k/\alpha-1} - d, \tag{13.4.9}$$

where

$$d = \begin{cases} \dfrac{k}{\alpha}\,(-z_0)^{k/\alpha-1} & \text{for Ritz Method,} \\ 0 & \text{for Bergman Kernel Method.} \end{cases}$$

Note that $(z-z_0)^{1/\alpha}$ is the dominant term in each of the asymptotic expansions (13.4.6)-(13.4.8). It appears that the mapping $z \mapsto z^{1/\alpha}$ that transforms an angle $\alpha\pi$ at $z_0 \in \Gamma$ into the angle π at the point $w_0 = f(z_0)$ will solve the corner problem. But this does not happen because if $1/\alpha$ is not an integer, a branch singularity always occurs at the corner z_0, and when α is an integer, the existence of the logarithm in (13.4.6) makes the corner z_0 a logarithmic branch point singularity even if $1/\alpha$ is an integer.

In the Ritz Method and the Bergman Kernel Method (§8.5 and §8.6) the minimum polynomial is constructed by taking the basis set as that of monomials z^j, $j = 0, 1, 2, \dots$. Then the singular basis function $\phi(z)$ associated with the corner singularity at $z_0 \in \Gamma$ in the above three cases of asymptotic expansions (13.4.6)-(13.4.8) has the form

$$\phi(z) = \begin{cases} \dfrac{d}{dz}\left\{(z-z_0)^{k+l/\alpha}\,(\log\,(z-z_0))^m\right\} & \text{in Case 1,} \\ \dfrac{d}{dz}\left\{(z-z_0)^{k+l/\alpha}\right\} & \text{in Case 2,} \\ \dfrac{d}{dz}\left\{(z-z_0)^{l/\alpha}\right\} & \text{in Case 3,} \end{cases} \tag{13.4.10}$$

which is used to augment the basis set in the RM and the BKM when determining the mapping function $f(z)$ from (8.5.25). It may be noted that the function $f(z)$, originally defined on D, can be extended by analytic continuation through a portion of its boundary into Ext (Γ). This procedure is used in §13.6 to investigate the nature and location of poles and pole-type singularities of $f(z)$ near the boundary Γ. The singularities of the Bergman kernel $K(z,a)$ in the region Ext (Γ) also affect the convergence of the polynomial series (8.5.20). These singularities are either poles of $K(z,a)$ that lie close to the boundary Γ or branch point singularities of the boundary itself. Their effect should always be taken into account when determining the mapping function for the exterior regions (see §13.7).

13.5 Singularity Behavior

In the neighborhood of a corner point $z_0 = \gamma\,(s_0)$ let the function $\gamma(s)$ have a Taylor series representation

$$\gamma(s) = \gamma\,(s_0) + \begin{cases} \sum\limits_{n=1}^{\infty} \dfrac{(s-s_0)^n}{n!}\,\gamma^{(n)}\left(s_0^-\right), & s < s_0, \\ \sum\limits_{n=1}^{\infty} \dfrac{(s-s_0)^n}{n!}\,\gamma^{(n)}\left(s_0^+\right), & s > s_0, \end{cases} \tag{13.5.1}$$

where $\gamma^{(n)}\left(s_0^\pm\right) = \lim\limits_{s\to s_0^\pm}\left\{\dfrac{d^n\gamma(s)}{ds^n}\right\}$. The boundary correspondence function $\phi(s)$ associated with the mapping function f is given by

$$\phi'(s) = i\,\frac{f'(\gamma(s))\,\overline{f(\gamma(s))}}{|f(\gamma(s))|^2}. \tag{13.5.2}$$

Then, by (13.4.3), the density function $\mu(s)$ is related to f by

$$\mu(s) = -\frac{\Im\left\{f'(\gamma(s))\,\overline{f(\gamma(s))}\right\}}{2\pi}, \tag{13.5.3}$$

where $|f(\gamma(s))| = 1$. Note that for the mapping function f_E this relation is, in view of (13.4.3), given by

$$\mu(s) = -\frac{\Im\left\{f_E'(\gamma(s))\,\overline{f_E(\gamma(s))}\right\}}{2\pi\,\hat{\gamma}}. \tag{13.5.4}$$

Hence, using (13.4.6)-(13.4.8), (13.5.1), and (13.5.3), a formal asymptotic expansion for $\mu(s)$ as $s \to s_0$ is given by

$$\mu(s) = \begin{cases} \sum_{j=1}^{\infty} a_j^-\,\psi_j(s-s_0), & s < s_0, \\ \sum_{j=1}^{\infty} a_j^+\,\psi_j(s-s_0), & s > s_0, \end{cases} \tag{13.5.5}$$

where the functions ψ_j depend on the value of α, $0 < \alpha < 2$. Then from (13.5.5) we conclude the following:

(a) If $1 < \alpha < 2$, i.e., if the corner z_0 is re-entrant, then $\mu(s)$ becomes unbounded at $s = s_0$.

(b) If $\dfrac{1}{1+q} < \alpha < 1$, where $q \ge 1$ is an integer, then $\mu(s)$ becomes unbounded at $s = s_0$.

(c) If $\alpha = \dfrac{1}{q}$, where $q \ge 1$ is an integer, then (13.5.5) does not involve rational powers of $(s - s_0)$. Since $a_1^- \ne a_1^+$, the function $\mu^{(q-1)}(s)$ has a jump discontinuity at $s = s_0$. Moreover, for some $j > 1$, one of the functions ψ_j in (13.5.5), obtained from the expansion (13.4.6), is a function of the form $\sigma^{2q-1}\log\sigma$, where σ stands for $(s - s_0)$ or $(s_0 - s)$. Thus, in general, the left and right $(2q-1)$th derivatives of $\mu(s)$ at $s = s_0$ become unbounded.

(d) In Case 3 of Theorem 13.2, without loss of generality, we take $\gamma(s)$ in the form

$$\gamma(s) = \gamma(s_0) + \begin{cases} (s_0 - s)\,e^{i\alpha\pi}, & s \le s_0, \\ s - s_0, & s \ge s_0. \end{cases} \tag{13.5.6}$$

Then, $\psi_j(\sigma) = \sigma^{-1+j/\alpha}$ for $j = 1, 2, \dots$, and $a_j^+ = (-1)^{j+1}\,a_j^-$ for $j = 1, 2, \dots$. In this case conclusions (a) and (b) remain the same, but (c) changes, viz., if $\alpha = 1/q$ ($q \ge 1$ an integer), then

(d.1) if q is odd, then $\mu(s)$ has no singularity at $s = s_0$; and

(d.2) if q is even, then $\mu^{(q-1)}(s)$, in general, has a jump discontinuity at $s = s_0$.

13.6 Pole-Type Singularities

Sometimes the function f (or f_E) has poles or pole-type singularities adjacent to the boundary Γ, which are located in the region D^* (or D). We will examine the nature and location of such singularities. This analysis, based on the work of Papamichael et al. [1986], will be confined to the function f. Thus, we will determine the nature and location of poles and pole-type singularities of f in D^*, which are obtained by considering an analytic continuation of f across the boundary Γ into D^*. The procedure expands the domain of f as much as possible provided the function f on the extended domain agrees with the original.

To determine the dominant poles of the mapping function $w = f(z)$ in $\text{Ext}\,(\Gamma)$, which are actually the poles of the analytic continuation of f across Γ into $\text{Ext}\,(\Gamma)$ that lie close to Γ, we can use the Schwarz reflection principle in the case when the boundary Γ consists of straight line segments and circular arcs. Then f has simple poles at the (finite) symmetric points of the origin with respect to the straight line segments and the circular arcs (see §2.6). But when the boundary Γ is more analytic than straight line segments and circular arcs, then in many cases the dominant poles of f can be determined by using a generalized symmetry principle as follows: Let the parametric equation of an analytic arc $\hat{\Gamma}$ of Γ be given by

$$z = \gamma(s), \quad s_1 < s < s_2. \tag{13.6.1}$$

Let G^* be a simply connected region in the complex ζ-plane, $\zeta = s + i\,t$, that satisfies the following two conditions:
$C1$. The function $z = \gamma(s)$ is univalent in G^*.
$C2$. The straight line $L = \{\zeta : \zeta = s + i\,t,\ s_1 < s < s_2,\ t = 0\}$ divides the region G^* into two partitions G_1 and G_2 which are symmetric to each other with respect to L and $G^* = G_1 \cup L \cup G_2$, and the image D_1 of G_1 under the transformation (13.6.1) is contained in D $(D_1 \subseteq D)$.

Obviously, G_1 and G_2, each subsets of G^*, are defined by

$$\begin{aligned} G_1 &= \{\zeta : \zeta \in G^* \text{ and } t > 0\}, \\ G_2 &= \{\zeta : \zeta \in G^* \text{ and } t < 0\}, \end{aligned} \qquad \text{or} \qquad \begin{aligned} G_1 &= \{\zeta : \zeta \in G^* \text{ and } t < 0\}, \\ G_2 &= \{\zeta : \zeta \in G^* \text{ and } t > 0\}. \end{aligned}$$

Under conditions (i) and (ii) above, the function (13.6.1) maps the region G^* conformally onto a region $D^* = D_1 \cup \hat{\Gamma} \cup D_2$ such that the regions G_1, G_2 and the straight line L are mapped onto the regions D_1, D_2 and the arc $\hat{\Gamma}$, respectively. Then the function

$$h(\zeta) = f\left(\gamma(\zeta)\right), \tag{13.6.2}$$

where $f(z)$ maps the region D onto the disk $|w| < R$, is univalent in the region $G_1 \cup L$, and $w = h(\zeta)$ maps the straight line L onto an arc of the circle $|w| = R$. Thus, by the reflection principle the function

$$H(\zeta) = \begin{cases} h(\zeta) & \zeta \in G_1 \cup L, \\ \dfrac{1}{\overline{\overline{h(\bar{\zeta})}}}, & \zeta \in G_2, \end{cases} \tag{13.6.3}$$

is meromorphic in G_2 and defines an analytic continuation of h across L into G_2. If η denotes the inverse of γ, then the function

$$F(z) = H\left(\eta(z)\right) = \begin{cases} f(z), & z \in D_1 \cup \hat{\Gamma}, \\ \dfrac{1}{\overline{\overline{f\left(\beta(z)\right)}}}, & z \in D_2, \end{cases} \tag{13.6.4}$$

where $\beta(z) = \gamma\left(\overline{\eta(z)}\right)$ is analytic in D_1 and meromorphic in D_2, defines an analytic continuation of f across $\hat{\Gamma}$ into D_2, where the points z and $\beta(z)$ are symmetric points with respect to the arc $\hat{\Gamma}$. Hence, we have the following result (Papamichael and Kokkinos [1981]):

Theorem 13.3. *The following cases hold:*
(a) If $0 \in D_1$, *then* $\gamma(z)$ *has exactly one zero in* G_1, *i.e., the function* $F(z)$ *has a simple pole at a point* $z_0 \in D_2$, *where* $z_0 = \gamma\left(\bar{\zeta}_0\right) = a(0)$ *is the inverse point of the origin with respect to the arc* $\hat{\Gamma}$.
(b) If $0 \in \partial D_1 \backslash \hat{\Gamma}$, *then* $\gamma(z)$ *has at least one zero* $\zeta_0 \in \partial G_1 \backslash L$, *and the function* $\gamma(z)$ *need not be one-to-one in the neighborhood of the points* ζ_0 *and* $\bar{\zeta}_0$.
(c) If $0 \in D_1 \cup \left(\partial D_1 \backslash \hat{\Gamma}\right)$, *then* F *has no poles in the region* $D_2 \cup \left(\partial D_2 \backslash \hat{\Gamma}\right)$.

To determine the behavior of $F(z)$ at the point $z_0 = \gamma(\underline{\zeta_0}) \in \partial D_2 \backslash \hat{\Gamma}$ in part (b) of the above theorem, let us assume that γ is analytic at ζ_0 and $\bar{\zeta}_0$. Then

$$\gamma(\zeta) = (\zeta - \zeta_0)^m \ \gamma_1(\zeta), \tag{13.6.5}$$

and

$$\gamma(\zeta) - \gamma(\bar{\zeta}_0) = (\zeta - \zeta_0)^n \ \gamma_1(\zeta), \tag{13.6.6}$$

where γ_1 and γ_2 are analytic and nonzero at the points ζ_0 and $\bar{\zeta}_0$, respectively. Then the mapping function is $f(z) = z\, f_1(z)$, $f_1(0) \neq 0$. Hence, from (13.6.4), for $z \in D_2$ the function $G(z) = \dfrac{1}{F(z)}$ can be written as

$$G(z) = \overline{\beta(z)}\, g_1(z), \quad g_1(z_0) \neq 0, \tag{13.6.7}$$

where g_1 is analytic at z_0. Since

$$\overline{\beta(z)} = (\eta(z) - \eta(z_0))^m \ a_1(z), \quad a_1(z_0) \neq 0, \tag{13.6.8}$$

we find that

$$\eta(z) - \eta(z_0) = (z - z_0)^{1/n}\, \eta_1(z), \quad \eta_1(z_0) \neq 0, \tag{13.6.9}$$

where a_1 and η_1 are analytic at z_0. Hence, from (13.6.8) and (13.6.9) we get

$$\overline{\beta(z)} = (z - z_0)^{m/n} \ (\eta_1(z))^m \ a_1(z),$$

and thus, from (13.6.7)

$$G(z) = (z - z_0)^{m/n}\, g_2(z), \quad g_2(z_0) \neq 0, \tag{13.6.10}$$

where g_2 is analytic at z_0. This leads to the following result.

Theorem 13.4. *The following cases hold:*
(a) If $m = n = 1$, *then* F *has a simple pole at* z_0.
(b) If $m = 2$, $n = 1$, *then* F *has a double pole at* z_0.

(c) If $m = 1$, $n = 2$, then F has a branch singularity of the form $(z - z_0)^{-1/2}$ at the point z_0.

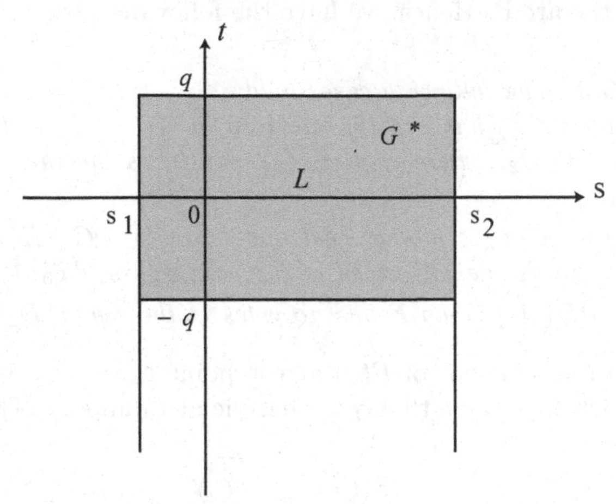

Figure 13.12 Region G^* .

Map 13.6. Let the arc $\hat{\Gamma}$ be defined by the ellipse $E : \dfrac{(x - x_c)^2}{a^2} + \dfrac{(y - y_c)^2}{b^2} = 1$, $a > b$, and let the parametric equation of $\hat{\Gamma}$ be $z = \gamma(s) = z_c + a\,e\,\cos(s - i\,q)$, $0 \le s_1 < s < s_2 < 2\pi$, where $z_c = x_c + i\,y_c$ is the center C , $e = \sqrt{1 - b^2/a^2}$ the eccentricity of the ellipse, $\cosh q = 1/e$, and $s_2 - s_1 < 2\pi$. Then the function $z = \gamma(\zeta)$, $\zeta = s + i\,t$, is univalent in the strip $\{\zeta : \zeta = s + i\,t,\ s_1 < s < s_2,\ -\infty < t < q\}$, and the region G^* is a symmetric subregion of the rectangle $\big\{\zeta : \zeta = s + i\,t,\ s_1 < s < s_2,\ -q < t < q\big\}$ (see Figure 13.12). Consider the case when G^* is the entire rectangle. Then the region $D^* = D_1 \cup \hat{\Gamma} \cup D_2$ can be determined by finding the images of the four sides of this rectangle under the transformation $z = \gamma(\zeta)$, $\zeta = s + i\,t$. Assuming that the regions G_1 and G_2 are defined by (13.6.1), the four typical regions D^* are presented in Figure 13.13(a)-(d) which correspond to the following four cases, respectively:

$$\text{(a)}\quad s_1 = 0,\ 0 < s_2 \le \frac{\pi}{2}; \qquad \text{(b)}\quad s_1 = 0,\ \frac{\pi}{2} < s_2 \le \pi;$$

$$\text{(c)}\quad s_1 = 0,\ \pi < s_2 \le \frac{3\pi}{2}; \qquad \text{(d)}\quad s_1 = 0,\ \frac{3\pi}{2} < s_2 < 2\pi.$$

In each Figure 13.13, the equations of the different arcs are

$$\begin{aligned}
\hat{\Gamma} &= \text{arc}\,\widehat{PQ} : \{z = z_c + a\,e\,\cos(s - i\,q),\quad 0 < s < s_2\}\,,\\
\hat{\Gamma}' &= \text{arc}\,\widehat{P'Q'} : \{z = \gamma(s - i\,q),\quad 0 < s < s_2\}\,,\\
\hat{\Gamma}^* &= \text{arc}\,\widehat{Q'R} : \{z = \gamma(s_2 + i\,t),\quad -q < t < q\}\,.
\end{aligned}$$

Arc $\hat{\Gamma}'$ is that of an ellipse E', and Γ^* that of a hyperbola orthogonal to both ellipses E and E' (the right branch of the hyperbola if $\cos s_2 > 0$ and the left branch if $\cos s_2 < 0$). In the case when $s_2 = \pi/2$ or $3\pi/2$, the hyperbola degenerates into a vertical straight line through the center C (R coincides with C and $\hat{\Gamma}$ becomes a part of the minor axis), whereas when $s = \pi$, it degenerates into the major axis, R coincides with the focus F_2, and $\hat{\Gamma}$ becomes a part of the major axis. The region D_1 is shaded in each figure, and D_2 is the region bounded by the arcs $\hat{\Gamma}$, $\hat{\Gamma}'$ and the subarc $\overset{\frown}{Q'Q}$ of Γ^*. Note that in figures (c) and (d) the region D_1 includes a cut on the major axis from F_2 to R (because in this case the mapping $z = \gamma(\zeta)$ yields a common image $z_c - a\,e\,\cos s$ of the points $(\pi \pm s) + i\,q$, $s > 0$).

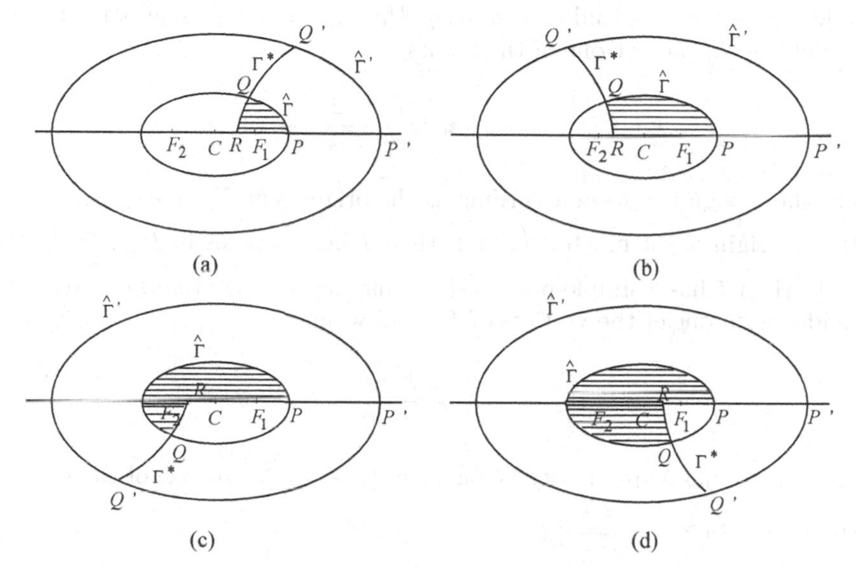

Figure 13.13 Different arcs.

The region D^* associated with any arc $\hat{\Gamma}$ with $0 \leq s_1 < s < 2\pi$ can be obtained from Figure 13.13 (a)-(d). For example, the region D^* associated with an arc $\hat{\Gamma}$ for $0 < s_1 < \pi/2$ and $\pi < s_2 < 3\pi/2$ is obtained by deleting the region of figure (a) from that of figure (c). The region D^* associated with an arc $\hat{\Gamma}$, which includes the two vertices $z_c \pm a$ of the ellipse E can also be obtained from these four figures. For example, if $-\pi/2 < s_1 < 0$ and $\pi < s_2 < 3\pi/2$, then the region D^* is the union of the region of figure (c) with the region obtained by reflecting the region of figure (a) on the major axis.

Now, using the results of Theorems 13.3 and 13.4, we conclude that $\gamma(\zeta)$ has exactly one zero in G_1 at the point $\zeta_0 = \cos^{-1}\left(-\dfrac{z_c}{a\,e}\right) + i\,q$, which means that the function f has a simple pole at the point

$$z_0 = \gamma\left(\bar{\zeta}_0\right) = z_c - \frac{(a^2 + b^2)\,\bar{z}_c - 2\,i\,a\,b\,\sqrt{a^2 - b^2 - \bar{z}_c^2}}{a^2 - b^2} \in D_2, \tag{13.6.11}$$

where the square root is chosen such that $0 < \arg\left\{\sqrt{a^2 - b^2 - \bar{z}_c^2}\right\} < \pi$.

If $0 \in \partial D_1 \backslash \hat{\Gamma}$, then the origin lies on the major axis between the foci F_1 and F_2, i.e., $-a\,e \leq x_c \leq a\,e$ and $y_c = 0$. Then, the following three situations arise:
(i) If the origin lies on a cut in the region D_1 but does not coincide with either focus of E, then there are two distinct values of $\cos^{-1}(-x_c/a\,e)$ in the interval (s_1, s_2), and associated with these two values there are two distinct zeros of $\gamma(\zeta)$ on the side $t = q$ of G_1. Hence, f has two simple poles at the two points

$$z_0 = \frac{-2\,b^2\,x_c \pm 2\,i\,a\,b\,\sqrt{a^2 - b^2 - x_c^2}}{a^2 - b^2} \in \hat{\Gamma}'. \tag{13.6.12}$$

(ii) If the origin does not lie on a cut of D_1 and does not coincide with either focus of E, then there is exactly one value of $\cos^{-1}(-x_c/a\,e)$ in the interval (s_1, s_2), and so $\gamma(\zeta)$ has a zero on the side $t = q$ of G_1. Thus, f has a simple pole at z_0 given by (13.6.12), where a proper sign is chosen so that z_0 lies on $\hat{\Gamma}'$.
(iii) If the origin coincides with either focus of E, i.e., $x_c = \pm a\,e$, $y_c = 0$, then $\gamma(\zeta)$ has a double zero at $\zeta_0 = i\,q$ and $\zeta_0 = \pi + i\,q$. Hence, f has a double pole at one of the vertices of the ellipse E', i.e., at one of the points

$$z_0 = \pm\frac{2\,b^2}{\sqrt{a^2 - b^2}}, \tag{13.6.13}$$

where the $\pm$ sign is chosen according as the origin is at F_1 or F_2.

If the origin is not in $D_1 \cup \left(\Gamma\backslash\hat{\Gamma}\right)$, then f has no poles in $D_2 \cup \hat{\Gamma}'$. If the origin lies in $\partial D_1\backslash\hat{\Gamma}$, then f has a simple pole at the point z_0 given by (13.6.11), except when the origin coincides with one of the vertices of E', i.e., when

$$x_c = \pm\frac{a^2 + b^2}{\sqrt{a^2 - b^2}}, \quad \text{and} \quad y_c = 0. \tag{13.6.14}$$

In this case f has a singularity of the form $(z - z_0)^{-1/2}$ at one of the foci of E, i.e., at one of the points $z_0 = \pm\dfrac{2\,b^2}{\sqrt{a^2 - b^2}}$.

Map 13.7. (a) Let the boundary Γ of the region D be the union of an elliptic curve Γ_1 and the straight line segment Γ_2, defined by

$$\Gamma_1 = \{z : z = 4\cos s - 2\,e + i\,b\sin s,\ -\pi/2 < s < \pi/2,\ 0 < b < 4\},$$
$$\Gamma_2 = \{z : z = x + i\,y,\ x = -2\,e,\ -b < y < b\},$$

where $e = \dfrac{\sqrt{16 - b^2}}{2}$ (see Figure 13.14).

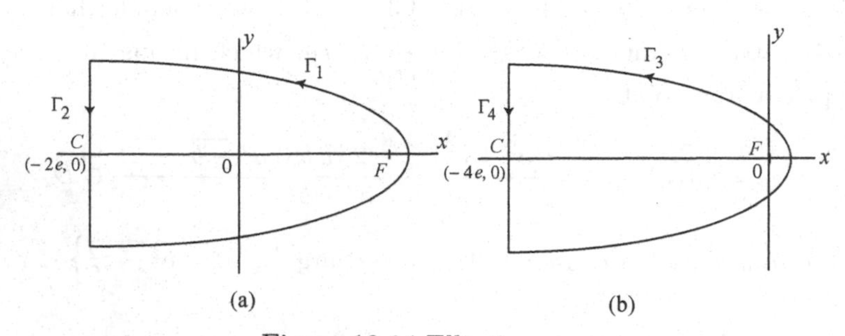

Figure 13.14 Elliptic curve.

There are two poles of f with respect to the curve Γ_1, and in view of (13.6.12) they are at

$$z_{1,2} = \frac{b^2 \pm 4\sqrt{3}\,b\,i}{\sqrt{16 - b^2}}.$$

There is one pole with respect to the line Γ_2 at $z_3 = -\sqrt{16 - b^2}$, which is the mirror image of 0 in Γ_2.

(b) If we translate the region D by $2e$ in the negative x direction, then the origin 0 coincides with the focus F_1, and the new region D' is bounded by arcs

$$\Gamma_3 = \{z : z = 4(\cos s - e) + i\, b \sin s,\ -\pi/2 < s < \pi/2\},$$
$$\Gamma_4 = \{z : z = x + i\, y,\ x = -4\, e,\ -b < y < b\}.$$

Then, in view of (13.6.13), the function f has a double pole with respect to the curve Γ_3 at $z_4 = \dfrac{2\, b^2}{\sqrt{16 - b^2}}$, and with respect to the line Γ_4 it has a simple pole at $z_5 = -2\sqrt{16 - b^2}$, which is the mirror image of O in Γ_4. Note that the boundary of the region D' is very close to the origin. In such a situation the mapping function f is connected to the mapping function f_1 of part (a) by

$$f(z) = \frac{|\alpha|}{\alpha} \frac{f_1(z) - \alpha}{1 - \bar{\alpha} f_1(z)}, \quad \alpha = f_1\left(\frac{ae}{2}\right).$$

Map 13.8. Let the region D be bounded by the straight line segments

$$\overline{AB} : \{z : z = x + i\, y,\ -2 < x < 2,\ y = -1/3\},$$
$$\overline{BC} : \{z : z = x + i\, y,\ x = 2,\ -1/3 < y < 1\},$$
$$\overline{AE} : \{z : z = x + i\, y,\ x = -2,\ -1/3 < y < 1\},$$

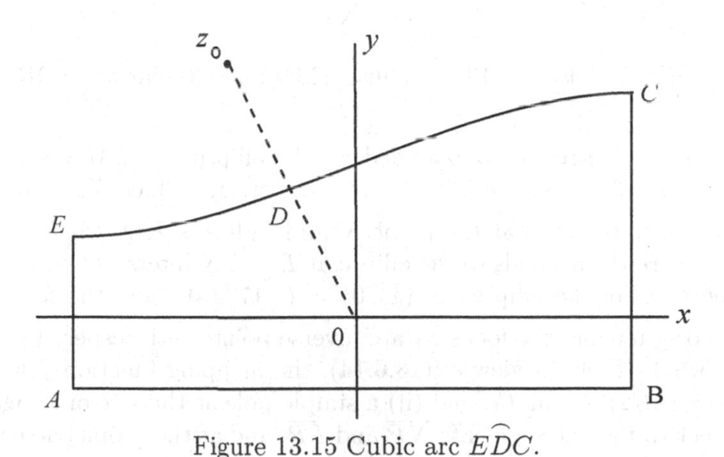

Figure 13.15 Cubic arc $E\widehat{D}C$.

and the cubic arc

$$E\widehat{D}C : \{z : z = \gamma(s),\ -2 < s < 2\},$$

where $\gamma(s) = s + i\left(\frac{2}{3} + \frac{1}{4} s - \frac{1}{48} s^3\right)$ (see Figure 13.15). The arc $C\widehat{D}E$ has a point of inflection at $x = 0$. The function $\gamma(\zeta)$ has a zero inside the boundary of the region D at the point

$$\zeta_0 = -0.160784962923 - 0.626680456065\, i.$$

Then

$$z_0 = \gamma\left(\bar{\zeta}_0\right) = -0.321569925846 + 1.25336091213\, i.$$

Also, since $c(\zeta_1) - c(\zeta_2) = (\zeta_1 - \zeta_2)\, R(\zeta_1, \zeta_2)$, where

$$R(\zeta_1, \zeta_2) = 1 + i\left(\frac{1}{4} - \frac{1}{48}\left(\zeta_1^2 + \zeta_1\,\zeta_2 + \zeta_2^2\right)\right),$$

and $R(\zeta_1, \zeta_2) \neq 0$ for all ζ_1 and ζ_2 in the rectangle $G = \{\zeta : \zeta = s + it, -2 < s < 2, -1 < t < 1\}$, the function $\gamma(\zeta)$ is one-to-one in G. Thus, there exists a simply connected region G^* that contains the points ζ_0 and $\bar{\zeta}_0$ and is such that the conditions $C1$ and $C2$, mentioned in the beginning of this section, are satisfied. Hence, in view of Theorem 13.3(a), the function f has a simple pole with respect to the arc $\overset{\frown}{CDE}$ at the point z_0.

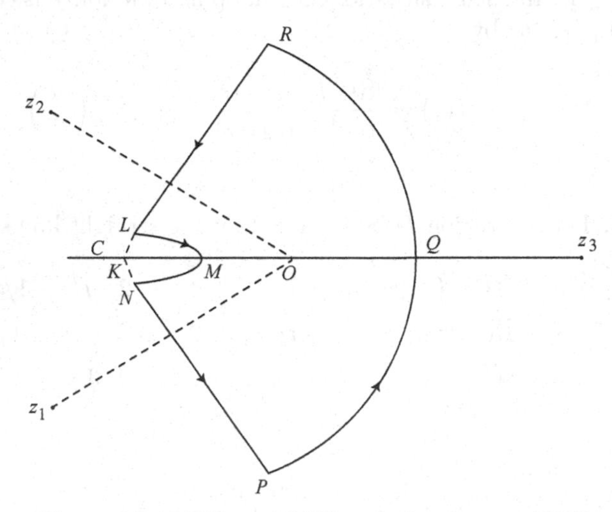

Figure 13.16 Ellipse LMN and circular arc PQR.

Map 13.9. The region D bounded by the elliptic arc $\overset{\frown}{LMN}$ which is defined by $z = 5\cos s - 17/2 + 3\,i\,\sin s$, $-\pi/5 < s < \pi/5$, the straight lines $\overline{NP}$ and $\overline{LR}$, and the circular arc $\overset{\frown}{PQR}$ whose center is at the point K and radius is KQ, where $Q = (7/2, 0)$ and K is the point where the normals to the ellipse at L and N intersect the x-axis. The coordinates of the center C of the ellipse are $(x_c, 0) = (-17/2, 0)$, and the focus F_1 is at $(-9/2, 0)$. Thus, the origin 0 and the focus F_1 are inverse points with respect to the elliptic arc $\overset{\frown}{LMN}$ (Figure 13.16). Then, in view of (13.6.14), the mapping function f has (i) a singularity of the type $(z + 9/2)^{-1/2}$ at F_1, and (ii) a simple pole at the mirror image z_1, z_2 of the origin with respect to the line segments $\overline{NP}$ and $\overline{LR}$, and at the geometric inverse z_3 of the origin with respect to the circular arc $\overset{\frown}{PQR}$.

13.7 Exterior Regions

In §10.3.3 we considered the case of the function $w = f_E(z)$ which maps the region Ext (D) univalently onto the region $U^* = \{|w| > 1\}$ such that $f_E(\infty) = 0$ and $\lim\limits_{z\to\infty} f_E'(z) > 0$. In this case, by using the inversion $z \mapsto z^{-1}$, we reduced the problem to that of the mapping of interior regions by the function $f(z)$.

In this section we are concerned with the following mapping problem: Assume that the origin lies inside a simply connected region D with the Jordan boundary Γ. Let a function $w = g(z)$ map the region Ext (Γ) conformally onto the disk $B(0, R) = \{|w| < R\}$ in the

w-plane. Also, the transformation $\zeta = 1/z$ maps the boundary Γ onto a Jordan contour Γ^* so that $z = \infty$ goes into $\zeta = 0$. Let $w = f(\zeta)$, $f(0) = 0$, $f'(0) = 1$, map the region D^*, bounded by the contour Γ^*, conformally onto the disk $B(0, R)$ (Figure 13.17). Thus, $w = g(z) = f(1/\zeta)$ maps the region Ext (Γ) conformally onto the disk $B(0, R)$ such that $g(\infty) = 0$. Hence, determining the mapping function g reduces to determining the interior mapping function f in such problems of exterior regions, which correspond to maps in §9.4. Also, the function $w = 1/g(z)$ maps the region Ext (Γ) conformally onto the exterior of the circle $|w| = 1/R$, and the quantity $d = 1/R$ is the transfinite diameter of the region $D \cup \Gamma$.

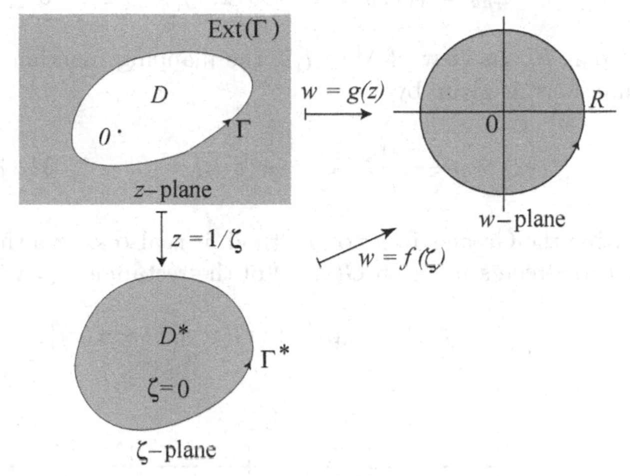

Figure 13.17 Ext (Γ) onto a disk.

Note that the function g is different from the function f_E studied earlier. We will use the RM and BKM to approximate the mapping function g. Since $f \in \mathcal{K}^1(D^*)$, the basis in RM is taken as $\{\phi_j(\zeta)\}$, as in §8.5, so that $\phi_1(0) = 1$ and $\phi_j(0) = 0$ for $j = 2, 3, \dots$, which leads to the complex linear system

$$\sum_{j=1}^{n} \langle \phi_j, \phi_i \rangle \, c_j = -\langle \phi_1, \phi_i \rangle, \quad i = 2, \dots, n, \tag{13.7.1}$$

which is solved for the unknowns c_j, $j = 2, \dots, n$. Thus, the nth RM approximations for the mapping function $f(\zeta)$ and the radius R are given by

$$f_n(\zeta) = \int_0^{\zeta} \Phi_n(t)\, dt, \quad R = \sqrt{\pi}\, \|\Phi_n\|, \tag{13.7.2}$$

where

$$\Phi_n(\zeta) = \phi_1(\zeta) + \sum_{j=2}^{n} \phi_j(\zeta), \tag{13.7.3}$$

is the nth approximation of $f'(\zeta)$.

In the RM and BKM (§8.5), since the basis set $\{\phi_j(\zeta)\}$ is a known complete set, first we approximate the Bergman kernel $K(\zeta, 0)$ of D^* by a finite Fourier sum. Then

$$f'(\zeta) = \frac{K(\zeta, 0)}{K(0, 0)}, \tag{13.7.4}$$

and $R = (\pi K(0,0))^{-1/2}$, as in (8.5.23) and (8.5.24). The details of the process are the same as in the five steps given above.

The basis set is taken as the set of monomials $\{\zeta^{j-1}\}$, $j = 1, 2, \dots$. Depending on the singularities of $K(\zeta, 0)$, the boundary singularities, and the poles of $f(\zeta)$, however, this basis is augmented by the functions $\phi(z)$ defined in (13.4.10).

Map 13.10. Consider mapping the rectangle

$$\Omega_{ab} = \{(x, y) : |x| < a/2, |y| < b/2\}, \quad a \geq b,$$

onto the unit disk U. In view of Map 7.7, the mapping function $f(z)$, known in terms of the elliptic functions, is given by

$$f(z) = \beta \frac{\zeta - \alpha}{\zeta - \bar{\alpha}}, \quad \zeta = \operatorname{sn}(z, k), \quad |\beta| = 1, \ \Im\{\alpha\} > 0.$$

Alternately, using the Green's function method, it is also known that the mapping function $f(z)$ is related to Green's function $G(z, z_0)$ of the rectangle Ω_{ab} with a pole at z_0 by

$$f(z) = \exp\{-2\pi G(z, z_0) + i H(z)\}, \tag{13.7.5}$$

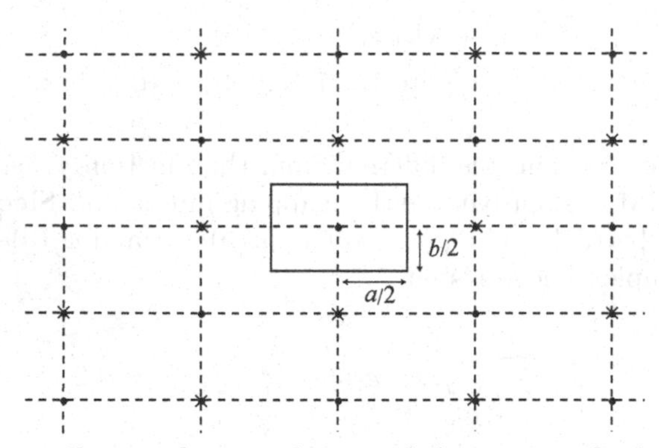

Figure 13.18 Rectangle onto the unit disk using method of images.

where $H(z)$ is the conjugate harmonic function of $G(z, z_0)$. The method of images can be used to express Green's function of Ω_{ab} as a double sum of logarithm functions (see Kythe [1996: 81]). In particular, at $z_0 = 0$

$$G(z, z_0) = \frac{1}{2\pi} \sum_{m,n=-\infty}^{\infty} (-1)^{m+n} \log \frac{1}{|z - z_{mn}|}, \tag{13.7.6}$$

where $z_{mn} = ma + i\,nb$ (Figure 13.18). Since the conjugate harmonic function of $\log|z - z_{mn}|$ is $\arg\{z - z_{mn}\}$, we find from (13.7.5) and (13.7.6) that the function $f(z)$ that maps the rectangle Ω_{ab} onto U such that $f(0) = 0$ is given by

$$f(z) = \exp\left\{ \sum_{m,n=-\infty}^{\infty} (-1)^{m+n} \log(z - z_{mn}) = \frac{\prod_{m+n=\text{even}} (z - z_{mn})}{\prod_{m+n=\text{odd}} (z - z_{mn})} \right\}. \tag{13.7.7}$$

As noted in Map 2.10, in the present case both f and the kernel function $K(z,0)$ have poles at all 'negative' images of the point $z_0 = 0$ with respect to the four sides of Ω_{ab} (these points are identified by an $\times$ in Figure 13.18). The poles at $z = \pm a$ and $z = \pm ib$ affect the convergence of the representation (8.5.10) of $K(z,0)$, even when $a = b$. But their effect is more significant the thinner the rectangle becomes, because in such cases ($b \ll a$) the distance of the poles at $\pm ib$ from the boundary of Ω_{ab} gets smaller compared with the dimensions of Ω_{ab}. Then the mapping function $f(z)$ from (13.7.7) is given by

$$f(z) = \frac{z}{(z^2 - a^2)(z^2 + b^2)}\, g(z), \tag{13.7.8}$$

where $g(z)$ is analytic in the region $\{(x,y) : |x/a| + |y/b| < 3\}$. Also, since from (8.5.25)

$$K(z,0) = \sqrt{\frac{K(0,0)}{\pi}}\, f'(z), \quad f'(0) = 0,$$

the set

$$\left\{\left(\frac{z}{z-a}\right)', \ \left(\frac{z}{z+a}\right)', \left(\frac{z}{z-ib}\right)', \left(\frac{z}{z+ib}\right)', \ z^j, \ j = 0, 1, \dots \right\},$$

where the prime denotes differentiation with respect to z, is best suited as the basis set for both BKM and VM.

Map 13.11. Consider the rectangle $\Omega_{a1} = \{(x,y) : |x| \le a, |y| \le 1\}$ (Figure 13.19).
CASE 1: ($a \neq 1$). Since the region has fourfold symmetry about 0, the odd powers of z do not appear in the polynomial representation (8.5.20) of the kernel function $K(z,0)$. Hence we take the basis set as $\{\phi_j^*(z)\} = \left\{z^{2(j-1)}\right\}_{j=1}^{N}$.
CASE 2: ($a = 1$). Since the region has eightfold symmetry about the origin and the polynomial representation of $K(z,0)$ has only powers of z that are multiples of 4, we take the basis set as $\{\phi_j^*(z)\} = \left\{z^{4(j-1)}\right\}_{j=1}^{N}$.

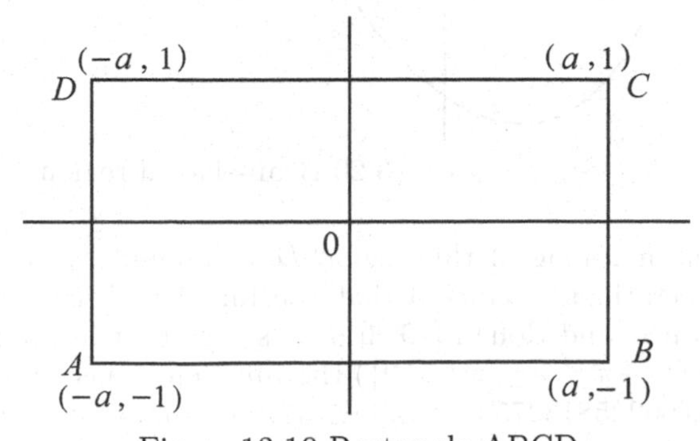

Figure 13.19 Rectangle ABCD.

The augmented basis (AB) is obtained by adding to the above orthonormal basis set the four singular functions that correspond to the four poles at $z = \pm 2a$ and $z = \pm 2i$. Because of the symmetry of the region, these four singular functions are combined into two functions

$\left(\dfrac{z}{z^2-4a^2}\right)'$ and $\left(\dfrac{z}{z^2+4}\right)'$ when $a \neq 1$. In the case when $a = 1$, these singular functions simplify to a single function $\left(\dfrac{z}{z^4-16}\right)'$. Hence the AB is given by

$$\phi_1(z) = \left(\frac{z}{z^2-4a^2}\right)', \quad \phi_2(z) = \left(\frac{z}{z^2+4}\right)',$$
$$\phi_{j+3} = z^{2j},\ j = 0, 1, \dots, \quad \text{when } a \neq 1;$$
$$\phi_1(z) = \left(\frac{z}{z^4-16}\right)', \quad \phi_{j+2} = z^{4j} \quad j = 0, 1, \dots, \quad \text{when } a = 1.$$

Map 13.12. Consider the bean-shaped region D bounded by the contour (Figure 13.20)

$$\Gamma : \Big\{ z : z = \gamma(s) = \frac{9}{4}\big[0.2 \cos s + 0.1 \cos 2s - 0.1$$
$$+ i\,(0.35 \sin s + 0.1 \sin 2s - 0.02 \sin 4s)\big], \quad -\pi \le s \le \pi \Big\}.$$

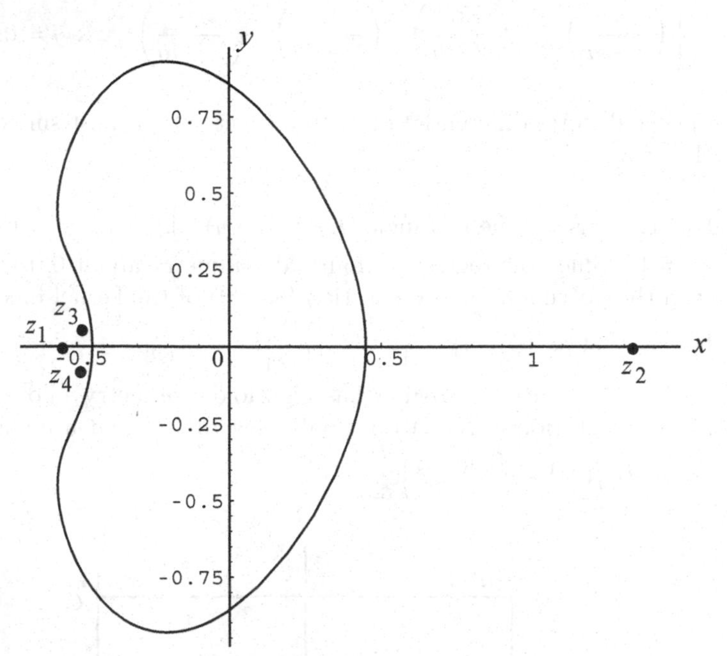

Figure 13.20 Bean-shaped region.

The conformal mapping of this region D was found by Reichel [1985] who, based on geometric considerations, predicted that the function f has a simple pole at $z \approx -0.61$. Papamichael, Warby and Hough [1983] have shown that in the neighborhood of the s-axis $(= \{\zeta : \zeta = s + it,\ -\pi \le s \le \pi\ t = 0\})$ the function f has (i) a simple pole at each of the points $z_1 = -0.650225813375$ and $z_2 = 1.311282520094$; and (ii) a singularity of the form $\sqrt{z - z_j}$, $j = 3, 4$, at the points $z_{3,4} = \pm 0.565672547402 \mp 0.068412683544\, i$. Hence, for the BKM with augmented basis (AB), we take

$$\psi(z) = \frac{d}{dz}\left\{ \frac{\sqrt{\sqrt{z - z_4} - \sqrt{z_3 - z_4}}}{z - z_1} \right\},$$

and the AB consists of the functions

$$\phi_1(z) = \left(\frac{z}{z-z_1}\right)', \qquad \phi_2(z) = \psi(z) + \psi(\bar z),$$
$$\phi_3(z) = \left(\frac{z}{z-z_2}\right)', \qquad \phi_{4+j}(z) = z^{j-1}, \quad j = 1, 2, \dots .$$

Example 13.3. We will compute the function that maps the unit circle $|w| = 1$ onto the family of squares

$$z\,\bar z + k\left(\frac{z^2 - \bar z^2}{4}\right)^2 = 1 \tag{13.7.9}$$

in the z–plane such that the point $w = 0$ goes into the point $z = 0$. Note that for $k = 1$, Eq (13.7.9) reduces to $(x^2-1)\,(y^2-1) = 0$ which represents the sides of the unit square. The square (13.7.9) is symmetric about the x and y axes and also about the lines $y = \pm x$. Hence,

$$z = \sum_{n=1}^{\infty} a_{4n-3}\, e^{i(4n-3)\theta}, \tag{13.7.10}$$

which gives (see Kythe [1998: 137])

$$\begin{aligned}
a_1^2 &= 1 - a_5^2 - a_9^2 - a_{13}^2 - \cdots ,\\
&\quad + \frac{k}{2a_1}\left[\frac{a_1^4}{4} + a_1^2 a_5^2 + \left(a_1 a_9 + \frac{a_5^2}{2}\right)^2 + (a_1 a_{13} + a_5 a_9)^2 + \cdots\right],\\
a_5 &= -\frac{a_5 a_9}{a_1} - \frac{a_9 a_{13}}{a_1} - \cdots + \frac{k}{2a_1}\left[-\frac{a_1^4}{8} + \frac{a_1^3 a_5}{2} + \left(a_1^2 a_9 + \frac{1}{2}\, a_1 a_5^3\right)\right.\\
&\quad \left. + (a_1 a_{13} + a_5 a_9)\left(a_1 a_9 + \frac{1}{2}\, a_1 a_5^2\right) + \cdots\right],\\
a_9 &= -\frac{a_5 a_{13}}{2} - \cdots + \frac{k}{2a_1}\left[-\frac{a_1^3 a_5}{2} + \frac{a_1^3 a_9}{2} + \frac{a_1^2 a_5^2}{4}\right.\\
&\quad \left. + \left(a_1^2 a_5 a_{13} + a_1 a_5^2 a_9\right) + \cdots\right],\\
a_{13} &= -\frac{a_5 a_{17}}{2} - \cdots + \frac{k}{2a_1}\left[-\frac{a_1^3 a_9}{2} - \frac{a_1^2 a_5^2}{2} - \frac{a_1^2 a_5^2}{2}\right.\\
&\quad \left. + \frac{1}{2}\left(a_1^3 a_{13} + a_1^2 a_5 a_9\right) + \cdots\right].
\end{aligned}$$

If we take the initial values as $a_1 = 1$, $a_5 = a_9 = \ldots = 0$, then computing up to the third successive approximations (see Table 3, Appendix E), we get the approximate mapping function as

$$\begin{aligned}
z &= \left(1 + \frac{k}{16} + \frac{3k^2}{256} + \frac{3k^2}{1024}\right) w - \left(\frac{k}{16} + \frac{7k^2}{256} + \frac{11k^2}{1024}\right) w^5\\
&\quad + \left(\frac{k^2}{64} + \frac{27k^3}{2048}\right) w^9 - \frac{11k^3}{2048}\, w^{13}.
\end{aligned} \tag{13.7.11}$$

For $k = 1$, this becomes

$$z = 1.077\, w - 0.1006\, w^5 + 0.0288\, w^9 - 0.0054\, w^{13},$$

which compares with the exact solution (4.5.31) with a maximum error of the order of 10^{-3}. ■

REFERENCES USED: Abramovici [1973], Carrier, Krook and Pearson [1966], Cooley and Tukey [1965], Cooley, Lewis and Welch [1970], Gaier [1964], Garrick [1936; 1949], Halsey [1979; 1982], James [1971], Kantorovich and Krylov [1958], Kober [1957], Kythe [1998], Piaggio and Strain [1947], Pennisi et al. [1963], Phillips [1943; 1966], Theodorsen [1931], Warschawski [1945].

14

Doubly Connected Regions

Some well-known numerical methods for approximating conformal mapping of doubly connected regions onto an annulus or the unit disk are presented. Numerical solutions are also confined to a limited class of regions where either one boundary is circular or axisymmetric. Most common methods use integral equations, iterations, polynomial approximations, and kernels. We will develop Symm's integral equations and the related orthonormal polynomial method. A dipole formulation that leads to the method of reduction of connectivity will be presented.

14.1 Annular Regions

Let $A(r_1, r_2)$, $r_1 < |z| < r_2$, denote an annular region in the z-plane.

Map 14.1. If the mapping of the annular region $A(r_1, r_2)$ onto a parallel strip defined by $\{\log r_1 < u < \log r_2\}$ in the w-plane is defined by the function $w = \log z$, then the annulus corresponds to an enumerable set of congruent rectangles, all with sides a and b such that $b = \log \dfrac{r_2}{r_1}$, and $a = 2\pi$. Since every doubly connected region can be mapped conformally onto an annulus of the type shown in Figure 14.1 and since two annuli with different ratios r_2/r_1 *cannot* be mapped conformally onto each other, the set of all doubly connected regions falls into classes of conformally equivalent regions where every class is characterized by the ratio r_2/r_1 of the radii belonging to that class. Hence, the ratio r_2/r_1, known as the *conformal invariant*, is related to the ratio a/b of the sides of the rectangles by

$$\frac{b}{a} = \frac{1}{2\pi} \log \frac{r_2}{r_1}, \quad \text{or} \quad \frac{r_2}{r_1} = e^{2\pi b/a}. \tag{14.1.1}$$

The linear transformation

$$\zeta = \frac{i - \sqrt{k}\, w}{1 + \sqrt{k}\, w}, \quad \text{or} \quad w = \frac{1}{i\sqrt{k}} \frac{\zeta - 1}{\zeta + 1}, \tag{14.1.2}$$

where k is defined in Map 7.7, maps the half-plane $\Im\{w\} \geq 0$ onto the circular region $|\zeta| \leq 1$ such that the four points corresponding to the points $w = \pm 1, \pm\dfrac{1}{k}$ are the vertices of a rectangle whose center is at $\zeta = 0$ (Figure 14.2). Then the angle ψ between the diagonals of the rectangle is given by

$$\tan\frac{\psi}{2} = \frac{2\sqrt{k}}{1-k}. \tag{14.1.3}$$

Combining the mapping (14.1.2) and Map 7.7 of the upper half-plane $\Im\{z\} > 0$ onto the rectangle, we obtain a conformal mapping of the rectangle onto the disk $|\zeta| < 1$. Thus, the angle ψ is another conformal invariant for doubly connected regions. A table of complete elliptic functions of the first kind for k from 0 to 1 and of k_1, $K(k)$, $K(k_1)$, a/b, r_2/r_1 and ψ (see Figure 14.2) is available in Andersen et al. [1962:165-166].

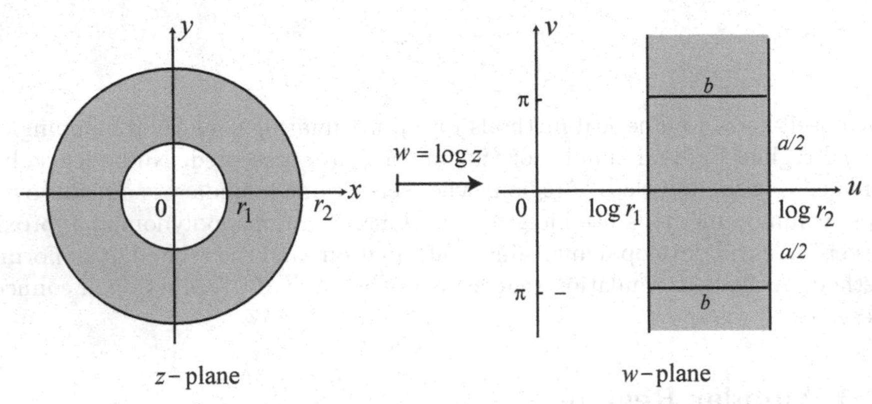

Figure 14.1 Annulus onto a set of enumerable congruent rectangles.

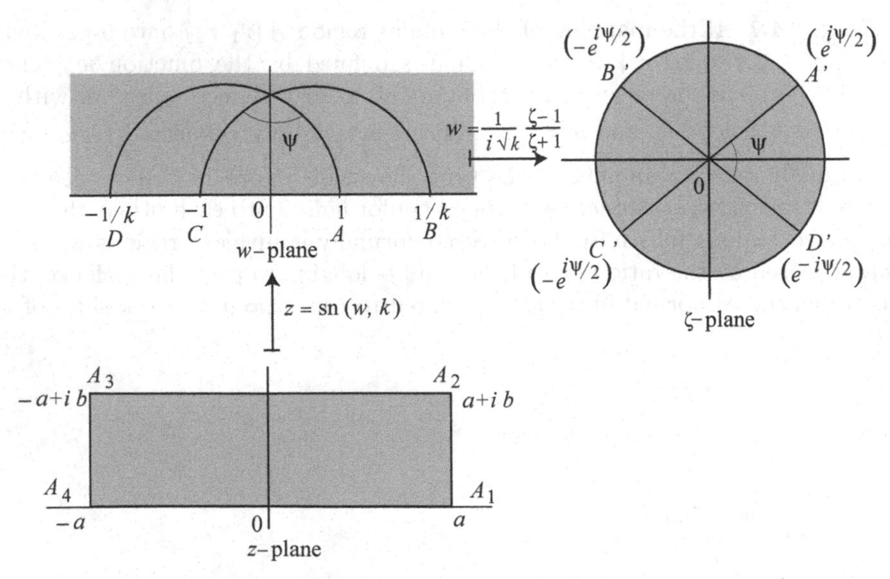

Figure 14.2 Upper half-plane onto the unit circle onto a rectangle.

Map 14.2. We know that the linear transformation $w = f(z) = \dfrac{z-a}{\bar{a}z-1}$, $|a| < 1$ maps the unit disk $B(0,1)$ onto itself. Moreover, this transformation always maps circles onto circles. We are interested in finding a particular value of a that will map the inner circle $C(c,c) \equiv |z-c| = c$ onto a circle of the form $|w| = R$. Let us choose a to be real and try to map the points 0 and $2c$ on the inner circle onto the points R and $-R$ on the circle $|w| = R$. Thus, we must have $f(0) = a = R$, and $f(2c) = \dfrac{2c - a}{2ca - 1} = -R$. Substituting the first into the second leads to the quadratic equation $ca^2 - a + c = 0$, which has two real solutions: $a = \dfrac{1 \pm \sqrt{1 - 4c^2}}{2c}$. Since $0 < c < \frac{1}{2}$, the solution with the plus sign gives $a > 1$, and therefore, it is rejected.

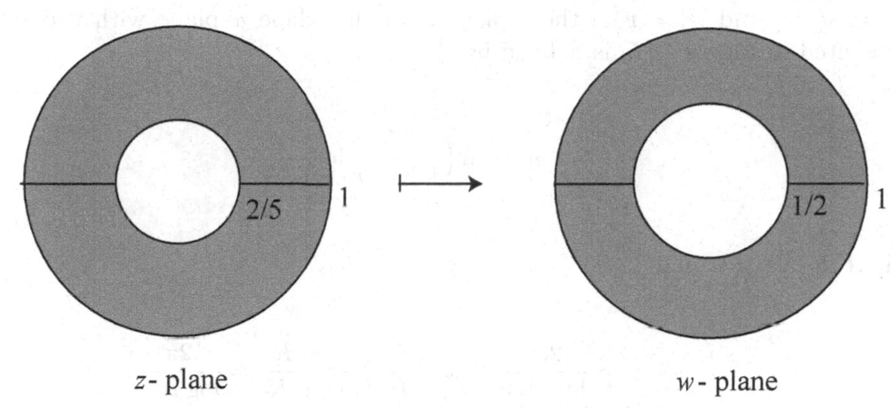

Figure 14.3 $A\left(\frac{2}{5}, 1\right)$ onto $A\left(\frac{1}{2}, 1\right)$.

Then the solution with the negative sign yields the required map

$$w = \frac{2cz - 1 + \sqrt{1 - 4c^2}}{\left(1 - \sqrt{1 - 4c^2}\right) z - 2c}. \tag{14.1.4}$$

In particular, taking $c = \frac{2}{5}$ gives $a = \frac{1}{2}$, which, when substituted into Eq (14.1.4), yields the transformation $w = \dfrac{2z - 1}{z - 2}$ that maps the annulus $A\left(\frac{2}{5}, 1\right)$ onto the annulus $A\left(\frac{1}{2}, 1\right)$, as shown in Figure 14.3.

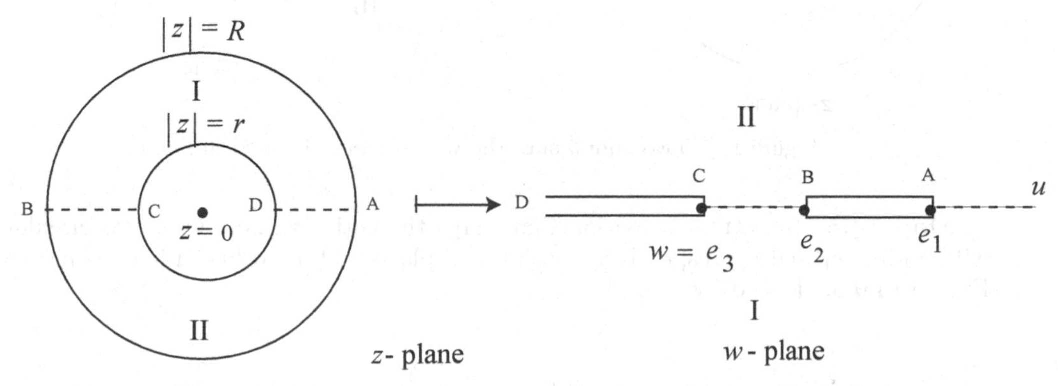

Figure 14.4 An annulus onto the whole plane with two slits in line.

Map 14.3. The transformation that maps an annulus in the z-plane, bounded by the circles $|z| = R$ and $|z| = r$, onto the whole w-plane with two slits in line, as presented in Figure 14.4, is defined by

$$w = \wp\Big(\log \frac{z}{r}\Big), \tag{14.1.5}$$

where $\omega_1 = \log \dfrac{R}{r}, R > r > 0$, and $\omega_3 = i\pi$. This transformation maps the points $z = R$ (A); r (D); $-R$ (B); and $-r$ (C) onto the points $w = e_1$ (A); ∞ (D); e_2 (B); and e_3 (C), respectively. It maps the circle $|z| = R$ onto the segment $e_2 \le u \le e_1$ of $v = 0$, counted twice; and the circle $|z| = r$ onto the half-line $-\infty < u \le e_3, v = 0$, counted twice.

Map 14.4. The transformation that maps the annuli, bounded by the circles $|z| = R$, $|z| = \sqrt{Rr}$, and $|z| = r$ in the z-plane onto the whole w-plane with two slits in line, as presented in Figure 14.5, is defined by

$$w = \operatorname{sn}\left(\frac{2K}{\log R/r} \log \frac{z}{\sqrt{Rr}}\right), \tag{14.1.6}$$

where $R > r > 0$, and

$$\tau = \frac{2i\pi}{\log R/r}, \quad k = \frac{\Theta_2(\tau)}{\Theta_3(\tau)}; \quad \frac{K'}{K} = \frac{2\pi}{\log R/r}, \tag{14.1.7}$$

and K, K' are defined as in §F.1.

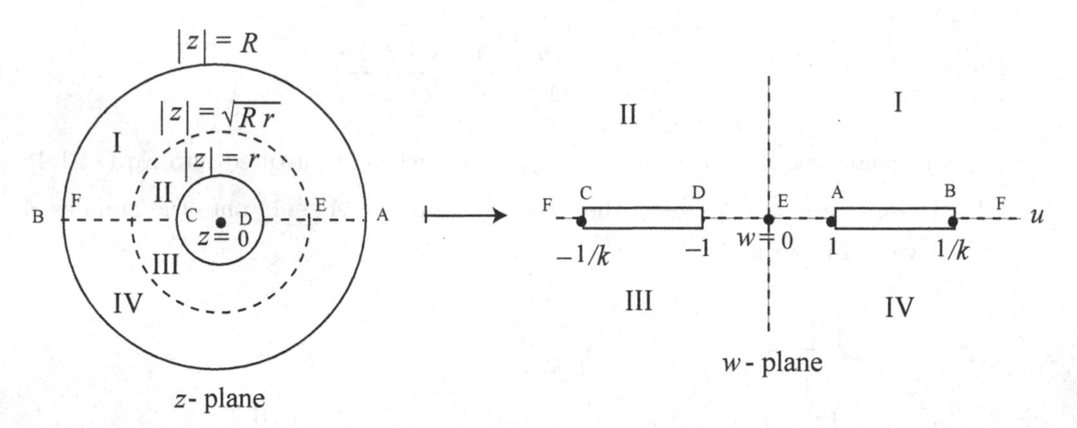

Figure 14.5 Two annuli onto the whole plane with two slits in line.

Map 14.5. The transformation that maps the entire z-plane with two circular holes with radius r_1 and r_2, respectively, onto the w-plane with two slits in line, as presented in Figure 14.6, is defined by

$$w = \wp\left(\log \frac{z+c}{z-c} + \frac{1}{2} \log \frac{a+c}{a-c}\right), \tag{14.1.8}$$

with

$$\omega_1 = \frac{1}{2} \log \frac{(b+c)(a+c)}{(b-c)(a-c)}, \quad \omega_3 = i\pi, \tag{14.1.9}$$

where $a > 0, b > 0, r_1 > 0, r_2 > 0$, but $a^2 - r_1^2 = b^2 - r_2^2$; and $c = \sqrt{a^2 - r_1^2} > 0$.

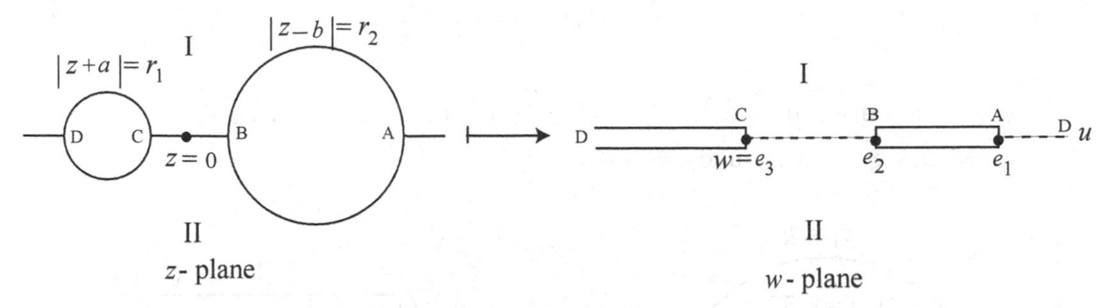

Figure 14.6 z-plane with two circular holes onto the whole plane with two slits in line.

This transformation maps the points $z = b + r_2$ (A); $b - r_2$ (B); $-a + r_1$ (C); and $-a - r_1$ (D) in the z-plane onto the points $w = e_1; e_2; e_3$; and ∞, respectively. It also maps the circle $|z - b| = r_2$ onto the segment $e_2 \leq u \leq e_1$ of $v = 0$, counted twice; and the circle $|z + a| = r_1$ onto the half-line $v = 0, -\infty < u \leq e_3$, counted twice.

Map 14.6. The transformation that maps the entire z-plane with two circular holes with radius r_1 and r_2, respectively, onto the w-plane with two finite slits in line, as presented in Figure 14.7, is defined by

$$w = \operatorname{sn}\left(\frac{K}{\lambda}\log\frac{z+c}{z-c} + \rho\right), \tag{14.1.10}$$

where

$$\lambda = \sqrt[4]{\frac{b+c}{b-c}\cdot\frac{a+c}{a-c}}, \quad \tau = \frac{i\pi}{\lambda}, \quad k = \frac{\Theta_2(\tau)}{\Theta_3(\tau)}, \quad \rho = \frac{K}{4\lambda}\log\left(\frac{a+c}{a-c}\cdot\frac{b-c}{b+c}\right) \tag{14.1.11}$$

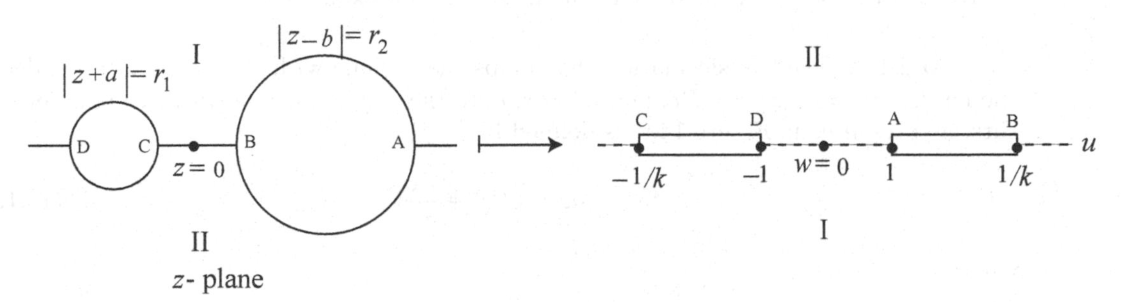

Figure 14.7 z-plane with two circular holes onto the whole plane with two finite slits in line.

This transformation maps the points $z = b + r_2$ (A); $b - r_2$ (B); $-a + r_1$ (C); $-a - r_1$ (D) onto the points $w = 1; 1/k; -1/k; -1$, respectively. It also maps the circle $|z - b| = r_2$ onto the segment $1 \leq u \leq 1/k$ of $v = 0$, counted twice; and the circle $|z + a| = r_1$ onto the segment $-1/k \leq u \leq -1$ of $v = 0$, counted twice.

Map 14.7. The transformation that maps the annuli in the z-plane, bounded by the circles $|z| = R$, $|z| = \sqrt{Rr}$, and $|z| = r$, onto the whole w-plane with two parallel finite slits, as presented in Figure 14.8, is defined by

$$w = -\zeta\Big(i\log\frac{z}{\sqrt{Rr}}\Big) + \frac{i\eta}{\pi}\log\frac{z}{\sqrt{Rr}}, \tag{14.1.12}$$

where $R > 0, r > 0$, $\omega_1 = \pi$, $\omega_3 = \frac{1}{2}\log\dfrac{R}{r}$; λ is the root of $\wp(\lambda + \omega_3) = -\dfrac{\eta}{\pi}$, $-\pi < \lambda < 0$; $\eta = \zeta(\omega_1)$, $\xi = \lambda + \omega_3$, and $\bar{\xi} = \lambda - \omega_3$.

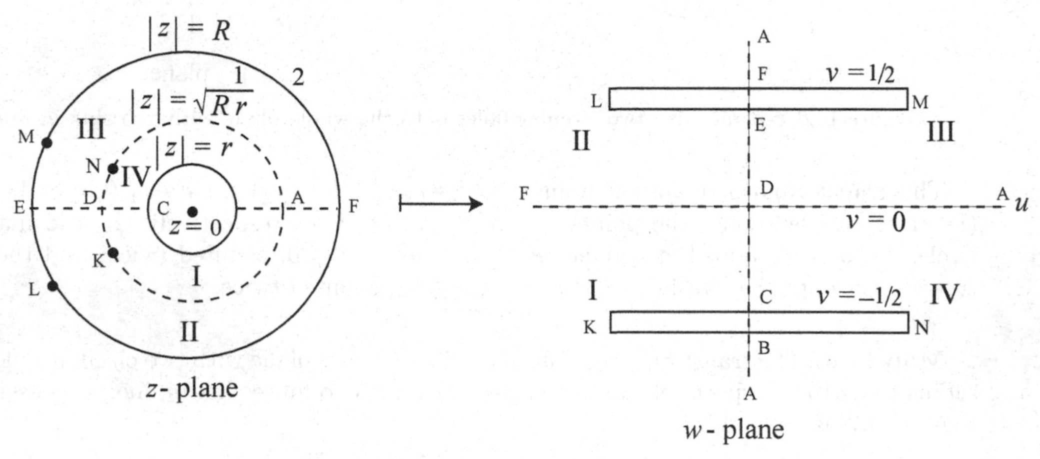

Figure 14.8 Annuli on the z-plane onto two parallel finite slits.

This transformation maps the points $z = R$ (F); $\sqrt{Rr}$ (A); r (B); $-r$ (C); $-\sqrt{Rr}$ (D); $-R$ (E); $r\,e^{i\lambda}$ (K); $R\,e^{i\lambda}$ (L); $r\,e^{-i\lambda}$ (N); $R\,e^{-i\lambda}$ (M) onto the points $w = i/2$ (F); ∞ (A); $-i/2$ (B); $-i/2$ (C); 0 (D); $i/2$ (E); $\zeta(\xi) - \dfrac{\eta\xi}{\pi}$ (K); $\zeta(\bar{\xi}) - \dfrac{\eta\bar{\xi}}{\pi}$ (L); $-\zeta(\bar{\xi}) + \dfrac{\eta\bar{\xi}}{\pi}$ (N); $-\zeta(\xi) + \dfrac{\eta\xi}{\pi}$ (M), respectively. It also maps the circles $|z| = R$ and $|z| = r$ onto the slits $v = -1/2$ (LM) and $v = 1/2$ (KN); and maps the circle $|z| = \sqrt{Rr}$ onto the line $v = 0$. Note that the location of the points K, L, M, N in the figure are not exact.

Map 14.8. The transformation that maps the z-plane with circular holes bounded by the circles $|z| = R$, $|z| = \sqrt{Rr}$, and $|z| = r$, onto the whole w-plane with two parallel finite slits, as presented in Figure 14.9, is defined by

$$w = \zeta(\chi) + \frac{\eta\chi}{\pi}, \tag{14.1.13}$$

where

$$\begin{aligned} &\chi = i\log\Big\{\frac{z+c}{z-c}\Big(\frac{a+c}{a-c}\cdot\frac{b-c}{b+c}\Big)^{1/4}\Big\};\\ &\omega_1 = \pi,\ \ \omega_3 = i\log\Big(\frac{a+c}{a-c}\cdot\frac{b+c}{b-c}\Big)^{1/4};\ \ \zeta(\omega_1) = \eta,\ \ \zeta(\omega_3) = \eta';\\ &p = c\frac{\mu+1}{\mu-1},\ \ q = \frac{2c\sqrt{\mu}}{|\mu-1|},\ \ \mu = \Big(\frac{a-c}{a+c}\cdot\frac{b+c}{b-c}\Big)^{1/2}. \end{aligned} \tag{14.1.14}$$

Again, $a > 0, b > 0, r_1 > 0, r_2 > 0$, but $a^2 - r_1^2 = b^2 - r_2^2$; $c = \sqrt{a^2 - r_1^2} > 0$; and λ is the root of $\wp(\lambda + \omega_3) = -\eta/\pi$, $-\pi < \lambda < 0$.

The point marked K represents $z = -a + r_1 e^{i\phi}$, where $\tan\phi = \dfrac{c\sin\lambda}{a\cos\lambda - r_1}$. The point marked N represents $z = -a + r_1 e^{-i\phi}$.

The point marked L represents $z = b + r_2 e^{i\psi}$, where $\tan\psi = \dfrac{c\sin\lambda}{r_2 - b\cos\lambda}$. The point marked M represents $z = b + r_2 e^{-i\psi}$.

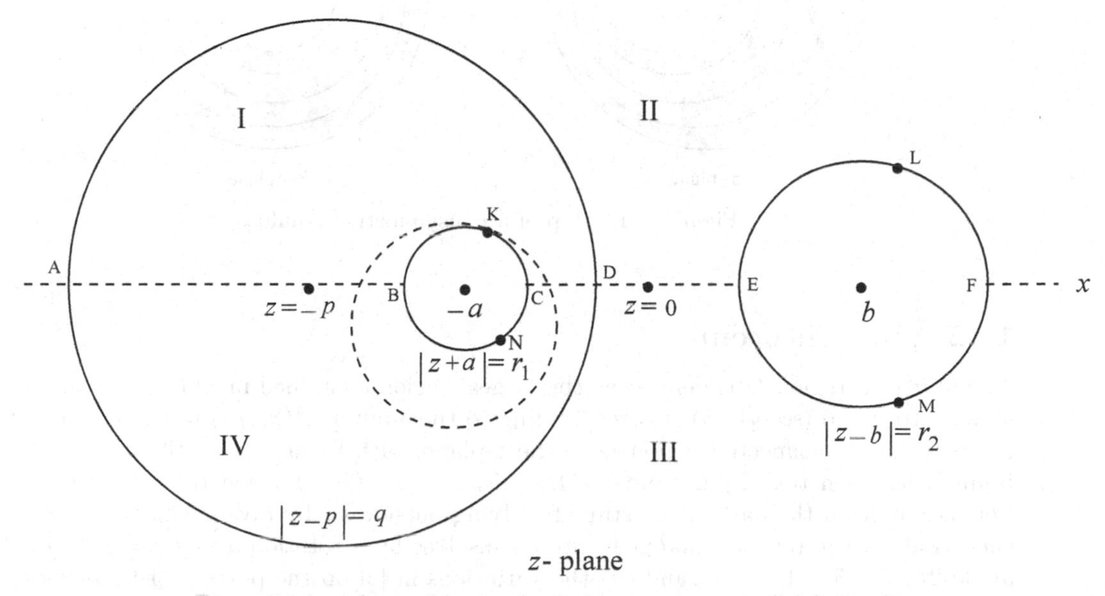

Figure 14.9 z-plane with circular holes onto two parallel finite slits.

Note that Figure 14.9 shows only the z-plane; the figure in the w-plane is that same as that in Figure 14.8.

This transformation maps (i) the circles $|z + a| = r_1; |z - b| = r_2$ onto the slit KN, and the slit LM, respectively; (ii) the circle $|z + p| = q$ onto the line $v = 0$. (iii) the segment $-a + r_1 \le x \le b - r_2$ of $y - 0$ onto the segment $-\frac{1}{2} \le v \le \frac{1}{2}$ of $u = 0$ (CE); (iv) the segment $-p - q < x \le -a + r_1$ of $y = 0$ onto the half-line $u = 0, -\infty, v \le -\frac{1}{2}$ (AB); (v) the half-lines $y = 0, x \ge b + r_2$, and $y = 0, -\infty < x \le -p - q$, together, onto the half-line $u = 0, \frac{1}{2} < v < \infty$ (FA); (vi) the circle γ, touching $|z + a| = r_1$ at K, exterior to $|z + a| = r_1$ and to $|z - b| = r_2$ onto the airfoil, surrounding the slit KN, cusp at K (not in figure).

If $a = b$, then $r_1 = r_2$, and $|z + p| = q$ is replaced by $x = 0$.

Map 14.9. The conformal mapping of a non-concentric annulus onto a symmetric annulus is presented in Figure 14.10.

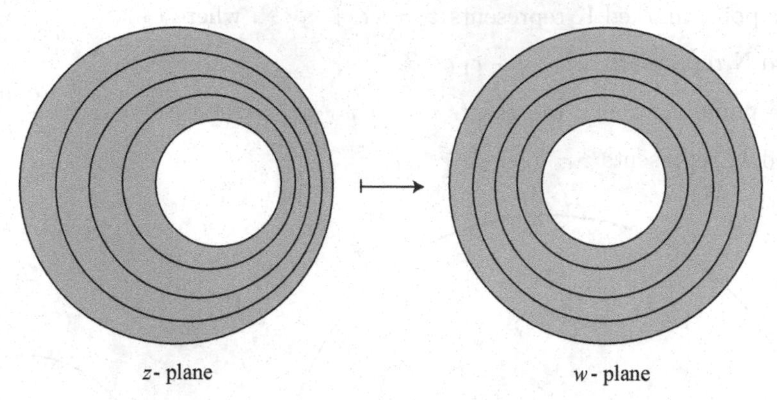

Figure 14.10 Map of a nonsymmetric annulus.

14.2 Area Theorem

The *star* of a region D is defined as the largest region contained in D that is starlike with respect to $z = 0$ (see §11.5). Corresponding to the annulus $A(\rho_1, \rho_2)$ in the w-plane, there exists a doubly connected region D in the z-plane, with Γ_1 and Γ_0 as the inner and outer boundaries such that Γ_1 lies outside the disk $|z| < r\,(1 - \delta_r)$, where $\delta_r \to 0$ as $r \to 0$. Let us partition the part of the star of D lying outside Γ_1 by rays emanating from $z = 0$ such that (i) the rays are mapped onto themselves by a rotation about $z = 0$ through an angle $2k\pi/n$, $k = 1, \dots, n$; and (ii) the variations in $|z|$ on the portion of Γ_1 between two consecutive rays (excluding the rays themselves) is less than a preassigned quantity $\varepsilon > 0$. Since Γ_1 is an analytic curve, such a system of rays exists. There are nm such rays, where m is an integer. We denote them in order by $l_1, l_2, \dots, l_{nm}$, where l_{mk+q} is obtained from l_q by a rotation through an angle $2k\pi/n$. Let r_k, $k = 1, 2, \dots, nm$, denote the largest distance from $z = 0$ to the part G_k of the star lying between the rays l_k and l_{k+1} and outside Γ_1 for $k = 1, 2, \dots, nm$, where $G_{nm+1} = G_1$ (Figure 14.11). The function $\zeta = \log\left(\frac{z}{r}\right)$ transforms this system of regions G_k into a system of regions H_k lying, respectively, in rectangles with sides of length $\log\left(\frac{r_k}{r}\right)$ and α_k, where $\sum_{k=1}^{nm} \alpha_k = 2\pi$ and $\alpha_{mk+j} \neq \alpha_j$.

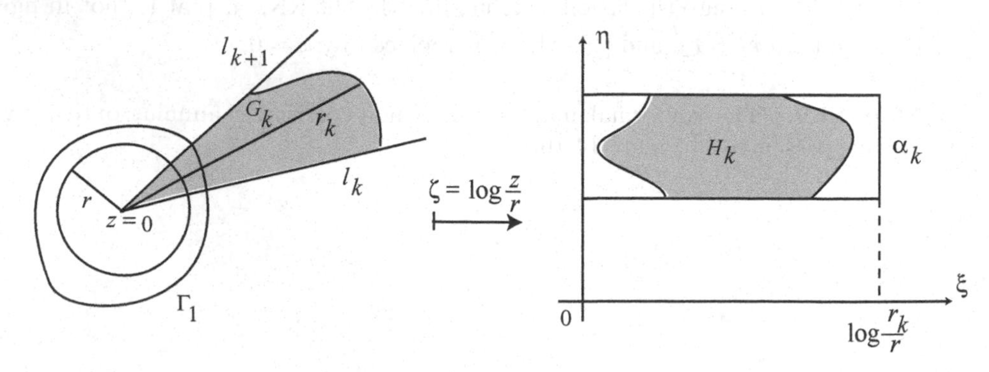

Figure 14.11 Regions G_k onto the regions H_k

If the region H_k is mapped onto a rectangle of sides a_k and b_k such that the boundary segments of H_k are mapped into the side of length b_k, then

$$\frac{a_k}{b_k} \geq \frac{\alpha_k}{\log \dfrac{r_k}{r}}, \tag{14.2.1}$$

which yields

$$\sum_{k=1}^{nm} \frac{a_k}{b_k} \geq \sum_{j=1}^{m} \sum_{k=0}^{n-1} \frac{\alpha_{mk+j}}{\log \dfrac{r_{mk+j}}{r}} = \sum_{j=1}^{m} \alpha_j \sum_{k=0}^{n-1} \frac{1}{\log \dfrac{r_{mk+j}}{r}}.$$

Since the system of regions G_k is the image of a system of strips contained in the annulus $\rho_1 < |w| < \rho_2$ under the mapping $w = f(z)$, we have

$$\frac{2\pi}{\log \dfrac{\rho_2}{\rho_1}} \geq \sum_{j=1}^{m} \alpha_j \cdot \min_j \sum_{k=0}^{n-1} \frac{1}{\log \dfrac{r_{mk+j}}{r}} = \frac{2\pi}{n} \cdot \min_j \sum_{k=0}^{n-1} \frac{1}{\log \dfrac{r_{mk+j}}{r}}. \tag{14.2.2}$$

Using the inequality $\dfrac{1}{n} \sum_{k=1}^{n} c_n \geq \sqrt[n]{\prod_{k=1}^{n} c_k}$, $c_k \geq 0$, twice, we find from (14.2.2) that

$$\sum_{k=0}^{n-1} \frac{1}{\log \dfrac{r_{mk+j}}{r}} \geq \frac{n}{\sqrt[n]{\prod_{k=1}^{n-1} \log \dfrac{r_{mk+j}}{r}}} \geq \frac{n^2}{\sum_{k=1}^{n-1} \log \dfrac{r_{mk+j}}{r}},$$

and hence, (14.2.2) yields

$$n \log \frac{\rho_2}{\rho_1} \leq \max_j \sum_{k=1}^{n-1} \log \frac{r_{mk+j}}{r},$$

or

$$M^n = \left(\frac{\rho_2}{\rho_1}\right)^n \leq \max_j \prod_{k=0}^{n-1} \frac{r_{mk+j}}{r}, \tag{14.2.3}$$

where M is the conformal modulus of the region $A(\rho_1, \rho_2)$.

Antonjuk [1958] has proved the following theorem for functions that are regular in an annulus:

Theorem 14.1. *Suppose that the function $w = f(z)$ is regular in the annulus $A(1, M) = \{1 < |z| < M$ and satisfies the conditions $|f(z) \geq 1$ and $\dfrac{1}{2i\pi} \displaystyle\int_C \frac{f'(z)}{f(z)}\, dz \geq 1$, where C is a contour in $A(1, M)$ which is not homologous to $z = 0$. Let A_f^* denote the star of a finite doubly connected Riemann surface A_f onto which the annulus $A(1, m)$ is mapped by the function $w = f(z)$ with respect to the system of rays emanating from the point $w = 0$. Then*

$$(P^* + \pi)\,(p^* + \pi) \geq \pi^2 M^4, \tag{14.2.4}$$

where P^* is the area of the star A_f^* and p^* the area of the preimage of A_f^*. The equality in (14.2.4) holds for functions of the form $f(z) = c\,z$, $|c| = 1$.

Map 14.10. The region $A(\rho, 1) = \{\rho < |\zeta| < 1\}$ is mapped conformally onto the unit disk $|t| < 1$, slit from $-L$ to $+L$ (Figure 14.12) by the function

$$t = L \operatorname{sn}\Big(\frac{2iK}{\pi} \log \frac{\zeta}{\rho} + K, k\Big), \quad k = L^2, \quad K = K(k) \tag{14.2.5}$$

(Nehari [1952:293-295]).

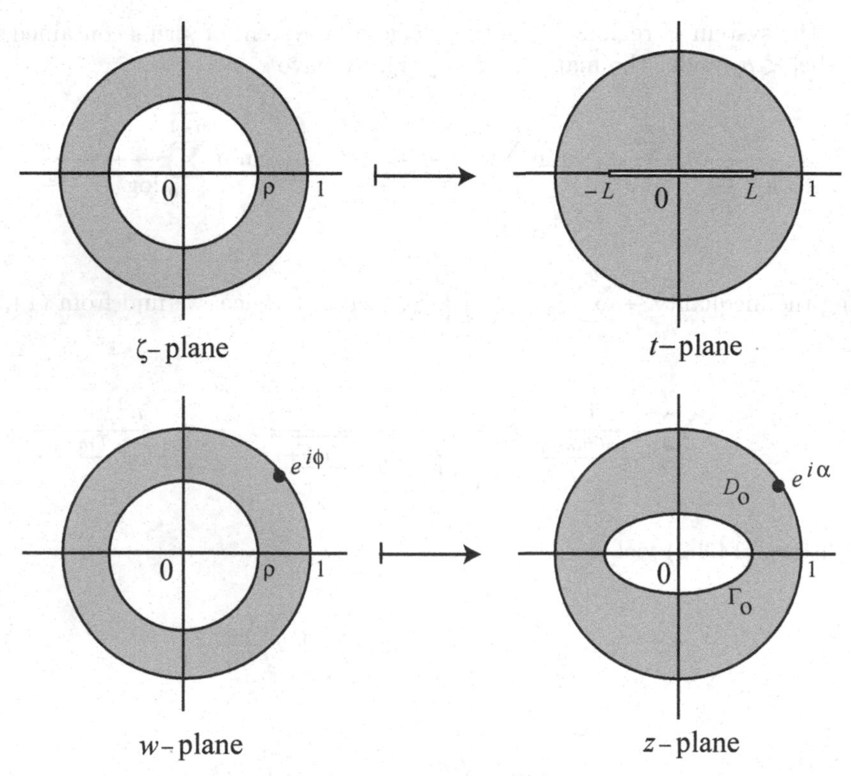

Figure 14.12 Annular region onto the slit unit disk and onto an annular region.

Let

$$\zeta = \rho/w, \quad t = \frac{L}{2}\left(z + z^{-1}\right), \quad |z| = 1. \tag{14.2.6}$$

Then the circle $|t| = 1$ is mapped onto the boundary Γ_0 in the z-plane by the mapping (14.2.6), where Γ_0 is in polar coordinates defined by

$$r = r(\theta) = \sqrt{\frac{1}{L^2} + \sin^2\theta} - \sqrt{\frac{1}{L^2} - \cos^2\theta}. \tag{14.2.7}$$

For example, if we take $k = L^2 = \sin 46°$, as in Gaier [1964:222], then the values of

$r_n = r\left(\frac{n\pi}{18}\right)$ for $n = 0, 1, 2, \dots, 9$ are given in Table 14.1.

Table 14.1 Values of r_n

n	r_n	n	r_n
0	0.554 435	5	0.417 632
1	0.543 464	6	0.395 150
2	0.515 529	7	0.379 357
3	0.480 595	8	0.370 041
4	0.446 599	9	0.366 967

We find that
(i) The modulus M of the region D_0 is given by Nehari [1952:294] as

$$M = \frac{1}{\rho} = \exp\left\{\frac{\pi}{4}\frac{K'(k)}{K(k)}\right\} = \frac{1}{\sqrt[4]{q(k)}} \approx 2.166187,$$

for $k = \sin 46°$ (for the function sn and q, see §10.4).
(ii) Let $w = e^{i\phi}$ be the map of $z = e^{i\alpha}$, $\alpha = \alpha(\phi)$. Then

$$\zeta \mapsto t \,:\, t = L\,\mathrm{sn}\Big((1 + 2\phi/\pi)\,K, k\Big), \tag{14.2.8}$$

$$z \mapsto t \,:\, t = L\,\cos\alpha, \tag{14.2.9}$$

where

$$\cos\alpha = \mathrm{sn}\Big((1 + 2\phi/\pi)\,K, k\Big) = \mathrm{sn}\Big((1 - 2\phi/\pi)\,K, k\Big).$$

Set $\alpha' = \pi/2 - \alpha$. Then $(1 - 2\phi/\pi)\,K = F(\alpha', k)$, where F is the hypergeometric function. In particular, let $\phi_n = \frac{n\pi}{18}$, $n = 1, 2, \dots, 8$, and set $r = 90 - 10\,n$. Then $F(\alpha', k) = rK/90$, $r = 80, 70, \dots, 10$, and $k = \sin 46°$. This yields the values of α'_n, and hence, of α_n.
(iii) Let $w = \rho\,e^{i\phi}$ map the z-plane onto the w-plane (Figure 14.12). We will determine $\beta = \beta(\phi) - \arg\{z\}$. First, we use the mapping

$$\zeta \mapsto t \,:\, t = L\,\mathrm{sn}\,(K + i\kappa + v, k)\,, \quad \kappa = \frac{2K\,\log M}{\pi}, \quad v = \frac{2K}{\pi}\,\phi. \tag{14.2.10}$$

Since $t = e^{i\psi}$ maps t onto $|t| = 1$, we have $1 = L\,\mathrm{sn}\,(K + i\,\kappa, k)$ for $\phi \neq 0$. Set $u = K + i\,\kappa$. Then $\mathrm{sn}\,u = 1/L$, $\mathrm{cn}\,u = \sqrt{1 - 1/L^2}$ which is purely imaginary, and $\mathrm{dn}\,u = \sqrt{1 - L^2}$ which is real. Also, $\mathrm{sn}\,v$, $\mathrm{cn}\,v$, and $\mathrm{dn}\,v$ are real. Hence, from (14.2.10), by equating real parts, we get

$$\cos\psi = \frac{\mathrm{cn}\,(v, k)\,\mathrm{dn}\,(v, k)}{1 - L^2\,\left[\mathrm{sn}\,(v, k)\right]^2}, \tag{14.2.11}$$

where $v = v_n = nK/9$, $n = 1, 2, \dots, 8$. Since, in view of (14.2.9), the mapping $z \mapsto t$ gives

$$\cos\psi = \frac{L}{2}\left(r + \frac{1}{r}\right)\cos\beta, \quad \sin\psi = \frac{L}{2}\left(r - \frac{1}{r}\right)\sin\beta,$$

we eliminate r, use (14.2.11), and obtain

$$\begin{aligned}\sin^2\beta &= \frac{-\left(1-L^2\right)+\sqrt{\left(1-L^2\right)^2+4L^2\sin^2\psi}}{2L^2} \\ &= \frac{1-l^2}{1-L^2\,\mathrm{sn}^2(v,k)}\,\mathrm{sn}^2(v,k).\end{aligned} \tag{14.2.12}$$

If we set $v_n = nK/9$, $n = 1, 2, \dots$, then we determine the values of $\beta_n = \beta(\phi_n)$ from the table given above.

Map 14.11. (Dirichlet problem for the annulus). Let two real-valued, 2π-periodic and continuous functions $u_1(\theta)$ and $u_2(\theta)$ be defined on the boundary of the annulus $A(r_1, r_2) = \{r_1 < |z| < r_2\}$. The Dirichlet problem for this region deals with determining a function $u(r, \theta)$ which is continuous in the closed region $A(r_1, r_2) \cup \Gamma_1 \cup \Gamma_0 = \{r_1 \le |z| \le r_2\}$, harmonic in $A(r_1, r_2)$, and takes the boundary value $u_1(\theta)$ on Γ_1 for $z = r_1 e^{i\theta}$ and $u_2(\theta)$ on Γ_0 for $z = r_2 e^{i\theta}$. The harmonic function $v(r, \theta)$, conjugate to $u(r, \theta)$, is in general not single-valued, and thus the function $f(z) = u + i\,v$ will have two summands. One is a single-valued function that can be expanded in a Laurent series for the annulus, and the other is $\log z$ with a real coefficient A. Thus,

$$u + i\,v = \sum_{n=-\infty}^{\infty} \gamma_n\, z^n + A\,\log z, \quad \gamma_n = \alpha_n + i\,\beta_n, \tag{14.2.13}$$

which, after separating into real and imaginary parts, yields

$$\begin{aligned}u &= \alpha_0 + \sum_{n=1}^{\infty} \left[\left(\alpha_n r^n + \alpha_{-n} r^{-n}\right)\cos n\theta - \left(\beta_n r^n - \beta_{-n} r^{-n}\right)\sin n\theta\right] \\ &\quad + A\,\log r, \\ v &= \beta_0 + \sum_{n=1}^{\infty} \left[\left(\beta_n r^n + \beta_{-n} r^{-n}\right)\cos n\theta + \left(\alpha_n r^n - \alpha_{-n} r^{-n}\right)\sin n\theta\right] + \theta.\end{aligned} \tag{14.2.14}$$

In the first equation in (14.2.14) we set $r = r_1$ and $r = r_2$. This gives us the Fourier series expansions for the functions $u_1(\theta)$ and $u_2(\theta)$, respectively, where the Fourier coefficients are given by

$$\begin{aligned}a_0^{(1)} &= \alpha_0 + A\,\log r_1 = \frac{1}{2\pi}\int_{-\pi}^{\pi} u_1(\theta)\,d\theta, \\ a_0^{(2)} &= \alpha_0 + A\,\log r_2 = \frac{1}{2\pi}\int_{-\pi}^{\pi} u_2(\theta)\,d\theta, \\ a_n^{(1)} &= \alpha_n\, r_1^n + \alpha_{-n}\, r_1^{-n} = \frac{1}{2\pi}\int_{-\pi}^{\pi} u_1(\theta)\,\cos n\theta\,d\theta, \\ a_n^{(2)} &= \alpha_n\, r_2^n + \alpha_{-n}\, r_2^{-n} = \frac{1}{2\pi}\int_{-\pi}^{\pi} u_2(\theta)\,\cos n\theta\,d\theta, \\ b_n^{(1)} &= \beta_{-n}\, r_1^{-n} - \beta_n\, r_1^n = \frac{1}{2\pi}\int_{-\pi}^{\pi} u_1(\theta)\,\sin n\theta\,d\theta, \\ b_n^{(2)} &= \beta_{-n}\, r_2^{-n} - \beta_n\, r_2^n = \frac{1}{2\pi}\int_{-\pi}^{\pi} u_2(\theta)\,\sin n\theta\,d\theta.\end{aligned}$$

Hence

$$A = \frac{a_0^{(2)} - a_0^{(1)}}{\log M}, \qquad \alpha_0 = \frac{a_0^{(1)} \log r_2 - a_0^{(1)} \log r_1}{\log M},$$
$$\alpha_n = \frac{a_n^{(1)} r_2^{-n} - a_n^{(2)} r_1^{-n}}{r_1^n r_2^{-n} - r_1^{-n} r_2^n}, \qquad \alpha_{-n} = \frac{a_n^{(1)} r_2^n - a_n^{(2)} r_1^n}{r_2^n r_1^{-n} - r_2^{-n} r_1^n}, \tag{14.2.15}$$
$$\beta_n = \frac{b_n^{(2)} r_1^{-n} - b_n^{(1)} r_2^{-n}}{r_1^n r_2^{-n} - r_1^{-n} r_2^n}, \qquad \beta_{-n} = \frac{b_n^{(1)} r_2^n - b_n^{(2)} r_1^n}{r_2^n r_1^{-n} - r_2^{-n} r_1^n}.$$

Map 14.12. (Neumann problem for the annulus). Let $F_1(\theta)$ and $F_2(\theta)$ denote the normal derivatives of the harmonic function $u(r, \theta)$ on the boundaries $\Gamma_1 = \{|z| = r_1\}$ and $\Gamma_0 = \{|z| = r_2\}$ of the annulus $A(r_1, r_2)$. Then the function $f(z) = u + i\, v$ has the same representation as in (14.2.13), where separating real and imaginary parts and satisfying the Neumann conditions on the boundaries Γ_1 and Γ_0, respectively, we get

$$\begin{aligned} F_1(\theta) &= \sum_{n=1}^{\infty} n\big[\left(\alpha_n r_1^{n-1} - \alpha_{-n} r_1^{n-1}\right) \cos n\theta \\ &\qquad - \left(\beta_n r_1^{n-1} - \beta_{-n} r_1^{n-1}\right) \sin n\theta\big] + \frac{A}{r_1}, \\ F_2(\theta) &= -\sum_{n=1}^{\infty} n\big[\left(\alpha_n r_2^{n-1} - \alpha_{-n} r_2^{n-1}\right) \cos n\theta \\ &\qquad - \left(\beta_n r_2^{n-1} - \beta_{-n} r_2^{n-1}\right) \sin n\theta\big] - \frac{A}{r_2}. \end{aligned} \tag{14.2.16}$$

The coefficient A is determined in two ways which are equal:

$$A = \frac{r_1}{2\pi} \int_{-\pi}^{\pi} F_1(\theta)\, d\theta = -\frac{r_2}{2\pi} \int_{-\pi}^{\pi} F_2(\theta)\, d\theta. \tag{14.2.17}$$

The other coefficients in the Fourier series expansions of $F_1(\theta)$ and $F_2(\theta)$ are given by

$$\begin{aligned} a_n^{(1)} &= n \left(\alpha_n r_1^{n-1} - \alpha_{-n} r_1^{-n-1}\right) = \frac{1}{\pi} \int_{-\pi}^{\pi} F_1(\theta) \cos n\theta\, d\theta, \\ a_n^{(2)} &= -n \left(\alpha_n r_2^{n-1} - \alpha_{-n} r_2^{-n-1}\right) = \frac{1}{\pi} \int_{-\pi}^{\pi} F_2(\theta) \cos n\theta\, d\theta, \\ b_n^{(1)} &= -n \left(\beta_n r_1^{n-1} + \beta_{-n} r_1^{-n-1}\right) = \frac{1}{\pi} \int_{-\pi}^{\pi} F_1(\theta) \sin n\theta\, d\theta, \\ b_n^{(1)} &= n \left(\beta_n r_2^{n-1} + \beta_{-n} r_2^{-n-1}\right) = \frac{1}{\pi} \int_{-\pi}^{\pi} F_2(\theta) \sin n\theta\, d\theta. \end{aligned} \tag{14.2.18}$$

Thus, after solving (14.2.18), we find that the coefficients $\alpha_{\pm n}$ and $\beta_{\pm n}$ in the series (14.2.16) are given by

$$\alpha_n = \frac{1}{n} \frac{a_n^{(1)} r_1 r_2^{-n} + a_n^{(2)} r_2 r_1^{-n}}{r_1^n r_2^{-n} - r_2^n r_1^{-n}}, \qquad \beta_n = -\frac{1}{n} \frac{b_n^{(1)} r_1 r_2^{-n} + b_n^{(2)} r_2 r_1^{-n}}{r_1^n r_2^{-n} - r_2^n r_1^{-n}},$$
$$\alpha_{-n} = \frac{1}{n} \frac{a_n^{(1)} r_1 r_2^n + a_n^{(2)} r_2 r_1^n}{r_1^n r_2^{-n} - r_2^n r_1^{-n}}, \qquad \beta_{-n} = \frac{1}{n} \frac{b_n^{(2)} r_2 r_1^n + b_n^{(1)} r_1 r_2^n}{r_1^n r_2^{-n} - r_2^n r_1^{-n}}. \tag{14.2.19}$$

14.3 Source Density

Let $w = f_\Omega(z)$ map conformally a finite, doubly connected region Ω, bounded by two Jordan contours Γ_1 and Γ_0, $\Gamma_1 \subset \Gamma_0$, onto the annulus $A(\rho, 1)$, with $M = 1/\rho$, $\rho < 1$, where ρ is initially unknown and is to be determined. Let us assume that the origin lies inside the region $\text{Int}\,(\Gamma_1)$. Then the function $f_\Omega(z)$, which is uniquely determined except for a rotation, can be represented in the form (as in §12.1)

$$f_\Omega(z) = e^{\log z + g(z) + i\,h(z)}, \tag{14.3.1}$$

where g and h are conjugate harmonic functions in Ω such that

$$\left|f_\Omega(z)\right| = \begin{cases} \rho, & \text{for } z \in \Gamma_1, \\ 1, & \text{for } z \in \Gamma_0. \end{cases} \tag{14.3.2}$$

Thus,

$$f_\Omega(z) = e^{\log|z| + g(x,y) + i\,[\arg\{z\} + h(x,y)]}, \quad z = x + i\,y, \tag{14.3.3}$$

where the boundary conditions (14.3.2) become

$$g(x, y) = \begin{cases} \log\rho - \log|z|, & z \in \Gamma_1, \\ -\log|z|, & z \in \Gamma_0. \end{cases} \tag{14.3.4}$$

As in §12.1, we will represent $g(x, y)$ as a single-layer logarithmic potential

$$g(x, y) = \int_{\Gamma = \Gamma_1 + \Gamma_0} \log|z - \zeta|\,\mu(\zeta)\,|d\zeta|, \tag{14.3.5}$$

where $\mu(\zeta)$ is the source density on Γ. Also,

$$h(x, y) = \int_\Gamma \arg\{z - \zeta\}\,\mu(\zeta)\,|d\zeta| + q, \tag{14.3.6}$$

where q is an arbitrary constant corresponding to an arbitrary rotation in the mapping function $f_\Omega(z)$ defined by (14.3.1). Then the boundary conditions (14.3.4) become

$$\int_\Gamma \log|z - \zeta|\,\mu(\zeta)\,|d\zeta| = \begin{cases} \log\rho - \log|z|, & z \in \Gamma_1, \\ -\log|z|, & z \in \Gamma_0, \end{cases} \tag{14.3.7}$$

and the condition (12.1.7) on the single-valuedness of h reduces to

$$\int_{\Gamma_1} \mu(\zeta)\,|d\zeta| = 0, \tag{14.3.8}$$

as in (12.1.13). Note that Eqs (14.3.7) and (14.3.8), known as Symm's integral equations, are coupled equations for $\mu(\zeta)$ and ρ and possess a unique solution (see Jawson [1963]). Once $\mu(\zeta)$ is determined, the functions g and h can be computed from (14.3.5) and (14.3.6), respectively, and hence, the mapping function $f_\Omega(z)$ from (14.3.1).

14.3.1 Numerical Computation of $\mu(\zeta)$, g, h, and f_Ω is carried out by the ONP method as in §12.2, i.e., by partitioning the boundary Γ into N sections $G_1, \dots, G_N$ and approximating $\mu(\zeta)$ by μ_j which is constant over each G_j, $j = 1, \dots, N$. Then Eq (14.3.7) is computed at each node $z_k = x_k + i\,y_k$, $j = 1, \dots, N$, together with Eq (14.3.8). This yields a system of $(N+1)$ linear equations in $(N+1)$ unknowns $\mu_1, \dots, \mu_N$ and $\log \rho$. The solution of this system leads to the approximate value of ρ. Then the functions g and h are approximated by finite terms corresponding to the integrals in (14.3.5) and (14.3.6), and the mapping function $f_\Omega(z)$ is finally obtained from (14.3.1). The method follows the same set of steps as in §12.2.

Let $\hat{f}_\Omega(z)$ and $\hat{\rho}$ denote the numerical approximations to $f_\Omega(z)$ and ρ. Then

$$\left|\hat{\rho} - \rho\right| \ll \max_z \left|\hat{f}_\Omega(z) - f_\Omega(z)\right|, \tag{14.3.9}$$

and, as Symm [1969] has noted, $\hat{\rho}$ is more accurate than $\hat{f}_\Omega(z)$. In fact, by the maximum modulus theorem, $\left|\hat{f}_\Omega(z) - f_\Omega(z)\right|$ takes its maximum value somewhere on the boundary $\Gamma = \Gamma_1 \cup \Gamma_0$, and, as in §10.2, this maximum value rarely exceeds $2\max_z \left|\hat{f}_\Omega(z) - f_\Omega(z)\right|$. But since

$$\left|\hat{f}_\Omega(z) - f_\Omega(z)\right| = \begin{cases} \left|\hat{\rho} - \rho\right|, & \text{at the nodes of } \Gamma_1, \\ 0, & \text{at the nodes of } \Gamma_0, \end{cases}$$

then, in view of (14.3.9), we should compute $\left|\hat{f}_\Omega(z)\right|$ at some point $z = Z_j$ between the nodes (which may be end points of G_j). Thus, the point Z_j is called the *internodal point* for G_j. The error E in $\left|\hat{f}_\Omega(z)\right|$ then is given by

$$E = \max_{Z_j} \left\{ \max_{z \in \Gamma_1} \left| \left|\hat{f}_\Omega(z)\right| - \rho \right|, \max_{z \in \Gamma_0} \left| \left|\hat{f}_\Omega(z)\right| - 1 \right| \right\}. \tag{14.3.10}$$

If the doubly connected regions are symmetric about one or both coordinate axes, then the total number of equations to be solved reduces from $(N+1)$ to $(N/2+1)$ or $(N/4+1)$, respectively. We will denote the approximate values of u, v, and $w = u + i\,v$ by $\hat{u}$, $\hat{v}$, and $\hat{w}$, respectively.

Map 14.13. Consider a pair of limaçons

$$\begin{aligned} \Gamma_1 &= \{x = a_1 \cos t + b_1 \cos 2t, y = a_1 \sin t + b_1 \sin 2t,\ a_1 > 0,\ b_1 > 0\}, \\ \Gamma_0 &= \{x = a_2 \cos t + b_2 \cos 2t, y = a_2 \sin t + b_2 \sin 2t,\ a_2 > 0,\ b_2 > 0\}, \end{aligned}$$

(see Figure 14.13 with $a_1 = 5$, $a_2 = 10$, $b_2 = 3$, and $b_1 = b_2/4$) where $t = 0\ (2\pi/N)\ 2\pi$ defines the distribution of nodes on each boundary and $N = 2(n-1)$. The values of a_1, a_2, b_1, b_2 are chosen such that $b_1/b_2 = (a_1/a_2)^2$ which ensures that the function

$$f_\Omega(z) = \frac{\sqrt{a_2^2 + 4b_2\, z} - a_2}{2b_2},$$

which maps Γ_0 onto the unit circle (see Muskhelishvili [1963: §48], who has determined analytic solutions for some doubly connected regions like Pascal's limaçons, epitrochoids, hypotrochoids, and elliptic rings), also maps Γ_1 onto a concentric circle of radius $\rho = a_1/a_2$, where $M = 1/\rho$. Because of symmetry about the x-axis, we take $t = 0(\pi/10)\,\pi$. The values of $\hat{u}$ and $\hat{v}$ are given below in Table 14.2.

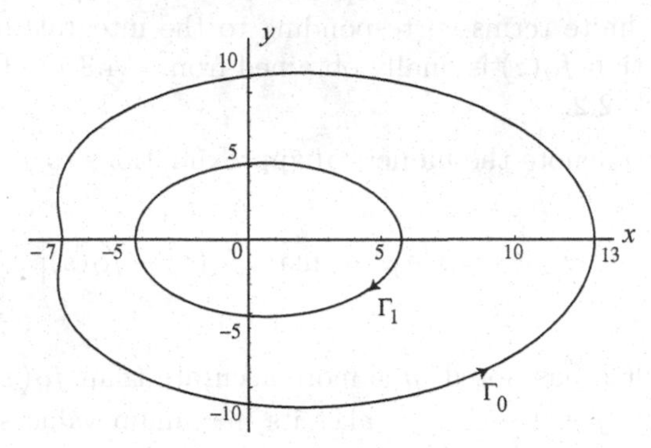

Figure 14.13 Limaçons.

Symm [1969] has taken $a_1 = 5$, $a_2 = 10$, $b_2 = 3$ and $b_1 = b_2/4$ and has shown that the error E increases as b_2 increases and that the boundary Γ_0 gradually changes from a circle ($b_2 = 0$) to a cardioid ($b = 5$). In each case E decreases as N increases. However, $\hat{\rho}$ varies very little, which indicates that even a crude partition of Γ is sufficient for a good approximation of ρ.

Table 14.2 Values of $\hat{u}$ and $\hat{v}$.

x	y	$\hat{u}$	$\hat{v}$
13.0	0.0	1.00000	−0.00006
11.9376	4.95353	0.95106	0.30902
9.01722	8.73102	0.80902	0.58778
4.95080	10.9433	0.58779	0.80902
0.66316	11.2739	0.30902	0.95106
−3.0	10.0	0.00002	1.00000
−5.51722	7.7472	−0.30902	0.95106
−6.8049	5.237	−0.58779	0.80902
−7.16312	3.02468	−0.80912	0.58779
−7.08351	1.32681	−0.95106	0.30902
−7.0	0.0	−0.99999	0.00002
5.75	0.0	0.49999	−0.00001
5.36205	1.98592	0.47552	0.15451
4.27685	3.65222	0.40452	0.29389
2.70716	4.75838	0.29389	0.40542
0.938322	5.19612	0.15451	0.47553
−0.75	5.0	0.00002	0.49999
−2.15185	4.31444	−0.15452	0.47553
−3.17069	3.33179	−0.29389	0.40452
−3.81322	2.22563	−0.36698	0.42726
−4.14852	1.10425	−0.47554	0.15452
−4.25	0.0	−0.49999	0.00003

14.4 Dipole Distribution

It is known that in a simply connected region D bounded by a Jordan contour $\Gamma : \{z = \gamma(s)\}$, where s denotes the arc length along Γ, $0 \le s \le L$, the dipole distribution density $\mu(s)$ satisfies the integral equation

$$\mu(s) = \frac{1}{\pi} \left[g(s) - \int_0^L \mu(t) \, \frac{\partial}{\partial n_t} \left(\log \frac{1}{r_{tz}} \right) dt \right] \tag{14.4.1}$$

where n_t is the inward normal at a point $\zeta = \gamma(t) \in \Gamma$, $0 \le t \le L$, $s \ne t$, and $r_{tz} = |z - \zeta|$, $z \in D$, and $g(s)$ denotes the boundary value of the potential function $u(z)$ on Γ (see Example 10.3 and Example 10.10). The following result is useful in numerical evaluation of conformal mapping: If the density $\mu(t)$ can be determined by solving Eq (14.4.1), then the Dirichlet problem and hence the problem of conformal mapping is reduced to quadratures.

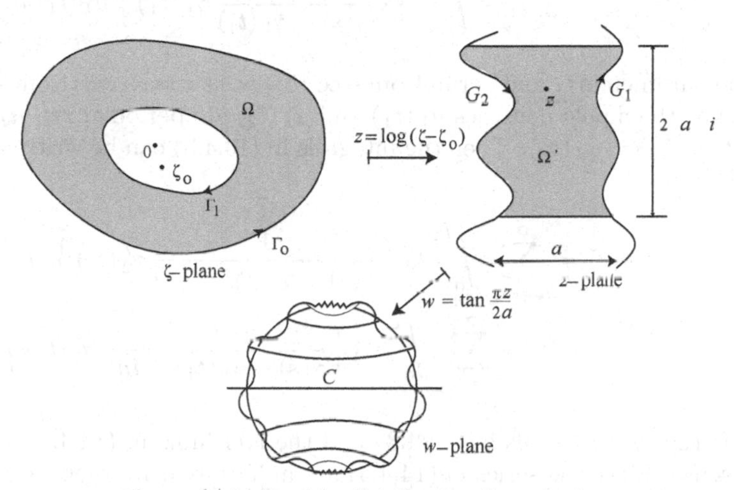

Figure 14.14 Double distribution formulation.

This dipole distribution formulation can be used for a doubly connected region Ω in the ζ-plane bounded by two Jordan contours Γ_1 and Γ_0, $\Gamma_1 \subset \Gamma_0$, and $0 \in \text{Int}\,(\Gamma_1)$, by transforming the region Ω into a simply connected region by the function $z = \log(\zeta - \zeta_0)$ which transforms Ω into an irregular strip Ω' with period $2i\pi$. Then the region Ω' is further transformed into an irregular circlelike region C in the w-plane by the function $w = \tan \dfrac{\pi z}{2a}$, where a is the mean-width of Ω'. Note that the region C may have infinitely many extrema ('humps') in the neighborhood of the two points corresponding to $\pm\infty$ (Figure 14.14). Thus, the boundary problem for the region Ω reduces to a boundary problem for the strip Ω', where the boundary values are $2i\pi$-periodic. Without loss of generality, we will consider the general case of the period ib, where b need not be 2π. Let the equations of the two boundaries G_1 and G_2 of the strip Ω' be $G_1 : z = \gamma(s_1)$ and $G_2 : z = \gamma(s_2)$, where s_1 and s_2 are the arc lengths on G_1 and G_2, respectively. If one period covers the arc lengths L_1 and L_2 such that $\gamma(s_1 + L_1) = \gamma(s_1) + ib$, and $\gamma(s_2 + L_2) = \gamma(s_2) + ib$, then the kernel in Eq (14.4.1) becomes

$$
\begin{aligned}
\frac{\partial}{\partial n_t}\left(\log \frac{1}{r_{tz}}\right) &= \Re\left\{\frac{\partial}{\partial n_t}\left(\log \frac{1}{\gamma(s)-\gamma(t)}\right)\right\} \\
&= \Re\left\{\frac{1}{\gamma(s)-\gamma(t)}\,\frac{\partial \gamma}{\partial n_t}\right\} = \Re\left\{\frac{i}{\gamma(s)-\gamma(t)}\,\frac{\partial \gamma}{\partial t}\right\} \\
&= -\Im\left\{\frac{1}{\gamma(s)-\gamma(t)}\,\gamma'(t)\right\},
\end{aligned} \tag{14.4.2}
$$

where n is the inward normal, $z=\gamma(s)\in\Omega'$, $\gamma(t)\in G_{1,2}$, and the potential at a point $z\in\Omega'$ is given by

$$
\begin{aligned}
u(z) = &\int_{-\infty}^{\infty} \Im\left\{\frac{1}{\gamma(s)-\gamma_2(t_2)}\,\gamma_2'(t_2)\right\}\mu_2(t_2)\,dt_2 \\
&- \int_{-\infty}^{\infty} \Im\left\{\frac{1}{\gamma(s)-\gamma_1(t_1)}\,\gamma_1'(t_1)\right\}\mu_1(t_1)\,dt_1,
\end{aligned} \tag{14.4.3}
$$

where the parameters t_1 and t_2 run from $-\infty$ to $+\infty$ as z traverses from $-i\infty$ to $+i\infty$. Let us assume that the dipole densities $\mu_1(t_1)$ and $\mu_2(t_2)$ are periodic, i.e., $\mu_1(t_1+L_1)=\mu_1(t_1)$, and $\mu_2(t_2+L_2)=\mu_2(t_2)$. Then the integrals in (14.4.3) can be written as sum of integrals, and we have

$$
\begin{aligned}
u(z) = &\sum_{n=-\infty}^{\infty}\int_0^{L_2} \Im\left\{\frac{1}{\gamma(s)-\gamma_2(t_2)-inb}\,\gamma_2'(t_2)\right\}dt_2 \\
&- \sum_{n=-\infty}^{\infty}\int_0^{L_1} \Im\left\{\frac{1}{\gamma(s)-\gamma_1(t_1)-inb}\,\gamma_1'(t_1)\right\}dt_1.
\end{aligned} \tag{14.4.4}
$$

Note that the two integrals in (14.4.3) and the two sums in (14.4.4) are convergent in the Cauchy sense. Since the series in (14.4.4) are uniformly convergent, we can interchange the integration and summation. Using formula (2.4.21) we find that

$$
\begin{aligned}
\sum_{n=-\infty}^{\infty}\frac{1}{\gamma(s)-\gamma(t)-inb} &= \frac{1}{\gamma(s)-\gamma(t)} \\
&\quad + \frac{1}{2ib}\sum_{n=1}^{\infty}\left\{\frac{1}{\frac{\gamma(s)-\gamma(t)}{ib}-n} + \frac{1}{\frac{\gamma(s)-\gamma(t)}{ib}+n}\right\} \\
&= \frac{\pi}{ib}\cot\left(\pi\,\frac{\gamma(s)-\gamma(t)}{ib}\right),
\end{aligned}
$$

and (14.4.4) becomes

$$
\begin{aligned}
u(z) = \pi &\int_0^{L_2} \Im\left\{\frac{1}{ib}\cot\left(\pi\,\frac{\gamma(s)-\gamma_2(t_2)}{ib}\right)\gamma_2'(t_2)\right\}\mu_2(t_2)\,dt_2 \\
- \pi &\int_0^{L_1} \Im\left\{\frac{1}{ib}\cot\left(\pi\,\frac{\gamma(s)-\gamma_1(t_1)}{ib}\right)\gamma_1'(t_1)\right\}\mu_1(t_1)\,dt_1.
\end{aligned} \tag{14.4.5}
$$

Let the point $z\in\Omega'$ approach a point z_s on the boundary of Ω'. Then, if $g(s)=u_+(z_s)$ is the prescribed boundary value, then $\lim\limits_{z\to z_s} u(z)=u_+(s_s)=u(z_s)+\pi\,g(s)$, and the integral equation for $\mu(s)$ is

$$\begin{aligned}\mu(s) = \frac{1}{\pi} g(s) + \int_0^{L_1} \Im\left\{\frac{1}{ib} \cot\left(\pi \frac{\gamma(s) - \gamma_1(t_1)}{ib}\right) \gamma_1'(t_1)\right\} \mu_1(t_1)\, dt_1 \\ - \int_0^{L_2} \Im\left\{\frac{1}{ib} \cot\left(\pi \frac{\gamma(s) - \gamma_2(t_2)}{ib}\right) \gamma_2'(t_2)\right\} \mu_2(t_2)\, dt_2.\end{aligned} \tag{14.4.6}$$

This integral equation can be solved by an iterative method, e.g., the one in §10.4. Although this method of *reduction of connectivity* seems especially suitable for those doubly connected regions that can be easily transformed into parallel strips, yet no numerical study has been done for it. In some special cases, however, the analysis becomes simpler, and it is presented in the following two case studies.

Map 14.14. (Andersen et al. [1962]). Consider the symmetric doubly connected region Ω which is transformed into the unit disk (or an annulus) by a chain of conformal maps f_1, f_2, f_3, f_4, f_5, f_6 and f_7, as shown in Figure 14.15.

Note that in the z-plane $z_0 \in \Gamma_0$ is chosen as a point on the axis of symmetry. The mapping goes from the z-plane through the z_1-plane, z_2-plane, z_3-plane, z_4-plane and finally to the unit disk in the w-plane. The conformal maps are as follows:

$$f_1 : z_1 = \log(z - z_0); \quad f_2 : z_2 = \tan\frac{\pi z_1}{2a};$$

$$f_3 : z_3 = \frac{(q+1)z_2 - 1 + q}{(q-1)z_2 + 1 + q};$$

$$f_4 : z_4 = \frac{1}{i\sqrt{k}} \frac{z_3 - 1}{z_3 + 1}; \quad f_5 : w = \frac{z_4 - i}{z_4 + i};$$

$$f_6 : z_6 = \int_0^{z_4} \frac{dt}{\sqrt{(1-t^2)(1-k^2t^2)}}; \quad f_7 : z_7 = \exp\left\{\frac{2\pi z_4}{ia}\right\}.$$

The five points marked as a, b, c, d, and e are traced through all of these mappings. The maps f_1 and f_2 are used in Figure 14.15; the map f_2 carries the points a and b into the points $e^{\pm i\alpha}$ and the points c and d into the points $e^{\pm i\beta}$, respectively, on the unit disk C_1, which are mapped by f_3 into the end points of two diameters of the unit disk C_2, where the angle ψ between these diameters is given by (14.1.3) and $k = \dfrac{\tan(\alpha/2)}{\tan(\beta/2)}$ (see §14.1); map f_4 carries the unit disk C_2 into the upper half-plane H, and finally f_5 maps the upper half-plane H onto the unit disk in the w-plane. Note in passing that f_6 maps the upper half-plane H onto the rectangle R which is mapped onto an annulus by f_7. Conversely, by using the maps f_7^{-1}, f_6^{-1}, and f_5^{1} in that order, an annulus can be mapped onto the unit disk.

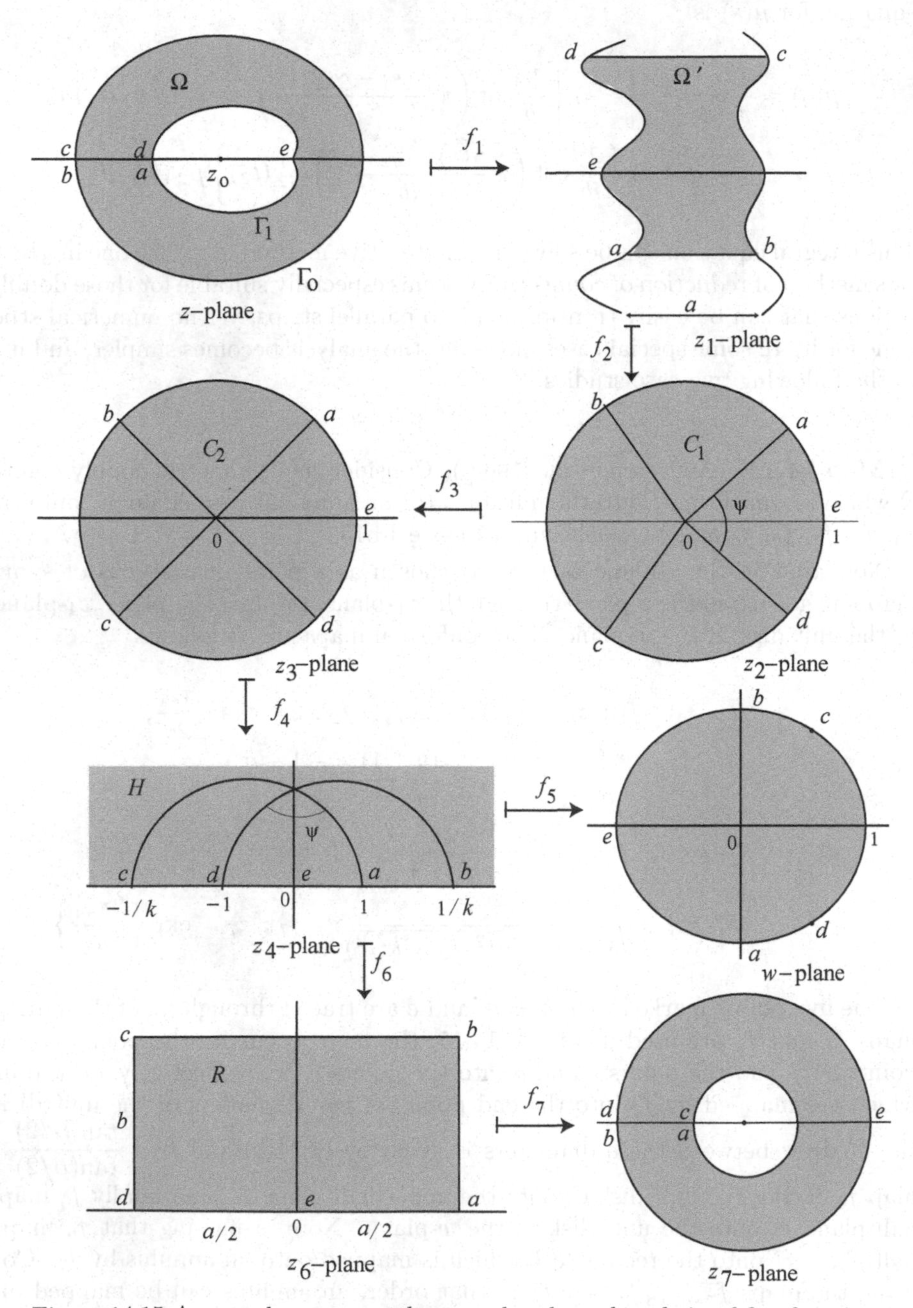

Figure 14.15 An annulus onto another annulus through a chain of five functions.

In this example the numerical computations are needed only for the mapping f_2 ($\Omega' \mapsto C_1$); other mappings are straightforward. The relation between the mapping f_2 ($\Omega' \mapsto C_1$) is provided by Gershgorin's integral equation (10.2.6) and the relation between their interior points by the Poisson integral (K.6). The mapping f_6 ($H \mapsto R$) is defined by the elliptic integral of the first kind (see §f.1.1).

Map 14.15. (Andersen et al. [1962]). In the nonsymmetric case, the horizontal lines in the z-plane need not go into straight horizontal lines in the w-plane, as shown in Figure 14.16. We will apply the above method of reduction of connectivity (from 2 to 1) to a periodic irregular strip Ω' onto a parallel strip S of width $\pi/2$. Then we can use the chain mapping of Map 14.14 to obtain the conformal mapping of a doubly connected region onto a unit disk or an annulus. This method involves the following steps:
1. Compute Green's function for the parallel strip S. This Green's function is not the usual Green's function with a logarithmic singularity; besides, it is Green's function with some unspecified period iq.
2. Use this Green's function to derive the integral equation for the boundary correspondence between the boundaries of the strip Ω' and S.

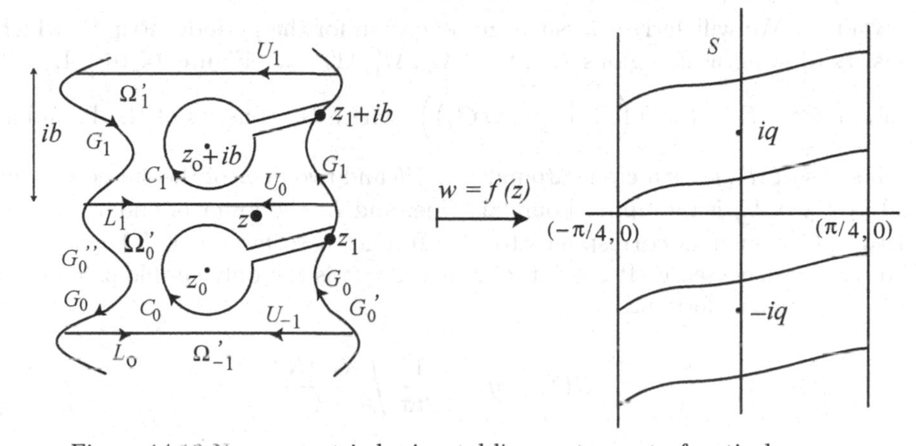

Figure 14.16 Nonsymmetric horizontal lines onto a set of vertical curves.

STEP 1. To compute Green's function, note that the function $w = \tan z$, where $z = 0$ goes into $w = 0$, maps the parallel strip $S : \left\{ -\frac{\pi}{4} \le \Re\{z\} \le \frac{\pi}{4} \right\}$ onto the unit circle $|w| \le 1$. The usual Green's function for the point $z = 0$ is $G(z, 0) = \log \tan z$. Then the iq-periodic Green's function is given by

$$G(z,q) = \sum_{n=-\infty}^{\infty} \log \tan(z + inq), \tag{14.4.7}$$

where the series is convergent in the Cauchy sense. This periodic Green's function can be represented in terms of elliptic theta functions as

$$\begin{aligned} G(z,q) &= \log \tan z + \sum_{n=1}^{\infty} \log\left[\tan(z+inq)\,\tan(z-inq)\right] \\ &= \log \tan z - \log\left\{ \prod_{n=1}^{\infty} \tanh(iz - nq)\,\tanh(iz+nq) \right\} \\ &= \log\left\{ \frac{1}{i}\,\frac{u-u^{-1}}{u+u^{-1}} \prod_{n=1}^{\infty} \frac{(1-h^{2n}u^2)\,(1-h^{2n}u^{-2})}{(1+h^{2n}u^2)\,(1+h^{2n}u^{-2})} \right\} \\ &= \log \frac{\vartheta_1(z,h)}{\vartheta_2(z,h)}, \end{aligned} \tag{14.4.8}$$

where we have set $u = e^{iz}$ and $h = e^{-q}$, and $\vartheta_{1,2}$ are elliptic theta functions such that $\vartheta_1(z + \pi/2, h) = \vartheta_2(z, h)$. These functions are numerically evaluated by the formulas (Abramowitz and Stegun [1972])

$$\begin{aligned} \vartheta_1(z, h) &= 2 \sum_{n=1}^{\infty} (-1)^{n-1}\, h^{(2n-1)^2/4} \sin(2n-1)z, \\ \vartheta_2(z, h) &= 2 \sum_{n=1}^{\infty} h^{(2n-1)^2/4} \cos(2n-1)z, \end{aligned} \tag{14.4.9}$$

where the series are convergent for $h < 1$, i.e., $q > 0$. If q is not too close to unity, the series converge very rapidly, and the computation of $\vartheta_{1,2}$ is straightforward.

STEP 2. We will derive Gershgorin's equation for the periodic strip Ω' which we assume consists of congruent regions $\dots, \Omega'_{-1}, \Omega'_0, \Omega'_1, \Omega'_2, \dots$ (Figure 14.16). Let C denote the contour $C = L_n \cup U_n \cup \left(\bigcup_{j=-n}^{n} (C_j \cup G_j) \right)$, where G_j consists of the boundary curves on both sides of Ω'_j, C_j is a circle around $z_0 + jib$ and two lines of the cut connecting this circle and $z_1 + jib$, U_j is the upper boundary line, and L_j the lower boundary line of Ω'_j (Figure 14.16). The point z_0 corresponds to $w = 0$, and z_1 to $w = \pi/4$. The variable ζ traverses C in the positive sense. If $z \in \operatorname{Int}(C)$, then $\zeta = z$ is the only simple pole of $f(z)$, in which case by Cauchy's formula

$$G(f(z), q) = \frac{1}{2i\pi} \int_C \frac{G(f(\zeta), q)}{\zeta - z}\, d\zeta.$$

Now, let the radii of the circles C_j tend to zero. Then the above integral along these circles vanishes because $G(f(\zeta), q) \approx \log(\zeta - z_0 - jib)$, and thus

$$\begin{aligned} G(f(z), q) &= \frac{1}{2i\pi} \int_{C \setminus \{C_j\}} \frac{G(f(\zeta), q)}{\zeta - z}\, d\zeta \\ &\quad + \frac{1}{2i\pi} \sum_{j=-n}^{n} \int_{z_1 + jib}^{z_0 + jib} \frac{G^+(f(\zeta), q) - G^-(f(\zeta), q)}{\zeta - z}\, d\zeta, \end{aligned}$$

where $G^+(f(\zeta), q) - G^-(f(\zeta), q) = 2i\pi$. Therefore,

$$G(f(z), q) = \frac{1}{2i\pi} \int_{C \setminus \{C_j\}} \frac{G(f(\zeta), q)}{\zeta - z}\, d\zeta + \sum_{j=-n}^{n} \log \frac{z_0 + jib - z}{z_1 + jib - z}.$$

Since $G_j = G_0 + jib$ and $G(f(\zeta + jib), q) = G(f(\zeta) = iq, q) = G(f(\zeta), q)$, we find that

$$\begin{aligned} G(f(z), q) &= \frac{1}{2i\pi} \int_{U_n \cup L_{-n}} \frac{G(f(\zeta), q)}{\zeta - z}\, d\zeta \\ &\quad + \frac{1}{2i\pi} \int_{G_0} G(f(\zeta), q) \sum_{j=-n}^{n} \frac{1}{\zeta + jib - z}\, d\zeta \\ &\quad + \sum_{j=-n}^{n} \log \frac{z_0 + jib - z}{z_1 + jib - z} \equiv I_1 + I_2 + I_3. \end{aligned} \tag{14.4.10}$$

Let $n \to \infty$. Then let $I_2 \to 0$. Using the formula (2.4.16), we have

$$\sum_{j=-n}^{n} \frac{1}{\zeta + jib - z} = \frac{1}{iL} \sum_{j=-n}^{n} \frac{1}{\dfrac{\zeta - z}{iL} + j} \to \frac{\pi}{iL} \cot \frac{\pi(\zeta - z)}{iL} \quad \text{as } n \to \infty,$$

and using formula (2.4.20), we find that, as $n \to \infty$,

$$\prod_{j=-n}^{n} \frac{z_0 + jib - z}{z_1 + jib - z} = \frac{z_0 - z}{z_1 - z} \prod_{j=1}^{n} \frac{1 - \left(\dfrac{z_0 - z}{jib}\right)^2}{1 - \left(\dfrac{z_1 - z}{jib}\right)^2} \to \frac{\sin \dfrac{z_0 - z}{jib}}{\sin \dfrac{z_1 - z}{jib}}.$$

Hence, as $n \to \infty$ in the Cauchy sense, we find from (14.4.1) that

$$\begin{aligned} I_2 + I_3 &= \frac{1}{2i\pi} \int_{G_0} G\left(f(\zeta), q\right) \frac{\pi}{iL} \cot \frac{\pi(\zeta - z)}{iL}\, d\zeta + \log \frac{\sin \dfrac{z_0 - z}{jib}}{\sin \dfrac{z_1 - z}{jib}} \\ &= G\left(f(z), q\right). \end{aligned} \tag{14.4.11}$$

Since $G\left(f(z), q\right)$ is purely imaginary, let $G\left(f(z), q\right) = i\, \Phi\left(f(z), q\right)$, where Φ is a real-valued function that defines the boundary correspondence and, by taking the imaginary part of (14.4.11), is defined by

$$\begin{aligned} \Phi\left(f(z), q\right) &= \frac{1}{2\pi} \int_{G_0} \Phi\left(f(\zeta), q\right) \Im\Big\{ \frac{\pi}{iL} \cot \frac{\pi(\zeta - z)}{iL}\, d\zeta \Big\} \\ &\quad + \arg \left\{ \frac{\sin \dfrac{z_0 - z}{jib}}{\sin \dfrac{z_1 - z}{jib}} \right\} \equiv J_1 + J_2. \end{aligned} \tag{14.4.12}$$

Now let the point z approach a point z_s on the boundary of G_0. Then

$$J_1 \to \frac{1}{2\pi} \int_{G_0} \Phi\left(f(\zeta), q\right) \Im\Big\{ \frac{\pi}{iL} \cot \frac{\pi(\zeta - z)}{iL}\, d\zeta \Big\} + \frac{1}{2} \Phi\left(f(z_s), q\right). \tag{14.4.13}$$

Hence, as $z \to z_s$, from (14.4.12) and (14.4.13) we obtain

$$\begin{aligned} \Phi\left(f(z_s), q\right) &= \frac{1}{\pi} \int_{G_0} \Phi\left(f(\zeta), q\right) \Im\Big\{ \frac{\pi}{iL} \cot \frac{\pi(\zeta - z)}{iL}\, d\zeta \Big\} \\ &\quad + 2 \arg \left\{ \frac{\sin \dfrac{z_0 - z}{jib}}{\sin \dfrac{z_1 - z}{jib}} \right\}, \end{aligned} \tag{14.4.14}$$

which is Gershgorin's integral equation for the periodic strip Ω'.

Note that Eq (14.4.14) can be simplified by taking the parametric equation of the curve G_0 as $\zeta = \zeta(t)$, where t is the arc length on G_0. Since G_0 has two parts, G_0' and G_0'', this parametric equation represents two equations. If we set $w(t) = f(\zeta(t))$, $z_s = \zeta(s)$, and $\Phi(f(\zeta(t)), q) = \phi(t, q)$, then Eq (14.4.14) becomes

$$\begin{aligned} \phi(s) = \frac{1}{\pi}\left(\int_{G_0'} + \int_{G_0''}\right) \phi(t,q)\, \Im\Big\{\frac{\pi}{iL}\cot\frac{\pi(\zeta(t)-\zeta(s))}{iL}\,\zeta'(t)\Big\}\,d\zeta \\ + 2\arg\left\{\frac{\sin\dfrac{z_0-\zeta(s)}{jib}}{\sin\dfrac{z_1-\zeta(s)}{jib}}\right\} = \arg\Big\{\frac{\vartheta_1(w(s),q)}{\vartheta_2(w(s),q)}\Big\}. \end{aligned} \tag{14.4.15}$$

Since $w(s) = \pm\frac{\pi}{4} + i\,y(s)$ for $\zeta(s) \in G_0$, we find from (14.4.9) that

$$\frac{\vartheta_1(w(s),q)}{\vartheta_2(w(s),q)} = \frac{\pm A(y(s)) + i\,B(y(s))}{A(y(s)) \mp i\,B(y(s))}, \tag{14.4.16}$$

where the upper and the lower sign is chosen according as $w(s) = \pm\frac{\pi}{4} + i\,y(s)$, and

$$\begin{aligned} A(y) &= \sum_{n=1}^{\infty} \sqrt{2}\,h^{n(n-1)}\cos\frac{(2n-1)\pi}{4}\cosh(2n-1)y, \\ B(y) &= \sum_{n=1}^{\infty} \sqrt{2}\,h^{n(n-1)}\sin\frac{(2n-1)\pi}{4}\sinh(2n-1)y = \frac{1}{i}\,A\left(y + \frac{i\pi}{2}\right). \end{aligned}$$

Hence the simplified form of Eq (14.4.14) is

$$\phi(s,q) = \arg\Big\{\frac{\vartheta_1(w(s),q)}{\vartheta_2(w(s),q)}\Big\} = \begin{cases} 2\arg\{A(y(s)) + i\,B(y(s))\} \text{ on } G_0', \\ \pi - 2\arg\{A(y(s)) + i\,B(y(s))\} \quad \text{on } G_0''. \end{cases} \tag{14.4.17}$$

This equation can be numerically computed by an iterative method.

Example 14.1. A doubly connected region Ω can be mapped univalently onto an annulus $A(1,R)$, $R > 1$, and the mapping is unique if a point z_0 on the boundary of Ω is transformed into a point w_0 on the boundary of $A(1,R)$. (Goluzin [1969: 208]; Wen [1992:97].) ■

Example 14.2. Consider coaxial ellipses

$$\Gamma_1 : \frac{x^2}{a_1^2} + \frac{y^2}{b_1^2} = 1, \quad \Gamma_0 : \frac{x^2}{a_2^2} + \frac{y^2}{b_2^2} = 1,$$

which are symmetric about both coordinate axes. Take the distribution $\alpha = 0\ (2\pi/N)\ 2\pi$ for the partition of the boundaries, and $N = 4(n-1)$. Note that if (i) $a_1 = 5$, $b_1 = 1$, $a_2 = 7$, $b_2 = 5$, (ii) $a_1 = 6$, $b_1 = 2$, $a_2 = 9$, $b_2 = 7$, or (iii) $a_1 = 7$, $b_1 = 2$, $a_2 = 9$, $b_2 = 6$, then the ellipses Γ_1 and Γ_0 are confocal, i.e., $a_1^2 - b_1^2 = a_2^2 - b_2^2$. In these cases there is an exact mapping function

$$f_\Omega(z) = \frac{z + \sqrt{z^2 - (a_2^2 - b_2^2)}}{a_2 + b_2},$$

with $\rho = \dfrac{a_1 + b_1}{a_2 + b_2}$. Also, E decreases as N increases. (Symm [1969].) ■

Example 14.3. Consider Cassini's ovals

$$\Gamma_1 = \left\{ \left[(x+b_1)^2 + y^2 \right] \left[(x-b_1)^2 - y^2 \right] = a_1^4 \right\},$$
$$\Gamma_0 = \left\{ \left[(x+b_2)^2 + y^2 \right] \left[(x-b_2)^2 - y^2 \right] = a_2^4 \right\},$$

(see Figure 14.1), which have symmetry about both coordinates axes. Partition Γ by taking equi-spaced points on the x-axis on each boundary with the same number of points on Γ_1 and Γ_0, and take $N = 4(n-1)$. Note that if (i) $a_1 = 2$, $b_1 = 1$, $a_2 = \sqrt[4]{2506} \approx 7.07389$, $b_2 = 7$, or (ii) $a_1 = 9$, $b_1 = 6$, $a_2 = \sqrt[4]{11116} \approx 10.26803$, $b_2 = 8$, then $\dfrac{a_1^4 - b_1^4}{a_2^4 - b_2^4} = \dfrac{b_1^2}{b_2^2}$, and in this case the exact mapping function is given by

$$f_\Omega(z) = \frac{a_2\, z}{\sqrt{a_2^4 - b_2^4 + b_2^2\, z^2}},$$

with $\rho = \dfrac{a_2 b_1}{a_1 b_2}$. Then E decreases as N increases (Symm [1969]). ■

14.5 Gaier's Variational Method

In Gaier's variational method we first determine the approximate function

$$H(z) = \frac{f'_\Omega(z)}{f_\Omega(z)} - \frac{1}{z} \tag{14.5.1}$$

by taking a finite series representation as

$$H(z) = \sum_{j=1}^{n} a_j\, \phi_j(z),$$

where $\{\phi_j(z)\}$ is the basis set of functions in $L^2(\Omega)$ which possess single-valued indefinite integrals in Ω. This set is augmented by adding appropriate singular functions to account for singularities on the boundary $\partial\Omega = \Gamma_1 \cup \Gamma_0$ and in $\mathrm{Int}\,(\Gamma_1) \cup \mathrm{Ext}\,(\Gamma_0)$. In fact, since the variational problem to minimize the integral

$$\|u\|^2 = \iint_\Omega |u(z)|^2\, dx\, dy, \quad u \in \mathcal{K}^1(\Omega), \tag{14.5.2}$$

(see §4.2) has a unique solution u_0 such that u_0 is orthogonal to $\mathcal{K}^0(\Omega)$, the function H is related to u_0 by

$$H(z) = \frac{u_0(z)}{\|u\|^2}. \tag{14.5.3}$$

Let us denote

$$A(z) = \log f_\Omega(z) - \log z. \tag{14.5.4}$$

Then $H(z) = A'(z)$, and for each function $\phi_j \in L^2(\Omega)$ which is continuous on $\partial\Omega$, we have, in view of Green's formula (2.3.11),

$$\langle \phi_j, H \rangle = \frac{1}{2i} \int_{\partial\Omega} \phi_j(z)\, \overline{A(z)}\, dz = i \int_{\partial\Omega} \phi_j(z)\, \log|z|\, dz. \tag{14.5.5}$$

The modulus $M = r_2/r_1$ of Ω is related to the function H by

$$\log M = \frac{1}{2\pi}\left\{\frac{1}{i}\int_{\partial\Omega}\frac{1}{z}\log|z|\,dz - \|H\|^2\right\}. \tag{14.5.6}$$

Gaier's variational method (VM) resembles the RM (§4.2) in many ways. Let the basis set $\{\phi_j(z)\} \in L^2(\Omega)$ be such that $\phi_1 \in \mathcal{K}^0(\Omega)$. Let $\mathcal{K}_n^m$, $m = 0, 1$, denote the n-dimensional counterparts of $\mathcal{K}^m(\Omega)$, i.e.,

$$\mathcal{K}_n^m(\Omega) = \mathcal{E}_n \cap \mathcal{K}^m(\Omega), \quad m = 0, 1, \tag{14.5.7}$$

where $\mathcal{E}_n = \text{span}\,(\phi_1, \phi_2, \dots, \phi_n)$. The set $\mathcal{K}_n^m$ is nonempty for $n = 1, 2, \dots$, and the n-dimensional problem corresponding to (14.5.2) is as follows:

Problem I_n^m: In the class $\mathcal{K}_n^1(\Omega)$ minimize $\|u\|$, defined by (14.5.2), over all $u \in \mathcal{K}_n^1(\Omega)$.

As in Problem I_n, the following results hold for the above problem:
(i) the problem I_n^m has a unique solution u_0;
(ii) the minimal function u_n is orthogonal to $\mathcal{K}_n^0(\Omega)$; and
(iii) the sequence $\{u_n\} \to u_0$ uniformly in Ω, i.e., in view of (14.5.3),

$$\frac{u_n(z)}{\|u\|^2} \to H(z), \tag{14.5.8}$$

almost uniformly in Ω (i.e., there is mean convergence in every compact subset of Ω). Hence, $\lim\limits_{n\to\infty} \|u_n - u_0\| = 0$. Let

$$h_j = \langle H, \phi_j\rangle, \quad j = 1, 2, \dots. \tag{14.5.9}$$

Note that $h_j \neq 0$. Also, since $\mathcal{K}_n^0(\Omega) = \{u \in \mathcal{E}_n : \langle H, u\rangle = 0\}$, the set $\{\bar{h}_1\,\phi_j(z) - \bar{h}_j\,\phi_1(z)\}$, $j = 2, 3, \dots, n$, is the basis of $\mathcal{K}^0(\Omega)$. Thus, if we take

$$u_n(z) = \sum_{j=1}^{n} c_j\,\phi_j(z), \tag{14.5.10}$$

then, since u_n is orthogonal to $\mathcal{K}^0(\Omega)$ and $\langle H, u_n\rangle = 1$, we obtain the linear $(n\times n)$ system of equations

$$\begin{gathered}\sum_{j=1}^{n}\{h_1\,\langle\phi_j, \phi_i\rangle - h_i\,\langle\phi_j, \phi_1\rangle\}\; c_j = 0, \quad j = 2, 3, \dots, n,\\ \sum_{j=1}^{n}\bar{h}_j\,c_j = 1,\end{gathered} \tag{14.5.11}$$

which determines the coefficients c_j. Then, in view of (13.4.8), the formula $H_n(z) = \dfrac{u_n(z)}{\|u_0]|^2}$ gives the nth approximation of the function $H(z) = A'(z)$ and the nth VM approximation of the mapping function $f_\Omega(z) \equiv f(z)$

$$f_n(z) = z\,e^{\int_\zeta^z H_n(t)\,dt}, \quad \zeta \in \overline{\Omega}. \tag{14.5.12}$$

Also, from (14.5.6)

$$M = \exp\left\{\frac{1}{2\pi}\left(\frac{1}{i}\int_{\partial\Omega}\frac{1}{z}\log|z|\,dz - \|H_n\|^2\right)\right\}, \tag{14.5.13}$$

which gives the nth VM approximation of the modulus M of the region Ω (M_n gives an upper bound to M).

Let us assume that the mapping function f_Ω is normalized so that the region Ω is mapped conformally onto the annulus $A(\rho, 1)$. Then the density function $\mu(s)$ (see §13.5) is related to the boundary correspondence function $\phi_\Omega(s)$ by

$$\phi'_\Omega(s) = 2\,\pi\,\gamma_1\,\mu(s), \tag{14.5.14}$$

where $\gamma_1 = \log M = -\log\rho$.

Let $\{\phi_j^*(z)\}$ denote the orthonormal basis of $L^2(\Omega)$. Then the function H has the Fourier series expansion

$$H(z) = \sum_{j=1}^{\infty}\beta_j\,\phi_j(z), \tag{14.5.15}$$

where the Fourier coefficients are given by $\beta_n = \langle\overline{\phi_j^*, H}\rangle$. Then the VM follows the same five-step procedure explained in §8.5.2, which leads to the nth approximation

$$H_n(z) = \sum_{j=1}^{n}\beta_j\,\phi_j^*(z), \quad \beta_j = \langle\phi_j^*, H\rangle, \quad j = 1, 2, \dots, n, \tag{14.5.16}$$

which, from (14.5.12), yields the nth approximation $f_n(z)$ of the mapping function $f_\Omega(z)$.

The basis set is taken as the set $\{z^j\}_{j=-\infty}^{\infty}$ which is a complete set in $L^2(\Omega)$. But the use of this set results in the same kind of problems as in the RM and BKM for simply connected regions. Due to the presence of singularities of the function H in the complement of Ω and corner points on the boundary, this basis set is augmented by adding singular functions related to each singular behavior. Thus, in the neighborhood of a branch point singularity at $z_j \in \partial\Omega$ the asymptotic expansion of the mapping function involves fractional powers of $(z - z_j)$. For a corner point $z_j \in \partial\Omega$ the asymptotic expansion of H augments the basis by singular functions of the form

$$\phi_j(z) = \begin{cases} \dfrac{1}{z^2}\left(\dfrac{1}{z} - \dfrac{1}{z_j}\right)^{r-1}, & \text{if } z_j \in \Gamma_1, \\ (z - z_j)^{r-1}, & \text{if } z_j \in \Gamma_0, \end{cases} \tag{14.5.17}$$

where $r = k + l/\alpha$, $\alpha = p/q > 0$, $k = 0, 1, 2, \dots$, and $1 \le l \le p$. A branch point singularity occurs if $p \neq 1$. If the arcs at a corner point are straight line segments, then the exponent $k + l/\alpha$ in (14.5.17) is replaced by l/α, $l = 1, 2, \dots$. Depending on the rational values of $k + l/\alpha$, the first few singular functions (14.5.17) are added to the basis set $\{z^j\}_{j=-\infty}^{\infty}$ to form the AB. Note that the singular functions in (14.5.17) for $z_j \in \Gamma_0$ are the same type as used for interior mappings and those in (14.5.17) for $z_j \in \Gamma_1$ are those used for exterior mappings of the simply connected regions discussed in Chapter 12. The function f_Ω may involve logarithmic terms if the asymptotic expansion is valid, but these logarithmic singularities can generally be ignored as they produce no serious computational problem.

Map 14.16. Consider the doubly connected region Ω bounded by a circle in a square, defined by

$$\Omega: \{(x,y): |x|<1, |y|<1\} \cap \{z: |z|>a, \quad a<1\}.$$

There are no corner singularities, so no AB is required. The region has eightfold symmetry about the origin, and thus, the basis set is taken as $z^{(-1)^{j+1}(2j+1)}$, $j=1,2,\dots$.

Map 14.17. Let $G_a = \{(x,y): |x|<a, |y|<a\}$ define a square region. Consider the doubly connected region Ω as a square in a square (square frame, Figure 14.17) defined by

$$\Omega = \{G_1 \cap \text{compl}\,(\bar{G}_a)\,,\ a<1\}.$$

Let z_j denote the four corners of the inner square. Then the singular functions associated with the branch point singularities at these corners are the functions $\phi_{rj}(z)$, $j=1,2,3,4$, where

$$r = k + \frac{2l}{3}, \quad k=0,1,\dots, \text{ and } 1 \le l \le 3. \tag{14.5.18}$$

Since the region Ω has eightfold symmetry about the origin, these four singular functions $\phi_{rj}(z)$ are combined into a single function

$$\tilde{\phi}_r(z) = \phi_{r1}(z) + \sum_{j=2}^{4} e^{i\theta_j}\,\phi_{rj}(z), \tag{14.5.19}$$

where the arguments θ_j are chosen such that

$$e^{i\pi/2}\,\tilde{\phi}_r\left(e^{i\pi/2}\,z\right) = \tilde{\phi}_r(z).$$

Since θ_j depend on the branches used in defining the functions $\phi_{rj}(z)$, one must be careful while constructing the singular functions of the form $\tilde{\phi}_r(z)$. The AB is

$$z^{(-1)^{j+1}(2j+1)},\ j=1,2,\dots;\quad \tilde{\phi}_r(z),\ r=\frac{2}{3},\frac{4}{3},\frac{5}{3},\frac{7}{3}.$$

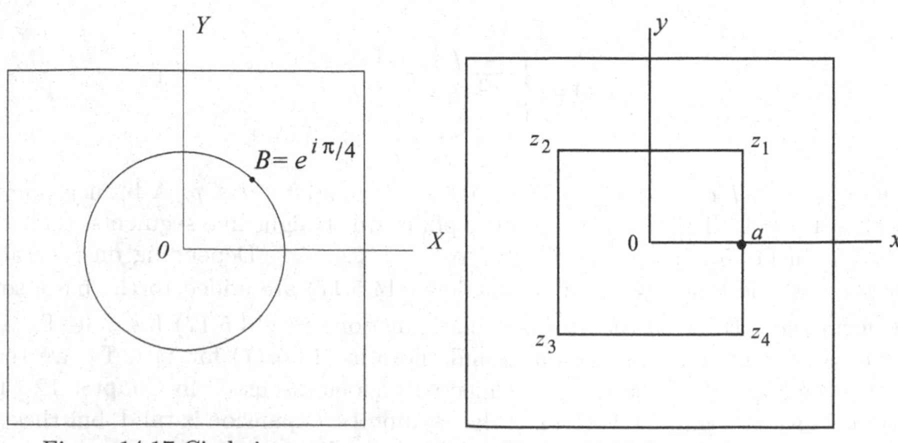

Figure 14.17 Circle in a square. Figure 14.18 Square in a square.

Map 14.18. Let

$$G_{ab} = \{(x,y) : |x| < a, |y| < b\} \cup \{|x| < b, |y| < a\}, \tag{14.5.20}$$

and

$$G_c = \{(x,y) : |x| < c, |y| < c\}.$$

Then consider the doubly connected region Ω which is a cross in a square (Figure 14.19), defined by

$$\Omega = G_c \cap \operatorname{compl}\left(\bar{G}_{ab}\right), \quad a < c,\, b < c.$$

Let the eight corners A, B, C, D, E, F, G, H of the cross-shaped region G_{ab} be denoted by z_j, $j = 1, 2, \dots, 8$, respectively. The singular functions associated with the branch point singularities at z_j, $j = 1, \dots, 8$, are given by

$$\phi_{rj}(z) = \frac{1}{z^2}\left(\frac{1}{z} - \frac{1}{z_j}\right)^{r-1},$$

as in (14.5.17), and r is defined by (14.5.18). In view of the symmetry these eight singular functions can be combined into two functions

$$\tilde{\phi}_{rj}(z) = \phi_{rj}(z) + \sum_{k=1}^{3} e^{i\theta_{2k+j}}\, \phi_{r,2k+j}(z), \quad j = 1, 2, \tag{14.5.21}$$

where the arguments θ_{2k+j} are chosen such that

$$e^{i\pi/2}\, \tilde{\phi}_{rj}\left(e^{i\pi/2} z\right) = \tilde{\phi}_{rj}(z), \quad j = 1, 2.$$

as in Map 14.17.

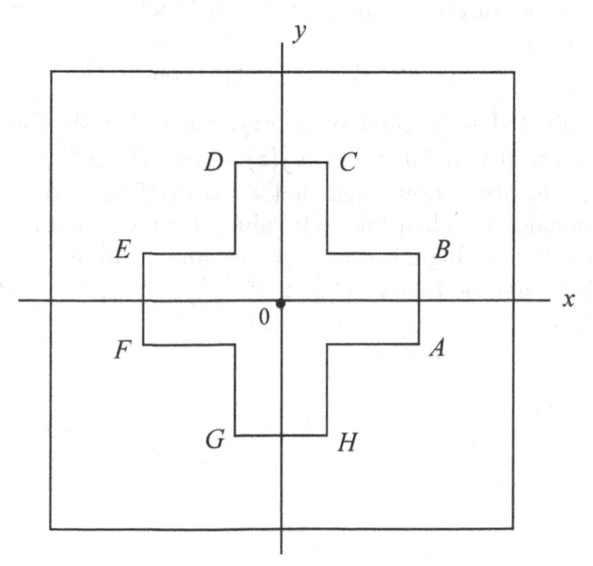

Figure 14.19 Cross in a square.

Example 14.4. Let α be rational. Show that the following first four functions ψ_j in the formal asymptotic expansion (13.5.5) for the density function $\mu(s)$ given by

$$\psi_1(\sigma) = \sigma^{-1+1/\alpha}, \quad 0 < \alpha < 2,$$

$$\psi_2(\sigma) = \begin{cases} \sigma^{1/\alpha}, & 0 < \alpha < 1, \\ \sigma \log \sigma, & \alpha = 1, \\ \sigma^{-1=2/\alpha}, & 1 < \alpha < 2, \end{cases}$$

$$\psi_3(\sigma) = \begin{cases} \sigma^{1+1/\alpha}, & 0 < \alpha < 1/2, \\ \sigma^3 \log \sigma, & \alpha = 1/2, \\ \sigma^{-1+2/\alpha}, & 1/2 < \alpha < 1, \\ \sigma^{1/\alpha}, & 1 \le \alpha < 2, \end{cases}$$

$$\psi_4(\sigma) = \begin{cases} \sigma^{2+1/\alpha}, & 0 < \alpha < 1/3, \\ \sigma^5 \log \sigma, & \alpha = 1/3, \\ \sigma^{-1+2/\alpha}, & 1/3 < \alpha \le 1/2, \\ \sigma^{1+1/\alpha}, & 1/2 < \alpha < 1, \\ \sigma^2 (\log \sigma)^2, & \alpha = 1, \\ \sigma^{-1+3/\alpha}, & 1 < \alpha < 2, \end{cases}$$

where, in particular, $a_j^{\pm}$, $j = 1, 2, 3$, satisfy the relations

$$\left.\begin{aligned} a_1^- &= \lambda^{1/\alpha}\, a_1^+, \quad 0 < \alpha < 2, \\ a_2^- &= -\lambda^{2/\alpha}\, a_2^+, \quad 1 \le \alpha < 2, \\ a_3^- &= -\lambda^{2/\alpha}\, a_3^+, \quad 1/2 \le \alpha < 1 \end{aligned}\right\}, \quad \text{and} \quad \lambda = \left| \frac{\gamma'\left(s_0^-\right)}{\gamma'\left(s_0^+\right)} \right|.$$

(Papamichael, Warby and Hough [1986].) ■

Example 14.5. Let $G_{ab} = \{(x, y) : |x| < a < 1, |y| < b < 1\}$ denote a rectangular region. Consider the doubly connected region Ω which is a rectangle in a circle (Figure 14.20) and defined by

$$\Omega = \{z : |z| < 1\} \cap \text{compl}\left(\bar{G}_{ab}\right).$$

If $a \ne b$, the region Ω has fourfold symmetry. Show that the four singular functions $\phi_{rj}(z)$ can be combined into two functions $\tilde{\phi}_{rj}(z) = \phi_{rj}(z) + e^{i\theta_j}$, $j = 1, 2$, where r is defined in (14.5.18), and θ_j are chosen such that $e^{i\pi}\, \tilde{\phi}_{rj}\left(e^{i\pi} z\right) = \tilde{\phi}_{rj}(z)$. If $a = b$, the region Ω has eightfold symmetry. Then for each value of r the four functions $\phi_{rj}(z)$, $j = 1, 2, 3, 4$, can be combined into a single function of the same form as (14.5.19), and in each case the monomial basis set can be taken as $\left\{z,\, z^{\pm(2j+1)}\right\}$, $j = 1, 2, \dots$ (Papamichael and Kokkinos [1984]). ■

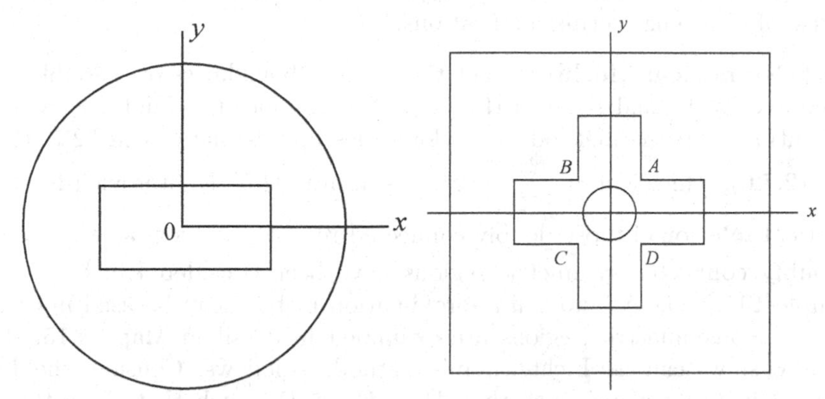

Figure 14.20 Rectangle in a circle. Figure 14.21 Cross in a square.

Example 14.6. Let G_{ab} be the cross-shaped region defined by (14.5.20). We will consider the circle-in-a-cross region Ω (Figure 14.21) defined by

$$\Omega = G_{ab} \cap \{z : |z| > c\}.$$

Let z_j, $j = 1, 2, 3, 4$, denote the four corners A, B, C, D of the outer boundary. Show that the singular functions associated with the branch point singularities at these points z_j are $\phi_{rj}(z) = (z - z_j)^{r-1}$, where r is defined in (14.5.18). Using symmetry it can be shown that these four functions can be combined into a single function $\tilde{\phi}_{rj}(z)$ of the form (14.5.21), and that the AB is formed by the monomial basis set $z^{(-1)^{j+1}(2j+1)}$ plus the functions $\tilde{\phi}_{rj}(z)$ with $r = 2/3, 4/3, 8/3, 10/3$. (Papamichael and Kokkinos [1984].) ■

14.6 Mapping onto Annulus

According to the Riemann mapping theorem, a doubly connected region can always be conformally mapped onto an annulus $A(r_1, r_2)$. Suppose, one of the radii, say r_1, of this annulus is chosen arbitrarily. Then the other radius r_2 must be obtained from the ratio r_2/r_1 which is the (fixed) conformal modus of the doubly connected region.

Note that the conformal ratio is related to capacitance, or conductance, of a potential field that is generated between the inner and outer boundaries.

The Schwarz-Christoffel transformation and the integral equation methods, like Lichtenstein, Garrick, Theodorsen and others discussed earlier, when applied to doubly connected regions do not display symmetry (see Nehari [1952: Ch 2], Gaier [1964; 1981], and Henrici [1986]). The integral equation method was, however, extended to doubly connected region by Symm [1969]. This method was used by Hough and Papamichael [1983] to examine the interior, exterior and doubly connected regions with boundaries with sharp corners by using the spline approximation. Moreover, Papamichael and Kokkinos [1984] examined the approximate transformation of boundaries with sharp corners using both a variational method and an orthonormalization method, and Gaier [1964] and Papamichael and Warby [1984] have discussed a connection between two numerical methods for doubly connected regions, one by Challis and Burley [1982] and the other by Garrick [1936].

A doubly connected region with one or more axis of symmetry can be examined by breaking it up into identical subregions (or mirror images), if possible, such that each subregion becomes simply connected. However, nonsymmetric, multiply connected regions are demanding and require more sophisticated programming methods.

The (numerical) conformal mapping of doubly connected regions has become important because of their engineering applications.

14.6.1 Numerical Methods. Let the annular boundaries of a doubly connected region be denoted by Γ_1 and Γ_2 such that $z = 0$ is interior of Γ_1 and $z = \infty$ is exterior to Γ_2. As mentioned in the method of Taylor series approximations in §2.7, the inverse of the series (2.7.6), which is $w = \sum_{j=-m}^{m} a_j z^j$, has been used by Kantorovich-Krylov [1964: 363] in numerical solution of those doubly connected regions that contain the point at infinity.

Doubly connected symmetric regions have been considered in §14.3, Map 14.14, and Example 14.2. The polynomial approximation method can be used in cases of symmetric regions. Nonsymmetric regions are examined in detail in Map 14.15. However, in the general case we can use Lichtenstein's method, as follows: Consider the doubly connected region D in the z-plane, such that $D = D_{z_1} \cap D_{z_2}$ such that Γ_1 is the outer boundary and Γ_2 the inner boundary of D_{z_1}, and let C be the annulus $D_{w_1} \cap D_{w_2}$, such that C_1 is the outer boundary and C_2 the inner boundary of D_{w_1} (Figure 14.22). Then, using the Lichtenstein-Gershgorin approximations (11.12.9) (for interior region) and (11.12.15) (for exterior region), we obtain a combination of these equations, as given in Muratov [1937], as

$$\Theta_1(s_1) = -2\beta(s_1) - \frac{1}{\pi}\int_{\Gamma_1} \frac{\Theta_1(t)\cos\alpha_s(t)}{r_{st}}\,dt + \frac{1}{\pi}\int_{\Gamma_2} \frac{\Theta_2(t)\cos\alpha_s(t)}{r_{st}}\,dt, \tag{14.6.1}$$

$$\Theta_2(s_2) = -2\beta(s_2) - \frac{1}{\pi}\int_{\Gamma_1} \frac{\Theta_1(t)\cos\alpha_s(t)}{r_{st}}\,dt + \frac{1}{\pi}\int_{\Gamma_2} \frac{\Theta_2(t)\cos\alpha_s(t)}{r_{st}}\,dt, \tag{14.6.2}$$

where s_1 and s_2 are the labels of s on Γ_1 and Γ_2, respectively.

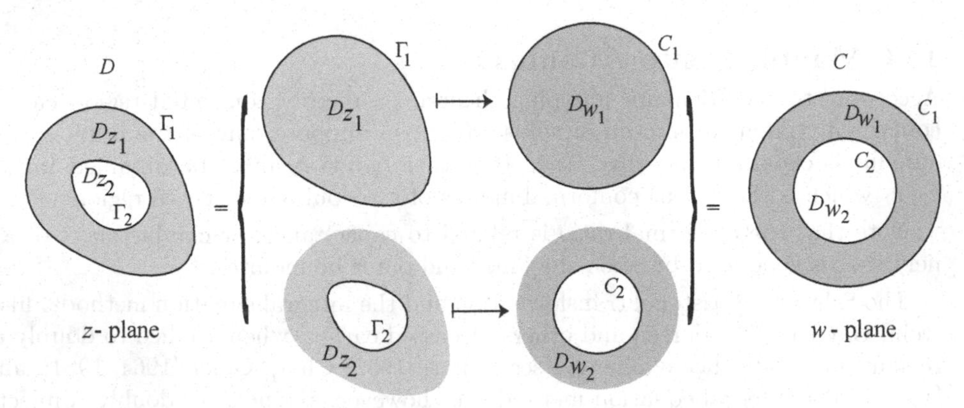

Figure 14.22 Mapping of doubly connected regions.

Next, we will consider the following question: If $D_{z_1} \cup \Gamma_1$ maps onto $D_{w_1} \cup C_1$, and $D_{z_2} \cup \Gamma_2$ maps onto $D_{w_2} \cup C_2$, then can we conclude that the doubly connected region D is mapped onto the annulus C in the w-plane? The answer is affirmative (see Figure 14.22), provided certain conditions are met, as explained in the following three major cases: (i) symmetry and circles, (ii) symmetric region with an inner circle, (iii) symmetric region with an outer circle, and (iv) other configurations in doubly connected regions.

Case (i). As in the Theodorsen-Garrick method (§11.13), when there are s number of symmetries in a simply connected region, then the mapping of its interior that includes $z = 0$, and of its exterior onto the w-plane can be approximated respectively by a polynomial of the form

$$w = \sum_{j=0}^{n} a_{1+js} z^{1+js}, \quad w = \sum_{j=0}^{n} a_{1-js} z^{1-js}. \tag{14.6.3}$$

We can combine the two polynomials to use for the mapping of a doubly connected region onto a circular annulus (see Figure 14.22).

Now, if one of the boundaries of the annular region in the w-plane is C_2 which is a circle of radius r_1, then the Lichtenstein-Gershgorin integral equation (14.6.2) reduces to

$$\Theta_2(s_2) = -2\beta(s_2) - \frac{1}{\pi} \int_{C_1} \frac{\Theta_1(\sigma_1) \cos(n_t, r)}{r} \, d\sigma_1 + \frac{1}{2\pi r_1} \int_{C_2} \Theta_2(\sigma_2) \, d\sigma_2. \tag{14.6.3}$$

If, in addition to the circle C_2 there are one or more axes of symmetry, Eq (14.6.3) becomes

$$\Theta_2(s_2) = -2\beta(s_2) + \pi - \frac{1}{\pi} \int_{C_1} \frac{\Theta_1(\sigma_1) \cos(n_t, r)}{r} \, d\sigma_1, \tag{14.6.4}$$

which means that if C_2 is a circle and the region has one or more axes of symmetry, then Eqs (14.6.2) and (14.6.4) must be used. Similarly, if C_1 is a circle and the region has one or more axes of symmetry, we use Eq (14.6.2) and

$$\Theta_1(s_1) = -2\beta(s_1) - \pi + \frac{1}{\pi} \int_{C_2} \frac{\Theta_2(\sigma_2) \cos(n_t, r)}{r} \, d\sigma_2. \tag{14.6.5}$$

Doubly connected regions presented in Figures 14.19 and 14.20 can be dealt by this method, so also a simplified cross-section of a solid propellant rocket grain, as shown in Figure 14.23, in which one of the boundaries in the z-plane is a circle.

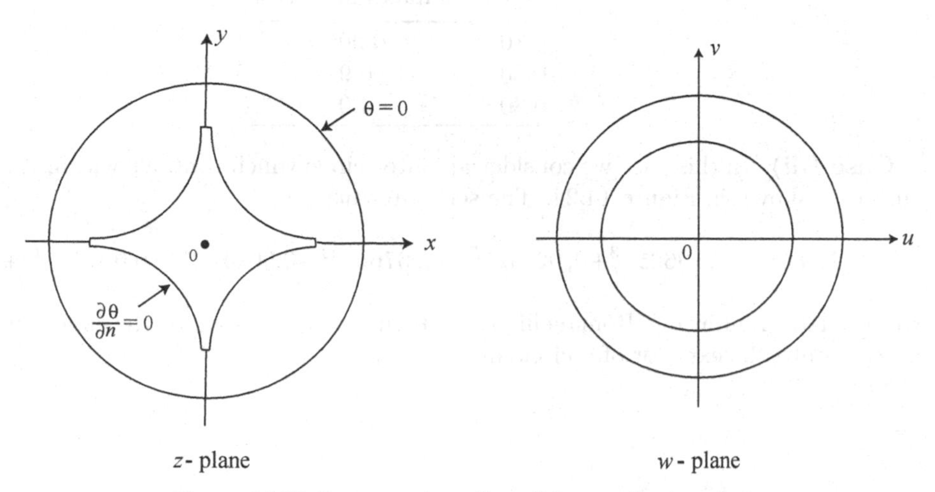

Figure 14.23 Cross-section of a solid propellant rocket.

The following Table 14.3 (Schinzinger and Laura [2003: 153]) gives the values of the coefficients to be used with the series expansion (14.6.3).

Table 14.3 Values of the coefficients.

Polygon	s	a_1	a_{1-s}	a_{1+2s}	a_{1+3s}	a_{1+4s}	a_{1+5s}	a_{1+6s}
Square	4	1.08	−0.108	0.04	−0.26	0.0174	−0.0127	0.00997
Pentagon	5	1.0526	−0.0704	0.0272	−0.0153	0.0103	−0.0079	0.0058
Hexagon	6	1.0376	−0.0496	0.0183	−0.0108	0.100	−0.0053	
Heptagon	7	1.0279	−0.0361	0.0119	−0.0060	0.0040	−0.0017	
Octagon	8	1.0219	−0.0282	0.0092	−0.0048	0.0030	−0.0022	0.0017

Case (ii). In this case the power series expansion of the form $w = \sum_{j=-m}^{m} a_j z^j$ can be used provided that the boundary Γ_1 has $z = 0$ as an interior point and the boundary Γ_2 has $z = \infty$ as an exterior point. Then we map the outer boundary Γ_1 onto the outer boundary C_1 of the annulus in the w-plane. The inner boundary Γ_2 will be mapped onto the inner circle C_2 in the w-plane, because the jth term in the above power series, that has the power of z as $(1 + js) > 0$, where s denotes the number of axes of symmetry, becomes extremely small. This holds for any point on and within the inner circle. For $j = 1$, we have $w = a_1 z$, so that the radius r_2 of the region Γ_1 becomes the radius $R_1 = a_1 r_1$ of the circle C_1. Example 11.5 is a representative example of this case, also the following example.

Example 14.7. Consider a square with outer boundary of size $2b$ with corners at $x + iy = \pm b \pm ib$, and with an inner boundary which is a circle of radius r_1 with center $(0, 0)$, i.e., concentric with the square. The following polynomial with axes of symmetry $s = 4$ is applicable in this case:

$$z = 1.08b\Big(w - \frac{1}{10}w^5 + \frac{1}{24}w^9 - \frac{5}{208}w^{13} + \cdots\Big), \tag{14.6.6}$$

with approximate errors given in Table 14.4.

Table 14.4 Approximate errors.

r_1/b	% maximum error
0.10	0.005
0.50	0.9
0.80	4.0

Case (iii). In this case we consider an outer circle which contains within it a concentric square as shown in Figure 14.24. The series expansion

$$w = a\big\{1.1804z - 0,1966z^{-3} + 0,0214z^{-7} - 0.0076z^{-11} + 0.0044z^{-15} - 0.0026z^{-19} + 0.0008z^{-23} \tag{14.6.7}$$

was used by Laura and Romanelli [1973] to map the exterior of the inner square of size $2a \times 2a$ onto the exterior of a circle in the w-plane.

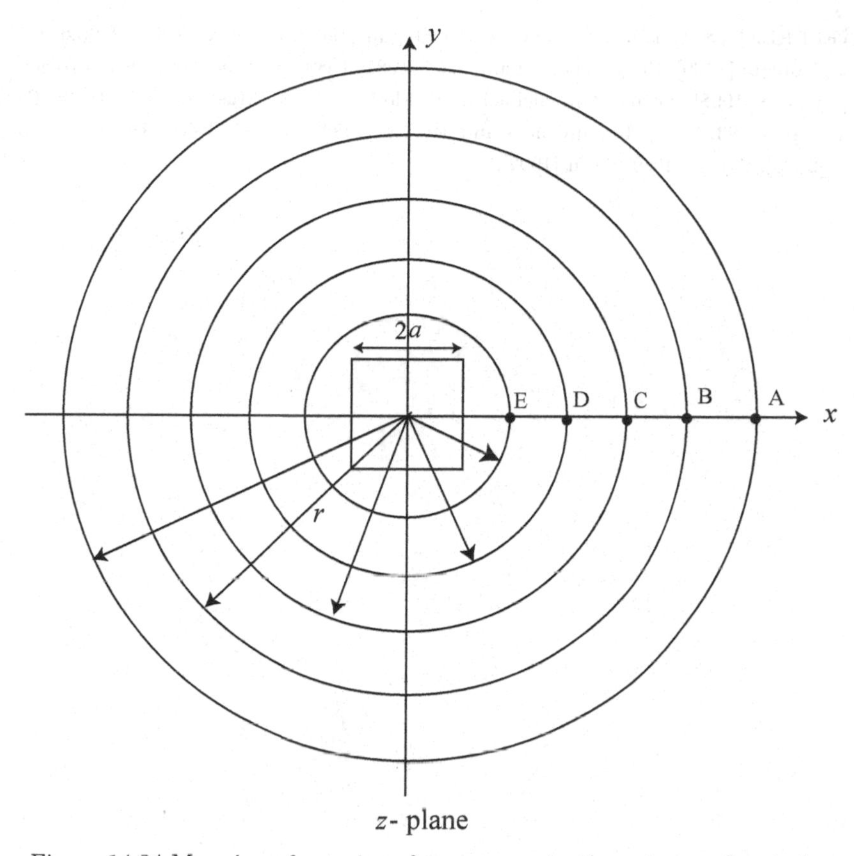

Figure 14.24 Mapping of exterior of a square onto the exterior of a circle.

It is shown that this transformation maps the square onto an inner circle while retaining the circular shape of the outer boundary as long as $r_2 > 2$, and the error due to extra terms in z raised a negative power becomes negligible, as seen from the following Table 14.5.

Table 14.5 Error due to extra terms.

Case	Ratio of circle to square $r/(2a)$	Maximum error (%)
A	5	0.03
B	4	0.06
C	3	0.02
D	2	1.1
E	1.5912	2.5

The mapping problems involving regions with slits have been mostly examined in earlier chapters The work of Nehari [1952] and Ellacott [1979] deal with circular, concentric slits; of von Koppenfels and Stallmann [1959] deal with parallel and radial slits; and of Papamichael and Kokkinos [1984] deal with regions of the shape of the cross.

REFERENCES USED: Abramowitz and Stegun [1968], Andersen et al. [1962], Copson [1975], Gaier [1964], Goluzin [1969], Hough and Papamichael [1981, 1983], Jawson [1963], Kantorovich and Krylov [1958; 1998], Kythe [1998], Levin, Papamichael and Sideridis [1978], Muslhelishvili [1963], Papamichael, Warby and Hough [1983; 1986], Papamichael and Warby [1984], Nehari [1952], Reichel [1985], Schinzinger and Laura [2003], Symm [1969], Wen [1992].

15

Multiply Connected Regions

The existence and uniqueness theorems for the conformal mappings of multiply connected regions onto canonical regions are discussed in Kythe [1998:358 ff]. We will first discuss a numerical method based on Mikhlin's integral equation formulation on the boundary, which is a Fredholm integral equation of the second kind and has a unique periodic solution. Then a numerical method, called Mayo's method, that uses a fast Poisson solver for the Laplacian (Mayo [1984]) is employed to determine the mapping function in the interior of the region which can be simply, doubly, or multiply connected, with accuracy even near the boundary. This method, in fact, computes the derivatives of the mapping function in the first application and the mapping function itself when applied twice. Most of the methods for conformal mapping compute the boundary correspondence function only.

15.1 Some Useful Results

We will define certain canonical regions, besides the unit disk and the annulus. A region with parallel cuts (slits) is understood to be a region obtained from the extended complex plane $\mathbb{C}_\infty$ by removing several mutually parallel line segments inclined at an angle θ to the positive real axis (we call them parallel finite cuts of inclination θ). By a region with spiral cuts we mean a region which is obtained by removing several logarithmic spirals from $\mathbb{C}_\infty$. Let α and c be real constants. The equation

$$\Im\{e^{-i\alpha}\log w\} = c \tag{15.1.1}$$

represents a logarithmic spiral in the w-plane with the origin as its asymptotic point, where α is the angle between the logarithmic spiral and a fixed ray emanating from the origin (α is known as the oblique angle of the spiral cuts). For $\alpha = 0$ the logarithmic spiral reduces to a ray $\arg\{w\} = c$ emanating from the origin, and for $\alpha = \pi/2$ it becomes a circle $|w| = e^c$ (unit circle for $c = 0$). The following theorem establishes the existence and uniqueness of the conformal mapping from a multiply connected region onto a region with parallel or spiral cuts. A result due to Hilbert [1909] is as follows:

Theorem 15.1. *Let Ω be a multiply connected region in the extended z-plane $\mathbb{C}_\infty$ and θ a real number. Then there exists a univalent meromorphic function $w = f_\theta(z)$ in Ω such* *that (i) it maps Ω conformally onto a region with parallel finite cuts of inclination θ in the extended w-plane; and (ii) it maps a given point $z = a$ into $w = \infty$, and in a neighborhood of $z = a$ the function $f_\theta(z)$ may be represented by a series of the form*

$$f_\theta(z) = \frac{1}{z-a} + a_1 (z-a) + \cdots , \tag{15.1.2}$$

or

$$f_\theta(z) = z + \frac{a_1}{z} + \cdots , \tag{15.1.3}$$

according to whether a is finite or not. Each of these functions is unique for the region Ω.

A similar theorem holds if the image of Ω has spiral cuts of oblique angle α in the w-plane. The mapping functions are the same as (15.1.2) and (15.1.3) which in this case are denoted by f_α instead of f_θ. A proof of this theorem can be found in Goluzin [1969:213] or Wen [1992:118]. Two results on the existence and uniqueness of conformal mapping of a multiply connected region Ω contained inside the unit disk $|z| < 1$ onto a region inside the unit disk $|w| < 1$ with concentric finite circular cuts and inside an annulus $r < |z| < 1$ are as follows:

Theorem 15.2. *Let Ω be a multiply connected region of connectivity $(n+1)$ inside the unit disk $|z| < 1$ where $\Gamma = |z| = 1$ is the boundary component of Ω and $0 \in \Omega$. Then there exists a unique, univalent analytic function $w = f(z)$ in Ω such that (i) it maps Ω conformally onto a region G inside the unit disk $|w| < 1$ which has n circular cuts centered at $w = 0$, and (ii) it maps the unit circle $|z| = 1$ conformally onto the unit circle $|w| = 1$ with $f(0) = 0$ and $f(1) = 1$.*

Theorem 15.3. *Let Ω be a multiply connected region of connectivity $n+1$ inside the annulus $r < |z| < 1$, where $\Gamma_0 = \{|z| = 1\}$ and $\Gamma_1 = \{|z| = r\}$ are the two boundary components of Ω. Then there exists a unique univalent analytic function $w = f(z)$ in Ω such that (i) it maps Ω conformally onto a region G in the w-plane formed by removing n concentric circular arcs centered at $w = 0$ from the annulus $\rho < |w| < 1$, where $0 < \rho < 1$, and (ii) it maps the unit circle Γ_0 conformally onto the unit circle $|w| = 1$, and the circle Γ_1 onto the circle $|w| < \rho$, with $f(1) = 1$.*

A region whose boundary consists of a finite union of circles is known as a *circular region*. We will consider the conformal mapping of a multiply connected region Ω onto a circular region G. This can be accomplished by a chain of two conformal mappings: (i) that of Ω onto a region G with parallel cuts, and (ii) that of G onto a circular region Δ. The former mapping is already established in Theorem 15.1. We need to discuss only the latter mapping. We will state the uniqueness theorem for conformal mappings onto circular regions; the existence of this mapping can be proved by the continuity method (see Wen [1992: 118]).

Lemma 15.1. *Let D be an $(n+1)$-connected circular region obtained by removing n disks from the unit disk $|z| < 1$, with $0 \in D$, and let Δ be an $(n+1)$-connected circular region obtained by removing n disks from the unit disk $|w| < 1$, with $0 \in \Delta$. If $w = f(z)$ maps D conformally onto Δ such that (i) $f(0) = 0$, $f(1) = 1$, and (ii) $f(\zeta_j) = \zeta_j$, $j = 1, 2, 3$, where ζ_j are three distinct points on $|z| = 1$, then $f(z) = z$.*

This lemma establishes the identity mapping; its proof is available in Wen [1992:118].

Theorem 15.4. *Let Ω be a multiply connected region in the extended z-plane. Then there exists at most one univalent meromorphic function $w = f(z)$ in Ω, which maps Ω conformally onto a circular region G in the w-plane, such that the point $z = \infty$ goes into $w = \infty$, and in a neighborhood of $z = \infty$ the function $f(z)$ has the series expansion (15.1.2).*

PROOF. Suppose that $w = f_1 z)$ and $w = f_2(z)$ are two univalent meromorphic functions in Ω, each of which satisfies the hypothesis of the theorem. Then the function $f_2\left(f_1^{-1}(w)\right)$, where $z = f_1^{-1}(w)$ denotes the inverse of $w = f_1(z)$, is univalent and meromorphic in the region G, maps G onto another circular region G' in the ζ-plane, maps the point $w = \infty$ into the point $\zeta = \infty$, and in the neighborhood of the point $w = \infty$ has the series expansion

$$f_2\left(f_1^{-1}(w)\right) = w + \frac{b_1}{w} + \cdots . \tag{15.1.4}$$

Now we must show that if $\zeta = F(w)$ is the function that maps the region G conformally onto G' such that $F(\infty) = \infty$, and has the series representation (15.1.4) in the neighborhood of the point $w = \infty$, then $F(w) = w$. In fact, by using linear transformations of w and ζ, we can map the regions G and G', respectively, onto circular regions D and Δ of Lemma 15.1. The univalent function obtained from these linear transformations satisfies conditions (i) and (ii) of this lemma and thus represents an identity mapping. Hence, $\zeta = F(w)$ is a linear transformation with $F(\infty) = \infty$, and therefore, $F(w) = aw + b$. But since $F(w)$ has an expansion of the form (15.1.4) in the neighborhood of $w = \infty$, we require that $a = 1$ and $b = 0$, i.e., $F(w) = w$. It proves that $f_2\left(f_1^{-1}(w)\right) = w$, and hence, $f_1(z) \equiv f_2(z)$. ■

The function $f_\theta(z)$, defined by (15.1.2) or (15.1.3), can be evaluated for arbitrary θ from the equation

$$f_\theta(z,a) = e^{i\theta}\left[\cos\theta\, f_0(z,a) - i\,\sin\theta\, f_{\pi/2}(z,a)\right]. \tag{15.1.5}$$

The difference $d(z)$ between the two sides of Eq (15.1.5) is regular in the region Ω, and $d(a) = 0$. Also, all values taken by $d(z)$ on any contour Γ_j lie on a circle (in the extended sense) $\Im\{e^{-i\theta} w\} = c$. The above equation also enables us to compute the function $f_\theta(z,a)$ for arbitrary θ if we know the functions $f_0(z,a)$ and $f_{\pi/2}(z,a)$. To get these relations in a symmetric form, we set

$$P(z,a) = \frac{1}{2}\left[f_{\pi/2}(z,a) - f_0(z,a)\right], \quad Q(z,a) = \frac{1}{2}\left[f_{\pi/2}(z,a) + f_0(z,a)\right]. \tag{15.1.6}$$

Since $\Im\{f_0(z,a)\} = \text{const}$, and $\Im\{f_{\pi/2}(z,a)\} = \text{const}$ on Γ_j, $j = 0,1,\dots,n$, for fixed $a \in \Omega$, i.e., since

$$f_0(z,a) = \overline{f_0(z,a)} + \text{const}, \quad f_{\pi/2}(z,a) = -\overline{f_{\pi/2}(z,a)} + \text{const},$$

we have on Γ_j, in view of (15.1.6),

$$P(z,a) = -\overline{Q(z,a)} + \overline{q_j(a)}, \tag{15.1.7}$$

where $q_j(a)$ are independent of $z \in \Gamma_j$. Now, for fixed $a, b \in \Omega$, we set

$$\begin{aligned} P(z,a,b) &= \frac{1}{2}\left[\log f_{\pi/2}(z,a,b) - \log f_0(z,a,b)\right], \\ Q(z,a,b) &= \frac{1}{2}\left[\log f_{\pi/2}(z,a,b) + \log f_0(z,a,b)\right]. \end{aligned} \tag{15.1.8}$$

Then

$$\begin{aligned} P'_z(z,a,b) &= \frac{d}{dz}P(z,a,b) = \overline{P(b,z)} - \overline{P(a,z)}, \\ Q'_z(z,a,b) &= \frac{d}{dz}P(z,a,b) = Q(b,z) - Q(a,z). \end{aligned} \tag{15.1.9}$$

Since $\Im\{\log f_0(z,a,b)\} = \text{const}$, and $\Im\{\log f_{\pi/2}(z,a,b)\} = \text{const}$ on Γ_j, we find from (15.1.8) that

$$P(z,a,b) = -\overline{Q(z,a,b)} + \overline{q_j(a,b)}, \tag{15.1.10}$$

where $q_j(a,b)$ are independent of $z \in \Gamma_j$. We also consider an integral on the entire boundary Γ of the region Ω in the positive direction:

$$\begin{aligned} I_1 &= \frac{1}{2i\pi}\int_\Gamma P(t,z)\,P'_t(t,a,b)\,dt = \sum_{j=0}^n \frac{1}{2i\pi}\int_{\Gamma_j} P(t,z)\,dP(t,a,b) \\ &= \sum_{j=0}^n \frac{1}{2i\pi}\int_{\Gamma_j}\Big[-\overline{Q(t,z)} + \overline{q_j(z)}\Big]\,d\Big(-\overline{Q(t,a,b)} + \overline{q_j(a,b)}\Big) \\ &= \sum_{j=0}^n \frac{1}{2i\pi}\int_{\Gamma_j}\overline{Q(t,z)}\,d\Big(\overline{Q(t,a,b)}\Big) \\ &= -\overline{\frac{1}{2i\pi}\int_\Gamma Q(t,z)\,Q'_t(t,a,b)\,dt}, \end{aligned}$$

because the function $Q(t,a,b)$ is single-valued on each contour Γ_j. Since $Q(t,z)$ and $Q(t,a,b)$ have simple poles in Ω at the point z and the points a and b, respectively, by using the residue theorem, we get $I_1 = -\overline{Q'_z(z,a,b)} - \overline{Q(a,z)} + \overline{Q(b,z)}$. But since $I_1 = 0$, we obtain the formula

$$Q'_z(z,a,b) = Q(b,z) - Q(a,z). \tag{15.1.11}$$

Similarly, if we consider the integral $I_2 = \dfrac{1}{2i\pi}\displaystyle\int_\Gamma Q(t,z)\,P'_t(t,a,b)\,dt$ and follow the above technique, we obtain the formula

$$P'_z(z,a,b) = \overline{P(b,z)} - \overline{P(a,z)}. \tag{15.1.12}$$

Note that the functions $P(z,a)$ and $Q(z,a)$ are themselves not necessarily analytic functions of a, as shown by taking the region Ω as $|z| > 1$. Then, in this case

$$P(z,a) = \frac{1}{1-|a|^2}\,\frac{z-a}{1-\bar a z}, \quad Q(z,a) = \frac{1}{1-|a|^2}\,\frac{1-\bar a z}{z-a}.$$

In the next section we will use the above formulas to solve the Dirichlet problem and construct Green's function for the multiply connected region Ω.

15.2 Dirichlet Problem

Let Ω be an (n+1)-connected region in the finite z-plane, bounded by $(n+1)$ Jordan contours Γ_j, $j = 0,1,\dots,n$ (Figure 15.1). Let $u_j(z)$, $j = 0,1,\dots,n$, be a set of harmonic functions that have the boundary values $u_j = \delta_{jk}$ on Γ_j, $k = 0,1,\dots,n$ (i.e., the boundary value is 1 on Γ_j and zero on Γ_k, $k \neq j$), and satisfy the relation $\displaystyle\sum_{j=0}^n u_j(z) = 1$ because the sum on the right side is a harmonic function and it must be 1 everywhere on the boundary $\Gamma = \partial\Omega$. The conjugate harmonic function $v_j(z)$, $j = 0,1,\dots,n$, in general, is not single-valued. Let $[\Delta v_j(z)]_{\Gamma_k}$ denote the increment in the function $v_j(z)$ as z traverses the contour Γ_k in the positive sense (marked by arrowheads in Figure 15.1), and suppose that

$$[\Delta v_j(z)]_{\Gamma_k} = 2\pi\, p_{k,j}, \quad k,j = 0,1,\dots,n, \tag{15.2.1}$$

where $p_{k,j}$ are constants. Then the set of analytic functions $w_j(z) = u_j(z) + i\, v_j(z)$, $j = 0, 1, \dots, n$, satisfies the following conditions:

$$\begin{aligned}&\text{(i) } w_j(z) = -\overline{w_j(z)} + c_{k,j} \text{ on } \Gamma_k, \text{ where } c_{k,j} \text{ is a constant;}\\ &\text{(ii) } [\Delta w_j(z)]_{\Gamma_k} = 2i\pi\, p_{k,j};\\ &\text{(iii) } w_j'(z) \text{ is regular in } \bar{\Omega}.\end{aligned} \tag{15.2.2}$$

Now consider the integrals

$$I_1 = \frac{1}{2i\pi} \int_\Gamma P(t,u)\, w_j'(t)\, dt, \quad I_2 = \frac{1}{2i\pi} \int_\Gamma P(t,u,v)\, w_j'(t)\, dt. \tag{15.2.3}$$

In view of Cauchy's theorem, each integral is equal to zero. We will use the formulas (15.2.2), (15.1.7), and (15.1.10) to obtain

$$\begin{aligned} I_1 &= \frac{1}{2i\pi} \int_\Gamma P(t,u)\, dw_j(t) = -\sum_{j=0}^{n} \frac{1}{2i\pi} \int_{\Gamma_j} \left(-\overline{Q(t,u)} + \overline{q_j(u)} \right) \overline{dw_j(t)} \\ &= \overline{\sum_{j=0}^{n} \frac{1}{2i\pi} \int_{\Gamma_j} (-Q(t,u) + q_j(u))\, dw_j(t)} \\ &= -\overline{\frac{1}{2i\pi} \int_\Gamma Q(t,u)\, w_j'(t)\, dt} + \overline{\sum_{j=0}^{n} q_j(u)\, p_{k,j}} \\ &= -\overline{w_j'(u)} + \overline{\sum_{j=0}^{n} q_j(u)\, p_{k,j}}, \end{aligned}$$

which gives

$$w_j'(u) = \sum_{j=0}^{n} q_j(u)\, p_{k,j}, \quad k = 0, 1, \dots, n. \tag{15.2.4}$$

Similarly, for the integral I_2 by using integration by parts we get

$$\begin{aligned} I_2 &= -\overline{\frac{1}{2i\pi} \int_\Gamma Q(t,u,v)\, w_j'(t)\, dt} + \overline{\sum_{j=0}^{n} q_j(u,v)\, p_{k,j}} \\ &= -\overline{w_j(u) - w_j(v)} + \overline{\sum_{j=0}^{n} q_j(u,v)\, p_{k,j}}, \end{aligned}$$

which yields

$$w_j(v) - w_j(u) = \sum_{j=0}^{n} q_j(u,v)\, p_{k,j}, \quad k = 0, 1, \dots, n. \tag{15.2.5}$$

Formulas (15.2.4) and (15.2.5) express the solution of the Dirichlet problem for a multiply connected region in terms of the functions $q_j(u)$ and $q_j(u,v)$ which define univalent mappings.

To find Green's function $G(z, z_0)$ for the region Ω, let the corresponding analytic function be denoted by $F(z, z_0)$ so that $G(z, z_0) = \Re\{F(z, z_0)\}$. The function $F(z, z_0)$ has a

logarithmic singularity at the point $z = z_0 \in \Omega$ and is not single-valued in Ω. The function $F(z, z_0)$ has the following properties:
(a) $F(z, z_0) = -\overline{F(z, z_0)}$+const on Γ_j;
(b) $[\Delta F(z, z_0)]_{\Gamma_j} = -2i\pi\, w_j(z_0)$, where w_j is defined in (15.2.2);
(c) $F'(z, z_0)$ is regular in Ω except at a simple pole $z = z_0$ with residue -1.
Now, consider the integrals

$$I_3 = \frac{1}{2i\pi} \int_\Gamma P(t, u)\, F'_z(t, z_0)\, dt,$$

and

$$I_4 = \frac{1}{2i\pi} \int_\Gamma P(t, u, v)\, F'_z(t, z_0)\, dt,$$

where $u, v \in \Omega$. Evaluating these integrals first by using the residue theorem and then using (15.1.7), (15.1.10) and the above mentioned property (b), we obtain

$$\begin{aligned} F'_z(u, z_0) &= Q(z_0, u) + \overline{P(z_0, u)} - \sum_{j=0}^{n} q_j(u)\, w_j(z_0), \\ F(v, z_0) - F(u, z_0) &= Q(z_0, u, v) + \overline{P(z_0, u, v)} - \sum_{j=0}^{n} q_j(u, v)\, w_j(z_0). \end{aligned} \tag{15.2.6}$$

After separating the real parts in (15.2.6), we find that

$$G(v, z_0) - G(u, z_0) = \log \left| f_{\pi/2}(z_0, u, v) \right| - \sum_{j=0}^{n} \log \rho_j(u, v) \cdot w_j(z_0), \tag{15.2.7}$$

where $\rho_j(u, v)$ are the radii of the circles on which lie the images of the contours Γ_j for $j = 0, 1, \dots, n$ under the mapping $w = f_{\pi/2}(z, u, v)$. Using the symmetry property of Green's functions and replacing z_0 by z in (15.2.7), we also get

$$G(v, z_0) - G(u, z_0) = \Re\left\{ f_{\pi/2}(z_0, u, v) \right\} - \sum_{j=0}^{n} \log \rho_j(u, v) \cdot w_j(z_0). \tag{15.2.8}$$

Together the two formulas (15.2.7) and (15.2.8) yield

$$F(z, v) - F(z, u) = \log f_{\pi/2}(z_0, u, v) - \sum_{j=0}^{n} \log \rho_j(u, v) \cdot w_j(z_0), \tag{15.2.9}$$

which is the analytic representation of functions that are meromorphic in the region Ω and real on its boundary Γ. These functions are known as *Schottky functions*.

15.3 Mikhlin's Integral Equation

As we have seen in §12.1, the function $w = f(z)$ that maps a simply connected region D with Jordan boundary Γ conformally onto the unit disk $|w| < 1$, such that a point $z_0 \in D$ goes into the point $w = 0$, can be written in the form $f(z) = (z - z_0)\, g(z)$, where $g(z)$ is analytic and nonzero in D. The function $F(z) = \log g(z)$ is also analytic in D. If $\zeta \in \Gamma$, then $|f(\zeta)| = |\zeta - z_0|\, |g(\zeta)| = 1$, and hence, $\Re\{F(\zeta)\} = \log|g(\zeta)| = -\log|\zeta - z_0|$. The problem of finding the function $F(z)$ reduces to solving the Laplace equation with the Dirichlet boundary value $-\log|\zeta - z_0|$, which determines $\Re\{F(z)\}$. This, followed by finding the conjugate harmonic function, leads to determining $g(z) = e^{F(z)}$ which finally yields the mapping function $f(z)$. This formulation is the same as in (12.1.1) in Symm's method.

In the case of a doubly connected region Ω in the z-plane bounded by the Jordan contours Γ_0 and Γ_1, $\Gamma_1 \subset \Gamma_0$, such that $0 \in \Gamma_1$, let the function $w = f_\Omega(z)$ map Ω conformally onto the annulus $A(\rho, 1)$. Since the function $f_\Omega(z)$ is bounded and nonzero in Ω, the function $\log f_\Omega(z)$ is nonsingular in Ω. Let

$$F_\Omega(z) = \log \frac{f_\Omega(z)}{z} = \log f_\Omega(z) - \log z, \tag{15.3.1}$$

where the function $F_\Omega(z)$ is single-valued and regular in Ω. If z traverses Γ_0 and Γ_1 in the positive direction, both $\arg\{f_\Omega(z)\}$ and $\arg\{z\}$ increase by 2π. Let $g(z) = \Re\{F_\Omega(z)\}$. Since the contours Γ_0 and Γ_1 are mapped onto circles, it is easy to find the boundary values of $g(z)$. Since $|f_\Omega(\zeta)| = \zeta$ for $\zeta \in \Gamma_1$ and $|f_\Omega(\zeta)| = 1$ for $\zeta \in \Gamma_0$, we find that

$$g(z) = \begin{cases} \log\rho - \log|\zeta|. & \zeta \in \Gamma_1, \\ -\log\zeta, & \zeta \in \Gamma_0, \end{cases} \tag{15.3.2}$$

which is the same as (14.2.4). Thus, the value of ρ (conformal modulus) is determined under the condition that the function $h(z)$, conjugate to $g(z)$, must be single-valued, i.e., it must satisfy the condition (14.2.8). The mapping function f_Ω is then determined from (14.2.1).

We will consider the mapping of a multiply connected region Ω onto the slit unit disk by the function $w = w(z)$. Let the region Ω be $(n+1)$-connected, $n \geq 2$, and bounded by Jordan contours Γ_j, $j = 0, 1, \dots, n$. We will assume that a point $z_0 \in \Omega$ is mapped into the point $w = 0$. As in the case of a simply connected region, the mapping function can be written as $w(z) = (z - z_0)\, g(z)$, where $g(z)$ is analytic and nonzero in Ω. Thus, if $W(z) = \log g(z)$, then $u(z) = \Re\{W(z)\} = -\log|\zeta - z_0|$, $\zeta \in \Gamma_0$. Now, suppose that the contours Γ_k, $k = 1, \dots, n$, are mapped onto the circles $|w| = \rho_k$, i.e., $|w(\zeta)| = \rho_k$ for $\zeta \in \Gamma_k$. Then

$$u(z) = \begin{cases} -\log|\zeta - z_0|, & \zeta \in \Gamma_0, \\ \log\rho_k - \log|\zeta - z_0|, & \zeta \in \Gamma_k,\ k = 1, \dots, n. \end{cases} \tag{15.3.3}$$

The problem of determining $u(z)$ reduces to solving a Dirichlet problem, and the mapping function $w(z)$ can be determined from $w(z) = e^{u(z)+i\,v(z)}$, where $v(z) = \Im\{W(z)\}$ must be single-valued.

As in §14.3, we will assume that the mapping problem can be solved in terms of the integral of a dipole density function $\mu(s)$ on the boundary $\Gamma \equiv \partial\Omega$, which is the real part of the Cauchy integral with density μ, i.e.,

$$u(t) = \Re\left\{\frac{1}{2i\pi}\int_\Gamma \frac{\mu(z)}{\zeta - z}\,dz\right\} = \frac{1}{2i\pi}\int_\Gamma \mu(s)\,\frac{\partial}{\partial n_s}\log r_{st}\,ds, \tag{15.3.4}$$

where $z = x + i\,y = \gamma(t)$, $r_{st} = |\zeta - z|$, and $\zeta = \gamma(s)$, $0 \leq s, t \leq L$ (as in (14.3.1)). Note that if Ω is simply connected, then we have an ordinary Dirichlet problem with boundary data $\mu(\zeta) = -\log|\zeta - z_0|$, $\zeta \in \Gamma$ and $z_0 \in \Omega$, and the density function $\mu(s)$ is obtained by solving the integral equation

$$\mu(t) + \frac{1}{\pi}\int_\Gamma \mu(s)\,\frac{\partial}{\partial n_s}\log r_{st}\,ds = -2\,\log|\zeta - z_0|. \tag{15.3.5}$$

But in the case of a multiply connected region, a modified Dirichlet problem is solved, where we must determine the density function $\mu(s)$ as well as the radii ρ_k which appear in (15.3.3). Let $\mu(t)$ denote the solution of the integral equation (Mikhlin [1957])

$$\mu(t) + \frac{1}{\pi} \int_\Gamma \left[\mu(s) \frac{\partial}{\partial n_s} \log r_{st} - \chi(s,t) \right] ds = -2 \log |\zeta - z_0|, \tag{15.3.6}$$

where

$$\chi(s,t) = \begin{cases} 1 & \text{if } s,\, t \text{ lie on the same contour,} \\ 0 & \text{otherwise.} \end{cases}$$

The radii (conformal moduli) ρ_k are given by

$$\rho_k = \frac{1}{\pi} \int_\Gamma \mu(s)\, ds. \tag{15.3.7}$$

Eq (15.3.6) is known as *Mikhlin's integral equation.* It is a Fredholm integral equation of the second kind whose kernel is given by

$$M(s,t) = \frac{\partial}{\partial n_s} \log r_{st} \in L^2[0, L],$$

which is bounded, because we have $M(s,t) = 0.5\, \kappa(s)$ for $s, t \in \Gamma$, where $\kappa(s)$ is the curvature of the contour at s.

The numerical solution of Eq (15.3.6) is obtained by using a Nyström method (quadrature) with the trapezoidal rule (see, Atkinson [1976], Delves and Mohamed [1985], Mayo [1986]). Eq (15.3.6) is then written in the quadrature as

$$\mu_n(t_i) + \frac{1}{\pi} \sum_j \left[\frac{\partial}{\partial n_{t_j}} \log r_{t_i,t_j} + \chi(t_i, t_j) \right] \mu_n(t_j)\, h = d(t_i), \tag{15.3.8}$$

where h is the mesh size, $d(t_i) = -2 \log |t_i - s_0|$, and $s_0 = \gamma(z_0)$. The mesh points are taken as equi-spaced points with respect to some boundary parameter, but the points used as nodes to compute the above quadrature are independent of the mesh points which are used for a fast Poisson solver (see next two sections). Since the trapezoidal rule is highly accurate on periodic regions, the accuracy of the solution of Mikhlin's equation is the same as that of the quadrature formula (15.3.8). This equation can also be solved by the methods developed in Chapters 9 and 11, and although the equation, in general, is not symmetric, it has positive real eigenvalues (Kellogg [1929]).

15.4 Mayo's Method

After the density function $\mu(s)$ is computed by the quadrature formula (15.3.8), we must still compute the Cauchy integral

$$W(t) = \frac{1}{2\pi} \int \frac{\mu(t)}{z - t}\, dz$$

at points in the interior of the region Ω. For this purpose a fast Poisson solver, developed by Mayo [1984], is used (see details in the next section). This is accomplished in the following two steps:

STEP 1. Compute

$$\Re\{W(t)\} = \frac{1}{2\pi}\int M(s,t)\,\mu(t)\,dt. \tag{15.4.1}$$

STEP 2. Compute $\Im\{W(t)\}$.

The details of step 1 are as follows: Embed the region Ω in a larger region R which is a rectangle with uniform mesh in both x and y directions. There exists a fast Poisson solver for the Laplacian in a rectangle (§15.5). Then the function $\Re\{W(t)\}$ defines another harmonic function $\hat{u}$ at points in $R\backslash\Omega$. The function $\hat{u}$ is a discontinuous extension of u from Ω into $R\backslash\Omega$. Define a function

$$U(t) = \begin{cases} u(t), & t \in \Omega, \\ \hat{u}(t), & t \in R\backslash\Omega. \end{cases}$$

Then we use the fast Poisson solver to compute an approximate solution of the discrete Laplace equation $\nabla^2 U = 0$ at all mesh points of R. Since both u and $\hat{u}$ are harmonic, we set the Laplacian to zero at those mesh points that have all four of their adjacent mesh points on the same side of the boundary. But to approximate the Laplacian at all remaining (irregular) mesh points, we take the following approach: Since both u and $\hat{u}$ are continuous along the normal direction but have a jump equal in magnitude to the density μ, we evaluate the jumps in the derivatives of u and $\hat{u}$ along the coordinate directions:

$$u_x - \hat{u}_x = \mu'(s)\,\frac{x'(s)}{x'(s)^2 + y'(s)^2}, \quad u_y - \hat{u}_y = \mu'(s)\,\frac{y'(s)}{x'(s)^2 + y'(s)^2}, \tag{15.4.2}$$

where the suffix indicates the variable of partial differentiation, and the prime denotes the derivative with respect to s. Note that these derivatives contain derivatives of μ and those of the boundary contours. Therefore, these jumps are used to approximate the discrete difference operators at irregular mesh points (see §15.5 for details).

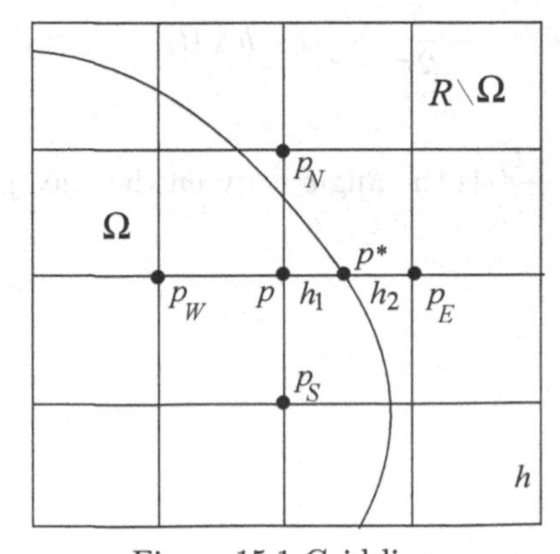

Figure 15.1 Grid lines.

As an example, suppose that a point p is inside Ω, but its adjacent neighbor to the right, p_E, is not inside Ω. Let p^* denote the point where the grid line between p and p_E cuts the boundary, and let $h_2 = |p^* - p_E|$ (Figure 15.1). Using the Taylor series at p and p_E, we find that

$$\begin{aligned}\hat{u}\left(p_E\right)-u(p) &= \left[\hat{u}(p^*)-u(p^*)\right]+h_2\left[\hat{u}_x(p^*)-u_x(p^*)\right]\\ &+\frac{h_2^2}{2}\left[\hat{u}_{xx}(p^*)-u_{xx}(p^*)\right]+h\,u_x(p)+\frac{h^2}{2}u_{xx}(p)+O(h^3)\\ &=\left\{\sum(3)\right\}+h\,u_x(p)+\frac{h^2}{2}u_{xx}(p)+O(h^3),\end{aligned} \tag{15.4.3}$$

where $\{\sum(3)\}$ denotes the sum of the first three terms on the right side, which are known quantities and can be expressed in terms of the solution of the integral equation and the boundary data, whereas the remaining terms account for the usual Taylor series. We also obtain the same kind of expression as (15.4.3) for the difference $U(p)-U\left(p_E\right)$ except that there may be no boundary term. Thus, an approximate solution of the discrete Laplacian of U can be computed as the sum of the four difference operators at all mesh points (details in the next section). The boundary data of U are obtained by approximating the values of the integral (15.4.1) at mesh points that lie at the edge of R.

In step 2, the values of $\Im\{W(t)\}=v(t)$ are easy to compute because, in view of the Cauchy-Riemann equations, we can express the discontinuities of v in terms of the discontinuities in u. Thus, the discrete Laplacian of v can be computed easily.

Once the discrete Laplacians of u and v are computed, we apply the fast Poisson solver twice to obtain the values of u and v at the mesh points. Thus, the solution of (15.3.8) has second-order accuracy in h.

Example 15.1. For Cassini's oval (Figure 12.1) where two different paths of the boundary are close to each other, the method described above in step 1 may fail to give very accurate results because the kernel becomes very large there. In this case (and others like it) the kernel is integrated exactly, i.e.,

$$\int_{t_{i-1}}^{t_{i+1}} M(s,t)\,ds=\int_{t_{i-1}}^{t_{i+1}}\frac{\partial\phi\left(s,t_j\right)}{\partial s}\,ds=\phi\left(t_{i+1},t_j\right)-\phi\left(t_{i-1},t_j\right),$$

where $\phi=\arg\{W(t)\}$. Then Eq (15.3.8) yields the system of equations

$$\mu\left(t_i\right)+\frac{1}{2\pi}\sum_j\hat{\phi}+h\,\chi\left(t_i,t_j\right)\,\mu\left(t-j\right)=d\left(t_i\right), \tag{15.4.4}$$

where $\hat{\phi}=\tan^{-1}\dfrac{t_{i+1}-t_j}{t_{i-1}-t_j}$ is the angle between the lines joining the points t_{i+1} and t_{i-1} to the point t_j.

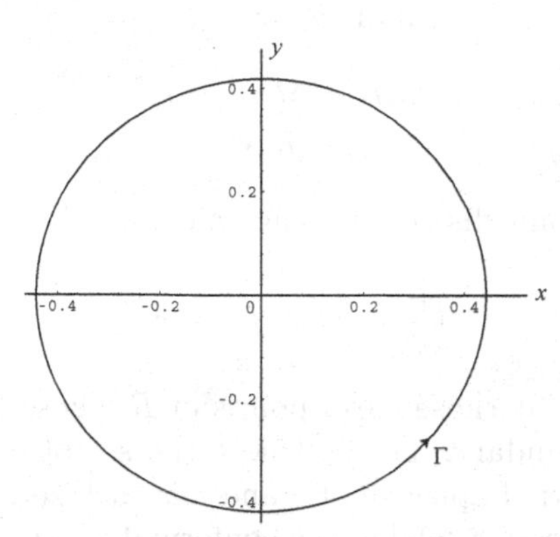

Figure 15.2 Cassini's oval.

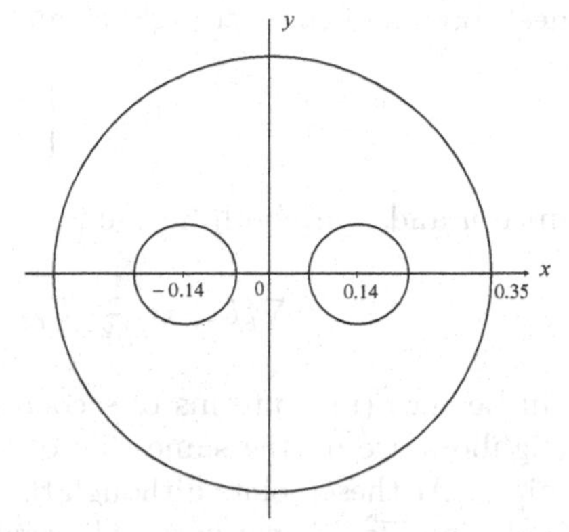

Figure 15.3 Triply connected region.

Mayo [1986] has considered the case of Cassini's oval

$$\Gamma = \{\, [(x+c)^2 + y^2]\, [(x-c)^2 + y^2] = a^4\},$$

whose polar equation is $r(\theta) = \sqrt{c^2 \cos 2\theta + \sqrt{a^4 - c^2 \sin^2 2\theta}}$, $x(\theta) = r(\theta)\cos(\theta)$, $y(\theta) = r(\theta)\sin(\theta)$, with $c = 0.1$ and $a = 0.43$ (Figure 15.2). Note that all Cassini's ovals in Figure 12.1 (Example 12.1) have $c = 1$.

The method described above is used for this contour with 90 mesh points equi-spaced with respect to θ. The exact mapping function is known (Symm [1966: 256]). The simply connected region $\mathrm{Int}\,(\Gamma)$ is embedded in the unit square. The maximum error found was 0.54×10^{-2} with $h = 1/32$, and 0.24×10^{-2} with $h = 1/64$. ■

Example 15.2. Consider the triply connected region bounded by the three circles $\Gamma_0 = \{|z| = 0.35\}$, $\Gamma_1 = \{|z - 0.14| = 0.08$, $\Gamma_2 = \{|z + 0.14| = 0.08\}$ (Figure 15.3). A total of 180 mesh points on the boundary Γ with mesh size $h = 1/128$ were taken to solve Eq (15.3.8). The region Ω and its image with the images of the grid lines, together with the unit circle and the slits, are given in Mayo [1986]. ■

15.5 Fast Poisson Solver

Now we will discuss the details of the fast Poisson solver for the Laplacian. Consider the integral equation (15.3.6) which we write as

$$\mu(t) + \frac{1}{\pi} \int_\Gamma M(s,t)\, \mu(s)\, ds = 2\, g(t), \quad t \in \Gamma, \tag{15.5.1}$$

where the kernel $M(s,t)$ is bounded and represents the normal derivative of Green's function for the Laplacian in the plane. In the region $R\backslash\Omega$ we define a harmonic function $\hat{u}$, by using the same formula as (15.3.4) in the form

$$\hat{u}(t) = \frac{1}{2\pi} \int_\Gamma M(s,t), \mu(s)\, ds. \tag{15.5.2}$$

The function $\hat{u}$ is a discontinuous extension of u in the region $R\backslash\Omega$. Let (x_i, y_j) denote the mesh points of the rectangle R, and let U be defined on R by

$$U_{ij} = \begin{cases} u\,(x_i, y_j)\,, & \text{if } (x_i, y_j) \in \Omega, \\ \hat{u}\,(x_i, y_j)\,, & \text{if } (x_i, y_j) \in R\backslash\Omega. \end{cases} \tag{15.5.3}$$

Since u and $\hat{u}$ are both harmonic, a five-point discrete Laplacian defined by

$$\nabla_h^2 U_{ij} = \frac{1}{h^2}\left[U_{i+1,j} + U_{i-1,j} + U_{i,j+1} + U_{i,j-1} - 4\,U_{ij}\right]$$

will be zero (up to terms of second order) at those mesh points of R whose four adjacent neighbors are on the same side of the boundary. Let S denote the set of irregular mesh points. At these points although the analytic Laplacian of u and $\hat{u}$ is each zero, the discrete Laplacian $\nabla_h^2 U$ is not zero. The central idea for solving the conformal mapping problem is to compute $\nabla_h^2 U$ at the points in S, evaluate values of $\hat{u}$ on ∂R, and then apply the fast Poisson solver on R.

Now, it is possible to approximate this discrete Laplacian without explicitly solving for u or $\hat{u}$ anywhere because we need only compute the jump discontinuities between u and $\hat{u}$ and those in their derivatives at the boundary in terms of the density μ. The procedure to compute these jump discontinuities is as follows: Since the discontinuity between the tangential derivatives of u and $\hat{u}$ at a point on the boundary is equal to the value of the density at that point, we have $u_s - \hat{u}_s = \mu'(s)$. Also, there is no discontinuity between their normal derivatives, i.e., $\dfrac{\partial u}{\partial n_s} = \dfrac{\partial \hat{u}}{\partial n_s}$. These two results and the knowledge of the direction of the contour lead to the formula (15.4.2) which computes the discontinuities between u_x and $\hat{u}_x$ and between u_y and $\hat{u}_y$. Higher order derivatives of u and $\hat{u}$ can be obtained by differentiating (15.4.2).

Since the function U is the real part of the Cauchy integral with the same density function μ, we will consider

$$W(t) = \frac{1}{2i\pi}\int_\Gamma \frac{\mu(\zeta)}{\zeta - z}\,d\zeta. \tag{15.5.4}$$

Since the kernel in Eq (15.5.1) is the real part of the kernel in (15.5.4), we have

$$\begin{aligned} \Re\left\{\frac{1}{2i\pi}\,\frac{d\zeta/ds}{\zeta - z}\right\} &= \frac{1}{2pi}\,\frac{y'(s)\,[x(s) - x(t)] - x'(s)\,[y(s) - y(t)]}{[x(s) - x(t)]^2 + [y(s) - y(t)]^2}\,ds \\ &= \frac{1}{2\pi}\,M(s,t)\,ds, \end{aligned} \tag{15.5.5}$$

where $\zeta = \gamma(s) = x(s) + i\,y(s)$ and $z = \gamma(t) = x(t) + i\,y(t)$. Thus, $u(z) = \Re\{W(t)\}$ for $z \in \Omega$, and $\hat{u}(z) = \Re\{W(z)\}$ for $z \in R\backslash\Omega$. Then the jump between u and $\hat{u}$ can be computed from the jump discontinuities of Cauchy integrals across the boundary contour (recall that Cauchy integrals are analytic functions). Thus, for example,

$$2\,W(z) = \begin{cases} \mu(z) + \dfrac{1}{i\pi}\displaystyle\int_\Gamma \frac{\mu(\zeta)}{\zeta - z}\,d\zeta, & \text{if } z \to \Gamma_-, \\ -\mu(z) + \dfrac{1}{i\pi}\displaystyle\int_\Gamma \frac{\mu(\zeta)}{\zeta - z}\,d\zeta, & \text{if } z \to \Gamma_+, \end{cases} \tag{15.5.6}$$

where $z \to \Gamma_\pm$ stands for whether z approaches from inside Ω or from outside Ω. Hence, there exists a discontinuity of magnitude $\mu(z)$ in W as z crosses Γ. Since $\mu(z)$ is a real function, we find that $u(z) - \hat{u}(z) = \mu(z)$ if $z \in \Gamma$.

The discontinuities in the first and second derivatives of $W(z)$ can be computed as follows: Since by integration

$$\frac{d}{dz}\,W(z) = \frac{1}{2i\pi}\int_\Gamma \frac{d}{dz}\,\frac{\mu(\zeta)}{\zeta - z}\,d\zeta = \frac{1}{2i\pi}\int_\Gamma \frac{\mu'(\zeta)}{\zeta - z}\,d\zeta,$$

we find that the derivative of a Cauchy integral with density μ is another Cauchy integral with density μ', and thus, $W'(z)$ has discontinuity of magnitude $\mu'(z)$ as z crosses Γ. Also, since $W(z)$ is analytic and $u_x(z) = \Re\{W'(z)\}$, we have

$$\begin{aligned} u_x(z) - \hat{u}_x(z) &= \Re\{\mu'(z)\} = \Re\left\{\frac{d\mu/ds}{dz/ds}\right\} = \frac{\mu'(s)\,x'(s)}{x'(s)^2 + y'(s)^2},\\ u_y(z) - \hat{u}_y(z) &= -\Im\{\mu'(z)\} = \frac{\mu'(s)\,y'(s)}{x'(s)^2 + y'(s)^2}, \end{aligned} \tag{15.5.7}$$

and since $\Re\{W''(z)\} = u_{xx}(z)$, we have

$$u_{xx}(z) - \hat{u}_{xx}(z) = \Re\{\mu''(z)\}, \quad u_{yy}(z) - \hat{u}_{yy}(z) = -\Im\{\mu''(z)\}.$$

These discontinuities can be used to approximate the discrete Laplacian at mesh points near the boundary.

An approximation of the discrete Laplacian of U at points of the set S can be computed as follows: If we consider a point p_E to the right of a point $p \in \Omega$ (Figure 15.1), we find that the difference $\hat{u}\,(p_E) - u(p)$ is given by (15.4.3), of which the first three terms $\{\sum(3)\}$ can be computed in terms of the density function μ and the distances of the irregular mesh points from the boundary. Now, if p_W is the mesh point to the left of $p \in \Omega$, then

$$U\,(p_W) - U(p) = \begin{cases} -h\,u_x(p) + \dfrac{h_2^2}{2}\,u_{xx}(p) + O\left(h^3\right), & \text{if } p_W \in \Omega,\\ \{\sum(3)\} - h\,u_x(p) + \dfrac{h_2^2}{2}\,u_{xx}(p) + O\left(h^3\right), & \text{if } p_W \notin \Omega. \end{cases} \tag{15.5.8}$$

Thus, in either case, from (15.4.3) and (15.5.8) we find that

$$U\,(p_W) + U\,(p_E) - 2\,U(p) = \left\{\sum(3)\right\} + h^2\,u_{xx}(p) + O\left(h^3\right).$$

Similarly, if p_N and p_S are points above and below p, respectively (Figure 15.1), then

$$U\,(p_N) + U\,(p_S) - 2\,U(p) = \left\{\sum(3)\right\} + h^2\,u_{yy}(p) + O\left(h^3\right).$$

Hence, $\nabla^2 u(p) = u_{xx}(p) + u_{yy}(p) = 0$ yields

$$h^2\,\nabla_h^2 U(p) = \left\{\sum(3)\right\} + O\left(h^3\right). \tag{15.5.9}$$

By an analogous argument, if p is in $R\backslash\Omega$, then $\nabla^2\hat{u}(p) = 0$ will also lead to formula (15.5.9). This formula gives second-order accuracy in approximating the discrete Laplacian of U at points of the set S. If we want to reach fourth-order accuracy in this approximation at points of S, we must use the fourth order Taylor series expansion. Then, for example, at the point p_E we will have

$$\begin{aligned}\hat{u}\,(p_E) - u(p) &= [\hat{u}(p^*) - u(p^*)] + h_2\,[\hat{u}_x(p^*) - u_x(p^*)] \\ &\quad + \frac{h_2^2}{2}\,[\hat{u}_{xx}(p^*) - u_{xx}(p^*)] + \frac{h_2^3}{6}\,[\hat{u}_{xxx}(p^*) - u_{xxx}(p^*)] \\ &\quad + h\,u_x(p) + \frac{h^2}{2}\,u_{xx}(p) + \frac{h^3}{6}\,u_{xxx}(p) + O\left(h^4\right) \\ &= \left\{\sum(4)\right\} + h\,u_x(p) + \frac{h^2}{2}\,u_{xx}(p) + \frac{h^3}{6}\,u_{xxx}(p) + O\left(h^4\right),\end{aligned}$$

where $\left\{\sum(4)\right\}$ is the sum of the first four terms on the right side in the above expression. Then

$$h^2\,\nabla_h^2 U(p) = \left\{\sum(4)\right\} + O\left(h^4\right). \tag{15.5.10}$$

This shows that if the solution of the integral equation is known almost accurately, we can compute an approximate solution with second order accuracy. Mayo's method solves Mikhlin's integral equation with machine accuracy because of the use of the trapezoidal rule in the quadrature formula with smooth boundary data. However, splines can be used to compute more accurate values for the derivatives of the density function. It has been found that in practice, second-order accuracy is sufficient to obtain an accurate solution.

The computational algorithm consists of the following steps:

STEP 1. Embed the region Ω in a rectangle R. This rectangle is chosen at least $3h$ distance away from Γ.

STEP 2. Find all irregular mesh points and their distances to the boundary in the x and y directions.

STEP 3. Solve the integral equation by using the quadrature formula (15.3.8), which replaces the integral equation by a sum at a set of boundary points. This yields a dense linear system of equations

$$\mu\,(t_i) + \sum w_i\,K(i,j)\,\mu\,(t_j) = 2\,g\,(t_i)\,, \quad i = 1, \dots, n, \tag{15.5.11}$$

where the points used as nodes are different from the mesh points. These nodes are chosen as equi-spaced points with respect to the parameter used on the boundary, and the trapezoidal rule is used for quadrature, in cases where the boundary data is not smooth, or for points near those boundary portions where the curvature is large, a Galerkin method with augmented bases containing singular points is needed (see §14.5). System (15.5.11) is solved by the Gaussian elimination method.

STEP 4. Interpolate the values of the density with a quintic spline which yields sixth order accuracy for values of the density at intermediate points.

STEP 5. Compute the discrete Laplacian at irregular points in the set S by using (15.4.3).

STEP 6. Compute the values of U at the edge of the grid.

STEP 7. Apply the fast Poisson solver.

STEP 8. Compute the derivatives u_x and u_y by (15.5.7).

STEP 9. Compute the conjugate function $v(z)$.

For details, see Mayo [1984; 1986].

REFERENCES USED: Atkinson [1976], Goluzin [1969], Kythe [1998], Mayo [1984; 1986], Mikhlin [1957], Symm [1966], Wen[1992].

Part 3: Applications

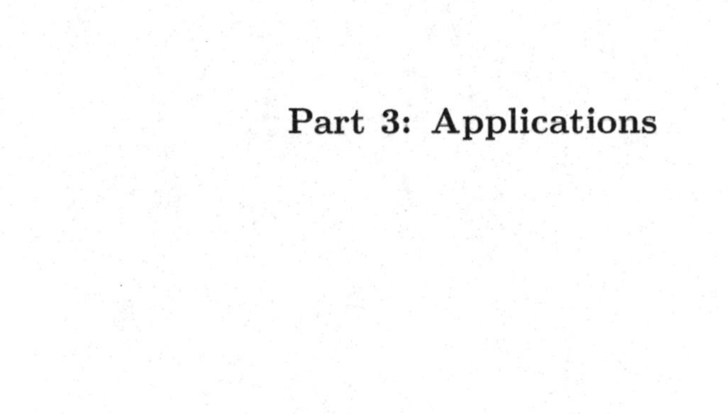

16

Grid Generation

Exact solutions of boundary value problems for simple regions, such as a circle, square or annulus, can be obtained with relative ease even in cases where the boundary conditions are rather complicated. Although Green's functions for such simple regions are known, the solution of a boundary value problem for regions with complex structures often becomes more difficult, even for a simple problem, such as the Dirichlet problem. One approach to solving these difficult problems is to conformally transform a given region into the simplest form. This will, however, result in change not only in the region and the associated boundary conditions but also in the governing differential equation. Grid generation methods using conformal mappings are presented for problems dealing with a cascade of blades, and inlet flow configurations.

16.1 Computational Region

Conformal mapping has been used to generate orthogonal boundary-fitted coordinates in solving various boundary value problems in simply connected regions. A useful work in this area is the book by Thompson, Warsi and Mastin [1985]. A grid is an integral part of finite difference or finite element methods. A discrete model becomes more efficient when it is constructed by using natural coordinate systems and maintaining a uniform connectivity pattern between grid nodes. These two requirements are met when the grid is obtained by coordinate transformations using conformal mapping methods so that the boundary of the physical region is represented by constant coordinate lines. Besides this adaptive feature, the conformal maps can be made to adapt to certain salient features, such as singularities. A grid generation methodology must be able to control the grid spacing effectively, especially near the boundary (see Tomamidis and Assanis [1991]).

Computational methods, like the finite differences or the finite elements, for solving boundary value problems are usually simple if the physical region has regular geometry over which a uniformly distributed grid can be imposed. However, if the region has arbitrary irregular geometry, such a region is first transformed into an associated computational region with regular geometry, like a rectangle or circle. In such cases the difficulty arises not only from the transformation of the governing equation(s) but also from the boundary conditions. The coordinate transformation and conformal boundary maps are generally used to transform an irregular physical region into the corresponding computational region. But such transformations and conformal mappings, in general, are very difficult to construct except in relatively simpler cases.

First, we will gather some transformation formulas from the physical (x, y)-region into the computational (ξ, η)-region.

16.1.1. Coordinate Transformations. To find the transformation from independent variables x, y of the physical plane into a set of independent variables ξ, η of the computational plane, let us assume that

$$\xi = \xi(x, y), \quad \eta = \eta(x, y), \tag{16.1.1}$$

or inversely,

$$x = x(\xi, \eta), \quad y = y(\xi, \eta). \tag{16.1.2}$$

The Jacobian J of the transformation is given by

$$J = \frac{\partial(x, y)}{\partial(\xi, \eta)} = \begin{vmatrix} x_\xi & y_\xi \\ x_\eta & y_\eta \end{vmatrix} = x_\xi\, y_\eta - x_\eta\, y_\xi \neq 0, \tag{16.1.3}$$

where the subscripts denote partial differentiation with respect to the indicated variable. If $u = u(x, y)$, then

$$\frac{\partial u}{\partial x} = \frac{1}{J}\left(y_\eta \frac{\partial u}{\partial \xi} - y_\xi \frac{\partial u}{\partial \eta}\right), \quad \frac{\partial u}{\partial y} = \frac{1}{J}\left(-x_\eta \frac{\partial u}{\partial \xi} + x_\xi \frac{\partial u}{\partial \eta}\right). \tag{16.1.4}$$

For the gradient, the transformation formulas for the conservative form are

$$u_x = \frac{1}{J}\left[(y_\eta\, u)_\xi - (y_\xi\, u)_\eta\right], \quad u_y = \frac{1}{J}\left[-(x_\eta\, u)_\xi + (x_\xi\, u)_\eta\right], \tag{16.1.5}$$

and for the nonconservative form are

$$u_x = \frac{1}{J}\left(y_\eta\, u_\xi - y_\xi\, u_\eta\right), \quad u_y = \frac{1}{J}\left(-x_\eta\, u_\xi + x_\xi\, u_\eta\right). \tag{16.1.6}$$

Let $\mathbf{u} = u_1\,\mathbf{i} + u_2\,\mathbf{j}$. Then for divergence the transformation formulas for the conservative form are

$$\nabla \cdot \mathbf{u} = \frac{1}{J}\left[(y_\eta\, u_1 - x_\eta\, u_2)_\xi + (-y_\xi\, u_1 + x_\xi\, u_2)_\eta\right], \tag{16.1.7}$$

and for the nonconservative form are

$$\nabla \cdot \mathbf{u} = \frac{1}{J}\left[y_\eta\, (u_1)_\xi - x_\eta\, (u_2)_\xi - y_\xi\, (u_1)_e ta + x_\xi\, (u_2)_\eta\right]. \tag{16.1.8}$$

Example 16.1. Using formulas (16.1.4), the equation of continuity

$$\frac{\partial u}{\partial x} + \frac{\partial v}{\partial y} = 0, \tag{16.1.9}$$

when transformed from (x, y)-coordinates of the physical plane into (ξ, η)-coordinates of the computational plane, becomes

$$y_\eta \frac{\partial u}{\partial \xi} - y_\xi \frac{\partial u}{\partial \eta} - x_\eta \frac{\partial v}{\partial \xi} + x_\xi \frac{\partial v}{\partial \eta} = 0. \blacksquare \tag{16.1.10}$$

Example 16.2. The Laplacian $\nabla^2 \equiv \dfrac{\partial^2}{\partial x^2} + \dfrac{\partial^2}{\partial y^2}$ is transformed into the following forms:

Conservative form:

$$\begin{aligned} J\nabla^2 u = & \left\{ \frac{1}{J} y_\eta \left[(y_\eta u)_\xi - \frac{1}{J} (y_\xi u)_\eta \right] - x_\eta \left[-(x_\eta u)_\xi + (x_\xi u)_\eta \right] \right\}_\xi \\ & + \left\{ -\frac{1}{J} y_\xi \left[(y_\eta u)_\xi + (y_\xi u)_\eta \right] + \frac{1}{J} x_\eta \left[-(x_\eta u)_\xi + (x_\xi u)_\eta \right] \right\}_\eta . \end{aligned}$$

Nonconservative form:

$$\begin{aligned} \nabla^2 u = & \frac{1}{J^2} \left[\left(x_\eta^2 + y_\eta^2\right) u_{\xi\xi} - 2\left(x_\xi x_\eta + y_\xi y_\eta\right) u_{\xi\eta} + \left(x_\xi^2 + y_\xi^2\right) u_{\eta\eta} \right] \\ & + \left[\left(\nabla^2 \xi\right) u_\xi + \left(\nabla^2 \eta\right) u_\eta \right]. \blacksquare \end{aligned}$$

Next, to present the basic concept of numerical grid generation for different boundary value problems, we will consider a very simple one-dimensional transformation and show how the computational region with uniformly distributed grids is obtained and how the governing equations are changed.

Example 16.3. Consider a plane steady-state boundary layer flow over a flat rectangular plate $\{0 \le x \le a,\ 0 \le y \le b\}$. If this problem is solved by the finite difference or finite element method, a rectangular grid is constructed over the physical region with more nodes concentrated near the wall (x-axis) where the gradients are assumed to be larger than elsewhere (see Figure 16.1). This grid is uniform along the x-axis but nonuniform along the y-axis.

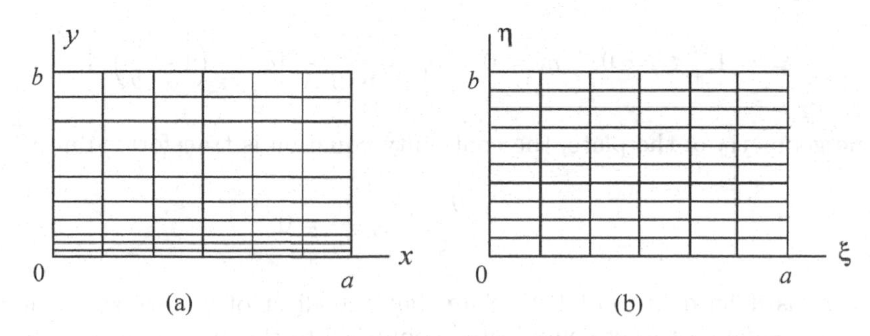

Figure 16.1 (a) Physical region. (b) Computational region.

In order to transform the grid in Figure 16.1(a) into the uniform grid of Figure 16.1(b), we use the coordinate transformation

$$\xi = x, \quad \eta = 1 - \frac{\ln \phi(y)}{\ln A}, \tag{16.1.11}$$

where

$$\phi(y)=\frac{\alpha+\left(1-\dfrac{y}{b}\right)}{\alpha-\left(1-\dfrac{y}{b}\right)},\quad A=\frac{\alpha+1}{\alpha-1},\quad 1<\alpha<\infty. \tag{16.1.12}$$

The inverse transformation is given by

$$x=\xi,\quad y=b\,\frac{(\alpha+1)-(\alpha-1)\,A^{1-\eta}}{1+A^{1-\eta}}. \tag{16.1.13}$$

This transformation (Roberts [1971]) makes the grid spacing uniform along the y-axis. The parameter α is known as the *stretching parameter*. The grid concentration as $\alpha\to 1$ is presented in Figure 16.2 where values of y are plotted for different values of α and η with $b=1$.

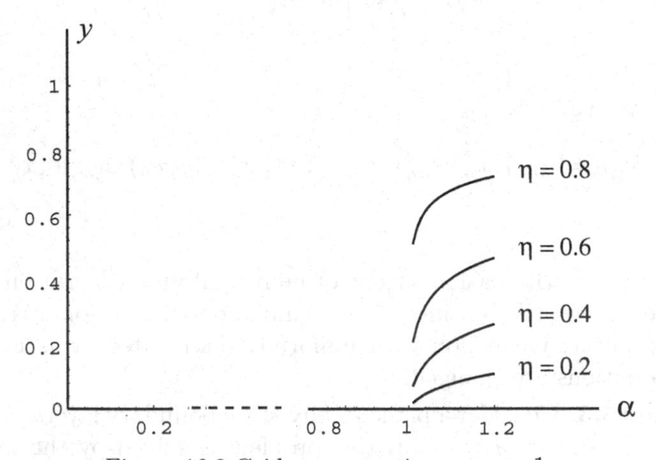

Figure 16.2 Grid concentration as $\alpha\to 1$.

Next, we will transform the differential equation from the physical region into the computational region by the transformation (16.1.11). For brevity, let us consider the equation of continuity (16.1.9). Since

$$\xi_x=1,\quad \xi_y=0,\quad \eta_x=0,\quad \eta_y=\frac{2\alpha}{b\,\ln A}\left[\alpha^1-\left(1-\frac{y}{b}\right)^2\right]^{-1} \tag{16.1.14}$$

for the geometry of the plate, the continuity equation is transformed into

$$\frac{\partial u}{\partial \xi}+\eta_y\,\frac{\partial v}{\partial \eta}=0, \tag{16.1.15}$$

where η_y is defined in (16.1.14). Note that the effect of uniformizing the grid makes the governing equation rather complicated compared to the original form. Moreover, after the problem is solved in the computational region, the solution is transformed back to the physical region by using (16.1.11). ■

16.1.2. Orthogonal Method. The use of conformal mappings to generate grids has some important limitations: (i) they are applicable to plane problems, (ii) they have no control over the interior grids, (iii) multiple-valued mapping functions are difficult to implement,

(iv) orthogonality is lost in arbitrary distribution of boundary points, (v) a very small change in the shape of the original boundary results in changes in the location of image boundary points, and (vi) finding a boundary map is in itself a difficult task.

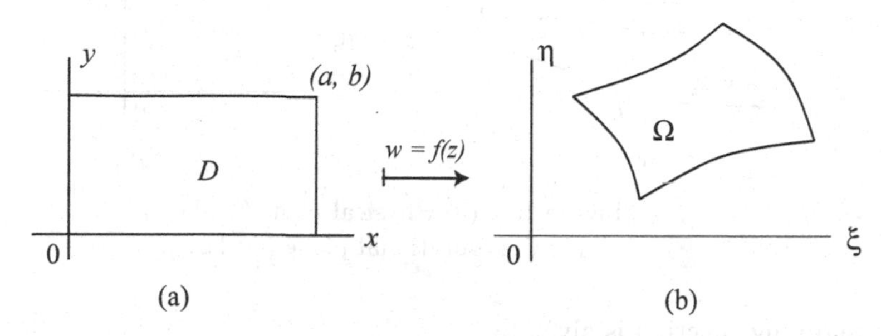

Figure 16.3 A rectangle onto an arbitrary 4-sided region.

A consequence of the Riemann mapping theorem is that a rectangle $R = \{0 \le x \le 1,\ 0 \le y \le b\}$ cannot be mapped univalently onto an arbitrary region D with four sides, as shown in Figure 16.3, unless the ratio a/b is restricted to a particular constant m, known as the *conformal module*. Then the mapping $w = f(z)$ can be constructed by solving the partial differential equation

$$m^2\, w_{\xi\xi} + w_{\eta\eta} = 0. \tag{16.1.16}$$

This yields a *quasiconformal* map (see Lehto and Virtaanen [1973]). Since m is domain-dependent and not known a priori, this approach is not feasible for grid generation. On the other hand, orthogonal transformations in the plane can be regarded as quasi-conformal mapping with a real dilation (Knupp and Steinberg [1993]).

16.2 Inlet Configurations

We will develop the orthogonal grid generation method by using conformal mapping. In this method the physical region is mapped onto the computational method by one- or two-step conformal maps. First, we will consider the grid generation for a case study involving inlet configuration in flow problems and determine the basic mappings by using both one-step and two-step methods.

Map 16.1. (One-step method). Consider the physical region in the z-plane in Figure 16.4(a). The boundary of this region, defined by $y = y_1(x)$ and $y = y_2(x)$, is given by two sets of data:

$$\begin{aligned} z_1^{(n)} &= x_1^{(n)} + i\, y_1^{(n)}, \quad 2 \le n \le n_1 - 1, \\ z_2^{(n)} &= x_2^{(n)} + i\, y_2^{(n)}, \quad 2 \le n \le n_2 - 1, \end{aligned} \tag{16.2.1}$$

where x_2 takes the minimum value $x_2^{n_3}$ at $n = n_3$, i.e., at the point E. This region is mapped onto the computational region in the $\zeta = (\xi, \eta)$-plane, which is a rectangular strip

$0 \le \eta \le 1$ (Figure 16.4).

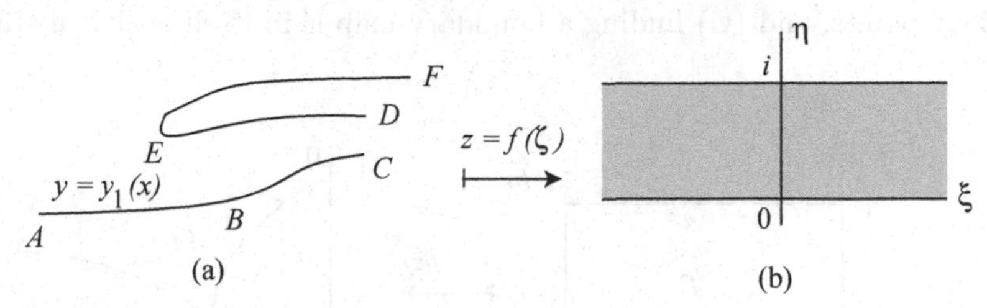

Figure 16.4 (a) Physical plane (z-plane).
(b) Computational plane (ζ-plane).

The mapping function is given by

$$\begin{aligned} z &= h\left[\zeta - \frac{1}{\pi}\left(1 + e^{-\pi\zeta}\right)\right] + A_1 + i\,B_1 \\ &\quad + \sum_{j=2}^{\kappa}\left[A_j \sin\frac{(j-1)\pi(\zeta - \xi_0)}{L} + i\,B_j \cos\frac{(j-1)\pi(\zeta - \xi_0)}{L}\right], \end{aligned} \tag{16.2.2}$$

where $h,\ A_j, B_j\,(j = 1, 2, \dots, \kappa)$, ξ_0 and L are yet unknown real constants to be determined. A particular case from (16.2.2) for $A_j = B_j = 0,\ j = 1, 2, \dots, \kappa$, is given by

$$z = h\left[\zeta - \frac{1}{\pi}\left(1 + e^{-\pi\zeta}\right)\right], \tag{16.2.3}$$

which maps the upper half-plane $y \ge 0$ with a cut at $y = h,\ x \ge 0$, onto the rectilinear strip $0 \le \eta \le i$ in the ζ-plane (Figure 16.4). If we rewrite (16.2.2) as

$$\begin{aligned} x + i\,y &= h\left[(\xi + i\,\eta) - \frac{1}{\pi}\left(1 + e^{-\pi(\xi + i\eta)}\right)\right] + A_1 + i\,B_1 \\ &\quad + \sum_{j=2}^{\kappa}\left[A_j \sin\frac{(j-1)\pi(\xi + i\,\eta - \xi_0)}{L} + i\,B_j \cos\frac{(j-1)\pi(\xi + i\,\eta - \xi_0)}{L}\right], \end{aligned} \tag{16.2.4}$$

set $\eta = 0$, and equate real and imaginary parts, we get

$$\begin{aligned} x = x_1(\xi) &= h\left[\xi - \frac{1}{\pi}\left(1 + e^{-\pi\xi}\right)\right] + A_1 \\ &\quad + \sum_{j=2}^{\kappa} A_j \sin\frac{(j-1)\pi(\xi - \xi_0)}{L}, \end{aligned} \tag{16.2.5}$$

and

$$y = y_1 = B_1 + \sum_{j=2}^{\kappa} B_j \cos\frac{(j-1)\pi(\xi + i\,\eta - \xi_0)}{L}. \tag{16.2.6}$$

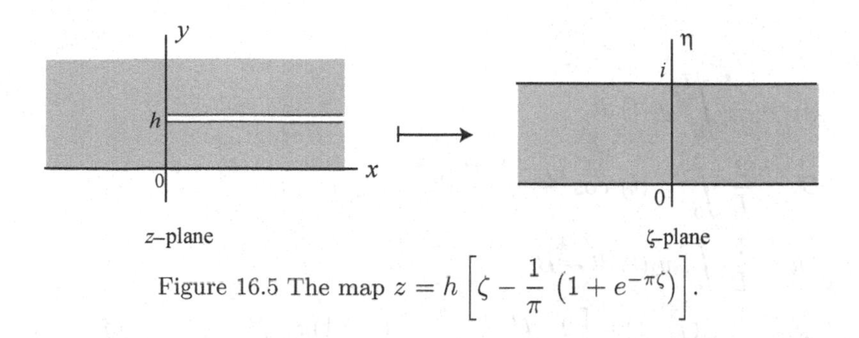

Figure 16.5 The map $z = h\left[\zeta - \frac{1}{\pi}\left(1 + e^{-\pi\zeta}\right)\right]$.

Again, if we set $\eta = 1$ in (16.2.4) and equate real and imaginary parts, we obtain

$$x + x_2(\xi) = h\left[\xi - \frac{1}{\pi}\left(1 + e^{-\pi\xi}\right)\right] + A_1 + \sum_{j=2}^{\kappa}\left[A_j \cosh\frac{(j-1)\pi}{L} + B_j \sinh\frac{(j-1)\pi}{L}\right] \sin\frac{(j-1)\pi(\xi - \xi_0)}{L}, \tag{16.2.7}$$

and

$$y = y_2 = h + B_1 + \sum_{j=2}^{\kappa}\left[A_j \sinh\frac{(j-1)\pi}{L} + B_j \cosh\frac{(j-1)\pi}{L}\right] \cos\frac{(j-1)\pi(\xi - \xi_0)}{L}. \tag{16.2.8}$$

Note that (16.2.6) and (16.2.8) imply that y_1 and y_2 are even, $2L$-periodic functions of $\xi - \xi_0 \equiv t$. The unknown constants are determined by the iterative method as follows: Assume that the nth approximations for h, A_j, B_j, ξ_0, and L are known. Let us denote them by $h^{(n)}$, $A_j^{(n)}$, $B_j^{(n)}$, $\xi_0^{(n)}$, and $L^{(n)}$. Then proceed as follows:

1. Substitute $h^{(n)}$, $A_j^{(n)}$, $B_j^{(n)}$, $\xi_0^{(n)}$, and $L^{(n)}$ into (16.2.5) and (16.2.7) and obtain the $(n+1)$th approximations for x_1 and x_2.
2. Find $\xi = \xi_{n_3}$, which makes $x_2(\xi)$ minimum, and determine A_1 so that $x_2\left(\xi_{n_3}\right) = x_2^{(n_3)}$.
3. Obtain the solutions of $x_1(\xi) = x_1^{(2)}$ and $x_2(\xi) = x_2^{(2)}$ and take the smaller value as ξ_1.
4. Obtain the solutions of $x_1(\xi) = x_2^{(n_1-1)}$ and $x_2(\xi) = x_2^{(n_2-1)}$, and take the smaller value as ξ_2.
5. Determine $z_1(\xi_1) = z_1^{(1)}$, $z_2(\xi_1) = z_2^{(1)}$, $z_1(\xi_2) = z_1^{(n_1)}$ and $z_2(\xi_2) = z_2^{(n_2)}$ by extrapolation.

Steps 1 through 5 determine $x_1(\xi)$ and $x_2(\xi)$ for the interval

$$\xi_1 \le \xi \le \xi_2 = \xi_1 + L. \tag{16.2.9}$$

6. Set $L = \xi_2 - \xi_1$ and $\xi_0 = \xi_1$.
7. Now the left sides of (16.2.6) and (16.2.8) for y_1 and y_2 are determined as functions of ξ on the interval (16.2.9) through the functions $x_1(\xi)$ and $x_2(\xi)$.
8. The constants h, A_j (for $j = 2, 3, \dots, \kappa$) and B_j (for $j = 1, 2, \dots, \kappa$) are determined by

Fourier analysis. Thus,

$$\begin{aligned} B_1 &= \frac{1}{L}\int_0^L y_1(t)\,dt,\\ B_j &= \frac{2}{L}\int_0^L y_1(t)\,\cos\frac{(j-1)\pi t}{L}\,dt, \quad 1\le j\le\kappa,\\ h &= \frac{1}{L}\int_0^L y_2(t)\,dt - B_1,\\ A_j &= \operatorname{csch}\frac{(j-1)\pi}{L}\left[\frac{2}{L}\int_0^L y_2(t)\,\cos\frac{(j-1)\pi t}{L}\,dt - B_j\,\cosh\frac{(j-1)\pi}{L}\right], \end{aligned} \tag{16.2.10}$$

for $2\le j\le\kappa$, where $t=\xi-\xi_0$.

9. All $(n+1)$th approximations, denoted generically by $\phi^{(n+1)}$, thus obtained are replaced by $(1-r)\,\phi^{(n)}+r\,\phi^{(n+1)}$, where r is a relaxation constant, usually 0.5. This assures the convergence of the successive approximations.

10. Repeat steps 1 through 9 until the desired convergence is achieved.

Note that the first approximation is given by $h=y_2^{(n_3)}$, $A_j=B_j=0\ (1\le j\le\kappa)$, $\xi_1=0$, and $L=x_1^{(n_1-1)}-x_1^{(2)}$. Since the hyperbolic functions are involved, double precision should be used in all computations.

Map 16.2. (Two-step method). Consider the same flow problem as in Map 16.1. The mapping is carried out as follows: First, map the region ABC'DEF in the z-plane onto the rectilinear strip $0\le\eta\le 1$ in the ζ-plane (see Figure 16.6) by

$$z = h\left[\zeta-\frac{1}{\pi}\left(1+e^{-\pi\zeta}\right)\right]+A_1+\sum_{k=2}\kappa_1 A_k\,\sin\frac{(k-1)\pi\,(\zeta-\xi_0)}{L_1}, \tag{16.2.11}$$

where the constants h, A_k, ξ_0 and L_1 are computed by an iterative method similar to that in Map 16.1. Note that the mapping (16.2.11), which is the mapping function for the inlet without center bodies, is obtained by taking $B_k=0\ (k\ge 1)$ in (16.2.2). The image of the arc BC in the ζ-plane is denoted by

$$\eta=\eta_1(\xi), \quad \xi_3\le\xi\le\xi_4. \tag{16.2.12}$$

Next, map the region ABCDEF in the ζ-plane onto the semi-infinite strip $u\ge 0$, $0\le v\le 1$, in the w-plane $(w=u+i\,v)$ by

$$\zeta=\frac{2}{\pi}\,\log\cosh\frac{\pi w}{2}+B_0+\sum_{k=1}^{\kappa_2}B_k\,\cos\frac{(2k-1)\pi(w-i)}{2L_2}, \tag{16.2.13}$$

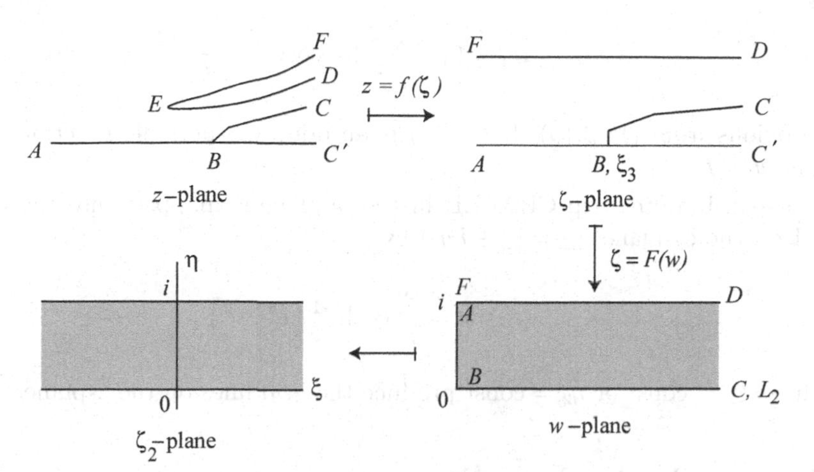

Figure 16.6 Two-step method.

such that the points $B\,(\zeta = \xi_3)$ and $C\,(\zeta = \xi_4 + i\,\eta_4)$ go into the points $w = 0$ and $w = L_2$, respectively. The unknown constants B_k $(0 \le k \le \kappa_2)$ and L_2 are determined as in §16.2.1. Thus,

$$
\begin{aligned}
\xi + i\,\eta &= \frac{2}{\pi} \log \cosh \frac{\pi(u + i\,v)}{2} \\
&\quad + B_0 \sum_{k=1}^{\kappa_2} B_k \cos \frac{(2k-1)\,(u - i(1-v))}{2\,L_2} \\
&= \frac{2}{\pi} \log \cosh \frac{\pi(u + i\,v)}{2} \\
&\quad + B_0 \sum_{k=1}^{\kappa_2} B_k \Big[\cos \frac{(2k-1)\,\pi u}{2\,L_2} \cosh \frac{(2k-1)\pi(1-v)}{2\,L_2} \\
&\quad + i \sin \frac{(2k-1)\,\pi u}{2\,L_2} \sinh \frac{(2k-1)\pi(1-v)}{2\,L_2} \Big],
\end{aligned}
\tag{16.2.14}
$$

which for $v = 0$ gives, after separating the real and imaginary parts,

$$
\xi = \frac{1}{\pi} \log \cosh^2 \frac{\pi u}{2} + B_0 + \sum_{k=1}^{\kappa_2} B_k \cosh \frac{(2k-1)\,\pi}{2\,L_2} \cos \frac{(2k-1)\,\pi u}{2\,L_2},
\tag{16.2.15}
$$

$$
\eta = \eta_1(\xi) = \sum_{k=1}^{\kappa_2} B_k \sinh \frac{(2k-1)\,\pi}{2\,L_2} \sin \frac{(2k-1)\,\pi u}{2\,L_2}.
\tag{16.2.16}
$$

Hence,

$$
B_k = \frac{2}{L_2} \operatorname{csch} \frac{(2k-1)\pi}{2\,L_2} \int_0^{L_2} \eta_1 \sin \frac{(2k-1)\pi u}{2\,L_2}\, du, \quad 1 \le k \le \kappa_2.
\tag{16.2.17}
$$

Also, at the point $B\,(w = 0)$ we have from (16.2.13)

$$
\xi_3 = B_0 + \sum_{k=1}^{\kappa_2} B_k \cosh \frac{(2k-1)\,\pi}{2\,L_2}.
$$

Thus,

$$B_0 = \xi_3 - \sum_{k=1}^{\kappa_2} B_k \cosh \frac{(2k-1)\pi}{2L_2}. \tag{16.2.18}$$

It is obvious from (16.2.16) that $\eta_1(u)$ is an odd, $4L_2$-periodic function symmetric about the line $u = L_2$.

The semi-infinite strip CBAFED in the w-plane is mapped onto the infinite strip $0 \le \eta_2 \le 1$ on the ζ_2-plane ($\zeta_2 = \xi_2 + i\,\eta_2$) by

$$w = \frac{2}{\pi} \cosh^{-1}\left(e^{\pi \xi_2/2}\right). \tag{16.2.19}$$

The lines $\xi_2 = \text{const}$ or $\eta_2 = \text{const}$ produce the grid lines on the z-plane.

16.3 Cascade Configurations

The Ives-Liutermoza method (Ives and Liutermoza [1977]) deals with the mapping problem that transforms the region exterior to a cascade of blades first onto a region exterior to a near circle. Then the second mapping transforms the interior of the near circle onto the unit disk. The success of the first mapping depends on the solidity of cascades; if it is low, the mapping yields an acceptable near circle, but if it increases, the image boundary degenerates into a peanut-shaped contour, in which case the second mapping will not work at all, and the Ives-Liutermoza method fails. To overcome this difficulty for the mapping problem of the cascade of blades, we will first determine the function that maps the region directly onto a rectangle for two periods of the cascade of blades, which can then be mapped onto the unit disk (see Figure 14.2 and 14.13). We will consider three cases described in the following case studies.

Map 16.3. First we will study the ordinary type of grids. Let two periods of the cascade of blades make a row in the y direction in the z-plane (physical plane $z = x + i\,y$, Figure 16.7(a)). The contour of one of the blades is defined by a set of data $z_n = x_n + i\,y_n$, $n = 1, \dots, N$, $z_N = z_1$, which are ordered clockwise, with $|x_n| \le 1$. Let h denote the pitch of the blades in the y-direction. The function that maps them onto an infinite strip in the ζ-plane (computational plane, Figure 16.7(b)) has the form

$$z = A_0 \left[\frac{h}{\pi} \log \operatorname{sn}(\zeta, k) - 1 + \sum_{j=1} MC_j \cos \frac{(j-1)\pi\zeta}{K'}\right], \tag{16.3.1}$$

where A_0 is a real parameter that finally approaches unity in an iterative scheme (see Step 3 of the algorithm given below), $C_j = A_j + i\,B_j$ and k, $0 < k < 1$, are constants to be determined, K and K' are complete elliptic integrals of the first kind with moduli k and $k' = \sqrt{1-k^2}$, respectively, and sn is one of the Jacobian elliptic functions (see §F.2). In particular when $A_0 = 1$, $k = e^{-2\pi/h}$, $A_j = 0 = B_j$ for $j \ge 1$, the function (16.3.1) represents the mapping of a cascade of flat plates of chord 2, pitch h, and zero stagger. If we set $\zeta = K + i\,\eta$ in (16.3.1) and separate the real and imaginary parts, we get

$$\begin{aligned} x = A_0 \Big[\frac{h}{\pi} \log \operatorname{dn}(\eta, k') - 1 + A_1 + \sum_{j=2}^{M} \Big\{ & A_j \cosh \frac{(j-1)\pi K}{K'} \cos \frac{(j-1)\pi\eta}{K'} \\ & - B_j \sinh \frac{(j-1)\pi K}{K'} \sin \frac{(j-1)\pi\eta}{K'} \Big\}\Big], \end{aligned} \tag{16.3.2}$$

$$y = A_0 \left[B_1 + \sum_{j=2}^{M} \left\{ A_j \sinh \frac{(j-1)\pi K}{K'} \sin \frac{(j-1)\pi\eta}{K'} + B_j \cosh \frac{(j-1)\pi K}{K'} \cos \frac{(j-1)\pi\eta}{K'} \right\} \right], \tag{16.3.3}$$

where dn is another Jacobian elliptic function.

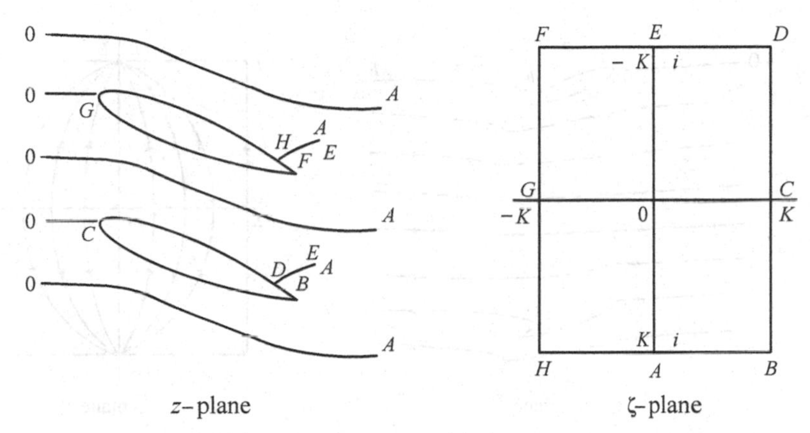

Figure 16.7 (a) Physical region, (b) Computational region.

The following algorithm for the iterative method, which starts with the data for the initial guess as that of the flat cascade, assumes that the nth approximations $A_j^{(n)}$, $B_j^{(n)}$ and $k^{(n)}$ are known. Then proceed as follows:

STEP 1. Substitute the known values of $A_j^{(n)}$, $B_j^{(n)}$ and $k^{(n)}$ into (16.3.2) and compute the values of $A_1^{(n+1)}$ such that $x_{\max} + x_{\min} = 0$. Then compute the constant $A_0^{(n+1)}$ such that $x_{\max} - x_{\min} = 2$. Use this data on the left side of (16.3.2) to yield the relation

$$\eta = \eta^{(n+1)}(x). \tag{16.3.4}$$

STEP 2. Choose a relaxation constant $\varepsilon_0\,(= 0.5)$, and replace $A_0^{(n+1)}$ by $(1-\varepsilon_0)\, A_0^{(n)} + \varepsilon_0\, A_0^{(n+1)}$, to yield a new value of $A_0^{(n+1)}$.

STEP 3. Use this value of $A_0^{(n+1)}$ to obtain $k^{(n+1)}$ from

$$\log k^{(n+1)} = A_0^{(n+1)} \log k^{(n)}. \tag{16.3.5}$$

In this process $A_0 \to 1$ as n increases.

STEP 4. Use (16.3.4) to write the profile of the blade in the form $y = y^{(n+1)}(\eta)$. Substitute this y on the left side of (16.3.3), and use Fourier series analysis on it to obtain

$$\begin{aligned} A_j^{(n+1)} &= \left[K'\, A + 0 \sinh \frac{(j-1)\pi K}{K'} \right]^{-1} \int_{-K'}^{K'} y \sin \frac{(j-1)\pi\eta}{K'}\, d\eta, \quad j \ge 2, \\ B_j^{(n+1)} &= \left[K'\, A + 0 \sinh \frac{(j-1)\pi K}{K'} \right]^{-1} \int_{-K'}^{K'} y \cos \frac{(j-1)\pi\eta}{K'}\, d\eta, \quad j \ge 1. \end{aligned} \tag{16.3.6}$$

STEP 5. Choose a relaxation constant $\varepsilon_1\,(= 0.1)$, and replace $A_j^{(n+1)}$ and $B_j^{(n+1)}$ by $(1-\varepsilon_1)\, A_j^{(n)} + \varepsilon_1\, A_j^{(n+1)}$ and $(1-\varepsilon_1)\, B_j^{(n)} + \varepsilon_1\, B_j^{(n+1)}$, respectively.

STEP 6. Repeat steps 1 through 5 until the required convergence is achieved.
Note that Steps 3 and 5 guarantee the convergence of successive approximations in this scheme. The grids drawn with solidity 1.58, $N = 42$, $M = 20$, $k = 4.1 \times 10^{-5}$, $K = 1.57$, and $K' = 11.5$ can be found in Inoue [1983], where the lines ξ =const surround the blades, and the lines η =const continue across the periodic boundaries OA and OE. The constant k and the ratio K/K' both decrease as the solidity of the blade increases. It has been observed that this method remains successful for solidity up to 0.29.

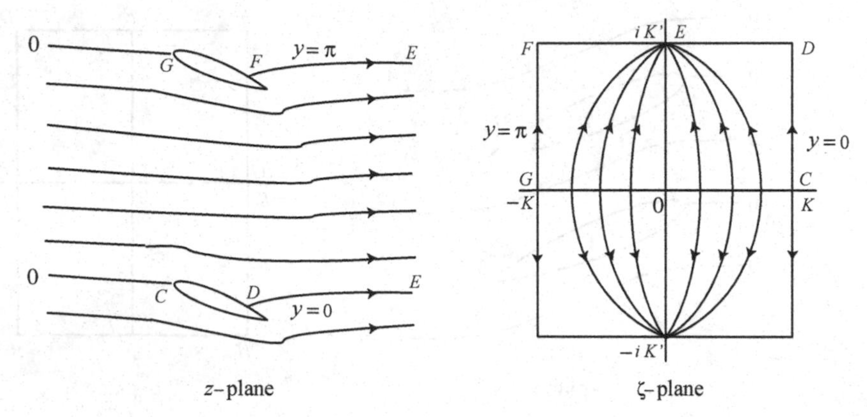

Figure 16.8 (a) Physical region. (b) Computational region.

Map 16.4. Another kind of grid is generated when the streamlines flow parallel to the x-axis in a special situation induced by the presence of sources and sinks in the physical plane. We will consider the problem of a cascade of two periodic blades. Let the streamlines of the flow through the cascade from left to right, as shown in Figure 16.8(a), represent a family of grid lines in the z-plane. Let $Z = X + i\,Y$ denote the complex velocity potential of the flow induced by the following distribution of sources and sinks in the ζ-plane (Figure 16.8(b)):

$$\begin{aligned} &\text{unit sinks at } \zeta = 2mK + 2niK'; \text{ and} \\ &\text{unit sources at } \zeta = 2mK + (2n-1)iK', \text{ where } m, n \text{ are integers.} \end{aligned} \tag{16.3.7}$$

Since the through-flow grid is based on (X, Y) coordinates, the flow induced by sources and sinks is defined by (see (7.2.13))

$$Z = \log \operatorname{sn}(\zeta, k), \tag{16.3.8}$$

or

$$\zeta = \int_0^{e^Z} \left[(1-t^2)(1-k^2t^2)\right]^{-1/2} dt. \tag{16.3.9}$$

Also, we have $Z(K) = 0$ and $Z(K + i\,K') = -\log k$, where $\zeta = K$ and $\zeta = K + i\,K'$ are the stagnation points of the flow. The through-flow grids for the cascade of Figure 16.8 can be drawn with the same data as above.

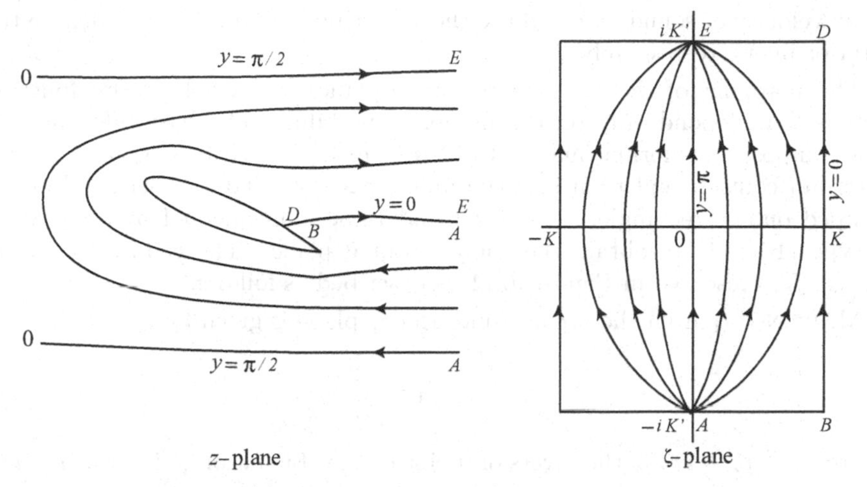

Figure 16.9 (a) Physical region. (b) Computational region.

Map 16.5. A different type of grid arises in the case when the streamlines of the flow that starts from infinity on the right encircles the cascade and returns to infinity to the right (Figure 16.9(a)). These streamlines represent the grid as a family of coordinate lines. The flow is induced by the distribution of sources and sinks in the ζ-plane which are as follows (Figure 16.9(b)):

$$\begin{aligned}&\text{unit sinks at } \zeta = 2mK + (4n+1)iK'; \text{ and}\\ &\text{unit sources at } \zeta = 2mK + (4n-1)iK', \text{ where } m, n \text{ are integers.}\end{aligned} \tag{16.3.10}$$

The complex velocity potential of this flow is defined by

$$Z = X + i\,Y = \log \operatorname{sn}\left(\frac{K'}{K}(\zeta + i\,K'), k_1\right), \tag{16.3.11}$$

where K_1 and K_1' are the complete elliptic functions of the first kind with moduli k_1 and $k_1' = \sqrt{1-k_1^2}$, respectively, such that $\dfrac{K_1'}{K_1} = \dfrac{2K'}{K}$ and $k_1 = \dfrac{1-k'}{1+k'}$ (see Landen transformation, Map 7.7). The inverse mapping of (16.3.11) is given by

$$\zeta = \frac{K}{K'}\int_0^{e^Z} \frac{dt}{\sqrt{(1-t^2)(1-k_1^2t^2)}} - i\,K'. \tag{16.3.12}$$

Note that $Z(K - i\,K') = 0$ and $Z(K + i\,K') = -\log k_1$, where $K \pm i\,K'$ are the stagnation points of the flow. The grid lines generated by this method for the above cascade can be drawn for the same data as in Map 16.3.

Map 16.6. The design of an airfoil and wing becomes significant in the transonic flow problem. The basic equations of fully developed flow potential ϕ around the configuration of an axisymmetric inlet of arbitrary geometry, shown in Figure 16.10, is given by

$$\left(a^2 - \phi_x^2\right)\phi_{xx} - 2\,\phi_x\phi_r\phi_{xr} + \left(a^2 - \phi_r^2\right)\phi_{rr} + \frac{a^2}{r}\,\phi_r = 0, \tag{16.3.13}$$

where (x, r) denote the coordinates along and normal to the centerline, respectively, and a is the velocity of sound. We will use the subscripts 'int' and 'ext' to denote the interior and exterior inlets, respectively.

The mapping of the physical region (z-plane) is carried out by functions with scale factors that depend only on the mapping modulus. The basic idea in the construction of a composite conformal map is to transform the physical boundary (inlet contour) into a Jordan contour and then into the unit circle using Fourier series. Finally the circle is mapped onto a rectangle, as in §14.1 and 14.3, supplemented by a coordinate stretching of type (16.1.11) to obtain the computational plane. The chain of conformal mappings $f_1, \dots, f_8$, presented in Figure 16.11, is described as follows:

MAPPING f_1 from the z-plane onto the z_1-plane is given by

$$z_1 = \frac{2\,r_\star}{z_\star - z},$$

where $z_\star = x_\star + i\,r_\star$ is the inversion point of the stagnation point $z = x + i\,r$.

MAPPING f_2 from the z_1-plane onto the z_2-plane is

$$z_2 = i\,\sqrt{i\,(z_1 + i)} + 1.$$

This separates the interior and exterior points at infinity and thus opens up the closed centerline in the z_1-plane. The square root of $z_1 + i$ is used with a cut starting at the branch point $z_1 = -i$.

MAPPING f_3 from the z_2-plane onto the z_3-plane is given by

$$z_3 = \frac{i\,z_2}{z_2 - 2}.$$

This bilinear transformation takes the point $z_2 = 2$ into ∞ and the centerline into the positive real axis in the z_3-plane. While approaching the interior infinity, the inside inlet in the z_3-plane tends to a line with a constant imaginary part for increasing positive values of the real part. This leads to a situation where the flow field in the z_3-plane 'opens up' as in the z_4-plane.

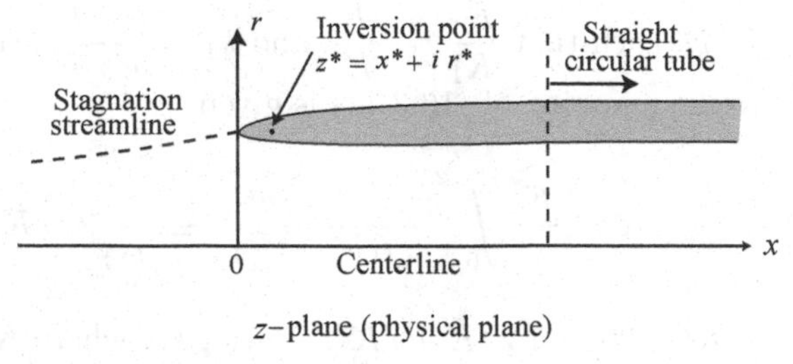

Figure 16.10 Interior radius in downstream in the inlet.

MAPPING f_4 from the z_3-plane to the z_4-plane is given by

$$z_4 = e^{c_3\,z_3},$$

where the constant c_3 is chosen such that $\Im\Big\{\lim_{z_3 \to \infty} c_3\,z_3\Big\} = \pi$ on the inlet interior side of the contour; thus $c_3 = \dfrac{\pi\,r_\star}{2\,r_{\text{int}}}$, where r_{int} denotes the interior radius for downstream in the inlet (see Fig 16.11 where the flow field contour is transformed into a Jordan contour without corners).

MAPPING f_5 from the z_4-plane onto the z_5-plane is given by

$$z_5 = -\frac{z_4 + (1 + i\, b_4)}{z_4 - (1 - i\, b_4)},$$

where the constant b_4 takes some suitable value between 0.1 and 1. Although the contour in the z_5-plane has a continuously varying tangent, it has curvature singularities at the two points at infinity.

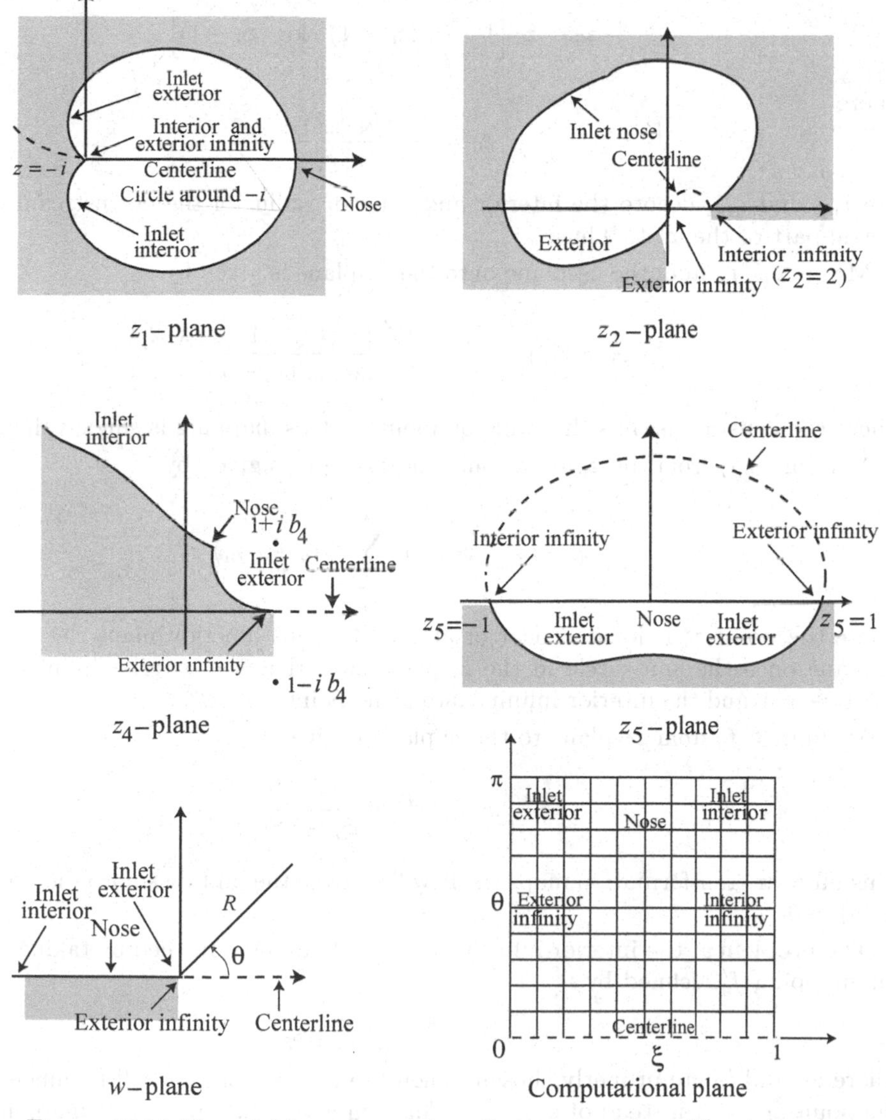

Figure 16.11 Flow field contour onto a Jordan contour without corners.

At this point we would like to use the Fourier series to transform the boundary into the unit circle. Since the exponential function is used in the mapping f_4, the far downstream region in the inlet interior is very dense around the interior infinity. Thus, the Fourier series mapping of the region in the z_5-plane will not be highly accurate in the neighborhood of the interior infinity which is at $z_5 = -1$. To avoid this, we use a Taylor series expansion of the mapping function f_5 about this point. Since the leading terms of this series do not contain the curvature at the interior infinity, the curvature singularity becomes negligible and can be neglected. We will further discuss the case of interior infinity at $z_5 = -1$ hereafter. At the exterior infinity, which is mapped into $z_5 = 1$, the curvature has a finite discontinuity, where, to compute the behavior of z near the exterior infinity, we use the transformation f_7, defined below, which removes the curvature singularity at $z_5 = 1$ before the boundary is mapped onto the unit circle.

A transformation of the type

$$z_5 \sim z_6 \left[1 + c_6 \, (z_6 - 1)^2 \log (z_6 - 1)\right],$$

where

$$c_6 = -\frac{2\, i\, b_4\, r_{\text{int}}\, r_{\text{ext}}}{\pi^2 r_\star^2},$$

and r_{int} and r_{ext} denote the interior and exterior radius of the downstream constant geometric part of the inlet. Thus,

MAPPING f_6 from the z_5-plane onto the z_6-plane is given by

$$z_5 = \left[1 + c_6 \, \frac{(z_6 - 1)^2 \, [\log (c_6 - 1) - i\, \pi/2]}{1 - i\, a_6\, c_6 \, (z_6 - 1)}\right],$$

where the constant a_6 has the value of about 5. This mapping is single-valued.

MAPPING f_7 from the z_6-plane onto the z_7-plane is given by

$$z_6 - z_6^* = z_7 \exp\left[\sum_{n=0}^{M} (\alpha_n + i\, \beta_n)\, z_7^n\right],$$

where the point z_6^* is located near or at $z_6 = 0$. This function maps the boundary in the z_6-plane onto the unit circle in the z_7-plane such that the exterior infinity goes into the point $z_7 = 1$ and the interior infinity into some point $z_7 = e^{i\theta_7}$.

MAPPING f_8 from z_7-plane to the w-plane is given by

$$w = -e^{i\theta_7/2}\, \frac{z_7 - 1}{z_7 - e^{i\theta_7}}.$$

This bilinear transformation maps the flow field from the unit disk onto the upper half-plane $\Im\{w\} > 0$.

The problem at the interior infinity at $z_5 = -1$ can be remedied by taking, instead of f_4, the mapping f_4', defined by

$$z_3' = z_3 - \frac{a_3}{z_3 + b_3},$$

where a_3 and b_3 are properly chosen. Then the new z_5-contour will be much 'fuller' below the point $z_5 = -1$ instead of $z_5 = -1$. This will affect the subsequent mappings f_5 and f_6, which will then be defined by

$$f_5' : \; z_5 = -\frac{z_4 - (a_4 + i\, b_4)}{z_4 - (a_4 - i\, b_4)},$$

with

$$c_6 = -\frac{2\, i\, b_4\, r_{\text{int}}\, r_{\text{ext}}}{\pi^2\, r_\star^2\, a_4\, (1 + a_3/b_3^2)},$$

where, in practice, we take $1 < a_3 < 2.5$, and $b_3 \approx 2$.

The Fourier coefficients in the mapping f_6 are computed from

$$\frac{dz_6}{dz_7} = \exp\left[\sum_{n=0}^{M} (\gamma_n + i\,\delta_n)\, z_7^n\right].$$

This increases the numerical accuracy over that obtained from differentiating the function f_6.

As a result of this chain of mappings the governing equation (16.3.13) is transformed into

$$\begin{aligned}
&\left(a^2 - q_1^2\right)\phi_{RR} - 2 q_1 q_2 \frac{1}{R}\,\phi_{R\theta} + \left(a^2 - q_2^2\right)\frac{1}{R^2}\,\phi_{\theta\theta} \\
&\quad + a^2 B\left(q_1 \frac{r_R}{r} + \frac{q_2}{R}\frac{r_\theta}{r} + B\left(a^2 + q_2^2\right)\frac{q_1}{R} + \left(q_1^2 + q_2^2\right)\left(q_1 B_R + \frac{q_2}{R}\, B_\theta\right)\right) = 0, \qquad (16.3.14)
\end{aligned}$$

where q_1 and q_2 are the velocity components in the R and θ direction, respectively, i.e.,

$$q_1 = \frac{1}{B}\,\phi_R, \quad q_2 = \frac{1}{RB}\,\phi_\theta.$$

Arlinger [1975] solved this problem by the finite difference method on the rectangle shown in Figure 16.11 (the computational plane).

A Fortran code (TOMCAT) for a method of automatic numerical generation of curvilinear coordinate system with grid lines coinciding with all boundaries of a multiply connected region is available (Thompson, Thames and Mastin [1977]). The computer code is independent of the boundary shapes and numbers, which are input data. The program has the following features: (i) automatic convergence controls activated by input parameters, if needed; (ii) a choice of several different types of initial guesses for the iterative process; (iii) gradual addition of coordinate system control; and (iv) general movement of the outer boundary out to its final position. Another program is available in Thompson et al. [1976] for computing the scale factors from the coordinates for use in solving partial differential equations on a coordinate system.

An adaptive grid scheme to solve the Poisson grid generation equations by methods related to Green's function, where the source terms are only position-dependent, uses the boundary element method and is given by Munipalli and Andersen [1996]. All of these schemes and programs solve the types of problems discussed in the above case studies.

Example 16.4. Note that under the mapping of a region D in the z-plane onto a region G in the w-plane by the function $w = f(z) = u(x, y) + i\, v(x, y)$, the Laplace equation for the function $u(x, y)$ is transformed into the Laplace equation for the function $U(\xi, \eta) = u\left(x(\xi, \eta), y(\xi, \eta)\right)$, i.e., the Laplacian ∇^2 satisfies the relation

$$\nabla_{xy}^2 = |f'(z)|^2\, \nabla_{\xi\eta}^2 = \frac{1}{|F'(w)|^2}\, \nabla_{\xi\eta}^2,$$

where $z = F(w)$ denotes the inverse mapping function (Sveshnikov and Tikhonov [1978: 194].) ■

REFERENCES USED: Arlinger [1975], Inoue [1983; 1985], Ives and Liutermoza [1977], Knupp and Steinberg [1993], Kythe [1998], Lehto and Virtaanen [1973], Munipalli and Andersen [1996], Özisik [1994], Roberts [1971], Sveshnikov and Tikhonov [1978], Thompson et al. [1977; 1985].

17

Field Theories

We will discuss mathematical models involving potential fields and related Laplace's, Poisson's, and other equations, which are encountered in different flow fields in continuum mechanics and physics. There are different methods to solve boundary value problems involving these equations, namely, analytic methods including Green's function, conformal mapping method, and numerical approximations using finite and boundary elements. In this chapter we will confine to the following two-dimensional equations: Laplace's, Poisson's, Helmholtz, biharmonic, and membrane equations, and provide their solutions in different domains in the (x, y)-plane, including related Green's functions; some examples are also provided. Conformal mapping methods will be discussed in subsequent chapters.

17.1 Mathematical Models

In many mathematical modeling formulations, partial derivatives are required to represent physical quantities. These derivatives always involve more than one independent variable, generally the space variables $x, y, \dots$ and the time variable t. Such formulations have one or more dependent variables, which are the unknown functions of the independent variables. The resulting equations are called *partial differential equations*, which, together with the initial and/or boundary conditions, represent physical phenomena and are known as the initial or boundary value problems.

The *order* of a partial differential equation is the same as the order of the highest partial derivative appearing in the equation. The partial derivatives $\dfrac{\partial u}{\partial x}, \dfrac{\partial u}{\partial y}, \dfrac{\partial^2 u}{\partial x^2}, \dfrac{\partial^2 u}{\partial y \partial x}$, and $\dfrac{\partial^2 u}{\partial y^2}$ are sometimes denoted by u_x, u_y, u_{xx}, u_{xy}, and u_{yy} (or p, q, r, s, and t), respectively. The most general first-order partial differential equation with two independent variables x and y has the form

$$F(x, y, u, p, q) = 0, \qquad p = u_x, \quad q = u_y. \tag{17.1.1}$$

The most general second-order partial differential equation is of the form

$$F(x, y, u, p, q, r, s, t) = 0, \qquad r = u_{xx}, \quad s = u_{xy}, \quad t = u_{yy}. \tag{17.1.2}$$

A partial differential equation is said to be *linear* if the unknown function u and all its partial derivatives appear in an algebraically linear form, i.e., of the first degree. For example, the equation

$$a_{11}\, u_{xx} + 2a_{12}\, u_{xy} + a_{22}\, u_{yy} + b_1\, u_x + b_2\, u_y + c_0\, u = f, \tag{17.1.3}$$

where the coefficients $a_{11}, a_{12}, a_{22}, b_1, b_2$, and c_0 and the function f are functions of x and y, is a second-order linear partial differential equation in the unknown $u(x, y)$.

An operator L is a linear differential operator iff $L(\alpha u + \beta v) = \alpha Lu + \beta Lv$, where α and β are scalars, and u and v are any functions with continuous partial derivatives of appropriate order. A partial differential equation of the form $Lu = 0$ is said to be *homogeneous*, whereas an equation of the form $Lu = g$, where $g \neq 0$ is a given function of the independent variables, is said to be *nonhomogeneous*. Thus, a linear homogeneous equation is such that whenever u is a solution of the equation, then cu is also a solution, where c is a constant. A function $u = \phi$ is said to be a *solution* of a partial differential equation if ϕ and its partial derivatives, when substituted for u and its partial derivatives occurring in the partial differential equation, reduce it to an identity in the independent variables. The *general* solution of a linear partial differential equation is a linear combination of all linearly independent solutions of the equation with as many arbitrary functions as the order of the equation; a partial differential equation of order k has k arbitrary functions. A *particular* solution of a partial differential equation is one that does not contain arbitrary functions or constants.

A partial differential equation is said to be *quasi-linear* if it is linear in all the highest-order derivatives of the dependent variable. For example, the most general form of a quasi-linear second-order equation is

$$A(x, y, u, p, q)\, u_{xx} + B(x, y, u, p, q)\, u_{xy} + C(x, y, u, p, q)\, u_{yy} + f(x, y, u, p, q) = 0. \tag{17.1.4}$$

17.2 Initial and Boundary Conditions

A partial differential equation subject to certain conditions in the form of initial or boundary conditions is known as an initial value or a boundary value problem. The initial conditions, also known as *Cauchy conditions*, are the values of the unknown function u and an appropriate number of its derivatives at the initial point.

The boundary conditions fall into the following three categories:

(i) *Dirichlet boundary conditions* (also known as boundary conditions of the first kind), when the values of the unknown function u are prescribed at each point of the boundary ∂D of a given domain D.

(ii) *Neumann boundary conditions* (also known as boundary conditions of the second kind), when the values of the normal derivatives of the unknown function u are prescribed at each point of the boundary ∂D.

(iii) *Robin boundary conditions* (also known as boundary conditions of the third kind, or mixed boundary conditions), when the values of a linear combination of the unknown function u and its normal derivative are prescribed at each point of the boundary ∂D.

Two types of boundary conditions remain unchanged by the transformation and problems with boundary conditions under conformal mappings: (i) The value of a harmonic function is constant on the boundary (one type of Dirichlet problem); and (ii) the normal derivatives of a harmonic function is zero on the boundary (one type of Neumann problem). The conditions are that $w = f(z)$ and its inverse are analytical. An example for these three categories is given in §19.1 (Example 19.1).

(iv) The *mixed boundary conditions* of the second-order partial differential equations are represented by

$$\left[\alpha u + \beta \frac{\partial u}{\partial n}\right]_{\partial D} = f(\mathbf{x})\Big|_{\partial D}, \tag{17.2.1}$$

where α and β are constants, $\frac{\partial u}{\partial n}$ denotes the normal derivative defined by

$$\frac{\partial u}{\partial n} = \mathbf{n} \cdot \nabla u = n_{x_1} \frac{\partial u}{\partial x_1} + \cdots + n_{x_n} \frac{\partial u}{\partial x_n},$$

where $\mathbf{x} = (x_1, \dots, x_n)$, and $\mathbf{n}$ is the outward normal to the boundary ∂D. A mixed boundary value problem can have a Dirichlet boundary condition on one part of the boundary, a Neumann boundary condition on another part, and a Robin boundary condition on still a third part of the boundary. It should be pointed out that Neumann boundary conditions do not lead to a unique solution of a boundary value problem. For example, if u is a solution of the Laplace equation $\nabla^2 u = 0$ on the rectangle of Figure 17.1, subject to the Neumann boundary conditions

$$\frac{\partial u}{\partial x}(x,0) = f_1(x), \qquad \frac{\partial u}{\partial y}(x,b) = f_2(x),$$
$$\frac{\partial u}{\partial x}(0,y) = g_1(y), \qquad \frac{\partial u}{\partial x}(a,y) = g_2(y),$$

where $f_{1,2}(x)$ and $g_{1,2}(y)$ are prescribed functions, then $w = u + c$, where c is a constant, is also a solution of this boundary value problem.

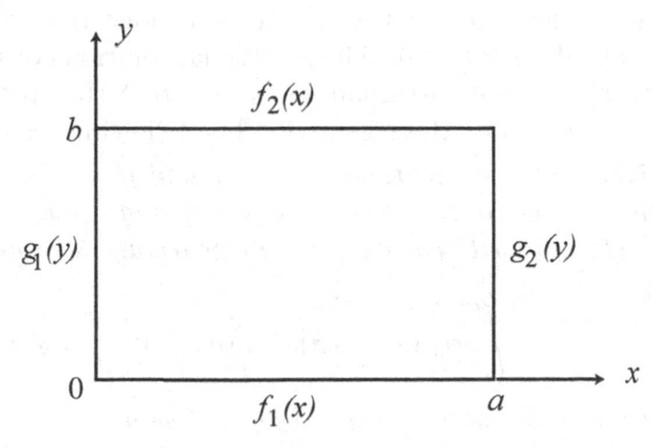

Figure 17.1 Laplace's equation in a rectangle under Neumann conditions.

(v) The *implicit conditions* arise in certain cases, especially when the partial differential equation represents a model of some real-life physical situation, certain restrictions are imposed by means of boundary conditions, which are implicit in nature. These restrictions are also known as *natural boundary conditions* as opposed to *essential boundary conditions*, which in the case of mixed boundary value problems take the form of initial conditions. These types of implicit conditions arise, for example, in the weak variational formulation of physical problems. An example for this category is given in §19.1 (Example 19.2).

Remember that $\nabla^2 u$ at a point (x, y) is a measure of the difference between the values of u at (x, y) and the average of the values of u in an infinitesimal neighborhood surrounding the point (x, y).

Example 17.1. Consider the two-dimensional Laplace's equation in polar coordinates, which is defined by

$$\nabla^2 u = \frac{1}{r}\frac{\partial}{\partial r}\left(r\frac{\partial u}{\partial r}\right) + \frac{1}{r^2}\frac{\partial^2 u}{\partial \theta^2} = 0.$$

Let the domain be the disk $0 \le r \le a$. This equation has the following two sets of solutions:
(i) $u_n(r, \theta) = r^n \left(A_n \sin n\theta + B_n \cos n\theta\right),\ n = 1, 2, \dots;$
(ii) $u_n(r, \theta) = \dfrac{1}{r^n}\left(C_n \sin n\theta + D_n \cos n\theta\right),\ n = 1, 2, \dots.$
Notice that the set (ii) has a singularity at $r = 0$, and, therefore, it is not acceptable for the given domain unless there is a source at the singularity. But the set (i) being bounded on the disk is acceptable. However, if the domain is the exterior $r > a$ of the above disk, then the set (ii) will be bounded and, therefore, acceptable for this domain. ■

(vi). The *periodic conditions* over an interval $[a, b]$ are of the type $y(a) = y(b), y'(a) = y'(b)$. It is encountered in the Sturm-Liouville problem consisting of the Sturm-Liouville equation

$$\frac{d}{dx}\left[p(x)\frac{dy}{dx}\right] + \left[q(x) + \lambda\, w(x)\right] y = 0, \tag{17.2.2}$$

which is a linear second-order ordinary differential equation defined on a given interval $a \le x \le b$ and satisfying the boundary conditions of the form

$$\begin{aligned} a_1\, y(a) + b_1\, y'(a) &= 0, \\ a_2\, y(b) + b_2\, y'(b) &= 0, \end{aligned} \tag{17.2.3}$$

where λ is a real parameter, and a_1, a_2, b_1, and b_2 are given real constants such that a_1 and b_1, or a_2 and b_2 are both not zero. It is obvious that the system (17.2.2)-(17.2.3) always has a trivial solution $y = 0$. The nontrivial solutions of this problem are called the *eigenfunctions* $\phi_n(x)$ and the corresponding values of λ the *eigenvalues* λ_n of the problem. The pair (ϕ_n, λ_n) is known as the *eigenpair*. The following result is useful:

Theorem 17.1. *Let the functions p, q, w, and p' in Eq (17.2.2) be real-valued and continuous on the interval $a \le x \le b$. Let $\phi_m(x)$ and $\phi_n(x)$ be the eigenfunctions of the problem (17.2.2)–(17.2.3) with corresponding eigenvalues λ_m and λ_n, respectively, such that $\lambda_m \ne \lambda_n$. Then*

$$\int_a^b \phi_m(x)\, \phi_n(x)\, w(x)\, dx = 0, \quad m \ne n, \tag{17.2.4}$$

i.e., the eigenfunctions ϕ_m and ϕ_n are orthogonal with respect to the weight function $w(x)$ on the interval $a \le x \le b$.

Proof of this theorem can be found in many textbooks on ordinary differential equations, e.g., Ross [1964], and Boyce and DiPrima [1962]. An example of this category is available in §19.1 (Example 19.3).

17.3 Classification of Second-Order Equations

If $f = 0$ in Eq (17.1.3), the most general form of a second-order homogeneous equation is

$$a_{11}\, u_{xx} + 2a_{12}\, u_{xy} + a_{22}\, u_{yy} + b_1\, u_x + b_2\, u_y + c_0\, u = 0. \tag{17.3.1}$$

To show a correspondence of this equation with an algebraic quadratic equation, we replace u_x by α, u_y by β, u_{xx} by α^2, u_{xy} by $\alpha\beta$, and u_{yy} by β^2. Then the left side of Eq (17.3.1) reduces to a second degree polynomial in α and β:

$$P(\alpha, \beta) = a_{11}\alpha^2 + 2a_{12}\alpha\beta + a_{22}\beta^2 + b_1\alpha + b_2\beta + c_0. \tag{17.3.2}$$

It is known from analytical geometry and algebra that the polynomial equation $P(\alpha, \beta) = 0$ represents a *hyperbola, parabola,* or *ellipse* according as its discriminant $a_{12}^2 - a_{11}a_{22}$ is positive, zero, or negative. Thus, Eq (17.3.1) is classified as hyperbolic, parabolic, or elliptic according as the quantity $a_{12}^2 - a_{11}a_{22}$ is positive, zero, or negative.

An alternative approach to classify the types of Eq (17.3.1) is based on the following theorem:

Theorem 17.2. *The relation $\phi(x, y) = C$ is a general integral of the ordinary differential equation*

$$a_{11}\, dy^2 - 2a_{12}\, dx\, dy + a_{22}\, dx^2 = 0 \tag{17.3.3}$$

iff $u = \phi(x, y)$ is a particular solution of the equation

$$a_{11}\, u_x^2 + 2a_{12}\, u_x\, u_y + a_{22}\, u_y^2 = 0. \tag{17.3.4}$$

PROOF. (Kythe et al. [2003:12]) Assume that the function $u = \phi(x, y)$ satisfies Eq (17.3.4). Then the equation

$$a_{11}\left(\frac{\phi_x}{\phi_y}\right)^2 - 2a_{12}\left(-\frac{\phi_x}{\phi_y}\right) + a_{22} = 0 \tag{17.3.5}$$

holds for all x, y in the domain of definition of $u = \phi(x, y)$ with $\phi_y \neq 0$. In order that the relation $\phi(x, y) = C$ is the general solution of Eq (17.3.3), we must show that the function y defined implicitly by $\phi(x, y) = C$ satisfies Eq (17.3.3). Suppose that $y = f(x, C)$ is such a function. Then

$$\frac{dy}{dx} = -\left[\frac{\phi_x(x, y)}{\phi_y(x, y)}\right]_{y=f(x,C)}.$$

Hence, in view of Eq (17.3.5), we have

$$\begin{aligned} a_{11}\left(\frac{dy}{dx}\right)^2 &- 2a_{12}\left(\frac{dy}{dx}\right) + a_{22} \\ &= \left[a_{11}\left(-\frac{\phi_x}{\phi_y}\right)^2 - 2a_{12}\left(-\frac{\phi_x}{\phi_y}\right) + a_{22}\right]_{y=f(x,C)} = 0. \end{aligned} \tag{17.3.6}$$

Thus, $y = f(x, C)$ satisfies Eq (17.3.4).

Conversely, let $\phi(x,y)=C$ be a general solution of Eq (17.3.3). We must show that for each point (x,y)

$$a_{11}\,\phi_x^2+2a_{12}\,\phi_x\,\phi_y+a_{22}\,\phi_y^2=0. \tag{17.3.7}$$

If we can show that Eq (17.3.7) is satisfied for an arbitrary point (x_0,y_0), then Eq (17.3.7) will be satisfied for all points. Since $\phi(x,y)$ represents a solution of Eq (17.3.3), we construct through (x_0,y_0) an integral of Eq (17.3.3), where we set $\phi(x_0,y_0)=C_0$, and consider the curve $y=f(x,C_0)$. For all points of this curve we have

$$\begin{aligned} a_{11}\left(\frac{dy}{dx}\right)^2 &- 2a_{12}\left(\frac{dy}{dx}\right)+a_{22} \\ &= \left[a_{11}\left(-\frac{\phi_x}{\phi_y}\right)^2-2a_{12}\left(-\frac{\phi_x}{\phi_y}\right)+a_{22}\right]_{y=f(x,C_0)}=0. \end{aligned}$$

If we set $x=x_0$ in this equation, we get

$$a_{11}\,\phi_x^2(x_0,y_0)+2a_{12}\,\phi_x(x_0,y_0)\,\phi_y(x_0,y_0)+a_{22}\,\phi_y^2(x_0,y_0)=0,$$

where $y_0=f(x_0,C_0)$. ■

Eq (17.3.3) or (17.3.4) is called the *characteristic equation* of the partial differential equation (17.1.3) or (17.3.1); the related integrals are called the *characteristics*.

Eq (17.3.6), regarded as a quadratic equation in dy/dx, yields two solutions:

$$\frac{dy}{dx}=\frac{a_{12}\pm\sqrt{a_{12}^2-a_{11}\,a_{22}}}{a_{11}}.$$

The expression under the radical sign determines the type of the differential equation (17.1.3) or (17.3.1). Thus, as before, Eq (17.1.3) or (17.3.1) is of the hyperbolic, parabolic, or elliptic type according as the quantity $a_{12}^2-a_{11}a_{22} \gtreqless 0$.

The following two-dimensional equations often appear in the study of physical problems.

1. *Heat equation* in R^1: $u_t=k\,u_{xx}$, where u denotes the temperature distribution and k the thermal diffusivity.

2. *Wave equation* in R^1: $u_{tt}=c^2u_{xx}$, where u represents the displacement, e.g., of a vibrating string from its equilibrium position, and c the wave speed.

3. *Laplace equation* in R^2: $\nabla^2u\equiv u_{xx}+u_{yy}=0$, where $\nabla^2=\nabla\cdot\nabla$ denotes the Laplacian.

4. *Poisson's equation* in R^n: $\nabla^2u=f$, also known as the nonhomogeneous Laplace equation in R^n; it arises in various field theories and electrostatics.

5. *Helmholtz equation* in R^3: $\left(\nabla^2+k^2\right)u=0$, which arises, e.g., in underwater scattering.

6. *Biharmonic equation* in R^3: $\nabla^4u\equiv\nabla^2(\nabla^2u)=0$; it arises in elastodynamics.

7. *Euler's equations* in R^3: $\mathbf{u}_t+(\mathbf{u}\cdot\nabla)\mathbf{u}+\dfrac{1}{\rho}\nabla p=0$, where $\mathbf{u}$ denotes the velocity field, and p the pressure.

8. *Navier-Stokes equations* in R^3: $\mathbf{u}_t+(\mathbf{u}\cdot\nabla)\mathbf{u}+\dfrac{1}{\rho}\nabla p=\nu\nabla^2\mathbf{u}$, where ν denotes the kinematic viscosity and ρ the density of the fluid.

9. *Maxwell's equations* in R^3: $\mathbf{E}_t - \nabla \times \mathbf{H} = 0$, $\mathbf{H}_t + \nabla \times \mathbf{E} = 0$, where $\mathbf{E}$ and $\mathbf{H}$ denote the electric and the magnetic field, respectively; they are a system of six equations in six unknowns.

Origins of these and other equations of mathematical physics are related to some interesting physical problems. We will derive some of them as examples, which will also bring out certain aspects of mathematical modeling of boundary value problems.

17.4 Superposition Principle

Let L denote a linear differential operator of any order and any kind. The superposition principle for homogeneous and nonhomogeneous linear differential equations is represented by the following two theorems:

Theorem 17.3. *Let $Lu = 0$ be a differential equation. Suppose u_1 and u_2 are two linearly independent solutions. Then $c_1u_1 + c_2u_2$ is also a solution.*

PROOF. By hypotheses $Lu_{1,2} = 0$. By definition $L(c_1u_1 + c_2u_2) = c_1Lu_1 + c_2Lu_2 = 0$. ■

Theorem 17.4. *If $Lu = \sum_1^n c_i f_i$ be a nonhomogeneous linear differential equation and if $Lg_k = f_k$, then $\sum_1^n c_i g_i$ is a solution of the above differential equation.*

PROOF. $L\sum_1^n c_i g_i = \sum_1^n c_i Lg_i = \sum_1^n c_i f_i$. Thus $\sum_1^n c_i g_i$ satisfies the differential equation. ■

It is obvious from these two theorems that if v is a solution of an equation $Lu = 0$ and if F a solution of $Lu = f$, then $v + F$ is also a solution of $Lu = f$. A generalized superposition principle is as follows:

Theorem 17.5. *If each of the functions u_i, $i = 1, 2, \dots$, is a solution of a linear homogeneous differential equation $L(u) = 0$, then the series $u = \sum_{i=1}^{\infty} C_i\, u_i$ is also a solution of this differential equation, provided that u and its derivatives appearing in $L(u)$ can be differentiated term-by-term.*

PROOF. If the derivatives of u appearing in $L(u) = 0$ can be differentiated term-by-term, we have

$$\frac{\partial^n u}{\partial x^m \partial t^{n-m}} = \sum_{i=1}^{\infty} C_i \frac{\partial^n u_i}{\partial x^m \partial t^{n-m}}, \tag{17.4.1}$$

and since the equation $L(u) = 0$ is linear and a convergent series can be differentiated term-by-term, we can write

$$L(u) = L\left(\sum_{i=1}^{\infty} C_i\, u_i\right) = \sum_{i=1}^{\infty} C_i\, L(u_i) = 0. \tag{17.4.2}$$

The sufficient condition for term-by-term differentiability is the uniform convergence of the series $\sum_{i=1}^{\infty} C_i \frac{\partial^n u_i}{\partial x^m \partial t^{n-m}}$. ■

17.5 Parabolic Equations

The *heat equation* with the constant coefficient,

$$u_t = k u_{xx} \tag{17.5.1}$$

(see §17.2) is encountered in the theory of heat and mass transfer, describing the one-dimensional unsteady thermal processes in quiescent media or solids, and one-dimensional unsteady mass exchange processes, with constant thermal diffusivity k. Let A, B and μ be arbitrary constants. Then the particular solutions of Eq (17.5.1) are:

$$u(x,t) = A(x^2 + 2kt) + B;$$
$$u(x,t) = A(x^3 + 6ktx) + B;$$

or in general:

$$u(x,t) = x^{2n} + \sum_{j=1}^{n} \frac{(2n)(2n-1)\dots(2n-2j+1)}{j!} (kt)^j x^{2n-2j};$$
$$u(x,t) = x^{2n+1} + \sum_{j=1}^{n} \frac{(2n+1)(2n)\dots(2n-2j+2)}{j!} (kt)^j x^{2n-2j+1}.$$

Also

$$u(x,t) = A\, e^{k\mu^2 t \pm \mu x} + B;$$
$$u(x,t) = A\, e^{-k\mu^2 t} \cos(\mu x) + B;$$
$$u(x,t) = A\, e^{-k\mu^2 t} \sin(\mu x) + B;$$
$$u(x,t) = A \frac{1}{t^{1/2}}\, e^{-x^2/(4kt)} + B;$$
$$u(x,t) = A \frac{1}{t^{3/2}}\, e^{-x^2/(4kt)} + B;$$
$$u(x,t) = A\, e^{-\mu x} \cos\left(\mu x - 2k\mu^2 t\right) + B;$$
$$u(x,t) = A\, e^{-\mu x} \sin\left(\mu x - 2k\mu^2 t\right) + B;$$
$$u(x,t) = A \operatorname{erf}\left(\frac{x}{2\sqrt{kt}}\right) + B;$$
$$u(x,t) = A \operatorname{erfc}\left(\frac{x}{2\sqrt{kt}}\right) + B;$$
$$u(x,t) = A\Big\{\sqrt{\frac{t}{\pi}} e^{-x^2/(4kt)} - \frac{x}{2\sqrt{k}} e^{x/(2\sqrt{kt})}\Big\} + B.$$

(See Carslaw and Jaeger [1959], Polyanin and Nazaikinskii [2016]).

The infinite series solution is (Carslaw and Jaeger [1959])

$$u(x,t) = f(x) + \sum_{n=1}^{\infty} \frac{(kt)^n}{n!}\, f_x^{2n)}(x),$$

where $f(x)$ is any infinitely differentiable function. This solution satisfies the initial condition $u(x,0) = f(x)$. The sum is finite if $f(x)$ is a polynomial.

The solutions involving arbitrary functions of time are (Carslaw and Jaeger [1959])

$$u(x,t) = g(t) + \sum_{n=1}^{\infty} \frac{1}{(2n)!k^n}\, x^{2n} g_t^{(n)}(t),$$
$$u(x,t) = xh(t) + x \sum_{n=1}^{\infty} \frac{1}{(2n+1)!k^n}\, x^{2n} h_t^{(n)}(t),$$

where $g(t)$ and $h(t)$ are infinitely differentiable functions. The sums are finite if $g(t)$ and $h(t)$ are polynomials. The first solution satisfies the boundary condition of the first kind $u(0,t) = g(t)$, and the second solution the boundary condition of the second kind $u_x(0,t) = h(t)$.

The fundamental solution is (Kythe [1995, Ch. 3])

$$u^*(x,t) = \frac{H(t)}{(4\pi kt)^{n/2}}\, e^{-|x|^2/(4kt)}.$$

17.6 Hyperbolic Equations

We will consider only the following types of hyperbolic equation:

17.6.1 Second-Order Hyperbolic Equations in One Space Variable. It is known as the *wave equation*, or the *equation of vibration of a string*, expressed as

$$\frac{\partial^2 u}{\partial t^2} = c^2 \frac{\partial^2 u}{\partial x^2}. \tag{17.6.1}$$

Its general solution is

$$u(x,t) = \phi(x+ct) + \psi(x-ct), \tag{17.6.2}$$

where $\phi(x)$ and $\psi(x)$ are arbitrary functions. The problem of a vibrating string of length l is defined by the wave equation (17.6.1) as the boundary value problem

$$\begin{aligned} &\frac{\partial^2 u}{\partial t^2} = c^2 \frac{\partial^2 u}{\partial x^2}, \quad 0 < x < l, \\ &u(0,t) = 0 = u(l,t), \quad t > 0, \\ &u(x,0) = f(x), \ u_t(x,0) = h(x), \quad 0 < x < l, \end{aligned} \tag{17.6.3}$$

where $f \in C^1(0,l)$ and $h(x)$ are given functions, and c is the wave velocity. The solution of this problem is

$$u(x,t) = \sum_{n=1}^{\infty} \Big[A_n \cos \frac{n\pi ct}{l} + B_n \sin \frac{n\pi ct}{l} \Big], \tag{17.6.4}$$

where

$$\begin{aligned} A_n &= \frac{2}{l} \int_0^l f(x) \sin \frac{n\pi x}{l}\, dx, \\ B_n &= \frac{2}{n\pi c} \int_0^l h(x) \sin \frac{n\pi x}{l}\, dx, \quad n = 1, 2, \dots \end{aligned}$$

The d'Alembert's solution is given by

$$u(x,t) = \frac{1}{2}\left[f(x+ct) + f(x-ct)\right] + \frac{1}{2}\left[g(x-ct) - g(x+ct)\right] = \phi(c-xt) + \psi(c+ct), \tag{17.6.5}$$

where

$$f(z) = \sum_{n=1}^{\infty} A_n \sin \frac{n\pi z}{l}, \quad g(z) = \sum_{n=1}^{\infty} B_n \cos \frac{n\pi z}{l}.$$

This signifies that at each point x of the string, the solution

$$u(x,t) = \sum_{n=1}^{\infty} \alpha_n \cos \frac{n\pi c}{l}(t+\delta_n) \sin \frac{n\pi x}{l}$$

describes a harmonic motion with amplitude $\alpha_n \sin \frac{n\pi x}{l}$, and each such motion of the string is called a *standing wave* with nodes at the points where $\sin \frac{n\pi x}{l} = 0$. These points remain fixed during the entire process of vibration. However, the string vibrates with maximum amplitudes α_n at the points where $\sin \frac{n\pi x}{l} = \pm 1$. Thus, for any t the structure of the standing wave is described by

$$u(x,t) = \sum_{n=1}^{\infty} C_n(t) \sin \frac{n\pi x}{l}, \quad C_n(t) = \alpha_n \cos \omega_n (t_{\delta_n}), \quad \omega_n = \frac{n\pi c}{l},$$

so that at times t when $\cos \omega (t+\delta_n) = \pm 1$, the velocity becomes zero and the displacement reaches its maximum value.

The fundamental solution $u^*(x,t)$ (Kythe [1995, Ch. 4]) is

$$u^*(x,t) = \frac{1}{2c} H(t - |x|) = \frac{1}{2c} H(t)[H(x+t) - H(x-t)].$$

If $u(x,t)$ is a solution of the wave equation (17.6.1), then the functions

$$\begin{aligned} u_1(x,t) &= A\, u(\pm\lambda x + C_1, \pm\lambda t + C_2), \\ u_2(x,t) &= A\, u\Big(\frac{x - \alpha t}{\sqrt{1-(\alpha/c)^2}}, \frac{t - \alpha c^{-2} x}{\sqrt{1-(\alpha/c)^2}}\Big), \\ u_3(x,t) &= A\, u\Big(\frac{x}{x^2 - c^2t^2}, \frac{t}{x^2 - c^2 t^2}\Big), \end{aligned}$$

are also solutions of the wave equation, where A, C_1, C_2, α, and λ are arbitrary constants. The function u_2 results from the invariance of the wave equation under the Lorenz transformation.

Example 17.2. (Goursat problem) If the boundary conditions are prescribed as

$$u = \begin{cases} f(x) & \text{for } x - ct = 0, 0 \le x \le a, \\ g(x) & \text{for } x + ct = 0, \ 0 \le x \le b, \end{cases}$$

where $f(0) = g(0)$, then the solution is

$$u(x,t) = f\Big(\frac{x+ct}{2}\Big) + g\Big(\frac{x-ct}{2}\Big) - f(0).$$

The domain of solution propagation is bounded by four lines: $x - ct = 0$, $x + ct = 0$, $x - ct = 2b$, and $x + ct = 2a$. ■

17.6.2 General Case. We will consider the wave equation of the form

$$\frac{\partial^2 u}{\partial t^2} = \frac{\partial^2 u}{\partial x^2} + \Phi(x,t). \tag{17.6.6}$$

In the domain $-\infty < x < \infty$, the Cauchy problem has the initial conditions

$$u = f(x) \text{ at } t = 0, \quad u_t = g(x) \text{ at } t = 0. \tag{17.6.7}$$

The solution is

$$u(x,t) = \frac{1}{2}\left[f(x-ct) + f(x+ct)\right] + \frac{1}{2c}\int_{x-ct}^{x+ct} g(s)\,ds + \frac{1}{2c}\iint_{x-c(t-\tau)}^{x+c(t-\tau)} \Phi(s,\tau)\,ds\,d\tau.$$

The *first boundary value problem* with the initial conditions (17.6.7) and the boundary condition $u = h(t)$ at $x = 0$ for all $x \in [0, \infty)$ has the solution

$$u(x,t) = u_1(x,t) + u_2(x,t),$$

where

$$u_1(x,t) = \begin{cases} \frac{1}{2}\left[f(x+ct) + f(x-ct)\right] + \dfrac{1}{2c}\int_{x-ct}^{x+ct} g(s)\,ds, & \text{if } t < x/c, \\ \frac{1}{2}\left[f(x+ct) - f(x-ct)\right] + \dfrac{1}{2c}\int_{x-ct}^{x+ct} g(s)\,ds + h\left(t - x.c\right), & \text{if } t > x/c, \end{cases}$$

$$u_2(x,t) = \begin{cases} \int_0^{t=x/c}\int_{x-c(t-\tau)}^{x+c(t-\tau)} \Phi(s,\tau)\,ds\,d\tau, & \text{if } t < x/c, \\ \int_0^{t-x/c}\int_{x-c(t-\tau)}^{x+c(t-\tau)} \Phi(s,\tau)\,ds\,d\tau + \int_{t-x(t/c}^{t}\int_{x-c(t-\tau)}^{x+c(t-\tau)} \Phi(s,\tau)\,ds\,d\tau, & \text{if } t > x/c. \end{cases}$$

The *second boundary problem* with the initial conditions (17.6.7) and the boundary condition $u_x = h(t)$ at $x = 0$ for all $x \in [0, \infty)$ has the solution

$$u(x,t) = u_1(x,t) + \frac{1}{2c}u_2(x,t),$$

where

$$u_1(x,t) = \begin{cases} \frac{1}{2}\left[f(x+ct) + f(x-ct)\right] + \dfrac{1}{2c}\int_{x-ct}^{x+ct} g(s)\,ds, & \text{if } t < x/c, \\ \frac{1}{2}\left[f(x+ct) + f(x-ct)\right] + \dfrac{1}{2c}\int_0^{x+ct} g(s)\,ds + \dfrac{1}{2c}\int_0^{ct-x} g(s)\,ds \\ \quad -c\int_0^{x+ct} h(s)\,ds, & \text{if } t > x/c, \end{cases}$$

$$u_2(x,t) = \begin{cases} \int_0^{t=x/c}\int_{x-c(t-\tau)}^{x+c(t-\tau)} \Phi(s,\tau)\,ds\,d\tau, & \text{if } t < x/c, \\ \int_0^{t-x/c}\int_0^{x+c(t-\tau)} \Phi(s,\tau)\,ds\,d\tau + \int_0^{x+c(t-\tau)}\int_0^{c(t-\tau)-x} \Phi(s,\tau)\,ds\,d\tau \\ \quad + \int_{t-x(t/c}^{t}\int_{x-c(t-\tau)}^{x+c(t-\tau)} \Phi(s,\tau)\,ds\,d\tau, & \text{if } t > x/c. \end{cases}$$

17.7 Elliptic Equations

Laplace's equation $\nabla^2 u = 0$ is often encountered in heat and mass transfer, fluid mechanics, elasticity, electrostatics, and other areas of continuum mechanics and physics. For example, in heat and mass transfer theory, this equation describes steady-state temperature distribution in the absence of heat sources and sinks in the region under study. A regular solution of Laplace's equation is called a harmonic function. Conversely, if $w = f(z) = u(x,y) + iv(x,y)$ is a conformal mapping, then each u and v is a harmonic function and thus satisfies Laplace's equation. The first boundary value problem for Laplace's equation is often referred to as the *Dirichlet problem* and the second boundary value problem as the *Neumann problem.*

Laplace's equation in two space variables in the rectangular coordinates system is written as

$$\frac{\partial^2 u}{\partial x^2} + \frac{\partial^2 u}{\partial y^2} = 0. \tag{17.7.1}$$

Its particular solutions are:

$$\begin{aligned}
u(x,y) &= Ax + By + C;\\
u(x,y) &= A\left(x^2 - y^2\right) + Bxy;\\
u(x,y) &= A\left(x^3 - 3xy^2\right) + B\left(3x^2y - y^3\right);\\
u(x,y) &= \frac{Ax + By}{x^2 + y^2} + C;\\
u(x,y) &= e^{\pm\mu x}\left(A\cos\mu y + B\sin\mu y\right);\\
u(x,y) &= \left(A\cos\mu x + B\sin\mu x\right)e^{\pm\mu y};\\
u(x,y) &= (A\sinh\mu x + B\cosh\mu x)(C\cos\mu y + D\sin\mu y);\\
u(x,y) &= (A\cos\mu x + B\sin\mu x)(C\sinh\mu y + D\cosh\mu y);\\
u(x,y) &= A\ln\left[(x - x_0)^2 + (y - y_0)^2\right] + B,
\end{aligned}$$

where A, B, C, D, x_0, y_0, and μ are arbitrary constants. The fundamental solution is

$$u^*(x,y) = \frac{1}{2\pi}\ln\frac{1}{r}, \quad r = \sqrt{x^2 + y^2}.$$

If $u(x,y)$ is a solution of Laplace's equation, then the functions

$$\begin{aligned}
u_1 &= Au\left(\pm\alpha x + C_1, \pm\alpha y + C_2\right),\\
u_2 &= Au(x\cos\beta + y\sin\beta, -x\sin\beta + y\cos\beta),\\
u_3 &= Au\Big(\frac{x}{x^2 + y^2}, \frac{y}{x^2 + y^2}\Big),
\end{aligned}$$

are also solutions everywhere they are defined; A, C_1, C_2, α, and β are arbitrary constants, and the sign of α in u_1 is taken independently of each other.

17.7.1 Boundary Value Problems for Laplace's Equation. Some specific features are:

(i) For outer boundary value problems on the plane, it is usually required to set the additional condition that the solution of Laplace's equation must be bounded at infinity.

(ii) The solution of the second boundary value problem is determined up to an arbitrary additive term.

(iii) Let the second boundary value problem in a closed domain $\bar{D}$ with piecewise smooth boundary ∂D be characterized by the condition $\frac{\partial u}{\partial n} = f(\mathbf{r})$ for $\mathbf{r} \in \partial D$, where $\frac{\partial u}{\partial n}$ is the derivative along the outward normal to ∂D. The necessary and sufficient condition for solvability of the problem is of the form: $\int_{\partial D} f(\mathbf{r})\, d\partial D = 0$.

Example 17.3. (First boundary value problem for $-\infty < x < \infty, 0 \le y < \infty$) The boundary condition on the half-plane is prescribed as $u(x,0) = f(x)$. The solution is

$$u(x,y) = \frac{1}{\pi}\int_{-\infty}^{\infty} \frac{y f(s)}{(x-s)^2 + y^2}\, ds = \frac{1}{\pi}\int_{-\pi/2}^{\pi/2} f(x + y\tan\theta)\, d\theta. \ \blacksquare$$

Example 17.4. (Second boundary value problem for $-\infty < x < \infty, 0 \le y < \infty$) The boundary condition on the half-plane is prescribed as $u_x(x,0) = f(x)$. The solution is

$$u(x,y) = \frac{1}{\pi}\int_{-\infty}^{\infty} f(s) \ln\sqrt{(x-s)^2 + y^2}\, ds + C,$$

where C is an arbitrary constant. ■

Example 17.5. The first boundary value problem in the first quadrant $0 \le x < \infty, 0 \le y < \infty$, with boundary conditions prescribed as $u(0,y) = f_1(y)$ and $u(x,0) = f_2(x)$, has the solution

$$\begin{aligned} u(x,y) &= \frac{4}{\pi} xy \int_0^{\infty} \frac{s f_1(s)}{[x^2 + (y-s)^2]\,[x^2 + (y+s)^2]}\, ds \\ &\quad + \frac{4}{\pi} xy \int_0^{\infty} \frac{s f_2(s)}{[(x-s)^2 + y^2]\,[(x+s)^2 + y^2]}\, ds. \ \blacksquare \end{aligned}$$

Example 17.6. The first boundary value problem in the infinite strip $0 < x < \infty, 0 \le y \le a$, with boundary conditions prescribed as $u(x,0) = h_1(x)$ and $u(x,a) = h_2(x)$, has the solution

$$\begin{aligned} u(x,y) &= \frac{1}{2a} \sin\left(\frac{\pi y}{a}\right) \int_{-\infty}^{\infty} \frac{h_1(s)}{\cosh[\pi(x-s)/a] - \cos(\pi y/a)}\, ds \\ &\quad + \frac{1}{2a} \sin\left(\frac{\pi y}{a}\right) \int_{-\infty}^{\infty} \frac{h_2(s)}{\cosh[\pi(x-s)/a] + \cos(\pi y/a)}\, ds. \end{aligned}$$

Similarly, the second boundary value problem in this infinite strip with boundary conditions prescribed as $u_x(x,0) = g_1(x)$ and $u_y(x,a) = g_2(x)$ has the solution

$$\begin{aligned} u(x,y) &= \frac{1}{2\pi}\int_{-\infty}^{\infty} g_1(s) \ln\{\cosh[\pi(x-s)/a] - \cos(\pi y/a)\}\, ds \\ &\quad - \frac{1}{2\pi}\int_{-\infty}^{\infty} g_2(s) \ln\{\cosh[\pi(x-s)/a] + \cos(\pi y/a)\}\, ds + C, \end{aligned}$$

where C is an arbitrary constant. ■

17.7.2 Cylindrical Polar Coordinates. The two-dimensional Laplace's equation in the cylindrical polar coordinates is

$$\nabla^2 u = \frac{\partial^2 u}{\partial r^2} + \frac{1}{r}\frac{\partial u}{\partial r} + \frac{1}{r^2}\frac{\partial^2 u}{\partial \theta^2}, \quad r = \sqrt{x^2 + y^2}. \tag{17.7.2}$$

Particular solutions are:

$$u(r) = A \ln r + B;$$

$$u(r,\theta) = \left(a\, r^m + \frac{B}{r^m}\right)(C\cos m\theta + D \sin m\theta),$$

where $m = 1, 2, \dots; A, B, C$, and D are arbitrary constants.

Example 17.7. The first boundary value problem in the domain $0 \le r \le R$ or $R \le r < \infty$ with the prescribed boundary condition $u(R, \theta) = f(\theta)$ has the solution

(a) for the inner problem $r \le R$:

$$u(r, \theta) = \frac{1}{2\pi} \int_0^{2\pi} f(s) \frac{R^2 - r^2}{r^2 - 2Rr\cos(\theta - s) + R^2}\, ds.$$

This solution is known as the *Poisson integral.*

(b) for the outer problem $r \ge R$:

$$u(r, \theta) = \frac{a_0}{2} \sum_{n=1}^{\infty} (a_n \cos n\theta + b_n \sin n\theta),$$

where

$$a_n = \frac{1}{\pi} \int_0^{2\pi} f(s) \cos(ns)\, ds, \quad n = 0, 1, 2, \dots,$$

$$b_n = \frac{1}{\pi} \int_0^{2\pi} f(s) \sin(ns)\, ds, \quad n = 1, 2, \dots.$$

The bounded solution for the outer problem $r \ge R$ is

$$u(r, \theta) = \frac{1}{2\pi} \int_0^{2\pi} f(s) \frac{r^2 - R^2}{r^2 - 2Rr\cos(\theta - s) + R^2}\, ds,$$

or in the series form

$$u(r, \theta) = \frac{a_0}{2} \sum_{n=1}^{\infty} \left(\frac{R}{r}\right)^n (a_n \cos n\theta + b_n \sin n\theta),$$

where the coefficients a_0, a_n, and b_n are the same as defined above. Note that in hydrodynamics and other applications, we sometimes encounter outer problems where we have to consider unbounded solutions as $r \to \infty$. ■

17.7.3 Transformation of Laplace's Equation. If a function $\phi(x, y)$ satisfies Laplace's equation at a point $z \in D$, it is analytic in D. Let $w = u + i\,v$ be the image of $z = x + i\,y$ under a conformal mapping. Then the partial derivatives $\dfrac{\partial u}{\partial x}, \dfrac{\partial u}{\partial v}, \dfrac{\partial v}{\partial x}$, and $\dfrac{\partial v}{\partial y}$ are defined at z. Since

$$\frac{\partial \phi(x, y)}{\partial x} = \frac{\partial \phi}{\partial u}\frac{\partial u}{\partial x} + \frac{\partial \phi}{\partial v}\frac{\partial v}{\partial x},$$

we have

$$\begin{aligned}\frac{\partial^2 \phi}{\partial x^2} &= \frac{\partial}{\partial x}\left[\frac{\partial \phi}{\partial u}\frac{\partial u}{\partial x} + \frac{\partial \phi}{\partial v}\frac{\partial v}{\partial x}\right] \\ &= \frac{\partial u}{\partial x}\frac{\partial}{\partial x}\left(\frac{\partial \phi}{\partial u}\right) + \frac{\partial \phi}{\partial u}\frac{\partial^2 u}{\partial x^2} + \frac{\partial v}{\partial x}\frac{\partial}{\partial x}\left(\frac{\partial \phi}{\partial v}\right) + \frac{\partial \phi}{\partial v}\frac{\partial^2 v}{\partial x^2}.\end{aligned} \tag{17.7.3}$$

Since

$$\frac{\partial}{\partial x}\Big(\frac{\partial \phi}{\partial u}\Big) = \frac{\partial}{\partial u}\Big(\frac{\partial \phi}{\partial u}\Big)\frac{\partial u}{\partial x} + \frac{\partial}{\partial v}\Big(\frac{\partial \phi}{\partial u}\Big)\frac{\partial v}{\partial x} = \frac{\partial^2 \phi}{\partial u^2}\frac{\partial u}{\partial x} + \frac{\partial^2 \phi}{\partial u \partial v}\frac{\partial v}{\partial x},$$

and

$$\frac{\partial}{\partial x}\Big(\frac{\partial \phi}{\partial v}\Big) = \frac{\partial}{\partial u}\Big(\frac{\partial \phi}{\partial v}\Big)\frac{\partial u}{\partial x} + \frac{\partial}{\partial v}\Big(\frac{\partial \phi}{\partial v}\Big)\frac{\partial v}{\partial x} = \frac{\partial^2 \phi}{\partial u \partial v}\frac{\partial u}{\partial x} + \frac{\partial^2 \phi}{\partial v^2}\frac{\partial v}{\partial x},$$

Eq (17.7.3) becomes

$$\frac{\partial^2 \phi}{\partial x^2} = \frac{\partial \phi}{\partial u}\frac{\partial^2 u}{\partial x^2} + \frac{\partial \phi}{\partial v}\frac{\partial^2 v}{\partial x^2} + \Big(\frac{\partial u}{\partial x}\Big)^2\frac{\partial^2 \phi}{\partial u^2} + \Big(\frac{\partial v}{\partial x}\Big)^2\frac{\partial^2 \phi}{\partial v^2} + 2\frac{\partial u}{\partial x}\frac{\partial v}{\partial x}\frac{\partial^2 \phi}{\partial u \partial v}. \tag{17.7.4}$$

If x is replaced by y, then

$$\frac{\partial^2 \phi}{\partial y^2} = \frac{\partial \phi}{\partial u}\frac{\partial^2 u}{\partial y^2} + \frac{\partial \phi}{\partial v}\frac{\partial^2 v}{\partial y^2} + \Big(\frac{\partial u}{\partial y}\Big)^2\frac{\partial^2 \phi}{\partial u^2} + \Big(\frac{\partial v}{\partial y}\Big)^2\frac{\partial^2 \phi}{\partial v^2} + 2\frac{\partial u}{\partial y}\frac{\partial v}{\partial y}\frac{\partial^2 \phi}{\partial u \partial v}. \tag{17.7.5}$$

Then combining (17.7.4) and (17.7.5) we obtain Laplace's equation

$$\begin{aligned}\frac{\partial^2 \phi}{\partial x^2} + \frac{\partial \phi}{\partial y} = 0 = & \frac{\partial \phi}{\partial u}\Big(\frac{\partial^2 u}{\partial x^2} + \frac{\partial^2 u}{\partial y^2}\Big) + \frac{\partial \phi}{\partial v}\Big(\frac{\partial^2 v}{\partial x^2} + \frac{\partial^2 v}{\partial y^2}\Big) \\ & + \frac{\partial^2 \phi}{\partial u^2}\Big[\Big(\frac{\partial u}{\partial x}\Big)^2 + \Big(\frac{\partial u}{\partial y}\Big)^2\Big] + \frac{\partial^2 \phi}{\partial v^2}\Big[\Big(\frac{\partial v}{\partial x}\Big)^2 + \Big(\frac{\partial v}{\partial y}\Big)^2\Big] \\ & + 2\frac{\partial^2 \phi}{\partial u \partial v}\Big[\frac{\partial u}{\partial x}\frac{\partial v}{\partial x} + \frac{\partial u}{\partial y}\frac{\partial v}{\partial y}\Big].\end{aligned} \tag{17.7.6}$$

Since $w = f(z)$ is analytic, both u and v satisfy Laplace's equation, i.e.,

$$\frac{\partial^2 u}{\partial x^2} + \frac{\partial^2 u}{\partial y^2} = \frac{\partial^2 v}{\partial x^2} + \frac{\partial^2 v}{\partial y^2} = 0.$$

Also, by the Cauchy-Riemann equations

$$\frac{\partial u}{\partial x}\frac{\partial v}{\partial x} + \frac{\partial u}{\partial y}\frac{\partial v}{\partial y} = 0, \quad \Big(\frac{\partial u}{\partial x}\Big)^2 + \Big(\frac{\partial u}{\partial x}\Big)^2 = \Big(\frac{\partial v}{\partial x}\Big)^2 + \Big(\frac{\partial v}{\partial x}\Big)^2.$$

Substituting these two equations into (17.7.6), we get

$$\Big[\Big(\frac{\partial v}{\partial x}\Big)^2 + \Big(\frac{\partial v}{\partial y}\Big)^2\Big]\Big[\Big(\frac{\partial u}{\partial x}\Big)^2 + \Big(\frac{\partial u}{\partial y}\Big)^2\Big] = \Big[\frac{\partial^2 \phi}{\partial u^2} + \frac{\partial^2 \phi}{\partial v^2}\Big]\Big[\Big(\frac{\partial v}{\partial x}\Big)^2 + \Big(\frac{\partial v}{\partial y}\Big)^2\Big] = 0, \tag{17.7.7}$$

where, in general, $\Big(\frac{\partial u}{\partial x}\Big)^2 + \Big(\frac{\partial u}{\partial y}\Big)^2 = \Big(\frac{\partial v}{\partial x}\Big)^2 + \Big(\frac{\partial v}{\partial y}\Big)^2 \neq 0$. Hence, Eq (17.7.7) leads to

$$\Big(\frac{\partial v}{\partial x}\Big)^2 + \Big(\frac{\partial v}{\partial y}\Big)^2 = \nabla^2_{u,v}\phi = 0. \tag{17.7.8}$$

17.8 Poisson's Equation

Poisson's equation

$$-\nabla^2 u(\mathbf{x}) = \Phi(\mathbf{x}), \quad \mathbf{x} = (x, y) \in \mathbb{R}^2, \tag{17.8.1}$$

is often encountered in heat and mass transfer theory, fluid mechanics, elasticity, electrostatics, and other areas of mechanics and physics. For example, it describes the steady-state temperature distribution in the presence of heat sources and sinks in the domain under study. Note that Laplace's equation is a special case of Poisson's equation with $\Phi = 0$.

Consider a finite domain $D \in \mathbb{R}^2$ with sufficiently smooth boundary. Let $\mathbf{x} = (x, y)$ and $\mathbf{x}' = (x', y')$, and $|\mathbf{x}-\mathbf{x}'| = (x-x')^2+(y-y')^2$. Then the solution of the *first boundary value problem* for Poisson's equation $-\nabla u(\mathbf{x}) = \Phi(\mathbf{x})$ in the domain D with the nonhomogeneous boundary condition $u = f(\mathbf{x})$ for $\mathbf{x} \in \partial D$ can be represented as

$$u(\mathbf{x}) = \int_D \Phi(\mathbf{x}')G(\mathbf{x}; \mathbf{x}')\, dx'\, dy' - \int_{\partial D} f(\mathbf{x}')\frac{\partial G}{\partial \mathbf{n}}\, dx'\, dy', \tag{17.8.2}$$

where $G(\mathbf{x}; \mathbf{x}')$ is the Green's function of the first boundary value problem, and $\dfrac{\partial G}{\partial \mathbf{n}}$ is the derivative of the Green's function with respect to $\mathbf{x}'$ along the outward normal $\mathbf{n}$ to the boundary ∂D. This Green's function G satisfies Laplace's equation in x, y in the domain D everywhere except the point (x', y'), at which G has a singularity of the form $\dfrac{1}{2\pi} \ln \dfrac{1}{|\mathbf{x}-\mathbf{x}'|}$, and satisfies the homogeneous boundary condition of the first kind with respect to x, y at the domain boundary, can be represented in the form

$$G(\mathbf{x}; \mathbf{x}') = \frac{1}{2\pi} \ln \frac{1}{|\mathbf{x}-\mathbf{x}'|} + g(\mathbf{x}, \mathbf{x}'), \tag{17.8.3}$$

where the auxiliary function $g(\mathbf{x}, \mathbf{x}')$ is determined by solving the first boundary value problem for Laplace's equation $\nabla^2 g = 0$ with the boundary condition $u\big|_{\partial D} = -\dfrac{1}{2\pi} \ln \dfrac{1}{|x-\mathbf{x}'|}$, where $\mathbf{x}'$ is treated as a two-dimensional free parameter. Note that $G(\mathbf{x}; \mathbf{x}') = G(\mathbf{x}', \mathbf{x})$. In polar coordinates we should set $\mathbf{x} = (r\cos\theta, r\sin\theta)$, $\mathbf{x}' = (r'\cos\theta', r'\sin\theta')$, $|\mathbf{x}-\mathbf{x}'|^2 = r^2 + r'^2 - 2rr'\cos(\theta-\theta')\, dr\, d\theta$.

The *second boundary value problem* for Poisson's equation (17.8.1) in the domain D is characterized by the boundary condition $\dfrac{\partial u}{\partial \mathbf{n}} = f(\mathbf{x})$ for $\mathbf{x} \in \partial D$. The solvability condition for this problem is

$$\int_D \Phi(\mathbf{x})\, dx\, dy + \int_{\partial D} f(\mathbf{x})\, dx'\, dy' = 0.$$

The solution of the second boundary value problem subject to this condition is

$$u(\mathbf{x}) = \int_D \Phi(\mathbf{x})G(\mathbf{x}; \mathbf{x}')\, dx\, dy + \int_{\partial D} f(\mathbf{x})G(\mathbf{x}; \mathbf{x}')\, dx'\, dy' + C, \tag{17.8.4}$$

where C is an arbitrary constant, and the Green's function G for this problem is determined by the following conditions: (i) the function G satisfies Laplace's equation in x, y in D everywhere except at the point (x', y'), at which at which G has a singularity of the form $\dfrac{1}{2\pi} \ln \dfrac{1}{|\mathbf{x}-\mathbf{x}'|}$, and (ii) satisfies the homogeneous boundary condition of the second kind with respect to x, y at the domain boundary, i.e., $\dfrac{\partial G}{\partial \mathbf{n}}\Big|_{\partial D} = 1/l$, where l is the length of the boundary. Green's function is unique up to an additive constant.

The *third boundary value problem* for Poisson's equation (17.8.1) in the domain D is characterized by the nonhomogeneous boundary condition $\dfrac{\partial u(\mathbf{x})}{\partial \mathbf{n}} + ku(\mathbf{x}) = f(\mathbf{x})$ for

$\mathbf{x} \in \partial D$, is given by formula (17.8.4) with $C = 0$, where $G = G(\mathbf{x}; \mathbf{x}')$ as its Green's function is determined by the following conditions: i) The function G satisfies Laplace's equation in x, y in D everywhere except at the point (x', y'), at which at which G has a singularity of the form $\frac{1}{2\pi} \ln \frac{1}{|\mathbf{x} - \mathbf{x}'|}$, and (ii) satisfies the homogeneous boundary condition of the third kind with respect to x, y at the domain boundary, i.e., $\left[\frac{\partial G}{\partial \mathbf{n}} + kG\right]_{\partial D} = 0$, where l is the length of the boundary. The Green's function can be represented by (7.8.3), where the auxiliary function g is obtained by solving Laplace's equation $\nabla^2 g(\mathbf{x}) = 0$ corresponding to the third boundary value problem. Note that Green's function is always symmetric with respect to its arguments, i.e., $G(\mathbf{x}; \mathbf{x}') = G(\mathbf{x}', \mathbf{x})$.

17.8.1 Particular Solutions of Poisson's Equation. We consider two cases of the right-hand side of Eq (17.8.1):

CASE 1. If $\Phi(\mathbf{x}) = \Phi(x, y) = \sum_{i=1}^{n} \sum_{j=1}^{n} a_{ij} \exp(B_i x + c_j y)$, then Poisson's equation has the solution of the form

$$u(x, y) = -\sum_{i=1}^{n} \sum_{j=1}^{n} \frac{a_{ij}}{b_i^2 + c_j^2} \exp(b_i x + c_j y).$$

CASE 2. If $\Phi(\mathbf{x}) = \Phi(x, y) = \sum_{i=1}^{n} \sum_{j=1}^{n} a_{ij} \sin(b_i x + p_i) \sin(c_j y + q_j)$, then Poisson's equation has the solution of the form

$$u(x, y) = -\sum_{i=1}^{n} \sum_{j=1}^{n} \frac{a_{ij}}{b_i^2 + c_j^2} \sin(b_i x + p_i) \sin(c_j y + q_j).$$

For different domains and boundary value problems the solutions are as follows.

(a) For the domain $-\infty < x, y < \infty$, the solution is

$$u(x, y) = \frac{1}{2\pi} \iint_{-\infty}^{\infty} \Phi(x', y') \ln \frac{1}{\sqrt{(x - x')^2 + (y - y')^2}}\, dx'\, dy'.$$

The first boundary value problem in this domain with the boundary condition $u(x, 0) = f(x)$ has the solution

$$u(x, y) = \frac{1}{\pi} \int_{-\infty}^{\infty} \frac{y f(x')\, dx'}{(x - x')^2 + y^2} + \frac{1}{2\pi} \iint_{-\infty}^{\infty} \Phi(x', y') \ln \frac{\sqrt{(x - x')^2 + (y + y')^2}}{\sqrt{(x - x')^2 + (y - y')^2}}\, dx'\, dy'.$$

The second boundary value problem in this domain with the boundary condition $u_y(x, 0) = f(x)$ has the solution

$$\begin{aligned} u(x, y) &= \frac{1}{\pi} \int_{-\infty}^{\infty} \ln \sqrt{(x - x')^2 + y^2}\, dx' \\ &+ \frac{1}{2\pi} \iint_{-\infty}^{\infty} \Phi(x', y') \Big[\ln \frac{1}{\sqrt{(x - x')^2 + (y + y')^2}} + \frac{1}{\sqrt{(x - x')^2 + (y - y')^2}} \Big]\, dx'\, dy' + C, \end{aligned}$$

where C is an arbitrary constant.

(b) For the horizontal strip $-\infty < x < \infty$, $0 \le y \le a$:

The first boundary value problem in this domain with the boundary conditions $u(x,0) = f_1(x)$ and $u(x,a) = f_2(x)$ has the solution

$$u(x,y) = \frac{1}{2a} \sin \frac{\pi y}{a} \int_{-\infty}^{\infty} \frac{f_1(s)}{\cosh[\pi(x-s)/a] - \cos(\pi y/a)}\, ds + \frac{1}{2a} \sin \frac{\pi y}{a} \int_{-\infty}^{\infty} \frac{f_2(s)}{\cosh[\pi(x-s)/a] + \cos(\pi y/a)}\, ds + \frac{1}{4\pi} \int_0^a \int_{-\infty}^{\infty} \frac{\cosh[\pi(x-s)/a] - \cos[\pi(y+y')/a]}{\cosh[\pi(x-s)/a] - \cos[\pi(y-y')/a]}\, ds.$$

The second boundary value problem in this domain with the boundary conditions $u_y(x,0) = f_1(x)$ and $u_y(x,a) = f_2(x)$ has the solution

$$u(x,y) = -\int_{-\infty}^{\infty} f_1(y')G(x,y;x',0)\,dy' + \int_{-\infty}^{\infty} f_2(y')G(x,y;x',a)\,dy' + \int_0^a \int_{-\infty}^{\infty} \Phi(x',y')G(x,y;x',y'),dx'\,dy' + C,$$

where the Green's function $G(x,y;x',y')$ is defined by

$$G(x,y;x',y') = \frac{1}{4\pi}\Big[\ln \frac{1}{\cosh[\pi(x-x')/a] - \cos[\pi(y-y')/a]} + \ln \frac{1}{\cosh[\pi(x-x')/a] - \cos[\pi(y+y')/a]}\Big]. \tag{17.8.5}$$

The mixed boundary value problem in this domain with the boundary conditions $u(x,0) = f_1(x)$ and $u_y(x,a) = f_2(x)$ has the solution

$$u(x,y) = \int_{-\infty}^{\infty} f_1(y')\Big[\frac{\partial}{\partial y'}G(x,y;x',y')\Big]_{y'=0} dy' + \int_{-\infty}^{\infty} f_2(y')G(x,y;x',a)\,dy' + \int_0^a \int_{-\infty}^{\infty} \Phi(x',y')G(x,y;x',y'),dx'\,dy' + C,$$

where

$$G(x,y;x',y') = \frac{1}{a}\sum_{n=0}^{\infty} \frac{1}{\mu_n} \exp(-\mu_n|x-x'|) \sin(\mu_n y) \sin(\mu_n y'),$$

and $\mu_n = \dfrac{(2n+1)\pi}{2a}$.

(c) For the semi-infinite strip in the first quadrant $0 \le x < \infty$, $0 \le y \le a$:

The first boundary value problem in this domain with the boundary conditions $u(x,0) = f_1(x)$, $u(x,0) = f_2(x)$, and $(x,a) = f_3(x)$ has the solution

$$u(x,y) = \int_0^a f_1(x')\Big[\frac{\partial}{\partial x'}G(x,x';y,y')\Big]_{x'=0} dy' + \int_0^a f_2(x')\Big[\frac{\partial}{\partial y'}G(x,x';y,y')\Big]_{y'=0} dy' \int_0^a f_3(x')\Big[\frac{\partial}{\partial y'}G(x,x';y,y')\Big]_{y'=a} dx' + \int_0^a \int_0^{\infty} \Phi(x',y')G(x,y;x',y'),dx'\,dy' + C,$$

where

$$G(x,y;x',y') = \frac{1}{4\pi}\Big[\ln\frac{\cosh[\pi(x-x')/a] - \cos[\pi(y+y')/a]}{\cosh[\pi(x-x')/a] - \cos[\pi(y-y')/a]} - \ln\frac{\cosh[\pi(x+x')/a] - \cos[\pi(y+y')/a]}{\cosh[\pi(x+x')/a] - \cos[\pi(y-y')/a]}\Big],$$

or in series form

$$G(x,y;x',y') = \frac{1}{a}\sum_{n=1}^{\infty}\frac{1}{q_n}\left[e^{-q_n|x-x'|} - e^{-q_n|x+x'|}\right]\sin(q_n y)\sin(q_n y'),$$

where $q_n = n\pi/a$.

(d) For the first quadrant $0 \le x < \infty$, $0 \le y < \infty$:

The first boundary value problem in this domain with the boundary conditions $u(0,y) = f_1(y)$, $u(x,0) = f_2(x)$, and $(x,a) = f_3(x)$ has the solution

$$u(x,y) = \frac{4xy}{\pi}\Big[\int_0^\infty \frac{y' f_1(y')\,dy'}{[x^2+(y-y')^2][x^2+(y+y')^2]} + \frac{x' f_2(x')\,dy'}{[(x-x')^2+y^2][(x+x')^2+y^2]}\Big] + \frac{1}{2\pi}\int_0^\infty\int_{-\infty}^\infty \Phi(x',y')\ln\frac{\sqrt{(x-x')^2+(y+y')^2}\sqrt{(x+x')^2+(y-y')^2}}{\sqrt{(x-x')^2+(y-y')^2}\sqrt{(x+x')^2+(y+y')^2}}.$$

17.8.2 Poisson's Equation in Polar Coordinates System. The two-dimensional Poisson's equation in polar coordinates is written as

$$\frac{1}{r}\frac{\partial}{\partial r}\Big(r\frac{\partial u}{\partial r}\Big) + \frac{1}{r^2}\frac{\partial^2 u}{\partial \theta^2} + \Phi(r,\theta) = 0, \quad r = \sqrt{x^2+y^2}. \tag{17.8.6}$$

The first boundary value problem in the circle $0 \le r \le R$, $0 \le \theta \le 2\pi$ with the boundary condition $u(R,\theta) = f(\theta)$ has the solution

$$u(r,\theta) = \frac{1}{2\pi}\int_0^{2\pi} f(\theta')\frac{R^2-r^2}{r^2-2Rr\cos(\theta-\theta')+R^2}\,d\theta' + \int_0^{2\pi}\int_0^R \Phi(r',\theta')G(r,\theta;r',\theta')\,r'\,dr'\,d\theta',$$

where

$$G(r,\theta;r',\theta') = \frac{1}{2\pi}\ln\frac{1}{|\mathbf{r}-\mathbf{r}'|} - \frac{1}{2\pi}\ln\frac{R}{r'|(R/r')^2\mathbf{r}'-\mathbf{r}|},$$

where $\mathbf{r} = (r,\theta), x = r\cos\theta, y = r\sin\theta; \mathbf{r}' = (x',y')), x' = r'\cos\theta', y' = r'\sin\theta'$. Then

$$G(r,\theta;r',\theta') = \frac{1}{4\pi}\ln\frac{r^2r'^2 - 2Rrr'\cos(\theta-\theta') + R^4}{R^2\left[r^2 - 2rr'\cos(\theta-\theta') + r'^2\right]}.$$

The first boundary value problem in the semi-circle $0 \le r \le R$, $0 \le \theta \le \pi$ with the boundary conditions $u(R,\theta) = f_1(\theta)$, $u(r,0) = f_2(r)$, and $u(r,\pi) = f_3(r)$ has the solution

$$u(r,\theta) = -R\int_0^\pi f_1(\theta')\Big[\frac{\partial}{\partial r'}G(r,\theta;r',\theta')\Big]_{r'=R}dr' + \int_0^R f_2(r')\frac{1}{r'}\Big[\frac{\partial}{\partial \theta'}G(r,\theta;r',\theta')\Big]_{\theta'=0}dr' - \int_0^R f_3(r')\frac{1}{r'}\Big[\frac{\partial}{\partial \theta'}G(r,\theta;r',\theta')\Big]_{\theta'=\pi}dr' + \int_0^\pi\int_0^R \Phi(r',\theta')G(r,\theta;r',\theta')\,r'\,dr'\,d\theta',$$

where

$$G(r,\theta;r',\theta') = \frac{1}{4\pi}\Big[\ln \frac{r^2r'^2 - 2Rrr'\cos(\theta-\theta') + R^4}{R^2\,[r^2 - 2rr'\cos(\theta-\theta') + r'^2]} - \ln \frac{r^2r'^2 - 2Rrr'\cos(\theta+\theta') + R^4}{R^2\,[r^2 - 2rr'\cos(\theta+\theta') + r'^2]}\Big].$$

The first boundary value problem in the circular sector $0 \le r \le R, 0 \le \theta \le \alpha$ with the boundary conditions $u(R,\theta) = f_1(\theta)$, $u(r,0) = f_2(r)$, and $u(r,\alpha) = f_3(r)$ has the solution

$$u(r,\theta) = -R\int_0^\alpha f_1(\theta')\Big[\frac{\partial}{\partial r'}G(r,\theta;r',\theta')\Big]_{r'=R} dr' + \int_0^R f_2(r')\frac{1}{r'}\Big[\frac{\partial}{\partial \theta'}G(r,\theta;r',\theta')\Big]_{\theta'=0} dr'$$
$$- \int_0^R f_3(r')\frac{1}{r'}\Big[\frac{\partial}{\partial \theta'}G(r,\theta;r',\theta')\Big]_{\theta'=\pi} dr' + \int_0^\alpha \int_0^R \Phi(r',\theta')G(r,\theta;r',\theta')\, r'\,dr'\,d\theta',$$

where

$$G(r,\theta;r',\theta') = \frac{1}{2\pi}\ln\Big\{\frac{|z^{\pi/\alpha} - \bar{\zeta}^{\pi/\alpha}||R^{2\pi/\alpha} - (\bar{\zeta}z)^{\pi/\alpha}|}{|z^{\pi/\alpha} - \zeta^{\pi/\alpha}||R^{2\pi/\alpha} - (\zeta z)^{\pi/\alpha}|}\Big\},$$

where $z = re^{i\theta}$, and $\zeta = r'e^{i\theta'}$.

17.8.3 Helmholtz Equation. Many problems related to steady-state oscillations (mechanical, acoustical, thermal, electromagnetic, and others) are described by the two-dimensional Helmholtz equation

$$-\left(\nabla^2 + k\right) u(\mathbf{x}) = \Phi(\mathbf{x}). \tag{17.8.7}$$

For $k < 0$, this equation describes mass transfer processes with first-order volume chemical reactions. Any elliptic equation with constant coefficients can reduce to the Helmholtz equation. This equation is called *homogeneous* if $\Phi = 0$ and *nonhomogeneous* if $\Phi \neq 0$.

A homogeneous boundary value problem is a boundary value problem for any Helmholtz equation with homogeneous boundary conditions; a particular solution of a homogeneous boundary value problem is the trivial solution $u \equiv 0$. The values k_n of the parameter k for which there are nontrivial solutions of the homogeneous boundary value problem are called *eigenvalues* and the corresponding solutions, $u = u_n$, are called the *eigenfunctions* of the boundary value problem, which have the following properties:

(i) There are infinitely many eigenvalues k_n; the set $\{k_n\}$ forms a discrete spectrum for the given boundary value problem.

(ii) All eigenvalues are positive, except for the eigenvalue $k_0 = 0$ which exists in the second boundary value problem, with the corresponding eigenfunction $u_0 = 0$; the eigenvalues occur in order of increasing magnitude so that $k_1 < k_2 < k_3 < \cdots$.

(iii) The eigenvalues $k_n \to \infty$ as $n \to \infty$. The following asymptotic estimate holds: $\lim\limits_{n\to\infty} \dfrac{n}{k_n} = \dfrac{S_2}{4\pi}$, where S_2 is the area of the two-dimensional domain under study.

(iv) The eigenfunctions $u_n = u_n(x,y)$ are defined up to a constant multiplier. In a domain D, any two eigenfunctions corresponding to different eigenvalues $k_n \neq k_m$ are orthogonal, i.e., $\int_D u_n\, u_m\, dD = 0$.

Any twice continuously differentiable functions $f = f(\mathbf{x})$ that satisfies the boundary conditions of a boundary value problem can be expanded into a uniformly convergent series

$$f(\mathbf{x}) = \sum_{n=1}^\infty f_n u_n, \quad \text{where } f_n = \frac{1}{\|u_n\|^2}\int_D f u_n\, dD, \quad \|u_n\|^2 = \int_D u_n^2\, dD.$$

If f is square-summable, then the series converges in the mean.

(v) The eigenvalues of the first boundary value problem do not increase if the domain is extended.

There are three possible cases for the nonhomogeneous Helmholtz equation with homogeneous boundary conditions:

CASE 1. If the parameter k is not equal to any one of the eigenvalues, then there exists the series solution

$$u = \sum_{n=1}^{\infty} \frac{A_n}{k_n - k} u_n, \quad A_n = \frac{1}{\|u_n\|^2} \int_D \Phi u_n \, dD, \quad \|u_n\|^2 = \int_D u_n^2 \, dD.$$

CASE 2. If k is equal to some eigenvalue $k = k_m$, then the solution of the nonhomogeneous problem exists only if the function Φ is orthogonal to u_n, i.e., $\int_D \Phi \, u_m \, dD = 0$. In this case the system is

$$u = \sum_{n=1}^{m-1} \frac{A_n}{k_n - k_m} u_n + \sum_{m+1}^{\infty} \frac{A_n}{k_n - k_m} u_n + C u_m, \quad A_n = \frac{1}{\|u_n\|^2} \int_D \Phi u_n \, dD, \quad \|u_n\|^2 = \int_D u_n^2 \, dD,$$

and C is an arbitrary constant.

CASE 3. If $k = k_m$ and $\int_D \Phi u_m \, dD \neq 0$, then the boundary value problem for the nonhomogeneous equation does not have a solution.

First boundary value problem. The solution of this problem for Eq (17.8.7) with the boundary condition $u(\mathbf{x}) = f(\mathbf{x})$ for $\mathbf{x} \in \partial D$ can be represented in the form

$$u(\mathbf{x}) = \int_D \Phi(\mathbf{x}) G(\mathbf{x}; \mathbf{x}') \, dx \, dy - \int_{\partial D} f(\mathbf{x}) \frac{\partial}{\partial \mathbf{n}} G(\mathbf{x}; \mathbf{x}') \, dx' \, dy',$$

where $\mathbf{x} = (x, y), \mathbf{x}' = (x', y') \in D$, and $\dfrac{\partial}{\partial \mathbf{n}}$ denotes the derivative along the outward normal to the boundary ∂D with respect to the variable x' and y'. The Green's function is given by

$$G(\mathbf{x}; \mathbf{x}') = \sum_{n=1}^{\infty} \frac{u_n(\mathbf{x}) u_n(\mathbf{x}')}{\|u_n\|^2 (k_n - k)}, \quad k \neq k_n, \tag{17.8.8}$$

where k_n and u_n are the eigenvalues and eigenfunctions of the first homogeneous boundary value problem.

Second boundary value problem. The solution of this problem for Eq (17.8.7) with the boundary condition $\dfrac{\partial u}{\partial \mathbf{n}}(\mathbf{x}) = f(\mathbf{x})$ for $\mathbf{x} \in \partial D$ can be represented in the form

$$u(\mathbf{x}) = \int_D \Phi(\mathbf{x}) G(\mathbf{x}; \mathbf{x}') \, dx \, dy + \int_{\partial D} f(\mathbf{x}) \frac{\partial}{\partial \mathbf{n}} G(\mathbf{x}; \mathbf{x}') \, dx' \, dy', \tag{17.8.9}$$

where the Green's function is given by

$$G(\mathbf{x}; \mathbf{x}') = -\frac{1}{k S_2} + \sum_{n=1}^{\infty} \frac{u_n(\mathbf{x}) u_n(\mathbf{x}')}{\|u_n\|^2 (k_n - k)}, \quad k \neq k_n,$$

where S_2 is the area of the domain D, and k_n and u_n are the positive eigenvalues and eigenfunctions of the second homogeneous boundary value problem.

Third boundary value problem. The solution of this problem for Eq (17.8.7) with the boundary condition $\frac{\partial u}{\partial \mathbf{n}}(\mathbf{x}) + ku(\mathbf{x}) = f(\mathbf{x})$ for $\mathbf{x} \in \partial D$ is given by (17.8.9), where the Green's function is defined by (17.8.8), where k_n and u_n are the positive eigenvalues and eigenfunctions of the third homogeneous boundary value problem.

Boundary conditions at infinity. Assuming that the function $\Phi(\mathbf{x})$ is finite and sufficiently rapidly decaying as $r \to \infty$, where $r = \sqrt{x^2 + y^2}$, the *boundary conditions at infinity in an infinite domain* are as follows:

(i) For $k < 0$, the vanishing condition of the solution at infinity is set as $\to 0$ as $r \to \infty$.

(ii) For $k > 0$, the radiation conditions, known as the *Sommerfeld conditions*,[1] at infinity are used, which are

$$\lim_{r\to\infty} \sqrt{r}\, u = \text{const}, \qquad \lim_{r\to\infty} \sqrt{r}\Big(\frac{\partial u}{\partial r} + i\sqrt{k}\, u\Big) = 0.$$

Particular solutions. The Helmholtz equation (17.8.7) has the following solutions:

$$\begin{aligned}
u(x, y) &= (Ax + B)(C \cos \mu y + D \sin \mu y), \quad k = \mu^2,\\
u(x, y) &= (Ax + B)(C \cosh \mu y + D \sinh \mu y), \quad k = -\mu^2,\\
u(x, y) &= (A \cos \mu x + B \sin \mu x)(Cy + D), \quad k = \mu^2,\\
u(x, y) &= (A \cosh \mu x + B \sinh \mu x)(Cy + D), \quad k = -\mu^2,\\
u(x, y) &= (A \cos \mu_1 x + B \sin \mu_1 x)(C \cos \mu_2 y + D \sin \mu_2 y), \quad k = \mu_1^2 + \mu_2^2,\\
u(x, y) &= (A \cos \mu_1 x + B \sin \mu_1 x)(C \cosh \mu_2 y + D \sinh \mu_2 y), \quad k = \mu_1^2 - \mu_2^2,\\
u(x, y) &= (A \cosh \mu_1 x + B \sinh \mu_1 x)(C \cos \mu_2 y + D \sin \mu_2 y), \quad k = -\mu_1^2 + \mu_2^2,\\
u(x, y) &= (A \cosh \mu_1 x + B \sinh \mu_1 x)(C \cosh \mu_2 y + D \sinh \mu_2 y), \quad k = -\mu_1^2 - \mu_2^2.
\end{aligned}$$

Fundamental solutions. They are

$$u^*(x, y) = \begin{cases} \dfrac{1}{2\pi} K_0(sr) & \text{if } k = -s^2 < 0,\\[2mm] \dfrac{i}{4} H_0^{(1)}(pr) & \text{if } k = p^2 > 0,\\[2mm] -\dfrac{i}{4} H_0^{(2)}(pr) & \text{if } k = p^2 > 0, \end{cases}$$

where $K_0(z)$ is the modified Bessel function of the second kind, $H_0^{(1)}$ and $H_0^{(2)}$ are the Hankel functions of the first and second kind of order zero. The leading term of the asymptotic expansion of these fundamental solutions, as $r \to 0$, is given by $\dfrac{1}{2\pi} \ln \dfrac{1}{r}$.

For different domains D, the solutions of the Helmholtz equation (17.8.7) are as follows:

[1] These conditions are: The sources must be sources, not sinks of energy. The energy which is radiated from sources must scatter to infinity; no energy may be radiated from infinity into the field.

(a) For the domain $-\infty < x, y < \infty$, the solution is

$$u(\mathbf{x}) = \begin{cases} \frac{1}{2\pi} \iint_{-\infty}^{\infty} \Phi(\mathbf{x}') K_0(sr), & \text{if } k = -s^2 < 0, \\ -\frac{i}{4} \iint_{-\infty}^{\infty} \Phi(\mathbf{x}') H_0^{(2)}(pr), & \text{if } k = p^2 > 0, \end{cases}$$

where $r = \sqrt{(x-x')^2 + (y-y')^2}$, and the Sommerfeld conditions are used to obtain the second solution.

(b) For the half-plane $-\infty < x < \infty, 0 \le y < \infty$, the solution of the first boundary value problem with the boundary condition $u(x, 0) = f(x)$ is

$$u(x,y) = \int_{-\infty}^{\infty} f(x') \Big[\frac{\partial}{\partial y'} G(x, y; x', y')\Big]_{y'=0} dy' + \int_0^{\infty} \int_{-\infty}^{\infty} \Phi(x', y') G(x, y; x', y')\, dx'\, dy',$$

where the Green's function is

$$G(x, y; x', y') = \begin{cases} \frac{1}{2\pi} [K_0(sr_1) - K_0(sr_2)] & \text{for } k = -s^2 < 0, \\ -\frac{i}{4} \Big[H_0^{(2)}(pr_1) - H_0^{(2)}(pr_2)\Big] & \text{for } k = p^2 > 0, \end{cases}$$

where $r_{1,2} = \sqrt{(x-x')^2 + (y \mp y')^2}$.

For this half planc, the solution of the second boundary value problem with the boundary condition $u_y(x, 0) = f(x)$ is

$$u(x, y) = -\int_{-\infty}^{\infty} f(x') G(x, y; x', 0)\, dy' + \int_0^{\infty} \int_{-\infty}^{\infty} \Phi(x', y') G(x, y; x', y')\, dx'\, dy',$$

where the Green's function is

$$G(x, y; x', y') = \begin{cases} \frac{1}{2\pi} [K_0(sr_1) + K_0(sr_2)] & \text{for } k = -s^2 < 0, \\ -\frac{i}{4} \Big[H_0^{(2)}(pr_1) + H_0^{(2)}(pr_2)\Big] & \text{for } k = p^2 > 0. \end{cases}$$

(c) For the first quadrant $0 \le x < \infty, 0 \le y < \infty$, the solution of the first boundary value problem with the boundary conditions $u(0, y) = f_1(y)$ and $u(x, 0) = f_2(x)$ is

$$u(x, y) = \int_0^{\infty} f_1(y') \Big[\frac{\partial}{\partial x'} G(x, y; x', y')\Big]_{x'=0} dy' + \int_0^{\infty} f_2(x') \Big[\frac{\partial}{\partial x'} G(x, y; x', y')\Big]_{y'=0} dy'$$
$$+ \int_0^{\infty} \int_0^{\infty} \Phi(x', y') G(x, y; x', y')\, dx'\, dy',$$

where the Green's function is

$$G(x, y; x', y') = \begin{cases} \frac{1}{2\pi} [K_0(sr_1) - K_0(sr_2) - K_0(sr_3) + K_0(sr_4)] & \text{for } k = -s^2 < 0, \\ -\frac{i}{4} \Big[H_0^{(2)}(pr_1) - H_0^{(2)}(pr_2) - H_0^{(2)}(pr_3) + H_0^{(2)}(rp_4)\Big] & \text{for } k = p^2 > 0, \end{cases}$$

where $r_{1,2} = \sqrt{(x-x')^2 + (y \mp y')^2}$, and $r_{3,4} = \sqrt{(x-x')^2 + (y \mp y')^2}$.

For this quadrant, the solution of the second boundary value problem with the boundary condition $u_x(0,y) = f(y)$, $u_y(x,0) = f_2(x)$ is

$$u(x,y) = -\int_0^\infty f_1(y')G(x,y;0,y')\,dy' - \int_0^\infty f_2(x')G(x,y;x',0)\,dx' + \int_0^\infty \int_0^\infty \Phi(x',y')G(x,y;x',y')\,dx'\,dy',$$

where the Green's function is

$$G(x,y;x',y') = \begin{cases} \frac{1}{2\pi}\left[K_0(sr_1) + K_0(sr_2) + K_0(sr_3) + K_0(sr_4)\right] & \text{for } k = -s^2 < 0, \\ -\frac{i}{4}\left[H_0^{(2)}(pr_1) + H_0^{(2)}(pr_2) + H_0^{(2)}(pr_3) + H_0^{(2)}(rp_4)\right] & \text{for } k = p^2 > 0, \end{cases}$$

(d) For the semi-infinite strip $-\infty < x < \infty, 0 \le y \le a$, the solution of the first boundary value problem with the boundary conditions $u(x,0) = f_1(x)$ and $u(x,a) = f_2(x)$ is

$$u(x,y) = \int_{-\infty}^\infty f_1(x')\Big[\frac{\partial}{\partial x'}G(x,y;x',y')\Big]_{y'=0} dy - \int_{-\infty}^\infty f_2(x')\Big[\frac{\partial}{\partial y'}G(x,y;x',y')\Big]_{y'=a} dx' + \int_0^\infty \int_{-\infty}^\infty \Phi(x',y')G(x,y;x',y')\,dx'\,dy',$$

where the Green's function is

$$G(x,y;x',y') = \frac{1}{2\pi}\sum_{n=-\infty}^{\infty}\left[K_0(sr_{n1}) - K_0(sr_{n2})\right],$$

where $r_{n1,n2} = \sqrt{(x-cx')^2 + (y \mp y' - 2na)^2}$, or

$$G(x,y;x',y') = \frac{1}{a}\sum_{n=1}^{\infty}\frac{1}{\beta_n}\exp\{-\beta_n|x-x'|\}\,\sin(q_n y)\sin(q_n y'),$$

where $\beta_n = \sqrt{q_n^2 - k}$, $q + -n = (\pi n)/a$.

For this semi-infinite strip, the solution of the second boundary value problem with the boundary condition $u_y(x,0) = f_1(x)$, $u_y(x,a) = f_2(x)$ is

$$u(x,y) = -\int_{-\infty}^\infty f_1(x')G(x,y;x',0)\,dy' + \int_{-\infty}^\infty f_2(x')G(x,y;x',0)\,dx' + \int_0^\infty \int_0^\infty \Phi(x',y')G(x,y;x',y')\,dx'\,dy',$$

where the Green's function is

$$G(x,y;x',y') = \frac{1}{2\pi}\sum_{n=-\infty}^{\infty}\left[K_0(sr_{n1}) + K_0(sr_{n2})\right],$$

or

$$G(x,y;x',y') = \frac{1}{2a}\sum_{n=1}^{\infty}\frac{1}{\beta_n}\exp\{-\beta_n|x-x'|\}\,\cos(q_n y)\cos(q_n y').$$

(e) For the rectangle $0 \le x \le a, 0 \le y \le b$, the solution of the first boundary value problem with the boundary conditions $u(0,y) = f_1(y)$, $u(a,y) = f_2(y)$, $u(x.0) = f_3(x)$, and $u(x,b) = f_4(x)$ is

$$\begin{aligned} u(x,y) &= \int_0^b f_1(y')\Big[\frac{\partial}{\partial x'}G(x,y;x',y')\Big]_{x'=0} dy - \int_0^b f_2(y')\Big[\frac{\partial}{\partial x'}G(x,y;x',y')\Big]_{x'=a} dy' \\ &= \int_0^a f_3(x')\Big[\frac{\partial}{\partial x'}G(x,y;x',y')\Big]_{y'=0} dx' - \int_0^a f_4(y')\Big[\frac{\partial}{\partial y'}G(x,y;x',y')\Big]_{y'=b} dy' \\ &+ \int_0^a \int_0^b \Phi(x',y')G(x,y;x',y')\,dx'\,dy', \quad \text{for } k \ne k_{mn}, \end{aligned}$$

where

$$k_{mn} = \pi^2\Big(\frac{n_2}{a^2} + \frac{m^2}{b^2}\Big), \quad n,m = 1,2,\dots,$$

$$u_{mn} = \sin\frac{n\pi x}{a}\sin\frac{m\pi y}{b}, \quad \|u_{mn}\|^2 = \frac{ab}{4},$$

are the eigenvalues, eigenfunctions, and the norm squared, respectively, and the Green's function is

$$G(x,y;x',y') = \frac{2}{a}\sum_{n=1}^{\infty}\frac{\sin(p_n x)\sin(p_n x')}{\beta_n\sinh(\beta_n b)}H_n(y,y') = \frac{2}{b}\sum_{n=1}^{\infty}\frac{\sin(q_m y)\sin(q_m y')}{\mu_m\sinh(\mu_m a)}Q_m(x,x'),$$

or

$$G(x,y;x',y') = \frac{4}{ab}\sum_{n=1}^{\infty}\sum_{m=1}^{\infty}\frac{\sin(p_n x)\sin(q_m y)\sin(p_n x')\sin(q_m y')}{p_n^2 + q_m^2 - k},$$

where

$$p_n = \frac{n\pi}{a}, ;q_m = \frac{m\pi}{b}, \quad \beta_n = \sqrt{p_n^2 - k}, \quad \mu_m = \sqrt{q_m^2 - k},$$

$$H_n(y,y') = \begin{cases} \sinh(\beta_n y')\sinh[\beta_n(b-y)] & \text{for } b \ge y > y' \ge 0, \\ \sinh(\beta_n y)\sinh[\beta_n(b-y')] & \text{for } b \ge y' > y \ge 0, \end{cases}$$

$$Q_m(x,x') = \begin{cases} \sinh(\mu_m x')\sinh[\mu_m(a-x)] & \text{for } a \ge x > x' \ge 0, \\ \sinh(\mu_m x)\sinh[\mu_m(a-x')] & \text{for } a \ge x' > x \ge 0, \end{cases}$$

For this rectangle, the solution of the second boundary value problem with the boundary condition $u_x(0,y) = f_1(y)$, $u_x(a,y) = f_2(y)$, $u_y(x,0) = f_3(x)$, and $u_y(x,b) = f_4(x)$ is

$$\begin{aligned} u(x,y) &= \int_0^a \int_0^b \Phi(x',y')G(x,y;x',y')\,dx'\,dy' \\ &- \int_0^b f_1(y')G(x,y;0,y')\,dy' + \int_0^b f_2(y')G(x,y;a,y')\,dy' \\ &- \int_0^a f_3(x')G(x,y;x',0)\,dx' + \int_0^a f_4(x')G(x,y;x',b)\,dx', \quad \text{for } k \ne k_{mn}. \end{aligned}$$

Polar coordinates system. In two-dimensional polar coordinates system, the Helmholtz equation (17.8.7) is written as

$$\frac{1}{r}\frac{\partial}{\partial r}\Big(r\frac{\partial u}{\partial r}\Big) + \frac{1}{r^2}\frac{\partial^2 u}{\partial\theta^2} + ku(r,\theta) = -\Phi(r,\theta), \quad r = \sqrt{x^2+y^2}.$$

Particular solutions of this equation for $\Phi \equiv 0$ are:

$$\begin{aligned}
u(r,\theta) &= [AJ_0(\mu r) + BY_0(\mu r)]\,(C\theta + D), \;\; k = \mu^2;\\
u(r,\theta) &= [AI_0(\mu r) + BK_0(\mu r)]\,(C\theta + D), \;\; k = -\mu^2;\\
u(r,\theta) &= [AJ_m(\mu r) + BY_m(\mu r)]\,(C\cos m\theta + D\sin m), \;\; k = \mu^2;\\
u(r,\theta) &= [AI_m(\mu r) + BK_m(\mu r)]\,(C\cos m\theta + D\sin m), \;\; k = -\mu^2,
\end{aligned}$$

where A, B, C, D are arbitrary constants, J_m and Y_m are Bessel functions, I_m and K_m are modified Bessel functions, and $m = 1, 2, \dots$.

17.8.4 Fourth-Order Stationary Equations. The biharmonic equation

$$\nabla^2\nabla^2 u \equiv \frac{\partial^4 u}{\partial x^4} + 2\frac{\partial^4 u}{\partial^2 x \partial^2 y} + \frac{\partial^4 u}{\partial y^4} = 0 \qquad (17.8.10)$$

is encountered in plane elasticity (where u is known as the *Airy stress function*); it is also used to describe slow flows of viscous incompressible fluids (in which case u is the *stream function*. This equation has particular solutions:

$$\begin{aligned}
u(x,y) &= Ax^3 + Bx^2y + Cxy^2 + Dy^3 + ax^2 + bxy + cy^2 + \alpha x + \beta y + \gamma,\\
u(x,y) &= (A\cosh\beta x + B\sinh\beta x + Cx\cosh\beta x + Dx\sinh\beta x)(a\cos\beta y + b\sin\beta y),\\
u(x,y) &= (A\cos\beta x + B\sin\beta x + Cx\cos\beta x + Dx\sin\beta x)(a\cosh\beta y + b\sinh\beta y),\\
u(x,y) &= Ar^2\ln r + Br^2 + C\ln r + D, \quad r = \sqrt{(x-a)^2 + (y-b)^2},\\
u(x,y) &= (Ax + By + c)(D\cosh\beta x + E\sinh\beta x)(a\cos\beta y + b\sin\beta y),\\
u(x,y) &= (Ax + By + c)(D\cosh\beta y + E\sinh\beta y)(a\cos\beta x + b\sin\beta x),\\
u(x,y) &= (x^2 + y^2)(D\cosh\beta x + E\sinh\beta x)(a\cos\beta y + b\sin\beta y),\\
u(x,y) &= (x^2 + y^2)(D\cosh\beta y + E\sinh\beta y)(a\cos\beta x + b\sin\beta x),
\end{aligned}$$

where $A, B, C, D, E, a, b, c, \alpha, \beta, \gamma$ are arbitrary constants. All solutions of Laplace's equation are also solutions of the biharmonic equation (17.8.10). Particular solutions of this equation in some orthogonal curvilinear coordinates system are as follows: (a) In polar coordinates (r, θ), where $x = r\cos\theta$ and $y = r\sin\theta$:

$$\begin{aligned}
u(r,\theta) &= \left(Ar^{2+k} + br^{2-k} + Cr^k + Dr^{-k}\right)\cos(k_n\theta),\\
u(r,\theta) &= \left(Ar^{2+k} + br^{2-k} + Cr^k + Dr^{-k}\right)\sin(k_n\theta),\\
u(r,\theta) &= Ar^2\ln r + Br^2 + C\ln r + D, \quad \text{at } k = 0.
\end{aligned}$$

(b) In bipolar coordinates (ξ, η), where $x = \dfrac{c\sinh\xi}{\cosh\xi - \cos\eta}, y = \dfrac{c\sin\eta}{\cosh\xi - \cos\eta}$:

$$\begin{aligned}
u(\xi,\eta) = \frac{a\cos k\eta + b\sin k\eta}{\cosh\xi - \cos\eta}&\\
\times \left[A\cosh(k+1)\xi + B\sinh(k+1)\xi + C\cosh(k-1)\xi + D\sinh(k-1)\xi\right].&
\end{aligned}$$

The fundamental solution is

$$u^*(x,y) = \frac{1}{8\pi} r^2 \ln r.$$

Other particular solutions in different domains are as follows:

(a) In the upper half-plane $-\infty < x < \infty, 0 \leq y < \infty$ with the boundary conditions $u(x,0) = 0$ and $u_y(x,0) = f(x)$, the solution is

$$u(x,y) = \int_{-\infty}^{\infty} f(x')G(x,x';y,0)\,dx', \quad G(x,y;0,0) = \frac{1}{\pi}\frac{y^2}{\pi^2+y^2}.$$

In the same domain with the boundary conditions $u_x(x,0) = f(x)$ and $u_y(x,0) = g(x)$, the solution is

$$u(x,y) = \frac{1}{\pi}\int_{-\infty}^{\infty} f(x')\Big[\arctan\Big(\frac{x-x'}{y}\Big) + \frac{(x-x')y}{(x-x')^2+y^2}\Big]\,dx' + \frac{y^2}{\pi}\int_{-\infty}^{\infty} \frac{g(x')}{(x-x')^2+y^2}\,dx' + C,$$

where C is an arbitrary constant.

17.9 Two-Dimensional Flows

In conformal mapping that maps a physical domain onto the model domain, we change the boundaries, the region, and generally the corresponding partial differential equation. Recall that conformal mappings do not solve physical problems by themselves; they help reduce the physical domain into the model domain in which the problem is easily solved. First, we select a suitable conformal map and then apply it to the physical domain, keeping in mind that the underlying physical process that initially provides the equation and the boundary conditions.

First, we will describe how Laplace's equation is transformed under a conformal map. Then we will show how Laplace's equation arises in certain potential field. Next, Poisson's equation, the wave equation, diffusion equation, and the equation for elastic deformation, together with their boundary conditions, are transformed under a conformal map.

A physical field is defined as the totality of a region in real space, every point of which is given by a set of coordinates, together with a single physical quantity ϕ, such as stress, strain, density, temperature, or electric potential, which may vary with time t.

17.9.1 Laplace's Equation under a Conformal Map. We will develop the relationship between the potential function $\phi(x,y)$ in the physical domain and the corresponding potential function $\psi(x,y)$ in the model domain, such that $\psi(u(x,y), v(x,y)) = \phi(x,y)$. Since the potential ϕ satisfies $\nabla^2\phi = 0$, we will show that $\nabla\psi = 0$ also. We have

Theorem 17.6. *Laplace's equation is invariant under conformal mapping.*

PROOF. Following Henrici [1974: Ch. 5], we set $\phi(x,y) = \psi(u,v)$. Then

$$\frac{\partial\phi}{\partial x} = \frac{\partial\psi}{\partial u}\frac{\partial u}{\partial x} + \frac{\partial\psi}{\partial v}\frac{\partial v}{\partial y}. \tag{17.9.1}$$

In the z and w planes, the complex gradients are $\nabla\phi = \dfrac{\partial\phi}{\partial x} + i\,\dfrac{\partial\phi}{\partial y}$ and $\nabla\psi = \dfrac{\partial\psi}{\partial u} + i\,\dfrac{\partial\psi}{\partial v}$.

Substituting the derivatives (17.9.1) and using the Cauchy-Riemann equations we get

$$
\begin{aligned}
\nabla\phi &= \Big(\frac{\partial\psi}{\partial u}\frac{\partial u}{\partial x} + \frac{\partial\psi}{\partial v}\frac{\partial v}{\partial y}\Big) + i\Big(\frac{\partial\psi}{\partial u}\frac{\partial u}{\partial x} + \frac{\partial\psi}{\partial v}\frac{\partial v}{\partial y}\Big) \\
&= \Big(\frac{\partial\psi}{\partial u}\frac{\partial u}{\partial x} + \frac{\partial\psi}{\partial v}\frac{\partial v}{\partial y}\Big) + i\Big(-\frac{\partial\psi}{\partial u}\frac{\partial u}{\partial x} + \frac{\partial\psi}{\partial v}\frac{\partial v}{\partial y}\Big) \\
&= \Big(\frac{\partial\psi}{\partial u} + i\frac{\partial\psi}{\partial v}\Big)\Big(\frac{\partial u}{\partial x} - i\frac{\partial v}{\partial x}\Big) \\
&= \nabla\psi\cdot\overline{f'(z)}.
\end{aligned} \tag{17.9.2}
$$

Next, we start with Eq (17.9.1), and get

$$
\frac{\partial^2\phi}{\partial x^2} = \frac{\partial^2\psi}{\partial u^2}\Big(\frac{\partial u}{\partial x}\Big)^2 + 2\frac{\partial^2\psi}{\partial u\partial v}\frac{\partial u}{\partial x}\frac{\partial v}{\partial x} + \frac{\partial^2\psi}{\partial v^2}\Big(\frac{\partial v}{\partial x}\Big)^2 + \frac{\partial\psi}{\partial u}\nabla^2 u + \frac{\partial\psi}{\partial v}\nabla^2 u. \tag{17.9.3}
$$

A similar equation for $\dfrac{\partial^2\phi}{\partial y^2}$ can be derived. Then adding these two equations, we get

$$
\begin{aligned}
\frac{\partial^2\phi}{\partial x^2} + \frac{\partial^2\phi}{\partial y^2} &= \frac{\partial^2\psi}{\partial u^2}\Big\{\Big(\frac{\partial u}{\partial x}\Big)^2 + \Big(\frac{\partial u}{\partial y}\Big)^2\Big\} + \frac{\partial^2\psi}{\partial v^2}\Big\{\Big(\frac{\partial v}{\partial x}\Big)^2 + \Big(\frac{\partial v}{\partial y}\Big)^2\Big\} \\
&\quad + 2\frac{\partial^2\psi}{\partial u\partial v}\Big\{\frac{\partial u}{\partial x}\frac{\partial v}{\partial x} + \frac{\partial u}{\partial y}\frac{\partial v}{\partial y}\Big\} + \frac{\partial\psi}{\partial v}\nabla^2 u + \frac{\partial\psi}{\partial v}\nabla^2 v \\
&= \frac{\partial^2\psi}{\partial u^2}|f'(z)|^2 + \frac{\partial^2\psi}{\partial v^2}|f'(z)|^2,
\end{aligned} \tag{17.9.4}
$$

thus yielding

$$
\nabla^2\psi(u,v) = \nabla^2\phi(x,y)|f'(z)|^2, \tag{17.9.5}
$$

where the remaining terms are zero because of the Cauchy-Riemann equations. Hence, $\nabla^2\psi = 0$ when $\nabla^2\phi = 0$, provided that $f'(z) \neq 0$, which is a condition satisfied by analytic functions at all points except their singular points. Note that $f'(z)$ in the above discussion can be replaced by $1/f'(w)$. ■

We have seen that potential field problems that include potential problems in two-dimensional steady-state fluid flows, heat flows, and electrostatics consists of two general types: the *Dirichlet boundary conditions*, which prescribe the potential on the boundary segment of the domain under study, and the *Neumann boundary conditions* which prescribe the potential gradient at the boundary segment, although some problems require a mix of both of these types.

In applications of conformal mappings, the physical plane in the z-plane is transformed into the mathematical *model plane* (w-plane). The Dirichlet conditions are directly transferable to the model plane since $w = f(z)$ is equivalent to the inverse mapping $z = f^{-1}(w)$. However, when transferring the Neumann condition we must apply the condition (17.9.2) concerning potential gradients, which requires the use of local derivatives unless the potential gradient at a boundary point is either zero (i.e., flow lines are parallel to the boundary) or a maximum (i.e., flow lines are normal to the boundary).

In the following chapters we will investigate the two-dimensional steady-state fluid flows, heat flows, and electrostatics under conformal transformations. The notation used in these flows is presented in Table 17.1.

Table 17.1

Fluid Flow	Heat Flow	Electrostatic
velocity potential $u(x, y)$	temperature $u(x, y)$	electrostatic potential $u(x, y)$
complex potential $f(z) = u(x, y) + iv(x, y)$	complex temperature $f(z) = u(x, y) + iv(x, y)$	complex potential $f(z) = u(x, y) + iv(x, y)$
streamlines $v(x, y) = c$	lines of heat flux $v(x, y) = c$	lines of force $v(x, y) = c$
equipotential lines $u(x, y) = c$	isotherms $u(x, y) = c$	equipotential lines $u(x, y) = c$
velocity vector $\mathbf{V} = \nabla u = \overline{f'(z)}$	heat flux $\mathbf{Q} = -k\nabla u = -k\overline{f'(z)}$	electric intensity $\mathbf{E} = -\nabla u = -\overline{f'(z)}$
at the boundary of an obstacle $v(x, y) = c$ $\dfrac{\partial u(x, y)}{\partial n} = 0$	along the boundary of insulation material $\dfrac{\partial u(x, y)}{\partial n} = 0$	at the boundary of a conductor $\dfrac{\partial u(x, y)}{\partial n} = 0$

Legend: k is the conductivity, and n is the outward normal to the boundary; $\mathbf{Q}$ is also called the *density of heat power flow*.

17.9.2 Sources and Sinks. Using the polar coordinates, the radial and tangential velocity components are defined by

$$V_r = \frac{\Lambda}{2\pi r}, \quad V_\theta = 0,$$

where Λ is a scaling factor known as the *source strength*. The volume flow rate per unit span $\dot{V}$ across a circle of radius r is given by

$$\dot{V} = \int_0^{2\pi} \mathbf{v} \cdot \mathbf{n}\, dA = \int_0^{2\pi} V_r\, r\, d\theta = \int_0^{2\pi} \frac{\Lambda}{2\pi r}\, r\, d\theta = \Lambda.$$

Thus, the source strength Λ specifies the rate of volume flow outward from the *source*. If $\Lambda < 0$, the flow is inward, and the flow is called a *sink*.

The origin $(0, 0)$ is called a *singular point*, or a *singularity*. As we approach this point, the magnitude of the radial velocity tends to infinity as $V_r \sim 1/r$. Hence, the flow at the singular point is not physically possible, although we can always use the source to represent actual flows.

We will consider three cases of location of sources and sinks.

(i) Consider the conformal map $w = z^2$ which maps the boundary C of a rectangular domain D in the z-plane onto the upper-half of the w-plane, (Figure 17.2)[2] where the rectangular coordinates system (x, y) is used to represent the source (I) and the sink (II)

[2] This kind of mapping can also be carried by Schwarz-Christoffel transformations (Chapter 7).

along the x- and y-axis, respectively, in the first quadrant of the z-plane and their images in the first and second quarter of the w-plane (Figure 16(a)). However, if the source (I) and sink (II) are taken along the y- and x-axis, respectively, as in Figure 16(b), then the polar coordinates system (r, θ) is used and the mapping is $w = 2i \log z$, which is a composite of $T = z^2$ from the z-plane onto the T-plane followed by the mapping $w = i \log T$ onto the w-plane.

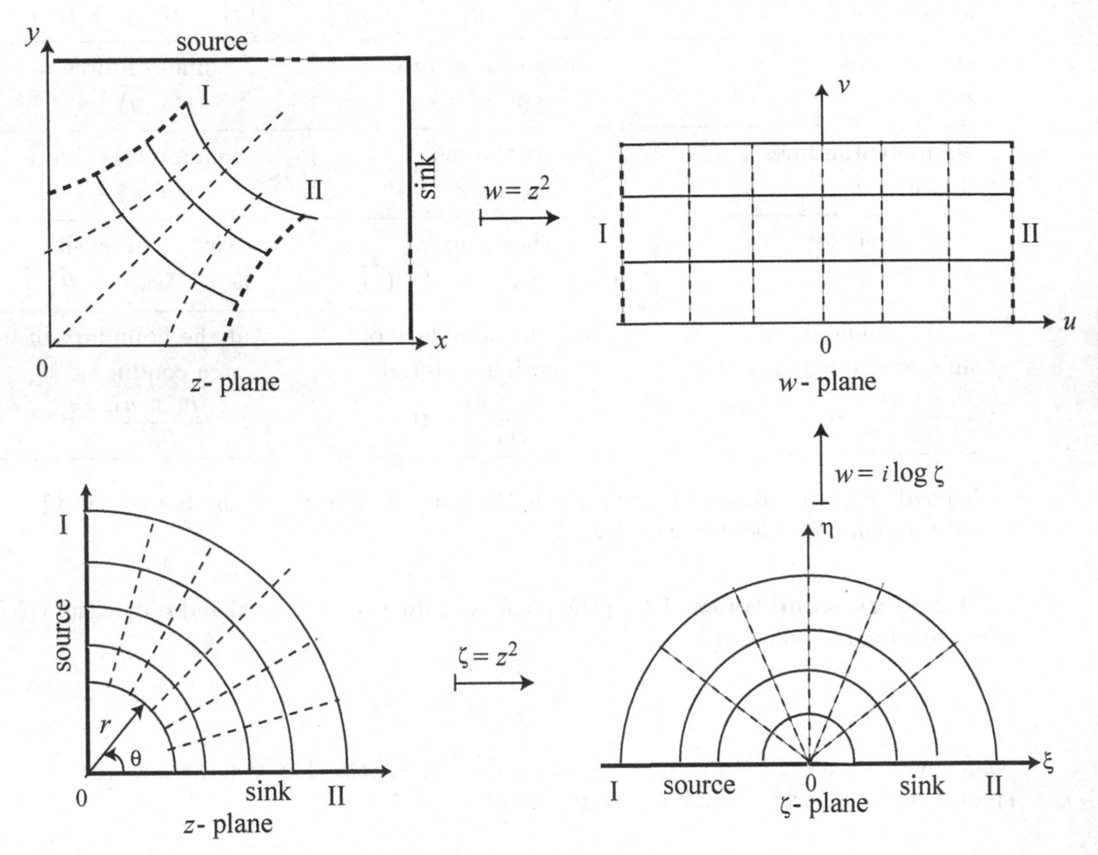

Figure 17.2 Sources and sinks along coordinate axes.

(ii) Consider the case where the entire rectangular boundary of the first quadrant, made up of x- and y-axes, in the z-plane represents the sink (II), and the source (I) is located at an interior point z_p in the first quadrant. The conformal transformation $w = z^2$ maps the sink boundary (II) on to the u-axis and point-source (I) at z_p onto the point w_p on the v-axis ($v \neq 0$) (Figure 17.3(b)), or the point-source (I) at an interior point w_p in the first quadrant of the w-plane and the sink (II) onto the point $\bar{w}_p$ in the fourth quadrant of the w-plane (Figure 17.3(c)). In this case, the field in the w-plane can be analyzed using a mirror image of the source (I). Since the source (I) and the sink (II) are now located on the vertical line in complex conjugate position, the ground plane is merely a line of symmetry as shown in Figure 17.3(c).

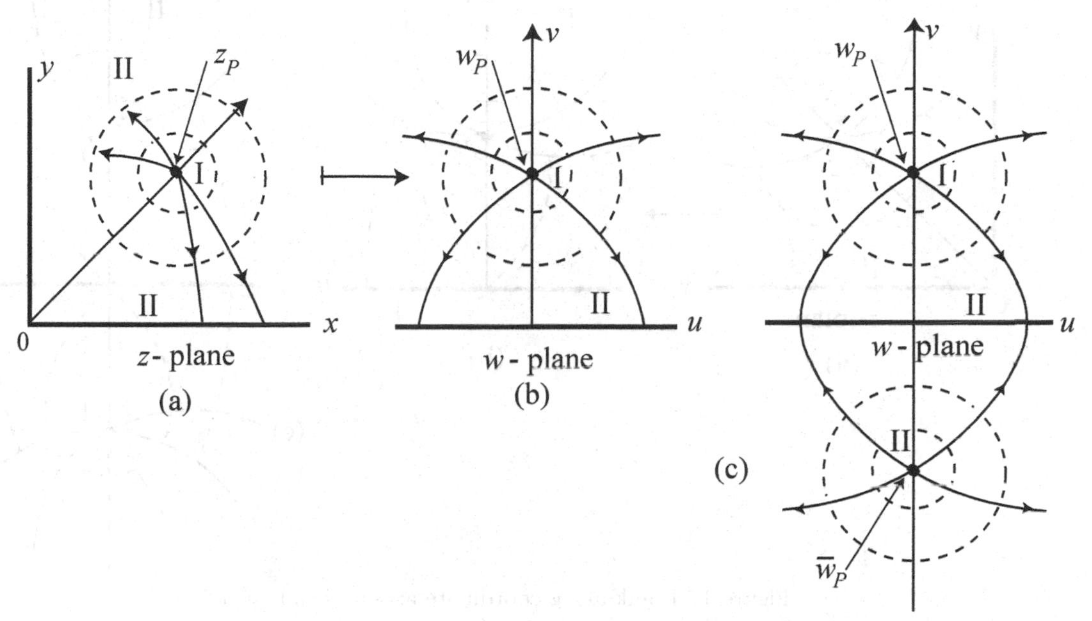

Figure 17.3 Sink along coordinates axes and sink at an interior point.

Then the potential distribution for a longitudinal source strength Q in a medium of permittivity k is given by the complex potential

$$\psi_c = \frac{Q}{2\pi k} \log \frac{z^2 - z_p^2}{z^2 - \bar{z}_p^2}. \tag{17.9.6}$$

For example, the potential due to two small parallel wires with charge $+Q$ and $-Q$, respectively, is given by the real part of Eq (17.9.6):

$$\psi = \frac{Q}{2\pi k} \log \frac{r_1}{r_2} = \frac{Q}{2\pi k} \log \frac{(u - u_p)^2 + (v - v_p)^2}{(u - u_p)^2 + (v - v_p)^2}. \tag{17.9.7}$$

(iii) In the case when the source (I) is located at both the x- and y-axis in the first quadrant of the z-plane, and the sink is located at the point at infinity, then the conformal mapping $w = z^2$ maps the source (I) onto the u-axis and the sink onto a point $v \neq 0$ (Figure 17.3(b)); also, the source (I) is mapped onto the u-axis and the sink is mapped onto a point $v \neq 0$ and its complex conjugate in the fourth quadrant (Figure 17.3(c)).

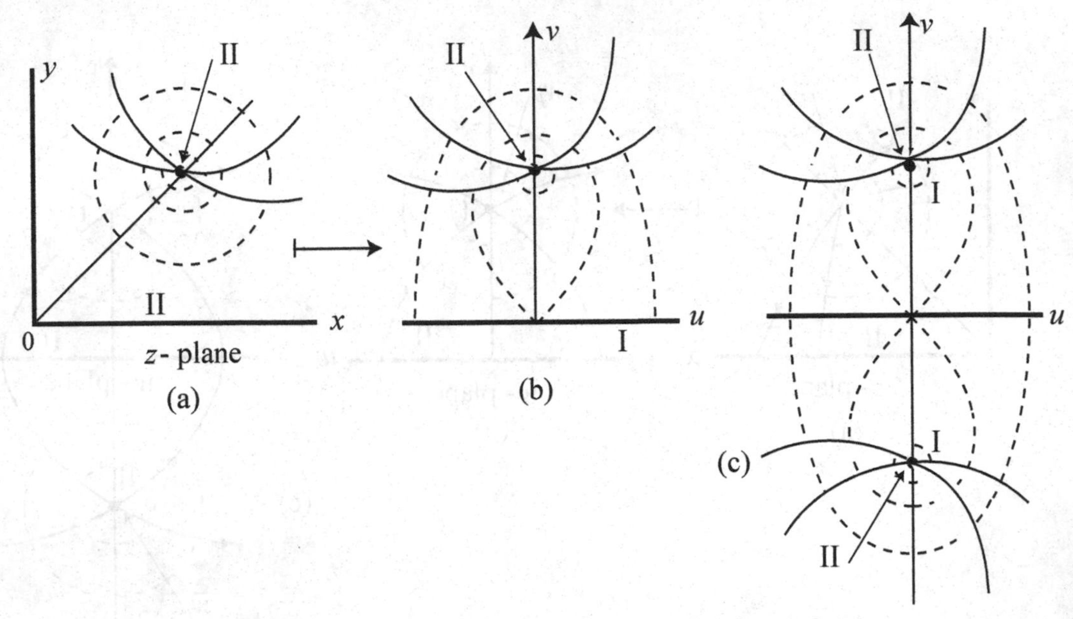

Figure 17.4 Sink along coordinate axes and sink at infinity.

REFERENCES USED: Boyce and DiPrima [1992], Carslaw and Jaeger [1984], Henrici [1974], Kythe [1995; 1996; 2011], Kythe et al. [2003]; Polyanin snd Nazaikinskii [2016], Ross [1964], Schinzinger and Laura [2003].

18

Fluid Flows

Certain applications of conformal mapping in the area of different types of two-dimensional fluid flows are discussed.

18.1 Viscous Laminar Flows

In laminar flows the fluid moves in layers, in contrast to the chaotic motion of a turbulent flow. We will consider internal laminar flows and solve the flow equation for velocity and pressure. Internal flows are laminar flows that occur within a channel or pipe. The fluid contained therein is of finite dimensions and confined by the channel or pipe walls. At the entry region the fluid develops a boundary layer near the channel or pipe walls, while the remaining central portion of the fluid may remain a uniform flow (Figure 18.1).

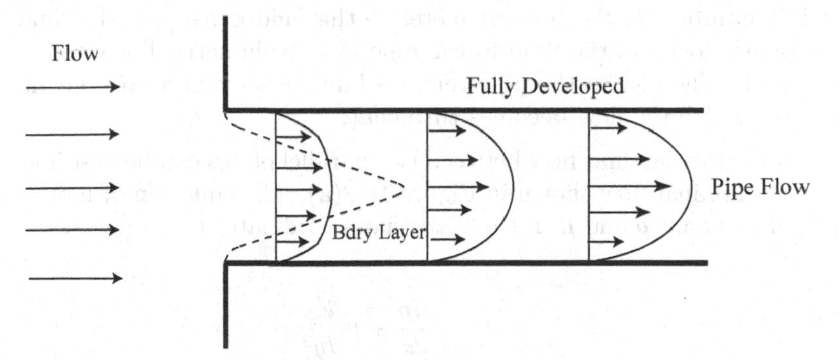

Figure 18.1 Internal pipe flow.

Recall that a uniform flow consisting of a velocity field $\mathbf{v} = u\mathbf{i} + v\mathbf{j}$ is constant. This velocity field is defined by free-stream speed V_0 and the flow angle α, such that

$$u = V_0 \cos\alpha, \quad v = V_0 \sin\alpha.$$

The corresponding potential and stream functions are

$$\phi(x,y) = V_0(x\cos\alpha + y\sin\alpha), \quad \psi(x,y) = V_0(y\cos\alpha - x\sin\alpha).$$

The uniform flow has zero divergence, i.e., $\nabla \cdot \mathbf{v} = 0$, since $\mathbf{v}$ is constant. This also means that $\phi(x,y)$ satisfies Laplace's equation, i.e., $\nabla^2\phi(x,y) = 0$. The uniform flow is *irrotational*, i.e., it has zero vorticity and zero circulation; thus,

$$\nabla \times \mathbf{v} = \mathbf{0},$$

or the equivalent irrotationality condition that $\psi(x,y)$ satisfies Laplace's equation, i.e., $\nabla^2\psi(x,y) = 0$.

Within the boundary layer, viscous stresses are very dominant, which slow down the fluid due to its friction with the walls. This slowdown propagates away from the walls; it happens thusly: as the fluid enters the channel, the fluid particles immediately next to the boundary to the walls are slowed down; these particles then interact viscously and slow down those in the second layer from the walls, and this process continues, resulting in the thickening of the boundary layers down the stream. Eventually the central core is eliminated and the velocity takes some average profile across the channel which is no longer affected by any edge effects that arise from the entrance region. This makes the flow independent of what happened at the channel entrance, and then we cannot solve the equation, for example, for velocity profile without including an entrance region in the calculations. At this stage the flow is said to be *fully developed*.

For example, in the case of pipe flow, Boussinesq [1867, 1868] has estimated that a flow can become fully developed over a distance $x_L \approx 0.03\,Re\,D$, where $Re = \rho V D/\mu$ is the Reynold's number, D the pipe diameter, ρ the fluid density, μ the fluid viscosity, and V the average velocity of the fluid in the pipe (V = volumetric flow-rate/cross-sectional area of the pipe). The flow in the pipe remains laminar so long as Re remains less than about 2300, after which the flow becomes turbulent.

A steady-state laminar flow between two parallel plates can be described, for example, by an internal laminar flow shown in Figure 18.2(a). The equation of motion for a Newtonian fluid with constant ρ and μ in the Cartesian coordinates (x,y) is

$$-\frac{dp}{dx} + \mu\frac{d^2v}{dy^2} = 0, \tag{18.1.1}$$

where v denotes the velocity in the x-direction (of the flow), and the body forces are assumed to be orthogonal to the direction of the flow, and thus, do not appear in Eq (18.1.1). Applying the no-slip boundary condition $v(x,h) = 0 = v(x,-h)$, and the condition that $dv/dy = 0$ at the midpoint of the distance $2h$ between the plates, we have

$$v(x,y) = \frac{1}{2\mu}\frac{dp}{dx}\left(y^2 - h^2\right), \tag{18.1.2}$$

which shows that the velocity profile is parabolic as in Figure 18.2(a). The flow is entirely driven by a pressure gradient. Thus, $v = 0$ if $dp/dx = 0$, and there is no flow.

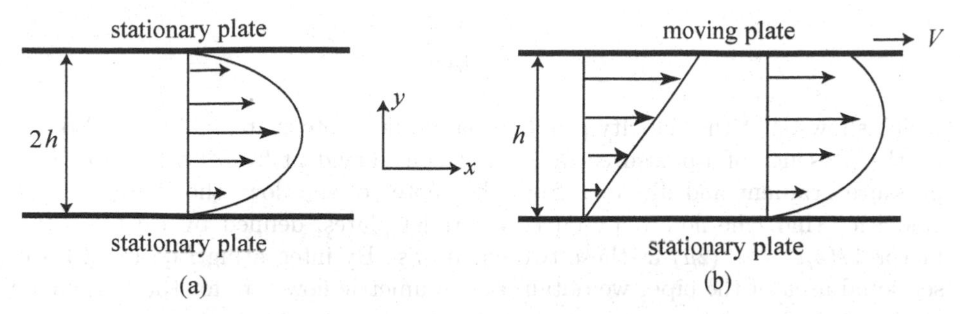

Figure 18.2 Flow between two plates: (a) stationary plates; (b) Couette flow.

The *Couette flow* is described as a steady-state laminar flow between two parallel plates where one plate is stationary and the other is moving, i.e., they are in relative motion (Figure 18.2(b)). Suppose that the upper plate moves at a speed V relative to the other. Although the pressure gradient is assumed to be constant, a nonzero pressure gradient in the direction of the flow may, in general, be present. In the absence of the body forces, the Newtonian fluid with constant ρ and μ is the same as Eq (18.1.1), but the new boundary conditions are $v(x, h) = V$ and $v(x, 0) = 0$. Thus, the velocity profile is given by

$$v(x, y) = \frac{1}{2\mu}\frac{dp}{dx}\left(y^2 - h^2 y\right) + \frac{Vy}{h}, \tag{18.1.3}$$

which shows that the velocity profile is a combination of parabolic and linear flows, where the parabolic part is entirely due to the presence of the pressure gradient, and the linear part is due to the motion of the upper plate moving at velocity V. Integrating Eq (8.1.3) across the gap between the plates, we find that the volumetric flow rate Q' per unit depth of the flow as

$$Q' = \int_0^h v\, dy = -\frac{h^3}{12\mu}\frac{dp}{dx} + \frac{1}{2}Vh. \tag{18.1.4}$$

The *Poiseuille flow* is a steady-state laminar flow in a cylindrical pipe, shown in Figure 18.3. The equation of motion for a Newtonian fluid with constant ρ and μ, with no body forces along the z-direction is defined in polar cylindrical coordinates by

$$-\frac{dp}{dz} + \frac{\mu}{r}\frac{d}{dr}\left(\frac{r dv_z}{dr}\right) = 0. \tag{18.1.5}$$

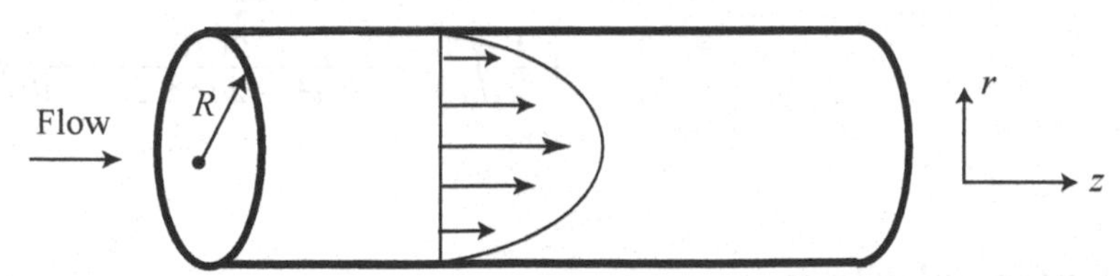

Figure 18.3 Poiseuille flow.

Assuming that the pressure gradient is constant, the boundary conditions are $v(R, z) = 0$ and $\frac{dv_z}{dr}(0, z) = 0$ since the flow is symmetric about the r-axis and therefore, the slope of

v_z with respect to r at $r = 0$ must be zero, thus giving the solution as

$$v_z = \frac{1}{4\mu}\frac{dp}{dz}\left(r^2 - z^2\right), \tag{18.1.6}$$

which shows that the velocity profile is parabolic. Note that this flow is also entirely driven by the presence of a pressure gradient. It is also evident from Eq (18.1.6) that for a given pressure gradient and distance from the center of the flow, the Poiseuille flow has lower velocities than the flow between two parallel plates, defined by Eq (18.1.2) because the factor $1/(4\mu) < 1/(2\mu)$ in these two equations. By integrating Eq (18.1.6) over the cross-sectional area of the pipe, we obtain the volumetric flow rate for the Poiseuille flow as

$$Q' = \int_0^R \int_0^{2\pi} v_z \, r \, d\theta \, dr = -\frac{\pi R^4}{8\mu}\frac{dp}{dz}. \tag{18.1.7}$$

Example 18.1. Consider the problem of a slow Stokes inflow of a viscous fluid into a half-plane through a slit of width $2a$ at a velocity U that makes an angle β with the normal to the boundary, where the angle is measured off from the normal direction counterclockwise. The stream function u is introduced by the relations $v_x = -\dfrac{\partial u}{\partial y}$ and $v_y = -\dfrac{\partial u}{\partial x}$, where v_x and v_y are the fluid velocity components. Then the problem is reduced to the special case of the above problem with

$$f(x) = \begin{cases} U\cos\beta & \text{for } |x| < a, \\ 0 & \text{for } |x| > a, \end{cases} \qquad g(x) = \begin{cases} U\sin\beta & \text{for } |x| < a, \\ 0 & \text{for } |x| > a, \end{cases}$$

and the solution is

$$\begin{aligned} u(x,y) &= \frac{U}{\pi}\left[(x-a)\cos\beta + y\sin\beta\right]\arctan\left(\frac{y}{x-a}\right) \\ &\quad - \frac{U}{\pi}\left[(x+a)\cos\beta + y\sin\beta\right]\arctan\left(\frac{y}{x+a}\right). \end{aligned}$$

On the disk $0 \le r \le a$, $0 \le \theta \le 2\pi$, with the boundary conditions (in polar coordinates) $u(a,\theta) = f(\theta)$, $u_r(a,\theta) = g(\theta)$, the solution is

$$\begin{aligned} u(r,\theta) = \frac{1}{2\pi a}(r^2 - a^2)\Big\{ &\int_0^{2\pi} \frac{[a - r\cos(\eta-\theta)]\, f(\eta)}{r^2 + a^2 - 2ar\cos(\eta-\theta)}\Big\} d\eta \\ &+ \frac{1}{2}\int_0^{2\pi} \frac{g(\eta)}{r^2 + a^2 - 2ar\cos(\eta-\theta)}\, d\eta\Big\}. \ \blacksquare \end{aligned}$$

18.2 External Flows

External flow occurs when the fluid flows over an object. There is always a perturbation in the flow because of its interaction with the object in its path. For example, a flow of fluid around a sphere, called the *creeping flow* (because of very small Re since the convective transfer of momentum is negligible compared to the viscous transfer of the momentum), presented in Figure 18.4, is an external flow. Although it is difficult to determine the general solution of flow around a sphere, an analytical solution of a creeping flow is possible.

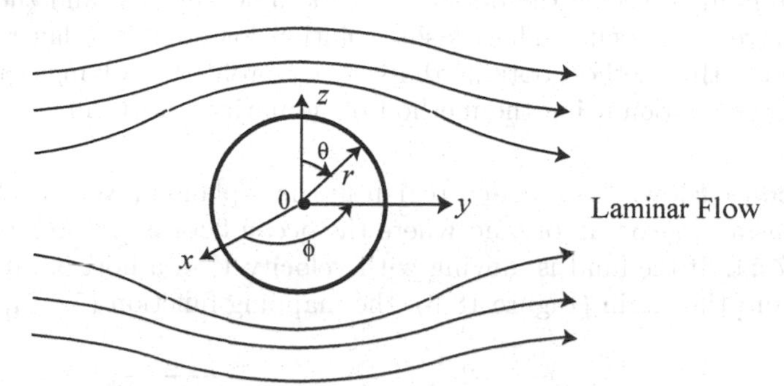

Figure 18.4 Flow around a sphere.

18.2.1 Creeping Flow around a Sphere. A well-known solution of the creeping flow around a sphere is that the total force F_0 on the sphere as a result of the fluid flowing around is is given by

$$F_0 = 6\pi\mu R V_0, \tag{18.2.1}$$

where μ is the fluid viscosity, R the radius of the sphere, and V_0 is the free stream velocity of the fluid flow. Note that Eq (18.2.1) consists of two parts, the first of which is known as the *form drag* F_p, given by $F_p = 2\pi\mu R V_0$, which arises because the normal stress on the backside (downstream) of the sphere is less than the normal stress on the front side (upstream) of the sphere. Note that the form drag arises from changes in the normal stress due to the motion of the fluid around the *form* of the body, and it does not arise from gravity or some other body force. The second part of Eq (18.2.1), known as *skin* or *friction* and denoted by F_f is given by $F_f = 4\pi\mu R V_0$, and it is due to the force exerted on the sphere by viscous shear stresses at the sphere-fluid interface.

18.2.2 Boundary Layer Flow over a Flat Plate. This flow is presented in Figure 18.5. The viscous effects may be negligible in the potential flow region, even if the fluid viscosity is considerable; this is because the velocity gradients are so small that there is practically no friction between different parts of the fluid.

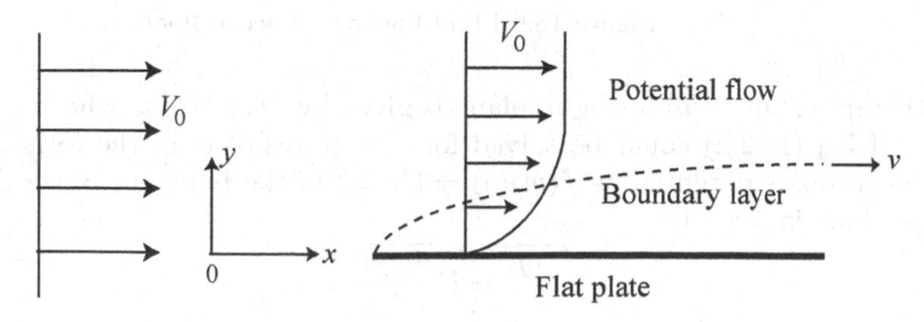

Figure 18.5 Boundary layer flow over a flat plate.

The precise location of the boundary layer is subject to convention, but the boundary layer thickness is defined by requiring that the fluid velocity $v \approx 0.99V_0$ at the boundary layer, where V_0 is the free-stream velocity. At distances far from the surface of the plate the flow is regarded as potential flow, whereas as closer to the plate the flow is regarded as boundary layer flow. If the pressure decreases in the direction of the flow, i.e., if $dp/dx < 0$, then the boundary layer thickness grows more slowly than the case when the pressure gradient is opposite to the direction of the flow, i.e., when $dp/dx > 0$. In this, and in such a case it may reverse the direction of the fluid velocity, and then a *separation* of the boundary layer will occur. The Blasius solution for boundary layer flow over a flat plate is well known; this method obtains the velocity profile by solving equations of momentum and mass conservation using the method of similarity transform.

18.2.3 Ocean Flow. The shaded region in the z-plane in Map 7.62 (Figure 7.55) can be visualized as a large ocean of fluid where the ocean floor is the step of height one as shown in Figure 7.54. If the fluid is moving with velocity V in a horizontal direction from left to right far from the origin (Figure 18.6), the mapping function (7.7.7), i.e.,

$$w = \frac{2i}{\pi}\left(\arcsin\sqrt{z} + \sqrt{z(1-z)}\right), \tag{18.2.2}$$

must be computed at $z = i$. Thus,

$$\begin{aligned} w &= \frac{2i}{\pi}\left(\arcsin\sqrt{i} + \sqrt{i(1-i)}\right) = \frac{2i}{\pi}\left[\arcsin\left(e^{i\pi/4}\right) + \sqrt{1+i}\right] \\ &\approx \frac{2i}{\pi}\left[(0.57 + 0.75\,i) + (1.1 + 0.46\,i\right] \approx -0.77 + 1.06\,i, \end{aligned}$$

where $\arcsin\left(e^{i\pi/4}\right) \approx 0.57 + 0.75\,i$, and $\sqrt{1+i} \approx 1.1 + 0.46\,i$. Hence, the point $z = i$ maps onto the point $w \approx -0.77 + 1.06\,i$.

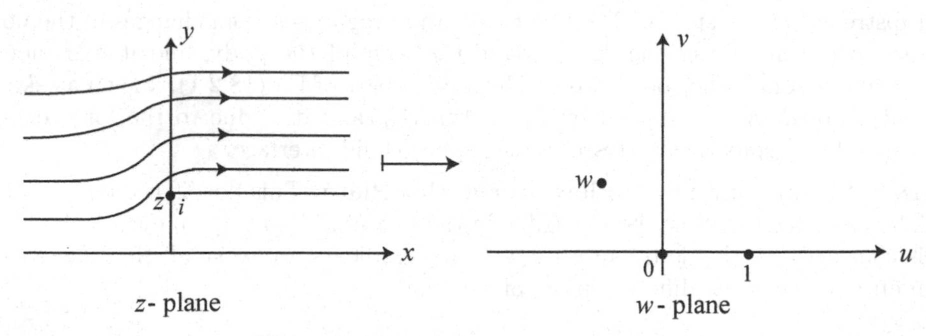

Figure 18.6 Fluid flow on an ocean floor.

The complex potential in the w-plane is given by $f(z) = U\,z$, where U is the velocity at infinity. If Eq (18.2.2) could be solved for z in terms of w in the form $z = g(w)$, we can write the complex potential as $f(g(w)) = Ug(w)$ in the w-plane. Since the velocity in the in the w-plane is given by

$$\overline{\frac{df}{dw}} = U\overline{\frac{dg(w)}{dw}} = U\overline{\frac{dz}{dw}}.$$

From (18.2.2) we have

$$\frac{dz}{dw} = \frac{\pi}{2i}\sqrt{\frac{z}{1-z}}. \tag{18.2.3}$$

To determine U such that when w is near ∞, the velocity is V. But when w is near ∞, z is also near ∞. Thus, since $\lim\limits_{z\to\infty}\sqrt{\dfrac{z}{1-z}} = \sqrt{-1} = i$, the velocity near infinity is equal to $\frac{\pi i}{2}\, i\, U = -\frac{\pi}{2} U$. Thus, we can take $U = -\frac{2V}{\pi}$, and then the velocity is $-V\, i \sqrt{\frac{\bar{z}}{1-\bar{z}}}$. Hence, the velocity at the point $w = -0.77 + 1.06\, i$, which is the image of the point $z = i$, is given by substituting the preimage $z = i$ of this point in the above expression, i.e.,

$$\text{velocity} = -V\, i \sqrt{\frac{-i}{1+i}} = V \sqrt{\frac{1+i}{2}} = V(1.18 + 0.86\, i),$$

and so the velocity vector has magnitude $1.46V$ and makes an angle of $\pi/5$ with the u-axis.

18.3 Ideal Fluid Flows

Consider a plane steady ideal fluid flow with the velocity vector field

$$\mathbf{v} = \begin{bmatrix} u(x,y) \\ v(x,y) \end{bmatrix} \quad \text{at the point } \mathbf{x} = \begin{bmatrix} x \\ y \end{bmatrix},$$

where $\mathbf{v}(\mathbf{x})$ denotes the instantaneous fluid velocity at the point $\mathbf{x} \in D \subset \mathbb{R}^2$. The flow is said to be *incompressible* (i.e., fluid volume does not change) iff

$$\nabla \cdot \mathbf{v} = \frac{\partial u}{\partial x} + \frac{\partial u}{\partial y} = 0. \tag{18.3.1}$$

The fluid is called *irrotational* (i.e., no vorticity, no circulation) iff

$$\nabla \times \mathbf{v} = \frac{\partial v}{\partial x} - \frac{\partial u}{\partial y} = 0. \tag{18.3.2}$$

A flow is called an *ideal fluid flow* if it is both incompressible and irrotational. Most liquids and gases behave like ideal flows.

Note that Eqs (18.3.1)-(18.3.2) are almost identical to the Cauchy-Riemann equations, except that these equations have a change in sign in front of the derivatives of v. This means that we can replace v by $-v$ in the Cauchy-Riemann equations and obtain Eqs (18.3.1)-(18.3.2).

Theorem 18.1. *The velocity vector field* $\mathbf{v} = [\,u(x,y) \quad v(x,y)\,]^T$ *induces an ideal fluid flow iff*

$$f(z) = u(x,y) - iv(x,y) \tag{18.3.3}$$

is a complex analytic function at $z = x + iy$.

Note that the components $u(x,y)$ and $-v(x,y)$ of the velocity vector field for the ideal fluid flows are necessarily harmonic conjugates. The function $f(z)$ defined by (18.3.3) is known as the *complex velocity* of the fluid flow. Under such a flow the fluid particles follow the trajectories $z(t) = x(t) + iy(t)$ obtained by integrating the system of ordinary differential equations

$$\frac{dx}{dt} = u(x,y), \ \frac{dy}{dt} - v(x,y), \tag{18.3.4}$$

or in complex form

$$\dot{z} \equiv \frac{dz}{dt} = \overline{f(z)}, \tag{18.3.5}$$

where the curves parameterized by the solution $z(t)$ are known as *streamlines* of the fluid flow. If the complex velocity $f(z_0)$ at a point z_0 vanishes, then the solution $z(t) = z_0$ of Eq (18.3.5) is constant, and the point z_0 is a *stagnation point* of the flow.

Example 18.2. $f(z) = 1$. The velocity vector $\mathbf{v} = [1 \quad 0]^T$. Then solving $\dot{z} = 1$, i.e., $\dot{x} = 1, \dot{y} = 0$, we get the solution $z(t) = t + z_0$, which represents a uniform horizontal flow by straight lines parallel to the real axis (see Figure 18.7(a)). ■

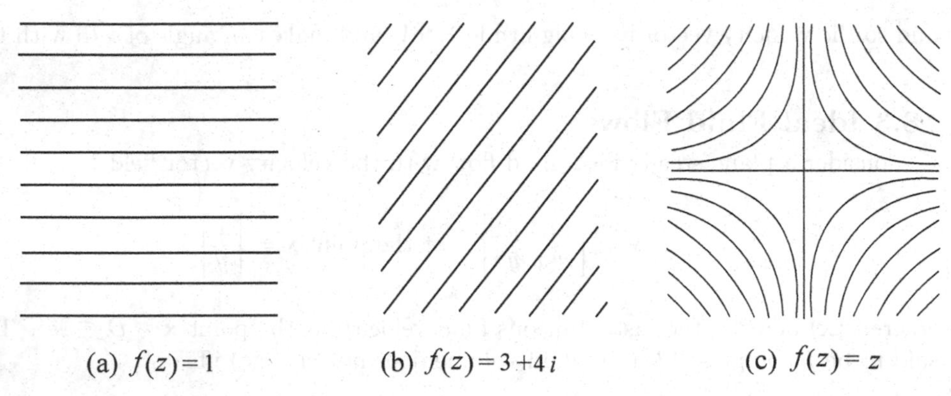

Figure 18.7 Complex fluid flows.

Example 18.3. $f(z) = c = a + ib$: We solve $\dot{z} = \bar{c} = a - ib$, to get $z(t) = \bar{c}t + z_0$. The streamlines are parallel to the straight lines inclined at an angle $\theta = \arg\{\bar{c}\} = \arg\{c\}$ to the real axis (see Figure 18.7(b) for $a = 3$ and $b = 4$). ■

Example 18.4. $f(z) = z = x + iy$: Solving $\dot{z} = \bar{z}$, i.e., $\dot{x} = x, \dot{y} = y$, we get $z(t) = x_0\, e^t + iy_o\, e^{-t}$, with stagnation point at $x = 0 = y$ (see Figure 18.7(c)). ■

Example 18.5. $f(z) = -iz = y - ix$: Solving $\dot{z} = i\bar{z}$, i.e., $\dot{x} = y, \dot{y} = x$, we get $z(t) = (x_0 \cosh t + y_0 \sinh t) + i\,(x_0 \sinh t + y_0 \cosh t)$. The flow moves along hyperbolas and rays $x^2 - y^2 = c^2$ (Figure 18.7(c) rotated by $\pi/4$). ■

Example 18.6. (Flow past a corner) A solid boundary in a fluid flow is characterized by the *no-flux condition* so that the fluid velocity $\mathbf{v}$ is tangent everywhere to the boundary. Thus, no fluid flows into or out of the solid boundary, and the flow will consist of streamlines and stagnation points. For example, the boundary of the first quadrant $Q = \{x > 0, y > 0\} \subset \mathbb{C}$ consists of the x and y axes and the origin. Since there are streamlines of the flow with complex velocity $f(z) = z = x + iy$, its restriction to Q represents an ideal flow past an inner right corner at the origin (Figure 18.8), where the individual fluid particles move along rectangular hyperbolas as they flow past the corner. ■

Figure 18.8 Flow inside a corner.

Now, suppose that the complex velocity $f(z)$ admits a complex anti-derivative, i.e., a complex analytic function $g(z)$ such that

$$g(z) = \phi(x,y) + i\psi(x,y), \quad \text{where } \frac{dg}{dz} = f(z). \tag{18.3.6}$$

Then from Cauchy-Riemann equations

$$\frac{dy}{dz} = \frac{\partial\phi}{\partial x} - i\frac{\partial\phi}{\partial y} = u - iv, \quad \text{or } \frac{\partial\phi}{\partial x} = u, \frac{\partial\phi}{\partial y} = v.$$

Thus, $\nabla\phi = \mathbf{v}$, and $\phi(x,y) = \Re\{g(z)\}$ defines a *velocity potential* of the fluid flow. That's why the function $g(z)$ is known as a *complex potential function* for the given velocity field. Thus, $\phi(x,y)$ is analytic, harmonic and satisfies Laplace's equation $\nabla^2\phi = 0$. Conversely, any harmonic function can be regarded as the potential for some fluid flow, where the real flow velocity is its gradient $\mathbf{v} = \nabla\phi$, which represents an ideal fluid flow. The harmonic function $\phi(x,y)$ is known as the *stream functions*; it satisfies Laplace's equation, and the potential and stream functions are related by the Cauchy-Riemann equations

$$\frac{\partial\phi}{\partial x} = u = \frac{\partial\psi}{\partial y}, \quad \frac{\partial\phi}{\partial y} = v = -\frac{\partial\phi}{\partial x}. \tag{18.3.7}$$

The level sets of the velocity potential, defined by $\{\phi(x,y) = c, c \in \mathbb{R}\}$, are known as *equipotential curves.* The velocity potential $\mathbf{v} = \nabla\phi \neq 0$[1] is normal to the equipotential curves. As mentioned earlier, $\mathbf{v} = \nabla\phi$ is also tangent to the level curves $\{\phi(x,y) = d\}$ of its harmonic conjugate function. But $\mathbf{v}$, being the velocity field, is tangent to the streamlines due to the fluid particles. Thus, these two systems of curves must be the same. Hence, the *level curves* of the streamlines are the streamlines of the flow, and the set of the equipotential curves $\{\phi = c\}$ and the set of the streamlines $\{\psi = d\}$ are two mutually orthogonal families of plane curves.

Example 18.7. (Flow around a disk) Consider the complex potential function

$$g(z) = z + \frac{1}{z} = \left(x + \frac{x}{x^2+y^2}\right) + i\left(y - \frac{y}{x^2+y^2}\right), \tag{18.3.8}$$

[1] $v = \nabla\phi \neq 0$ means that the flow is away from a stagnation point.

where the real and imaginary parts are solutions of two-dimensional Laplace's equation. The corresponding complex flow field is

$$f(z) = \frac{dg}{dz} = 1 - \frac{1}{z^2} = 1 - \frac{x^2 - y^2}{(x^2 + y^2)^2} + i\frac{2xy}{(x^2 + y^2)^2}. \tag{18.3.9}$$

The equipotential curves and streamlines are plotted in Figure 18.9.

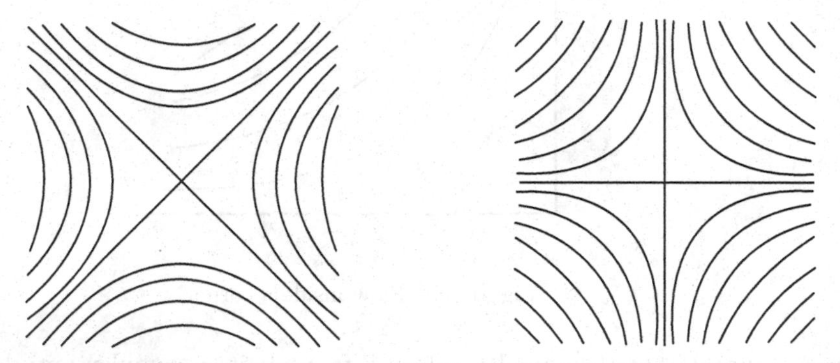

Figure 18.9 Equipotential lines and streamlines.

The points $z = \pm 1$ are stagnation points of the flow, and $z = 0$ is a singularity. Thus, the fluid moving along the positive real axis approach the stagnation point $z = -1$ as $t \to \infty$. Note that the streamlines $\psi(x, y) = y - \dfrac{y}{x^2 + y^2} = d$ become horizontal at large distances, and thus far away from the origin; the flow is similar to a uniform horizontal flow, from left to right, with unit complex velocity $f(z) \equiv 1$.

The level curves for $d = 0$ consist of the unit circle $|z| = 1$ at the real axis. In particular, the unit circle consists of two semicircular streamlines together with two stagnation points. The flow velocity $\mathbf{v} = \nabla\phi$ is everywhere tangent to the unit circle, and thus satisfies the no-flux condition $\mathbf{v} \cdot \mathbf{n} = 0$ along the boundary $|z| = 1$ (see Figure 18.10). ■

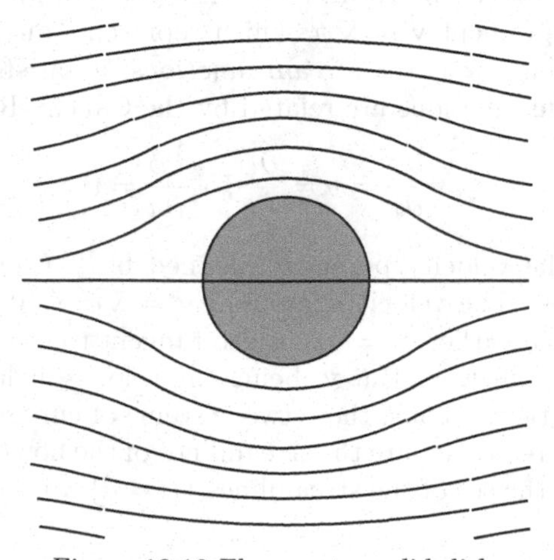

Figure 18.10 Flow past a solid disk.

Electromagnetic potential. In the case when $\phi(x, y)$ represents an electromagnetic potential function, the level curves of the harmonic conjugate $\psi(x, y)$ are the paths followed by charged particles under the electromagnetic force field $\mathbf{v} = \nabla\phi$.

Temperature distribution. If $\phi(x,y)$ represents the equilibrium temperature distribution in a plane, its level curves are the isotherms (curves of constant temperature) while the level lines of its harmonic conjugate $\psi(x,y)$ represent the curves along which heat energy flows.

Membrane. If $\phi(x,y)$ represents the height of a deformed membrane, then its level curves are contour lines of elevation, and the level curves of its harmonic conjugate $\psi(x,y)$ are the curves of steepest descent, as if a stream of water flows down the membrane ignoring any inertial effects of the flow.

18.3.1 Complex Potential Flow. For two-dimensional flows, the complex potential, denoted by $f(z) = u(x,y) + i\,v(x,y)$, due to incompressibility and irrotationality, satisfies the equation

$$\frac{df}{dz} = u - i\,v = q\,e^{-i\theta},$$

where $f(z)$ is an analytic function of $z = x + i\,y$ in the region D of the z-plane occupied by the flow. We will use the *log-hodograph* conformal transformation

$$\zeta = \log\left\{U_s \frac{dz}{df}\right\} = \log\frac{U_s}{q} = i\,\theta, \tag{18.3.10}$$

where U_s is the velocity at the wall.

18.4 Potential Flows

The integral equation

$$\Phi(s) = \int_\Gamma N(s,t)\,\Phi(t)\,dt + 2\log\left|\frac{P-\gamma(s)}{Q-\gamma(s)}\right|, \quad s \neq 0, \tag{18.4.1}$$

derived in §10.2.3, is known as *Carrier's integral equation*. This equation describes the problem of the potential flow of an inviscid fluid past a periodic array of airfoils of arbitrary shape (more on this problem in Chapter 13). The research developed by Joukowski to determine the force exerted by a flow on a body around which it is flowing eventually led to the theoretical foundation for practical aircraft construction, and the methods of conformal mapping played an important role in modern aviation. During the 1930s, Theodorsen's iterative method became a pioneer in transforming the exterior of the unit circle onto the exterior of an almost circular contour. But in the potential flow analysis of airfoils, James's method developed in 1971 turns out to be more successful for all types of contours which do not have corner singularity. These methods are discussed in Chapter 13, where it is noted (§13.2) that every multi-element potential flow always uses the single-element mapping methods at some stage during the computations.

18.4.1 Potential Flow of Ideal Fluids. Let the x- and y-components of velocity of a fluid particle be denoted by V_x and V_y, respectively, such that its magnitude is $V = \left(V_x^2 + V_y^2\right)^{1/2}$. Then using Bernoulli's equation for ideal fluid flow at constant level, we get $P/\rho + V^2/2 =$ const, where P is the pressure and ρ is the mass density. The velocity components can also be expressed in terms of a velocity potential $u(x,y)$ by $V_x = \dfrac{\partial u}{\partial x}$ and $V_y = \dfrac{\partial u}{\partial y}$, and we obtain the *complex potential* as

$$f(z) = u(x,y) + i\,v(x,y), \tag{18.4.2}$$

where $v(x, y)$ is the *stream function* of the flow, which, by using the Cauchy-Riemann equations reduces to

$$\frac{\partial v}{\partial x} = -\frac{\partial u}{\partial y} = -V_y, \quad \frac{\partial v}{\partial y} = \frac{\partial u}{\partial x} = V_x. \tag{18.4.3}$$

The lines $u(x, y) = c$, where c is a constant, are known as *equipotential lines*, whereas the lines $v(x, y) = c$ are the *streamlines*. Note that the difference between the values of v at two points in the z-plane is numerically equal to the rate of mass flow at unit mass density ($\rho = 1$) across a curve joining these two points. Thus, the complex potential function $f(z)$ can be used to express both the potential and stream functions in terms of the variable z. A similar complex potential function also exists in the w-plane. The derivative of $f(z)$, known as the *complex velocity*, defined by Eq (18.4.2), is

$$f'(z) = \frac{df}{dz} = \frac{\partial u}{\partial x} + i, \frac{\partial v}{\partial x} = V_x - i\, V_y = V. \tag{18.4.4}$$

Let $w = f(z) = u + i\, v$ map a domain D in the z-plane onto a domain G in the w-plane. Then the equipotential lines and streamlines generated by a flow are mapped conformally on the w-plane, and the potential flow in the w-plane is given by

$$\frac{df}{dw} = \frac{df}{dz}\frac{dz}{dw} = V_u - i\; V_v. \tag{18.4.5}$$

Thus, Eqs (18.4.4) and (18.4.5) yield

$$V_u - i\, V_v = \frac{V_x - i\, V_y}{f'(z)}, \quad \text{where} \quad \sqrt{V_u^2 + V_v^2} = \frac{\sqrt{V_x^2 + V_y^2}}{|f'(z)|}, \tag{18.4.6}$$

which shows that the absolute velocity at a point in the w-plane is the ratio between the absolute velocity at the corresponding point in the z-plane and $|f'(z)|$. Moreover, the equipotential lines $(x, y) = c$ and the streamlines $v(x, y) = c$ are orthogonal to each other, and they together determine the flow in the w-plane under a conformal transformation.

Example 18.8. The streamlines around a circular cylinder placed normal to a flow can be mapped onto parallel streamlines past a horizontal strip (Figure 18.11(a)) by the conformal mapping

$$w = V_0 \left(z + \frac{a^2}{z} \right), \quad \text{or} \quad z = \frac{w}{2} \pm \sqrt{\frac{w^2}{4} - a^2}, \tag{18.4.7}$$

(see Map 4.33). The real and imaginary parts of the complex potential $w = f(z)$ in polar cylindrical coordinates are given by

$$u(r, \theta) = V_0 \frac{r^2 + a^2}{r} \cos\theta, \quad v(r, \theta) = V_0 \frac{r^2 - a^2}{r} \sin\theta, \tag{18.4.8}$$

from which we obtain the streamlines that satisfy $v_0(r, \theta) = 0$, and these streamlines are represented by the circle $r = a$ and the real axis $\theta = 0, \pi$. ■

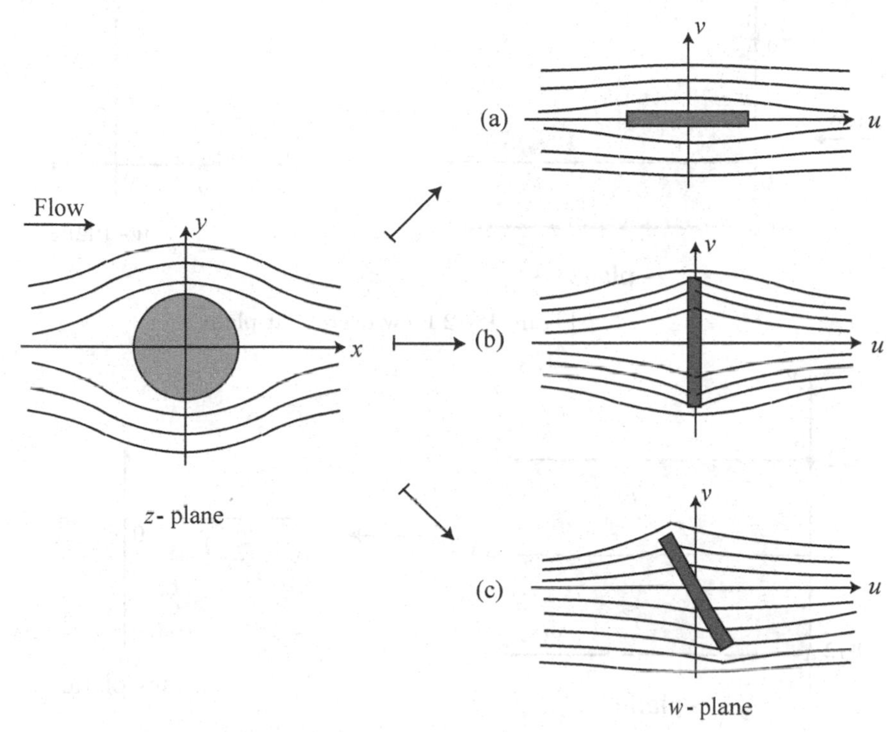

Figure 18.11 Flow around a circular cylinder in the z-plane, mapped onto (a) flow past horizontal plate, (b) flow past a vertical plate, and (c) flow past a slant plate.

For other cases of flow around a cylinder in the absence of rotational flow, see Figure (18.11(b)), which represents a flow past a vertical plate under the conformal mapping $w = z - \dfrac{a^2}{z}$ (Map 4.34), and Figure 18.11(c)), which represents a flow past a slant plate, inclined at an angle α, under the conformal mapping $w = z\, e^{i\alpha} - \dfrac{a^2}{z\, e^{i\alpha}}$ (Map 4.34). The conformal mappings for more general shapes can be found in Woods [1961], Davis [1979], and Dias [1986]. Another work on cavity flows by Chaudhrey and Schinzinger [1992] on mapping of curve segments by direct methods is also useful.

18.4.2 Flow over a Plate. Map 5.46 conformally maps the semi-strip in the first quadrant of the z-plane onto the upper half-plane $\Im\{w\} > 0$. We will consider a particular case when a semi-strip in the right-half of the z-plane is mapped onto the upper-half of the w-plane (Figure 18.12). This mapping is given by

$$w = \frac{1}{2}(b + c) + \frac{1}{2}(b - c)\cosh\left\{\frac{i\pi}{z_1 - z_0}(z - z_0)\right\}. \tag{18.4.9}$$

If we take $b = 1$ and $c = -1$, then the mapping (18.41) reduces to $w = -i \sinh z$ (Map 5.70), which maps the horizontal region $-\pi/2 \leq y \leq \pi/2$, $x \geq 0$ in the z-plane onto the lower-half $\Im\{w\} < 0$ of the w-plane (Figure 18.13).

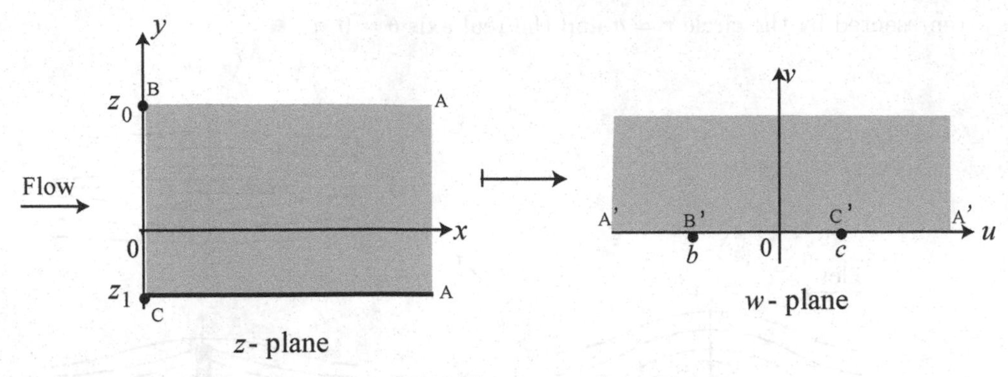

Figure 18.12 Flow over a flat plate.

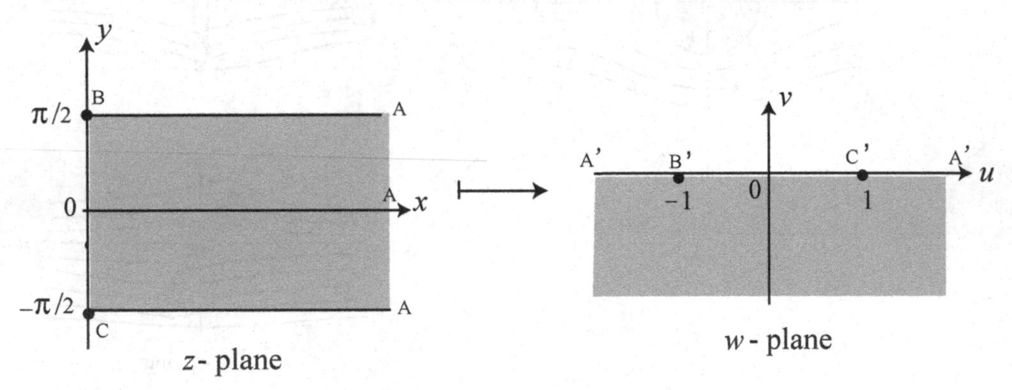

Figure 18.13 Flow over a flat plate.

18.4.3 Potential Flow at Separation. The separation must be smooth, otherwise $U_s = 0$ at separation which is not consistent with the velocity of the free-stream velocity. A smooth separation and a wedge flow at separation are presented in Figure 18.14. The potential near separation with no curvature at the wall of separation is defined by $\zeta \approx -ik\sqrt{z}$. For different cases of k, the potential flows are shown in Figure 18.15, where case (a) represents a sharp edge, $k > 0$; case (b) represents a smooth wall, $k > 0$; case (c) smooth wall, $k < 0$; and case (d) represents the *Villat condition* where the cavity pressure is the lowest with $U_s = U_\infty$. Note that the cases (a) and (b) present favorable pressure gradient, while case (c) an adverse pressure gradient. Figure 18.16(a) represents the Villat condition for subcritical flow, where the angle $\theta = 55°$, and (b) represents the supercritical flow with angle $\theta = 120°$.

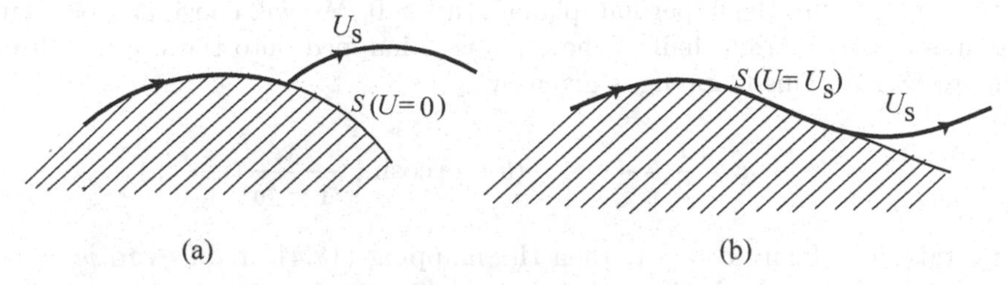

Figure 18.14 (a) Wedge flow, and (b) smooth flow at separation.

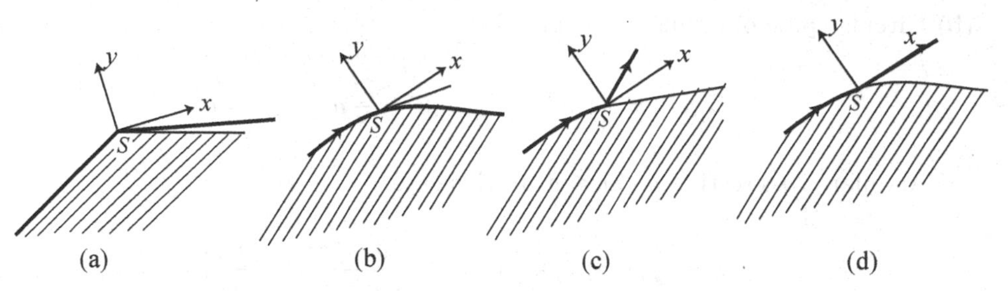

Figure 18.15 Different cases of k.

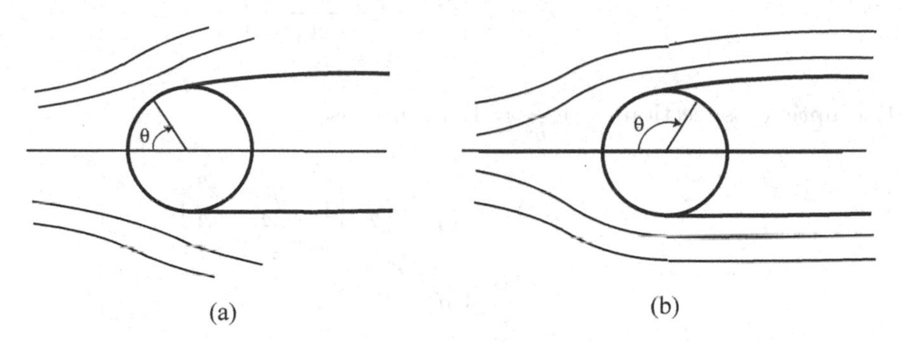

Figure 18.16 Villat condition: (a) subcritical flow; (b) supercritical flow.

18.5 Boundary Layer Flows

The presence of the convection terms in the momentum equations for the Navier-Stokes fluid flows introduces nonlinearity in the equations and makes it very difficult to solve them. There are, however, certain important classes of fluid flows in which these convection terms do not exist. They are known as the boundary layer flows, and we will present some problems where the boundary of the flow regions are cylindrical surfaces with generators along the x-axis. These flows are of the following two types: (i) Steady-state flows through different pipes of uniform cross-section with constant pressure gradient, and (ii) unsteady flows generated by the motion of a solid boundary along the x-axis.

Example 18.9. The first type of fluid flows are governed by the Poisson's equation

$$\frac{\partial^2 u}{\partial y^2}+\frac{\partial^2 u}{\partial z^2}=\frac{1}{\mu}\frac{dp}{dx}, \tag{18.5.1}$$

subject to the boundary condition $u=0$ at the walls of the pipe, where u is the velocity component of the flow along the x-axis, μ the coefficient of viscosity, and dp/dx the constant pressure gradient. The general analytical solution of Eq (18.5.1) subject to the given boundary condition is

$$u(y,z)=-f(y,z)\,\frac{1}{\mu}\frac{dp}{dx}, \tag{18.5.2}$$

where the function $f(y,z)$ depends on the shape of the cross-section of the pipe and is such that $\nabla^2 f(y,z)=-1$. The volume flux across the cross-section of the pipe is determined by $\dfrac{C}{\mu}\dfrac{dp}{dx}$, where C is a constant that depends on the volume flux per unit area of the cross-section. Some special cases are as follows:

(a) Two-dimensional channel $-a\le z\le a$: In this case

$$f(y,z)=\frac{1}{2}\left(a^2-z^2\right),\quad C=\frac{2}{3}a^3.$$

(b) Circular pipe of radius a: In this case

$$f(y,z) = \frac{1}{4}\left(a^2 - r^2\right), \quad C = \frac{\pi}{8}\,a^4, \quad r^2 = y^2 + z^2.$$

(c) Annular cross-section $a \le r \le b$: In this case

$$\begin{aligned} f(y,z) &= \frac{1}{4}\left[a^2 - r^2 + \frac{b^2 - a^2}{\ln(b/a)}\,\ln\left(\frac{r}{a}\right)\right], \\ C &= \frac{\pi}{8}\left[b^4 - a^4 - \frac{(b^2-a^2)^2}{\ln(b/a)}\right]. \end{aligned}$$

(d) Elliptic cross-section $\dfrac{y^2}{a^2} + \dfrac{z^2}{b^2} = 1$: In this case

$$\begin{aligned} f(y,z) &= \frac{a^2b^2}{2\left(a^2+b^2\right)}\left(1 - \frac{y^2}{a^2} - \frac{z^2}{b^2}\right), \\ C &= \frac{\pi}{4}\,\frac{a^3b^3}{a^2+b^2}. \end{aligned}$$

(e) Rectangular cross-section $|y| \le a$, $|z| \le b$: In this case

$$\begin{aligned} f(y,z) = \frac{a^2}{2} - \frac{y^2}{2} \\ - 2a^2\left(\frac{2}{\pi}\right)^3 \sum_{n=0}^{\infty} \frac{(-1)^n}{(2n+1)^3}\,\frac{\cosh(2n+1)(\pi z/2a)}{\cosh(2n+1)(\pi b/2a)}\,\cos\frac{(2n+1)\pi y}{2a}, \\ C = \frac{4}{3}\,ba^3 - 8a^4\left(\frac{2}{\pi}\right)^5 \sum_{n=0}^{\infty} \frac{1}{(2n+1)^5}\,\tanh\frac{(2n+1)\pi b}{2a}. \end{aligned}$$

In particular, if $a = b$, then $C = 0.562308\,a^4$.

If the cross-section is none of the above regions, the soap film method is used to experimentally compute f and C. We will not discuss this method here, but interested readers can see the experimental technique in Taylor [1937]. ■

Example 18.10. This is an example of the second type of flows. Consider an infinite plate that is moved in its own plane with a constant velocity U at time $t = 0$ in a fluid which is initially at rest. If the plate lies in the (x,y)-plane, then the velocity distribution satisfies the equation

$$\frac{\partial u}{\partial t} = \nu\,\frac{\partial^2 u}{\partial z^2}, \tag{18.5.3}$$

where ν is the kinematic viscosity of the fluid. The initial and boundary conditions are $u(z,0) = 0$, and $u(0,t) = U$ for $t > 0$. The solution is

$$u = U\,\operatorname{erfc}\left(\frac{z}{2\sqrt{\nu t}}\right). \tag{18.5.4}$$

A distinction between this boundary layer problem and the similar heat conduction problem is that there is no true propagation velocity in the latter case. But by defining the boundary layer thickness as the distance in which the value of u drops to a certain preassigned fraction of U, we find from (18.5.4) that the boundary layer thickness is proportional to $\sqrt{\nu t}$, and it grows at a rate proportional to $\sqrt{\nu/t}$. ■

Besides the above examples, there are many situations in which the solutions of the heat conduction equation are applicable to the problems of fluid flows. Some examples are as follows.

Example 18.11. In the cylindrical polar coordinates (r, θ, z), let the velocity components of a fluid flow be denoted by u_r, u_θ, and u_z. Consider the case of the flow for which $u_r = 0 = u_z$, $u_\theta = u_\theta(r, t)$, and the pressure $p =$ const. Then u_θ satisfies the equation

$$\frac{\partial u_\theta}{\partial t} = \nu \left(\frac{\partial^2 u_\theta}{\partial r^2} + \frac{1}{r} \frac{\partial u_\theta}{\partial r} - \frac{u_\theta}{r^2} \right), \tag{18.5.5}$$

and the vorticity $\omega_z = \frac{1}{r} \frac{\partial}{\partial r} (r u_\theta)$ satisfies the diffusion equation

$$\frac{\partial \omega_z}{\partial t} = \nu \left(\frac{\partial^2 \omega_z}{\partial r^2} + \frac{1}{r} \frac{\partial \omega_z}{\partial r} \right). \tag{18.5.6}$$

A solution of Eq (18.5.6) is

$$\omega_z = \frac{\Gamma}{4\pi\nu t} e^{-r^2/4\nu t}, \tag{18.5.7}$$

where Γ is the initial value of the circulation about the origin. This solution describes the dissolution of a vortex filament concentrated at the origin at time $t = 0$. ■

Example 18.12. The diffusion equation (18.5.5) also describes the motion of a fluid contained inside or outside an infinite cylinder that is rotating about its axis. The vorticity ω_z is initially concentrated at the surface of the cylinder, but it spreads outside into the fluid such that $\omega_z \to 0$ as $t \to \infty$, and then $u_\theta = A/r$, where A is a constant. However, inside the cylinder, the vorticity w_z approaches a constant value which is equal to twice the angular velocity of the cylinder, and the fluid tends to rotate like a rigid body. ■

Example 18.13. Consider the case of a steady flow parallel to an infinite plate on which the nonzero component of velocity assumes a prescribed nonzero value. This represents the steady-state flow far downstream of the leading edge of a semi-infinite plate. If there is no suction in the plate, the boundary layer grows indefinitely downstream such that at a finite distance the velocity eventually becomes zero. But if there is suction in the plate, the boundary layer eventually stops growing and we obtain the 'asymptotic suction profile.' Let the x- and z-axis lie along and perpendicular to the plate, and the corresponding components of velocity be u and w, respectively, and let the pressure p be independent of x. Then the equation of continuity $\frac{\partial u}{\partial x} + \frac{\partial w}{\partial z} = 0$ implies that $w =$ const$= -W$, say, at the plate. Thus, the governing equation of the flow is

$$-W \frac{du}{dz} = \nu \frac{d^2 u}{dz^2}, \tag{18.5.8}$$

with $p =$ const throughout the flow, subject to the boundary conditions $u = 0$ at $z = 0$, and $\lim_{z \to \infty} u(z) = U < +\infty$, where U is the mean-stream velocity of the flow. Hence, the solution of this problem is

$$u = U \left(1 - e^{-Wz/\nu} \right). \tag{18.5.9}$$

Note that as $\nu \to 0$, the velocity approaches the mean stream velocity which gets concentrated inside a boundary layer of the plate. ■

Example 18.14. Consider the circulating flow around a rotating circular cylinder with suction. Let an infinite cylinder that is immersed in a fluid at rest be suddenly rotated about its axis with constant angular velocity. Then the vorticity is given by $\omega_z = \frac{1}{r}\frac{\partial}{\partial r}(r u_\theta)$ (see Example 18.11). We know that initially this vorticity is concentrated at the surface of the cylinder, but it spreads out until $\omega_z = 0$ everywhere. Finally, the steady-state solution is given by

$$u_\theta = \frac{\Gamma_1}{2\pi r}, \tag{18.5.10}$$

where Γ_1 is the circulation around the cylinder. The solution (18.5.10) is known as the *asymptotic suction profile.*

Now, if there is suction throughout the surface of the cylinder, the velocity will eventually attain its steady-state value when the outward diffusion is balanced by the convection of vorticity toward the cylinder. Let a denote the radius of the cylinder, and let $-V$ be the component of the suction velocity perpendicular to the surface of the cylinder. Then the radial velocity u_r for the flow is given by

$$u_r = -\frac{aV}{r}.$$

This equation satisfies the equation of continuity $\operatorname{div}\cdot \mathbf{v} = 0$. Since the rate of diffusion of vorticity across a circle $r = a$ is given by $-2\pi a\nu r\frac{\partial \omega_z}{\partial r}$, and the rate of convection is $2\pi a r u_r \omega_z$, we find that ω_z satisfies the first-order equation

$$\frac{\partial \omega_z}{\partial r} + R\frac{\omega_z}{r} = 0, \tag{18.5.11}$$

where $R = aV/\nu$. The solution of Eq (18.5.11) is

$$\omega_z = \frac{1}{r}\frac{\partial}{\partial r}(r u_\theta) = A\left(\frac{a}{r}\right)^R, \tag{18.5.12}$$

where A is the value of ω_z at the cylinder, i.e., $A = \omega_z(a)$. Since the circulation $\Gamma = 2\pi r u_\theta$, we find from (18.5.12) that

$$\Gamma = \begin{cases} \Gamma_1 - \dfrac{2\pi a^2}{R-2} A\left(\dfrac{a}{r}\right)^{R-2} & \text{if } R \neq 2, \\ \Gamma_1 + 2\pi a^2 A \ln \dfrac{r}{a} & \text{if } R = 2. \end{cases}$$

Thus, for $R \le 2$ the solution with finite circulation at $r = \infty$ is $\Gamma = \Gamma_1$, where $\omega_z = 0$ and $u_\theta = \Gamma_1/2\pi r$, as given by (18.5.10). However, for $R > 2$, the value of circulation at infinity is Γ_1, and A can be adjusted so as to give any prescribed value of circulation at the cylinder. Hence, to maintain different values of circulation at the cylinder and at infinity the suction velocity V must be such that $V > \frac{2\nu}{a}$. ■

18.5.1 Presence of Circulation. In Examples 18.11 and 18.14 we noticed the presence of circulation in the fluid flow. The property of lift or propulsion is present in the design of an airplane wing and hydrofoils, the sails of yachts and boats, and the blades of turbines and propellers. The application of conformal mapping is appropriate in all such types of fluid flows that include circulation, since without circulation such flows would not be feasible. A flow induced by concentrated source of strength S at the origin $z = 0$, defined by $w = S \log z = \phi + i\psi$, where $\phi = S \ln r$ and $\psi = S\theta$ $(z = r\, e^{i\theta})$, is shown in Figure 18.17(a). The velocity is radical everywhere, and $\dfrac{d\phi}{dr} = \dfrac{S}{r}$. In Figure 18.17(b), there is a vortex situated at the origin $z = 0$, and there is a flow defined by $w = \dfrac{i\Gamma}{2\pi} \log z$, where Γ is the (clockwise) circulation. Notice that the equipotential lines and the streamlines in Figures 18.17(a) and 18.17(b) are the reverse of each other.

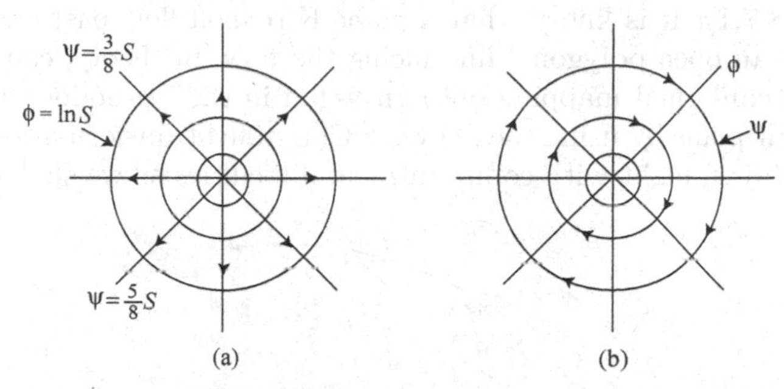

Figure 18.17 Potential field of (a) a concentrated source and (b) a vortex.

Now, since the flow around a circular cylinder is given by Eq (18.4.7) (Figure 18.14), the flow in the presence of clockwise circulation around the same cylinder will be defined by

$$w = V_0\Big(z + \frac{a^2}{z}\Big) + \frac{i\,\Gamma}{2\pi} \log z, \tag{18.5.13}$$

where the second term in Eq (18.5.13) is the vortex term added to Eq (18.4.7). Thus, the combined flows presented in Figure 18.18 are composed of Figure 18.14 in the z-plane and Figure 18.14(a), and shows the effect of superposition of irrotational flow and a vortex with circulation Γ. Also, Figure 18.18(a), (b), and (c) represent the three cases when $\Gamma \lesseqgtr 4\pi a V_0$, respectively.

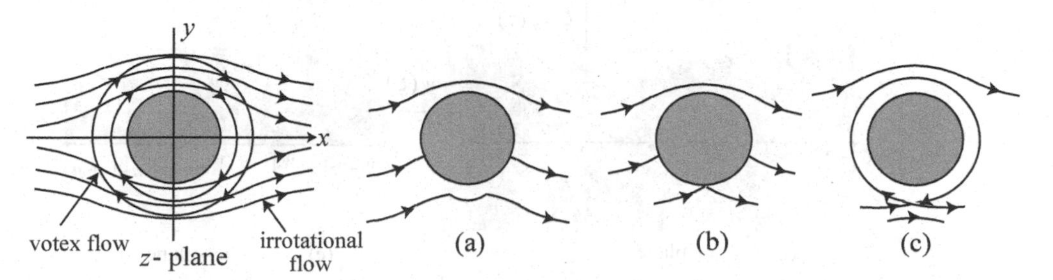

Figure 18.18 Superposition of irrotational flow and a vortex with circulation.

Notice that the lift on a cylinder with circulation depends on the pressure distribution along the length of the cylinder. In such cases, the Bernoulli equation (pressure $p = p_0 V_0^2/2$, see §18.4.1) leads to the determination of the force

$$F = F_y + iF_x = -\frac{\rho}{2} \int_C \Big(\frac{dw}{dz}\Big)^2 dz, \tag{18.5.14}$$

whence we get

$$F_x = 0, \quad F_y = \rho V_0 \Gamma, \tag{18.5.15}$$

where the drag force F_x is obviously zero, since the viscosity and boundary layer effects are ignored, and F_y gives the lift. Woods [1961: 187] has shown by plotting Eq (18.5.13) that there exists a jump of width Γ along the real axis.

18.6 Kirchhoff's Flow Problem

The classical Kirchhoff flow problem (Kirchhoff [1869]) deals with the flow of an ideal incompressible fluid past an obstacle and around a stationary wake bounded by free streamlines (see §18.7.1). It is known that a plane Kirchhoff flow past a solid polygonal obstacle composed of an open polygonal line facing the flow, in theory, can be determined by constructing its conformal mapping onto an n-gon in the log-hodograph plane and then onto the upper half-plane by using the Schwarz-Christoffel transformation. In practice, however, this approach is fraught with computational difficulties as we shall soon see.

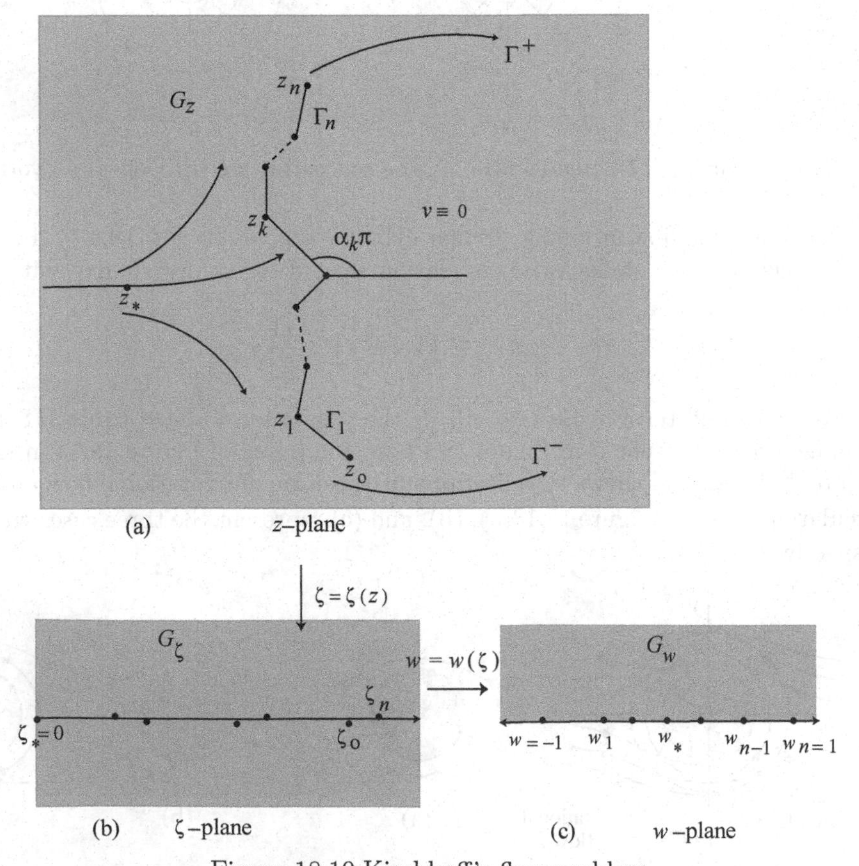

Figure 18.19 Kirchhoff's flow problem.

The geometry of a plane Kirchhoff flow is as follows: A solid obstacle with a open polygonal boundary Γ, composed of n straight line segments $\Gamma_k = (z_{k-1}, z_k)$, $k = 1, \dots, n$, lies in the region G_z in the z-plane (physical plane) as shown in Figure 18.19(a). The ideal incompressible fluid flow, flowing past the obstacle Γ, is assumed to be irrotational. Let the complex velocity be denoted by $v(z)$ and normalized by $v(\infty) = 1$. The flow divides between an upper and a lower part at an unspecified stagnation point z_* where the upper flow passes over z_n and the lower over z_0. Then the flow continues smoothly forward past z_n and z_0 with finite acceleration around a wake in which $v \equiv 0$. The two streamlines, Γ^+ and Γ^- denote the curves of discontinuity which separate the wake from the rest of the flow, and the stream function is zero on $\Gamma^\pm$. The shape of $\Gamma^\pm$ is not known but must be determined by using the condition that $|v(z)| = 1$ along these straight line segments. Note that this condition follows from Bernoulli's equation ($p + 0.5\,|v| = \text{const}$) and the fact that the pressure must remain constant throughout the wake and continuous on $\Gamma^\pm$.

Thus, the Kirchhoff flow problem can be stated as follows: Given the obstacle Γ in the physical region G_z in the z-plane, determine the velocity field $v(z)$, the streamlines $\Gamma^\pm$, and the location of the stagnation point z_* for the above flow. Also compute the associated lift and drag coefficients. A conformal mapping solution of this problem can be stated as follows: Let τ denote the hodograph (or conjugate velocity) plane so that the complex conjugate velocity is defined by $\tau(z) = \bar{v}(z)$. Since the flow is incompressible and irrotational, the velocity $v = \nabla\phi$, where $\phi(z)$ is the real part of the complex velocity potential $\zeta(z) = \phi(z) + i\,\psi(z)$ such that $\nabla^2\phi = 0$ and $\psi(z)$ is the stream function. Thus,

$$\tau(z) = \frac{d\zeta}{dz}. \tag{18.6.1}$$

The function $\zeta(z)$, regular in G_z, maps the region G_z conformally onto a slit region G_ζ in the ζ-plane, where the slit begins at $\zeta_* = \zeta\,(z_*)$ at which point the flow separates to go around the polygonal obstacle (Figure 18.19(b)). Without loss of generality we take $\zeta_* = 0$. Let a new complex variable w be defined by

$$w = \sqrt{\frac{2\zeta}{W}} + w_*, \quad \zeta = \frac{W}{2}\,(w - w_*)^2, \tag{18.6.2}$$

where W is real and $w_* \in (-1, 1)$ (Figure 18.19(b) in which $w_k = w(\zeta_k)$ is marked), G_w is the upper half-plane where $[-1, 1]$ is the image of Γ, and $(1, \infty)$ and $-\infty, -1)$ that of Γ^+ and Γ^-, respectively. We shall discuss two methods.

(i) Classical Hodograph Method. In this method the upper half-plane G_w in the w-plane is mapped onto the hodograph region $G_\tau = \tau\,(G_\zeta)$ in the τ-plane. The technique leads to a solution, at least theoretically, for the general Kirchhoff flow problem, because although G_z is unknown due to the presence of unknown free streamlines, the region G_τ is almost known in the following sense: Since the flow must be tangential on the solid boundary Γ, we know $\arg\{\tau(z)\}$ depending on whether the point on Γ is downstream of z_* toward z_0 or upstream toward z_n, respectively, i.e.,

$$\arg\{\tau(z)\} = \begin{cases} (1 - \alpha_k)\,\pi & \text{for } z \in \Gamma_k,\ w < w_*, \\ -\alpha_k\pi & \text{for } z \in \Gamma_k,\ w > w_*. \end{cases} \tag{18.6.3}$$

Also

$$|\tau(z)| = 1 \quad \text{for } z \in \Gamma^\pm. \tag{18.6.4}$$

The region G_τ is 'gearlike', bounded by circular arcs and subsets of rays passing through the origin. By introducing the log-hodograph variable

$$\Omega(z) = -\log \tau(z), \tag{18.6.5}$$

the region G_τ is mapped onto a region G_Ω which is bounded by vertical and horizontal line segments. Then we can use a Schwarz-Christoffel transformation to map G_w onto G_Ω. This method establishes a relation between τ and ζ and then integrates (18.6.1) to obtain ζ and τ as functions of z.

However, in practice only a few simple cases involving a flat plate (Kirchhoff [1869]; for the classical case see §18.6.1) and certain wedges (Birkhoff and Zarantonello [1957]; Gurevich [1965]; Robertson [1965] ; and Elcrat [1982]) have been solved by this method because the complexity of the conformal mappings grows as the number of sides of the polygonal obstacle increases. Then the resulting parameter problem inherent in the Schwarz-Christoffel integrals must be determined numerically which is not an easy task. Another difficulty stems from the fact that although the vertices in the w-plane are known, those in the Ω-plane must be computed by integrating (18.6.1), which can be very time consuming. Finally, the most serious difficulty with this method is that, in general, G_Ω is a Riemann surface with slits or branch points of unknown dimensions rather than just a polygon.

(ii) Elcrat-Trefethen Method. This method, developed by Elcrat and Trefethen [1986], for computing flows past an arbitrary obstacle with a high degree of accuracy, uses a modified Schwarz-Christoffel integral to map the upper half-plane $\Im\{w\} > 0$ directly onto the physical domain G_z rather than onto the log-hodograph plane. The Schwarz-Christoffel formula (7.1.2) is modified as follows: Let the polygonal line $\Gamma = \bigcup_{k=1}^{n} \Gamma_k$ in the z-plane (the obstacle in the physical plane) have vertices z_k, sides $\Gamma_k = (z_{k-1}, z_k)$, interior angles $\alpha_k \pi$ (counterclockwise), $k = 1, \dots, n$, and exterior angles μ_k, where $\mu_k = \alpha_{k-1} - \alpha_k$ for $k = 1, \dots, n-1$, and $\mu_n = \alpha_1 - \alpha_n + 2$, thus $\sum_{k=1}^{n} \mu_k = 2$. Let $z = F(w)$ be the conformal mapping of the upper half-plane $\Im\{w\} > 0$ onto the G_z such that the point $w = \infty$ goes into a point on Γ_1. Let $W_k = w(\Gamma_k)$ denote the intervals (w_{k-1}, w_k), where $w_k = F(z_k)$. Then $\arg\{\frac{dz}{dw}\}$ has a constant value α_π on each W_k and a jump of $\mu_k \pi$ at w_k, i.e.,

$$\arg\{\frac{dz}{dw}\} = \alpha_1 \pi \quad \text{for } w \in (w_n, \infty), \tag{18.6.6}$$

$$\Delta \arg\{\frac{dz}{dw}\} = \mu_k \pi \quad \text{at } w = w_k. \tag{18.6.7}$$

Let $g_k(w) = (w - w_k)^{-\mu_k}$ denote the factors in the formula (2.3.5), where the branch of $g_k(w)$ is chosen such that $g_k(w) > 0$ for $w > w_k$. Then $\arg\{g_k(w)\} = \text{const}$ except for a jump of $\mu_k \pi$ at w_k. It means that the function $g_k(w)$ maps $\Im\{w\} > 0$ onto the sector bounded by the rays $e^{-i\mu_k\pi}\,\mathbb{R}^+$ and $\mathbb{R}^+$ in the z-plane (Figure 18.20(a)). Hence,

$$\frac{dz}{dw} = C\, e^{i\alpha_1\pi} \prod_{k=1}^{n} g_k(w) = C\, e^{i\alpha_1\pi} \prod_{k=1}^{n} (w - w_k)^{-\mu_k}, \quad C > 0,$$

and the modified Schwarz-Christoffel formula is

$$z = F(w) = C_0 + C\, e^{i\alpha_1\pi} \int^{w} (\zeta - w_k)^{-\mu_k}\, d\zeta, \tag{18.6.8}$$

where C_0 is a complex constant.

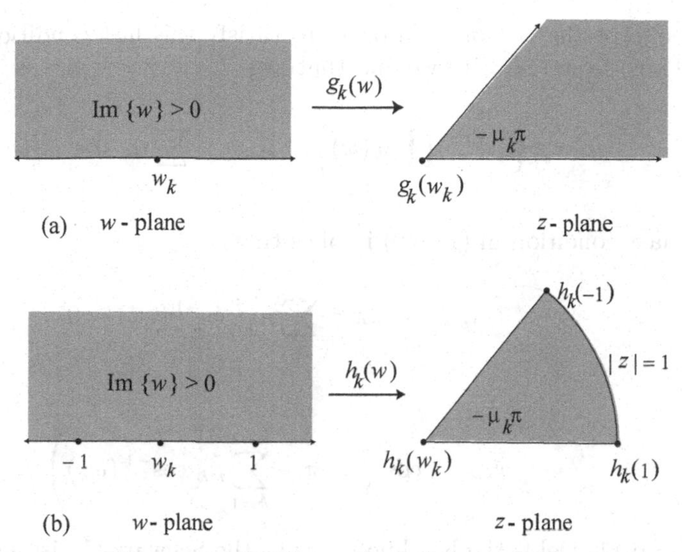

Figure 18.20 Elcrat-Trefethen method.

Now, consider the function $z = F_1(\zeta)$ which maps the slit region G_ζ in the ζ-plane onto the region G_z in the z-plane (Figure 18.19). Since we know $\arg\{\frac{dz}{d\zeta}\}$ for $w \in [-1, 1]$ and $\left|\frac{dz}{d\zeta}\right|$ elsewhere, we have

$$
\begin{aligned}
&\arg\{\frac{dz}{d\zeta}\} = \alpha_k\,\pi \quad \text{for } w = w_k, \\
&\Delta\,\arg\{\frac{dz}{d\zeta}\} = \mu_k\pi \quad \text{at } w = w_k \text{ for } k = 1, \dots, n-1, \\
&\left|\frac{dz}{d\zeta}\right| = 1 \quad \text{for } |w| > 1, \\
&\arg\{\frac{dz}{d\zeta}\} = 0 \quad \text{at } w = \infty.
\end{aligned}
\tag{18.6.9}
$$

Thus, the Kirchhoff flow problem is a modification of the Schwarz-Christoffel problem (18.6.6) in the sense that a constant-modulus condition, instead of the constant-argument condition, is applied over the boundary. Define a function $h_k(w)$ by

$$
h_k(w) = \left[\frac{w - w_k}{1 - w_k\,w + \sqrt{(1-w^2)\,(1-w_k^2)}}\right]^{-\mu_k}, \tag{18.6.10}
$$

where the branch is chosen such that $h_k(w) > 0$ for $w \in (w_k, 1)$. The function h_k has singularity at w_k (like the function g_k) and also at $w = \pm 1$, and it maps the half-plane $\Im\{w\} > 0$ onto the closed circular sector bounded by the rays $e^{-i\mu_k\pi}\,\mathbb{R}^+$, $\mathbb{R}^+$ and the unit circle $|z| = 1$ (Figure 18.20(b)). Since $\left|h_k(w)\right| = 1$ for $|w| > 1$, we have

$$
\frac{dz}{d\zeta} = \frac{1}{\tau} = e^{i\alpha_n\,\pi}\prod_{*} h_k(w), \tag{18.6.11}
$$

where $\prod_{*}$ denotes the product over $k = 1, \dots, n-1$, ($\sum_{*}$ defined analogously) and w_* is the preimage of z_*. Note that the function defined by (18.6.11) satisfies all the conditions in (18.6.9) except the last one. In order to satisfy this last condition we must choose w_* properly. Thus, from (18.6.10) we find that $\arg\{h_k(w)\} = -\mu_k \cos^{-1}(-w_k)$, and hence,

$$\arg\Big\{e^{i\alpha_n \pi} \prod_* h_k(w)\Big\} = \alpha_n \pi - \sum_* \mu_k \cos^{-1}(w_k).$$

Hence, the last condition in (18.6.9) implies that

$$\alpha_n \pi = \alpha_n \pi - \sum_* \mu_k \cos^{-1}(w_k) = 0,$$

and thus,

$$w_* = -\cos\left(\alpha_n \pi - \sum_{k=1}^{n-1} \mu_k \cos^{-1}(w_k)\right). \tag{18.6.12}$$

Then Eq (18.6.11) yields the Kirchhoff flow as the Schwarz-Christoffel integral

$$\begin{aligned} z &= C + e^{i\alpha_n \pi} \int^{\zeta} \prod_* h_k(t)\, dt \\ &= C + e^{i\alpha_n \pi}\, W \int^{w} (t - w_*) \prod_* h_k(t)\, dt \\ &= C + W\, e^{i\alpha_n \pi} \int^{w} \Big(1 - w_* t + \sqrt{(1-t^2)(1-w_*^2)}\Big) \\ &\times \prod_{k=1}^{n-1} \left(\frac{t - w_k}{1 - w_* t + \sqrt{(1-t^2)(1-w_*^2)}}\right)^{-\mu_k} dt, \end{aligned} \tag{18.6.13}$$

where we replaced the integration with respect to ζ by that with respect to w by setting $d\zeta = W\,(w - w_*)\,dw$ and in the last step canceling the common factor $(t - w_*)$. The above formula basically matches the formula derived by Monakov [1983: 185, Eq (5)], where he erroneously takes $w_* = 0$. A Fortran package for the Kirchhoff flow problem, containing the files scpack.exe and kirch1.exe, can be obtained from the second author in Elcrat and Trefethen [1986].

18.6.1 Original Kirchhoff's Flow Problem. The original Kirchhoff's flow problem deals with an irrotational flow of a weightless, ideal, incompressible fluid past a flat plate AB with separation of the jet, such that the modulus of flow velocity is equal to the modulus of the approaching stream v_0 on the surfaces of the jets AD and BD (Figure 18.21). The region of constant pressure behind the plate extends to infinity and the plate AB is perpendicular to the flow. This problem deals with the determination of the function $\zeta(w)$ such that $v_0 \dfrac{dz}{dw} = \zeta(w)$. Show that the function

$$\zeta(w) = -\sqrt{\frac{u_0}{w}} - \sqrt{\frac{u_0}{w} - 1},$$

where u_0 is a real constant maps the w-plane cut along the positive real axis from C to $+\infty$ onto the upper half-plane $\Im\{\zeta\} > 0$ from which a semicircle of unit radius is removed

(Figure 18.22). Use the following conformal maps: $\tau = \dfrac{\zeta - 1}{\zeta + 1}$, $\tau_1 = \tau^2$, $t = \sqrt{\dfrac{w}{u_0}}$, and $t_1 = \dfrac{1+t}{1-t}$ (Gurevich [1965: 15-20]).

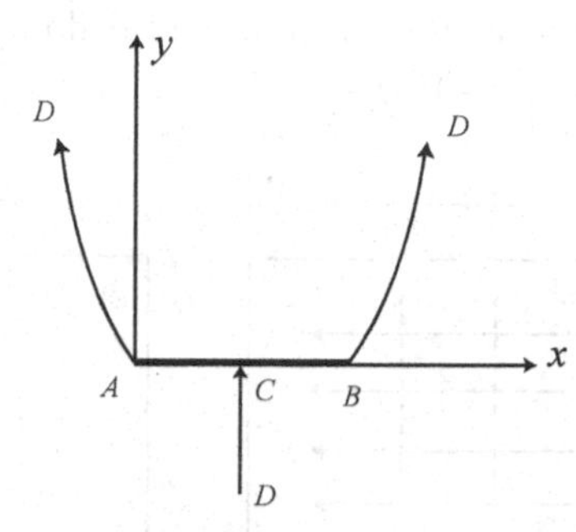

Figure 18.21 Modulus of the flow velocity.

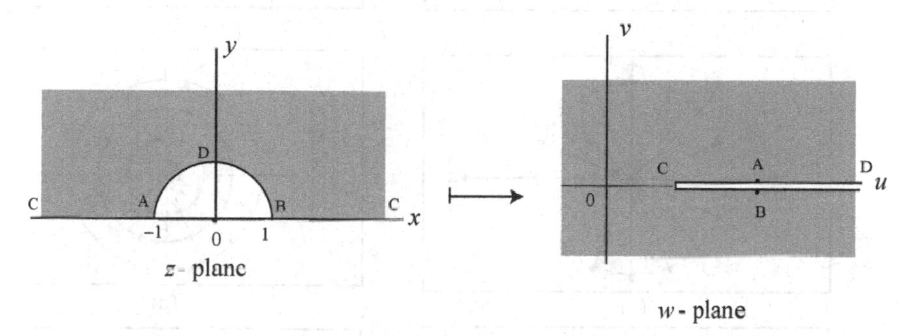

Figure 18.22 Conformal map.

18.7 Streamlines in Fluid Flows

Consider a two-dimensional steady-state flow of an ideal fluid, which is incompressible (i.e., of constant density) and non-viscous. Then the velocity of the flow can be derived from a potential (harmonic), and any force the flow creates on an obstacle must be perpendicular to the obstacle's surface. Remember that such a fluid never exists in nature, but it is used to approximate certain physical situations.

Let $f(z) = u(x,y) + i\,v(x,y)$ be an analytic function, where $u(x,y)$ is called the *velocity potential*, and $v(x,y)$ the *stream function*. The equations of the streamlines are $xy = c$, where c is a (constant) parameter along any streamline. Now, we will discuss the following types of fluid flows with their streamlines and velocity potential, which are presented in Figure 18.23.

(i) A uniform horizontal flow from left to right ($w = z$) (Figure 18.23(a));

(ii) flow around a corner (Map: $w = z^2$) (Figure 18.23(b));

(iii) radial flow with source at $z = 0$, fluid emerging from the origin (Map 5.20: $w = \log z$) (Figure 18.23(c));

(iv) whirlpool or vortex at $z = 0$ (conformal map: $w = i \log z$) (Figure 18.23(d));

(v) source at $z = a$ and sink at $z = b$, i.e., fluid emerges at $z = a$ and absorbed at $z = b$ (conformal map: $w = \log \dfrac{z-a}{z-b}$) (Figure 18.23(e));

(vi) dipole at $z = 0$, the dipole consisting of a large source and sink by an infinitesimal distance (conformal map: $w = \frac{1}{z}$) (Figure 18.23(f));

(vii) cylinder of radius a in a uniform flow from left to right (conformal map: $w = z + \frac{a^2}{z}$) (Figure 18.23(g)); and

(viii) source at $z = 0$ in a uniform horizontal flow from left to right (conformal map: $w = Vz + m \log z$) (Figure 18.23(h)).

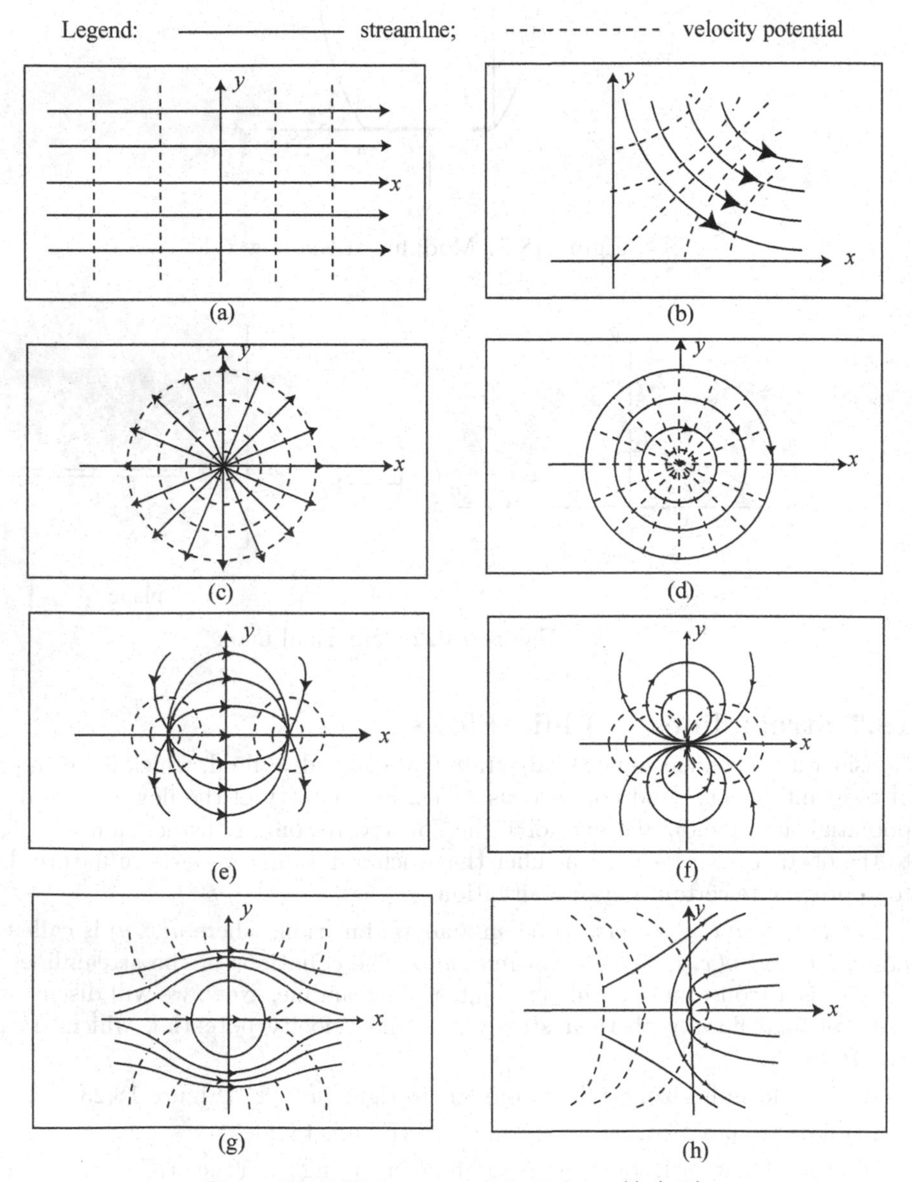

Figure 18.23 Eight types of fluid flows (i)-(viii).

Example 18.15. Since the streamlines are defined by $v =$ const, they are obtained by equating the imaginary part of the map $w = z + \dfrac{a^2}{z}$ to the constant c. This imaginary part is given by

$$w = x + iy + \frac{a^2}{x+iy} = x + iy + \frac{a^2(x-iy)}{x^2+y^2} = x + \frac{a^2}{x+y^2} + i\Big(y - \frac{a^2 y}{x^2+y^2}\Big).$$

Hence the streamlines are $v(x,y) = y - \dfrac{a^2 y}{x^2+y^2} = c$. ■

If the fluid is non-viscous, there is no friction between the fluid and the boundary, any streamline can be regarded as the boundary along which the fluid must flow. Consider the case of a right-angled corner, as shown in Figure 18.24. The fluid continuously enters the shaded region through the edge where the two walls meet Figure 18.24(a)). Once the steady state is attained, the streamlines can be described as follows: If we use two streamlines along the positive real and imaginary axes as the boundaries of the fluid, we find that the streamlines in the first quadrant are simply the straight lines emerging from the origin (Figure 18.24(b)).

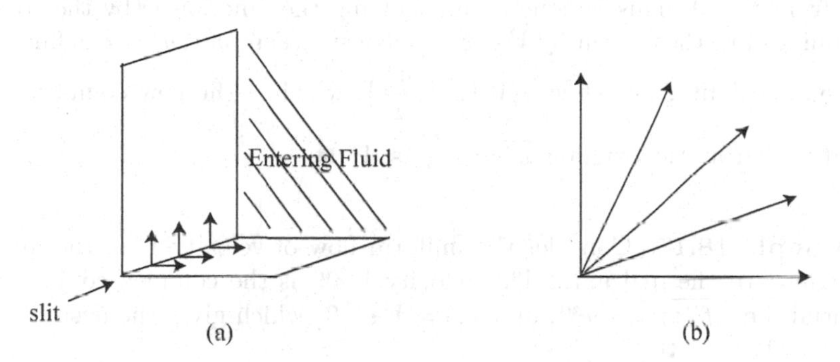

Figure 18.24 Flow in a right-angled corner.

18.7.1 Stagnation Point. At each point we can describe the velocity of the fluid flow by means of a vector tangent to a streamline. If there is no well-defined tangent at a point, then the velocity must be zero there. These points are called *stagnation points* of the fluid. Consider the analytic function $w = f(z) = z + \dfrac{a^2}{z}$ of Example 18.15. For this function the stagnation points are at $z + \pm a$, because the streamline along the real axis splits at these points to form the circle $|z| = a$ as shown in Figure 18.23(g). The stagnation points occur where $f'(z) = 0$. Thus, $f'(z) = 1 - \dfrac{a^2}{z} = 0$ yields $z = \pm a$. Also, along the real axis, $f(z)$ itself is real and thus, $v(x,y) = 0$ is the streamline along the real axis.

18.7.2 Velocity Potential in Fluid Flows. The streamlines of a fluid flow are found, as the level lines of the imaginary part of a given analytic function. The velocity of the flow is a vector along the tangent to the streamlines. Recall that the gradient $\nabla u = u_x + i\, u_y$ is at each point (x,y) orthogonal to the level curve $u =$const through that point, and the modulus of the gradient, $|\nabla u| = \sqrt{u_x^2 + u_y^2}$ is the directional derivative du/ds in the direction of the most rapid increase in u. In fact, the gradient ∇u describes the *velocity vector* of the flow, which using the Cauchy-Riemann equation $u_y = -u_x$ becomes $\nabla u = u_x - u_y$. Since the derivative $\dfrac{\partial f}{\partial x} = \dfrac{\partial(u+iv)}{\partial x} = u_x + i\, v_x$, we compare it with the above result for ∇u and find that the velocity vector is $\overline{f'(z)}$.

We have found that the analytic function $f(z) = u + i\,v$ satisfies all the features of an ideal fluid with the following properties:

(i) The stream function is $v(x, y)$, and the streamlines are given by $v(x, y) = c$, where c is an arbitrary constant; and

(ii) the velocity potential is given by $u(x, y)$, where the equipotential lines are given by $u(x, y) = c$, and the velocity vector by $\overline{f'(z)}$. The function $f(z)$ is called the *complex potential* of the fluid flow.

Example 18.16. To find the complex potential, velocity vector, and the stream function of the fluid flow when a cylindrical pipe of radius a is placed in a uniform horizontal flow with velocity V far from the pipe, consider a horizontal flow past this pipe described by the function $w = z + \dfrac{a^2}{z}$ (Figure 18.23(g)). The velocity of the flow is described by $\overline{w'} = 1 - \dfrac{a^2}{\bar{z}^2}$. When z is large (i.e., away from the pipe), the velocity is nearly 1. So we must adjust the above mapping function w so that it has the same streamlines, but the velocity V is far away from $z = 0$. This is done by multiplying the function w by the constant V, and the streamlines are then given by $V\,v(x, y)$ =const, where $v(x, y) = c$ defines the same type of streamlines. Thus, we set $w = V\left(z + \dfrac{a^2}{z}\right)$, which is the new complex potential, and the velocity far from the origin is $\overline{w'} = \dfrac{Va^2}{\bar{z}^2} \approx V$. ■

Example 18.17. Consider the uniform flow of velocity V in the direction inclined at an angle α to the real axis. The velocity $V\,e^{i\alpha}$ is the complex conjugate of the complex potential, i.e., $\overline{f'(z)} = V\,e^{i\alpha}$, or $f'(z) = V\,e^{-i\alpha}$, which gives the desired complex potential as $f(z) = V\,e^{i\alpha}\,z$. ■

Example 18.18. To find the points where the speed of the flow described in Example 18.16 is a maximum, consider a source of fluid at the origin, defined by $f(z) = m \log z$, where m is called the *strength* of the source. Figure 18.25 shows a cylinder of fluid of radius r and height 1 at time t.

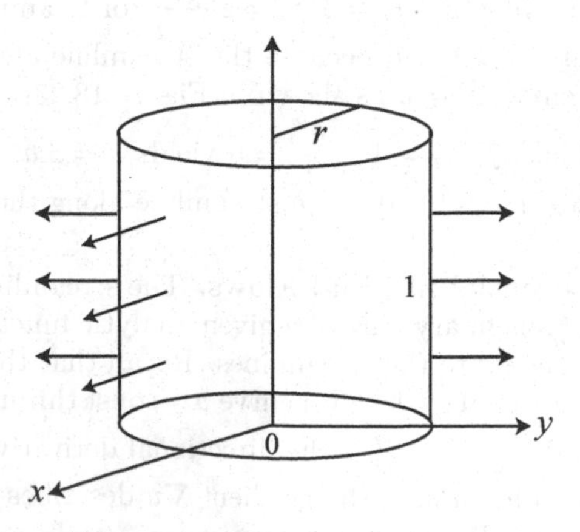

Figure 18.25 Flow past a cylinder.

A short time later at $t+dt$, the radius of the cylinder increases to $r+dr$. Since the velocity vector has magnitude $|f'(z)| = |m/z| = m/r$, and since the speed is described by dr/dt, we get $dr/dt = m/r$. Thus, the rate at which the volume increases is given by

$$\frac{d(\text{volume})}{dt} = \frac{d(\pi r^2)}{dt} = 2\pi r \frac{dr}{dt} = 2\pi r \frac{m}{r} = 2\pi m.$$

Thus, the complex potential $m \log z$ describes a source of strength m at the origin in which a fluid emerges at the rate $2\pi m$ units of volume per unit time for each unit of length perpendicular to the z-plane. ■

Example 18.19. Consider the flow described by $w = m\log(\sin z)$. Note that near the origin, $\sin z \approx z$, and thus $w \approx m \log z$ for very small z, and this is the source of strength m at the origin. At $z = \pi$, $\sin z \approx \pi - z$, and we have

$$w = m\log(\sin z) \approx m\log(\pi - z) = m\log(-1)(z-\pi) = m\log(-1) + m\log(z-\pi) = m\log(z-\pi),$$

where the term $\log(-1)$, being a constant added to the complex potential, is dropped because it does not alter the physical flow in any way. The term $m\log(z-\pi)$ is a source of strength m at $z = \pi$. Thus, continuing this way and knowing that $\sin z$ has zeros at all integral multiples of π, we have sources of strength m at the points $z = n\pi$, where $n = 0, \pm 1, \pm 2, \ldots$. The streamlines of $w = m\log(\sin z) = m\log|\sin z| + i\,m\arg\{\sin z\}$ are the curves $\arg\{\sin z\} = c$, where c is an arbitrary constant. These curves and the equipotential lines $|\sin z| = \text{const}$ are presented in Figure 18.26. The mapping $\sin z$ is described in Map 5.9(v), and presented in Figure 5.10(e); also, see Maps 5.27 and 5.28. ■

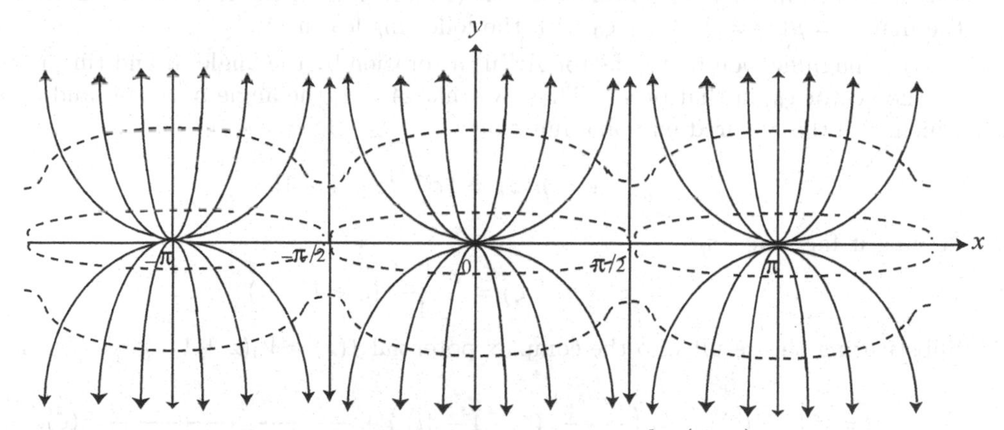

Figure 18.26 Flow defined by $w = m\log(\sin z)$.

Besides using an analytic function as the complex potential to describe the fluid flow, we can also use properties of the mapping function $M(z)$ that maps each point of the z-plane onto a point in the w-plane.

18.8 Conformal Mapping of Flow Patterns

Fluid flows can also be described by using mapping properties of analytic functions by transforming $f(z)$ into a new flow pattern. Suppose that $f(z)$ describes the flow past a solid disk (Figure 18.27(a)) from left to right in the z-plane which we assume is made of a flexible material. Now consider another analytic function $\zeta = \mu(z)$, where $\zeta = \xi + i\eta$, which maps each point of the z-plane onto a new point on the ζ-plane. This mapping can be thought of as a distortion of the flexible surface into a new shape shown in Figure 18.27(b), where the streamlines drawn on the ζ-plane have also been distorted. We can solve $\zeta = \mu(z)$ for z in terms of ζ so that $z = \mu^{-1}(\zeta)$. Then the complex potential describing the new flow pattern is given by $f(z) = f\left(\mu^{-1}(\zeta)\right)$. In the following examples we will see how the mapping properties of the function $\zeta = \mu(z)$ can generate new models of different fluid flows.

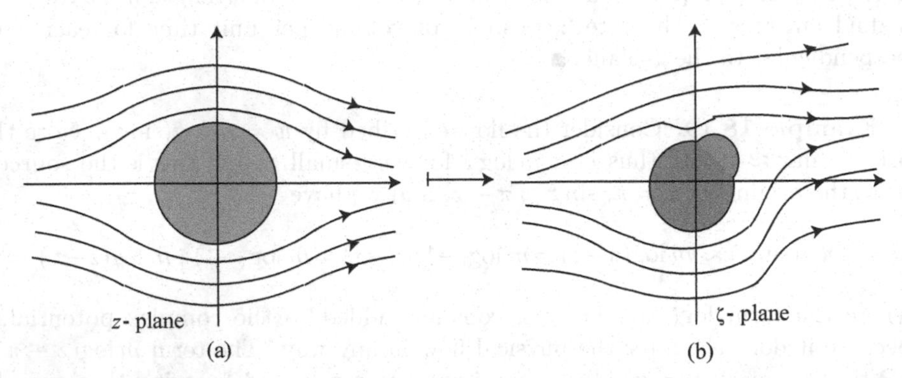

Figure 18.27 (a) Flow past a solid disk. (b) Distorted flow.

Example 18.20. In Example 18.16 the complex potential $f(z) = V_0\left(z + \dfrac{1}{z}\right)$ describes a uniform horizontal flow around a circular cylindrical obstacle of radius 1 about the origin. If the flow is in the direction $\pi/4$ of the velocity V far from the origin around a cylindrical obstacle of radius 4 with center at $4+4i$ (Figure 18.28), the flow can be described under the map $\zeta = \mu(z) = A\,e^{i\alpha}z + \zeta_0$ with the following features:

(i) a magnification by the factor A; (ii) a rotation by the angle α; and (iii) a translation by the vector ζ_0, in that order. Thus, we take $A = 4$, the angle $\alpha = \pi/4$, and $\zeta_0 = 4+4i$, which give the required mapping function

$$\zeta = \mu(z) = 4\,e^{i\pi/4}z + 4 + 4i.$$

Solving it for z we get

$$z = \mu^{-1}(\zeta) = \frac{e^{-i\pi/4}}{4}\,(\zeta - 4 - 4i)\,.$$

Substituting this result into the complex potential $f(z) = V_0(z + 1/z)$, we obtain

$$f(z) = f\left(\mu^{-1}(\zeta)\right) = V_0\Big(\frac{e^{-i\pi/4}}{4}\,(\zeta - 4 - 4i)\Big) + \frac{4}{e^{-i\pi/4}\,(\zeta - 4 - 4i)} \equiv F(\zeta). \tag{18.8.1}$$

Now, we choose V_0 such that the speed of the flow remains V far from $\zeta = 0$. Note that for large $|\zeta|$, we find from Eq (18.8.1) that

$$F(\zeta) \approx \frac{V_0\,e^{-i\pi/4}\,(\zeta - 4 - 4i)}{4},$$

and so the velocity is approximately $\overline{F'(\zeta)} \approx \frac{1}{4}V_0\,e^{i\pi/4}$. Thus, the speed $V = V_0/4$, and so we choose $V_0 = 4V$. Then from (18.8.1) the complex potential is

$$V\Big[e^{-i\pi/4}\,(\zeta - 4 - 4i) + \frac{16\,e^{i\pi/4}}{\zeta - 4 - 4i}\Big]. \tag{18.8.2}$$

Since the function $\mu(z)$ represents a conformal map, we must have $\mu'(z) \neq 0$ in the interior of the region of the flow. The streamlines are already presented in Figure 18.23 and Figure 18.26. ■

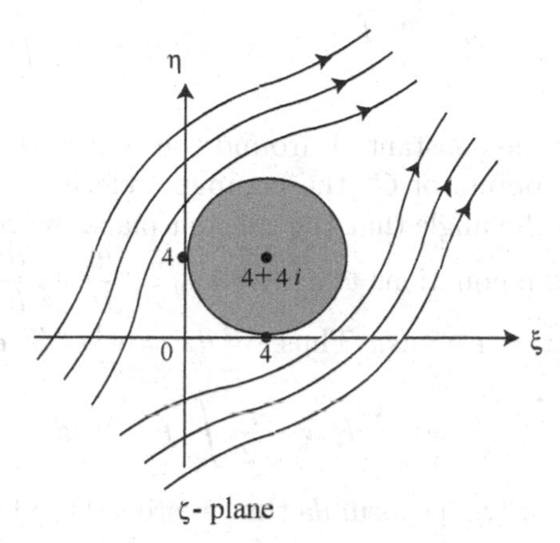

Figure 18.28 Flow around a cylinder.

18.9 Joukowski Maps

Consider the irrotational plane flow of an ideal fluid around an infinite plate. Let the (x, y)-plane intersect the plate along the segment $-a \leq x \leq a$ and the velocity vector of the flow lie in the (x, y)-plane and have a prescribed complex value w_∞. Since the Joukowski function (Map 6.5)

$$z = F(\zeta) = \frac{a}{2}\left(\zeta + \frac{1}{\zeta}\right) \tag{18.9.1}$$

conformally maps the exterior of the unit circle in the ζ-plane onto the z-plane cut along the segment $-a \leq x \leq a$, we have $F(\infty) = \infty$, and $F'(\infty) = a/2$. Hence, the flow problem reduces to that of an irrotational flow around a circular cylinder of unit radius in the ζ-plane with complex velocity $W_\infty = \dfrac{a}{2}\, w_\infty$ at infinity. The complex potential of this latter problem is given by

$$\psi(\zeta) = \frac{a}{2}\left(\bar{w}_\infty\, \zeta + \frac{w_\infty}{\zeta}\right). \tag{18.9.2}$$

If we set $\zeta = \dfrac{z + \sqrt{z^2 - a^2}}{a}$ and $\dfrac{1}{\zeta} = \dfrac{z - \sqrt{z^2 - a^2}}{a}$, which are obtained from (18.9.1) such that $\sqrt{z^2 - a^2} > 0$ for $z = x > a$, and separate w_∞ into real and imaginary parts $(w_\infty = (v_x)_\infty + i\,(v_y)_\infty)$, where the vector $\mathbf{v}_\infty = (v_x)\,\mathbf{i} + (v_y)\,\mathbf{j}$ and v_x and v_y are related to $w = u + iv$ by the Cauchy-Riemann equations $v_x = \dfrac{\partial u}{\partial x} = \dfrac{\partial v}{\partial y}$ and $v_y = \dfrac{\partial u}{\partial y} = -\dfrac{\partial v}{\partial x}$, then the complex potential of the original problem from (18.9.2) is given by

$$f(z) = (v_x)_\infty\, z - i\,(v_y)_\infty\, \sqrt{z^2 - a^2}. \tag{18.9.3}$$

Now, since the force of pressure acting on an element ds of any contour C is proportional to the hydrodynamic pressure p at the given point of flux and is directed along the inward normal $\mathbf{n}$ such that $-d\mathbf{n} = \mathbf{j}\,dx - \mathbf{i}\,dy$, we find that the components of the force acting on the contour C are given by $R = R_x + i\,R_y$, where $R_x = -\displaystyle\int_C p\,dy$ and $R_y = \displaystyle\int_C p\,dx$. Since

$p = A - \dfrac{\rho v^2}{2}$ from Bernoulli's integral, where A is a constant and ρ is the fluid density, we obtain the pressure force with which the flow acts on the plate as

$$R = \frac{\rho}{2} \int_C \mathbf{v}^2 \,(dx - i\,dy) = -\frac{\rho}{2} \int_C \mathbf{v}^2\, \overline{dz}, \tag{18.9.4}$$

where the integral of the constant A around the contour C is zero. Since the velocity acts tangentially to C at points of C, the complex velocity w of the flow is related to $\mathbf{v}$ by $w = \mathbf{v}\, e^{i\phi}$, where ϕ is the angle that the tangent makes with the x-axis. Then, since in view of the Cauchy-Riemann equations $w = v_x + i\, v_y = \dfrac{\partial u}{\partial x} + i\, \dfrac{\partial u}{\partial y} = \dfrac{\partial u}{\partial x} - i\, \dfrac{\partial v}{\partial x} = \overline{f'(z)}$, we have $\mathbf{v}\, e^{i\phi} = f'(z)$, and $\overline{dz} = e^{-i\phi}\, ds$. Thus, $\mathbf{v}^2\, dz = \mathbf{v}^2\, e^{-2i\phi}\, e^{i\phi}\, ds = f'^2(z)\, dz$, and (18.9.4) becomes

$$R = -\frac{\rho}{2} \int_C f'^2(z)\, dz, \tag{18.9.5}$$

which is known as *Chaplygin's formula* that expresses the force exerted by a flow on a body around which it flows in terms of the derivative of the complex potential.

If we use the formula

$$f(z) = \bar{w}_\infty\, z + \frac{\Gamma_\infty}{2i\pi} \log z + \sum_{n=0}^{\infty} \frac{c_n}{z^n},$$

which defines the complex potential in the neighborhood of the point at infinity, where Γ_∞ is the circulation of the flow at infinity, we find that

$$f'(z) = \bar{w}_\infty\, z + \frac{\Gamma_\infty}{2i\pi}\, \frac{1}{z} + \sum_{n=2}^{\infty} \frac{c_n}{z^n},$$

$$f'^2(z) = \frac{\bar{w}_\infty\, \Gamma_\infty}{i\pi}\, \frac{1}{z} + \sum_{n=2}^{\infty} \frac{b_n}{z^n},$$

and thus, $\displaystyle\int_C f'^2(z)\, dz = 2\bar{w}_\infty\, \Gamma_\infty$. Then (18.9.5) yields $R = \rho\, (v_y)_\infty\, \Gamma_\infty - \rho\, (v_x)_\infty\, \Gamma_\infty$, or

$$|R| = \rho\, |\mathbf{v}_\infty|\, |\Gamma_\infty|, \tag{18.9.6}$$

known as *Joukowski's theorem* on lifting force, which states that the force of pressure of an irrotational flow with velocity $\mathbf{v}_\infty$ at infinity and flowing around a contour C with circulation Γ is given by the formula $R = \rho\, |\mathbf{v}_\infty|\, |\Gamma|$. The direction of the force is obtained by rotating the vector $\mathbf{v}_\infty$ through an angle $\pi/2$ in the direction of the circulation.

The Joukowski map is defined by $w = \frac{1}{2}(z + 1/z)$. Since $dw/dz = \frac{1}{2}\left(1 - 1/z^2\right) = 0$ iff $z = \pm 1$, the Joukowski map is conformal everywhere except at the critical points ± 1 and at the singularity $z = 0$ where it is not defined. Take a point $z = e^{i\theta}$ on the unit circle. Then $w = \frac{1}{2}\left(e^{i\theta} + e^{-i\theta}\right) = \cos\theta$ lies on the real u-axis with $-1 \le u \le 1$, which means that the Joukowski map reduces the unit circle in the z-plane to the real line-segment $[-1, 1]$ in the w-plane.

The inverse map $z = w \pm \sqrt{w^2 - 1}$ shows that every $w \neq \pm 1$ is the image of two different points z. If w is not on the critical line-segment $[-1, 1]$, then a point w with the minus sign lies inside the unit circle and the one with the plus sign lies outside the unit circle, and if $-1 < w < 1$, then both points lie on the unit circle and a common vertical line. Thus, the Joukowski map is a one-to-one conformal mapping from the exterior of the unit circle $|z| > 1$ onto $\mathbb{C}\backslash[-1, 1]$ which is the exterior of the unit line segment $[-1, 1]$. Also, the Joukowski map transforms the concentric circles $|z| = r \neq 1$ onto ellipses with foci at ± 1 in the w-plane (see Figure 18.29).

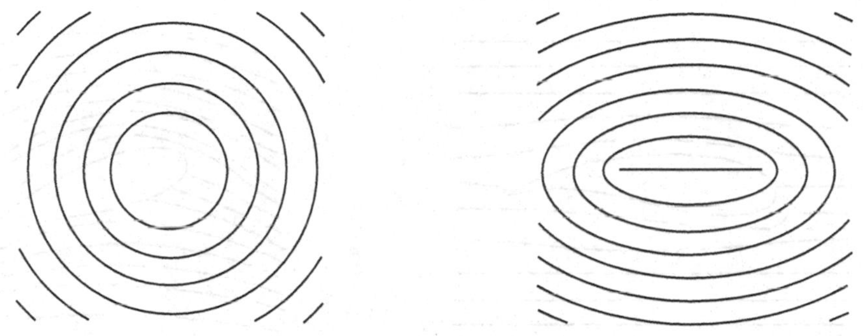

Figure 18.29 Joukowski map.

Under the Joukowski map, the image curves take different shapes depending on the values of the center and radius of the circle in the z-plane; some of these images in the w-plane are shown in Figure 18.30 for the pairs (center, radius).

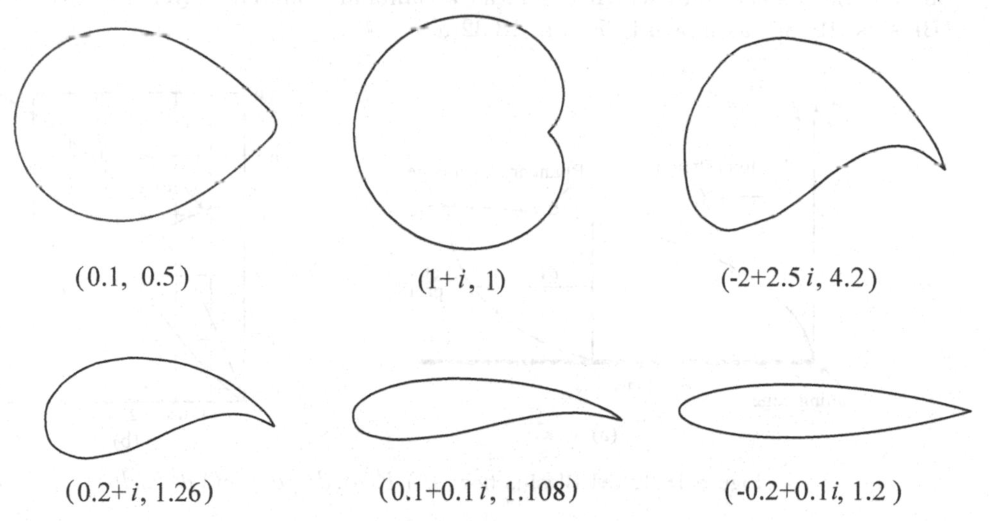

Figure 18.30 Airfoils for different centers and radii.

If the circle passes through the singular point $z = -1$, then its image is not smooth and has a cusp at $w = 1$, and some of the images in Figure 18.29 are shaped like the cross-section through an idealized airfoil (plane wing).

Example 18.21. (Tilted plate) In the case of a tilted plate in a uniform horizontal fluid flow, the cross-section is the line segment $z(t) = t\,e^{i\phi}$, $-1 \leq t \leq 1$, which is obtained by rotating the horizontal line-segment $[-1, 1]$ through an angle $-\phi$ (see Figure 18.31), in which the flow is shown as going from left to right and ϕ is known as the *attack angle* to the flow relative to the flow). The resulting flow is to tilt the segment by the angle $-\phi$ while rotating the streamlines which become asymptotically horizontal. Thus, the complex potential of the flow is of the form

$$\chi(z) = e^{i\phi}\left(z\cos\phi - i\sin\sqrt{z^2 - e^{-2i\phi}}\right).$$

The corresponding streamlines of the flow are shown in Figure 18.31. ■

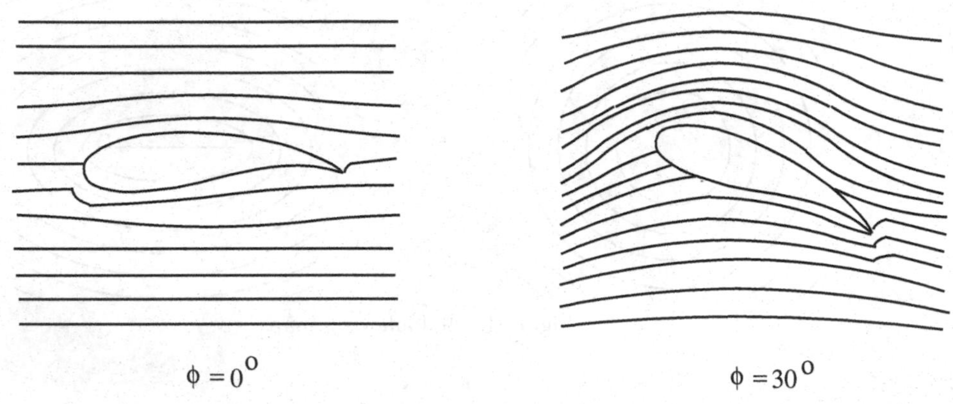

Figure 18.31 Flow past a titled airfoil.

A nonzero circulation around a body creates lift, although the precise relation depends on the Blasius theorem, which is related to laminar boundary layer flow over a flat plate (Blasius [1908]), as shown in Figure 18.32(a).

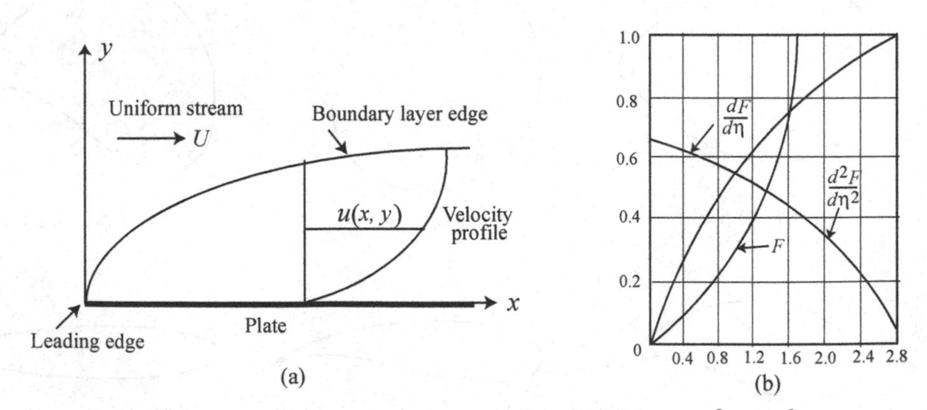

Figure 18.32 (a) Blasius flow; (b) $F(\eta), dF/d\eta$, and $d^2F/d\eta^2$.

The laminar boundary layer equations for this problem are

$$\frac{\partial u}{\partial s} + \frac{\partial v}{\partial n} = 0, \quad u\frac{\partial u}{\partial s} + v\frac{\partial u}{\partial n} = \nu\frac{\partial^2 u}{\partial n^2},$$

where ν is the viscosity of the fluid. Using the stream function ψ, given by $u = \dfrac{\partial\psi}{\partial n}$ and $v = -\dfrac{\partial\psi}{\partial s}$, the governing equation, known as the *Blasius equation*, is

$$\frac{\partial\psi}{\partial n}\frac{\partial^2\psi}{\partial s\partial n} - \frac{\partial\psi}{\partial s}\frac{\partial^2\psi}{\partial n^2} = \nu\frac{\partial^3\psi}{\partial n^3}, \tag{18.9.7}$$

subject to the following boundary conditions: (i) $\psi = 0$ on the solid surface, $n = 0$ since $v = 0$; (ii) $\dfrac{\partial \psi}{\partial n} = 0$ on the solid surface, $n = 0$ since $u = 0$; and (iii) $\dfrac{\partial \psi}{\partial n} \to U$ as $n \to \infty$ since $u \to U$. Following Blasius [1908], the Blasius solution for a steady, planar, laminar boundary layer with zero pressure gradient (i.e., $dU/ds = 0$) is

$$\psi = \sqrt{4\nu U s}\, F(\eta), \quad \eta = \left(\frac{U}{4\nu s}\right)^{1/2} n, \tag{18.9.8}$$

and the velocities $u(s,n)$ and $v(s,n)$ are given by

$$u = U\frac{dF}{d\eta}, \quad v = \left(\frac{\nu U}{s}\right)^{1/2}\left(F - \eta\frac{dF}{d\eta}\right). \tag{18.9.9}$$

The forms of (η), $dF/d\eta$, and $d^2F/d\eta^2$ are shown in Figure 18.32(b). For the Blasius theorem see §18.10.1.

18.10 Airfoils

The fluid motion around an airfoil can be obtained from the flow past an off-center circle, as shown above in Figure 18.30. First, consider an affine map $w = az + b$. It moves the unit disk $|z| \le 1$ to the disk $|w - b| \le |a|$ (with center b and radius $|a|$). The angular component of a creates a rotation, so that the streamlines around the new disk will be inclined at an angle $\phi = \arg\{a\}$ with the horizontal, although the boundary circle will pass through $w = 1$ so long $|a| = |1 - b|$. Now, using the Joukowski transformation $\zeta = \frac{1}{2}(w + 1/w) = \frac{1}{2}\left(az + b + \dfrac{1}{az + b}\right)$, which maps the new disk $|w - b| \le |a|$ to the airfoil, the complex potential for the flow past the airfoil is obtained by substituting the inverse map

$$z = \frac{w - b}{a} = \frac{\zeta - b + \sqrt{\zeta^2 - 1}}{a}$$

into the disk potential (18.3.6), we get

$$\Theta(\zeta) = \frac{\zeta - b + \sqrt{\zeta^2 - 1}}{a} + \frac{a\left(\zeta - b - \sqrt{\zeta^2 - 1}\right)}{b^2 + 1 - 2b\zeta}. \tag{18.10.1}$$

Finally, replacing ζ by $e^{i\phi}\zeta$ in (18.10.1), the streamlines become asymptotically horizontal. ■

Note that potential flows do not produce lift, so an airplane with such a wing would not fly! The question is then, how do the birds and commercial airplanes fly? The answer lies in the phenomenon of circulation.

The integral of the complex velocity $f(z)$ along a curve C is given by

$$\int f(z)\, dz = \int_C (u\, dx + v\, dy) - i \int_C (v\, dx - u\, dy) = \int_C \mathbf{v} \cdot d\mathbf{x} - i \int_C \mathbf{v} \cdot \mathbf{n}\, ds,$$

such that its real part denotes the *circular integral*, while its imaginary part is minus *flux integral*, which can also be defined by $\int_C \mathbf{v} \times d\mathbf{x}$. If the complex velocity admits a single-valued complex potential in a simply connected domain

$$\chi(z) = \phi(z) - i\psi(z), \quad \chi'(z) = f(z), \tag{18.10.2}$$

then the complex integral is independent of path over a curve connecting a to b, which yields $\int_C f(z)\,dz = \chi(b) - \chi(a)$ by the fundamental theorem. Thus, the circulation and flux integral for an ideal fluid flow are also path independent, so that the circulation integral is given by

$$\int_C \mathbf{v}\cdot d\mathbf{x} = \int_C \nabla\phi\cdot d\mathbf{x} = \phi(b) - \phi(a),$$

and the flux integral by

$$\int_C \mathbf{v}\times d\mathbf{x} = \int_C \nabla\psi\cdot d\mathbf{x} = \psi(b) - \psi(a),$$

which is the difference in the values of the stream function at the endpoint of C and the 'flux potential' for the flow depends only on the endpoints. If C is a closed contour and $\chi(z)$ is analytic within C, then$-\int{}_C\mathbf{v}\cdot d\mathbf{v} = 0 =-\int{}_C\mathbf{v}\times d\mathbf{x}$, implying that there is no circulation or flux within any closed curve. Thus, lift on a body is possible only under a nonzero circulation around it. It is governed by the Blasius theorem, which is as follows.

18.10.1 Blasius Theorem. Consider a steady two-dimensional harmonic flow with velocity $(U, 0)$ at infinity, past a cylinder of radius a centered at the origin, is not unique. However, such a harmonic flow may be written as

$$w = U(z + a^2/z) + \frac{iT}{2\pi}\log z, \tag{18.10.3}$$

where the principal branch of the logarithm function is taken. Now, we seek a conformal map from the z-plane onto the w-plane to obtain a complex potential $w(z)$, such that streamlines map onto streamlines.

Example 18.22. The map $z(\zeta) = \zeta + \dfrac{b^2}{\zeta}$ (Map 6.8) maps the circle of radius $a > b$ in the ζ-plane onto the ellipse of semi-major axis $\dfrac{a^2+b^2}{a}$ and the semi-minor (y-axis) $\dfrac{a^2-b^2}{a}$ in the z-plane, while the exterior is mapped onto the exterior. A uniform flow with velocity $(U, 0)$ at infinity, past the circular cylinder $|\zeta| = a$, has the complex potential $\zeta(z) = U(\zeta + a^2/\zeta)$. Inverting the map and requiring that $\zeta \approx z$ for large $|z|$ gives $\zeta = \frac{1}{2}\left(z + \sqrt{z^2 - 4b^2}\right)$. Then $w(z) = w(\zeta(z))$ is the complex potential for uniform flow past the ellipse. Note that $dz/d\zeta \to 1$ as $z \to \infty$. Thus, infinity maps by the identity and so the uniform flow imposed on the circular cylinder is also imposed on the ellipse.

Moreover, the same mapping $w = f(z) = z + \dfrac{a^2}{z}$ describes a flow with stagnation points at $z = \pm a$. The streamlines along the real axis split at these points to form the circle $|z| = a$, which can be proved as follows: the stagnation point occurs where $f'(z) = 0$. Since

$$f'(z) = 1 - \frac{a^2}{z^2} = 0,$$

we get $z = \pm a$ as the stagnation points. Along the real axis, $f(z)$ itself is real, and thus, $v(x, y) = 0$ is a streamline along the real axis. Since

$$v(x, y) = y\Big(1 - \frac{a^2}{x^2 + y^2}\Big),$$

we get $v(x, y) = 0$ either when $y = 0$ (real axis), or when $x^2 + y^2 = a^2$ (circle of radius a). ■

Theorem 18.2. (Circle theorem) *Let a harmonic flow have complex potential $f(z)$, analytic in the domain $|z| \leq a$. If the circular cylinder of radius a is placed at the origin, then the new complex potential is $w(z) = f(z) + \overline{f(a^2/z)}$.*

PROOF. We need to verify that the surface of the cylinder is a streamline. Note that on the circle $a^2/\bar{z} = z$, we have $w(z) = f(z) + \overline{f(z)}$, which means that the stream function $\psi = 0$, i.e., the circle is a streamline. Next, note that the added term is an analytic function of z if it is not singular at z, since if $f(z)$ is analytic at z, so is $\overline{f(\bar{z})}$. Since f is analytic in $|z| \leq a$, the function $f(a^2/z)$ is analytic in $|z| \geq a$, and so also for the function $\overline{f(\bar{z})}$. Hence, the only singularity of $w(z)$ in $|z| > a$ are those of $f(z)$. ■

Example 18.23. If a cylinder of radius a is placed in a uniform flow, then $f = Uz$ and $w = Uz + U\,\overline{(a^2/z)}$. If a cylinder is placed in the flow of a point source at $b > a$ on the x-axis, then $f(z) = \dfrac{Q}{2\pi}\log(z - b)$ and

$$w(z) = \frac{Q}{2\pi}\Big(\log(z - b) + \overline{\log(a^2/\bar{z} - b)} = \frac{Q}{2\pi}\left(\log(z - b) + \log(z - a^2/z) - \log z\right) + C, \tag{18.10.4}$$

where Q is the source strength and C is a constant. This result helps us verify that the imaginary part of w is constant when $z = a\,e^{i\theta}$. Note that the image system of the source, with singularities within the circle, consists of a source of strength Q at the image point a^2/b, and a source of strength $-Q$ at the origin. ■

In fluid dynamics the calculation of the force exerted by the fluid on a rigid body is important. In a 2-D steady harmonic flow the calculation is done by using the complex potential.

Theorem 18.3. (Blasius theorem) *Let a steady uniform flow past a fixed two-dimensional body with boundary C be a harmonic flow with velocity potential $w(z)$. Then, if no external body forces are present, the force (X, Y) exerted by the fluid on the body is given by*

$$X - iY = \frac{i\rho}{2}\oint_C \left(\frac{\partial w}{\partial z}\right)^2 dz, \tag{18.10.5}$$

where the integral is taken around the boundary in counter-clockwise sense.

The formula (18.10.5), due to Blasius, reduces the force calculation to a complex contour integration.

PROOF. Recall that $dX - i\,dY = p(-dy - i\,dx) = -i\,p\,d\bar{z}$, where p denotes pressure. Also, by Bernoulli's theorem for steady ideal flows we have

$$p = -\frac{\rho}{2}\left|\frac{\partial w}{\partial z}\right|^2 + C, \tag{18.10.6}$$

and thus,

$$X - i\,Y = \frac{i\rho}{2}\oint_C \frac{\partial w}{\partial z}\overline{\frac{\partial w}{\partial z}}\,d\bar{z}. \tag{18.10.7}$$

But since the boundary C is a streamline, so that $d\psi = 0$ there, and so on C we have $\overline{\dfrac{\partial w}{\partial z}}\,d\bar{z} = d\bar{w} = \dfrac{\partial w}{\partial z}\,dz$, and substituting this into (18.10.7), the result follows. ■

18.10.2 Boundary Layer Flow. We will derive the Blasius solution for a laminar flow over a flat plate under the assumption that the flow is governed by the equation for a steady 2-D Newtonian fluid with negligible body forces, i.e., by

$$\frac{\partial u}{\partial x} + \frac{\partial u}{\partial y} = 0 \quad \text{(conservation of mass)}, \tag{18.10.8}$$

$$\rho\Big(u\frac{\partial u}{\partial x} + \nu\frac{\partial u}{\partial y}\Big) = v\frac{\partial^2 u}{\partial y^2} \quad \text{(momentum balance in the } x\text{-direction)}, \tag{18.10.9}$$

subject to the boundary conditions $u(x,0) = v(x,0) = 0$ and $u(x,\delta) = U$. Using the following change of variables

$$\eta = \sqrt{\frac{U}{vx}} = \sqrt{Re_x}\,\frac{y}{x},\ f(\eta) = \frac{\psi}{\sqrt{vxU}},\ \frac{u}{U} = \frac{\partial f}{\partial \eta} = f'(\eta),$$

the above equations reduce to

$$f''' + \frac{1}{2} f f'' = 0, \tag{18.10.10}$$

subject to the new boundary conditions $f(x,0) = f'(x,0) = 0$ and $\lim_{y\to\infty} f'(x,y) = 1$. Then, taking $\eta = 5$ so that $\dfrac{u}{U} = 0.9$, the Blasius solutions are as follows:

Disturbance Thickness: $\dfrac{\delta}{x} = \dfrac{5}{\sqrt{Re}}$,

Displacement Thickness: $\dfrac{\delta''}{x} = \Big(\dfrac{\delta''}{x}\Big)\Big(\dfrac{\delta}{x}\Big) = \dfrac{1.72}{\sqrt{Re}}$,

Momentum Thickness: $\dfrac{\theta}{x} = \Big(\dfrac{\theta}{\delta}\Big)\Big(\dfrac{\delta}{x}\Big) = \dfrac{0.665}{\sqrt{Re}}$,

Wall Shear Stress: $\tau_w = \mu\dfrac{\partial u}{\partial y}\Big|_{y=0} = \mu U\Big[\dfrac{\partial f'}{\partial \eta}\dfrac{\partial \eta}{\partial y}\Big]_{y=0} = \mu U f'_{\eta=0}\sqrt{\dfrac{U}{vx}} = \dfrac{0.332\rho U^2}{\sqrt{Re}}$,

Friction Coefficient: $c_f = \dfrac{\tau_w}{\frac{1}{2}\rho U^2} = \dfrac{0.664}{\sqrt{Re}}$,

Drag Coefficient for Friction: $C_{D,f} = \dfrac{1}{A}\displaystyle\int c_f\, dA = \dfrac{1}{l}\int_{x=0}^{l} \dfrac{0.664}{\sqrt{Re}}\, dx = \dfrac{1.328}{\sqrt{Re}}$.

The numerical data for the Blasius solution for the laminar flow over a flat plate is presented in Table 18.1.

Table 18.1 Blasius solution for the laminar flow over a flat plate.

$\eta = y\sqrt{\frac{U}{\nu x}}$	$f(\eta)$	$f'(\eta) = \frac{u}{U}$	$f''(\eta)$
0.0	0.0000	0.0000	0.3321
0.5	0.0415	0.1659	0.3309
1.0	0.1656	0.3298	0.3230
1.5	0.3701	0.4868	0.3026
2.0	0.6500	0.6298	0.2668
2.5	0.9964	0.7513	0.2174
3.0	1.3969	0.8461	0.1614
3.5	1.8378	0.9131	0.1078
4.0	2.3059	0.9555	0.0642
4.5	2.7903	0.9795	0.0340
5.0	3.2834	0.9916	0.0159
5.5	3.7807	0.9969	0.0066
6.0	4.2798	0.9990	0.0024
6.5	4.7795	0.9997	0.0008
7.0	5.2794	0.9999	0.0002
7.5	5.7794	1.0000	0.0001
8.0	6.2794	1.0000	0.0000

The dimensionless velocity profile for the Blasius solution for laminar flow over a flat plate is presented in Figure 18.33.

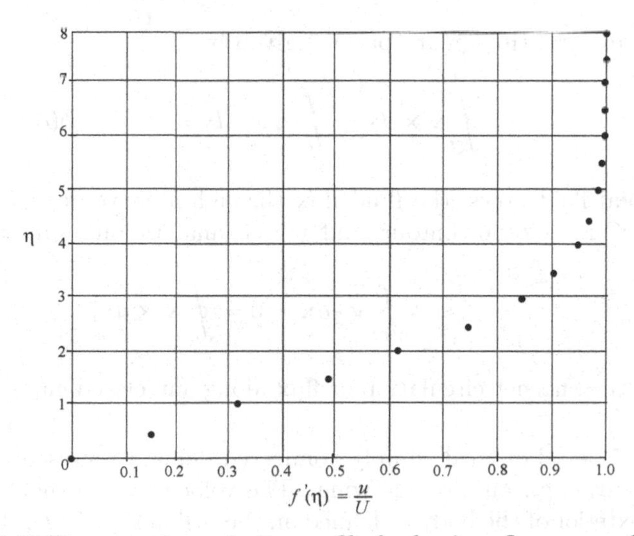

Figure 18.33 Dimensionless velocity profile for laminar flow over a flat plate.

Note that the magnitude of the velocity ratio v/u is given by $(Re_s)^{-1/2}$, where $Re_s = Us/\nu$ is the Reynolds number based on U and the distance s along the surface measured from the leading edge. Since the initial condition for the boundary layer is that the ratio u/U must be small, it requires that Re_s must be large. This will be true if $s \gg \nu/U$. Thus, the boundary layer approximation is valid except in a small region close to the leading edge whose length is ν/U, which has negligible effect and therefore, can be ignored.

18.10.3 Circulation. Consider a steady-state flow of an incompressible irrotational flow. Let $f(z) = u(x,y) + i\,v(x,y)$ denote the complex velocity corresponding to the real velocity vector $\mathbf{v} = (u(x,y), v(x,y))^T$ at the point $(x,y)^T$. The complex velocity $f(z)$ along a curve C is given in terms of two real line integrals, by

$$\int_C f(z)\,dz = \int_C (u + i\,v)(dx + i\,dy) = \underbrace{\int_C (u\,dx + v\,dy)}_{\text{circulation integral}} + i\underbrace{\int_C (v\,dx - u\,dy)}_{\text{flux integral}}. \tag{18.10.11}$$

Note that the circulation integral (real part) is defined by $\int_C \mathbf{v}\cdot d\mathbf{x}$, while the flux integral is defined by $\int_C \mathbf{v}\cdot\mathbf{n}\,ds = \int_C \mathbf{v}\times d\mathbf{x}$.

If the complex velocity admits a single-valued complex potential $\chi(z) = \phi(z) + i\,\psi(z)$, where $\chi'(z) = f(z)$, then the complex integral is independent of the path, and thus, by the fundamental theorem, for any curve C connecting α to β

$$\int_c f(z)\,dz = \chi(\beta) - \chi(\alpha), \tag{18.10.12}$$

Then the circulation integral (real part) is given by

$$\int_C \mathbf{v}\cdot d\mathbf{x} = \int_C \nabla\phi\cdot d\mathbf{x} = \phi(\beta) - \phi(\alpha), \tag{18.10.13}$$

and the flux integral (imaginary part) is given by

$$\int_C \mathbf{v}\times d\mathbf{x} = \int_C \nabla\psi\cdot d\mathbf{x} = \psi(\beta) - \psi(\alpha). \tag{18.10.14}$$

Thus, for ideal fluid flows, the fluid flux through a curve depends only on its endpoints. Moreover, if C is a closed contour, and $\chi(z)$ is analytic on its interior, then

$$\oint_C \mathbf{v}\cdot d\mathbf{x} = 0 = \oint_C \mathbf{v}\times d\mathbf{x}, \tag{18.10.15}$$

and there is then no net circulation or flux along any closed curve.

Let $D \subset \mathbb{C}$ be a bounded, simply connected domain representing the cross-section of a cylindrical body, e.g., an airplane wing. The velocity vector field $\mathbf{v}$ of a steady-state flow around the exterior of the body is defined on the region $\Omega = \mathbb{C}\backslash\bar{D}$. There is no flux boundary condition $\mathbf{v}\cdot\mathbf{n} = 0$ on ∂D indicates that there is no fluid flow across the boundary of the solid body. The resulting circulation of the fluid around the body is given by $\oint_C \mathbf{v}\cdot d\mathbf{x}$, where $C \in \Omega$ is any simply closed contour encircling the body. By Cauchy's theorem, the value of this integral does not depend on the choice of C. However, if the corresponding complex velocity $f(z)$ admits a single-valued complex potential in Ω, then from Eq (18.10.15) we know that the circulation integral is zero, and thus, the body will not experience any lift.

The only way to get lift is through a single-valued complex velocity with a non-zero circulation integral. This demands that its complex potential should be multiple-valued. There is only one function that has such a property, and it is the complex logarithm

$$\lambda(z) = \log(az + b), \quad \text{with derivative } \lambda'(z) = \frac{a}{az + b},$$

which is single-valued away from the singularity at $z = -b/a$ (see Map 5.20b). Thus, we should introduce the family of complex potentials

$$\chi_\gamma(z) = z + \frac{1}{z} + i\,\gamma \log z. \tag{18.10.16}$$

The corresponding complex velocity

$$f_\gamma(z) = \frac{\partial \chi_\gamma}{\partial z} = 1 - \frac{1}{z^2} + \frac{i\,\gamma}{z} \tag{18.10.17}$$

remains asymptotically 1 at large distances. On the unit circle $z = e^{i\theta}$. Thus,

$$f_\gamma\left(e^{i\theta}\right) = \frac{1}{2} - \frac{1}{2}\,e^{-2i\theta} + i\,\gamma\, e^{-i\theta} = (\sin\theta + \gamma) i\, e^{-i\theta}$$

is a real multiple of the complex tangent vector $i\,e^{-i\theta} = \sin\theta - i\cos\theta$, and hence, its normal velocity or flux vanishes iff γ is real. By Cauchy's theorem, if C is a curve going once around the disk in a counter-clockwise direction, then

$$\oint_C f_\gamma(z)\,dz = \oint_C \left(1 - \frac{1}{z^2} + \frac{i\gamma}{z}\right) dz = -2\pi\gamma. \tag{18.10.18}$$

Hence, if $\gamma \neq 0$, the circulation integral is non-zero, and there is a net lift on the cylinder. From the plots of the streamlines for the flow corresponding to a few values of γ in Figure 18.31, we notice that the asymmetry of the streamlines accounts for the lift experienced by the disk. In particular, for $|\gamma| \leq 2$, the stagnation points have moved from ± 1 to

$$z_\pm = \pm\sqrt{1 - \tfrac{1}{4}\gamma^2} - \frac{1}{2} i\,\gamma. \tag{18.10.19}$$

When we compose the modified potentials (18.10.16) with the Joukowski transformation (Map 6.1), we obtain a complex potential for flow around the airfoil which is image of the unit disk. The conformal map does not affect the value of the complex integrals, and hence, for any $\gamma \neq 0$, there is nonzero circulation around the airfoil under this modified fluid flow, and the airplane will definitely fly.

Notice that the value $-2\pi\gamma$ in (18.10.18) is arbitrary for the circulation integral, and this will lead to an arbitrary amount of lift. However, Kutta [1911] hypothesized that 'Nature chooses the constant γ so as to keep the velocity of the flow at the trailing edge of the airfoil finite'. This requires that the trailing edge $w = 1$ of the Joukowski map must be a stagnation point. The corresponding point on the unit circle is

$$z = \frac{1 - \beta}{\alpha} = e^{i(\psi - \phi)}, \tag{18.10.20}$$

where $\phi = \arg\{\alpha\}$ and $\psi = \arg\{1 - \beta\}$, since we must have $|\alpha| = |1 - \beta|$ so that the image of the unit circle goes through $w = 1$. Equating Eq (18.10.20) to Eq (18.10.19), we obtain the *Kutta's formula*

$$\gamma = 2\sin(\phi - \psi), \tag{18.10.21}$$

which produces the corresponding circulation via (18.10.18). As long as the attack angle ϕ is moderate, the resulting flow and lift will remain in a fairly good agreement with the experimental data. For more details, see Lamb [1945], Batchelor [1967], Henrici [1974], and Keener [1988].

18.10.4 Lift. Returning to the problem of lift on a body, the Blasius theorem shows that such a lift requires a nonzero circulation around it. Let $D \subset \mathbb{C}$ be a compact, simply connected domain representing the cross-section of a cylindrical body, such as an airplane wing. The velocity field $\mathbf{v}$ of a steady flow around the exterior of the wing is defined on the domain $G = \mathbb{C}\backslash\bar{D}$, and the no flux boundary conditions $\mathbf{v} \cdot \mathbf{n} = 0$ on $\partial G = \partial D$ show that there is no fluid flowing across this boundary. Thus, there is no circulation of the fluid across this boundary, since this circulation is given by $\oint_C \mathbf{v} \cdot d\mathbf{x}$, where $C \subset G$ is any close contour around the body (wing), i.e.,

$$\oint_C \mathbf{v} \cdot \mathbf{n} = 0 = \oint_C \mathbf{v} \times \mathbf{n} = 0. \tag{18.10.22}$$

However, if the associated complex velocity $f(z)$ admits a single-valued complex potential in G, then in view of (18.10.22) where the circulation integrals are zero. There will be *no* lift. Moreover, since the stream lines of the flow are symmetric above and below the disk, as in Figure 18.10, there cannot be any vertical force either. Thus, any airplane with zero circulation integral will not fly.

However, airplanes do fly. So the lift is only possible if we consider a single-valued complex velocity with zero circulation integral, but require that its complex potential be multiple-valued. The only function that accomplishes it is the complex logarithm $w(z) = \log(az+b)$, with the single-valued derivative $w'(z) = \dfrac{a}{az+b}$ which is away from the singularity at $z = -b/a$ (see Map 5.20a). Thus, introducing the family of complex potentials

$$\chi_\gamma(z) = z + \frac{1}{z} + i \log z, \tag{18.10.23}$$

the corresponding complex velocity

$$f_\gamma(z) = \frac{d\chi_\gamma}{dz} = 1 - \frac{1}{z^2} + \frac{i\gamma}{z} \sim 1,$$

that is, the velocity remains 1 asymptotically at large distances. Notice that on the unit circle $z = e^{i\theta}$, we have

$$f_\gamma\left(e^{i\theta}\right) = \frac{1}{2}\left(1 - e^{-2i\theta}\right) - i\gamma,\ e^{i\theta} = (\sin\theta + \gamma)\, ie^{-i\theta},$$

which is a real multiple of the complex tangent vector $ie^{-i\theta} = \sin\theta + i\cos\theta$, and its normal velocity or flux vanishes iff γ is real. If C is a curve going once counter-clockwise around the disk, then, using Cauchy's theorem 2.1 and Cauchy's formula (2.4.10), we get

$$\oint_C f_\gamma(z)\, dz = \oint_C \left(1 - \frac{1}{z^2} + \frac{i\gamma}{z}\right) dz = -2\pi\gamma. \tag{18.10.24}$$

This means that if $\gamma \neq 0$, the circulation integral is not zero, and the body (wing) has a lift, as shown in Figure 18.34, which shows streamlines for the lift of the disk. In particular, if $|\gamma| \leq 2$, the stagnation points move from ± 1 to

$$z_\pm = \pm\sqrt{1 - \tfrac{1}{4}\gamma^2} - \frac{1}{2} i\gamma. \tag{18.10.25}$$

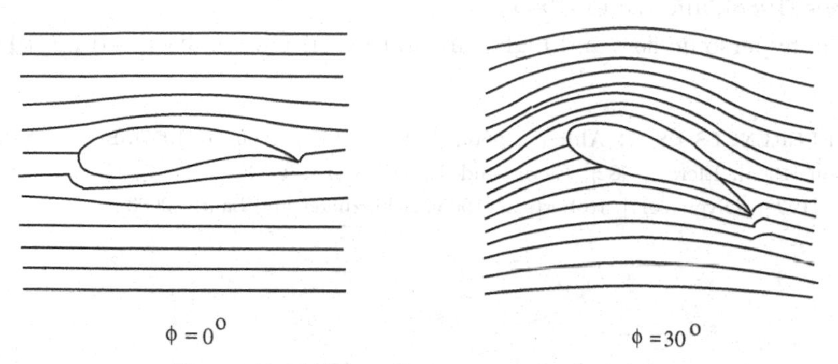

Figure 18.34 Kutta flow past a tilted airfoil.

The composite mapping of the Map 6.2 and the modified potential (18.10.23) yields a complex potential around the airfoil so that the circulation integral has an arbitrary value $-2\pi\gamma$, and thus an arbitrary amount of lift. The question as to which of these possible values correspond to the true physical situation follows from the Kutta hypothesis (Kutta [1902]), according to which the constant γ must be chosen so that the velocity of the flow at the trailing edge of the airfoil stays finite, thereby requiring that the trailing edge $\zeta = 1$ must be a stagnation point. In view of the Joukowski Map 6.2, the trailing edge corresponds to $w = 1$, and hence, under this map the corresponding point on the unit circle is

$$z = \frac{1-b}{a} - e^{i(\psi-\phi)}, \tag{18.10.26}$$

where $\phi = \arg\{a\}$ and $\psi = \arg\{1-b\}$, since we require that $|a| = |1-b|$ so that the image of the unit circle goes through the point $w = 1$. Also, equating (18.10.26) to (18.10.25), we obtain the *Kutta formula*

$$\gamma = 2\sin(\phi - \psi), \tag{18.10.27}$$

which produces the corresponding circulation controlled by Eq (18.10.24). Note that in Eq (18.10.27), the angle ϕ is known as the *attack angle*. Some flows of the airfoil are presented in Figure 18.35, which shows that as long as the attack angle ϕ is moderate, the resulting flow and lift remain in a fairly good agreement with the experimental data.

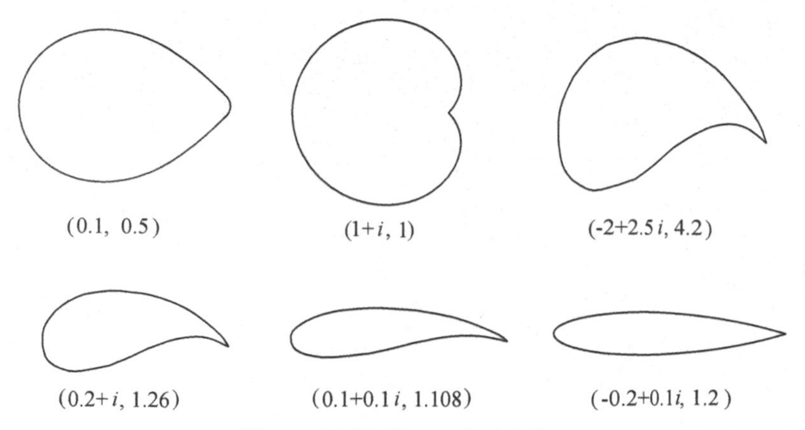

Figure 18.35 Flows of airfoils.

For more information and further development, see Batchelor [1967], Henrici [1974], Keener [1988], and Lamb [1945].

The hypersonic flow and high temperature effects are discussed in §19.6.

REFERENCES USED: Ahlfors [1966], Birkoff and Zarantonello [1957], Boas [1987], Carrier, Krook and Pearson [1966], Elcrat [1982], Elcrat and Trefethen [1986], Gaier [1964], Kantorovich and Krylov [1958], Kythe [1998], Olver [2017], Robertson [1965], Schinzinger and Laura [2003].

19

Heat Transfer

We will discuss steady heat transfer problems and their solutions both by analytical and conformal mapping methods. The conformal transformations that are used consist of $z = \sin w$, $w = -i\dfrac{z-1}{z+1}$, and $T = z^2$, $w = i \log T = 2i \log z$. These transformations help solve some steady and transient heat conduction problems.

19.1 Heat Flow

The properties of heat flows are summarized in Table 17.1 at the end of §17.9.1. Accordingly, in two-dimensional heat flow problems, the potential is the temperature denoted by $u(x, y)$; the lines of heat flux are $v(x, y) = c$ (c constant), thus giving the complex temperature $f(z) = u(x, y) + iv(x, y)$; a line of equal temperature is an isotherm $u(x, y) = c$; and the heat flux is defined by $\mathbf{Q}$, also known as the density of heat power flow. Thus, the equations governing the heat flow in a region D are

$$\begin{aligned} &\mathbf{Q} = k\mathbf{U}, \\ &\mathbf{U} = -\nabla u, \\ &\nabla^2 u = \begin{cases} 0 & \text{for the region } D \text{ without heat generation,} \\ -h/k & \text{for the region } D \text{ with a uniformly distributed heat source,} \end{cases} \end{aligned} \tag{19.1.1}$$

where $h = J_c^2 \rho$ watts per unit volume caused by current flow.

The *heat equation* with the constant coefficient,

$$u_t = k u_{xx} \tag{19.1.2}$$

with its particular solutions, has been introduced in §17.2. Some examples are given below.

In heat transfer problems there are three categories of boundary conditions, namely, Dirichlet, Neumann, and Robin, discussed in §17.2; they are also known as the isothermal, adiabatic, and outer heat conduction boundary conditions, respectively. The last category includes the effects of convection, radiation, and heat conduction into the surrounding medium, which is usually regarded as negligible.

The following problems are examples of each category:

$$\begin{aligned} u_t &= k\, u_{xx}, \quad 0 < x < l,\ t > 0, \\ u(x,0) &= f(x), \quad u_t(x,0) = g(x),\ 0 < x < l, \\ u(0,t) &= T_1(t), \quad u(l,t) = T_2(t),\ t > 0; \end{aligned} \tag{19.1.3}$$

$$\begin{aligned} u_t &= k\, u_{xx}, \quad 0 < x < l,\ t > 0, \\ u(x,0) &= f(x), \quad u_t(x,0) = g(x),\ 0 < x < l, \\ u_x(0,t) &= T_3(t), \quad u_x(l,t) = T_4(t),\ t > 0; \end{aligned} \tag{19.1.4}$$

$$\begin{aligned} u_t &= k\, u_{xx}, \quad 0 < x < l,\ t > 0, \\ u(x,0) &= f(x), \quad u_t(x,0) = g(x),\ 0 < x < l, \\ &\left.\begin{aligned} u(0,t) + \alpha\, u_x(0,t) = 0, \\ u(l,t) + \beta\, u_x(l,t) = 0, \end{aligned}\right\} \quad t > 0. \end{aligned} \tag{19.1.5}$$

An example of the category (iv) (implicit conditions) discussed in §17.2, is as follows.

Example 19.1. Consider the one-dimensional heat equation

$$\frac{\partial u}{\partial t} = k\,\frac{\partial^2 u}{\partial x^2}, \quad -\infty < x < \infty, \quad t > 0,$$

where k denotes the thermal diffusivity. This equation has the following sets of solutions:
(i) $u_0(x,t) = A\,x + B$, where A and B are arbitrary constants; and
(ii) $u_\lambda(x,t) = \cosh(\lambda x)\, e^{k\lambda^2 t}$, where $\lambda \ge 0$ is a real parameter.
Note that for any fixed t, $\lim\limits_{x\to\pm\infty} u_\lambda(x,t) = +\infty$. Since these solutions are not bounded as $x \to \pm\infty$, they may not be acceptable for physical problems. If we require the solution u to be bounded at infinity in both x and t, we must impose the following conditions (natural conditions):

$$\lim_{x\to\pm\infty} u(x,t) < \infty, \quad \text{and} \quad \lim_{t\to\infty} u(x,t) < \infty,$$

as well as the initial condition $u(x,0) = f(x)$, $-\infty < x < \infty$, where $f(x)$ is a preassigned bounded function for all x, e.g., $f(x) = e^{-x^2}$. ■

Example 19.2. (Ring-shaped heat conductor) As an example of periodic boundary conditions discussed in §17.2, consider a heat conductor of unit length such that periodic boundary conditions are prescribed at its two ends $x = \pm 1/2$, i.e., the heat pole at $x = 0$ repeats periodically (see Figure 19.1(a)).[1]

Because of this periodicity, the temperature u and all its derivatives coincide at the end points $x = \pm 1/2$ and there is no jump discontinuity at these end points. This is achieved by bending the conductor rod into a ring such that its two ends coincide. Although this ring is drawn circular in the above figure, its shape is immaterial for a linear heat conduction problem. However, the lateral surface of the ring is assumed to be adiabatically closed. The initial temperature is taken as $u(x,0) = f(x)$, where $f(x)$ is an arbitrary function that

[1] If the length of the conductor is l, then introduce a dimensionless coordinate $x' = x/l$ and write x instead of x'.

is symmetric with respect to $x = 0$. Then the Fourier expansion of $f(x)$ is a cosine series satisfying the periodicity condition at the end points, i.e.,

$$f(x) = \sum_{n=0}^{\infty} A_n \cos 2\pi nx, \quad A_n = 2 \int_{-1/2}^{1/2} f(x) \cos 2\pi nx \, dx. \tag{19.1.6}$$

The general solution of the heat equation $u_t = k\, u_{xx}$ is given by

$$u(x,t) = \sum_{n=0}^{\infty} A_n \, e^{-4\pi^2 n^2 kt} \cos 2\pi nx. \tag{19.1.7}$$

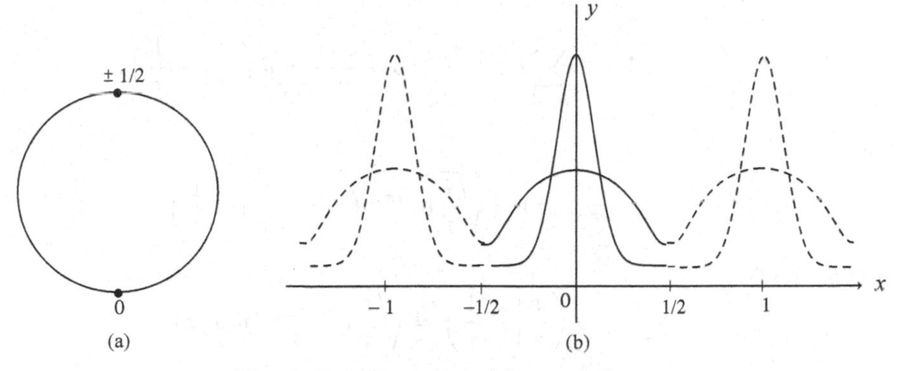

Figure 19.1 Ring-shaped heat conductor.

Let $f(x)$ denote the unit source, i.e., $f(x) = \delta(x)$, which means that $f(x) = 0$ for $x \neq 0$ and $\int_{-1/2}^{1/2} f(x)\, dx = 1$. Then the coefficients A_n in (19.1.6) are given by $A_0 = 1$, $A_n = 2$ for $n \geq 1$. In this case the solution $u(x,t)$, given by (19.1.7), becomes the theta-function $\vartheta(x|\, t)$. Hence, the solution of this problem is

$$\vartheta(x|\, t) = 1 + 2 \sum_{n=1}^{\infty} e^{-4\pi^2 n^2 kt} \cos 2\pi nx, \tag{19.1.8}$$

or, taking $4i\pi kt = \tau$,

$$\vartheta(x|\, \tau) = 1 + 2 \sum_{n=1}^{\infty} e^{i\pi\tau n^2} \cos 2\pi nx. \tag{19.1.9}$$

This series converges faster for large kt, and represents the later phases of exponential damping due to the unit source. The curves sketched in Fig. 19.1(b) for the solution (19.1.8) show the behavior for large kt (where $kt > 1$) by flat curves and the behavior for small kt ($kt < 1$) by steep curves. However, to understand the temperature distribution we go back to the Fourier series (19.1.7). From the heat source $U_0(x,t)$ given in the ring we find that at points $x = n$, where $n = \pm 1, \pm 2, \dots$, the identical heat sources are

$$U_n(x,t) = \frac{1}{\sqrt{4\pi kt}}\, e^{-(x-n)^2/4kt}. \tag{19.1.10}$$

Now, consider the series

$$u(x,t) = \sum_{n=-\infty}^{+\infty} U_n(x,t) = \frac{1}{\sqrt{4\pi kt}} \sum_{n=-\infty}^{\infty} e^{-(x-n)^2/4kt}, \tag{19.1.11}$$

which for small values of kt converges rapidly. In this series the dominant terms are U_0 and possibly U_{-1} and U_1; other terms have no effect because of the factor $e^{n^2/kt}$ in (19.1.10). Therefore, the representation (9.2.15) is a good complement of (19.1.8). If we rewrite the series (19.1.11) as

$$u(x,t) = \frac{1}{\sqrt{4\pi kt}}\, e^{-x^2/4kt} \left[1 + 2\sum_{n=1}^{+\infty} e^{-n^2/4kt} \cos\frac{inx}{2kt}\right],$$

where we have used the identity $\cos ix = \left(e^x + e^{-x}\right)/2$, then in terms of τ the terms in the square brackets become

$$1 + 2\sum_{n=1}^{+\infty} e^{-i\pi n^2/\tau} \cos\frac{2\pi nx}{\tau} = \vartheta\left(\frac{x}{\tau}\,\Big|\, -\frac{1}{\tau}\right).$$

Hence,

$$\vartheta(x|\,\tau) = \sqrt{\frac{i}{\tau}}\, e^{-i\pi x^2/\tau} \cdot \vartheta\left(\frac{x}{\tau}\,\Big|\, -\frac{1}{\tau}\right),$$

or, conversely,

$$\vartheta\left(\frac{x}{\tau}\,\Big|\, -\frac{1}{\tau}\right) = \sqrt{\frac{\tau}{i}}\, e^{i\pi x^2/\tau} \cdot \vartheta(x|\,\tau), \tag{19.1.12}$$

which is known as the *transformation formula* for the theta-function. ∎

Example 19.3. Consider the one-dimensional heat conduction equation

$$\frac{\partial u}{\partial t} = k\frac{\partial^2 u}{\partial x^2}, \quad 0 < x < l,$$

subject to the boundary and initial conditions

$$u(0,t) = 0 = u(l,t), \quad t > 0, \quad u(x,0) = f(x), \quad 0 < x < l, \tag{19.1.13}$$

where $f \in C^1$ is a prescribed function. In physical terms, this problem represents the heat conduction in a rod when its ends are maintained at zero temperature while the initial temperature u at any point of the rod is prescribed as $f(x)$. Using the method of separation of variables and the prescribed conditions, the solution is (Kythe et al. [2003: 128])

$$u(x,t) = \sum_{n=1}^{\infty} C_n \sin\frac{n\pi x}{l} e^{-kn^2\pi^2 t/l^2}, \tag{19.1.14}$$

where

$$C_n = \frac{2}{l}\int_0^l f(x) \sin\frac{n\pi x}{l}\, dx, \; n = 1, 2, \dots . \; ∎$$

An interesting situation, considered in the next example, arises if the function $f(x)$ is zero in the initial condition (19.6.3), and the boundary conditions are nonhomogeneous.

Example 19.4. Consider the dimensionless partial differential equation governing the plane wall transient heat conduction

$$u_t = u_{xx}, \quad 0 < x < 1, \tag{19.1.15}$$

with the boundary and initial conditions

$$u(0,t) = 1, \quad u(1,t) = 0, \quad t > 0; \quad u(x,0) = 0, \quad 0 < x < 1. \tag{19.1.16}$$

Since the nonhomogeneous boundary condition in (19.1.16) does not allow us to compute the eigenfunctions, as in Example 19.3, we proceed as follows: First, we find a particular solution of the problem, which satisfies only the boundary conditions. Although there is more than one way to determine the particular solution, we can, for example, take the steady-state case, where the equation becomes $\tilde{u}_{xx} = 0$, which, after integrating twice, has the general solution

$$\tilde{u}(x) = c_1 x + c_2,$$

which, using the boundary conditions $\tilde{u}(0) = 1$, $\tilde{u}(1) = 0$ gives the steady-state solution as

$$\tilde{u}(x) = 1 - x.$$

Next, we formulate a homogeneous problem by writing $u(x,t)$ as a sum of the steady-state solution $\tilde{u}(x)$ and a transient term $v(x,t)$, i.e., $u(x,t) = \tilde{u}(x) + v(x,t)$, or

$$v(x,t) = u(x,t) - \tilde{u}(x). \tag{19.1.17}$$

Hence, the problem reduces to finding $v(x,t)$. If we substitute v from (19.1.17) into (19.1.15), we get

$$v_t = v_{xx}, \tag{19.1.18}$$

where the boundary conditions (19.1.16) and the initial condition (19.6.8) reduce to

$$\begin{aligned} v(0,t) &= u(0,t) - \tilde{u}(0) = 0, \\ v(1,t) &= u(1,t) - \tilde{u}(1) = 0, \\ v(x,0) &= u(x,0) - \tilde{u}(x) = x - 1. \end{aligned}$$

Then its general solution from (19.1.14) is given by

$$v(x,t) = \sum_{n=1}^{\infty} C_n \, e^{-n^2\pi^2 t} \sin n\pi x,$$

and the coefficients C_n are determined from (19.6.5) as

$$C_n = 2 \int_0^1 (x-1) \sin n\pi x \, dx = -\frac{2}{n\pi}.$$

Thus,

$$v(x,t) = -\frac{2}{\pi} \sum_{n=1}^{\infty} \frac{1}{n} \, e^{-n^2\pi^2 t} \sin n\pi x,$$

and finally from (19.1.17)

$$u(x,t) = 1 - x - \frac{2}{\pi} \sum_{n=1}^{\infty} \frac{1}{n} \, e^{-n^2\pi^2 t} \sin n\pi x.$$

In general, if the thickness of the plate is l, the solution is

$$u(x,t) = 1 - \frac{x}{l} - \frac{2}{\pi} \sum_{n=1}^{\infty} \frac{1}{n} e^{-n^2\pi^2 t/l^2} \sin \frac{n\pi x}{l}.$$

The solution for the half-space is derived by letting $l \to \infty$. Since

$$\lim_{l\to\infty} u(x,t) = 1 - \frac{2}{\pi} \sum_{n=1}^{\infty} \frac{l}{n\pi} e^{-n^2\pi^2 t/l^2} \sin \frac{n\pi x}{l} \cdot \frac{\pi}{l},$$

let $n\pi/l = \xi$ and $\pi/l = d\xi$. Then

$$\lim_{l\to\infty} u(x,t) = 1 - \frac{2}{\pi} \int_0^{\infty} \frac{1}{\xi} e^{-\xi^2 t} \sin \xi t \, d\xi = 1 - \operatorname{erf}\left(\frac{x}{2\sqrt{t}}\right) = \operatorname{erfc}\left(\frac{x}{2\sqrt{t}}\right). \blacksquare$$

Example 19.5. (Heat conduction equation in $R^1 \times R^+$) Consider a laterally insulated rod of uniform cross section with area A and constant density ρ, constant specific heat c, and constant thermal conductivity a. We assume that the temperature $u(x,t)$ is a function of x and t only, $t > 0$, and use the law of conservation of energy to derive the heat conduction equation. Consider a segment PQ of the rod, with coordinates x and $x + \Delta x$ (Figure 19.2). Let R denote the rate at which the heat is accumulating on the segment PQ. Then, assuming that there are no heat sources or sinks in the rod, the rate R is given by

$$R = \int_x^{x+\Delta x} \frac{\partial c\rho A u(\xi,t)}{\partial t} \, d\xi.$$

Note that R can also be evaluated as the total flux across the boundaries of the segment PQ, which gives $R = aA[u_x(x+\Delta x,t) - u_x(x,t)]$. Now, using the mean-value theorem for integrals, we have

$$c\rho A \, u_t(x + h\Delta x, t) \, \Delta x = aA[u_x(x+\Delta x,t) - u_x(x,t)], \quad 0 < h < 1.$$

After dividing both sides by $c\rho A\Delta x$ and taking the limit as $\Delta x \to 0$, we get

$$u_t(x,t) = k \, u_{xx}(x,t). \blacksquare$$

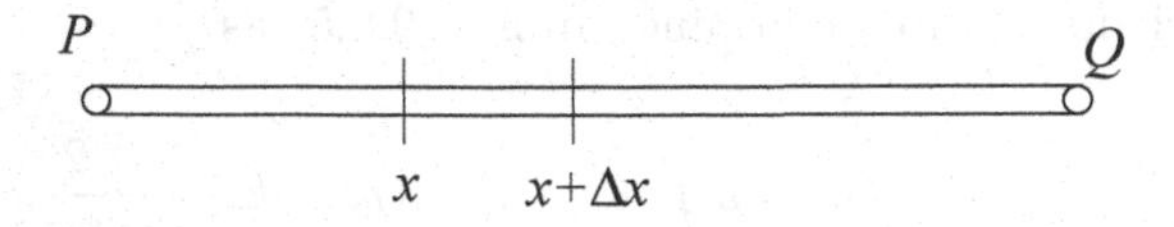

Figure 19.2 Segment PQ on a thin uniform rod.

19.1.1 Method of Separation of Variables. Some examples of steady-state heat conduction, or potential, problems, using the method of separation of variables, are given below.

Example 19.6. We consider the steady-state heat conduction (or potential) problem for the rectangle $R \; \{0 < x < a, \; 0 < y < b\}$:

$$u_{xx} + u_{yy} = 0, \quad x, y \in R, \tag{19.1.19}$$

subject to the Dirichlet boundary conditions

$$u(0,y) = 0 = u(a,y),\ u(x,0) = 0,\ u(x,b) = f(x). \tag{19.1.20}$$

Physically, this problem arises if three edges of a thin isotropic rectangular plate are insulated and maintained at zero temperature, while the fourth edge is subjected to a variable temperature $f(x)$ until the steady-state conditions are attained throughout R. Then the steady-state value of $u(x,y)$ represents the distribution of temperature in the interior of the plate. Using the method of separation of variables, the solution is (see Kythe et al. [2003: 130])

$$u(x,y) = \sum_{n=1}^{\infty} C_n \sin\frac{n\pi x}{a} \sinh\frac{n\pi y}{a}, \tag{19.1.21}$$

where

$$C_n \sinh\frac{n\pi b}{a} = \frac{2}{a}\int_0^a f(x)\sin\frac{n\pi x}{a}\,dx, \quad n = 1,2,\dots.$$

In particular, if $f(x) = f_0 = \text{const}$, then

$$C_n \sinh\frac{n\pi b}{a} = \frac{2f_0[1-(-1)^n]}{n\pi}.$$

Thus, from (19.1.21), we have

$$u(x,y) = \frac{2f_0}{\pi}\sum_{n=1}^{\infty}\frac{1-(-1)^n}{n}\,\frac{\sin(n\pi x/a)\sinh(n\pi y/a)}{\sinh(n\pi b/a)}. \blacksquare$$

Example 19.7. Consider the steady-state heat conduction or potential problem

$$u_{xx} + u_{yy} = 0, \quad 0 < x < \pi, \quad 0 < y < 1, \tag{19.1.22}$$

subject to the mixed boundary conditions $u(x,0) = u_0 \cos x$, $u(x,1) = u_0 \sin^2 x$, $u_x(0,y) = 0 = u_x(\pi,y)$. Using the separation of variables method, the solution is given by (see Kythe et al. [2003: 132])

$$\begin{aligned} u(x,y) &= \frac{u_0}{2}y + u_0\left[\cosh y - \frac{\cosh 1 \sinh y}{\sinh 1}\right]\cos x - u_0\frac{\sinh 2y}{2\sinh 2}\cos 2x \\ &= u_0\left[\frac{1}{2}y + \frac{\sinh(1-y)}{\sinh 1}\cos x - \frac{\sinh 2y}{2\sinh 2}\cos 2x\right]. \blacksquare \end{aligned} \tag{19.1.23}$$

19.2 Heat Transfer

We will derive the heat transfer equation for a uniform isotropic body. Let $u(x,y,z,t)$ denote the temperature of a uniform isotropic body at a point (x,y,z) and time t. If different parts of the body are at different temperatures, then heat transfer takes place within the body. Consider a small surface element δS of a surface S drawn inside the body. Under the assumption that the amount of heat δQ passing through the element δS in time δt is proportional to $\delta S\,\delta t$, and the normal derivative is $\dfrac{\partial u}{\partial n}$, we get

$$\delta Q = -a\frac{\partial u}{\partial n}\,\delta S\,\delta t = -a\,\delta S\,\delta t\,\nabla_n u, \tag{19.2.1}$$

where a is the thermal conductivity of the body, which depends only on the coordinates (x, y, z) of points in the body but is independent of the direction of the normal to the surface S, and ∇_n denotes the gradient in the direction of the outward normal to the surface element δS. Let Q denote the *heat flux* which is the amount of heat passing through the unit surface area per unit time. Then Eq (19.2.1) implies that

$$Q = -a \frac{\partial u}{\partial n}. \tag{19.2.2}$$

Now, consider an arbitrary volume V bounded by a smooth surface S. Then, in view of Eq (19.2.2), the amount of heat entering through the surface S in the time interval $[t_1, t_2]$ is given by

$$Q_1 = -\int_{t_1}^{t_2} dt \iint_S a(x, y, z) \frac{\partial u}{\partial n}\, dS = \int_{t_1}^{t_2} dt \iiint_V \nabla \cdot (a \nabla u)\, dV, \tag{19.2.3}$$

by divergence theorem, where n is the inward normal to the surface S. Let δV denote a volume element. The amount of heat required to change the temperature of this volume element by $\delta u = u(x, y, z, t + \delta t) - u(x, y, z, t)$ in time δt is

$$\delta Q_2 = [u(x, y, z, t + \delta t) - u(x, y, z, t)]\; c(x, y, z)\rho(x, y, z)\, \delta V,$$

where $c(x, y, z)$ and $\rho(x, y, z)$ are the specific heat and density of the body, respectively. Integrating (19.2.3) we find that the amount of heat required to change the temperature of the volume V by δu is given by

$$Q_2 = \iiint_V [u(x, y, z, t + \delta t) - u(x, y, z, t)]\, c\rho\, dV = \int_{t_1}^{t_2} dt \iiint_V c\rho \frac{\partial u}{\partial t}\, dV.$$

Next, we assume that the body contains heat sources, and let $g(x, y, z, t)$ denote the density of such heat sources. Then the amount of heat released by or absorbed in V in the time interval $[t_1, t_2]$ is

$$Q_3 = \int_{t_1}^{t_2} dt \iiint_V g(x, y, z, t)\, dV. \tag{19.2.4}$$

Since $Q_2 = Q_1 + Q_3$, we find from (19.2.3)-(19.2.4) that

$$\int_{t_1}^{t_2} dt \iiint_V \left[c\rho \frac{\partial u}{\partial t} - \nabla \cdot (a \nabla u) - g(x, y, z, t) \right] dV = 0,$$

or, since the volume V and the time interval $[t_1, t_2]$ are arbitrary, we get

$$\begin{aligned} c\rho \frac{\partial u}{\partial t} &= \nabla \cdot (a \nabla u) + g(x, y, z, t) \\ &= \frac{\partial}{\partial x}\left(a \frac{\partial u}{\partial x} \right) + \frac{\partial}{\partial y}\left(a \frac{\partial u}{\partial y} \right) + \frac{\partial}{\partial z}\left(a \frac{\partial u}{\partial z} \right) + g(x, y, z, t), \end{aligned} \tag{19.2.5}$$

which is the *heat conduction equation* for a uniform isotropic body. If c, ρ, and a are constant, Eq (19.2.5) becomes

$$\frac{\partial u}{\partial t} = k \left(\frac{\partial^2 u}{\partial x^2} + \frac{\partial^2 u}{\partial y^2} + \frac{\partial^2 u}{\partial z^2} \right) + f(x, y, z, t), \tag{19.2.6}$$

where $k = a/c\rho$ is known as the *thermal diffusivity*, and $f = g/c\rho$ denotes the heat source (sink) function. In the absence of heat sources or sinks (i.e., when $g(x, y, z, t) = 0$), Eq (19.2.6) reduces to the *homogeneous heat conduction equation*

$$\frac{\partial u}{\partial t} = k\,\nabla^2 u = k\left(\frac{\partial^2 u}{\partial x^2} + \frac{\partial^2 u}{\partial y^2} + \frac{\partial^2 u}{\partial z^2}\right). \tag{19.2.7}$$

In the case when the temperature distribution throughout the body reaches the steady state, i.e., when the temperature becomes independent of time, Eq (19.2.7) reduces to the *Laplace equation*

$$\nabla^2 u = \frac{\partial^2 u}{\partial x^2} + \frac{\partial^2 u}{\partial y^2} + \frac{\partial^2 u}{\partial z^2} = 0. \blacksquare \tag{19.2.8}$$

19.2.1 Ill-Posed Problems. Consider a cylindrical material that extends to infinity in a direction perpendicular to the z-plane. The temperature on the boundary of this cylinder is known. The problem is to find the temperature inside the cylinder. Assuming that the temperature is not changing with time, let $u(x, y)$ denote the temperature at a point (x, y). Recall that the function $u(x, y)$, which is the real part of $f(z) = u(x, y) + iv(x, y)$, satisfies Laplace's equation. The function $f(z)$ is known as the *complex temperature*. The problem to solve is to determine the appropriate function $f(z)$ such that its real part $u(x, y)$ matches the given temperature on the boundary of the cylinder.

If $f(z)$ denotes the complex temperature, then its real part $u(x, y) = c$, where c is a constant, denotes the level lines of the temperature, called the *isotherms*, and the level lines of $v(x, y) = c$ are called the *lines of flux*, which are the lines along which heat flows.

Note that the solution of a well-posed flow problem always exists and is unique. We will now discuss the problem of ill-posed boundary value problems.

The existence of the solution of a boundary value problem is generally difficult to establish. One of the methods to show that a solution exists is to construct it. We have already constructed solutions of different types of boundary value problems. For example, the existence of the solution of the Dirichlet problem $\nabla^2 u = f(\mathbf{x})$ in a domain Ω, such that $u = g(\mathbf{x}_s)$ on the boundary $\partial\Omega$, where $\mathbf{x}_s$ is an arbitrary point on the boundary, is established in Kythe et al. [2003: 232] in terms of the Green's function. To understand what an ill-posed problem means, we will first note that a well-posed problem is the one for which a unique solution exists, and this solution depends continuously not only on the prescribed initial and boundary conditions but also on the coefficients in the equations, or boundary conditions, parameters, and geometry as well. A problem that is not well-posed is said to be *ill-posed*. For example, the Neumann problem $\nabla^2 u = F(\mathbf{x})$ in a domain Ω, such that $\dfrac{\partial u}{\partial n} = g(\mathbf{x}_s)$ on the boundary $\partial\Omega$, where $\mathbf{x}_s$ is an arbitrary point on the boundary, is solved in Kythe et al. [2003: 232]. It is shown that by the divergence theorem (see Appendix A)

$$\iiint_\Omega F(\mathbf{x})\,d\Omega = \iiint_\Omega \nabla^2 u\,d\Omega = \iint_S \frac{\partial u}{\partial n}\,dS = \iint_S g(\mathbf{x}_s)\,dS.$$

Thus, if the consistency condition $\displaystyle\iiint_\Omega F(\mathbf{x})\,d\Omega = \iint_S g(\mathbf{x}_s)\,dS$ is not satisfied, the Neumann problem cannot have a solution. Examples of ill-posed problems arise, *inter alia*, in the areas of heat conduction, forced vibrations, bifurcations, Cauchy problems for hyperbolic systems, and singular perturbations. We will not discuss all these problems, but solve some specific problems only.

Example 19.8. The initial value problem for the Laplace equation $u_{xx} + u_{yy} = 0$, $x > 0$, $-\infty < y < \infty$, subject to the initial conditions $u(0, y) = f(y)$, $u_x(0, y) = g(y)$, $-\infty < y < \infty$, is ill-posed in the sense that the solution does not depend continuously on the data functions f and g. The details are as follows: For $f = f_1 = 0$ and $g = g_1 = 0$ we have the solution $u_1 = 0$. Also, for $f = f_2 = 0$ and $g = g_2 = \dfrac{\sin ny}{n}$ the solution is $u_2 = \dfrac{1}{n^2} \sin ny \sinh nx$. Since the data functions f_1 and f_2 are the same, and since $\lim_{n\to\infty} |g_1 - g_2| = \lim_{n\to\infty} \left| -\dfrac{\sin ny}{n} \right| = 0$ uniformly for all $y \in R^1$, we conclude that the pairs of the data functions f_1, g_1 and f_2, g_2 can be made arbitrarily close by choosing n sufficiently large. We will compare the solutions u_1 and u_2 at $y = \pi/2$ for an arbitrary small, fixed value of $x > 0$ and odd positive values of n. Then

$$\lim_{n\to\infty} \left| u_1\left(x, \frac{\pi}{2}\right) - u_2\left(x, \frac{\pi}{2}\right) \right| = \lim_{n\to\infty} \frac{1}{n^2} \sinh nx = \lim_{n\to\infty} \frac{e^{nx} - e^{-nx}}{2n^2} = \infty.$$

Hence, by choosing n sufficiently large, the maximum difference between the data functions can be made arbitrarily small, but the maximum difference between the corresponding solutions becomes arbitrarily large. Thus, this initial value problem, and all such problems for elliptic equations, are, in general, ill-posed. ■

Example 19.9. Consider the backward heat equation $v_t = v_{xx}$, $0 < x < 1$, $0 < t < T$, subject to the boundary and initial conditions $v(0, t) = 0 = v(1, t)$ for $0 < t < T$, and $v(x, T) = f(x)$ for $0 < x < 1$. Note that the initial condition of the forward condition has been replaced by a terminal condition, which specifies the state at the final time $t = T$. The initial value problem for the backward heat equation is completely identical to the terminal problem for the forward equation. The problem is to determine the previous states $v(x, t)$ for $t < T$, which have resulted up to the state $f(x)$ at time T. This problem has no solution for arbitrary $f(x)$. Even if the solution exists, it does not depend continuously on the data. To see this, let $v_0(x)$ denote the initial state $v(x, 0)$. Then the function

$$v(x, t) = \sum_{n=1}^{\infty} C_n \, e^{-n^2\pi^2 t} \sin n\pi x, \quad 0 < x < 1, \; 0 < t < T,$$

where

$$C_n = 2 \int_0^1 v_0(x) \sin n\pi x \, dx, \quad n = 1, 2, \dots,$$

will be the solution of the problem provided the coefficients C_n are such that

$$f(x) = \sum_{n=1}^{\infty} C_n \, e^{-n^2\pi^2 T} \sin n\pi x \, dx, \quad 0 < x < 1. \tag{19.2.9}$$

However, the series in (19.2.9) converges uniformly to a function in the class $C^\infty(0, 1)$ irrespective of what value C_n has. Hence, no solution exists when $f(x) \notin C^\infty(0, 1)$. Moreover, even if $f(x)$ is in the class $C^\infty(0, 1)$, the solution does not depend continuously on the data. We will consider two special cases of $f(x)$.

CASE 1. $f(x) = \dfrac{\sin N\pi x}{N}$, where N is an integer. Then the unique solution of the problem is

$$v(x, t) = \frac{1}{N} e^{N^2\pi^2 (T-t)} \sin N\pi x, \quad 0 < x < 1, 0 < t < T.$$

Note that $|f(x)| \to 0$ as $N \to \infty$, which means that this data function differs by an arbitrarily small quantity from the data function $f \equiv 0$, which yields the solution $v \equiv 0$. On the other hand, $|v(x,t)| \to \infty$ for $0 < t < T$ as $N \to \infty$, which means that the solution does not depend continuously on the data.

CASE 2. $f(x) = x$. Then

$$C_n\, e^{-n^2\pi^2 T} = 2\int_0^1 x\,\sin n\pi x\,dx = (-1)^{n+1}\,\frac{2}{n}.$$

Thus,

$$v_0(x) = \sum_{n=1}^{\infty} (-1)^{n+1}\,\frac{2}{n}\,e^{n^2\pi^2 T}\,\sin n\pi x\,dx.$$

This series is divergent, and hence, $v_0(x) = v(x,0)$ cannot be determined. Moreover, the solution does not remain close to x for $x \neq 0$. For another case, it can be shown that the above problem is ill-posed for $f(x) = x(1-x), 0 < x < 1$. ■

Example 19.10. To show that the backward heat conduction problem in a finite rod is ill-posed, let the ends of the rod of length l be kept at zero temperature subject to the condition $v(x,1) = g(x)$, which is known, and assume that there are no sources/sinks. The problem reduces to finding the initial state $v(x,0) = f(x)$. Since

$$v(x,t) = \sum_{n=1}^{\infty} C_n\,\sin\frac{n\pi x}{l}\,e^{-n^2\pi^2 t/l^2},$$

where

$$C_n = \frac{2}{l}\int_0^l f(\xi)\,\sin\frac{n\pi\xi}{l}\,d\xi,$$

we get

$$v(x,t) = \frac{2}{l}\sum_{n=1}^{\infty}\Big\{\int_0^l f(\xi)\,\sin\frac{n\pi\xi}{l}\,d\xi\Big\}\,\sin\frac{n\pi x}{l}\,e^{-n^2\pi^2 t/l^2}.$$

At $t = 1$, we get

$$g(x) = v(x,1) = \frac{2}{l}\sum_{n=1}^{\infty}\Big\{\int_0^l f(\xi)\,\sin\frac{n\pi\xi}{l}\,d\xi\Big\}\,\sin\frac{n\pi x}{l}\,e^{-n^2\pi^2/l^2}. \tag{19.2.10}$$

Let the Fourier sine series for $g(x)$ be given by

$$g(x) = \sum_{n=1}^{\infty} G_n\,\sin\frac{n\pi x}{l}, \tag{19.2.11}$$

where

$$G_n = \frac{2}{l}\int_0^l g(\xi)\,\sin\frac{n\pi\xi}{l}\,d\xi.$$

Comparing (19.2.10) and (19.2.11), we find that

$$G_n = \frac{2}{l}\,e^{-n^2\pi^2/l^2}\int_0^l f(\xi)\,\sin\frac{n\pi\xi}{l}\,d\xi,$$

which, being an integral equation of the first kind, is known to be ill-posed (Kythe and Puri [2002, Ch. 11]). ■.

In heat transfer problems there are other features that impose certain restrictions so as to make a problem well-posed. These features include boundedness of the region and the insulation at certain boundaries. They are explained by the following example.

Example 19.11. Consider a vertical slab with its left face maintained at the constant temperature 20° and the right face at 90° in Fahrenheit or centigrade (Figure 19.3(a)). First, we find a harmonic function $u(x, y)$ with no singularities or discontinuities inside the slab $(0 \leq x \leq 1)$, such that $u(0, y) = 20$ and $u(1, y) = 90$. Since the boundary conditions do not depend on y, the function u is independent of y, and we can take $u = A + B$. Applying the boundary conditions we find that $A = 70$ and $B = 20$, so that temperature distribution in the slab is given by $u(x, y) = 70x + 20$, and thus, the complex temperature is $f(z) = 70z + 20$. Notice that if this were an electrostatic problem, then $u(x, y)$ defines the potential inside a channel bounded by conducting plates at $x = 0$ and at $x = 1$, with the left plate kept at potential 20 and the right at potential 90.

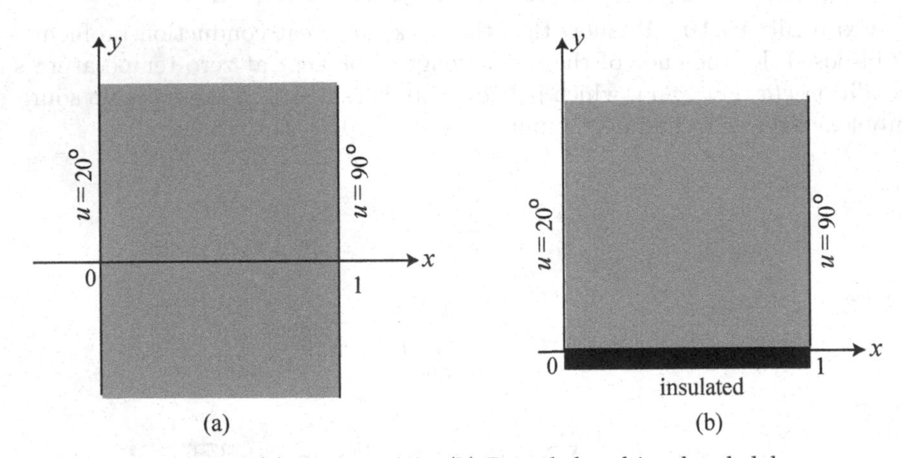

Figure 19.3 (a) Vertical slab. (b) Bounded and insulated slab.

If $f(z) = u(x, y) + iv(x, y)$ defines the complex temperature, then the level lines of temperature $\Re\{f(z)\} = u(x, y) = c$ (constant) are called the *isotherms*, and the level lines of $\Im\{f(z)\} = v(x, y) = c$ are called the *lines of flux*. Note that heat flows along the lines of flux.

However, there exists another solution to this heat conduction problem. Consider the function

$$\sin \pi z = \sin \pi x \cosh \pi y + i \cos \pi x \sinh \pi y.$$

The real part of this function is harmonic, and zero for $x = 0$ and $x = 1$. Thus, the function

$$U(x, y) = 70x + 20 + \sin \pi x \cosh \pi y$$

is another solution of the problem, and this heat conduction problem is ill-posed.

Now, what happens to the temperature across the top and the bottom of the slab? Since the slab is infinite in the y-direction, there is really no top or bottom of the slab. However, since $\lim_{y \to \pm\infty} \cosh y = \infty$, the solution $U(x, y)$ approaches infinity for large y, i.e., there exists a great heat source both at the top and the bottom of the slab. If we require that the solution $u(x, y)$ be *bounded* throughout the slab, the other solution $U(x, y)$ would have been eliminated, thus giving a unique solution. This demands that we add the restriction that the temperature be bounded when the material (or region) extends to infinity.

Moreover, since the presence of insulation at a boundary, or a part of it, stops the flow of heat in or out of that boundary, we can also specify that a portion of the boundary be *insulated*. The insulated segment of the boundary coincides with a line of flux $v(x, y) = c$. This means that $\dfrac{\partial u(x,y)}{\partial n} = 0$, where n is normal to the insulated boundary surface. ■

Example 19.12. To find the temperature inside the bounded slab (Figure 19.3(b)), the solution $u(x, y) = 70x + 20$ found in Example 19.11 is the required temperature on the left and right boundaries. Also the insulated segment on the x-axis is a line of flux $y = 0$. The derivative along the outward normal to the insulated boundary is $\dfrac{\partial u}{\partial y} = 0$. ■

Example 19.13. Solve the two-dimensional steady heat conduction problem for the quadrant $x, y > 0$ with thermal conductivity a if the side $y = 0$ is maintained at zero temperature, while the other side $x = 0$ is thermally insulated except for the region $0 < y < b$ through which heat flows with constant density q (Figure 19.4).

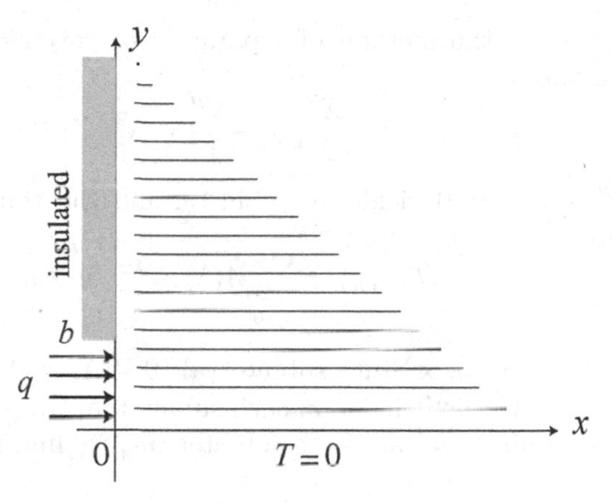

Fig. 19.4 Heat conduction in the first quadrant with a vertical insulated wall.

Thus, we solve Laplace's equation

$$\frac{\partial^2 T}{\partial x^2} + \frac{\partial^2 T}{\partial y^2} = 0, \quad 0 < x < \infty,\ 0 < y < \infty, \tag{19.2.12}$$

subject to the boundary conditions

$$T(x, 0) = 0, \quad \left.\frac{\partial T}{\partial x}\right|_{x=0} = f(y) = \begin{cases} -\dfrac{q}{a}, & 0 < y < b, \\ 0, & b < y < \infty. \end{cases}$$

Applying the Fourier sine transform with respect to y, we get

$$\frac{d^2\tilde{T}}{dx^2} - \alpha^2 \tilde{T} = 0,$$

which has the solution $\tilde{T}(x,\alpha) = \sqrt{\frac{\pi}{2}}\,B(\alpha)\,e^{-\alpha x}$. On inversion this gives

$$T(x,y) = \int_0^\infty B(\alpha)\,e^{-\alpha x}\sin\alpha y\,d\alpha, \tag{19.2.13}$$

where the coefficient $B(\alpha)$ is determined from the boundary condition

$$\frac{\partial T}{\partial x}\Big|_{x=0} = f(y) = \int_0^\infty \alpha\,B(\alpha)\sin\alpha y\,d\alpha, \quad 0 < y < \infty,$$

which gives

$$B(\alpha) = -\frac{2}{\pi\alpha}\int_0^\infty f(y)\sin\alpha y\,dy = \frac{2q}{\pi a}\,\frac{1-\cos\alpha b}{\alpha^2}.$$

Hence, the solution of the problem is

$$T(x,y) = \frac{2q}{\pi a}\int_0^\infty \frac{1-\cos\alpha b}{\alpha^2}\,e^{-\alpha x}\sin\alpha y\,d\alpha. \tag{19.2.14}$$

Alternatively, we use the method of separation of variables, assuming that $T(x,y) = X(x)Y(y)$, and obtain

$$\frac{X''}{X} = -\frac{Y''}{Y} = \lambda^2.$$

The equation $Y'' + \lambda^2 Y = 0$ yields $y_\lambda = \sin\lambda y$, and the equation $X'' - \lambda^2 X = 0$ yields $X_\lambda = e^{-\lambda x}$. Hence,

$$T(x,y) = \sum_{\lambda>0} A(\lambda)\,e^{-\lambda x}\sin\lambda y. \tag{19.2.15}$$

We partition the interval $[0,\infty)$ into subintervals $0 < \lambda_1 < \lambda_2 < \dots$. Let $A(\lambda)\,e^{\lambda x}\sin\lambda y$ have an average value $A(\lambda)\,e^{\lambda x}\sin\lambda y$ on each subinterval $(\lambda_{i-1},\lambda_i)$, $i = 1,2,\dots$, of uniform length $d\lambda$. Then summing over all these subintervals, we find from (19.2.15)

$$T(x,y) = \int_0^\infty A(\lambda)\,e^{-\lambda x}\sin\lambda y\,d\lambda,$$

which is the same as (19.2.13). Thus, the solution is given by (19.2.14) with α replaced by λ. ■

19.3 Conformal Transformations

We will consider the conformal mapping $z = \sin\zeta$ (See Map 5.9(v), Map 5.27, and Map 5.28.) The transformations from the z-plane onto the ζ-plane involve

$$\begin{aligned} x &= \sin\xi\cosh\eta, \\ y &= \cos\xi\sinh\eta, \end{aligned} \tag{19.3.1}$$

with the inverse transformations

$$\begin{aligned} \xi &= \arcsin\left[\frac{1}{2}\left(\sqrt{(x+1)^2+y^2} - \sqrt{(x-1)^2+y^2}\right)\right], \\ \eta &= \cosh^{-1}\left[\frac{1}{2}\left(\sqrt{(x+1)^2+y^2} + \sqrt{(x-1)^2+y^2}\right)\right], \end{aligned} \tag{19.3.2}$$

where the range of the arcsin function is $[-\pi/2, \pi/2]$.

Example 19.14. To find the temperatures inside the shaded region in the z-plane shown in Figure 19.5 with the indicated boundary conditions, we use Example 19.11, and use the first two mappings in Figure 19.6.

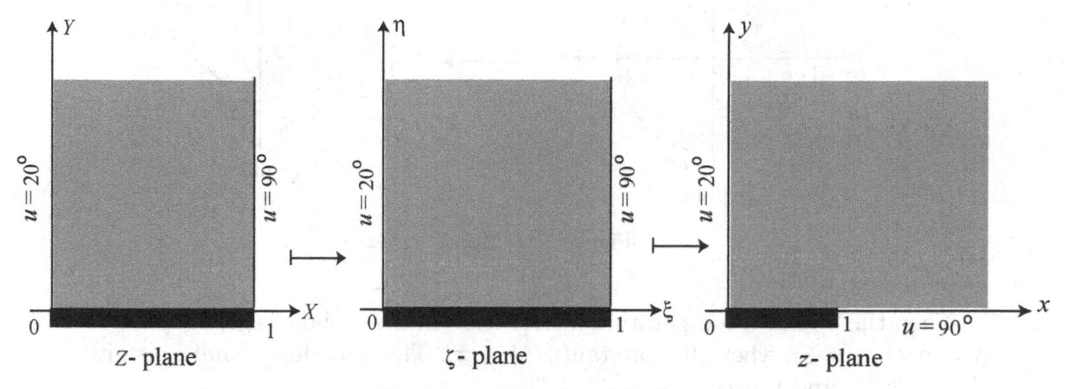

Figure 19.5 Map $z = \sin\zeta$.

Note that the figures in the Z- and ζ-planes correspond to the problem of Example 19.11. Thus, the solution in the w-plane is $u = 70X + 20$, where x is replaced by X as $Z = X + iY$. The first transformation onto the ζ-plane is the magnification $\zeta = \pi Z/2$, or $Z = 2\zeta/\pi$. Hence,

$$Z = X + iY = \frac{2\xi}{\pi} + i\frac{2\eta}{\pi},$$

which yields $X = 2\xi/\pi$. Then the solution for u is

$$u = \frac{140\xi}{\pi} + 20.$$

The second mapping $z = \sin\zeta$ uses Eq (19.3.2) to transform this value of temperature u to the required value

$$u = \frac{140}{\pi} \arcsin\left[\frac{1}{2}\left(\sqrt{(x+1)^2 + y^2} - \sqrt{(x-1)^2 + y^2}\right)\right] + 20. \blacksquare$$

Example 19.15. Consider the case of temperature distribution throughout the upper and lower semicircular regions of the unit disk with the given boundary temperature

$$u = \begin{cases} 20^\circ & \text{for } 0 < \theta < \pi, \\ 0^\circ & \text{for } -\pi < \theta < 0. \end{cases}$$

To find the temperature at the points $z = 0$ and $z = 1/2$, we use the above mapping function $\zeta = i\dfrac{1-z}{1+z}$ which maps region $|z| \leq 1$ onto the upper half of the ζ-plane (see Map 3.25), such that the upper semicircular boundary is mapped onto the positive ξ-axis while the lower boundary is mapped onto the negative ξ-axis (see Figure 19.6).

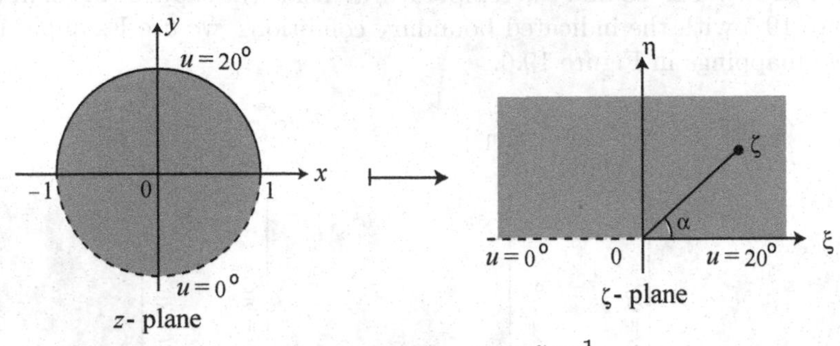

Figure 19.6 Map $\zeta = -i\dfrac{z-1}{z+1}$.

Then the bounded temperature distribution in the ζ-plane is given by $u = A\alpha + B$, where $\alpha = \arctan(\eta/\xi)$, where $0 \le \arctan(\eta/\xi) \le \pi$. The boundary conditions give $B = 20$ and $A = -20/\pi$, and hence,

$$u = \frac{20}{\pi} \arctan \eta\xi + 20, \tag{19.3.3}$$

using which in the mapping function $\zeta = -i\dfrac{z-1}{z+1}$ we obtain

$$\xi + i\eta = -i\frac{x_1 + iy}{x+1+iy} = -i\frac{x^2+y^2-1+2iy}{(x+1)^2+y^2}.$$

Hence, equating real and imaginary parts, we have

$$\xi = \frac{2y}{(x+1)^2+y^2}, \quad \eta = \frac{1-x^2-y^2}{(x+1)^2+y^2}. \tag{19.3.4}$$

Substituting (19.3.4) into (19.3.3) we get the temperature distribution

$$u(x,y) = -\frac{20}{\pi} \arctan \frac{1-x^2-y^2}{2y} + 20, \quad 0 \le \arctan(\cdots) \le \pi.$$

Then the temperature at $z = 0$ is

$$u(0,0) = -\frac{20}{\pi} \arctan(\infty) + 20 = -\frac{20}{\pi}\frac{\pi/2}{+}20 = 10°,$$

since $\tan(\pi/2) = \infty$; and the temperature at (0,1/2) is

$$u(0,1/2) = -\frac{20}{\pi} \arctan(3/4) + 20 \approx 16.9°. \blacksquare$$

Example 19.16 To find the temperature inside the semicircular region, shown in Figure 19.7, with the given boundary values, we use the transformation $\zeta = i\dfrac{1-z}{1+z}$ to map the given upper semicircular region onto the first quadrant of the ζ-plane (see Map 3.25), such that the lower boundary of the z-plane maps onto the positive η-axis and the upper boundary maps onto the positive ξ-axis (Figure 19.7).

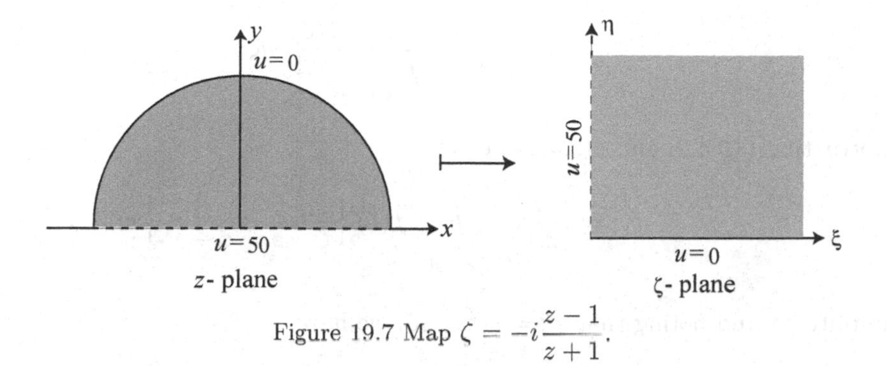

Figure 19.7 Map $\zeta = -i\dfrac{z-1}{z+1}$.

Then the solution in the ζ-plane is

$$u(\xi, \eta) = \frac{100}{\pi} \arctan \frac{\eta}{\xi},$$

which, using Eq (19.3.4), becomes

$$u(x, y) = \frac{100}{\pi} \arctan \frac{1 - x^2 - y^2}{2y},$$

where the domain of the arctan function is $[0, \pi/2]$. ■

19.4 Poisson's Integral Formulas

There are two Poisson's integral formulas which are unique solutions of the Dirichlet problem. Given on the boundary of a simply connected domain, these formulas determine the values of the function inside the domain.

19.4.1 First Poisson's Integral Formula. If u is defined and continuous on the disk $D(0, r) = \{z : |z| < r\}$, then for $\rho < r$

$$u\left(\rho\, e^{i\phi}\right) = \frac{r^2 - \rho^2}{2\pi} \int_0^{2\pi} \frac{u\left(re^{i\theta}\right)}{r^2 - 2r\rho\cos(\theta - \phi) + \rho^2}\, d\theta. \tag{19.4.1}$$

PROOF. $D(0, r)$ is simply connected; u is harmonic on $D(0, r)$. Then there exists an analytic function $f(z)$ in $D(0, r)$ such that

$$u(z) = \Re\{f(z)\}, \quad |z| < r.$$

By Cauchy's integral formula

$$f(z) = \frac{1}{2i\pi} \int_{\Gamma_s : |z| = s|r} \frac{f(\zeta)}{\zeta - z}\, d\zeta \tag{19.4.2}$$

for all z within Γ_s, where $\rho \in \Gamma_s$. (Recall that the symmetric point of z with respect to the circle $|z - a| + R$ is $z^* = \dfrac{R^2}{\bar{z} - \bar{a}} + a$.) Then the symmetric point z^* of z with respect to the circle Γ_s is given by $z^* = \dfrac{s^2}{\bar{z}}$. ($z^*$ is also called the reflection of z in the circle $|z| = s$.) Note that if z is within Γ_s, then z^* is given by

$$\frac{1}{2i\pi} \int_{\Gamma_s:|z|=s|r} \frac{f(\zeta)}{\zeta - z^*}\, d\zeta. \tag{19.4.3}$$

Subtracting (19.4.2) and (19.4.3) we get

$$f(z) = \frac{1}{2i\pi} \int_{\Gamma_s:|z|=s|r} f(\zeta) \left\{ \frac{1}{\zeta - z} - \frac{1}{\zeta - z^*} \right\} d\zeta.$$

Simplifying and noting that $s^2 = |\zeta|^2 = \zeta\bar{\zeta}$, we have

$$\begin{aligned} \frac{1}{\zeta - z} - \frac{1}{\zeta - z^*} &= \frac{1}{\zeta - z} - \frac{1}{\zeta - \dfrac{s^2}{\bar{z}}} = \frac{1}{\zeta - z} - \frac{\bar{z}}{\zeta\bar{z} - \zeta\bar{\zeta}} \\ &= \frac{1}{\zeta - z} - \frac{\bar{z}}{\zeta(\bar{z} - \bar{\zeta})} = \frac{\zeta(\bar{z} - \bar{\zeta}) + \bar{z}(\zeta - z)}{\zeta(\zeta - z)(\bar{z} - \bar{\zeta})} \\ &= \frac{|\zeta|^2 - |z|^2}{\zeta|\zeta - z|^2}. \end{aligned}$$

Hence,

$$f(z) = \frac{1}{2i\pi} \int_{\Gamma_s:|z|=s|r} f(\zeta) \frac{|\zeta|^2 - |z|^2}{\zeta|\zeta - z|^2}\, d\zeta.$$

Now, let $z = \rho e^{i\phi}$, $\rho < s < r$ and $\zeta = s e^{i\theta}$. Then

$$f(z) = \frac{1}{2i\pi} \int_0^{2\pi} \frac{f(se^{i\theta})\,(s^2 - \rho^2)}{se^{i\theta}\left|se^{i\theta} - \rho e^{i\phi}\right|^2}\, ise^{i\theta}\, d\zeta.$$

But

$$\begin{aligned} \left|se^{i\theta} - \rho e^{i\phi}\right|^2 &= \left|s\cos\theta - \rho\cos\phi + i\,(s\sin\theta - \rho\sin\phi)\right|^2 \\ &= (s\cos\theta - \rho\cos\phi)^2 + (s\sin\theta - \rho\sin\phi)^2 \\ &= s^2 - 2s\rho\sin\theta\cos\phi + \rho^2. \end{aligned}$$

Then

$$u(\rho e^{i\phi}) = \Re\{f(\rho e^{i\phi})\} = \frac{1}{2\pi} \int_0^{2\pi} \frac{u(se^{i\theta})\,(s^2 - \rho^2)}{s^2 - 2s\rho\sin\theta\cos\phi + \rho^2}\, d\theta.$$

Let $s \to r$ through a sequence of radii $\{\rho_n\} \to r$. Then

$$u(\rho e^{i\phi}) = \frac{1}{2\pi} \int_0^{2\pi} \frac{u(re^{i\theta})\,(r^2 - \rho^2)}{r^2 - 2r\rho\sin\theta\cos\phi + \rho^2}\, d\theta, \quad \rho < r. \tag{19.4.4}$$

19.4.2 Poisson's Formula for the Unit Disk. If the domain is the unit circle $|z| \le 1$, $z = r\,e^{i\theta}$, then

$$u(r,\theta) = \frac{1 - r^2}{2\pi} \int_0^{2\pi} \frac{u(1,\phi)}{1 - 2r\cos(\theta - \phi) + r^2}\, d\phi, \tag{19.4.5}$$

where the point $r, \theta)$ is inside the unit circle, and $u(1,\phi)$ is a point on the boundary $|z| = 1$. The function $u(1,\phi)$ is assumed known (see Figure 19.8(a), where the length between the fixed point P and any point Q is equal to $|\overline{PQ}|^2 = 1 - 2r\cos(\theta - \phi) + r^2$.) Then, formula (19.4.5) can also be written as

$$u(r,\theta) = \frac{1 - r^2}{2\pi} \int_0^{2\pi} \frac{u(1,\phi)}{|\overline{PQ}|^2}\, d\phi, \tag{19.4.6}$$

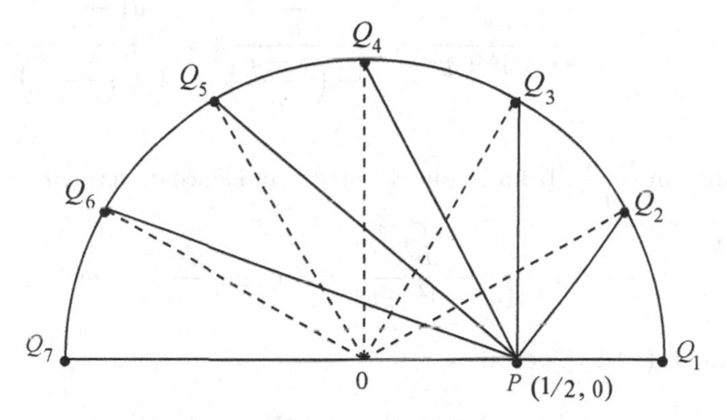

Figure 19.8 (a) Unit disk, (b) z-plane.

19.4.3 Second Poisson's Integral Formula. This is Poisson's integral representation for the Dirichlet problem in the half-plane. To find the solution of the Laplace equation $u_{xx} + u_{yy} = 0$ in the domain $|x| < \infty$ and $y \geq 0$, subject to the conditions that $u \to 0$ as $|x| \to \infty$ or as $y \to \infty$, and $u(x, 0) = \delta(x)$, we apply the Fourier transform to the partial differential equation with respect to x and get

$$\tilde{u}_{yy} - \alpha^2 \tilde{u} = 0,$$

whose solution is

$$\tilde{u} = Ae^{-|\alpha| y}.$$

Applying the boundary condition at $y = 0$ in the transform domain, we get $\tilde{u}(\alpha, 0) = A = \dfrac{1}{\sqrt{2\pi}}$. Hence, $\tilde{u} = \dfrac{1}{\sqrt{2\pi}} e^{-|\alpha| y}$. On inverting, we obtain

$$u(x,y) = \frac{1}{\pi} \frac{y}{x^2 + y^2}.$$

Now, we use the convolution theorem[2] for the Fourier transform (Kythe et al. [2003: 189]) to obtain the solution to the problem with arbitrary condition $u(x, 0) = f(x)$. Then the solution is

$$u(x,y) = \frac{1}{\pi} \int_{-\infty}^{\infty} u(s,0) \frac{y\, ds}{(s-x)^2 + y^2}, \tag{19.4.7}$$

which is known as the Poisson integral representation for the Dirichlet problem in the half-plane. The values of the harmonic function $u = u(s, 0)$ are assumed known on the real axis.

[2] The convolution of $f(t)$ and $g(t)$ over $(-\infty, \infty)$ is defined by $f \star g = \dfrac{1}{\sqrt{2\pi}} \int_{-\infty}^{\infty} f(s) g(x - s)\, ds = \dfrac{1}{\sqrt{2\pi}} \int_{-\infty}^{\infty} f(x - s) g(s)\, ds$.

Formula (19.4.7) allows us to compute $u(x,y)$ at any point (x,y) in the upper half-plane $y>0$ by simply knowing the value of u on the x-axis (see Figure 19.8(b)).

To get the geometrical significance of formula (19.4.7), note that the point P at (x,y) is fixed while the point Q moves along the x-axis, i.e., we can regard x and y as constants while s is treated as a variable to perform the integration. Then the integrand in (19.4.7), without the factor $u(s,0)$, can be rewritten as

$$\frac{y\,ds}{(s-x)^2+y^2} = \frac{\dfrac{ds}{y}}{1+\left(\dfrac{s-x}{y}\right)^2} = \frac{d\left(\dfrac{s-x}{y}\right)}{1+\left(\dfrac{s-x}{y}\right)^2}.$$

Since $\theta = \arctan\dfrac{s-x}{y}$ (from Figure 19.8(b)), and since $d\arctan u = \dfrac{du}{1+u^2}$, we can rewrite

$$\frac{y\,ds}{(s-x)^2+y^2} = d\arctan\frac{s-x}{y} = d\theta.$$

Hence, formula (19.4.7) becomes

$$u(x,y) = \frac{1}{\pi}\int_{-\pi/2}^{\pi/2} u(s,0)\,d\theta. \tag{19.4.8}$$

Example 19.17. Consider the case of heat flow when the temperature at different points on the boundary of the unit disk is as given in Table 19.1 and Figure 19.9. Use Poisson's formula (19.4.6) to determine the temperature at $z=1/2$.

Table 19.1 Temperature on the unit circle.

angle ϕ in radians	0	$\pi/6$	$\pi/3$	$\pi/2$	$2\pi/3$	$3\pi/4$	π	$\cdots$	0
temperature $u(1,\phi)$	0	10	20	30	20	10	0	$\cdots$	0

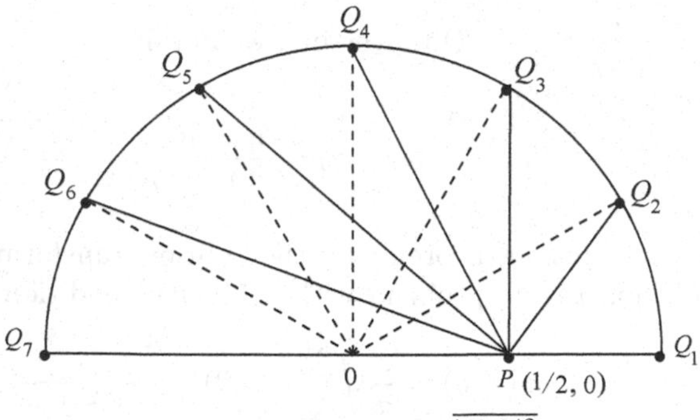

Figure 19.9 Distances $\overline{|PQ|^2}$.

We will estimate the temperature $u(1/2,0)$ by using the Riemann sum, so that formula (19.4.6) becomes

$$u(r,\theta) \approx \frac{1-r^2}{2\pi}\sum_{n=2}^{6}\frac{u(1,\phi_n)}{\overline{|PQ_n|^2}}\,\frac{\pi}{6}, \tag{19.4.9}$$

where we have taken only the non-zero values of $u(1, \phi)$ which are at the points $Q_2, \dots, Q_6$ (Figure 19.9). Instead of considering the entire unit circle, we have presented only the upper semicircle, so as to avoid considering points with zero temperature. To estimate the distances $\overline{|PQ|^2}$ we have used the laws of cosines for a triangle. Thus,

$$\overline{|PQ_2|^2} = (1)^2 + (1/2)^2 - 2 \cdot 1 \cdot (1/2) \cos(\pi/6) \approx 0.384; \text{ and similarly}$$
$$\overline{|PQ_3|^2} = 0.75; \ \overline{|PQ_4|^2} = 1.25; \ \overline{|PQ_5|^2} = 1.75; \ \overline{|PQ_6|^2} \approx 2.116.$$

Hence,

$$\begin{aligned} u(1/2, 0) &\approx \frac{0.75}{2\pi} \frac{\pi}{6} \Big[\frac{10}{0.384} + \frac{20}{0.75} + \frac{30}{1.25} + \frac{20}{1.75} + \frac{10}{2.116}\Big] \\ &\approx 0.625 \big[2.604 + 2.667 + 2.4 + 1.142 + 0.473\big] \approx 6.8. \ \blacksquare \end{aligned}$$

Example 19.18. Estimate the temperature at the $z = i$ when the temperatures on the x-axis are determined as

$$u(x, 0) = \begin{cases} 0 & \text{for } x < 1, \\ 20x & \text{for } 1 \le x \le 3, \\ 0 & \text{for } x > 3. \end{cases}$$

Using Poisson's formula (19.4.8), we have for this problem

$$u(0, 1) = \frac{1}{\pi} \int_{\pi/4}^{\arctan(3)} 20s \, d\theta = \frac{1}{\pi} \int_{\pi/4}^{\arctan(3)} 20s \, d(\arctan s).$$

This integral holds because $u(s, 0) = 0$ for values of θ outside the range $\pi/4 \le \theta \le \arctan(3)$ (see Figure 19.10).

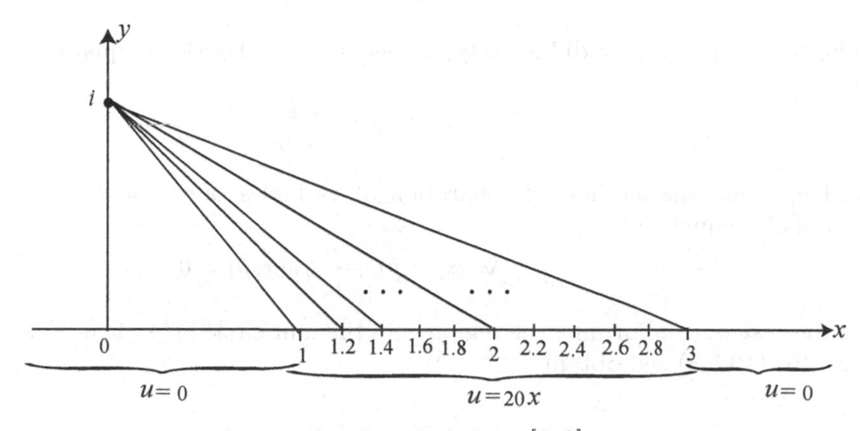

Figure 19.10 Subintervals of $[1, 3]$.

To evaluate this integral we will use the Riemann sum. Thus, we divide the interval $1 \le x \le 3$ into 10 equal subintervals of length 0.2 each. The endpoints of the intervals are $1, 1.2, 1.4, \dots, 2.8, 3$. We choose the value of u at the midpoint of each interval, i.e., at the points $x = 1.1, 1.3, 1.5, \dots, 2.9$. Then we compute

$$u(0, 1) \approx \frac{1}{\pi} \sum_{n=1}^{10} 20x_n \, \Delta\theta_n,$$

where $\Delta\theta_n = \arctan(x_n + 0.1) - \arctan(x_n - 0.1)$. The computation is presented in Table 19.2, where the angles are in radians.

Table 19.2.

n	x_n	$20x_n$	$\Delta\theta_n$	$20x_n\Delta\theta_n/\pi$
1	1.1	22	0.0289	0.6358
2	1.3	26	0.0239	0.6214
3	1.5	30	0.0194	0.5820
4	1.7	34	0.0167	0.5678
5	1.9	38	0.0139	0.5282
6	2.1	42	0.0117	0.4914
7	2.3	46	0.0100	0.4600
8	2.5	50	0.0089	0.4450
9	2.7	54	0.0077	0.4158
10	2.9	58	0.0067	0.3868

The last column adds up to 6.136. Hence, $u(0,1) \approx 6.136°$. ■

19.5 Diffusion Equation

Since Laplace's equation is invariant under conformal mapping, we do not need a different notation other than u, say, to express the potential in the physical as well as the model planes. Thus, using Eq (17.9.5), we have

$$\nabla^2\phi(x,y) = \nabla^2\psi(u,v)|f'(z)|^2 = \nabla^2\psi(u,v)|dw/dz|^2 = \nabla^2\psi(u,v)\,|dz/dw|^{-2}. \tag{19.5.1}$$

The basic equation for diffusion type phenomena is Fourier's equation which is expressed as

$$\nabla^2\phi = \frac{1}{c^2}\frac{\partial\phi}{\partial t}, \tag{19.5.2}$$

which, using the method of separation of variables $\phi(xy) = \phi_0(x,y)\,e^{\alpha t}$, becomes the Helmholtz equation

$$\nabla^2\phi_0(x,y) + \frac{\alpha}{c^2}\phi_0(x,y) = 0. \tag{19.5.3}$$

In the case of transforming the region onto the unit circle $|w| < 1$, we substitute Eq (19.5.1) into Eq (19.5.3) and obtain

$$\nabla^2\psi_0(w,\bar{w}) + \frac{\alpha}{c^2}\,|dz/dw|^2\psi_0(w,\bar{w}) = 0. \tag{19.5.4}$$

Some examples, using different methods, are as follows:

Example 19.19. To find the solution of the diffusion equation $u_t = k\,u_{xx}$, $0 < x < \pi$, $t > 0$, subject to the boundary conditions $u(0,t) = 1 - e^{-t}$ and $u(\pi,t) = 0$ for $t \geq 0$, and the initial condition $u(x,0) = 0$ for $0 < x < \pi$, we apply the Laplace transform method, which yields

$$\frac{d^2\bar{u}}{dx^2} = \frac{s}{k}\,\bar{u},$$

with $\bar{u}(0,s) = \dfrac{1}{s(s+1)}$, and $\bar{u}(\pi,s) = 0$. The solution in the Laplace domain is

$$\bar{u}(x,s) = \frac{1}{s(s+1)} \frac{\sinh\sqrt{s/k}\,(\pi - x)}{\sinh\sqrt{s/k\pi}}.$$

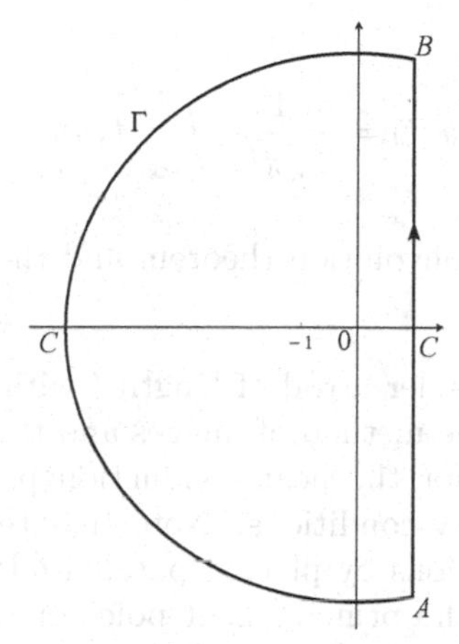

Figure 19.11 Contour for integration.

The inversion formula gives

$$u(x,t) = \frac{1}{2\pi i}\int_{c-i\infty}^{c+i\infty} \frac{e^{st}\sinh\sqrt{s/k}(\pi - x)}{s(s+1)\sinh\sqrt{s/k}\,\pi}\,ds,$$

where c is any positive constant. Assuming that k is not of the form n^{-2}, the integrand has simple poles at $s = 0, -1, -kn^2$, $n = 1, 2, \cdots$. The contour is shown in Fig. 19.11, where the left semicircle, with $\Re\{s\} = c$, is defined as the limit of a sequence of semicircles Γ_n that cross the negative s axis between the poles at $-kn^2$ and $-k(n+1)^2$. The limit of the integrand around Γ_n is zero as $n \to \infty$. The residue at the pole $s = 0$ is $(\pi - x)/\pi$, and at the pole $s = -1$ is

$$-\,e^{-t}\,\frac{\sin[(\pi - x)/\sqrt{k}]}{\sin(\pi/\sqrt{k})}.$$

The residue at $s = -kn^2$ is given by

$$\lim_{s\to kn^2} \frac{s+n^2}{s(s+1)}\,e^{st}\,\frac{\sinh\sqrt{s/k}(\pi - x)}{\sinh\sqrt{s/k}\,\pi} = \frac{2\sin nx}{n\pi(kn^2-1)}\,e^{-kn^2 t}.$$

Hence,

$$u(x,t) = \frac{\pi - x}{\pi} - e^{-t}\,\frac{\sin[(\pi - x)/\sqrt{k}]}{\sin(\pi/\sqrt{k})} + \frac{2}{\pi}\sum_{n=1}^{\infty} \frac{\sin nx}{n(kn^2-1)}\,e^{-kn^2 t}.$$

Note that $u \to \dfrac{\pi - x}{\pi}$ as $t \to \infty$, which gives the steady-state temperature in the interval $0 \le x \le \pi$. ■

Example 19.20. To find the general solution of the diffusion equation $u_t = k u_{xx}$ under the nonhomogeneous initial condition $u(x,0) = f(x)$, subject to the boundary conditions that $\lim_{x\to\pm\infty} f(x), u(x,t) \to 0$, we apply Fourier transform and get

$$\tilde{u}_t + k\alpha^2 \tilde{u} = 0,$$

with the initial condition $\tilde{u}(\alpha, 0) = \tilde{f}(\alpha)$, where α is the variable of the transform. The solution, after applying the initial condition, is

$$\tilde{u}(\alpha, t) = \tilde{f}(\alpha) e^{-k\alpha^2 t},$$

which, on inversion, yields

$$u(x,t) = \frac{1}{2\sqrt{\pi k t}} \int_{-\infty}^{\infty} f(s) e^{-(x-s)^2/(4kt)}\, ds,$$

where we have used the convolution theorem and the Fourier transform of e^{-kx^2} which is $\dfrac{1}{\sqrt{2\pi k}} e^{-\alpha^2/(4k)}$. ■

Example 19.21. Consider a rod of length l with periodic boundary conditions at the end points. We will use the method of images and the result of Example 19.2 to determine Green's function $G(x,t)$ for the heat conduction problem in the following four cases of different types of boundary conditions. Note that this problem is regarded as that of an infinite sequence of reflections by placing parallel mirrors at the end points (as an optical example); then not only the primary heat pole (marked in Figures 19.12(a)-(d)) but also all its images are reflected at both ends of the rod.

(a) $u = 0$ at both $x = 0$ and $x = l$.

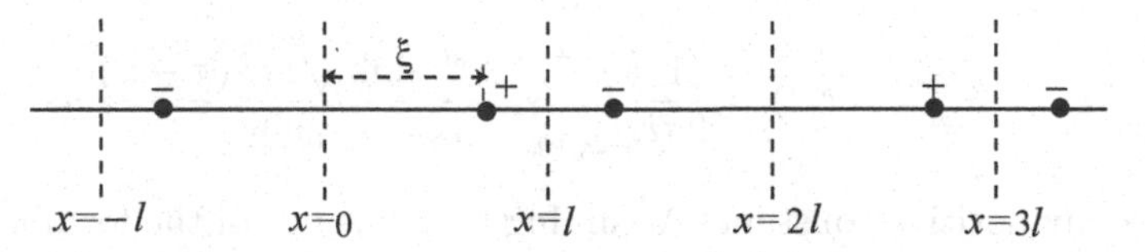

Fig. 19.12(a) Heat conduction in a rod.

$$\text{Solution:} \quad f(x) = \sum B_n \sin \frac{n\pi x}{l}, \quad B_n = \frac{2}{l} \int_0^l f(x) \sin \frac{n\pi x}{l}\, dx,$$

$$G(x,t) = \vartheta\left(\frac{x-\xi}{2l} \Big| \tau\right) - \vartheta\left(\frac{x+\xi}{2l} \Big| \tau\right).$$

(b) $\dfrac{\partial u}{\partial x} = 0$ at both $x = 0$ and $x = l$.

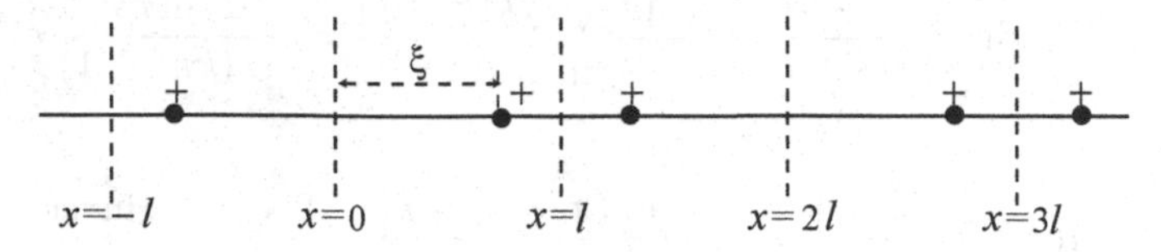

Figure 19.12(b) Heat conduction in a rod.

$$\text{Solution:} \quad f(x) = \sum A_n \cos \frac{n\pi x}{l},$$

$$A_n = \frac{2}{l} \int_0^l f(x) \cos \frac{n\pi x}{l}\, dx, \quad A_0 = \frac{1}{l} \int_0^l f(x)\, dx,$$

$$G(x,t) = \vartheta\left(\frac{x-\xi}{2l} \Big| \tau\right) + \vartheta\left(\frac{x+\xi}{2l} \Big| \tau\right).$$

(c) $u = 0$ at $x = 0$, and $\dfrac{\partial u}{\partial x} = 0$ at $x = l$.

ξ + + − −
x=−l x=0 x=l x=2l x=3l

Figure 19.12(c) Heat conduction in a rod.

Solution $f(x) = \sum B_n \sin \dfrac{(n+1/2)\pi x}{l}$,

$$B_n = \frac{2}{l}\int_0^l f(x) \sin \frac{(n+1/2)\pi x}{l}\, dx,$$

$$G(x,t) = \vartheta\left(\frac{x-\xi}{4l}\,\Big|\,\tau\right) - \vartheta\left(\frac{x+\xi}{4l}\,\Big|\,\tau\right) + \vartheta\left(\frac{x+\xi-2l}{4l}\,\Big|\,\tau\right) - \vartheta\left(\frac{x+\xi-2l}{4l}\,\Big|\,\tau\right).$$

(d) $\dfrac{\partial u}{\partial x} = 0$ at $x = 0$, and $u = 0$ at $x = l$

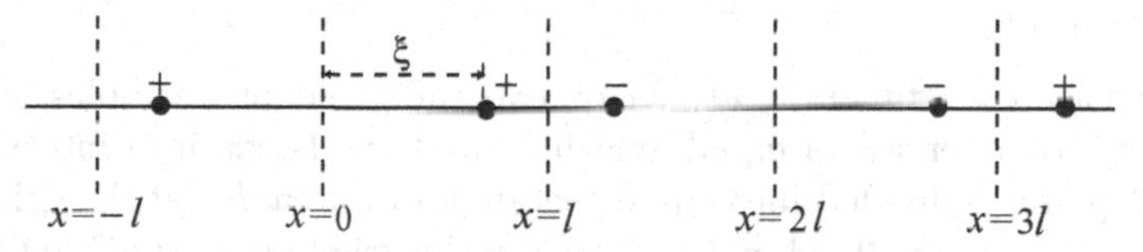

Figure 19.12(d) Heat conduction in a rod.

The solution is

$$f(x) = \sum A_n \cos \frac{(n+1/2)\pi x}{l}, \quad A_n = \frac{2}{l}\int_0^l f(x) \cos \frac{(n+1/2)\pi x}{l}\, dx,$$

$$G(x,t) = \vartheta\left(\frac{x-\xi}{4l}\,\Big|\,\tau\right) + \vartheta\left(\frac{x+\xi}{4l}\,\Big|\,\tau\right) - \vartheta\left(\frac{x+\xi-2l}{4l}\,\Big|\,\tau\right) - \vartheta\left(\frac{x+\xi-2l}{4l}\,\Big|\,\tau\right). \blacksquare$$

19.5.1 Finite Difference Method. Consider the one-dimensional diffusion equation

$$u_t = k\, u_{xx}, \tag{19.5.5}$$

where k denotes the thermal diffusivity. For the grid $(x_i, t_j) = (ih, jl)$, we will discuss the following three finite difference schemes:

(a) Forward Difference (Explicit Scheme):

$$\frac{U_{i,j+1} - U_{i,j}}{l} = k\, \frac{U_{i+1,j} - 2U_{i,j} + U_{i-1,j}}{h^2},$$

or

$$U_{i,j+1} = \left(1 + r\, \delta_x^2\right) U_{i,j}, \tag{19.5.6}$$

where $r = k\,l/h^2$.

(b) Backward Difference (Implicit Scheme):

$$\frac{U_{i,j+1} - U_{i,j}}{l} = k\,\frac{U_{i+1,j+1} - 2U_{i,j+1} + U_{i-1,j+1}}{h^2},$$

or

$$\left(1 - r\,\delta_x^2\right) U_{i,j+1} = U_{i,j}. \tag{19.5.7}$$

(c) Crank-Nicolson (Implicit Scheme):

$$\frac{U_{i,j+1} - U_{i,j}}{l} = k\,\frac{\delta_x^2 U_{i,j} + \delta_x^2 U_{i,j+1}}{h^2},$$

or

$$\left(1 - \frac{r}{2}\delta_x^2\right) U_{i,j+1} = \left(1 + \frac{r}{2}\delta_x^2\right) U_{i,j}. \tag{19.5.8}$$

Note that the Crank-Nicolson scheme is derived by averaging the finite differences at the points (i,j) and $(i,j+1)$ (see Exercise 12.2). Also, the above forward difference scheme (19.5.6) is conditionally stable iff $r < 1/2$, but the other two schemes (19.5.7) and (19.5.8) are always stable.

In the case of a function $u(x,t)$ of two independent variables x and t, we partition the x-axis into intervals of equal length h, and the t-axis into intervals of equal length l. The (x,t)-plane is divided into equal rectangles of area hl by the grid lines parallel to Ot, defined by $x_i = ih$, $i = 0, \pm 1, \pm 2, \dots$, and by the grid lines parallel to Ox, defined by $y_j = jl$, $j = 0, \pm 1, \pm 2, \dots$ (Fig. 19.13). We will use the following notation: Let $u_P = u(ih, jl) = u_{i,j}$ denote the value of the function $u(x,t)$ at a mesh point (node) $P\,(ih, jl)$. Then, in view of formula (12.6), we have the following three central difference schemes:

$$\begin{aligned} \left.\frac{\partial^2 u}{\partial x^2}\right|_P = \left.\frac{\partial^2 u}{\partial x^2}\right|_{i,j} &= \frac{u\left((i+1)h, jl\right) - 2u(ih, jl) + u\left((i-1)j, jl\right)}{h^2} \\ &= \frac{U_{i+1,j} - 2U_{i,j} + U_{i-1,j}}{h^2} \equiv \frac{\delta_x^2\, U_{i,j}}{h^2}, \quad i = 1, 2, \dots, n-1, \end{aligned} \tag{19.5.9}$$

with a truncation error $-\dfrac{h^2}{12} u_{xxxx}(\bar{x}, t)$, where $x_{i-1} < \bar{x} < x_i$;

$$\left.\frac{\partial^2 u}{\partial t^2}\right|_P = \left.\frac{\partial^2 u}{\partial t^2}\right|_{i,j} = \frac{U_{i,j+1} - 2U_{i,j} + U_{i,j-1}}{l^2} \equiv \frac{\delta_t^2\, U_{i,j}}{l^2}, \quad j = 1, 2, \dots, m-1, \tag{19.5.10}$$

with a truncation error $-\dfrac{l^2}{12} u_{tttt}(x, t')$, where $t_{j-1} < t' < t_j$; and

$$\left.\frac{\partial^2 u}{\partial x \partial t}\right|_P = \frac{U_{i+1,j+1} - U_{i+1,j-1} - U_{i-1,j+1} + U_{i-1,j-1}}{4hl}, \tag{19.5.11}$$

with a truncation error $-\frac{h^2}{6}u_{xxxt}(\bar{x},\bar{t}) - \frac{l^2}{6}u_{xttt}(x',t')$.

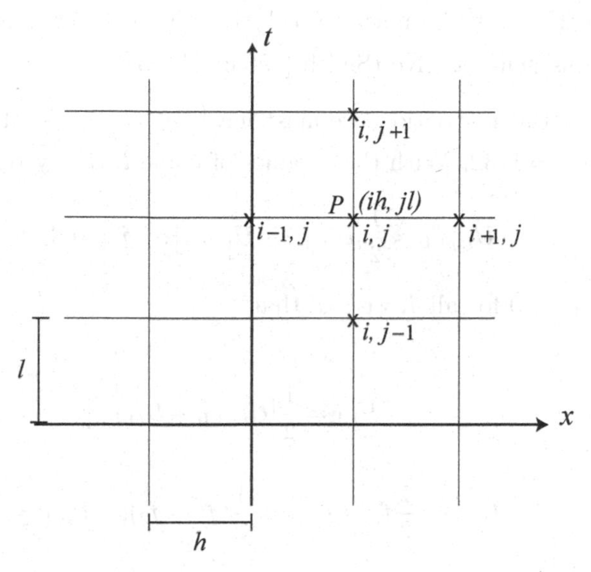

Figure 19.13 Grid lines.

Example 19.22. Consider the boundary value problem

$$\begin{aligned} u_t &= u_{xx}, \quad 0<x<1, \quad t>0, \\ u(0,t) &= 0 = u(1,t), \quad \text{for } t>0, \\ u(x,0) &= f(x), \quad \text{for } 0<x<1. \end{aligned} \tag{19.5.12}$$

Since the space derivative is of second order, we will use the central difference scheme (19.5.9), where $U_{i,j} = u(x_i, t_j)$ denotes the temperature at a point x_i at time t_j. For the time derivative we use the forward difference scheme

$$u_t = \frac{U_{i,j+1} - U_{i,j}}{l},$$

where $l = t_{j+1} - t_j$. Then Eq (19.5.12) is approximated by

$$\begin{aligned} &r\left[U_{i+1,j} - 2U_{i,j} + U_{i-1,j}\right] = U_{i,j+1} - U_{i,j}, \\ &i = 1,2,\dots,n-1, \quad \text{and} \quad j = 0,1,2,\dots, \end{aligned} \tag{19.5.13}$$

where $r = l/h^2$, and the boundary and initial conditions become

$$\begin{aligned} U_{0,j} &= 0 = U_{n,j}, \quad \text{for } j = 1,2,\dots, \\ U_{i,0} &= f(x_i) = f_i, \quad \text{for } i = 0,1,2,\dots,n. \end{aligned}$$

After rearranging the terms in Eq (19.5.13), we get

$$U_{i,j+1} = r\,U_{i-1,j} + (1-2r)\,U_{i,j} + r\,U_{i+1,j}. \tag{19.5.14}$$

This difference equation allows us to compute $U_{i,j+1}$ from the values of U that are computed for earlier times. Note that the value $U_{i,0} = f_i$ is a prescribed value of u at time $t = 0$ and $x = x_i$.

As an example, we take $n = 4$, i.e., $h = 1/4$. The value of r in Eq (19.5.14) must be chosen properly so that the solution remains stable. It has been determined that for a stable solution the value of r must be such that the coefficients of u on the right side of Eq (19.5.14) remains non-negative (Smith [1985: Ch. 3]). Hence, we must have $0 < r \le \dfrac{1}{2}$. Then, in view of this restriction, we must have $l \le rh^2 = \dfrac{1}{32}$. In the marginal case when $r = 1/2$, we get $l = 1/32$. With these values of r and l, the system (19.5.14) reduces to

$$U_{i,j+1} = \frac{1}{2}\big[U_{i-1,j} + U_{i+1,j}\big], \quad j = 0, 1, 2, \dots .$$

Since $U_{0,j} = U_{4,j} = 0$ for all j, we find that
for $j = 0$:

$$U_{i,1} = \frac{1}{2}\big[U_{i-1,0} + U_{i+1,0}\big],$$

or, successively,

$$U_{1,1} = \frac{1}{2}f_2, \quad U_{2,1} = \frac{1}{2}(f_1 + f_2), \quad U_{3,1} = \frac{1}{2}f_2;$$

for $j = 1$:

$$U_{1,2} = \frac{1}{2}U_{2,1}, \quad U_{2,2} = \frac{1}{2}(U_{1,1} + U_{3,1}), \quad U_{3,2} = \frac{1}{2}U_{2,1},$$

and so on. The values of $U_{i,j}$ for $f(x) = \cos \pi x$ are listed below in Table 19.3 for some successive values of t.

Table 19.3 Values of $U_{i,j}$ for $f(x) = \cos \pi x$.

t	$x = 0$	$x = 0.25$	$x = 0.5$	$x = 0.75$	$x = 1$
0	1	$1/\sqrt{2} \approx 0.7071$	0	$-1/\sqrt{2} \approx -0.7071$	-1
1/32	0	0.5	0	-0.5	0
1/16	0	0	0	0	0
3/32	0	0	0	0	0
1/8	0	0	0	0	0

Notice that the values average out in the outer columns. This will happen if $r = 1/2$ is chosen. The solution for problem (19.5.12) for different values of t with $r = 0.1$ is presented in the following table Table 19.4.

Table 19.4 Solution of problem (19.5.12).

t	$x = 0$	$x = 0.25$	$x = 0.5$	$x = 0.75$	$x = 1$
0	1	$1/\sqrt{2} \approx 0.7071$	0	$-1/\sqrt{2} \approx -0.7071$	-1
1/32	0	0.665685	0	-0.665685	0
1/16	0	0.532548	0	-0.532548	0
3/32	0	0.426039	0	-0.426039	0
1/8	0	0.340831	0	-0.340831	0 ■

19.5.2 Steady-State Temperature. In this case the temperature at the point (x, y) must be equal to the temperatures at points in its neighborhood. If the temperature was lower than the average temperature, heat would flow toward the point (x, y) and cause the temperature there to rise, thus destroying the steady state.

If we expand the function u into a Taylor's series about the point (x, y), we get

$$\begin{aligned} u(x+h, y+k) &= \sum_{n=0}^{\infty}\sum_{m=0}^{\infty} \frac{\partial^{m+n}u(x,y)}{\partial x^m \partial y^n}\frac{h}{m!}\frac{k^n}{n!} \\ &= u(x.y) + u_x(x.y)h + u_y(x,y)k + u_{xx}(x,y)\frac{h^2}{2} + u_{xy}(x.y)hk + u_{yy}(x,y)\frac{k^2}{2} \\ &+ u_{xxx}(x,y)\frac{h^3}{3!} + \cdots . \end{aligned} \tag{19.5.15}$$

Thus, the difference between u at (x, y) and an adjacent point $(x + h, y + k)$ is

$$u(x+h, y+k) - u(x,y) = u_x h u_y k + u_{xx}\frac{h^2}{2} + u_{xy}hk + u_{yy}\frac{k^2}{2} + u_{xxx}\frac{h^3}{3!} + \cdots . \tag{19.5.16}$$

Then the average of the differences (19.5.16) over the infinitesimal square surrounding the point (x, y) of side 2ε is

$$\frac{1}{4\varepsilon^2}\int_{-\varepsilon}^{\varepsilon}\int_{-\varepsilon}^{\varepsilon} \{u(x+h, y+k) - u(x,y)\}\, dk\, dh. \tag{19.5.17}$$

If we replace the integrand in (19.5.17) by the right side of (19.5.16) and integrate, we find that the first and second terms give

$$\frac{u_x(x,y)}{4\varepsilon^2}\int_{-\varepsilon}^{\varepsilon}\int_{-\varepsilon}^{\varepsilon} h\, dh\, dk = 0 - \frac{u_y(x,y)}{4\varepsilon^2}\int_{-\varepsilon}^{\varepsilon}\int_{-\varepsilon}^{\varepsilon} k\, dk\, dh,$$

but the third term gives

$$\frac{u_{xx}}{4\varepsilon^2}\frac{u_x(x,y)}{4\varepsilon^2}\int_{-\varepsilon}^{\varepsilon}\int_{-\varepsilon}^{\varepsilon}\frac{h^2}{2}\, dh\, dk = \frac{u_{xx}}{2\varepsilon}\frac{h^3}{6}\Big|_{-\varepsilon}^{\varepsilon} = \frac{u_{xx}\varepsilon^2}{6} \neq 0.$$

The fourth term vanishes like the first two terms, but the fifth term does not vanish and gives $\dfrac{u_{yy}\varepsilon^2}{6}$. The remaining terms will involve higher powers of ε and will henceforth be ignored. Thus, combining these values we have

$$\frac{1}{4\varepsilon^2}\int_{-\varepsilon}^{\varepsilon}\int_{-\varepsilon}^{\varepsilon} \{u(x+h, y+k) - u(x,y)\}\, dk\, dh\, dk\, dh \approx \frac{\varepsilon^2}{6}\nabla^2 u. \tag{19.5.18}$$

19.6 High Temperature Effects

For incompressible flows, the Bernoulli equation gives a stagnation pressure, i.e., static and dynamic pressure, p_0, as $p_0 = p + \frac{1}{2}\rho v^2$. For compressible flows, the stagnation pressure is defined by

$$p_0 = p\Big(1 + \frac{\gamma - 1}{2}M^2\Big)^{\gamma/(\gamma-1)}, \tag{19.6.1}$$

where M is the Mach number, and γ is the specific heat of the fluid. Expanding Eq (19.6.1) in a Taylor's series expansion, we obtain

$$
\begin{aligned}
p_0 &= \Big[1 + \frac{\gamma}{\gamma-1}\frac{\gamma-1}{2}M^2 + \frac{\gamma}{2(\gamma-1)}\Big(\frac{\gamma}{\gamma-1}-1\Big)\Big(\frac{\gamma-1}{2}M^2\Big)^2 + \cdots\Big] \\
&= p\Big[1 + \frac{\gamma}{2}M^2 + \frac{\gamma}{2}\Big(\frac{M^2}{2}\Big)^2 + \cdots\Big] \\
&= p + \frac{\rho v^2}{2} + \frac{1}{2}\rho v^2 \frac{M^2}{4} + \cdots, \quad \text{using } M^2 = \frac{\rho v^2}{\gamma p},
\end{aligned}
\tag{19.6.2}
$$

where the first term is related to the Bernoulli equation. Note that the higher terms are negligible for small M (< 0.3), since $M^2/4 < (0.3)^2/4 = 0.0225$.

Static properties are those properties which are observed if the observer is moving with the supersonic flow, and they are always defined with respect to the observer. On the other hand, the *stagnation properties* are always defined by conditions at a point, and represent the static properties that are measured if the fluid (i.e., air) is made to stop at that point isentropically, typically at the nose of an aircraft.

19.6.1 Hypersonic Flow and High Temperature Effects. According to van Dyke, a hypersonic flow past a body at high Mach number is essentially nonlinear. Let τ denote the thickness-to-body ratio of a body. Then for thin bodies, the product $M\tau$ is of order 1. The thin region between the shock and the body is called the *shock layer*. In thin bodies the shock layer is very close to the body. Shock curvature implies that shock strength varies for different streamlines (stagnation pressure and velocity gradients).

The *hypersonic tunnel* for air propulsion can be described as follows:

At 80,000 feet and dynamic pressure greater than 2 psf, the sonic boom is generated; also at 120,000 feet and dynamic pressure less than 1 psf. For other details, see http://www.onera.fr/conferences/ramjet-scramjet-pde/images/hypersonic-funnel.gif.

For a typical re-entry case of a supersonic flight, very little deceleration is needed until the jet reaches denser air, so as to avoid large fluctuations in aerodynamic loads and landing point. The altitude z of different types of atmosphere that are encountered in supersonic flights are as follows:

$0 < z < 10$ km: Troposphere;
$10 < z < 50$ km: Stratosphere;
$50 < z < 80$ km: Mesosphere;
$z > 80$ km: Ionosphere (contains ions and free electrons);
also,
$60 < z < 85$ km: NO^+;
$85 < z < 140$ km: NO^+, and O_2^+;
$140 < z < 200$ km: NO^+, O_2^+, and O^+;
$z > 200$ km: presence of N^+ and O^+.

A simple model for the variation of density with altitude is defined as

$$
dp = -\rho g\, dz, \quad p = \frac{\rho \hat{R} T}{\hat{M}}, \tag{19.6.3}
$$

where ρ is the fluid density, and the temperature $T = \text{const}$ for the isothermal case (which is not practical). If we neglect dissociation and ionization, and assume that the molecular weight remains constant, then

$$\frac{dp}{p} \approx -\frac{g\hat{M}}{\hat{R}T}\,dz, \quad \rho \approx \rho_0 \ln\left\{\frac{g\hat{M}z}{\hat{R}T}\right\}. \tag{19.6.4}$$

19.6.2 Shock Layer. The presence of vorticity in the shock layer is defined by *Crocco's theorem*, which states that

$$T\nabla s = \nabla h_0 = \mathbf{u} \times \boldsymbol{\omega}, \tag{19.6.5}$$

where T is temperature, s is elevation of the point above a reference plane, h_0 is the height of the inlet, $\mathbf{u}$ is the flow velocity vector, and $\boldsymbol{\omega}$ is the vorticity vector. The thin boundary layer merges with shock waves to produce a *merged shock-viscous layer*, and various effects, such as temperature, density, and pressure, arise in the flow field so that specific heats and mean molecular weight may not remain constant. In most supersonic flights, except in the case of hypervelocity projectiles, low-density flow occurs at very high altitudes. The *Knudsen number* is defined as $K = \lambda/L$, where λ is the mean free path and L is the characteristic length. Note that for $z < 60$ km, the mean free path $L < 1$. However, for $z > 120$ the continuum assumption may not hold. Details about the various effects at supersonic flows are presented in Figure 19.14. Other points of interest in this figure are: At the location marked 1, the Mach number $M \gg 1$, and between the locations 1 and 2 the shock wave is close to the body, creating a thin shock layer. At location marked 2, the vorticity interaction begins, and the boundary layer is δ; and finally at location marked 3, viscous interactions start and the boundary layer at this location is $\delta \propto M^2/\sqrt{Re}$, where Re is the Reynolds number.

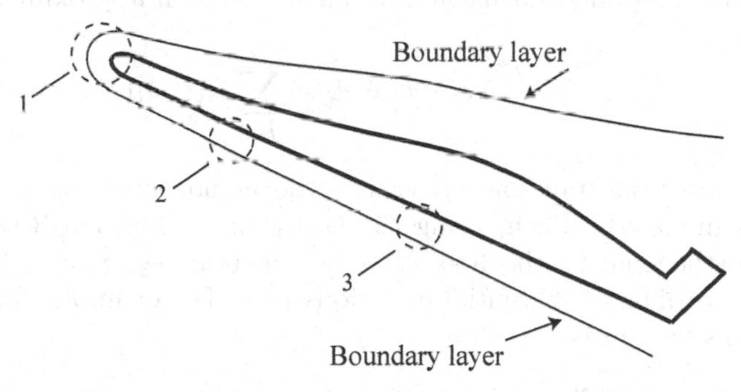

Figure 19.14 Various effects of supersonic flows.

19.6.3 Flow through a Constriction. The Bernoulli equation can be regarded as a statement about the conservation of energy principle appropriate to fluid flows. The qualitative behavior, usually called the *Bernoulli effect*, is the lowering of fluid pressure where the flow velocity is increased. This lowering of pressure may be seen counter-intuitive, but it seems less when pressure is considered as energy density. In the case of high-velocity steady-state flows through a constriction, kinetic energy must increase at the expense of pressure energy, i.e., the energy per unit volume before and after the fluid passes through a constriction is defined by

$$\underbrace{p_1 + \frac{1}{2}\rho v_1^2 + \rho g h_1}_{\text{energy per unit volume before}} = \underbrace{p_2 + \frac{1}{2}\rho v_2^2 + \rho g h_2}_{\text{energy per unit volume before}}, \tag{19.6.6}$$

where h_1 is the height of the inlet, h_2 is the height of the constriction, and $v_{1,2}$ represent the flow velocity, $p_{1,2}$ the pressure energy, $\frac{1}{2}\rho v_{1,2}^2$ the kinetic energy per unit volume, and $\rho g h_{1,2}$ the potential energy per unit volume before and after the flow through the constriction. The Bernoulli effect is the reduction in pressure which occurs when the fluid velocity is increased and internal pressure is decreased for a flow through a constriction, such that finally $p_2 < p_1$ and $v_2 > v_1$, as shown in Figure 19.15, where p_1 is the inlet pressure, p_2 is the pressure at the constriction, v_1 is the effective fluid velocity, and v_2 is the fluid velocity at the constriction.

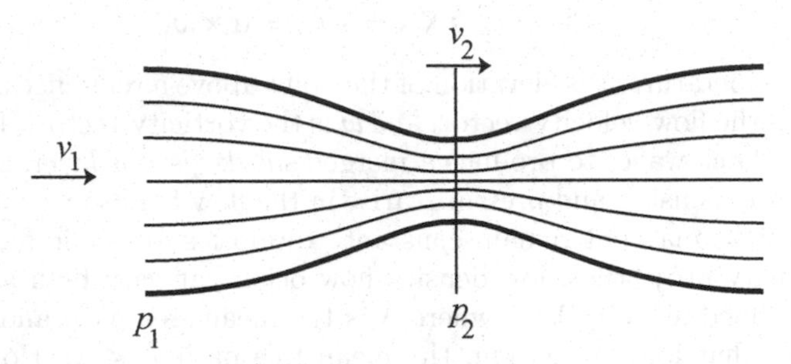

Figure 19.15 Supersonic flow through a constriction.

For more details, see Anderson [2007], and [2016].

19.7 Transient Problems

For time-dependent problems the semi-discrete formulation is used to choose the basis functions. Thus, for one-dimensional problems the Nth approximate solution is taken as

$$\tilde{u}_N(x,t) = \phi_0 + \sum_{j=1}^{N} c_j(t)\,\phi_j(x), \tag{19.7.1}$$

where, as before, the functions ϕ_j satisfy the homogeneous boundary conditions and ϕ_0 is chosen as in (9.55). Then, using the Galerkin or Rayleigh-Ritz method such that the residual is orthogonal to the first N basis functions ϕ_i, $i = 1, 2, \cdots, N$, we obtain the N first-order ordinary differential equations in t. For example, for the diffusion equation $u_t = \nabla^2 u$, this system is

$$\sum_{j=1}^{N} \dot{c}_j(t)\,\langle \phi_j, \phi_i\rangle = \sum_{j=1}^{N} c_j(t)\,\langle \phi_j, \phi_i\rangle + \langle \phi_j, \phi_0\rangle,$$

where the dot denotes the time derivative. The initial conditions for this system are subject to another Galerkin approximation such that its residual $R = u(x,0) - \tilde{u}_N(x,0)$ is orthogonal to the first N basis functions ϕ_j. This yields the system of N algebraic equations

$$\sum_{j=1}^{N} c_j(0)\,\langle \phi_j, \phi_i\rangle = \langle \phi_j, u(x,0) - \phi_0(r)\rangle,$$

which is generally solved for the unknowns $c_i(0)$ by numerical methods.

Example 19.23. Consider the heat conduction equation

$$\frac{\partial u}{\partial t} = \frac{\partial^2 u}{\partial r^2} + \frac{1}{r}\frac{\partial u}{\partial r}, \quad 0 < r < 1,$$

subject to the boundary conditions $u(1,t) = 0 = u_r(0,t)$, and the initial condition $u(r,0) = \ln r$. For a general formulation, we will first consider the annular region $a < r < 1$ $(a > 0)$, and then let $a \to 0$. Now the boundary conditions become $u(1,t) = 0 = u_r(a,t)$. For the first-order approximation, we take the basis function as $\phi_1(r) = c_0 + c_1 r + c_2 r^2$. To determine the coefficients c_0, c_1, and c_2, we require that ϕ_1 satisfies the above boundary conditions. Thus, $\phi_1(1) = c_0 + c_1 + c_2 = 0$, and $\dfrac{\partial \phi_1(a)}{\partial r} = c_1 + 2c_2 a = 0$. By solving these two equations in terms of c_0, we find that $c_1 = 2ac_0/(1-2a)$, and $c_2 = -c_0/(1-2a)$. If we take $c_0 = 1-2a$, then $c_1 = 2a$, and $c_2 = -1$, and the basis function becomes $\phi_1(r) = 1 - 2a + 2ar - r^2$, or $\phi_1(r) = 1 - b + br - r^2$, with $b = 2a$. This suggests that for the Nth approximation we should choose the basis functions as

$$\phi_j(r) = 1 - b_j + b_j r - r^{j+1}, \quad b_j = (j+1)a^j, \quad j = 1, 2, \cdots, N,$$

with $\phi_0 = 0$. The Nth-order approximate solution is then taken in the semi-discrete form as

$$\tilde{u}_N(r,t) = \sum_{j=1}^{N} c_j(t)\,\phi_j(r).$$

The residual is given by

$$\sum_{j=1}^{N} \left\{ \dot{c}_j(t)\,\phi_j + c_j(t) \left[(j+1)^2 r^{j-1} - \frac{b_j}{r} \right] \right\}.$$

Then for the Galerkin method, as $a \to 0$, we have

$$\begin{aligned} 0 &= \sum_{j=1}^{N} \int_0^1 \left\{ \dot{c}_j(t)\,\phi_j + c_j(t) \left[(j+1)^2 r^{j-1} - \frac{b_j}{r} \right] \right\} \left(1 - b_i + b_i r - r^{i+1}\right\}\, r\, dr \\ &= \sum_{j=1}^{N} \int_0^1 \left\{ \dot{c}_j(t)\, f(i,j) + c_j(t)\, g(i,j) \right\}, \end{aligned}$$

for $i = 1, 2, \cdots, N$, where

$$\begin{aligned} f(i,j) &= \frac{(1-b_i)(1-b_j)}{2} + \frac{b_i + b_j - b_i b_j}{3} + \frac{b_i b_j}{4} - \frac{1-b_j}{i+3} \\ &\quad - \frac{1-b_i}{j+3} - \frac{b_j}{i+4} - \frac{b_i}{j+4} + \frac{1}{i+j+4}, \\ g(i,j) &= (j+1)^2 \left[\frac{1-b_i}{j+1} + \frac{b_i}{j+2} - \frac{1}{i+j+2} \right] - b_j + \frac{b_i b_j}{2} + \frac{b_j}{i+2}. \end{aligned}$$

The initial condition is, in general, satisfied approximately. This is accomplished by requiring that the residual

$$R = \sum_{j=1}^{N} c_j(0)\,\phi_j(r) - \ln r$$

be orthogonal to the basis functions $\phi_i(r)$, i.e., $\langle \phi_i, R \rangle = 0$ for $i = 1, 2, \cdots, N$. This means that

$$\lim_{\varepsilon \to 0} \int_\varepsilon^1 R\,\phi_i(r)\, r\, dr = 0 \quad \text{for } i = 1, 2, \cdots, N,$$

since R has a logarithmic singularity at $r = 0$. After evaluating this improper integral, we obtain a system of N algebraic equations:

$$\sum_{j=1}^{N} c_j(0) f(i,j) = \frac{1}{4} + \frac{5}{36} b_j + \frac{1}{(i+3)^2}, \quad i = 1, 2, \cdots, N,$$

which is solved for the unknowns $c_j(0)$. ■

Note that there are other weighted residual methods, like the collocation method, least-squares method, and the method of moments, which are sometimes used, but we will not discuss them. Interested readers can find detailed information on these methods in Connor and Brebbia [1973], Davies [1980], Kantorovitch and Krylov [1958], Kythe [2003], and Reddy [1984].

Example 19.24. The geodesics[3] for the problems (a) on the xy-plane, take $I = \int ds = \int \sqrt{1 + y'^2}\, dx$; (b) on the xy-plane, take $I = \int ds = \int \sqrt{1 + r^2 (d\theta/dr)^2}\, dr$; and (c) on the cylinder $x^2 + y^2 = a^2$, $-\infty < z < \infty$, take $x = a \cos t$, $y = a \sin t$, and $I = \int ds = \int \sqrt{a^2 + (dz/dt)^2}\, dt$ are: (a) straight lines $y = c_1 x + c_2$; (b) straight lines $r \cos(\theta - c_1) = c_2$; and (c) $z = c_1 t + c_2$. ■

REFERENCES USED: Carslaw and Jaeger [1959], Kythe [1998], Kythe et al. [2003], Kantorovitch and Krylov [1958], Schinzinger and Laura [2003], Smith [1985].

[3] A geodesic is the shortest distance between two points on the surface of the Earth, which is the segment of a great circle.

20

Vibrations and Acoustics

We will study the vibrations of strings and other media, and present related topics from the science of acoustics. The case of vibration of a string has been mentioned in §17.3 and an example presented in §17.6.1. Vibrations are defined by hyperbolic equations which are extensively discussed in Chapters 17, 18, and 19. The propagation and dispersion of waves has also been mentioned in previous chapters. We will study some aspects of vibrating membranes and acoustics as to how they are related to conformal mapping. In engineering studies the notation $w = u + i\,v$ has been used for transverse deflections. However, since we use w for the model plane (w-plane) in conformal mappings, we will, therefore, use ϕ for the potential function to represent deflection.

20.1 Wave Propagation and Dispersion

We discuss three important aspects of the hyperbolic equation, namely, wave propagation, wave dispersion, and damped waves.

20.1.1 Wave Propagation. The solution of the problem of a vibrating string, defined by the one-dimensional wave equation

$$\frac{\partial^2 u}{\partial t^2} = c^2 \frac{\partial^2 u}{\partial x^2}, \quad 0 < x < l,$$

and subject to the boundary conditions $u(0,t) = 0 = t(l,t)$, $t > 0$, and the initial conditions $u(x,0) = f(x)$, $u_t(x,0) = g'(x)$, is given by

$$\begin{aligned} u(x,t) &= \frac{1}{2}\sum_{n=1}^{\infty} A_n \Big\{ \sin\frac{n\pi(x+ct)}{l} + \sin\frac{n\pi(x-ct)}{l} \Big\} \\ &\quad + \frac{1}{2}\sum_{n=1}^{\infty} B_n \Big\{ \cos\frac{n\pi(x-ct)}{l} - \cos\frac{n\pi(x+ct)}{l} \Big\} \\ &= \frac{1}{2}\left[f(x+ct) + f(x-ct) \right] + \frac{1}{2}\left[-g(x+ct) + g(x-ct) \right], \end{aligned} \tag{20.1.1}$$

which can be represented as $u(x,t) = u_1(x,t) + u_2(x,t)$, where

$$u_1(x,t) = \frac{f(x+ct)+f(x-ct)}{2}, \quad u_2(x,t) = \frac{g(x+ct)-g(x-ct)}{2c}. \tag{20.1.2}$$

The function $u_1(x,t)$ represents the path of the propagation of the initial displacement without the initial velocity, i.e., for $g'(x) = 0$. The other function $u_2(x,t)$ contains the initial velocity (initial impulse) with zero initial displacement. Geometrically, the function $u(x,t)$ represents a surface in the (u,x,t)-space (Figure 20.1(a)). The intersection of this surface by a plane $t = t_0$ is described analytically by $u = u(x, t_0)$, which exhibits the profile of the string at time t_0. However, the intersection of the surface $u(x,t)$ by a plane $x = x_0$ is given by $u = u(x_0, t)$, which represents the path of the motion of the point x_0.

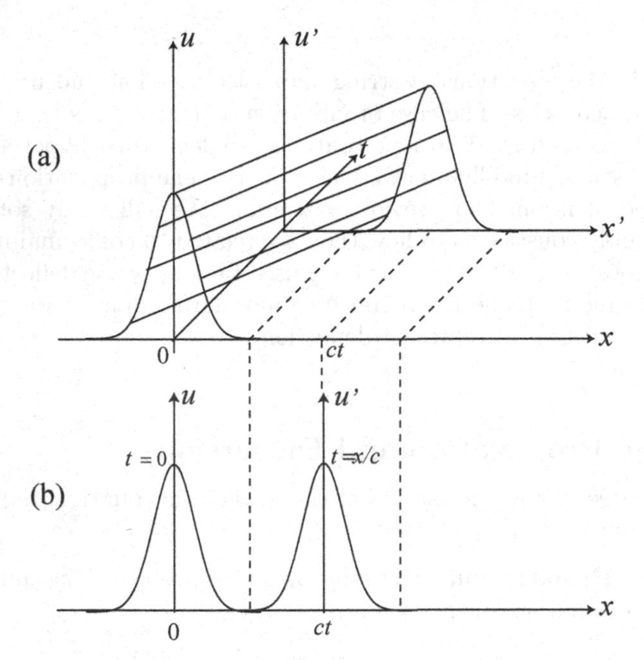

Figure 20.1 Profile of propagating waves.

The function $u = f(x-ct)$ represents a *propagating wave*. The structure of a propagating wave at different times t is described as follows: Assume that an observer moves parallel to the x-axis with velocity c (Figure 20.1(b)). If the observer is at the initial time $t = 0$ at the position $x = 0$, then he (she) has moved along the path toward the right until some time t. Let a new coordinate system, defined by $x' = x - ct$, $t' = t$, move along with the observer. Then $u(x,t) = f(x-ct)$ is defined in this new coordinate system by

$$u(x', t') = f(x'),$$

which means that the observer sees one and the same profile $f(x')$ during the entire time t. Thus, $f(x-ct)$ represents a fixed profile $f(x')$, which moves to the right with velocity c (hence, a propagating wave). In other words, in the (x,t)-plane the function $u = f(x-ct)$ remains constant on the line $x - ct =$const.

The other function $f(x+ct)$ similarly represents a wave propagating toward the left with velocity c. For this wave we have a similar explanation. Thus, the initial form of both waves is characterized by the function $f(x')/2$, which is equal to one half of the original displacement.

Example 20.1. The one-dimensional wave equation $u_{tt} = c^2 u_{xx}$, subject to the initial conditions $u(x,0) = f(x)$, $u_t(x,0) = g'(x)$, has the solution

$$u(x,t) = \phi(x-ct) + \psi(x+ct), \quad x \in R^1, \tag{20.1.3}$$

where ϕ represents a disturbance (wave) traveling in the positive x direction with velocity c, while ψ is a disturbance traveling in the negative x direction with the same velocity. Since at $x=0$, Eq (20.1.3) gives $\phi(-ct) + \psi(ct) = 0$, we find that

$$u(x,t) = \phi(x-ct) - \phi(-x-ct).$$

The initial conditions give

$$\begin{aligned} f(x) &= \phi(x) + \psi(x) = \phi(x) - \phi(-x), \\ g'(x) &= -c\,\phi'(x) + c\,\psi'(x) = -c\,\phi'(x) + c\,\phi'(-x). \end{aligned} \tag{20.1.4}$$

If we integrate the second equation in (20.1.4), with x_0 as an arbitrary point in R^1, we obtain

$$g(x) = -c\big[\phi(x) + \phi(-x)\big] + 2c\phi\,(x_0) + g\,(x_0)\,,$$

which, when added to the first equation in (20.1.4), gives

$$2\phi(x) = f(x) - \frac{1}{c}\,g(x) + A, \quad \text{where} \quad A = -2\phi\,(x_0) + \frac{1}{c}\,g\,(x_0)\,. \tag{20.1.5}$$

Substituting (20.1.5) into (20.1.3), we find the solution as

$$\begin{aligned} u(x,t) &= \frac{1}{2}\Big[f(x-ct) + f(x+ct) + \frac{1}{c}\Big(\int_{x_0}^{x+ct} - \int_{x_0}^{x-ct}\Big)\,g'(s)\,ds\Big] \\ &= \frac{1}{2}\Big[f(x-ct) + f(x+ct) + \frac{1}{c}\int_{x-ct}^{x+ct} g'(s)\,ds\Big] \\ &= \frac{1}{2}\Big[f(x-ct) + f(x+ct) + \frac{1}{c}\,g(x+ct) - \frac{1}{c}\,g(x-ct)\Big], \end{aligned} \tag{20.1.6}$$

which is valid only for all $x \in R^1$.

A geometrical interpretation of Eq (20.1.6) is obvious from Figure 20.2, where we draw through the point P in the (x,t)-plane two lines PA and PB, which satisfy the equation $x \pm ct$ =const. The disturbance at the point P at the time T is caused by those initial data that lie on the segment $AB = (x_1, x_2)$. In optics, the triangle APB is called the *retrograde light cone*. If the initial velocities are zero, i.e., if $g=0$, then

$$u(x,t) = \frac{1}{2}\Big[f(x-ct) + f(x+ct)\Big]. \tag{20.1.7}$$

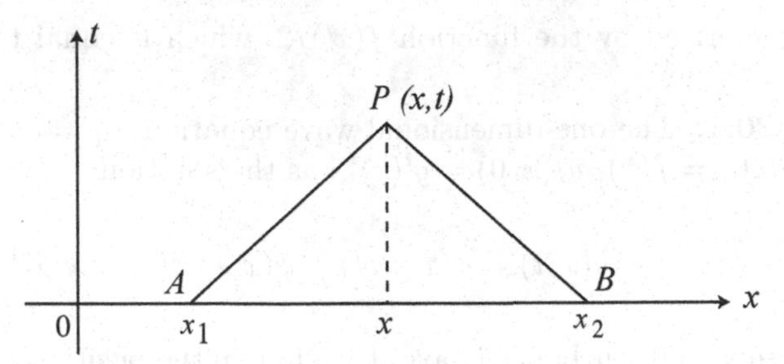

Figure 20.2 Huygens principle.

This represents the *Huygens principle*, which states that each point of an advancing wavefront becomes a new source of secondary waves such that the envelope tangent to all these secondary waves forms a new wavefront. The secondary waves have the same frequency and speed as the primary advancing waves. Thus, the disturbance at x at time t, originating from the sources at A and B at time zero, needs the time $\tau = (x_2 - x)/c = (x - x_1)/c$ to reach the point x. ■

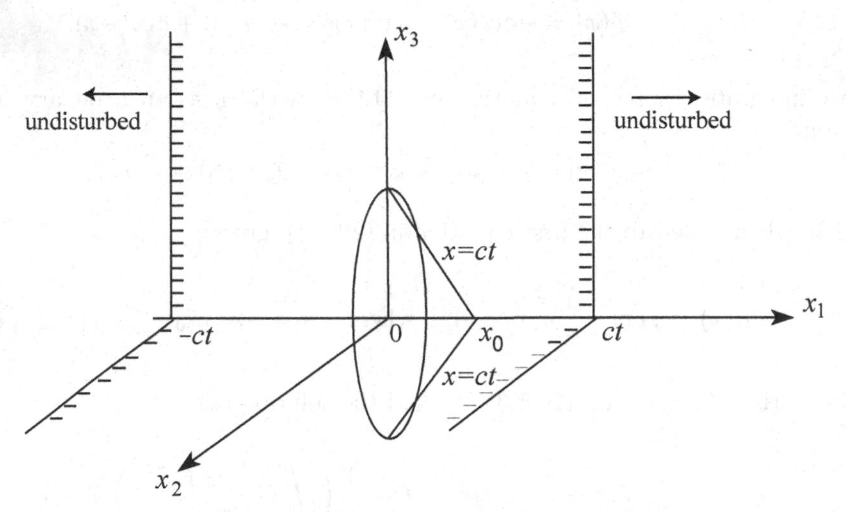

Figure 20.3 Wave propagation in R^1.

In $\mathbb{R}^1$, the Green's functions for the wave operator is

$$G_1(x,t;x',t') = \frac{1}{2c} H\left[c(t-t') - (|x-x'|)\right],$$

which shows that the wave originating instantaneously at a point source $\delta(x,t)$ at time $t > 0$ covers the interval $-ct \leq x \leq ct$, where there exist two edges defined by $x = \pm ct$ that move forward with velocity c. This wave is observed behind the front edge and has amplitude $1/2c$. Hence, the wave diffusion occurs in this case. A three-dimensional representation of Green's function in $\mathbb{R}^1$ is shown in Figure 20.3. It can be viewed as that of a wave starting at the point source and propagating as a plane wave $|x| \leq ct$ whose front edge $|x| = ct$ moves with the velocity c perpendicular to the plane $x = 0$. There does not exist a rear edge of the wave in this case.

In $\mathbb{R}^2$, the Green's function defined by

$$G_2(r; x', t') = \frac{H\left(c(t-t') - r\right)}{2\pi c\sqrt{c^2(t-t')^2 - r^2}}, \quad r^2 = (x-x')^2 + (y-y')^2,$$

with $(\mathbf{x}', t') = (\mathbf{0}, 0)$, shows that the disturbance originates instantaneously at the point source $\delta(\mathbf{x}, t)$, and at time $t > 0$ it occupies the entire circle $|\mathbf{x}| \le ct$ (see Figure 20.4). The wavefront at $|\mathbf{x}| = ct$ propagates throughout the plane with velocity c, but wave propagation exists behind the front edge at all subsequent times, and the wave has no rear edge. The wave diffusion occurs in this case, and the Huygens principle does not apply.

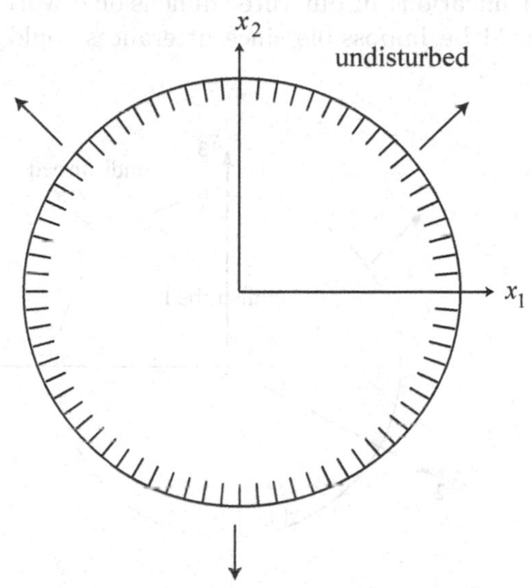

Figure 20.4 Wave propagation in R^2.

In $\mathbb{R}^3$, the Green's function

$$G_3(x, y, z, t; x', y', z', t') = \frac{1}{4\pi c}\delta\Big(t - t' - \frac{1}{c}\sqrt{(x-x')^2 + (y-y')^2 + (z-z')^2}\Big),$$

where δ is the Dirac delta function, with $(\mathbf{x}', t') = (\mathbf{0}, 0)$, implies that the disturbance originating at a point source $\delta(\mathbf{x}, t)$ at time $t > 0$ occupies a spherical surface of radius ct and center at the origin. The wave propagates as a spherical wave with wavefront at $|\mathbf{x}| = ct$ and velocity c, and after the wave has passed there is no disturbance (see Figure 20.5). The Huygens principle applies in this case. The amplitude of the wave decays like r^{-1} as the radius increases. For the derivation of these three Green's functions, see Kythe et al. [2003: 229ff]. The graphs of these three Green's functions are shown in Figure 20.5.

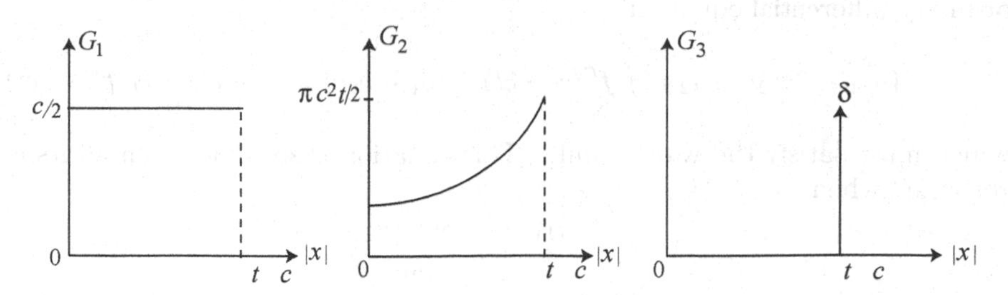

Figure 20.5 Green's functions G_1, G_2, and G_3.

There is a significant difference between the two- and three-dimensional cases. If a stone is dropped in a calm shallow pond, the leading water wave spreads out in a circular form with its radius increasing uniformly with time, but the water contained by this wave continues to move after its passage. This is because of the Heaviside function $H(t-t')$ in the Green's function solution $G_2(r; x', t')$ given above, which leaves a wake behind it. On the other hand, in the three-dimensional case, if a shot fired suddenly at time $t = t'$ in still air is heard only on expanding spherical surfaces with center at the firing gun and radius $c|t-t'|$, where c is the velocity of sound. However, the air does not continue to reverberate after the passage of this wave. This is because of the presence of the Dirac delta function in the solution, which represents a sharp bang and no tail effect. The Huygens principle accounts for the simplicity of communications in our three-dimensional world. If it were two dimensional, communication would be impossible since utterances could be hardly distinguished from one another.

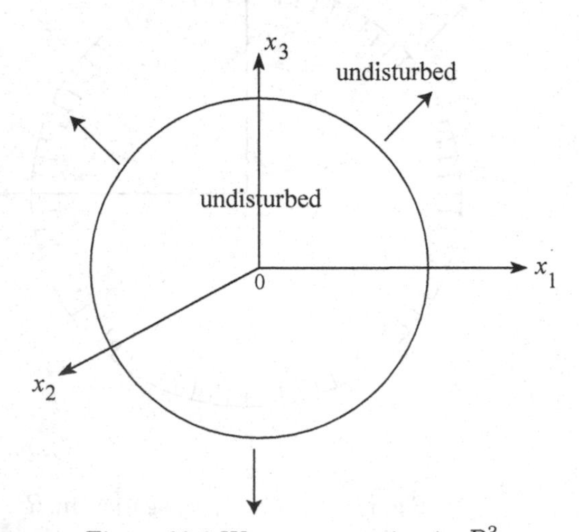

Figure 20.6 Wave propagation in R^3.

20.1.2 Wave Dispersion. The one-dimensional wave equation $u_{tt} = c^2 u_{xx}$ has solutions of arbitrary form given by (20.1.3). Consider the general form (17.3.1) of the homogeneous, linear partial differential equation of the second order with constant coefficients. Then the problem of determining which solutions are of arbitrary form (20.1.3) reduces to that of constructing relationships among the coefficients in Eq (17.3.1), which guarantee the solution of this equation to be of the form

$$u(x,t) = f(x-ct), \tag{20.1.8}$$

where f is an arbitrary function. Substituting (20.1.8) into (17.3.1) we get the linear ordinary differential equation

$$\left(a_{11} - 2a_{12}c + a_{22}c^2\right) f''(x-ct) + (b_1 - b_2 c)\, f'(x-ct) + c_0\, f(x-ct) = 0,$$

which must satisfy the wave profile. This equation is solvable when all its coefficients are zero, i.e., when

$$\begin{aligned} a_{11} - 2a_{12}c + a_{22}c^2 &= 0,\\ b_1 - b_2 c &= 0,\\ c_0 &= 0. \end{aligned} \tag{20.1.9}$$

If the differential equation (17.3.1) is hyperbolic, then it is necessary and sufficient that the conditions (20.1.9) must be satisfied. The first equation in (20.1.9) yields the wave velocity

$$c = \frac{a_{12} \pm \sqrt{a_{12}^2 - a_{11}a_{22}}}{a_{22}}.$$

Thus, there exist two velocities of wave propagation for hyperbolic differential equations for which $a_{12}^2 - a_{11}a_{22} > 0$. Then, for both values of c we find from the remaining two equations in (20.1.9) that $b_1 = b_2 = c_0 = 0$. Therefore, a solution of the form (20.1.8) for a propagating wave without decay or growth is possible only for an equation of the form

$$a_{11}u_{xx} + 2a_{12}u_{xt} + a_{22}u_{tt} = 0. \tag{20.1.10}$$

If $a_{22} \neq 0$, then Eq (20.1.10) represents a wave equation in a moving coordinate system. To see this, we use the characteristic coordinates $\xi,\ \eta = x - \beta_{1,2}\, t$, where

$$\beta_{1,2} = \frac{a_{11}}{a_{12} \pm \sqrt{a_{12}^2 - a_{11}a_{22}}} = \frac{a_{12} \mp \sqrt{a_{12}^2 - a_{11}a_{22}}}{a_{22}} = c_{2,1},$$

provided $a_{12}^2 - a_{11}a_{22} > 0$. Then Eq (20.1.10) is transformed into

$$\begin{aligned} a_{11}\left[u_{\xi\xi} + 2u_{\xi\eta} + u_{\eta\eta}\right] - 2a_{12}\left[\beta_1 u_{\xi\xi} + (\beta_1 + \beta_2)\, u_{\xi\eta} + \beta_2 u_{\eta\eta}\right] \\ + a_{22}\left[\beta_1^2 u_{\xi\xi} + 2\beta_1\beta_2 u_{\xi\eta} + \beta_2^2 u_{\eta\eta}\right] = 0. \end{aligned}$$

Note that the coefficients of both $u_{\xi\xi}$ and $u_{\eta\eta}$ vanish. Thus, Eq (20.1.10) reduces to $u_{\xi\eta} = 0$, which is the wave equation with solutions

$$u = f(\xi) + g(\eta) = f\left(x - \beta_1 t\right) + g\left(x - \beta_2 t\right).$$

This solution represents two waves with speed β_1 and β_2, respectively.

For elliptic equations (i.e., when $a_{12}^2 - a_{11}a_{22} < 0$), we do not get solutions that represent waves with real velocities.

For parabolic equations (i.e., when $a_{12}^2 - a_{11}a_{22} = 0$), Eq (20.1.10) becomes

$$\left(\alpha\,\frac{\partial}{\partial x} + \beta\,\frac{\partial}{\partial t}\right)\left(\alpha\,\frac{\partial}{\partial x} + \beta\,\frac{\partial}{\partial t}\right) u = 0, \tag{20.1.11}$$

where $\alpha = \sqrt{a_{11}}$, $\beta = \sqrt{a_{22}}$, and $\alpha\beta = a_{12}$, and its solution is given by

$$u(x,t) = f(\beta x - \alpha t) + x\, g(\beta x - \alpha t). \tag{20.1.12}$$

Hence, solutions in the form of propagating waves exist in this case.

In physics, the propagating wave is denoted by $u = f\,(x - t/c)$, and the concept of wave dispersion is slightly different. Thus, in physics, a harmonic wave is represented by $u(x,t) = e^{i(\omega t - \nu x)}$, where ω is the frequency, $\nu = 2\pi/\lambda$ the wave number, and λ the wave length. Then the velocity by which the phase $\delta = \omega t - \nu x$ of the wave moves is known as the phase velocity and is equal to ω/ν. A wave dispersion occurs when the phase velocity of a harmonic wave depends on the frequency, and the harmonic signals are displayed relative to each other, thus creating a distorted signal.

20.1.3 Damped Waves. Eq (17.3.1) admits solutions in the form of damped waves, which are given by

$$u(x,t) = \mu(t)\, f(x - ct), \quad \text{or} \quad u(x,t) = \mu(x)\, g(x - ct). \tag{20.1.13}$$

We will consider the first form of the waves; the other form can be analyzed similarly. Substituting (20.1.13) into (17.3.1) we get

$$\begin{gathered}\left(a_{11} - 2a_{12}c + a_{22}c^2\right) \mu f'' + \left[(b_1 - b_2 c)\, \mu + 2\,(a_{12} - a_{22}c)\, \mu' \right] f' \\ \left(c_0 \mu + b_2 \mu' + a_{22}\mu''\right) f = 0.\end{gathered}$$

Also, since f is arbitrary, the coefficients of f, f' and f'' must be equal to zero:

$$\begin{aligned} &a_{11} - 2a_{12}c + a_{22}c^2 = 0, \\ &(b_1 - b_2 c)\, \mu + 2\,(a_{12} - a_{22}c)\, \mu' = 0, \\ &c_0 \mu + b_2 \mu' + a_{22}\mu'' = 0. \end{aligned} \tag{20.1.14}$$

Since the function $\mu(t)$ satisfies an ordinary differential equation with constant coefficients, it has the form $\mu(t) = e^{\pm \kappa t}$. We consider the case when $\mu(t) = e^{-\kappa t}$. Thus, substituting it into Eqs (20.1.14), we get

$$\begin{aligned} &a_{11} - 2a_{12}c + a_{22}c^2 = 0, \\ &(b_1 - b_2 c)\, \kappa + 2\,(a_{12} - a_{22}c) = 0 \\ &a_{22}\kappa^2 - b_2\kappa + c_0 = 0. \end{aligned} \tag{20.1.15}$$

If we eliminate c and κ from the system (20.1.15), we obtain a condition of compatibility of these three equations. The first equation shows that only the hyperbolic differential equation admits damped waves as solutions. The damping coefficient κ is obtained from the second equation as $\kappa = \dfrac{b_2 \pm \sqrt{b_2^2 - 4c_0 a_{22}}}{2a_{22}}$. If we substitute this value of κ into the third equation, we obtain the following relation for the coefficients:

$$4\left(a_{12}^2 - a_{11}a_{22}\right) c_0 + \left(a_{11}b_2^2 - 2a_{12}b_1 b_2 + a_{22}b_1^2\right) = 0. \tag{20.1.16}$$

If this relation is satisfied, then the solutions of Eq (17.3.1) exist in the form of damped waves.

Example 20.2. Consider the telegraph equation

$$u_{xx} - CL\, u_{tt} + (CR + Lg)\, u_t + gR\, u = 0, \tag{20.1.17}$$

where C is the capacity, L the induction coefficient, R the resistance, and g the loss coefficient of a conductor through which an electric current passes. This equation is hyperbolic ($a_{12}^2 - a_{11}a_{22} = CL > 0$). Also, it does not have a solution as a propagating wave if g and R are both nonzero, because then the relation (20.1.16) is satisfied. The velocity c of the damped wave is given by the first equation in (20.1.15) as $c = 1/\sqrt{CL}$; the damping coefficient κ by the second equation in (20.1.15) as $\kappa = (CR + Lg)/2CL$; and the third equation yields $4CLgR - (CR + Lg)^2 = -(CR - Lg)^2 = 0$, i.e., $CR = Lg$, which is the compatibility condition for Eq (20.1.15). Using these values, the damped wave is given by

$$u(x,t) = e^{-\kappa t}\, f(x - ct), \quad \kappa = \frac{R}{L} = \frac{g}{C}, \quad c = \sqrt{\frac{1}{CL}}.$$

A physical interpretation of this analysis is as follows: It is important for a telegraphic communication over large distances that no damping of the wave should occur in a cable. This requires that either R or g be small, or L or C be large. Thus, the propagation of an undamped signal should result in an undisturbed reproduction of this signal at the receiving end, because wave dispersion, if it occurs, always impairs the quality of reception independent of the quality of telegraphic equipment and cables. A similar situation also exists in the acoustical effects in telephones. ■

20.1.4 Thermal Waves. As we have seen above, every sinusoidal wave has its source in some kind of periodic disturbance. The kinematical aspects of wave motion are important. Like other kinds of waves, thermal or heat waves possess similar kinematics. We will discuss the problem of propagation of thermal waves in the earth.

Example 20.3. It is well known that the changes in the earth's surface occur daily (day-night cycle) as well as annually (summer-winter cycle). Disregarding the earth's inhomogeneity, we will assume that the periodic temperature distribution in the earth is homogeneous. Since the effect of the initial temperature becomes small after several temperature variations, we have to solve a steady-state boundary value problem without initial conditions. Thus, we find a bounded solution $u(y,t)$ for the following problem:

$$u_t = k\, u_{yy}, \quad 0 \le y < \infty, \quad -\infty < t, \tag{20.1.18}$$

$$u(0,t) = A \cos \omega t. \tag{20.1.19}$$

Note that the range $0 \le y < \infty$ represents the medium from the earth's surface to its interior. We assume the solution of the form $u(y,t) = A\, e^{\alpha y + \beta t}$, where α and β are constants to be determined, and A is a preassigned constant. Then substituting this solution into Eq (20.1.18) and using the complex form of the boundary condition $u(0,t) = A\, e^{i\omega t}$, we find that $\beta = i\omega$, and $\alpha^2 = \omega/k = i\omega/k$, which gives $\alpha = \pm(1+i)\sqrt{\omega/2k}$. Hence,

$$u(y,t) = A\, e^{\pm y\sqrt{\omega/2k} + i\left[\pm y\sqrt{\omega/2k} + \omega t\right]}. \tag{20.1.20}$$

The bounded solution is obtained by choosing the minus sign. Thus, taking the real part of (20.1.20), we obtain the required solution as

$$u(y,t) = A\, e^{-y\sqrt{\omega/2k}} \cos\left(y\sqrt{\omega/2k} - \omega t\right). \tag{20.1.21}$$

It is known that when the temperature of the earth's surface changes periodically over a long period of time, the temperature fluctuations in its interior develop with the same period. The structure of the thermal waves is as follows:

(i) The amplitude of these waves, given by $A\, e^{-y\sqrt{\omega/2k}}$, decreases exponentially with the depth y. Thus, an increase in depth results in a decay of the amplitude.

(ii) The temperature fluctuations in the earth occur with a phase lag of $y/\sqrt{2k\omega}$, which denotes the time that lapses between the temperature maximum and minimum inside the earth, and the corresponding time point on the surface is proportional to the depth y.

(iii) The change in the temperature amplitude is given by $e^{-y\sqrt{\omega/2k}}$, which means that the depth of penetration of the temperature is smaller when the period $2\pi/\omega$ is smaller. Thus, for two distinct temperature distributions at time t_1 and t_2, the corresponding depths y_1 and y_2 at which the relative temperature change is the same are related by

$$y_2 = y_1 \sqrt{t_2/t_1}.$$

For example, a comparison between the daily ($t_1 = 1$) and the annual variation with $t_2 = 365\, t_1$ yields $y_2 = \sqrt{365}\, y_1 \approx 19.104\, y_1$, i.e., the depth of penetration of the annual temperature distribution with the same amplitude as on the surface is about 19 times larger than the depth of penetration of the daily temperature distribution. ■

20.2 Vibrations of Strings

We will discuss the vibrations of a string and the membrane motion as related to acoustics. In the case of a membrane, we will determine the upper bound of its fundamental frequency. First, we present the case of vibrations of a string and that of a circular membrane. Consider a stretched string of length l that is fixed at both ends. It is assumed that (i) the string is thin and flexible, i.e., it offers no resistance to change of form except a change in length, and (ii) the tension T in the string is much larger than the force due to gravity acting on it so that the latter can be neglected. Let the string in its equilibrium state be situated along the x-axis. Let $u(x, t)$ denote the displacement of the string at time t from its equilibrium position. The shape of the string at a fixed t is represented in Figure 20.7.

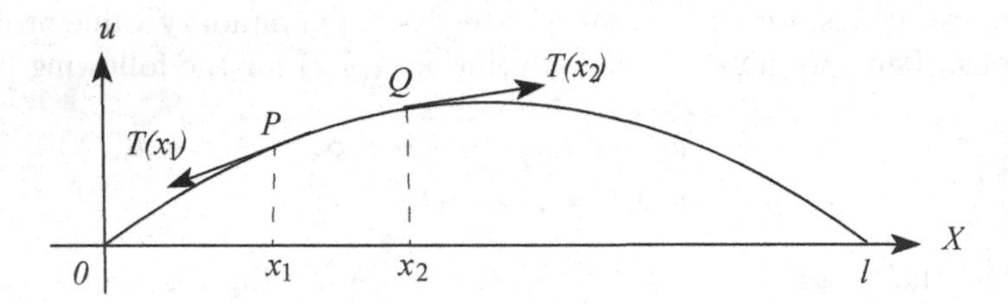

Figure 20.7 Vibrations of a string.

We assume that the vibrations are small, which implies that the displacement $u(x, t)$ and its derivative u_x are small enough so that their squares and products can be neglected. As a result of vibrations, let a segment (x_1, x_2) of the string be deformed into the segment PQ. Then at time t the length of the arc $\widehat{PQ}$ is given by

$$\int_{x_1}^{x_2} \sqrt{1 + u_x^2}\, dx \approx x_2 - x_1, \tag{20.2.1}$$

which simply means that under small vibrations the length of the segment of the string does not change. By Hooke's law, the tension T at each point in the string is independent of t, i.e., during the motion of the string, any change in T can be neglected in comparison with the tension in equilibrium. We will now show that the tension T is also independent of x. In fact, it is evident from Figure 20.7 that the x-component of the resulting tension at the points P and Q must be in equilibrium, i.e.,

$$T(x_1) \cos \alpha(x_1) - T(x_2) \cos \alpha(x_2) = 0,$$

where $\alpha(x)$ denotes the angle between the tangent at a point x and the positive x-axis at time t. Since the vibrations are small, we have

$$\cos \alpha(x) = \frac{1}{\sqrt{1 + \tan^2 \alpha(x)}} = \frac{1}{\sqrt{1 + u_x^2}} \approx 1, \tag{20.2.2}$$

which implies that $T(x_1) \approx T(x_2)$. Since x_1 and x_2 are arbitrary, the magnitude of T is independent of x. Hence, if T_0 denotes the tension at equilibrium and T the tension in the vibrating string, then $T \approx T_0$ for all x and t.

Now, the sum of the components of tension $T(x_1)$ at P and $T(x_2)$ at Q along the u-axis is given by

$$\begin{aligned} &T_0\left[\sin\alpha(x_2) - \sin\alpha(x_1)\right] \\ &= T_0\left[\frac{\tan\alpha(x_2)}{\sqrt{1+\tan^2\alpha(x_2)}} - \frac{\tan\alpha(x_1)}{\sqrt{1+\tan^2\alpha(x_1)}}\right] \\ &= T_0\left[\frac{u_{x_2}}{\sqrt{1+u_{x_2}^2}} - \frac{u_{x_1}}{\sqrt{1+u_{x_1}^2}}\right] \quad \text{by using (20.2.2)} \\ &\approx T_0\left[\frac{\partial u}{\partial x_2} - \frac{\partial u}{\partial x_1}\right] = T_0\left[\left.\frac{\partial u}{\partial x}\right|_{x=x_2} - \left.\frac{\partial u}{\partial x}\right|_{x=x_1}\right] \\ &= T_0\int_{x_1}^{x_2}\frac{\partial^2 u}{\partial x^2}\,dx. \end{aligned} \tag{20.2.3}$$

Let $g(x,t)$ denote the external force per unit length acting on the string along the u-axis. Then the component of $g(x,t)$ acting on the segment $\widehat{PQ}$ along the u-axis is given by

$$\int_{x_1}^{x_2} g(x,t)\,dx. \tag{20.2.4}$$

Let $\rho(x)$ be the density of the string. Then the inertial force on the segment $\widehat{PQ}$ is

$$-\int_{x_1}^{x_2}\rho(x)\,\frac{\partial^2 u}{\partial t^2}\,dx. \tag{20.2.5}$$

Hence, the sum of the components (20.2.3), (20.2.4), and (20.2.5) must be zero, i.e.,

$$\int_{x_1}^{x_2}\left[T_0\frac{\partial^2 u}{\partial x^2} + g(x,t) - \rho(x)\frac{\partial^2 u}{\partial t^2}\right]\,dx = 0. \tag{20.2.6}$$

Since x_1 and x_2 are arbitrary, it follows from (20.2.6) that the integrand must be zero, which gives

$$\rho(x)\frac{\partial^2 u}{\partial t^2} = T_0\frac{\partial^2 u}{\partial x^2} + g(x,t). \tag{20.2.7}$$

This represents the partial differential equation for the vibrations of the string. If $\rho = \text{const}$, then (20.2.7) reduces to

$$\frac{\partial^2 u}{\partial t^2} = c^2\frac{\partial^2 u}{\partial x^2} + f(x,t), \tag{20.2.8}$$

where $c = \sqrt{T_0/\rho}$, and $f(x,t) = g(x,t)/\rho$. In the absence of external forces, Eq (20.2.8) becomes

$$\frac{\partial^2 u}{\partial t^2} = c^2\frac{\partial^2 u}{\partial x^2}, \tag{20.2.9}$$

which is the wave equation for free vibrations (oscillations) of the string. ■

Next, we consider the nonhomogeneous wave equation

$$u_{tt} = c^2 u_{xx} + f(x,t), \quad 0 < x < l, \tag{20.2.10}$$

with the homogeneous (Dirichlet) boundary conditions $u(0,t) = 0 = u(l,t)$, $t > 0$, and the initial conditions $u(x,0) = g(x)$, $u_t(x,0) = h(x)$, $0 \le x \le l$. If we use the method of separation of variables, the Dirichlet boundary conditions suggest that we assume a Fourier series solution of the form

$$u(x,t) = \sum_{n=1}^{\infty} u_n(t) \sin \frac{n\pi x}{l},$$

where t is regarded as a parameter. The functions f, g, h are written in terms of the Fourier series as

$$\begin{aligned} f(x,t) &= \sum_{n=1}^{\infty} f_n(t) \sin \frac{n\pi x}{l}, \quad \text{where} \quad f_n(t) = \frac{2}{l} \int_0^l f(\xi,t) \sin \frac{n\pi\xi}{l}\, d\xi; \\ g(x) &= \sum_{n=1}^{\infty} g_n \sin \frac{n\pi x}{l}, \qquad \text{where} \quad g_n = \frac{2}{l} \int_0^l g(\xi) \sin \frac{n\pi\xi}{l}\, d\xi; \\ h(x) &= \sum_{n=1}^{\infty} h_n \sin \frac{n\pi x}{l}, \qquad \text{where} \quad h_n = \frac{2}{l} \int_0^l h(\xi) \sin \frac{n\pi\xi}{l}\, d\xi. \end{aligned} \tag{20.2.11}$$

After substituting (20.2.11) into (20.2.10), we get

$$\sum_{n=1}^{\infty} \left\{ \ddot{u}_n(t) + \frac{n^2\pi^2c^2}{l^2} u_n(t) - f_n(t) \right\} \sin \frac{n\pi x}{l} = 0,$$

where $\ddot{u} = d^2u/dt^2$. This relation is satisfied if all the coefficients of the series are zero, i.e., if

$$\ddot{u}_n(t) + \frac{n^2\pi^2c^2}{l^2} u_n(t) = f_n(t). \tag{20.2.12}$$

The solution $u_n(t)$ of this ordinary differential equation with constant coefficients is easily obtained under the initial conditions

$$\begin{aligned} u(x,0) &= g(x) = \sum_{n=1}^{\infty} u_n(0) \sin \frac{n\pi x}{l} = \sum_{n=1}^{\infty} g_n \sin \frac{n\pi x}{l}, \\ u_t(x,0) &= h(x) = \sum_{n=1}^{\infty} \dot{u}_n(0) \sin \frac{n\pi x}{l} = \sum_{n=1}^{\infty} h_n \sin \frac{n\pi x}{l}. \end{aligned}$$

Thus, $u_n(0) = g_n$, and $\dot{u}_n(0) = h_n$. Now, we define the solutions $u_n(t)$ in the form

$$u_n(t) = u_n^{(1)}(t) + u_n^{(2)}(t),$$

where $u_n^{(1)}(t)$ is a particular solution of Eq (20.2.12), which, using the variation of parameters method, is given by

$$u_n^{(1)}(t) = \frac{1}{n\pi c} \int_0^t \sin \frac{n\pi c(t-\tau)}{l}\, f_n(\tau)\, d\tau,$$

represents the solution of the nonhomogeneous equation with the homogeneous initial conditions, and

$$u_n^{(2)}(t) = g_n \cos \frac{n\pi ct}{l} + \frac{lh_n}{n\pi c} \sin \frac{n\pi ct}{l}$$

is the solution of the homogeneous equation with the prescribed initial conditions. Hence,

$$\begin{aligned} u(x,t) &= \sum_{n=1}^{\infty} \left[u_n^{(1)}(t) + u_n^{(2)}(t) \right] \\ &= \sum_{n=1}^{\infty} \frac{1}{n\pi c} \int_0^t \sin \frac{n\pi c(t-\tau)}{l} \sin \frac{n\pi x}{l} f_n(\tau)\, d\tau \\ &\quad + \sum_{n=1}^{\infty} \left(g_n \cos \frac{n\pi ct}{l} + \frac{lh_n}{n\pi c} \sin \frac{n\pi ct)}{l} \right) \sin \frac{n\pi x}{l}. \end{aligned} \tag{20.2.13}$$

Note that the second term is the solution of the corresponding problem with $f = 0$ (representing a freely vibrating string with prescribed initial conditions. The first term represents the forced vibrations of the string under the influence of an external force. Delillo [1994b] discusses how the geometry of the region affects the conditions and accuracy of this method

20.3 Vibrations of Membranes

Suppose that a membrane that is a perfectly flexible, thin, stretched sheet occupies a region D in the xy-plane in its equilibrium state. Further, let the membrane be subjected to a uniform tension T applied on its boundary ∂D perpendicular to the lateral surface. This means that the force acting on an element ds of the boundary ∂D is equal to $T\,ds$. We will examine the transverse oscillations of the membrane, which move perpendicular to the xy-plane at each point in the direction of the u-axis. Thus, the displacement u at a point $(x,y) \in D$ is a function of x, y and t. Assuming that the oscillations are small, i.e., the functions u, u_x, and u_y are so small that their squares and products can be neglected, let $A \in D$ denote an arbitrary area of the membrane, which is bounded by the curve L and lies in a state of equilibrium in the xy-plane. After the membrane is displaced from its equilibrium position, let the area A be deformed into an area A' bounded by a curve L' (see Figure 20.7), which at time t is defined by

$$A' = \iint_A \sqrt{1 + u_x^2 + u_y^2}\, dx\, dy \approx \iint_A dx\, dy = A.$$

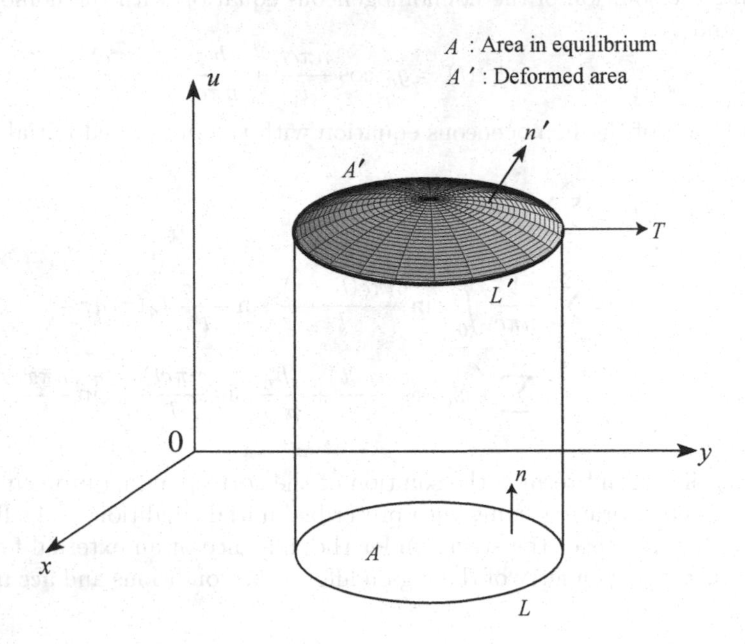

Figure 20.8 Vibrations of a membrane.

Thus, we can neglect the change in A during the oscillations, and the tension in the membrane remains constant and equal to its initial value T.

Note that the tension T which is perpendicular to the boundary L' acts at all points in the tangent plane to the surface area A'. Let ds' denote an element of the boundary L'. Then the tension acting on this element is $T\,ds'$, and $\frac{\partial u}{\partial n} = \cos\alpha$, where α is the angle between the tension T and the u-axis, and n is the outward normal to the boundary L. The component of the tension acting on the element ds' in the direction of the u-axis is $T\frac{\partial u}{\partial n}\,ds'$. Hence, the component of the resultant force acting on the boundary L' along the u-axis is

$$T\int_{L'} \frac{\partial u}{\partial n}\,ds' \approx T\int_{L} \frac{\partial u}{\partial n}\,ds = \iint_A \left(\frac{\partial^2 u}{\partial x^2} + \frac{\partial^2 u}{\partial y^2}\right) dx\,dy, \tag{20.3.1}$$

by Green's identity, where, in view of small oscillations, we have taken $ds' \approx ds$, and replaced L' by L. Let $g(x, y, t)$ denote an external force per unit area acting on the membrane along the u-axis. Then the total force acting on the area A' is given by

$$\iint_A g(x, y, t)\,dx\,dy. \tag{20.3.2}$$

Let $\rho(x, y, t)$ be the surface density of the membrane. Thus, the inertial force at all times t is

$$\iint_A \rho(x, y, t)\frac{\partial^2 u}{\partial t^2}\,dx\,dy. \tag{20.3.3}$$

Since the sum of the inertial force and the total force is equal and opposite to the resultant of the tension on the boundary L', we find from (20.3.1)-(20.3.3) that

$$\iint_A \left[\rho(x, y, t)\frac{\partial^2 u}{\partial t^2} - T\left(\frac{\partial^2 u}{\partial x^2} + \frac{\partial^2 u}{\partial y^2}\right) - g(x, y, t)\right] dx\,dy = 0,$$

or, since A is arbitrary,

$$\rho(x,y,t)\frac{\partial^2 u}{\partial t^2} = T\left(\frac{\partial^2 u}{\partial x^2} + \frac{\partial^2 u}{\partial y^2}\right) + g(x,y,t). \tag{20.3.4}$$

This is the partial differential equation for small oscillations of a membrane. If the density $\rho = \text{const}$, then Eq (20.3.4) in the absence of external forces reduces to

$$\frac{\partial^2 u}{\partial t^2} = c^2\left(\frac{\partial^2 u}{\partial x^2} + \frac{\partial^2 u}{\partial y^2}\right), \quad c = \sqrt{T/\rho}. \blacksquare \tag{20.3.5}$$

To justify Eq (20.3.5), let the membrane's equilibrium position lie in the (x,y)-plane. Assuming that the membrane has negligible bending resistance, ignoring the gravitational body forces, and assuming that constant tension force is applied uniformly to the membrane in all directions, the transverse deflection measure $u(x,y,t)$ will be in the z-direction. A $dx\,dy$ element of the membrane as viewed along the y-axis is shown in Figure 20.9, in which (a) represents $\dfrac{du}{dx} + \dfrac{\partial}{\partial x}\Big(\dfrac{\partial u}{\partial x}\Big)\,dx$. A similar view occurs along the x-axis.

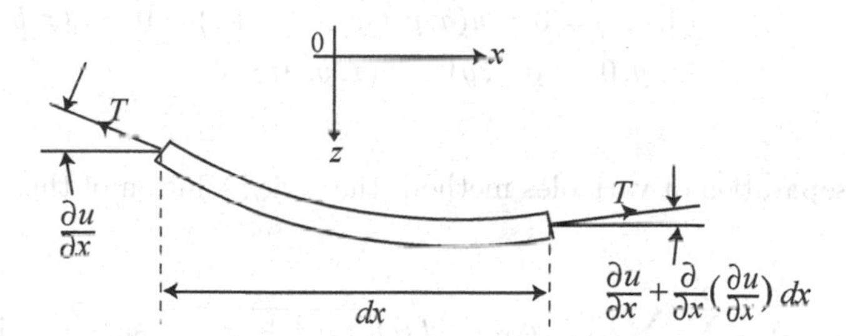

Figure 20.9 View of the $(dx\,dy)$ element of the membrane.

Using Newton's law in the z-direction, we have

$$T\,dy\,\frac{\partial^2 u}{\partial x^2}\,dx + T\,dx\,\frac{\partial^2 u}{\partial y^2}\,dy = \rho\,dx\,dy\frac{\partial^2 u}{\partial t^2},$$

which, after dividing by $dx\,dy$, gives

$$\frac{\partial^2 u}{\partial x^2} + \frac{\partial^2 u}{\partial y^2} = \frac{1}{c^2}\frac{\partial^2 u}{\partial t^2}, \quad c = \sqrt{T/\rho},$$

which is the wave equation (20.3.5). Note that T denotes the tension per unit length and ρ the mass per unit area.

20.3.1 Force at the Boundary of a Membrane. We will discuss transverse vibrations of an elastic membrane in $\mathbb{R}^2$ in the following cases:

CASE 1. Consider a rectangular membrane, shown in Figure 20.10.

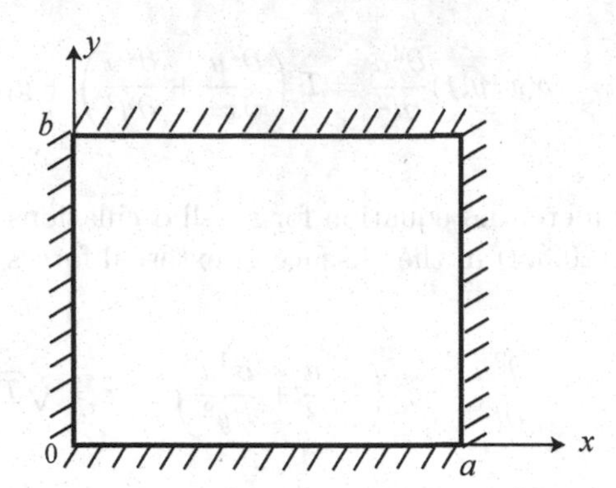

Figure 20.10 Rectangular membrane.

The boundary value problem in $\mathbb{R}^2$ is

$$\begin{aligned} &\frac{\partial^2 u}{\partial x^2}+\frac{\partial^2 u}{\partial y^2}=\frac{1}{c^2}\frac{\partial^2 u}{\partial t^2}, \\ &u(0,y,t)=0=u(a,y,t), \quad u(x,0,t)=0=u(x,b,t), \\ &u(x,y,0)=g(x,y), \quad \dot{u}(x,y,0)=0. \end{aligned} \tag{20.3.6}$$

Using the separation of variables method, the series solution of this problem is

$$u(x,y,t)=\sum_{m=1}^{\infty}\sum_{n=1}^{\infty} A_{mn}\cos\left(\pi ct\sqrt{m^2/a^2+n^2/b^2}\right)\sin\frac{m\pi x}{a}\sin\frac{n\pi y}{b}, \tag{20.3.7}$$

where

$$A_{mn}=\frac{4}{ab}\int_0^a\int_0^b g(\xi,\eta)\sin\frac{m\pi\xi}{a}\sin\frac{n\pi\eta}{b}. \tag{20.3.8}$$

The frequency of vibrations f is given by

$$f=\frac{\omega}{2\pi}=\frac{c}{2}\sqrt{m^2/a^2+n^2/b^2}. \tag{20.3.9}$$

Note that in the case of a square ($a=b$), the frequency is not unique with each mode shape, i.e., $\omega_{mn}=\omega_{nm}$.

CASE 2. Consider a membrane of arbitrary shape clamped along its boundary and governed by Eq (20.3.5) subject to the Dirichlet boundary condition $u\left(B(x,y)=0;t\right)=0$, where $B(x,y)=0$ is a functional relation which defines the boundary of the membrane (Figure 20.11). Recall that the Dirichlet problem is also known as the first boundary value problem.

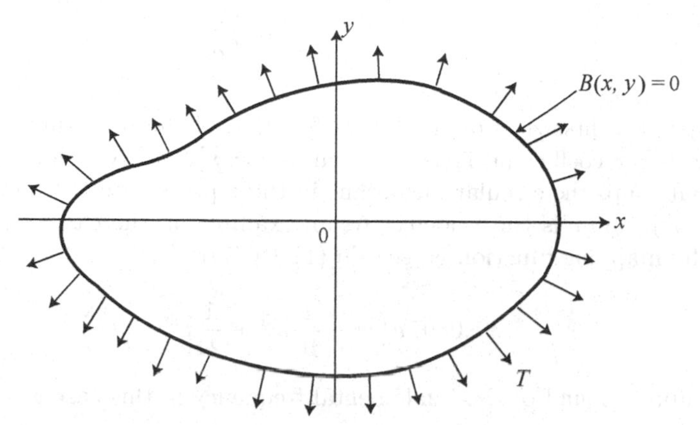

Figure 20.11 Membrane of arbitrary shape.

This membrane has become a valid mathematical model for various technological and scientific problems ranging from physiological systems to machine and transducer elements. A survey of all such technical and scientific references for determining the variational modes of membranes can be found in Mazumdar [1975].

Case 3. Consider the case of an axisymmetric membrane of infinite extent with some prescribed initial displacements. The initial value problem is formulated in polar cylindrical coordinates as follows:

$$\begin{gathered}\frac{\partial^2 u}{\partial r^2}+\frac{1}{r}\frac{\partial u}{\partial r}=\frac{1}{c^2}\frac{\partial^2 u}{\partial t^2},\\ u(r,0)g(r),\quad u_t(t,0)=0.\end{gathered}\tag{20.3.10}$$

Applying the zero-order Hankel transform over the variable r to Eq (20.3.10), we get

$$\frac{d^2\hat{u}}{dt^2}+c^2\lambda^2\hat{u}=0,\quad \hat{u}(\lambda,0)=\hat{g}(\lambda),\quad \hat{u}_t(\lambda,0)=0.$$

This equation has the solution

$$\hat{u}(\lambda,t)=\hat{g}(\lambda)\cos c\lambda t,$$

which on inversion gives

$$u(r,t)=\int_0^\infty \lambda\hat{g}(\lambda)\cos c\lambda t J_0(\lambda r)\,d\lambda.\tag{20.3.11}$$

If the initial condition is $u(r,0)=g(r)=\begin{cases}\dfrac{1}{\pi a^2} & \text{for } r<a,\\ 0 & \text{for } r>a,\end{cases}$, then the solution becomes

$$u(r,t)=\begin{cases}-\dfrac{1}{2\pi}\dfrac{ct}{r^{1/2}\,(c^2t^2-r^2)^{3/2}}, & \text{if } 0<r<ct,\\ 0 & \text{if } ct<r<\infty.\end{cases}\tag{20.3.12}$$

The particular case when $a\to 0$ corresponds to a plucked membrane.

20.3.2 Fundamental Frequency of a Membrane. The fundamental circular frequency ω_{11} of a membrane satisfies the inequality

$$\omega_{11} \le \frac{\alpha_{00}}{c_1}\, c, \tag{20.3.13}$$

where α_{00} is the first zero of the Bessel function of the first kind and zero order ($\alpha_{00} = 2.4048$), c_1 is the coefficient of the w^1-term in $z = \sum c_n w^n$, which is the conformal transformation that maps the circular membrane in the z-plane onto the unit circle in the w-plane (see (20.3.4)), and c is the velocity. As an example, in the case of a square membrane of side $2a$, the mapping function is (see Eq (11.10.11))

$$1.0807\, a\Big(w - \frac{1}{10} w^5 + \frac{1}{24} w^9 - \cdots\Big),$$

thus, the upper bound of the fundamental frequency in this case is

$$\omega_{11} \le \frac{2.4048}{1.0807}\frac{c}{a} = 2.2267\frac{c}{a}.$$

The exact solution, obtained by using the method of separation of variables, is

$$\omega_{11} = \frac{\pi c}{a}\sqrt{m^2 + n^2}\Big|_{m=n=1} = 2.2214\,\frac{c}{a}.$$

In the case of a star-shaped membrane, whose boundary is defined in polar coordinates by $R(t) = 1 + \lambda \cos 4t$, $|\lambda| < 1$, the coefficient $c_1 = \left(1 - \frac{9}{4}\lambda^2\right)$ (Laura [1964]). Hence, we obtain the upper bound of the fundamental frequency in this case as

$$\omega_{11} \le \frac{2.4048}{1 - \dfrac{9}{4}\lambda^2}\, c.$$

20.3.3 Free Vibrations of Membranes. In the case of normal modes of harmonic motion, we substitute $u(x, y, t) = W(x, y)\, e^{i\omega t}$ in Eq (20.2.14), then we obtain

$$\frac{\partial^2 W}{\partial x^2} + \frac{\partial^2 W}{\partial y^2} + \Big(\frac{\omega}{c}\Big)^2 W = 0, \tag{20.3.14}$$

which, using the fact that $z = x + iy$ and $\bar z = x - iy$, becomes

$$4\frac{\partial^2 W}{\partial z\, \partial \bar z} + \Big(\frac{\omega}{c}\Big)^2 W = 0, \tag{20.3.15}$$

If $z = f(w)$ is the analytic function that conformally maps the unit circle $|w| = 1$ onto the physical domain in the z-plane, the Maxwell's equation (20.3.15) can be written as

$$4\frac{\partial^2 W}{\partial w\, \partial \bar w} + \Big(\frac{\omega}{c}\Big)^2 |f'(w)|^2 W = 0, \tag{20.3.16}$$

Notice that the above transformation has resulted in a complicated differential equation but simpler transformed domain which is the unit circle. Thus, the boundary conditions are

$$W(w, \bar w)\Big|_{|w|=1} = 0. \tag{20.3.17}$$

Note that if the original domain is doubly connected, then it is transformed onto an annulus in the w-plane, but the boundary condition on the annulus still remains simple.

We can use any of the weighted residual methods, like collocation, Galerkin, least squares and others, to determine an approximated solution of the boundary value problem (20.3.16)-(20.3.17). For example, using the Galerkin method (see §24.2)), let

$$W(w,\bar{w}) \approx W_a(w,\bar{w}) = \sum_{n=1}^{N} B_n f_n(w,\bar{w}), \tag{20.3.18}$$

where each function f_n satisfies the boundary condition (20.3.17). Then substituting Eq (20.3.18) into Eq (20.3.16), we obtain a residual function $\varepsilon(w,\bar{w})$ which is not necessarily zero since $W_a(w,\bar{w})$ is not an exact solution of Eq (20.3.16). This residual function must be orthogonal to each function f_n on the physical domain D, i.e.,

$$\int_D \varepsilon(w,\bar{w}) f_n(w,\bar{w})\, dD = 0 \quad \text{for } n = 1, 2, \dots, N. \tag{20.3.19}$$

This orthogonality condition yields a homogeneous system of N algebraic equations in the N unknown B_n, which finally gives a determinant equation whose roots ω_{ij}/c will give the natural frequency parameters for this boundary value problem. For details of the Galerkin method, see Kythe [1995: 33ff], Kythe and Wei [2004: 9ff], and Sideridis [1984].

Example 20.4. As an application, consider the case of a square membrane of side $2a$. The mapping function is (7.1.12), or

$$z = a\,\frac{\int_0^w (1+t^4)^{-1/2}\, dt}{\int_0^1 (1+t^4)^{-1/2}\, dt}, \tag{20.3.20}$$

which in series form is the expression (11.10.11). Since the boundary in the w-plane is the unit circle, we can disregard the Θ-dependence in the w-plane in the first approximation. Thus, Eq (20.3.18) becomes

$$W(w,\bar{w}) \approx W_a(w,\bar{w}) = \sum_{n=1}^{N} B_n \left(1 - (r^2)^n\right), \quad r^2 = |w|^2. \tag{20.3.21}$$

For $N = 1$ (one-term approximation), we get $\omega_{11}\, a/c = 2.26$, while for $N = 2$ (two-term approximation) $\omega_{11} a/c = 2.222$; the exact eigenvalue is $\omega_{11} a/c = 2.2214$.

In the case of a regular hexagonal membrane, Laura and Faulstich [1965] determined the approximate mapping function as

$$z = a\left(1 - 037\, w - 0.0496\, w^7 + 0.0183\, w^{13} - \cdots\right), \tag{20.3.22}$$

where a denotes the apothem of the hexagon. For $N = 1$, we have $\omega_{11} a/c = 2.36$; for $N = 2$ we have $\omega_{11} a/c = 2.317$, while the Szegö upper bound gives $\omega_{11} a/c < \dfrac{2.4048}{1.0376} = 2.318$. ■

Example 20.5. Consider the boundary problem of a vibrating circular membrane of radius a, where the modes of vibrations are axisymmetric. The fundamental frequency coefficient is given by $\Omega_{00} = a\sqrt{\rho/T}\,\omega_{00} \equiv \alpha_{00}$, where ρ is the area density of the membrane,

T the applied radial tension per unit length, ω_{00} the fundamental circular frequency, and α_{00} is the first zero of the Bessel function of the first kind and zero order ($\alpha_{00} = 2.4048$). We will use the Galerkin or Rayleigh-Ritz method to determine the fundamental frequency coefficient Ω_{00} for this vibrating membrane. The fundamental mode shape is defined by

$$W(r) \approx W_a(r) = A\left(a^2 - r^2\right).$$

Since this value is higher than the exact eigenvalue (by 5%), Rayleigh [1894] suggested to change this expression to

$$W(r) \approx W_a(r) = A\left(a^\gamma - r^\gamma\right). \tag{20.3.23}$$

Then the Galerkin method is applied to obtain $\Omega_{00} = \Omega_{00}(\gamma)$. Since this value is an upper bound, we must have $\dfrac{\partial \Omega_{00}}{\partial \gamma} = 0$, which optimizes the calculated value of the fundamental frequency coefficient as $\Omega_{00} = 2.41$. This method, now known as the Rayleigh-Ritz method (see §24.3), can be found in Schmidt [1981; 1983]. ■

Example 20.6. Consider a square membrane of side a with rounded corners (Figure 20.12) in the z-plane which is mapped onto the unit disk in the w-plane by the conformal transformation

$$z = aL\left(w + \eta w^{1+n}\right), \tag{20.3.24}$$

where $L = 25/24$, $\eta = -(1/25)$, and $n = 4$. Substituting Eqs (20.3.23) and (20.3.24) into Eq (20.3.15) and applying the Galerkin variational method, we get

$$\omega_{11}\frac{a}{c} = \frac{1}{L}\left\{\frac{\gamma/2}{\frac{1}{2} - \dfrac{2}{\gamma+2} + \dfrac{1}{2\gamma+2} + \eta^2(1+n)^2\left(\dfrac{1}{2n+2} - \dfrac{2}{2n+\gamma+2} + \dfrac{1}{2n+2\gamma+2}\right)}\right\}^{1/2}. \tag{20.3.25}$$

Then for the above values of L, n and η, the minimum of the expression in (20.3.25) occurs at $\gamma = 1.40$, which gives $\omega_{11}\, a/c = 2.3.16$, which is very close to the eigenvalue $\Omega_{11} = 2.308$ obtained by Irie et al. [1983], where the details of this method can be found. ■

Example 20.7. Consider the epitrochoidal region shown in Figure 20.12. We use the mapping $z = a\left(w + \frac{1}{6}w^5\right)$ (Map 20.1) and using the same method as in the above example, we get $\omega_{11}\, a/c = 2.393$, which compares very well with the result obtained by Baltrukonis et al. [1965] placing it in the range $2.384 \le \omega_{11}\, a/c \le 2.385$. ■

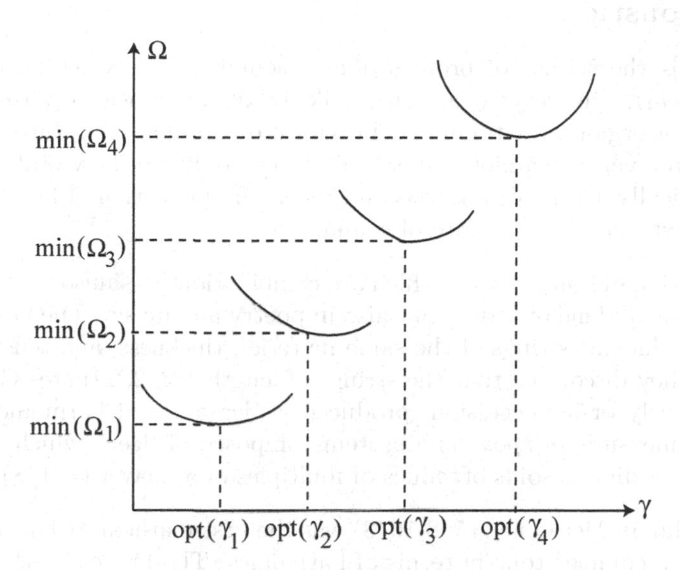

Figure 20.12 Epitrochoidal region.

Example 20.8. We will consider transverse vibrations of composite membranes, and in particular determine the fundamental frequencies of vibration for a class of nonhomogeneous membranes shown in Figure 20.13. The details can be found in Laura et al [1985], where the conformal mapping method of Case 2 of §20.2.3 is used. Schinzinger and Laura [2003: 338] have compared the results obtained from the following three methods: conformal mapping, Fourier expansion collocation, and the finite elements, for the square and the hexagonal membranes.

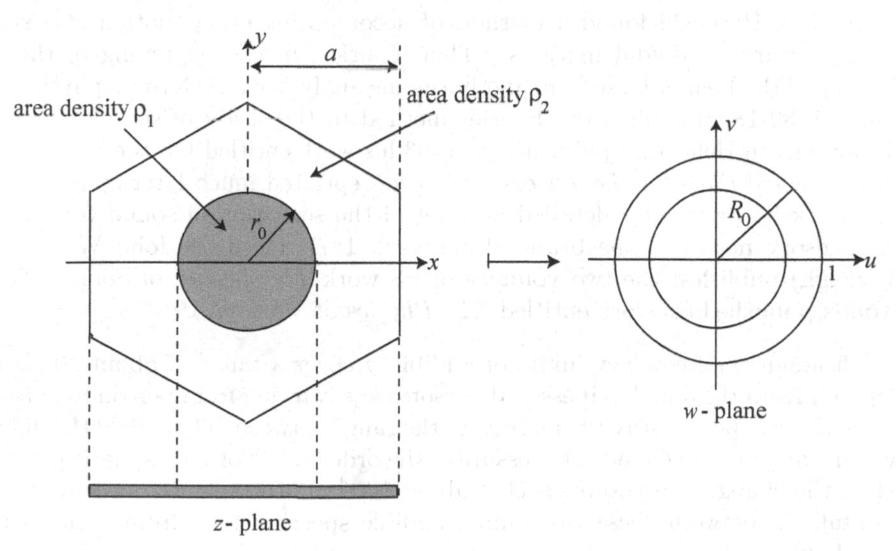

Figure 20.13 Nonhomogeneous membrane.

20.4 Acoustics

Acoustics is the science of propagation of sound waves, which surround us and interact with our hearing in everyday activities, like talking, and hearing speech, poetry, and music. Our hearing organs are stimulated by vibrations propagating through an elastic medium like air. However, we cannot 'see' sound; we can only see the vibrations on an oscilloscope. Mathematically, harmonic analysis and repeated application of fast Fourier transform help clarify the structure and nature of sound.

Modulation of sound waves, which are combinations of sinusoids, is a useful tool not only in processing all kind of waves, but also in poetry and music. The Greeks produced musical sounds by plucking strings of the same material, thickness, and tension, but with different lengths. They discovered that the strings of length $1, 3/2, 2$ (notes C, G, C), when plucked simultaneously or in succession, produced a pleasant and harmonious sound. The Latin meter (or measure, $\mu\epsilon'\tau\rho o\nu$) is a system composed of 'feet', which are divided into seven types of periodic sinusoids of values of multiples of a quaver ($= 1/8$).

Abbé Martin Mersenne (1588-1648) was the first mathematician to publish a qualitative analysis of a complex tone in terms of harmonics. The French mathematician and philosopher Pierre Gassendi was the first to measure the speed of sound in 1635 at 478 m/s, an incorrect result. The correct speed was measured under the Royal Academy of Sciences in Paris in 1750 to be 332 m/s at 0°C; subsequently this speed in water was determined to be 1,435 m/s at 8°C.

Every sound is characterized by a precise number of vibrations per second. This fact was announced by both Galileo and Hooke. But in the 17th century Leonard Euler, Daniel Bernoulli, Jean le Rond d'Alembert, and Joseph Louis Lagrange all investigated the vibrating string, and d'Alembert provided the solution of the wave equation.

In 1753, Bernoulli found a method of decomposing every motion of a string as a sum of elementary sinusoidal motions. Then Fourier, in the beginning of the 18th century, developed the Fourier transform and harmonic analysis. The German physicist Georg Simon Ohm (1789-1854) applied the Fourier method to the study of sound. Later the physicist Hermann von Helmholtz published in 1863 his work entitled *On the Sensation of Tone as a Physiological Basis for the Theory of Music*, reprinted much later as Helmholtz [1954]. In this work he provided a detailed account of the sensation of sound from the ear through the sensory nerves to the brain. Finally, in 1877 and 1878 John William Stuart (Lord Rayleigh) published the two volumes of his work *The Theory of Sound*. Then Alexander Woods published his work entitled *The Physics of Music* in 1913.

Although the frequency limits of audibility cover a range of about 20 Hz to 20,000 Hz, depending on the tone loudness and person's age, varying from person to person, the human ear with its super sensitivity can hear in the range between 500 to 4000 Hz and responds very well in this range to change of pressure of the order 10^{-10} of atmospheric pressure. However, when the change in pressure reaches about 10^{-4} of atmospheric pressure, hearing becomes painful. In between these two limits, audible speech ranges from minimum audibility to very loud noise.

20.4.1 Harmonic Analysis. Sound is detected as a plane wave by a microphone, or an oscilloscope, or the human (animal) ear. A simple sound, generally produced by a tuning fork struck close to the tip of the fork, as a single tone, agitates the molecules of the air in its vicinity, which are immediately subjected to repeated compressions followed by expansions.

The motion of each molecule over time t is governed by

$$y = A \sin 2\pi f t, \tag{20.4.1}$$

where f is the frequency, or number of oscillations (cycles) per second, A is the amplitude of the oscillation, i.e., A is the maximum displacement of the molecule, and the oscillations take place in the direction of propagation. Thus, this sound is a longitudinal compression wave.

The hearing starts around 1×10^{-12} Watts/meter^2. The ear drum is a wonderful receptor of sound waves. Even at and near the lowest perceptible sound intensities, the eardrum vibrates much less distance (i.e., with much less amplitude) than the diameter of a hydrogen atom, which is twice the Bohr radius ($a_0 = 5.29 \times 10^{-12}$ m $= 5.29$ pm). It is so amazing to learn that if the energy in a single 1-Watt night light were converted into acoustical energy and distributed in equal amount to every person in the world, it would still be audible to every person with normal hearing.

The intensity (loudness) of sound is proportional to the square of the amplitude, which is measured in 'decibels' (dB), the one-tenth part of a 'bel'. The unit 'bel' was chosen by electric engineers at the Bell Telephone Company in 1920s to honor the inventor of the telephone, Alexander Graham Bell (1847-1922), from his last name slightly shortened. Let I denote the sound intensity, and let I_0 correspond to the minimum audibility level for intact hearing. The value $I_0 = 0$ dB was fixed by international agreement. It corresponds to a pressure of 20 μPa, where 1 atmospheric pressure is equal to 10^5 pascals. The sound intensity I is then compared to $I_0 > 0$, as the ratio I/I_0 which is a rational number (no units). Then $\log_{10}\{I/I_0\}$ is the measure of sound intensity I in 'bels'. This implies that we can hear sound loudness over 14 orders of magnitude, i.e., the jet noise at 140 dB on a runway has a loudness of 10^{14} dB greater than threshold which is 1×10^{-12} Watts/meter^2. Thus, our ears can measure loudness over a very large range. Table 20.1 provides the decibel levels for typical sounds.

Table 20.1 Decibel Levels for Typical Sounds

Source	dB	Intensity
Threshold	0	1×10^{-12}
Breathing	20	1×10^{-10}
Whispering	40	1×10^{-8}
Soft talking	60	1×10^{-6}
Loud talking	80	1×10^{-4}
Yelling	100	1×10^{-2}
Loud concert	120	1
Jet takeoff	140	100

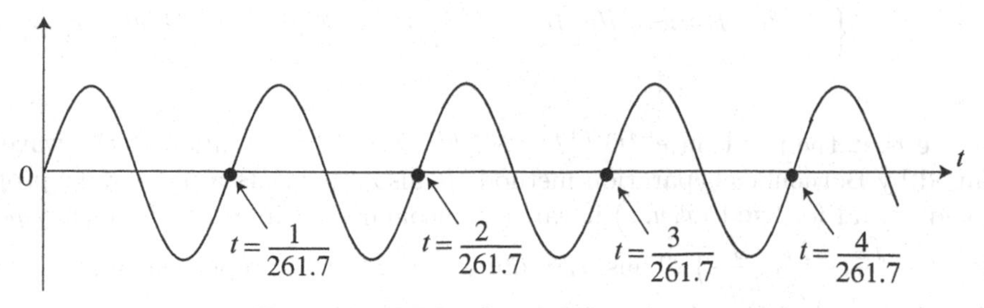

Figure 20.14 Waveform of a single tone.

In view of formula (20.4.1), if the single tone of a tuning fork is converted to an electric current and fed into an oscilloscope, the result is a graph of a single sinusoid with frequency f, shown in Figure 20.14. The value of $f = 261.7$ cycles per second, which is the *first harmonic*, is called the *fundamental*, since the frequency of the whole sound wave is 261.7 Hz. The higher harmonics, or *overtones*, are integral multiples of the fundamental. The waveform is then decomposed as the sum of harmonics with appropriate amplitudes. Thus, harmonic analysis of sound analyzes the sound spectrum.

20.4.2 Huygens Principle. Consider the Cauchy problem for the three-dimensional wave equation

$$u_{tt} = c^2 \left(u_{xx} + u_{yy} + u_{zz}\right), \ -\infty < x, y, z < \infty, \ t > 0, \tag{20.4.2}$$

subject to the initial conditions $u(x, y, z, 0) = 0$, $u_t(x, y, z, 0) = g(x, y, z)$, $-\infty < x, y, z < \infty$, where c is the wave speed. Assuming that g is absolutely integrable, and u and its first and second partial derivatives are absolutely integrable in x, y, z for each t, and using the Fourier transform, we find that

$$\begin{aligned} &\frac{\partial^2 U}{\partial t^2} + c^2 (f_x^2 + f_y^2 + f_z^2)\, U = 0, \\ &U(f_x, f_y, f_z, 0) = 0, \\ &\frac{\partial U}{\partial t}(f_x, f_y, f_z, 0) = G(f_x, f_y, f_z). \end{aligned}$$

The solution of this problem is

$$U(f_x, f_y, f_z, t) = G(f_x, f_y, f_z) \frac{\sin \rho\, ct}{\rho\, c}, \tag{20.4.3}$$

where $\rho = \sqrt{f_x^2 + f_y^2 + f_z^2}$. Then the formal solution of the Cauchy problem can be written as

$$u(x, y, z, t) = \int_{f_x^2+f_y^2+f_z^2<t^2} G(f_x, f_y, f_z)\, \frac{\sin \rho\, ct}{\rho\, c}\, e^{-i\,(f_x x + f_y y + f_z z)}\, df_x\, df_y\, df_z, \tag{20.4.4}$$

Since $\sin(\rho c t) = \dfrac{1}{2i}\left(e^{i\,\rho c t} - e^{-i\,\rho c t}\right)$, the solution (20.4.4) becomes

$$\begin{aligned} u(x, y, z, t) = \frac{1}{2} &\int_{f_x^2+f_y^2+f_z^2<t^2} \frac{G(f_x, f_y, f_z)}{\rho c} \times \\ &\left\{ e^{-i\,\rho(ct - f_x x/\rho - f_y y/\rho - f_z z/\rho)} - e^{-i\,\rho(ct + f_x x/\rho + f_y y/\rho + f_z z/\rho)} \right\} df_x\, df_y\, df_z. \end{aligned} \tag{20.4.5}$$

Notice that the function $e^{-i\,\rho(ct + f_x x/\rho + f_y y/\rho + f_z z/\rho)}$ is a solution of the wave equation obtained by Bernoulli's separation method. It also represents a plane wave propagating with speed c, and for fixed (x, y, z) it varies sinusoidally in time with frequency ρc in the direction $\left(-\dfrac{f_x}{\rho}, -\dfrac{f_y}{\rho}, -\dfrac{f_z}{\rho}\right)$. Thus, the solution u in (20.4.5) represents a plane wave in various directions and with various frequencies.

To obtain the solution of the Cauchy problem (20.4.2) in a sphere of radius r, we introduce the spherical coordinates by setting $f_x - x = r\sin\theta\cos\phi$, $f_y - y = r\sin\theta\sin\phi$, $f_z - z = r\cos\theta$. Then the solution becomes (Weinberger [1965: 335])

$$u(x,y,z,t) = \frac{1}{2}\int_0^{2\pi}\int_0^{\pi} f(x+ct\sin\theta\cos\phi, y+ct\sin\theta\sin\phi, z+ct\cos\theta)\,\sin\theta\,d\phi\,d\theta. \quad (20.4.6)$$

Note that this solution does not depend on the behavior of the function f both outside as well as inside the above spherical domain. It depends only on the values of f on the surface of this sphere. This fact provides another insight into the Huygens principle: if 'a signal concentrated in the neighborhood of a point P at time zero is concentrated at time $t > 0$ near a sphere of radius ct centered at P, then a listener at a distance d from a musical instrument hears exactly what has been played at time $t - d/c$, rather than a mixture of all the notes played up to that time'.

20.4.3 Audio Signals. The multiplication of two audio signals is usually carried out by slowly varying signals. Assume that neither of the two audio signals is slowly varying. Let us first consider the product of two sinusoids $x_1 = \cos(\alpha_1 n + \phi_1)$ and $x_2 = \cos(\alpha_2 n + \phi_2)$:

$$\begin{aligned} x_1 x_2 &= \cos(\alpha_1 n + \phi_1)\cos(\alpha_2 n + \phi_2) \\ &= \frac{1}{2}\left[\cos\left((\alpha_1 + \alpha_2)n + (\phi_1 + \phi_2)\right) + \cos\left((|\alpha_1 - \alpha_2|)n + (\phi_1 - \phi_2)\right)\right]. \end{aligned} \quad (20.4.7)$$

The two components in (20.4.7) are called the *sidebands*.

The shifting of the component frequencies of a sound is called *ring modulation*. A very simple form is shown in Figure 20.15, in which an oscillator provides a carrier signal which is simply multiplied by the input (called the *modulating signal*).

In general, the term 'ring modulation' is used to mean multiplication of any two signals, although we have considered sinusoidal signals in the above definition. Eq (20.4.7) implies that in terms of two phases $\phi_{1,2}$ and the same amplitude a we have

$$2a\cos(\alpha n + \phi_1)\cos(\alpha n + \phi_2) = a\cos\left(2\alpha n + (\phi_1 + \phi_2)\right) + a\cos(\phi_1 - \phi_2). \quad (20.4.8)$$

The second term on the right side in (20.4.8) has zero frequency; its amplitude depends on the relative phases of the two sinusoids, ranging from $+a$ to $-a$ as the phase difference varies from 0 to π radians.

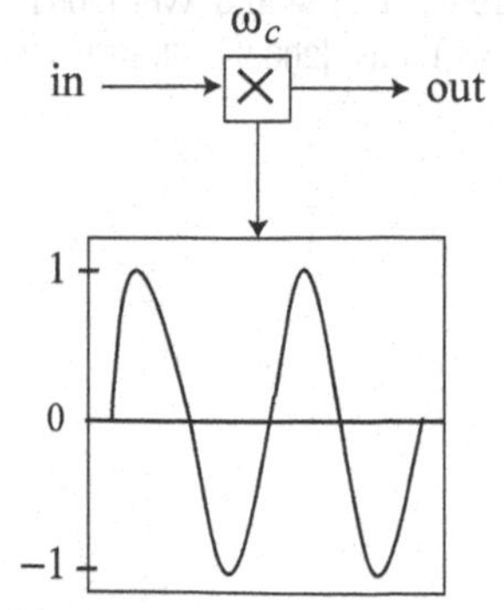

Figure 20.15 Ring modulation.

In the case when a signal consists of more than one partial each, the signal of frequency α_1 can be replaced by a sum of finitely many sinusoids, such as $a_1 \cos(\alpha_1 n) + \cdots + a_k \cos(\alpha_k n)$. If this signal is multiplied by another signal of frequency β, we obtain partials at frequencies $\alpha_1 + \beta$, $|\alpha_1 - \beta|$, ... , $\alpha_k + \beta$, $|\alpha_k - \beta|$. The resulting spectrum is the original spectrum together with its reflection about the vertical axis. This composite spectrum is then shifted right by the carrier frequency ω_c. Finally, if any component of the shifted spectrum is still left on the vertical axis, it is reflected back to make its frequency positive.

The multiplication of two audio signals, assumed to be slowly varying, is defined by the trigonometric identity (20.4.8), valid for $\phi_1 - \phi_2 > 0$. If $\phi_1 - \phi_2 < 0$, simply interchange the two sinusoids, so that $\phi_1 - \phi_2$ becomes positive. The formula (20.4.8) means that the multiplication of two sinusoids yields a sum of two terms, one involving the sum of the two frequencies and the other their difference. It also provides a method for ring modulation (shifting the component frequencies in waveshaping), as shown in Figure 20.15, in which an oscillator provides a carrier signal ω_c which is then multiplied by the input. Note that ring modulation generally implies multiplication of any two signals. In the case of audio signals it always uses a carrier signal ω_c.

20.4.4 Acoustic Waveguides. These waveguides, even with complicated cross-section, have been analyzed using conformal mapping techniques by Kashin and Merkulov [1966], Laura [1967], Hine [1971], Laura and Maurizi [1971], and Laura, Romanelli and Maurizi [1972]. Three-dimensional axisymmetric problems related to diffraction and radiation for a general class of bodies of revolution were developed by Pond [1970] who used the conformal mapping of the region outside the meridian profile of the body onto the region outside the unit circle. The numerical analysis of the boundary value problem in the model space was done by the Galerkin method. Berger [1978] obtained a numerical solution for the transient vibrations of an arbitrary shell of revolution surrounded by an acoustic medium, where the region outside the shell is conformally mapped onto the region outside the unit circle. The scattering of acoustic waves has been studied using conformal mapping techniques by Bowman et al. [1969: 37]. A recent work by Dozier [1984] deals with underwater acoustic signals from ocean surfaces where a sequence of conformal mappings is used to locally flatten successive segments of the surface which is assumed piecewise linear and frozen in time.

Finite element methods for vibration analysis including (i) free vibrations of an elastic rod, of an elastic beam and of an elastic plate, (ii) axial vibrations of a plastic rod, and (iii) eigenvalue problems for the hyperbolic, parabolic, and Helmholtz equations are discussed in Kythe and Wei [2004: ch 13].

REFERENCES USED: Baltrukonis et al. [1965], Bowman et al. [1969], Hine [1971], Kashin and Merkulov [1966], Kythe [1995; 1998], Kythe and Wei [2004], Kythe and Kythe [2012], Laura [1964, 1967], Mazumdar [1975], Schinzinger and Laura [2003], Schmidt [1981; 1983], Sideridis [1984], Weinberger [1965].

21

Electromagnetic Field

Electromagnetism is a branch of physics involving the study of the electromagnetic force, a type of physical interaction that occurs between electrically charged particles. The electromagnetic force usually exhibits electromagnetic fields such as electric fields, magnetic fields and light. There are many mathematical descriptions of the electromagnetic field. In classical electrodynamics, electric fields are described as electric potential and electric current. Maxwell's equations describe how electric and magnetic fields are generated and altered by each other and by charges and currents.

We will discuss electrostatic field, electric field, electric potential, electric capacitors, AC circuits, and Laplace's and Poisson's equations, and use conformal mapping techniques to solve some problems related to these topics.

21.1 Electromagnetic Field

Gauss's law states that the electric flux out of any closed surface is proportional to the total charge enclosed within the surface. In applying Gauss's law to the electric field at a point charge, it becomes consistent with Coulomb's law which states that the electric charge acting at a point charge q_1 as a result of the presence of a second point charge q_2 is given by $F = \dfrac{kq_1q_2}{r^2} = \dfrac{q_1q_2}{4\pi\varepsilon r^2}$, where F is the electric force. Using Gauss's law, we obtain formulas for the (radial) electric field and the potential outside an infinitely long wire with a line charge q:

$$E(r) = \frac{q}{2\pi\varepsilon r}, \quad \text{and} \quad v(r) = \text{const} - \frac{q}{2\pi\varepsilon\, r}. \tag{21.1.1}$$

Maxwell's equations are a set of four differential equations that form the theoretical basis for describing classical electromagnetism. They are based on the following four laws: (i) GAUSS'S LAW: Electric charges produce an electric field, and the electric flux across a closed surface is proportional to the charge enclosed; (ii) GAUSS'S LAW FOR MAGNETISM: There are no magnetic monopoles. The magnetic flux and Faraday's law quantitative across a closed surface are zero; (iii) FARADAY'S LAW: Time-varying magnetic fields produce an electric field; and (iv) AMPÈRE'S LAW: Steady currents and time-varying electric fields (this is due to Maxwell's correction) produce a magnetic field.

Let q and $\mathbf{v}$ be respectively the electric charge and velocity of a particle. Then the Lorenz law defines the electric field $\mathbf{E}$ and magnetic field $\mathbf{B}$ by specifying the total electromagnetic force $\mathbf{F}$ as $\mathbf{F} = q\mathbf{E} + q\mathbf{v} \times \mathbf{B}$, i.e., it contains one part of the electromagnetic force that arises from interaction of a moving charge ($q\mathbf{v}$ as the magnetic field) and the electric field.

In integral form, these four laws are expressed as follows:

Gauss's law states that the total charge contained within a closed surface is proportional to the total electric flux (the sum of the normal component of the field) across the surface:

$$\int_S \mathbf{E} \cdot d\mathbf{n} = \frac{1}{\varepsilon_0} \int \rho \, dV,$$

where ε_0 is the electric constant, and ρ is the charge density.

Gauss's law for magnetism states that although magnetic dipoles can produce an analogous magnetic flux, which carries a similar mathematical form, the total magnetic charge over all space must sum to zero:

$$\int_S \mathbf{B} \cdot d\mathbf{a} = 0.$$

Faraday's law states that the electric and magnetic fields become intertwined when the fields undergo time evolution. Faraday discovered in the 1820s that a change in magnetic flux produces an electric field over a closed loop:

$$\int_{\text{loop}} \mathbf{E} \cdot dS = -\frac{d}{dt} \int_S \mathbf{B} \cdot d\mathbf{a},$$

with the orientation of the loop defined according to the right-hand rule, and the negative sign reflects Lenz's law.

Ampère's law suggests that steady current across a surface leads to a magnetic field (expressed in terms of the flux). Also, Maxwell determined that rapid changes in the electric flux $(d/dt)\mathbf{E} \cdot d\mathbf{a}$ can also lead to changes in magnetic flux. Thus

$$\int_{\text{loop}} \mathbf{B} \cdot dS = \mu_0 \int_S \mathbf{j} \cdot d\mathbf{a} + \mu_0 \varepsilon_0 \frac{d}{dt} \mathbf{E} \cdot d\mathbf{a},$$

where $\mathbf{j}$ is the current density.

In summary,

$$\begin{aligned}
&\text{Gauss's law}: \quad \int_S \mathbf{E} \cdot d\mathbf{a} = \frac{1}{\varepsilon_0} \int \rho \, dV, \\
&\text{Gauss's law for magnetism}: \quad \int_S \mathbf{B} \cdot d\mathbf{a} = 0, \\
&\text{Faraday's law}: \quad \int_{\text{loop}} \mathbf{E} \cdot ds = -\frac{d}{dt} \int_S \mathbf{B} \cdot d\mathbf{a}, \\
&\text{Ampère's law}: \quad \int_{\text{loop}} \mathbf{B} \cdot ds = \mu_0 \int_S \mathbf{j} \cdot d\mathbf{a} + \varepsilon.
\end{aligned} \tag{21.1.2}$$

In differential form:

$$
\begin{aligned}
&\text{Gauss's law:}\quad \nabla\cdot\mathbf{E} = \frac{\rho}{\varepsilon_0},\\
&\text{Gauss's law for magnetism:}\quad \nabla\cdot\mathbf{B} = 0,\\
&\text{Faraday's law:}\quad \nabla\times\mathbf{E} = -\frac{d\mathbf{B}}{dt},\\
&\text{Ampère's law:}\quad \nabla\times\mathbf{B} = \mu_0\mathbf{j} + \mu_0\varepsilon_0\frac{\partial\mathbf{E}}{\partial t}.
\end{aligned}
\tag{21.1.3}
$$

21.2 Electrostatic Field

Let the analytic function $f(z) = u(x,y) + i\,v(x,y)$ define the two-dimensional electromagnetic field. Consider several parallel cylindrical metal conductors, which are very long so that the electric field established is identical in any plane orthogonal to these cylinders. Suppose electric charges are placed on each such conductor. Then an electrostatic potential $u(x,y)$ is established in free space surrounding the conductors, with the following properties: (i) $\nabla^2 u(x,y) = 0$ in free space, i.e., Laplace's equation $\dfrac{\partial^2 u}{\partial x^2} + \dfrac{\partial^2 u}{\partial y^2} = 0$ holds; and (ii) $u = \text{const}$ on the surface of each conductor; and (ii) the function $f(z)$ is called the *complex electrostatic potential* or simple *complex potential.*

Example 21.1. Let the positive and negative halves of the real axis each represent infinite conducting plates separated from each other by insulation at the origin. Suppose the right plate $x > 0, y = 0$ is kept at potential $u = 3$, while the left plate $x < 0, y = 0$ is kept at potential $u = -4$. The electrostatic potential u in the upper half-plane $y > 0$ is determined by considering the imaginary part of the function $f(z) = \log z$, which is $\arg\{z\} = \theta$, where $z = \log|z| + i\arg\{z\} = \log|z| + i\,\theta$. Moreover, θ takes the value 0 on the positive real axis and π on the negative real axis. Thus, we must find the constants A and B such that $u = A\theta + B$ is the desired potential. Since $\theta = 0$ when $u = 3$, we find that $B = 3$. Also, when $\theta = \pi$ when $u = -4$, we find that $A = -7/\pi$. Then the desired potential is $u = -\dfrac{7}{\pi}\theta + 3 = -\dfrac{7}{\pi} + \arctan\dfrac{y}{x} + 3$, where $0 \le \arctan\dfrac{y}{x} \le \pi$. Thus, the electrostatic potential is the real part of the complex potential $f(z) = \dfrac{7}{\pi}\,i\log z + 3$. ■

Example 21.2. Three conductors on the x-axis, separated by insulation at $x = \pm 1$, are charged with electrostatic potentials

$$
u(x,y) = \begin{cases} -12 & \text{for } -\infty < x < -1,\\ 9 & \text{for } -1 < x < 1,\\ 7 & \text{for } 1 < x < \infty, \end{cases}
$$

as shown in Figure 21.1(a).

To determine the electrostatic potential in the upper half-plane $y > 0$, assume that $u = A\theta_1 + B\theta_2 + C$, where A, B, C are constants to be determined, and θ_1 and θ_2 are the angles at $x = -1$ and $x = 1$, respectively, as shown in Figure 21.1(b). Since u is harmonic because θ_1 is the imaginary part of $\log(z-1)$ and θ_2 the imaginary part of $\log(z+1)$. Using the boundary conditions, we have: (i) $u = 7$ when $\theta_1 = 0 = \theta_2$; (ii) $u = 9$ when $\theta_1 = \pi$ and $\theta_2 = 0$; and (iii) $u = -12$ when $\theta_1 = \pi = \theta_2$, we find that $C = 7, A = 2/\pi$, and $B = -19/\pi$. Thus,

$$
u = \frac{2}{\pi}\theta_1 - \frac{19}{\pi}\theta_2 + 7 = \frac{2}{\pi}\arctan\frac{y}{x-1} - \frac{19}{\pi}\arctan\frac{y}{x+1} + 7,
$$

where $0 \leq \theta_1, \theta_2 \leq \pi$. This electrostatic potential is the real part of the complex potential

$$f(z) = -\frac{2i}{\pi}\log(z-1) + \frac{19i}{\pi}\log(z+1) + 7. \blacksquare$$

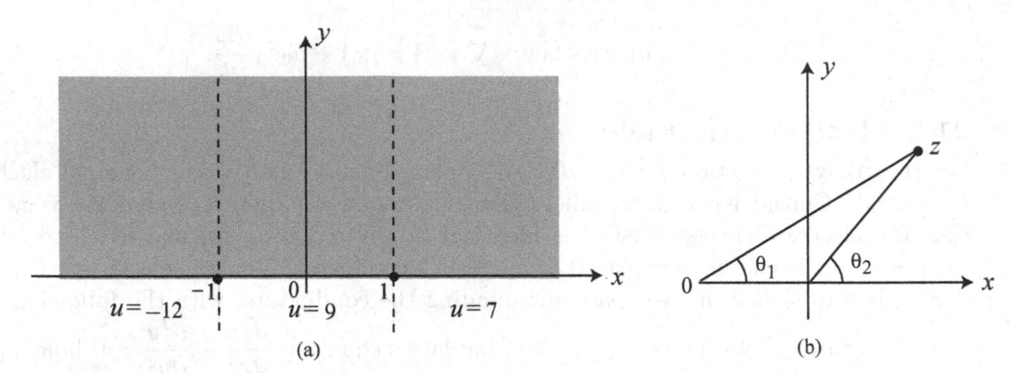

Figure 21.1 Three charged conductors.

21.2.1 Electric Potential. In view of Example 18.18, the function $f(z) = m\log(z - z_0)$ describes the source of fluid flow at a point z_0. This situation can be perceived as an infinitely long line piercing the complex z-plane orthogonally at the point z_0 with fluid rushing out radially in all directions, where the quantity of fluid from each linear foot of the line source is $2\pi m$ units of volume per second.

In the case of an electrostatic field, these line sources of fluid flow become the line sources of the electric potential. Imagine a straight wire conductor orthogonal to the z-plane at the origin, on which an electric charge of density q units of charge per linear foot of distance on the wire is evenly distributed. Then the electric potential generated by this wire is $u = -2q\log r$, and the complex potential is

$$f(z) = 2q\log z. \tag{21.2.1}$$

PROOF. In general, a line charge q per unit length is subjected to the force $q\mathbf{E}$ per unit length, where $\mathbf{E}$ is the *electric field intensity vector*, defined by the equations

$$\mathbf{E} = -\nabla u, \quad \text{and} \quad \mathbf{E} = -\overline{f'(z)}. \blacksquare \tag{21.2.2}$$

Example 21.3. Let a line of charge q_1 per unit length be at $z = 0$ and another charge q_2 per unit length be at $z = 1$. Then from (21.2.1) the complex potential is given by the sum of the potentials due to the sources at $z = 0$ and $z = 1$, i.e., by

$$f(z) = -2q_1\log z - 2q_2\log(z-1). \tag{21.2.3}$$

Since the potential $u = \Re\{f(z)\}$, we have

$$\begin{aligned} u &= -2q_1\ln|z| - 2q_2\ln|z-1| = -2q_1\ln\sqrt{x^2+y^2} - 2q_2\ln\sqrt{(x-1)^2+y^2} \\ &= -q1\ln\sqrt{x^2+y^2} - q_2\ln\sqrt{(x-1)^2-y^2}. \end{aligned}$$

From (21.2.2) we get the electric field intensity vector as $\mathbf{E} = \dfrac{2q_1}{\bar{z}} + \dfrac{2q_2}{\bar{z}-1}$.

As we have seen, the electrostatic potential is defined by $\Re\{f(z)\} = u$. But what does $\Im\{f(z)\} = v$ define? Going back to the fluid flow problems, we recall that $v = c$, where c is constant, define the streamlines, and the velocity vectors are tangent to these lines. In the electrostatic case, the lines $v = c$ are called the *lines of force* and the vector $\mathbf{E}$ is tangent to these lines.

For example, if $q_1 = q_2 = q$ in Example 21.3, the complex potential (21.2.3) becomes $f(z) = -2q \log|z(z-1)| - 2qi \arg\{z(z-1)\}$. Since $\arg\{z(z-1)\} = \arctan \dfrac{2xy - y}{x^2 - y^2 - x}$, the imaginary part of this complex potential is given by

$$v(x,y) = -2q \arctan \frac{2xy - y}{x^2 - y^2 - x},$$

and thus, the lines of force, defined by $v = c$, are given by

$$\frac{2xy - y}{x^2 - y^2 - x} = \tan\Big(-\frac{c}{2q}\Big) = C,$$

where C is an arbitrary constant. The equipotential lines $u =$ const and the lines of force $v =$ const at the points $z = \pm 1$ are presented in Figure 21.2. ■

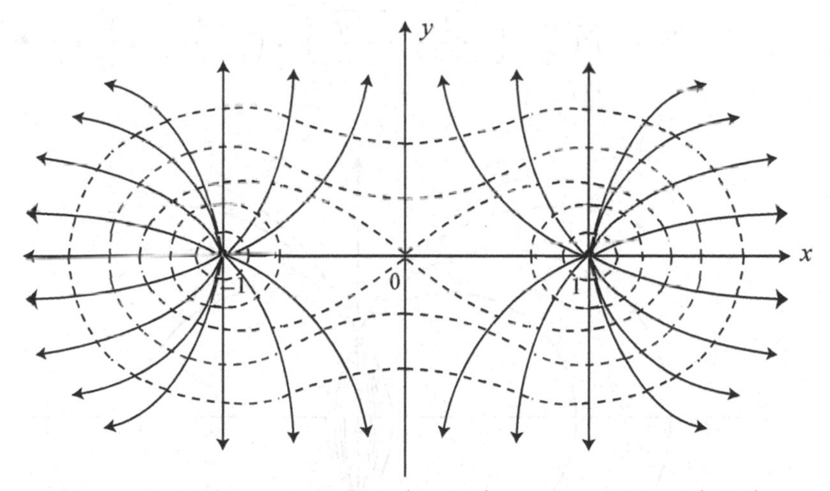

Figure 21.2 Equipotential lines (dashed) and lines of force (solid).

Example 21.4. Let the line charge q unit per length be placed at the points $z = n\pi$, $n = 0, \pm 1, \pm 2, \dots$. This case is similar to the fluid flow problem defined by the function $f(z) = m \log(z - z_0)$, where m is replaced by $-2q$. Thus, the complex potential is $f(z) = -2q \log(\sin z)$. ■

Example 21.5. Consider a line of charge q per unit length located at the point $z = 1$, and let the real axis be a ground conductor, i.e., it has potential $u = 0$. Recall that the line charge in this case is $-2q \log|z - 1|$ at $z = i$, but it will not make the potential zero on the real axis. Let us place a charge $-q$ at $z = -i$. Then the potential on the real axis will be zero. Thus, the electrostatic potential and the complex potential are given, respectively, by

$$u(x,y) = -2q \log|z-i| + 2q \log|z+i|, \text{ and } f(z) = -2q \log(z-i) + 2q \log(z+i) = 2q \log \frac{z+i}{z-i}.$$

The charge at $z = -i$ is called the *image* of the charge at $z = i$. Now, using the method of images (see Kythe [1998: 341]), since

$$\frac{z+i}{z-i} = \frac{x^2+y^2-1+2i\,x}{x^2+(y-1)^2},$$

the complex potential is given by

$$\begin{aligned} f(z) &= 12\log\frac{|z+i|}{|z-i|} + 2qi\,\arg\left\{\frac{z+i}{z-i}\right\} \\ &= 2q\log\frac{\sqrt{x^2+(y+1)^2}}{\sqrt{x^2+(y-1)^2}} + 2qi\,\arctan\frac{2x}{x^2+y^2-1} \\ &= q\log\frac{x^2+(y+1)^2}{x^2+(y-1)^2} + 2qi\,\arctan\frac{2x}{x^2+y^2-1}. \end{aligned}$$

Notice that the electrostatic potential $u(x, y) = \Re\{f(z)\}$, and the lines of force are $v(x, y) = \Im\{f(z)\} = c$, or $\dfrac{2x}{x^2+y^2-1} = C$, which can be rewritten as $(x-C)^2+y^2 = C^2+1$, and they represent the circles with centers $(C, 0)$ on the real axis passing through the line charges at $z = \pm i$ (Figure 21.3). The electric intensity vector $\mathbf{E}$ is given by

$$\mathbf{E} = -\overline{f'(z)} = \overline{\left(-\frac{2q}{z-i} + \frac{2q}{z+i}\right)} = \overline{\left(\frac{4qi}{z^2+1}\right)} = \frac{1-4q}{\bar{z}^2+1}. \blacksquare$$

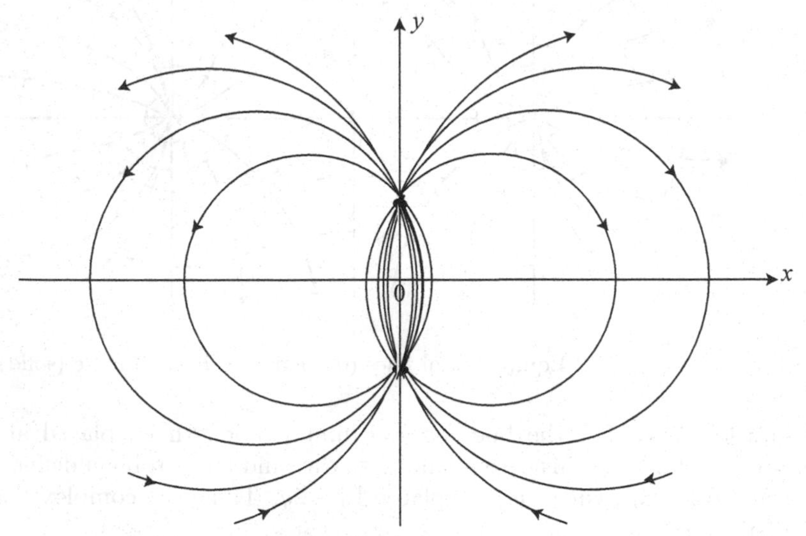

Figure 21.3 Lines of force.

The conformal transformation $z = e^{\zeta}$ maps the upper-half of the z-plane onto the infinite strip (channel) $0 \le \Im\{\zeta\} \le \pi$ in the ζ-plane (see Map 5.5). Let $\zeta = \xi + i\eta$. Now, it is known that under this mapping the real axis on the ζ-plane, $\eta = 0, -\infty < \xi < \infty$, is mapped onto the positive real axis of the z-plane and that the line $\zeta = \xi + i\pi$, $-\infty < \xi < \infty$, is mapped onto the negative real axis of the z-plane, and also the line segment $\zeta = i\eta$, $0 \le \eta \le \pi$ is mapped onto the upper-half of the unit circle on the z-plane (see Map 5.3 and Map 5.6).

Example 21.6. Consider a channel of width π bound on both sides by grounded conductors. Then under the mapping $z = e^{\zeta}$, a line of charge q at the point $\zeta = i\pi/2$ is mapped onto $z = 1$; the lower boundary of the channel is mapped onto positive x-axis and the upper boundary onto the negative x-axis. To solve this problem, consider a line source of charge at $z = i$, where x-axis is at zero potential. In Example 21.5, the complex potential is given by $f(z) = 2q \log \dfrac{z+i}{z-i}$. If we set $z + e^{\zeta}$, we get the complex potential as

$$f(z) = 2q \log \frac{e^{\zeta}+i}{e^{\zeta}-i}. \tag{21.2.4}$$

To find the potential $u(\xi, \eta)$ and the lines of force $v(\xi, \eta) = C$, notice that

$$\frac{e^{\zeta}+i}{e^{\zeta}-i} = \frac{e^{\xi}e^{i\eta}+i}{e^{\xi}e^{i\eta}-i} = \frac{e^{2\xi}-1+2i\,e^{\xi}\cos\eta}{e^{2\xi}-1+2i\,e^{\xi}\sin\eta}.$$

Then its real part is

$$\left|\frac{e^{\zeta}+i}{e^{\zeta}-i}\right| = \frac{\sqrt{\left(e^{2\xi}-1\right)^2+4e^{2\xi}\cos^2\eta}}{e^{2\xi}+1-2e^{\xi}\sin\eta},$$

and the imaginary part is

$$\arg\left\{\frac{e^{\zeta}+i}{e^{\zeta}-i}\right\} = \arctan\frac{2e^{\xi}\cos\eta}{e^{2\xi}-1}.$$

Combining these two parts, we find that the complex electrostatic potential is given by

$$u(\xi, \eta) = q \log \frac{\left(e^{2\xi}-1\right)^2+4e^{2\xi}\cos^2\eta}{\left(e^{2\xi}+1-2e^{\xi}\sin\eta\right)^2},$$

and the lines of force are given by

$$v(\xi, \eta) = 2q \arctan\frac{2e^{\xi}\cos\eta}{e^{2\xi}-1} = c,$$

or by

$$\frac{2e^{\xi}\cos\eta}{e^{2\xi}-1} = C,$$

where C is an arbitrary constant. ■

Example 21.7. A line of charge q per unit length is located at $z = 1/2$ inside the unit circle which is a ground conductor (Figure 21.4). The problem is to find the complex and electrostatic potentials in $|z| \leq 1$.

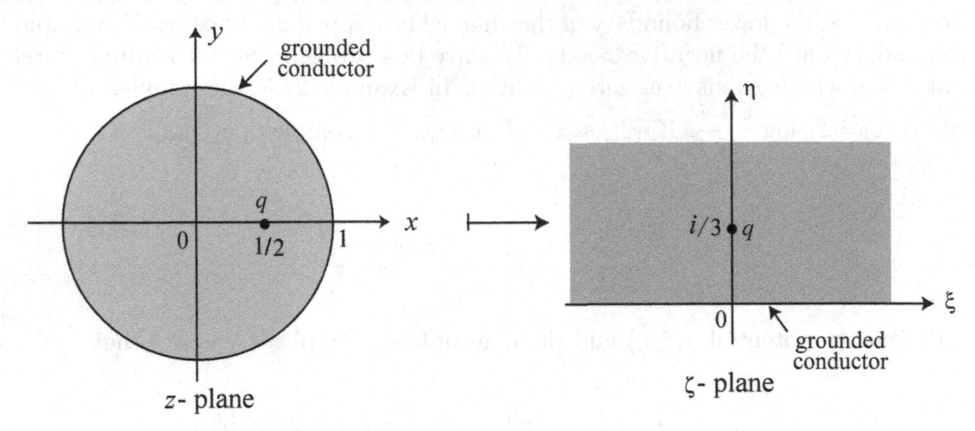

Figure 21.4 Unit circle as ground conductor.

We will use the transformation $\zeta = -i\dfrac{z-1}{z+1}$ (see Map 3.25), which maps the conducting boundary $|z| = 1$ onto the real ζ-axis and the line of charge at $z = 1/2$ onto the point $\zeta = i/3$, as shown in Figure 21.4. Then, placing an image line of charge $-q$ per unit length at $\zeta = -i/3$, the complex potential in the ζ-plane is given by

$$f(\zeta) = -2q \log(\zeta - i/3) + 2q \log(\zeta + i/3) = 2q \log \frac{\zeta + i/3}{\zeta - i/3}.$$

Using the above bilinear transformation this complex potential is mapped onto the unit circle $|z| = 1$ as

$$f(z) = 2q \log \frac{-i\dfrac{z-1}{z+1} + \dfrac{i}{3}}{-i\dfrac{z-1}{z+1} - \dfrac{i}{3}} = 2q \log \frac{2z-1}{z-2}.$$

The electrostatic potential $u = \Re\{f(z)\}$ is then given by

$$u(x, y) = 2q \log \left| \frac{2z-1}{z-2} \right| = q \log \frac{(2x-1)^2 + 4y^2}{(x-2)^2 + y^2}. \ \blacksquare$$

In electrostatic problems, the streamlines are parallel to the electric field with electric intensity $\mathbf{E}$. We have

$$\mathbf{E}(z) = E_x(x, y) + i\, E_y(x, y) = -\overline{\frac{df^*(z)}{dz}},$$

or

$$E_x(x, y) = -\frac{\partial u(x, y)}{\partial x}, \quad E_y(x, y) = -\frac{\partial u(x, y)}{\partial y}.$$

Example 21.8. The potential from a line charge, of charge density q along the z-axis is $u(r) = -2q \log r = -2q\, \Re\{\log z\}$. The electric field can be calculated in two methods:

METHOD 1. we have

$$-\overline{\left[\frac{d}{dz}\left[(-2q\log z\right]\right]} = 2q\frac{1}{\bar{z}} = 2q\frac{z}{|z|^2} = 2q\left[\frac{x}{x^2+y^2} + i\frac{y}{x^2+y^2}\right],$$

$$E_x = 2q\frac{x}{x^2+y^2}, \quad E_y = 2q\frac{y}{x^2+y^2}.$$

METHOD 2. We have

$$E_x(x,y) = -\frac{\partial u(x,y)}{\partial x} = -\left[-2q\frac{\partial \log\sqrt{x^2+y^2}}{\partial x}\right] = 2q\frac{x}{x^2+y^2},$$

$$E_y(x,y) = -\frac{\partial u(x,y)}{\partial y} = -\left[-2q\frac{\partial \log\sqrt{x^2+y^2}}{\partial y}\right] = 2q\frac{y}{x^2+y^2}. \blacksquare$$

Example 21.9. Consider two infinite parallel flat plats, separated by a distance and maintained at zero potential. A line of charge q per unit length is located between the two plates at a distance a from the lower plate. The problem is to find the electrostatic potential in the region D of the z-plane. The transformation $w = e^{\pi z/d}$ maps the shaded region of the z-plane onto the upper half of the w-plane (see Map 5.10 also), such that the point $z = ia$ is mapped onto the point $w_0 = e^{i\pi a/d}$; the points on the lower plate $z = x$ and on the upper plate $z = x + id$ are mapped onto the real u-axis for $u > 0$ and $u < 0$, respectively. i.e., the points A, B, C, D, E, F are mapped onto A', B', C', D', E', F', respectively.

Consider the line of charge q at w_0 and a line of charge $-q$ at $\bar{w}_0$. The associated complex potential is given by

$$F(w) = -2\log(w - w_0) + 2q\log(w - \bar{w}_0) = 2q\log\frac{w - \bar{w}_0}{w - w_0}.$$

The real part is zero on the real axis $u = 0$, which satisfies the boundary condition on the plates. Hence, the electrostatic potential at any point in the shaded region of the z-plane is given by

$$\Re\{F(w)\} = 2q\log\frac{w - e^{-iv}}{w - e^{iv}}, \quad v = \pi a/d. \blacksquare \tag{21.2.5}$$

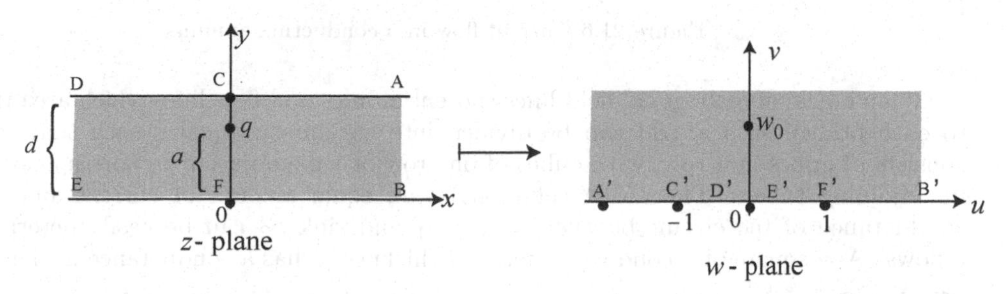

Figure 21.5 Problem of two infinite parallel flat plates.

21.3 Electric Field

Let $\mathbf{j}$ denote the current density vector, V the voltage, $\mathbf{E}$ the electric field intensity, which is the gradient of the voltage, but opposite in sign, to $\mathbf{E}$, and σ the conductivity which is the reciprocal of resistivity. In the absence of any sources in the medium, these quantities are related as follows:

$$\mathbf{j} = \sigma\,\mathbf{E}, \tag{21.3.1}$$

$$\mathbf{E} = -\nabla V, \tag{21.3.2}$$

$$\nabla^2 V = 0. \tag{21.3.3}$$

Any 'flux tube', a pipe of uniform depth, between two adjacent flowlines carries a fixed total current throughout its entire length from electrode to electrode in a source-free region. Only the current density changes as the tube narrows or widens. No current passes through the 'walls' of such a tube. Assuming that conduction takes place in a conductor of uniform thickness or depth in $\mathbb{R}^3$ along the z-axis, i.e., normal to the (x, y)-plane, the flux 'tubes' appearing in the (x, y)-plane are indicated by the outlines of their side walls as shown in Figure 21.6(a) or (b).

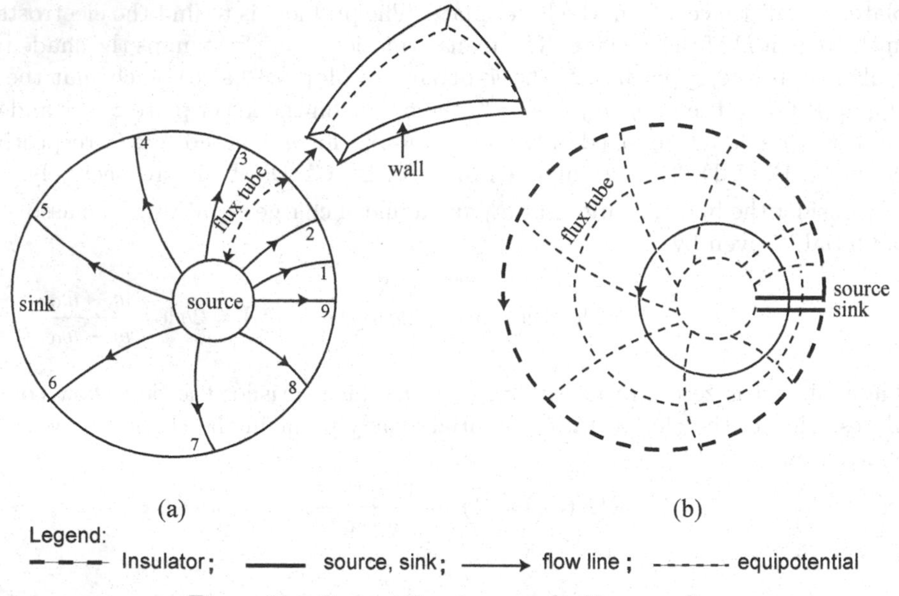

Figure 21.6 Current flow in a conducting annulus.

Given a grid of orthogonal field lines (potential lines and flow lines which are orthogonal to each other), such a grid can be divided into curvilinear squares such that each tube consists of cubes in a row, with cubes of one row or tube adjacent to corresponding cubes in adjoining tubes. Then each tube carries an equal amount of current and the total conductance of the circuit between source S_1 and sink S_2 can be easily determined, as follows: Any square of a conducting plate of thickness d has a conductance in the direction of value $G = \sigma\,\dfrac{ad}{b}$, where ad is the cross-sectional area of length b. Since, by definition, $a = b$ for any square of width a and length b, including a curvilinear square, we get $G = \sigma d$, which is independent of the size of the square and of the direction of flow as long as it occurs parallel to either side a or side b and uniformly enters and leaves the boundaries at opposite ends. The advantage of dividing the given electric field into squares is now obvious (see Figure 21.7). The current flow in Figure 21.6(a) is mostly radial, while that in Figure 21.6(b) is mostly circular.

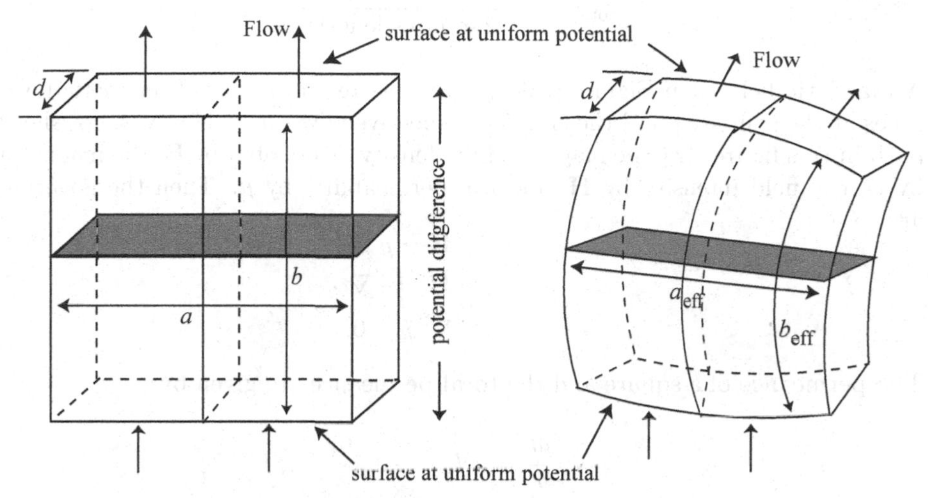

Figure 21.7 Current flow through 'squares' in the (x, y)-plane.

Note that the total conductance of the configuration in Figure 21.6(a) is that of 9 flux tubes side-by-side, having a total width equivalent to nine squares, and a total length equivalent to two squares. Thus, the conductance of the path is $(9/2)\sigma d$, and

$$G_{\text{total}} = \sigma d \, \frac{9\,\text{squares widthwise}}{2\,\text{squares lengthwise}} = 4.5\,\sigma d. \tag{21.3.4}$$

The ratio 4.5 is called the *modulus* or *shape factor* of the device in Figure 21.6(a), as defined in Schinzinger and Laura [2003:163]. This device can be scaled up or down, but as long as the ratio of the squares widthwise to lengthwise and the thickness d remain unchanged, the conductance stays at $4.5\,\sigma d$. In practice, however, other matters, such as limits imposed by heat dissipation, mechanical strength, or electrical breakdown across the electrodes, are taken into account.

In Figure 21.6(b), the total conductance $G_{\text{total}} = \dfrac{1}{4.5}\sigma d$, where $1/4.5$ is its *modulus* or *shape factor*.

If the rectangular surface is square in Figure 21.7, then $a = b$, which gives $a_{\text{eff}} = b_{\text{eff}}$. Also, the shaded part of this figure represents the cross-sectional area of flow, which is ad or $a_{\text{eff}}d$.

Figure 21.6 also represents the electric field between two charge carrying electrodes. The flow quantity is the dielectric flux density or the displacement vector, denoted by $\mathbf{D}$, the potential is the electric potential ϕ, and the negative of the potential gradient is the electric field intensity $\mathbf{E}$. For a medium characterized by permittivity ε, we have

$$\begin{aligned} \mathbf{D} &= \varepsilon\,\mathbf{E}, \\ \mathbf{E} &= -\nabla\phi, \\ \nabla^2\phi &= 0. \end{aligned} \tag{21.3.5}$$

We will use Figure 21.7 to establish the capacitance of a square which comes to

$$C = \varepsilon\,\frac{ad}{b} = \varepsilon\,d. \tag{21.3.6}$$

Also, as in Eq (21.3.4), the total capacitance for Figure 21.6 is given by

$$C_{\text{total}} = \varepsilon\, d\, \frac{9\,\text{squares widthwise}}{2\,\text{squares lengthwise}} = 4.5\, \varepsilon\, d. \tag{21.3.7}$$

A magnetic field as in Figure 21.6(a) exists because of a pair of magnetic north and south poles at the source S_1 and the sink S_2, respectively; we have, however, not shown the return path in this figure. Let the magnetic flux density be denoted by $\mathbf{B}$, the magnetomotive force by ϕ, the field intensity by $\mathbf{H}$, and the permeability by μ. Then the governing equations are

$$\begin{aligned} \mathbf{B} &= \mu\, \mathbf{H}, \\ \mathbf{H} &= -\nabla\phi, \\ \nabla^2\phi &= 0. \end{aligned} \tag{21.3.8}$$

The permeance of a square and the total permeance are given by

$$P = \mu\, \frac{ad}{b} = \mu\, d. \tag{21.3.9}$$

$$P_{\text{total}} = \mu\, d\, \frac{9\,\text{squares widthwise}}{2\,\text{squares lengthwise}} = 4.5\, \mu\, d. \tag{21.3.10}$$

A magnetic field is commonly caused not by a permanent magnet but by an electric current or an energized winding. Figure 21.6(a) may be regarded as representing the cross-section of a coaxial cable, with the source S_1 carrying current into the plane of the figure and the sink S_2 carrying return current in the opposite direction, both S_1 and S_2 representing perfectly conducting thin walls or 'current sheets' in cross-section. Then the magnetic flux lines due to the current in the cable would be directed tangentially through the annulus, i.e., orthogonal to the direction shown in Figure 21.6(a). Figure 21.6(b) may also be used as a model by assuming that S_1 and S_2 represent the source and sink, respectively, of a 'magnetomotive force' or scalar magnetic potential measured in Ampere-turns. Thus, for a path between S_1 and S_2 in this electric model, the total permeance is given by

$$P_{\text{total}} = \frac{2}{9}\, \mu\, d, \tag{21.3.11}$$

and the inductance L, with $N = 1$ turn, by

$$L = N^2 P_{\text{total}} = \frac{2}{9}\, \mu\, d. \tag{21.3.12}$$

Finally, the characteristic impedance of the coaxial cable in the lossless case (i.e., with no resistance) is given by (Schinzinger and Laura [2003: 165])

$$Z_0 = \sqrt{L/C} = \frac{2}{9}\, \sqrt{\mu/\varepsilon}. \tag{21.3.13}$$

21.3.1 Hall Effect. The classical Hall effect, discovered by Edwin Hall in 1879, is the production of a potential difference across an electrical conductor when a magnetic field is applied in a direction perpendicular to that of flow of current. As defined in Trefethen and Williams [1986], the Hall effect is the production of a voltage difference (known as the *Hall voltage*) across an electrical conductor, transverse to an electric current in the conductor and to an applied magnetic field perpendicular to the current. The Hall effect devices are often used as magnetometers, i.e., to measure magnetic fields, or inspect materials, such as tubing or pipelines. The measurement of this voltage difference is important in semiconductor electronics. Its sign reveals whether the material being tested carries current primarily by electron (type n) or by 'holes' (type p).

Let D denote a planar polygonal domain (as in Problem O, see §8.8, and Trefethen and Williams [1986]), which can be regarded as having been cut from a thin material of uniform electrical conductivity (Figure 21.8). On each boundary arc Γ_k, k in an index set $\Sigma \subseteq \{1, \dots, n\}$, a fixed constant voltage $u = u_k$ is applied, while the remainder of the boundary is insulated. Assuming that Σ contains no adjacent pairs $k, k+1$ except with $u_k = u_{k+1}$, let $K \leq n/2$, $K \geq 1$, be the number of disjoint components of $\bigcup_{k \in \Sigma} \bar{\Gamma}_k$, that is, the number of separate voltages applied on the boundary ∂D.

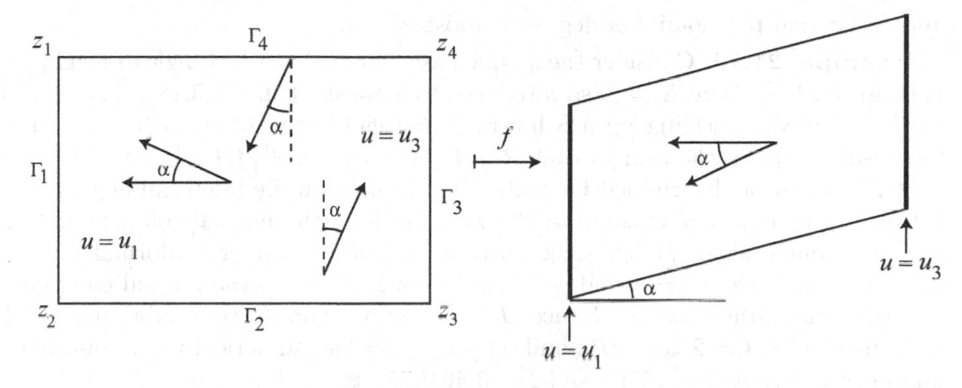

Figure 21.8 Schwarz-Christoffel transformations.

If a constant magnetic field $\mathbf{B}$ is applied perpendicular to D, the resulting steady-state voltage $u(z)$, electric field $\mathbf{E}(z)$, and current density $\mathbf{j}(z)$ are as in §21.1, defined by

$$\mathbf{E} = -\nabla u, \quad \mathbf{j} = \mathbf{E} + \mathbf{j} \times \mathbf{B}. \tag{21.3.14}$$

If the two vectors $\mathbf{E}$ and $\mathbf{j}$ are interpreted as complex scalars and if $E = |\mathbf{E}|$, $j = |\mathbf{j}|$, and $B = |\mathbf{B}|$, then Eq (21.3.14) becomes $E = (1 - iB)j$. Thus, $\mathbf{E}$ and $\mathbf{j}$ form a constant angle $\alpha = \arctan B$ at every point $z \in D$, i.e,

$$\frac{j}{E} = \frac{|\mathbf{j}|}{|\mathbf{E}|} = \cos\alpha, \quad \arg\{j\} - \arg\{E\} = \alpha, \tag{21.3.15}$$

where α is called the *Hall angle*. Without loss of generality, we assume that $|\alpha| < \pi/2$. Then the problem of determining u is of the form of Problem O: On each side with $k \in \Sigma$, the voltage u is constant, which implies that it satisfies the oblique derivative boundary conditions with $\theta_k = \pi/2$. On each side with $k \notin \Sigma$, there is no current through the boundary, i.e., $\mathbf{j}$ is tangent to Γ_k. Then using Eqs (21.3.14) and (21.3.15), ∇u makes an angle α clockwise from Γ_k in this case, which implies that the directional derivative of u is zero at an angle clockwise from the normal to Γ_k, i.e., $\theta_k = \alpha$. Thus, an oblique derivative boundary condition is satisfied on every side of ∂D. Moreover, we have two more conditions: the values $u = u_k$ on Γ_k are specified for $k \in \Sigma$, and any physically meaningful solution must be continuous. This leads to the following *general Hall effect problem*:

Problem O$_\mathrm{H}$. (Trefethen and Williams [1986]) Let D, Σ, u_k, and α be given as defined above. Find a solution to the problem O subject to the homogeneous oblique derivative boundary conditions (8.8.1) with

$$\theta_k = \begin{cases} \pi/2 & \text{for } k \in \Sigma, \\ \alpha & \text{for } k \notin \Sigma, \ |\alpha| < \pi/2, \end{cases} \tag{21.3.16}$$

together with the additional conditions

$$u \in C(\bar{D}), \quad u(z) = u_k \quad \text{for all } z \in \Gamma_k, k \in \Sigma. \tag{21.3.17}$$

The conditions (21.3.17) ensure a unique solution in the form $u = \Re\{\psi \circ \phi\}$, where

$$\psi(t) = C_3 + e^{i\theta_{k_m}-1} \int_{t_\phi}^{t} p(t') \prod_{k=1}^{K} (t' - q_k)^{\alpha/\pi - \frac{1}{2}} (t' - r_k)^{-\alpha/\pi - \frac{1}{2}} \, dt', \tag{21.3.18}$$

and p is a real polynomial of degree at most $K - 1$.

Example 21.10. Consider the simple case where D is a rectangle of height 1 and length L (Figure 21.9). Here $K = 2$, so p reduces to a constant in (21.3.18). The constant C_3 can be determined by setting up a Schwarz-Christoffel integral (21.3.18), then shift and scale the resulting parallelogram so that $\Re\{f(\Gamma_1)\} = u_1$ and $\Re\{f(\Gamma_3)\} = u_3$. Then, the height h of $f(D)$ can be determined by evaluating an integral, i.e., determining the total current I from Γ_3 to Γ_1, for I is equal to the magnitude of the normal component of $\mathbf{j}$ along Γ_1, or equivalently along $f(\Gamma_1)$, since current is invariant under conformal mapping. Using (21.3.15), we have $|\mathbf{E}| = 1$ and $|\mathbf{j}| = \cos\alpha$ along $f(\Gamma_1)$, so the normal component of $\mathbf{j}$ has constant magnitude $\cos^2\alpha$. Hence, $I = h\cos^2\alpha$. The Fortran subroutine SCPACK has been used with $L = 2, \alpha = \pi/6$, and $u_3 - u_1 = 1$. The numerical values obtained for height and current are $h = 0.615567$ and $I = 0.461675$. ■

In the case when D is a more general polynomial with $K \geq 3$ prescribed voltages, the situation changes qualitatively because now Eq (21.3.18) contains a nontrivial polynomial p, which introduces slits and/or branch points in $f(D)$. We will consider an example.

Example 21.11. Consider a 6-gon shown on the left in Figure 21.9 with vertices at $-1 + i$, -1, 0, $\frac{1}{2} - \frac{i}{2}$, $2 - \frac{i}{2}$, $2 + \frac{i}{2}$.

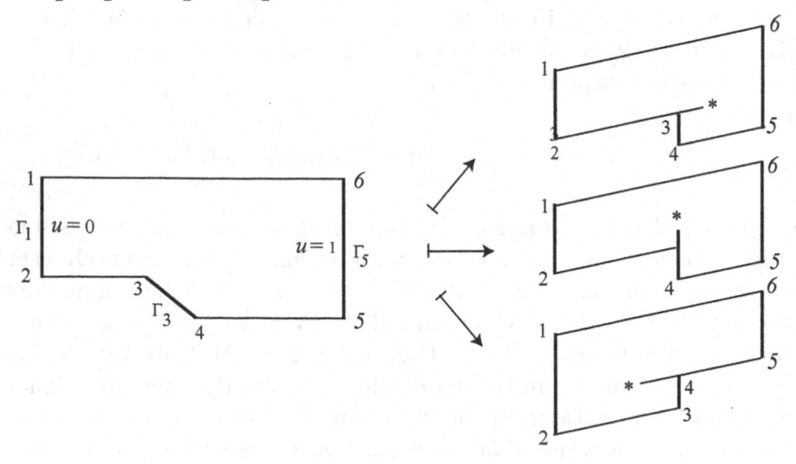

Figure 21.9 A 6-gon.

In this problem, voltages are prescribed as follows: $u_1 = 0, u_3 = U$, and $u_5 = 1$ on the sides Γ_1, Γ_3 and Γ_5, while the sides Γ_2, Γ_4 and Γ_6 are insulated. Let $p(t) = c(t - t_*)$, where

c is a constant, so that t_* marks the 'extra vertex' corresponding to the end of the slit. For arbitrary U, t_* must lie anywhere on $\mathbb{R}$, but let us assume $0 < U < 1$. Then there are exactly three possibilities for $f(D)$ as presented in Figure 21.9. In each case one of the sides of $f(D)$ protrudes beyond the vertex at which it might be expected to stop, then doubles back, so that $f(D)$ is now not a 6-gon but a 7-gon with vertices $w_1, \dots, w_6$ and w_*. These three cases differ in the order in which the vertices and prevertices appear along the boundary as follows:

$$\begin{aligned}
&(\mathrm{a}): \; w_1, w_2, w_*, w_3, w_4, w_5, w_6;\\
&(\mathrm{b}): \; w_1, w_2, w_3, w_*, w_4, w_5, w_6;\\
&(\mathrm{c}): \; w_1, w_2, w_3, w_3, w_4, w_*, w_5, w_6.
\end{aligned}$$

They also differ physically in each case, i.e., (a) current flows into Γ_3 only; current flows both into and out of Γ_3; and (c) current flows out of Γ_3 only. However, it might happen that $f(D)$ has no slit in it. If U is specified as in Figure 21.9, nothing is known a priori as to which of the cases (a)-(c) will occur, nor do we know the correct slit length or any other vertical dimensions of $f(D)$, although we know its horizontal (i.e., oblique) dimensions. This is why the determination of ψ is a generalized parameter problem, which is linear. For a practical illustration, let $K > 2$, $U = \frac{1}{2}$, and $\alpha - \pi/6$, and the numerical solution is determined, which turns out to be of type (b). The results are: total current 0.449016 out of Γ_5; 0.166622 into Γ_3; 0.045959 out of Γ_3, and 0.328354 into Γ_1.

The cases (a) and (c) can occur, as is shown in Figure 21.10 with a 3×3 array of numerically determined boundary polygons $f(\partial D)$ with $\alpha = -\pi/6, 0, \pi/6$ and $U = 0.1, 0.5, 0.9$. The case (a) is obtained with $U = 0.1$, (b) with $U = 0.5$, and (c) with $U = 0.9$. ■

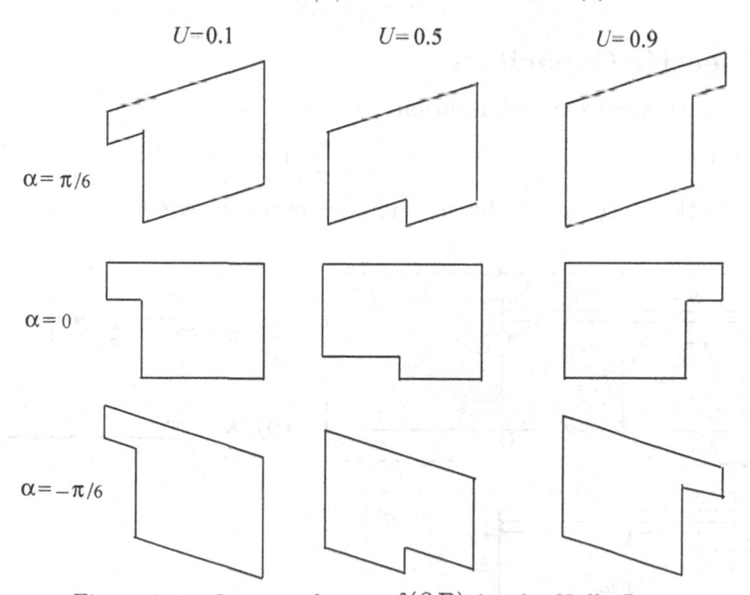

Figure 21.10 Image polygons $f(\partial D)$ for the Hall effect.

21.4 Electromagnetic Waves

Electromagnetic waves are generated by the electromagnetic radiation (EMR) as a form of energy emitted and absorbed by charged particles. This energy behaves like a wave as it travels through space. It has both electric and magnetic components, which are always in a fixed ratio of intensity with each other, oscillate in phase perpendicular to each other, and are orthogonal to the direction of energy and wave propagation. These properties are presented in Figure 21.11, where the solid curves are in the vertical $(\mathbf{k}, \mathbf{E})$-plane which corresponds to the electric field $\mathbf{E}$, while the dotted curves are in the horizontal $(\mathbf{k}, \mathbf{B})$-plane which corresponds to the magnetic field $\mathbf{B}$. Electromagnetic waves can be regarded as a self-propagating transverse oscillating wave of electric and magnetic fields. In vacuum, electromagnetic radiation propagates at the speed of light.

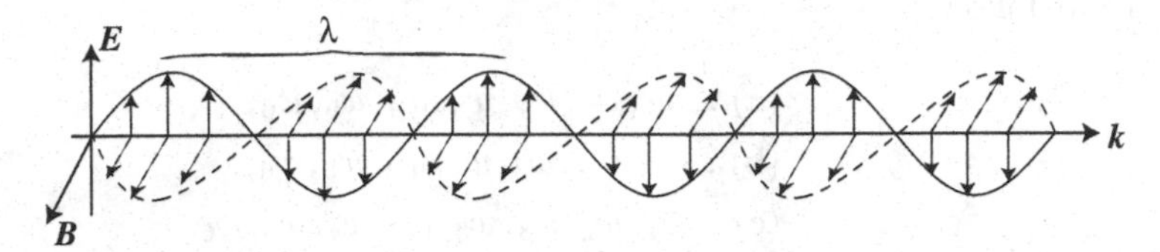

Figure 21.11 Components of electromagnetic radiation.

The classification of electromagnetic radiation is based on the frequency of its waves. It is governed by the laws of electrodynamics (known as Maxwell's equations). Electric and magnetic fields obey the superposition principle. Since they are vector fields, all electric and magnetic field vectors add together by laws of vector addition.

Electromagnetic radiation exhibits both wave properties and particle properties simultaneously. This is known as the *wave-particle duality*. As a wave, it is transverse, so that the oscillations of the wave are perpendicular to the direction of the energy transfer. The electric and magnetic parts of the field remain in a fixed ratio and satisfy Maxwell's equations involving both $\mathbf{E}$ and $\mathbf{B}$ fields.

21.5 Electric Capacitors

We present Maxwell's transformation

Map 21.1a. $$z = \frac{a}{\pi}\left(1 + w + w^2\right)$$

that shows the fringing field in the end-zone of an electric capacitor.

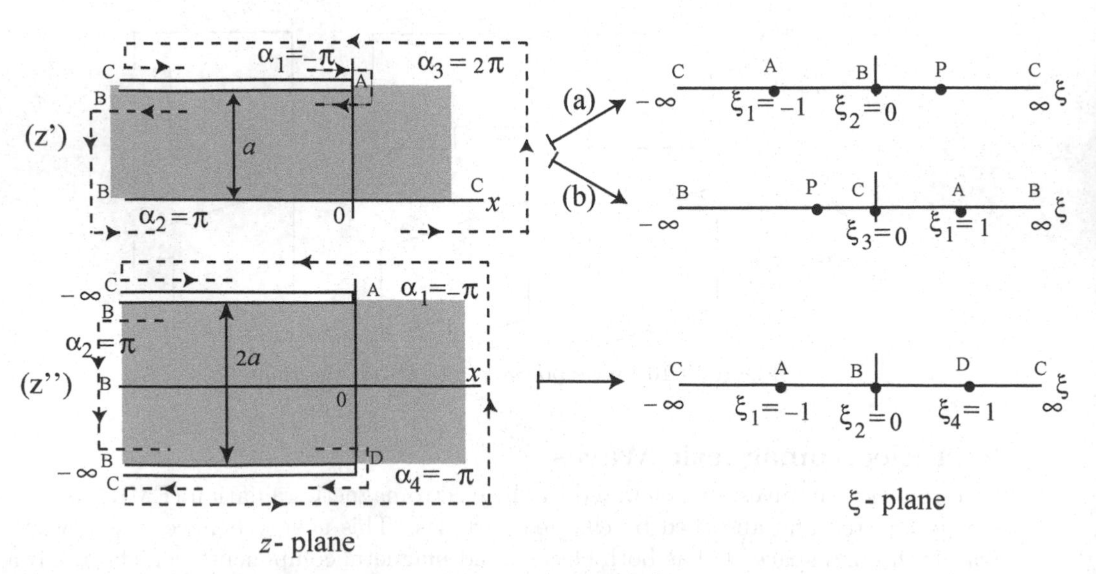

Figure 21.12 One-quarter (z') and one-half (z'') of an electric capacitor.

We will derive the results using Schwarz-Christoffel transformation, where the freedom of choosing points is significant. Thus, we can select points in the ζ-plane that correspond to the vertices in the z-plane. The following three cases are analyzed, each with a different set of points. The part (z') in Figure 21.12 represents one-quarter of the capacitor, while the part (z'') represents one-half of an electric capacitor.

We start with (z'): Recall that $\zeta = \xi + i\eta$. The point C is at $\xi = \infty$, point A is at $\xi_1 = -1$, and point B is at $\xi_2 = 0$. Thus, we have $\alpha_1 = -\pi$ and $\alpha_2 = \pi$, and the transformation is given by

$$\begin{aligned} z(\xi) &= C_1 \int \frac{d\xi}{(\xi+1)^{-1}(\xi-0)} + C_2 \\ &= C_1 \int \frac{\xi+1}{\xi}\, d\xi + C_2 = C_1\left(\xi + \log\xi\right) + C_2. \end{aligned} \tag{21.5.1}$$

At the point B, the boundary condition $\xi = 0$, $z = -\infty \pm ia$ is not easily applicable. However, the integration constant C_1 can be obtained by using the infinitesimal changes as we track through an arc around ξ_2 with $\xi = r\, e^{i\theta}$ and $r \to 0$ in the ξ-plane. Then we get

$$\begin{aligned} &\int_{-\infty+ia}^{-\infty+i0} dz = C_1 \int_0^{0+} \frac{\xi+1}{\xi}\, d\xi = C_1 \int_0^{0+} \left(1 + \frac{1}{\xi}\right) d\zeta \\ &-ia = \lim_{\xi\to 0} C_1 \int_\pi^0 \left(1 + \frac{1}{r}\, e^{i\theta}\right) ire^{i\theta}\, d\theta = -iC_1\pi, \end{aligned} \tag{21.5.2}$$

which gives $C+1 = a/\pi$.

At the point A, set $\xi = -1, z = 0 + ia$. Then

$$ia = \frac{a}{\pi}\big[-1 + \log(-1)\big] + C_2 = \frac{a}{\pi}\,(-1 + ia) + C_2,$$

which gives $C_2 = a/\pi$. Hence $z = \dfrac{a}{\pi}\,(1 + \zeta + \log\zeta)$, which, using $w = \log\zeta$, gives

$$z = \frac{a}{\pi}\,(1 + w + e^w)\,, \tag{21.5.3}$$

or

$$x = \frac{a}{\pi}\,(1 + u + e^u\cos v)\,, \quad y = \frac{a}{\pi}\,(e^u \sin v + v)\,. \tag{21.5.4}$$

Thus, $v = 0$ corresponds to $y = 0$, and $x = \frac{a}{\pi}\,(1 + e^u + u)$ shows that even close points on the u-axis correspond to distant points on the x-axis. The line $v = \pi$ corresponds to $y = a$ and $x = \frac{a}{\pi}\,(1 - e^u + u)$. The points A, B, C correspond to similar points in Figure 21.13,

part (z') and (a).

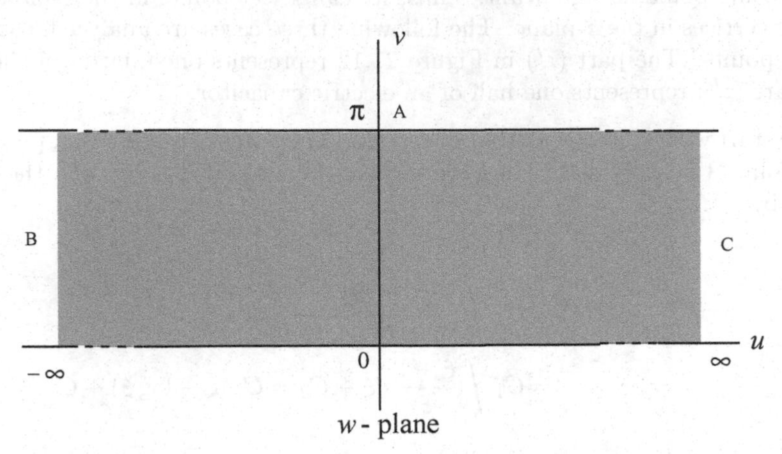

Figure 21.13 One-quarter (z') and one-half (z'') of an electric capacitor.

Map 21.1b. We again start with the part (z'), but with a different selection of the three points $\xi_2 = \pm\infty$ (point B), $\xi_3 = 0$ (point C), and $\xi_1 = 1$ (point A). Then, since $\alpha_1 = \pi, \alpha_3 = 2\pi$, we obtain the integral equation

$$z(\xi) = C_1 \int \frac{d\xi}{(\xi+0)^2 (\xi-1)^{-1}} + C_2$$
$$= C_1 \int \frac{\xi - 1}{\xi^2}\, d\xi + C_2 = C_1\Big(\log \xi + \frac{1}{\xi}\Big) + C_2. \tag{21.5.5}$$

Now, at the point B the change in z is from $0 + ia$ to $0 + i0$ and the angle θ changes from 0 to π as we traverse the ξ-axis. Thus, with $r \to \infty$ at B, we get

$$\int_{-\infty+ia}^{-\infty+i0} dz = C_1 \int_{0-}^{0+} \frac{\xi-1}{\xi^2}\, d\xi,$$

or

$$-ia = \lim C_1 \int_0^{\pi} \frac{re^{i\theta}-1}{r^2 e^{2i\theta}}\, ire^{i\theta}\, d\theta,$$

which gives $C_1 = -a/\pi$.

At point A, we have $\xi = 1$ and $z = ia$, $ia = -\frac{a}{\pi}\big[\log(1) = 1\big] + C_2$, which gives $C_2 = ia + a/\pi$. Hence,

$$z(i) = -\frac{a}{\pi}\big[\log\xi + \tfrac{1}{\xi}\big] + ia + \tfrac{a}{\pi} = \frac{a}{\pi}\Big(1 - \frac{1}{\xi} - \log\xi\Big) + ia,$$

which, with $w = \log\zeta$, we get

$$z(w) = \frac{a}{\pi}\big[1 - e^{w} - w\big] + ia. \tag{21.5.6}$$

If we use the transformation $w \mapsto w^*$, we find the same result as above, but with w^* instead of w, i.e., $w^* = i\pi - w$, thus

$$z(w^*) = \frac{a}{\pi}\big[1 + e^{-w^*} + w^*\big]. \tag{21.5.7}$$

Map 21.1c. We start with (z''), and find that it yields similar results as obtained in Maps 21.1a and 21.1b if we use the symmetry on the ξ-axis, and the transformation $w = 2\log\zeta$.

21.5.1 Capacitance between Two Conductors. Let q be the electrical charge on either conductor and Δu be the difference between one conductor and the other. Then the *capacitance* C is defined by

$$C = \frac{|q|}{|\Delta u|}.$$

In two-dimensional problems, the capacitance is calculated per unit length, denoted by c, and its cross-section is typically displayed in the complex plane. In this case, let q_L denote the charge per unit length of one of the conductors. Then

$$c = \frac{|q_L|}{\Delta u|}.$$

Theorem 21.1. *The electrical charge per unit length on a conductor that belongs to a charged two-dimensional configuration of conductors is*

$$q_L = \frac{\varepsilon}{4\pi}\Delta v(x,y), \tag{21.5.8}$$

where ε is the permeability (or dielectric constant) of the surrounding material, and $\Delta v(x,y)$ is the decrement (initial value minus final value) of the stream function as we proceed in the positive direction once around the boundary of the cross-section of the conductor in the complex plane.

The stream function $v(x,y) \equiv v(z)$ is a multiple-valued function defined using a branch cut. Thus, this function does not return to its original value after encircling the conductor, which means that $\Delta v(z) \neq 0$. Then

$$c = \frac{\varepsilon}{4\pi}\frac{|\Delta v|}{|\Delta u|}. \tag{21.5.9}$$

Theorem 21.2. *The capacitance of a two-dimensional system of conductors is not affected by a conformal transformation of its cross-section.*

21.5.2 Problem of Two Cylinders. The problem is to find the potential outside two conducting cylinders, each with radius r_0 and their centers separated by a distance d, and kept at the potential $\pm V_0/2$ (Figure 21.14(a)). The solution is constructed based on the potential of two line charges (Figure 21.14(b)), where the potential is given by

$$v(x,y) = \frac{q}{2\pi\varepsilon}\log\frac{r_2}{r_1} = \frac{q}{2\pi\varepsilon}\log\frac{\sqrt{(x+a)^2+y^2}}{\sqrt{(x-a)^2+y^2}}, \tag{21.5.10}$$

where q and a are parameters related to V_0, r_0, and d.

To determine the surface of the equipotential surfaces, consider the points on the surface with constant potential equal to the potential $v(x_P, y_P) \equiv v_P$ at the point P (Figure 21.14(b)), which satisfies

$$r_2 = r_1\, e^u, \quad \text{where} \quad u = \frac{V_P}{\left(\dfrac{q}{2\pi\varepsilon}\right)}. \tag{21.5.11}$$

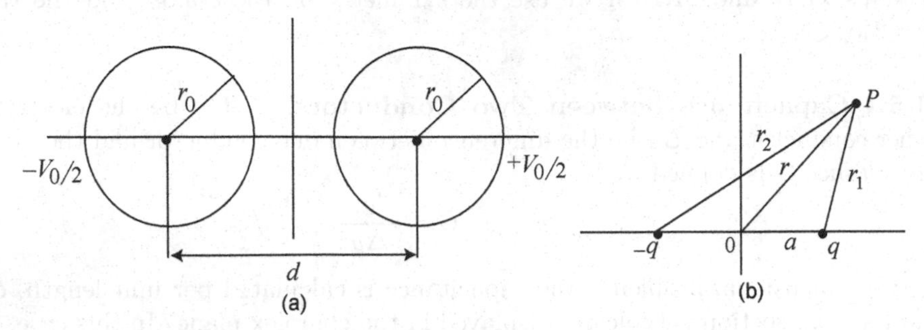

Figure 21.14 Problem of two conducting cylinders.

In this form, it is hard to determine what the surface looks like. But in terms of the coordinates (x, y), the surface is a circle (i.e., a cylinder in $\mathbb{R}^3$) defined by

$$(x - a \coth u)^2 + y^2 = \frac{a^2}{\sinh^2 u}, \tag{21.5.12}$$

where $a = r_0 \sinh u$, $d = 2a \coth u = 2r_0 \cosh u$, $q/(2\pi\varepsilon) = V_0/(2u)$, where $u = \cosh^{-1} \frac{d}{2r_0}$. Hence, the two conducting cylinders are held at the potential $\pm V_0/2$, where

$$v(x, y) = \frac{V_0}{2} \frac{\log(r_2/r_1)}{u}. \tag{21.5.13}$$

Example 21.12. Consider an equipotential surface (Figure 21.14(a)) with the potential $u = 0.8$. Then from (21.1.3) we find that this surface corresponds to the potential $V(x, y) = V_P = (0.8q)/(2\pi\varepsilon)$, and if, for example, the charge is $q = 2\pi\varepsilon$ volts, the equipotential surface $u = 0.8$ would correspond to the surface of 0.8 volts. The plot, if desired, can be obtained by first calculating the radius of the circle $a/\sinh(0.8) \approx 1.13a$ and translating the circle away from the origin by $a \coth(0.8) \approx 1.5a$. ■

In practical problems we can replace the parameters V_P, a, and q by $V_0/2, r_0$, and d, respectively. Thus,

$$a = r_0 \sinh u, \quad d = 2a \coth u = 2r_0 \cosh u. \tag{21.5.14}$$

Since u is defined in terms of q, Eq (21.1.3) determines

$$\frac{q}{2\pi\varepsilon} = \frac{V_)}{2u}, \quad = \cosh^{-1} \frac{d}{2r_0}. \tag{21.5.15}$$

Hence, the solution of the problem of two conducting cylinders held at potential $\pm V + 0/2$ is

$$V(x, y) = \frac{V_0}{2} \frac{\log(r_2/r_1)}{u}, \tag{21.5.16}$$

where $r_1 = \sqrt{(x-a)^2 + y^2}$, $r_2 = \sqrt{(x+a)^2 + y^2}$, and a and u are determined from (21.5.14) and (21.5.15).

21.6 AC Circuits

Complex numbers can be used to analyze a circuit containing an AC generator, resistors, capacitors, and inductance coils in the same way as they are used to analyze a DC circuit containing a battery and resistors. There exists a *phase difference* between the current through a device and the voltage across the device. The AC voltage $V(t)$ and current $I(t)$ are said to be *sinusoidal*, i.e., both these quantities vary with time according to

$$V(t) = V_p \sin(2\pi \mathfrak{f} t + \phi_V), \quad I(t) = I_p \sin(2\pi \mathfrak{f} t + \phi_I), \tag{21.6.1}$$

where $\mathfrak{f}$ is the frequency of the AC generator, and ϕ_V and ϕ_I are the *phase angles*, and the subscript p denotes the peak or maximum voltage or current. The phase difference between the voltage and current is $\phi_V - \phi_I$.

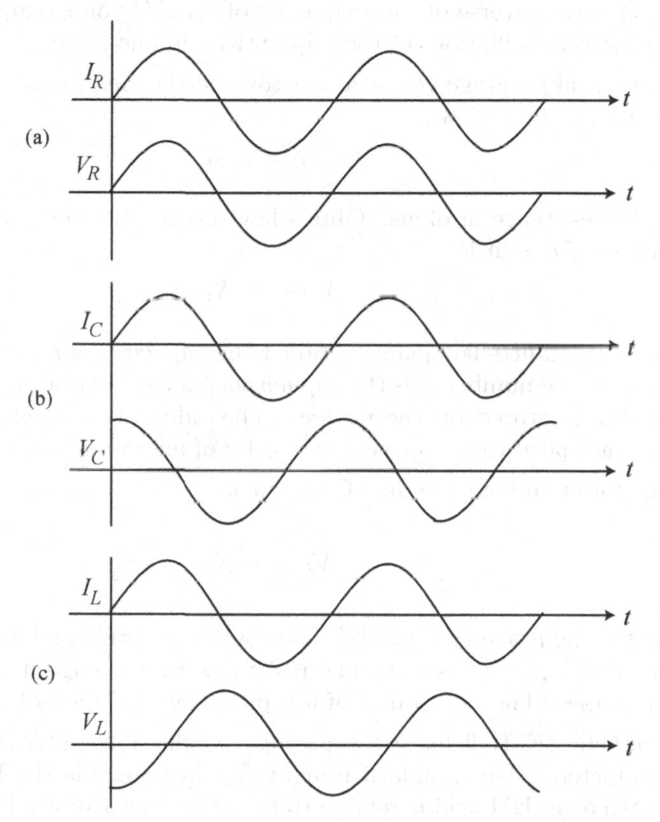

Figure 21.15 Conformal and isogonal mappings.

In an AC circuit, the current I_R through a resistor is in phase with V_R which is the voltage across the resistor so that $\phi_V - \phi_I = 0$. This implies that when one of these quantities (e.g., the current) is at a maximum, the other is zero. Also, the current and voltage are 90° *out of phase* with each other for a capacitor and for an inductor. This means that for either device, the voltage becomes zero at the instant when the current is a maximum. In the case of a capacitor, the voltage is 90° behind the current at every instant, which means that $\phi_V - \phi_I = -90°$ for a capacitor. In the case of an inductor the voltage is 90° ahead of the current at every instant, so that $\phi_V - \phi_I = 90°$. These phase relations are shown in Figure 21.15(a)–(c), where (a) has voltage in phase with the current through a resistor; (b) has voltage 90° behind the current through a capacitor; and (c) has voltage 90° ahead of the current through an inductor; the subscripts R, C, and L represent the resistor, capacitor, and inductor, respectively. For more details, see Serway and Jewett [2004: 1034-1042].

Since there is an equal amount of positive or negative voltage or current, the average voltage or current over one period is zero. Thus, average voltage and current do not define the peak or maximum values V_p and I_p. Instead, these average values, called *root mean squared* or *rms* values, are determined by taking the averages of the squares of the voltage and current over one cycle. For example, the rms of voltage is given by

$$V_{\text{rms}} = \sqrt{\frac{1}{T}\int_0^T [V(t)]^2\,dt} = \sqrt{\frac{1}{T}V_p^2\int_0^t \sin^2(2\pi\mathfrak{f}t)\,dt} = \frac{V_p}{2\sqrt{2}}, \tag{21.6.2}$$

where $T = 1/\mathfrak{f}$ is the inverse of the frequency of the AC voltage and current, i.e., T is the time it takes for the oscillations of these quantities in one cycle.

The rms (or peak) voltage across a resistor and the rms (or peak) current through the resistor is given by *Ohm's law*:

$$V_R = I_R R, \tag{21.6.3}$$

where R is the resistance in ohms. Ohm's law relating the rms voltage and current for a capacitor in an AC circuit is

$$V_C = I_C X_C, \tag{21.6.4}$$

where X_C is a resistance-like quantity called the *capacitance reactance*, defined by $X_C = 1/(2\pi\mathfrak{f}C)$. The device number C is the *capacitance* of a capacitor; it measures the maximum charge that can be stored on the device. The value of C is expressed in farads. The capacitance of a typical capacitor is of the order of microfarads (μfd).

Ohm's law for an inductor in an AC circuit is

$$V_L = I_L X_L, \tag{21.6.5}$$

where X_L is the *inductive reactance* defined by $X_L = 2\pi\mathfrak{f}L$, and L is called *inductance* of the inductor, and it measures the coil's resistance to a change in the current through it, measured in henries. The inductance of a typical coil is of the order of millihenries (mH).

Electromagnetic (EM) fields are typically generated by alternating currents (AC) in electrical conductors. The standard unit of EM frequency is the Hertz, abbreviated Hz. The wavelength of an EM field is related to the frequency f of the EM wave, related by the formula: wavelength $= 300/f$ meters.

21.6.1 Two-Dimensional Phase Diagrams. These are used to describe the phase difference between the voltage across and the current through the above-mentioned devices.

(i) An RCL AC series circuit containing a resistor, a capacitor, and an inductor in series, is shown in Figure 21.16(a). The same current flows through each device in such a circuit, and the sum of the potential differences across the device is the potential difference across the generator. Notice that the voltage across the resistor is in phase with the current through the three devices. Hence, V_R is also along the positive horizontal axis; the voltage across the capacitor is 90° behind the current through it, so V_C is along the negative vertical axis; V_L is ahead of the current by 90°, and V_L is along the positive vertical axis (Figure

21.16(b), where the length of the lines, with arrowhead, represent the rms or peak values of V_R, V_L, and V_C).

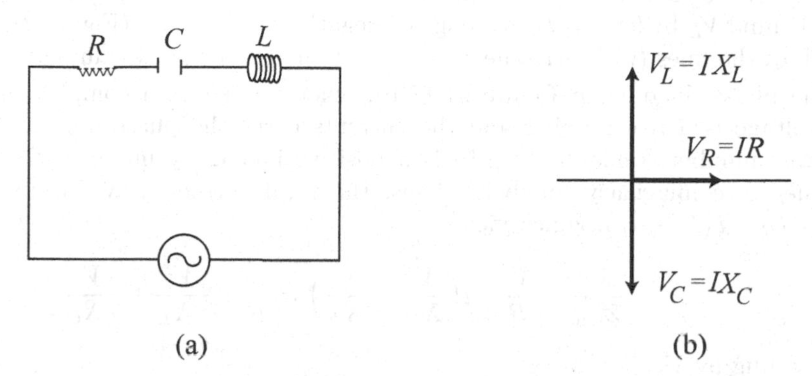

Figure 21.16 (a) An RCL AC series circuit. (b) Phase diagram.

The phase diagram in Figure 21.16(b) when viewed as a complex plane, can represent the current and voltage as complex numbers, where the current is taken along the real axis and therefore a real number. Thus, $V_R \to V_R$ is a real number, while $V_C \to -i\,V_C$ is a negative imaginary number, so also is $V_L \to i\,V_L$. Each of these reactances acts as a resistor in a DC circuit such that the reactance is a measure of how the device 'impedes' the flow of current. In the case of AC circuits, the complex number that represents the reactance of a device is called the *complex impedance*.

Addition of voltages across the device in a series circuit gives the net voltage $V_{\text{net}} = V_R + i\,V_L - i\,V_C$. If the voltages are expressed in terms of the common current and various impedances, we get $IZ_{\text{net}} = IR + I\,(i\,X_L - i\,X_C)$, which yields the net impedance

$$Z_{\text{net}} = R + i\,X_L + (-i\,X_C) = R + i\,(X_L - X_c)\,. \tag{21.6.6}$$

(ii) An AC parallel circuit containing a resistor, capacitor, and inductor in parallel, is shown in Figure 21.17(a).

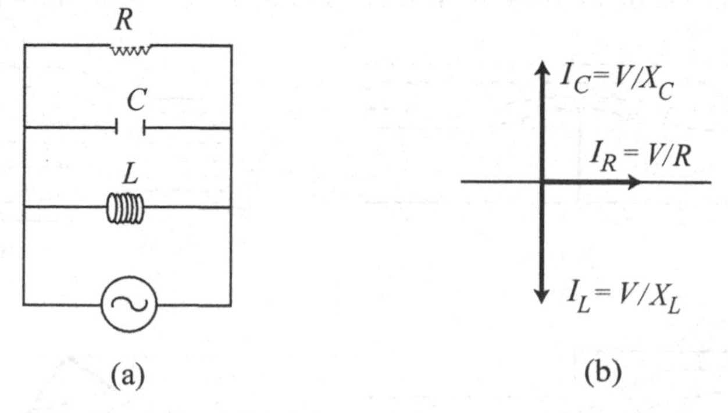

Figure 21.17 (a) An RCL AC parallel circuit. (b) Phase diagram.

The potential differences across these three devices are the same, and the sum of the currents through the devices adds to the current drawn from the generator. Since the voltage across all three devices is the same, the line representing this voltage is taken along the positive horizontal axis and the currents are represented by lines drawn at appropriate phase angles relative to the voltage. Thus, the current through the resistor is in phase with the voltage across it; so I_R is along the positive horizontal axis. The current through the capacitor is 90° ahead of the voltage across it, so I_C is along the positive vertical axis. Since I_L is behind V_L by 90°, so I_L is along the negative vertical axis (Figure 21.17(b), where the length of the lines (with arrowhead) represent the *rms* or peak values of I_R, I_L, and I_C).

The phase diagram in Figure 21.17(b), when viewed as a complex plane, shows that the voltage is a real number and the currents carry the phase angles. Thus, $I_R \to I_R$ is a real number, while $I_C \to i\, I_C$ is a positive imaginary number whereas $I_L \to -i\, I_L$ is a negative imaginary number. Thus, the total current drawn from the generator is $I_{\text{net}} = I_R + i\,(I_C - I_L)$, which yields

$$\frac{V}{Z_{\text{net}}} = \frac{V}{R} + i\Big(\frac{V}{X_C} - \frac{V}{X_L}\Big) = \frac{V}{R} + \frac{V}{i\,X_L} - \frac{V}{i\,X_C},$$

or, dividing by V,

$$\frac{1}{Z_{\text{net}}} = \frac{1}{R} + \frac{1}{i\,X_L} + \frac{1}{i\,X_C}. \tag{21.6.7}$$

Older Terminology. It is still used for the *electromotive force* or *emf*, where E is used to represent voltage. Some of the mnemonics are as follows:

ELI : voltage (E) across the inductor (L) is 90° ahead of current (I) through it.
ICE : current (I) through a capacitor (C) is 90° ahead of voltage (E) across it.
EIR : voltage (E) across a resistor (R) and current (I) are in phase.

21.7 Laplace's Equation

One basic feature of conformal mapping is that it does not solve problems, but it may reduce hard problem to easier ones. The amount of work that must be done to solve easier problems varies considerably with the application. In some cases the original problem may be reduced to a model problem whose solution is known exactly. This is the case in the fluid flow problems of Figure 21.18, in which a crooked channel may be mapped onto an infinite straight channel of constant width. If the simpler methods fail, a solution of the model domain may be found by finite difference or finite element methods. For problems involving Poisson's equation or more complicated equations, this procedure is normally necessary.

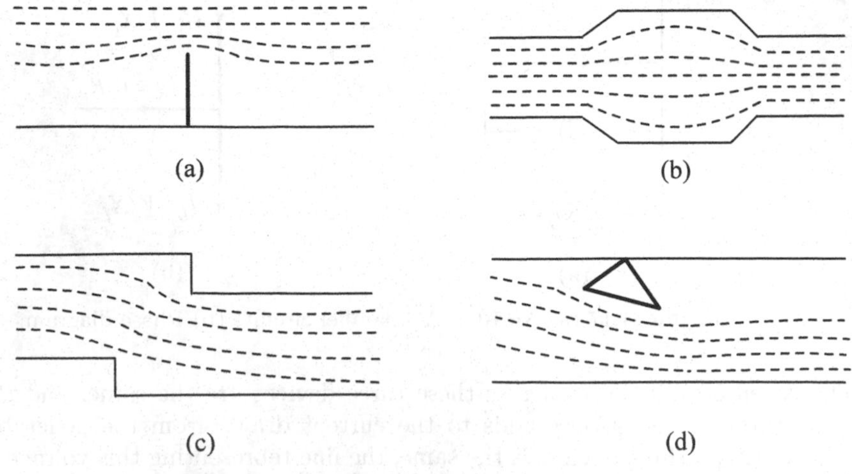

Figure 21.18 Different channels for fluid flows.

If a problem involving Laplace's equation with either Dirichlet or Neumann conditions can be mapped conformally to a disk, then Poisson's formula or Dini's formula (see Kantorovich and Krylov [1958]) provides integral representations of the solution at each interior point. Such integrals are readily computed to high accuracy. However, the main disadvantage of this approach is that a new integral may be evaluated for each point at which the solution is desired.

If the solution is required at many points in the region of interest, then it is probably more efficient to solve Laplace's equation by a trigonometric expansion of the form $a_0 + \sum_{k=1}^{m} r^k \left(a_k \sin k\theta + b_k \cos k\theta\right)$, where the coefficients a_k and b_k are chosen so as to fit the boundary conditions closely. The only disadvantage in this method is that convergence may be slow if the boundary conditions are not smooth.

Example 21.13. (Non-coaxial cable) Consider the problem of determining the electrostatic potential inside a non-coaxial cylindrical cable with prescribed constant potential value on the two bounding cylinders. Assume, e.g., the larger cylinder has radius 1 and is centered at the origin, while the smaller cylinder has radius 2/5 and is centered at $z = 2/5$. The resulting electrostatic potential is independent of the longitudinal coordinates, and thus it can be regarded as a planar potential in the annular region between two circles that represent the cross-section of the cylinders. The required potential, therefore, satisfies the Dirichlet boundary value problem $\nabla^2\phi = 0$ for $|z| < 1$ and $|z - 2/5| > 2/5$, given that $\phi = a$ when $z = 1$, and $\phi = b$ when $|z - 2/5| = 2/5$. The bilinear transformation $w = \dfrac{2z-1}{z-2}$ maps the annular region in the z-plane onto the annulus $A_{1/2,1} = \{\frac{1}{2} < |w| < 1\}$, which is the cross-section of the cable. The transformed potential $\Phi(u, v)$ satisfies the Dirichlet boundary conditions $\Phi = a$ when $|u| = \frac{1}{2}$, and $\Phi = b$ when $|u| = 1$. Thus, the potential Φ is given by

$$\Phi(u, v) = \alpha \log |u| + \beta,$$

where α, β are constant. For the above dimensions of the two cylinders, the particular potential function

$$\Phi(u, v) = \frac{b-a}{\log 2} \log |u| + b = \frac{b-a}{2\log 2} \log(u^2 + v^2) + b$$

satisfies the prescribed boundary conditions. Hence the required non-coaxial electrostatic potential is obtained by composition with the above conformal map, and is given by

$$\phi(x, y) = \frac{b-a}{\log 2} \log \left| \frac{2z-1}{z-2} \right| = b = \frac{b-a}{2\log 2} \log \left(\frac{(2x-1)^2 + y^2}{(x-2)^2 + y^2} \right) + b. \blacksquare$$

21.7.1 Electric Potential and Field between Two Infinite Sheets. This is a problem in which the original problem is reduced to a model problem whose solution is known exactly. Figure 21.19(a) represents an infinite region bounded by a straight boundary fixed at potential $\phi = 0$ and one jagged boundary fixed at potential $\phi = 2$. This may be regarded as a simple electrostatic problem. The main question to be answered computationally is: what are the voltage ϕ and the electric field $E = -\nabla\phi$ at a given point, either within the field or on the boundary? The procedure begins by mapping the given region onto the unit disk using a Schwarz-Christoffel transformation, like Map 7.4; then analytically onto an infinite straight channel, as in the four cases of Figure 21.18. In the straight channel, ϕ and

E are known trivially, and the data thus obtained may be transferred to the problem region through a conformal map that connects the two regions. The details are simple enough, so they are left out. Figure 21.19(b) shows $|E|$ as a function of z on the upper and lower boundaries of the original region. Detailed behavior of ϕ, $|E|$, and $\arg\{\phi\}/\pi$ at three points near $3 + 1.5i$ are presented in Table 21.1.

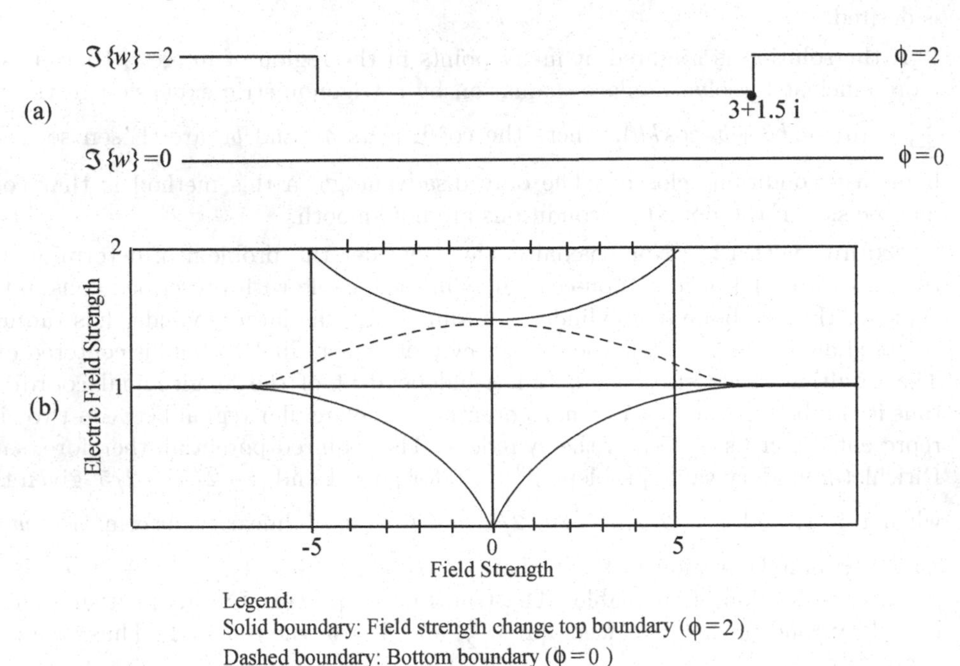

Figure 21.19 Electric potential and field between two infinite sheets.

The computed potential ϕ, field strength $|E|$, and $\arg\{E\}/\pi$ at three points near $3+1.5i$ are presented in Table 21.1.

Table 21.1 (Trefethen [1979: 41]).

w	ϕ	$\lvert E\rvert$	$\arg\{E\}/\pi$
$3.1 + 1.4i$	1.7564	1.3082	−0.3823
$3.01 + 1.49i$	1.9486	2.4403	−0.2833
$3.01 + 1.499i$	1.9889	5.2137	−0.2572
$3.01 + 1.50i$	2.0000	−−	−0.2500

21.7.2 Poisson's Equation. A 7-sided polygon, including a 16×32 finite-difference grid in the unit disk, is presented in Figure 21.20(a), and a (r, θ) grid for $r = 4$ and $\theta = 8$ on the unit disk in the w-plane is shown in Figure 21.20(b). On this region we will solve Poisson's equation

$$\nabla^2 \phi(x, y) = \frac{1}{5} \sin 2x \left(1 - 2(y + 1)^\alpha\right), \tag{21.7.1}$$

subject to the Dirichlet boundary conditions

$$\phi(x, y) = \rho(x, y) = \frac{1}{10} \sin 2x(y + 1)^2 \tag{21.7.2}$$

on the boundary. The procedure is as follows: We map the region onto the disk and solve a transformed problem in the disk in polar coordinates using a second-order fast finite-difference solver (PWSPLR, by Swarztrauber and Sweet [1975]), and obtain $\rho(x, y)$, which is the correct solution in the interior as well as on the boundary. Thus, the accuracy of the numerical solution is easily determined.

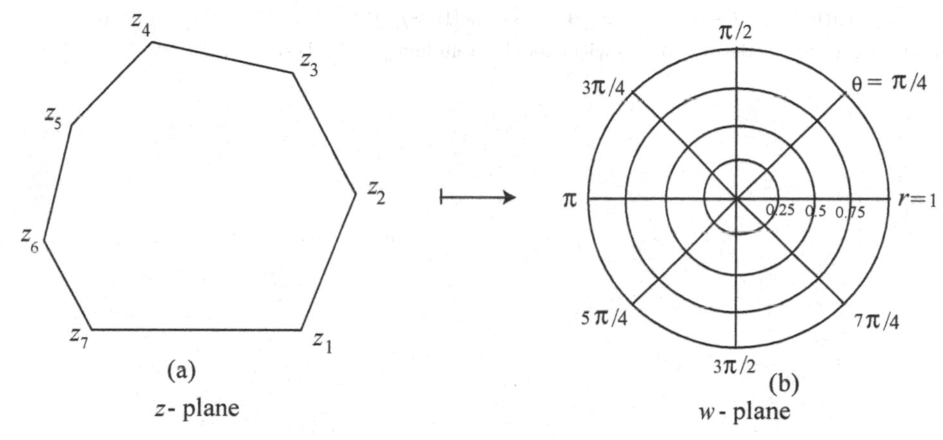

Figure 21.20 A 7-sided polygon with a 4×8 finite-difference grid in the unit disk.

Computed results for four different grids are presented in Table 21.2.

Table 21.2 (Trefethen [1979: 43]).

Grid ($r \times \theta$)	Transformation and setup time	Past Position solver time	Max. Error	RMS Error
4×8	1.3 secs.	< 0.01 secs.	0.132	0.0309
8×16	2 secs.	0.01 secs.	0.055	0.0085
16×32	5 secs.	0.03 secs.	0.031	0.0037
32×64	16 secs.	0.15 secs.	0.026	0.0012

To summarize, the main suggestions for solving Laplace's and Poisson's equations using Schwarz-Christoffel transformations with better accuracy are as follows:

(i) Use Map 7.4 (unit disk) rather than the upper half-plane as the domain model; it provides better numerical scaling.

(ii) Use complex contour integrals interior to the model domain rather than along the boundary; it makes the treatment of unbounded polygons easy.

(iii) Use compound Gauss-Jacobi quadrature in complex arithmetic to accurately evaluate the Schwarz-Christoffel integrals.

(iv) Formulate the parameter problem as a constrained nonlinear system in $N - 1$ variables.

(v) Eliminate constraints in the nonlinear system by a simple variable transformation.

(vi) Solve the system by a packaged nonlinear solver, that requires no initial estimate.

(vii) Once the parameter problem is solved, compute a reliable estimate of the accuracy of further computations.

(viii) If accurate computation of the inverse mapping in two steps is required, use a packaged o.d.e. solver and a packaged complex root-finder.

A program listing is available in Trefethen [1979: 46-55].

REFERENCES USED: Kober [1957], Kythe [1998/2012; 2015], Mader [1991], Purcell [2013], Schinzinger and Laura [2003], Serway and Jewitt [2004], Trefethen [1979], Trefethen and Williams [1986].

22

Transmission Lines and Waveguides

Applications of conformal mapping to transmission systems cover a wide range of electrical engineering problems. For example, the power system engineer needs to know to what extent the nonlinear and frequency dependent characteristics of conductors and soils will affect the propagation of surges on very high-voltage lines. Communication engineers and their circuit designers cover the areas of maintaining low loss and low distortion transmission on lines, waveguides, or microstrip configurations. The problems of dielectric breakdown, power dissipation, and nonionizing electromagnetic radiation are important, and conformal mapping helps the analysis and synthesis of transmission systems.

We will use conformal mapping to determine the modulus, impedance, propagation constant, or other characteristics of transmission lines, microstrip lines, and waveguides. Microstrip and coplanar waveguide are chosen for the analysis, which included three configurations of each transmission line geometry: a reference with no additional thin film material, one with the thin film on top of the conductors, and one with the thin film beneath the conductors but on top of the transmission line substrate.

Conformal mapping techniques are used in the analysis and synthesis of transmission lines, microstrip lines, and waveguides. Many applications and recent developments in transmission technology also require calculation of secondary effects, nonuniformities, and new configurations in conformal transformations. Such applications demand that the systems engineer should know to what extent the nonlinear frequently dependent characteristics of conductors affect the propagation of surges on ultra high-voltage lines. The communication engineers and integrated circuit designers are interested in low loss, low distortion transmission on lines, waveguides or microstrip lines.

We will discuss basic Maxwell's equations and their transformation from the physical space onto the model space, and their application in transmission lines and waveguides.

22.1 Maxwell's Equations

Following Schinzinger and Laura [2003: 195]), consider a potential field of the kind presented in Figure 22.1, in which the physical field in the z-plane is transformed into the image plane $w = u + iv$. Let $u = U$ and $v = V$ describe particular field lines representing flow and potential. Consider an infinitesimally curvilinear square in the z-plane formed by the intersection of the field lines U, $U + du$, V, and $V + dv$. Under a conformal mapping $w = f(z)$, all these four sides and both diagonals are magnified or contracted by the same factor M (or m) defined in Map 3.1. This factor is the same in all directions at a point, but its magnitude generally depends on the location of the point.

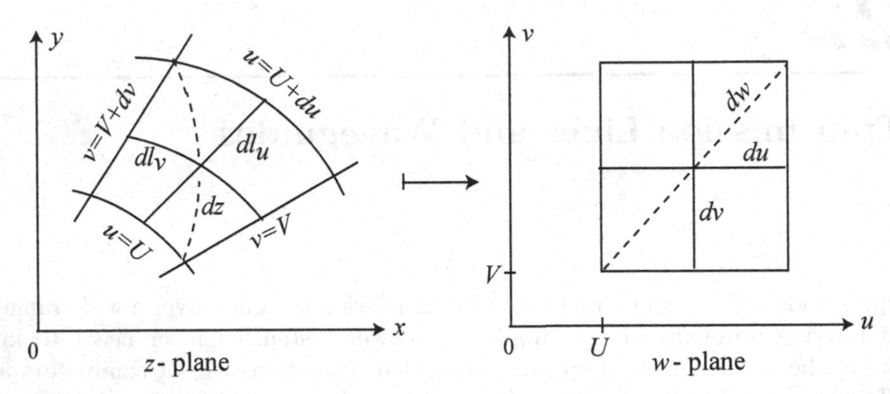

Figure 22.1 Conformal map of a curvilinear system.

Using Figure 22.1, we find that the relations holding at the level of different lengths are as follows:

$$\begin{aligned} dl_u &\leftrightarrow du & dl_v &\leftrightarrow dv, \\ dl_u &= M_u(x,y)\,du & dl_v &= M_v(x,y)\,dv, \\ dl_u &= |dz/\sqrt{2}| & dl_v &= |dz/\sqrt{2}|. \end{aligned} \tag{22.1.1}$$

Since conformal mapping produces uniform scaling, we have $M_u = dl_u/du, M_v = dl_v/dv$, which gives

$$M_u = M_v = M = |dz/dw| = 1/|f'(z)|. \tag{22.1.2}$$

In polar coordinates, we have

$$\begin{aligned} dl_r &= M_r\,dr, & dl_\theta &= M_\theta r\,d\theta, \\ M_r &= dl_r/dr, & M_\theta &= dl_\theta/(r\,d\theta), \end{aligned} \tag{22.1.3}$$

and $M_r = M_\theta = M = |dz/dw| = 1/|f'(z)|$ as above in (22.1.2). The inverse relations are expressed in terms of the scale factor $m = 1/M$. Thus,

$$\begin{aligned} du &= m_u\,dl_u & dv &= m_v\,dl_v, \\ m_u &= du/dl_u, & m_v &= dv/dl_v, \end{aligned} \tag{22.1.4}$$

so that

$$m_u = m_v = m + |dw/dz| = |f'(z)|. \tag{22.1.5}$$

In polar coordinates, we have

$$\begin{aligned} dr &= m_r\,dl_r, & r\,d\theta &= m_\theta\,dl_\theta, \\ m_r &= dr/dl_r, & m_\theta &= r\,dl_\theta/dl_\theta, \end{aligned} \tag{22.1.6}$$

and $m_r = m_\theta = m = |dw/dz| = |f'(z)|$ as above in (22.1.5). The scale factor m or M can be obtained from dz/dw and dw/dz using different methods, such as (i) analytically by differentiation, (ii) graphically by field plots, and (iii) experimentally from current-field analogs. See Meinke [1949], Piloty [1949], and Flachenecker and Lange [1967].

In the cases of field lines U and V we have

$$|\nabla U_z| = \Big[\Big(\frac{\partial U}{\partial l_u}\Big)^2 + \Big(\frac{\partial U}{\partial l_v}\Big)^2\Big]^{1/2},$$

thus,

$$\begin{aligned}|\nabla U_w| &= \Big[\Big(\frac{\partial U}{\partial u}\Big)^2 + \Big(\frac{\partial U}{\partial v}\Big)^2\Big]^{1/2} \\ &= \Big[\Big(\frac{\partial U}{\partial l_u} M_u^2\Big)^2 + \Big(\frac{\partial U}{\partial l_v} M_v^2\Big)^2\Big]^{1/2}.\end{aligned}$$

Since $M_u = M_v = M = 1/m$, we get

$$|\nabla U_w| = M|\nabla U_z|, \quad \text{or} \quad |\nabla U_z| = m\,|\nabla U_w|. \tag{22.1.7}$$

Maxwell's equations (21.1.3) for electric field intensity $\mathbf{E}$, or displacement density $\mathbf{D} = \varepsilon\mathbf{E}$, (or dielectric constant) for the surrounding material, and magnetic field strength $\mathbf{H}$ (or flux density $\mathbf{B} = \mu\mathbf{H}$) in a uniform source-free medium of permeability μ, permittivity ε, and conductivity σ become

$$\operatorname{curl}\mathbf{E} = -\mu\frac{\partial \mathbf{H}}{\partial t}, \tag{22.1.8}$$

$$\operatorname{curl}\mathbf{H} = \sigma\mathbf{E} + \varepsilon\frac{\partial \mathbf{E}}{\partial t}, \tag{22.1.9}$$

$$\div\,\mathbf{D} = 0, \tag{22.1.10}$$

$$\div\,\mathbf{B} = 0. \tag{22.1.11}$$

Let the physical space be (x, y, ζ),[†] $z = x + iy$[‡] and the model (transformed) space be (u, v, ζ), $w = u + iv$, $w = w(z)$. The scale factors m and $1/M$ are defined as $M = |dz/dw| = 1/m$. Then the transformations are

$$z \mapsto w, \quad \zeta_z = \zeta_w = \zeta, \quad |dz/dw| = M_u = M_v = M, \quad M_\zeta = 1.$$

The vector components of $\mathbf{E}$ and $\mathbf{H}$ are

$$\begin{aligned}&E_x,\ E_y,\ E_\zeta \quad \text{where} \quad E_u = ME_x,\ E_v = ME_y,\ E_\zeta(\text{same, unchanged});\\ &H_x,\ H_y,\ H_\zeta \quad \text{where} \quad H_u = MH_x,\ E_v = MH_y,\ H_\zeta(\text{same, unchanged}).\end{aligned} \tag{22.1.12}$$

Then Maxwell's equation (22.1.8) becomes

$$\operatorname{curl}\mathbf{E} = -\mu\frac{\partial \mathbf{H}}{\partial t} = -i\,\omega\mu\mathbf{H}. \tag{22.1.13}$$

[†] x, y, ζ are the three axes in a rectangular coordinate system in $\mathbb{R}^3$. Note that ζ used here is a real quantity, and does not mean $\zeta = \xi + i\eta$ used in the ζ-plane in conformal mappings.

[‡] Although $j = \sqrt{-1}$ is used for the imaginary unity in electrical engineering, we will keep our notation $i = \sqrt{-1}$ for the imaginary unity, and denote the current by J.

such that

$$\frac{\partial E_\zeta}{\partial y} - \frac{\partial E_y}{\partial \zeta} = -i\,\omega\mu H_x, \quad \frac{\partial E_\zeta}{\partial v} - \frac{\partial E_v}{\partial \zeta} = -i\,\omega\mu H_u, \tag{22.1.13a}$$

$$\frac{\partial E_x}{\partial \zeta} - \frac{\partial E_\zeta}{\partial x} = -i\,\omega\mu H_y, \quad \frac{\partial E_u}{\partial v} - \frac{\partial E_\zeta}{\partial u} = -i\,\omega\mu H_v, \tag{22.1.13b}$$

$$\frac{\partial E_y}{\partial x} - \frac{\partial E_x}{\partial y} = -i\,\omega\mu H_\zeta, \quad \frac{\partial E_v}{\partial u} - \frac{\partial E_u}{\partial v} = -i\,\omega\mu M^2 H_\zeta. \tag{22.1.13c}$$

Eq (22.1.9) becomes

$$\operatorname{curl}\mathbf{H} = \sigma\mathbf{E} + \varepsilon\frac{\partial \mathbf{E}}{\partial t} = (\sigma + i\,\omega\varepsilon)\mathbf{E}. \tag{22.1.14}$$

such that

$$\frac{\partial H_\zeta}{\partial y} - \frac{\partial H_y}{\partial \zeta} = (\sigma + i\,\omega\varepsilon)E_x, \quad \frac{\partial H_\zeta}{\partial v} - \frac{\partial H_v}{\partial \zeta} = i\,(\sigma + i\,\omega\varepsilon)E_u, \tag{22.1.14a}$$

$$\frac{\partial H_x}{\partial \zeta} - \frac{\partial H_\zeta}{\partial x} = (\sigma + i\,\omega\varepsilon)E_y, \quad \frac{\partial H_u}{\partial \zeta} - \frac{\partial H_\zeta}{\partial u} = (\sigma + i\,\omega\varepsilon)E_v, \tag{22.1.14b}$$

$$\frac{\partial H_y}{\partial x} - \frac{\partial H_x}{\partial y} = (\sigma + i\,\omega\varepsilon)E_\zeta, \quad \frac{\partial H_v}{\partial u} - \frac{\partial H_u}{\partial v} = (\sigma + i\,\omega\varepsilon)M^2 E_\zeta. \tag{22.1.14c}$$

Eq (22.1.10) becomes

$$\div\mathbf{D} = 0, \tag{22.1.15}$$

such that

$$\varepsilon\Big(\frac{\partial E_x}{\partial x} + \frac{\partial E_y}{\partial y} + \frac{\partial E_\zeta}{\partial \zeta}\Big) = 0, \quad \varepsilon\Big(\frac{\partial E_u}{\partial u} + \frac{\partial E_v}{\partial v} + M^2\frac{\partial E_\zeta}{\partial \zeta}\Big) = 0. \tag{22.1.15a}$$

Eq (22.1.11) becomes

$$\div\mathbf{B} = 0, \tag{22.1.16}$$

such that

$$\mu\Big(\frac{\partial H_x}{\partial x} + \frac{\partial H_y}{\partial y} + \frac{\partial H_\zeta}{\partial \zeta}\Big) = 0, \quad \mu\Big(\frac{\partial H_u}{\partial u} + \frac{\partial H_v}{\partial v} + M^2\frac{\partial H_\zeta}{\partial \zeta}\Big) = 0. \tag{22.1.16a}$$

Also,

$$E_y = E_{ym}\, e^{i\,\omega t}, \quad \frac{\partial E_y}{\partial t} = i\,\omega E_y, \tag{22.1.17}$$

where ω is the frequency in radians per second. The scale factor m or $1/M$ is defined by (22.1.3) with $M_\zeta = 1$. However, in the above transformations for Maxwell's equation, the longitudinal axis ζ is not involved.

According to experimental data, the vector fields $\mathbf{E}, \mathbf{H}, \mathbf{D}, \mathbf{B}$, and the current density $\mathbf{j}$ are not independent and therefore, they should be supplemented with the material equations (constitutive relations) of the medium. For isotropic linear media, these equations are

$$\mathbf{D} = \varepsilon\mathbf{E}, \quad \mathbf{B} = \sigma\mathbf{H}, \quad \mathbf{j} = \lambda\mathbf{E}, \tag{22.1.18}$$

where ε and μ are the dielectric permittivity and the magnetic permeability of the medium. For a perfect dielectric, i.e., for a medium that does not conduct electric current at all, we should set $\lambda = 0$.

The general solution of Maxwell's equations (22.1.8)–(22.1.11) is

$$\mathbf{H} = -\operatorname{curl}\operatorname{curl}\boldsymbol{\Psi}, \quad \mathbf{E} = \mu \operatorname{curl}\boldsymbol{\Psi}_t, \tag{22.1.19}$$

where the vector function $\boldsymbol{\Psi}$ satisfies the equation

$$\mu\,(\varepsilon\boldsymbol{\Psi}_{tt} + \sigma\boldsymbol{\Psi}_t) - \nabla^2\boldsymbol{\Psi} = \mathbf{0}. \tag{22.1.20}$$

By successively eliminating the vectors $\mathbf{H}$ and $\mathbf{E}$ from the system (22.1.8)-(22.1.11), we obtain two independent overdetermined subsystems:

(a) for the magnetic field intensity,

$$\begin{aligned} &\mu\,(\varepsilon\mathbf{H}_{tt} + \sigma\mathbf{H}_t) - \nabla^2\mathbf{H} = \mathbf{0}, \\ &\div\,\mathbf{H} = 0, \end{aligned} \tag{22.1.21}$$

(b) for the electric field intensity,

$$\begin{aligned} &\mu\,(\varepsilon\mathbf{E}_{tt} + \sigma\mathbf{E}_t) - \nabla^2\mathbf{E} + \nabla^2(\div\mathbf{E}) = \mathbf{0}, \\ &\div\,\mathbf{E} = 0. \end{aligned} \tag{22.1.22}$$

22.2 Wave Propagation

There are two types of wave formulations: (a) a transverse electromagnetic (TEM wave), and (b) a transmission magnetic (TM)wave.

Example 22.1. For a TEM wave propagation in the ζ direction, the electric and magnetic fields are confined to the transverse (x, y)-plane. Thus, $E_\zeta = 0, H_\zeta = 0$ in both physical space and the model space. For the sake of simplicity, we will assume that the wave is polarized vertically with respect to $\mathbf{E}$ so that $E_x = 0 = H_y$ as shown in Figure 22.2. Then Eqs (22.1.13a) and (22.1.14b) are reduced, respectively, to

$$\frac{\partial E_y}{\partial \zeta} = -i\,\omega\mu H_x, \quad \frac{\partial H_x}{\partial \zeta} = (\sigma + i\,\omega\varepsilon)E_y, \tag{22.2.1}$$

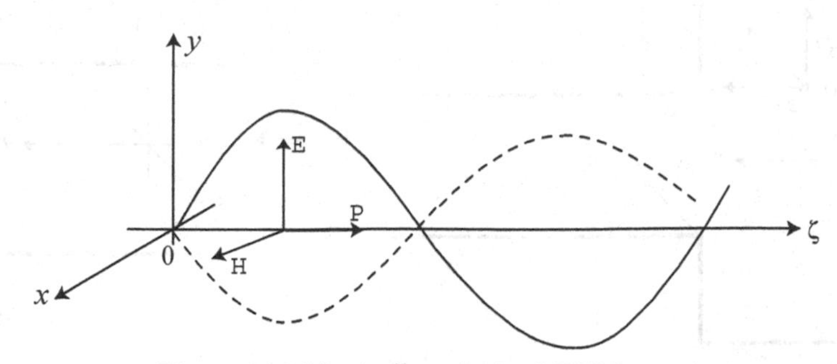

Figure 22.2 Vertically polarized TEM wave.

After further differentiation with respect to ζ and t, and then combining the equations, we obtain the Helmholtz equations

$$
\begin{aligned}
\frac{\partial^2 E_y}{\partial \zeta^2} &= -i\,\omega\mu(\sigma + i\,\omega\varepsilon)E_y, \\
\frac{\partial^2 H_x}{\partial \zeta^2} &= -i\,\omega\mu(\sigma + i\,\omega\varepsilon)H_x.
\end{aligned}
\tag{22.2.2}
$$

The wave propagation along the direction of increasing ζ can be expressed in terms of a complex propagation constant Γ which modifies the wave's amplitude and phase according to the following relations:

$$
E_y = E_{ym}\, e^{i\omega t}\, e^{-\Gamma\zeta}, \quad H_x = H_{xm}\, e^{i\omega t}\, e^{-\Gamma\zeta}. \tag{22.2.3}
$$

Then Eqs (22.2.2) can be written as

$$
\begin{aligned}
\Gamma^2 E_y &= -i\,\omega\mu(\sigma + i\,\omega\varepsilon)E_y, \\
\Gamma^2 H_x &= -i\,\omega\mu(\sigma + i\,\omega\varepsilon)H_x,
\end{aligned}
\tag{22.2.4}
$$

which gives Γ as

$$
\Gamma^2 = -i\,\omega\mu(\sigma + i\,\omega\varepsilon), \quad \text{or} \quad \Gamma = \omega\Big[\mu\varepsilon\Big(1 - \frac{\sigma}{\omega\varepsilon}\Big)\Big]^{1/2}, \tag{22.2.5}
$$

whence the ratio $E_y/H_x \equiv \eta$, known as *intrinsic impedance*, is given by

$$
\eta = \left[\frac{i\,\omega\mu}{\sigma + i\,\omega\varepsilon}\right]^{1/2}. \tag{22.2.6}
$$

Under conformal mapping from the physical z-plane onto the model w-plane, we find that the wave equations are identical to Eqs (22.2.4), except that E_y and H_x are replaced by E_v/M and H_u/M, respectively; but the factor $1/M$ cancels out. Hence, the TEM wave can as well be analyzed in the w-plane provided an appropriate conformal map is selected.

(b) In the case of a transverse mode (TM) wave propagating along the x-axis, the transmission channel is not uniform in the direction of propagation. For example, we may have a parallel plate transmission line which undergoes a step change in plate separation. However, this configuration can be treated with conformal mapping techniques. Let the (y, ζ)-plane be the transverse plane, with propagation along the x-axis (Figure 22.3). Then $H_x = 0$. Also, assume that $E_\zeta = 0 = H_y$. Then E_x components will appear in the vicinity of the change in plate separation.

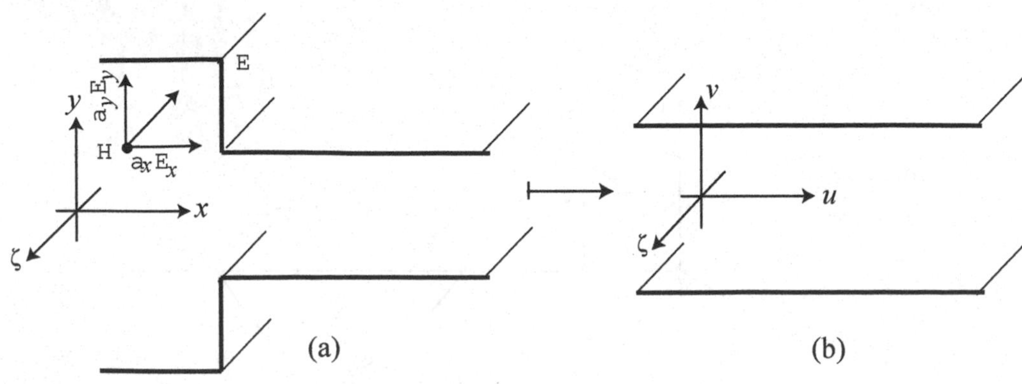

Figure 22.3 TM wave.

By differentiating Eq (22.1.13c) with respect to x and Eq (22.1.15) with respect to y, then adding the equations and using Eq (22.1.14b), we obtain the wave equation in terms of E_y as

$$\frac{\partial^2 E_y}{\partial y^2} + \frac{\partial^2 E_y}{\partial \zeta^2} = -i\,\omega\mu(\sigma + i\,\omega\varepsilon)E_y. \tag{22.2.7}$$

which is equal to $\Gamma^2 E_y$, as in Eq (22.2.4). Other results can be similarly obtained. ■

Note that the scale factor M that occurs in the above equations in the model plane is significant. Suppose that an appropriate conformal transformation is determined which maps the physical (x, y)-plane onto the model (u, v)-plane and that no transformation is required for the third dimension ζ. The same procedure we have used above to derive the equation for E_y will permit us to derive the wave equation for E_v. Thus, from Eqs (22.1.13c), (22.1.15) and (22.1.14b) we find that

$$\frac{\partial^2 E)v}{\partial \zeta^2} + \frac{\partial^2 E_v}{\partial v^2} = i\,\omega\mu m^2(\sigma + i\,\omega\varepsilon)E_v. \tag{22.2.8}$$

Similarly, using Eqs (22.1. 14a), (22.1.14b) and (22.1.13c) we get

$$\frac{\partial^2 H_\zeta}{\partial u^2} + \frac{\partial^2 H_\zeta}{\partial v^2} = i\,\omega\mu m^2(\sigma + i\,\omega\varepsilon)H_\zeta. \tag{22.2.9}$$

22.3 Transmission Lines

A transmission line, e.g., coaxial cables and twisted pair cables, is a pair of electrical conductors carrying an electrical signal from one place to another. The two conductors have inductance per unit length, which is calculated from their size and shape. They have capacitance per unit length, which is calculated from the dielectric constant of the insulation. In the early days of cable making, there would be current leaking through the insulation, but in modern cables, such leakage is negligible. The electrical resistance of the conductors, however, is significant because it increases with frequency. The magnetic fields generated by high-frequency currents drive all currents to the outer edge of the conductor that carries them, so the higher the frequency, the thinner the layer of metal available to carry the current, and the higher the effective resistance of the cable. We will state the two well-known Kirchhoff's laws and then discuss solutions of the equations that govern a transmission line.

There are two theorems that are stated below for ready reference.

(i) **Kirchhoff's Current Law** (KCL theorem), which states that the total current or charge entering a junction or node is exactly equal to the charge leaving the node, i.e., the algebraic sum of all currents entering and leaving a node must be equal to zero. This law is also known as the *conservation of charge.*

(ii) **Kirchhoff's Voltage Law** (KVL theorem), which states that in any closed loop network the total charge around the loop is equal to the sum of all the voltage drops within the same loop, which is equal to zero, i.e., the algebraic sum of all voltages within the loop must be zero. This law is also known as the *conservation of energy.*

A transmission line is a two-port network connecting a generator circuit at the sending end to a load at the receiving line. The length of a transmission line is very important in transmission line analysis. Common types of transmission lines are: (i) coaxial line; (ii) two-wire line; (iii) parallel-line line; (iv) microstrip line; and (v) waveguide.

The following distributed parameters are used to characterize the circuit properties of a transmission line;

R: resistance per unit length (Ω/m);
L: inductance per unit length (H/m);
G: conductance per unit length (S/m);
C: capacitance per unit length (F/m); and
Δz: increment of length (m).

These parameters are related to the physical properties of the material filling the space between two wires, i.e.,

$$LC = \mu\varepsilon, \quad \frac{G}{C} = \frac{\sigma}{\varepsilon}, \tag{22.3.1}$$

where μ, ε, and σ are the permittivity, permeability, and conductivity, respectively, of the surrounding medium.

For the *coaxial* and *two-wire* transmission lines, the above distributed parameters are related to the physical properties and geometrical dimensions as given in Table 22.1.

Table 22.1 Coaxial and Two-Wire Transmission Lines.

Parameter	Coaxial	Two-Wire
R	$\frac{R_s}{2\pi}\left(\frac{1}{a}+\frac{1}{b}\right)$	$\frac{R}{\pi a}$
L	$\frac{\mu}{2\alpha}\ln(b/a)$	$\frac{\mu}{\pi}\ln\left[(d)/(2a)+\sqrt{(d/2a)^2-1}\right]$
G	$\frac{2\pi\sigma}{\ln(b/a)}$	$\frac{\pi\sigma}{\ln\left[(d)/(2a)+\sqrt{(d/2a)^2-1}\right]}$
C	$\frac{2\pi\varepsilon}{\ln(b/a)}$	$\frac{\pi\varepsilon}{\ln\left[(d)/(2a)+\sqrt{(d/2a)^2-1}\right]}$

where R_s denotes the surface resistivity of the conductors.

Example 22.2. Consider a short section Δz of a transmission line shown in Figure 22.4. Using KVL and KCL circuit theorems, we obtain the following differential equations for this section of the transmission line:

$$v(z,t) - R\Delta z\, J(z,t) - L\Delta z \frac{\partial J(z,t)}{\partial t} - v(z+\Delta z,t) = 0,$$
$$J(z,t) - G\Delta z\, v(z+\Delta z,t) - C\Delta z\, \frac{\partial v(z+\Delta z,t}{\partial t} - J(z+\Delta z,t) = 0, \tag{22.3.2}$$

where $J(z,t)$ denotes current. Let $\Delta z \to 0$. Then we obtain a system of two coupled equations

$$-\frac{\partial v(z,t)}{\partial z} = R\, J(z,t) + L\frac{\partial J(z,t)}{\partial t},$$
$$-\frac{\partial J(z,t)}{\partial z} = Gv(z,t) + C\frac{\partial v(z,t)}{\partial t}. \tag{22.3.3}$$

In the case of sinusoidally varying voltages and currents, we use the *phasor forms*

$$v(z,t) = \Re\{V(z)\,e^{i\omega t}\}, \quad J(z,t) = \Re\{I(z)\,e^{i\omega t}\}, \tag{22.3.4}$$

where $V(z)$ and $I(z)$ are called *phasors* of $v(z,t)$ and $J(z,t)$. The coupled equations (22.3.3) can be written in terms of these phasors as

$$-\frac{\partial V(z)}{\partial z} = (R+i\omega L)I(z), \quad -\frac{\partial I(z)}{\partial z} = (G+i\omega C)V(z), \tag{22.3.5}$$

which after decoupling, yield

$$\frac{\partial^2 V(z)}{\partial z^2} = \gamma^2 V(z), \quad \frac{\partial^2 I(z)}{\partial z^2} = \gamma^2 I(z), \tag{22.3.6}$$

where $\gamma = \alpha + i\beta = \sqrt{(R+i\omega L)(G+i\omega C)}$, is called the *complex propagation constant*, with real part α known as the *attenuation constant* (Np/m) and the imaginary part β known as the *phase constant* (rad$/m$), and two quantities are generally functions of ω.

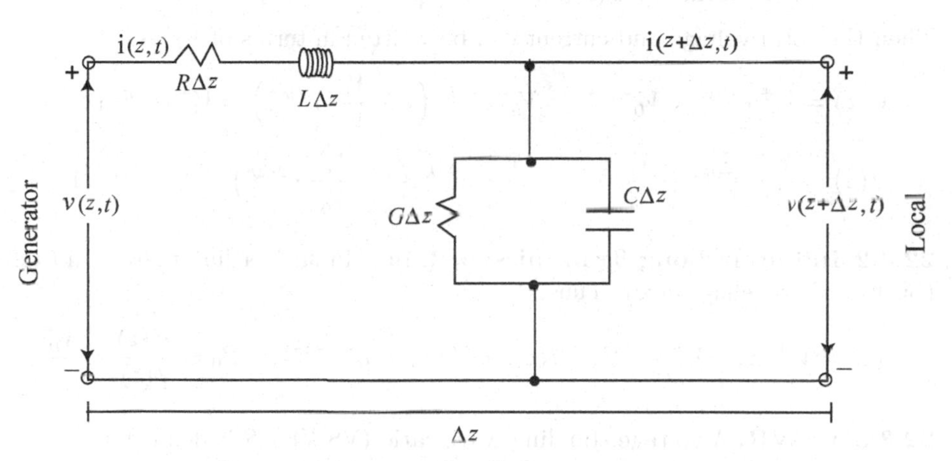

Figure 22.4 Section Δz of a transmission line..

The solutions of the transmission line equations (22.3.6) are

$$V(z) = V^+(z) + V^-(z) = V_0^+\,e^{-\gamma z} + V_0^-\,e^{\gamma z}, \quad I(z) = I^+(z) + I^-(z) = I_0^+\,e^{-\gamma z} + I_0^-\,e^{\gamma z}, \tag{22.3.7}$$

where V_0^+ and I_0^+ are the wave amplitudes of the forward traveling wave at $z=0$ and V_0^- and I_0^- the wave amplitudes of the backward traveling wave. Note that these amplitudes are generally complex numbers.

The solutions (22.3.7) imply that

$$\frac{V_0^+}{I_0^+} = -\frac{V_0^-}{I_0^-} = \frac{R+i\omega L}{\gamma}. \tag{22.3.8}$$

This is known a the *characteristic impedance* Z_0, given by

$$Z_0 = \frac{R+i\omega L}{\gamma} = \frac{\gamma}{G+i\omega C} = \sqrt{\frac{R+i\omega L}{G+i\omega C}}. \tag{22.3.9}$$

Note that Z_0 and γ are the two important parameters of a transmission line; they depend on the distributed parameters R, L, G, C of the line itself and ω, but not the length of the line. ■

22.3.1 Lossless Transmission Lines. For lossless transmission lines, $R = G = 0$, $\alpha = 0$, $\beta = \omega\sqrt{LC} = \omega\sqrt{\mu\varepsilon}$. Then

$$\text{phase velocity } u_p = \frac{\omega}{\beta} = \frac{1}{\sqrt{LC}} = \frac{1}{\sqrt{\mu\varepsilon}},$$

$$\text{complex propagation constant } \gamma = i\beta = i\omega\sqrt{\mu\varepsilon} = 2\pi f i\sqrt{\mu\varepsilon} = i\frac{2\pi}{\lambda} = ki,$$

$$\text{wavelength along the transmission line } \lambda = \frac{u+p}{f} = \frac{1}{f\sqrt{\mu\varepsilon}} = \frac{\omega}{f\beta} = \frac{2\pi}{\beta} = \frac{1}{f\sqrt{LC}},$$

$$\text{characteristic impedance } Z_0 = \sqrt{\frac{R+j\omega L}{G+i\omega C}} = \sqrt{\frac{L}{C}}.$$

Let Γ_L define a reflection coefficient at $z = 0$ as the ratio

$$\Gamma_L = \frac{\text{reflected voltage at } z=0}{\text{incident voltage at } z=0} = \frac{V_0^- \, e^{ik(0)}}{V_0^+ \, e^{ik(0)}} = \frac{V+0^-}{V_0^+} = |\Gamma_L|\, e^{i\theta_L}. \tag{22.3.10}$$

Then the total voltage and current can be written in terms of Γ_L as

$$V(z) = V_0^+ \, e^{-ikz} + V_0^- \, e^{ikz} = V_0^+ \, e^{-ikz}\Big(1 + \frac{V_0^-}{V_0^+}\, e^{2ikz}\Big) = V_0^+ \, e^{-ikz}\left(1 + \Gamma_L \, e^{2ikz}\right);$$

$$I(z) = \frac{V_0^+}{Z_0}\, e^{-ikz} - \frac{V_0^-}{Z_0}\, e^{ikz} = \frac{V_0^+}{Z_0}\, e^{-ikz}\Big(1 - \frac{V_0^-}{V_0^+}\, e^{2ikz}\Big) = I_0^+ \, e^{-ikz}\left(1 - \Gamma_L \, e^{2ikz}\right). \tag{22.3.11}$$

22.3.2 Infinitely Long Transmission Line. In such a line there is no reflected wave (backward traveling wave). Thus,

$$V(z) = V^+(z) = V_0^+ \, e^{-ikz}, \quad I(z) = I^+(z) = I_0^+ \, e^{-ikz}, \quad Z_0 = \frac{V(z)}{I(z)} = \frac{V_0^+}{I_0^+}, \quad \Gamma_L = 0. \tag{22.3.12}$$

22.3.3 VSWR. A voltage standing wave ratio (VSWR) S is defined as

$$S = \frac{|V_0^+|\,(1+|\Gamma_L|)}{|V_0^+|\,(1-|\Gamma_L|)} = \frac{1+|\Gamma_L|}{1-|\Gamma_L|}, \tag{22.3.13}$$

which is a dimensionless quantity, and thus,

$$|\Gamma_L| = \frac{S-1}{S+1}. \tag{22.3.14}$$

The special terminations of transmission lines are presented in Table 22.2.

Table 22.2 Special Terminations.

Γ_L	S	Z_L
0	1	Z_0 (matched)
-1	∞	0 (short-circuited)
1	∞	∞ (open-circuited)

At high frequencies, the wavelength is much smaller than the circuit size, which results in different phases at different locations in the circuit.

22.3.4 Parallel Wire Transmission Lines. There are two geometries involved in transmission lines: the parallel wire line, and the circular coaxial line, and both can conveniently provide reference images in conformal mapping of more complicated transmission line configurations. These two basic lines, (a) parallel wires and (b) circular coaxial cable, are presented in Figure 22.5, with their inductance L and capacitance C defined in §22.3, and the characteristic impedance Z_0 defined by Eq (22.3.9).

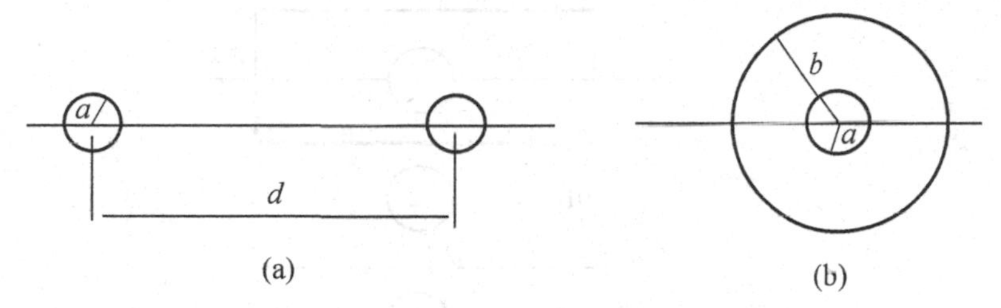

Figure 22.5 Two basic transmission lines.

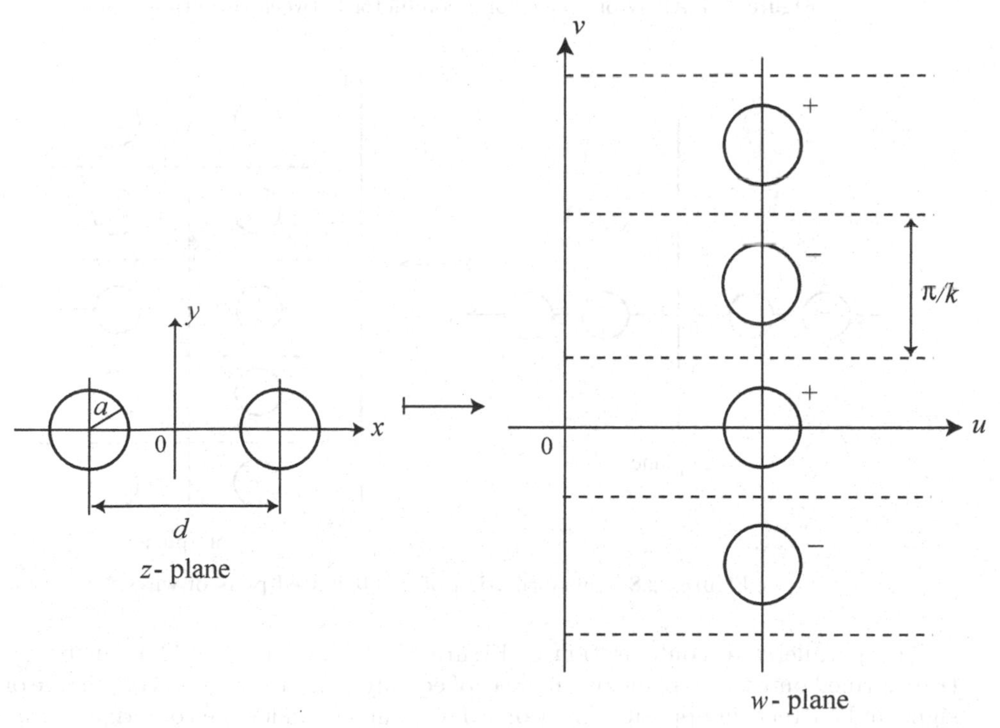

Figure 22.6 Transformation $w = k^{-1} \log z$.

In the case (a), it is assumed at $a \ll d$, the magnetic field inside the wire is neglected, and the line is lossless in air. Then

$$L = \frac{\mu}{\pi} \ln(d/a), \quad C = \frac{\varepsilon\pi}{\ln(d/a)}, \quad Z_0 = 120 \ln(d/a). \tag{22.3.15}$$

In the case (b), for both thin cylindrical shells, or thicker conductors at high frequency, and for lossless line in air, we have

$$L = \frac{2\mu}{\pi}\ln(b/a), \quad C = \frac{2\varepsilon\pi}{\ln(b/a)}, \quad Z_0 = 60\ln(b/a). \tag{22.3.16}$$

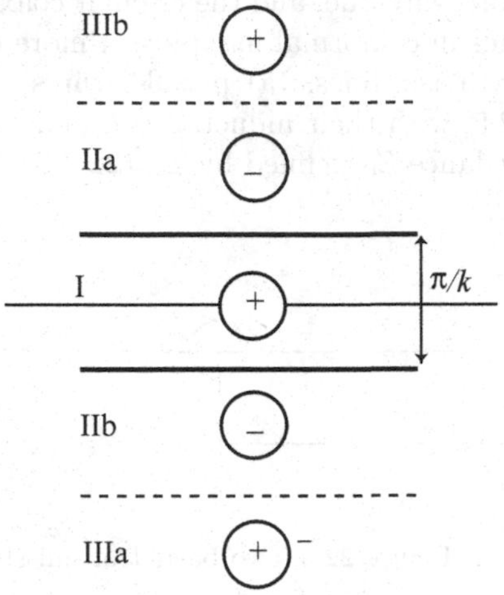

IIIb
+
IIa
−
I
+
π/k
IIb
−
IIIa
+

Figure 22.7 Array of images of a conductor between two ground plates.

y
+
−
0
+
−
x
z- plane
v
0
u
w - plane

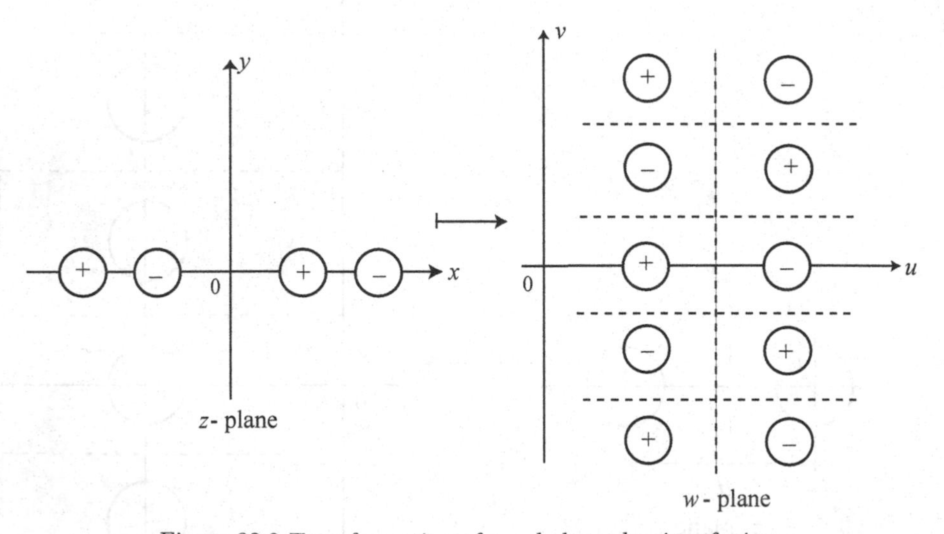

Figure 22.8 Transformation of two balanced pairs of wires.

The parallel wire configuration of Figure 22.5(a) and Figure 22.6 can be conformally transformed onto the w-plane by an array of equally spaced wires carrying charge of opposite signs and currents in opposite directions. For example, under the conformal transformation

$$w = k^{-1}\log z, \quad z = e^{kw}, \text{ where } u = k^{-1}\log(r_z),\ v = \Theta_z/k, \tag{22.3.17}$$

the distances between adjacent wires become $\Delta v = h = \pi/k$, since a wire is encountered with each rotation $\Delta\Theta_z = \pi$ in the z-plane.

The array of wires is the result of producing the image of a wire, or of several wires, between parallel ground plates of infinite extent. For example, consider Figure 22.7, in which, suppose, a positively charged wire is placed half-way between two grounded plates located at $v = \pm\pi/(2k)$. Among other wires beyond the plates, IIa and IIb represent the images of the original wire I; and IIIa is the image of IIa with respect to the upper ground plate; there would follow a IVa as image of IIIa, and a IVb as image of IIIb, and so on. The array thus produced by the conformal mapping allows us to interject the ground plates at lines of symmetry without altering the electric or magnetic fields.

As a result of conformal mapping, the characteristic impedance of a single wire between two parallel plates behaving like return conductors can be determined by taking the radius a small, which can be considered in the z-plane as representing an incremental displacement from the center of the wire, i.e., $a = |dz| = dr_z$. Since $|dz/dw| = k|z|$, the displacement $|dw|$ in the w-plane will be $|dw| = |dx|/|kz| = a/|kz|$. For small radii, z is almost constant for all points around the periphery of each wire at $|z| = d/2$, so that the corresponding radius in the w-plane becomes $b = |dw| = 2a/(kd)$ for small b.

Note that the characteristic impedance Z_0 of the pair of wires I and II is the same as that of, say, the wires I and IIa in the w-plane. However, the capacitance between wire I and the ground plate in the w-plane is twice that of the capacitance between I and II in the z-plane, because the inductances follow the inverse of that ratio. Hence, the characteristic impedance of the single-wire-between-parallel-plates configurations is one half of the parallel wire configuration, i.e.,

$$Z_0 = 60 \ln \frac{d}{a} = 60 \ln \frac{2h}{\pi b}. \tag{22.3.18}$$

Also, the array of two pairs of parallel wires symmetrically arranged on the x-axis conformally mapped onto the w-plane is shown in Figure 22.8, where the electric and magnetic fields have equipotential lines that are indicated by broken lines.

Based on Frankel [1942, 1977], a selection of five cross-sections of shielded wires are presented in Figure 22.9, where it is assumed that the conductors are lossless and thin; $\mu = \mu_0$, $\varepsilon = \varepsilon_0$; and wire radii b are small compared to other dimensions.

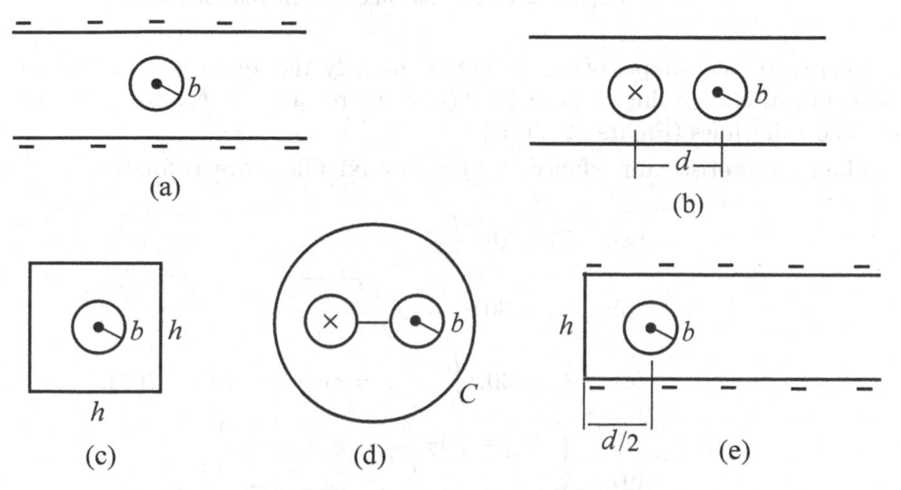

Figure 22.9 Different cross-section of shielded wires.

The characteristic impedances obtained by conformal mapping for these five selections are as follows:

$$\begin{aligned}&(a):\ Z_0 = 60\ln\frac{2h}{\pi b};\quad (b):\ Z_0 = 120\ln\left[\frac{2h}{\pi b}\tanh\frac{\pi d}{2h}\right];\quad (c):\ Z_0 = 60\ln\frac{1.08h}{\pi b};\\ &(d):\ Z_0 = 120\ln\left[\frac{H}{b}\frac{4c^2 - H^2}{4c^2 + H^2}\right];\quad (e):\ Z_0 = 60\ln\ln\left[\frac{2h}{\pi b}\tanh\frac{\pi d}{2h}\right],\end{aligned} \tag{22.3.19}$$

and if $\varepsilon = \varepsilon_r\varepsilon_0$, then divide Z_0 by $\sqrt{\varepsilon_r}$.

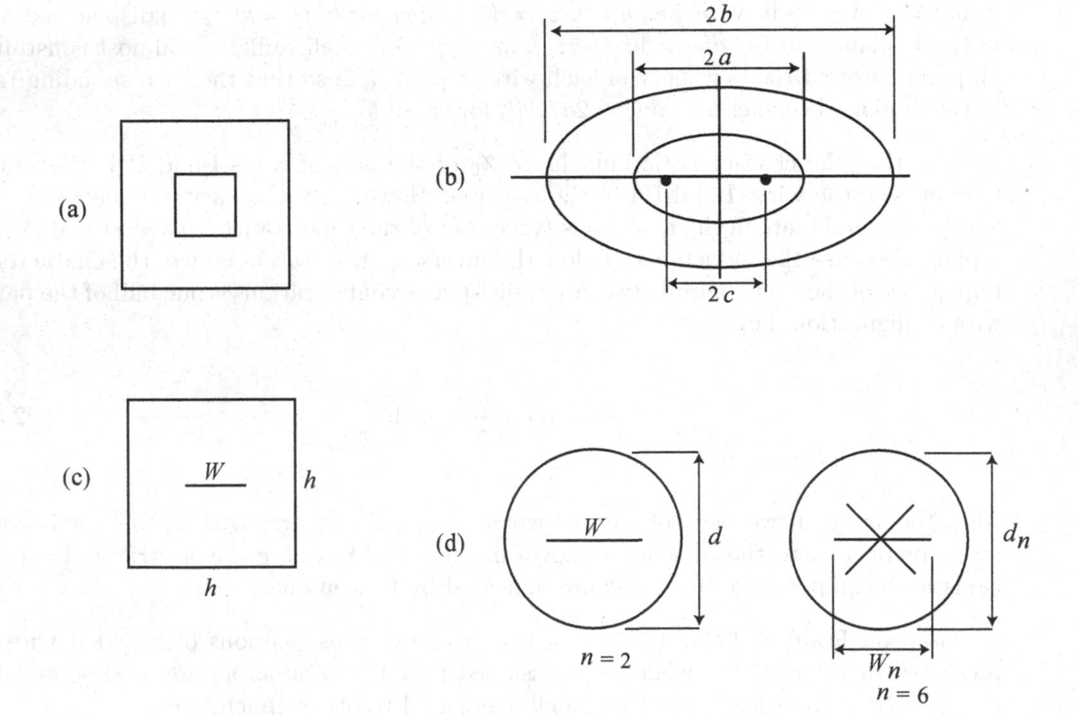

Figure 22.10 Coaxial lines of different shapes.

There are other shapes of coaxial cables, namely, the square coaxial line (Figure 22.10(a)); the confocal elliptic line (Figure 22.10(b)); the rectangular coaxial strip (Figure 22.10(c)); and the n-fin lines (Figure 22.10(d)).

The characteristic impedances for these coaxial lines are as follows:

$$\begin{aligned}&(a):\quad Z_0 = 15\pi\frac{K}{K'};\\ &(b):\quad Z_0 = 60\ln\frac{b+\sqrt{b^2-c^2}}{a+\sqrt{a^2-c^2}};\\ &(c):\quad Z_0 = 30\pi\frac{K}{K'},\quad k = \mathrm{cn}(1.854W/h; 0.707);\\ &(d):\quad \begin{cases} Z_{02} = 30\pi\dfrac{K}{K'}, & k = \dfrac{d^2-W^2}{d^2+W^2};\\ Z_{0n} = \dfrac{2}{n}Z_{02}, & \dfrac{W_n}{d_n} = \dfrac{W}{d},\end{cases}\end{aligned} \tag{22.3.20}$$

where the value for Z_0 for (a) is from Schinzinger and Laura [2003], for (b) and (d) from

Harrington [1961], for (c) from Oberhettinger and Magnus [1949]. Note that coaxial configurations can serve as waveguides for acoustic and electromagnetic waves.

22.3.5 Transmission with Lossy Conductors. We will consider two applications: (i) a cable enclosed in a metal pipe (Figure 22.11(a)), and (ii) a wire above ground (Figure 22.11(b)), which are conformally mapped onto a two conductor coaxial line (Figure 22.11(c)). This application has been studied by Stratton [1941], and Schinzinger and Ametani [1978].

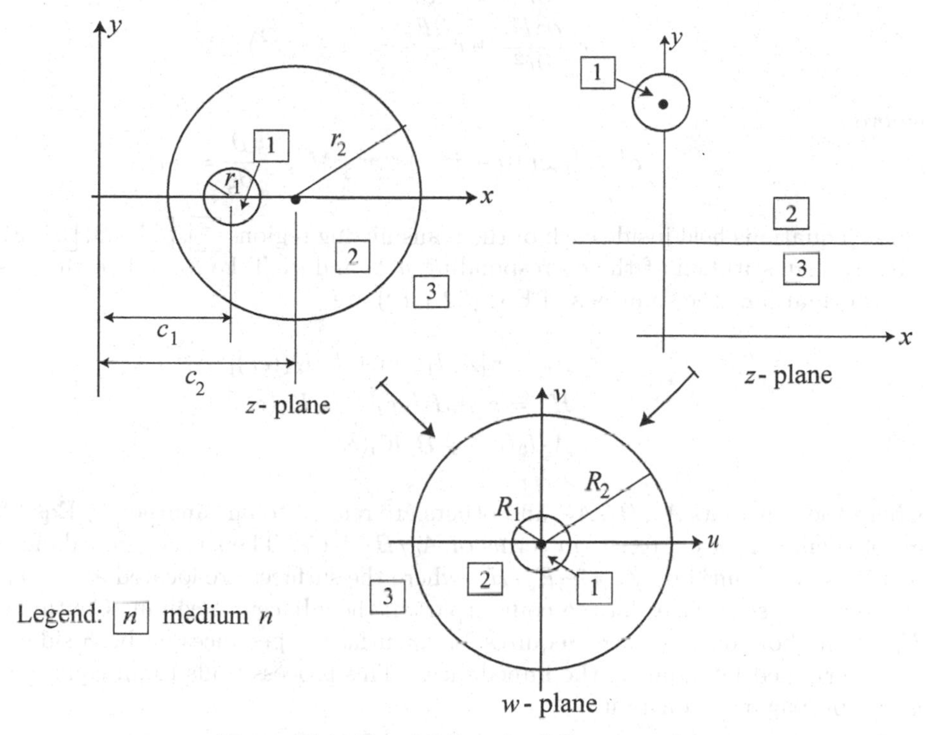

Figure 22.11 Coaxial lines in different media.

The series impedance is

$$Z = Z_1 + Z_2 + Z_3, \quad \text{where}$$

$$Z_1 = (i+i)\frac{1}{2\pi r_1 \delta_1 \sigma_1}\frac{I_0(\gamma_1 r_1)}{I_1(\gamma_1 r_1)} = \text{'inner' impedance of interior conductor;} \tag{22.3.21}$$

$$Z_2 = i\omega \frac{\mu_2}{2\pi} \ln \frac{r_2}{r_1} = \text{inductive reactance due to annular field;} \tag{22.3.22}$$

$$Z_3 = (i+i)\frac{1}{2\pi r_2 \delta_3 \sigma_3}\frac{K_0(\gamma_3 r_3)}{K_1(\gamma_3 r_3)} = \text{'inner' impedance of outer conductor;} \tag{22.3.23}$$

and

$$\delta = 2/\sqrt{\omega\mu\sigma} \quad \text{depth of penetration;}$$

$$\gamma = \begin{cases} i\,\omega\mu(\sigma + i\,\mu\sigma) - r^2 & \text{for intrinsic propagation constant} \\ \sqrt{i\omega\mu\sigma} & \text{for conductors} \end{cases};$$

I_0 and K_0 are the modified Bessel functions of the first and second kind, respectively. Moreover, the shunt admittance $Y = i\,\omega 2\pi/\ln(r_2/r_1)$, which is the capacitive admittance of the annulus; and the propagation constant $\Gamma = \sqrt{ZY} = Z/Z_0$.

In the polar coordinate system where $z = r\,e^{i\theta}$, Maxwell's equations are

$$\begin{aligned} r^2\frac{\partial^2 E_\varsigma}{\partial r^2} + r\frac{\partial E_\varsigma}{\partial r} &= c^2 r^2 E_\varsigma, \\ r^2\frac{\partial^2 H_\theta}{\partial r^2} + r\frac{\partial H_\theta}{\partial r} &= \left(1 + c^2 r^2\right) H_\theta, \\ r^2\frac{\partial^2 E_r}{\partial r^2} + r\frac{\partial E_r}{\partial r} &= \left(1 + c^2 r^2\right), \end{aligned} \tag{22.3.24}$$

where

$$c^2 = \left[i\,\omega\mu(\sigma + i\,\omega\varepsilon) - r^2\right] M^2, \quad \frac{\partial}{\partial\varsigma} = -r.$$

These equations hold inside each of the transmitting regions $\boxed{1}$, $\boxed{2}$, and $\boxed{3}$ (Figure 22.11), with the substitution of the corresponding μ, ε, and σ. Take $M = 1$ in the absence of any transformation. The solutions of Eqs (22.1.19) are

$$\begin{aligned} E_r &= r\left[A_r I_1(cr) + B_r K_1(cr)\right], \\ H_\theta &= r\left[A_\theta I_1(cr) + B_\theta K_1(cr)\right]; \\ &A_\varsigma I_0(cr) + B_\varsigma K_0(cr), \end{aligned} \tag{22.3.25}$$

where the constants A_r, B_r, A_ς, and others are related to one another by Eqs (22.3.15), and all of them can be expressed in terms of $A_\varsigma/B_\varsigma \equiv C_\varsigma$. The surface impedance in the three media is determined by $Z_s = -E_\varsigma/H_\theta$, where the surfaces are located at $r = r_1$ and $r = r_2$. However, these surfaces have a counterpart in the adjacent medium. Continuity of E_ς and H_θ at the boundary surfaces requires equal surface impedances on both sides, and $+\varsigma$ can be determined by equating the impedances. This process leads to an equation for Γ which is the propagation constant.

The series impedance Z can be determined from $Z = rz_0 = rV'J$, where

$$V = \int_{r_1}^{r_2} E_r\,dr, \quad J = 2\pi r H_\theta(r),$$

where these quantities assume propagation in the TEM mode, i/e., E_ς is neglected. Since the values of E_ς are relatively very small, the computed results are very close to the experimental values.

22.3.6 Pipe-Enclosed Cable. Consider the cable shown in an off-center position in Figure (22.10(a)). A bilinear transformation of the form (see Map 3.15)

$$w = \frac{z - a}{z + a},$$

where

$$s^2 = c_1^2 - r_1^2, \quad c_{1,2} = \left(r_2^2 - r_1^2 \mp b^2\right)/(2b), \quad R_{1,2} = r_{1,2}/\left(c_{1,2} + s\right),$$

will map this cable in the z-plane onto two concentric circles in the w-plane (Figure 22.10(c)).

22.3.7 Homogeneous Strip Lines. These lines consist of two or more parallel thin conductors, with rounded edges, arranged face-to-face or side-by-side. In the case of a single homogeneous strip lines, the dielectric is assumed to be air (of permittivity $\varepsilon = \varepsilon_0$). The characteristic impedance is

$$Z_0 = \frac{\mu_0 \varepsilon_0}{C} = \frac{\varepsilon_0}{C}\sqrt{\mu_0/\varepsilon_0} = 120\pi\varepsilon_0/C.$$

Any conformal transformation to the geometry of the strip lines should be applied to a repeatable segment of the strip line cross-section. This will allow one to obtain the total impedance of the strip line as a parallel combination of Z_0 of the segment.

Example 22.3. Consider the cases of two plates, (a) and (b), in Figure 22.12, where the ground plate in (b) is regarded as a plane of symmetry halfway between the strips of (a). We use the Schwarz-Christoffel transformation and obtain, using points in the T-plane, the integral and its general solution

$$z = Ck \int_0^1 \frac{t^2 - \lambda^2}{[(1-t^2)(1-k^2t^2)]^{1/2}}\, dt = \frac{C}{k}\left[(1-k^2\lambda^2)F(k,t) - E(k,t)\right], \tag{22.3.26}$$

where F and E are elliptic integrals of the first and second kind, respectively. The constants C, k, and λ are determined from the function relationships which direct the solution by numerical methods; thus, for example, for $W = h = 2h_a = 1$, we get $k = 0.14337, \lambda = 3.8716, C = -0.30491$ (Durrand [1966:317]).

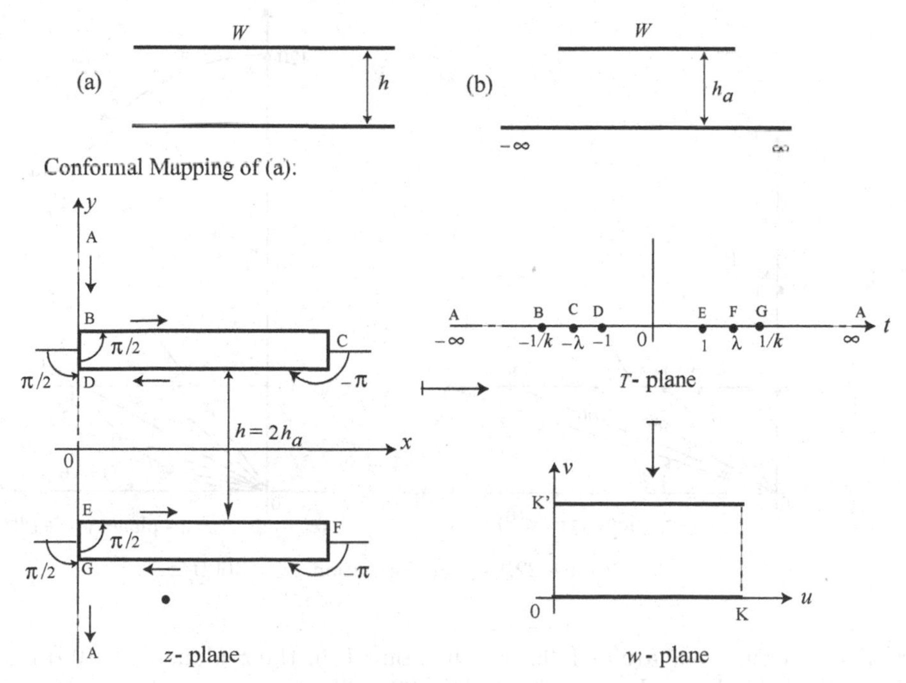

Figure 22.12 Symmetrical segments of two plates.

Using the logarithmic derivative of the theta function $\theta_1$and its parameter K'/K, Schneider [1969] mapped the twin plates in the Z-plane directly onto the w-plane, and obtained

$$z = -h\frac{K}{\pi}\frac{\partial}{\partial \omega} \ln \theta_1(\omega, K'/K).$$

In case (b), since $\dfrac{W}{h} = \dfrac{2}{\pi}\dfrac{\partial}{\partial\xi}\ln\theta_4(\xi, K'/K)$ and $\text{dn}^2(2K\xi) = E/K$, we get the characteristic impedance $Z_0 = 60\pi K'/K$, where after substituting $h = 2h_a$ and $Z_0 = 120\pi K'/K$, we will find Z_0 for the case (a). ■

Other cases of homogeneous (single dielectric) strip lines, namely, balanced, even and odd mode coupled and slotted, are discussed in Schinzinger and Laura [2003: 292ff], using the Schwarz-Christoffel transformations.

The selection of inhomogeneous strip lines with two dielectrics and quasi-TEM mode is presented in a tabular form in Schinzinger and Laura [2003: 304-305]; it describes microstrip lines, shielded striplines, and coplanar waveguides in the z-plane. Using the Schwarz-Christoffel transformations, these case are studied there and the characteristic impedances Z_0 are determined in each case.

22.4 Conformal Mapping and Electric Transmission

An application of the transformation $w = k/z$ (Map 3.10) is used in electric transmission through a single wire, and discussed in the following example.

Example 22.4. (Schinzinger and Laura [2003: 40]) It is an application to a single wire transmission line with ground return situated at $x = 0, y = 60$, and running parallel to the ground plane at $y = 30$. The cross-section of this geometry is shown on the z-plane in Figure 22.13.

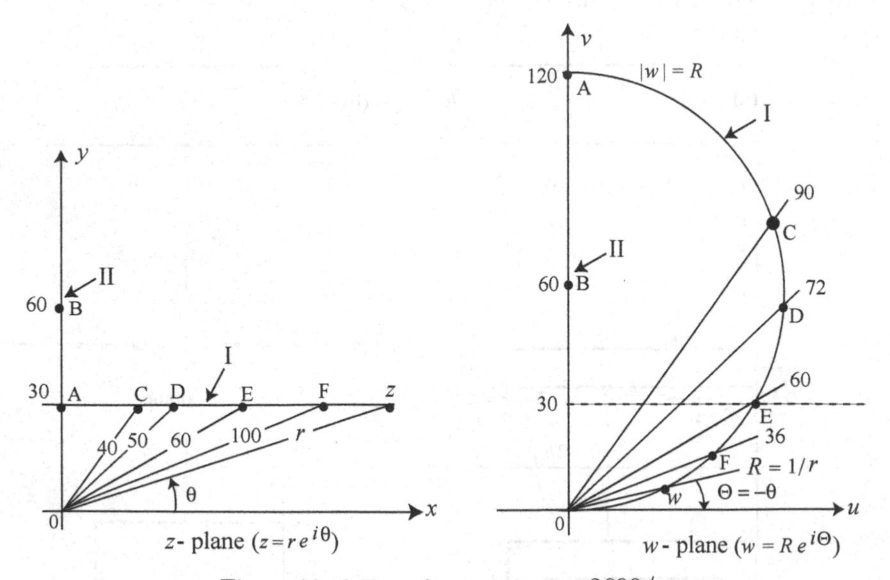

Figure 22.13 Transformation $w = 3600/z$.

The computations are as follows: The point F in the z-plane ($r = 100$) is mapped onto the point F in the w-plane at $R = 3600/100 = 36$ at an angle which is negative of that in the z-plane, since the positive direction of Θ is taken clockwise in the w-plane. In the same manner, the points A, B, C, and D in the z-plane are mapped onto the same points in the w-plane: Point E in the z-plane has $R = 3600/60 = 60$; point D has $R = 3200/50 = 72$; point C has $R = 3600/40 = 90$; point A has $R = 3600/120 = 30$, so that the point B is at $R = 60$. Thus, in the w-plane $R = 3600/r$ and $\Theta = -\theta$. These results are shown on the w-plane, where $w = Re^{i\Theta}, R = 1/r$. All the surface points on the wire are at $z = (0 + 60\,i) + R\,e^{i\Theta}$.

Since $R = k/r^2 = 3600/60^2 = 1$, the magnification of the transformation is $m = 1$ for the entire wire cross-section, implying that the wire itself will not change in size if its radius is small enough. Finally, note that it is an isogonal mapping and not a conformal one, i.e., the magnitude of an angle will be preserved but not its sense of rotation. This problem is solved in Schinzinger and Laura [2003: 40-43], with $k = 6400$.

We will reformulate this application for the transformation of the wire under the assumption that the cross-section of the wire is a circle of radius R and centered at the point B (Figure 22.14). Then the points on the surface of the wire are given by

$$R = |z - ic| = |k/w - ic| = \left|\frac{k}{u+iv} - ic\right| = \left|\frac{k + cv - icu}{u+iv}\right|,$$

thus, giving

$$R^2 = \frac{(k+cv)^2 + c^2u^2}{u^2+v^2}, \tag{22.4.1}$$

which can be rearranged as

$$u^2 + \left(v + \frac{kc}{c^2 - R^2}\right)^2 = \frac{k^2R^2}{(c^2 - R^2)^2}. \tag{22.4.2}$$

The image of this circle in the w-plane is a circle with radius $\frac{kR}{c^2 - R^2}$, centered at $(u, v) = \left(0, -\frac{kc}{c^2 - R^2}\right)$. If we let $k = c^2 - R^2$, the center of the image of the wire will be at $(0, -c)$. If R is very small, we may have $k = c^2$, and the wire in the w-plane will have radius $\frac{R}{1 - R^2/c^2} = \frac{c^2R}{c^2 - R^2} = \frac{kR}{c^2 - R^2} = R$.

The transformation of the ground plane $y = g$ is obtained from $z = k/w$, which gives

$$z = x + iy = \frac{k}{u+iv} = \frac{k(u - iv)}{u^2+v^2}.$$

Thus,

$$y = -\frac{v}{u^2+v^2}, \tag{22.4.3}$$

which is rearranged with $y = g$ as

$$u^2 + \left(v + \frac{k}{2g}\right)^2 = \left(\frac{k}{2g}\right)^2. \tag{22.4.4}$$

This is another circle centered at $v = -k/(2g)$.

If we want the centers of the circle of the wire (22.4.2) and of the ground plane (22.4.4) to coincide, the origin of the z-plane must be placed such that $\frac{kc}{R^2 - c^2} = -\frac{k}{2g}$, or

$$g = \frac{c}{2} - \frac{R^2}{2c}. \tag{22.4.5}$$

We can describe the physical model of this transmission in terms of the height h of the wire above ground, i.e., we take $h = c - g$. Then

$$h = c - g = c - \left(\frac{c}{2} - \frac{R^2}{2c}\right) = \frac{c^2 + R^2}{2c}. \tag{22.4.6}$$

Since Eq (22.4.6) is $c^2 - 2hc + R^2 = 0$, its roots are $c = h \pm \sqrt{h^2 - R^2}$. Hence, the proper value of c is $c = h + \sqrt{h^2 - R^2}$, which for a wire of very small radius ($R \ll h$), gives $c = 2h$. The general transformation with $k = c^2 - R^2$ is presented in Figure 22.14, where, for given R and h ($R \ll h$), we have $c = h + \sqrt{h^2 - R^2} \approx 2h$, $g = c - h \approx h$, and $k = c^2 - R^2 \approx c^2$. If we take $c = 60$ and a very small R, we have $k = 3600$, which takes us back to Example 22.4. ■

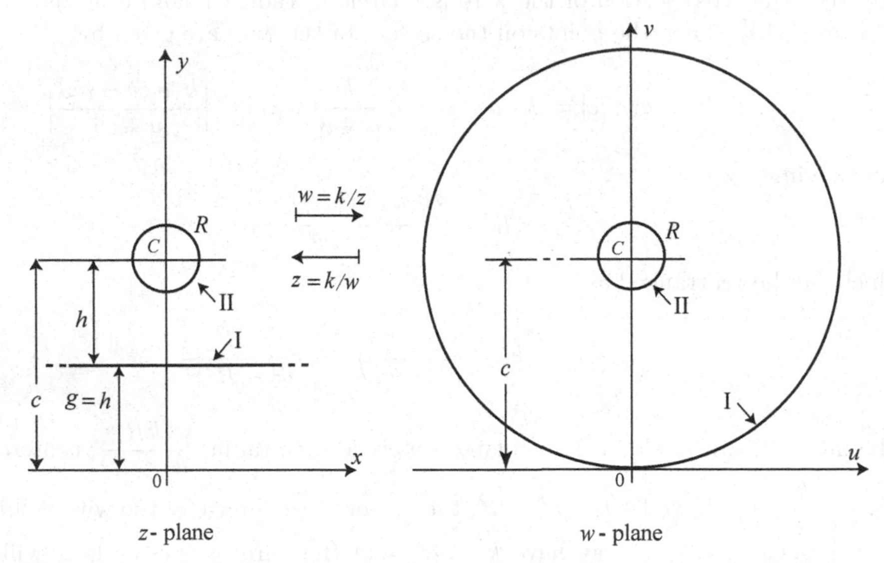

Figure 22.14 Transformation of a circle and a line into two concentric circles.

22.5 Conformal Mapping and Waveguides

While discussing waveguides for electromagnetic wave propagation at microwave and radio wave frequencies, the term *waveguide* in electromagnetics and communication engineering refers to any linear structure that conveys electromagnetic waves between its endpoints. However, the original and most common meaning is a hollow metal pipe used to carry radio waves. This type of waveguide is used as a transmission line mostly at microwave frequencies, for such purposes as connecting microwave transmitters and receivers to their antennas, in equipment such as microwave ovens, radar sets, satellite communications, and microwave radio links.

A dielectric waveguide uses a solid dielectric rod rather than a hollow pipe. An optical fiber is a dielectric guide designed to work at optical frequencies. Transmission lines such as the microstrip, coplanar waveguide, stripline or coaxial cable may be considered to be waveguide.

The electromagnetic wave in a metal-pipe waveguide may be imagined as traveling down the guide in a zig-zag path, being repeatedly reflected between opposite walls of the guide. In the particular case of a rectangular waveguide, it is possible to base an exact analysis on this point of view. Propagation in a dielectric waveguide may be viewed in the same way, with the waves confined to the dielectric by total internal reflection at its surface. Some structures, such as non-radioactive dielectric waveguides and the Goubau line, use both metal walls and dielectric surfaces to confine the wave.

Depending on the frequency, waveguides can be constructed from either conductive or dielectric materials. Generally, the lower the frequency to be passed the larger the waveguide is. For example, the natural waveguide the earth forms given by the dimensions between the conductive ionosphere and the ground as well as the circumference at the median altitude of the Earth is resonant at 7.83 Hz (known as *Schumann resonance*). On the other hand, waveguides used in extremely high frequency (EHF) communications can be less than a millimeter in width.

Electromagnetic waveguides are analyzed by solving Maxwell's equations, or their reduced form, the electromagnetic wave equation, with boundary conditions determined by the properties of the material and their interfaces. These equations have multiple solutions, or modes, which are eigenfunctions of the system of equations. Each mode (solution) is characterized by a cutoff frequency below which the mode cannot exist in the guide. Waveguide propagation modes depend on the operating wavelength and polarization and the shape and size of the guide. The longitudinal mode of a waveguide is a particular standing wave pattern formed by waves confined in the cavity. The transverse modes are classified into different types: (i) transverse electric (TE) modes which have no electric field in the direction of propagation; (ii) transverse magnetic (TM) modes which have no magnetic field in the direction of propagation; (iii) transverse electromagnetic (TEM) modes which have no electric and magnetic fields in the direction of propagation; and (iv) hybrid modes which have both electric and magnetic field components in the direction of propagation.

In hollow waveguides (single conductor), TEM waves are not possible, since Maxell's equations require that the electric field must then have zero divergence and zero curl and be equal to zero at the boundaries, resulting in a zero field, or equivalently $\nabla^2\Phi = 0$ with boundary conditions guaranteeing only the trivial solution. This contrasts with two-conductor transmission lines used at low frequencies; coaxial cable, parallel wire line and stripline, in which TEM mode is possible. Moreover, the propagating mode (i.e., TE and TM) inside the waveguide can be mathematically expressed as the superposition of TEM waves (Chakravorty [2015]).

The mode with the lowest cutoff frequency is called the *dominant mode* of the guide. It is common to choose the size of the guide such that only this one mode can exist in the frequency band of operation. In rectangular and circular (hollow pipe) waveguides, the dominant modes are designated the $\text{TE}_{1,0)}$ mode and $\text{TE}_{1,1}$ modes, respectively (Modi and Balanis [2016]).

Waveguides in practice act like equivalents of cables for high-frequency systems. For such applications, it is desired to operate waveguides with only one mode propagating through the waveguide. With rectangular waveguides, it is possible to design the waveguide such that the frequency band over which only one mode propagates is as high as $2:1$ (i.e., the ratio of the upper band edge to lower band edge is 2). The relation between the waveguide dimensions and the lowest frequency is: If W is the greater of the two dimensions, then the longest wavelength that will propagate is $\lambda = 2W$ and the lowest frequency is thus $f = c/\lambda = c/(2W)$. With circular waveguides, the highest possible bandwidth allowing only a single mode to propagate is only $1,3601 : 1$. (Modi and Balanis [2016]). Since rectangular waveguides have a much larger bandwidth over which only a single mode can propagate, there are standards for rectangular waveguides, but for circular ones. A list of standard waveguides is available in Fuller [1969]. For dielectric waveguides, see Rana [2005].

22.6 Helmholtz Equation and Rib-Shaped Waveguide

The Helmholtz equation, also known as the *scalar wave equation*, deals with the electromagnetic distribution under certain approximation in a given system. We will find a solution of the Helmholtz equation in complex geometries using conformal mapping and the homotopy perturbation method (He [1999]).

In the optical system if the refractive index contrast is small and the field is independent of polarization, we use the scalar wave equation

$$\nabla^2 \phi(\mathbf{r}) + \left(n_r^2 k_0^2 - \beta^2\right) \phi(\mathbf{r}) = 0, \tag{22.6.1}$$

where ϕ is the electromagnetic field, $\mathbf{r}$ is the position vector, n_r is the refractive index, k_0 is the vacuum wave number, and β is the propagation constant. Eq (22.6.1) can be solved using different methods, such as Green's function method (see Kythe et al. [2003: 234], and Kythe [2011: 219]), the discrete spectral-index method (Ng and Stern [1998]), and the beam propagation method (Shih and Chao [2008]).

The model field $\phi(x, y)$ for an anti-resonant reflecting optical waveguide (ARROW) structure is described by the Helmholtz equation (22.6.1), which is rewritten as

$$\frac{\partial^2 \phi(x,y)}{\partial x^2} + \frac{\partial^2 \phi(x,y)}{\partial y^2} + \lambda^2 \phi(x,y) = 0, \quad \lambda^2 = n_r^2 k_0^2 - \beta^2. \tag{22.6.2}$$

We will solve this equation using the anti-resonance boundary condition $\phi(\Gamma) = 0$, where Γ is the boundary. This boundary condition is valid at a single wavelength; however, it provides a very good approximation for well-confined modes at wavelengths close to anti-resonance.

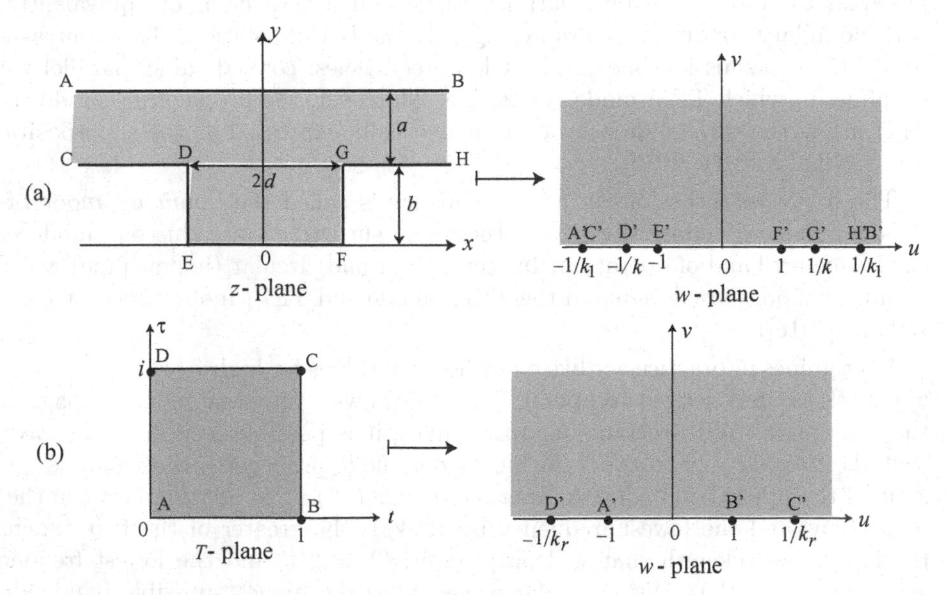

Figure 22.15 ARROW waveguide cross-section.

The conformal map is the Schwarz-Christoffel transformation that maps the half-plane $\Im\{w\} > 0$ onto the polygon in the z-plane (see Figure 22.15(a)), defined as the inverse mapping of in Map 7.1 by

$$z(w) = A \int_0^w \prod_{j=1}^n (w - a_j)^{-\alpha_j} \, dw, \tag{22.6.3}$$

where A is a scaling factor, n is the number of sides of the polygon, and $\alpha_j \pi$ are the external angles at the vertices a_k of the polygon for $k = 1, 2, \dots, n$.

The next conformal mapping of a unit square in the T-plane onto the upper half-plane $\Im\{w\} > 0$ (Figure 22.15(b)) is obtained from the Jacobi elliptic integral of the first kind and is given by (see Map 7.7)

$$\begin{aligned} T(w) - T(0) &= C_{\text{sq}} \int_0^w \frac{dw}{\sqrt{1-w^2}\sqrt{1-k_r^2 w^2}} = \frac{1}{2K(k_r)} \int_0^w \frac{d\theta}{\sqrt{1-\theta^2}\sqrt{1-k_r^2\theta^2}} \\ &= \frac{\operatorname{arcsn}(w, k_r)}{2K(k_r)}, \end{aligned} \tag{22.6.4}$$

where $T(0) = \frac{1}{2}$, $K(k_r)$ is the complete elliptic integral of the first kind of modulus k_r, $\operatorname{arcsn}(w, k_r)$ is the inverse of the Jacobi elliptic sine amplitude $\operatorname{sn}(z, k_r)$, both of modulus k_r, and the scaling factor $C_{\text{sq}} = [2K(k_r)]^{-1}$ is determined by the length of the base line AB (Figure 22.15(b)) of the unit square, since $T_{\text{B}} - T_{\text{A}} = C_{\text{sq}}\, 2K(k_r) = 1$. The modulus k_r controls the aspect ratio of the rectangle since $T_{\text{D}} - T_{\text{A}} = C_{\text{sq}}\, (i\, K'(k_r)) = i\, K'(k)/ [2K(k_r)] = i$, where $K'(k_r) = K(k_r') = K\left(\sqrt{1-k_r^2}\right)$. These results imply that $k_r \approx 0.17157$ is needed for an aspect ratio of 1. From Eq (22.6.4) we get the inverse map as

$$w(T) = \operatorname{sn}\left[(2T-1)\, K(k_r), k_r\right]. \tag{22.6.5}$$

Next, the mapping (22.6.3) from the upper half-plane $\Im\{w\} > 0$ onto the rib-shaped waveguide in the z-plane (Figure 22.15(a)), is given by

$$z(w) = C \int_0^w \frac{\sqrt{1-k^2w^2}}{(1-k_1^2w^2)\sqrt{1-w^2}}\, dw = \frac{2ak_1}{\pi} \frac{\sqrt{1-k_1^2}}{\sqrt{k^2-k_1^2}} \int_0^w \frac{\sqrt{1-k^2\theta^2}}{(1-k^2-k_1^2)\sqrt{1-\theta^2}}\, d\theta, \tag{22.6.6}$$

where $w = 0$ is mapped onto $z = 0$. The scaling factor C is determined by requiring that the integral must increase by $\Delta z = a\, i$ when w passes $1/k_1$ between the points H and B. According to Gibbs [1958], this integral can be expressed in terms of the Jacobi elliptic integral of the third kind, $\Pi(\zeta, \alpha, k)$, in terms of two parameters ζ and α (see Appendix F). Then defining ζ from $w = \operatorname{sn}(\zeta)$ and α from $k_1 = k \operatorname{sn}(\alpha)$, and using a change of variables $\vartheta = \operatorname{sn}(\theta)$, which gives $d\vartheta = \operatorname{cn}(\theta)\, d\theta$, the integral (22.6.6) becomes

$$z(\zeta) = \frac{2a}{\pi} \frac{\operatorname{sn}(\alpha)\operatorname{dn}(\alpha)}{\operatorname{cn}(\alpha)} \int_0^\zeta \frac{1-k^2\operatorname{sn}^2\vartheta}{1-k_1^2\operatorname{sn}^2\vartheta}\, d\vartheta = \frac{2a}{\pi}\Big(\frac{\operatorname{sn}(\alpha)\operatorname{dn}(\alpha)}{\operatorname{cn}(\alpha)}\, \zeta - \Pi(\zeta, \alpha)\Big), \tag{22.6.7}$$

where all Jacobi elliptic functions are of modulus k, and the constants k and α are determined by the aspect ratios of the rib-shaped structure by considering mapping of the point G (Figure 22.15(a)) and separating real and imaginary parts, which gives

$$\frac{d}{a} = \frac{2K(k)}{\pi}\Big(\frac{\operatorname{sn}(\alpha)\operatorname{dn}(\alpha)}{\operatorname{cn}(\alpha)} - Z(\alpha, k)\Big), \tag{22.6.8}$$

$$\frac{b}{a} = \frac{2K'(k)}{\pi}\Big(\frac{\operatorname{sn}(\alpha)\operatorname{dn}(\alpha)}{\operatorname{cn}(\alpha)} - Z(\alpha, k)\Big) - \frac{\alpha}{K(k)}, \tag{22.6.9}$$

where $Z(\alpha, k)$ is the Jacobi zeta function. Then using Eqs (22.6.7), (22.6.8) and (22.6.9), the rib-shaped waveguide of different dimensions d, a, b can be easily mapped onto the upper half-plane $\Im\{w\} > 0$ (Figure 22.15(a)), which is then mapped onto the unit square (Figure 22.15(b)), although the Helmholtz equation becomes nonlinear under these two conformal maps. To see this, assume that the potential in the physical z-plane $\phi(z) = \phi(x, y)$ and the potential in the model T-plane $\psi(T) = \psi(\xi, \eta)$ are related by $\psi(\xi(x,y), \eta(x,y)) = \phi(x,y)$. Then the Laplace operator becomes

$$\nabla^2_{x,y}\phi(x,y) = \nabla^2_{\xi,\eta}\psi(\xi,\eta)\Big|\frac{dT}{dz}\Big|^2,$$

and thus, the Helmholtz equation to be solved in the mapped domain, with $T = \xi + i\eta$, becomes

$$\frac{\partial^2\psi(\xi,\eta)}{\partial\xi^2} + \frac{\partial^2\psi(\xi,\eta)}{\partial\eta^2} = \Big|\frac{dT}{dz}\Big|^{-2}\lambda^2\psi(\xi,\eta) = 0, \tag{22.6.10}$$

where for the conformal transformation from the rib-shaped waveguide onto the unit square,

$$\frac{dT}{dz} = \frac{dT}{dw}\frac{dw}{dz} = \frac{\pi}{4aK(k_r)}\frac{\mathrm{cn}(\alpha)}{\mathrm{sn}(\alpha)\,\mathrm{dn}(\alpha)}\frac{1-k_1^2w^2}{\sqrt{1-k_r^2w^2}\sqrt{1-k^2w^2}}. \tag{22.6.11}$$

22.6.1 Homotopy Perturbation Method. The perturbation methods provide approximate solutions for boundary value and initial value problems. These methods are used when such problems contain a small parameter, say ε, and the solution for $\varepsilon = 0$ is known. This parameter occurs, in general, in a partial differential equation of the form

$$L(u) + \varepsilon N(u) - g(\mathbf{r}) = 0, \quad \mathbf{r} \in D, \tag{22.6.12}$$

with boundary conditions $B\left(u, \dfrac{\partial u}{\partial n}\right) = 0$ on the boundary $\Gamma \equiv \partial D$, where L is a linear partial differential operator, N is either a nonlinear or a linear differential operator, and $g(\mathbf{r})$ is an analytic function, which makes the solution of Eq (22.6.12) difficult. If, by taking $\varepsilon = 0$, Eq (22.6.12) reduces to an ordinary differential equation, then the perturbation method fails. Another kind of perturbation problems arises by perturbing the boundary. In this case the parameter ε will appear in the boundary conditions.

Following Reck et al. [2011], and using the concept of homotopy,[1] we can set up a homotopic equation of the form

$$H(v,p) = (1-p)\,[L(v) - L(u_0)] + p\,[A(v) - g(\mathbf{r})] = 0, \quad p \in [0,1], \tag{22.6.13}$$

where $A(v)$ is a general differential operator, which can be separated into a linear and a nonlinear operator L and N, respectively, and u_0 is an initial approximation satisfying the boundary conditions, and $u = \sum\limits_{n=0}^{\infty} v_n p^n$. Then the solution of Eq (22.6.12) is $u = \lim\limits_{p\to 1} v = \sum\limits_{n=0}^{\infty} v_n$. For Eq (22.6.10) the following homotopy can be constructed:

$$H = (1-p)\Big(\frac{\partial^2\psi}{\partial\xi^2} + \frac{\partial^2\pi}{\partial\eta^2} - \frac{\partial^2\psi_0}{\partial\xi^2} - \frac{\partial^2\psi_0}{\partial\eta^2}\Big) + p\Big(\frac{\partial^2\psi}{\partial\xi^2} + \frac{\partial^2\psi}{\partial\eta^2} + \Big|\frac{dT}{dz}\Big|^{-2}\lambda^2\psi\Big) = 0, \tag{22.6.14}$$

[1] In topology, homotopy is defined as a continuous transformation of one continuous function to another if one can be 'continuously deformed' into the other.

which reduces to Eq (22.6.10) for $p = 1$. By equating terms of identical powers of p we obtain the following set of equations:

$$
\begin{aligned}
p^0 &: \quad \psi_0, \\
p^1 &: \quad \frac{\partial^2 \psi_0}{\partial \xi^2} + \frac{\partial^2 \psi_0}{\partial \eta^2} + \left|\frac{dT}{dz}\right|^{-2} \lambda^2 \psi_0 + \frac{\partial^2 \psi_1}{\partial \xi^2} + \frac{\partial^2 \psi_1}{\partial \eta^2} = 0, \\
p^2 &: \quad \left|\frac{dT}{dz}\right|^{-2} \lambda^2 \psi_1 + \frac{\partial^2 \psi_2}{\partial \xi^2} + \frac{\partial^2 \psi_2}{\partial \eta^2} = 0, \\
p^3 &: \quad \left|\frac{dT}{dz}\right|^{-2} \lambda^2 \psi_2 + \frac{\partial^2 \psi_3}{\partial \xi^2} + \frac{\partial^2 \psi_3}{\partial \eta^2} = 0, \\
&\vdots \\
p^n &: \quad \left|\frac{dT}{dz}\right|^{-2} \lambda^2 \psi_{n-1} + \frac{\partial^2 \psi_n}{\partial \xi^2} + \frac{\partial^2 \psi_n}{\partial \eta^2} = 0,
\end{aligned}
\tag{22.6.15}
$$

which has the solution is $\psi = \sum\limits_{n=0}^{\infty} \psi_n p^n$ for $p \to 1$. Since ψ_0 is the initial guess and known, we are left with the problem of solving an infinite set of Poisson's equations of the form

$$\nabla^2_{\xi,\eta} \psi_n(\xi, \eta) = h_n(\xi, \eta), \tag{22.6.16}$$

where $h_n(\xi, \eta)$ is the source term. Since the conform mapping transformed the rib-shaped waveguide onto an $s \times t$ rectangle (here the unit square (Figure 22.15)), the solution of Poisson's equations can be expressed as a two-dimensional Fourier series, i.e.,

$$\psi_n(\xi, \eta) = \sum_{j=1}^{\infty} \sum_{m=1}^{\infty} E_{mj} \sin\left(\frac{m\pi}{s}\xi\right) \sin\left(\frac{j\pi}{t}\eta\right), \tag{22.6.17}$$

where the coefficients E_{mj} are determined by substituting Eq (22.6.17) into Poisson's equation and using the orthogonality relations

$$\int_0^s \sin\left(\frac{m\pi\xi}{s}\right) \sin\left(\frac{q\pi\xi}{s}\right) d\xi = \frac{s}{2}\delta_{mq}, \quad \text{and} \quad \int_0^t \sin\left(\frac{j\pi\eta}{t}\right) \sin\left(\frac{r\pi\eta}{t}\right) d\eta = \frac{t}{2}\delta_{jr}.$$

Then

$$E_{mj} = -\frac{4}{st\kappa_{mj}} \int_0^s \int_0^t h_n(\xi, \eta) \sin\left(\frac{m\pi\xi}{s}\right) \sin\left(\frac{j\pi\eta}{t}\right) d\xi\, d\eta, \tag{22.6.18}$$

where the coefficients κ_{mj} are given by

$$\kappa_{mj} = \left(\frac{m\pi}{s}\right)^2 + \left(\frac{j\pi}{t}\right)^2, \quad m, j = 1, 2, \dots.$$

Thus, the solution of the nonlinear Helmholtz equation (22.6.10) is obtained in the form of an infinite series of solutions for Poisson's equations, where the choice of the initial guess ψ_0 that satisfies the boundary conditions is the first eigenfunction for the Helmholtz eigenvalue problem (22.6.2), i.e.,

$$\psi_0(\xi, \eta) = \sin\left(\frac{\pi\xi}{s}\right) \sin\left(\frac{\pi\eta}{t}\right). \tag{22.6.19}$$

For the unit square we take $s = t = 1$. Now, substituting Eqs (22.6.19) and (22.6.17) into Eq (22.6.15), the first term of the homotopy solution is obtained with the source function

$$h_1(\xi,\eta) = -\left(\nabla^2\psi_0 + |dT/dz|^{-2}\lambda^2\psi_0\right) = -\left(|dT/dz|^{-2}\lambda^2 - 2\pi^2\right)\psi_0,$$

which gives

$$\psi_1(\xi,\eta) = \sum_{j=1}^{\infty}\sum_{m=1}^{\infty} \frac{4\sin(m\pi\xi)\sin(j\pi\eta)}{\kappa_{mj}} \times \iint_0^1 \left(\left|\frac{dT}{dz}\right|^{-2}\lambda^2 - 2\pi^2\right)\sin(\pi\xi)\sin(\pi\eta)\sin(m\pi\xi)\sin(j\pi\eta)\,d\xi\,d\eta. \tag{22.6.20}$$

In general, by repeated application of Eq (22.6.17) into Eq (22.6.15) we find that for $n > 1$

$$\psi_n(\xi,\eta) = \sum_{j=1}^{\infty}\sum_{m=1}^{\infty} \frac{4\sin(m\pi\xi)\sin(j\pi\eta)}{\kappa_{mj}} \int_0^1\int_0^1 \left(\left|\frac{dT}{dz}\right|^{-2}\lambda^2 - 2\pi^2\right)\sin(m\pi\xi)\sin(j\pi\eta)\,d\xi\,d\eta, \tag{22.6.21}$$

where

$$\lambda^2 = -\frac{\int_0^1 \psi\nabla^2(\xi,\eta)\psi\,d\Gamma}{\int_0^1 |dT/dz|^{-2}\psi^2\,d\Gamma}, \tag{22.6.22}$$

which can be calculated using both the initial guess ψ_0 and the above solutions including higher order terms.

Another example is that of a half coaxial waveguide with outer radius r_a and inner radius r_b corresponding to the physical plane $z = r\,e^{i\theta}$ with $r_b \leq r \geq r_a$ and $0 \leq \theta \leq 2\pi$ (see Map 5.20). In this case the mapping function $T(z) = \log(z/r_b) = \ln(r/r_b) + i\theta$, or $z(T) = r_b\,e^T$, maps the physical plane onto the rectangle $0 \leq \xi \leq \ln(r_a/r_b)$ and $0 \leq \eta \leq \pi$, with the Jacobian $|dT/dz|^{-2} = r_b^2 e^{2\xi}$.

The general Helmholtz equation has been solved using a combination of conformal mapping and the homotopy perturbation for ARROW structures of the rib-shaped and half coaxial waveguides. This method can also be used for other waveguide structures so long as a zero boundary condition is used for these waveguides.

The results for the eigenvalue λ^2 for the second mode in a rib-shaped waveguide with aspect ratios $b/a = 1, 2$ and $d/(2a) = 1$ as a function of homotopy perturbation with the number of Fourier terms as parameters are presented in Table 22.3. These results are compared to the result from FEM computed using MATLAB. The HPM result $\lambda^2 = 13.99$ matches with the FEM for 12 or more Fourier terms at homotopy order 6. The performance of these methods is compared in Table 22.4.

Table 22.3 Eigenvalues λ^2 (Reck et al. [2011]).

Mode	λ^2_{HPM}	λ^2_{FEM}	ε	b/a	$d/(2a)$
1	12.16	12.13	0.29%	1	1
1	6.788	6.789	0.14%	2	1
2	19.06	18.84	1.78%	1	1
2	13.99	13.96	0.75%	2	1

Comparison with the FEM method is provided in Table 22.4.

Table 22.4 Comparison of HPM and FEM Methods (Reck et al. [2011])

Points	Memory [MB]		CPU Time [s]	
	FEM	HPM	FEM	HPM
12417	40	7	0.9	1.1
56000	65	30	4.4	4.9
109857	173	45	13.3	9.9
143265	208	59	16.5	12.8
172257	254	73	22.8	16.0

22.7 Coplanar Waveguides

Coplanar waveguides (CPWs), used extensively in microwave and transmission lines, have wide range of applications. The structure of a CPW consists of a center conductor and two ground planes printed on the same surface of a dielectric slot. This structure has a very high frequency response, provides immediate access to adjust power plane, low conduction and dispersion loss, and generates elliptical polarized magnetic fields with two modes of propagation, TEM and non-TEM. We will obtain closed-form design equations using conformal mapping and analytical results in terms of complete elliptic integral, and provide approximate formula for its computation.

There are two main types of coplanar lines, as follows.

(i) The CPW which is composed of a median metallic strip separated by two narrow slits from an infinite ground plate, as shown in Figure 22.16(a). The dimensions of a CPW are: the central strip of width $2d$, and slots of width s. This structure is symmetric along a vertical plane located on the middle of the central strip.

(ii) The coplanar slot (CPS) consists of two strips running side-by-side shown in Figure 22.16(b); it is complementary of the topology of the CPW.

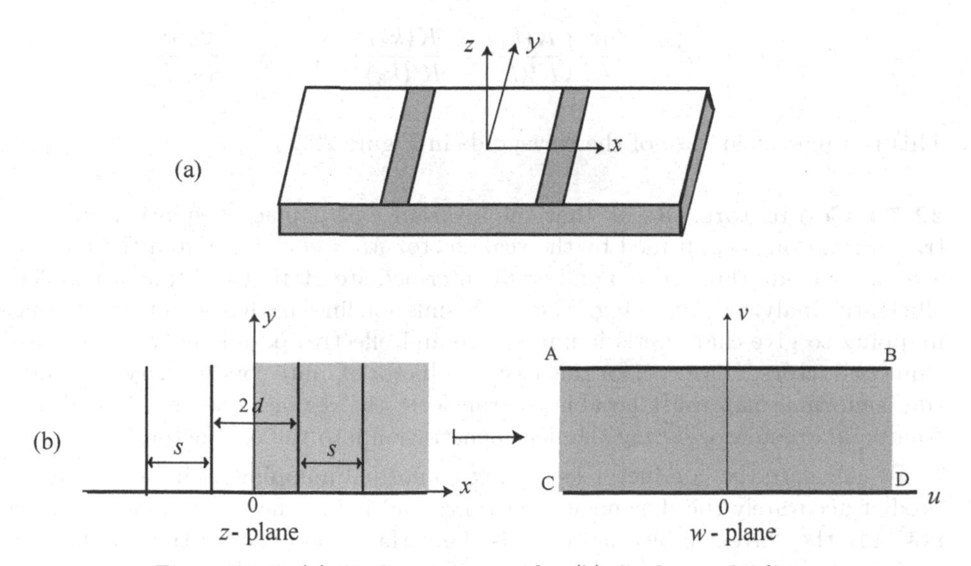

Figure 22.16 (a) Coplanar waveguide. (b) Coplanar slot line.

In view of the symmetry along the vertical line (Figure 22.16(b)), the quasi-static analysis of a CPW in the z-plane (shaded part) conformally mapped onto a parallel strip (capacitor ABCD, Figure 22.16(c)) in the w-plane (model plane) is given by the Schwarz-Christoffel transformation

$$w = \int_{z_0}^{z} \frac{dz}{\sqrt{(z-d)(z-d-s)}}, \tag{22.7.1}$$

where the capacitor ABCD is assumed to have a magnetic wall so that BC and AD also become magnetic walls with no resulting fringing field. The capacitances per unit length $C_{\rm top}$ of the dielectric field on the top and $C_{\rm bot}$ on the bottom of the dielectric substrate are given by

$$C_{\rm top} = 2\varepsilon_0 \frac{K(k_1)}{K'(k_1)}, \quad C_{\rm bot} = 2\varepsilon_0 \varepsilon_r \frac{K(k_1)}{K'(k_1)}, \tag{22.7.2}$$

where $K(k)$ and $K'(k)$ are the complete elliptic integral of the first and second kind, $K' = \sqrt{1-K^2}$, and $k_1 = d/2(d+s)$, where the accuracy of the above formulas ranges between 10^{-5} and $3 \cdot 10^{-6}$. Thus, the total line capacitance is $C_{\rm top} + C_{\rm bot}$, and the effective permittivity and the impedance are, respectively,

$$\varepsilon_{re} = \frac{\varepsilon_r + 1}{2}, \quad Z = \frac{30\pi}{\sqrt{\varepsilon_{re}}} \frac{K'(K_1)}{K(k_1)}. \tag{22.7.3}$$

However, in practice, the substrate has a finite thickness h (Figure 22.16(c)). Then the effective permittivity becomes

$$\varepsilon_{re} = 1 + \frac{\varepsilon_r - 1}{2} \frac{K(k_2)}{K(k_2)} \frac{K'(k_1)}{K(k_1)}, \quad k_2 = \frac{2\sqrt{k_1}}{1+k_1}. \tag{22.7.4}$$

Finally, for a CPW over a dielectric of finite thickness and backed by an infinite ground plate, the quasi-TEM wave is a hybrid between the microstrip and true CPW mode, and so $\varepsilon_{re} = 1 + q\,(\varepsilon_r - 1)$, where q is known as the *filling factor*, and its impedance is

$$Z = \frac{60\pi}{\sqrt{\varepsilon_{re}}} \Big[\frac{K(k_1)}{K'(k_1)} + \frac{K(k_3)}{K'(k_3)}\Big]^{-1}, \quad k_3 = \frac{2\sqrt{k_2}}{1+k_2}. \tag{22.7.5}$$

This is a particular case of the waveguide in Figure 22.15.

22.7.1 Conductors. Recall that the invariance of Laplace's equation under a conformal transformation is confirmed by the scale factor $M = |dw/dz|$. Often these transformations are carried out through a number of intermediate steps to obtain a simpler and more efficiently analyzable mapping. Many transmission lines have been analyzed using conformal mapping to give characteristic impedance and effective permittivity in the case of lossless thin conductors. However, for the lossy conductor of finite conductivity and finite thickness, the conformal map must be able to transform the regions both inside and outside of the conductor when considering the field penetration into the conductor.

To calculate the conductor loss using conformal mapping techniques and to frequently predict accurately the dependent resistance and inductance, the model should account for not only the current crowding towards the surface and edges of the conductor due to skin and proximity effects, but also due to uniform current distribution at low frequency.

Example 22.5. Consider the case of two rectangular conductors, A and B, in the z-plane, which are mapped into infinitesimally thin coplanar strips A' and B' in the intermediate T-plane by the map $T = f(z)$, $T = t + i\,\tau$, and then mapped into parallel plates A" and B" onto the w-plane by a map $w = g(T)$ (Figure 22.17.)

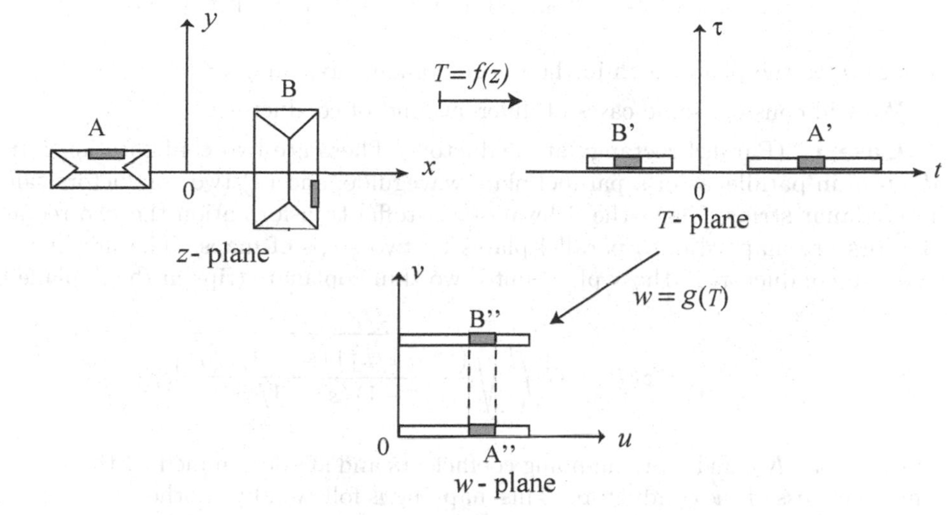

Figure 22.17 Mapping of two rectangular conductors A and B.

A point (x, y) on the surface of the conductor in the z-plane with corresponding effective internal impedance (EII) of $Z_{\mathrm{E}}(x, y)$ maps onto a point in the w-plane at (u, v_{top}) on the top plate. The EII is scaled in the w-plane by

$$M(u, v_{\mathrm{top}})Z_{\mathrm{E}}(x, y) = M(u, v)Z_{\mathrm{E}}\Big(\Re\{f\,(g^{-1}(u, v_{\mathrm{top}}))\,\}, \Im\{f\,(g^{-1}(u, v_{\mathrm{top}}))\,\}\Big), \tag{22.7.6}$$

where M is a scaling factor, $g^{-1}(u, v)$ is the inverse of the mapping function, and (u, v_{top}) is a point on the top plate, and the subscript E denotes the EII. Let dZ_{top} denote the differential series impedance per unit length of the top plate due to a differential width du. Then

$$dZ_{\mathrm{top}} = \frac{M(u, v_{\mathrm{top}})Z_{\mathrm{E}}(u, v_{\mathrm{top}})}{du},$$

where Z_{E} is the EII at some point (x, y) in the z-plane corresponding to a given point (u, v) in the w-plane. Using the same method, the differential series impedance per unit length dZ_{bot} of the bottom plate due to a differential width du is given by

$$dZ_{\mathrm{bot}} = \frac{M(u, v_{\mathrm{bot}})Z_{\mathrm{E}}(u, v_{\mathrm{bot}})}{du},$$

where (u, v_{bot}) is a point on the bottom plate. Assuming a uniform magnetic field between the two plates and using the transverse resonance technique, the inductance due to a differential width du and a separation $|v_{\mathrm{top}} - v_{\mathrm{bot}}|$ is given by

$$dL = \frac{\mu_0\,|v_{\mathrm{top}} - v_{\mathrm{bot}}|}{du},$$

where μ_0 is the permeability of free space. Next, the total differential series impedance per unit length is given by

$$dZ_{\mathrm{total}} = dZ_{\mathrm{top}} + dZ_{\mathrm{bot}} + i\omega\, dL. \tag{22.7.7}$$

Finally, the total series impedance per unit length $Z(\omega)$ for the transmission line can be approximated by the parallel combination of each differential impedance. Hence,

$$Z(\omega) = \Big[\int_0^{u_0} \frac{du}{j\omega\mu_0|v_{\text{top}} - v_{\text{bot}}| + Z_{\text{E}}(u, v_{\text{top}}) + Z_{\text{E}}(u, v_{\text{bot}})}\Big]^{-1}, \tag{22.7.8}$$

where μ_0 is the plate width in the image domain (w-plane).

We will consider some cases of different kind of conductors.

Case 1. (Parallel rectangular conductors) There are two configurations: (i) Two conductors in parallel as in a parallel plate waveguide, and (ii) two conductors side-by-side as in coplanar strips. Using the Schwarz-Christoffel transformation the two rectangular conductors are mapped onto parallel plates by two steps of maps. The first map transforms the two conductors in the z-plane onto two thin coplanar strips in the T-plane by

$$z(T) = C\int_0^t \sqrt{\frac{(s^2 - 1/k_1^2)\,(s^2 - 1/k_2^2)}{(s^2-1)\,(s^2 - 1/k^2)}}\,ds, \tag{22.7.9}$$

where C, k_1, K_2, and k are mapping coefficients and are determined by the separation width and thickness of the conductors. This mapping is followed by another map from the T-plane onto the w-plane, defined by

$$w(T) = \int_0^t \frac{ds}{\sqrt{(s^2-1)\,(s^2-1/k^2)}}, \tag{22.7.10}$$

which transforms the coplanar strips in the T-plane onto the parallel plates in the w-plane. The scale factor from the z-plane onto the T-plane is

$$M(T) = \left|\frac{dT}{dz}\right| = \left|\frac{1}{C}\sqrt{\frac{(T^2-1)\,(T^2-1/k^2)}{(T^2-1k_1^2)\,(T^2-1/k_2^2)}}\right|,$$

the scale factor from the T-plane to the w-plane is

$$N(T) = \left|\frac{dT}{dw}\right| = \sqrt{(T^2-1)\,(T^2-1/k^2)},$$

and the series impedance per unit length, $Z(\omega)$, is given by

$$Z(\omega) = \Big[\int_1^{1/k} \frac{ds}{j\omega\mu_0|v_{\text{top}} - v_{\text{bot}}|\,N(s) + 2Z_{\text{E}}(s)M(s)}\Big]^{-1}.$$

Case 2. V-groove conductor-backed coplanar waveguide (VGCPW) This coplanar waveguide has its backside ground plate made V-shaped in order to reduce current crowding and, therefore, reduce the conductor loss. However, no accurate model for the actual conductor loss in this structure has been studied. Only the attenuation constant is evaluated using conformal mapping. To determine if the V-groove reduces loss, design constraints are first taken into consideration. A design is shown in Figure 22.18(a), where the angle corresponds to an anisotropically etched groove in (100) Si. The coplanar waveguide (CPW) gap

$(b-a)$ is a function of the V-groove distance d for a constant $Z_0 = 50\Omega$, with $d = 16\,\mu$m for a 50Ω V-groove microstrip line (i.e., no CPW ground loss) is presented in Figure 22.18(b).

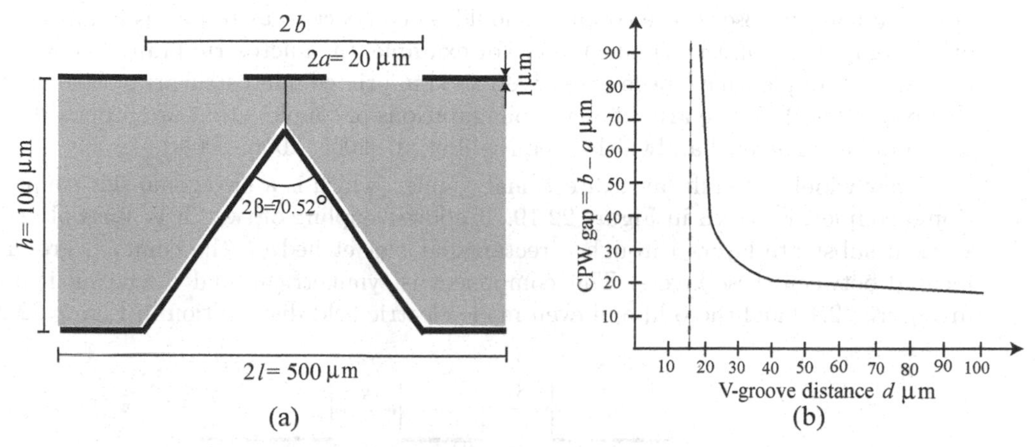

Figure 22.18 (a) V-groove conductor-based coplanar waveguide, (b) gap $(b-a)$ vs. d.

Other cases, like that of the normal coplanar waveguide (CPW) and microstrip lines of normal structure and V-shaped ground, are treated in the same way. It is observed that the conformal mapping method combined with the EII is very efficient in computing the conductor loss for the transmission line mode. This method consists of two parts: (i) The EII is assigned on the surface of the conductor, which represents the internal behavior of the conductor, and (ii) the conformal map is found for given geometries from the list of Schwarz-Christoffel transformations.

When numerically calculating the hyperelliptic integrals, the integration and parameter evaluation in Schwarz-Christoffel transformation is carried out by first dividing each side of the conductor into 10 segments and then 24-point Gaussian quadrature is used in each interval. However, to compute the integral having singular points with a tolerable accuracy, a large number of points is inevitable. In this case the Gauss-Chebyshev and Gauss-Jacobi quadrature formulas consider singular points in the integral properly, and therefore, appear to be a good choice in the hyperelliptic integrals.

In hyperelliptic integrals of Schwarz-Christoffel transformations, the mapping coefficient should be *a priori* known before computing the series impedance. Various iterative optimization schemes can be used to find the mapping coefficients. It is found that Powell's method (or Powell's conjugate direction method), which is an algorithm for finding a local minimum of a real-valued function is very effective; it is a direct search method and one of the least-square methods with constraints (Powell [1964]). There are other direct methods, such as the Hooke and Jeeves method for unconstrained optimization without using derivative (Hooke and Jeeves [1961]), and others, which can be adopted to coefficient evaluation often quite efficiently. There are gradient methods as well as direct search methods, e.g., the steepest descent method, Newton's method, and Newton-Raphson method.

Although the conformal mapping technique has been quite successful in evaluating the conductor loss, it does not work for multi-conductor (more than two) transmission lines, because it does not result in a simple parallel plate. For conformal mapping of multilayer microstrip lines, see Svacina [1992].

22.7.2 CPW Directional Couplers. In modern wireless communication systems the design of high-performance and low-cost directional couplers is very important. These couplers constitute the basic components used in applications of microwave integrated circuits, to combine or divide RF signals, antenna feeds, balanced mixers, modulators and others. For practical purposes these couplers should be compact so as to be easily integrated with other components in the same circuit. For example, the microstrip branch line couplers or hybrid ring couplers have been extensively used in printed microstrip array feeding networks (Nguyen [1995]). There are different configurations of couplers that are proposed when the couplers have narrow bandwidths (see Nedil et al. [2005; 2006; 2008]).

A new wideband multilayer directional coupler, which is a hexagonal slot-coupled directional coupler, is shown in Figure 22.19. It allows coupling of two CPW lines placed in two stacked substrate layers through a rectangular slot etched on the common ground plane located between these layers. This component is symmetrical, and the layout is presented in Figure 22.19 and the odd and even-mode electric field distribution in Figure 22.20.

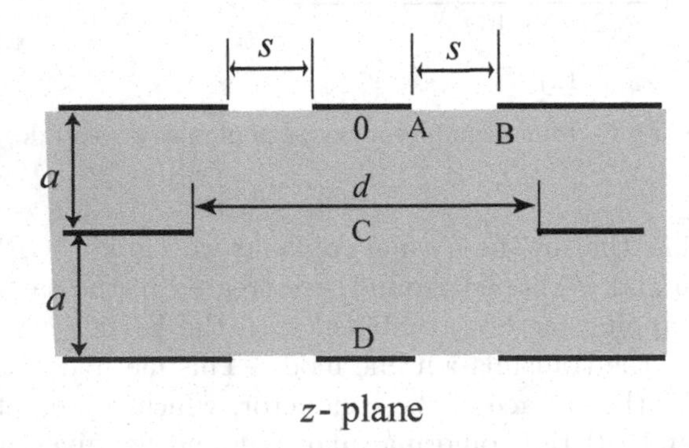

Figure 22.19 Broadside directional slot-coupled coupler.

The odd and even-mode configurations are considered in Figure 22.20 under conformal mapping, where it is assumed that these configurations have infinitely wide ground planes, and all conductors are assumed to be perfectly conducting and with zero thickness. This structure supports both odd and even modes, and the odd and even-mode impedances, Z_{odd} and Z_{even} are calculated using conformal methods which determine the coupling capacitance per unit length.

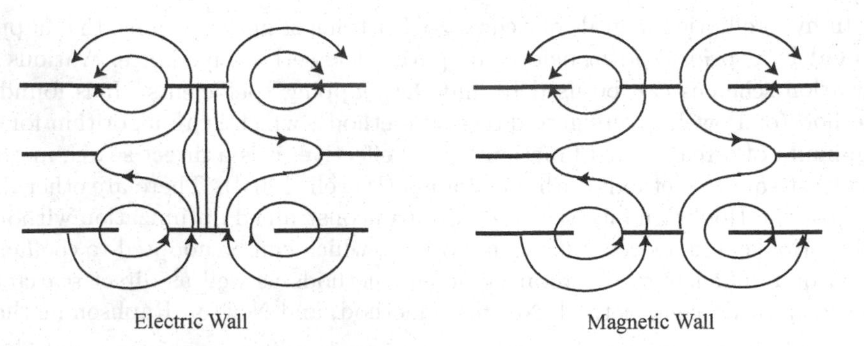

Figure 22.20 Odd- and even-mode electric fields.

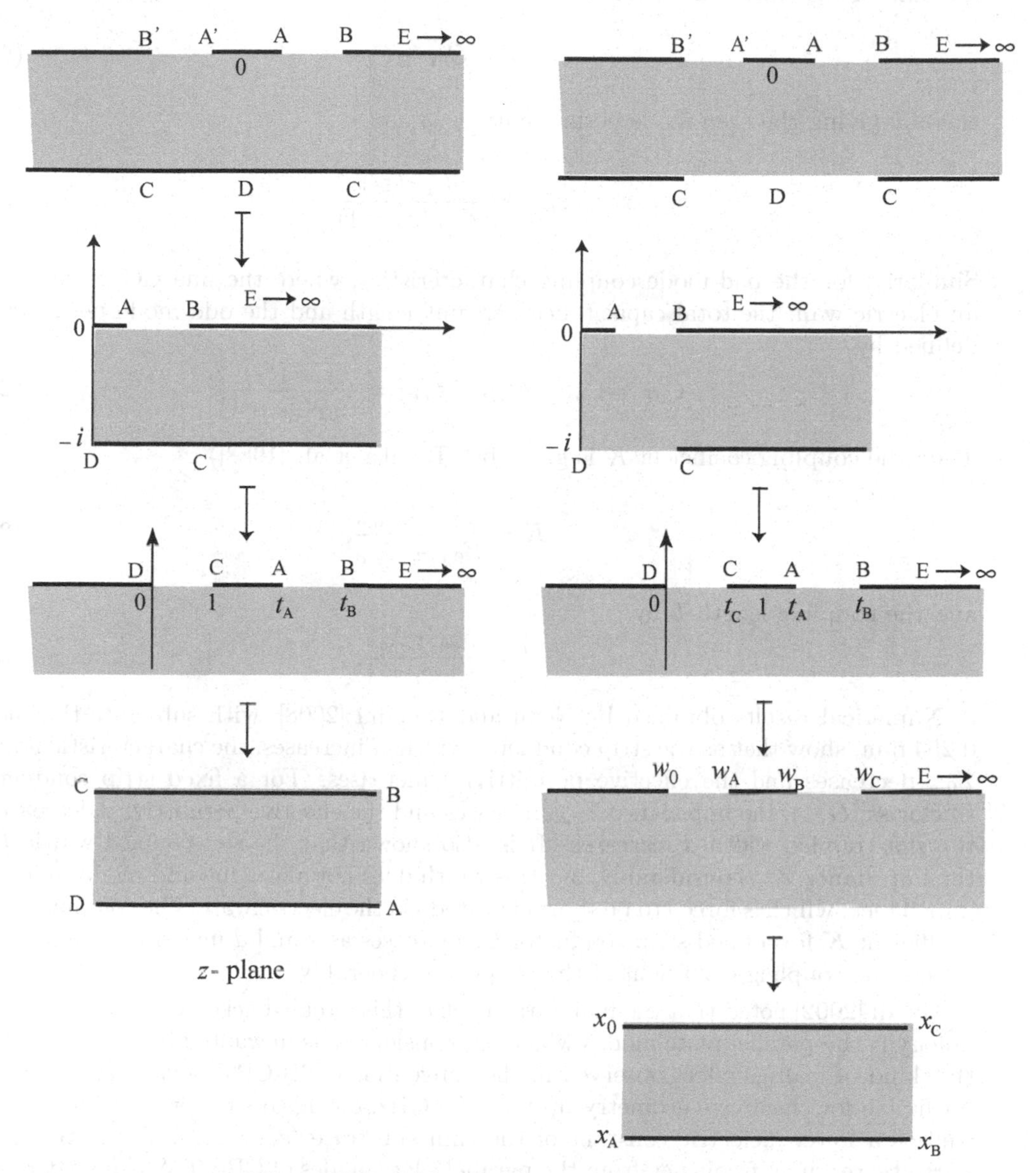

Figure 22.21 Odd and even-mode configurations.

This structure (Figure 22.21(a)) supports both fundamental (even and odd) modes, with their respective coplanar impedances Z_{even} and Z_{odd} per unit length, which are calculated using conformal mapping techniques. These modes, shown in Figure 22.21(b), can be isolated by assuming an electric wall for the odd mode and a magnetic wall for the even modes (as shown on the left and right side). The odd mode is obtained when the currents have equal amplitudes but opposite phases, but the even mode propagates when equal currents, in amplitude and phase, flow on the two coupled lines. For each mode the overall capacitance per unit length is the sum of the coupling capacitance for the air and the dielectric regions. Using the sequence of conformal mappings shown in Figure 22.21 (which can be identified as Maps 3.9 and 6.1), with the line CC' regarded as the magnetic line, the total even-mode capacitance C_{eT} per unit length (the suffix e for even mode) is defined as the sum of the capacitances for the even mode, denoted by C_{e1} and C_{e2},[†] i.e.,

$$C_{eT} = C_{e1} + C_{e2}, \tag{22.7.11}$$

thereby giving the even-mode permittivity $\varepsilon_{e,\text{eff}}$ as

$$\varepsilon_{e,\text{eff}} = \frac{C_{eT}(\varepsilon_r)}{C_{eT}(\varepsilon_r = 1)}. \tag{22.7.12}$$

Similarly, for the odd-mode coupling characteristics, where the line CC' is considered as an electric wall, the total capacitance per unit length and the odd-mode permittivity are defined by

$$C_{oT} = C_{o1} + C_{o2}, \quad \varepsilon_{o,\text{eff}} = \frac{C_{oT}(\varepsilon_r)}{C_{oT}(\varepsilon_r = 1)}. \tag{22.7.13}$$

Then the coupling coefficient K is given by (Tanaka et al. [1988])

$$K = \frac{Z_{0,e} - Z_{0,o}}{Z_{0,e} + Z_{0,o}}, \tag{22.7.14}$$

and the coupling length L by

$$L = \frac{\lambda_{ge} + \lambda_{go}}{8}. \tag{22.7.15}$$

Numerical results obtained by Nedil and Denidni [2008], with substrate thickness $a = 0.254$ mm, show that as the strip conductor width G increases, the characteristic impedance $Z_{0,o}$ decreases and the effective permittivity increases. For a fixed strip conductor and thickness (G, a), the impedance $Z_{0,e}$ increases and the effective permittivity decreases when the slot-coupled width d increases. It is also shown that the slot-coupled width d affects the impedance $Z_{0,o}$ considerably, but this width does not affect the odd-mode characteristic impedance, which is forced to be short circuited via the electric wall. The computed coupling coefficient K for a fixed strip conductor G, increases as s and d increase. The parameter d affects the coupling coefficient of the coupler considerably.

Haydl [2002] noted that the main drawback of this circuit-backed CPW (CB-CPW) technology is the parallel plate modes which are considered as unwanted bulk modes. However, this kind of leakage effect observed in the conventional CB-CPW geometry becomes quite negligible for the above geometry up to 18.33 GHz, owing to smaller lateral dimensions as well as a lower dielectric constant of the thin substrate ($\varepsilon_r = 2.2$). Thus, the minimum parasitic resonant frequency from the parallel-plate modes of CB-CPW can be derived from

$$F_{mn} = \frac{c}{2\sqrt{\varepsilon_r}} \sqrt{\left(\frac{m}{d_g}\right)^2 + \left(\frac{n}{L_g}\right)^2}, \tag{22.7.16}$$

which is based on the rectangular patch theorem (Haydl [2002]), where c is the velocity of light, ε_r is the relative permittivity, and d_g (= 7 mm) and L_g (= 30 mm) are the width and length of the ground in the coupler shown in Figure 22.19.

22.7.3 Cell Simulation. The design and simulation of measurement cells can be completed using a combination of LINPAR transmission line software (Djordjevic et al. [1999]) and

[†] The suffix e is for even mode and o for odd mode.

MATLAB. As a file manager, MATLAB acts as a creator and editor of the structure-defining data file, LINPAR executioner, and data processor. LINPAR uses a two-dimensional special technique to perform numerical computation of the transmission line structures, to provide the primary transmission line matrices, and inductance, capacitance, resistance and conductance per unit length, and to produce primary transmission line parameters. The LINPAR analysis is quasi-static where bound charges in a vacuum replace the dielectric materials, and free charges replace the conductors. The boundary conditions for the electrostatic potential and the normal component of the electric field derive a set of integral equations for the charge distribution.

LINPAR is applicable to both the microstrip and the CPW lines as a multi-layered and multi-conductor planar structure. Thus, LINPAR analysis of multi-layered structures with N conductor s results in an $N \times N$ matrix with elements x_{ij}, $i, j = 1, 2, \ldots, n$, for each transmission line parameter. In the case of the CPW structure, it results in a 3×3 matrix for each parameter. In each case, the element x_{22} corresponding to the transmission line parameter for the center conductor of the planar structure is very significant. Details of the application of LINPAR can be found in Skidmore [2012: 24ff].

22.8 Nonuniform Waveguides

There are three examples of nonuniform waveguides shown in Figure 22.22: (a) the cross-sectional configuration which is difficult to analyze in the (x, y)-plane (the waveguide a.1 is from Meinke et al. [1963], and waveguide a.2 is from Choi et al. [1988]); (b) the waveguide with a bend, that destroys longitudinal uniformity in the (x, ζ)-plane; and (c) the cross-section varies along the length of the guide. However, using conformal transformations, these and similar cases can be mapped onto familiar prismatic or longitudinal waveguides shown in Figure 22.23.

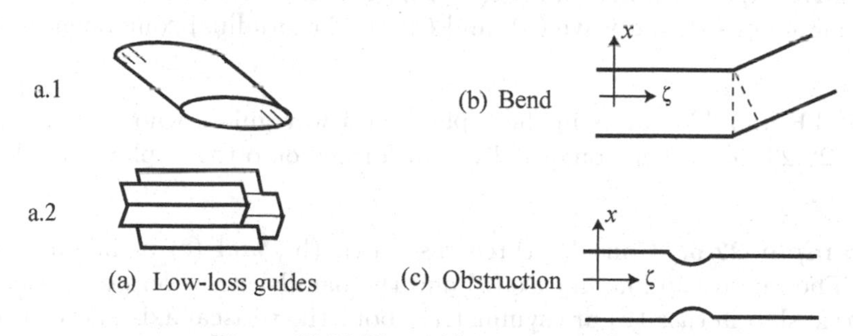

Figure 22.22 Nonuniform waveguides.

Figure 22.22 Nonuniform waveguides.

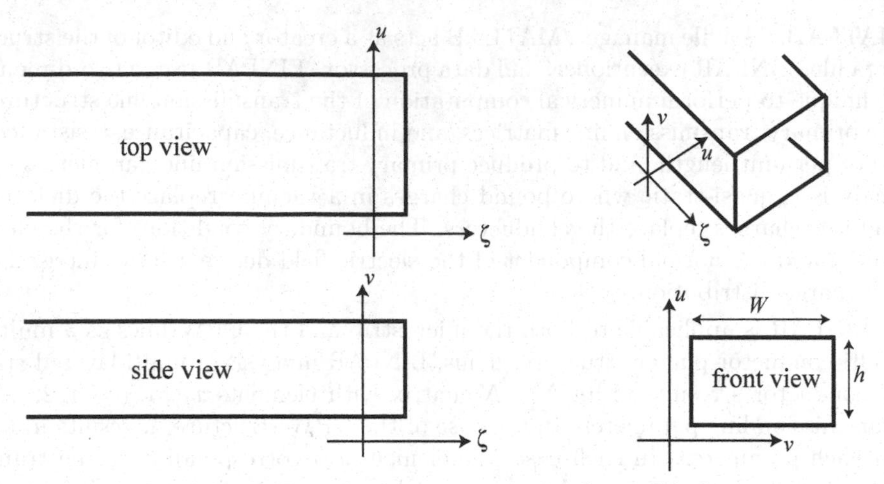

Figure 22.23 Rectangular waveguides in the model plane.

The hollow waveguide consisting of a single conductor cannot support the transverse electromagnetic mode (TEM), although mathematically one can imagine the physically acceptable modes as consisting of superposed, oblique TEM components. The transverse electric (TE) and transverse magnetic (TM) modes can propagate at various levels. For example, the fundamental TE mode (TE_0) has its electric field vector normal to those walls which are the closest. The *field intensity* varies as $E \sin(\pi u/W)$, i.e., it peaks at the center and tapers off towards the walls $u = 0, W$ where the electric field must vanish. The longest wavelength that can be accommodated is the *cutoff wavelength* $\lambda_c = 2W$. Besides λ_c, two other wavelengths will determine the propagation of the wave: they are (i) the wavelength of the signal or exciting wave (λ), and (ii) the longitudinal component in the waveguide.

The TE and TM waves in the z-plane in a waveguide, whose cross-section is shown in Figure 22.23, have been conformally transformed onto the w-plane (model plane).

Example 22.6. Consider three cases (a), (b), and (c) of inhomogeneous microstrip lines. The microstrip line in case (a) and the parallel strips opposing each other across the dielectric slab in case (b) are symmetric about the vertical axis and can be analyzed using the Schwarz-Christoffel transformation as in Example 22.3.

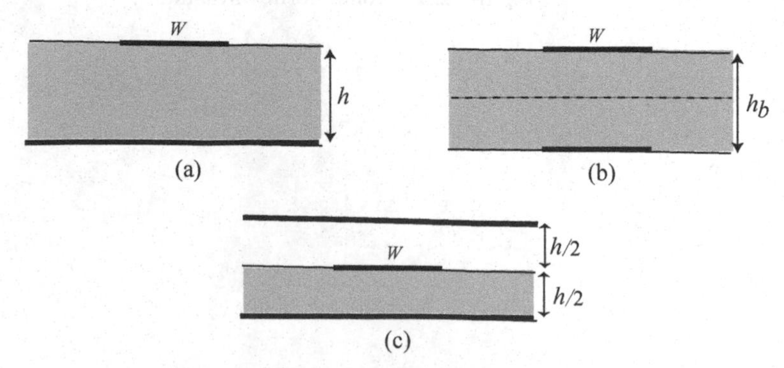

Figure 22.24 Inhomogeneous microstrip line.

In cases (a) and (b), the characteristic impedance Z_0 is given by (Wheeler [1965; 1978])

$$Z_0 = \frac{30}{\sqrt{\varepsilon_{\text{eff}}}} \ln \left[1 + A^2 \left(B + \sqrt{B^2 + (1+\varepsilon_r)\pi^2/(2B^2\varepsilon_r)}\right)\right], \tag{22.8.1}$$

where

$$A = 4h/W, \quad B = \frac{14 + 8/\varepsilon_r}{11}, \quad \varepsilon_{\text{eff}} = \frac{1}{2}(1+\varepsilon_r).$$

In case (c) which represents the shielded stripline, the characteristic impedance Z_0 is given by (Homentcovshi et al. [1988])

$$Z_0 = \frac{30\pi}{\sqrt{\varepsilon_{\text{eff}}}} \frac{K'(k)}{K(k)}, \tag{22.8.2}$$

where ε_{eff} is the same as above, and $k = \tanh(\pi W/(2h))$. ∎

Example 22.7. Consider the case of an inhomogeneous microstrip line shown in Figure 22.25, where its first quadrant in the z-plane is transformed onto the T-plane by mapping the conductor boundaries neglecting the line separating the two dielectrics, which is then mapped onto the w-plane to obtain the rectangle EGBD.

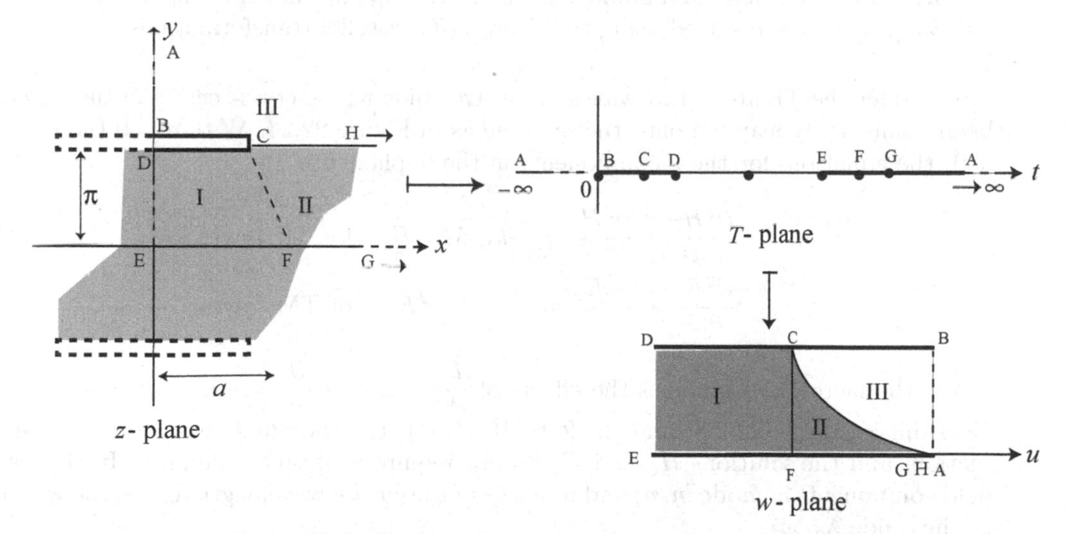

Figure 22.25 Conformal mapping of the microstrip line.

This mapping is carried out in the same manner as in Example 22.3 using the Schwarz-Christoffel transformation. The horizontal line CH in the z-plane corresponds to the border between region III with $\varepsilon = \varepsilon_0$ and region II with $\varepsilon = \varepsilon_r \varepsilon_0$. The line CH is mapped onto the curved line in the w-plane.

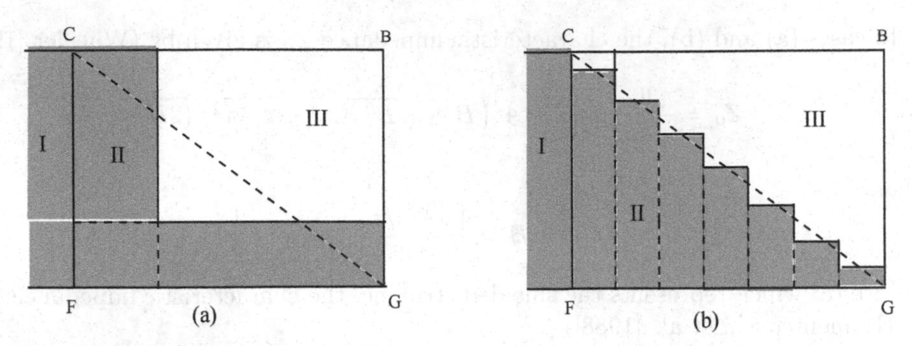

Figure 22.26 Two approximate methods.

A couple of numerical methods with results that approximate the test values are: (a) the one by Wheeler [1965], and (b) the other by Joshi et al. [1980]. Wheeler's method uses different techniques to represent region II, for example, he forms an equivalent lab in parallel with I and a horizontal slab in series with III, all depending on the proportions of the cross-section (Figure 22.26(a)). In the other method by Joshi et al. [1980], region II is divided into several vertical rectangular slabs as shown in Figure 22.26(b), where the capacitance (or modulus) between FH and CB consists of many series-parallel combinations of elemental capacitors. ■

There are other cases of inhomogeneous microstrip lines in Schinzinger and Laura [2003: 301-308]. They are resolved using the Schwarz-Christoffel transformations.

Consider the TE and TEM waves in a wave guide whose cross-section in the z-plane has been conformally mapped onto the w-plane as in Figure 22.23. With $\sigma = 0$ (no conductivity), the equations for the ζ-components in the w-plane are

$$\frac{\partial^2 H_\zeta}{\partial u^2} + \frac{\partial^2 H_\zeta}{\partial v^2} = -k_W M^{-2} H_\zeta \quad \text{for TE waves,} \tag{22.8.3}$$

$$\frac{\partial^2 E_\zeta}{\partial u^2} + \frac{\partial^2 E_\zeta}{\partial v^2} = -k_W M^{-2} E_\zeta \quad \text{for TM waves,} \tag{22.8.4}$$

where the factor k_W combines the effects of $\dfrac{\partial}{\partial t} = i\omega$ and $\dfrac{\partial}{\partial z} = -2\pi i/\lambda_g$. These equations are similar to Eqs (22.2.8) and (22.2.9). If $M = 1$, the waveguide has a rectangular cross-section, and the solutions H_ζ and E_ζ do not require conformal mapping. In this case the field components in mode m, n, and in terms of the cut-off wavelength λ_c and the wavelength in the guide λ_g, are

$$H_\zeta = A_{mn} \cos(k_m u) \cos(k_n v), \quad E_\zeta = B_{mn} \sin(k_m u) \sin(k_n v), \tag{22.8.5}$$

and the impedances are

$$\begin{aligned} Z_{\text{TE}} &= \frac{E_u}{H_v}\bigg|_{\text{TM}} = \omega^2 \mu\varepsilon \left[1 - \frac{\lambda^2}{\lambda_c^2}\right]^{1/2}, \\ Z_{\text{TM}} &= \frac{E_u}{H_v}\bigg|_{\text{TE}} = \omega^2 \mu\varepsilon \left[1 - \frac{\lambda^2}{\lambda_c^2}\right]^{-1/2}, \end{aligned} \tag{22.8.6}$$

where the wave numbers are:

$$k_m = m\pi/W, k_n = n\pi/h; \quad k_c^2 = k + m^2 + k_n^2 = m^2\pi^2/W + n^2\pi^2/h^2; \quad k_W^2 = \omega^2\mu\varepsilon - (2\pi/\lambda_g);$$

the cut-offs frequencies are $\lambda_c = 2\pi/k_c$, $\quad f_c = ck_c/(2\pi)$, $\quad c = [\mu\varepsilon]^{-1/2}$;
the wavelength is $\lambda_g = \lambda \left[1 - (\lambda/\lambda_c)^2\right]^{-1/2}$;
the attenuation is $\Gamma = 2\pi i/\lambda_g$ (ideal walls, $\sigma = 0$);
and the mode indices are m and n.† When the rectangular cross-section is a result of conformal mapping, the scale factor $M(u,v) = |dz/dw|$ appears in the above results.

22.8.1 Conformal Mapping of the Transverse Section of the Waveguide. Let the transformation $w = w(z)$ map the transverse cross-section of the waveguide in the z-plane onto a rectangle in the w-plane, and let $M(u,v) = |dz/dw|$. We will discuss the TE waves; the treatment of TM waves is similar. For $M = 1$, the solution is given by (22.8.5). However, for $M \neq 1$, let

$$H_\zeta = \sum_{m=0}^{\infty}\sum_{n=0}^{\infty} A_{mn} \cos\frac{m\pi u}{W} \cos\frac{n\pi v}{h}, \tag{22.8.7}$$

where the coefficients A_{mn} that satisfy the first of the Eqs (22.8.5) are obtained by expressing $1/M^2(u,v)$ as a Fourier series (Meinke and Gundlach [1962])

$$M^{-2}(u,v) = \sum_{m=0}^{\infty}\sum_{n=0}^{\infty} C_{mn} \cos\frac{m\pi u}{W} \cos\frac{n\pi v}{h}, \tag{22.8.8}$$

where

$$C_{mn} = \frac{p}{Wh}\int_{u=0}^{W}\int_{v=0}^{h} M^{-2}(u,v) \cos\frac{m\pi u}{W} \cos\frac{n\pi v}{h}, \tag{22.8.9}$$

where

$$p = \begin{cases} 1 & \text{for } m = 0 \text{ and } n = 0, \\ 2 & \text{for } m = 0 \text{ or } n = 0, \\ 4 & \text{for } m > 0 \text{ and } n > 0. \end{cases}$$

Note that C_{00} is the ratio of the cross-section area in the physical plane and the image area in the model plane. The conformal mapping can be adjusted by taking $C_{00} = 1$. To determine the other coefficients C_{mn}, we can use graphical Fourier series when integration becomes cumbersome. Further, substituting H_ζ from Eq (22.8.5) and M^{-2} from Eq (22.8.8), Meinke et al. [1963] obtained

$$A_{mn}\left[\left(\frac{m\pi}{W}\right)^2 + \left(\frac{n\pi}{h}\right)^2\right] = \left(\frac{2\pi}{\lambda_c}\right)^2 \left[A_{mn}\left(1 + \frac{1}{2}C_{0,2n} + \frac{1}{2}C_{2m,0} + \frac{1}{4}C_{2m,2n} + \cdots\right)\right], \tag{22.8.10}$$

where the factor $(2\pi/\lambda_c)^2$ is due to the propagation constant Γ. The cut-off wavelength is found from Eq (22.8.6) as

$$\lambda_c = 2\left[\frac{m^2}{W^2} + \frac{n^2}{h^2}\right]\left\{1 + \frac{1}{2}C_{0,2n} + \frac{1}{2}C_{2m,0} + \frac{1}{4}C_{2m,2n} + \cdots\right\}^{-1/2}. \tag{22.8.11}$$

The wave equations (22.8.5) can be solved by other methods. For example, Laura and Chi [1964] and Schinzinger and Laura [2003: 314] used the collocation method and applied it

† Here m does not mean $1/M$.

to the vane-like 'inductations' in circular waveguides to force the modes into a predetermined pattern; and in their 1980 paper they used the conformal transformation $w = A(z + pz^2)$ (Map 4.18), where A represents the scale factor and p a shape factor, thereby obtaining a cardioid in the limit when $p = \frac{1}{2}$. The wave numbers computed by these methods were in agreement up to $p = \frac{1}{2}$.

LITERATURE USED. Bowman, Senior and Uslenghi [1969], Brown [1967], Djordjevic, Bazdar, Sarkar and Harrington [1999], Durrand [1966], Flachenecker and Lange [1967], Foster and Anderson [1974], Haydl [2002], He [1999], Homentcovshi et al. [1988], Joshi [1980], Kumar, Saxena, Kapoor, Kala and Pant [2012], Ng and Stern [1998], Oberhettinger and Magnus [1949], Nedil, Denidni and Talbi [2005], Nedil and Denidni [2006; 2008], Nguyen [1995], Reck, Thomson and Hansen [2011], Schinzinger and Laura [2003], Shih and Chao [2008], Skidmore [2012], Tanaka, Tunoda and Aikawa [1988], Wandell [1991], Wen [1970], Wheeler [1964; 1965; 1978].

23

Elastic Medium

We will analyze linear elastic continua under the assumption that they undergo small strains. The linear theory of elasticity is based on the following two basic assumptions: (i) The material is subject to an infinitesimal strain and the stress is expressed as a linear function of strain, and (ii) any variation in the orientation of this material due to displacements is negligible. These assumptions lead to small strain and equilibrium equations under an undeformed geometry. The linearity assumption is an attempt to simplify the mathematical aspect of the behavior of solids. Although we assume that the material properties are linear, the deformations in a body may not be completely linear. For example, under certain loads, various materials exhibit plastic deformation while others creep with time, or they may crack, in which case the stresses are redistributed.

We will use conformal mapping methods to solve problems in elastic media related to the plane stresses and to torsion of prismatic rods. The classical theory of elasticity can be found in Sneddon and Berry [1958], Sokolnikoff [1956], and Green and Zerna [1968] for use of conformal mapping. Application of transformations to stress concentration around holes is available in Savin [1961].

23.1 Stress and Strain

We will express small strains and related stresses with respect to a right-hand rectangular coordinate system. Thus, in a Cartesian system where the coordinates are denoted by $\mathbf{x} = (x_1, x_2)$, consider an infinitesimal element (Figure 23.1). The stress vector is defined by

$$\boldsymbol{\sigma} = \begin{bmatrix} \sigma_{11} & \sigma_{12} \\ \sigma_{21} & \sigma_{22} \end{bmatrix}. \tag{23.1.1}$$

Note that if the coordinate system is taken as (x, y), instead of (x_1, x_2), then the normal stresses σ_{11}, σ_{22} are denoted by σ_x, σ_y respectively, and the shearing stresses σ_{12}, σ_{21} by τ_{xy}, τ_{yx} respectively. The equilibrium of the infinitesimal element implies that $\sigma_{12} = \sigma_{21}$.

Thus, we need to consider only two independent components of the shearing stress.

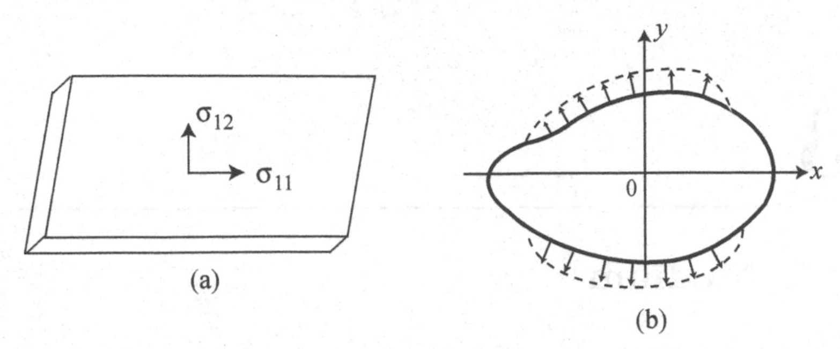

Figure 23.1 (a) Plane stresses on an infinitesimal element, (b) nonuniform loading.

Corresponding to these stresses, the normal and shearing strains are defined as follows:

$$\begin{aligned}&\text{Normal strains:}\quad \sigma_{ii} = u_{i,i}, \quad i = 1, 2;\\ &\text{Shearing strains:}\quad \tau_{ij} = \frac{1}{2}\left(u_{i,j} + u_{j,i}\right), \quad i, j = 1, 2 \quad (i \neq j),\end{aligned} \tag{23.1.2}$$

where (u_1, u_2) are translations along the (x_1, x_2) directions respectively. As before, only three of the above shearing strains are independent.

To determine the normal stresses $\sigma_x(x, y)$ and $\sigma_y(x, y)$ and the shear stress $\tau_{xy}(x, y)$, we must first obtain the strains $\varepsilon_x, \varepsilon_y$ and γ_{xy}, which will finally provide the displacements $u(x, y)$ and $v(x, y)$. Mushkhelishvili [1963] has shown that this problem can be stated in terms of the stress function $U(x, y)$ such that

$$\sigma_x = \frac{\partial^2 U}{\partial x^2}, \quad \sigma_y = \frac{\partial^2 U}{\partial y^2}, \quad \tau_{xy} = -\frac{\partial^2 U}{\partial x \partial y}, \tag{23.1.3}$$

where $U(x, y)$ is a solution of the biharmonic equation

$$\nabla^4 U = 0, \tag{23.1.4}$$

such that the governing boundary conditions are satisfied.

The biharmonic equation (23.1.4) is encountered in plane elasticity problems where U is known as the *Airy stress function.* It is also used to describe slow flows of viscous incompressible flows, in which case U is the stream function. The particular solutions of Eq (23.1.4) are:

$$\begin{aligned}
U(x, y) &= Ax^3 + Bx^2 y + Cxy^2 + ax^2 + bxy + cy^2 + \alpha x + \beta y + q,\\
U(x, y) &= (A\cosh\beta x + B\sinh\beta y + Cx\cosh\beta x + Dx\sinh\beta x)(a\cos\beta y + b\sin\beta y),\\
U(x, y) &= (A\cos\beta x + B\sin\beta y + Cx\cos\beta x + Dx\sin\beta x)(a\cosh\beta y + b\sinh\beta y),\\
U(x, y) &= Ar^2\ln r + Br^2 + C\ln r + D, \quad r = \sqrt{(x-a)^2 + (y-b)^2},\\
U(x, y) &= (Ax + By + C)(D\cosh\beta x + E\sinh\beta x)(a\cos\beta x + b\sin\beta x),\\
U(x, y) &= (x^2 + y^2)(D\cosh\beta x + E\sinh\beta x)(a\cos\beta y + b\sin\beta y),\\
U(x, y) &= (x^2 + y^2)(D\cosh\beta y + E\sinh\beta)(a\cos\beta x + b\sin\beta x),
\end{aligned} \tag{23.1.5}$$

where $A, B, C, D, E, a, b, c, \alpha, \beta$, and q are arbitrary constants. In polar coordinates the solutions of Eq (23.1.4) are:

$$\begin{aligned} U(r,\theta) &= \left(Ar^{2+\lambda} + Br^{2-\lambda} + Cr^{\lambda} + Dr^{-\lambda}\right)\cos(\lambda\theta), \\ U(r,\theta) &= \left(Ar^{2+\lambda} + Br^{2-\lambda} + Cr^{\lambda} + Dr^{-\lambda}\right)\sin(\lambda\theta), \\ U(r,\theta) &= Ar^2 \ln r + Br^2 + C\ln r + D \quad \text{at } \lambda = 0, \end{aligned} \tag{23.1.6}$$

where A, B, C, D, and λ are arbitrary constants. The fundamental solution of Eq (23.1.4) is

$$U^*(x,y) = \frac{1}{8\pi} r^2 \ln r, \quad r = \sqrt{x^2 + y^2}. \tag{23.1.7}$$

Various forms of the general solution of Eq (23.1.4) in terms of harmonic functions are:

$$\begin{aligned} U(x,y) &= xu_1(x,y) + u_2(x,y), \\ U(x,y) &= yu_1(x,y) + u_2(x,y), \\ U(x,y) &= (x^2+y^2)u_1(x,y) + u_2(x,y), \end{aligned} \tag{23.1.8}$$

where u_1 and u_2 are arbitrary harmonic functions satisfying Laplace's equation $\nabla^2 u_k = 0$ $(k = 1, 2)$. The complex form of the representation of the general solution is

$$U(x,y) = \Re\{\bar{z}\, f(z) + g(z)\}, \tag{23.1.9}$$

where $f(z)$ and $g(z)$ are arbitrary analytic functions of $z = x + iy$.

Example 23.1. (Boundary value problem for the upper half-plane $-\infty < x < \infty, 0 \leq y < \infty$) (a) The boundary conditions are $U(x,0) = 0$ and $U_y(x, 0 = f(x)$, and the solution is

$$U(x,y) = \int_{-\infty}^{\infty} f(\xi) G(x-\xi, y)\, d\xi, \quad G(x,y) = \frac{1}{\pi} \frac{y^2}{x^2+y^2}. \tag{23.1.10}$$

(b) The boundary conditions are $U_x(x,0) = f(x)$ and $U_y(x,0) = g(x)$, and the solution is

$$U(x,y) = \frac{1}{\pi} \int_{-\infty}^{\infty} f(\xi) \Big[\arctan\left(\frac{x-\xi}{y}\right) + \frac{y(x-\xi)}{(x-\xi)^2+y^2} \Big] + \frac{y^2}{\pi} \int_{-\infty}^{\infty} \frac{g(\xi)\, d\xi}{(x-\xi)^2+y^2} + C, \tag{23.1.11}$$

where C is an arbitrary constant. ■

Example 23.2. (Boundary value problem for a disk $0 \leq r \leq a, 0 \leq \theta \leq 2\pi$) The boundary conditions are $U(a, 0) = f(\theta)$, $U_r(a, 0) = g(\theta)$, and the solution is

$$U(r,\theta) = \frac{1}{2\pi a}(r^2 - a^2) \Big[\frac{(a - r\cos(s-\theta))\, f(s)\, ds}{(r^2 + a^2 - 2ar\cos(s-\theta))^2} - \frac{1}{2} \int_0^{2\pi} \frac{g(s)\, ds}{(r^2+a^2-2ar\cos(s-\theta))^2} \Big]. \tag{23.1.12}$$

23.2 Stress Function

We will discuss the case of an infinite plate. However, we first analyze the stress equation (23.1.4) more closely. The solution of this stress equation can be expressed in the z-plane in terms of analytic functions $\phi(z)$ and $\chi(z)$; thus,

$$U(x,y) = \Re\{\bar{z}\phi(z) + \chi(z)\}, \tag{23.2.1}$$

which can also be written as

$$U(x,y) = \frac{1}{2}\left[\bar{z}\phi(z) + z\bar{\phi}(\bar{z}) + \chi(z) + \bar{\chi}(\bar{z})\right]. \tag{23.2.2}$$

From (23.1.3) and (23.2.1) we get

$$\begin{aligned} \sigma_x &= \Re\{2\phi'(z) - \bar{z}\phi''(z) - \chi''(z)\}, \\ \sigma_y &= \Re\{2\phi'(z) + \bar{z}\phi''(z) + \chi''(z)\}, \\ \gamma_{xy} &= -\Im\{\bar{z}\phi''(z) + \chi''(z)\}, \end{aligned} \tag{23.2.3}$$

where the first two equations (23.2.3) give

$$\begin{aligned} \sigma_x + \sigma_y &= 4\Re\{\phi'(z)\}, \\ \sigma_y - \sigma_x &= 2\Re\{\bar{z}\phi'(z) + \chi''(z)\}, \end{aligned} \tag{23.2.4}$$

Similarly,

$$\begin{aligned} \sigma_y - \sigma_x - 2i\,\tau_{xy} &= 2\left[z\phi''(\bar{z}) + \bar{\chi}''(\bar{z})\right], \\ \sigma_y - \sigma_x + 2i\,\tau_{xy} &= 2\left[\bar{z}\phi''(\bar{z}) + \bar{\chi}''(\bar{z})\right]. \end{aligned} \tag{23.2.5}$$

The boundary conditions are derived as follows: Using the boundary element shown in Figure 23.2, the components of the resultant force per unit area at a point C are given by

$$p_x = \sigma_x \cos\alpha + \tau_{xy}\sin\alpha, \quad p_y = \sigma_y \sin\alpha + \tau_{xy}\cos\alpha, \tag{23.2.6}$$

where $\cos\alpha = \dfrac{dy}{ds}$, $\sin\alpha = -\dfrac{dx}{ds}$, $ds^2 = dx^2 + dy^2$.

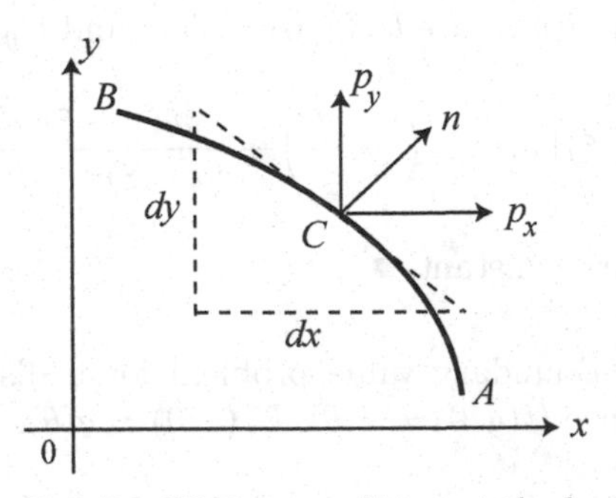

Figure 23.2 Equilibrium condition at the boundary.

Substituting (23.1.3) into (23.2.6), we get

$$\begin{aligned} p_x &= \frac{\partial^2 U}{\partial y^2}\frac{dy}{ds} + \frac{\partial^2 U}{\partial x \partial y}\frac{dx}{ds} \\ &= \frac{\partial}{\partial y}\Big(\frac{\partial U}{\partial y}\Big)\frac{dy}{ds} + \frac{\partial}{\partial x}\Big(\frac{\partial U}{\partial y}\Big)\frac{dx}{ds} = \frac{d}{ds}\Big(\frac{\partial U}{\partial y}\Big), \\ p_y &= -\frac{\partial^2 U}{\partial x^2}\frac{dx}{ds} - \frac{\partial^2 U}{\partial x \partial y}\frac{dy}{ds} = -\frac{d}{ds}\Big(\frac{\partial U}{\partial x}\Big). \end{aligned} \tag{23.2.7}$$

Hence, the components of the total resultant force due to the curved boundary element AB are given by

$$\begin{aligned} F_x &= \int_A^B \frac{d}{ds}\Big(\frac{\partial U}{\partial y}\Big)\, ds = \frac{\partial U}{\partial y}\Big|_A^B, \\ F_y &= -\int_A^B \frac{d}{ds}\Big(\frac{\partial U}{\partial x}\Big)\, ds = -\frac{\partial U}{\partial x}\Big|_A^B. \end{aligned} \tag{23.2.8}$$

Thus, from Eqs (23.2.1) and (23.2.8) we obtain

$$F_x + i\, F_y = -i\left[\phi(z) + z\bar{\phi}'(\bar{z}) + \bar{\chi}'(\bar{z})\right] = 0. \tag{23.2.9}$$

Note that in the case of stress-free boundary, we have

$$\phi(z) + z\bar{\phi}'(\bar{z}) + \bar{\chi}'(\bar{z}) = 0. \tag{23.2.10}$$

23.3 Infinite Plate and Conformal Mapping

Consider the case of an infinite plate with a hole of arbitrary size and a prescribed loading on both boundaries. Using Mushkhelishvili [1963], we can conclude from Eq (23.2.4) that both $\phi'(z)$ and $\chi''(z)$ must be finite at infinity, and they can, therefore, be expanded in infinite series of the form

$$\frac{d\phi(z)}{dz} = \sum_{n=0}^{\infty} C_n z^{-n}, \quad \frac{d^2\chi(z)}{dz^2} = \sum_{n=0}^{\infty} D_n z^{-n}. \tag{23.3.1}$$

We will now use conformal mapping, which should transform the hole of arbitrary shape in the physical plane onto a circular hole in the model plane (w-plane). Let such a transformation be $w = w(z)$, where $z = x + i\,y = r\,e^{i\theta}$ and $w = u + i\,v = R\,e^{i\Theta}$, where the scale factor (magnification) is $m = |dw/dz|$, or $m^{-1} = M = |dz/dw|$. Thus, in the w-plane we have

$$\frac{d\phi(w)}{dw} = \sum_{n=0}^{\infty} A_n w^{-n}, \quad \frac{d^2\chi(w)}{dw^2} = \sum_{n=0}^{\infty} B_n w^{-n}, \tag{23.3.2}$$

which upon integration give

$$\begin{aligned} \phi(w) &= A_0 w + A_1 \log w + \sum_{n=2}^{\infty} \frac{A_n w^{1-n}}{1-n}, \\ \frac{d\chi}{dw} &= B_0 w + B_1 \log w + \sum_{n=2}^{\infty} \frac{B_n w^{1-n}}{1-n}. \end{aligned} \tag{23.3.3}$$

Since the components of the displacement u and v must be single-valued, we must have $A_1 = B_1 = 0$. Hence, Eqs (23.3.3) can be expressed as

$$\phi = Aw + \sum_{n=1}^{\infty} a_n w^{-n}, \quad \frac{d\chi}{dw} = Bw + \sum_{n=1}^{\infty} b_n w^{-n}. \tag{23.3.4}$$

By change of variables we have

$$\phi'(z) = \frac{d\phi}{dz} = \frac{d\phi}{dw}\frac{dw}{dz} = \frac{\frac{d\phi}{dw}}{\frac{dz}{dw}}, \quad \chi'(z) = \frac{d\chi}{dz} = \frac{d\chi}{dw}\frac{dw}{dz} = \frac{\frac{d\chi}{dw}}{\frac{dz}{dw}},$$

which after substituting into Eq (23.2.10) gives

$$\frac{dz}{dw}\phi(w) + z(w)\frac{d\bar{\phi}(\bar{w})}{d\bar{w}} + \frac{d\bar{\chi}}{d\bar{w}} = 0, \quad r = 1, \tag{23.3.5}$$

or equivalently,

$$\frac{dz}{dw} + \bar{\phi}(\bar{w}) + \bar{z}(\bar{w})\frac{d\phi(w)}{dw} + \frac{d\chi}{dw} = 0, \quad r = 1. \tag{23.3.6}$$

Also, from the first equation in (23.3.4) we get

$$\sigma_x + \sigma_y = 4\Re\{\phi'(z)\} = 4\Re\Big\{\frac{d\phi}{dw}\Big/\frac{dz}{dw}\Big\}, \tag{23.3.7}$$

and similarly after some lengthy calculation we get

$$\sigma_x - \sigma_y - 2i\,\tau_{xy} = \frac{2}{\left(\frac{d\bar{z}}{d\bar{w}}\right)^3}\Big[z(w)\bar{\chi}''(\bar{w})\frac{d\bar{z}}{d\bar{w}} - z(w)\bar{\phi}'(\bar{w})\frac{d^2\bar{z}}{d\bar{w}^2} + \bar{\chi}''(\bar{w})\frac{d\bar{z}}{d\bar{w}} - \bar{\chi}'(\bar{w})\frac{d^2\bar{z}}{d\bar{w}^2}\Big], \tag{23.3.8}$$

or

$$\sigma_x - \sigma_y - 2i\,\tau_{xy} = \frac{2}{\left(\frac{d\bar{z}}{d\bar{w}}\right)^3}\Big[\bar{z}(\bar{w})\chi''(w)\frac{dz}{dw} - \bar{z}(\bar{w})\phi'(w)\frac{d^2z}{dw^2} + \chi''(w)\frac{dz}{dw} - \chi'(w)\frac{d^2z}{dw^2}\Big], \tag{23.3.9}$$

Next, we use the conformal transformation

$$z = f(w) = M\sum_{n=0}^{\infty} a_n w^{1-n}, \quad a_0 = 1, \tag{23.3.10}$$

where M is the scale factor. Then, using the boundary conditions at infinity

$$\sigma_x\big|_{x,y\to\infty} = \sigma_x(\infty), \quad \sigma_y\big|_{x,y\to\infty} = \sigma_y(\infty),$$

we find from (23.3.7) that

$$\sigma_x(\infty) + \sigma_y(\infty) = 4\Re\Big[\frac{d\phi}{dw}/\frac{dz}{dw}\Big]_\infty = 4\frac{A}{M}, \tag{23.3.11}$$

which gives

$$A = \frac{M}{4}\left[\sigma_x(\infty) + \sigma_y(\infty)\right]. \tag{23.3.12}$$

Similarly, $(\sigma_x - \sigma_y)\big|_\infty$ can be determined from the second equation in (23.2.4). Further, since $\phi''(z)\big|_\infty = 0$, we have

$$(\sigma_x - \sigma_y|)\big|_\infty = 2\Re\{\chi''(z)\}. \tag{23.3.13}$$

However, since

$$2\Re\left[\chi''(z)\right]_\infty = \Re\Big[2\frac{d}{dw}\Big(\frac{d\chi}{dw}\,\frac{dz}{dw}\Big)\frac{dw}{dz}\Big]_\infty = \Re\Big\{\frac{2B}{M^2}\Big\}, \tag{23.3.14}$$

we finally get

$$\frac{1}{2}M^2\left[\sigma_y(\infty) - \sigma_x(\infty)\right] = \Re\{B\}.$$

Using the shear stress we can similarly show that $M^2\tau_{xy}(\infty) = \Im\{B\}$. Moreover, if $\sigma_x(\infty) = \sigma_y(\infty) = T_1$ and $\tau_{xy}(\infty) = 0$, then

$$A = \frac{1}{2}MT_1, \quad \text{and } \Re\{B\} = 0 = \Im\{B\}. \tag{23.3.15}$$

Next, we determine the stress field in the infinite plate with a circular hole subject to the boundary conditions $\sigma_x(\infty) = \sigma_y(\infty) = T_1$ in the case when the hole undergoes *circumferentially periodic disturbances* produced by the boring process (Figure 23.3).

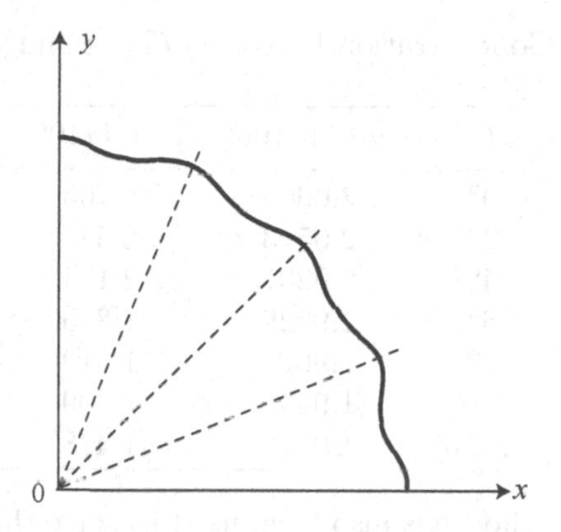

Figure 23.3 Hole with periodic disturbances.

Assuming that there are n periods around the circle and the amplitude of radial disturbance is $\mu = \max\{\delta r/r\}$, let the region be conformally transformed onto the w-plane with a circular hole by the following mapping:

Map 23.1.

$$z = z(w) = M\left(w + \mu w^{1-n}\right), \quad 0 \le \mu \le \frac{1}{n-1}. \tag{23.3.16}$$

Then in view of (23.3.15), we have

$$A = \frac{1}{2}MT_1, \quad \text{and } B = 0.$$

Example 23.3. Let the number of circumferential periods $n = 16$, and the radial disturbance $\mu = \Delta r/r = 1/100$. Thus, there are 16 axes of symmetry, and substituting (23.3.4) into (23.3.7) we get

$$\phi(w) = Aw + a_{15}w^{-15}, \tag{23.3.17}$$

$$\chi'(w) = Bw + b_1 w^{-1}, \tag{23.3.18}$$

where

$$a_{15} = -\frac{1}{2}MT_1\mu, \quad A = \frac{1}{2}MT_1, \quad b_1 = -\frac{MT_1}{1-7\mu}\left[1 - 7\mu + 15\mu(1-7\mu)\right], \quad B = 0.$$

It is obvious that the stress distribution along the boundary of the hole will attain their maximum value there. This point should be taken into consideration in design lest a crack might develop at this boundary.

Finally, in the w-plane the value of $\sigma_x + \sigma_y$, defined by Eq (23.3.7), becomes for $r = 1$

$$\sigma_x + \sigma_y \Big|_{r=1} \equiv \sigma_T = 4\Re\Big[\frac{d\phi}{dw}\Big/\frac{dz}{dw}\Big]_{r=1} = 2T_1 \frac{1 - 225\mu^2}{1 - 30\mu\cos 16\theta + 225\mu^2}, \tag{23.319}$$

where $\mu = \max\{\Delta r/r$ is the perturbation. The stress concentration factor σ_T/T_1 for the stressed infinite plate with a hole is defined by Eq (23.3.4) for $n = 16$. The values of this stress concentration factor are presented in Table 23.1 for $\mu = 1/1000, 1/100$, and $1/20$. Note that if T_1 is applied in one direction only, then the stress concentration factors will be considerably higher.

Table 23.1 Stress Concentration Factor σ_T/T_1 (Schinzinger and Laura [2003: 385]).

θ	$\mu = 1/1000$	$\mu = 1/100$	$\mu = 1/20$
$0°$	2.0609	2.706	14.000
$2°$	2.0513	2.545	3.000
$4°$	2.0257	2.194	0.967
$6°$	1.9929	1.855	0.509
$8°$	1.9629	1.619	0.352
$10°$	1.9443	1.499	0.294
$11°15'$	1.9409	1.478	0.286

A finite element method has also been used to solve this problem; see Laura, Reyes and Rossi [1974].

Finite element methods for plane elasticity and Stokes equations are discussed in Kythe and Wei [2004: Chapters 11, 12].

REFERENCES USED: Kythe and Wei [2004], Schinzinger and Laura [2003], Mushkhelishvili [1963], Sneddon and Berry [1958], Sokolnofoff [1956].

24

Finite Element Method

The pioneers in the development of the finite element method include Courant [1943], Prager and Synge [1947], Schoenberg [1948], Pólya and Szegö [1951], Hersch [1955], and Weinberger [1965]. Courant's work on the torsion problem is regarded as a classic; it defined piecewise linear polynomials over a triangulated region. Prager and Synge found approximate solutions for plane elasticity problems based on the concept of function space. Schoenberg developed the theory of splines, and used piecewise polynomials (interpolation functions) for approximation.

The finite element method is sometimes used in cases of irregular boundaries. It provides fast and economical solutions. The only disadvantage of this method is in the processing time for mesh generation. There are good programs already available, and they are mentioned toward the end of this chapter. Conformal mapping techniques are often used, especially in generating meshes and in infinite boundaries.

Recall that the conformal transformations map a region D in the z-plane (physical plane) onto a region G in the w-plane (model plane). If the domain $\Omega \subset G$ is regarded as a region in the model plane, then the function u (and v) must be replaced by a different notation, say ϕ and ψ, since we have used $w = u + iv$.

We will discuss the weak variational form, the Galerkin method, and the Rayleigh-Ritz method, followed by linear three-node triangular elements, single dependent variable problems and their local weak formulation, stiffness matrix and load vector, boundary integrals, applications to fluid flows, heat conduction, torsion, vibrations of elastic rods, electric circuits and capacitance, and other topics.

24.1 Weak Variational Form

The weak variation formulation of boundary value problems is derived from the fact that variational methods for finding approximate solutions of boundary value problems, viz., Galerkin, Rayleigh-Ritz, collocation, or other weighted residual methods, are based on weak variational statements of the boundary value problems.

Consider a general second-order boundary value problem

$$\frac{\partial F}{\partial u} - \frac{\partial}{\partial x}\Big(\frac{\partial F}{\partial p}\Big) - \frac{\partial}{\partial y}\Big(\frac{\partial F}{\partial q}\Big) = 0 \quad \text{in } \Omega, \tag{24.1.1}$$

subject to the boundary conditions

$$u = u_0 \quad \text{on } \Gamma_1, \quad \frac{\partial F}{\partial p} n_x + \frac{\partial F}{\partial q} n_y = q_0 \quad \text{on } \Gamma_2, \tag{24.1.2}$$

where $F = F(x, y, u, p, q)$, $p = u_x$, $q = u_y$, and n_x, n_y are the direction cosines of the unit vector $\mathbf{n}$ normal to the boundary $\partial\Omega \equiv \Gamma = \Gamma_1 \cup \Gamma_2$ of a two-dimensional region Ω such that $\Gamma_1 \cap \Gamma_2 = \emptyset$.

For example, a special case of (24.1.1) is when F is defined as

$$F = \frac{1}{2}\Big[k_1\Big(\frac{\partial u}{\partial x}\Big)^2 + k_2\Big(\frac{\partial u}{\partial y}\Big)^2\Big] - f\,u.$$

This equation arises in heat conduction problems in a two-dimensional region with k_1, k_2 as thermal conductivities in the x, y directions, and f being the heat source (or sink). If $k_1 = k_2 = 1$, then we get Poisson's equation $-\nabla^2 u = f$ with appropriate boundary conditions.

The weak variational formulation for the boundary value problem (24.1.1)-(24.1.2) is defined by

$$\iint_\Omega \Big[w\frac{\partial F}{\partial u} + \frac{\partial w}{\partial x}\frac{\partial F}{\partial p} + \frac{\partial w}{\partial y}\frac{\partial F}{\partial q}\Big]\, dx\, dy - \int_{\Gamma_2} w\, q_0\, ds = 0. \tag{24.1.3}$$

Details of derivation of the form (24.1.3) can be found, e.g., in Kythe and Wei [2004: 3-5]. The form (24.1.3) can be written in terms of the bilinear and linear differential forms as

$$b(w, u) = l(w), \tag{24.1.4}$$

where

$$b(w,u) = \iint_\Omega \Big[\frac{\partial w}{\partial x}\frac{\partial F}{\partial p} + \frac{\partial w}{\partial y}\frac{\partial F}{\partial q}\Big]\, dx\, dy, \quad l(w) = -\iint_\Omega w\frac{\partial F}{\partial u}\, dx\, dy + \int_{\Gamma_2} w q_0\, ds. \tag{24.1.5}$$

Formula (24.1.4) defines the weak variational form for Eq (24.1.1) subject to the boundary conditions (24.1.2). The quadratic functional associated with this variational form is given by

$$I(u) = \frac{1}{2} b(u, u) - l(u). \tag{24.1.6}$$

24.2 Galerkin Method

Consider the boundary value problem

$$L\,u = f \qquad \text{in } \Omega, \tag{24.2.1}$$

subject to the boundary conditions

$$u = g \qquad \text{on } \Gamma_1, \quad \frac{\partial u}{\partial n} + k\,u = h \qquad \text{on } \Gamma_2, \tag{24.2.2}$$

where $\Gamma = \Gamma_1 \cup \Gamma_2$ is the boundary of the region Ω. Let us choose an approximate solution $\tilde{u}$ of the form

$$\tilde{u} = \sum_{i=1}^{N} c_i \, \phi_i. \tag{24.2.3}$$

An approximate solution does not, in general, satisfy the system (24.2.1)-(24.2.2). The residual (error) associated with an approximate solution is defined by

$$r(\tilde{u}) \equiv L\,\tilde{u} - f = L\Big(\sum_{i=1}^{N} c_i \, \phi_i\Big) - f. \tag{24.2.4}$$

Note that if u_0 is an exact solution of (24.2.1)-(24.2.2), then $r(u_0) = 0$. The Galerkin method requires that the residual must be orthogonal with respect to the basis functions ϕ_i (also called the *trial functions*) used in (24.2.3), i.e., $\iint_\Omega r\,(\tilde{u})\;\phi_i\,dx\,dy = 0$ for $1, \dots, N$. Hence,

$$\iint_\Omega \big[L(\tilde{u}) - f\big]\,\phi_i \,dx\,dy = 0, \qquad i = 1, \dots, N, \tag{24.2.5a}$$

or

$$\sum_{j=1}^{n} c_j \iint_\Omega \phi_i \, L\phi_j \,dx\,dy = \iint_\Omega f\phi_i \,dx\,dy,$$

which in the matrix form is written as

$$\mathbf{A}\,\mathbf{c} = \mathbf{b}, \tag{24.2.5b}$$

where

$$A_{ij} = \iint_\Omega \phi_i \, L\phi_j \,dx\,dy, \qquad b_i = \iint_\Omega f\phi_i \,dx\,dy. \tag{24.2.5c}$$

In the examples given below, we choose different values of N for the trial function $\tilde{u}$. There is some guidance from geometry for such choices. However, the larger the N is, the better the approximation becomes.

Example 24.1. Consider Poisson's equation

$$-\nabla^2 u \equiv -\left(\frac{\partial^2 u}{\partial x^2} + \frac{\partial^2 u}{\partial y^2}\right) = c, \quad 0 < x < a, \quad 0 < y < b,$$

such that $u = 0$ at $x = 0, a$ and $y = 0, b$. First, we choose the first-order approximate solution as

$$\tilde{u}_1^{(1)} = \alpha\, xy(x-a)(y-b).$$

Note that this choice satisfies all four Dirichlet boundary conditions. The Galerkin equation (24.2.5a) gives

$$\int_0^b \int_0^a \big[-2\alpha(y^2 - by + x^2 - ax) - c\big]\, xy(x-a)(y-b)\,dx\,dy = 0$$

which simplifies to $\frac{\alpha}{90}\big[a^3b^3(a^2+b^2)\big] - \frac{a^3b^3c}{36} = 0$. Thus, $\alpha = \frac{5c}{2\left(a^2+b^2\right)}$. Hence,

$$\tilde{u}_1^{(1)} = \frac{5c}{2(a^2+b^2)}\, xy(x-a)(y-b). \blacksquare$$

Example 24.2. We will use the Galerkin method to solve the eigenvalue problem $\nabla^2 u + \lambda u = 0$ in the polar coordinates for $0 < r < a$.

To solve $\dfrac{1}{r}\dfrac{d}{dr}\Big(r\dfrac{du}{dr}\Big) + \lambda u = 0, \quad 0 < r < a$, take $\phi_j(r) = \cos\dfrac{j\pi r}{2a}$. For the first approximation, we have $\phi_1 = \cos\dfrac{\pi r}{2a}$, and $\tilde{u}_1 = \alpha_1 \cos\dfrac{\pi r}{2a}$, which leads to

$$2\pi \int_0^a \Big\{\frac{1}{r}\frac{d}{dr}\Big[\frac{r\pi}{2a}\Big(-\sin\frac{\pi r}{2a}\Big)\Big]\alpha_1 + \lambda\alpha_1 \cos\frac{\pi r}{2a}\Big\}\, r\, dr = 0.$$

This gives the equation for the eigenvalue λ as

$$\frac{\pi^2}{4}\Big(\frac{1}{2} + \frac{2}{\pi^2}\Big) - \lambda a^2\Big(\frac{1}{2} - \frac{2}{\pi^2}\Big) = 0.$$

Hence,

$$\lambda_1 = \frac{\pi^2(\pi^2+4)}{4a^2(\pi^2-4)} \approx \frac{5.8304}{a^2}.$$

The exact value is $\lambda_1 = \dfrac{5.779}{a^2}$. For the second-order approximation $\tilde{u}_2 = \alpha_1 \cos\dfrac{\pi r}{2a} + \alpha_2 \cos\dfrac{3\pi r}{2a}$, which gives $\lambda_2 = \dfrac{5.792}{a^2}$. Note that the disadvantage of Eq (24.2.5) is that it requires ϕ_j to satisfy the differential requirement of L. ■

24.3 Rayleigh-Ritz Method

Consider Poisson's equation $-\nabla^2 u = f$, with the homogeneous boundary conditions $u = 0$ on Γ_1 and $\partial u/\partial n = 0$ on Γ_2. Then, the weak variational form leads to

$$I(u) = \iint_\Omega \Big\{\frac{1}{2}\big|\nabla u\big|^2 - fu\Big\}\, dx\, dy = 0, \tag{24.3.1}$$

where $\big|\nabla u\big|^2 = u_x^2 + u_y^2$. A generalization of the result in (24.3.1) for the case of the system $Lu = f$ with the above homogeneous boundary conditions, where L is a linear self-adjoint and positive definite operator, leads to the functional

$$I(u) = \frac{1}{2}\iint_\Omega \{uLu - 2fu\}\, dx\, dy. \tag{24.3.2}$$

Theorem 24.1. *If the operator L is self-adjoint and positive definite, then the unique solution of $Lu = f$ with homogeneous boundary conditions occurs at a minimum value of $I(u)$.*

An application of Theorem 24.1 is the Rayleigh-Ritz method, where we find the direct solution of the variational problem for the system $Lu = f$ by constructing minimizing sequences and securing the approximate solutions by a limiting process based on such sequences. Thus, we choose a complete set of linearly independent basis (test) functions ϕ_i, $i = 1, \cdots$, and then approximate the exact solution u_0 by taking the approximate solution $\tilde{u}$ in the form

$$\tilde{u} = \sum_{i=1}^{n} c_i \phi_i, \tag{24.3.3}$$

where the constants c_i are chosen such that the functional $I(\tilde{u})$ is minimized at each stage. If $\tilde{u} \to u_0$ as $n \to \infty$, then the method yields a convergent solution. At each stage the method reduces the problem to that of solving a set of linear algebraic equations. The details for the boundary value problem $-\nabla^2 u = f$ with homogeneous boundary conditions are as follows: Using (24.3.3) in the functional (24.3.1) we get

$$\begin{aligned} I(\tilde{u}) = I(c_1, \cdots, c_n) &= \iint_\Omega \left\{ \left(\frac{\partial \tilde{u}}{\partial x}\right)^2 + \left(\frac{\partial \tilde{u}}{\partial y}\right)^2 - 2\tilde{u} f \right\} dx\, dy \\ &= \iint_\Omega \left\{ \left(\sum c_i \frac{\partial \phi_i}{\partial x}\right)^2 + \left(\sum c_i \frac{\partial \phi_i}{\partial y}\right)^2 - 2f \sum c_i \phi_i \right\} dx\, dy. \end{aligned}$$

Thus,

$$\begin{aligned} I(c_i) = c_i^2 &\iint_\Omega \left\{ \left(\frac{\partial \phi_i}{\partial x}\right)^2 + \frac{\partial \phi_i}{\partial y}\right)^2 \right\} dx\, dy \\ &+ 2 \sum_{i \neq j} c_i c_j \iint_\Omega \left(\frac{\partial \phi_i}{\partial x} \frac{\partial \phi_j}{\partial x} + \frac{\partial \phi_i}{\partial y} \frac{\partial \phi_j}{\partial y} \right) dx\, dy - 2c_i \iint_\Omega \phi_i f\, dx\, dy. \end{aligned}$$

Hence,

$$\frac{\partial I}{\partial c_i} = 2A_{ii} c_i + 2 \sum_{i \neq j} A_{ij}\, c_j - 2h_i, \tag{24.3.4}$$

and

$$A_{ij} = \iint_\Omega \left(\frac{\partial \phi_i}{\partial x} \frac{\partial \phi_j}{\partial x} + \frac{\partial \phi_i}{\partial y} \frac{\partial \phi_j}{\partial y} \right) dx\, dy, \quad h_i = \iint_\Omega \phi_i f\, dx\, dy.$$

Now, if we choose c_i such that $I(c_i)$ is a minimum (i.e., $\partial I/\partial c_i = 0$), then from (24.3.4) we get

$$\sum_{j=1}^{n} A_{ij} c_i = h_i, \quad i = 1, \cdots, n, \tag{24.3.5}$$

which in the matrix notation is

$$\mathbf{A}\,\mathbf{c} = \mathbf{h}, \tag{24.3.6}$$

where the vector $\mathbf{c} = [c_1, \cdots, c_n]^T$. Note that (24.3.6) is a system of linear algebraic equations to be solved for the unknown parameter c_i, and $\mathbf{A}$ is nonsingular if L is positive definite.

The Rayleigh-Ritz method is alternatively also developed by solving for u the equation (24.1.4), where we require that w satisfy the homogeneous essential conditions only. Then this problem is equivalent to minimizing the functional (24.1.6). In other words, we find an approximate solution of (24.1.4) in the form

$$u_n = \sum_{j=1}^{n} c_j \phi_j + \phi_0, \tag{24.3.7}$$

where the functions ϕ_j, $j = 1, \ldots, n$, satisfy the homogeneous boundary conditions while the function ϕ_0 satisfies the nonhomogeneous boundary condition, and the coefficients c_j

are chosen such that Eq (24.1.4) is true for $w = \phi_i$, $i = 1, \cdots, n$, i.e., $b(\phi_i, u_n) = l(\phi_i)$, or $b\Big(\phi_i, \sum_{j=1}^n c_j\phi_j + \phi_0\Big) = l(\phi_i)$ for $i = 1, \cdots, n$. Thus,

$$\sum_{j=1}^{n} c_j b(\phi_i, \phi_j) = l(\phi_i) - b(\phi_i, \phi_0), \quad i = 1, \cdots, n. \tag{24.3.8}$$

This is a system of n linear algebraic equations in n unknowns c_j and has a unique solution if the coefficient matrix in (24.3.8) is nonsingular and thus has an inverse.

The functions ϕ_i must satisfy the following requirements: (i) ϕ_i must be well defined such that $b(\phi_i, \phi_j) \neq 0$, (ii) ϕ_i must satisfy at least the essential homogeneous boundary condition, (iii) the set $\{\phi_i\}_{i=1}^n$ must be linearly independent, and (iv) the set $\{\phi_i\}_{i=1}^n$ must be complete. The term ϕ_0 in the representation (24.3.7) is dropped if all boundary conditions are homogeneous.

Example 24.3. Consider the Stokes flow of an ideal fluid in a channel of width b with zero vertical velocity component ($v = 0$). Then $u = u(y)$ is the velocity component in the y direction (Figure 24.1). The equation governing the Stokes flow is

$$\frac{\partial p}{\partial x} = \mu \frac{d^2 u}{dy^2}, \tag{24.3.9}$$

with the boundary conditions $u(0) = 0 = u(b)$. Since $\dfrac{\partial p}{\partial y} = 0$, Eq (24.3.9) implies that $\mu \dfrac{d^2 u}{dy^2}$ is constant. Hence, $\dfrac{\partial p}{\partial x} = \text{const}$, say, f_0. Then, this problem has the exact solution

$$u(y) = -\frac{f_0}{2\mu}\left(by - y^2\right).$$

This exact solution suggests that we consider an approximate solution of the form $\tilde{u} = \alpha \sin \dfrac{\pi y}{b}$ with $\phi(y) = \sin \dfrac{\pi y}{b}$. Then, using the Galerkin method, the residual is

$$r\,(\tilde{u}) = \mu \frac{d^2 \tilde{u}}{dy^2} - f_0 = -\mu\,\alpha \frac{\pi^2}{b^2} \sin \frac{\pi y}{b} - f_0.$$

Hence, solving $\displaystyle\int_0^b r\,(\tilde{u})\,\phi(u)\,dy = 0$, we find that

$$\int_0^b \Big[-\mu\,\alpha \frac{\pi^2}{b^2} \sin \frac{\pi y}{b} - f_0 \Big] \sin \frac{\pi y}{b}\,dy = 0,$$

which gives $\alpha = -\dfrac{4 f_0 b^2}{\mu \pi^3}$. Note that the Rayleigh-Ritz method also gives the same result; for details, see Kythe and Wei [2004]. Next, we take $f_0 = 1 = \mu = 1$, and $b = 1$. Then $\alpha = -4/\pi^3$. Since the flow is symmetric about the line $x = 0.5$, the values of the exact and approximate solutions of u are compared in the following table.

y	u_{exact}	u_{approx}
0.0	0.0	0.0
0.1	−0.045	−0.039865
0.2	−0.08	−0.075827
0.3	−0.105	−0.104368
0.4	−0.12	−0.122692
0.5	−0.125	−0.129006 ■

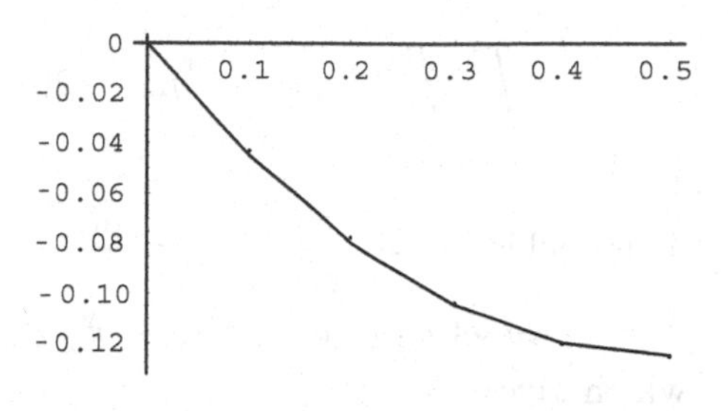

Figure 24.1 Stokes flow.

Example 24.4. Consider the boundary value problem $-k\Big(\dfrac{\partial^2 T}{\partial x^2}+\dfrac{\partial^2 T}{\partial y^2}\Big)=f$ in the region Ω with boundary conditions as shown in Figure 24.2. The following boundary conditions are prescribed: $kT_x = q_0(y)$ on HA; $kT_x = -\beta\,(T-T_\infty)$ on BC; $T=T_0(x)$ on AB, and $\partial T/\partial n = q_0 = 0$ on $CDEFGH$ (insulated), where k is the thermal conductivity of the material of the region Ω, β and T_∞ are ambient quantities, and $\partial T/\partial n = -\partial T/\partial x = -T_x$ on HA (two-dimensional heat conduction). The boundary conditions on $C_1 = AB$ (prescribed temperature T_0): $n_x = 0$, $n_y = -1$; on $C_2 = BC$ (convective boundary, T_∞): $n_x = 1$, $n_y = 0$; on $C_3 = CDEFGH$ (insulated boundary): $q = \partial T/\partial n = 0$; and on $C_4 = HA$ (prescribed conduction $q_0(y)$): $n_x = -1$, $n_y = 0$. The weak variational forms for this problem are

$$
\begin{aligned}
b(w,T) &= \iint_\Omega k\Big(\frac{dw}{dx}\frac{dT}{dx}+\frac{dw}{dy}\frac{dT}{dy}\Big)\,dx\,dy + \beta\int_0^b w(a,y)T(a,y)\,dy,\\
l(w) &= -\int_0^b w(0,y)q_0(y)\,dy + \beta T_\infty\int_0^b w(a,y)\,dy. \blacksquare
\end{aligned}
$$

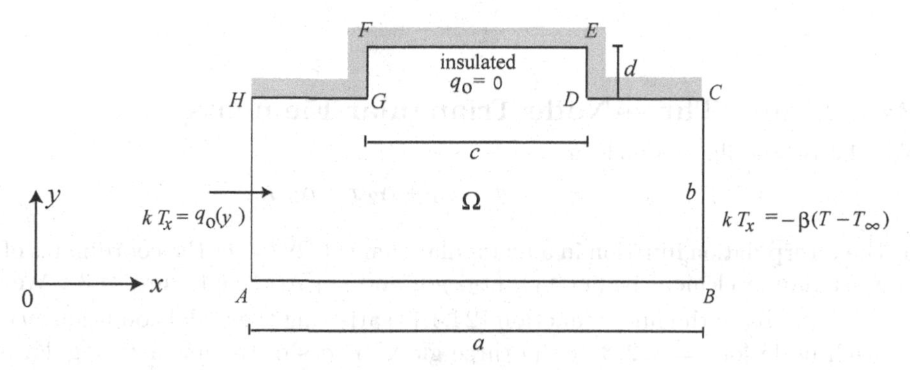

Figure 24.2 Region Ω.

Example 24.5. We will use the Galerkin method to solve Poisson's equation $\nabla^2 u = 2$, subject to the Dirichlet boundary condition $u=0$ along the boundary of the square $\{-a \le x,\ y \le a\}$, we use the basis functions $\phi(x,y)=(a^2-x^2)(a^2-y^2)$, and consider the approximate solution

$$
\tilde u_N(x,y) = (a^2-x^2)(a^2-y^2)(A_1 + A_2x^2 + A_3y^3 + \cdots + A_n x^{2i}y^{2j}).
$$

Then, for $N = 1$, we have

$$\int_{-a}^{a}\int_{-a}^{a} [-2(a^2 - y^2)A_1 - 2(a^2 - x^2)A_1 + 2](a^2 - x^2)(a^2 - y^2)\, dx\, dy = 0.$$

This yields $A_1 = \frac{5}{8a^2}$, $\tilde{u}_1 = \frac{5(a^2 - x^2)(a^2 - y^2)}{8a^2}$. We must have $A_2 = A_3$. Then for $N = 3$, we set $\tilde{u}_2 = (a^2-x^2)(a^2-y^2)[A_1+A_2(x^2+y^2)]$, where $A_1 = \frac{1295}{1416a^2}$, $A_2 = \frac{525}{4432a^4}$, which gives

$$\tilde{u}_2(x, y) = \frac{35}{4432a^2}(a^2 - x^2)(a^2 - y^2)\Big[74 + \frac{15}{a^2}(x^2 + y^2)\Big].$$

Alternately, if we choose the basis functions as $\phi_{jk} = \cos\frac{j\pi x}{2a}\cos\frac{k\pi y}{a}$, where j, k are odd, then

$$\tilde{u}_N = \sum_{\substack{j,k=1\\ j,k \text{ odd}}} \alpha_{jk} \cos\frac{j\pi x}{a}\cos\frac{k\pi y}{2a},$$

which leads to

$$\int_{-a}^{a}\int_{-a}^{a} \Big[\sum_{j,k} \alpha_{jk}\Big(\frac{j^2\pi^2}{4a^2} + \frac{k^2\pi^2}{4a^2}\Big)\cos\frac{j\pi x}{a}\cos\frac{k\pi xy}{2a}\Big]\cos\frac{m\pi x}{2a}\cos\frac{k\pi y}{2a}\, dx\, dy = 0.$$

Hence, for $j = m$ and $k = n$, we have

$$\alpha_{jk} = \frac{128a^2(-1)^{j+k-2)/2}}{jk(j^2 + k^2)\pi^4}. \blacksquare$$

24.4 Linear Three-Node Triangular Elements

We choose the linear function

$$u = \alpha_1 + \alpha_2 x + \alpha_3 y, \tag{24.4.1}$$

as the interpolation function in a triangular element $\Omega^{(e)}$. Let the coordinates of the nodes of this triangular element be (x_1, y_1), (x_2, y_2), and (x_3, y_3) (see Figure 24.3). We will solve for α_1, α_2, and α_3 in the linear function (24.4.1) satisfying the nodal conditions $u(x_i, y_i) = u_i^{(e)}$ at each node for $i = 1, 2, 3$ at the three global nodes of the linear triangular element $\Omega^{(e)}$. Then

at node 1: $u_1^{(e)} = \alpha_1 + \alpha_2 x_1 + \alpha_3 y_1$; at node 2: $u_2^{(e)} = \alpha_1 + \alpha_2 x_2 + \alpha_3 y_2$;
at node 3: $u_3^{(e)} = \alpha_1 + \alpha_2 x_3 + \alpha_3 y_3$.

Thus, we solve

$$\begin{Bmatrix} u_1^{(e)} \\ u_2^{(e)} \\ u_3^{(e)} \end{Bmatrix} = \begin{bmatrix} 1 & x_1 & y_1 \\ 1 & x_2 & y_2 \\ 1 & x_3 & y_3 \end{bmatrix} \begin{Bmatrix} \alpha_1 \\ \alpha_2 \\ \alpha_3 \end{Bmatrix},$$

and obtain

$$\alpha_1 = \frac{1}{2|\Omega^{(e)}|}\left[u_1^{(e)}\left(x_2 y_3 - x_3 y_2\right) + u_2^{(e)}\left(x_3 y_1 - x_1 y_3\right) + u_3^{(e)}\left(x_1 y_2 - x_2 y_1\right)\right],$$
$$\alpha_2 = \frac{1}{2|\Omega^{(e)}|}\left[u_1^{(e)}\left(y_2 - y_3\right) + u_2^{(e)}\left(y_3 - y_1\right) + u_3^{(e)}\left(y_1 - y_2\right)\right],$$
$$\alpha_3 = \frac{1}{2|\Omega^{(e)}|}\left[u_1^{(e)}\left(x_3 - x_2\right) + u_2^{(e)}\left(x_1 - x_3\right) + u_3^{(e)}\left(x_2 - x_1\right)\right],$$

where

$$|\Omega^{(e)}| = \text{area of the element } \Omega^{(e)} = \frac{1}{2}\begin{bmatrix} 1 & x_1 & y_1 \\ 1 & x_2 & y_2 \\ 1 & x_3 & y_3 \end{bmatrix}.$$

The area of a triangular element $\Omega^{(e)}$ is also denoted by $A^{(e)}$.

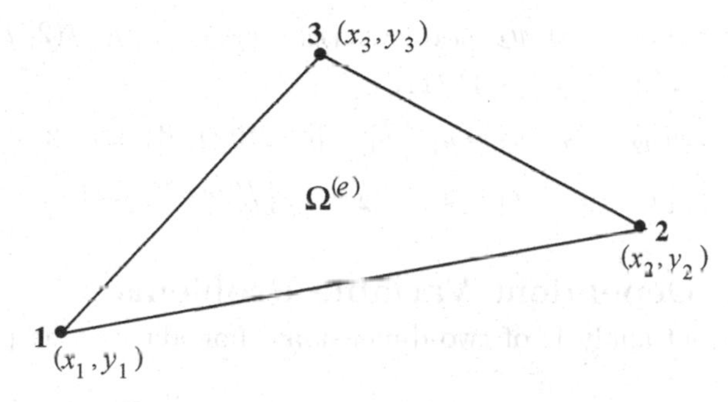

Figure 24.3 3-Node triangular element

Hence, since $u = \alpha_1 + \alpha_2\, x + \alpha_3\, y$, we get

$$u = \sum_{i=1}^{3} u_i^{(e)} \phi_i^{(e)}, \tag{24.4.2}$$

where

$$\phi_i^{(e)} = a_i^{(e)} + b_i^{(e)}\, x + c_i^{(e)}\, y, \tag{24.4.3}$$

and

$$\left.\begin{aligned} a_i^{(e)} &= \frac{x_j y_k - x_k y_j}{2|\Omega^{(e)}|}, \\ b_i^{(e)} &= \frac{y_j - y_k}{2|\Omega^{(e)}|}, \\ c_i^{(e)} &= \frac{x_k - x_j}{2|\Omega^{(e)}|}, \end{aligned}\right\} \quad \begin{aligned} i &= 1,2,3 \\ j &= 2,3,1 \\ k &= 3,1,2. \end{aligned} \tag{24.4.4}$$

The functions $\phi_i^{(e)}$ satisfy the conditions $\phi_i^{(e)}\left(x_j^{(e)}\right) = \delta_{ij}$, and $\sum_{i=1^N} \phi_i^{(e)}(x) = 1$, where δ_{ij} denotes the Kronecker delta, which is equal to 1 if $i = j$ and zero otherwise. Note that $a_1^{(e)} + a_2^{(e)} + a_3^{(e)} = 1$.

Example 24.6. Consider the linear triangular element $\Omega^{(e)}$ shown in Figure 24.3. Let $(x_1, y_1) = (2,2)$, $(x_2, y_2) = (5,3)$, and $(x_3, y_3) = (3,6)$. Then we have $2|\Omega^{(e)}| = 11$, and from (24.4.4) we get

$$a_1^{(e)} = \frac{x_2y_3 - x_3y_2}{2|\Omega^{(e)}|} = \frac{21}{11}, \quad a_2^{(e)} = \frac{x_3y_1 - x_1y_3}{2|\Omega^{(e)}|} = -\frac{6}{11}, \quad a_3^{(e)} = \frac{x_1y_2 - x_2y_1}{2|\Omega^{(e)}|} = -\frac{4}{11},$$
$$b_1^{(e)} = \frac{y_2 - y_3}{2|\Omega^{(e)}|} = -\frac{3}{11}, \quad b_2^{(e)} = \frac{y_3 - y_1}{2|\Omega^{(e)}|} = \frac{4}{11}, \quad b_3^{(e)} = \frac{y_1 - y_2}{2|\Omega^{(e)}|}1 = -\frac{1}{11},$$
$$c_1^{(e)} = \frac{x_3 - x_2}{2|\Omega^{(e)}|} = -\frac{2}{11}, \quad c_2^{(e)} = \frac{x_1 - x_3}{2|\Omega^{(e)}|} = -\frac{1}{11}, \quad c_3^{(e)} = \frac{x_2 - x_1}{2|\Omega^{(e)}|} = \frac{3}{11}.$$

Then, from (24.4.3) we obtain the interpolation functions

$$\phi_1^{(e)} = \frac{1}{11}\Big(21 - 3x - 2y\Big), \quad \phi_2^{(e)} = \frac{1}{11}\Big(-6 + 4x - y\Big), \quad \phi_3^{(e)} = \frac{1}{11}\Big(-4 - x + 3y\Big).$$

The values of $a_i^{(e)}$, $b_i^{(e)}$ and $c_i^{(e)}$, $i = 1, 2, 3$, can also be written in matrix form as

$$\begin{aligned}
\mathbf{a}^{(e)} &= [\,x_2y_3 - x_3y_2 \quad x_3y_1 - x_1y_3 \quad x_1y_2 - x_2y_1\,]/(2|\Omega^{(e)}|) \\
&= [\,21 \quad -6 \quad -4\,]/11, \\
\mathbf{b}^{(e)} &= [\,y_2 - y_3 \quad y_3 - y_1 \quad y_1 - y_2\,]/(2|\Omega^{(e)}|) = [\,-3 \quad 4 \quad -1\,]/11, \\
\mathbf{c}^{(e)} &= [\,x_3 - x_2 \quad x_1 - x_3 \quad x_2 - x_1\,]/(2|\Omega^{(e)}|) = [\,-2 \quad -1 \quad 3\,]/11. \blacksquare
\end{aligned}$$

24.5 Single Dependent Variable Problems

The finite element analysis of two-dimensional boundary value problems involves the following steps:

1. The boundary value problem is defined in a given domain Ω by a second-order partial differential equation which is subject to prescribed boundary and initial values, and
2. the boundary $\partial\Omega$ of the domain Ω is a closed curve in most problems.

Thus, the finite elements for the domain Ω are two-dimensional figures, like triangles, rectangles, or quadrilaterals. A mesh of these elements covers the given domain, and the solution of the boundary value problem is approximated over this finite element mesh. Obviously, such a solution contains the discretization as well as approximation errors; the former error is because of the approximation of the domain, and the latter because of the approximation of the numerical solution.

We consider the general second-order equation

$$-\frac{\partial G_1}{\partial x} - \frac{\partial G_2}{\partial y} + c\,u - f = 0, \quad \text{in } \Omega, \tag{24.5.1}$$

where c and f are known functions of x and y, subject to the prescribed boundary conditions: $u = \hat{u}$ on Γ_1, and $-(G_1\,n_x + G_2\,n_y) = q_n$ on Γ_2, where $\Gamma_1 \cup \Gamma_2 = \partial\Omega$ and $\Gamma_1 \cap \Gamma_2 = \emptyset$, and

$$G_1 \equiv a_{11}\,\frac{\partial u}{\partial x} + a_{12}\,\frac{\partial u}{\partial y}, \quad G_2 \equiv a_{21}\,\frac{\partial u}{\partial x} + a_{22}\,\frac{\partial u}{\partial y}, \tag{24.5.2}$$

with a_{ij} $(i, j = 1, 2)$, as known functions of x and y. Note that if $a_{11} = a = a_{22}$, $a_{12} = 0 = a_{21}$, and $c = 0$, then Eq (24.5.1) reduces to Poisson's equation

$$-\frac{\partial}{\partial x}\Big(a\,\frac{\partial u}{\partial x}\Big) - \frac{\partial}{\partial y}\Big(a\,\frac{\partial u}{\partial y}\Big) = f \quad \text{in } \Omega. \tag{24.5.3}$$

A mesh of quadrilateral elements in the region Ω is shown in Figure 24.4. This mesh consists of different geometric figures of triangular, rectangular, or quadrilateral shapes. A typical element is denoted by $\Omega^{(e)}$, and the discretization error is represented as the portions of the region (shaded in Figure 24.4) between its boundary $\Gamma \equiv \partial\Omega$ and the boundaries of the elements that lie toward the boundary Γ.

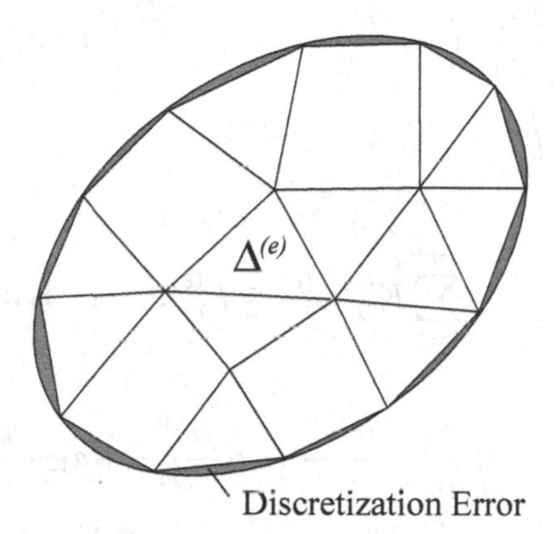

Figure 24.4 Finite elements.

24.5.1 Local Weak Formulation. Let the essential boundary conditions be prescribed as $u = u_0$ and the natural boundary condition as $-\big(G_1\, n_x + G_2\, n_y\big) = q_n$. Let w denote the test function. Then the weak variational form for an element $\Omega^{(e)}$ is

$$
\begin{aligned}
0 = \iint_{\Omega^{(e)}} \Big[\frac{\partial w}{\partial x}\Big(a_{11}\,\frac{\partial u}{\partial x} + a_{12}\,\frac{\partial u}{\partial y}\Big) + \frac{\partial w}{\partial y}\Big(a_{21}\,\frac{\partial u}{\partial x} + a_{22}\,\frac{\partial u}{\partial y}\Big) \\
+ c\,wu - wf\Big]\, dx\, dy - \int_{\Gamma^{(e)}\cap\Gamma_2} wq_n\, ds,
\end{aligned}
\tag{24.5.4}
$$

where $\Gamma^{(e)} = \partial\Omega$ is the boundary of $\Omega^{(e)}$. The bilinear and the linear forms are given by

$$
\begin{aligned}
b(w,u) &= \iint_{\Omega^{(e)}} \Big(\frac{\partial w}{\partial x}\, G_1 + \frac{\partial w}{\partial y}\, G_2 + c\,wu\Big)\, dx\, dy, \\
l(w) &= \iint_{\Omega^{(e)}} wf\, dx\, dy + \int_{\Gamma^{(e)}\cap\Gamma_2} wq_n\, ds.
\end{aligned}
\tag{24.5.5}
$$

24.5.2 Finite Element Equation. Let the function u be approximated by

$$
u \approx \sum_{i=1}^{n} u_i^{(e)}\, \phi_i^{(e)},
\tag{24.5.6}
$$

where $u_i^{(e)} = u\,(x_i, y_i)$, and $\phi_i^{(e)}$ are the linear interpolation functions such that $\phi_i^{(e)}\,(x_j, y_j) = \delta_{ij}$. Note that Eq (24.5.4) holds for any test function w. The natural choice for w is the n interpolation shape functions $\phi_i^{(e)}$, $i = 1, \dots, n$, and each choice of w yields an algebraic relation between $u_i^{(e)}$. The ith algebraic equation is known as the finite element equation, which is obtained by substituting (24.5.6) for u and $\phi_j^{(e)}$ for w in Eq (24.5.4), i.e.,

$$
\begin{aligned}
0 = \sum_{i=1}^{n} u_i^{(e)} \Big\{ \iint_{\Omega^{(e)}} \Big[\frac{\partial \phi_j^{(e)}}{\partial x} \Big(a_{11} \frac{\partial \phi_i^{(e)}}{\partial x} + a_{12} \frac{\partial \phi_i^{(e)}}{\partial y} \Big) \\
+ \frac{\partial \phi_j^{(e)}}{\partial y} \Big(a_{21} \frac{\partial \phi_i^{(e)}}{\partial x} + a_{22} \frac{\partial \phi_i^{(e)}}{\partial y} \Big) + c\, \phi_i^{(e)} \phi_j^{(e)} - f\, \phi_j \Big] \, dx\, dy \Big\} \\
- \int_{\Gamma^{(e)}} \phi_j^{(e)} q_n \, ds, \quad j = 1, \dots, n,
\end{aligned}
$$

which we write as

$$
\sum_{i=1}^{n} K_{ij}^{(e)} u_i^{(e)} = F_j^{(e)}, \quad \text{or} \quad \mathbf{K}\mathbf{u} = \mathbf{F}, \tag{24.5.7}
$$

where

$$
\begin{aligned}
K_{ij}^{(e)} = \iint_{\Omega^{(e)}} \Big[\frac{\partial \phi_j^{(e)}}{\partial x} \Big(a_{11} \frac{\partial \phi_i^{(e)}}{\partial x} + a_{12} \frac{\partial \phi_i^{(e)}}{\partial y} \Big) \\
+ \frac{\partial \phi_j^{(e)}}{\partial y} \Big(a_{21} \frac{\partial \phi_i^{(e)}}{\partial x} + a_{22} \frac{\partial \phi_i^{(e)}}{\partial y} \Big) + c\, \phi_i^{(e)} \phi_j^{(e)} \Big] \, dx\, dy, \\
F_j^{(e)} = \iint_{\Omega^{(e)}} f\, \phi_j \, dx\, dy + \int_{\Gamma^{(e)} \cap \Gamma_2} \phi_j^{(e)} q_n \, ds.
\end{aligned} \tag{24.5.8}
$$

Note that $K_{ij}^{(e)} = K_{ji}^{(e)}$ only if $a_{12} = a_{21}$, and $\mathbf{F}^{(e)} = \mathbf{f}^{(e)} + \mathbf{Q}^{(e)}$.

Before we solve the system (24.5.7), we will derive formulas for the evaluation of the stiffness matrix and the force vector for triangular and rectangular elements, impose boundary conditions, compute the boundary integrals, and discuss the assembly of local matrices.

24.5.3 Evaluation of Stiffness Matrix and Load Vector. We assume that the coefficients a_{ij} and c, and the function f are constant. Then the matrix $\mathbf{K}$ and the vector $\mathbf{f}$ are evaluated as described below.

(a) For a Triangular Element $\Omega^{(e)}$, the matrix $\mathbf{K}^{(e)}$ is composed of four double integrals

$$
\begin{aligned}
H_{ij}^{11} &= \iint_{\Omega^{(e)}} \frac{\partial \phi_i^{(e)}}{\partial x} \frac{\partial \phi_j^{(e)}}{\partial x} \, dx\, dy, \qquad H_{ij}^{12} = \iint_{\Omega^{(e)}} \frac{\partial \phi_i^{(e)}}{\partial x} \frac{\partial \phi_j^{(e)}}{\partial y} \, dx\, dy, \\
H_{ij}^{21} &= \iint_{\Omega^{(e)}} \frac{\partial \phi_i^{(e)}}{\partial y} \frac{\partial \phi_j^{(e)}}{\partial y} \, dx\, dy, \qquad H_{ij} = \iint_{\Omega^{(e)}} \phi_i^{(e)} \phi_j^{(e)} \, dx\, dy.
\end{aligned} \tag{24.5.9}
$$

Thus,

$$
\mathbf{K}^{(e)} = a_{11} \mathbf{H}^{11} + a_{12} \mathbf{H}^{12} + a_{21} \left(\mathbf{H}^{12}\right)^T + a_{22} \mathbf{H}^{22} + c\, \mathbf{H}. \tag{24.5.10}
$$

The vector $\mathbf{f}^{(e)}$ is defined by

$$
f_j^{(e)} = \iint_{\Omega^{(e)}} f\, \phi_j \, dx\, dy. \tag{24.5.11}
$$

Note that the integrals in (24.5.10) and (24.5.11) are of the type

$$
I_{mn} = \iint_{\Omega^{(e)}} x^m y^n \, dx\, dy. \tag{24.5.12}
$$

The integrals I_{mn} for $m, n = 0, 1, 2$ have the following values:

$$I_{00} = A^{(e)} \equiv |\Omega^{(e)}| \quad \text{(area of the element } \Omega^{(e)}\text{)}, \quad I_{10} = A^{(e)}\,\hat{x}, \quad \hat{x} = \frac{1}{3}\sum_{k=1}^{3} x_k,$$

$$I_{01} = A^{(e)}\,\hat{y}, \quad \hat{y} = \frac{1}{3}\sum_{k=1}^{3} y_k, \quad I_{11} = \frac{A^{(e)}}{12}\Big(\sum_{k=1}^{3} x_k y_k + 9\,\hat{x}\hat{y}\Big), \tag{24.5.13}$$

$$I_{20} = \frac{A^{(e)}}{12}\Big(\sum_{k=1}^{3} x_k^2 + 9\,\hat{x}^2\Big), \quad I_{02} = \frac{A^{(e)}}{12}\Big(\sum_{k=1}^{3} y_k^2 + 9\,\hat{y}^2\Big).$$

Then, using the results in (24.4.3)-(24.4.4), we get $\dfrac{\partial \phi_i^{(e)}}{\partial x} = b_i^{(e)}$, and $\dfrac{\partial \phi_i^{(e)}}{\partial y} = c_i^{(e)}$, which, in view of formulas (24.5.9), yield

$$\begin{aligned} H_{ij}^{11} &= A^{(e)}\,b_i^{(e)} b_j^{(e)}, \quad H_{ij}^{12} = A^{(e)}\,b_i^{(e)} c_j^{(e)}, \quad H_{ij}^{22} = A^{(e)}\,c_i^{(e)} c_j^{(e)}, \\ H_{ij} &= A^{(e)}\Big[a_i^{(e)} a_j^{(e)} + \Big(a_i^{(e)} b_j^{(e)} + a_j^{(e)} b_i^{(e)}\Big)\,\hat{x} + \Big(a_i^{(e)} c_j^{(e)} + c_i^{(e)} a_j^{(e)}\Big)\,\hat{y}\Big] \\ &\qquad + b_i^{(e)} b_j^{(e)}\,I_{20} + \Big(b_i^{(e)} c_j^{(e)} + b_j^{(e)} c_i^{(e)}\Big)\,I_{11} + c_i^{(e)} c_j^{(e)}\,I_{02}, \\ f_j^{(e)} &= \frac{f^{(e)} A^{(e)}}{3}, \quad Q_j^{(e)} = \frac{q_n A^{(e)}}{3}. \end{aligned} \tag{24.5.14}$$

The values of $\mathbf{K}^{(e)}$ and $\mathbf{f}^{(e)}$ are then evaluated for each element $\Omega^{(e)}$ from the data (coordinates) of the nodes.

(b) For a Rectangular Element $\Omega^{(e)} = \big\{(x, y) : 0 \le x \le a, 0 \le y \le b\big\}$, let a_{mn}, c and f have the constant values $a_{mn}^{(e)}$, $c^{(e)}$ and $f^{(e)}$, respectively, for $m, n = 1, 2$. Then, evaluating the double integrals, we get

$$\mathbf{H}^{11} = \frac{b}{6a}\begin{bmatrix} 2 & -2 & -1 & 1 \\ -2 & 2 & 1 & -1 \\ -1 & 1 & 2 & -2 \\ 1 & -1 & -2 & 2 \end{bmatrix}, \quad \mathbf{H}^{12} = \frac{1}{4}\begin{bmatrix} 1 & 1 & -1 & -1 \\ -1 & -1 & 1 & 1 \\ -1 & -1 & 1 & -1 \\ 1 & 1 & -1 & -1 \end{bmatrix},$$

$$\mathbf{H}^{22} = \frac{a}{6b}\begin{bmatrix} 2 & 1 & -1 & -2 \\ 1 & 2 & -2 & -1 \\ -1 & -2 & 2 & 1 \\ -2 & -1 & 1 & 2 \end{bmatrix}, \quad \mathbf{H} = \frac{ab}{36}\begin{bmatrix} 4 & 2 & 1 & 2 \\ 2 & 4 & 2 & 1 \\ 1 & 2 & 4 & 2 \\ 2 & 1 & 2 & 4 \end{bmatrix}, \tag{24.5.15}$$

$$\mathbf{f}^{(e)} = \frac{abf^{(e)}}{4}\,[1 \quad 1 \quad 1 \quad 1]^T.$$

24.5.4 Evaluation of Boundary Integrals. This is an important aspect in the process of finite elements method. Consider the boundary integral

$$Q_j^{(e)} = \int_{\partial\Omega^{(e)}} \phi_j^{(e)} q_n^{(e)}\,ds, \tag{24.5.16}$$

where $q_n^{(e)} = q_n^{(e)}(s)$ is prescribed, and s is measured along the boundary $\partial\Omega^{(e)}$. Note that in the case of two adjacent triangular elements e and e' (see Figure 24.5) the function $q_n^{(e)}$

cancels $q_n^{(e')}$ on the interface of these two elements. Also, in Figure 24.5 the function $q_n^{(e)}$ along the side $k\,l$ of the element e cancels $q_n^{(e')}$ along the side $m\,n$ of the element e', where the sides $k\,l$ and $m\,n$ represent the same interface between the elements e and e'. This situation can be regarded as the equilibrium state of the internal forces, known as *interface continuity*.

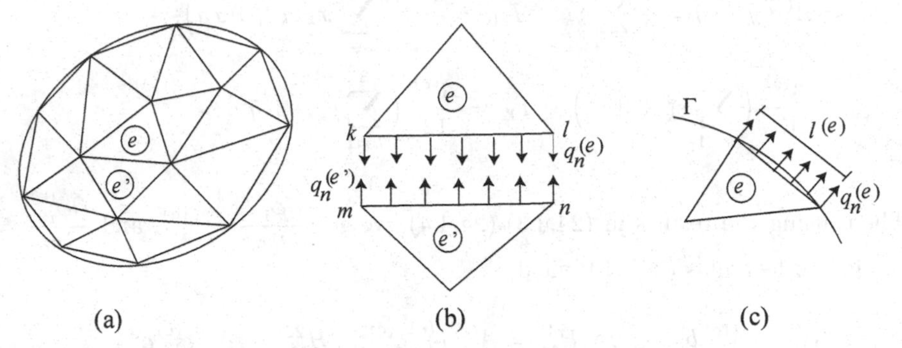

Figure 24.5 Interface continuity and boundary elements.

Now, if an element $\Omega^{(e)}$ falls on the boundary of the region, then the function $q_n^{(e)}(s)$ is either known, or it can be computed if not prescribed. In the latter case the primary variable must be prescribed on that side of the element $\Omega^{(e)}$. Again, this boundary consists of linear one-dimensional elements. Hence, to evaluate the boundary integrals we compute the line integrals (24.5.16).

24.5.5 Assembly of Element Matrices. This is carried in the same way as in one-dimensional problems. For example, consider a mesh of two elements shown in Figure 24.6.

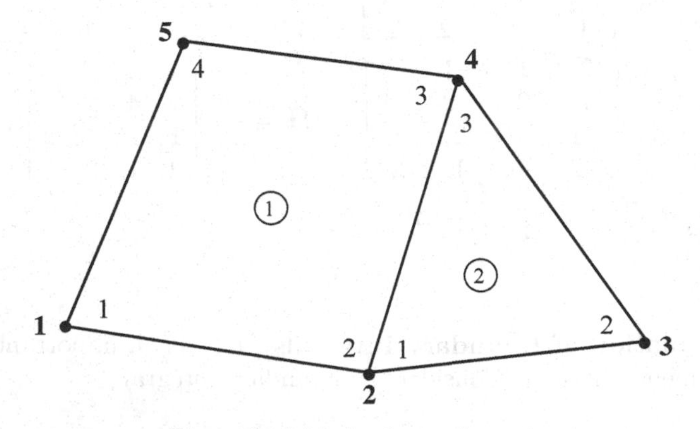

Figure 24.6 Mesh of two elements.

Let K_{ij} and $K_{ij}^{(e)}$ denote the global and local coefficient matrices, respectively. Then the following relations hold for the stiffness matrix $\mathbf{K}$:

Global	$\longrightarrow$ Local
K_{11}	$K_{11}^{(1)}$
$K_{12} = K_{21}$	$K_{12}^{(1)} = K_{21}^{(1)}$
$K_{13} = K_{31}$	0
$K_{14} = K_{41}$	$K_{13}^{(1)} = K_{31}^{(1)}$
$K_{15} = K_{51}$	$K_{14}^{(1)} = K_{41}^{(1)}$
K_{22}	$K_{22}^{(1)} = K_{11}^{(2)}$
$K_{23} = K_{32}$	$K_{12}^{(2)} = K_{21}^{(2)}$
$K_{24} = K_{42}$	$K_{23}^{(1)} + K_{13}^{(2)} = K_{32}^{(1)} + K_{31}^{(2)}$
$K_{25} = K_{52}$	$K_{24}^{(1)} = K_{42}^{(1)}$
K_{33}	$K_{22}^{(2)}$
$K_{34} = K_{43}$	$K_{23}^{(2)} = K_{32}^{(2)}$
$K_{35} = K_{53}$	0
K_{44}	$K_{33}^{(1)} + K_{33}^{(2)}$
$K_{45} = K_{54}$	$K_{34}^{(1)} = K_{43}^{(1)}$
K_{55}	$K_{44}^{(1)}$

The above relations between the global and local nodes can also be obtained from the connectivity matrix $\mathbf{C}$ for the mesh shown in Figure 24.6. This matrix is

$$\mathbf{C} = \begin{bmatrix} 1 & \mathbf{1} & \mathbf{2} & \mathbf{4} & \mathbf{5} \\ 2 & \mathbf{2} & \mathbf{3} & \mathbf{4} & \end{bmatrix},$$

where the first column gives the element number, and the bold face numbers refer to the global nodes. The vector $\mathbf{F} = \mathbf{f} + \mathbf{Q}$ can be similarly written. Thus, we have

$$\mathbf{K} = \begin{bmatrix} K_{11} & K_{12} & K_{13} & K_{14} & K_{15} \\ K_{21} & K_{22} & K_{23} & K_{24} & K_{25} \\ K_{31} & K_{32} & K_{33} & K_{34} & K_{35} \\ K_{41} & K_{42} & K_{43} & K_{44} & K_{45} \\ K_{51} & K_{52} & K_{53} & K_{54} & K_{55} \end{bmatrix},$$

$$\mathbf{F} = [\,F_1 \quad F_2 \quad F_3 \quad F_4 \quad F_5\,]^T = \begin{Bmatrix} f_1^{(1)} \\ f_2^{(1)} + f_1^{(2)} \\ f_2^{(2)} \\ f_3^{(1)} + f_3^{(2)} \\ f_4^{(1)} \end{Bmatrix} + \begin{Bmatrix} Q_1^{(1)} \\ Q_2^{(1)} + Q_1^{(2)} \\ Q_2^{(2)} \\ Q_3^{(1)} + Q_3^{(2)} \\ Q_4^{(1)} \end{Bmatrix}.$$

The extended form of the matrix $\mathbf{K}$ in terms of the local functions $K_{ij}^{(e)}$ can be written by replacing the global forms by the respective local forms given in the above relations. This is left as an exercise.

The primary variables are given by: $U_1 = u_1^{(1)}$, $U_2 = u_2^{(1)} = u_1^{(2)}$, $U_3 = u_2^{(2)}$, $U_4 = u_3^{(1)} = u_3^{(2)}$, $U_5 = u_4^{(1)}$. This enables us to write the finite element equation $\mathbf{Ku} = \mathbf{F}$.

Example 24.7. Consider the mesh of elements shown in Figure 24.7.

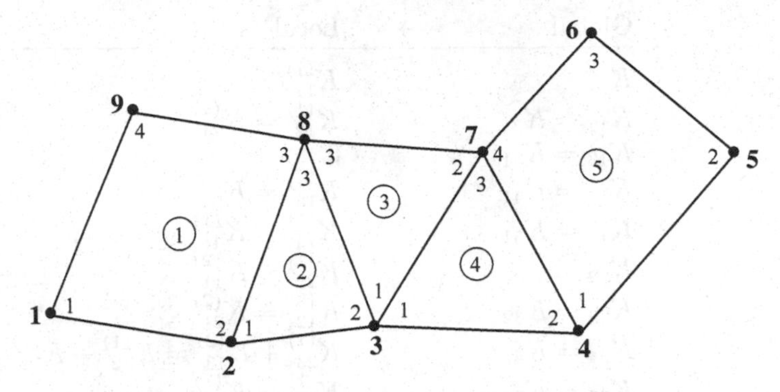

Figure 24.7 Mesh of triangular and rectangular elements.

The connectivity matrix $\mathbf{C}$ is given by

$$\mathbf{C} = \begin{bmatrix} 1 & \mathbf{1} & \mathbf{2} & \mathbf{8} & \mathbf{9} \\ 2 & \mathbf{2} & \mathbf{3} & \mathbf{8} & \\ 3 & \mathbf{3} & \mathbf{7} & \mathbf{8} & \\ 4 & \mathbf{3} & \mathbf{4} & \mathbf{7} & \\ 5 & \mathbf{4} & \mathbf{5} & \mathbf{6} & \mathbf{7} \end{bmatrix}.$$

After establishing the correspondence between the global and local nodes, details of which are left as an exercise, we obtain

$$\mathbf{K} = \begin{bmatrix} K_{11} & K_{12} & K_{13} & \dots & K_{18} & K_{19} \\ K_{21} & K_{22} & K_{23} & \dots & K_{28} & K_{29} \\ \dots & \dots & \dots & \dots & \dots & \dots \\ K_{91} & K_{92} & K_{93} & \dots & K_{98} & K_{99} \end{bmatrix},$$

$$\mathbf{F} = [\,F_1 \quad F_2 \quad \dots \quad F_8 \quad F_9\,]^T = \left\{ \begin{matrix} f_1^{(1)} \\ f_2^{(1)} + f_1^{(2)} \\ f_2^{(2)} + f_1^{(3)} + f_1^{(4)} \\ f_2^{(4)} + f_1^{(5)} \\ f_2^{(5)} + f_1^{(2)} \\ f_3^{(5)} \\ f_4^{(5)} + f_3^{(4)} + f_2^{(3)} \\ f_3^{(3)} + f_3^{(2)} + f_3^{(1)} \\ f_4^{(1)} \end{matrix} \right\} + \left\{ \begin{matrix} Q_1^{(1)} \\ Q_2^{(1)} + Q_1^{(2)} \\ Q_2^{(2)} + Q_1^{(3)} + Q_1^{(4)} \\ Q_2^{(4)} + Q_1^{(5)} \\ Q_2^{(5)} + Q_1^{(2)} \\ Q_3^{(5)} \\ Q_4^{(5)} + Q_3^{(4)} + Q_2^{(3)} \\ Q_3^{(3)} + Q_3^{(2)} + Q_3^{(1)} \\ Q_4^{(1)} \end{matrix} \right\}.$$

The extended form of the matrix $\mathbf{K}$ in terms of the local functions $K_{ij}^{(e)}$ can be written by replacing the global forms by the respective local forms given in the above relations. This is left as an exercise to the reader since the size of the matrix is too large to present here.

The primary variables are given by: $U_1 = u_1^{(1)}$, $U_2 = u_2^{(1)} = u_1^{(2)}$, $U_3 = u_2^{(2)} = u_1^{(3)} = u_1^{(4)}$, $U_4 = u_2^{(4)} = u_1^{(5)}$, $U_5 = u_2^{(5)}$, $U_6 = u_3^{(5)}$, $U_7 = u_4^{(5)} = u_3^{(4)} = u_2^{(3)}$, $U_8 = u_3^{(3)} = u_3^{(2)} = u_3^{(1)}$, $U_9 = u_4^{(1)}$. This enables us to write the finite element equation $\mathbf{Ku} = \mathbf{F}$. ■

Example 24.8. We will solve Poisson's equation $-\nabla^2 u = 2$ over the triangular region shown in Figure 24.8, subject to the boundary conditions

$$u(x,y) = 5 - 1.5\,y + 2.5\,y^2 \quad \text{on the boundary joining the nodes } \mathbf{1} \text{ and } \mathbf{6},$$
$$-\frac{\partial u}{\partial y} = 0 \quad \text{for } y = 0, \text{ and} \quad \frac{\partial u}{\partial x} = 0 \quad \text{for } x = 0.$$

We will use a uniform mesh of four equivalent triangular elements. The local nodes are chosen such that all four triangles are identical in both geometry and orientation. This reduces the numerical computation significantly in that we compute the required quantities only for one (the first) element. The numbering of the six global nodes is arbitrary. Note that there are no discretization errors in the problem.

Using formulas (24.4.3)-(24.4.4) we find for the element $\Omega^{(1)}$ that $A^{(1)} = 3/16$, and

$$b_1 = -4/3, \quad b_2 = 4/3, \quad b_3 = 0, \quad c_1 = 0, \quad c_2 = -2, \quad c_3 = 2.$$

Then, after using the formulas (24.5.9) and (24.5.14), we get

$$K_{ij}^{(e)} = A^{(e)}\left[b_i^{(e)}\, b_j^{(e)} + c_i^{(e)}\, c_j^{(e)}\right], \quad f_j^{(e)} = \frac{f\,A^{(e)}}{3},$$

which yields

$$\mathbf{K}^{(1)} = \begin{bmatrix} 1/3 & -1/3 & 0 \\ -1/3 & 13/12 & -3/4 \\ 0 & -3/4 & 3/4 \end{bmatrix}, \quad \mathbf{F}^{(1)} = \frac{8}{9}\begin{Bmatrix} 1 \\ 1 \\ 1 \end{Bmatrix} + \begin{Bmatrix} Q_1^{(1)} \\ Q_2^{(1)} \\ Q_3^{(1)} \end{Bmatrix}, \tag{24.5.17}$$

where $Q_j^{(1)} = Q_{11}^{(1)} + Q_{21}^{(1)} + Q_{13}^{(1)}$.

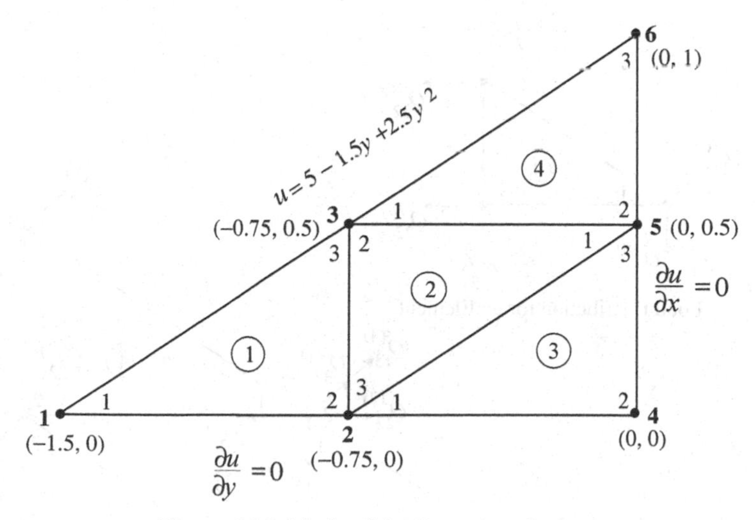

Figure 24.8 Mesh of 4 triangular elements.

The connectivity matrix of the finite element domain is given by

$$\mathbf{C} = \begin{bmatrix} 1 & \mathbf{1} & \mathbf{2} & \mathbf{3} \\ 2 & \mathbf{2} & \mathbf{5} & \mathbf{3} \\ 3 & \mathbf{2} & \mathbf{4} & \mathbf{5} \\ 4 & \mathbf{3} & \mathbf{5} & \mathbf{6} \end{bmatrix}.$$

Thus, the connectivity matrix **K** and the load vector **F** are defined by

$$\mathbf{K} = \begin{bmatrix} K_{11}^{(1)} & K_{12}^{(1)} & K_{13}^{(1)} & 0 & 0 & 0 \\ & K_{22}^{(1)} + K_{33}^{(2)} + K_{11}^{(3)} & K_{23}^{(1)} + K_{32}^{(2)} & K_{12}^{(3)} & K_{31}^{(2)} + K_{13}^{(3)} & 0 \\ & & K_{33}^{(1)} + K_{22}^{(2)} + K_{11}^{(1)} & 0 & K_{21}^{(2)} + K_{12}^{(4)} & K_{13}^{(4)} \\ & & & K_{22}^{(3)} & K_{23}^{(3)} & 0 \\ & \text{sym} & & & K_{11}^{(2)} + K_{33}^{(3)} + K_{22}^{(4)} & K_{32}^{(4)} \\ & & & & & K_{33}^{(4)} \end{bmatrix},$$

$$\mathbf{F} = \begin{Bmatrix} F_1^{(1)} \\ F_2^{(1)} + F_3^{(2)} + F_1^{(3)} \\ F_3^{(1)} + F_2^{(2)} + F_1^{(4)} \\ F_2^{(2)} \\ F_1^{(2)} + F_3^{(3)} + F_2^{(4)} \\ F_3^{(4)} \end{Bmatrix}.$$

Note that $U_1 = 5$, $U_3 = 39/8$, and $U_6 = 6$.

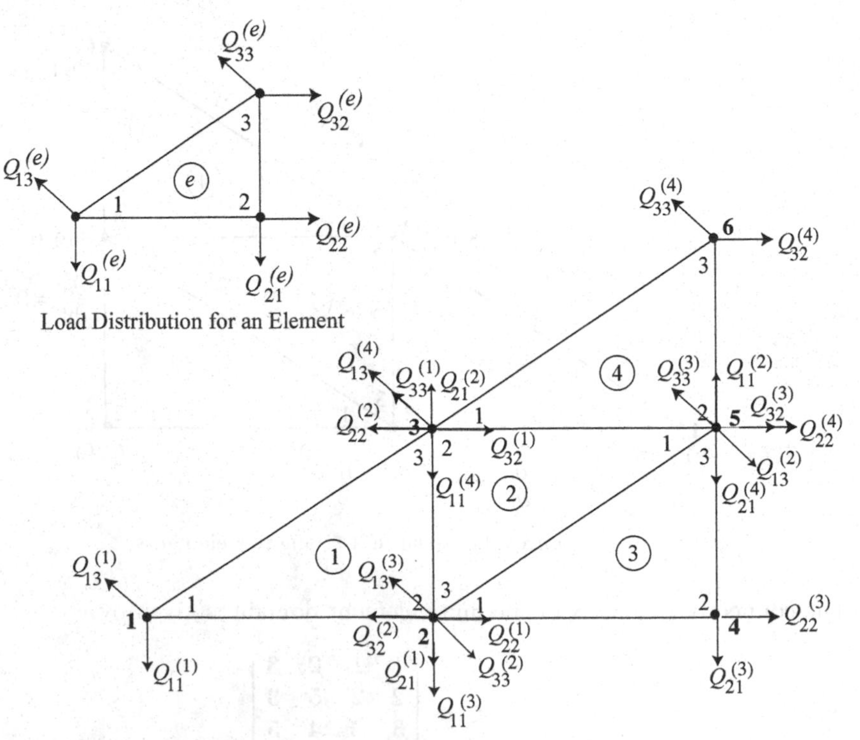

Figure 24.9 Resolution of the vector **Q**.

Since

$$Q_1^{(1)} = Q_{13}^{(1)}, \quad Q_3^{(1)} + Q_2^{(2)} + Q_1^{(4)} = Q_{33}^{(1)} + Q_{13}^{(4)}, \quad Q_3^{(4)} = Q_{33}^{4)},$$

(see Figure 24.9 for the resolution of the vector $\mathbf{Q}$), we use the values from (24.5.17), which holds for all four elements, i.e., $\mathbf{K}^{(1)} = \mathbf{K}^{(2)} = \mathbf{K}^{(3)} = \mathbf{K}^{(4)}$ and $\mathbf{F}^{(1)} = \mathbf{F}^{(2)} = \mathbf{F}^{(3)} = \mathbf{F}^{(4)}$,[†] and using the prescribed boundary conditions solve the system

$$\begin{bmatrix} 1/3 & -1/3 & 0 & 0 & 0 & 0 \\ -1/3 & 13/6 & -3/2 & -1/3 & 0 & 0 \\ 0 & -3/2 & 13/6 & 0 & -2/3 & 0 \\ 0 & 0 & 0 & 13/12 & -3/4 & 0 \\ 0 & 0 & -2/3 & -3/4 & 13/6 & -3/4 \\ 0 & 0 & 0 & 0 & -3/4 & 3/4 \end{bmatrix} \begin{Bmatrix} U_1 = 5 \\ U_2 \\ U_3 = 39/8 \\ U_4 \\ U_5 \\ U_6 = 6 \end{Bmatrix}$$

$$= \frac{8}{9} \begin{Bmatrix} 1 \\ 3 \\ 3 \\ 1 \\ 3 \\ 1 \end{Bmatrix} + \begin{Bmatrix} Q_1^{(1)} \\ Q_2^{(1)} + Q_3^{(2)} + Q_1^{(3)} = 0 \\ Q_3^{(1)} + Q_2^{(2)} + Q_1^{(4)} \\ Q_2^{(2)} = 0 \\ Q_1^{(2)} + Q_3^{(3)} + Q_2^{(4)} = 0 \\ Q_3^{(4)} \end{Bmatrix}.$$

The values of the unknown quantities are

$$U_2 = 4.67601, \quad U_4 = 5.45655, \quad U_5 = 6.6965.$$

These values then give

$$Q_{13}^{(1)} = -2.55867, \quad Q_{33}^{(1)} + Q_{13}^{(4)} = -12.5112, \quad Q_{33}^{(4)} = -1.41126. \blacksquare$$

Example 24.9. We use two linear triangular elements over the unit square (Figure 24.10) and solve Poisson's equation

$$\frac{\partial^2 u}{\partial x^2} + \frac{\partial^2 u}{\partial y^2} = 0, \quad (x, y) \in \Omega = [0, 1] \times [0, 1], \tag{24.5.18a}$$

subject to the boundary conditions

$$\begin{aligned} &u(1, y) = 1, && \frac{\partial u}{\partial x}(0, y) = 0, \quad 0 \le y \le 1, \\ &\frac{\partial u}{\partial y}(x, 0) = 0 && \frac{\partial u}{\partial y}(x, 1) = 1 - u(x, 1), \quad 0 \le x \le 1. \end{aligned} \tag{24.5.18b}$$

For a linear triangular element $\Omega^{(e)}$ the stiffness matrix is defined by

$$K_{ij}^{(e)} = A^{(e)} \left[b_i^{(e)} b_j^{(e)} + c_i^{(e)} c_j^{(e)} \right]. \tag{24.5.19}$$

The connectivity matrix is

$$\mathbf{C} = \begin{bmatrix} \mathbf{1} & \mathbf{4} & \mathbf{2} \\ \mathbf{3} & \mathbf{4} & \mathbf{1} \end{bmatrix},$$

[†] If the local nodes are numbered counterclockwise in a manner different from that in Figure 24.8, the stiffness matrix and the force vector must be computed for each element separately before their assembly and the solution of the system (24.5.7).

where we have dropped the element numbers. For the element $\Omega^{(1)}$, we have $A^{(1)} = 0.5$, $b_1^{(1)} = 0$, $b_2^{(1)} = 1$, $b_3^{(1)} = -1$, $c_1^{(1)} = -1$, $c_2^{(1)} = 0$, $c_3^{(1)} = 1$. Then $K_{11}^{(1)} = 0.5$, $K_{12}^{(1)} = 0 = K_{21}^{(1)}$, $K_{13}^{(1)} = -0.5 = K_{31}^{(1)}$, $K_{22}^{(1)} = 0.5$, $K_{23}^{(1)} = -0.5 = K_{32}^{(1)}$, and $K_{33}^{(1)} = 1$, which yield

$$\mathbf{K}^{(1)} = \begin{bmatrix} 0.5 & 0 & -0.5 \\ 0 & 0.5 & -0.5 \\ -0.5 & -0.5 & 1 \end{bmatrix}.$$

Similarly, for the element $\Omega^{(2)}$, we have $A^{(2)} = 0.5$, $b_1^{(2)} = 1$, $b_2^{(2)} = 0$, $b_3^{(2)} = -2$, $c_1^{(2)} = -1$, $c_2^{(2)} = 1$, $c_3^{(2)} = 0$. Then $K_{11}^{(2)} = 1$, $K_{12}^{(2)} = -0.5 = K_{21}^{(2)} = K_{13}^{(2)} = K_{31}^{(2)}$, $K_{22}^{(2)} = 0.5$, $K_{23}^{(2)} = 0 = K_{32}^{(2)}$, and $K_{33}^{(2)} = 0.5$, which yield

$$\mathbf{K}^{(2)} = \begin{bmatrix} 1 & -0.5 & -0.5 \\ -0.5 & 0.5 & 0 \\ -0.5 & 0 & 0.5 \end{bmatrix}.$$

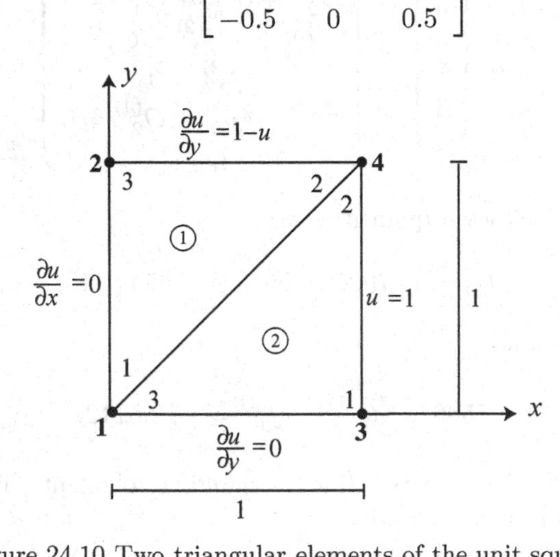

Figure 24.10 Two triangular elements of the unit square.

Hence, after assembly the stiffness matrix is given by

$$\mathbf{K} = \begin{bmatrix} 1 & -0.5 & -0.5 & 0 \\ -0.5 & 1 & 0 & -0.5 \\ -0.5 & 0 & 1 & -0.5 \\ 0 & -0.5 & -0.5 & 1 \end{bmatrix},$$

and the force vector $\mathbf{f} = \mathbf{0}$.

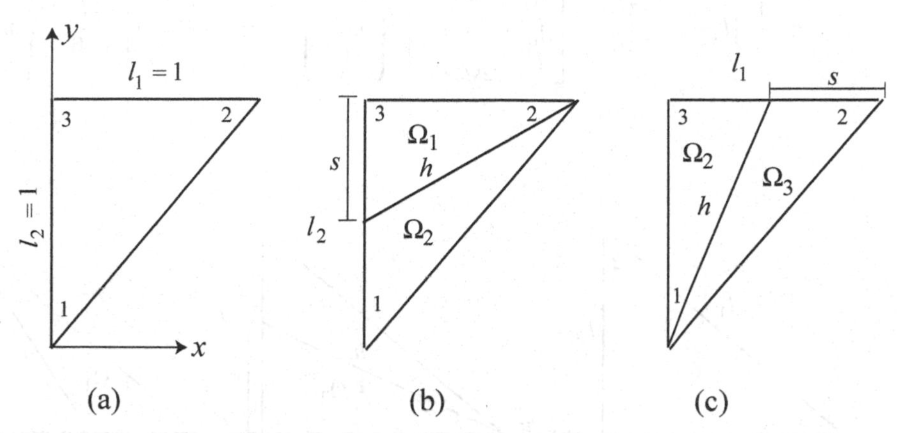

Figure 24.11 2-Element triangular mesh.

Now, before we solve the equation $\mathbf{Ku} = \mathbf{Q}$, we will evaluate the vector $\mathbf{Q}$. Since $\phi_1 = 0$ on l_1 and l_3, and $\phi_1 = \dfrac{sh/2}{l_2 h/2} = \dfrac{s}{l_2}$ on l_2 (see Figure 24.11b), and $\phi_2 = 0$ on l_2 and l_3, and $\phi_1 = 0$, $\phi_2 = \dfrac{(l_1 - s)\,h/2}{l_1 h/2} = \dfrac{l_1 - s}{l_1}$, $\phi_3 = \dfrac{sh/2}{l_1 h/2} = \dfrac{s}{L} - 1$ on l_1, we get

$$
\begin{aligned}
Q_1^{(1)} &= \int_0^{l_3} \phi_1\, q_n\, ds + \int_0^{l_1} \phi_1\, q_n\, ds + \int_0^{l_2} \phi_1\, q_n\, ds = 0 + 0 + \int_0^{l_2} \phi_1\, q_n\, ds = \int_0^{l_2} \frac{s}{l_2}\, q_1^{(1)}\, ds; \\
Q_2^{(1)} &= \int_0^{l_3} \phi_2\, q_n\, ds + \int_0^{l_1} \phi_2\, q_n\, ds + \int_0^{l_2} \phi_2\, q_n\, ds = 0 + \int_0^{l_1} \phi_2\, q_n\, ds + 0 = \int_0^{l_1} \phi_2\,(1-u)\, ds \\
&= \int_0^{l_1} \frac{l_1 - s}{l_1}\, ds - \sum_{i-1}^{3} \int_0^{l_1} \phi_2\, \phi_i\, ds\, u_i, \text{ where } u = \sum_{i=1}^{3} \phi_i\, u_i, \\
&= \frac{1}{l_1}\left[l_1 s - s^2/2\right]_0^{l_1} - \begin{bmatrix} \int_0^{l_1} \phi_2\phi_1\, ds & \int_0^{l_1} \phi_2\phi_2\, ds & \int_0^{l_1} \phi_2\phi_2\, ds \end{bmatrix} \begin{Bmatrix} u_1^{(1)} \\ u_2^{(1)} \\ u_3^{(1)} \end{Bmatrix} \\
&= \frac{l_1}{2} - \begin{bmatrix} 0 & \dfrac{l_1}{3} & \dfrac{l_1}{6} \end{bmatrix} \begin{Bmatrix} u_1^{(1)} \\ u_2^{(1)} \\ u_3^{(1)} \end{Bmatrix}.
\end{aligned}
$$

Similarly, since $\phi_1 = 0$, $\phi_2 = \dfrac{l_1 - s}{l_1}$, $\phi_3 = \dfrac{s}{l_1}$ on l_1, and $\phi_1 = \dfrac{sh/2}{l_2 h/2} = \dfrac{s}{l_2}$, $\phi_2 = 0$, $\phi_3 = \dfrac{(l_2 - s)\,h/2}{l_2 h/2} = \dfrac{l_2 - s}{l_2}$ (Figure 24.11c), we get

$$
\begin{aligned}
Q_3^{(1)} &= \int_0^{l_3} \phi_3\, q_n\, ds + \int_0^{l_1} \phi_3\, q_n\, ds + \int_0^{l_2} \phi_3\, q_n\, ds = 0 + \int_0^{l_1} \phi_3\, q_n\, ds + \int_0^{l_2} \phi_3\, q_n\, ds \\
&= \int_0^{l_1} \phi_3\,(1-u)\, ds + \int_0^{l_2} \phi_3 q_1^{(1)}\, ds \\
&= \int_0^{l_2} \frac{l_2 - s}{l_2}\, q_1^{(1)}\, ds + \int_0^{l_1} \frac{s}{l_1}\, ds + \sum_{i=1}^{3} \int_0^{l_1} \phi_3\, \phi_i\, ds\, u_i, \text{ where } u = \sum_{i=1}^{3} \phi_i\, u_i,
\end{aligned}
$$

$$
= \frac{1}{l_1}\left[l_1 s - \frac{s^2}{2}\right]_0^{l_1} - \begin{bmatrix} \int_0^{l_1} \phi_2\phi_1\, ds \\ \int_0^{l_1} \phi_2\phi_2\, ds \\ \int_0^{l_1} \phi_2\phi_3\, ds \end{bmatrix} \begin{Bmatrix} u_1^{(1)} \\ u_2^{(1)} \\ u_3^{(1)} \end{Bmatrix} = \frac{l_1}{2} - \begin{bmatrix} 0 & \frac{l_1}{3} & \frac{l_1}{6} \end{bmatrix} \begin{Bmatrix} u_1^{(1)} \\ u_2^{(1)} \\ u_3^{(1)} \end{Bmatrix}.
$$

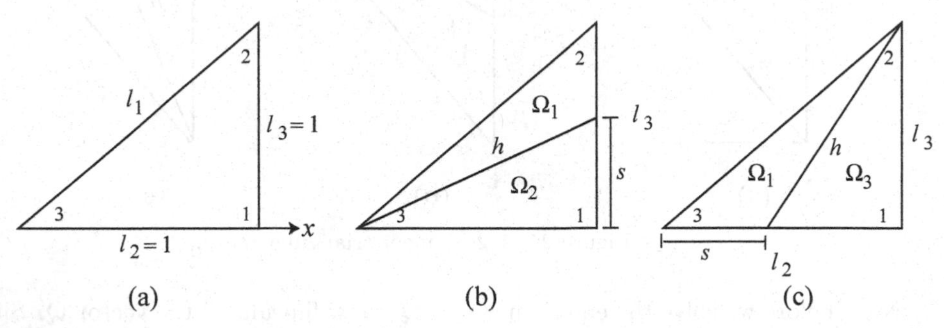

Figure 24.12 2-Element triangular mesh.

Hence,

$$
\begin{aligned}
\mathbf{Q}^{(1)} &= \begin{Bmatrix} \int_0^{l_2} \frac{s}{l_2}\, q_1^{(1)}\, ds \\ l_1/2 \\ \int_0^{l_2} \frac{l_2 - s}{l_2}\, q_1^{(1)}\, ds + \frac{l_1}{2} \end{Bmatrix} - \begin{bmatrix} 0 & 0 & 0 \\ 0 & l_1/3 & l_1/6 \\ 0 & l_1/6 & l_1/3 \end{bmatrix} \begin{Bmatrix} u_1^{(1)} \\ u_2^{(1)} \\ u_3^{(1)} \end{Bmatrix} \\
&= \begin{Bmatrix} \int_0^{1} \frac{s}{l_1}\, q_1^{(1)}\, ds \\ 0.5 \\ \int_0^{1} (1-s)\, q_1^{(1)}\, ds + 0.5 \end{Bmatrix} - \begin{bmatrix} 0 & 0 & 0 \\ 0 & 1/3 & 1/6 \\ 0 & 1/6 & 1/3 \end{bmatrix} \begin{Bmatrix} u_1^{(1)} \\ u_2^{(1)} \\ u_3^{(1)} \end{Bmatrix},
\end{aligned}
\tag{24.5.20}
$$

where $l_1 = l_2 = 1$.

Now, for the element 2, refer to Figure 24.12(a-c), and note that $\phi_1 = 0$ on l_1; $\phi_1 = \frac{sh/2}{l_2 h/2} = \frac{s}{l_2}$, $\phi_2 = 0$, and $\phi_3 = \frac{(l_2 - s)\,h/2}{l_2 h/2} = \frac{l_2 - s}{l_2}$ on l_2 (Figure 24.12c); and $\phi_1 = \frac{(l_3 - s)\,h/2}{l_3 h/2} = \frac{l_3 - s}{l_3}$, $\phi_2 = \frac{sh/2}{l_3 h/2} = \frac{s}{l_3}$, and $\phi_3 = 0$ on l_3 (Figure 24.12b). Thus, we have

$$
\begin{aligned}
Q_1^{(2)} &= \int_0^{l_3} \phi_1\, q_n\, ds + \int_0^{l_1} \phi_1\, q_n\, ds + \int_0^{l_2} \phi_1\, q_n\, ds = \int_0^{l_3} \phi_1\, q_1^{(2)}\, ds + 0 + \int_0^{l_2} \phi_1\, q_2^{(2)}\, ds \\
&= \int_0^{l_3} \frac{l_3 - s}{l_3}\, q_1^{(2)}\, ds + \int_0^{l_2} \frac{s}{l_2}\, q_2^{(2)}\, ds; \\
Q_2^{(2)} &= \int_0^{l_3} \phi_2\, q_n\, ds = \int_0^{l_3} \frac{s}{l_3}\, q_1^{(2)}\, ds; \\
Q_3^{(2)} &= \int_0^{l_2} \phi_3\, q_n\, ds = \int_0^{l_2} \frac{l_2 - s}{l_2}\, q_2^{(2)}\, ds.
\end{aligned}
$$

Thus,

$$\mathbf{Q}^{(2)} = \begin{Bmatrix} \int_0^{l_3} \frac{l_3 - s}{l_3}\, q_1^{(2)}\, ds + \int_0^{l_2} \frac{s}{l_2}\, q_2^{(2)}\, ds \\ \int_0^{l_3} \frac{s}{l_3}\, q_1^{(2)}\, ds \\ \int_0^{l_2} \frac{l_2 - s}{l_2}\, q_2^{(2)}\, ds \end{Bmatrix} - \begin{bmatrix} 0 & 0 & 0 \\ 0 & 0 & 0 \\ 0 & 0 & 0 \end{bmatrix} \begin{Bmatrix} u_1^{(1)} \\ u_2^{(1)} \\ u_3^{(1)} \end{Bmatrix}$$
$$= \begin{Bmatrix} \int_0^1 (1-s)\, q_1^{(2)}\, ds + \int_0^1 s q_2^{(2)}\, ds \\ \int_0^1 s q_1^{(2)}\, ds \\ \int_0^1 (1-s)\, q_2^{(2)}\, ds \end{Bmatrix} - \begin{bmatrix} 0 & 0 & 0 \\ 0 & 0 & 0 \\ 0 & 0 & 0 \end{bmatrix} \begin{Bmatrix} u_1^{(1)} \\ u_2^{(1)} \\ u_3^{(1)} \end{Bmatrix}, \tag{24.5.21}$$

where $l_2 = l_3 = 1$. Then, combining (24.5.20) and (24.5.21) we obtain

$$\mathbf{Q} = \begin{Bmatrix} \int_0^1 s q_1^{(1)}\, ds + \int_0^1 (1-s)\, q_2^{(2)}\, ds \\ \int_0^1 (1-s) q_1^{(1)}\, ds + \frac{1}{2} \\ \int_0^1 (1-s)\, q_1^{(2)}\, ds + \int_0^1 s q_2^{(2)}\, ds \\ \int_0^1 s q_1^{(2)}\, ds + \frac{1}{2} \end{Bmatrix} - \begin{bmatrix} 0 & 0 & 0 & 0 \\ 0 & 1/3 & 0 & 1/6 \\ 0 & 0 & 0 & 0 \\ 0 & 1/6 & 0 & 1/3 \end{bmatrix} \begin{Bmatrix} U_1 \\ U_2 \\ U_3 \\ U_4 \end{Bmatrix} \tag{24.5.22}$$
$$= \mathbf{Q}_b - \mathbf{K}_b\, \mathbf{u}.$$

Thus, since $\mathbf{K}\mathbf{u} = \mathbf{Q} = \mathbf{Q}_b - \mathbf{K}_b\, \mathbf{u}$, we get

$$\left(\mathbf{K} + \mathbf{K}_b\right) \mathbf{u} = \mathbf{Q}_b, \tag{24.5.23}$$

where

$$\mathbf{K} + \mathbf{K}_b = \begin{bmatrix} 1 & -0.5 & -0.5 & 0 \\ -0.5 & 1 & 0 & -0.5 \\ -0.5 & 0 & 1 & -0.5 \\ 0 & -0.5 & -0.5 & 1 \end{bmatrix} + \begin{bmatrix} 0 & 0 & & 0 \\ 0 & 1/3 & 0 & 1/6 \\ 0 & 0 & 0 & 0 \\ 0 & 1/6 & 0 & 1/3 \end{bmatrix} \begin{bmatrix} 1 & -0.5 & -0.5 & 0 \\ -0.5 & 4/3 & 0 & -1/3 \\ -0.5 & 0 & 1 & -0.5 \\ 0 & -1/3 & -0.5 & 4/3 \end{bmatrix}.$$

Then, from (24.5.23) we obtain

$$\begin{bmatrix} 1 & -0.5 & -0.5 & 0 \\ -0.5 & 4/3 & 0 & -1/3 \\ -0.5 & 0 & 1 & -0.5 \\ 0 & -1/3 & -0.5 & 4/3 \end{bmatrix} \begin{Bmatrix} U_1 \\ U_2 \\ U_3 \\ U_4 \end{Bmatrix} = \begin{Bmatrix} \int_0^1 s q_1^{(1)}\, ds + \int_0^1 (1-s)\, q_2^{(2)}\, ds \\ \int_0^1 (1-s) q_1^{(1)}\, ds + 0.5 \\ \int_0^1 (1-s)\, q_1^{(2)}\, ds + \int_0^1 s q_2^{(2)}\, ds \\ \int_0^1 s q_1^{(2)}\, ds + 0.5 \end{Bmatrix}$$
$$= \begin{Bmatrix} 0 \\ 0.5 \\ \int_0^1 (1-s)\, q_1^{(2)}\, ds \\ \int_0^1 s q_1^{(2)}\, ds + 0.5 \end{Bmatrix}.$$

After applying the essential boundary conditions $U_1 = U_4 = 1$, this system simplifies to

$$\begin{bmatrix} 1 & -0.5 \\ -0.5 & 4/3 \end{bmatrix} \begin{Bmatrix} U_1 \\ U_2 \end{Bmatrix} = \begin{Bmatrix} 0.5 \\ 5/6 \end{Bmatrix},$$

which yields $U_1 = 1 = U_2$, and $q_1^{(2)} = 0$. ■

Example 24.10. We will solve Laplace's equation $\nabla^2 u = 0$ to compute the two-dimensional electrostatic field due to the electrodes shown in Figure 24.13.

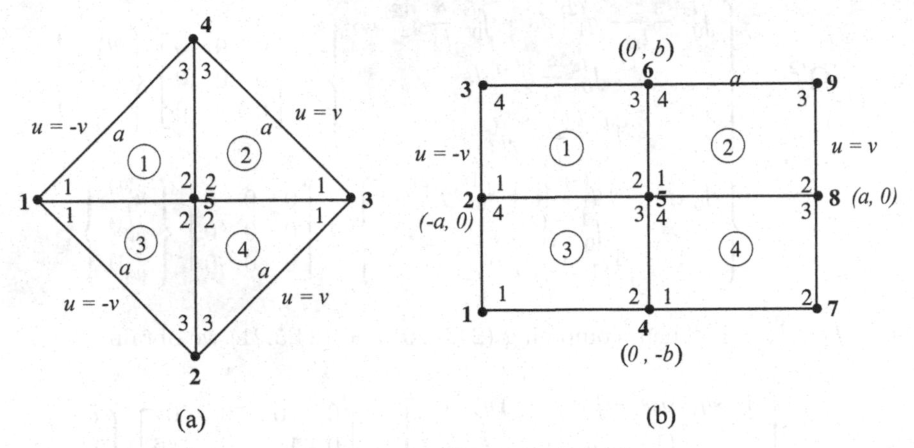

Figure 24.13 Electrostatic field.

Part (a). Four Identical Triangular Elements. We consider the four identical triangular elements (Figure 24.13a), and compute the stiffness matrix and the force vector for the element $\Omega^{(1)}$. Thus, from (24.4.3)-(24.4.4) we have $A^{(1)} = a^2/4$, and

$$
\begin{aligned}
b_1^{(1)} &= -\sqrt{2}/a, \quad b_2^{(1)} = \sqrt{2}/a, \quad b_3^{(1)} = 0, \\
c_1^{(1)} &= 0, \quad c_2^{(1)} = -\sqrt{2}/a, \quad c_3^{(1)} = \sqrt{2}/a,
\end{aligned}
$$

which leads to

$$
\mathbf{K}^{(1)} = \frac{1}{2}\begin{bmatrix} 1 & -1 & 0 \\ -1 & 2 & -1 \\ 0 & -1 & 1 \end{bmatrix}, \quad \mathbf{F}^{(1)} = \{\mathbf{0}\} + \begin{Bmatrix} Q_1^{(1)} \\ Q_2^{(1)} \\ Q_3^{(1)} \end{Bmatrix}. \tag{24.5.24}
$$

Note that by taking the numbering of the local nodes as in Figure 24.13(a), the above results hold for all four $\mathbf{K}^{(e)}$ and $\mathbf{F}^{(e)}$ for $e = 1, 2, 3, 4$. Since the connectivity matrix is

$$
\mathbf{C} = \begin{bmatrix} \mathbf{1} & \mathbf{5} & \mathbf{4} \\ \mathbf{3} & \mathbf{5} & \mathbf{4} \\ \mathbf{1} & \mathbf{5} & \mathbf{2} \\ \mathbf{3} & \mathbf{5} & \mathbf{2} \end{bmatrix}.
$$

The stiffness matrix $\mathbf{K}$ and the force vector $\mathbf{F}$ are given by

$$
\mathbf{K} = \begin{bmatrix} K_{11}^{(1)} + K_{11}^{(3)} & K_{12}^{(3)} & 0 & K_{13}^{(1)} & K_{12}^{(1)} + K_{12}^{(3)} \\ & K_{33}^{(3)} + K_{33}^{(4)} & K_{31}^{(4)} & 0 & K_{32}^{(3)} + K_{32}^{(4)} \\ & & K_{11}^{(2)} + K_{11}^{(4)} & K_{13}^{(2)} & K_{12}^{(2)} + K_{12}^{(4)} \\ & \text{sym} & & K_{33}^{(1)} + K_{33}^{(2)} & K_{32}^{(1)} + K_{32}^{(2)} \\ & & & & K_{22}^{(1)} + K_{22}^{(2)} + K_{22}^{(3)} + K_{22}^{(4)} \end{bmatrix},
$$

$$
\mathbf{F} = [-v \quad -v \quad v \quad v \quad 0]^T.
$$

Hence, we solve the system (taking $a = 1$)

$$\begin{bmatrix} 1 & 0 & 0 & 0 & -1 \\ 0 & 1 & 0 & 0 & -1 \\ 0 & 0 & 1 & 0 & -1 \\ 0 & 0 & 0 & 1 & -1 \\ -1 & -1 & -1 & -1 & 4 \end{bmatrix} \begin{Bmatrix} U_1 \\ U_2 \\ U_3 \\ U_4 \\ U_5 \end{Bmatrix} = \begin{Bmatrix} -v \\ -v \\ v \\ v \\ 0 \end{Bmatrix},$$

which gives $U_1 = -v$, $U_2 = -v$, $U_3 = v$, $U_4 = v$, and $U_5 = 0$. The exact solution is given by (see Lebedev et al. [1965, Problem 271]

$$u(x,y) = \frac{v\sqrt{2}}{a}(x+y) - \frac{4v}{\pi}\sum_{n=1}^{\infty}\Big\{(-1)^n \sinh\frac{n\pi\left[a-\sqrt{2}\,(y-x)\right]}{2a} + \sinh\frac{n\pi\left[a+\sqrt{2}\,(y-x)\right]}{2a}\Big\}\frac{\sin\frac{n\pi\left[a-\sqrt{2}\,(y+x)\right]}{2a}}{n\sinh n\pi},$$

where $u(x,y)$ denotes the electrostatic potential. The above results match with those obtained from this exact solution.

Part (b). Four Identical Rectangular Elements. We consider the four identical triangular elements (Figure 24.13b), and compute the stiffness matrix and the force vector for the element 1. This is left as an exercise (Exercise 6.6).

The exact solution is given by (Lebedev et al. [1965, Problem 271]

$$u(x,y) = v\left[\frac{x}{a} + \frac{2}{\pi}\sum_{n=1}^{\infty}\frac{\cosh(n\pi y/a)}{\cosh(n\pi b/a)}\frac{\sin(n\pi x/a)}{n}\right],$$

where $u(x,y)$ denotes the electrostatic potential. The above results match with those obtained from this exact solution. ■

Example 24.11. Find the distribution of d-c current in a thin rectangular sheet, if the current is applied by electrodes at the points $x = -a$, $y = 0$ and $x = a$, $y - 0$ (see Figure 24.14). This problem is equivalent to solving Laplace's equation

$$-\Big(\frac{\partial^2 u}{\partial x^2} + \frac{\partial^2 u}{\partial y^2}\Big) = 0 \quad \text{in } \Omega = \big\{(x,y) : -a \le x \le a, -b \le y \le b\big\}, \tag{24.5.25}$$

subject to the boundary conditions

$$\frac{\partial u}{\partial x}\Big|_{y=\pm b} = 0. \tag{24.5.26}$$

$$\frac{\partial u}{\partial x}\Big|_{y=\pm a} = f(y) = \begin{cases} -\dfrac{J}{2\varepsilon kh} \equiv A & \text{if } |y| < \varepsilon, \\ 0 \quad \text{if } |y| > \varepsilon. \end{cases} \tag{24.5.27}$$

We will use the following values: Total current $J = 1.198 \times 10^3$ A; conductivity $k = 0.599 \times 10^8\,\Omega^{-1}$ m^{-1}; thickness of the sheet $h = 0.1$ cm $= 10^{-3}$ m; $\varepsilon = 0.1$ cm $= 10^{-3}$ m;

$a = 5$ cm, $b = 3$ cm, and $c = 0.1$ cm. Then $A = -10A \cdot \Omega \cdot m^{-1}$.

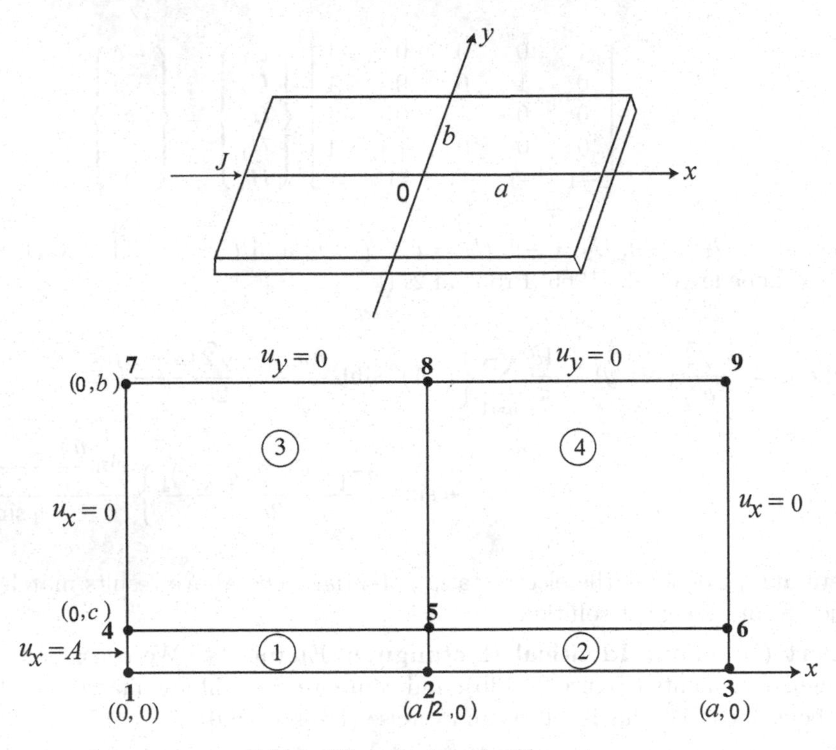

Mesh of Four Rectangular Elements

Figure 24.14 Distribution of d-c current.

Since there exists a bi-axial symmetry, we will model the first quadrant of the domain for finite element analysis. To discretize this quadrant we use a 2×2 mesh of four rectangular elements, of which the element 1 is identical to the element 2, and the element 3 is identical to the element 4. There is no discretization error in this model.

We find that for the element 1 and 2, each with sides 0.025 m and 0.001 m, the stiffness matrix is given by

$$\mathbf{K}^{(1)} = \begin{bmatrix} \frac{626}{75} & \frac{623}{150} & -\frac{313}{75} & -\frac{1249}{150} \\ \frac{623}{150} & \frac{626}{75} & -\frac{1249}{150} & -\frac{313}{75} \\ -\frac{313}{75} & -\frac{1249}{150} & \frac{626}{75} & \frac{623}{150} \\ -\frac{1249}{150} & -\frac{313}{75} & \frac{623}{150} & \frac{626}{75} \end{bmatrix} = \mathbf{K}^{(2)}.$$

Similarly, for the elements 3 and 4, each with sides 2.5 cm and 2.9 cm, we have

$$\mathbf{K}^{(3)} = \begin{bmatrix} \frac{1466}{2175} & -\frac{1057}{4350} & -\frac{733}{2175} & -\frac{409}{4350} \\ -\frac{1057}{4350} & \frac{1466}{2175} & -\frac{409}{4350} & -\frac{733}{2175} \\ -\frac{733}{2175} & -\frac{409}{4350} & \frac{1466}{2175} & -\frac{1057}{4350} \\ -\frac{409}{4350} & -\frac{733}{2175} & -\frac{1057}{4350} & \frac{1466}{2175} \end{bmatrix} = \mathbf{K}^{(4)}.$$

The connectivity matrix is

$$\mathbf{C} = \begin{bmatrix} 1 & 2 & 5 & 4 \\ 2 & 3 & 6 & 5 \\ 4 & 5 & 8 & 7 \\ 5 & 6 & 9 & 8 \end{bmatrix}.$$

Thus, after assembly the stiffness matrix is given by

$$\mathbf{K} = \begin{bmatrix} \frac{626}{75} & \frac{623}{150} & 0 & -\frac{1249}{150} & -\frac{313}{75} & 0 & 0 & 0 & 0 \\ \frac{623}{150} & \frac{1252}{75} & -\frac{313}{75} & -\frac{313}{75} & -\frac{1249}{75} & -\frac{313}{75} & 0 & 0 & 0 \\ 0 & -\frac{313}{75} & \frac{626}{75} & 0 & -\frac{313}{75} & -\frac{1249}{150} & 0 & 0 & 0 \\ -\frac{1249}{150} & -\frac{313}{75} & 0 & \frac{1308}{145} & \frac{567}{145} & 0 & -\frac{409}{4350} & -\frac{733}{2175} & 0 \\ -\frac{313}{75} & -\frac{1249}{75} & -\frac{313}{75} & \frac{567}{145} & \frac{2616}{145} & \frac{567}{145} & -\frac{737}{2175} & -\frac{409}{2175} & -\frac{737}{2175} \\ 0 & -\frac{313}{75} & -\frac{1249}{150} & 0 & \frac{567}{145} & \frac{1308}{145} & 0 & -\frac{733}{2175} & -\frac{409}{2175} \\ 0 & 0 & 0 & -\frac{409}{4350} & -\frac{733}{2175} & 0 & \frac{1466}{2175} & -\frac{1057}{4350} & 0 \\ 0 & 0 & 0 & -\frac{733}{2175} & -\frac{409}{2175} & -\frac{733}{2175} & -\frac{1057}{4350} & \frac{2932}{2175} & -\frac{1057}{4350} \\ 0 & 0 & 0 & 0 & -\frac{733}{2175} & -\frac{409}{2175} & 0 & -\frac{1057}{4350} & \frac{1466}{2175} \end{bmatrix}.$$

The vector $\mathbf{f} = \mathbf{0}$, and thus,

$$\mathbf{F} = \mathbf{Q} = \begin{Bmatrix} Q_1^{(1)} \\ Q_2^{(1)} + Q_1^{(2)} \\ Q_2^{(2)} \\ Q_4^{(1)} + Q_1^{(3)} \\ Q_3^{(1)} + Q_4^{(2)} + Q_2^{(3)} + Q_1^{(4)} \\ Q_3^{(2)} + Q_2^{(4)} \\ Q_4^{(3)} \\ Q_3^{(3)} + Q_4^{(4)} \\ Q_3^{(4)} \end{Bmatrix}.$$

Since $Q_1^{(1)} = 0$, $Q_2^{(1)} + Q_1^{(2)} = 0$, $Q_2^{(2)} = \dfrac{0.001\,A}{2} = \dfrac{0.001 \times (-10)}{2} = -0.005$, $Q_4^{(1)} + Q_1^{(3)} = 0$, $Q_3^{(1)} + Q_4^{(2)} + Q_2^{(3)} + Q_1^{(4)} = 0$, $Q_3^{(2)} + Q_2^{(4)} = 0$, $Q_4^{(3)} = 0$, $Q_3^{(3)} + Q_4^{(4)} = 0$, and $Q_3^{(4)} = 0$, we solve the system $\mathbf{KU} = \mathbf{F}$, and obtain $U_1 = 4.31058 \times 10^{-4}$, $U_2 = 5.64686 \times 10^{-4}$, $U_3 = 2.99468 \times 10^{-5}$, $U_4 = 4.45957 \times 10^{-4}$, $U_5 = 5.34319 \times 10^{-4}$, $U_6 = 7.96766 \times 10^{-5}$, $U_7 = 4.60917 \times 10^{-4}$, $U_8 = 3.649 \times 10^{-4}$, and $U_9 = 4.20937 \times 10^{-4}$. The exact solution, given by

$$u(x, y) = -\frac{J}{2kh}\left[\frac{x}{b} + \frac{2}{\pi}\sum_{n=1}^{\infty} \frac{\sinh(n\pi x/b)}{n\cosh(n\pi a/b)} \cos\frac{n\pi y}{b}\right] + \text{const.}$$

(see Lebedev et al. [1965, Problem 190]), yields results which differ from the finite element solution, because of the unknown constant in the exact solution. ∎

Example 24.12. Consider Laplace's equation $-\nabla^2 u = 0$ on a rectangle $\Omega = \{(x, y) : 0 < x < \pi,\ 0 < y < 1\}$ such that $u(x, 0) = \sin x$, $u(x, 1) = 1 + \sin 2x$, $u(0, y) = y = u(\pi, y)$, by considering a mesh of 2×2 linear triangular elements, as shown in Figure 24.15.

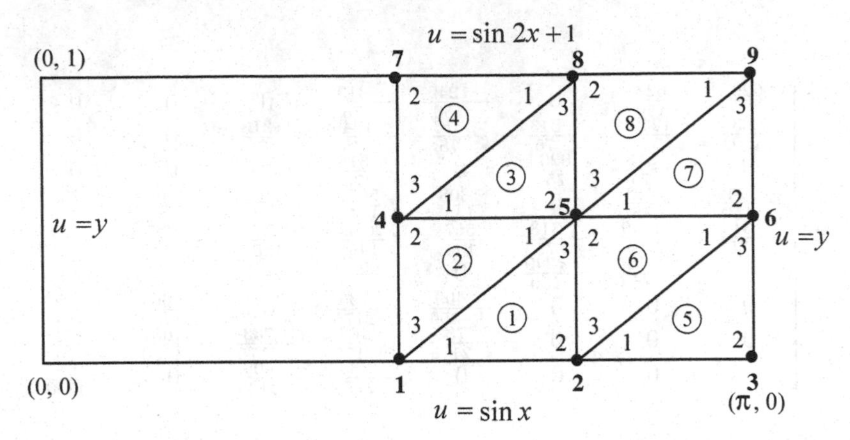

Figure 24.15 Heat conduction in a rectangle.

We find that

$$\mathbf{K}^{(e)} = \frac{1}{\pi} \begin{bmatrix} \pi^2/4+1 & -1 & -\pi^2/4 \\ -1 & 1 & 0 \\ -\pi^2/4 & 0 & \pi^2/4 \end{bmatrix}.$$

Note that $U_1 = 1$, $U_2 = 1/\sqrt{2}$, $U_3 = 0$, $U_6 = 1/2$, $U_7 = 1$, $U_8 = 0$, and $U_9 = 1$. Also, $f = 0$. Then, solve

$$\begin{bmatrix} K_{44}^{(e)} & K_{45}^{(e)} \\ K_{54}^{(e)} & K_{55}^{(e)} \end{bmatrix} \begin{bmatrix} U_4 \\ U_5 \end{bmatrix} = \begin{Bmatrix} K_{41}^{(e)} U_1 + K_{47}^{(e)} U_7 \\ K_{52}^{(e)} U_2 + K_{56}^{(e)} U_6 \end{Bmatrix}.$$

We have $U_4 = 0.324782$, $U_5 = 0.252299$. The exact solution is given by

$$u(x,y) = y + \Big(\cosh y - \frac{\cosh 1}{\sinh 1} \sinh y\Big) \sin x + \frac{\sinh 2y}{\sinh 2} \sin 2x,$$

which gives $U_4 = u(\pi/2, 1/2) = 0.943409$, $U_5 = u(\pi, 1/2) = 0.489511$. ■

Example 24.13. Consider Laplace's equation $-\nabla^2 u = 0$ on a rectangle $\Omega = \{(x,y) : 0 < x < \pi,\ 0 < y < 1\}$ such that $u(x,0) = \cos x$, $u(x,1) = \sin^2 x$, $u_x(0,y) = 0 = u_x(\pi,y)$, by considering a mesh of 2×2 linear triangular elements, as shown in Figure 24.15. Using the matrix $\mathbf{K}^{(e)}$ in Example 24.12, and noting that $U_1 = 0$, $U_2 = -1/\sqrt{2}$, $U_3 = -1$, $U_7 = 1$, $U_8 = 1/2$, and $U_9 = 0$, we obtain $U_4 = 0.140261$, $U_5 = -0.0273159$, and $U_6 = -0.30838$. The exact solution is given by (see Kythe et al. [2003:132-133])

$$u(x,y) = y + \Big(\cosh y - \frac{\cosh 1}{\sinh 1} \sinh y\Big) \cos x - \frac{\sinh 2y}{\sinh 2} \cos 2x,$$

which yields $u(\pi/2, 0.5) = U_4 = 0.574027$, $u(3\pi/4.0.5) = U_5 = -0.0635378$, $u(\pi, 0.5) = U_6 = -0.517437$. ■

Example 24.14. Consider Laplace's equation $-\nabla^2 u = 0$ in a half-strip (see Figure 24.16) subject to the boundary conditions

$$u(0,y) = f(y), \qquad \lim_{x \to +\infty} u(x,y) = u_\infty,$$
$$u_y(x,0) = 0, \qquad u_y(x,b) + \beta[u(x,b) - u_\infty] = 0,$$

where β is the film coefficient. Take $f(y) = 50\ (1 + y)$, and $u_\infty = 20°$ C. ■

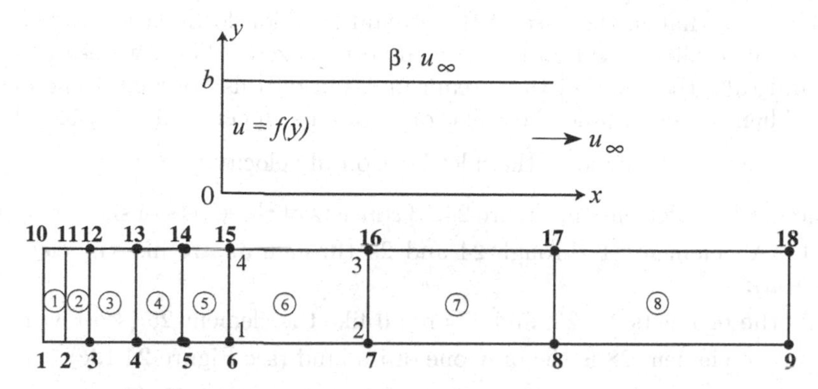

Figure 24.16 Rectangular mesh of the finite half-strip.

24.6 Fluid Flows

We will study two dimensional steady-state fluid flows of ideal fluids. These flows are also known as the potential flows, which are governed by Poisson's equation. We will solve the flow problem around an elliptic cylinder and that of a partially filled circular cylinder.

Example 24.15. (Flow around an elliptical cylinder) Formulate and solve the problem of irrotational flow of an ideal fluid around an elliptical cylinder using (a) the stream function ψ, and (b) the velocity potential ϕ. The flow in the case (a) is governed by Laplace's equation $\nabla^2\psi = 0$, and in the case (b) by $-\nabla^2\phi = 0$. We will, therefore, solve the equation $-\nabla^2 u = 0$ in a rectangular region Ω which is 8×4 m^2, where u is either the stream function ψ or the velocity potential ϕ. Thus, in the former case the velocity components $\mathbf{u} = (u_1, u_2)$ are given by $u_1 = \dfrac{\partial\psi}{\partial y}, u_2 = -\dfrac{\partial\psi}{\partial x}$, whereas in the latter case by $u_1 = -\dfrac{\partial\phi}{\partial x}, u_2 = -\dfrac{\partial\phi}{\partial y}$. In either case the constant term in the solution u does not affect the velocity field.

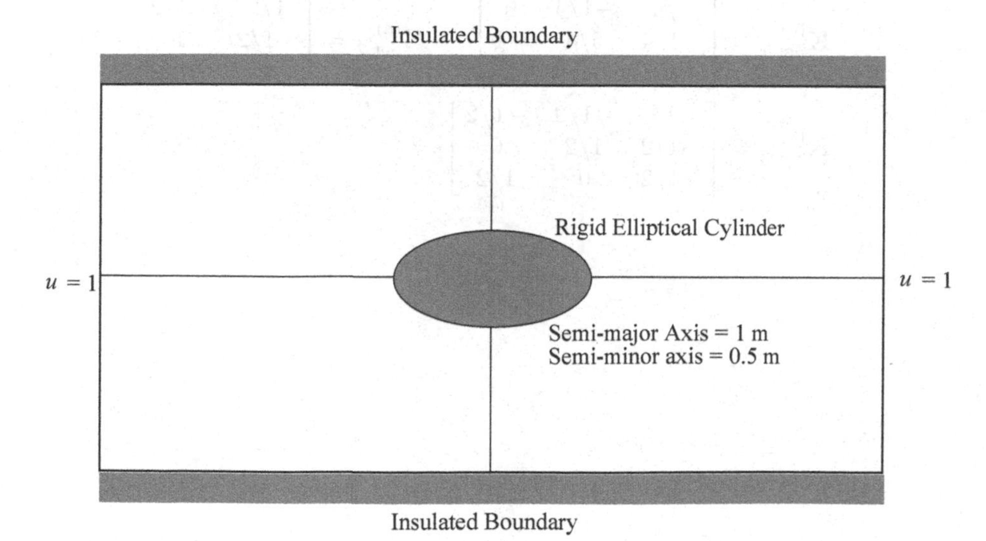

Figure 24.17 Flow around an elliptical cylinder.

Since the problem is symmetric about both horizontal and vertical centerlines, we will consider only the top left quadrant. A mesh of 32 triangular elements is chosen (Figure 24.18). Note that in the case of the stream function formulation the velocity component orthogonal to the horizontal line of symmetry is zero. Thus, we use this line as a stream line, and take the value of the stream function on this horizontal line of symmetry to be zero. Then we determine the value of ψ on the upper wall by using the condition that $\dfrac{\partial \psi}{\partial y} = U_0$, where U_0 denotes the inlet horizontal velocity.

The mesh of elements in Figure 24.18 consists of three sets of similar triangular elements:

SET 1. All elements 1 through 24 and 29 through 32 are like the element 1 (see Figure 24.19a);

SET 2. the elements 25, 26, and 27 are all like the element 25 (see Figure 24.19b); and

SET 3. the element 28 is the only one of its kind (see Figure 24.19c).

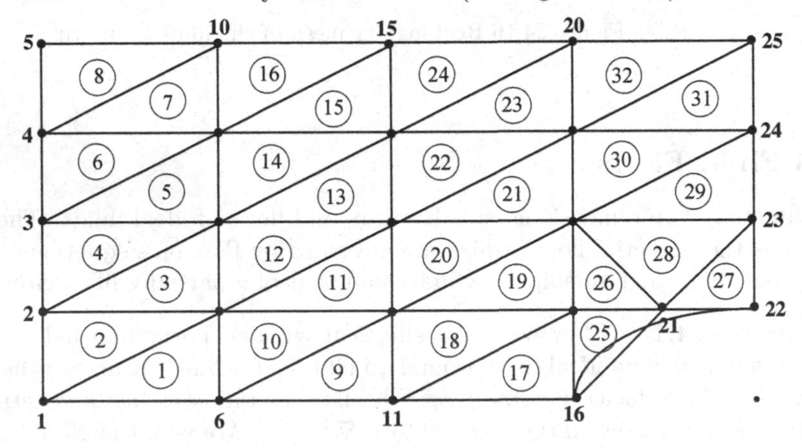

Figure 24.18 Mesh of 32 triangular elements.

We use formula (24.5.9) to compute the stiffness matrices for the first two sets of triangular elements, and formulas (24.4.3)-(24.4.4) for the third set. They are given by

$$\mathbf{K}^{(e)}_{\text{Set 1}} = \begin{bmatrix} 1/4 & -1/4 & 0 \\ -1/4 & 5/4 & -1 \\ 0 & -1 & 1 \end{bmatrix}, \quad \mathbf{K}^{(e)}_{\text{Set 2}} = \begin{bmatrix} 1/2 & -1/2 & 0 \\ -1/2 & 1 & -1/2 \\ 0 & -1/2 & 1/2 \end{bmatrix},$$

$$\mathbf{K}^{(e)}_{\text{Set 3}} = \begin{bmatrix} 1 & -1/2 & -1/2 \\ -1/2 & 1/2 & 0 \\ -1/2 & 0 & 1/2 \end{bmatrix}.$$

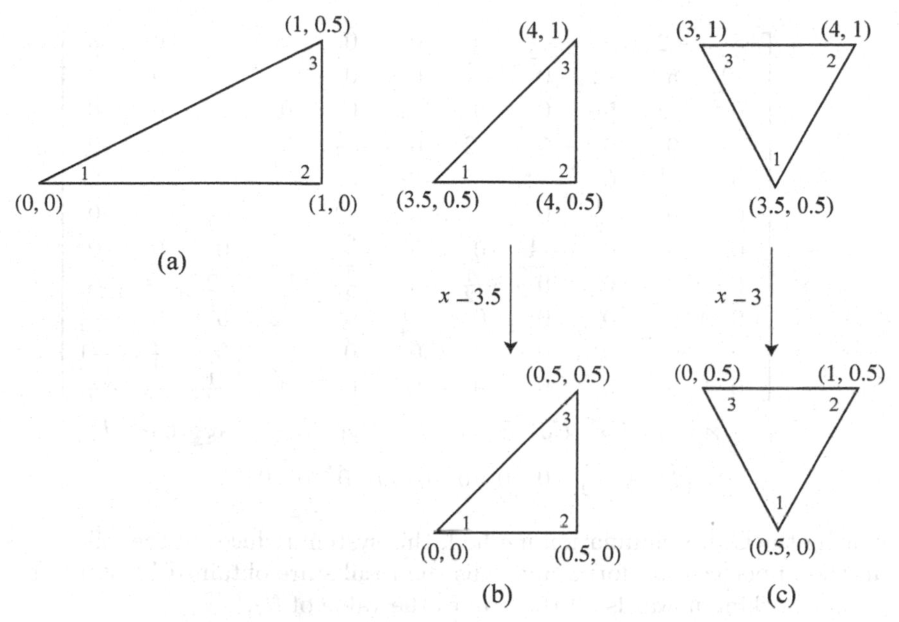

Figure 24.19 Three sets of triangular elements.

The connectivity matrix for this mesh is

$$\mathbf{C} = \begin{bmatrix} e & i & j & k & & e & i & j & k & & e & i & j & k \\ 1 & 1 & 6 & 7 & & 12 & 13 & 8 & 7 & & 23 & 14 & 19 & 20 \\ 2 & 7 & 2 & 1 & & 13 & 8 & 13 & 14 & & 24 & 20 & 15 & 14 \\ 3 & 2 & 7 & 8 & & 14 & 14 & 9 & 8 & & 25 & 21 & 17 & 16 \\ 4 & 8 & 3 & 2 & & 15 & 9 & 14 & 15 & & 26 & 17 & 21 & 18 \\ 5 & 3 & 8 & 9 & & 16 & 15 & 10 & 9 & & 27 & 21 & 22 & 23 \\ 6 & 9 & 4 & 3 & & 17 & 11 & 16 & 17 & & 28 & 21 & 23 & 18 \\ 7 & 4 & 9 & 10 & & 18 & 17 & 12 & 11 & & 29 & 18 & 23 & 24 \\ 8 & 10 & 5 & 4 & & 19 & 12 & 17 & 18 & & 30 & 24 & 19 & 18 \\ 9 & 6 & 11 & 12 & & 20 & 18 & 13 & 12 & & 31 & 19 & 24 & 23 \\ 10 & 12 & 7 & 6 & & 21 & 13 & 18 & 19 & & 32 & 25 & 21 & 19 \\ 11 & 7 & 12 & 13 & & 22 & 19 & 14 & 13 & & & & & \end{bmatrix}.$$

Then the assembled stiffness matrix is given by $\mathbf{K}$, which is given below in the assembled equation. Since $U_1 = U_6 = U_{11} = U_{16} = U_5 = U_{10} = U_{15} = U_{21} = U_{22} = U_{25} = 0$, and $U_{21} = U_3 = U_4 = 1$, the assembled equation $\mathbf{KU} = \mathbf{Q}$ simplifies to the following 11×11 system

$$
\begin{bmatrix}
5 & -2 & 0 & -\frac{1}{2} & 0 & 0 & 0 & 0 & 0 & 0 & 0 \\
-2 & 5 & -2 & 0 & -\frac{1}{2} & 0 & 0 & 0 & 0 & 0 & 0 \\
0 & -2 & 5 & 0 & 0 & -\frac{1}{2} & 0 & 0 & 0 & 0 & 0 \\
-\frac{1}{2} & 0 & 0 & 5 & -2 & 0 & -\frac{1}{2} & 0 & 0 & 0 & 0 \\
0 & -\frac{1}{2} & 0 & -2 & 5 & -2 & 0 & -\frac{1}{2} & 0 & 0 & 0 \\
0 & 0 & -\frac{1}{2} & 0 & -2 & 5 & -2 & 0 & -\frac{1}{2} & 0 & 0 \\
0 & 0 & 0 & -\frac{1}{2} & 0 & 0 & \frac{9}{2} & -\frac{3}{2} & 0 & 0 & 0 \\
0 & 0 & 0 & 0 & -\frac{1}{2} & 0 & \frac{3}{2} & \frac{19}{4} & -2 & -\frac{1}{4} & 0 \\
0 & 0 & 0 & 0 & 0 & -\frac{1}{4} & 0 & -2 & 5 & 0 & -\frac{1}{4} \\
0 & 0 & 0 & 0 & 0 & 0 & 0 & -\frac{1}{4} & 0 & \frac{9}{4} & -1 \\
0 & 0 & 0 & 0 & 0 & 0 & 0 & 0 & -\frac{1}{2} & -1 & \frac{5}{2}
\end{bmatrix} \times
$$

$$
\times \, [U_7 \;\; U_8 \;\; U_9 \;\; U_{12} \;\; U_{13} \;\; U_{14} \;\; U_{17} \;\; U_{18} \;\; U_{19} \;\; U_{23} \;\; U_{24}]^T
$$

$$
= [\tfrac{1}{2} \;\; \tfrac{1}{2} \;\; \tfrac{1}{2} \;\; 0 \;\; 0 \;\; 0 \;\; 0 \;\; 0 \;\; 0 \;\; 0 \;\; 0]^T .
$$

By using the Gauss elimination method, this system reduces to the following system, which is in the upper echelon form, and thus the results are obtained by starting at the value of U_{24} and working upwards all the way to the value of U_7.

$$
\begin{bmatrix}
1 & -0.4 & 0 & -0.1 & 0 & 0 & 0 \\
0 & 1 & -0.47619 & -0.47619 & -0.119 & 0 & 0 \\
0 & 0 & 1 & -0.0235 & -0.0588 & -0.1235 & 0 \\
0 & 0 & 0 & 1 & -0.4109 & -0.00239 & -0.10125 \\
0 & 0 & 0 & 0 & 1 & -0.497 & -0.0502 \\
0 & 0 & 0 & 0 & 0 & 1 & -0.0263 \\
0 & 0 & 0 & 0 & 0 & 0 & 1 \\
0 & 0 & 0 & 0 & 0 & 0 & 0 \\
0 & 0 & 0 & 0 & 0 & 0 & 0 \\
0 & 0 & 0 & 0 & 0 & 0 & 0 \\
0 & 0 & 0 & 0 & 0 & 0 & 0
\end{bmatrix}
$$

$$
\begin{bmatrix}
0 & 0 & 0 & 0 \\
0 & 0 & 0 & 0 \\
0 & 0 & 0 & 0 \\
-0.122 & 0 & 0 & 0 \\
-0.063 & -0.1273 & 0 & 0 \\
-0.345 & -0.0029 & 0 & 0 \\
1 & -0.4913 & -0.0603 & 0 \\
0 & 1 & -0.0312 & -0.127 \\
0 & 0 & 1 & -0.455 \\
0 & 0 & 0 & 1
\end{bmatrix}
\begin{Bmatrix}
U_7 \\ U_8 \\ U_9 \\ U_{12} \\ U_{13} \\ U_{14} \\ U_{17} \\ U_{18} \\ U_{19} \\ U_{23} \\ U_{24}
\end{Bmatrix}
=
\begin{Bmatrix}
0.1 \\ 0.16667 \\ 0.20588 \\ 0.02084 \\ 0.04267 \\ 0.04838 \\ 0.00545 \\ 0.01006 \\ 0.01137 \\ 0.00175 \\ 0.00378
\end{Bmatrix} .
$$

Solving this upper echelon form, we get

$$
\begin{aligned}
&U_7 = 0.21756, \quad U_8 = 0.2811185, \quad U_9 = 0.217569, \quad U_{12} = 0.0511291, \\
&U_{13} = 0.070667, \quad U_{14} = 0.05122, \quad U_{17} = 0.011065, \quad U_{18} = 0.016152, \\
&U_{19} = 0.011961, \quad U_{23} = 0.003476, \quad U_{24} = 003783.
\end{aligned}
$$

For the stream function ψ, we find that $\psi_7 = 0.8246008$, $\psi_8 = 0.7735452$, $\psi_9 = 0.824699$, $\psi_{12} = 0.8059803$, $\psi_{13} = 0.7942585$, $\psi_{14} = 0.8090609$, $\psi_{17} = 0.7835873$, $\psi_{18} = 0.8113408$, $\psi_{19} = 0.8275934$, $\psi_{23} = 0.7848073$, and $\psi_{24} = 0.6837628$.

The contours of the velocity potential and of the stream function are presented in Figure 24.20 and Figure 28.21, respectively. ■

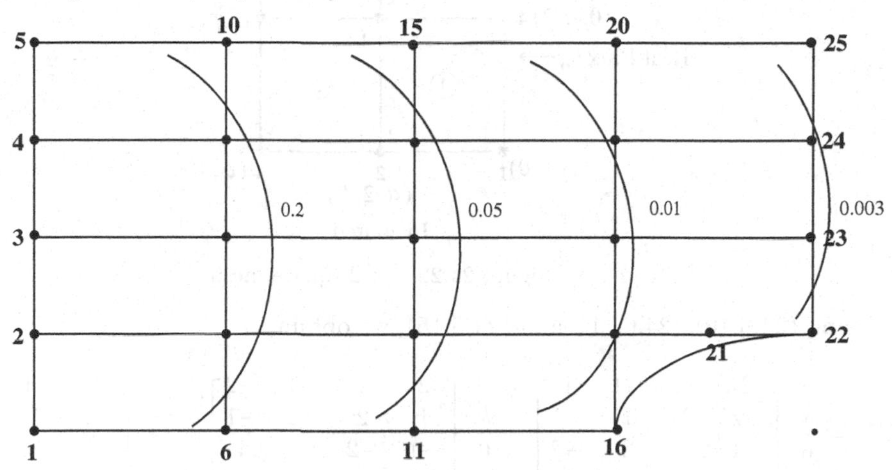

Figure 24.20 Contours of the velocity potential.

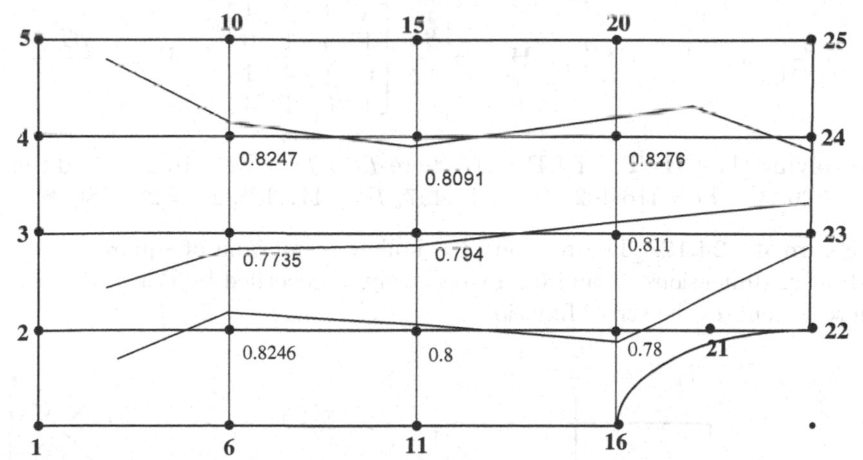

Figure 24.21 Contours of the stream function.

Example 24.16. Consider the steady-state heat transfer problem with a prescribed convection coefficient β, thermal conductivity k, and an internal heat generation f_0, in a square region of side a. This region is insulated both at the top and the bottom; a uniform heat flux q_0 acts on the left side; and the right side is kept at a prescribed temperature T_0. The boundary conditions are shown in Figure 24.22, where T_∞ denotes the ambient temperature. Take a mesh of 2×2 linear square elements, and compute the temperature distribution at the global nodes. Use the following data: $a = 0.002$ m; $k = 30$ W/(m·°C),

$\beta = 80$ W/(m$^2\cdot°$)C; $T_\infty = 10$°C; $T_0 = 100$°C, $f_0 = 10^7$ W/m^3, and $q_0 = 3 \times 10^5$ W/m^2.

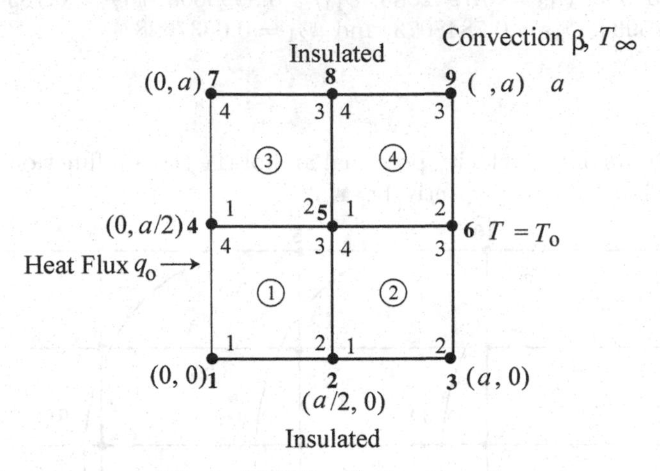

Figure 24.22 2 × 2 square mesh.

Using (24.6.10), (24.6.11), and (24.6.15), we obtain

$$\mathbf{K}^{(e)} = \frac{k}{6}\begin{bmatrix} 2 & -2 & -1 & 1 \\ -2 & 2 & 1 & -1 \\ -1 & 1 & 2 & -2 \\ 1 & -1 & -2 & 2 \end{bmatrix} + \frac{k}{6}\begin{bmatrix} 2 & 1 & -1 & -2 \\ 1 & 2 & -2 & -1 \\ -1 & -2 & 2 & 1 \\ -2 & -1 & 1 & 2 \end{bmatrix} = \frac{k}{6}\begin{bmatrix} 4 & -1 & -2 & -1 \\ -1 & 4 & -1 & -2 \\ -2 & -1 & 4 & -1 \\ -1 & -2 & -1 & 4 \end{bmatrix},$$

$$\mathbf{f}^{(e)} = \frac{f_0 a^2}{16}\,[1 \;\; 1 \;\; 1 \;\; 1]^T, \quad \mathbf{H}^{(e)} = \frac{\beta a}{12}\begin{bmatrix} 4 & 1 & 0 & 1 \\ 1 & 4 & 1 & 0 \\ 0 & 1 & 4 & 1 \\ 1 & 0 & 1 & 4 \end{bmatrix}, \quad \mathbf{P}^{(e)} = \frac{\beta T_\infty a}{2}\,[1 \;\; 1 \;\; 1 \;\; 1]^T,$$

and solving $(\mathbf{K}+\mathbf{H})\;\mathbf{T} = \mathbf{f}+\mathbf{P}+\mathbf{Q}$, where $T_3 = T_6 = T_9 = 100$, we find that $T_1 = 450.357$, $T_2 = 326.768$, $T_4 = 416.402$, $T_5 = 311.457$, $T_7 = 441.355$, $T_8 = 288.786$. ■

Example 24.17. The torsion of a hollow membrane of square cross section of inner and outer dimensions $2a$ and $6a$, respectively, is governed by Poisson's equation $-\nabla^2 u - 2$, where u denotes the stress function.

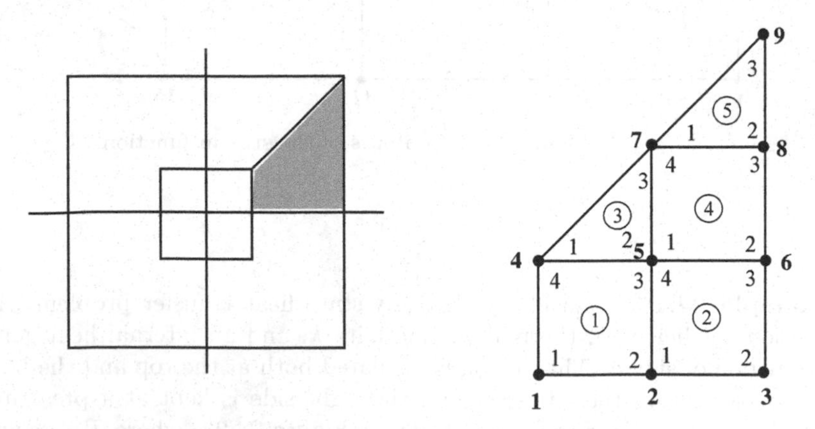

Figure 24.23 Mesh of elements for the shaded region.

The boundary conditions are $u = 2r$ on the outer boundary and $u = 3r^2$ on the inner boundary, where r denotes the ratio of the outer and inner dimensions of the squares. Take $a = 1$, and compute the stress function at the global nodes of the mesh of elements shown in Figure 24.23.

HINT. For the triangular elements, use (I.1)-(I.2) given in Appendix I, which give

$$\mathbf{K}^{(e)} = \frac{1}{2}\begin{bmatrix} 1 & -1 & 0 \\ -1 & 2 & -1 \\ 0 & -1 & 1 \end{bmatrix}, \quad \mathbf{f}^{(e)} = \frac{1}{3}\,[1 \;\; 1 \;\; 1]^T.$$

For the rectangular elements ($a = b = 1$), using (I.5) we have

$$\mathbf{K}^{(e)} = \frac{1}{6}\begin{bmatrix} 4 & -1 & -2 & -1 \\ -1 & 4 & -1 & -2 \\ -2 & -1 & 4 & -1 \\ -1 & -2 & -1 & 4 \end{bmatrix}, \quad \mathbf{f}^{(e)} = \frac{1}{16}\,[1 \;\; 1 \;\; 1 \;\; 1]^T.$$

Use the boundary conditions $U_1 = U_4 = 27$, $U_3 = U_6 = U_8 = U_9 = 6$.

ANS. $U_2 = 15.8651$, $U_5 = 13.4852$, $U_7 = 5.91974$. ■

Example 24.18. Use the Prandtl theory of torsion, governed by

$$-\Big(\frac{\partial^2 u}{\partial x^2} + \frac{\partial^2 u}{\partial y^2}\Big) + 2g\theta = 0,$$

where $u(x, y)$ is the stress function, g the shear modulus (N/cm^2), and θ is the angle of twist per unit length (rad/cm). Let Ω denote the cross section of an elliptical membrane being twisted, and compute the stress function u at the global nodes marked in Figure 24.24, where the semi-major axis is $a = 3$ and the semi-minor axis $b = 2$, and $g\,\theta = 2$.

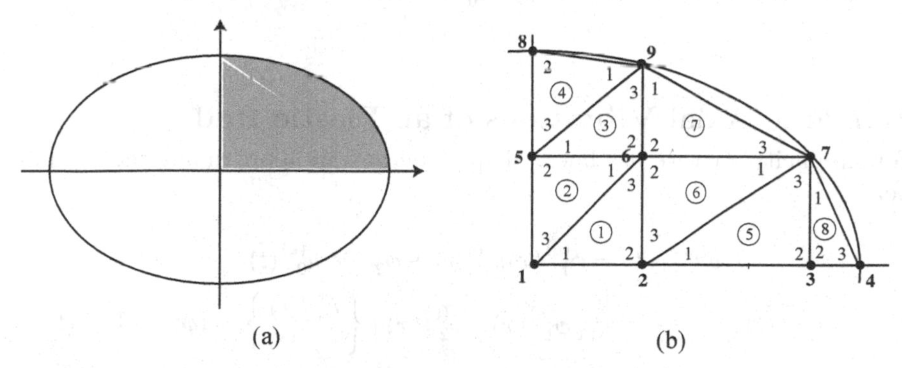

Figure 24.24 Mesh of elements for the shaded region.

Computing the stiffness matrices and the force vectors for the elements 1, 2, 3, 5, 6, 7, 8, and (24.4.3)-(24.4.4) for the element 4, we obtain $U_1 = 4.32553$, $U_2 = 2.11027$, $U_3 = 4.24104$, $U_5 = 3.18775$, $U_6 = 4.36778$. Compare these values with the exact solution

$$u(x, y) = \frac{G\theta a^2 b^2}{a^2 + b^2}\Big(1 - \frac{x^2}{a^2} - \frac{y^2}{b^2}\Big). \;■$$

Example 24.19. Consider the two-dimensional steady heat conduction problem for the quadrant $x, y > 0$

$$\frac{\partial^2 T}{\partial x^2} + \frac{\partial^2 T}{\partial y^2} = 0, \quad 0 < x < \infty,\; 0 < y < \infty,$$

where k is the thermal conductivity, if the side $y = 0$ is maintained at zero temperature, while the other side $x = 0$ is thermally insulated except for the region $0 < y < b$ through which heat flows with constant density q (Figure 24.25), that is, the boundary conditions are

$$T(x,0) = 0, \quad \frac{\partial T}{\partial x}\bigg|_{x=0} = f(y) = \begin{cases} -\dfrac{q}{a}, & 0 < y < b, \\ 0, & b < y < \infty. \end{cases}$$

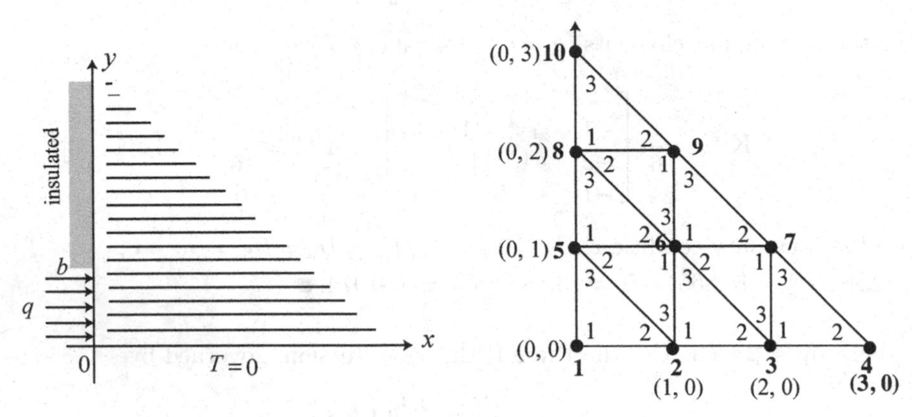

Figure 24.25 Mesh of triangular elements.

The exact solution is (Kythe et al. [2003: 280])

$$T(x,y) = \frac{2q}{\pi a}\int_0^\infty \frac{1-\cos\alpha b}{\alpha^2}\, e^{-\alpha x}\,\sin\alpha y\, d\alpha. \blacksquare \tag{24.6.1}$$

24.7 Free Axial Vibrations of an Elastic Rod

For simplicity, first we use linear shape functions to approximate the axial displacement u. Let

$$\begin{aligned} u^{(e)}(x,t) &= \phi_1^{(e)}(x)u_1^{(e)}(t) + \phi_2^{(e)}(x)u_2^{(e)}(t) \\ &= \begin{bmatrix} \phi_1^{(e)}(x) & \phi_2^{(e)}(x) \end{bmatrix} \begin{Bmatrix} \dot u_1^{(e)}(t) \\ \dot u_2^{(e)}(t) \end{Bmatrix} = (\boldsymbol{\phi}^{(e)})^T \mathbf{U}^{(e)}(t) \end{aligned}$$

be the linear interpolation function in x. Here the nodal displacement vector $\mathbf{U}^{(e)}(t) = \begin{Bmatrix} u_1^{(e)}(t) \\ u_2^{(e)}(t) \end{Bmatrix}$ is assumed to be a function of time t. Then, using the method of separation of variables, we have

$$\begin{aligned} (\dot u^{(e)})^2 &= \begin{bmatrix} \dot u_1^{(e)}(t) & \dot u_2^{(e)}(t) \end{bmatrix} \left(\begin{Bmatrix} \phi_1^{(e)}(x) \\ \phi_2^{(e)}(x) \end{Bmatrix} \begin{bmatrix} \phi_1^{(e)}(x) & \phi_2^{(e)}(x) \end{bmatrix} \right) \begin{Bmatrix} \dot u_1^{(e)}(t) \\ \dot u_2^{(e)}(t) \end{Bmatrix} \\ &= \begin{bmatrix} \dot u_1^{(e)}(t) & \dot u_2^{(e)}(t) \end{bmatrix} \begin{bmatrix} \phi_1^{(e)}(x)\phi_1^{(e)}(x) & \phi_1^{(e)}(x)\phi_2^{(e)}(x) \\ \phi_2^{(e)}(x)\phi_1^{(e)}(x) & \phi_2^{(e)}(x)\phi_2^{(e)}(x) \end{bmatrix} \begin{Bmatrix} \dot u_1^{(e)}(t) \\ \dot u_2^{(e)}(t) \end{Bmatrix}. \end{aligned}$$

The kinetic energy for an element is given by

$$E_k^{(e)} = \frac{1}{2}\int_{x_1^{(e)}}^{x_2^{(e)}} \rho^{(e)}A^{(e)}\left(\dot{u}^{(e)}\right)^2 dx = \frac{\rho^{(e)}A^{(e)}l^{(e)}}{6}\left[\dot{u}_1^{(e)}(t) \quad \dot{u}_2^{(e)}(t)\right]\begin{bmatrix}2 & 1\\ 1 & 2\end{bmatrix}\begin{Bmatrix}\dot{u}_1^{(e)}(t)\\ \dot{u}_2^{(e)}(t)\end{Bmatrix}$$

$$= \frac{1}{2}\left(\dot{\mathbf{U}}^{(e)}(t)\right)^T\mathbf{M}^{(e)}\dot{\mathbf{U}}(t)^{(e)},$$

where

$$\mathbf{M}^{(e)} = \frac{\rho^{(e)}A^{(e)}l^{(e)}}{3}\begin{bmatrix}2 & 1\\ 1 & 2\end{bmatrix}.$$

Similarly, the local elastic potential energy is given by

$$E_p^{(e)} = \frac{1}{2}(\mathbf{U}^{(e)})^T\mathbf{K}^{(e)}\mathbf{U}^{(e)},$$

where

$$\mathbf{K}^{(e)} = \frac{E^{(e)}A^{(e)}}{2}\int_{x_1^{(e)}}^{x_2^{(e)}}\begin{bmatrix}\frac{d\phi_1^{(e)}}{dx}\frac{d\phi_1^{(e)}}{dx} & \frac{d\phi_1^{(e)}}{dx}\frac{d\phi_2^{(e)}}{dx}\\ \frac{d\phi_2^{(e)}}{dx}\frac{d\phi_1^{(e)}}{dx} & \frac{d\phi_2^{(e)}}{dx}\frac{d\phi_2^{(e)}}{dx}\end{bmatrix}dx = \frac{E^{(e)}A^{(e)}}{l^{(e)}}\begin{bmatrix}1 & -1\\ -1 & 1\end{bmatrix},$$

and the work for the element is given by

$$W^{(e)} = \int_{x_1^{(e)}}^{x_2^{(e)}} f\left[u_1^{(e)}(t) \quad u_2^{(e)}(t)\right]\begin{Bmatrix}\phi_1^{(e)}(x)\\ \phi_2^{(e)}(x)\end{Bmatrix}dx = (\mathbf{U}^{(e)})^T\mathbf{F}^{(e)},$$

where

$$\mathbf{F}^{(e)} = \frac{f^{(e)}l^{(e)}}{2}\begin{Bmatrix}1\\ 1\end{Bmatrix}.$$

Therefore, the local energy is

$$I^{(e)}\left(\mathbf{U}^{(e)}\right) = \frac{1}{2}(\dot{\mathbf{U}}^{(e)})^T(t)\mathbf{M}^{(e)}\dot{\mathbf{U}}(t)^{(e)} + \frac{1}{2}(\mathbf{U}^{(e)})^T(t)\mathbf{K}^{(e)}\mathbf{U}(t)^{(e)} - (\mathbf{U}^{(e)})^T\mathbf{F}^{(e)},$$

and the total energy is given by

$$I(\mathbf{U}) = \sum_{e=1}^{NE} I^{(e)}\left(\mathbf{U}^{(e)}\right) = \frac{1}{2}(\dot{\mathbf{U}})^T(t)\mathbf{M}\dot{\mathbf{U}}(t) + \frac{1}{2}\mathbf{U}^T(t)\mathbf{K}\mathbf{U}(t) - \mathbf{U}^T\mathbf{F},$$

where $\mathbf{U}$ is the corresponding global nodal displacement vector, and $\mathbf{M}$, $\mathbf{K}$, $\mathbf{F}$ are the global matrices obtained after assembly of the local matrices. By Hamilton's principle, we need to solve for $\mathbf{U}$ from the global system

$$\mathbf{M}\ddot{\mathbf{U}}(t) + \mathbf{K}\mathbf{U}(t) = \mathbf{F},$$

that satisfies the appropriate initial and boundary conditions. A problem of interest in vibration is to determine solutions of the form

$$\mathbf{U}(t) = \mathbf{X}\sin\omega t = \begin{Bmatrix}X_1\\ X_2\\ \vdots\\ X_N\end{Bmatrix}\sin\omega t,$$

where $\mathbf{X}$ is independent of time in the corresponding homogeneous problem

$$\mathbf{M}\ddot{\mathbf{U}}(t) + \mathbf{K}\mathbf{U}(t) = \mathbf{0},$$

which is equivalent to solving for ω and $\mathbf{X}$ from

$$(\mathbf{K} - \omega^2\mathbf{M})\,\mathbf{X} = \mathbf{0},$$

where $\mathbf{X}$ also satisfies appropriate conditions translated from that of $\mathbf{U}$. A solution to the last equation is a pair of eigenvalue and eigenvector $(\omega^2, \mathbf{X})$, where $\mathbf{X}$ is known as the free vibration mode and the associated ω the corresponding frequency.

Example 24.20. Compute the axial modes of vibration of an elastic rod with Young's modulus $E = 29 \times 10^6$ lb/in^2, density $\rho = 0.29$ lb/in^3, cross-sectional area $A = 1$ in^2, and length $L = 100$ in, by using two linear finite elements (Figure 24.26). Assume that one end of the rod at $x = 0$ is fixed and the other end is free of stress, i.e., $E\dfrac{du}{dx}(L) = 0$.

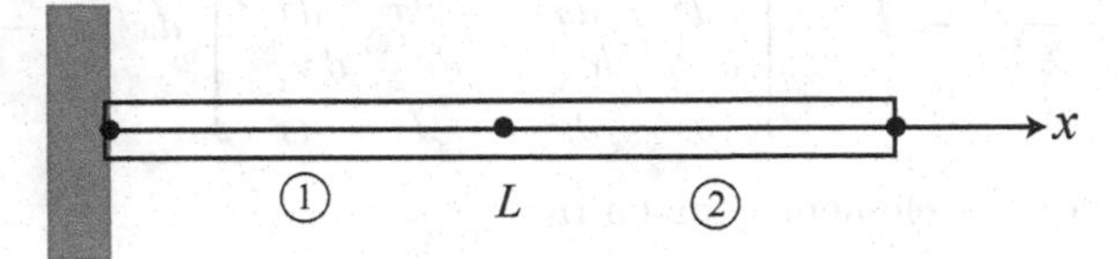

Figure 24.26 Mesh of 2 linear elements.

We have

$$\mathbf{M}^{(e)} = \frac{0.29 \cdot 50}{3}\begin{bmatrix} 2 & 1 \\ 1 & 2 \end{bmatrix} = \frac{145}{30}\begin{bmatrix} 2 & 1 \\ 1 & 2 \end{bmatrix},$$

$$\mathbf{K}^{(e)} = \frac{29 \times 10^6}{50}\begin{bmatrix} 1 & -1 \\ -1 & 1 \end{bmatrix} = 5.8 \times 10^5\begin{bmatrix} 1 & -1 \\ -1 & 1 \end{bmatrix}, \qquad e = 1, 2.$$

The global system is

$$\left(1.2 \times 10^5\begin{bmatrix} 1 & -1 & 0 \\ -1 & 2 & -1 \\ 0 & -1 & 1 \end{bmatrix} - \omega^2\begin{bmatrix} 2 & 1 & 0 \\ 1 & 4 & 1 \\ 0 & 1 & 2 \end{bmatrix}\right)\begin{Bmatrix} X_1 \\ X_2 \\ X_3 \end{Bmatrix} = \begin{Bmatrix} 0 \\ 0 \\ 0 \end{Bmatrix}.$$

Applying the boundary condition $U_1 = u(0) = 0$ gives $X_1 = 0$ for the fixed left end of the rod. Then

$$\left(1.2 \times 10^5\begin{bmatrix} 2 & -1 \\ -1 & 1 \end{bmatrix} - \omega^2\begin{bmatrix} 4 & 1 \\ 1 & 2 \end{bmatrix}\right)\begin{Bmatrix} X_2 \\ X_3 \end{Bmatrix} = \begin{Bmatrix} 0 \\ 0 \end{Bmatrix}.$$

The solutions are

$$\omega_1 = 398.0578, \qquad \begin{Bmatrix} X_2^{(1)} \\ X_3^1 \end{Bmatrix} = \begin{Bmatrix} 0.5774 \\ -0.8165 \end{Bmatrix},$$

and

$$\omega_2 = 113.9298, \qquad \begin{Bmatrix} X_2^{(2)} \\ X_3^2 \end{Bmatrix} = \begin{Bmatrix} 0.5774 \\ 0.8165 \end{Bmatrix}.$$

In general, if $\mathbf{F} \neq 0$, the global system

$$\mathbf{M}\ddot{\mathbf{U}}(t) + \mathbf{K}\mathbf{U}(t) = \mathbf{F}$$

is said to be nonhomogeneous, and it can be solved by numerical schemes based on finite difference and/or finite element methods. The choice of the numerical schemes may depend on the form of **F**. ■

24.8 Electric Potential

The finite element method can be used for an approximate solution of axisymmetric exterior-field problems by truncating the unbounded domain, obtained by conformal mapping from the original, unbounded domain. Although the proposed method presented below with four examples can be used for both bounded and unbounded axisymmetric fields, it is best suited to unbounded field problems.

24.8.1 Axisymmetric Exterior-Field Problems. The FE method is used to find an approximate solution of axisymmetric exterior-field problems by truncating the unbounded domain at a sufficiently large distance and imposing conditions on the terminating boundaries, and then conformally transforming it so that the domain in the model plane is bounded. This method has been used by McDonald and Wexler [1972], Bettess [1981], and Medina and Taylor[1983].

Let a three-dimensional axisymmetric scalar field described in the polar cylindrical coordinates (r, θ, z), with the z-axis as the axis of symmetry, be defined by the Euler-Lagrange equation

$$\frac{1}{r}\left[\nabla\cdot(r\boldsymbol{\kappa}\cdot\nabla\phi)\right]=g(z,r),\quad \boldsymbol{\kappa}=\begin{bmatrix}\kappa_{11}(r,z) & \kappa_{12}(r,z)\\ \kappa_{21}(r,z) & \kappa_{22}(r,z)\end{bmatrix}, \tag{24.8.1}$$

where the matrix $\boldsymbol{\kappa}$ is positive-definite at all points (r, z), and the function $g(r, z)$ defines the source space distribution. Eq (24.8.1) can be written as

$$\frac{1}{r}\left\{\frac{\partial}{\partial z}\left(\kappa_{11}\frac{\partial\phi}{\partial z}+\kappa_{12}\frac{\partial\phi}{\partial r}\right)\right]+\frac{\partial}{\partial r}\left[r\left(\kappa_{21}\frac{\partial\phi}{\partial z}+\kappa_{22}\frac{\partial\phi}{\partial r}\right]\right\}=g(z,r). \tag{24.8.2}$$

Consider the domain D in the azimuthal (z, r)-plane with boundary C (Figure 24.27).

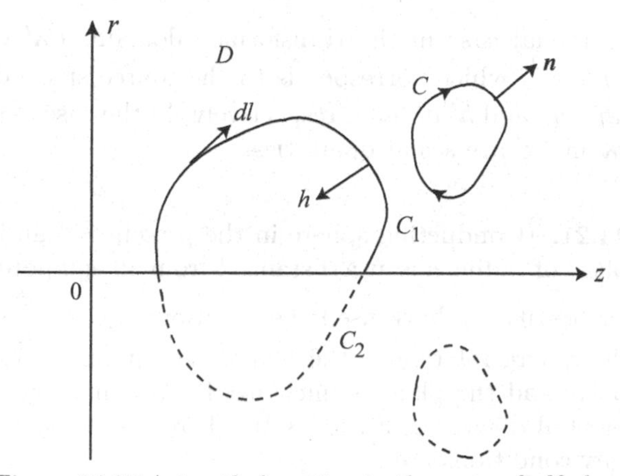

Figure 24.27 Azimuthal section in the upper half-plane

The boundary conditions are of mixed type, i.e.,

$$\phi\Big|_{C_1}=\phi+0(s),\quad s\in C_1,\quad\text{and}\quad [(r\boldsymbol{\kappa}\cdot\nabla\phi)\cdot\mathbf{n}]+\sigma(s)\phi(s)=h(s),\ \text{ on } C_2, \tag{24.8.3}$$

where $\sigma(s) \geq 0$, and $\mathbf{n}$ denotes the outward unit normal to the boundary $C = C_1 + C_2$. The solution of this general problem, available in Davies [1980], is

$$F = \frac{1}{2} \int_D [(\nabla \phi) \cdot (r\boldsymbol{\kappa} \cdot \nabla \phi + 2r\phi g] \, dz \, dr + \frac{1}{2} \int_{C_2} (r\phi^2 - 2h\phi) \, dl. \tag{24.8.4}$$

Let the domain $D + C_1 + C_2$ in the (z, r)-plane be conformally mapped onto the bounded domain $G + \Gamma_1 + \Gamma_2$ in the $w = (u, v)$-plane by

$$z + i\, r \equiv z(u, v) + i\, r(u, v) = f(w), \quad w = u + iv. \tag{24.8.5}$$

Then using the Cauchy-Riemann equations, Eq (24.8.4) can be written as

$$F = \frac{1}{2} \int_G \Big[(\widetilde{\nabla} \Phi) \cdot (R\boldsymbol{\kappa}' \cdot \widetilde{\nabla} \Phi + 2R\Phi g' \Big|\frac{df}{dw}\Big|\Big] \, dz \, dr + \frac{1}{2} \int_{\Gamma_2} (R\Phi^2 - 2h'\Phi) \Big|\frac{df}{dw}\Big| \, dl', \tag{24.8.6}$$

where $\widetilde{\nabla} = a_u \dfrac{\partial}{\partial u} + a_v \dfrac{\partial}{\partial v}$, and Φ, R, g', σ' and h' denote the functions ϕ, r, g, σ and h in terms of u and v, respectively. The tensor $\boldsymbol{\kappa}' = T\boldsymbol{\kappa}T^{-1}$, where

$$T = \begin{bmatrix} \dfrac{\partial u}{\partial z} & \dfrac{\partial u}{\partial r} \\ \dfrac{\partial v}{\partial z} & \dfrac{\partial v}{\partial r} \end{bmatrix}. \tag{24.8.7}$$

In the two-dimensional case in the transformed domain, $R\boldsymbol{\kappa}'$ can be interpreted as representing $Rg' \big|df/dw\big|^2$, which corresponds to the source space distribution, with σ and h replaced by $\sigma' \big|df/dw\big|$ and $h' \big|df/dw\big|$, respectively. In the case of isotropic and homogeneous medium, both $\boldsymbol{\kappa}$ and $\boldsymbol{\kappa}'$ are scalar quantities.

Example 24.21. (Conducting sphere in the presence of an infinite ground plane) The center of the sphere of radius a is at a distance b from an equipotential infinite plane (Figure 24.28, where for brevity we have used the notation: $\phi_{\mathbf{n}} = \dfrac{\partial \phi}{\partial \mathbf{n}}$). An electric potential ϕ_0 is applied to the sphere relative to the plane. The infinitely extended dielectric medium between the sphere and the plane is unchanged and is homogeneous with permittivity ε_0. The electric potential ϕ satisfies Eq (24.8.2), where $\kappa_{11} = \kappa_{22} = \varepsilon_0$ and $\kappa_{12} = \kappa_{21} = g = 0$, and the boundary conditions are

$$\phi = \begin{cases} \phi_0 & \text{for } (z-b)^2 + r^2 = a^2, \\ 0 & \text{at } z = 0, \\ 0 & \text{at infinity;} \end{cases} \qquad \text{and} \quad \frac{\partial \phi}{\partial r} \equiv \phi_r = 0 \quad \text{at } r = 0. \tag{24.8.8}$$

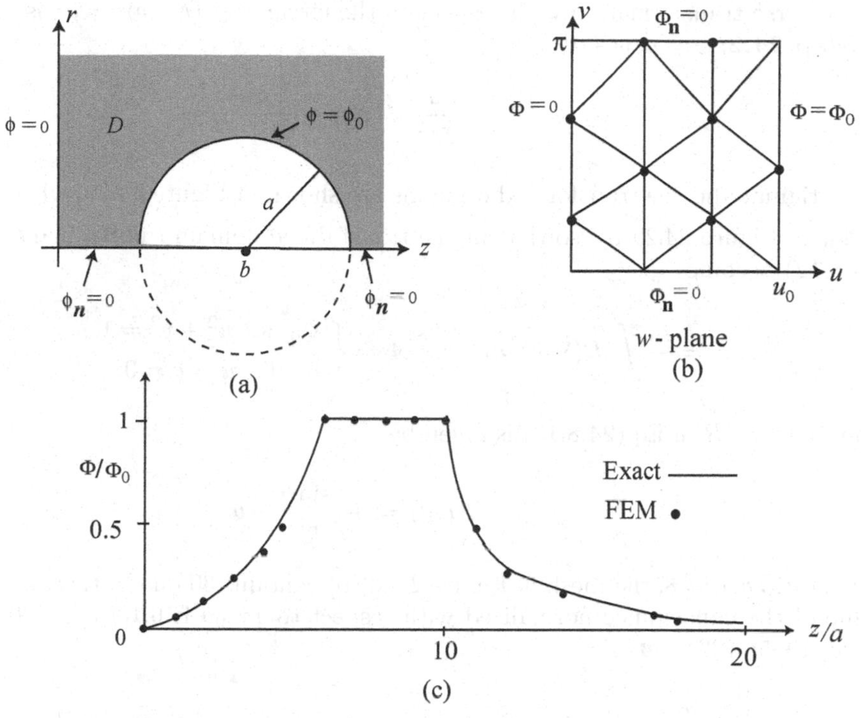

Figure 24.28 (a) Azimuthal section for sphere-to-plane, (b) FE mesh.

Use the conformal map **Map 24.1:** $z = f(w) = c\tanh(w/2)$, which maps the domain D onto the rectangular domain G (Figure 24.28(b)), where $c = a\sinh u_0$ and $u_0 = \cosh^{-1}(b/a)$, we obtain

$$z = \frac{c\sinh u}{\cosh u + \cos v}, \quad r = \frac{c\sin v}{\cosh u + \cos v}. \tag{24.8.9}$$

Then the transformed functional in Eq (24.8.6) becomes

$$F = \frac{\varepsilon_0}{2}\int_G \left(\frac{c\sin v}{\cosh u + \cos v}\right)\left(\widetilde{\nabla}\Phi\right)^2 du\,dv, \quad \text{where} \quad \Phi = \begin{cases} \Phi_0 & \text{for } u = u_0, \\ 0 & \text{for } u = 0. \end{cases} \tag{24.8.10}$$

The normalized electrostatic capacitance C of this system is obtained from $F_{\min}$ of the functional (24.8.10), which, according to Morse and Feshbach [1953], is $C = \dfrac{F_{\min}}{\varepsilon_0 a\phi_0^2}$. The numerical solution for $b/a = 8$ is shown in Figure 24.28(c). ■

Example 24.22. (Capacitance of a torus in free space) As in Example 24.21, this is also inverse transformation. The capacitance of a torus, shown in Figure 24.29, is calculated from the minimum value of the functional

$$F = \frac{\varepsilon_0}{2}\int_D r(\nabla\phi)^2\,dz\,dr, \tag{24.8.11}$$

subject to the constraints

$$\phi = \begin{cases} \phi_0 & \text{for } z^2 + (r-b)^2 = a^2, \\ 0 & \text{at infinity.} \end{cases}$$

The inverse transformation with respect to the circle $z^2 + (r - b)^2 = a^2$ is given by
Map 24.2: $z = a/w + ib$,
which yields

$$z = \frac{au}{u+v^2}, \quad r = -\frac{av}{u^2 + v^2}{}_b. \tag{24.8.12}$$

The original and the transformed domains are shown in Figure 24.29, where $\dfrac{\partial \Phi}{\partial \mathbf{n}}$ is written as $\Phi_{\mathbf{n}}$ in Figure 24.29(b), so that in the transformed domain (model plane) Eqs (24.8.11)-(24.8.12) become

$$F = \frac{\varepsilon_0}{2} \int_G R(\widetilde{\nabla}\Phi)^2 \, du \, dv, \quad \Phi = \begin{cases} \Phi_0 & \text{for } u^2 + v^2 = 1, \\ 0 & \text{for } u = v = 0. \end{cases} \tag{24.8.13}$$

The function R in Eq (24.8.13) is given by

$$R(r, \theta) = -\frac{a \sin\theta}{r} + b.$$

For a ratio $b/a = 8$, the mesh in Figure 24.29(b) contains 30 quadratic triangles. The exact value of the capacitance normalized with respect to $4\pi\varepsilon_0 a$ is 6.1007 ± 0.00092 (Wong and Ciric [1985: 132]). ■

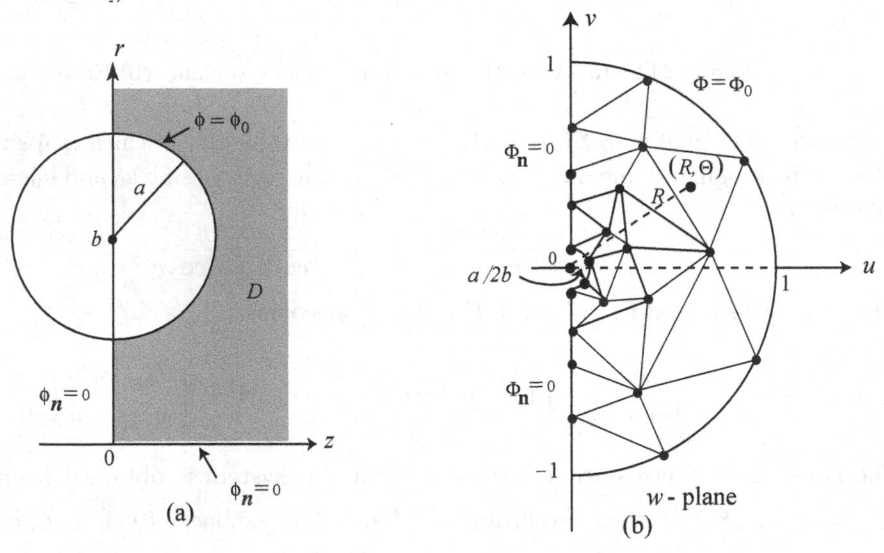

Figure 24.29 (a) Azimuthal section of a torus, (b) FE mesh.

Example 24.23. (Conducting sphere in a nonhomogeneous dielectric) Consider a conducting sphere of radius a in an infinitely extended nonhomogeneous dielectric, with the electric potential ϕ_0 with respect to the zero potential at infinity. Let the variation of permittivity ε of the dielectric be $\varepsilon = \varepsilon_0\Big(\dfrac{10a}{\rho} + 1\Big)$, where ρ denotes the radial distance from the center of the sphere. Using the inverse conformal transformation $z = a/w$, the physical domain D is mapped onto a bounded domain G (Figure 24.30), so that the transformed functional in Eq (24.8.6) is

$$F = -\frac{\varepsilon_0 a}{2} \int_G \frac{v}{u^2 + v^2} \left(10\sqrt{u^2 + v^2} + 1\right) (\widetilde{\nabla}\Phi) \, du \, dv. \tag{24.8.14}$$

The FEM solution for normalized potential distribution in the radial direction is presented in Figure 24.30(c): I with 25 quadratic elements. The potential distribution so obtained yields electrostatic capacitance $C = F_{\min}/\left(\varepsilon + 0a\Phi_0^2\right)$ which is 68% above the exact value of $10/(\ln 11)$.

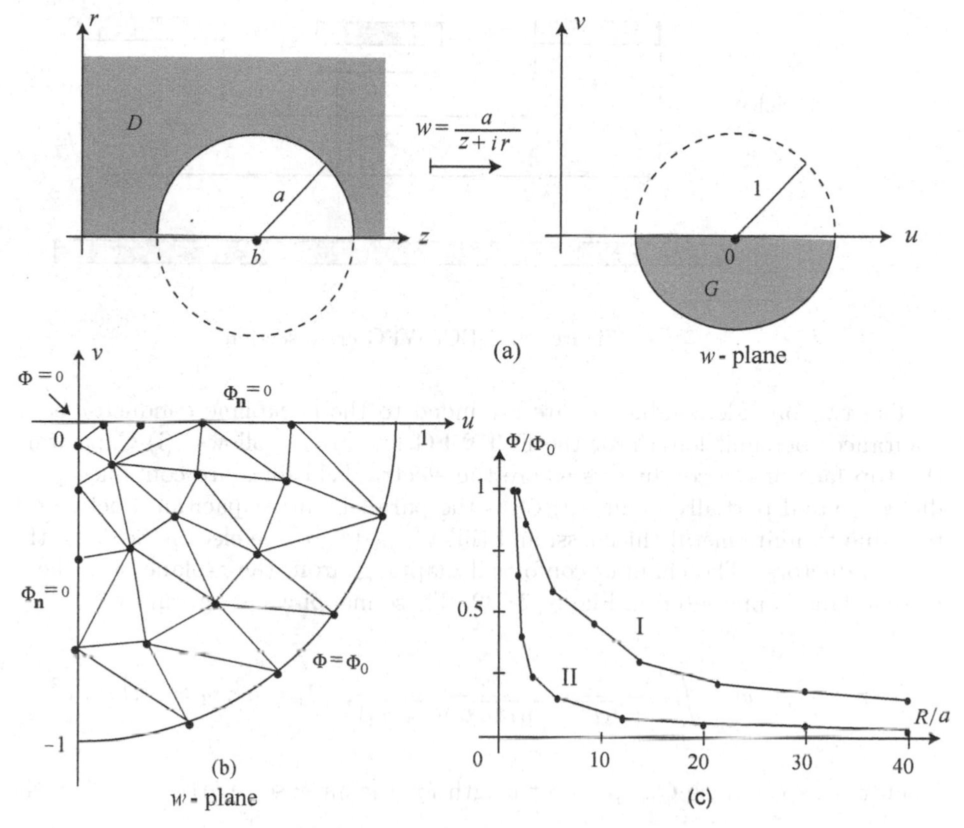

Figure 24.30 (a) Azimuthal section of a sphere, (b) FE mesh.

Note that Figure 24.30(c) has two curves marked I and II; the first one refers to Example 24.23 and the second one to Example 24.24 (which follows). ■

Example 24.24. (Charged sphere with external source distribution) Consider a conducting sphere of radius a in an infinitely extended homogeneous dielectric with a source distribution outside the sphere, given by $g = 1/\rho^4$, where ρ is the distance from the center of the sphere. Using the same conformal map as in Example 24.23, the normalized potential distribution is shown in Figure 24.30(c). It is found that within a radial distance of 11 times the radius of the sphere, the minimum percentage error in the nodal potential is 1.04%. ■

24.9 Waveguide

An analytic expression for an effective dielectric constant and impedance of a coplanar waveguide with finite grounds and embedded in a dielectric is developed by Jessie and Larson [2001]. Consider a closed-form CPW structure with finite grounds, buried in a dielectric, conductor backed, and with finite conductor thickness (ECPWFG) shown in Figure 24.31, where the dielectric height extends from the grounded plane at the bottom to above the conductors, which is $h_1 + t + h_2$, and the conductor pitch is $b + a$.

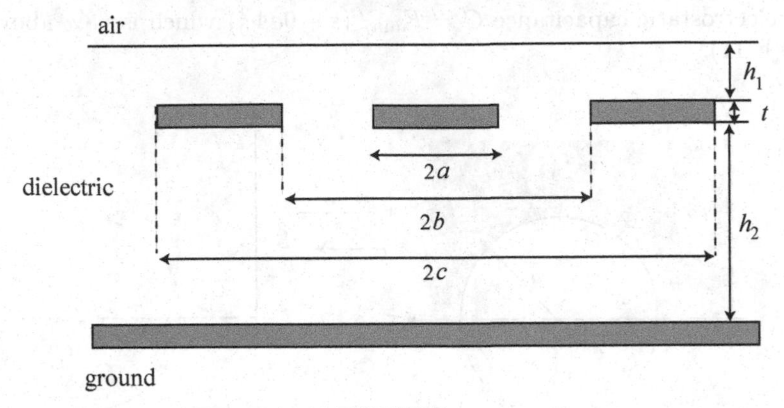

Figure 24.31 ECPWFG cross-section.

The two outside conductors are grounded to the backplane conductor below. The capacitances per unit length for the ECPWFG line are as follows: (i) Capacitance C_1 is for the top face of the conductors where the electric field lines are contained partially in the dielectric and partially in air; (ii) C_2 is the parallel plate capacitance between the conductors due to finite metal thickness; and (iii) C_3 is due to the electric fields on the bottom of the conductors. The chain of conformal mappings from the z-plane onto the t-plane onto the w-plane is presented in Figure 24.32. These mappings are given by

$$\tau = z^2, \quad w = \int_{t_i}^{t_j} \frac{dt}{\sqrt{t(t-t_1)(t-t_2)(t-t_3)}}, \quad t_1 = a^2, t_2 = b^2, t_3 = c^2. \tag{24.9.1}$$

Then the capacitance $C_{1\mathrm{A}}$ per unit length for the air case (marked by A) in the z-plane is given by

$$C_{1\mathrm{A}} = 2\varepsilon_0 \frac{K(k_{1\mathrm{A}})}{K'(k_{1\mathrm{A}})}, \quad k_{1\mathrm{A}} = \frac{a}{b}\sqrt{\frac{1-b^2/c^2}{1-a^2/c^2}}, \tag{24.9.2}$$

where K and K' are solutions to the complete elliptic integrals of the first kind and its complement, respectively.

Since the formulation with dielectric present above the conductors is similar to that derived with the expression of the dielectric, the thickness h_2 is transformed into unity offset from the origin. The transformation onto the τ-plane for the configuration with $\varepsilon \neq 1$ dielectric above the interface is given by

$$t = \cosh^2\left(\frac{\pi z}{2h_2}\right). \tag{24.9.3}$$

Then the capacitance $C_{1\mathrm{D}}$ per unit length for the dielectric case (marked by D) is

$$C_{1\mathrm{D}} = 2\varepsilon_0\,(\varepsilon_r - 1)\,\frac{K(k_{1\mathrm{D}})}{K'(k_{1\mathrm{D}})}, \quad k_{1\mathrm{D}} = \frac{\sinh\left(\frac{\pi a}{2h_2}\right)}{\sinh\left(\frac{\pi b}{2h_2}\right)}\sqrt{\frac{1-\frac{\sinh^2(\pi b/(2h_2))}{\sinh^2(\pi c/(2h_2))}}{1-\frac{\sinh^2(\pi a/(2h_2))}{\sinh^2(\pi c/(2h_2))}}}. \tag{24.9.4}$$

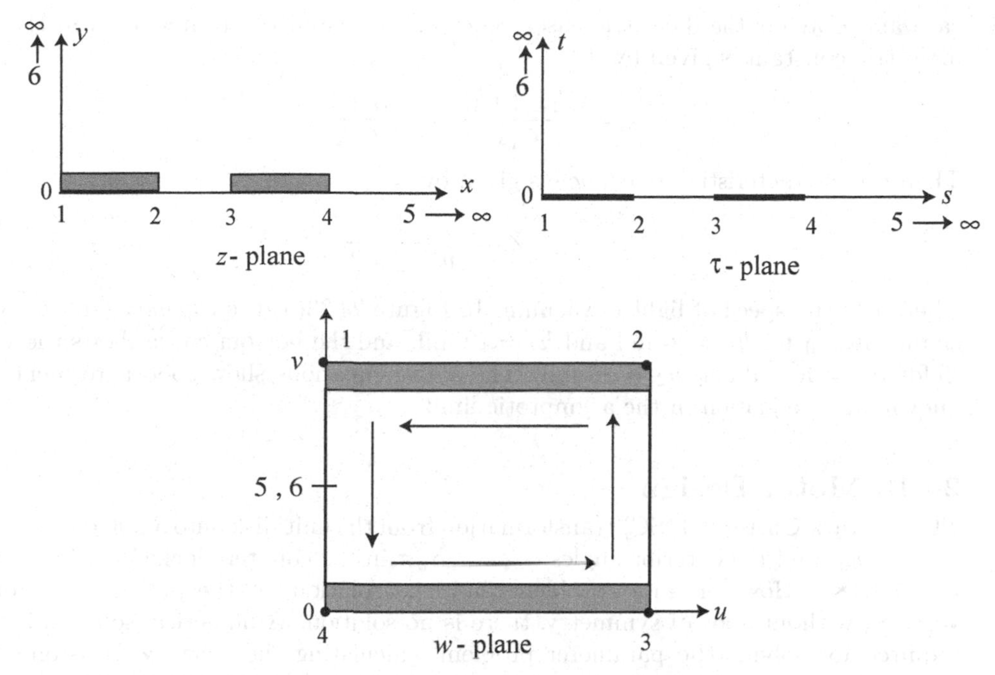

Figure 24.32 Conformal mappings of top face of ECPWFG without dielectric.

Thus, the total capacitance above the dielectrics is $C_{1\mathrm{A}} + C_{1\mathrm{D}}$. The capacitance C_3 per unit length in the w-plane can be derived in a similar manner. The transformations for the configuration with the only dielectric below the interface is the same as Eqs (24.9.3)-(24.9.4) with h_2 replaced by h_1 and ε_r replaced by $\varepsilon_r - 1$. The numerical results are plotted in Figure 24.33 for $a = 10$ mil, $b = 18$ mil, and c varied from 20 to 40 mil in the graphs (a), and for $a = 10$ mil, b varied from 12 to 32 mil, and $c = b + 2$ mil in the graphs (b). The CPW curve is denoted by the solid line and the ECPWFG curve by the dotted line.

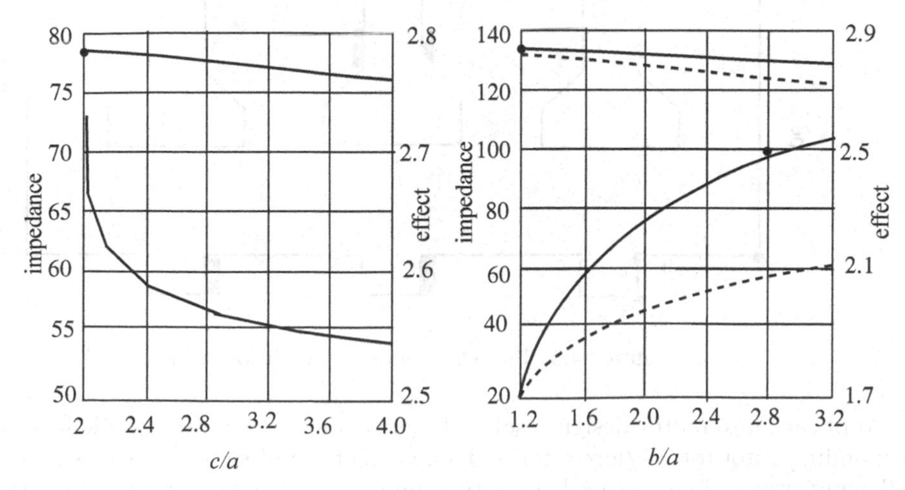

Figure 24.33 Graphs (a) and (b).

The capacitance C_2 due to finite metal thickness is evaluated using the classic parallel plate capacitor formula, because the fringing fields at the top and bottom faces of the conductors are contained in the above derivations. They are denoted by $C_{2\mathrm{A}}$ for the air case and C_{2D} for the dielectric case. Since all capacitances are now known, the effective dielectric constant is given by

$$\varepsilon_{\text{eff}} = \frac{C_{1A} + C_{1D} + C_{3D} + C_{2D}}{C_{1A} + C_{3A} + C_{2A}} \equiv \frac{C}{C^*}. \tag{24.9.5}$$

Then the characteristic impedance is given by

$$Z_0 = \frac{1}{v\sqrt{\varepsilon_{\text{eff}}}\, C^*}, \tag{24.9.6}$$

where v is the speed of light in vacuum. In Figure 24.33(a) the top curve shows ECPWFG change in ε_{eff} for $h_1 = 40$ mil and $h_2 = 10$ mil, and the bottom curve shows the values for $Z_0$for $h_1 = 40$ mil and $h_2 = 28$ mil. The above equations show good agreement with the known CPW equation in the asymptotic limit.

24.10 Motor Design

The Schwarz-Christoffel (SC) transformation from the unit disk onto a polygon with vertices $w_1, \dots, w_n$ and the exterior angles $\alpha_\pi, \dots, \alpha_n\pi$ in the counter-clockwise order is given by Eq (7.1.18). However, most problems have no solution for the prevertices. For $n > 3$ vertices, without a lot of symmetry, there is no solution. A numerical solution is therefore required for solving the parameter problem, calculating the Schwarz-Christoffel integral, and inverting the conformal mapping. Some historical milestones are:

1820s: Gauss provided the idea of conformal mapping
1867-1890: Schwarz and Christoffel discover the SC formula
1900-01: F. W. Carter uses SC mapping for the field between poles
1980: Trefethen - SCPACK Fortran program
1998: Driscoll and Vavavis: CRDT algorithm for multiply elongated regions

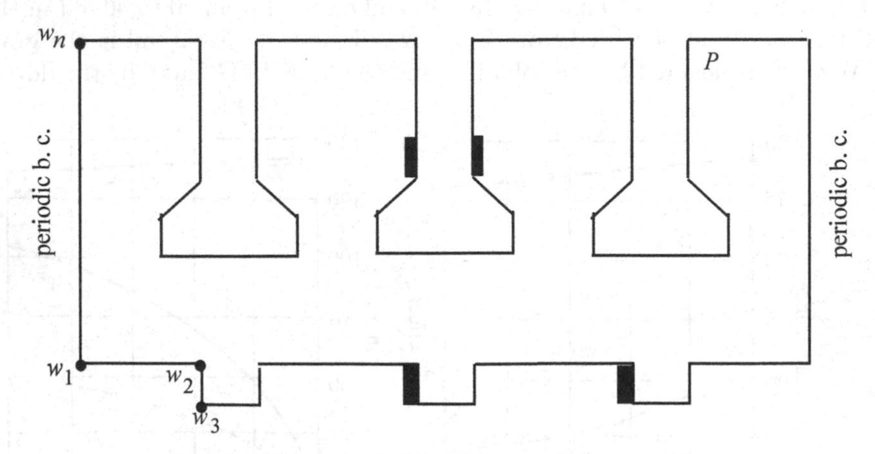

Figure 24.34 Periodic boundary conditions at sides.

Application to motor design resolves to calculate the electromagnetic fields and the corresponding rotor torque/forces for a given geometry and set of materials and sources. The following assumptions are needed: (i) two-dimensional machine cross-section, (ii) the air gap is a polygon, without curves, with n sides, (iii) the magnetics are linear, (iv) the boundary condition at polygonal edges is periodic, and (v) the sources are finite, discrete currents, as shown in Figure 24.34. The Schwarz-Christoffel transformations needed are presented in Figure 24.35, and their application to motor design in Figure 24.36.

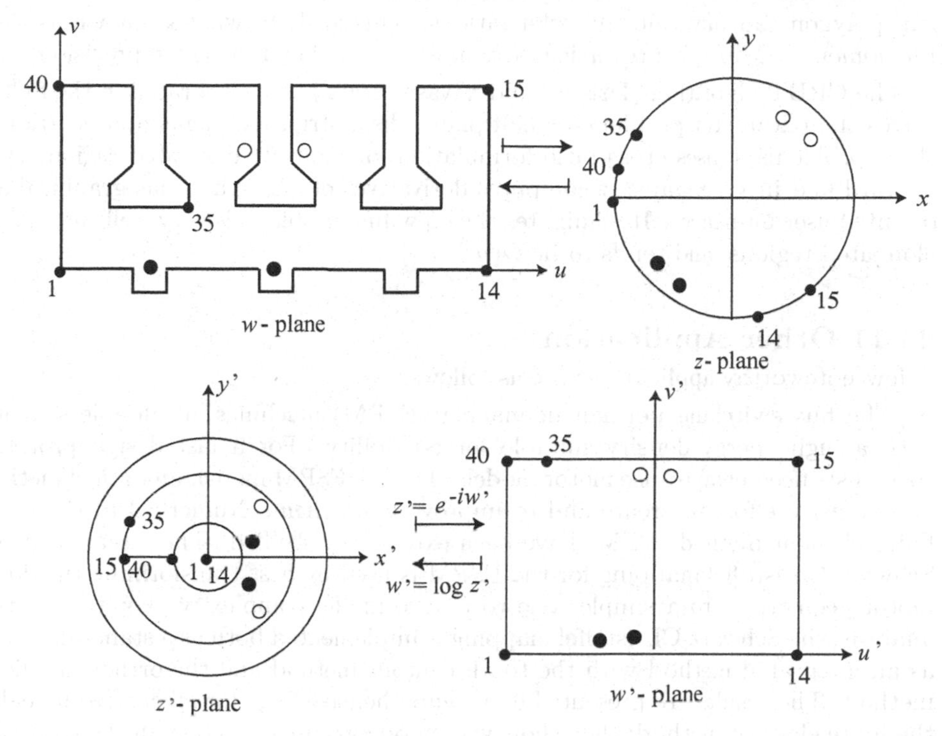

Figure 24.35 Conformal transformation.

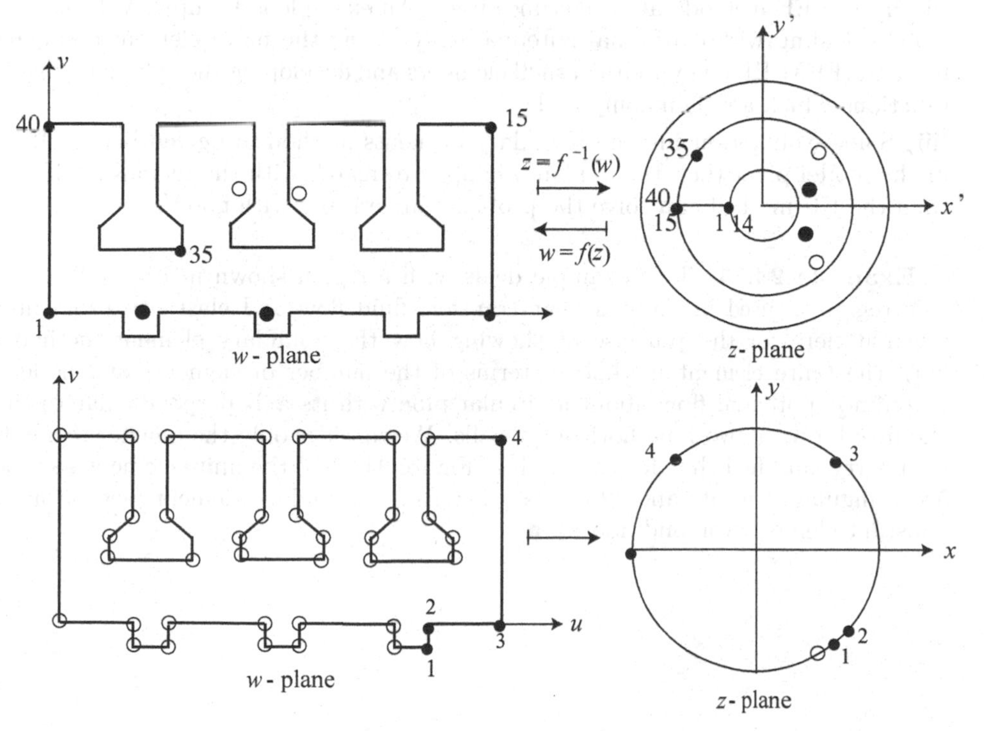

Figure 24.36 Application to motor design.

Notice that three prevertices in Figure 24.35 can be placed arbitrarily, and the motor gap polygon can have multiple elongations. This leads to what is known as the *crowding phenomenon.* Multiple prevertices are indistinguishable in machine precision.

The CRDT algorithm (Driscoll and Vivasis [1998]) is incorporated in the SC Toolbox; it solves the parameter problem for half-plane, disk, strip, rectangle, and exterior maps. For these problems it uses cross ratio formulation for multiply elongated regions. It computes forward and inverse maps; it computes derivatives of maps; and has graphical and object-oriented user interfaces. It eliminates the crowding problem, is very well suited for multiply-elongated regions, and tends to be $O(n^3)$.

24.11 Other Applications

A few noteworthy applications are as follows:

(i) The flux-switching permanent magnet (FSPM) machines are double salient machines with a high energy density suitable for e-mobility. For a fast design process, machine specialists need easy-to-use motor models. For the FSPM model, analytical methods require an extensive effort to create and to improve upon them. Numerical methods such as the finite element method (FEM) have been extensively studied. The research shows that the Schwarz-Christoffel mapping for the FSPM is used by first transforming the double salient motor geometry into a simpler one to reduce model complexity. For the electromagnetic analysis, the Schwarz-Christoffel mapping is implemented both as a stand-alone method and as an integrated method with the tooth contour method and the orthogonal field diagram method. The results are presented in a comprehensive form for all created models including the finite element method; they show very good agreement among all the methods reported by Ilhan et al. [2010, 2012], and Gysen et al. [2010].

(ii) Hybrid FE methods are sometimes used. An example is Kempel, Volakis, Woo and Yu [1992], dealing with conformal antenna arrays using the finite element-boundary integral method (FEM-BI) on cylindrical shell elements and developing the cylinder's dyadic Green's functions which are then computed.

(iii) Some examples using the boundary elements method are given below. If the regions in the model plane (i.e., the w-plane) happen to match with the regions in these examples, then the BE method can solve the problem numerically with good results.

Example 24.25. This example deals with a region shown in Figure 24.37. Although this region is used in various heat transfer, fluid flow and elasticity problems, we have given it here for the purpose of showing how the boundary element method compares with the finite element method in terms of the number of elements and nodes. Imagine a confined potential flow about a circular pipe with its axis perpendicular to the plane of the flow between two long horizontal walls. We consider only the quarter region because of symmetry, and find that for the mesh in Figure 24.37(b) the finite element method requires 58 triangular elements and 42 nodes, whereas the boundary element method needs only 24 constant elements (or mid-nodes). ■

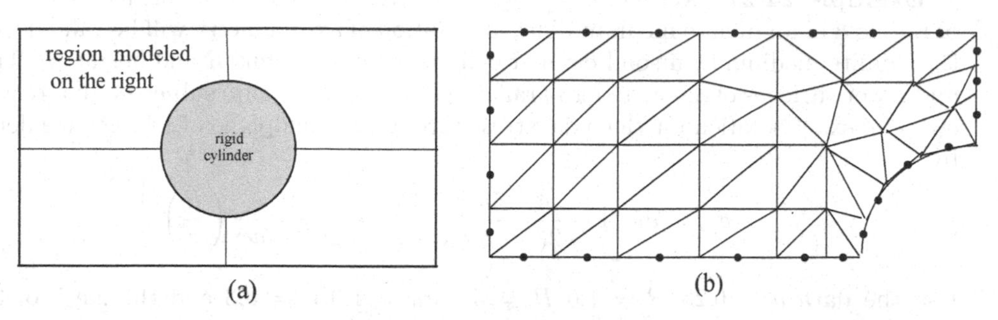

Figure 24.37 Model region.

Example 24.26. Consider the problem of a hollow circular pipe of radii $a = 10$ and $b = 15$ units, respectively, under an internal pressure $p = 100$ (see Figure 24.38). The other data is: $\mu = 80,000$, and $\nu = 0.25$. Because of axial symmetry, the input file is created with constant elements as in Figure 24.38(b). For the plane stress case, the displacement is given by

$$u(r) = \frac{pa^2}{E(b^2 - a^2} \left[(1-\nu) r + (1+\nu) \frac{b^2}{r^2} \right], \quad a \le r \le b,$$

and the stress by

$$\sigma(r) = \frac{pa^2}{E(b^2 - a^2)} \left(1 - \frac{b^2}{r^2} \right), \quad \theta(r) = \frac{pa^2}{E(b^2 - a^2)} \left(1 - \frac{b^2}{r^2} \right), \quad \text{for } a \le r \le b.$$

Both circumferential and radial results for displacements and stresses compare very well with the exact solutions. However, the boundary element results for the stresses in the vicinity of the boundary do not match with the exact solutions; but this was expected. It is found that the boundary element results are, in general, correct for those interior points which lie at a distance more than half an element length away from the boundary.

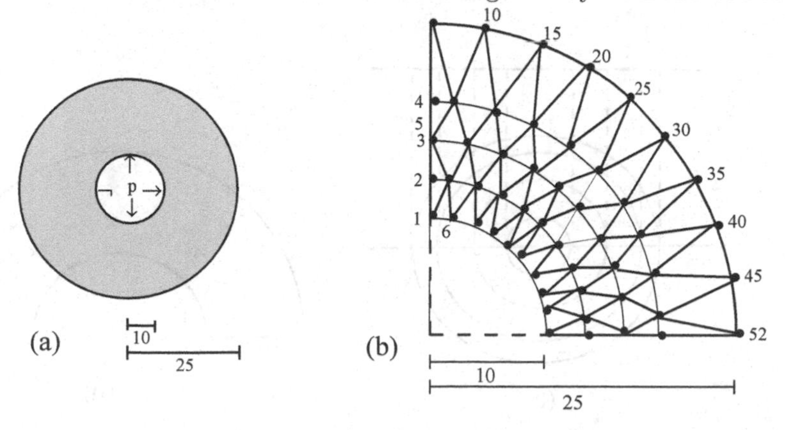

Figure 24.38 A hollow circular pipe.

If this problem is solved with the finite element method of Figure 24.38(b), with 52 nodes and 76 triangular elements, the results do not agree with the exact solutions. This means that if constant strains are used in linear elasticity, the resulting finite elements computed at the center of each element will produce poor results. This method should therefore be avoided. ■

Example 24.27. Consider the problem of stress concentration due to a spherical cavity of radius a in an infinite medium. The assumption of axisymmetry will be valid if we replace the infinite medium by a solid circular cylinder of large radius R and height H. The upper and lower surfaces of the cylinder parallel to the plane $z = 0$ are subjected to a tensile stress σ_0. The exact solution for the axial stress through the midplane of a large cylinder is given by

$$\sigma_{zz} = \sigma_0 \left[1 + \frac{4 - 5\nu}{2(7 - 5\nu} \left(\frac{R}{r} \right)^3 + \frac{9}{2(7 - 5\nu)} \left(\frac{R}{r} \right)^5 \right].$$

Use the data $a = 0.25, R = 1.0, H = 4.0, \sigma_0 = 1.0, \nu = 0.3$ and the mesh of 32 linear elements shown in Figure 24.39. ■

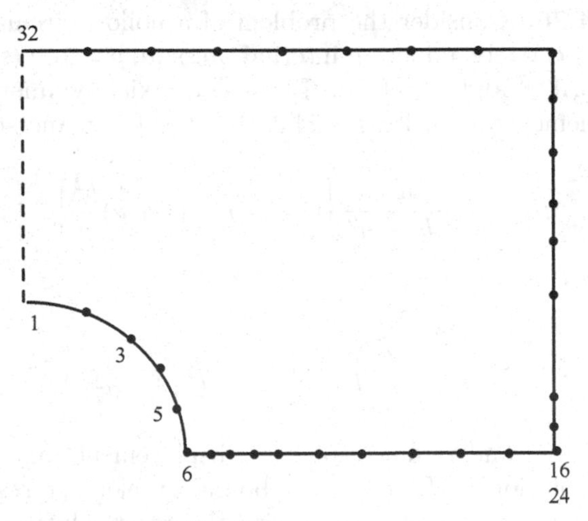

Figure 24.39 Mesh of 32 linear elements.

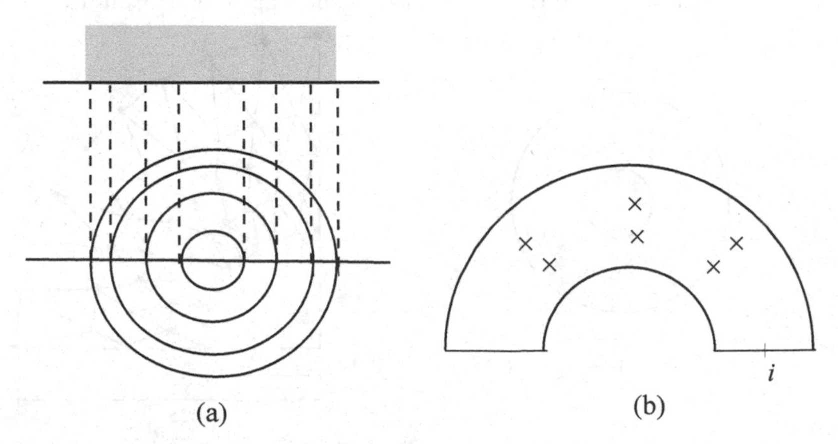

Figure 24.40 Concrete circular structure.

Example 24.28. In order to investigate the impedance of a rigid foundation in a concrete circular structure (Figure 24.40), we are interested in computing the stresses when subjected to torsional, vertical, horizontal, or rocking displacements. The boundary conditions are:

For torsional displacements: $u_r = 0 = u_z, u_\theta = u_\theta^0(r); p_r = 0 = p_z, p_\theta = p_\theta^0(r)$.

For vertical displacements: $u_r = u_r^0(r), u_\theta = 0, u_z = u_z^0; p - r = p_r^0(r), p_\theta = 0, p_z = p_z^0(r)$.

For horizontal and rocking displacements: $u_r = u_r^0 \cos\theta$, $u_\theta = -u_\theta^0 \sin\theta$, $u_z = u_z^0 \cos\theta$; $p_r = p_r^0 \sin\theta, p_\theta = -p_\theta^0 \sin\theta, p_z = -p_z \cos\theta$.

One-half of the symmetric regions between any two concentric rings can be discretized (Figure 24.40(b)), using constant, linear or quadratic elements $\tilde{S}_j, j = 1, \dots, N$. If we choose constant elements, then the values of $\mathbf{u}$ and $\mathbf{p}$ are constant, say, $\mathbf{u}^0$ and $\mathbf{p}^0$ respectively, on each boundary element $\tilde{S}_j$. If we further take the point i such that $\theta = 0$, then from (25.6.45) and (25.6.46) we obtain the following system of algebraic equations for the case of Figure 24.40(b) when $0 \le \theta \le \pi$:

$$\begin{aligned} \frac{1}{2}\begin{Bmatrix} u_r^0 \\ u_\theta^0 \\ u_z^0 \end{Bmatrix} + 2\iint_S [A]\left[u_r^0(\mathbf{x}') \quad u_\theta^0(\mathbf{x}') \quad u_z^0(\mathbf{x}')\right]^T dS \\ = 2\iint_S [B]\left[p_r^0(\mathbf{x}') \quad p_\theta^0(\mathbf{x}') \quad p_z^0(\mathbf{x}')\right]^T dS \end{aligned} \tag{24.11.1}$$

where $[A]$ and $[B]$ are 3×3 matrices given by

$$[A] = \begin{bmatrix} p_{11}\cos\theta + p_{12}\sin\theta & 0 & p_{13} \\ 0 & -p_{21}\sin\theta + p_{22}\cos\theta & 0 \\ p_{31}\cos\theta + p_{32}\sin\theta & 0 & p_{33} \end{bmatrix}, \tag{24.11.2}$$

$$[B] = \begin{bmatrix} u_{11}\cos\theta + u_{12}\sin\theta & 0 & u_{13} \\ 0 & -u_{21}\sin\theta + u_{22}\cos\theta & 0 \\ u_{31}\cos\theta + u_{32}\sin\theta & 0 & u_{33} \end{bmatrix}, \tag{24.11.3}$$

integration is performed over the half-ring $0 \le \theta \le \pi$ (hence the factor 2), and $dS = r\,dr\,d\theta$. ■

(iv) Conformal mapping techniques for consumer products are described by Upton, Haddad and Sørensen [2007]. It analyses the practical noise source identification by introducing reliable sound intensity measurement techniques with associated mapping as described in Fahy [1989]. Such mappings are based on intensity mapping methods, near-field acoustical holography (NAH) and beamforming. They produce results that are mapped over flat planes, which are mapped over a conformal surface corresponding to that of the measured object. The method used in this area is the boundary-element modeling (IBEM). Upton et al. [2007] describe a conformal mapping method that is based on NAH, and uses the statistically optimized NAH (SONAH) algorithm that allows the use of a small, hand-held microphone array, and positional detectors in the array that allow the use of the so-called *patch* holography. In this method, the acoustic quantities on the mapping surface are calculated by using a transfer matrix defined in such a way that all propagating waves and a weighted set of evanescent waves are projected with optimal average accuracy. Note that SONAH differs from NAH by avoiding spatial Fourier transforms, since the processing is carried out directly in the spatial domain. The basic theory of SONAH includes a description of phenomena such as spatial aliasing and wave-number domain leakage and a set of formulas for the estimation of error level of the method, and for visualizing the regions of the SONAH predictions for some typical microphone array geometries. This algorithm also investigates the sensitivity of the inherent error level distribution to changes in its parameters. The main advantage of SONAH as compared with NAH is that the usual requirement of a measurement aperture that extends beyond the source can be relaxed. Both NAH and

SONAH are based on the assumption that all sources are on the one side of the measurement plane whereas the other side is source-free (see, e.g., Jacobson and Jaud [2007], and Hald [2009], and references given on these articles). For the superposition method applied to near-field acoustical holography, see Sarkissian [2005].

(v) On conformal mapping methods for interfacial dynamics, see Crowdy [1999; 2000; 2003].

(vi) Spontaneous branching of discharge channels is frequently observed. Meulenbroek et al. [2003] have proposed a new branching mechanism based on simulations of a simple continuous discharge model in high fields. They present analytical results for such streamers in the Lozansky-Firsov limit (see Meulenbroek et al. [2003, 2004] for rationalization of streamer branching by conformal mapping techniques), where they can be modeled as moving equipotential ionization fronts. This model can be analyzed by conformal mapping techniques, which allow the reduction of the dynamical problem to finite sets of nonlinear ordinary differential equations. The solutions illustrate that branching is generic for the intricate head dynamics of streamers in the Lozansky-Firsov-limit. There are two transitions: the first, known as the avalanche-to-streamer transition, is a classical concept; it occurs when the space charge cannot be ignored. In the streamer phase, the interior of the ionized channel is screened from the externally applied field, and the field at the active head is enhanced. This field enhancement makes the streamer propagate more rapidly than the avalanche. The second transition deals with the case where ionizing gradients are much steeper and the field enhancement much stronger, which often results in the streamer splitting spontaneously according to the old concept of Lozansky and Firsov [1973, 1975] who found the simple parabolic front shape solution for the 'ideally conducting' streamer. The conformal mapping and numerical results are available in Ebert et al. [2003].

REFERENCES USED: Davies [1980], Driscoll and Vivasis [1998], Ebert et al. [2003], Gysen, Ilhan, Meesen, Paulides and Lomonova [2010], Gysen, Ilhan, Motoasca, Pauldes, et al. [2010], Kempel, Volakis, Woo and Yu [1992], Kythe [1995], Kythe et al. [2003], Kythe and Wei [2004], Meulenbroek et al. [2003], Morse and Feshbach [1953], Upton, Haddad and Sørensen [2007].

25

Computer Programs and Resources

25.1 Numerical Methods

This part contains applications of conformal transformations to the flow problems of ideal incompressible fluids, heat transfer problems, eletromagnetic and electric transmissions, and elastic problems.

The conformal mapping technique transforms the physical domain into the model domain and solves the problem in the model domain, mostly by analytical methods. However, the results need to be computed often to provide additional and usable information about the problem thus solved. There are different numerical methods to use, including free-hand plotting, which, although very fast and capable of handling irregular boundaries, may not be very reliable. However, it can be used when approximate solutions suffice the enquiry; it can also be used to check if the solutions from other methods are reliable. Another useful method involves experimentation in a laboratory, which provides data in many cases; it is worthwhile to use with recurrent problems so that the same laboratory setup can be used. Other mathematically oriented methods are as follows:

(i) Finite difference method. It is easily programmable using grids that are simple to use, and it does not require large computer memory. However, since it requires that the entire domain be computed, difficulties arise if the boundary of the domain is irregular. This method works well with small and special programs.

(ii) Finite element method. It handles the irregular boundaries very well by using triangular elements. The solutions are fast and economical, although it needs more processing time for mesh generation. Many good programs are available.

(iii) Boundary element method. It is useful in problems with one dimension less than those computed by finite difference and finite element methods. Boundaries are very easily handled, and in problems involving Laplace's equation it can compute points one at a time. The main disadvantage is generally processing, involving mesh generation, although it is reduced for the elliptic cases.

(iv) Source simulation method. It handles both interior and exterior domains very well. But it requires problem-specific programming or interactive operation. It is usually advantageous in cases when sources can be simulated as charges in the form of uniformly charged line segments or sets of individual charges.

(v) MONTE CARLO METHOD. It is simple and easily programmable, but it is limited to interior problems. Most of the time it is not very efficient. It may be used when accuracy is not demanded and when only isolated points of the domain are of interest. In order to integrate a function over a complicated domain D, this method picks random points over some simple domain D' which is a superset of D, checks whether each point is within D, and estimates the area (or volume) of D as that of D' multiplied by the fraction of points falling within D'. The Mathematica implementation is: `NINtegrate[f, ..., Method -> MonteCarlo]`

An estimate of the uncertainty produced by this technique is given by

$$\int f\,dV \approx V\langle f\rangle \pm \frac{\langle f^2\rangle - \langle f\rangle^2}{N}.$$

For more details, see Ueberhaber [1997], and Weinzierl [2000].

25.2 Software

Many programs and codes that may be required for computations are available in the public domain. Other sources for the software for the methods discussed above are as follows.

(i) For the finite difference method, besides algorithms for forward difference (explicit scheme), backward difference (implicit scheme), and the Crank-Nicolson implicit scheme, a Mathematica program, and solutions for certain examples involving second-order partial differential equations, such as one-dimensional steady-state heat conduction initial and boundary value problems, wave equation with the Neumann initial conditions, and wave equation boundary value problems, Poisson's equation with the Dirichlet boundary value problem, and Laplace's equation on the quarter circular region $x^2 + y^2 < 1$, $y > 0$, are available in Kythe et al. [2003].

(ii) For the finite element method, the theory, algorithms and Fortran programs are available in Kythe and Wei [2004].

(iii) For boundary element methods, the basic theory with many useful algorithms for solving boundary value problems, and a C program are available in Kythe [1995]. This C-program works only with the GNU C-compiler. A very useful summary and the said C code are also available in Kythe and Schäfferkotter [2005: 547ff].

For conformal mapping in general, there are algorithms and a few Mathematica codes available in Kythe [1998].

Various useful algorithms and computer programs are included in the Bibliography. This list is, however, not comprehensive.

25.2.1 Finite Difference Approximations. Fortran subprograms (subroutines) for the solution of partial differential equations are available in Swarztrauber and Sweet [1975]. The description of these subroutines are as follows:

(i) Subroutine PWSCRT. It solves the standard five-point finite difference approximation to the Helmholtz two- and three-dimensional equations $\dfrac{\partial^2 u}{\partial x^2} + \dfrac{\partial^2 u}{\partial y^2} + \lambda u = f(x, y)$, and

$\dfrac{\partial^2 u}{\partial x^2} + \dfrac{\partial^2 u}{\partial y^2} + \dfrac{\partial^2 u}{\partial z^2} = f(x,y,z)$. In the two-dimensional equation, the solution may sometimes fail for $\lambda > 0$. The subroutine for the three-dimensional equation may be used with a fast Fourier transform (FFT) routine to solve the three-dimensional Poisson's equations. A two-dimensional boundary value problem is presented that uses the PWSCRT subroutine to solve the problem

$$\frac{\partial^2 u}{\partial x^2} + \frac{\partial^2 u}{\partial y^2} - 4u = \left[2 - (4+\pi^2/4)x^2\right] \cos\frac{\pi}{2}(y+1)$$

on the rectangle $0 < x < 2, -1 < y < 3$, subject to the boundary conditions

$$u(0,y) = 0, \ \frac{\partial u}{\partial x}(2,y) = 4\cos\frac{\pi}{2}(y+1), \ \ -1 < y < 3, \quad \text{and } u \text{ periodic in } y.$$

The numerical solution is obtained by the finite difference approximation, and it uses the program XAMPLE, where the x-interval $0 \le x \le 2$ is divided into 100 panels and the y-interval $-1 \le y \le 3$ into 80 panels. This solution has four-digit accuracy as compared with the exact solution $u(x,y) = x^2 \cos\dfrac{\pi}{2}(y+1)$.

Subroutine (ii) PWSPLR. It provides a finite difference approximation to the Helmholtz equation, which in polar coordinates is

$$\frac{1}{r}\frac{\partial}{\partial x}\left(r\frac{\partial u}{\partial r}\right) + \frac{1}{r^2}\frac{\partial^2 u}{\partial \theta^2} + \lambda u = f(r,\theta). \tag{25.2.1}$$

If the domain is a portion of the disk $A < r < B$, $C < \theta < D$, and if the boundary data is in the form of a linear system of equations, then the problem is solved by the subroutine POIS (see Subroutine (vi) given below). A grid of points (r_i, θ_j) is defined by selecting integers N and M such that

$$\begin{aligned} r_i &= A + (i-1)\Delta r, \quad i = 1,2,\dots,N+1, \\ \theta_j &= C + (j-1)\delta\theta, \quad j = 1,2,\dots,N+1, \end{aligned} \tag{25.2.2}$$

where $\Delta r = (B-A)/M$ and $\theta_j = (D-c)/N$. If each derivative of Eq (25.2.1) is approximated by a centered difference approximation and the approximation to $u\,(r_i,\theta_j)$ is denoted by $u_{i,j}$, then for $r_i \neq 0$ we get

$$\begin{aligned} &\frac{1}{\Delta r^2 r_i}\left[\left(r_i + \frac{1}{2}\Delta r\right)(u_{i+1,j} - u_{i,j}) - \left(r_i - \frac{1}{2}\Delta r\right)(u_{i,j} - u_{i-1,j})\right] \\ &\quad + \frac{1}{\Delta\theta\, r_i^2}\left[u_{i,j-1} - 2u_{i,j} - u_{i,j+1}\right] + \lambda u_{i,j} = f\,(r_i,\theta_j), \end{aligned}$$

or

$$\begin{aligned} &\frac{r_i^{-1/2\Delta r}}{\Delta r^2 r_i} u_{i-1,j} - \frac{2}{\Delta r^2} u_{i,j} + \frac{r_i^{1/2\Delta r}}{\Delta r^2 r_i} u_{i+1,j} \\ &\quad + \left(\frac{1}{\Delta\theta\, r_i}\right)^2 [u_{i,j-1} - 2u_{i,j} + u_{i,j+1}] + \lambda u_{i,j} = f\,(r_i,\theta_j), \end{aligned} \tag{25.2.3}$$

which becomes the basic equation used to determine the unknowns $u_{i,j}$, where near the boundary this equation is modified by the boundary conditions. For this purpose, let us first consider the case when $A > 0$ so that the origin is not included in the disk. Then at $r = A$, two different types of boundary conditions are described:

(a) Given a function g such that $u(A,\theta) = g(\theta)$, $C \le \theta \le D$, we have

$$u_{i,j} = g(\theta_j), \quad = 1, 2, \dots, N+2, \tag{25.2.4}$$

and the i-index of the unknowns $u_{i,j}$ begins with $i = 2$. Then incorporating Eq (25.2.4) into Eq (25.2.3) for $i = 2$, we get

$$-\frac{2}{\Delta r^2} u_{2,j} + \frac{r_2^{1/2\Delta r}}{\Delta r^2 r_2} u_{3,j} + \left(\frac{1}{\Delta\theta\, r_2}\right)^2 [u_{2,j-1} - 2u_{2,j} + u_{2,j+1}] + \lambda u_{2,j}$$
$$= f(r_2, \theta_j) - \frac{r_2^{-1/2\Delta r}}{\Delta r^2 r_2} g(\theta_j).$$

(b) Given a function $h(\theta)$ such that

$$\frac{\partial u}{\partial r}(A,\theta) = h(\theta), \tag{25.2.5}$$

the solution is unknown at $r = A$, and the i-index of the unknowns $u_{i,j}$ begins with $i = 1$. Assuming that Eq (25.2.1) holds at $r = A$ and that the unknown $u_{1,j}$ is defined by Eq (25.2.3) with $i = 1$, we have

$$\begin{aligned} &\frac{r_1^{-1/2\Delta r}}{\Delta r^2 r_1} u_{1-1,j} - \frac{2}{\Delta r^2} u_{1,j} + \frac{r_1^{1/2\Delta r}}{\Delta r^2 r_1} u_{2,j} \\ &\quad + \left(\frac{1}{\Delta\theta\, r_1}\right)^2 [u_{1,j-1} - 2u_{1,j} + u_{1,j+1}] + \lambda u_{1,j} = f(r_1, \theta_j). \end{aligned} \tag{25.2.6}$$

This assumption requires that a virtual point (r_0, θ_j) just lying outside the boundary of the disk be introduced. Then the unknowns $u_{0,j}$ are eliminated from Eq (25.2.6) by approximating Eq (25.2.5) with a second-order central difference to get

$$u_{0,j} = u_{2,j} - 2\Delta r\, h(\theta_j). \tag{25.2.7}$$

Combining Eq (25.2.6) with Eq (25.2.7), we obtain the defining equation for ui, j as

$$-\frac{2}{\Delta r^2} u_{1,j} + \frac{2}{\Delta r^2} u_{2,j} + \left(\frac{1}{\Delta\theta\, r_1}\right)^2 [u_{1,j-1} - 2u_{1,j} + u_{1,j+1}] + \lambda u_{1,j}$$
$$= f(A, \theta_j) + \frac{2r_1 - \Delta r}{\Delta r\, r_1} h(\theta_j).$$

The same boundary conditions may be specified at $r = B$. Two boundary conditions may be specified at $\theta = C, D$, but a third type is also possible, which is defined as follows: Assuming that $u(r, C+\theta) = u(r, D+\theta)$, which leads to the conditions

$$u_{i,0} = u_{i,N}, \quad \text{and} \quad u_{i,N+1} = u_{i,1}, \tag{25.2.8}$$

and also assuming again that Eq (25.2.1) holds at $r = A$, we see that $u_{i,0}, u_{i,1}, \dots, u_{i,N+1}$ are unknowns, but because of Eq (25.2.8) a complete nonredundant set of unknowns is $u_{i,1}, u_{i,2}, \dots, u_{i,N}$. The defining equation for $u_{i,j}$, i.e., Eq (25.2.6), is modified using Eq (25.2.8), to give

$$\frac{r_i^{-1/2\Delta r}}{\Delta r^2 r_i} u_{i-1,j} - \frac{2}{\Delta r^2} u_{i,j} + \frac{r_1^{1/2\Delta r}}{\Delta r^2 r_i} u_{i+1,j}$$
$$+ \left(\frac{1}{\Delta\theta\, r_i}\right)^2 [u_{1,j-1} - 2u_{i,j} + u_{i,2}] + \lambda u_{i,1} = f(r_i, C).$$

Similarly, the equation for $u_{i,N}$ becomes

$$\frac{r_i^{-1/2\Delta r}}{\Delta r^2 r_i} u_{i-1,N} - \frac{2}{\Delta r^2} u_{i,N} + \frac{r_1^{1/2\Delta r}}{\Delta r^2 r_i} u_{i+1,N} + \left(\frac{1}{\Delta\theta\, r_i}\right)^2 [u_{1,N-1} - 2u_{i,N} + u_{i,2}] + \lambda u_{i,1} = f(r_i, \theta_N).$$

When periodicity in θ is assumed, it is not necessary that $D - C = 2\pi$. However, it is likely that $k(D-c) = 2\pi$ for some integer k which may correspond to the solution for the kth wave number. This permits solutions for high wave numbers with less computation.

When $A = 0$, the origin is included in the region of the disk. There is no change in the possible boundary conditions on the θ-boundaries, but the above boundary condition (b) may not be prescribed at $r = A$; instead, the above condition (a) may be prescribed, or we may prescribe the following condition:

For solution at $r = A = 0$, the solution is unspecified and Eq (25.2.3) cannot be used, and the equation for $u_{1,1}$ is obtained by multiplying Eq (25.2.1) by r over the sector of the disk $0 \le r \le \frac{1}{2}\Delta r$, $C \le \theta \le D$, which gives

$$\frac{1}{2}\Delta r \int_C^D \frac{\partial u}{\partial r}\left(\frac{1}{2}\Delta r, \theta\right) d\theta + \int_0^{\frac{1}{2}\Delta r} \frac{1}{r}\left[\frac{\partial u}{\partial \theta}(r, D) - \frac{\partial u}{\partial \theta}(r, C)\right] dr + \int_0^{\frac{1}{2}\Delta r}\int_C^D r(\lambda u - r)\, d\theta\, dr = 0 \equiv I_1 + I_2 + I_3 = 0. \tag{25.2.9}$$

Then

$$I_1 = \frac{1}{\Delta r}[u(r_2, \theta) - u(0, 0)]] \approx \frac{\Delta\theta}{2}\Big[\sum_{j=2}^{N} u_{2.j} + \frac{1}{2}u_{2,1} + \frac{1}{2}u_{2,N+1} - N u_{1,1}\Big]. \tag{25.2.10}$$

Integral I_2 is zero when periodicity in θ is specified. Suppose that $\frac{\partial u}{\partial \theta}$ is specified on the two boundaries. Then $\frac{\partial u}{\partial \theta} = 0$, since

$$\frac{\partial u}{\partial \theta}(r, \theta) = r\Big[-\sin\theta \frac{\partial u}{\partial x} + \cos\theta \frac{\partial u}{\partial y}\Big].$$

Hence, using a Taylor series, a linear approximation to $\frac{\partial u}{\partial \theta}$ near $r = 0$ is

$$\frac{\partial u}{\partial \theta}(r, \theta) = \frac{\partial u}{\partial \theta}(0, 0) + r\frac{\frac{\partial u}{\partial \theta}(\Delta r, \theta) - \frac{\partial u}{\partial \theta}(0, v)}{\Delta r} = \frac{r}{\Delta r}\frac{\partial u}{\partial \theta}(\Delta r, \theta).$$

Using this formula we get

$$I_2 = \int_0^{\frac{1}{2}\Delta r} \frac{1}{\Delta r}\Big[\frac{\partial u}{\partial \theta}(\Delta r, D) - \frac{\partial u}{\partial \theta}(\Delta r, C)\Big] dr = \frac{1}{2}\Big[\frac{\partial u}{\partial \theta}(r_2, \theta_{N+1}) - \frac{\partial u}{\partial \theta}(r_2, \theta_1)\Big], \tag{25.2.11}$$

where the derivatives on the right-hand side of Eq (25.2.11) are input data. Finally, a second-order approximation to I_3 is obtained by replacing $(\lambda u - f)(r, \theta)$ by $(\lambda u - f)(0, 0)$ and integrating, which yields

$$I_3 \approx \frac{\Delta r^2}{8}(D - C)\left[\lambda u_{1,1} - f(0,0)\right] = \frac{N\Delta\theta\,\delta r^2}{8}\left[\lambda u_{1,1} - f(0,0)\right]. \tag{25.2.12}$$

Thus, combining Eqs (25.2.10), (25.2.11) and (25.2.12), and dividing by $\frac{1}{8}\Delta\theta\,\Delta r^2$, we obtain formula:

$$\begin{aligned} \Big(\lambda - \frac{4}{\Delta r^2}\Big)u_{1,1} + \frac{4}{\Delta r^2}\sum_{j=2}^{N}\Big\{u_{2,j} + \frac{1}{2}u_{2,1} + \frac{1}{2}u_{2,N+1}\Big\} \\ = f(0,0) - \frac{4}{N\,\Delta\theta\,\Delta r^2}\big[\partial D(2) - \partial C(2)\big], \end{aligned} \tag{25.2.13}$$

where ∂D and ∂C denote the boundaries D and C. By examining the input parameters, the subroutine PWSPLR determines which $u_{i,j}$ are unknowns, then sets up the equations defining them, and incorporates the given boundary data. The final system of equations can then be solved by the subroutine POIS (see below Subroutine (vi)).

(iii) Subroutine PWSCYL. It provides a finite difference approximation to the Helmholtz equation in cylindrical coordinates defined by

$$\frac{1}{r}\frac{\partial}{\partial r}\Big(r\frac{\partial u}{\partial r}\Big) + \frac{\partial^2 u}{\partial z^2} + \frac{1}{r^2}u = f(r, z).$$

The details can be found in Swarztrauber and Sweet [1975: 37-59].

(iv) Subroutine PWSCSP. It provides a finite difference approximation to the Helmholtz equation in spherical coordinates, assuming axisymmetry (i,e., no dependence on longitudes z), which is defined by

$$\frac{1}{r^2}\frac{\partial}{\partial r}\Big(r^2\frac{\partial u}{\partial r}\Big) + \frac{1}{r^2\sin\theta}\frac{\partial}{\partial\theta}\Big(\sin\theta\frac{\partial u}{\partial\theta}\Big) + \frac{1}{r^2\sin^2\theta}u = f(,\theta),$$

where θ is the colatitude, and r the radial coordinate. The details can be found in Swarztrauber and Sweet [1975: 60-80].

(v) Subroutine PWSSSP. It provides a finite difference approximation to the Helmholtz equation in spherical coordinates and on the surface of the unit sphere. The details can be found in Swarztrauber and Sweet [1975: 81-95].

(vi) Subroutine POIS. It solves the linear system of equations

$$\begin{aligned} &A(I) * X(I-1, J) + B(I) * X(I, J) \\ &\quad + C(I) * [X(I+1, J) + X(I, J-1) - 2 * X(I, J) + X(I, J+1)] = Y(I, J), \\ &\quad \text{for } I = 1, 2, \ldots, N;\ J = 1, 2, \ldots, N. \end{aligned}$$

The details are given in Swarztrauber and Sweet [1975: 96-114].

(vii) Subroutine BLKTRI. It solves the system of linear equations of the form

$$\begin{aligned} &AN(J) * X(I, J-1) + AM(I) * X(I-1, J) + (BN(J) + BM(I)) * X(I, J) \\ &\quad + CN(J) * X(I, J+1) + CM(I) * X(I+1, J) = Y(I, J), \\ &\quad \text{for } I = 1, 2, \ldots, N, \text{ and } J = 1, 2, \ldots, N, \end{aligned}$$

where $I \pm 1$ is evaluated modulo M and $J \pm 1$ modulo N, i.e.,

$$X(I,0) = X(I,N); \;\; X(I,N+1) = X(I,1); \;\; X(0,J) = X(M,J); \;\; X(M+1,J) = X(1,J).$$

These equations usually result from discretization of separable elliptic equations. Boundary conditions may be Dirichlet, Neumann, or periodic. The details are given in Swarztrauber and Sweet [1975: 115-131].

The least-square solution of singular linear systems of equations is given in Swarztrauber and Sweet [1975: 133-137].

25.3 Computer Codes

The following computer codes are available in Kythe and Wei [2004: 351-391]:

Computer codes in Mathematica are available to compute solutions of the matrix equation **KU** = **F**, and to plot the solutions.

Algorithms for the following equations are available in Zwillinger [2002: 771-776]:
Algorithm for Poisson's equation finite-difference

$$\frac{\partial^2 u}{\partial x^2} + \frac{\partial^2 u}{\partial y^2} = f(x,y), \quad a \leqq x \leqq y, \, c \leqq y \leqq d,$$

subject to the boundary condition $u(a,y) = u(b,y) = h(x,y)$, $c \leqq y \leqq d$, and $u(x,c) = u(x,d) = g(x,y)$, $a \leqq x \leqq y$.
Algorithm for heat or diffusion equation finite-difference

$$\frac{\partial u}{\partial t} = k \frac{\partial^2 u}{\partial x^2}, \quad 0 < x < l, \, t > 0,$$

subject to the initial and boundary conditions $u(0,t) = 0 = u(l,t)$, $0 \leqq x \leqq l$ for $t > 0$, and $u(x,0) = f(x)$ for $0 \leqq x \leqq l$. This algorithm uses the Crank-Nicolson method.
Algorithm for the wave equation

$$\frac{\partial u}{\partial t} - a^2 \frac{\partial^2 u}{\partial x^2}(x,t) = 0, \quad 0 < x < l, \, t > 0.$$

Software Content information is as follows: Fortran: http://www.fortran.com; MATLAB: http://www.mathworks.com; Mathematica: http://www.wolfram.com.

Electronic math resources are available at: http//www.mathworld.wolfram.com; http://www.google.com; and http://dir.yahoo.com/science/math. For Mathematica and MATLAB subroutines, see Polyanin and Nazaikinskii [2016: 1277-1398].

SCPACK Version 2 and the User's guide are available in Trefethen [1983, 1989b].

Mathematica subroutines for numerical solutions of Fredholm and Volterra integral equations of the first and second kinds, with examples, are available in Kythe and Puri [2002].

Simulative investigations of manufacturing processes and subsequent structural analysis for composite components operate with different software packages, e.g., those from ESI Group, SIMULIA Abaqus and MSC.Nastran. Most interfaces between simulation and analysis in Laminate Modeler tool[1] is based on CATIA data form. The software Beta-CAE

[1] Anaglyph Ltd. MSC.Laminate Modeler, www.anaglyph.co.uk

ANSA[2] is used in an internal approach for data transfer into a crash simulation. The e-Xtream Engineering tools[3] convert Moldflow data onto an imported mesh from ANSYS, ABAQUS or PAM-Crash. ANSY[4] software converts data sets from the FiberSIM tool only on an internal mesh.

For airfoil problems, the following references are useful: Abbott et al. [1959], Arafeh and Schinzinger [1978], Babinsky [2003], Bertin and Cummings [2009], Buzbee et al. [1971], Clancy [1975], Clancy et al. [1971], Cooley and Tuckey [1965], Croom and Holmes [1985], Croom et al. [1970], Davis and Rabinowitz [1961], De Rivas [1972], Dorr [1970], Gutknecht [1981; 1983], Henrici 1979], Hockney [1970], Holmes et al. [1984], Houghton and Carpenter [2003], Hu [1998], Hurt [1960], Kantorovitch [1934], Khamayseh and Mastin [1996], Linz [1985], Longuet-Higgins and Cocket [1976], Morris [2009], Morris and Rusk [2003], Phillips [2010], Swarztrauber [1974], Swarztrauber and Sweet [1973; 1975], Sweet [1973; 1974], Traub [2016], Trefethen [1979; 1980; 1983; 1989a, b].

Hu [1998] has published Algorithm 785: a software package for computing Schwarz-Christoffel conformal transformation for doubly connected polygonal regions. This package fully describes the mathematical, numerical, and practical perspectives. It solves the so-called accessory parameter problem associated with the mapping function as well as evaluates forward and inverse maps. The robustness of the package is reflected by the flexibility in choosing the accuracy of the parameters to be computed, the speed of computation, the ability of mapping 'difficult' regions, and being user friendly. Several examples are presented to demonstrate the capabilities of the package.

For Aerospaceweb's information on Thin Foil Theory, see (http://www.aerospaceweb.org/question/aerodynamics/q0136.shtml). NASA Publications [2011] are also available.

Bazant and Crowdy [2005] have considered moving free boundary value problems formulated using conformal mappings. Recent developments in the material sciences include models of void electro-migration in metals, brittle fracture, and viscous sintering. Conformal mapping dynamics has also been used for stochastic problems, such as diffusion-limited aggregation and dielectric breakdown leading to new developments in fractal pattern formulation. They also discuss conformal mapping methods for interfacial dynamics.

Lastly, Upton, Haddad and Sørensen [2007] have discussed conformal mapping techniques for consumer products using NAH and SONAH methods. Also see §24.11(iv). For finite element methods in electrical engineering problems, consult Salon [1995], Silvestor and Ferrari [1996], Bianchi [2005], and Jin [2014].

REFERENCES USED: Bazant and Crowdy [2005], Buzzbee, Dorr, George and Golub [1971], Buzzbee, Golub and Nielson [1971], de Rivas [1972], Dorr [1970], Hockney [1970], Kythe and Puri [2002], Kythe et al. [2003], Kythe and Wei [2004], Swarztrauber [1974; 1974; 1975], Swarztrauber and Sweet [1973; 1975], Sweet [1973; 1974].

[2] BETA CAE Systems USA Inc., *ANSA v13.1.3 Edition. http://www.beta-cae.gr*, 2013.
[3] e-Xtream Engineering. *DIGIMAP Release Notes.* 4th ed, 2012.
[4] ANSYS Documentaion, 14th ed., 2013.

A

Green's Identities

Let D be a finite domain in R^n bounded by a piecewise smooth, orientable surface ∂D, and let w and F be scalar functions and $\mathbf{G}$ a vector function in the class $C^0(D)$. Then

$$\text{Gradient theorem:} \quad \int_D \nabla F \, dD = \oint_{\partial D} \mathbf{n} F \, dS,$$

$$\text{Divergence theorem:} \quad \int_D \nabla \cdot \mathbf{G} \, dD = \oint_{\partial D} \mathbf{n} \cdot \mathbf{G} \, dS,$$

$$\text{Stokes's theorem:} \quad \int_D \nabla \times \mathbf{G} \, dD = \oint_{\partial D} \mathbf{G} \cdot \mathbf{t} \, dS,$$

where $\mathbf{n}$ is the outward normal to the surface ∂D, $\mathbf{t}$ is the tangent vector at a point on ∂D, $\oint$ denotes the surface or line integral, and dS (or ds) denotes the surface (or line) element depending on the dimension of D. The divergence theorem in the above form is also known as the Gauss theorem. Stokes's theorem in $\mathbb{R}^2$ is a generalization of *Green's theorem* which states that if $\mathbf{G} = (G_1, G_2)$ is a continuously differentiable vector field defined on a region containing $D \cup \partial D \subset \mathbb{R}^2$ such that ∂D is a Jordan contour, then

$$\int_D \left(\frac{\partial G_2}{\partial x_1} - \frac{\partial G_1}{\partial x_2} \right) dx_1 \, dx_2 = \oint_{\partial D} G_1 \, dx_1 + G_2 \, dx_2. \tag{A.1}$$

Let the functions $M(x, y)$ and $N(x, y)$, where $(x, y) \in D$, be the components of the vector $\mathbf{G}$. Then, by the divergence theorem

$$\begin{aligned} \int_D \left(\frac{\partial M}{\partial x} + \frac{\partial N}{\partial y} \right) dx \, dy &= \oint_\Gamma \left[M \cos(\mathbf{n}, x) + N \cos(\mathbf{n}, y) \right] ds, \\ &= \oint_\Gamma M \, dx + N \, dy, \end{aligned} \tag{A.2}$$

with the direction cosines $\cos(\mathbf{n}, x)$ and $\cos(\mathbf{n}, y)$, where $\Gamma = \partial D$. If we take $M = f \, g_x$ and $N = f \, g_y$, then (A.2) yields

$$\int_D \left(\frac{\partial f}{\partial x} \frac{\partial g}{\partial x} + \frac{\partial f}{\partial y} \frac{\partial g}{\partial y} \right) dx \, dy = \int_\Gamma f \frac{\partial g}{\partial n} \, ds - \int_D f \nabla^2 g \, dx \, dy, \tag{A.3}$$

which is known as *Green's first identity*. Moreover, if we interchange f and g in (A.2), we get

$$\int_D \left(\frac{\partial f}{\partial x}\frac{\partial g}{\partial x} + \frac{\partial f}{\partial y}\frac{\partial g}{\partial y} \right) dx\,dy = \int_\Gamma g \frac{\partial f}{\partial n}\, ds - \int_D g\nabla^2 f\, dx\,dy. \tag{A.4}$$

If we subtract (A.3) from (A.4), we obtain *Green's second identity*:

$$\int_D \left(f\nabla^2 g - g\nabla^2 f \right) dx\,dy = \int_\Gamma \left(f\frac{\partial g}{\partial n} - g\frac{\partial f}{\partial n} \right) ds, \tag{A.5}$$

which is also known as *Green's reciprocity theorem*. Note that Green's identities are valid even if the domain D is bounded by finitely many closed curves. In that case, however, the line integrals must be evaluated over all paths that make the boundary of D. If f and g are real and harmonic in $D \subset \mathbb{R}$, then from (A.5)

$$\int_\Gamma \left(f\frac{\partial g}{\partial n} - g\frac{\partial f}{\partial n} \right) ds = 0. \tag{A.6}$$

Let D be a simply connected region in the complex plane $\mathbb{C}$ with boundary Γ. Let z_0 be any point inside D, and let D be the region obtained by indenting a disk $B(z_0, \varepsilon)$ from D, where $\varepsilon > 0$ is small (Figure A.1 (a)). Then ∂D consists of the contour Γ together with the contour $\partial B(z_0, \varepsilon) = \Gamma_\varepsilon$.

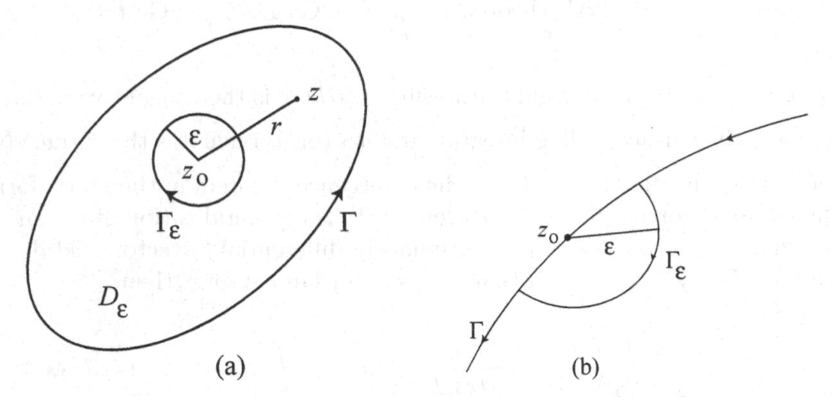

Figure A.1 Simply connected region and its boundary.

If we set $f = u$ and $g = \log r$ in (A.6), where $z \in D$ and $r = |z - z_0|$, then, since $\dfrac{\partial}{\partial n} = -\dfrac{\partial}{\partial r}$ on Γ_ε, we get

$$\int_\Gamma \left(u\frac{\partial}{\partial n}(\log r) - (\log r)\frac{\partial u}{\partial n} \right) ds - \int_{\Gamma_\varepsilon} \left(\frac{u}{r} - (\log r)\frac{\partial u}{\partial r} \right) ds = 0. \tag{A.7}$$

Now, let $\varepsilon \to 0$ in (A.7). Then, since

$$\lim_{\varepsilon\to 0} \int_{\Gamma_\varepsilon} \frac{u}{r}\, ds = \lim_{\varepsilon\to 0} \int_0^{2\pi} u(z_0 + \varepsilon\theta)\, \frac{1}{\varepsilon}\, \varepsilon\, d\theta = 0,$$

$$\lim_{\varepsilon\to 0} \int_{\Gamma_\varepsilon} \log r\, \frac{\partial u}{\partial r}\, ds = \lim_{\varepsilon\to 0} \int_0^{2\pi} \log\varepsilon\, \frac{\partial u}{\partial \varepsilon}\, \varepsilon\, d\theta = 0,$$

we obtain

$$2\pi\, u(z_0) = \int_\Gamma \left[u\, \frac{\partial}{\partial n}(\log r) - (\log r)\, \frac{\partial u}{\partial n} \right] ds, \tag{A.8}$$

which is known as *Green's third identity*. Note that Eq (A.8) gives the value of a harmonic function u at an interior point z_0 in terms of the boundary values of u and $\frac{\partial u}{\partial n}$. If the contour Γ has no corners and if the point z_0 is on the boundary Γ, then instead of the whole disk $B(z_0, \varepsilon)$ we consider a half disk at the point z_0 deleted from D (Figure A.1(b)), and Green's third identity becomes

$$\pi\, u(z_0) = \text{p.v.} \int_\Gamma \left[u\, \frac{\partial}{\partial n}(\log r) - (\log r)\, \frac{\partial u}{\partial n} \right] ds, \tag{A.9}$$

where p.v. denotes the principal value of the integral, i.e., it is the limit, as $r \to 0$, of the integral over the contour Γ obtained by deleting that part of Γ which lies within the circle of radius ε and center z_0.

Note that if the point z_0 lies on the boundary Γ, a consequence of Green's third identity (A.8) is

$$u(z_0) = \frac{1}{2\pi} \int_\Gamma u \frac{\partial G}{\partial n}\, ds. \tag{A.10}$$

REFERENCES USED: Carrier, Krook and Pearson [1966], Kythe [1996], Kythe, Puri and Schäferkotter [1997], Nehari [1952], Koppenfels [1959], Wayland [1970].

B

Cauchy's P.-V. Integrals

B.1 Numerical Evaluation

We will derive approximate formulas to compute Cauchy principal-value (p.-v.) integrals. We will consider the Cauchy p.-v. integral

$$F(t_0) = \frac{1}{i\pi} \,-\!\!\!\!\!\!\int_\Gamma \frac{f(t)}{t - t_0}\, dt, \quad t_0 \in \Gamma, \tag{B.1}$$

where Γ is a smooth, open arc or a Jordan contour and $f \in H^1(\Gamma)$, i.e., f satisfies the Hölder condition on Γ with $\alpha = 1$ (see §2.2.1). We can rewrite (B.1) as

$$F(t_0) = \frac{1}{i\pi} \,-\!\!\!\!\!\!\int_\Gamma \frac{f(t) - f(t_0)}{t - t_0}\, dt + \frac{f(t_0)}{i\pi} \,-\!\!\!\!\!\!\int_\Gamma \frac{dt}{t - t_0} = F_1(t_0) + F_2(t_0). \tag{B.2}$$

The integral $F_1(t_0)$ is an improper integral whereas $F_2(t_0)$ may be evaluated directly. Therefore, the problem reduces to evaluating the improper integral

$$F_1(t_0) = \frac{1}{i\pi} \int_\Gamma f_1(t)\, dt, \quad f_1(t) = \frac{f(t) - f(t_0)}{t - t_0}. \tag{B.3}$$

B.1.1 Method of Separation of Singularities. If the independent variable t in (B.3) is changed to the arc length parameter of Γ or a real parameter in the parametric equation of Γ, then certain classical results can be applied to the improper integral (B.3). In fact, if $f(t) \in H^1$, then $f_1(t)$ is bounded, and (B.3) becomes a proper integral. In particular, if Γ is a straight line segment on the real axis, then the variable t is real. If Γ is a circle or a circular arc with a fixed radius r, then the substitution $t = r\, e^{i\theta}$ also transforms (B.3) into an integral of the real variable θ. The method of separation of singularities is very easy to use when Γ is a straight line segment $[a, b]$ of the real axis. The following result can be proved by induction.

Theorem B.1. *Let* $G(t) = \begin{cases} \dfrac{g(t) - g(t_0)}{t - t_0}, & \text{if } t \neq t_0, \\ g'(t_0), & \text{if } t = t_0, \end{cases}$ *where* $g(t) \in C^{n+1}[a, b]$, $n \geq 0$, *and* $a \leq t_0 \leq b$. *Then*

$$G^{(k)}(t) = \begin{cases} \dfrac{g^{(k+1)}(\tau_k)}{k+1}, & \text{if } t \neq t_0, \\ \dfrac{g^{(k+1)}(t_0)}{k+1}, & \text{if } t = t_0, \end{cases} \qquad t_0 < \tau_k < t. \tag{B.4}$$

In the case when

$$F(t_0) = \frac{1}{i\pi} \rlap{-}\int_\Gamma \frac{f(t)}{t - t_0}\, w(t)\, dt, \quad t_0 \in \Gamma, \tag{B.5}$$

where $w(t) \geq 0$ is a weight function which may have integrable singularities at certain points, e.g., at the end points, then the above theorem can be used, provided (B.5) is written as

$$F(t_0) = \frac{1}{i\pi} \rlap{-}\int_\Gamma f_1(t)\, w(t)\, dt + \frac{f(t_0)}{i\pi} \rlap{-}\int_\Gamma \frac{w(t)}{t - t_0}\, w(t)\, dt, \quad t_0 \in \Gamma, \tag{B.6}$$

in which the second integral may be evaluated exactly. The above results are due to Ivanov [1968] and Pukhteev [1980].

B.1.2 Gauss-Chebyshev Type Quadrature. As an application of the above result we will find an approximate value of the Cauchy p.-v. integral

$$I(x) = \frac{1}{\pi} \rlap{-}\int_{-1}^{1} \frac{g(t)}{t - x} \frac{dt}{\sqrt{1 - t^2}}, \quad -1 < x < 1, \tag{B.7}$$

with weight function $w(t) = \dfrac{1}{\sqrt{1 - t^2}}$, where $g(x)$ is sufficiently smooth. First, note that if $g(t) \equiv 1$, the value of the integral (B.7) is zero. In fact, it can be proved by the contour integration method that

$$\frac{1}{\pi} \int_{-1}^{1} \frac{dt}{(t - x)\sqrt{1 - t^2}} = 0, \quad -1 < x < 1, \tag{B.8}$$

by taking a closed contour Γ^- containing the line segment $-1 \leq x \leq 1$, and using the residue theorem to obtain $\dfrac{1}{2i\pi} \displaystyle\int_{\Gamma^-} \dfrac{d\zeta}{\zeta - z)\sqrt{1 - \zeta^2}} = \dfrac{1}{\sqrt{1 - z^2}}$, where Γ^- is clockwise and $\sqrt{1 - \zeta^2}$ is the analytic continuation of the positive real-valued function $\sqrt{1 - x^2}$ along the upper side of $-1 < x < 1$. As Γ shrinks to $-1 \leq x \leq 1$, we get

$$\frac{1}{i\pi} \int_{-1}^{1} \frac{dt}{(t - z)\sqrt{1 - t^2}} = \frac{1}{\sqrt{1 - z^2}}, \quad z \notin [-1, 1].$$

Then (B.7) follows as $z \to x$ from the upper side (or lower side) and using Plemelj formula (2.4.19). Thus, we can rewrite (B.7) as

$$I(x) = \frac{1}{\pi} \int_{-1}^{1} \frac{g(t) - g(x)}{t - x} \frac{dt}{\sqrt{1 - t^2}}, \quad -1 < x < 1. \tag{B.9}$$

Now, if $f \in C^{2n}[-1, 1]$, then we have the following Gauss-Chebyshev quadrature formula (see Abramowitz and Stegun [1972])

$$I \equiv I[f] = \frac{1}{\pi} \int_{-1}^{1} \frac{f(t)}{\sqrt{1 - t^2}}\, dt \approx \frac{1}{n} \sum_{k=1}^{n} f(t_k), \tag{B.10}$$

where t_k, $k = 1, 2, \dots, n$, are the zeros of the Chebyshev polynomial $T_n(x) = \cos\left(n \cos^{-1} x\right)$ of the first kind and degree n, i.e., $t_k = \cos \dfrac{(2k-1)\pi}{2n}$, $k = 1, 2, \dots, n$, with the remainder R_n after n terms defined by

$$R_n = \frac{1}{(2n)!\, 2^{2n-1}} f^{2n}(\tau), \quad -1 < \tau < 1. \tag{B.11}$$

If we set $f(t) = \dfrac{g(t) - g(x)}{t - x}$ in (B.10), we obtain the quadrature formula

$$I(x) \approx \frac{1}{n} \sum_{k=1}^{n} \frac{g(t_k)}{t_k - x} + \frac{g(x)}{n} \sum_{k=1}^{n} \frac{1}{x - t_k}, \quad -1 < x < 1, \quad x \neq t_k. \tag{B.12}$$

However, since $\displaystyle\sum_{k=1}^{n} \frac{1}{x - t_k} = \frac{T_n'(x)}{T_n(x)}$ and $T_n'(x) = n\, U_{n-1}(x)$, where $U_{n-1}(x) = \dfrac{\sin\left(n \cos^{-1} x\right)}{\sqrt{1 - x^2}}$ is the Chebyshev polynomial of the second kind and degree $n - 1$, we find that

$$\sum_{k=1}^{n} \frac{1}{x - t_k} = \frac{n\, U_{n-1}(x)}{T_n(x)},$$

and formula (B.12) becomes

$$I(x) \approx \frac{1}{n} \sum_{k=1}^{n} \frac{g(t_k)}{t_k - x} + g(x) \frac{U_{n-1}(x)}{T_n(x)}, \quad -1 < x < 1, \quad x \neq t_k, \tag{B.13}$$

which is a Gauss-Chebyshev type quadrature formula. In particular, if $x = x_j$ is taken as a zero of $U_{n-1}(x)$, i.e., if $x_j = \cos \dfrac{j\pi}{n}$, $j = 1, 2, \dots, n-1$, then formula (B.13) simplifies to

$$I(x_j) \approx \frac{1}{n} \sum_{k=1}^{n} \frac{g(t_k)}{t_k - x_j}, \quad j = 1, 2, \dots, n-1. \tag{B.14}$$

This quadrature formula is useful in solving the following integral equation of the first kind:

$$\frac{1}{\pi} \int_{-1}^{1} \frac{\phi(t)}{t - x}\, dt + \int_{-1}^{1} k(x,t)\, \phi(t)\, dt = f(x), \quad -1 < x < 1, \tag{B.15}$$

where $k(x) \in H^0\left([-1,1] \times [-1,1]\right)$.

B.1.3 Approximation by Arcwise Linear Functions. The arc length parameter s can be used for approximating (B.1) in the case of simple contours (of integration). If the contour is complicated, we will first approximate the function $f(t)$ by an arcwise linear function and estimate error.

Let AB denote an open, smooth arc, and let $f(t) \in H^\alpha$, $0 < \alpha \leq 1$, be defined on AB. We partition AB by $A = t_1, \dots, t_n = B$, and let Γ_j denote a partition (t_j, t_{j+1}). Let $y_j = f(t_j)$, $\Delta y_j = y_{j+1} - y_j$, and $D_j = \Delta y_j / \Delta t_j$. We will define arcwise linear functions

$L_j(t)$ on AB such that $L_j(t) = y_j + D_j\,(t - t_j)$ for $t \in \Gamma_j$, $j = 1, \dots, N$. Then, if the error is denoted by $e_j(t)$, we have

$$\begin{aligned} |e_j(t)| &= |f(t) - L_j(t)| \le |f(t) - y_j| + |D_j|\,|t - t_j| \\ &\le C\,|t - t_j|^\alpha + \frac{C}{|\Delta(t_j)|^{1-\alpha}}\,|t - t_j| \le C\,\delta^\alpha, \end{aligned} \tag{B.16}$$

where C is a constant and $\delta = \max_j |\Delta y_j|$. This inequality implies that the error $e_j(t)$ becomes very small if $\{t_j\}$ is very dense, i.e., if δ is very small.

For example, let us consider the error when the integral in (B.1) is replaced by $\int_\Gamma \frac{L_j(t)}{t-\tau}\,dt$. Then $\int_\Gamma \frac{e_j(t)}{t-\tau}\,dt = \int_\Gamma \frac{e_j(t) - e_j(\tau)}{t-\tau}\,dt + e_j(\tau)\int_\Gamma \frac{dt}{t-\tau} = I_1(\tau) + I_2(\tau)$, and, in view of (B.16), we have

$$|I_2(\tau)| \le C\,|e_j(\tau)| \le C\,\delta^\alpha. \tag{B.17}$$

Now, if t and τ are in the same Γ_j, we have

$$|e_j(t) - e_j(\tau)| \le C\,|t - \tau|^\alpha \le C\,\delta^{\alpha-\varepsilon}\,|t - \tau|^\varepsilon, \tag{B.18}$$

where ε, $0 < \varepsilon < \alpha$, is very small. If $\tau \in \Gamma_{j-1}$ and $t \in \Gamma_k$, $j \le k$, then, since $e_j(\tau_j) = e_j(\tau_k) = 0$, we find that

$$\begin{aligned} |e_j(t) - e_j(\tau)| &\le |e_j(t)| + |e_j(\tau)| \le |e_j(t) - e_j(\tau_k)| + |e_j(\tau) - e_j(\tau_j)| \\ &\le C\,\left(|t - \tau_k|^\alpha + |t - \tau_j|^\alpha\right) \le C\,\delta^{\alpha-\varepsilon}\,\left(|t - \tau_k|^\varepsilon + |t - \tau_j|^\varepsilon\right) \\ &\le C\,\delta^{\alpha-\varepsilon}\,\left(|\tau_k\,t|^\varepsilon + |\tau_j\,t|^\varepsilon\right) \le C\,\delta^{\alpha-\varepsilon}\,\left(|\tau\,t|^\varepsilon\right) \le C\,\delta^{\alpha-\varepsilon}\,\left(|t - \tau|^\varepsilon\right). \end{aligned}$$

Thus, (B.18) always holds for some constant C, and

$$|I_1(t)| \le C\,\delta^{\alpha-\varepsilon}\int_\Gamma |t - \tau|^{-1+\varepsilon}\,|dt| \le C\,\delta^{\alpha-\varepsilon}\int_\Gamma s^{-1+\varepsilon}\,ds \le C_\varepsilon\,\delta^{\alpha-\varepsilon}, \tag{B.19}$$

where C_ε is a constant that depends on ε. In general, $C_\varepsilon \to +\infty$ as $\varepsilon \to 0$. Hence, the inequalities (B.18) and (B.19) yield

$$\left|\int_\Gamma \frac{f(t)}{t-\tau}\,dt - \frac{L_j(t)}{t-\tau}\,dt\right| \le C_\varepsilon\,\delta^{\alpha-\varepsilon}, \quad 0 < \varepsilon < \alpha. \tag{B.20}$$

Since ε is arbitrary, we find that for very small δ the left side of (B.20) becomes arbitrarily small. Thus, $\sum_{j=1}^{N} L_j(t)$ approximates the kernel $f(t)$ in Cauchy's p.-v. integral (B.1), and

$$\int_\Gamma \frac{f(t)}{t-\tau}\,dt \approx \sum_{j=1}^{N} \int_{\Gamma_j} \frac{L_j(t)}{t-\tau}\,dt, \tag{B.21}$$

which can be easily computed. For more on this topic, see the references cited below.

REFERENCES USED: Ivanov [1968], Lu [1984; 1994], Pukhteev [1980].

C

Riemann Mapping Theorem

A proof of the Riemann mapping theorem is presented. Associated with this proof is the Riemann-Hilbert problem, which is related to two boundary value problems of the theory of analytic functions, known as the Hilbert and the Riemann problems. A close relationship exists between the Hilbert problem and the theory of singular integral equations. Although the latter may be developed to a large extent without the former, it is this relationship that makes the latter theory simple and clear. For example, the iterative method for solving Theodorsen's integral equation, as outlined in Chapter 11, is based on a certain Riemann-Hilbert problem. In fact, Theodorsen's problem is a linearized version of a singular integral equation of the second kind.

C.1 Theorem and Proof

The Riemann mapping theorem (§2.6, Theorem 2.15) is stated as follows: Let $D \subset \mathbb{C}$ be a simply connected region. Then there exists a bijective conformal map $f : D \mapsto U$, where U is the open unit disk. Moreover, the map f is unique provided that $f(z_0) = 0$ and $f'(z_0) > 0$ for $z_0 \in D$.

Ahlfors [1953:172] has hailed this theorem as "one of the most important theorems of complex analysis." We will first provide a short historical development of this theorem before proving it. Riemann [1851: 40] stated that "two simply connected plane surfaces can always be mapped onto each other such that each point of the one corresponds to a unique point of the other in a continuous manner and the correspondence is conformal; moreover, the correspondence between an arbitrary interior point and an arbitrary boundary point of the one and the other may be given arbitrarily, but when this is done the correspondence is determined completely." This assertion came to be known as the *Riemann mapping theorem*, in which the explicit requirement that the boundary has at least two points is necessary to rule out the cases where the domain is either the complex plane or the sphere; these are the counter-examples provided by Riemann himself. The uniqueness of the mapping f was first specified by conditions on it only at an interior point as it was made explicit by Carathéodory [1913a,b; 1914] and Koebe [1913].

As Gray [1994] has mentioned, Riemann's own proof used the Dirichlet principle, and for this reason it was not accepted, specially in view of Hilbert's first paper [1905] on the Dirichlet principle, in which he stated the problem as follows: Suppose a boundary curve and a function on this curve are given. Let S be the part of the plane bounded by this curve. Then the function $f(x, y)$ is taken for which the value of the integral

$$L(\alpha, \beta) = \int_S \left[\left(\frac{\partial \alpha}{\partial x} - \frac{\partial \beta}{\partial y} \right)^2 + \left(\frac{\partial \alpha}{\partial y} + f \frac{\partial \beta}{\partial x} \right)^2 \right] dS \tag{C.1.1}$$

is finite, where α and β are two arbitrary real functions of x and y, and S is the surface over which the integral is taken. The function f is necessarily harmonic. Hilbert claimed that such considerations had led Riemann to his proof of the existence of functions with the given boundary values. However, Weierstrass was first to show that this approach was not reliable. According to Hilbert, the Dirichlet principle had "fallen into disrepute, and only Brill and Noether [1894] continued to hope that it could be resurrected, perhaps in a modified form."

If we vary α in the integral (C.1.1) by a continuous function, or by one discontinuous only at a single point, the integral $L(\alpha, \beta)$ attains a minimal value, and this minimum is attained by a unique function if we exclude the points of discontinuity. Such unique minimizing function is harmonic. In Riemann's proof, a function λ is constructed such that it vanishes on the boundary, may be discontinuous at isolated points, and for which the integral

$$L(\lambda) = \int \left[\left(\frac{\partial \alpha}{\partial x} \right)^2 + \left(\frac{\partial \lambda}{\partial y} \right)^2 \right] dT \tag{C.1.2}$$

is finite. The integral $L(\lambda)$ was called the *Dirichlet integral* later by Hilbert. Let $\alpha + \lambda = w$. Then Riemann considered the integral

$$\Omega = \int \left[\left(\frac{\partial \alpha}{\partial w} - \frac{\partial \beta}{\partial y} \right)^2 + \left(\frac{\partial w}{\partial y} + \frac{\partial \beta}{\partial x} \right)^2 \right] dT, \tag{C.1.3}$$

and wrote: "In the totality of these functions, λ represents a connected domain closed in itself, in which each function can be transformed continuously into every other, and a function cannot approach indefinitely close to one which is discontinuous along a curve with $L(\lambda)$ becoming infinite." Thus, for each λ, the integral Ω only becomes infinite with L, which depends continuously on λ and can never be less than zero. Hence, Ω has at least one minimum. The uniqueness follows directly from functions of the form $u + h\lambda$ near to a minimum u.

After this brief historical background we will provide a complete proof of the Riemann mapping theorem, which requires a prior knowledge of the maximum principle (Theorem 2.10) and Schwarz's lemma (Lemma 2.1, §2.4). There are other proofs of the Riemann mapping theorem, available in Gray [1994], and Rudin [1976].

PROOF. This proof takes into account Riemann's own approach and is based on the existence of a solution to the Dirichlet problem in any domain. We will use real variables, so that $z = x + i\,y$ is written as (x, y), and $f(z) = u + i\,v$ is written as $f(u, v)$. Consider the boundary value problem in a domain $D \subset \mathbb{C}$:

$$\frac{\partial^2 u}{\partial x^2} + \frac{\partial^2 u}{\partial y^2} = 0, \quad u(x, y) = \log |x + i\,y - a|, \tag{C.1.4}$$

where $x + i\,y \in \partial D$, and $a \in D$. Thus, $\log |x + i\,y - a|$ is bounded on ∂D. In view of the existence theorem, there exists a unique function u satisfying the boundary value problem

(C.1.4). Since the boundary value of u is bounded, u remains bounded on the interior of D. The function u satisfies the Laplace equation in (C.1.4). By the regularity theorem for the Laplace equation, u is differentiable and harmonic, thus analytic (synonymously, holomorphic or regular) on D, and the curl of this vector field is given by $(-u_y, u_x)$. Since D is simply connected, by Poincaré's lemma,[1] there exists a function v such that this vector field is $\text{grad}\, v = v_x + i\, v_y$. Since v is only determined up to an additive constant, we impose the condition $v(a) = 0$. This leads to the Cauchy-Riemann equations (2.1.6), and $u + i\, v$ is analytic on D.

Now, define three functions p, q, and f as

$$\begin{aligned} p(z) &= u(z) - \log|z-a|, \\ q(z) &= v(z) - \arg|z-a|, \\ f(z) &= e^{-p(z)-i\,q(z)} = (z-a)\, e^{-u(z)-i\,v(z)}. \end{aligned} \tag{C.1.5}$$

Note that the function p is single-valued while q is multiple-valued with the branch point at a, yet both satisfy the Laplace equation in $D\backslash\{a\}$. Also, the value of q increases by $2\pi i$ after one winding around the point a. However, the function f is single-valued since the exponentiation operation cancels the multiple-valued property of q. Moreover, f is analytic on D.

Notice that $p(z) = 0$ for $z \in \partial D$. We will show that $p(z) \geq 0$ whenever $z \in D\backslash\{a\}$. Since $\log|z-a| \to -\infty$ as $z \to a$, and $p(a)$ is finite, there exists an $\varepsilon > 0$ such that $p(z) > 0$ when $0 < |z-a| \leq \varepsilon$. Consider the region $G = \{z \in D : |z| > \varepsilon\}$. For a point $z \in \partial G$, either $|z-a| = \varepsilon$ or z must lie on ∂D. In either case we get $p(z) \geq 0$, which, in view of the maximum principle (Theorem 2.10), implies that $p(z) \geq 0$ for all $z \in G$. We already have $p(z) \geq 0$ whenever $0 < |z-a| < \varepsilon$, so $p(z) \geq 0$ whenever $z \in D\backslash\{a\}$. Since $|f(z) = e^{-p(z)} \leq 1$, we have $|f(z)| \leq 1$ when $z \in D$. Also, f(a)=0 and $f'(a)$ is real and positive because $f'(a) = e^{-u(a)-iv(a)} = e^{-u(a)} \in \mathbb{R}$.

To show that f is bijective, consider the level sets of p. For the sake of simplicity, we exclude those points where $f' = 0$. Then for every real number $r > 0$, define $A(r) = \{z \in D : |f(z) \geq e^{-r}$ and $f'(z) = 0\}$. We must show that $A(r)$ is finite. Note that the set $\{z \in D : |f(z)| \geq e^{-r}\}$ is compact. Hence, if $A(r)$ were infinite, it would have an accumulation point. But since $f(z) = 0$ whenever $z \in A(r)$ and f is analytic, this would imply that $f(z) \equiv 0$, which is a contradiction. Hence, $A(r)$ is finite.

Next, choose r such that $f'(z) \neq 0$ whenever $|z| = r$. Let $C(r)$ denote the level set $C(r) = \{z : p(z) = r\}$. We will show that $C(r)$ is smooth and homeomorphic[2] to a circle. Since $C(r)$ is a level set of a continuous function on a compact set, $C(r)$ is compact. Let w be a point on $C(r)$. By assumption, $f'(w) \neq 0$. Thus, by the inverse function theorem, there exists a neighborhood E of w on which f is invertible, i.e., f^{-1} exists, and is an analytic function. Since $z \in C(r)$ if and only if $|f(z)| = e^{-r}$, it follows that $C(r) \cap E$ is the image of an arc of the circle $\{z : |z| = e^{-r}\}$ under the mapping f^{-1}. Also, f^{-1} is an analytic function which is differentiable also. Hence, $C(r) \cap E$ is bijectively mapped onto a line segment. This is true for every point $z \in C(r)$, and thus $C(r)$ is a compact one-dimensional manifold. Therefore, it must be either a circle or a finite union of circles. Suppose $C(r)$ is a union of more than one circle. Then by Jordan theorem (§2.2), each of these circles divides

[1] This lemma states that a closed form on a starshape set is exact; it is a generalization of the fundamental theorem of calculus.

[2] Two curves are homeomorphic if they can be deformed into each other by a continuous, bijective and invertible mapping.

the complex plane into two components, one the interior and the other the exterior, with $C(r)$ as a common boundary. If there were two circles which together comprise $C(r)$, one of these circles would have to lie inside the other and there would be an open set Q which would have the two circles as its boundary. However, since $p(z)$ is assumed to be constant on both circles and takes the same value on both, the maximum principle would imply that p is constant in the region Q, which in turn would imply that $f = \text{const}$ on D. But this is impossible. Hence $C(r)$ consists of only one circle.

Next, we must show that a lies on $\text{Int}(C(r))$. Since the winding number of $C(r)$ about the point a is 1, the phase of $q(z)$ will increase by 2π after traversing $C(r)$ once. Since f^{-1} is analytic, $C(r)$ is not only homeomorphic to a circle; however, it is a smooth curve, and therefore, it has a tangent and a normal. The normal and tangential derivatives are obtained from the Cauchy-Riemann equations which in this case are $\dfrac{\partial p}{\partial t} = \dfrac{\partial q}{\partial n}$ and $\dfrac{\partial p}{\partial n} = -\dfrac{\partial q}{\partial t}$. Since $p(x) = r$ for $z \in C(r)$ and $p(x) \geq r$ for $x \in \text{Int}(C(r))$, so by the Cauchy-Riemann equations, both $\dfrac{\partial p}{\partial t} > 0$ and $\dfrac{\partial p}{\partial n} > 0$. This implies that $\dfrac{\partial q}{\partial t} \leq 0$, which is impossible because otherwise all the derivatives of p and q would vanish, i.e., f' would be zero, contrary to the hypothesis. Hence, q is a monotonically decreasing function on $C(r)$ and $\arg\{q\}$ decreases by 2π upon traversing $C(r)$, which together imply that the function e^{-iq} is a bijection from $C(r)$ onto the unit circle E. Hence, f is a bijection from $C(r)$ onto the circle of radius e^{-r}.

Finally, we must show that f' cannot have any zeros inside $C(r)$. Consider the points z for which $f'(z) = 0$, and choose r such that $f'(z) \neq 0$ whenever $|z| = r$. Note that $C(r)$ is a smooth closed curve, and since $f'(z) = -f(z)\big(u'(z) + i\,v'(z)\big)$. We use the fact that the argument of a product is the sum of the arguments of the factors. Consider f' and $u' + i\,v'$ separately. By argument principle (§2.4.4), f has only one simple zero, located at a, inside $C(r)$, and so $\arg\{f\}$ will increase by 2π after traversing $C(r)$ once. On the other hand, $\arg\{u' + i\,v'\}$ is determined by using the fact that the argument of the derivative of an analytic function is the same along any direction chosen to determine it. So we choose the normal direction, and find that $\arg\{u' + i\,v'\}$ changes by -2π after traversing $C(r)$ once. Hence, $\arg\{f'\}$ remains the same after once traversing $C(r)$. Since f' is analytic inside $C(r)$, by the argument principle, f' cannot have any zero inside $C(r)$. This proof is also available in Kythe [2016:29-32]. ■

C.2 Homogeneous Hilbert Problem

Consider the n-tuply connected region of Figure C.1, with the boundary $\Gamma = \cup_{k=1}^{n}\Gamma_k$. The *homogeneous Hilbert problem* states: Find a sectionally analytic function $\Phi(z)$ of finite degree at infinity such that

$$\Phi^+(t) = G(t)\,\Phi^-(t) \quad \text{on } \Gamma, \tag{C.2.1}$$

where $\Phi^+(t)$ and $\Phi^-(t)$ are limiting values from the right and the left at a point $t \in \Gamma$ (if t is an end point, then $\Phi^+(t) = \Phi^-(t) = \Phi(t)$), and $G(t)$, defined on Γ, satisfies the Hölder condition and $G(t) \neq 0$ at the point $t \in \Gamma$. If we take the logarithm on both sides of (C.2.1), we get

$$[\log \Phi(t)]^+ - [\log \Phi(t)]^- = \log G(t).$$

As t moves along the contours Γ_k in the positive sense, i.e., counterclockwise for $k = 0$ and clockwise for $k = 1, \dots, n$ (Figure C.1), $\log G(t)$ increases by integral multiples of $2i\pi$. Thus,

$$\frac{1}{2i\pi}\,[\log \Phi(t)]_{\Gamma_k} = \frac{1}{2\pi}\,[\arg\{G(t)\}]_{\Gamma_k} = \lambda_k, \quad k = 0, 1, \dots, n,$$

where λ_k are integers (positive, negative, or zero), and $[\arg\{G(t)\}]_{\Gamma_k}$ denotes the increment of $\arg\{G(t)\}$ as it goes around the contour Γ_k. The sum

$$\kappa = \sum_{k=0}^{n} \lambda_k = \frac{1}{2i\pi} [\log \Phi(t)]_\Gamma = \frac{1}{2\pi} [\arg\{G(t)\}]_\Gamma = \lambda_k$$

is known as the *index* of the Hilbert problem and also as the *index* of the function $G(t)$ given on Γ. The index κ is an integer.

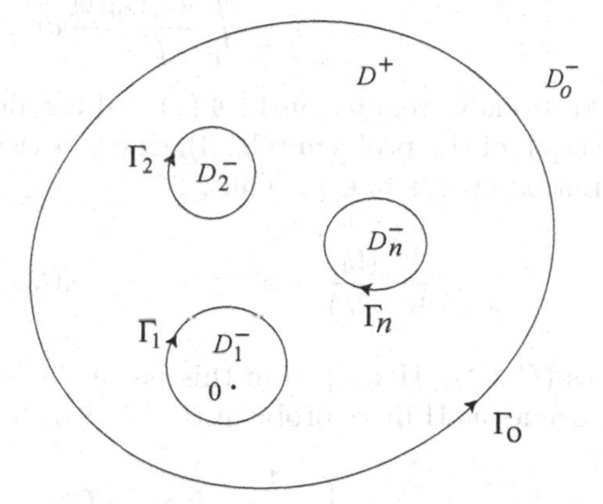

Figure C.1 n-tuply connected domain.

Let $a_1, \dots, a_n$ be arbitrary fixed points in the regions $D_1^-, \dots, D_n^-$, and let the origin of the coordinates be in the region D^+. Let

$$p(z) = \begin{cases} (z-a_1)^{-\alpha_1} \cdots (z-a_n)^{-\alpha_n} & \text{if } \Gamma = \cup_{k=1}^{n} \Gamma_k, \\ 1 \quad \text{if } \Gamma = \Gamma_0. \end{cases}$$

Let

$$G_0(t) = t^{-\kappa}\, p(t)\, G(t). \tag{C.2.2}$$

Then, after traversing the contours $\Gamma_0, \Gamma_1, \dots, \Gamma_n$, the function $\arg\{G_0(t)\}$ returns to its initial value, and therefore $\log G_0(t)$ is a well-defined single-valued and continuous function on Γ and satisfies the Hölder condition there with an arbitrarily fixed branch on each contour Γ_k.

To determine the solution $\Phi(z)$ of the homogeneous Hilbert problem (C.2.1), we introduce a new unknown function

$$\Psi(z) = \begin{cases} p(z)\, \Phi(z) & \text{in } D^+ \\ z^k\, \Phi(z) & \text{in } D^-, \end{cases}$$

which is regular except possibly at $z = \infty$. Then the condition (C.2.1) can be written as

$$\Psi^+(t) = G_0(t)\, \Psi^-(t). \tag{C.2.3}$$

First, we will find the *fundamental solution* for the homogeneous Hilbert problem. By taking the logarithm on both sides of (C.2.3), we formally get

$$\log \Psi^+(t) - \log \Psi^-(t) = \log G_0(t). \tag{C.2.4}$$

If we assume that $\log \Psi(z)$ is single-valued, sectionally regular on Γ, and zero at $z = \infty$, then

$$\log \Psi(z) = \frac{1}{2i\pi} \int_\Gamma \frac{\log G_0(t)}{t-z}\, dt,$$

thus, by using the Plemelj formulas (2.4.19) we find that

$$\Psi(z) = e^{g(z)}, \tag{C.2.5}$$

where

$$g(z) = \frac{1}{2i\pi} \int_\Gamma \frac{\log G_0(t)}{t-z}\, dt. \tag{C.2.6}$$

Obviously, $\Psi(z)$ is sectionally regular on Γ, $\Psi(z) \neq 0$ for all finite z, and $\Psi(\infty) = 1$. It is also a particular solution of the problem (C.2.3), since, in view of (C.2.6), $g^+(t_0) - g^-(t_0) = \log G_0(t_0)$ for an arbitrary point $t_0 \in \Gamma$. Thus,

$$\frac{\Psi^+(t_0)}{\Psi^-(t_0)} = e^{\log G_0(t_0)} = G_0(t_0),$$

which is the same as (C.2.4). Hence, from this particular solution we obtain a particular solution of the homogeneous Hilbert problem (C.2.1) which we represent as

$$Z(z) = \begin{cases} \dfrac{1}{p(z)} & \text{for } z \in D^+ \\ z^{-\kappa}\, e^{g(z)} & \text{for } z \in D^-, \end{cases} \tag{C.2.7}$$

where $g(z)$ is defined by (C.2.5). This particular solution $Z(z)$ is called a *fundamental solution* of the homogeneous Hilbert problem (C.2.1) because it vanishes nowhere in any finite part of the z-plane if $\kappa > 0$ and $Z(\infty) = 0$. This also holds for the boundary values $Z^+(t)$ and $Z^-(t)$. Note that $Z(z)$ is of degree $(-\kappa)$ at $z = \infty$. By using the Plemelj formulas (2.4.19), we get

$$g^+(t_0) = \frac{1}{2} \log G_0(t_0) + g(t_0), \quad g^-(t_0) = -\frac{1}{2} \log G_0(t_0) + g(t_0),$$

and then, in view of (C.2.7), the boundary values $Z^+(t)$ and $Z^-(t)$ are given by

$$Z^+(t_0) = \frac{e^{g(t_0)}\sqrt{G_0(t_0)}}{p(t_0)} = \frac{e^{g(t_0)}\sqrt{G(t_0)}}{t_0^{\kappa/2}\sqrt{p(t_0)}},$$
$$Z^-(t_0) = \frac{e^{g(t_0)}}{t_0^{\kappa}\sqrt{G_0(t_0)}} = \frac{e^{g(t_0)}}{t_0^{\kappa/2}\sqrt{p(t_0)G_0(t_0)}},$$

where we have used (C.2.2) which also gives

$$\frac{\sqrt{G(t)}}{t^{\kappa/2}\sqrt{p(t)}} = \frac{\sqrt{G_0(t)}}{p(t)}.$$

Theorem C.1. *All solutions of the homogeneous Hilbert problem (C.2.1) which have finite degree at infinity are given by*

$$\Phi(z) = Z(z)\, P(z), \tag{C.2.8}$$

where $P(z)$ is an arbitrary polynomial. PROOF. Let $\Phi(z)$ be any solution. Then by (C.2.1)

$$\Phi^+(t) = G(t)\,\Phi^-(t), \quad \text{and} \quad Z^+(t) = G(t)\,Z^-(t).$$

Since $Z^+(t) \neq 0$ and $Z^-(t) \neq 0$, we find that

$$\frac{F^+(t)}{Z^+(t)} = \frac{F^-(t)}{Z^-(t)} = G(t),$$

which implies that the function $\dfrac{\Phi(z)}{Z(z)}$ is regular in the entire z-plane, and has finite degree at $z = \infty$. Thus, it is a polynomial, which proves the theorem. ■

Some of the consequences of this theorem are as follows:

(i) The limiting values $\Phi^+(t)$ and $\Phi^-(t)$ of any solution $\Phi(z)$ of the homogeneous Hilbert problem satisfy the Hölder condition because the function $G(t)$ does so.

(ii) If the polynomial $P(z)$ is of degree n, then the degree of the solution $\Phi(z)$, given by (C.2.8), at $z = \infty$ is $(n - \kappa)$, i.e., the degree of this solution is not less than the degree $(-\kappa)$ of $Z(z)$. The degree of $\Phi(z)$ and $Z(z)$ at $z = \infty$ are the same only if $n = 0$, i.e., if $P(z) \equiv \text{const} \neq 0$. Hence, the index $(-\kappa)$ is the lowest possible degree of a solution of the homogeneous Hilbert problem.

(iii) The fundamental solution $Z(z)$ has the following properties:

(a) $Z(z)$ does not vanish in any finite part of the z-plane;

(b) $Z(z)$ has the lowest possible degree $(-\kappa)$ at $z = \infty$; and

(c) $Z(z)$ is a factor of every solution of the problem (C.2.1).

(iv) An application of the solution (C.2.1) produces the following result: If $\kappa \leq 0$, the homogeneous Hilbert problem has no solution vanishing at $z = \infty$, except the trivial solution $\Phi(z) \equiv 0$. If $\kappa > 0$, it has exactly κ linearly independent solutions $\Phi(z), z\Phi(z), \dots, z^{\kappa-1}\Phi(z)$, each of which vanishes at $z = \infty$. In fact, since the polynomial $P(z)$ in (C.2.8) is of degree at most $(\kappa - 1)$, all solutions of (C.2.1) that vanish at $z = \infty$ must have the form $\Phi(z) = Z(z)\,p_{\kappa-1}(z)$, where

$$p_{\kappa-1}(z) = c_0 + c_1 z + \cdots + c_{\kappa-1} z^{\kappa-1},$$

and $c_0, c_1, \dots, c_{\kappa-1}$ are arbitrary constants.

(v) If the contour Γ_0 is at infinity, then the solution (C.2.8) holds only for $\lambda_k = 0$, i.e., for $\kappa = 0$.

(vi) The homogeneous Hilbert problems that correspond to the conditions

$$\Phi^+(t) = G(t)\,\Phi^-(t) \quad \text{and} \quad \Psi^+(t) = [G(t)]^{-1}\,\Psi^-(t),$$

are known as *adjoint* problems of one another. Hence, if the former problem has index κ and fundamental solution $Z(z)$, then the latter has index $(-\kappa)$ and fundamental solution $[Z(z)]^{-1}$.

C.3 Nonhomogeneous Hilbert Problem

A generalization of the homogeneous Hilbert problem is the nonhomogeneous problem which states: Find a sectionally analytic function $\Phi(z)$ of finite degree at infinity such that

$$\Phi^+(t) = G(t)\,\Phi^-(t) + g(t) \quad \text{on } \Gamma, \tag{C.3.1}$$

where $G(t)$ and $g(t)$, defined on Γ ($t \in \Gamma$), satisfy the Hölder condition on Γ and $G(t) \neq 0$ on Γ.

The nonhomogeneous Hilbert problem can be easily solved by using the results of the previous section. Let $Z(z)$ be the fundamental solution for the homogeneous problem, defined by (C.2.7). Then this is also the fundamental solution for (C.3.1) with $g(t) \equiv 0$, in which case Eq (C.3.1) yields

$$G(t) = \frac{Z^+(t)}{Z^-(t)},$$

which when substituted in (C.3.1) gives

$$\frac{\Phi^+(t)}{Z^+(t)} - \frac{\Phi^-(t)}{Z^-(t)} = \frac{g(t)}{Z^+(t)}.$$

Since the function $\dfrac{\Phi(z)}{Z(z)}$ has finite degree at infinity, we get

$$\frac{\Phi(z)}{Z(z)} = \frac{1}{2i\pi} \int_\Gamma \frac{g(t)}{Z^+(t)} \frac{dt}{t-z} + P(z),$$

where $P(z)$ is an arbitrary polynomial. Hence, the *general solution* of the nonhomogeneous Hilbert problem (C.3.1) is given by

$$\Phi(z) = \frac{Z(z)}{2i\pi} \int_\Gamma \frac{g(t)}{Z^+(t)} \frac{dt}{t-z} + Z(z)\, P(z). \tag{C.3.2}$$

The function $Z(z)$ which is the fundamental solution for the corresponding homogeneous problem is called the *fundamental function* for the nonhomogeneous Hilbert problem. The index of this problem is also κ.

We will examine the solution (C.3.2) when $\Phi(\infty) = 0$. Then the degree of $Z(z)$ at infinity is $(-\kappa)$, and the solution (C.3.2) will vanish at infinity for $\kappa \geq 0$ iff the degree of $P(z)$ is $\leq (\kappa - 1)$. Hence, for $\kappa = 0$ it suffices to take $P(z) \equiv 0$. For $\kappa < 0$, obviously $P(z) \equiv 0$, and the coefficients of $z^{-1}, z^{-2}, \dots, z^{-\kappa}$ must be zero in the expansion

$$\frac{1}{2i\pi} \int_\Gamma \frac{g(t)}{Z^+(t)(t-z)}\, dt = -\sum_{j=0}^{\infty} \frac{z^{-(j+1)}}{2i\pi} \int_\Gamma \frac{t^j\, g(t)}{Z^+(t)}\, dt,$$

i.e.,

$$\frac{1}{2i\pi} \int_\Gamma \frac{t^j\, g(t)}{Z^+(t)}\, dt = 0, \quad \text{for } j = 0, 1, \dots, -\kappa - 1,$$

which is a necessary and sufficient condition for the solution to vanish at infinity for $\kappa < 0$. Hence, we have proved the following:

Theorem C.2. *If $\kappa \geq 0$, the general solution of the nonhomogeneous Hilbert problem (C.3.1) that vanishes at infinity is given by*

$$\Phi(z) = \frac{Z(z)}{2i\pi} \int_\Gamma \frac{g(t)}{Z^+(t)} \frac{dt}{t-z} + Z(z)\, P_{\kappa-1}(z), \tag{C.3.3}$$

where $P_{\kappa-1}(z)$ are arbitrary polynomials of degree $\leq (\kappa - 1)$ and $P_{\kappa-1}(z) = 0$ for $\kappa = 0$. If $\kappa < 0$, then the solution is given by

$$\Phi(z) = \frac{Z(z)}{2i\pi} \int_\Gamma \frac{g(t)}{Z^+(t)} \frac{dt}{t - z}, \tag{C.3.4}$$

provided conditions (C.3.3) are satisfied. Note that for $\kappa = 0$ there is a unique solution that vanishes at infinity. For $\kappa < 0$ there is a unique solution vanishing at infinity, if such a solution exists, but for $\kappa > 0$ there is an unlimited number of solutions, and the general solution (C.3.2) contains κ arbitrary constants.

In view of the Schwarz reflection and symmetry principles, the Hilbert problems can be applied to the upper half-plane D^+ or the unit disk. The general theory of the Riemann-Hilbert problem is presented in the next section. The Riemann-Hilbert problem is a linearized form of a singular integral equation of the second kind.

C.4 Riemann-Hilbert Problem

A generalization of the Hilbert boundary problem is the Riemann-Hilbert problem. This problem deals with determining a function $\Phi(z) = u + iv$ which is regular in D^+, continuous on $D^+ \cup \Gamma$, and satisfies the boundary condition

$$\Re\{a + ib\}\, \Phi^+ \equiv a\,u - b\,v = c \quad \text{on } \Gamma, \tag{C.4.1}$$

where $a(t), b(t), c(t)$ are real continuous functions defined for $t \in \Gamma$, satisfy the Hölder condition, and are such that $a^2 + b^2 \neq 0$ everywhere on Γ. Before we solve this problem, note that if

$$\Phi_k(z) = u_k + iv_k, \quad k = 1, \dots, n,$$

is any particular solution of the homogeneous problem

$$a\,u - b\,v = 0 \quad \text{on } \Gamma,$$

then any linear combination

$$\Phi(z) = \sum_{k=1}^{n} C_k\, \Phi_k(z)$$

is also a solution of the homogeneous problem, where C_k, $k = 1, \dots, n$, are real constants, and the functions $\Phi_k(z)$ are linearly independent. Now we will determine the solution of problem (C.4.1) for the circle.

C.4.1 Riemann-Hilbert Problem for the Unit Disk. Let U^+ denote the unit disk with boundary Γ ($|z| = 1$). Then boundary condition (C.4.1) becomes

$$2\,\Re\{a + ib\}\, \Phi^+(t) = (a + ib)\, \Phi^+(t) + (a - ib)\, \overline{\Phi^+(t)} = 2c \quad \text{on } \Gamma. \tag{C.4.2}$$

Let the solution $\Phi(z)$ be a sectionally analytic function on Γ such that it can be extended in U^+ by the function $\Phi^*(z)$, defined by (2.2.12), i.e.,

$$\Phi^*(z) = \bar{\Phi}\left(\frac{1}{z}\right) = \Phi(z) \quad \text{for } |z| \neq 1. \tag{C.4.3}$$

The function $\Phi^*(z)$ is bounded at $z = \infty$. Boundary condition (C.4.2) becomes

$$(a + ib)\,\Phi^+(t) + (a - ib)\,\Phi^-(t) = 2c, \tag{C.4.4}$$

or,

$$\Phi^+(t) = G(t)\,\Phi^-(t) + g(t), \tag{C.4.5}$$

where

$$G(t) = -\frac{a - ib}{a + ib}, \quad g(t) = \frac{2c}{a + ib}. \tag{C.4.6}$$

Thus, the Riemann-Hilbert problem (C.4.2) reduces to a solution of the Hilbert problem (C.4.4). However, if $\Phi(z)$ is any solution of the Hilbert problem (C.4.4), it may not be the solution of the original Riemann-Hilbert problem (C.4.1) because it may fail to satisfy condition (C.4.3). But we can always construct a solution of the problem (C.4.2) by using the function $\Phi(z)$. Note that if $\Phi(z)$ satisfies (C.4.2), then taking conjugates in (C.4.2) and using the fact that $\Phi^-(t) = \bar{\Phi}^-(1/t)$, we get

$$(a - ib)\,\Phi^{*-}(t) + (a + ib)\,\Phi^{*+}(t) = 2c,$$

which shows that $\Phi(z)$ is also the solution of the Hilbert problem (C.4.4). Let

$$\Omega(z) = \frac{1}{2}\left[\Phi(z) + \Phi^*(z)\right]. \tag{C.4.7}$$

Then $\Omega(z)$ is the solution of the Riemann-Hilbert problem (C.4.2), because $\Phi(z) = \Phi^*(z) = \frac{1}{2}\,[\Phi(z) + \Phi^*(z)]$.

We will determine the complete set of solutions of the Riemann-Hilbert problem (C.4.2). First, we consider homogeneous problem (C.4.2) for $c([\Phi(z) + \Phi^*(z)]) \equiv 0$. Let κ be the index of the function $G([\Phi(z) + \Phi^*(z)])$, i.e.,

$$\begin{aligned}\kappa &= \frac{1}{2i\pi}\,[\log G(t)]_\Gamma = \frac{1}{2i\pi}\left[\log\frac{a - ib}{a + ib}\right]_\Gamma = \frac{1}{2\pi}\,[\arg\{a - ib\} - \arg\{a + ib\}]_\Gamma \\ &= \frac{1}{\pi}\,[\arg\{a - ib\}]_\Gamma\,.\end{aligned}$$

Thus, κ is an even integer since $a(t)$ and $b(t)$ are continuous functions. The number κ is the index of the Riemann-Hilbert problem (C.4.2).

Let $Z(z)$ be the fundamental solution of the homogeneous Hilbert problem (C.4.5), i.e.,

$$Z(z) = \begin{cases} C\,e^{g(z)} & \text{for } |z| < 1, \\ C\,z^{-\kappa}\,e^{g(z)} & \text{for } |z| > 1, \end{cases}$$

where $C \neq 0$ is an arbitrary constant and

$$g(z) = \frac{1}{2i\pi}\int_\Gamma \frac{\log\left[t^{-\kappa}\,G(t)\right]}{t - z}\,dt = \frac{1}{2\pi}\int_\Gamma \frac{h(t)}{t - z}\,dt, \tag{C.4.8}$$

where

$$h(t) = \arg\left\{-\,t^{-\kappa}\,\frac{a - ib}{a + ib}\right\}$$

is a real-valued continuous function defined on Γ. Using (2.2.12) we get

$$g^*(z) = \frac{1}{2\pi} \int_\Gamma \frac{h(t)}{t-z}\, dt - i\alpha = g(z) - i\alpha,$$

where α is a real constant defined by

$$\alpha = \frac{1}{2i\pi} \int_\Gamma \frac{h(t)}{t}\, dt = \frac{1}{2\pi} \int_0^{2\pi} h(t)\, d\theta, \quad t = e^{i\theta}.$$

Hence,

$$Z^*(z) = \begin{cases} \bar{C}\, e^{g^*(z)} = \bar{C}\, e^{g(z)-i\alpha} & \text{for } |z| > 1, \\ \bar{C}\, z^\kappa\, e^{g(z)-i\alpha} & \text{for } |z| < 1, \end{cases}$$

or, for all $z \notin \Gamma$,

$$Z^*(z) = \frac{\bar{C}}{C}\, z^\kappa\, e^{-i\alpha}\, Z(z),$$

or, taking $\bar{C}/C = e^{i\alpha}$, we get

$$Z^*(z) = z^\kappa\, Z(z). \tag{C.4.9}$$

Now we will discuss two cases:

CASE 1. If $\kappa \geq 0$, the homogeneous Hilbert problem (C.4.4) for $c(t) \equiv 0$ has a nonzero solution bounded at infinity and is given by

$$\Phi(z) = P(z)\, Z(z), \tag{C.4.10}$$

where

$$P(z) = C_0\, z^\kappa + C_1\, z^{\kappa-1} + \cdots + C_\kappa$$

is an arbitrary polynomial of degree $\leq \kappa - 1$. Thus (C.4.10) is a solution of the Riemann-Hilbert problem (C.4.2) iff $\Phi^*(z) = \Phi(z)$, i.e., $P^*(z)Z^*(z) = P(z)Z(z)$, where $P^*(z) = \bar{P}(1/z)$ and $Z^*(z) = z^\kappa\, Z(z)$. Since

$$z^\kappa\, \bar{P}\left(\frac{1}{z}\right) = \bar{C}_0 + \bar{C}_1\, z + \cdots + \bar{C}_\kappa\, z^\kappa = C_0 z^\kappa + C_1\, z^{\kappa-1} + \cdots + C_\kappa = P(z), \tag{C.4.11}$$

i.e., $C_\kappa = \bar{C}_{\kappa-n}$, $n - 0, 1, \ldots, \kappa$, we can set $C_n = A_n + iB_n$, $n = 0, 1, \ldots, \kappa/2$, where A_n, B_n are real numbers with $B_{\kappa/2} = 0$. Then $C_m = A_{\kappa-m} - iB_{\kappa-m}$ for $m = \kappa/2 + 1, \ldots, \kappa$. Thus, there are in all $(\kappa + 1)$ arbitrary real constants. We will denote them by $D_0, D_1, \ldots, D_\kappa$. Then the formal general solution of the homogeneous Riemann-Hilbert problem (C.4.2) is given by

$$\Phi(z) = \sum_{k=0}^{\kappa} D_k\, \Phi_k(z),$$

where Φ_k $(k = 0, 1, \ldots, \kappa)$ are linearly independent solutions of the same problem.

CASE 2. If $\kappa \leq -2$, the homogeneous Hilbert problem (C.4.4) has no nonzero solution bounded at infinity. Hence, there is only the trivial (zero) solution of the homogeneous Riemann-Hilbert problem (C.4.4).

Theorem C.3. *For $\kappa \ge 0$ the homogeneous Riemann-Hilbert problem (C.4.4) has exactly $(\kappa+1)$ linearly independent solutions. Its general solution is given by (C.4.10), where $Z(z)$ is the fundamental solution of the Hilbert problem (C.4.4) subject to condition (C.4.9). For $\kappa \le -2$, the homogeneous Riemann-Hilbert problem (C.4.2) has only the trivial solution $\Phi(z) = 0$.*

Now we will consider the nonhomogeneous Riemann-Hilbert problem (C.4.2). We can construct its general solution, provided that we find only one particular solution. Then the general solution is the sum of this particular solution and the general solution of the homogeneous problem. However, the problem of finding a particular solution of the Riemann-Hilbert problem (C.4.4) is equivalent to finding any particular solution of the Hilbert problem (C.4.4) that is bounded at $z = \infty$, because, in view of (C.4.7), it will provide us with a particular solution of the Riemann-Hilbert problem (C.4.2). Hence,

Theorem C.4. *For $\kappa \ge 0$ there always exists a solution for the nonhomogeneous Riemann-Hilbert problem (C.4.2). For $\kappa \le -2$, this problem has a solution iff the following conditions are satisfied:*

$$\int_0^{2\pi} e^{i(n+\kappa/2}\,\Omega(\theta)\,c(\theta)\,d\theta = 0, \quad n = 1, 2, \dots, -\kappa - 1, \tag{C.4.12}$$

where

$$\Omega(\theta) = \frac{1}{\sqrt{a^2(\theta) + b^2(\theta)}} \exp\left\{ -\frac{1}{4\pi} \int_0^{2\pi} h(\psi) \cot\frac{\psi - \theta}{2}\, d\psi \right\}.$$

Note that conditions (C.4.12) are equivalent to the $(-\kappa - 1)$ conditions

$$\int_0^{2\pi} \Omega(\theta)\,c(\theta)\,\cos n\theta\,d\theta = 0, \quad n = 0, 1, \dots, -\frac{\kappa}{2} - 1,$$

$$\int_0^{2\pi} \Omega(\theta)\,c(\theta)\,\sin n\theta\,d\theta = 0, \quad n = 1, \dots, -\frac{\kappa}{2} - 1.$$

The solutions of the Hilbert problem (C.4.4) do not vanish at infinity, although they are bounded there. Thus, conditions (C.4.12) become

$$\int_\Gamma \frac{t^n\,g(t)}{Z^+(t)}\,dt = 0, \quad n = 0, 1, \dots, -\kappa - 1,$$

or

$$\int_\Gamma \frac{t^n\,c(t)}{[a(t) + ib(t)]\,Z^+(t)}\,dt = 0, \quad n = 0, 1, \dots, -\kappa - 2. \tag{C.4.13}$$

Since, in view of (C.4.8),

$$g^+(t_0) = \frac{i}{2}\,h(t_0) + \frac{1}{2\pi} \int_\Gamma \frac{h(t)}{t - t_0}\,dt,$$

we set $t = e^{i\theta}$, $t_0 = e^{i\theta_0}$. Then

$$g^+(t_0) = \frac{i}{2}\,h(t_0) + \frac{1}{4\pi} \int_0^{2\pi} h(t) \cot\frac{\theta - \theta_0}{2}\,d\theta + \frac{i}{4\pi} \int_0^{2\pi} h(t)\,d\theta.$$

In view of (C.4.11) the last term in the above expression is equal to $i\alpha/2$. Hence, $Z(z) = e^{g(z)-i\alpha/2}$ for $|z| < 1$, and

$$e^{ih(t_0)} = -t_0^{\kappa}\,\frac{a(t_0) - ib(t_0)}{a(t_0) + ib(t_0)},$$

which yields

$$Z^+(t_0) = \pm t^{\kappa/2} \sqrt{-\frac{a(t_0) - ib(t_0)}{a(t_0) + ib(t_0)}}\, \exp\left\{\frac{1}{4\pi}\int_0^{2\pi} h(t)\,\cot\frac{\theta-\theta_0}{2}\,d\theta\right\}.$$

This proves the theorem. ■

Some particular cases of Theorem C.4 are as follows:

1. If $\kappa \le -2$ and (C.4.13) is satisfied, then the Hilbert problem (C.4.4) has a unique solution which, in view of (C.3.4) and (C.4.6), is given by

$$\Phi(z) = \frac{Z(z)}{i\pi}\int_\Gamma \frac{c}{(a+ib)\,Z^+(t)\,(t-z)}\,dt.$$

Since this solution is unique, it is also the solution of problem (C.4.2).

2. For $\kappa \ge 0$, the formula

$$Q(z) = \frac{Z(z)}{i\pi}\int_\Gamma \frac{c}{(a+ib)\,Z^+(t)\,(t-z)}\,dt,$$

gives a particular solution of the problem (C.4.4). Then a particular solution of problem (C.4.2), as in (C.4.7), is given by

$$\Phi(z) = \frac{1}{2}\left[Q^+(z) + Q^*(z)\right], \tag{C.4.14}$$

where the function $Q^*(z)$ is defined as follows: Since, in view of (38.18), $Z^*(z) = z^{\kappa}\,Z(z)$, $\overline{Z^+(t)} = Z^{*-}(t) = t^{\kappa}\,Z^-(t)$, and $(a-ib)\,Z^-(t) = -(a+ib)\,Z^-(t)$, we have

$$\begin{aligned} Q^*(z) &= Z^*(z)\left\{-\frac{1}{i\pi}\int_\Gamma \frac{c}{(a+ib)\,\overline{Z^+(t)}\,(t-z)}\,dt + \frac{1}{i\pi}\int_\Gamma \frac{c}{(a-ib)\,\overline{Z^+(t)}}\,dt\right\} \\ &= z^{\kappa}\,Z(z)\left\{\frac{1}{i\pi}\int_\Gamma \frac{c\,t^{\kappa}}{(a+ib)\,Z^+(t)\,(t-z)}\,dt - \frac{1}{i\pi}\int_\Gamma \frac{c\,t^{\kappa}}{(a+ib)\,t\,Z^+(t)}\,dt\right\}. \end{aligned}$$

Substituting this in (C.4.14), we get a particular solution for the Hilbert problem (C.4.4) for $\kappa \ge 0$ as

$$\begin{aligned} \Phi(z) = &\frac{Z(z)}{2i\pi}\left\{\int_\Gamma \frac{c\,dt}{(a+ib)\,Z^+(t)\,(t-z)} + z^{\kappa}\int_\Gamma \frac{c\,t^{-\kappa}\,dt}{(a+ib)\,Z^+(t)\,(t-z)}\right\} \\ &- \frac{z^{\kappa}\,Z(z)}{2i\pi}\int_\Gamma \frac{c\,t^{-\kappa}\,dt}{(a+ib)\,t\,Z^+(t)}. \end{aligned} \tag{C.4.15}$$

3. For $\kappa = 0$, solution (C.4.15) simplifies to

$$\Phi(z) = \frac{Z(z)}{i\pi}\int_\Gamma \frac{c\,dt}{(a+ib)\,Z^+(t)\,(t-z)} - \frac{Z(z)}{2i\pi}\int_\Gamma \frac{c\,dt}{(a+ib)\,t\,Z^+(t)}. \tag{C.4.16}$$

Example C.1. (Dirichlet problem for the unit disk) We will discuss this problem for the unit disk U^+, i.e., we will find a function u which is harmonic in U^+, continuous on $U^+ \cup \Gamma$, and satisfies the boundary condition

$$u = f(t), \quad t \in \Gamma,$$

where $f(t)$ is a real-valued continuous function given on Γ. This problem is a special case of the Riemann-Hilbert problem with $a = 1$, $b = 0$, $c = f(t)$. Then the corresponding Hilbert problem (C.4.4) for $c \equiv 0$ reduces to

$$\Phi^+(t) + \Phi^-(t) = 0. \tag{C.4.17}$$

This corresponds to case 1 of §C.2 with index $\kappa = 0$. Then the fundamental solution for the problem (C.4.17) is given by

$$Z(z) = \begin{cases} A & \text{for } z \in U^+, \\ -A & \text{for } z \in U^-, \end{cases}$$

where A is an arbitrary constant. To satisfy the condition (C.4.9) we require that $Z^*(z) = Z(z)$, and thus it suffices to take $A = i$, which yields

$$Z(z) = \begin{cases} i & \text{for } z \in U^+, \\ -i & \text{for } z \in U^-, \end{cases}$$

Hence, from Theorem C.3 the general solution of the corresponding Riemann-Hilbert problem is given by $\Phi(z) = C\,i$, where C is an arbitrary real constant. Then from (C.4.16) the general solution of the nonhomogeneous Cauchy problem is given by

$$\Phi(z) = \frac{1}{i\pi}\int_\Gamma \frac{f(t)}{t-z}\,dt - \frac{1}{2i\pi}\int_\Gamma \frac{f(t)}{t}\,dt - i\,C = \frac{1}{2i\pi}\int_\Gamma \frac{f(t)}{t}\,\frac{t+z}{t-z}\,dt + i\,C. \tag{C.4.18}$$

This solution is known as the *Schwarz formula*. Note that this solution has been obtained by applying the results of §C.2, which assume that the function $f(t)$ satisfies the Hölder condition. But the solution (C.4.18) is also valid if instead of this condition only continuity of the function $f(t)$ is assumed. ■

The Riemann-Hilbert problem for the half-plane can be reduced to that for the unit disk and the fact that conformal mapping of a half-plane onto the unit disk leads to an inversion.

REFERENCES USED: Ahlfors [1953], Kantorovich and Krylov [1958], Kythe [1996; 2016], Muskhelishvili [1992], Sveshnikov and Tikhonov [1974], Wegmann [1986].

D

Gudermannian

D.1 Gudermannian.

The Gudermannian function $\mathrm{gd}(x)$ is related to the circular functions and hyperbolic functions. It is defined as

$$\mathrm{gd}(x) = \int_0^x \frac{1}{\cosh t}\, dt, \quad -\infty < x < \infty.$$

Other alternative definitions are:

$$\begin{aligned}\mathrm{gd}(x) &= \arcsin(\tanh x) = \arctan(\sinh x) = \mathrm{arccsc}((\coth x) \\ &= \mathrm{sgn}(x) \cdot \arccos(\mathrm{sech}\, x) = \mathrm{sgn}(x) \cdot \mathrm{arcsec}(\cosh x) \\ &= 2 \arctan\left[\tanh(x/2)\right] = 2\arctan\left(e^x\right) - \tfrac{\pi}{2}.\end{aligned}$$

Some identities are:

$$\begin{aligned}&\sin(\mathrm{gd} x) = \tanh x; \quad \csc(\mathrm{gd} x) = \coth x; \\ &\cos(\mathrm{gd} x) = \mathrm{sech}\, x; \quad \sec(\mathrm{gd} x) = \cosh x; \\ &\tan(\mathrm{gd} x) = \sinh x; \quad \cot(\mathrm{gd} x) = \mathrm{csch}\, x; \\ &\tan\left(\frac{1}{2}\mathrm{gd} x\right) = \tanh\left(\frac{1}{2}x\right).\end{aligned}$$

The inverse $\mathrm{gd}^{-1}x$ is defined as

$$\begin{aligned}\mathrm{gd}^{-1}(x) &= \int_0^x \frac{1}{\cos t}\, dt, \quad -\tfrac{\pi}{2} < x < \tfrac{\pi}{2} \\ &= \ln\left|\frac{1+\sin x}{\cos x}\right| = \frac{1}{2}\ln\left|\frac{1+\sin x}{1-\sin x}\right| \\ &= \ln\left|\tan x + \sec x\right| = \ln\left|\tan\left(\tfrac{\pi}{4} + \frac{1}{2}x\right)\right| \\ &= \mathrm{arctanh}(\sin x) = \mathrm{arcsinh}(\tan x) = \mathrm{arccoth}(\csc x) = \mathrm{arcsinh}(\cot x) \\ &= \mathrm{sgn}(x)\,\mathrm{arcsinh}(\sec x) = \mathrm{sgn}(x)\,\mathrm{arcsinh}(\cos x) \\ &= -i\,\mathrm{gd}(ix).\end{aligned}$$

Some identities for the inverse are:

$$\begin{aligned}
\sinh\left(\mathrm{gd}^{-1}(x)\right) &= \tan x; &\quad \operatorname{csch}\left(\mathrm{gd}^{-1}(x)\right) &= \cot x;\\
\cosh\left(\mathrm{gd}^{-1}(x)\right) &= \sec x; &\quad \operatorname{sech}\left(\mathrm{gd}^{-1}(x)\right) &= \csc x\\
\tanh\left(\mathrm{gd}^{-1}(x)\right) &= \sin x; &\quad \coth\left(\mathrm{gd}^{-1}(x)\right) &= \csc x.
\end{aligned}$$

The derivatives are

$$\frac{d}{dx}\left(\mathrm{gd}\,(x)\right) = \operatorname{sech} x; \quad \frac{d}{dx}\left(\mathrm{gd}^{-1}(x)\right) = \sec x.$$

REFERENCES USED: Carrier, Krook and Pearson [1966], Kythe [1996], Olver et al. [2010], Weisstein [2003].

E

Tables

The successive approximations are presented in tabular form. Note that the desired level of successive approximation is obtained by first choosing an initial guess. In all tables given below the initial guess for all coefficients is taken as zero. Then substituting these values into the right side of the respective equations, the first approximation is determined. Then the values of the first approximation are substituted into the right side of the same equations, and the second approximation is obtained. This process is continued successively until the desired approximation is attained.

TABLE E.1.

Coefficient	α_3	α_5	α_7	α_9	α_{11}
Initial Guess	0	0	0	0	0
1st Approx.	$-\lambda^2$	0	0	0	0
2nd Approx.	$-\lambda^2$	λ^4	0	0	0
3rd Approx.	$-\lambda^2 - \lambda^6$	λ^4	$-\lambda^6$	0	0
4th Approx.	$-\lambda^2 - \lambda^6$	$\lambda^4 + 3\lambda^8$	$-\lambda^6$	λ^8	0
5th Approx.	$-\lambda^2 - \lambda^6 - 4\lambda^{10}$	$\lambda^4 + 3\lambda^8$	$-\lambda^6 - 5\lambda^{10}$	λ^8	λ^{10}

TABLE E.2.

Coefficient	A_1	A_2	A_3	A_4	A_5
Initial Guess	0	0	0	0	0
1st Approx.	$-\frac{\lambda}{2}$	0	0	0	0
2nd Approx.	$-\frac{\lambda}{2}$	$\frac{\lambda^2}{2}$	0	0	0
3rd Approx.	$-\frac{\lambda}{2}+\frac{\lambda^3}{4}$	$\frac{\lambda^2}{2}$	$-\frac{5\lambda^3}{8}$	0	0
4th Approx.	$-\frac{\lambda}{2}+\frac{\lambda^3}{4}$	$\frac{\lambda^2}{2}-\frac{9\lambda^4}{16}$	$-\frac{5\lambda^3}{8}$	$\frac{7\lambda^4}{8}$	0
5th Approx.	$-\frac{\lambda}{2}+\frac{\lambda^3}{4}-\frac{3\lambda^5}{32}$	$\frac{\lambda^2}{2}-\frac{9\lambda^4}{16}$	$-\frac{5\lambda^3}{8}+\frac{9\lambda^5}{8}$	$\frac{7\lambda^4}{8}$	$-\frac{21\lambda^5}{16}$

TABLE E.3.

Coefficient	a_1	a_5	a_9	a_{13}
Initial Guess	1	0	0	0
1st Approx.	$1+\frac{k}{16}$	$-\frac{k}{16}$	0	0
2nd Approx.	$1+\frac{k}{16}+\frac{3k^2}{256}$	$-\frac{k}{16}-\frac{7k^2}{256}$	$\frac{k^2}{64}$	0
3rd Approx.	$1+\frac{k}{16}+\frac{3k^2}{256}+\frac{3k^3}{1024}$	$-\frac{k}{16}-\frac{7k^2}{256}-\frac{11k^3}{1024}$	$\frac{k^2}{64}+\frac{27k^3}{2048}$	$-\frac{11k^3}{2048}$

TABLE E.4.

Coefficient	a_3	a_5	a_7	a_9
Initial Guess	0	0	0	0
1st Approx.	$-\lambda$	0	0	0
2nd Approx.	$-\lambda$	λ^2	0	0
3rd Approx.	$-\lambda+5\lambda^3$	λ^2	$-\lambda^3$	0
4th Approx.	$-\lambda+5\lambda^3$	$\lambda^2-11\lambda^3$	$-\lambda^3$	λ^4

Table E.5.

Coefficient	a_1	a_5	a_9
Initial Guess	1	0	0
1st Approx.	$1+\frac{\lambda}{16}$	$-\frac{-\lambda}{16}$	0
2nd Approx.	$1+\frac{\lambda}{16}+\frac{7\lambda^2}{256}$	$-\frac{\lambda}{16}-\frac{7\lambda^2}{256}$	0
3rd Approx.	$1+\frac{\lambda}{16}+\frac{7\lambda^2}{256}+\frac{9\lambda^3}{1024}$	$-\frac{\lambda}{16}-\frac{7\lambda^2}{256}-\frac{19\lambda^3}{2048}$	$\frac{\lambda^3}{2048}$

References used: Kythe [1998].

F

Elliptic Functions

F.1 Elliptic Integrals

It is known that if $u = \arcsin x$, then $u' = \dfrac{1}{\sqrt{1-x^2}}$, and $(u')^2 = \dfrac{1}{1-x^2}$. So we define

$$u = F(k, x) = \int_0^x \frac{dx'}{\sqrt{1-x'^2}} = \arcsin x.$$

The three kinds of elliptic integrals are as follows:

F.1.1 Incomplete Elliptic Integral of the First Kind is defined by

$$u = F(k, \phi) = \int_0^\phi \frac{d\theta}{\sqrt{1-k^2\sin^2\theta}} = \int_0^x \frac{dv}{\sqrt{(1-v^2)(1-k^2v^2)}},$$

where $\phi = \operatorname{am} u$ is the amplitude of u, $x = \sin\phi$, and $0 < k < 1$ (here and below).

F.1.2. Complete Elliptic Integral of the First Kind is defined by

$$\begin{aligned} K(k) = F(k, \pi/2) &= \int_0^{\pi/2} \frac{d\theta}{\sqrt{1-k^2\sin^2\theta}} = \int_0^1 \frac{dv}{\sqrt{(1-v^2)(1-k^2v^2)}} \\ &= \frac{\pi}{2}\left\{1 + \left(\frac{1}{2}\right)^2 k^2 + \left(\frac{1\cdot 3}{2\cdot 4}\right)^2 k^4 + \cdots + \left[\frac{(2n-1)!!}{(2n)!!}\right]^2 k^{2n} + \cdots\right\}. \end{aligned}$$

F.1.3 Incomplete Elliptic Integral of the Second Kind is defined by

$$E(k, \phi) = \int_0^\phi \sqrt{1-k^2\sin^2\theta}\, d\theta = \int_0^x \frac{\sqrt{1-k^2v^2}}{\sqrt{(1-v^2)}}.$$

F.1.4. Complete Elliptic Integral of the Second Kind is defined by

$$\begin{aligned} E(k) = E(k, \pi/2) &= \int_0^{\pi/2} \sqrt{1-k^2\sin^2\theta}\, d\theta = \int_0^1 \frac{\sqrt{1-k^2v^2}}{\sqrt{(1-v^2)}}\, dv \\ &= \frac{\pi}{2}\left\{1 + \left(\frac{1}{2}\right)^2 k^2 - \left(\frac{1\cdot 3}{2\cdot 4}\right)^2 \frac{k^4}{3} - \cdots - \left[\frac{(2n-1)!!}{(2n)!!}\right]^2 \frac{k^{2n}}{2n-1} - \cdots\right\}. \end{aligned}$$

F.1.5 Incomplete Elliptic Integral of the Third Kind is defined by

$$\Pi(k,n,\phi)=\int_0^\phi \frac{d\theta}{(1+n\sin^2\theta)\sqrt{1-k^2\sin^2\theta}}=\int_0^x \frac{dv}{(1+nv^2)\sqrt{(1-v^2)(1-k^2v^2)}},$$

where n is the *characteristic parameter.*

F.1.6. Complete Elliptic Integral of the Third Kind is defined by

$$\Pi(k,n,\pi/2)=\int_0^{\pi/2} \frac{d\theta}{(1+nv^2)\sqrt{1-k^2\sin^2\theta}}=\int_0^1 \frac{dv}{(1+nv^2)\sqrt{(1-v^2)(1-k^2v^2)}}.$$

Notation: The argument k is the *elliptic module*, ($k^2<1$), and

$$k'=\sqrt{1-k^2},\quad K'(k)=K(k'),\quad E'(k)=E(k'),$$

where k' is the *complementary module.* The following properties hold:

$$\begin{aligned}&K(-k)=K(k),\quad K(k)=K'(k');\\ &E(k)=E'(k'),\quad E(-k)=E(k);\\ &E(k)K'(k)+E'(k)K(k)-K(k)K'(k)=\frac{\pi}{2}.\end{aligned}$$

Conversion formulas for complete elliptic integrals:

$$\begin{aligned}K\Big(\frac{1-k'}{1+k'}\Big)&=\frac{1+k'}{2}\,K(k);\\ E\Big(\frac{1-k'}{1+k'}\Big)&=\frac{1+k'}{2}\,[Ek+k'K(k)]\,;\\ K\Big(\frac{2\sqrt{k}}{1+k}\Big)&=(1+k)K(k);\\ E\Big(\frac{2\sqrt{k}}{1+k}\Big)&=\frac{1}{1+k}\,\big[2E(k)-(k')^2K(k)\big]\,.\end{aligned}$$

Differentiation formulas and differential equations:

$$\frac{dK(k)}{dk}=\frac{E(k)}{k(k')^2}-\frac{K(k)}{k},\quad \frac{dE(k)}{dk}=\frac{E(k)+K(k)}{k}.$$

The functions $K(k)$ and $K'(k)$ satisfy the second-order linear ordinary differential equation

$$\frac{d}{dk}\Big[k\left(1-k^2\right)\frac{dK}{dk}\Big]-kK=0.$$

The functions $E(k)$ and $E'(k)$ satisfy the second-order linear ordinary differential equation

$$\left(1-k^2\right)\frac{d}{dk}\Big(k\frac{dE}{dk}\Big)+kE=0.$$

The complete elliptic integrals satisfy

$$\begin{aligned}K(k)&=F\Big(\frac{k,\pi}{2}\Big),\quad E(k)=E\Big(\frac{k,\pi}{2}\Big),\\ K'(k)&=F\Big(\frac{k',\pi}{2}\Big),\quad E'(k)=E\Big(k',\frac{\pi}{2}\Big).\end{aligned}$$

Properties of complete elliptic integrals:

$$F(k,-\phi) = -F(k,\phi), \quad F(k, n\pi \pm \phi) = 2nK(k) \pm F(k,\phi),$$
$$E(k,-\phi) = -E(k,\phi), \quad E(k, n\pi \pm \phi) = 2nE(k) \pm E(k,\phi).$$

F.1.2 Landen's Transformation (Gauss's Transformation). Set $\tan\phi = \dfrac{\sin 2\phi_1}{k+\cos 2\phi_1}$, or $k\sin\phi = \sin(2\phi_1 - \phi)$. Then

$$F(k,\phi) = \int_0^{\phi} \frac{d\theta}{\sqrt{1-k^2\sin^2\theta}} = \frac{2}{1+k}\int_0^{\phi_1} \frac{d\theta_1}{\sqrt{1-k_1^2\sin^2\theta_1}},$$

where $k_1 = \dfrac{2\sqrt{k}}{1+k}$. By successive applications, the sequences $k_1, k_2, \dots$ and $\phi_1, \phi_2, \dots$ are obtained such that $k < k_1 < k_2 < \cdots < 1$ where $\lim_{n\to\infty} k_n = 1$. Then

$$F(k,\phi) = \sqrt{\frac{k_1 k_2 \dots}{k}} \int_0^{\Phi} \frac{d\theta}{\sqrt{1-\sin^2\theta}} = \sqrt{\frac{k_1 k_2 \dots}{k}} \ln\tan\left(\frac{\pi}{4} + \frac{\Phi}{2}\right),$$

where

$$k_1 = \frac{2\sqrt{k}}{1+k}, \quad k_2 = \frac{2\sqrt{k_1}}{1+k_1}, \quad \dots, \text{and } \Phi = \lim_{n\to\infty} \phi_n.$$

F.2 Jacobi's Elliptic Functions

Jacob's elliptic functions are defined as

$$\sin\phi = \sin\operatorname{am}(u,k) = \operatorname{sn}(u,k),$$
$$\cos\phi = \cos\operatorname{am}(u,k) = \operatorname{cn}(u,k),$$
$$\sqrt{1-k^2\sin^2\phi} = \sqrt{1-k^2\sin^2\operatorname{am}(u,k)} = \operatorname{dn}(u,k).$$

The functions sn, cn, dn are read as sine amplitude, cosine amplitude, and delta amplitude, respectively. Note that

$$\operatorname{sn}(u,0) = \sin u; \quad \operatorname{cn}(u,0) = \cos u; \quad \operatorname{dn}(u,0) = 1;$$
$$\operatorname{sn}(u,1) = \tanh u; \quad \operatorname{cn}(u,1) = \operatorname{sech} u; \quad \operatorname{dn}(u,1) = \operatorname{sech} u;$$
$$\operatorname{sn}\left(\frac{K}{2},k\right) = \frac{1}{\sqrt{1+k'}}, \quad \operatorname{cn}\left(\frac{K}{2},k\right) = \frac{k'}{\sqrt{1+k'}}, \quad \operatorname{dn}\left(\frac{K}{2},k\right) = \sqrt{k'};$$
$$\operatorname{sn}(K,k) = 1, \quad \operatorname{cn}(K,k) = 0; \quad \operatorname{dn}(K,k) = k'.$$

Let $x = \operatorname{sn} u$, $\operatorname{sn}' u = \dfrac{d\operatorname{sn} u}{du} = \operatorname{cn} u \operatorname{dn} u$; then $\operatorname{sn} u$ satisfies the first-order differential equation

$$\left(\frac{dx}{du}\right)^2 = \left(1-x^2\right)\left(1-k^2x^2\right).$$

Let $y = \operatorname{cn} u$, $\operatorname{cn}' u = \dfrac{d \operatorname{cn} u}{du} = -\operatorname{sn} u \operatorname{dn} u$; then $\operatorname{cn} u$ satisfies the first-order differential equation

$$\left(\frac{dy}{du}\right)^2 = \left(1 - y^2\right)\left(1 - k^2 + k^2 y^2\right).$$

Let $z = \operatorname{dn} u$, $\operatorname{dn}' u = \dfrac{d \operatorname{dn} u}{du} = -k^2 \operatorname{sn} u \operatorname{dn} u$; then $\operatorname{cn} u$ satisfies the first-order differential equation

$$\left(\frac{dz}{du}\right)^2 = \left(1 - z^2\right)\left(z^2 - (1 - k^2)\right).$$

The elliptic functions $\operatorname{sn} u, \operatorname{cn} u, \operatorname{dn} u$ also satisfy the algebraic differential equations of second order, as follows:

(i) If $x = \operatorname{sn} u$, then

$$x'' = \frac{d^2 x}{du^2} = -\left(1 - k^2\right) x + 2k^2 x^3.$$

(i) If $y = \operatorname{cn} u$, then

$$y'' = \frac{d^2 y}{du^2} = -\left(1 - 2k^2\right) y - 2k^2 y^3.$$

(iii) (i) If $z = \operatorname{dn} u$, then

$$z'' = \frac{d^2 z}{du^2} = \left(2 - k^2\right) z - 2z^3.$$

The relation among $\operatorname{sn} u, \operatorname{cn} u$, and $\operatorname{dn} u$ is as follows:

$$\begin{aligned} x &= \sin(\operatorname{am} u) = \operatorname{sn} u, \\ \sqrt{1 - x^2} &= \cos(\operatorname{am} u) = \operatorname{cn} u, \\ \sqrt{1 - k^2 x^2} &= \sqrt{1 - k^2 \operatorname{sn}^2 u} = \operatorname{dn} u. \end{aligned}$$

The factor $(i - k^2)$ is known as the *complementary elliptic modulus.*

The following identities hold:

$$\operatorname{sn}(-u) = -\operatorname{sn} u, \quad \operatorname{cn}(-u) = \operatorname{cn} u, \quad \operatorname{dn}(-u) = \operatorname{dn} u;$$

$$\operatorname{sn}^2 u + \operatorname{cn}^2 u = 1; \quad \operatorname{dn}^2 u + k^2 \operatorname{sn}^2 u = 1; \quad \operatorname{dn}^2 u - k^2 \operatorname{cn}^2 u = k'^2;$$

$$\operatorname{sn}^2 u = \frac{1 - \operatorname{cn} 2u}{1 + \operatorname{dn} 2u}; \quad \operatorname{cn}^2 u = \frac{\operatorname{dn} 2u + \operatorname{cn} 2u}{1 + \operatorname{dn} 2u}; \quad \operatorname{dn}^2 = \frac{1 - k^2 + \operatorname{dn} 2u + k^2 \operatorname{cn} u}{1 + \operatorname{dn} 2u};$$

$$\sqrt{\frac{1 - \operatorname{cn} 2u}{1 + \operatorname{cn} 2u}} = \frac{\operatorname{sn} u \operatorname{dn} u}{\operatorname{cn} u}; \quad \sqrt{\frac{1 - \operatorname{dn} 2u}{1 + \operatorname{dn} 2u}} = \frac{k \operatorname{sn} u \operatorname{cn} u}{\operatorname{dn} u}.$$

Additional formulas are:

$$\begin{aligned} \operatorname{sn}(u + v) &= \frac{\operatorname{sn} u \operatorname{cn} v \operatorname{dn} v + \operatorname{cn} u \operatorname{sn} v \operatorname{dn} v}{1 - k^2 \operatorname{sn}^2 u \operatorname{sn}^2 v}, \\ \operatorname{cn}(u + v) &= \frac{\operatorname{cn} u \operatorname{cn} v - \operatorname{sn} u \operatorname{sn} v \operatorname{dn} u \operatorname{dn} v}{1 - k^2 \operatorname{sn}^2 u \operatorname{sn}^2 v}, \\ \operatorname{dn}(u + v) &= \frac{\operatorname{dn} u \operatorname{dn} v - k^2 \operatorname{sn} u \operatorname{sn} v \operatorname{cn} u \operatorname{cn} v}{1 - k^2 \operatorname{sn}^2 u \operatorname{sn}^2 v}. \end{aligned}$$

Catalan's constant is

$$\frac{1}{2}\int_0^1 K\,dk = \frac{1}{2}\int_{k=0}^1 \int_{\theta=0}^{\pi/2} \frac{d\theta\,dk}{\sqrt{1-k^2\sin^2\theta}} = \frac{1}{1^2} - \frac{1}{3^2} + \frac{1}{5^2} + \cdots \approx 0.915965594.$$

F.2.1 Periods. Let

$$K = \int_0^{\pi/2} \frac{d\theta}{\sqrt{1-k^2\sin^2\theta}}, \quad \text{and} \quad K' = \int_0^{\pi/2} \frac{d\theta}{\sqrt{1-k'^2\sin^2\theta}},$$

where $k' = \sqrt{1-k^2}$. The periods of Jacobi's elliptic functions are

$$\begin{aligned} &\operatorname{sn} u \text{ has periods } 4K \text{ and } 2iK'; \\ &\operatorname{cn} u \text{ has periods } 4K \text{ and } 2K + 2iK'; \\ &\operatorname{dn} u \text{ has periods } 2K \text{ and } 4iK'. \end{aligned}$$

F.2.2 Special Values:

$$\operatorname{sn} 0 = 0, \quad \operatorname{cn} 0 = 1, \quad \operatorname{dn} 0 = 1, \quad \operatorname{am} 0 = 0.$$

F.2.3 Legendre's Relations:

$$EK' + E'K - KK' = \pi/2,$$

where

$$\begin{aligned} E &= \int_0^{\pi/2} \sqrt{1-k^2\sin^2\theta}\,d\theta, & K &= \int_0^{\pi/2} \frac{d\theta}{\sqrt{1-k^2\sin^2\theta}}, \\ E' &= \int_0^{\pi/2} \sqrt{1-k'^2\sin^2\theta}\,d\theta, & K' &= \int_0^{\pi/2} \frac{d\theta}{\sqrt{1-k'^2\sin^2\theta}}. \end{aligned}$$

F.2.4 Jacobi Integrals of the Third Kind $\Pi(z,\alpha,k)$ is the definite integral

$$\Pi(z,\alpha,k) = k^2 \operatorname{sn}(\alpha,k)\operatorname{cn}(\alpha,k)\operatorname{dn}(\alpha,k) \int_0^1 \frac{\sqrt{1-k^2t^2}}{\sqrt{1-t^2}}\,dt.$$

F.2.5 Jacobi Zeta Function $Z(u,k) = Z(u)$ is related to the incomplete elliptic integrals of the first and second kind by

$$Z(u,k) = E(u,k) - F(u,k)\frac{E(k)}{K(k)}.$$

F.3 Weierstrass Elliptic Function

The Weierstrass elliptic function $\wp(z)$ is defined by

$$\wp(z) = \wp\left(z \middle| \omega_1, \omega_2\right) = \frac{1}{z^2} + \sum_{m,n} \left[\frac{1}{(z - 2m\omega_1 - 2n\omega_2)^2} - \frac{1}{(2m\omega_1 + 2n\omega_2)^2} \right],$$

where the summation is taken over all integers m and n, except $m = n = 0$. This function is a complex, double period function of a complex variable z with periods ω_1 and ω_2, i.e.,

$$\begin{aligned} &\wp(-z) = \wp(z), \\ &\wp\left(z + 2m\omega_1 + 2n\omega_2\right) = \wp(z), \end{aligned}$$

where $m, n = 0, \pm 1, \pm 2 \dots$, and $\Im\{\omega_2/\omega_1\} \neq 0$. The above series representation for $\wp(z)$ converges everywhere except for second-order poles located at $z_{mn} = 2m\omega_1 + 2n\omega_2$. Note that

$$\wp\left(z_1 + z_2\right) == -\wp(z_1) - \wp(z_2) + \frac{1}{4}\left[\frac{\wp'(z_1) - \wp'(z_2)}{\wp(z_1) - \wp(z_2)}\right]^2.$$

The Weierstrass elliptic function $\wp(z)$ is also defined by

$$\wp(z) = \frac{1}{z^2} - \sum_{k=2}^{\infty} g_k z^{2k-2},$$

where $g_2 = \dfrac{g_2}{20}, g_3 = \dfrac{g_2}{28}$, and

$$g_k = \frac{3}{(2k+1)(k-3)} \sum_{m=2}^{k-2} g_m g_{m-k}, \; k \geq 4.$$

The parameters g_2 and g_3 are known as the *invariants*.

The inverse function of $w = \wp(z) = \wp\left(z; g_2, g_3\right)$ is defined as

$$z = \int_{-\infty}^{w} \frac{ds}{\sqrt{4\left(s - e_1\right)\left(s - e_2\right)\left(s - e_3\right)}} = \int_{-\infty}^{w} \frac{ds}{\sqrt{4s^3 - g_2 s - g_3}}.$$

The periods of $\wp(z)$ are written as $2\omega_1, 2\omega_2$, and $\omega_3 = -\omega_1 - \omega_2$; thus, $\wp(\omega_1) = e_1, \wp(\omega_2) = e_2$, and $\wp(\omega_3) = e_3$.

Other notations are:

(i) $\theta_0(z|\tau) = \sum\limits_{n=-\infty}^{\infty} (-1)^n q^{n^2} \cos 2n\pi z$, where $q = e^{i\pi\tau}$. Similar notation are for $\theta_1(z|\tau), \theta_2(z|\tau)$, and $\theta_3(z|\tau)$; and

(ii) $\Theta_0(\tau) = \theta_0(0|\tau), \Theta_2(\tau) = \theta_2(0|\tau)$, and $\Theta_3(\tau) = \theta_3(0|\tau)$.

F.4 Jacobi's Theta Functions

Let $q = e^{in\tau}, |q| < 1$, where q is the nome and τ is the half-period. Then the Jacobi theta function is expressed as $\vartheta_n(z,q)$ in terms of the nome, or $\vartheta_n(z|\tau)$ in terms of the half-period, $n = 1, \dots, 4$. The function $\vartheta_n(z,q)$ is defined successively by

$$\vartheta_1(z,q) = \sum_{n=-\infty}^{\infty} (-1)^{n-1/2} q^{(n+1/2)^2} e^{(2n+1)iz},$$

$$\vartheta_2(z,q) = \sum_{n=-\infty}^{\infty} q^{(n+1/2)^2} e^{(2n+1)iz},$$

$$\vartheta_3(z,q) = \sum_{n=-\infty}^{\infty} q^{n^2} e^{2niz},$$

$$\vartheta_4(z,q) = \sum_{n=-\infty}^{\infty} (-1)^n q^{n^2} e^{2niz}.$$

Also (Whittaker and Watson [1990:463-464]),

$$\begin{aligned}
\vartheta_1(z,q) &= 2\sum_{n=0}^{\infty}(-1)^{n-1/2} q^{(n+1/2)^2} \sin\left[(2n+1)iz\right] \\
&= 2q^{1/4}\sum_{n=0}^{\infty}(-1)^n q^{n(n+1)} \sin\left[(2n+1)z\right], \\
\vartheta_2(z,q) &= \sum_{n=0}^{\infty} q^{(n+1/2)^2} e^{(2n+1)iz} \\
&= 2q^{1/4}\sum_{n=0}^{\infty}(-1)^n q^{n(n+1)} \cos\left[(2n+1)z\right], \\
\vartheta_3(z,q) &= 1 + 2\sum_{n=0}^{\infty} q^{n^2} \cos(2nz), \\
\vartheta_4(z,q) &= 1 + 2\sum_{n=0}^{\infty}(-1)^n q^{n^2} \cos(2nz).
\end{aligned}$$

Also,

$$\begin{aligned}
\vartheta_1(z,q) &= 2q^{1/4} \sin z - 2q^{9/4}\sin(3z) + 2q^{25/4}\sin(5z) + \cdots, \\
\vartheta_2(z,q) &= 2q^{1/4} \cos z + 2q^{9/4}\cos(3z) + 2q^{25/4}\cos(5z) + \cdots, \\
\vartheta_3(z,q) &= 1 + 2q\cos(2z) + 2q^4\cos(4z) + 2q^9\cos(6z) + \cdots, \\
\vartheta_4(z,q) &= 1 - 2q\cos(2z) + 2q^4\cos(4z) - 2q^9\cos(6z) + \cdots.
\end{aligned}$$

See Weisstein [2003].

F.4.1 Connection with Jacobi elliptic functions.

$$\operatorname{sn}(w) = \frac{\vartheta_3(0)\vartheta_1(v)}{\vartheta_2(0)\vartheta_4(v)}, \quad \operatorname{cn}(w) = \frac{\vartheta_4(0)\vartheta_2(v)}{\vartheta_2(0)\vartheta_4(v)}, \quad \operatorname{dn}(w) = \frac{\vartheta_4(0)\vartheta_3(v)}{\vartheta_3(0)\vartheta_4(v)}, \quad w = 2Kv.$$

The parameters are related by

$$k = \frac{\vartheta_2^2(0)}{\vartheta_3^2(0)}, \quad k' = \frac{\vartheta_4^2(0)}{\vartheta_3^2(0)}, \quad K = \frac{\pi}{2}\vartheta_3^2(0), \quad K' = -i\tau K.$$

F.5 Modular Function

A function f is said to be *modular* (or *elliptic modular*) if it satisfies the following conditions: (i) f is meromorphic in the upper half-plane; (ii) $f(A\tau) = f(\tau)$ for every matrix A in the modular group Gamma[1]; and (iii) the Laurent series of f has the form

$$f(\tau) = \sum_{n=-m}^{m} a(n)\, e^{2\pi i n \tau}.$$

The elliptic lambda function

$$\lambda(\tau) = \lambda(q) = h^2)q) - \frac{\vartheta_2^4(0,q)}{\vartheta_3^4(0,q)},$$

where z, q and r are the same as in §F.3, is a λ-modular function defined on the upper half-plane, and $\vartheta_i(z,q)$ are the Jacobi's theta functions. It satisfies the functional equations

$$\lambda(\tau + z) = \lambda(\tau), \quad \lambda\Big(\frac{\tau}{2\tau+1}\Big) = \lambda(\tau).$$

F.6 Weierstrass Zeta Function

The Weierstrass zeta function $\zeta(z)$ is defined by

$$\frac{d\zeta(z)}{dz} = -\wp(z),$$

such that $\lim_{z\to 0} |\wp(z) = z^{-1}| = 0$. Then

$$\zeta(z) - z^{-1} = -\int_0^z \left[\wp(z) - z^{-2}\right] dz = -\sum{}' \int_0^z \left[(z - \Omega_{mn})^{-2} - \Omega_{mn}^{-2}\right] dz,$$

where $\sum'$ means that the term $m = n = 0$ is omitted from the sum. Thus $\zeta(z)$ is an odd function. Integrating $\wp(z + 2\omega_1) = \wp(z)$ gives

$$\zeta(z + 2\omega_1) = \zeta(z) = 2\eta_1.$$

Let $z = \omega_1$, which yields $\zeta(-\omega_1) + 2\eta_1 = -\zeta(\omega_1) + 2\eta_1$, or $\eta_1 = \zeta(\omega_1)$. Similarly, $\eta_2 = \zeta(-\omega_2)$. From Whittaker and Watson [1990:444], we have

$$\eta_1\omega_2 - \eta_2\omega_1 = \frac{1}{2} i\pi.$$

---

[1] It is defined as follows: Of all Möbius transformations of the form $\tau' = \dfrac{a\tau + b}{c\tau + d}$, $ad - bc = 1$, the λ-group is a subgroup with a and d odd, and b and c even. This group can also be represented by the 2×2 matrix $A = \begin{bmatrix} a & b \\ c & d \end{bmatrix}$, $\det A = 1$.

If $x+y+z=0$, then (Whittaker and Watson [1990:446])

$$[\zeta(x)+\zeta(y)+\zeta(z)]^2+\zeta'(zx)+\zeta'(y)+\zeta'(z)=0.$$

Also,

$$2\frac{\begin{vmatrix}1 & \wp(x) & \wp^2(x)\\ 1 & \wp(y) & \wp^2(y)\\ 1 & \wp(z) & \wp^2(z)\end{vmatrix}}{\begin{vmatrix}1 & \wp(x) & \wp'(x)\\ 1 & \wp(y) & \wp'(y)\\ 1 & \wp(z) & \wp'(z)\end{vmatrix}}=\zeta(x+y+z)-\zeta(x)-\zeta(y)-\zeta(z).$$

The series expansion of $\zeta(z)$ is

$$\zeta(z)=z^{-1}-\sum_{k=2}^{\infty}\frac{g_k z^{2k-1}}{2k-1},$$

where g_2, g_3, and g_k are defined as above. See Abramowitz and Stegun [1972:635; 627-671].

F.7 Weierstrass Sigma Function

The Weierstrass sigma function $\sigma(z)$ is related to the Weierstrass zeta function $\zeta(z)$ by

$$\frac{d}{dz}\log\sigma(z)=\zeta(z),$$

such that $\lim_{z\to\infty}\frac{\sigma(z)}{z}=1$. Then

$$\sigma(z)=z\prod_{m,n=-\infty}^{\infty\;\prime}\left[\left(1-\frac{z}{\Omega_{mn}}\right)\exp\left(\frac{z}{\Omega_{mn}}+\frac{z^2}{2\Omega_{mn}^2}\right)\right],$$

where $'$ means that the term with $m=n=0$ is omitted from the product. Also,

$$\sigma(z+2\omega_1)=e^{-2\eta_1(z+\omega_1)}\,\sigma(z),$$
$$\sigma(z+2\omega_2)=-e^{-2\eta_2(z+\omega_2)}\,\sigma(z),$$

and

$$\sigma_j(z)=\frac{e^{-\eta_j(z)\,\sigma(z+\omega_j)}}{\sigma(\omega_j)},\quad j=1,2,\dots.$$

The function $\sigma(z)$ can be expressed in terms of the Jacobi theta function using

$$\sigma(z|\omega_1,\omega_2)=\frac{2\omega_1}{\pi\vartheta_1}\exp\left(\frac{-\nu^2\vartheta_1''}{6\vartheta_1'}\right)\vartheta_1\left(\nu\Big|\frac{\omega_2}{\omega_1}\right),$$

where $\nu=\frac{\pi z}{2\omega_1}$, $\eta_1=-\frac{\pi^2\vartheta_1'''}{12\omega_1\vartheta_1'}$, and $\eta_2=-\frac{\pi^2\omega_2\vartheta_1'''}{12\omega_1^2\vartheta_1'}-\frac{i\pi}{2\omega_1}$. Also,

$$\sigma(z)=\sum_{m,n=0}^{\infty}a_{mn}\left(\frac{1}{2}g_2\right)^m(2g_3)^n\frac{z^{4m+6n+1}}{(4m+6n+1)!},$$

where $a_{00} = 1$ and $a_{mn} = 0$ for either subscript negative; the other values are given by the recurrence relation

$$a_{mn} = 3(m+1)a_{m+1,n+1} + \frac{16}{3}(n+1)a_{m-2n+1} - \frac{1}{3}(2m+3n-1)(4m+6n+1)a_{m-1,n}.$$

See Abramowitz and Stegun [1972:635-636], and Whittaker and Watson [1990:446].

REFERENCES USED: Abramowitz and Stegun [1972], Whittaker and Watson [1990].

G

Gauss-Jacobi Rule

G.1 Gauss-Jacobi Rule

The Jacobi polynomials $P_n^{(\alpha,\beta)}(x)$ are orthogonal polynomials with the weight function $w(x) = (1-x)^\alpha(1+x)^\beta$, $\alpha > -1$, $\beta > -1$. The Gauss-Jacobi rule (also known as the *Mehler quadrature formula*) is defined by

$$\int_{-1}^{1} (1-x)^\alpha (1+x)^\beta f(x)\,dx = \sum_{i=1}^{n} w_i\, f(x_i) + E, \tag{G.1}$$

where

$$w_i = -\frac{2n+\alpha+\beta+2}{n+\alpha+\beta+1}\,\frac{\Gamma(n+\alpha+1)\Gamma(n+\beta+1)}{(n+1)!\,\Gamma(n+\alpha+\beta+1)}\,\frac{2^{\alpha+\beta}}{P_n^{\prime(\alpha,\beta)}(x)\,P_{n+1}^{(\alpha,\beta)}(x)}, \tag{G.2}$$

and the error term E_n in the n-point rule is

$$E_n = \frac{\Gamma(n+\alpha+1)\Gamma(n+\beta+1)\Gamma(n+\alpha+\beta+1)}{(2n+\alpha+\beta+1)\,[\Gamma(2n+\alpha+\beta+1)]^2} \times \frac{n!\,2^{2n+\alpha+\beta+1}}{(2n)!}\,f^{(2n)}(\xi), \quad \xi \in (-1,1). \tag{G.3}$$

The Gauss-Legendre rule is a special case of formula (G.2) with $\alpha = \beta = 0$. The Gauss-Chebyshev rule is another special case with $\alpha = \beta = -1/2$.

For integrands with the Jacobi weight function $w(x) = (1-x)^\alpha(1+x)^\beta$, Piessens and Branders [1973] use the formulas

$$\int_{-1}^{1} (1-x)^\alpha (1+x)^\beta g(x)\,dx \approx \sum_{k=0}^{N} b_k\, G_k(\alpha,\beta) + E_N^1, \tag{G.4}$$

$$\int_{-1}^{1} (1-x)^\alpha (1+x)^\beta \ln\left(\frac{1+x}{2}\right) g(x)\,dx \approx \sum_{k=0}^{N} b_k\, I_k(\alpha,\beta) + E_N^2, \tag{G.5}$$

where $g(x)$ is assumed to have a rapidly convergent Chebyshev series expansion

$$g(x) = {\sum_{k=0}^{\infty}}' a_k\, T_k(x),$$

and

$$b_k = \frac{2}{N} {\sum_{m=0}^{N}}'' g\left(x_m\right) T_k\left(x_m\right), \quad x_m = \cos\frac{m\pi}{N},$$

$$G_n(\alpha,\beta) = 2^{\alpha+\beta+1}\frac{\Gamma(\alpha+1)\Gamma(\beta+1)}{\Gamma(\alpha+\beta+2)}\, {}_3F_2\left[\begin{matrix} n, -n, \alpha+1 \\ \frac{1}{2}, \alpha+\beta+2 \end{matrix}; 1\right],$$

$$E_k^1 \approx a_{k+1}\big(G_{k+1}(\alpha,\beta) - G_{k-1}(\alpha,\beta)\big),$$

$$E_k^2 \approx a_{k+1}\big(I_{k+1}(\alpha,\beta) - I_{k-1}(\alpha,\beta)\big),$$

and $I_n(\alpha,\beta)$ are obtained from the recurrence relation

$$\begin{aligned}(\alpha+\beta+n+2)\, I_{n+1}(\alpha,\beta) + 2(\alpha-\beta)\, I_n(\alpha,\beta) + (\alpha+\beta-n+2)\, I_{n-1}(\alpha,\beta) \\ = 2G_n(\alpha,\beta) - G_{n-1}(\alpha,\beta) - G_{n+1}(\alpha,\beta).\end{aligned}$$

Example G.1. Compute $\displaystyle\int_{-1}^{1} (1-x)^{1/2}(1+x)^{1/2}\, g_i(x)\, dx$, $i = 1, 2$, where $g_1(x) = e^{-2x}$, and $g_2(x) = (x+1.01)^{-3/2}$. The results are given in Table 3.2.1, due to Piessens and Branders [1973].

Table G.1 Exact Values and Absolute Errors.

	Exact	N	Gauss-Jacobi	Formula (G.4)-(G.5)
$g_1(x)$	2.1643953820	5	7.9(−7)	1.8(−6)
		6	6.0(−9)	7.2(−8)
$g_2(x)$	4.2785147337	10	2.7(−1)	2.5(−1)
		20	2.3(−2)	2.6(−1)
		50	8.1(−6)	9.3(−6)
		100	$< 5.0(-11)$	$< 5.0(-11)$

The value of these two integrals are $\frac{\pi}{2}\, I_1(2) \approx 2.4985665285$ and 5.7963478953, respectively (by Mathematica), which casts doubt about the accuracy of the above method.

G.2 Piessens' Method

Following Piessens [1972], we assume that $\bar{f}(s)$ can be represented as

$$\bar{f}(s) = s^{-a} \sum_{k=0}^{\infty} c_k\, P_k^{(\alpha,\beta)}\left(1 - bs^{-1}\right), \tag{G.6}$$

where $P_k^{(\alpha,\beta)}(\cdot)$ denotes the Jacobi polynomials of degree k, and a, α, β and b are free parameters; their choice will be discussed later. The coefficients c_k are given by

$$c_k = \frac{1}{h_k}\int_{-1}^{1} (1-u)^{\alpha}(1+u)^{\beta}\, P_k^{(\alpha,\beta)}(u)\, \psi(u)\, du,$$

where

$$h_k = \int_{-1}^{1} (1-u)^\alpha (1+u)^\beta \left[P_k^{(\alpha,\beta)}(u)\right]^2 du,$$

$$\psi(u) = \left(\frac{b}{1-u}\right)^a \bar{f}\left(\frac{b}{1-u}\right).$$

Inverting the series (G.6) term-by-term, we find that

$$f(t) = \frac{t^{a-1}}{\Gamma(a)} \sum_{k=0}^{\infty} c_k \, \frac{(\alpha+1)_k}{k!} \, \phi_k\left(\frac{bt}{2}\right),$$

where $(\alpha+1)_k$ is the Pochhammer's symbol with $(\alpha+1)_0 = 1$, and $\phi_k(x)$ is a polynomial of degree k defined by

$$\phi_k(x) = {}_2F_2 \left[\begin{matrix} -k, & k+\alpha+\beta+1 \\ \alpha+1, & a \end{matrix}; x\right].$$

To evaluate numerical values of c_k, we note that if the series (G.6) is truncated after $M+1$ terms, we obtain

$$f(t) \approx \frac{t^{a-1}}{\Gamma(a)} \sum_{k=0}^{M} \frac{(2k+\alpha+\beta+1)\,\Gamma(k+\alpha+\beta+1)}{2^{\alpha+\beta+1}\Gamma(\alpha+1)\Gamma(k+\beta+1)} \, \phi_k\left(\frac{bt}{2}\right) \sum_{j=1}^{N} V_j \, P_k^{(\alpha,\beta)}(u_j), \tag{G.7}$$

where N is the order of the Gauss-Jacobi quadrature formula, u_j are the nodes and $V_j = w_j \, \psi(u_j)$ the weights, w_j being the weights of the Gauss-Jacobi quadrature. In the special case when $\alpha = \beta = -1/2$, formula (G.7) becomes

$$f(t) \approx \frac{t^{a-1}}{\Gamma(a)} \sum_{k=0}^{M}{}' \, c_k \, \phi_k\left(\frac{bt}{2}\right), \tag{G.8}$$

where

$$\phi_k(x) = {}_2F_2 \left[\begin{matrix} -k, & k \\ \frac{1}{2}, & a \end{matrix}; x\right],$$

and the coefficients c_k are obtained by using Clenshaw's method for the computation of Chebyshev coefficients as

$$c_k \approx \frac{2}{N} \sum_{m=0}^{N}{}'' \psi\left(\cos\frac{m\pi}{N}\right) \cos\left(\frac{mk\pi}{N}\right) \tag{G.9}$$

or

$$c_k \approx \frac{2}{N+1} \sum_{m=0}^{N}{}'' \psi\left(\cos\frac{(2m+1)\pi}{2(N+1)}\right) \cos\left(\frac{(2m+1)k\pi}{2(N+1)}\right), \quad k \le N. \tag{G.10}$$

In formula (G.9) we need the value of $\lim_{s\to\infty} s\bar{f}(s)$. If this limit is not known, we use formula (G.10).

The polynomials $\phi_k(x)$ in formula (G.8) become very large as k increases, and they have alternating signs. For example, some of these polynomials are

$$\begin{aligned}
\phi_0(x) &= 1, \\
\phi_1(x) &= 1 - \frac{2}{a}\,x, \\
\phi_2(x) &= 1 - \frac{8}{a}\,x + \frac{8}{a(a+1)}\,x^2, \\
\phi_3(x) &= 1 - \frac{18}{a}\,x + \frac{48}{a(a+1)}\,x^2 - \frac{32}{a(a+1)(a+2)}\,x^3, \\
\phi_4(x) &= 1 - \frac{32}{a}\,x + \frac{160}{a(a+1)}\,x^2 - \frac{256}{a(a+1)(a+2)}\,x^3 \\
&\quad + \frac{128}{a(a+1)(a+2)(a+3)}\,x^4.
\end{aligned}$$

In general, using Fasenmyer's technique (see Rainville [1960]), $\phi_k(x)$ for formula (G.7) can be determined from the recurrence formula

$$\phi_k(x) = (A + B\,x)\,\phi_{k-1}(x) + (C + D\,x)\,\phi_{k-2}(x) + E\,\phi_{k-3}(x), \quad k \geq 3,$$

where

$$\begin{aligned}
A &= \frac{(2n+\alpha+\beta)(2n+\alpha+\beta-1)(a+n-1)(a+n-2)(n-1)}{(\alpha+n)(a+n-1)(2n+\alpha+\beta-3)(n+\alpha+\beta)} \\
&\qquad - \frac{n(2n+\alpha+\beta-1)}{n+\alpha+\beta}, \\
B &= \frac{(2n+\alpha+\beta)(2n+\alpha+\beta-1)}{(\alpha+n)(a+n-1)(n+\alpha+\beta)}, \\
C &= -1 - A - E, \\
D &= B\,\frac{n-1}{n+\alpha+\beta-1}, \\
E &= D\,\frac{(n-2)(n+\beta-2)(a+1-n-\alpha-\beta)}{(2n+\alpha+\beta-3)(2n+\alpha+\beta-4)}, \\
\phi_0(x) &= 1, \\
\phi_1(x) &= 1 - \frac{\alpha+\beta+2}{a(\alpha+1)}\,x, \\
\phi_2(x) &= 1 - \frac{2(\alpha+\beta+3)}{a(\alpha+1)}\,x + \frac{(\alpha+\beta+3)(\alpha+\beta+4)}{a(a+1)(\alpha+1)(\alpha+2)}\,x^2.
\end{aligned}$$

The parameter a must be such that $\bar{f}(s) \to s^{-a}$ as $s \to \infty$. However, there may be functions $\bar{f}(s)$ for which such a does not exist.

Generally it is convenient to take $\alpha = \beta = -0.5$. This choice simplifies the calculations considerably. However, these values of α and β are not suitable when the Laplace transform is known in a small interval on the real line. In that case formula (G.9) must be used and the value of N must be so low that it satisfies the condition $\cos \dfrac{\pi}{2(N+1)} \leq 1 - \dfrac{b}{A}$, where $[0, A]$ is the interval in which $\bar{f}(s)$ is known. This restricts the number of coefficients c_k that can be calculated. The problem can be avoided by taking α large and $\beta \approx 1$.

The value of b is related to the interval of convergence on the real line for the series in (G.6). The minimum value of $\Re\{s\}$ for which the series in (G.6) is convergent is $b/2$.

Piessens computes the polynomials $\phi_k(x)$ and gives a generating formula for these polynomials. Hypergeometric functions are available in Mathematica; for details see `piessens.nb`. Finally, the error is given by

$$E(t) \approx \frac{t^{a-1}}{\Gamma(a)}\, c_{M+1}\phi_{M+1}\Big(\frac{bt}{2}\Big). \tag{G.11}$$

Formulas (G.7) and (G.8) are compared with those by Salzer [1958] and Luke [1969], where Salzer approximates the Laplace transform $\bar{f}(s)$ by an interpolating function $\bar{f}(s) \approx s^{-a}\,Q_N(1/s)$, where $Q_n(x)$ is a polynomial of degree N; this formula uses equally spaced interpolation points $s_k = k$. The approximating function is then inverted exactly. If N is large, this inversion loses accuracy. Luke's method is a generalization of methods of Erdélyi [1943], Lanczos [1956], Miller and Guy [1966], and is also related to the method of Bellman, Kalaba and Lockett [1966] (see §4.3.1). In this method the original function $f(t)$ is obtained as a series of shifted Jacobi polynomials

$$f(t) \approx \left(1 - 2^{-\lambda t}\right)^{\alpha} e^{-bt} \sum_{k=0}^{N} a_k\, P_k^{(\alpha,\beta)}\left(2e^{-\lambda t} - 1\right),$$

where α, β, λ and b are free parameters, and

$$\begin{aligned} a_k = \frac{\lambda\,(2k+\alpha+\beta+1)}{\Gamma(k+\alpha+1)} \sum_{m=0}^{k} (-1)^m \binom{k}{m} \frac{\Gamma(2k-m+\alpha+\beta+1)}{\Gamma(k-m+\beta+1)} \\ \times F\left(\lambda\,(k-m+\beta+1) - b\right). \end{aligned}$$

Example G.2. Piessens [1971] considers two examples: (i) $\bar{f}(s) = \dfrac{1}{\sqrt{s^1+1}}$, for which $f(t) = J_0(t)$, and (ii) $\bar{f}(s) = \dfrac{e^{-(1/s)}}{\sqrt{s}}$, for which $f(t) = \dfrac{\cos(2\sqrt{t})}{\sqrt{\pi t}}$, and compares the results for these two examples with those of Salzer, Luke, and the exact solution.

G.3. Gauss-Jacobi Quadrature

Erdogan and Gupta [1972], Erdogan, Gupta and Cook [1973], and Krenk [1975] show that by a proper choice of collocation points, both CSK1 and CSK2 can be solved using the Gauss-Jacobi quadrature rule. Consider a CSK2 in its general form

$$a_i\,\phi_i(x) + \frac{b_i}{\pi} -\!\!\!\!\!\!\int {}^{1}_{-1}\phi_i(s)\,\frac{ds}{s-x} + \lambda \sum_{i=1}^{N} \int_{-1}^{1} k_{ij}(x,s)\phi(s)\,ds = f_i(x), \quad |x| < 1, \tag{G.12}$$

for $j = 1, \dots, N$, where a_i and b_i are real constants, the kernels $k_{ij}(x,s) \in H\left([-1,1]\right)$, and the free terms $f_i(x)$ are known functions. The unknown functions $\phi_i(x)$ or their first derivatives have integrable singularities at the endpoints $x = \pm 1$. A general closed-form solution of Eq (G.12) is not known. However, for a numerical solution based on the Gauss-Jacobi quadrature rule, Erdogan, Gupta and Cook [1973] have found a group of fundamental functions defined by

$$w_i(x) = (1-x)^{\alpha_j}(1+x)^{\beta_j}, \tag{G.13}$$

where

$$\alpha_j = \frac{1}{2i\pi} \log \frac{a_j - ib_j}{a_j + ib_j} + N_j,$$
$$\beta_j = -\frac{1}{2i\pi} \log \frac{a_j - ib_j}{a_j + ib_j} + M_j,$$

N_j and M_j being integers for $j = 1, \dots, N$, and for each of the N equations in (G.12) the *index* κ_j of the integral operators K_{ij} is defined by

$$\kappa_j = -(\alpha_j + \beta_j) = -(N_j + M_j), \quad j = 1, \dots, N.$$

Since we have assumed that ϕ_i or their first derivatives have integrable singularities at the endpoints, the index must be $-1, 0, 1$ (see Mushkhelishvili [1992]). The numerical solution of Eq (G.12) is given by

$$\phi_i(x) = g_i(x)\, w_i(x),$$

where

$$g_i(x) = \sum_{i=1}^{\infty} c_{ij}\, P_j^{(\alpha,\beta)}(x). \tag{G.14}$$

Here $P_n^{(\alpha,\beta)}(x)$ are Jacobi polynomials of degree n with indices α and β, and c_{ij} are constants to be determined. The general scheme is to truncate the series (G.14) for $i = 1, \dots, n$, and determine methods to compute the unknown coefficients c_{ij}. We discuss this problem below for a simple case of Eq (G.12), but the method can be easily extended to Eq (G.12).

G.4 Solution by Jacobi Polynomials

For the sake of simplicity we consider a special case of Eq (G.12) of the form

$$a\,\phi(x) + \frac{b}{\pi} -\!\!\!\!\!\!\int_{-1}^{1} \phi(s) \frac{ds}{s - x} + \lambda \int_{-1}^{1} k(x, s)\phi(s)\, ds = f(x), \quad |x| < 1, \tag{G.15}$$

for which, by using the orthogonality properties of the Jacobi polynomials, Erdogan, Gupta and Cook [1973] derive an infinite system of linear algebraic equations

$$\phi(x) = \sum_{n=0}^{\infty} c_n\, w(x)\, P_n^{(\alpha,\beta)}(x), \tag{G.16}$$

where $w(x) = (1 - x)^{\alpha}(1 + x)^{\beta}$ and c_n, $n = 0, 1, \dots$, are constants to be determined. Before substituting (G.16) into Eq (G.15), note that for the index $\kappa = (-1, 0, 1)$ we have (see Tricomi [1957], Szegö [1939])

$$\begin{aligned} \frac{1}{\pi} -\!\!\!\!\!\!\int_{-1}^{1} w(s)\, P_n^{(\alpha,\beta)}(s) \frac{ds}{s - x} &= \cot(\pi\alpha)\, w(x)\, P_n^{(\alpha,\beta)}(x) \\ &\quad - \frac{2^{\alpha+\beta}\, \Gamma(\alpha)\Gamma(n + \beta + 1)}{\pi\Gamma(n + a + \beta + 1)} F\Big(n + 1, -n - \alpha - \beta; 1 - \alpha; \frac{1 - x}{2}\Big), \\ -1 < x < 1, \quad \Re\{\alpha\} > -1, &\ \Re\{\alpha\} \neq 0, 1, \dots, \quad \Re\{\beta\} > -1, \\ \cot(\pi\alpha) = \cot \pi \Big[\frac{1}{2i\pi} \log\Big(\frac{a - ib}{a + ib}\Big) + N\Big] &= -\frac{a}{b}, \quad \alpha + \beta = -\kappa, \end{aligned} \tag{G.17}$$

and

$$P_{n-\kappa}^{(-\alpha,-\beta)}(x) = \frac{\Gamma(n-\kappa-\alpha+1)}{\Gamma(1-\alpha)\,\Gamma(n-\kappa+1)}\, F\Big(n+1, -n+\kappa; 1-\alpha; \frac{1-x}{2}\Big). \tag{G.18}$$

Hence, combining (G.17) and (G.18), we get

$$\begin{aligned} a\,w(x)\,P_n^{(\alpha,\beta)}(x) + \frac{b}{\pi} - \int {}_{-1}^{1} w(s)\,P_n^{(\alpha,\beta)}(s)\,\frac{ds}{s-x} \\ = -2^{-\kappa}\, b\, \frac{\Gamma(\alpha)\Gamma(1-\alpha)}{\pi}\, P_{n-\kappa}^{(-\alpha,-\beta)}(x), \quad |x|<1. \end{aligned} \tag{G.19}$$

Then, substituting (G.16) into Eq (G.15) and using (G.19), we obtain an infinite system of algebraic equations

$$\sum_{n=0}^{\infty} c_n \Big[-\frac{2^{-\kappa}\, b}{\sin(\pi\alpha)}\, P_{n-\kappa}^{(-\alpha,-\beta)}(x) + h_n(x) \Big] = f(x), \tag{G.20}$$

where

$$h_n(x) = \lambda \int_{-1}^{1} w(s)\, P_n^{(\alpha,\beta)}(s)\, k(x,s)\, ds, \quad |x|<1.$$

Now, we use the orthogonality relations

$$\int_{-1}^{1} P_n^{(\alpha,\beta)}(s)\, P_m^{(\alpha,\beta)}(s)\, w(s)\, ds = \begin{cases} 0 & \text{if } n \neq m, \\ \theta_m^{(\alpha,\beta)} & \text{if } n = m, \end{cases}$$

for $m = 0, 1, 2, \dots$, where

$$\theta_m^{(\alpha,\beta)} = \frac{2^{\alpha+\beta+1}}{2m+\alpha+\beta+1}\, \frac{\Gamma(m+\alpha+1)\Gamma(m+\beta+1)}{m!\,\Gamma(m+\alpha+\beta+1)}, \quad m = 0, 1, 2, \dots,$$

and

$$\theta_0^{(\alpha,\beta)} = \int_{-1}^{1} w(s)\, ds = \frac{2^{\alpha+\beta+1}\,\Gamma(\alpha+1)\Gamma(\beta+1)}{\Gamma(\alpha+\beta+2)}.$$

Then we truncate the series (G.20) to obtain

$$-\frac{2^{-\kappa}\, b}{\sin(\pi\alpha)}\, \theta_m(-\alpha,-\beta)\, c_{m+\kappa} + \sum_{m=0}^{N} d_{nm}\, c_n = F_m, \quad m = 0, 1, \dots, N, \tag{G.21}$$

where

$$\begin{aligned} d_{nm} &= \int_{-1}^{1} P_m^{(-\alpha,-\beta)}(x)\, w(-\alpha,-\beta,x)\, h_n(x)\, dx, \\ F_m &= \int_{-1}^{1} P_m^{(-\alpha,-\beta)}(x)\, w(-\alpha,-\beta,x)\, f(x)\, dx, \\ w(-\alpha,-\beta,x) &= (1-x)^{-\alpha}(1+x)^{-\beta} = w^{-1}(x). \end{aligned} \tag{G.22}$$

There are three cases to consider.

CASE 1. $\kappa = -1$: Note that the first term of the series (G.20) is equal to a constant multiplied by $c_0\, P_1^{(-\alpha,-\beta)}(x)$. Hence, in solving Eq (G.15) we can take $c_{-1} = 0$. Also, since $P_0^{(-\alpha,-\beta)}(x) = 1$, it can be seen from Eqs (G.21) and (G.22) that the first equation obtained from (G.21) for $m = 0$ is equivalent to the consistency condition

$$\int_{-1}^{1} \left[f(x) - \int_{-1}^{1} k(x,s)\phi(s)\, ds \right] \frac{ds}{w(s)} = 0.$$

Thus, Eqs (G.21) give $(N+1)$ linear equations to compute the unknown constants $c_0, \dots, c_N$.

CASE 2. $\kappa = 0$: This case does not need any additional conditions, and Eqs (G.21) give the unique solution for $c_0, \dots, c_N$.

CASE 3. $\kappa = 1$: In this case there are $(N+2)$ unknown constants $c_0, \dots, c_{N+1}$ but only $(N+1)$ equations given by (G.21). Thus, we need one more equation, which is provided by the equilibrium or compatibility condition $\displaystyle\int_{-1}^{1} \phi(s)\, ds = A$, which, after its substitution into (G.16) and using the orthogonality condition, reduces to

$$c_0\, \theta_0(\alpha, \beta) = A.$$

Then Eqs (G.21) together with this condition are solved to compute the $(N+2)$ constants $c_0, \dots, c_{N+1}$.

REFERENCES USED: Abramowitz and Stegun [1972].

H

Orthogonal Polynomials

Functions analytic in a region D can always be represented by Taylor series only if D is a circular disk. If the boundary Γ of D is not a circle, such a series representation is not possible. Therefore, we must find a sequence of functions which depend only on the region D so that any analytic function in D can be expanded in the form of a series involving functions from this sequence. It turns out that all such functions are certain polynomials of a special form which are either orthogonal on the boundary Γ or in the region D. They not only provide a series expansion for analytic functions on D but also play a significant role in conformal mapping.

H.1 Polynomials Orthogonal on the Boundary

We will analyze the structure of polynomials that are orthogonal on a contour. Let Γ be an arbitrary rectifiable curve (not necessarily closed) of length l. By using the Gram-Schmidt orthogonalization process (see Gaier [1964:132] for an algorithm for this process) we construct a sequence of polynomials $\{P_0(z), P_1(z), \dots, P_n(z), \dots\}$ with the following properties:
(i) $P_n(z)$ is a polynomial of degree n in z;
(ii) the coefficient of z^n in $P_n(z)$ is positive; and
(iii) the polynomials $P_n(z)$ are orthonormal (orthogonal and normalized) along the curve Γ, i.e.,

$$\frac{1}{l}\int_\Gamma P_n(z)\,\overline{P_m(z)}\,ds = \delta_{nm}, \tag{H.1}$$

where δ_{nm} is the Kronecker delta. We will introduce the constants

$$h_{pq} = \frac{1}{l}\int_\Gamma z^p\,\bar{z}^q\,ds. \tag{H.2}$$

Note that $ds = |dz|$ and $h_{pq} = \bar{h}_{qp}$. Consider the positive-definite Hermitian quadratic forms

$$H_n(t) = \sum_{p,q=0}^{n} h_{pq}\,t_p\,\bar{t}_q = \frac{1}{l}\int_\Gamma \left|t_0 + t_1 z + \cdots, + t_n z^n\right|^2 ds, \tag{H.3}$$

with determinants D_n defined by

$$D_0 = 1, \quad D_n = \begin{vmatrix} h_{00} & h_{10} & \cdots & h_{n0} \\ h_{01} & h_{11} & \cdots & h_{n1} \\ \cdots & \cdots & \cdots & \cdots \\ h_{0n} & h_{1n} & \cdots & h_{nn} \end{vmatrix}. \tag{H.4}$$

Then the Szegö polynomials $\sigma_n(z)$ are represented by

$$\sigma_n(z) = \frac{1}{\sqrt{D_{n-1}\, D_n}} \begin{vmatrix} h_{00} & h_{10} & \cdots & h_{n0} \\ h_{01} & h_{11} & \cdots & h_{n1} \\ \cdots & \cdots & \cdots & \cdots \\ h_{0\,n-1} & h_{1\,n-1} & \cdots & h_{n\,n-1} \\ 1 & z & \cdots & z^n \end{vmatrix}. \tag{H.5}$$

It can be verified that these polynomials possess the above three properties. As regards the question of expansion of an arbitrary analytic function in a series involving Szegö polynomials, the following result due to Smirnov [1928] holds:

Theorem H.1. *Suppose that a function $f(z)$ is analytic inside a region D bounded by a Jordan curve Γ, has almost everywhere boundary values on Γ, and can be represented in terms of these boundary values by Cauchy integrals. Then $f(z)$ can be expanded in a series involving Szegö polynomials:*

$$f(z) = \sum_{n=0}^{\infty} A_n\, \sigma_n(z), \tag{H.6}$$

which is uniformly convergent everywhere within Γ, and the coefficients A_n are determined by

$$A_n = \frac{1}{l} \int_\Gamma f(z)\, \overline{\sigma_n(z)}\, ds. \tag{H.7}$$

For a proof of this theorem see Smirnov [1928].

As an application of Szegö polynomials to conformal mapping, let $w = F(z)$ map the region D in the z-plane onto the disk $|w| < R$ such that a point $a \in D$ goes into the origin $w = 0$, $F(a) = 0$, and $F'(a) = 1$. In view of Theorem 4.3.1, out of all functions $F(z)$ analytic on D and normalized at a by $F(a) = 1$, the function $\sqrt{f'(z)}$ minimizes the integral (4.3.1), i.e.,

$$I = \frac{1}{l} \int_\Gamma |F(z)|^2\, ds \quad \text{in the class } \mathcal{L}^1. \tag{H.8}$$

Using the series expansion (H.6) for the function $F(z)$ in terms of Szegö polynomials $\sigma_n(z)$, and using $F(a) = 1$, we find that

$$\sum_{j=0}^{\infty} A_j\, \sigma_j(a) = 1. \tag{H.9}$$

Then, from (H.8)

$$\frac{1}{l} \int_\Gamma F(z)\, \overline{F(z)}\, ds = \sum_{j=0}^{\infty} A_j\, \bar{A}_j. \tag{H.10}$$

Then the system of coefficients corresponding to the function $F(z) = \sqrt{f'(z)}$ attains the minimum value for the sum in (H.10) such that the condition (H.9) is satisfied. We will denote these coefficients by δ_j, and set $A_j = \delta_j + \varepsilon\, \eta_j$. Since, in view of (H.9), the condition

$$\sum_{j=0}^{\infty} \delta_j\, \sigma_j(a) = 1 \tag{H.11}$$

still holds, the numbers η_j must be such that

$$\sum_{j=0}^{\infty} \eta_j\, \sigma_j(a) = 0. \tag{H.12}$$

Then, from (H.10)

$$\begin{aligned}\frac{1}{l}\int_\Gamma F(z)\,\overline{F(z)}\,ds &= \sum_{j=0}^{\infty} \delta_j\,\bar\delta_j + \varepsilon \sum_{j=0}^{\infty} \eta_j\,\bar\delta_j \\ &\quad + \bar\varepsilon \sum_{j=0}^{\infty} \bar\eta_j\,\delta_j + |\varepsilon|^2 \sum_{j=0}^{\infty} \eta_j\,\bar\eta_j.\end{aligned} \tag{H.13}$$

Since the expression on the right side in (H.13) must be less than $\sum_{j=0}^{\infty} \delta_j\,\bar\delta_j$, which is the minimum value of the integral (H.8), it is necessary and sufficient that the coefficients of ε and $\bar\varepsilon$ vanish for all η_j subject to the condition (H.12), i.e.,

$$\sum_{j=0}^{\infty} \eta_j\,\bar\delta_j = 0 = \sum_{j=0}^{\infty} \bar\eta_j\,\delta_j. \tag{H.14}$$

Now, from (H.12) we get $\eta_0 = -\sum_{j=0}^{\infty} \eta_j\,\sigma_j(a)$, since $\sigma_0(z) = 1$, and

$$\sum_{j=0}^{\infty} \eta_j\left[\bar\delta_j - \bar\delta_0\,\sigma_j(a)\right] = 0, \tag{H.15}$$

which is valid for arbitrary $\eta_1, \eta_2, \dots$ only if

$$\bar\delta_j = \bar\delta_0\,\sigma_j(a). \tag{H.16}$$

Substituting the values of δ_j from (H.16) and (H.11) we get

$$\delta_0 \sum_{j=0}^{\infty} \sigma_j(a)\,\overline{\sigma_j(a)} = 1. \tag{H.17}$$

Set

$$S(z,a) = \sum_{j=0}^{\infty} \overline{\sigma_j(a)}\,\sigma_j(z). \tag{H.18}$$

Then we have

$$\begin{aligned}\delta_0 &= \bar{\delta}_0 = \frac{1}{S(a,a)} = \frac{\overline{\sigma_0(a)}}{S(a,a)},\\ \delta_j &= \frac{\overline{\sigma_j(a)}}{S(a,a)},\\ \sqrt{f'(z)} &= \frac{1}{S(a,a)}\sum_{j=0}^{\infty}\overline{\sigma_j(a)}\,\sigma_j(z) = \frac{S(z,a)}{S(a,a)},\\ f(z) &= \frac{1}{S^2(a,a)}\int_a^z S^2(z,a)\,dz,\end{aligned} \tag{H.19}$$

where $S(z,a)$ is the Szegö kernel. In order to derive an approximate formula for $f(z)$, we will assume that only n Szegö polynomials $\sigma_j(z)$ are known. Then

$$S_n(z,a) \approx \sum_{j=0}^{n}\overline{\sigma_j(a)}\;\sigma_j(z), \tag{H.20}$$

and

$$f(z) \approx \frac{1}{S(a,a)}\int_a^z S^2(z,a)\,dz. \tag{H.21}$$

The radius R of the disk $|w| < R$ is given exactly by

$$\begin{aligned}R &= \frac{1}{2\pi}\int_\Gamma |f'(z)|^2\,ds = \frac{1}{2\pi\,S^2(a,a)}\int_a^z S(z,a)\,\overline{S(z,a)}\,ds\\ &= \frac{1}{2\pi\,S^2(a,a)}\sum_{j=0}^{\infty}\overline{\sigma_j(a)}\,\sigma_j(a) = \frac{l}{2\pi\,S(a,a)}.\end{aligned} \tag{H.22}$$

Then obviously the function $g(z)$ that maps the region D onto the unit disk U is given by

$$g(z) = \frac{2\pi}{l\,S^2(a,a)}\int_a^z S^2(z,a)\,dz. \tag{H.23}$$

Example H.1. We will determine the mapping function $F(z)$ that maps the square $\{-1 \le x,\, y \le 1\}$ onto the disk $|w| \le R$. Since $z = x+i$ on AB, $z = x-i$ on DC, $z = 1+iy$ on CB, and $z = -1+iy$ on DA, the numbers

$$\begin{aligned}h_{pq} = \frac{1}{8}\Big\{\int_{-1}^{1} &\left[(x+i)^p(x-i)^q + (x-i)^p(x+i)^q\right]dx\\ &+ \left[(1+iy)^p(1-iy)^q + (-1+iy)^p(-1-iy)^q\right]dy\Big\}\end{aligned} \tag{H.24}$$

are computed in `cs422.nb` for $p, q = 0, 1, \dots, 8$ (see Notes, at the end). If Mathematica is not used, then (H.24) can be written as

$$\begin{aligned} h_{pq} &= \frac{1}{8}\int_{-1}^{1}\Big\{(x+i)^p(x-i)^q+(x-i)^p(x+i)^q \\ &\qquad + i^{p-q}\left[(x+i)^p(x-i)^q+(x-i)^p(x+i)^q\right]\Big\}\,dx \\ &= \frac{1+i^{p-q}}{8}\int_{-1}^{1}\left[(x+i)^p(x-i)^q+(x-i)^p(x+i)^q\right]dx \\ &= \frac{1+i^{p-q}}{4}\,\Re\Big\{\int_{-1}^{1}(x+i)^p(x-i)^q\,dx\Big\} \\ &= \begin{cases} 0, & \text{if } p-q\neq 4k, \\ \frac{1}{2}\int_{-1}^{1}(x^2+1)^q\,\Re\left\{(x+i)^{p-q}\right\}dx, & \text{if } p-q=4k, \end{cases} \end{aligned} \tag{H.25}$$

and then the numbers h_{pq} can be evaluated with the same values as in (H.25). Now, from (H.5),

$$\sigma_0(z)=1,\quad \sigma_1(z)=\frac{\sqrt{3}}{2}\,z,\quad \sigma_2(z)=\frac{1}{2}\sqrt{\frac{15}{7}}\,z^2,$$

$$\sigma_3(z)=\frac{1}{4}\sqrt{\frac{35}{6}}\,z^3,\quad \sigma_4(z)=\frac{3}{16}\sqrt{\frac{7}{22}}\,(4+5z^4),$$

$$\sigma_5(z)=\frac{3}{8}\sqrt{\frac{11}{379}}\,z(8+7z^4),\quad \sigma_6(z)=\frac{1}{128}\sqrt{\frac{429}{3941}}\,z^2(220+147z^4),$$

$$\sigma_7(z)=\frac{1}{8}\sqrt{\frac{65}{96222}}\,z^3(182+99z^4).$$

Since all $\sigma_j(0)$ are zero except for $j=1,4$, we find from (H.18) that

$$S(z,0)=\sum_{n=0}^{\infty}\overline{\sigma_n(0)}\,\sigma_n(z)\approx 1+\frac{63}{1408}\,(4+5z^4),$$

$S(0,0)=\dfrac{415}{352}$, and thus, from (H.19)

$$\begin{aligned} f(z) &\approx \frac{1}{S^2(0,0)}\int_0^z S^2(z,0)\,dz = z+\frac{63}{830}\,z^5+\frac{441}{110224}\,z^9 \\ &\approx z+0.0759036\,z^5+0.004\,z^9, \end{aligned} \tag{H.26}$$

which can be compared with (4.2.19). Let $z=\phi(w)$ be the inverse function of $w=f(z)$ such that $\phi(w)$ maps the circle $|w|=R$ onto the given square, and $\phi(0)=0$, $\phi'(0)=1$. By using the Schwarz-Christoffel transformation analogous to Map 7.4, the function $z=\phi(w)$ is represented by the elliptic integral

$$z=\int_0^w\frac{d\zeta}{\sqrt{1+k^4\zeta^4}}=w-\frac{k^4}{10}w^5+\frac{k^8}{24}w^9+\cdots, \tag{H.27}$$

where

$$k=\int_0^1\frac{d\zeta}{\sqrt{1+\zeta^4}}\approx 0.927037. \tag{H.28}$$

On inversion, (H.27) yields

$$w = f(z) = z + \frac{k^4}{10} z^5 + \frac{k^8}{120} z^9 + \frac{11k^{12}}{15600} z^{13} + \cdots, \tag{H.29}$$

(see, e.g., Gaier [1964:148].) A comparison of (H.27) and (H.30) shows that $\frac{k^4}{10} = \frac{63}{830}$, or $k = \sqrt[4]{\frac{63}{83}} \approx 0.933395$, which, after comparing with the value of k in (H.29) shows that the polynomial approximation of $f(z)$ has an error of 0.636%. This means that the polynomial $f(z)$ maps the boundary of the square onto some curve that does not quite coincide with the circle $|w| = R$. In order to determine the closeness of this curve to the circle $|w| = R$, we evaluate $|f(1)|$ and $|f(1+i)|$, which are given by $|f(1)| = 1.0799036$, and $|f(1+i)| = 1.075368896$, which shows that the radius of the circle onto which the square is mapped by the approximate polygon lies between these two values. However, from (H.22) we find that $R = 1/k \approx 1.078705$, which gives a maximum error of at most 0.5% of the value of R.

The polynomial that maps the given square onto the unit disk U can be determined from (H.23). The exact solution is given by the elliptic integral

$$\begin{aligned} z &= \frac{\int_0^w (1+t^4)^{-1/2}\,dt}{\int_0^1 (1+t^4)^{-1/2}\,dt} \\ &\approx 1.08 \left(w - \frac{1}{10} w^5 + \frac{1}{24} w^9 - \frac{5}{208} w^{11} + \cdots \right). \blacksquare \end{aligned} \tag{H.30}$$

The exact mapping function is known in terms of Jacobian elliptic functions (see Map 7.4).

H.2 Polynomials Orthogonal to a Region

Let D be, as before, a simply connected region with a Jordan boundary Γ and area A. Using the Schmidt orthogonalization process, we construct a system of polynomials $\{Q_0(z), Q_1(z), \dots, Q_n(z), \dots\}$, with the following properties:
(i) $Q_n(z)$ is a polynomial of degree n in z;
(ii) the coefficient of z^n in $Q_n(z)$ is positive; and
(iii) the polynomials $Q_n(z)$ are orthonormal (orthogonal and normalized) along the curve Γ, i.e.,

$$\frac{1}{A} \iint_D Q_n(z)\,\overline{Q_m(z)}\,dx\,dy = \delta_{nm}. \tag{H.31}$$

These properties are similar to those in §H.1, except that the line integral is now replaced by the surface integral. We introduce the constants

$$\gamma_{pq} = \frac{1}{A} \iint_D z^p\,\bar{z}^q\,dx\,dy, \tag{H.32}$$

and, analogous to (H.4), define the determinants Δ_n by

$$\Delta_0 = 1, \quad \Delta_n = \begin{vmatrix} \gamma_{00} & \gamma_{10} & \cdots & \gamma_{n0} \\ \gamma_{01} & \gamma_{11} & \cdots & \gamma_{n1} \\ \cdots & \cdots & \cdots & \cdots \\ \gamma_{0n} & \gamma_{1n} & \cdots & \gamma_{nn} \end{vmatrix}. \tag{H.33}$$

Then the polynomials

$$\Pi_n(z) = \frac{1}{\sqrt{\Delta_{n-1}\,\Delta_n}} \begin{vmatrix} \gamma_{00} & \gamma_{10} & \cdots & \gamma_{n0} \\ \gamma_{01} & \gamma_{11} & \cdots & \gamma_{n1} \\ \cdots & \cdots & \cdots & \cdots \\ \gamma_{0n} & \gamma_{1n} & \cdots & \gamma_{nn} \\ 1 & z & \cdots & z^n \end{vmatrix} \tag{H.34}$$

are orthogonal in the region D, and form a complete closed system. Any function $f(z)$ analytic on D such that the integral

$$\iint_D |f(z)|^2\,dx\,dy < +\infty$$

can be uniquely expanded in a series involving the polynomials $\Pi_n(z)$, i.e.,

$$f(z) = \sum_{n=0}^{\infty} \alpha_n\,\Pi_n(z), \tag{H.35}$$

where the coefficients α_n are determined by

$$\alpha_n = \frac{1}{A} \iint_D f(z)\,\overline{\Pi_n(z)}\,dx\,dy. \tag{H.36}$$

As an application, note that, in view of §4.1, the function $F(z) = f_0(z) \in \mathcal{K}^1$ maps the region D conformally onto the disk $|w| < R$ such that a point $a \in D$ goes into $w = 0$ and $f_0'(a) = 1$. Out of all analytic functions $F(z) \in \mathcal{K}^1$ with $F(a) = 0$ and $F'(a) = 1$, the function $f_0(z)$ gives the minimum for the integral (4.1.2). Analogous to Theorem H.1 the function $f_0(z)$ can be represented in a series expansion involving the polynomials $\Pi_j(z)$ as

$$f_0(z) = \frac{1}{K(a,a)} \int_a^z K(z,a)\,dz, \tag{H.37}$$

where, as in (4.1.5),

$$K(z,a) = \sum_{j=0}^{\infty} \overline{\Pi_j(a)}\,\Pi_j(z). \tag{H.38}$$

Then the area of the circle $|w| = R$ is given by

$$\begin{aligned} \pi R^2 &= \iint_D |F'(z)|^2\,dx\,dy \\ &= \frac{A}{K^2(a,a)} \sum_{j=0}^{\infty} \overline{\Pi_j(a)}\,\Pi_j(z) = \frac{A}{K(a,a)}, \end{aligned} \tag{H.39}$$

whence

$$R = \sqrt{\frac{A}{\pi\,K(a,a)}}, \tag{H.40}$$

and the mapping function is determined by

$$f_0(z) = \sqrt{\frac{A}{\pi\,K(a,a)}} \int_a^z K(a,a)\,dz. \tag{H.41}$$

MATHEMATICA CODE: CS442.NB:

```
A[j_, k_] := Integrate[ Integrate[ (x+ I*y)^j * (x-I*y)^k,
{x, -1,1}], {y, -1,1}];
MatA = Table[A[j,k], {j,1,8}, {k,1,8}];
MatrixForm[MatA];
B=Table[A[j,0], {j,1,8}];
c=LinearSolve[MatA,- B];
(* These are the coefficients of phi_8[z] *)
phi8[z_] := 1 + c . Table[z^i, {i, 8}];
phi8[z];
(* The mapping function is given by f'[z]=phi8[z] *)
f[z_] := Integrate[phi8[t], {t, 0,z}];
f[z]
```

REFERENCES USED: Gaier [1964], Goluzin [1957; 1969], Kantorovich and Krylov [1958], Kythe [1998], Nehari [1952]

I

Special Finite Elements

Some special triangular and rectangular elements lead to different stiffness matrices and force vectors. The following three cases are valid for the Laplacian $-\nabla^2$ on a right-angled triangular element with sides a and b, $a \geq b$, and the location of the local nodes 1, 2, and 3, such that the local node 1 is at the origin.

1. For a right-angled linear triangular element $\Omega^{(e)}$ with base a and altitude b, if the local nodes 1, 2 and 3 are at $(0,0)$, $(a,0)$ and (a,b), respectively, and the local node 2 is at the right angle (see Fig. I.1a), then

$$\mathbf{K}^{(e)} = \frac{1}{2ab} \begin{bmatrix} b^2 & -b^2 & 0 \\ -b^2 & a^2+b^2 & -a^2 \\ 0 & a^2 & a^2 \end{bmatrix}, \tag{I.1}$$

and

$$\mathbf{f}^{(e)} = \frac{f_0 ab}{6} \begin{Bmatrix} 1 \\ 1 \\ 1 \end{Bmatrix}. \tag{I.2}$$

2. For a right-angled linear triangular element $\Omega^{(e)}$ with base a and altitude b, if the local node 1 is at the right angle and the nodes 2 and 3 are numbered counter-clockwise (see Fig. I.1b), then

$$\mathbf{K}^{(e)} = \frac{1}{2ab} \begin{bmatrix} a^2+b^2 & -b^2 & -a^2 \\ -b^2 & b^2 & 0 \\ -a^2 & 0 & a^2 \end{bmatrix}, \tag{I.3}$$

and the force vector $\mathbf{f}^{(e)}$ is the same as in (I.2).

3. For a right-angled linear triangular element $\Omega^{(e)}$ with base a and altitude b, if the local node 3 is at the right angle and the nodes 1 and 2 are numbered counter-clockwise (see Fig. I.1c), then

$$\mathbf{K}^{(e)} = \frac{1}{2ab} \begin{bmatrix} a^2 & 0 & -a^2 \\ 0 & b^2 & -b^2 \\ -a^2 & -b^2 & a^2+b^2 \end{bmatrix}, \tag{I.4}$$

and the force vector $\mathbf{f}^{(e)}$ is the same as in (I.2).

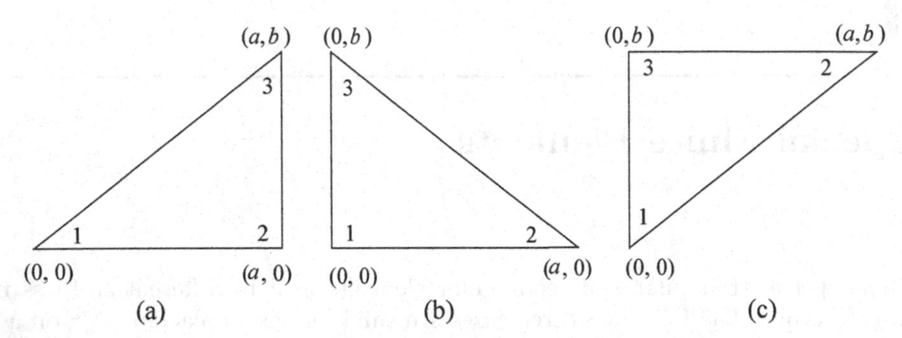

Fig. I.1. Three Cases of a Right Triangle.

4. For a 4-node bilinear square element $\Omega^{(e)}$ of side a, the stiffness matrix and the force vector for the Laplacian $-\nabla^2$ are given by

$$\mathbf{K}^{(e)} = \frac{1}{6} \begin{bmatrix} 4 & -1 & -2 & -1 \\ -1 & 4 & -1 & -2 \\ -2 & -1 & 4 & -1 \\ -1 & -2 & -1 & 4 \end{bmatrix}, \quad \mathbf{f}^{(e)} = \frac{f^{(e)}a^2}{4} \begin{Bmatrix} 1 \\ 1 \\ 1 \\ 1 \end{Bmatrix}. \tag{I.5}$$

5. For heat transfer problems with convective conductance β, the following elements are mostly used, with the respective stiffness matrices.

(5a) For a 3-node linear triangular element:

$$\begin{aligned} \mathbf{S}^{(e)} &= \frac{\beta_{12}^{(e)} l_{12}^{(e)}}{6} \begin{bmatrix} 2 & 1 & 0 \\ 1 & 2 & 0 \\ 0 & 0 & 0 \end{bmatrix} \\ &\quad + \frac{\beta_{23}^{(e)} l_{23}^{(e)}}{6} \begin{bmatrix} 0 & 0 & 0 \\ 0 & 2 & 1 \\ 0 & 1 & 2 \end{bmatrix} + \frac{\beta_{31}^{(e)} l_{31}^{(e)}}{6} \begin{bmatrix} 2 & 0 & 1 \\ 0 & 0 & 0 \\ 1 & 0 & 2 \end{bmatrix}, \end{aligned} \tag{I.6}$$

$$\begin{aligned} \mathbf{P}^{(e)} &= \frac{\beta_{12}^{(e)} T_\infty^{12} l_{12}^{(e)}}{2} \begin{Bmatrix} 1 \\ 1 \\ 0 \end{Bmatrix} \\ &\quad + \frac{\beta_{23}^{(e)} T_\infty^{23} l_{23}^{(e)}}{2} \begin{Bmatrix} 0 \\ 1 \\ 1 \end{Bmatrix} + \frac{\beta_{31}^{(e)} T_\infty^{31} l_{31}^{(e)}}{2} \begin{Bmatrix} 1 \\ 0 \\ 1 \end{Bmatrix}. \end{aligned} \tag{I.7}$$

(5b) For a 4-node bilinear rectangular element:

$$\mathbf{S}^{(e)} = \frac{\beta_{12}^{(e)} l_{12}^{(e)}}{6} \begin{bmatrix} 2 & 1 & 0 & 0 \\ 1 & 2 & 0 & 0 \\ 0 & 0 & 0 & 0 \\ 0 & 0 & 0 & 0 \end{bmatrix} + \frac{\beta_{23}^{(e)} l_{23}^{(e)}}{6} \begin{bmatrix} 0 & 0 & 0 & 0 \\ 0 & 2 & 1 & 0 \\ 0 & 1 & 2 & 0 \\ 0 & 0 & 0 & 0 \end{bmatrix}$$

$$+\frac{\beta_{34}^{(e)} l_{34}^{(e)}}{6}\begin{bmatrix}0&0&0&0\\0&0&0&0\\0&0&2&1\\0&0&1&2\end{bmatrix}+\frac{\beta_{41}^{(e)} l_{41}^{(e)}}{6}\begin{bmatrix}2&0&0&1\\0&0&0&0\\0&0&0&0\\1&0&0&2\end{bmatrix}, \tag{I.8}$$

$$\mathbf{P}^{(e)}=\frac{\beta_{12}^{(e)}T_\infty^{12}l_{12}^{(e)}}{2}\begin{Bmatrix}1\\1\\0\\0\end{Bmatrix}+\frac{\beta_{23}^{(e)}T_\infty^{23}l_{23}^{(e)}}{2}\begin{Bmatrix}0\\1\\1\\0\end{Bmatrix}$$

$$+\frac{\beta_{34}^{(e)}T_\infty^{34}l_{34}^{(e)}}{2}\begin{Bmatrix}0\\0\\1\\1\end{Bmatrix}+\frac{\beta_{41}^{(e)}T_\infty^{41}l_{41}^{(e)}}{2}\begin{Bmatrix}1\\0\\0\\1\end{Bmatrix}. \tag{I.9}$$

(5c) For a 6-node quadratic triangular element:

$$\mathbf{S}^{(e)}=\frac{\beta_{13}^{(e)} l_{13}^{(e)}}{30}\begin{bmatrix}4&2&-1&0&0&0\\1&16&2&0&0&0\\-1&2&4&0&0&0\\0&0&0&0&0&0\\0&0&0&0&0&0\\0&0&0&0&0&0\end{bmatrix}+\frac{\beta_{35}^{(e)} l_{35}^{(c)}}{30}\begin{bmatrix}0&0&0&0&0&0\\0&0&0&0&0&0\\0&0&4&2&-1&0\\0&0&2&16&2&0\\0&0&-1&2&4&0\\0&0&0&0&0&0\end{bmatrix}$$

$$+\frac{\beta_{51}^{(e)} l_{51}^{(e)}}{30}\begin{bmatrix}4&0&0&0&-1&2\\0&0&0&0&0&0\\0&0&0&0&0&0\\0&0&0&0&0&0\\-1&0&0&0&4&2\\2&0&0&0&2&16\end{bmatrix}, \tag{I.10}$$

$$\mathbf{P}^{(e)}=\frac{\beta_{13}^{(e)}T_\infty^{13}l_{13}^{(o)}}{6}\begin{Bmatrix}1\\4\\1\\0\\0\\0\end{Bmatrix}+\frac{\beta_{35}^{(e)}T_\infty^{35}l_{35}^{(e)}}{6}\begin{Bmatrix}0\\0\\1\\4\\1\\0\end{Bmatrix}+\frac{\beta_{51}^{(e)}T_\infty^{51}l_{51}^{(e)}}{6}\begin{Bmatrix}1\\0\\0\\0\\1\\4\end{Bmatrix}. \tag{I.11}$$

REFERENCES USED: Kythe and Wei [2004].

J

Schwarz Formula

We will consider specially the case when D is a circle with center at the origin and radius R. In this case we carry out the mapping onto the unit disk by

$$w = \frac{z}{R}, \tag{J.1}$$

and Green's function $G(z, z_0) = \log \frac{w - w_0}{1 - w\bar{w}_0}$ becomes

$$G(z, z_0) = \log \frac{R(z - z_0)}{R^2 - z\bar{z}_0}. \tag{J.2}$$

Then

$$d\,G(z, z_0) = \Big(\frac{1}{z - z_0} + \frac{\bar{z}_0}{R^2 - z\bar{z}_0}\Big)\, dz. \tag{J.3}$$

Since $|z|^2 = z\bar{z} = R^2$ on the boundary of the circle and $dz = iz\, d\theta$, so

$$d\,G(z, z_0) = i\,\Big(\frac{z}{z - z_0} + \frac{\bar{z}}{\bar{z} - \bar{z}_0} - 1\Big)\, d\theta. \tag{J.4}$$

Setting $z = R\,e^{i\theta}$ and $z_0 = \rho\, e^{i\phi}$, we find that

$$d\,G(z, z_0) = i\,\frac{R^2 - \rho^2}{R^2 + \rho^2 - 2R\rho\cos(\theta - \phi)}\, d\theta. \tag{J.5}$$

If we substitute (J.5) in (A.10), we obtain the Poisson integral

$$u(\rho\, e^{i\phi}) = \frac{1}{2\pi}\int_0^{2\pi} u(R\, e^{i\phi})\frac{R^2 - \rho^2}{R^2 + \rho^2 - 2R\rho\cos(\theta - \phi)} d\theta. \tag{J.6}$$

By a similar integral representation we can determine the harmonic function $v(z)$ which is conjugate to $u(z)$. In view of (2.5.2)

$$v(z) - v(0) = \int_0^z \frac{\partial u}{\partial n}\, ds. \tag{J.7}$$

When we apply this operation on (J.6) and follow through the corresponding integrations and differentiations, we get

$$v(\rho e^{i\phi}) - v(0) = \frac{1}{2\pi} \int_0^{2\pi} u(R\, e^{i\phi}) \int_0^{\rho e^{i\phi}} \frac{\partial}{\partial u} \left[\frac{R^2 + \rho^2}{R^2 + \rho^2 - 2R\rho \cos(\theta - \phi)} \right] ds\, d\theta, \quad \text{(J.8)}$$

where the inner integral is taken on an arbitrary path that lies entirely in the interior of the circle. Note that

$$\begin{aligned} \frac{R^2 - \rho^2}{R^2 + \rho^2 - 2R\rho \cos(\theta - \phi)} &= \frac{z}{z - z_0} + \frac{\bar{z}}{\bar{z} - \bar{z_0}} - 1 \\ &= \Re \left\{ \frac{2z}{z - z_0} - 1 \right\} = \Re \left\{ \frac{z + z_0}{z - z_0} \right\}. \end{aligned} \quad \text{(J.9)}$$

Thus,

$$\Im \left\{ \frac{z + z_0}{z - z_0} \right\} = \frac{-2R\rho \sin(\theta - \phi)}{R^2 + \rho^2 - 2R\rho \cos(\theta - \phi)}, \quad \text{(J.10)}$$

and hence

$$v(\rho e^{i\phi}) = v(0) - \frac{1}{2\pi} \int_0^{2\pi} u(R\, e^{i\theta}) \frac{2R \sin(\theta - \phi)}{R^2 + \rho^2 - 2R\rho \cos(\theta - \phi)} d\theta. \quad \text{(J.11)}$$

If we combine (J.6) and (J.11), we obtain the Schwarz formula:

$$w(\rho e^{i\phi}) = i\, v(0) + \frac{1}{2\pi} \int_0^{2\pi} u(R\, e^{i\phi}) \frac{Re^{i\theta} + \rho e^{i\phi}}{Re^{i\theta} - \rho e^{i\phi}}\, d\theta, \quad \text{(J.12)}$$

which allows us to determine the value of a complex potential function $f(z) = u(z) + i\, v(z)$ in a circle with prescribed boundary values $u(z)$ and $v(0)$.

REFERENCES USED: Kythe [1998].

Bibliography

(Note: First author or single author is cited with last name first.)

Abbott, Ira H., and Albert E. von Doenhoff. 1959. *Theory of Wing Sections.* Dover, New York.

Abbott, M. B., and D. R. Bosco. 1990. *Computational Fluid Dynamics.* New York: Wiley.

Abel, N. H. 1824. *Mémoire sur les équation algébriques.* Christina.

Abramovici, F. 1973. The accurate calculation of Fourier integrals by the fast Fourier transform technique. *J. Comp. Phys.* 11: 28-37.

Abramowitz M., and I. A. Stegun (eds.) 1968/1972. *Handbook of Mathematical Functions.* Dover, New York.

Agur, E. E., and J. Vlachopoulos. 1981. Heat transfer to molten polymer flow in tubes *J. Appl. Polym. Sci.* 26: 765-773.

Ahlberg, E. N., and J. Walsh. 1967. *The Theory of Splines and Their Applications.* Academic Press, New York.

Ahlfors, L. V. 1930. Untersuchungen zur Theorie der konformen Abbildung und der granzen Funktionen. *Act Soc. Sci. Fenn. A*, 1-40.

———. 1935. Zur Theorie der Überlagerungsflächen. *Acta Math.* 65: 157-194.

———. 1952. Remarks on the Neumann-Poincaré integral equation. *Pacific J. Math.* 2: 271-280.

———. 1953/1966. *Complex Analysis.* New York: McGraw-Hill.

Akduman, I., and R. Kress. 2002. Acoustic response of a non-circular cylindrical enclosure using conformal mapping. *Inverse Problems.* 18: 1659-1672.

Akhiezer, N. I. 1990. *Elements of the Theory of Elliptic Functions.* AMS Translation of Mathematical Monographs. Vol. 79. Providence, RI: American Mathematical Society.

Alenicyn, Ju. E. (Ю. Е. Аленисын). 1964. Conformal mapping of a multiply connected domain onto many-sheeted canonical surfaces. *Izv. Akad. Nauk SSSR Ser. Mat.* 28: 607-644 (Russian).

———. 1965. Conformal mapping of multiply connected domains onto surfaces of several sheets with rectilinear cuts. *Izv. Akad. Nauk SSSR Ser. Mat.* 29: 887-902 (Russian).

Alexander, Greg. 1997. *NACA Airfoil Series.* Aerospaceweb.org (http://www.aerospaceweb.org/ question/ airfoils/q0041.shtml). Updated May 25, 2018.

Amir-Moez, Ali R. 1967. Conformal transformation charts used by electrical engineers.

Mathematics Magazine. 40: 268-270.

Andersen, Chr., S. E. Christiansen, O. Møller, and H. Tornehave. 1962. Conformal mapping, Chap. 3 in *Selected Numerical Methods* (C. Gram, ed.) Kopenhagen: Regnecentralen.

Anderson, David, and Scott Eberhardt. 1988. *Understanding Flight*, 2nd ed. Prentice-Hall.

Anderson, John D. 2007. *Fundamentals of Aerodynamics*, 4th ed. McGraw-Hill, New York.

———. 2016. Some reflections on the history of fluid dynamics, in R. W. Johnson's *Handbook of Fluid Dynamics*. 2nd ed. Boca Raton, FL: CRC Press.

Anderson, J. M., K. F. Barth, and D. A. Brannen. 1977. Research problems in complex analysis. *Bull. London Math. Soc.* 9: 129-162.

Antonjuk, G. K. (Г. К. Антонюк). 1958. On the covering of areas for functions regular in an annulus. *Vestnik Leningrad Univ.* 13: 45-65 (Russian).

Apostel, T. M. 1967-1969. *Calculus*. Blaisdell, Waltham, MA.

Apostolatos, Theocharis A. 2003. Hodograph: A useful geometrical tool for solving some difficult problems in dynamics. *Am. J. Physics*. 71: 261-266.

Arafeh, Samir A., and R. Schinzinger. 1978. Estimation algorithms for large scale power systems. *IEEE Trans. on Power and Apparatus*. PAS-98: 1680-1688.

Arbenz, K. 1958. *Integralgleichungen für einige Randwertprobleme für Gebiete mit Ecken* Dissertation, ETH Zurich.

Arlinger, B. G. 1975. Calculation of transonic flow around axisymmetric inlets. *AIAA J.* 13: 1614-1621.

Asmar, Nakhlé H., and Gregory C. Jones. 2002. *Applied Complex Analysis with Partial Differential Equations*. Prentice-Hall.

Atkinson, K. E. 1976. *A Survey of Numerical Methods for the Solution of Fredholm Integral Equations of the Second Kind*. SIAM, Philadelphia.

———. 1997. *The Numerical Solution of Integral Equations of the Second Kind*. Cambridge University Press.

Babinsky, Holger. November 2003. *How do wings work?* (http://www.iop.org/EJ/article/0031-9120/38/6/001/pe3_6_001.pdf) (PDF), Physics Education.

Babuska, I. 1971. Error bounds for the finite element method. *Numer. Math.* 16: 322-333.

Bairstow, L., and A. Berry. 1919. Two-dimensional solutions of Poisson's and Laplace's equations. *Proc. Royal Soc. London*, Ser. A 95: 457-475.

Baker, C. T. H. 1978. *The Numerical Treatment of Integral Equations*. Oxford: Clarendon Press.

Baker, G. R., D. I. Meiron, and S. A. Orszag. 1980. Vortex simulations of the Rayleigh-Taylor instability. *Phys. Fluids*. 23: 1485-1490.

Baker, M. 1964. *The Principles and Applications of Variational Methods*. Cambridge, MA: MIT Press.

Balanis, C. A.1989. *Advanced Engineering Electromagnetics*. New York: John Wiley.

Baltrukonis, M. Chi, and P. A. A. Laura. 1965. Axial shear vibrations of star shaped bars: Kohn-Kato bounds. *Eighth Midwestern Mech. Conf., Developments in Mech.* Vol. 4: 449-467.

Banin, A. M. (A. M. Банин). 1943. Approximate conformal transformation applied to a plane parallel flow past an arbitrary shape. *PMM* (Прикладная Математика и Мечаника, Акад. Наук СССР, Отделение Теч. Наук, Инст. Мечаники 7: 131-140 (Russian).

Banerjee, P. K., and R. Butterfield. 1981. *Boundary Element Methods in Engineering Science*. New York: McGraw-Hill.

Barnard, R. W., and K. Pearce. 1986. Rounding corners of gearlike domains and the omitted area problem. *J. Comput. Appl. Math.* 14: 217-226; also in *Numerical Conformal Mapping* (L. N. Trefethen, ed.), 217-226. North-Holland, Amsterdam.

Batchelor, G. K. 1967. *An Introduction to Fluid Dynamics.* Cambridge University Press.

Bateman, H. 1959. *Partial Differential Equations of Mathematical Physics.* Cambridge University Press.

Bazant, Martin Z., and Darren Crowdy. 2005. Conformal mapping methods for interfacial dynamics, in *Handbook of Material Modeling*, (S. Yip et al. eds.) Vol, I, Ch. 4, Art. 4.10. Springer Science and Business Media, New York.

Beardon, Alan F. 1987. On Fornberg's numerical method for conformal mapping. *American Mathematical Monthly.* 94: 48-53.

Becker, M. 1964. *The Principles and Applications of Variational Methods.* Cambridge, MA: MIT Press.

Bell, S. R. 1981. Biharmonic mappings and the $\bar{\delta}$-problem. *Ann. Math.* 114: 103-112.

Bellman, R. 1970. *Introduction to Matrix Analysis*, 2nd ed. New York: McGraw-Hill.

———, H. Kagiwada, and R. E. Kalaba. 1965. Identification of linear systems via numerical inversion of Laplace transforms. *IEEE Trans. Automatic Control.* AC-10: 111-112.

———, Kalaba, R. E., and J. A. Lockett. 1966. *Numerical Inversion of the Laplace Transform: Applications to Biology, Economics, Engineering and Physics.* American Elsevier, New York.

Berezin, I. S., and N. P. Zhidkov. (И. С. Березин и Н. П. Жидков. 1965. *Computing Methods*, Vol. 2. Addison-Wesley, Reading, MA, and Pergamon Press, Oxford, UK; translation of Методы вычислении, Физматгиз, Москва (Russian).

Berger, B. S. 1978. Transient motion of an elastic shell of revolution in an acoustic medium. *J. Appl. Mech.*. 100-1: 149-152.

Bergman, S. 1922. Über die Entwicklung der harmonischen Funktionen der Ebene und des Raumes nach Orthogonalfunktionen. *Math. Annalen.* 86: 237-271; Thesis, Berlin 1921.

———. 1923-24. Über Bestimmung der Verzweigungspunkte eines hyperelliptischen Integrals aus seinen Periodizitätsmoduln mit Anwendungen auf die Theorie des Transformators. *Math. Zeit.* 19: 8-25.

———. 1925. Über die Berechnung des magnetischen Feldes in einem Einphasen-Transformator. *ZAMM.* 5: 319-331.

———. 1947. Punch-card machine methods applied to the solution of the torsion problem. *Quart. Appl. Math.* 5: 69-81.

———. 1950. The kernel function and conformal mapping. *AMS Math. Surveys.* 5: American Mathematical Society.

———, and M. Schiffer. 1948. Kernel functions in the theory of partial differential equations of elliptic type. *Duke Math. J.* 15: 535-566.

———, and M. Schiffer. 1949. Kernel functions and conformal mapping, I, II. *Bull. Am. Math. Soc.* 55: 515.

———, and M. Schiffer. 1951. Kernel functions and conformal mapping. *Compositio Math.* 8: 205-249.

Bergström, H. 1958. An approximation of the analytic function mapping a given domain inside or outside the unit circle. *Mém. Publ. Soc. Sci. Arts Lettr. Hainaut, Volume hors Série.* 193-198.

Berrut, J.-P. 1976. *Numerische Lösung der Symmschen Integralgleichung durch Fourier-Methoden.* Master's Thesis, ETH, Zurich, 1976.

———. 1985. *Über Integralgleichungen und Fourier-Methoden zur numerischen konformen Abbildung.* Doctoral Thesis, ETH, Zurich, 1985.

———. 1986. A Fredholm integral equation of the second kind for conformal mapping.

J. Comput. Appl. Math. 14: 99-110; also in *Numerical Conformal Mapping* (L. N. Trefethen, ed.) North-Holland, Amsterdam, 99-110.

Bertin, John J., and Russel M. Cummings. 2009. in *Aerodynamics for Engineering Students* (Butterworth Heinmann, ed.) 5th ed. p. 199.

Bettess, Pete. 1981. Operation counts for boundary integral and finite element methods. *Intern. J. for Numer. Meth. Eng.* 17: 306-308.

Betz, A. 1964. *Konforme Abbildung.* Springer-Verlag, Berlin.

Bianchi, Nicola. 2005. *Electrical Machine Analysis Using Finite Elements.* Boca Raton, FL. Taylor and Francis/CRC.

Bickley, W. C. 1929. Two-dimensional potential problems concerning a single closed boundary. *Phil. Trans.* A 228: 235.

———. 1930. The effect of rotation upon the lift and moment of a Joukowski aerofoil. *Proc. Royal Soc.* A 127: 186.

———. 1934. Two-dimensional potential problem for the space outside a rectangle. *Proc. Lond. Math. Soc.*, Ser. 2. 37: 82-105.

Bieberbach, L. 1880. *Einführing in die konforme Abbidung.* English translstion: *Conformal Mapping,* by F. Steinhardt. 2010. AMS Chelsea Publishing, American Mathematical Society, RI.

———. 1914. Zur Theorie und Praxis der konformen Abbildung. *Rend. del Circolo mat. Palermo.* 38: 98-112.

———. 1916. Über die Koeffizienten derjenigen Potenzreihen, welche eine schlichte Abbildung des Einheitakreises vermitten. *Sitzungber. Preuss. Akad. Wissen. Phys-Math.* 138: 940-955.

———. 1924. Über die konforme Kreisabbildung nahezu kreisförmig Bereiche. *Sitzungsbereichte der Preuss. Akad. Wissen.* 181-188.

Binns, K. J. 1971. Numerical methods of conformal mapping. Short Note. *Proc. IEE, London.* 118: 909-910.

———, and P. J. Lawrenson. 1973. *Analysis and Computation of Electric and Magnetic Field Problems*, 2nd ed. Pergamon Press, New York.

Bird, R. B., R. Armstrong, and O. Hassager. 1976. *Dynamics of Polymeric Fluids*, Vol. 1. Fluid Mechanics. New York: Wiley, New York.

Birkoff, G., and D. Young. 1950. Numerical quadrature of analytic and harmonic functions. *J. Math. Phys.* 29: 217-221.

———, and G.-C. Rota. 1962. *Ordinary Differential Equations.* Blaisdell, Waltham, MA.

———, D. Young, and E. H. Zarantonello. 1951. Numerical methods in conformal mapping. *Am. Math. Soc., Proc. Fourth Symposium Appl. Math.* 117-140. McGraw-Hill, New York, 1953.

———, and E. H. Zarantonello. 1957. *Jets, Wakes and Cavities.* Academic Press, New York.

———, D. M. Young, and E. H. Zarantonello. 1951. Effective conformal transformation of smooth simply connected domains. *Proc. Nat. Acad. Sci. USA.* 37: 411-414.

Bisshopp, F. 1983. Numerical conformal mapping and analytic continuation. *Quart. Appl. Math.* 41: 125-142.

Bjørstad, Peter, and Eric Grusse. 1987. Conformal mapping of circular polygons. *SIAM J. Sci. Stat. Comput.* 8-1: 19-32.

Blasius, H. 1908. Grenzschichten in Flüssigkeiten mit kleiner Reibung. *Z. Math. Phys.* 56:137; 60: 397398.

Blaskett, D. R., and H. Schwerdtfeger. 1945-46. A formula for the solution of an arbitrary analytic function. *Quart. Appl. Math.* 3: 266-268.

Bloch, A. 1925. Les Théoremes de M. Valiron sur les fonctions entiéres et la théorie de

l'unifirmisation. *Ann. Fac. Sci. Univ. Toulouse*, III. 17: 1-22.
Boas, R. P. 1987. *Invitation to Complex Analysis.* Random House. New York.
Boger, D. V., A. Cabelli, and A. L. Halmos. 1975. The behavior of a power-law fluid flowing through a sudden expansion. *A.I.Ch.E. Journal.* 21: 540-549.
Bonk, Mario, and Pekka Koskela. 2002. Electrostatic imaging via conformal mapping. *American Journal of Mathematics.* 124: 1247-1287.
Borre, K. 2001. *Plane Networks and Their Applications.* Boston: Birkhäuser.
Boussinesq, J. 1867. Théorie des exeérience de M. Poiseule sur l'écoulement des liqides dans les tubes capilaires. *Comptes Rendus de l'Académie des Scs.* 65: 46-48.
———. 1868. Mémoire sur l.influence des frottements dans les mouvements réguliers des fluides. *JMPA.* 13:377-424.
Bowman, F. 1933. Notes on two-dimensional electric field problems, notes 1 and 2. *Proc. London Math. Soc.*, 2nd series. 39: 205-215.
———. 1936. Notes on two-dimensional electric field problems, note 5. *Proc. London Math. Soc.*, 2nd series. 41: 271-277.
———, T. B. A. Senior, and P. L. E. Uslenghi. 1969. *Electromagnetic and Acoustic Scattering by Simple Shapes.* Amsterdam: North-Holland Publ. Co.
Boyce, W. E., and R. C. DiPrima. 1962. *Elementary Differential Equations*, 5th ed.; 7th ed., 2001. New York: Wiley.
Bradfield, K. N. E., S. G. Hooker, and R. V. Southwell. 1937. Some Applications of Conformal Transformation to Problems in Hydrodynamics. *Proceedings of the Royal Society of London.* Series A, Mathematical and Physical Sciences. 159, No. 898: 315-346.
Brass, H. 1982. Zur numerischen Berecknung der konjugierten Funktion, in *Numerical Methods of Approximation Theory* (L. Collatz, et al., eds.), Vol. 6, 43-62. Bassel: Birkhäuser.
Brebbia, C. A. 1980. *The Boundary Element Method for Engineers.* London: Peachtree Press.
Brennen, D. A., and J. C. Clunie (eds.) 1980. *Contemporary Complex Analysis.* Academic Press, New York.
Brill, A., and M. Noether. 1894. Bericht über die Entwicklung der Theorie der analytischen Funktionen in älterer and neuerer Zeit. *Jahresbericht der Deutschen Mathematiker-Vereningung.* 3: 107-566.
Broman, Arne. 1948. Identities in the Theory of Conformal Mapping. *Proceedings of the National Academy of Sciences of the United States of America.* 34: 605-610.
Bronstein, M. 1997. *Symbolic Integration I: Transcendental Functions.* Springer-Verlag, New York.
Brown, J. 1967. *Electromagnetic Wave Theory.* URSI Symposium. Pergamon Press.
Bruch, John C. 1975. A note on conformality. *Computers and Graphics.* 1: 361-374.
———, and Roger C. Wood. 1972. Problem 70-5, Conformal Mapping of a Cross Slit Strip. *IEEE Trans. on Ed.* E-15: 73-80.
Brunner, H., and P. J. van der Houwen. 1986. *The Numerical Solution of Volterra Equations.* Amsterdam: North-Holland.
Brychkov, Y. A. 2008. *Handbook of Special Functions: Derivatives, Integrals, Series and Other Formulas.* Boca Raton, FL: CRC Press.
Burkhardt, H. 1912. *Einführung in die Theorie der analytischen Funktionen.* Leipzig.
———. 1920. *Elliptische Funktionen.* Leipzig.
Burington, R. S. 1940. On the Circles of Curvature of the Images of Circles under a Conformal Map. *American Mathematical Monthly.* 47: 362-373.
Burnside, W. 1891. On functions determined from their discontinuities, and a certain form

of boundary condition. *Proc. Royal Soc. London*, Ser. A. 22: 346-358.

Buzbee, B. L., F. W. Dorr, J. A. George, and G. H. Golub. 1971. On direct methods for solving Poisson's equations. *SIAM J. Numer. Anal.* 7: 627-656.

———, G. H. Golub, and C. W. Nielson. 1971. The direct solution of the discrete Poisson equation on irregular regions. *SIAM J. Numer. Anal.* 8: 722-736.

Carathéodory, C. 1913a. Über die gegenseitige Beziehung der Ränder bei der konformen Abbildung des Inneren einer Jordanschen Kurve auf einen Kreis. *Mathematische Annalen.* 73: 305-320.

———. 1913b. Über die Begrenzung einfach zusammenhängender Gebiete. *Mathematische Annalen.* 73: 323-370.

———. 1914. Elementarer Beweis für den Fundamentalsatz der konformen Abbildung. *Schwarz-Festschrift, J. Springer, Berlin.* 19-41.

———. 1916. *Über das Neumann-Poincarésche Problem für ein Gebiet mit Ecken.* Dissertation, Uppsala.

———. 1932. *Conformal Representation.* Cambridge Tracts in Math. and Math. Phys., No. 28, London; Cambridge University Press, Cambridge, 1969.

Carleman, T. 1922. Sur la résolution de certaines équations intégrales. *Archiv för mathematik astronomi och fysik.* 16: 181-196.

Carnaham, B., H. A. Luther, and J. O. Wilkes. 1969. *Applied Numerical Methods.* New York: Wiley.

Carrier, G. F. 1947. On a conformal mapping technique. *Quart. Appl. Math.* 5: 101-104.

———, M. Krook, and C. E. Pearson. 1966. *Functions of a Complex Variable: Theory and Technique.* McGraw-Hill, New York.

———, and C. Pearson. 1988. *Partial Differential Equations*, 2nd ed. New York: Academic Press.

Carslaw, H. S., and J. C. Jaeger. 1959. *Conduction of Heat in Solids*, 2nd ed. New York: Oxford University Press.

Carter, P. W. 1926. The magnetic field of the dynamo-elastic machine. *J. Inst. Elect. Eng.* 64: 1115.

Casarella, M. J., P. A. A. Laura, and M. Chi. 1966. On the approximate solution of flow and heat transfer through non-circular conduits with uniform wall temperature. *Brit. J. Appl. Phys.* 18: 1327-1335.

———, and N. Ferragut. 1971. On the approximate solution of flow and heat transfer through non-circular conduits with uniform wall temperature and heat generation. *Nucl. Engg. Design.* 16: 387-398.

Caylay, A. 1895. *An Elementary Treatise on Elliptic Functions.* 2nd ed. London.

Cerimele, B. J. 1970. A Uniqueness Result in Conformal Mapping. *SIAM Review.* 12: 145.

Chakravarthy, S., and D. Anderson. 1979. Numerical conformal mapping. *Math. Comp.* 33: 953-969.

Chakravorty, Pragnan. 2015. Analysis of rectangular waveguides: An intuitive approach. *IETE Journal of Education.* 55: 76-80

Challis, N. V., and D. M. Burley. 1982. A numerical method for conformal mapping. *IMA J. Numer. Anal.* 2: 169-181.

Chaplygin, S. A. (С. А. Чаплыгин). 1933. A theory of grid wing (1914), in *Collected Papers*, Vol. 2. Moscow: Gostekhizdat, Moscow (Russian).

Chaudhrey, Maqsood A., and Ronald Schinzinger. 1992. Cavitational flows and global injectivity of conformal maps. *COMPEL.* 11: 263-275.

Chebyshev, P. L. (П. Л. Чебышев). 1896. On geographic maps construction (1895), in *Full Collection of Works*, Vol. 5. St. Petersbourg: Acad. Sci. Publ. (Russian).

Chhabra, R. P., and J. F. Richardson. 1999. *Non-Newtonian Flow in the Process Industries:*

Fundamentals and Engineering Applications. Butterworth-Heinemann.

Choi, Y. M., D. J. Harris, and K. F. Tsang. 1988. Theoretical and experimental characteristics of single V-groove guide for X-band and 100 GHz operation. *IEEE Trans. Microwave Th. and Techn.* MTT-36: 715-723.

Christoffel, E. B. 1867-1868. Sul problema della temperature stazionaire e la reppresentazione di una data superficie. *Annali di Matematica,* (2) 1: 89-104.

Churchill, R. V. 1960. *Complex Variables and Applications.* New York: McGraw-Hill.

———. 1972. *Operational Methods,* 3rd ed. New York: McGraw-Hill.

———, and J. W. Brown. 1978. *Fourier Series and Boundary Value Problems,* 3rd ed. New York: McGraw-Hill.

———, and J. W. Brown. 1978. *Fourier Series and Boundary Value Problems.* New York: McGraw-Hill.

Clancy, L. J. 1975. *Aerodynamics.* Pitman Publishing Ltd., London.

———, G. H. Golub, and C. W. Nielson. 1971. On direct methods for solving Poisson's equations. *SIAM J. Numer. Anal.* 7: 627-656.

Clough, R. W. 1960. The finite element method in plane stress analysis. *Proc. 2nd ASCE Conf. Electronic Comput.,* Pittsburgh, PA, Sept. 1960, 345-378.

Cockroft, F. W. 1927. The effect of curved boundaries on the distribution of electrical stress round conductors. *J. Inst. Elect. Eng.* 66: 385.

Cohen, Harold. 2007. *Complex Analysis with Applications in Science and Engineering.* 2nd ed. New York: Springer.

Collin, Robert E. 2001. *Foundations for Microwave Engineering.* 2nd Ed. New York: John Wiley.

Connor, J. J., and C. A. Brebbia. 1973. *Finite Element Techniques for Structural Engineers.* London: Butterworths.

Cooley, J. W., and J. W. Tukey. 1965. An algorithm for the machine calculation of complex Fourier series. *Math. Comp.* 19: 297-301.

———, P. A. W. Lewis, and P. D. Welch. 1970. The fast Fourier transform algorithm: Programming considerations in the calculation of sine, cosine, and Laplace transforms. *J. Sound Vibration.* 12: 315-337.

Coon, A. R., N. I. M. Gould, and Ph. L. Toint. 2000. *Test-Region Methods.* MPS/SIAM Series on Optimization. SIAM, Philadelphia.

Copson, E. T. 1975. *Partial Differential Equations.* Cambridge University Press, London.

Courant, R. 1943. Variational methods for the solution of problems of equilibrium and vibrations. *Bull. Amer. Math. Soc.* 49: 1-23.

———. 1950. *Dirichlet's Principle, Conformal Mapping and Minimal Surfaces.* New York: Interscience.

———. 1964, 1965. *Differential and Integral Calculus,* Vol. 1, 2. New York: Interscience.

———, and D. Hilbert. 1963, 1965. *Methods of Mathematical Physics,* Vol. 1, 2. New York: Interscience.

Crocker, M. J. (ed). 1998. *Encyclopedia of Acoustics.* John Wiley, New York.

Crowdy, D. 1999. A note on viscous sintering and quadrature identities. *Eur. J. Appl. Math.* 10: 623634.

———. 2000. Hele-Shaw flows and water waves. *J. Fluid Mech.* 409: 223-242.

———. 2003. Viscous sintering of unimodal and bimodal cylindrical packings with shrinking pores. *Eur. J. Appl. Math.* 14: 421-445.

———, and J. Marshall. 2004. Constructing multiply-connected quadrature domains. *SIAM J. Appl. Math.* 64: 1334-1359.

Croom, C. C., and B. J. Holmes. 1985-04-01. *Flight evaluation of an insect contamination protection system for laminar flow wings.*

(https://ntrs.nasa.gov/search.jsp?R=19850067951).

———, P. A. W. Lewis, and P. D. Welch. 1970. The fast Fourier transform algorithm: Programming considerations in the calculation of sine, cosine, and Laplace transforms. *J. Sound Vibration.* 12: 315-337.

D'Angelo, John P. 1984. Length of ray-images under conformal maps. *American Mathematical Monthly.* 91: 413-414.

Daeppen, H. 1988. *Die Schwarz-Christoffel-Abbildung für zweifach zusfimmenhangende Gebietemit Anwendungen.* Ph.D. thesis, ETH, Zurich.

Dai, Y. H., et al. 1998. Testing different nonlinear conjugate gradient methods. *Research Report, Institute of Computational Mathematics and Scientific/Engineering Computing.* Chinese Academy of Sciences.

———, and Y. Yuan. 1999. *Nonlinear Conjugate Gradient Methods.* Shanghai Scientific and Technology Publisher.

———, and Y. Yuan. 2000. A nonlinear conjugate gradient with a strong global convergence. *SIAM J. Optimiz.* 10: 177-182.

Davies, A. J. 1980. *The Finite Element Method.* Oxford: Clarendon Press.

Davies, B. 1978. *Integral Transforms and Their Applications.* New York: Springer-Verlag.

Davis, H. F. 1963. *Fourier Series and Orthogonal Functions.* Boston: Allyn and Bacon.

Davis, P., and P. Rabinowitz. 1956. Numerical experiments in potential theory using orthonormal functions. *J. Washington Acad. Sci.* 46: 12-17.

———, and P. Rabinowitz. 1961. Advances in orthonormalizing computation, in *Advances in Computers.* (ed. Franz L. Alt), Vol. 2, 55-133.

———, and P. Rabinowitz. 1975. *Methods of Numerical Integration.* Academic Press.

Davis, R. T. 1979. Numerical methods for coordinate generation based on Schwarz-Christoffel transformations, in *A Collection of Papers*, AIAA Computational Fluid Dynamics Conference, Amer. Inst. of Aeronautics and Astronautics; Paper # 79-1463: 180-194.

Davis, R. T. 1979. Numerical methods for coordinate generation based on Schwarz-Christoffel transformations. *AIAA Paper 79-1463.* Williamsburg, VA.

Davy, N. 1944. The field between equal semi-infinite rectangular electrodes or magnetic pole-pieces. *Phil. Mag.* (7) 35: 819.

Daymond, S. D., and J. Hodgekinson. 1939. A type of aerofoil. *Quarterly J. Math.* 10: 136.

Dean, W. R. 1944. Note on the shearing motion of a fluid past a projection. *Proc. Camb. Phil. Soc.* 40: 214.

de Branges, L. 1985. A proof of the Bieberbach conjecture. *Acta Math.* 154: 137-152.

De Cicco, John. 1942. Regions and their "patterns" in conformal mapping. *National Mathematics Magazine.* 16: 275-279.

———. 1946. Geometry of scale curves in conformal maps. *American Journal of Mathematics.* 68: 137-146.

de Cristoforis, Massimo Lanza. 1991. Conformal image warping. *Transactions of the American Mathematical Society.* 323: 509-527.

de Rivas, E.K. 1972. On the use of non-uniform grids in finite difference equations. *J. Comput. Phys.* 10: 202-210.

Delillo, Thomas K. 1994a. Upper bound for distortion of capacity under conformal mapping. *SIAM Journal on Numerical Analysis.* 31: 788-812.

———. 1994b. The accuracy of numerical conformal mapping methods: A survey and results. *SIAM J. on Numer. Anal.* 31:788-812.

Delves, L. M. 1977. A fast method for the solution of Fredholm integral equation. *J. Inst. Math. Appl.* 20: 173-182.

———, and J. Walsh (eds.) 1974. *Numerical Solution of Integral Equations.* Clarendon

Press, Oxford.

———, and J. L. Mohamed. 1985. *Computational Methods for Integral Equations.* Cambridge University Press, Cambridge.

Deresiewicz, H. 1961. Thermal stress in a plate due to disturbance of uniform heat flow in a hole of general shape. *J. Appl. Mech., Trans. ASME.* 28: 147-149.

Devaney, R. L., and L. Keen. 1988. Dynamics of maps with constant Schwarzian derivative, in *Complex Analysis* (I. Laine, S. Rickman and T. Sorvali, eds.), Proc. Joensuu 1987. Lecture Notes in Math., Vol. 1351. 92-100. Berlin: Springer-Verlag.

Dias, Frederic. 1986. *On the Use of Schwarz-Christoffel Transformation for the Numerical Solution of Potential Flow Problems.* Ph.D. Dissertation, University of Wisconsin, Madison.

———, Alan R. Elcrat, and Lloyd N. Tefethen. 1987. Ideal jet flow in two dimensions. Numerical Analysis Report 87-1, Dept. of Mathematics, MIT, Cambridge.

Diaz, K. P. 1987. The Szegö kernel as a singular integral kernel on a family of weakly pseudoconvex domains. *Trans. Amer. Math. Soc.* 304: 141-170.

Djordjevic, A. R., M. B. Bazdar, T. K. Sarkar, and R. F. Harrington. 1999. *LINPAR for Windows.* Boston: Artech House.

Dorr, F.W. 1970. The direct solution of the discrete Poisson equation on a rectangle. *SIAM Review.* 12: 248-263.

Douglas, J. 1931. Solution of the problem of Plateau. *Trans. Am. Math. Soc.* 33: 263-321.

Dozier, L. B. 1984. A numerical treatment of rough surface scattering for the parabolic wave equation. *J. Acoustical Society of America.* 75: 1415-1432.

Driscoll, Tobin A. 1996. *Algorithm 756: A MATLAB Toolbox for Schwarz-Christoffel Mapping.* ACM Transactions on Mathematical Software (TOMS). 22 n.2: 168-186.

———, and Stephen A. Vavasis. 1998. Mapping drug distribution patterns in solid tumors: Toward conformal chemotherapy for local tumor control. *SIAM J. Sci. Comput.* 19: 1783-1803.

———, and Lloyd N. Trefethen. 2002. *Schwarz-Christoffel Mapping.* Cambridge Monographs on Applied and Computational Mathematics. Cambridge, UK: Cambridge University Press.

Dryfus, L. D. 1924. Über die Anwendung der konforme Abbildung zur Berechnung der Durchschlags und Überschlagserscheinungen zwischen kantigen Konstructionsteilen unter Oel. *Archiv für Elektrotechnik.* 13: 23-145.

Duff, G. D. F. 1956. *Partial Differential Equations.* University of Toronto Press.

Duffy, D. G. 1994. *Transform Methods for Solving Partial Differential Equations.* CRC Press, Boca Raton, FL.

Durrand, Emile. 1966. *Electrostatique, Tome 2: Problemes Generaux Conducteurs.* Paris: Mason.

Ebert, U., B. Meulenbrook, C. Montijn, A. Rocco, and W. Hundsdorfer. 2003. Spontaneous branching of anode-directed discharge streamers: Conformal analysis and numerical results. Defense Technical Information Center, Compilation Part Notice ADP014946. *Proc. 26th Intern. Conf. on Phenomena in Ionized Gases*, held in Greifswald, Germany on 15-20 July 2003. vol. 4: 25-27.

Elcrat, A. R. 1982. Separated flow past a plate with spoiler. *SIAM J. Math. Anal.* 13: 632-639.

———, and L. N. Trefethen. 1986. Classical free-streamline flow over a polygonal obstacle. *J. Comput. Appl. Math.* 14: 251-265; also in *Numerical Conformal Mapping* (L. N. Trefethen, ed.) North-Holland, Amsterdam, 1986: 251-265.

Ellacott, S. W. 1978. A technique for approximate conformal mapping, in *Multivariate Approximation* (D. Handscomb, ed.) 301-314. London: Academic Press.

———. 1979. On the approximate conformal mapping of multiply connected domains. *Numer. Math.* 33: 437-446.

Embree, Mark, and Lloyd N. Trefethen. 1999. Numerical conformal mapping using cross-ratios and Delaunay triangulation. *SIAM Rev.* 41: 745-761.

Emerson, D. T. 1997. The work of Jagdis Chandra Bose: 100 years of MM-wave research *IEEE Trans. on Microwave Theory and Research.* 45: 2267-2273.

England, A. H. 1971. *Complex Variable Methods in Elasticity.* Wiley, London.

Epstein, B. 1948. A method for the solution of the Dirichlet problem for certain types of domains. *Quart. Appl. Math.* 6: 301-317.

———. 1962. *Partial Differential Equations.* McGraw-Hill, New York.

Erdélyi, A. 1943. Inversion formulae for the Laplace transformation. *Phil. Mag.* 34: 533-537.

———. 1943. Note on an inversion formula for the Laplace transformation. *J. London Math. Soc.* 18: 72-77.

———, W. Magnus, F. Oberhettinger, and F. G. Tricomi. 1954. *Tables of Integral Transforms,* Vol. 1. New York: McGraw-Hill.

Éĭdus, D. M. (Д. М. Éйдус). 1966. On some boundary-value problems in infinite regions. *Amer. Math. Soc. Transl,* (2). 53: 139-166; translation of Некоторие краваые задачи в бесконечных областях, Изв. Акад. Наук СССР, Ser. Мат. 27 (1963), 1055-1080. (Russian).

Erdogan, F., and G. D. Gupta. 1972. On the numerical solution of singular integral equations. *Quart. Appl. Math.* 30: 525.

———, G. D. Gupta, and T. S. Cook. 1973. Numerical solution of singular integral equations, in *Methods of Analysis and Solution of Crack Problems* (G. C. Sih, ed.), Mechanics of Fracture, Vol. 1: 368-425. Noordhoff, Leyden.

Erkama, T. 1988. Rational Riemann maps, in *Complex Analysis* (I. Laine, S. Rickman and T. Sorvali, eds.) *Proc. Joensuu 1987,* Lecture Notes in Math # 1351, 101-109. Berlin: Springer-Verlag.

Euler, L. 1775. On representations of a spherical surface on the plane, in *Collected Works,* Series I, Vol. 28, 248-275.

Evans, D. J. 1974. *Software for Numerical Methods.* New York: Academic Press.

Fahy, F. J. 1989. *Sound Intensity.* 2nd ed. E. & F. N., Spon.

Farlow, S. J. 1962. *Partial Differential Equations for Scientists and Engineers.* Wiley, New York.

Fastook, J. L. 1993. The finite-element method for solving conservation equations in glaciology. *Computational Science and Engineering.* 1: 55-67.

Ferrand, J. 1945. Sur la déformation analytique d'un domaine. *Comptes Rendus de l'Acad. de Scs. de Paris.* 221: 132-134.

Ferrari, C. 1933. Sulla transformazione conforme di duo cherchi in due profili alari. *Mem. della Accad. delle Sci, di Torino* (22) 67, no. 2.

Fitzgerald, C. H. 1985. The Bieberbach Conjecture: Retrospective. *Notices of the American Mathematical Society,* 2-5.

———, and Ch. Pommerenke. 1985. The de Branges theorem on univalent functions. *Trans. Amer. Math. Soc.* 290: 683-690.

Flachenecker, G, and K. Lange. 1967. Conformal transformation applied to relative loads in waveguides, in J. Brown, *Electromagnetic Wave Theory,* URSI Symposium, Pergamon Press.

Fletcher, C. A. J. 1988. *Computational Techniques for Fluid Dynamics.* Vol. II, Berlin: Springer-Verlag.

Fletcher, R., and M. J. D. Powell. 1963. A rapidly convergent descent method for minimization. *Comput. J.* 6: 163-168.

Florence, A. L., and J. N. Goodier. 1960. Thermal stress due to disturbance of uniform heat flow by an insulated ovaloid hole. *J. Appl. Mech., Trans. ASME.* 27: 635-639.

Flores, A. F., J. C. Gottifredi, G.V. Morales and O. D. Quiroga. 1991. Heat transfer to power-Law fluids flowing in tubes and flat ducts with viscous heat generation. *Chemical Engineering Science.* 46: 1385-1392.

Floryan, J. M. 1985. Conformal-mapping-based coordinate generation method for channel flows. *J. Comput. Phys.* 58: 229-245.

———, and C. Zemach. 1987. Schwarz-Christoffel mappings: A general approach. *J. Comput. Phys.* 72: 347-371.

Fock, V. 1929. Über die konforme Abbildung eines Kreisvierecks mit verschwindenden Winkeln. *J. Reine angew. Math.* 161: 137-151.

Fornberg, B. 1980. A numerical method for conformal mapping. *SIAM J. Sci. Comput.* 1: 386-400.

———. 1984. A numerical method for conformal mapping of doubly connected regions. *SIAM J. Sci. Stat. Comp.* 5: 771-783.

Forray, M. J. 1972. *Variational Calculus in Science and Engineering.* Academic Press, New York.

Forsyth, A. R. 1918. *Theory of Functions of a Complex Variable.* 3rd ed. Cambridge.

Foster, K., and R. Anderson. 1974. Transmission line properties by conformal mapping. *Proc. IEE,* 121-5: 337-339.

Fox, E. N., and J. McNamee. 1948. The two-dimensional potential problem of seepage into a cofferdam. *Philosophical Mag.* (7). 39: 165-203.

Frankel, Sidney. 1942. Characteristic impedance of parallel wires in rectangular troughs. *Proc. IRE.* 30: 182-190.

———. 1977. *Multiconductor Transmission Line Analysis.* Artech House.

Frances, G., and N. Hanges. 1995. Explicit formulas for the Szegö kernel on certain weakly pseudoconvex domains. *Proc. Amer. Math. Soc.* 123: 3161-3168.

Franklin, P. 1944. *Methods of Advanced Calculus.* McGraw-Hill, New York.

Frazer, R. A. 1926. On the motion of circular cylinders in a viscous fluid. *Phil. Trans.* 225: 93.

Friedlander, F. G. 1982. *An Introduction to the Theory of Distributions.* Cambridge: Cambridge Univ. Press.

Freiberg, M. S. 1951. *A new method for the effective determination of conformal maps.* Ph. D. Thesis, Univ. of Minnesota, 1951.

Friedrichs, K. O. 1962. A finite difference scheme for the Neumann and the Dirichlet problems. *NYO-9760.* Courant Institute of Mathematical Science, New York University, New York.

Froberg, C. E. 1969. *Introduction to Numerical Analysis.* Reading, MA: Addison-Wesley.

Fuller, Baden. 1969. *Microwaves.* New York: Pergamon Press.

Fulks, W. 1993. *Complex Variables: An Introduction.* New York; Marcel Dekker, Inc.

Gaier, D. 1964. *Konstruktive Methoden der konformen Abbildung.* Springer Tracts in Natural Philosophy, Vol. 3. Springer-Verlag, Berlin.

———. 1976. Integralgleichung erster Art und konforme Abbildung. *Math. Zeit.* 147:113-129.

———. 1980. *Vorlesungen über Approximation in Komplexen.* Birkhäuser Verlag, Basel.

———. 1983. Numerical aspects in conformal mapping, in *Computational Aspects of Complex Analysis.* Reidel, Dordrecht, Boston 51-78.

———. 1983. Numerical methods in conformal mapping, in *Computational Aspects of Complex Analysis,* H. Werner et al. (eds), 51-78. D. Reidel Publ.

———, and O. Hübner. 1976. Schnelle Auswertung von Ax bei Matrizen A zyklischer Bauart, Toeplitz- und Hankel-Matrizen. *Mitt. Math. Sem. Giessen.* 121: 27-38.

———, and N. Papamichael, 1987. On the comparison of two numerical methods for conformal mapping. *IMA J. Numer. Anal.* 71: 261-282.

Gakhov, F. D. (Ф. Д. Гахов). 1937. On the Riemann boundary problem. *Matem. Sbornik.* 2(44): 165-170.

Gandy, R. W. G., and R. V. Southwell. 1940. Relaxation methods applied to engineering problems. V. Conformal transformation of a region in plane space. *Phil. Trans. Royal Soc., London*, Ser. A , 238: 453-475.

Garlick, A. R. 1983. The use of distorting grids and flux splitting to model axisymmetric adiabatic explosions. *J. Comp. Phys.* 52: 427-447.

Garrick, I. E. 1936. Potential flow about arbitrary biplane wing sections. *Report 542, NACA*, 1936.

———. 1949. Conformal mapping in aerodynamics, with emphasis on the method of successive conjugates. *Symposium on the Construction and Applications of Conformal Maps, National Bureau of Standards, Appl. Math. Ser.* 18: 137-147.

———. 1952. Conformal mapping in aerodynamics, with emphasis on the method of successive conjugates, in *Construction and Applications of Conformal Maps.* Proc. of a Symposium held at the UCLA in 1949; National Bureau of Standards, Applied Math. Series, 137-147.

Gauss, C. F. 1822. Allgemeine Auflösung der Aufgabe die Thiele einer gegebener Fläche so abzubilden, daßdie Abbildung dem abgebildeten in den kleinsten Thielen ähnlich wird. *Werke*, Vol. IV: 189-216.

———. 1827. Disquisitiones generales circa superficies, in *Collected Works*, Vol. IV, 219-258. Original and a translation in P. Dombroski, *Astérisque*, 62. Math. de France, Paris 1979.

Gautschi, W. 1977. The condition of orthogonal polynomials. *Math. Comp.* 26: 923-924.

———. 1978. Questions of numerical condition related to polynomials, in *Recent Advances in Numerical Analysis* (C. de Boor and G. Golub, eds.) Academic Press, New York.

———. 1979. Condition of polynomials in power form. *Math. Comp.* 33: 343-352.

Gehring, F. W., and W. K. Hayman. 1963. An inequality in the theory of conformal mapping. *J. Math Pures Appl.*l 41: 353-361.

Gelfand, I. M., and G. E. Shilov. 1964. *Generalized Functions and Operations*, Vol. I (Translation from Russian). Academic Press, New York.

Gerabedian, P. R. 1964. *Partial Differential Equations.* Wiley, New York.

———, and M. M. Schiffer. 1949. Identities in the theory of conformal mapping. *Trans. Am. Math. Soc.* 65: 187-238.

———, E. McLeod, Jr., and Martin Vitousek. 1954. Studies in the Conformal Mapping of Riemann Surfaces. *American Mathematical Monthly.* Part 2: Proceedings of the Symposium on Special Topics in Applied Mathematics. (Aug.- Sep., 1954). 61: 8-10.

Gershgorin, S. I. (С. И. Гершгорин). 1933. On the conformal mapping of a simply connected region onto a disc. *Matem. Sbornik.* 40: 48-58 (Russian).

Gibbs, W. J. 1958. *Conformal Transformations in Electrical Engineering.* London: Chapman and Hall.

Gilbarg, D. 1949. A generalization of the Schwarz-Christoffel transformation. *Proc. Nat. Acad. Sci.* 35: 609-611.

———. 1960. Jets and Cavities, in *Handbuch der Physik*, Vol 9. Springer-Verlag, Berlin, 311-445.

Gladshteyn, I. S., and I. M. Ryzhik. 2007. *Tables of Integrals, Series and Products.* (Alan Jeffrey and Daniel Zwillinger, eds.), 7th ed. New York: Academic Press.

Glauert, H. 1924. A method of calculating the characteristics of a tapered wing. *Report and Memoranda of the Aeronaut. Res. Comm.* # 824.

———. 1924. A theory of thin aerofoils. *Report and Memoranda of the Aeronaut. Res. Comm.* # 910.

———. 1929. The force and moment of an oscillating aerofoil. *Report and Memoranda of the Aeronaut. Res. Comm.* # 1242.

———. 1929/1948. *The Elements of Aerofoil and Airscrew Theory.* Cambridge University Press, London.

Golub, G., and J. Welsch. 1969. Calculation of Gaussian quadrature rules. *Math. Comp.* 23: 221-230.

Goluzin, G. M. (Г. М. Голузин). 1934. The solution of fundamental plane problems of mathematical physics in the case of Laplace's equation and multiply connected regions bounded by circles. *Matem. Sbornik.* 41: 246-276 (Russian).

———. 1937. On conformal mapping of doubly connected regions bounded by rectilinear and circular polygons, in *Conformal Mapping of Simply and Doubly Connected Regions.* ONTI, Moscow, 90-97 (Russian).

———. 1937. Conformal mapping of multiply connected regions on a slit plane by a method of functional equations, in *Conformal Mapping of Simply and Doubly Connected Regions.* ONTI, Moscow, 98-110 (Russian).

———. 1939. Iterationsprozesse für konforme Abbildungen mehrfach zusammenhängender Bereiche. *Matem. Sbornik*, New Series, 6 (41): 377-382 (Russian).

———. 1957. *Geometrische Funktionentheorie.* Deutscher Verlag der Wissenschaften, Berlin; German translation of Геометрическая Теория Функций Комплексного Переменного, Москва-Ленинград, 1952 (Russian).

———. 1969. *Geometric Theory of Functions of a Complex Variable.* Transl. Math. Monographs Vol. 26, American Mathematical Society, Providence, RI; English translation of Геометрическая Теория Функций Комплексного Переменного, Наука, Москва, 1966 (Russian).

Göhre, W. 1930. Das elektrostatische Field zweier Kondensatorformen. *ZAMM,* 10: 547.

Goodman, A. W. 1950. On the Schwarz-Christoffel transformation and p-valent functions. *Trans, Am. Math. Soc.* 68: 204-223.

———. 1960. Conformal mapping onto certain curvilinear polygons. *Univ. Nac. Tucuman.* A 13: 20-26.

Goursat, E. 1884. Démostration du théorème de Cauchy. *Acta Math.* 4: 197-200.

Grafarend, Erik W., and Friedrich W. Krumm. 2006. *Map Projections.* Berlin: Springer-Verlag.

Gragg, W. B., and R. A. Tapia. 1974. Optimal error bounds for the Newton-Kantorovich theorem. *SIAM J. Num. Anal.* 11: 10-13.

Gram, C. (ed.) 1962. *Selected Numerical Methods.* Regnecentralen, Kopenhagen.

Gray, J. 1994. On the history of the Riemann mapping theorem. *Rendiconti del Circolo Matematico di Palermo.* Serie II, Supplemento. 34: 47-94.

Gredshteyn, I. S., and I. W. Ryzhik. 1965. *Tables of Integrals, Series and Products.* Academic Press, New York.

Green, A. E., and W. Zerna. 1968. *Theoretical Elasticity.* Oxford: The Clarendon Press.

Green, S. L. 1953. *The Theory and Use of the Complex Variable.* Isaac Pitman, London; first ed. 1939.

Greenberg, M. D. 1971. *Application of Green's Functions in Science and Engineering.* Prentice-Hall, Englewood Cliffs, NJ.

Greenhill, G. 1910. Theory of a stream-line past a plane barrier. *R. & H.* 19.

Greiner, P., and E. Stein. 1978. On the solvability of some differential operators. *Proc.*

Internat. Conf., Cortona, Italy, 1976-1977. pp. 106-165.

Griffiths, David J. 1995. *Introduction to Quantum Mechanics.* Upper Saddle River, NJ: Prentice Hall.

Gronwall, T. H. 1914. Some remarks on conformal representation. *Ann. Math.* 16: 72-76.

Grötzsch, H. 1931. Zur konformen Abbildung mehrfach zusammenhängender, schlichter Bereiche (Iterationsverfahren). *Berichte Verhand. Sächsichen Akad. Wissen. zu Leipzig, Math.-Phys. Klasse.* 83: 67-76.

Grunsky, H. 1932. Neue Abschätzungen zur konformen Abbild ein- und mehrfach zusammenhängender Berichte. *Schr. Math. Seminars Inst. Angew. Math. Univ. Berlin.* 1: 93-140.

Gu, Xianfeng David, and Shing-Tung Yau. 2008. *Computational Conformal Geometry.* Advanced Lectures in Mathematics series, Vol. 3. Somerville, MA: International Press.

Gu, Y. Wang, T. F. Chan, P. M. Thompson, and S.-T. Yau. 2004. Genus zero surface conformal mapping and its application to brain surface mapping. *IEEE Trans. on Medical Imaging.* 23: 949-958.

Gupta, K. C., R. Garg, L. Bahl, and P. Bhartia. 1992. *Microwave and Slotlines.* New York: Addison-Wesley.

Gurevich, M. I. (М. И. Гуревич). 1965. *Theory of Jets in Ideal Fluids.* Academic Press, New York; translation of Теория Стрие Идал'ное Жидкости, Госидарственное Издалел'ство Физ.-Мат. Лит., Москва, 1961 (Russian).

Gutknecht, M. H. 1977. Existence of a solution to the discrete Theodorsen equation for conformal mapping. *Math. Comp.* 31: 478-480.

———. 1979. Fast algorithms for the conjugate periodic function. *Computing.* 22: 79-91.

———. 1981. Solving Theodorsen's integral equation for conformal maps with the fast Fourier transform and various nonlinear iterative methods. *Num. Math.* 36: 405-429.

———. 1983. On the computation of the conjugate trigonometric rational function and on a related splitting problem. *SIAM J. Numer. Anal.* 20: 1198-1205.

———. 1983. Numerical experiments on solving Theodorsen's integral equation for conformal maps with the fast Fourier transform and various nonlinear iterative methods. *SIAM J. Sci. Stat. Comput.* 4: 1-30.

———. 1986. Numerical conformal mapping methods based on function conjugation. *J. Comput. Appl. Math.* 14: 31-77; also in *Numerical Conformal Mapping* (L. N. Trefethen, ed.), North-Holland, Amsterdam, 1986, 31-77.

Gysen, B., E. Ilhan, K. Meesen, J. Paulides, and E. Lomonova. 2010. Modeling of flux switching permanent magnet machines with Fourier analysis. *Magnetics IEEE Trans. on 2010.* 46():1599-1502.

Haberman, R. 1987. *Elementary Applied Partial Differential Equations,* 2nd ed. Prentice-Hall, Englewood Cliffs, NJ.

Hackbusch, W. 1985. *Multigrid Methods and Applications.* Berlin: Springer-Verlag.

Hadamard, J. 1908. Mémoire sur le problème d'analyse relatif à l'équalibre de plaques élastiques encastrées. *Mémoires Acad. Sci.* 33, No. 4.

Hageman, L. A., and D. M. Young. 1981. *Applied Iterative Methods.* Academic Press, New York.

Hall, R. R. 1976. The length of ray images under starlike mappings. *Mathematika.* 23: 147-150.

Hald, J. 2009. Basic theory and properties of statistically optimized near-field acoustical holography. *J. Acoust. Soc. Am.* 125: 2105-2120. doi: 10.1121/1.3079773.

Halley, E. 1695. An easy demonstration of the analogy of the logarithmic tangents to the Meridian line, or sum of the secants, with various methods for computing the same to the utmost exactness. *Philosophical Transactions.* 19: 202-214.

———. 1980. A conformal mapping inequality for starlike functions of order 1/2 *Bull. London Math. Soc.* 12: 119-126.

Halliday, David, and Robert Resnick. 2014. *Fundamentals of Physics*, 10th ed. John Wiley, New York.

Halsey, N. D. 1979. Potential flow analysis of multielement airfoils using conformal mapping. *AIAA J.* 17: 1281-1288.

———. 1982. Comparison of the convergence characteristics of two conformal mapping methods. *AIAA J.* 20: 724-726.

Hansen, J. H. 1934. The practical application of conformal representation. Thesis, Liverpool, October 1934.

Harrington, Roger. 1961. *Time-Harmonic Electromagnetic Fields.* New York: McGraw-Hill.

Hartmann, M., and G. Opfer. 1986. Uniform approximation as a numerical tool for constructing conformal maps. *J. Comput. Appl. Math.* 14: 193-206; also in *Numerical Conformal Mapping* (L. N. Trefethen, ed.), North-Holland, Amsterdam, 193-206.

Häuser, J., and C. Taylor (eds.) 1986. *Numerical Grid Generation in Computational Fluid Dynamics.* Swansea, UK: Pineridge Press.

Haydl, W. H. 2002. On the use of vias in conductor-backed coplanar circuits. *IEEE Trans. Microwave Theory Tech.* 6, no. 50: 1571-1577.

Hayes, J. K., D. K. Kahaner, and R. G. Kellner. 1972. An improved method for numerical conformal mapping. *Math. Comput.* 26: 327-334.

———. 1975. A numerical comparison of integral equations of the first and second kind for conformal mapping. *Math. Comp.* 29: 512-521.

Harvánek, Zdeněk, Petr Beneš, and Stanislv Klusáček. 2014. Comparison of patch holography methods for confined space. Inter-Noise 2014, Melbourne, Australia. p1023.pdf. (www.acoustics.asn.au).

He, J. 1999. Homotopy perturbation method. *Comput. Methods Appl. Mech. Eng.* 178: 257-262.

Hecht, K. T. 2000. *Quantum Mechanics.* New York: Springer-Verlag.

Heinhold, J., and R. Albrecht. 1954. Zur Praxis der konformen Abbildung. *Rendiconti Circulo Mat. Palermo.* 3: 130-148.

Heins, A. E. 1974. The generalized radiation problem, in *Topics in Analysis* (A. Dodd and B. Eckmann, eds.), Colloquium on Math. Anal., Jyväskylä. Lecture Notes in Math., Vol. 419, 166-177. Springer-Verlag, Berlin.

Heins, Maurice. 1949. Conformal mapping and convergence of a power series. *The Annals of Mathematics.* 2nd Ser. 50: 686-690.

———. 1953. Deduction of Cardano's formula by conformal mapping. *Proceedings of the National Academy of Sciences of the United States of America.* 39: 322-324.

———. 1954. Another remark on "Some Problems in Conformal Mapping." *Proceedings of the National Academy of Sciences of the United States of America.* 40: 302-305.

———. 1958. On integrating factors and on conformal mappings. *Transactions of the American Mathematical Society.* 89: 267-276.

Helmholtz, H. 1954. *On the Sensation of Tone.* New York: Dover.

Henrici, P. 1974. *Applied and Computational Complex Analysis*, vol. I. Wiley, New York.

———. 1976. Einige Anwendungen der schnellen Fouriertransformation, in *Moderne Methoden der Numerischen Mathematik* (J. Albrecht and L. Collatz, eds.), Birkhäuser, Basel 111-124.

———. 1979. Fast Fourier methods in computational complex analysis. *SIAM Rev.* 21: 481-527.

———. 1979. Barycentric formulas for trigonometric interpolation. *Numer. Math.* 33:

225-234.

———. 1986. *Applied and Computational Complex Analysis*, Vol. III, Wiley, New York.

Hersh, J. 1955. Equations differentielles et functions de cellules. *Comptes Rendus Acad. Sci Paris.* 240:1602-1604.

Herglotz, G. 1917. Über die Nullstellen der hypergeometrischen Funktion. *Berichte Verhand. Sächs. Akad. Wissen. Leipzig, Math.-Phys. Klasse.* 69: 510-534.

Hesthaven, J. S., and T. Warburton. 2008. *Nodal Discontinuous Galerkin Methods: Algorithms, Analysis, and Applications.* Vol. 54 of Text in Applied Mathematics. New York: Springer.

Hilbert, D. 1901/1935. Das Dirichletsche Prinzip (1901), in *Gesammelte Abhandlungen*, Vol.. 3, Springer-Verlag, Berlin, 1935.

———. 1905. Über das Diricletsche Prinzip, *Journal für Mathematik.* 129: 63-67; *Gesammelte Abhandlungen.* 3: 10-14.

———. 1909/1935. Zur Theorie der konformen Abbildung (1909), in *Gesammelte Abhandlungen*, Vol. 3, Springer-Verlag, Berlin, 1935.

———. 1924. *Grundzüge einer allgemeinen Theorie der linearen Integralgleichungen*, 2nd ed. Leipzig-Berlin, 83-94.

Hildebrand, H. B. 1965. *Methods of Applied Mathematics*, 2nd ed. Prentice-Hall, Englewood Cliffs, NJ.

Hille, E. 1976. *Ordinary Differential Equations in the Complex Domain.* Wiley, New York.

Hinch, E. J. 1991. *Perturbation Methods.* Cambridge University Press.

Hine, M. J. 1971. Eigenvalues for a uniform fluid waveguide with an eccentric-annulus cross-section. *Journal of Sound and Vibration.* 15: 295-305.

Hiroyasu Izeki. 1996. Computational conformal mapping for surface grid generation. *Transactions of the American Mathematical Society.* 348: 4939-4964.

Hockney, R.W. 1970. The potential calculation and some applications. *Meth. in Comput. Phys.* 9: 135-211.

Hodgkinson, J. 1930. Conformal representation by means of Lamé functions. *J. London Math. Soc.* 5: 296-306.

———, and E. Poole. 1924. The conformal representation of the area of a plane bounded by two straight or circular slits. *Proc. London Math. Soc.* 23: 396-422.

Hoffman K. 1975. *Analysis in Euclidean Space.* Englewood Cliffs, NJ: Prentice-Hall.

Hooke, R., and T. A. Jeeves. 1961. Direct search solution of numerical and statistical problems. *J. Assoc. Comput. Math.* 8: 212-229.

Höhndorf, F. 1926. Verfahren zur Berechnung des Auftriebes gegebener Tragflächen-Profile. *ZAMM.* 6: 265-283.

Hockney, R. 1970. The potential calculation and some applications. *Methods in Comput. Phys.* 9: 135-211.

Hoffman, K. 1975. *Analysis in Euclidean Space.* Prentice-Hall, Englewood Cliffs, NJ.

Hoidn, H.-P. 1982. Osculation methods for the conformal mapping of doubly connected regions. *ZAMP.* 33: 640-652.

———. 1986. A reparameterisation method to determine conformal maps. *J. Comput. Appl. Math.* 14: 155-161; also in *Numerical Conformal Mapping* (L. N. Trefethen, ed.), North-Holland, Amsterdam,1986, 155-161.

Holmes, B. J., C. J., Obra, and L. P. Yip. 1984-06-01. *Natural laminar flow experiments on modern airplane surfaces* (https://ntrs.nasa.gov/search.jsp?R=19840018592).

Holzmüller, G. 1882. *Einfürung in die Theorie der isogonalen Verwandtschaften und der konforme Abbildungen.* Leipzig.

Homentkovshi, D, A. Manolescu, A. M. Manolescu, and C. Burileanu. 1988. An analytical solution for the coupled stripline-like microstrip line problem. *IEEE Trans. Microwave*

Theory and Techn. 36: 1002-1007.

Hooke, R., and T. A. Jeeves. 1961. Direct search solution of numerical and statistical problems. *J. Assoc. Comput. Mach.* 8: 212-229.

Hough, D. M., and N. Papamichael. 1981. The use of splines and singular functions in an integral equation method for conformal mapping. *Numer. Math.* 37: 133-147.

———. 1983. An integral equation method for the numerical conformal mapping of interior, exterior and doubly-connected domains. *Numer. Math.* 41: 287-307.

Houghton, E. L., and P. W. Carpenter. 2003. *Aerodynamics for Engineering Students.* Butterworth Heinmann, ed., 5th ed. p. 17-18. ISBN 978-0-7506-5111-3.

Householder, A. S. 1970. *The Numerical Treatment of a Single Nonlinear Equation.* New York: McGraw-Hill.

Howe, D. 1973. The application of numerical methods to the conformal transformation of polygonal boundaries. *J. Inst. Math. Applics.* 12: 125-136.

Hromadka, T. V., and C. Lai, 1987. *The Complex Variable Boundary Element Method in Engineering Analysis.* Berlin: Springer-Verlag.

Hsiao, G. C., P. Kopp, and W. L. Wendland. 1980. A Galerkin collocation method for some integral equations of the first kind. *Computing.* 25: 89-130.

Hu, Chenglie. 1998. Algorithm 785: A software package for computing Schwarz-Christoffel conformal transformation for doubly connected polygonal regions. *ACM Trans. on Math. Software* (TOMS). 24: 317-333. New York: ACM Press.

Hübner, O. 1979. Zur Numerik der Theodorsenschen Integralgleichung in der konformen Abbildung. *Mitt. Math. Sem. Giessen.* 140: 1-32.

———.1982 . Über die Anzahl der Lösungen der diskreten Theodorsen-Gleichung. *Numer. Math.* 39: 195-204.

———. 1986 . The Newton method for solving the Theodorsen integral equation. *J. Comput. Appl. Math.* 14: 19-30; also in *Numerical Conformal Mapping* (L. N. Trefethen, ed.), North-Holland, Amsterdam, 1986, 19-30.

Humi, K. M., and W. B. Miller. 1992. *Boundary Value Problems and Partial Differential Equations.* PWS-KENT Publishing Company, Boston, MA.

Hurt, H. H., Jr. (January 1965). 1960. *Aerodynamics for Naval Aviators.* U. S. Government Printing Office, Washington, DC: U.S. Navy, Aviation Training Division. pp. 21-22, NAVWEPS 00-80T-80.

Hurwitz, A., and R. Courant. 1925. *Vorlesungen über allgemeine Funktionentheorie und elliptische Funktionen.* 2nd ed. Berlin.

Ibragimov, N. H. (H. X. Ибрагимов). 1976. Huygens' principle. *Amer. Math. Soc. Transl.* (2), Vol. 104: 141-152; translation of Принцип Гюйгенса,, 159-170 (Russian).

Ilhan, E., J. Paulides, and L. Lomonova. 2010. Tooth contour method implementation for the flux-switching PM machines. In *2010 XIX Intern. Conf. on Electrical Machines (ICEM).* Rome, Sept. 2010: 6-8.

———, E., B. Gysen, J. Paulides, and E. Lomonova. 2010. Analytical hybrid model for flux switching permanent magnet machines. *IEEE Trans. on 2010 XIX Intern. Conf. on Electrical Machines.* 46(6): 1762-1765.

———, Motoasca, E. T., J. J. Pauldes, et al. 2012. Conformal mapping: Schwarz-Christoffel method for flux-switching PM machines. *Math Sci.* 6: 37. https://doi:org 101186/2251-7456-6-37.

Inoue, K. 1983. Grid generation for cascades using conformal mapping. *J. Comput. Phys.* 52: 130-140.

———. 1985. Grid generation for inlet configurations using conformal mapping. *J. Comput. Phys.* 58: 146-154.

Irie, T., G. Yamada, and M. Sonoda. 1983. Natural frequencies of square membrane and square plate with rounded corners. *Journal of Sound and Vibration.* 86-2: 249-255.

Isaacson, E., and H. B. Keller. 1966. *Analysis of Numerical Methods.* New York: John Wiley.

Isola, Dario. 2005. *Joukowski Airfoil Transformation.* MATLAB Central. http://www.mathworks.com/matlabcentral/fileexchange/loadFile.do?object Id=6670.

Ivanov, V. I, and M. K. Trubetskov. 1994. *Handbook of Conformal Mapping with Computer-Aided Visualization.* CRC Press, Boca Raton, FL.

———, and M. K. Trubetskov. 1995. *The Accuracy of Numerical Conformal Mapping Methods: A Survey of Examples and Results.* CRC Press, Boca Raton, FL.

———, and V. Yu. Papov. 2002. *Conformal Mappings and Their Applications* (Конформные отображения и их приложения). Moscow: Editorial URSS (Russian).

Ivanov, V. V. (В. В. Иванов). 1968. *Theory of Approximate Methods and Applications to Numerical Solution of Singular Integral Equations.* Kiev (Russian).

Ives, D. C. 1976. A modern look at conformal mapping including multiply connected regions. *AIAA J.* 14: 1006-1011.

———, and J. F. Liutermoza. 1977. Analysis of transonic cascade flow conformal mapping and relaxation techniques. *AIAA J.* 15: 647-652.

Iwaniec, Kari and Tadeusz, and Gaven Martin Astala. 2009. *Elliptic Partial Differential Equations and Quasiconformal Mappings in the Plane.* Princeton University Press, Princeton, NJ.

Jacobson, Finn, and Virginia Jaud. 2007. Statistically optimized near field acoustic holography using an array of pressure velocity probes. *J. Acoust. Soc. Am.* 121: 1550; doi: 10.1121/1.2434245.

James, R. M. 1971. A new look at two-dimensional incompressible airfoil theory. *Report J0918/01*, Douglas Aircraft Co., Long Beach, CA, May 1971.

Janna, W. 1993. *Introduction to Fluid Mechanics.* Boston: PWS Publishing Company.

Jawson, M. A. 1963. Integral equation methods in potential theory. I. *Proc. Roy. Soc. A.* 275: 23-32.

———, and G. T. Symm. 1977. *Integral Equation Methods in Potential Theory and Elastostatics.* Academic Press, London.

Jeffrey, A. 1992. *Complex Analysis and Applications.* CRC Press, Boca Raton, FL.

Jeffreys, H. 1928. On aerofoils of small thickness. *Proc, Royal Soc.* A 121.

———. 1958. On the conformal mapping of nearly-circular domains. *Transactions of the American Mathematical Society.* 88: 207-213.

———, and B. S. Jeffreys. 1946. *Methods of Mathematical Physics.* Cambridge University Press, Cambridge.

Jeltsch, R. 1969. Numerische konforme Abbildung mit Hilfe der Formel von Cisotti. *Diplomarbeit, ETH, Zurich.*

Jenkins, James A. 1952. On conformal mapping of regions bounded by smooth curves. *Proceedings of the American Mathematical Society.* 3: 147-151.

———. 1958. *Univalent Functions and Conformal Mapping.* Springer.

———. 1953. Studies in the conformal mapping of Riemann surfaces, I. *Proceedings of the American Mathematical Society.* 4: 978-981.

———. 1969. A procedure for conformal mapping of triply-connected domains. *Proceedings of the American Mathematical Society.* 22: 324-325.

Jessie, D., and L. Larson. 2001. Conformal mapping for buried CPW with finite grounds. *Electronics Letters.* 37: 1521-1523.

Jin, Jian-Ming. 2014. *The Finite Element Method in Electromagnetics.* 3rd ed. Wiley, New York.

John, F. 1982. *Partial Differential Equations.* New York: Springer-Verlag.

Joshi, K. K., J. S. Rao, and B. N. Das. 1980. Analysis of inhomogeneously filled stripline and microstripline. *IEEE Proc.* 127-H: 11-14.

Joukowski, N. E. (Н. Е. Жуковский).1880. Modification of Kirchhoff's method for determination of fluid motion in two dimensions with constant velocity. (1890); in *Collected Papers*, Vol. 2, Gostekhizdat, Moscow, 1950 (Russian).

———. 1910-1912. Über die Konturen des Tragflächen der Drachenflieger, in *Collected Papers*, Vol. 5. Gl. red. av. lit., Moscow (Russian).

———. 1911. Theoretical foundation of aeronautics (1911), in *Collected Papers*, Vol. 6, Gostekhizdat, Moscow, 1950 (Russian).

———. 1915. Vorticity theory of the propeller (1915), in *Collected Papers*, Vol. 4, Gostekhizdat, Moscow, 1950 (Russian).

Julia, G. 1926. Sur le représentation conforme des aires simplement convexes. *Comptes Rendus Acad. Sci. Paris.* 182: 1314-1316.

———. 1927. Sur une série de polynômes liée à la représentation conforme des aires simplement convexes. *Comptes Rendus Acad. Sci. Paris.* 183: 10-12.

———. 1931. Dévelopment en série de polynômes ou de fonctions rationelles de la fonction qui fournit la représentation conforme d'une aire simplement convexe sur un cercle. *Annales de l'École normale Sup.* 44: 289-316.

Kacimov, A. R. 2000. Green's functions for multiply connected domains via conformal mapping. *Journal of Engineering Mathematics.* 37: 397-400(4).

Kakaç, S., and Y. Yener. 1993. *Heat Conduction*, 3rd Ed. Washington, DC: Taylor and Francis.

Kaloni, P. N. 1967. Fluctuating flow of an elastico-viscous fluid past a porous flat plate. *Phys. Fluids.* 10: 1344-1346.

Kanekal S., A. Sahai, H. M. Kim, R. E. Jones, and D. Brown. 1997. A deformation of flat conformal structures. *European Journal of Cancer.* 33: 179-180(2).

Kantorovich, L. V. (Л. В. Канторович). 1933. Sur quelques méthodes de la détermination de la fonction qui effectue une représentation conforme. *Bull. Acad. Sci. URSS.* 7: 229-235.

———. 1933. Sur la représentation conforme. *Matem. Sbornik.* 40: 294-325.

———. 1934. Quelques rectifications à mon mémoire "Sur la représentation conforme." *Matem. Sbornik.* 41: 179-182.

———. 1934. On the approximate computation of some types of definite integrals and other applications of the method of removing singularities. *Matem. Sb.* (2), 41: 235-245 (Russian).

———. 1934. Sur la représentation conforme des domaines multiconvexes. *Comptes Rendus Acad. Sci. URSS.* 2: 441-445.

———. 1937. Conformal mapping of a circle on a simply connected region, in *Conformal Mapping of Simply and Doubly Connected Regions.* Gostekhizdat, Leningrad-Moscow (Russian).

———, and V. I. Krylov (В. И. Крылов). 1936. *Methods for the Approximate Solution of Partial Differential Equations.* Gostekhizdat, Leningrad-Moscow,1936 (Russian).

———, and V. I. Krylov. 1958. *Approximate Methods for Higher Analysis.* New York: Interscience.

———, and V. I. Krylov. 1964. *Approximate Methods for Higher Analysis.* First Russian edition 1936; English Translation 1941 (ed. C. D. Benster); Interscience, New York, 1958; Groningen: Noordhoff.

Kanwal, R. 1983. *Generalized Functions: Theory and Technique.* New York: Academic Press.

Karunakaran, V. 1983. Comment on: Potential flow analysis of multielement airfoils using conformal mapping. *Proceedings of the American Mathematical Society.* 87: 289-294.

Kashin, V. A., and V. V. Merkulov. 1966. Determination of the eigenvalues fir waveguide with complex cross section. *Soviet Phys.: Acoustics.* 11: 285-287.

Kasner, Edward. 1940. On the use of conformal mapping in shaping wing profiles. *Transactions of the American Mathematical Society.* 48: 50-62.

———, and John De Cicco. 1944. On Mr. L. Kantorovitch's method in the theory of conformal mapping and on the application of that method to aerodynamics. *Proceedings of the National Academy of Sciences of the United States of America.* 30: 162-164 (Spanish).

———, and John De Cicco. 1945. Scale curves in conformal maps. *American Journal of Mathematics.* 67: 157-166.

Kassab, A. J., R. T. Bailey, and C. K. Hsieh. 1993. CVBEM in simply- and multiply-connected domains for the solution of heat conduction problems, in *Advances in Boundary Element Methods* (M. H. Aliabadi and C. A. Brebbia, eds.), Computational Mechanics Publications, Southampton, and Elsevier Applied Science, London.

Katz, Joseph and Allen Plotkin. 1991. *Low Speed Aerodynamics*, 2nd ed. McGraw-Hill.

Kehren, E. 1932. Die Anwendung der konformen Abbildung in der Elektrostatik. *Ann. der Physik.* (5), 14: 367.

Keldyš M. V. (М. В. Келдыш). 1939. Conformal mappings of multiply connected domains on canonical domains. *Uspekhi Matem. Nauk.* 6: 90-119; translation of Конформые отображенния многоцвязных областей на канонические области, Успехи мат. наук, 6 (1939), 90-119 (Russian).

Keener, J. P. 1988. *Principles of Applied Mathematics. Transformations and Approximations.* Addison-Wesley, New York.

Kellogg, O. D. 1929/1953. *Foundations of Potential Theory.* Springer-Verlag, Berlin, 1929; Dover, New York, 1953.

Kempel, Leo C., John L. Volakis, Alex C. Woo, and C. Long Yu. 1992. A finite element-boundary integral method for conformal antenna arrays on a circular cylinder. *NASA-CR-190610*, Final Report 027723-6-F, July 1992. Dept. Electrical Engg. and Computer Sc., University of Michigan, An Arbor, MI.

Kershaw, D. 1974. Singular integrals and boundary value problems, in *Numerical Solutions of Integral Equations* (L. M. Delves and J. Walsh, eds.) 258-274. Oxford: Clarendon Press.

Kerzman, N., and E. M. Stein. 1978. The Cauchy kernel, the Szegö kernel, and the Riemann mapping function. *Math. Ann.* 236: 85-93.

———, and M. R. Trummer. 1986. Numerical conformal mapping via the Sgezö kernel. *J. Comput. Appl. Math.* 14: 111-123; also in *Numerical Conformal Mapping* (L. N. Trefethen, ed.), North-Holland, Amsterdam, 1986, 111-123.

Kevorkian, J. 1990. *Partial Differential Equations.* Belmont: Wadsworth & Brooks/Cole.

Khajalia, G. (Г. Хаjалия). 1940. Sur la théorie de la représentation conforme des domaines doublement convexes. *Matem. Sbornik*, New Series, 8: 97-106.

Khamayseh, Ahmed, and C. Wayne Mastin. 1996. Handbook of conformal mapping with computer-aided visualization. With 1 IBM-PC floppy disk (5.25 inch; HD). *J. Comput. Phys.* 123: 394-401.

Khellaf, K., and G. Lauriat. 1996. A new analytical solution for heat transfer in the entrance region of ducts: Hydrodynamically developed flows of power-law fluids with constant wall temperature. *Int. J. Heat Mass Transfer.* 40: 3443-3447.

Khvedelidze, B. V. (Б. В. Хведелидзе). 1941. On the Poincaré boundary value problem of the theory of logarithmic potential for multiply connected regions. *Soobshcheniya AN Gruz. SSR* (Сообшшения АН Груз. ССР). 2: 571-578, 865-872 (Russian).

Kirchhoff, G. R. 1869. Zur Theorie freier Flüssigkeitsstrahlen. *J. Reine Angew. Math.* 70: 289-298.

———. 1882. *Gesammelte Abhandlungen.* Leopzig.

Knupp, P., and S. Steinberg. 1993. *Fundamentals of Grid Generation.* CRC Press, Boca Raton, FL.

Kober, H. 1945-1948. *Dictionary of Conformal Representations.* Admiralty Computing Service, Dept. of Scientific Research and Experiment, Admiralty, London, Parts 1-5.

———. 1952/1957. *Dictionary of Conformal Representations.* Dover, New York, 1952; 2nd ed., Dover, New York, 1957.

Koebe, P. 1913. Ränderzuordung bei konformer Abbildung. *Göttinger Nachrichten.* 286-288.

———. 1915. Abhandlungen zur Theorie der konformen Abbildung. I. Die Kreisabbildung des allgemeisnten einfach und zweifach zusammenhängenden schlichten Bereichs und die Ränderzuordnung bei konformer Abbildung. *J. Reine Angew. Math.* 145: 177-223.

———. 1918. Abhandlungen zur Theorie der konformen Abbildung. IV. Abbildung mehrfach zusammenhängender schlichter Bereiche auf Schlichtbereiche. *Acta Math.* 41: 304-344.

Kohl, E. 1930. Beitrag zur Lösung des ebenen Spannungsprobleme. *ZAMM.* 10: 141.

Komatu, Y. 1943. Untersuchungen über konforme Abbildung von zweifach zusammenhängender Gebieten. *Proc. Phys.-Math. Soc. Japan,* (3) 25: 1-42.

———, and M. Ozawa. 1951. Conformal mapping of multiply connected domains. I. *Kodai Math. Sem. Rep.* 81-95.

———, and M. Ozawa. 1952. Conformal mapping of multiply connected domains. II. *Kodai Math. Sem. Rep.* 39-44.

Kondrat'ev, V. A., and O, A. Oleinik. 1983. Boundary-value problems for partial differential equations in non-smooth domains. *Russian Math. Surveys.* 38:1-86.

Korn, G. A. 1961. *Mathematical Handbook for Scientists and Engineers. Definitions, Theorems, and Formulas.* New York: McGraw-Hill.

Kranz, S. G. 1990. *Complex Analysis: The Geometric Viewpoint.* The Mathematical Association of America.

Kreiszig, E. 1978. *Introductory Functional Analysis with Applications.* Wiley, New York.

Krenk, S. 1975. On quadrature formulas for singular integral equations of the first and second kind. *Quart. Appl. Math.* (Oct. 1975) 225-232.

———. 1978. Quadrature formulas of closed type for the solution of singular integral equations. *J. Inst. Math. Appl.* 22: 99-107.

Krylov, V. I. (В. И. Крылов). 1937. Concerning a method of constructing a function which maps a region conformally on a circle, in *Conformal Mapping of Simply and Doubly Connected Regions.* Gostekhizdat, Leningrad-Moscow (Russian).

———. 1938. An application of integral equations to the proof of certain theorems for conformal mapping. *Matem. Sbornik.* 4: 9-30 (Russian).

———, and N. Bogolyubov. 1929. Sur la solution approchée du problème de Dirichlet. *Comptes Rendus Acad. Sci. URSS.* 283-288.

———, V. V. Lugin, and L. A. Yanavich. 1963. *Tables of Numerical Integration for* $\int_0^1 x^\beta (1-x)^\alpha f(x)\,dx$. Akademiya Nauk Belorusskoi S. S. R., Minsk. (Russian)

Kufarev, P. P. 1935-1937. Über das zweifachzusammenhängende Minimalgebiet. *Bull. de l'Institut de Math. et Mécan. Univ. Kouybycheff de Tomsk.* 1: 228-236.

———. 1947. On a method of numerical determination of the parameters in the Schwarz-Christoffel integral. *Doklady Akad. Nauk SSSR* (new Series). 57: 535-537 (Russian).

———. 1950. On conformal mapping of complementary regions. *Doklady Akad. Nauk SSSR.* 73: 881-884.

Kulisch, U. 1963. Ein Iterationsverfahren zur konformen Abbildung des Einheitskreises auf ein Stern. *ZAMM.* 43: 403-410.

Kulshrestha, P. K. 1973. Distortion of spiral-like mappings. *Proc. Royal Irish Acad.* 73, Sec. A: 1-5.

———. 1973. Generalized convexity in conformal mappings. *J. Math. Anal. Appl.* 43: 441-449.

Kumar, Mukesh, R. Saxena, A. Kapoor, P. Kala, and R. Pant. 2012. Theoretical characterization of coplanar waveguides using conformal mapping. *International Journal of Advanced Research in Computer Science and Electrical Engineering.* 1: 48-51.

Kutta, W. H. 1902. Lifting forces in flowing fluids. A part of his Ph. D. thesis, 1894.

———. 1911. Die ebene Zirkulationsströmungen. *Berichte Bayer. Akad. d. Wissen.* 57-64.

Kythe, Dave K., and P. K. Kythe. 2012. *Algebraic and Stochastic Coding Theory.* Boca Raton, FL: CRC Press.

Kythe, P. K. 1995. *An Introduction to Boundary Element Methods.* CRC Press, Boca Raton, FL.

———. 1996. *Fundamental Solutions for Differential Operators and Applications.* Boston: Birkhäuser.

———. 1998/2012. *Computational Conformal Mapping.* Boston: Birkhäuser; ebook, 2012. Springer Math & Business, New York.

———. 2011. *Green's Functions and Linear Differential Equations: Theory, Applications, and Computation.* Taylor & Francis Group/CRC Press.

———. 2015. *Sinusoids: Theory and Technological Applications.* Boca Raton, FL : CRC Press.

———. 2016. *Complex Analysis: Conformal Inequalities, and the Bieberbach Conjecture.* Boca Raton: CRC Press.

———, and P. Puri. 1983. Wave structure in oscillatory Couette flow of a dusty gas. *Acta Mech.* 46: 127-135

———, and J. H. Abbott. 1986. Propagation and interference of waves in oscillatory heat conduction in composite media. *Rivista di Mat. Univ. Parma.* 12 : 227-236.

———, and P. Puri. 1987. Unsteady MHD free-convection flows with time-dependent heating in a rotating medium. *Astrophysics and Space Science.* 135 : 219-228.

———, P. Puri, and M. R. Schäferkotter. 1997. *Partial Differential Equations and Mathematica.* CRC Press.

———, and P. Puri. 2002. *Computational Methods for Linear Integral Equations.* Boston: Birkhäuser.

———, P. Puri, and Michael R. Schäferkotter. 2003. *Partial Differential Equations and Boundary Value Problems with Mathematica*, 2nd ed. Chapman & Hall/ CRC, Boca Raton, FL.

———, and Dongming Wei. 2004. *An Introduction to Linear and Nonlinear Finite Element Analysis: A Computational Approach.* Birkhäauser, Boston.

———, and M. R. Schäferkotter. 2005. *Handbook of Computational Methods for Integration.* Chapman & Hall/CRC.

Labus, J. 1927. Berechnung des elektrischen Feldes von Hochspannungstransformatoren mit Hilfe der konformen Abbildung, wenn mehere Wicklungen vershiedenen Potentials vorhanden sind. *Archiv für Elektrotechnik.* 19: 82-103.

Lamb, H. 1932/1945. *Hydrodynamics.* Cambridge Univ. Press; Dover, New York.

Lambert, J. H. 1772. *Anmerkungen und Zusätze zur Entwerfung der Land- und Himmelskarten.* English translation: Notes and Comments on the Composition of Terrestrial and Celestial Maps, Ann Arbor, University of Michigan, 1972.

Lanczos, C. 1961. *Linear Differential Operators.* New York: Van Nostrand.

Landau, E. 1926. Einige Bemerkungen über schlichte Abbildung. *Jahresbericht Deut. Math.-Vereinigung.* 34: 239-243.

Laura, P. A. A. 1964. On the determination of the natural frequency of a star-shaped membrane. *Journal of the Royal Astronautical Society.* April 1964.

———. 1967. Calculations of eigenvalues for uniform fluid waveguide with complicated cross section. *J. Acoust. Soc. Am.* 42: 21-26.

———. 1975. A survey of modern applications of the method of conformal mapping. *Revista de la Unión Mathematica Argentina.* 27: 167-179 .

———, and M. Chi. 1964. Approximate method for the study of heat conduction in bars of arbitrary cross section. *J. Heat Transfer, Trans. ASME 86, Ser. C*, 466-467.

———, and M. Chi. 1965. An application of conformal mapping to a three-dimensional unsteady heat conduction problem. *Aeronaut. Quart.* 16: 221-230.

———, and A. J. Faulstich. 1965. An application of conformal mapping to the determination of the natural frequency of membranes of regular polygonal shape. *Proc. Ninth Midwestern Mechanics Conference.* University of Wisconsin, Madison, WI. 155-163.

———, and A. J. Faulstich. 1968. Unsteady heat conduction in plates of polygonal shape. *Int. J. Heat Mass Transfer.* 11: 297-303.

———, and P. A. Shahady. 1966. Longitudinal vibrations of a solid propellant rocket motor *Proc. Third Southeastern Conf. on Theor. Appl. Mech.* Pergamon Press, New York, 623-633.

———, and E. Romanelli. 1973. Determination of eigenvalues in a class of waveguides of doubly connected cross section. *J. Sound and Vibration.* 26: 395-400.

———, and M. J. Maurizi. 1971. Comments on 'Eigenvalues for a uniform fluid waveguide with an eccentric-annulus cross-section.' *Journal of Sound and Vibration.* 18: 445-447.

———, E. Romanelli, and M. J. Maurizi. 1972. On the analysis of waveguides of doubly-connected cross section by the method of conformal mapping. *J. Sound Vibration.* 20: 27-38.

———, J. A. Reyes, and R. E. Rossi. 1974. Numerical experiments on the determination of stress concentration factors. *Strain.* April 1974: 58-63.

———, L. C. Nava, and V. H. Cortinez. 1985. A modification of the Galerkin method and the solution of the Helmholtz equation in regions of complicated boundary shape. *Journal of the Acoustical Society of America.* 77.

Lavrent'ev, M. A. (M. A. Лаврентьев). 1934. *Theory of Conformal Mappings* (К Теории Конформбых Отображений). Труды физ.-мат. и-та. им. Б. А. Стеклова. (Russian).

———. 1936. Boundary problems in the theory of univalent functions. *Amer. Math. Soc. Transl.* (2) 32: 1-35; translation of О некоторих граничхых задачах в теории однолистных функций, Mat. Sb., 1 (43): 815-844 (Russian).

———. 1947. *Conformal Mappings with Applications to Some Problems of Mechanics.* Gostekhizdat, Moscow-Leningrad. (Russian).

———. 1934/1984. On the theory of conformal mappings. *Amer. Math. Soc. Transl.* (2) 122: 1-69; Trudy Fiz.-Mat. Inst. Steklova, 5 (1934), 159-245 (Russian).

Lawrenson, P. J., and S. K. Gupta. 1968. Conformal transformation employing direct-sketch techniques of minimization. *Proc. IEE, London.* 115:427-431.

Lawrentjew, M. A., and B. W. Schabat (М. А. Лаврентьев у Б . Щ. Шабат). 1967. *Methoden der komplexen Funktiontheorie.* VEB Deutscher Verlag der Wissenschaften, Berlin (German translation).

Leach, W. Marshall, Jr. 1999. *Introduction to Electroacoustics and Audio Amplifier Design*, 2nd ed. Kendall/Hunt Publishing Co., Dubuque, IA.

Lebedev, N. N. 1955. On the theory of conformal mappings of a circle onto nonoverlapping regions. *Doklady Akad. Nauk SSSR.* 103: 553-555 (Russian).

———, I. P. Skalskaya, and Y. S. Uflyand. 1965. *Worked Problems in Applied Mathematics.* New York: Dover.

Lehman, R. S. 1957. Development of the mapping function at an analytic corner. *Pacific J. Math.* 7: 1437-1449.

Lehto, O., and K. I. Virtaanen. 1973. *Quasiconformal Mappings in the Plane.* Springer-Verlag, Berlin.

Leja, F. 1934. Une méthode de construction de la fonction de Green appartenant à un domain plan quelconque. *Comptes Rendus Acad. Sci. Paris.* 198: 231-234.

———. 1935. Construction de la fonction analytique effectuant la représentation conforme d'un domaine plan quelconque sur le cercle. *Math. Ann.* 111: 501-504.

———. 1936. Sur une suite de polynômes et la représentation conforme d'un domain plan quelconque sur le cercle. *Annales Soc. Polonaise Math.* 14: 116-134.

Leont'ev, A. F. 1988. Representation of functions in convex domains by generalized exponential series. *Amer. Math. Soc. Transl.* (2) 140: 121-130;

Levi, M. 2007. Riemann mapping Theorem by steepest descent. *American Math. Monthly.* 114: 245-251.

Levin, D., N. Papamichael, and A. Sideridis. 1978. The Bergman kernel method for an improved conformal mapping of simply-connected domains. *J. Inst. Math. Appl.* 22: 171-187.

Levinson, N., and R. M. Redheffer. 1970. *Complex Variables.* Holden-Day, San Francisco.

Lewent, L. 1925. *Conformal Representation.* Translated by R. Jones and D. H. Williams. London.

Lewy, H. 1950. Developments at the confluence of analytic boundary conditions. *Univ. Calif. Publ. Math.* 1: 247-280.

Lichtenstein, L. 1911. Über die konforme Abbildung ebener analytischer Gebiete mit Ecken. *J. Reine Angew. Math.* 140: 100-119.

———. 1917. Zur konformen Abbildung einfach zusammenhängender schlichter Gebiete. *Archiv Math. Phys.* 25: 179-180.

Liebmann, H. 1918. Die angenäherte Ermittlung harmonischer Funktionen und konformer Abbildungen (nach Ideen von Boltzmann und Jacobi). *Sitzungberichte math.-phys. Klasse Akad. Wissen. München.* 385-416.

Linz, D. P. 1985. On the approximate computation of certain strongly singular integrals. *Computing.* 35: 345-353.

Liu, T. J., H. M. Lin, and C. N. Hong. 1988. Comparison of two numerical methods for the solution of non-Newtonian flow in ducts. *Int. J. Numer. Methods Fluids.* 8: 845-861.

———, S. Wen, and J. Tsou. 1994. Three-dimensional finite element analysis of polymeric fluid flow in an extrusion die. Part I: Entrance effect. *Polymer Engineering Science.* 34: 827-834.

Logan, D. L. 1997. *A First Course in the Finite Element Method.* PWS Publishers.

Longuet-Higgins, M. S., and E. D. Cokelet. 1976. The deformation of steep surface waves on water, I. A numerical method of computation. *Proc. Roy. Soc. London, Ser. A.* 350: 1-26.

Love, A. E. H. 1892. On the theory of discontinuous fluid motions in two dimensions. *Proc. Camb. Phil. Soc.* 7: 175.

Löwner, K. 1923. Untersuchungen über schlichte konforme Abbildungen des Einheitskreises, I. *Math. Ann.* 89: 103-121.

Lozansky, E. D., and O. B. Firsov. 1973. The theory of the initial stages of streamer propagation. *J. Phys. D: Appl. Phys.* 6:976.

———(Лозанский, Е. Д., и О. Б. Фирсов). 1975. Теория искры (Theory of the Spark). Fizmatgiz, Moskva. (Russian).

Lu, J. 1984. A class of quadrature formulas of Chebyshev type for singular integrals. *J. Math. Anal. Appl.* (2) 100: 416-435.

Luke, Y. L. 1969. *The Special Functions and Their Approximations*, Vol. 2 New York: Academic Press, 255-269.

Lurye, A. J. 1927. Zur Schwarz-Christoffelschen Formel. *Annales Inst. Polytech. Nom M. I. Kalinin, Leningrad.* 30: 113-120 (Russian).

Lyusternik, L. A. 1926. Über einige Anwendungen der direkten Methoden in der Variationsrechnung. *Matem. Sbornik.* 33: 173-201 (Russian).

———. 1947. Remarks on the numerical solution of boundary problems for Laplace's equation and the calculation of characteristic values by the method of networks. *Trudy Mathem. Inst. V. A. Steklova.* 20: 49-64 (Russian).

Mackie, A. G. 1989. *Boundary Value Problems.* Scottish Academic Press., Edinburgh.

MacCluer, C. R. 1994. *Boundary Value Problems and Orthogonal Expansions.* New York: IEEE Press.

Mader, R. E. 1991. *Programming in Mathematica*, 2nd ed. Addison-Wesley, Redwood City, CA.

Maiti, M. 1968. A note on the integral equation methods in potential theory. *Quart. Appl. Math.* 25: 480-484.

Malentiev, P. B. 1937. Approximate conformal mapping, in *Conformal Mapping of Simply and Doubly Connected Regions*, Gostekhizdat, Leningrad-Moscow (Russian).

Makarov, N. G. 1986. Conformal Mapping of Regions Bounded by Curvilinear Polygons. *Proceedings of the American Mathematical Society.* 96:233-236.

Mangler, W. 1930. Two remarks on the Schwarz-Christoffel transformation. *ZAMM.* 18: 251.

Marčenko, A. R. 1935. Sur la représentation conforme. *Comptes Rendus (Doklady) Acad. Sci. URSS* 1: 289-290.

Marchman, James F. 2012. *Computer Programs for AOE 3014.* Virginia Polytechnic: Aerospace and Ocean Engineering Department. http://www.aoe.vt.edu/marchman /software/.

Markushevich, A. I. (А. И. Маркушевич). 1950. *Theory of Analytic Functions* (Теория аналитических функций). Gostekhizdat, Moskva-Leningrad (Russian).

Marsden, J. E., and M. J. Hoffman. 1987. *Basic Complex Analysis*, 2nd ed. W. H. Freeman and Company, New York.

Mathews, John H., and Kurtis K. Fink. 2004. *Numerical Methods Using Matlab*, 4th ed. Prentice-Hall,

———, John J., and Russell W. Howell. 2008. *Dictionary of Conformal Mapping*, Part 1. California State University Fullerton. http://math.fullerton.edu/mathews/c2003/ConformalMappDictionary.1.html.

Mathis, H. F. 1963. Algebraic Structure and Conformal Mapping. *Mathematics Magazine.* 36: 25-30.

Mayo, A. 1984. The fast solution of Poisson's and the biharmonic equations on irregular regions. *SIAM J. Numer. Anal.* 21: 285-299.

———. 1986. Rapid methods for the conformal mapping of multiply connected regions. *J. Comput. Appl. Math.* 14: 143-153; also in *Numerical Conformal Mapping* (L. N. Trefethen, ed.) North-Holland, Amsterdam, 143-153.

Mazumdar, J. 1975. A review of approximate methods for determining the variational modes of membranes. *The Shock and Vibration Design.* 7: 6ff.

McDonald, B. H., and A. Wexler. 1972. Finite-element solution of unbounded field problems. *IEEE Trans. on Microwave Theory and Techniques.* 20: 841-847.

McOwen R. 1996. *Partial Differential Equations.* Prentice-Hall, Englewood Cliffs, NJ.

Medina, Francisco, and Robert Taylor. 1983. Finite element techniques for problems of unbounded domains. *Intern. J. for Numer. Meths. in Eng.* 19. https://doi.org/10.1002/nme.162019808.

Meinke, Hans H. 1949. Ein allgemeines Losungsverfahren für inhomogene zylinder-symmetrische Wellenfelder. *Zeit. für angew. Physik.* 1: 509-516.

———, and F. W. Gundlach. 1962. *Taschenbuch der Hochfrequenztechnik.* Springer-Verlag, Berlin.

———, K. P. Lange, and J. F. Ruger. 1963. TE and TM-waves in waveguides of very general cross-sections. *Proc. IEEE.* 51: 1436-1443.

Meiron, D. I., S. A. Orszag, and M. Israeli. 1981. Applications of numerical conformal mapping. *J. Comp. Phys.* 40: 345-360.

Melent'ev, P. V. (П. В. Мелентьев). 1937. Несколъко нобых методов и приёмов приближенных вычислений (Several New Methods and Devices for Approximate Computations). ОНТИ, Ленинград-Москва (Russian).

Mengoni, O. 1936. Die konforme Abbildung gewisser Polyeder auf die Kugel. *Monatshefte für Math. Phys.* 44: 159-185.

Menikoff, R., and C. Zemach. 1980. Methods for numerical conformal mapping. *J. Comp. Phys.* 36: 366-410.

———, 1983. Rayleigh-Taylor instability and the use of conformal maps for ideal fluid flow. *J. Comput. Phys.* 51: 28-64.

Mercator, G. 1569. *Nova et acuta orbis terrae descripto ad usum navigantium emendate accomodata.* (A new and enlarged description of the earth with corrections for use in navigation).

Messiha, S. A. S. 1966. Laminar boundary layers in oscillatory flow along an infinite flat plate with variable suction. *Proc. Camb. Phil. Soc.* 62: 329-337.

Meulenbroek, Bernard, A. Rocco, and U. M. Ebert. 2003. Stremer branching rationalized by conformal mapping. *MAS Report E0311*, Sept 30, 2003. Cemtrum voor Wiskunde en Informe, National Research Institute for Mechanics and Computer Science, Netherlands Organization for Scientific Research, ISSN 1386-3703.

———, Andrea Rocco, and Ute Ebert. 2004. Steamer branching rationalized by conformal mapping techniques. *Phys. Rev.* E 69; doi.org/10.1103/PhysRevE.69

Meyer, Eva-Suzanne. 1979. Praktische Vefahren zur konformen Abbildungen Garadenpolygonen. Dissertation, Universität Hannover.

Mikhlin, S. G. (С. Г. Михлин). 1957. *Integral Equations and Their Applications.* Pergamon, London.

———. 1964. *Variational Methods in Mathematical Physics.* Pergamon Press, New York.

Milin, I. M. (И. М. Милин). 1977. *Univalent Functions and Orthogonal Systems.* Translations of Mathematical Monographs, Vol. 49. American Mathematical Society, Providence, RI.

Miller, M., and W. T. Guy, Jr. 1966. Numerical inversion of the Laplace transform by the use of Jacobi polynomials. *SIAM J. Numer. Anal.* 3: 624-635.

Minda, C. David. 1979. Conformality and semiconformality of a function holomorphic in the disk. *American Mathematical Monthly.* 86: 684-686.

Miser, Hugh J. 1942. On conformal mapping of infinite strips. *National Mathematics Magazine.* 16: 333-337.

Modi, A. Y., and C. A. Balanis. 2016. PEC-PMC baffle inside circular cross section waveguide for reduction of cutoff frequency. *IEEE Microwave and Wireless Components Letters.* 26: 171-173.

Monakov, V. N. 1983. *Boundary-Value Problems with Free Boundaries for Elliptic Systems of Equations.* American Mathematical Society, Providence, RI.

Morris, Wallace J., II. 2009. *A universal prediction of stall onset for airfoils at a wide range of Reynolds number flows.* (http://adsabs.harvard.edu/abs/2009PhDThesis146M).

———, and Zvi Rusk. October 2003. Stall onset aerofoils at low to moderately high Reynolds number flows. *Journal of Fluid Mechanics.* 733: 439-472. available at (https://www.cambridge.org/core/journal-of-fluid-mechnics/article/stall-onset-on-aerofoils-at-low-moderately-high-reynolds-number-flows/ 648F9A27BAEEBE84CF381225519749BC).

Morse, P. M., and K. U. Ingard. 1968. *Theoretical Acoustics.* McGraw-Hill, New York.

Morse, P., and H. Feshbach. 1953. *Methods of Theoretical Physics,* Vol. I, II. McGraw-Hill, New York.

Müller, M. 1938. Zur konforme Abbildung angenähert kreisförmiger Gebiete. *Math. Zeit.* 43: 628-636.

Müller, W. 1924. Zur Konstruction von Traflächen-Profilen. *ZAMM.* 4: 213.

———. 1926. Stromlinien und Kraftlinien in der konformen Abbildung. *ZAMM.* 6: 284.

———. 1927. Zylinder in einer unstetigen Potentialströmung. *ZAMM.* 7: 13.

Munipalli, R., and D. A. Andersen. 1996. An adaptive grid scheme using the boundary element method. *J. Comput. Phys.* 127: 452-463.

Muratov, M. I. (М. И. Муратов). 1937. Conformal mapping of a half-plane on a region close to it, in *Conformal Mapping of Simply and Doubly Connected Regions.* Gostekhizdat, Leningrad-Moscow (Russian).

Muskhelishvili, N. I. (Н. И. Мусхелишвили). 1953/1992. *Singular Integral Equations.* Noordhoff, Leiden, 1953; 2nd ed., Dover, New York, 1992; translation of Сингулярные Интегральные Уравненя, Москва, 1946 (Russian).

———. 1963. *Some Basic Problems of the Mathematical Theory of Elasticity.* Noordhoff, Groningen.

NASA Publications: *Left from flow turning.* http://www.grc.nasa.gov/WWW/K-12/ airplane/right2.html. Archived. http://web.archive.org/web/20110705131635/http://www.grc.nasa.gov/WWW/K-12/ airplane/right2.html.

Nash, S. G., and A. Sofer. 1996. *Linear and Nonlinear Programming.* New York: McGraw-Hill.

National Aeronautics and Space Administration. 2011. *Conformal Mapping: Joukowski Transformation.* htttp://www.grc.nasa.gov/WWW/K-12/airplane/map.html.

Naidu, P. S., and K. O. Westphal. 1966. Some theoretical considerations of ion optics of the mass spectrometer ion source - I . *Brit. J. Appl. Phys.* 17: 645-651.

Nedil, M., and T. A. Denidni. 2006. Quasi-static analysis of a new wideband directional coupler using CPW multilayer technology. *IEEE, MTT-S Int. Micro. Symp. Dig.,* San Francisco, June 2006. 1133-1136.

———, and T. A. Denidni. 2008. Analysis and design of an ultra wideband directional coupler. *Progress in Electromagnetics Research.* 1: 291-305.

———, T. A. Denidni, and L. Talbi. 2005. CPW multilayer slot-coupled directional coupler. *Electron. Lett.* 12, No. 41: 45-46.

Needham, T. 1997. *Visual Complex Analysis.* Clarendon Press, Oxford.

Nehari, Z. 1949. The Schwarzian derivative and schlicht functions. *Bull. Am. Math. Soc.* 55: 545-551.

———. 1949. The kernel function and canonical conformal maps. *Duke Math. J.* 16: 165-178.

———. 1952. *Conformal Mapping.* Dover Publications, New York.

———, and Vikramaditya Singh. 1956. On the Conformal Mapping of Multiply Connected Regions. *Proceedings of the American Mathematical Society.* 7: 370-378.

Neumann, C. 1877. *Untersuchungen über das logarithmische und Newtonsche Potential.*Leipzig.

Nevanlinna, R. 1925. Zur Theorie der meromorphen Funktionen. *Acta Math.* 46: 1-99.

———. 1939. Über das alternierende Verfahren von Schwarz. *J. Reine Angew. Math.* 180: 121-128.

———, and V. Paatero. 1964/1969. *Introduction to Complex Analysis.* Addison-Wesley Publishing Co., Reading, MA ; *Einführung in die Funktionentheorie.* Birkhäuser, Basel, 1964.

Neville, E. H. 1944. *Jacobian Elliptic Functions.* Oxford.

Ng, W., and M. Stern. 1998. Analysis of multiple-rib waveguide structures by the discrete spectral-index method. *Proc. IEEE Conference on Optoelectronics, 1998.* 365-371.

Nguyen, C. 1995. Investigation of hybrid modes in broadside-coupled coplanar waveguide for microwave and millimeter-wave integrated circuits. *IEEE Antenna Propagat. Symposium,* Montral, Canada, July 1995. 18-23.

Niethammer, W. 1966. Iterationsverfahren bei der konformen Abbildung. *Computing.* 1: 146-153.

Noether, F. 1921. Über eine Klasse singulärer Integralgleichungen. *Math. Ann.* 82: 42-63.

Novinger, W. P.1975. A numerical comparison of integral equations of the first and second kind for conformal mapping. *American Mathematical Monthly.* 82: 279-282.

Nyström, E. J. 1930. Über die praktische Auflösung von Integralgleichungen mit Anwendungen auf Randwertaufgaben. *Acta Math.* 54: 185-204.

Oberhettinger, F. 1990. *Tables of Fourier Transforms and Fourier Transforms of Distributions.* Springer-Verlag, Berlin.

———, and W. Magnus. 1949. *Anwendung der elliptischen Funktionen in Physik und Technik.* Berlin: Springer-Verlag.

Oberkampf, W. L., and S. C. Goh. 1974. *Computational Mechanics.* Lecture Notes in Math., No. 461, 569-580. Springer-Verlag, Berlin.

O'Brien, V. 1981. Conformal mappings for internal viscous flow problems. *J. Comput. Phys* 44: 220-226.

Okano D., H. Ogata, and K. Amano. 2003. Conformal metrics and size of the boundary. *Journal of Computational and Applied Mathematics.* 152: 441-450(10).

Oliner, Arthur. 2006. The evolution of electromagnetic waveguides: From hollow metallic guides to microwave integrated circuits, in Sarkar et al. *History of Wireless,* Chapter 16. New York: John Wiley.

Olver, F. W. J. 1974. *Asymptotics and Special Functions.* Academic Press, New York.

———, D. W. Lozier, R. F. Boisvert, and C. W. Clark (eds.). 2010. *NIST Handbook of Mathematical Analysis.* Cambridge University Press, Cambridge.

Olver, Peter J. 2017. *Complex Analysis and Conformal Mapping.* Lecture Notes, cml/pdf. available at www.users.notes.umn.edu.

Opfer, G. 1979. New extremal properties for constructing conformal mappings. *Numer. Math.* 32: 423-429.

———. 1980. Conformal mappings onto prescribed regions via optimization techniques. *Numer. Math.* 35: 189-200.

———. 1982. Solving complex approximation problems by semiinfinite-finite optimization techniques: A study on convergence. *Numer. Math.* 39: 411-420.

Ortega, J. M. 1972. *Numerical Analysis.* Academic Press, New York.

———, and W. C. Rheinboldt. 1970. *Iterative Solution of Nonlinear Equations in Several Variables.* Academic Press, New York.

Osserman, Robert. 2004. *Mathematical Mapping from Mercator to the Millennium.* Chapter 18, 233-257.

Ostrowski, A. 1929. Mathematische Miszellen XV. Zur konformen Abbildung einfachzusammenhängender Gebiete. *Jahresbericht Deut. Math.-Vereiningung.* 38: 168-182.

———. 1930. Über konforme Abbildungen annährend kreisförmiger Gebiete. *Jahresbericht Deut. Math.-Verieinigung.* 39: 78-81.

———. 1952. On a discontinuous analogue of Theodorsen's and Garrick's model. Symposium on the construction and application of conformal maps. *National Bureau of Standards Appl. Math. Ser.* 18: 165-174.

———. 1952. On the convergence of Theodorsen's and Garrick's method of conformal mapping. *National Bureau of Standards, Appl. Math.* Ser. 18: 149-164.

———. 1955. Conformal mapping of a special ellipse on the unit circle, in *Experiments in the Computation of Conformal Maps Appl. Math. Ser., National Bureau of Standards,* Vol. 42.

Özisik, M. N. 1994. *Finite Difference Methods in Heat Transfer.* CRC Press, Boca Raton, FL.

———. 1980. *Heat Conduction.* Wiley, New York.

Page, W. M. 1912. Two-dimensional problems in electrostatics and hydrodynamics. *Proc. London Math. Soc.* II, 11: 321.

Panton, Ronald L. 1996. *Incompressible Flow,* 3rd ed. Cambridge University Press, Van Dyke.

Papamichael, N., and J. Whitman. 1973. A cubic spline technique for the one-dimensional heat conduction equation. *J. Inst. Math. Appl.* 11: 111-113.

———, and C. A. Kokkinos. 1981. Two numerical methods for the conformal mapping of simply-connected domains. *Comput. Meth. Appl. Mech. Engg.* 28: 285-307.

———, and C. A. Kokkinos. 1982. Numerical conformal mapping of exterior domains. *Comput. Meth. Appl. Mech. Engg.* 31: 189-203.

———, and C. A. Kokkinos. 1984. The use of singular functions for the approximate conformal mapping of doubly-connected domains. *SIAM J. Sci. Stat. Comput.* 5: 93-106.

———, and M. K. Warby. 1984. Pole-type singularities and the numerical conformal mapping of doubly-connected domains. *J. Comput. Appl. Math.* 10: 93-106.

———, M. K. Warby, and D. M. Hough. 1983. The determination of the poles of the mapping function and their use in numerical conformal mapping. *J. Comput. Appl. Math.* 9: 155-166.

———, M. K. Warby, and D. M. Hough. 1986. The treatment of corner and pole-type singularities in numerical conformal mapping techniques. *J. Comput. Appl. Math.* 14: 163-191; in *Numerical Conformal Mapping* (L. N. Trefethen, ed.), North-Holland, Amsterdam, 1986, 163-191.

———, and Nikos Stylianopoulos. 2010. *Numerical Conformal Mapping.* Hackensack, NJ: World Scientific.

Patankar, S. V. 1980. *Numerical Heat Transfer and Fluid Flow.* Hemisphere Publishing Corporation.

Payne, L. E. 1975. *Improperly Posed Problems in Partial Differential Equations.* SIAM, Philadelphia, PA.

Pennisi, L. L., L. I. Gordon, and S. Lasher. 1963. *Elements of Complex Variables.* Holt, Rinehart and Winston, New York.

Perlmutter, M,, and R. Siegel. 1963. Effect of specularly reflecting grey surface on thermal radiation through a tube and from its heated wall. *ASME J. Heat Transfer.* 85: 55-62.

Perring, W. G. A. 1927. The theoretical pressure distribution around Joukowski aerofoils.

Reports and Memoranda of the Aeronaut. Res. Comm. # 1106.

Petrovskii, I. G. 1967. *Partial Differential Equations.* W. B. Saunders Co., Philadelphia.

Petyt, M. 1998. *Introduction to Finite Element Vibration Analysis.* Cambridge University Press.

Phillips, E. G. 1943. *Functions of a Complex Variable with Applications.* Interscience, New York.

———. 1966. *Some Topics in Complex Analysis.* Pergamon Press, Oxford.

Phillips, H. B., and N. Wiener. 1923. Nets and the Dirichlet problem. *J. Math. Phys., MIT.* 2: 105-124.

Phillips, Warren F. 2010. *Mechanics of Flight,* 2nd ed. Wiley, New York.

Piaggio, H. T. H., and M. N. Strain. 1947. The conformal transformation $Z = (lz^2+2mz+n)/(pz^2+2qz+r)$. *J. London Math. Soc.* 22: 165-167.

Picard, É. 1879. Sur une propiétée des fonctions entiŕes. *C. R. Acad. Sci. Paris.* 88: 1024-1027.

———. 1927. *Leçons aux quelques types simples d'équations aux dérivées partielles.* Paris.

Piele, Donald T., Morris W. Firebaugh, and Robert Manulik. 1977. An elementary approach to the problem of extending conformal maps to the boundary. *American Mathematical Monthly.* 84: 677-692.

Pierce, A. D. 1989. *Acoustics: An Introduction to Its Physical Principles and Applications.* The Acoustical Society of America, New York.

Pierpont, J. 1959. *Functions of a Complex Variable.* New York: Dover.

Piessens, R. 1972a. Improved methods for Fourier coefficients of a function given at a set of arbitrary points. *Electron. Letters.* 8: 250-251.

———. 1972b. A new numerical method for the inversion of the Laplace transform. *J. Inst. Maths. Appl.* 10: 185-192.

———. 1971. Gaussian quadrature formulas for the numerical integration of Bromwich's integral and the inversion of the Laplace transform. *J. Engg. Math.* 5: 1-9.

———. 1971. Calculation of Fourier coefficients of a function given at a set of arbitrary points. *Electron. Letters.* 7: 681-682.

———, and M. Branders. 1973. The evaluation and application of some modified moments. *BIT* 13: 443-450.

Piloty, R. 1949. Die Anwendung der konformen Abbildung auf die Feldgleichungen in homogenen Rechteckrohren. *Zeitschrift für angewandte Physik.* 1: 441-448.

Plemelj, J. 1908. Ein Ergänzungssatz zur Cauchyschen Integraldarstellung analytischer Funktionen. *Monatshefte für Math. Phys.* 19: 205-210.

Polyanin, Andrei D., and Vladimir E. Nazaikinskii. 2016. *Handbook of Linear Partial Differential Equations for Engineers and Scientists.* 2nd ed. Boca Raton, FL: CRC Press.

Polubarinova-Kochina, P. Ya. 1962. *Theory of Groundwater Movement.* Translated from Russian by J. M. R. de Wiest. Princeton Univ. Press.

Pólya, G., and G. Szegö. 1925. *Aufgaben und Lehrsätze aus der Analysis.* Berlin.

———, and G. Szegö. 1951. *Isoperimetric Inequalities in Mathematical Physics.* Princeton University.

Pommerenke, Ch. 1985. The Bieberbach conjecture. *Mathematical Intelligencer.* 7(2): 23-25.

———. 1992a. *Boundary Behaviour of Conformal Maps.* Berlin: Springer-Verlag.

———. 1992b. *Boundary Behaviour of Conformal Maps.* Grundlehren der Mathematischen Wissenschaften, Vol. 299. Berlin: Springer-Verlag.

Pond, H. L. 1970. Sound radiation from a general class of bodies of revolution. *Naval Underwater Systems Center, New London,* NUSC Report No. NL-3031.

Porter, D., and D. S. G. Stering. 1993. *Integral Equations.* Cambridge: Cambridge University Press.

Powell, M. J. D. 1964. An efficient method for finding the minimum of a function of several variables calculating derivatives. *Computer J.* 7: 155-162.

———. 1968. A Fortran subroutine for solving systems of non-linear algebraic equations. *Tech. Report AERE-R 5947.* Harwell, UK.

———. 1989. A tolerant algorithm for linearly constrained optimization calculations. *Math. Programming.* 45: 547-566.

———. 1994. A direct search optimization method that models the objective and constraint functions by linear interpolation, in *Advances in Optimization and Numerical Analysis* (S. Gomez and J.-P. Hennart, eds). Kluwer Academic, Dordrecht, 51-67.

———. 2002. 'UOBYQA': Unconstrained optimization by quadratic approximation. *Math. Programming.* 92: 555-582.

———. 2006. The NEWUOA software for unconstrained optimization without derivatives, in *Large-Scale Nonlinear Optimization,* (G. Di Pillo and M. Roma, eds). Springer-Verlag, Berlin, 255-297.

Pozar, D. M. 2001. *Microwave and RF Design of Wireless Systems.* New York: John Wiley.

———. 2005. *Microwave Engineering.* 3rd ed. New York: John Wiley.

Prager, W., and J. L. Synge. 1947. Approximations in elasticity based on the concept of function space. *Quart. Appli. Math.* 5: 241-269.

Privaloff, I. 1916. Sur les fonctions conjuguées. *Bull. Soc. Math. France.* 44: 100-103.

Privalov, I. I. (И. И. Привалов). 1934. On the boundary problem in the theory of analytic functions. *Matem. Sbornik, N. S.* 41: 519-526.

———. 1956. *Randeigenschaften Analytischer Funktionen,* 2nd ed. Deutscher Verlag der Wissenschaften, Berlin (Russian).

Protter M., and H. Weinberger. 1984. *Maximum Principles in Differential Equations.* Springer-Verlag, Berlin.

Pukhteev, G. N. (Г. Н. Пухтеев). 1980. *Exact Methods for Calculation of Cauchy-type Integrals.* Nauka, Novosibirsk.

Purcell, E. M. 2013. *Electricity and Magnetism.* 3rd ed. Cambridge University Press.

Puri, P., and P. K. Kythe. 1988. Some inverse Laplace Transforms of the exponential form. *ZAMP.* 39: 150-156.

———, and P. K. Kythe. 1989. Waves in unsteady viscoelastic free-convection periodic flows in rotating mediums. *Developments in Mechanics.* 15: 425-426.

———, and P. K. Kythe. 1995. Nonclassical thermal effects In Stokes Second problems. *Acta Mechanica.* 112: 1-9

———, and P. K. Kythe. 1997. Discontinuities in velocity gradients and temperature in the Stokes' first problem with nonclassical heat conduction. *Quarterly of Applied Mathematics.* LV, #1: 167-176.

Rabinowitz, P. 1966. Numerical experiments in conformal mapping by the method of orthogonal polynomials. *J. Assoc. Comp. Mach.* 13: 296-303. Radon, J. 1919. Über die Randwertaufgaben beim logarithmischen Potential. *Sitz.-Ber. Wien. Akad. Wiss.*, Abt. IIa 128: 1123-1167.

Raineville, E. D. 1960. *Special Functions.* New York: Macmillan & Co.

Ramo, Simon, John R. Whinnery, and Theodore Van Duzer. 1994. *Fields and Waves in Communication Electronics.* New York: John Wiley. 321-324.

Rana, Farhan. 2005. Lecture 26: Dielectric slab waveguides. Class Notes ECE 3-3: Electromagnetic Fields and Waves. Electrical Engineering Dept., Cornell University.

Rapoport, I. 1940. Sur le Problème plan inverse de la théorie du potential. *Comptes Rendus Acad. Sci. URSS*, New Series. 28: 305-307.

Rayleigh, J. W. S. 1894. *Theory of Sound.* Vol.1. 2nd ed. London: Macmillan; reprinted 1945 by Dover, New York. 112-113.

Reck, Kasper, Erik V. Thomsen, and Ole Hansen. 2011. Solving the Helmholtz equation in conformal mapped ARROW structures using homotopy perturbation method. *Optics Express.* 19: 1808-1823; doi: 10.1364/OE.19.001808.

Reddy, J. N. 1984/1993. *An Introduction to the Finite Element Method.* McGraw-Hill, New York.

Reich, E., and S. E. Warschawski. 1960. On canonical conformal maps of regions of arbitrary connectivity. *Pacific J. Math.* 10: 965-989.

Reichel, L. 1985. On polynomial approximation in the complex plane with application to conformal mapping. *Math. Comp.* 44: 425-433.

———. 1986. A fast method for solving certain integral equations of the first kind with application to conformal mapping. *J. Comput. Appl. Math.* 14: 125-142; also in *Numerical Conformal Mapping* (L. N. Trefethen, ed.), North-Holland, Amsterdam, 125-142.

———. 1988. Curvature, circles, and conformal maps. *SIAM Journal on Numerical Analysis.* 25: 1359-1368.

Reischel, L. 1987. Parallel iterative methods for the solution of Fredholm integral equations of the second kind, in *Hypercube Multiprocessors.* (M. T. Heath, ed.) SIAM, Philadelphia, PA, pp. 520-529.

———. 1989. Fast solution methods for Fredholm equations of the second kind. *Numer. Math.* 57: 719-736.

Ren, Y., B. Zhang, and H. Qiao. 1999. A simple Taylor-series expansion method for a class of second kind integral equations. *J. Compu. Appl. Math.* 110: 15-24.

Richardson, M. K. 1965. *A numerical method for the conformal mapping of finite doubly connected regions with application to the torsion problem for hollow bars*, Ph. D. Thesis, University of Alabama.

Richardson, S. 1989. Finite-element methods for conformal mappings. *SIAM Review.* 31: 484-485.

Rici, John R. 1983. *Numerical Methods, Software, and Analysis.* IMSL Reference Edition. New York: McGraw-Hill.

Richter, G. R. 1978. Numerical solution of integral equations of the first kind with nonsmooth kernels. *SIAM J. Numer. Anal.* 15: 511-522.

Rickey, F. V., and P. M. Tuchinsky. 1980. An application of geography to mathematics. *Mat. Magazine.* 53: 162-166.

Riemann, B. 1851. Grundlagen für eine allgemeine Theorie der Funktionen einer veränderlichen komplexen Grösse (1851), in *Collected Works*, Dover, New York, 1953.

Riesz, F. and M. 1916/1923. Über die Randwerte einer analytischen Funktion. *Comptes Rendus du Quatrième Congrés des Mathématiciens Scandinaves à Stockholm*, 1916, 27-44; *Math. Zeitschrift.* 18: 95.

Roach, G. F. 1970/1982. *Green's Functions: Introductory Theory and Applications.* Van Nostrand Reinhold, London; 1982. *Green's Functions.* 2nd ed. Cambridge: Cambridge University Press.

Roberts, G. O. 1971. Computational meshes for boundary layer problems. *Proc. Second Int. Conf. Num. Methods Fluid Dynamics*, Lecture Notes in Physics, 8. Springer-Verlag, New York, 117-177.

Robertson, J. A. 1965. *Hydrodynamics in Theory and Application.* Prentice-Hall, Englewood Cliffs, NJ.

Rodin, Burton. 1985. Mapping theorems in complex analysis. *Proceedings of the American Mathematical Society.* 94: 297-300.

Rosenblatt, A. 1936. Sur la représentation conforme de domaines plans. *Comptes Rendus Acad. Sci. Paris.* 202: 1398-1400.

———. 1936. Sur la représentation conforme de domaines bornés par des courbes générales. *Comptes Rendus Acad. Sci. Paris.* 202: 1832-1834.

———. 1943. On Mr. L. Kantorovič's method in the theory of conformal mapping and on the application of that method to aerodynamics. *Actas de la Academia Nacional de Ciencias Exactas, Fisicas y Naturales de Lima.* 6: 199-219.

———. 1943. General comparison of conformal and equilong geometries. *Acta Acad. Ci. Lima.* 6: 199-219.

———, and S. Turski. 1936. Sur la représentation conforme de domaines plans. *Comptes Rendus Acad. Sci. Paris.* 202: 899-901.

Rosenhead, L. 1931. The lift on a flat plate between parallel walls. *Proc. Royal Soc.* A 132: 127.

———. Aerofoil in a wind tunnel. *Proc. Royal Soc.* A 140: 583.

———, and S. D. Daymond. 1937. The lateral force on a keel and rudder. *Proc. Camb. Phil. Soc.* 33: 62.

———, and B. Davidson. 1941. Wind tunnel correction for a circular open jet tunnel with a reflection plate. *Proc. Royal Soc.* A 177: 366.

Ross, S. L. 1964. *Differential Equations.* Waltham: Blaisdell.

Rossing, T. D., F. R. Moore, and P. A. Walter. 2002. *The Science of Sound*, 3rd ed. Addison-Wesley, San Francisco.

Rothe, H. 1908. Über das Grundtheorem und die Obertheoreme der automorphen Funktionen im Falle der Hermite-Laméschen Gleichung mit vier singulären Punkten. *Monatshefte Math. Phys.* 19: 258-288.

———, F. Ollendorff, and K. Pohlhausen. 1933. *Theory of Functions Applied to Engineering Problems.* M.I.T. Press.

Royden, H. L. 1952. A modification of the Neumann-Poincaré method for multiply-connected regions. *Pacific J. Math.* 2: 353-394.

Rubenstein, Z. 1969. *A Course in Ordinary and Partial Differential Equations.* Academic Press, New York.

Rudin, W. 1976. *Principles of Mathematical Analysis.*, 3rd ed. McGraw-Hill, New York.

———. 1987. *Real and Complex Analysis*, 3rd ed. McGraw-Hill, New York.

Ruehr, Otto G. 2002. Analytical-numerical treatment of the one-phase Stephan problem with constant applied heat flux, in *Integral Methods in Science and Engineering* (P. Schiavone, C. Constanda and A. Mioduchowski, eds.) Boston: Birhäuser.

Saff, E. B., and A. D. Snider. 1976. *Fundamentals of Complex Analysis for Mathematics, Science and Engineering.* Prentice-Hall, Englewood Cliffs, NJ.

Saffman, P. G. 1974. The structure and decay of trailing vortices. *Arch. Mech. (Archivum Mechaniki Stosowanej).* 26: 423-439.

Sagan, Hans. 1989. *Boundary and Eigenvalue Problems in Mathematical Physics.* New York: Dover.

Salzer, H. E.1958. Tables for the numerical calculation of inverse Laplace transforms.*J. Math. Phys.* 37: 89-108.

Sansone, G., and J. Gerretsen. 1960. *Lectures on the Theory of a Complex Variable.* Noordhoff, Groningen.

Salon, Sheppard J. 1995. *Finite Element Analysis of Electrical Machines.* Springer Science+Business Media. New York.

Sarkissian, A. 2005. Method of superposition applied to patch near-field acoustic holography. *J. Acoust. Soc. Am.* 118:671-678.

Sato, Risaburo, and Tetsuo Ikeda. 1979. Line constants, in *Microwave Filters and Circuits,*

Ch. 5, (Matsumoto Akio, ed.). Academic Press.

Savin, G. N. 1961. *Concentration around Holes.* Translated from Russian original of 1951 by E. Gross. New York: Pergamon Press.

Schaginyan, A. L. (А. Л. Шагинян). 1944. Sur les polynômes extrémaux qui présentent l'approximation d'une fonction rélisant la représentation conforme d'un domaine sur un cercle. *Comptes Rendus (Doklady) de l'Académie des Sciences de l'URSS*, New Series. 45: 50-52.

Schiffer, M. 1946. The kernel function of an orthonormal system. *Duke Math. J.* 13: 529-540.

———. 1946. Hadamard's formula and variation of domain functions. *Am. J. Math.* 68: 417-448.

———. 1948. An application of orthonormal functions in the theory of conformal mapping. *Am. J. Math.* 70: 147-156.

Schinzinger, Ronald, and Patricio A. A. Laura. 1991/2003. *Conformal Mapping: Methods and Applications*, New York: Dover; Elsevier Science Publishers, 1991.

———, and Akihiro Ametani. 1978. Surge propagation characteristics of pipe enclosed underground cables. *IEEE Trans. Power App. and Sys.* PAS-97: 1680-1688.

Schleiff, M. 1968. Über Näheungsverfahren zur Lösung einer singulären linearen Integrodifferentialgleichun. *ZAMM.* 48: 477-483.

Schneider, M. V. 1969. Microstrip lines for microwave integrated circuits. *The Bell System Technical J.* 48:1421-1444.

Schoenberg, I. J. 1948. Contributions to the problem of approximation of equidistant data by analytic functions: Part A: On the problem of smoothing or graduation. A first class of analytic approximation formulae. *Quart. Appl. Math.* 4: 45-99.

Schmidt, R. 1981. A variant of the Rayleigh-Ritz method. *Journal of the Industrial Mathematical Society.* 31: 37-56.

———. 1983. Techniques for estimating natural frequencies. *Journal of Engineering Mechanics.* 109: 654-657.

Schücker, T. 1991. *Distributions, Fourier Transforms and Some of Their Applications to Physics.* World Scientific, Singapore.

Schwarz, H. A. 1869. Über einige Abbildungsaufgaben. *J. Reine Angew. Math.* 70: 105-120; also in *Gesammelte Mathematische Abhandlung*, II. Springer-Verlag, Berlin, 1890, 65-83.

———. 1869. Notizia sulla rappresentazione conforme di un'area ellitica sopra un'area circolare, in *Gesammelte Mathematische Abhandlung*, II. Springer-Verlag, Berlin, 102.

———. 1872. Über diejenigen Fälle, in welchen die Gaussische hypergeometrische Reihe eine algebraische Funktion ihres vierten Elementes darstellt, in *Gesammelte Mathematische Abhandlung*, II, Springer-Verlag, Berlin, 1890, 211.

———. 1890. Über die Integration der partiellen Glechung $\dfrac{\partial^2 u}{\partial x^2}+\dfrac{\partial^2 u}{\partial y^2}$ unter vorgeschriebenen Bedingungen. *M. B. Preußische Akademie*, 767-795; in *Werke 2*, 144-171.

———. 1893. *Formeln und Lehrsätze zum Gebrauch der elliptischen Funktionen.* Springer-Verlag, Berlin.

———. 1890. *Gesammelte Abhandlungen.* Berlin.

Scott, Jefferey A. 2003. *Lift Coefficient and Thin Airfoil Theory.* 10 August 2003. http://www.aerospaceweb.org/question/aerodynamics/q0136.shtml.

Seidel, W. 1931. Über die Ränderzuordnung bei konformen Abbildungen. *Math. Ann.* 104: 182-243.

———. 1952. Bibliography of numerical methods in conformal mapping. *National Bureau of Standards Appl. Math.* Ser. 8, 269-280.

Segerlind, L. J. 1984. *Applied Finite Element Analysis*, 2nd ed. New York: Wiley.
Serway, R., and J. Jewitt. 2004. *Physics for Scientists and Engineers.* 6th ed. Belmont, CA: Brooks/Cole-Thompson Learning.
Shampine, Lawrence F., and Richard C. Allen, Jr. 1973. *Numerical Computing: An Introduction.* Philadelphia, PA: W. B. Saunders.
Sheil-Small, T. 1969. Some conformal mapping inequalities for starlike and convex functions. *J. London Math. Soc.* (2) 1: 577-587.
Shiffman, M. 1939. The plateau problem for non-relative minima. *Annals of Math.* 40: 834-854.
Shih, C. T., and S. Chao. 2008. Simplified numerical method for analyzing TE-like modes in a three-dimensional circularly bent dielectric rib waveguide by solving two one-dimensional eigenvalue equations. *J. Opt. Soc. Am.* B25: 1031-1037.
Shilov, Georgi E. 1973. *Elementary Real and Complex Analysis.* New York: Dover.
Sideridis, A. B. 1984. A numerical solution of the membrane eigenvalue problem. *Computing.* 32: 167-176.
Siegel, R., M. E. Goldstein, and J. M. Savino. 1970. Conformal mapping procedure for transient and steady state two-dimensional solidification. *Fourth Intern. Heat Transfer Conf., Versailles, France, 1970*, in *Heat Transfer*, Elsevier, Amsterdam.
Siegel, R., and J. R. Howell. 1992. *Thermal Radiation Heat Transfer.* 3rd ed. Hemisphere Publishing Co., Washington, DC.
Silverman, R. A. 1967. *Introductory Complex Analysis.* Prentice-Hall, Englewood Cliffs, NJ.
Sivestor, Peter, P, and Ronald L. Ferrari. 1996. *Finite Elements for Electrical Engineers.* 3rd ed. Cambridge University Press.
Singh, V. 1960. An integral equation associated with the Szegö kernel function. *Proc. London Math. Soc.* 10: 376-394.
Skidmore, Scott M. 2012. *Analysis and Optimization of Broadband Measurement Cells for the Characterization of Dielectric Polymer Films.* M. S. Thesis, Graduate School, University of South Florida.
Smirnov, V. (B. Смирнов). 1928. Sur la theéorie des polynomes orthogonaux à une variable complexe. *J. Soc. Phys.-math. Léningrade.* 2: 155-179.
———. 1932. Über die Ränderzuordnung bei konformen Abbildungen. *Math. Ann.* 107: 313-323.
———. 1964. *A Course in Higher Mathematics*, Vol. IV: Integral Equations and Partial Differential Equations. Pergamon Press, London.
Smith, G. D. 1985. *Numerical Solutions of Partial Differential Equations: Finite Difference Methods*, 3rd ed. Oxford: Clarendon Press.
Smythe, W. R. 1989. *Static and Dynamic Electricity*, 3rd ed. Boca Raton, FL: Taylor & Francis.
Sneddon, I. N. 1957. *Partial Differential Equations.* McGraw-Hill, New York.
———, and D. S. Berry. 1958. *The Classical Theory of Elasticity*, in *Handbuch der Physik*, S. Flugge (ed.) Berlin: Springer-Verlag.
———. 1978. *Fourier Transforms and Their Applications.* Springer-Verlag, Berlin.
Sommerfeld, A. 1964. *Partial Differential Equations in Physics*, Vol. VI. Academic Press, New York.
Sokolnikoff, I. S. 1956. *Mathematical Theory of Elasticity.* 2nd ed. New York: McGraw-Hill.
Southwell, R. V. 1946. *Relaxation Methods in Theoretical Physics.* Oxford University Press, Oxford.
Sparrow, E. M., and A. Haji-Sheikh. 1966. Flow and heat transfer in ducts of arbitrary

shape with arbitrary thermal boundary conditions. *J. Heat Transfer, Trans. ASME.* 88: 351-356.

———, and R. D. Cess. 1978. *Radiation Heat Transfer.* McGraw-Hill-Hemisphere.

Specht, E. 1951. Estimates of the mapping function and its derivatives in conformal mapping. *Trans. Amer. Math. Soc.* 53: 183-196.

Squires, W. 1975. Computer implementation of Schwarz-Christoffel method for generating two-dimensional glow grids. *J. Franklin Inst.* 299: 315-321.

Sridhar, K., and R. Davis. 1985. A Schwarz-Christoffel method for generating two-dimensional flow grid. *J. Fluid Eng.* 58: 330-337.

Stakgold, I. 1968. *Boundary Value Problems of Mathematical Physics*, Vol. II Macmillan, New York.

———. 1970. *Green's Functions and Boundary Value Problems.* New York: Wiley.

———. 1979. *Green's Functions and Boundary Value Problems.* Wiley, New York.

Steffe, J. F. 1992. *Rheological Methods in Food Process Engineering.* Freeman Press.

Stein, E. M. 1972. *Boundary Behavior of Holomorphic Functions of Several Complex Variables.* Princeton Univ. Press, Princeton, NJ.

Stein, N. P. (Н. П. Штеин). 1937. The determination of parameters in the Schwarz-Christoffel formula, in *Conformal Mapping of Simply and Doubly Connected Regions*, Gostechizdat, Leningrad-Moscow (Russian).

Stephenson, K. 1987/88. Concerning the gross star theorem, in *Complex Analysis* (I. Laine, S. Rickman and T. Sorvali, eds.) *Proc. Joensuu* 1987; Lecture Notes in Math # 1351, 1988. 328-338. Springer-Verlag, Berlin.

Stiefel, E. 1956. On solving Fredholm integral equations. Applications to conformal mapping and variational problems of potential theory. *J. Soc. Indust. Appl. Math.* 4: 63-85.

Stratton, J. A. 1941. *Electromagnetic Theory.* New York: McGraw-Hill.

Street, R. L. 1973. *The Analysis and Solution of Partial Differential Equations.* Brooks/Cole, Monterey, CA.

Streeter, V. L. 1966. *Fluid Mechanics.* McGraw-Hill, New York.

Svacina, J. 1992. Analysis of multilayer microstrip lines by a conformal mapping method. *IEEE Trans. Microwave Theory Tech.* 40: 769-772.

Sveshnikov, A. G., and A. N. Tikhonov (А. Г. Свешников и А. Н. Тихонов. 1978. *The Theory of Functions of a Complex Variable.* Mir Publishers, Moscow; translation of Теория Функций Комплексной Переменной, Наука, Мос- ква, 1974 (Russian).

Swarztrauber, Paul N. 1974. The direct solution of the discrete Poisson equation on the surface of a sphere. *J. Comput. Phys.* 15: 46-54.

———. 1974. A direct method for the discrete solution of separable elliptic equations. *SIAM J. Numer. Anal.* 11: 1136-1150.

———, and R. A. Sweet. 1973. The direct solution of the discrete Poisson equation on a disk. *SIAM J. Numer. Anal.* 10: 900-907.

Sweet, R.A. 1973. Direct methods for the solution of Poisson's equation on a staggered grid. *J. Comput. Phys.* 12: 422-428.

———. 1974. A generalized cyclic reduction algorithm. *SIAM J. Numer. Anal.* 11: 506-520.

———, and Ronald Sweet. 1975. *Efficient FORTRAN Subprograms for the Solution of Elliptic Partial Differential Equations.* NCAR Technical Note, NCAR-TN/IA-109. Boulder, CO: National Center for Atmospheric Research.

Symm, G. T. 1963. Integral equation methods in potential theory. II. *Proc. Roy. Soc.* A 275: 33-46.

———. 1966. An integral equation method in conformal mapping. *Numer. Math.* 9:

250-258.

———. 1967. Numerical mapping of exterior domains. *Numer. Math.* 10: 437-445.

———. 1969. Conformal mapping of doubly-connected domains. *Numer. Math.* 13: 448-457.

———. 1980. The Robin problem for Laplace's equation, in *New Developments in Boundary Element Methods* (C. A. Brebbia, ed.), CML Publications, Southampton.

Szegö, G. 1921. Über orthogonale Polynome, die zu einer gegeben Kurve der komplexen Ebene gehören. *Math. Zeit.* 9: 218-270.

———. 1939. *Orthogonal Polynomials*, AMS Colloquium Publications 23, American Mathematical Society, New York.

———. 1950. Conformal mapping of the interior of an ellipse onto a circle. *Amer. Math. Monthly.* 57: 474-478.

Tammi, O. 1974. On Green's inequalities, in *Topics in Analysis* (A. Dodd and B. Eckmann, eds.), Colloquium on Math. Anal., Jyväskylä 1970; Lecture Notes in Math., Vol. 419, 370-375. Springer-Verlag, Berlin.

Tanaka, T., K. Tsunoda, and M. Aikawa. 1988. Slot-coupled directional couplers between double-sided substrate microstrip lines and their applications. *IEEE Trans. Microwave Theory Tech.* 12: 1752-1757.

Taylor, G. I. 1937. *The Determination of Stresses by Means of Soap Films. The Mechanical Properties of Fluids.* Blackie & Son, Ltd., London, Glasgow, 136.

Teichmüller, O. 1938. Untersuchungen über konforme und quasikonforme abbildung. *Deutsche Math.* 3: 621-678.

Thamburaj, P., and J. Q. Sun. 2001. Note on a paper by Sinha and Odgaard: Application of conformal mapping to diverging open channel flow. *Journal of Sound and Vibration.* 241: 283-295(13).

Theodorsen, T. 1931. Theory of wing sections of arbitrary shape. *National Advisory Committee on Aeronautics*, Tech. Rep. 411.

———, and I. E. Garrick. 1933. General potential theory of arbitrary wing sections. *National Advisory Committee on Aeronautics*, Tech. Rep. 452.

Thompson, J. F. 1952. *Numerical Grid Generation.* North-Holland, Amsterdam.

———. 1982. *Elliptic Grid Generation.* Elsevier Publishing Co., Inc., New York.

———, F. C. Thames, and C. W. Mastin. 1976. Boundary-fitted coordinate system for solution of partial differential equations on fields containing any number of arbitrary two-dimensional bodies. *NASA, CR-2729.*

———, F. C. Thames, and C. W. Mastin. 1977. TOMCAT: A code for numerical generation of boundary-fitted curvilinear coordinate system on fields containing any number of arbitrary two-dimensional bodies. *J. Comput. Phys.* 24: 274-302.

———, Z. A. Warsi, and C. W. Mastin. 1985. *Numerical Grid Generation: Foundations and Applications.* North-Holland, New York.

Truckembrodt, E. 1980. *Fluidmechanik*, vol. 2. Springer-Verlag.

Thurman, Robert E. 1994. An inverse problem for circle packing and conformal mapping. *Transactions of the American Mathematical Society.* 346:605-616.

Timman, R. 1951. The direct and the inverse problem of airfoil theory. A method to obtain numerical solutions. *Nat. Luchtv. Labor. Amsterdam*, Report F.16.

Timoshenko, S. P., and N. Godier. 1951. *Theory of Elasticity.* New York: McGraw-Hill.

———, Young, D. H., and W. Weaver, Jr. 1074. *Vibration Problems in Engineering.* New York: Wiley.

Titchmarsh, E. C. 1968. *Theory of Functions.* Oxford University Press, Oxford.

Todd, J. (ed.) 1955. *Experiments in the Computation of Conformal Maps.* U.S. National Bureau of Standards, Appl. Math. Ser. 42, U.S. Govt. Printing Office.

———, and S. E. Warschawski. 1955. On the solution of the Lichtenstein-Gershgorin integral equation in conformal mapping, II, Computation experiments. *National Bureau Standards Appl. Math. Series.* 42: 31-44.

Tomamidis, P., and D. N. Assanis. 1991. Generation of orthogonal grids with control of spacing. *J. Compt. Phys.* 94: 437-453.

Tong , P., and J. N. Rossettos. 1977. *Finite-Element Method.* Cambridge: The MIT Press.

Traub, Lance W., 2016-03-24. Semi-empirical prediction of airfoil hysteresis. *Aerospace.* 3:9. doi:10.3390/aeospace3020009 (https://doi.org/10.3390%2Faerospace3020009); (http://www.mdpi.com/2226-4310/3/2/9).

Trefethen, L. N. 1979. *Numerical Computation of the Schwarz-Christoffel Transformation.* Tech. Report STAN-CS-79-710, March 1979, Computer Science Department, Stanford University.

———. 1980. Numerical computation of the Schwarz-Christoffel transformation. *SIAM J. Sci. Stat. Comput.* 1: 82-102.

———. 1983. *SCPACK Version 2 User's Guide*, Internal Report 24, Institute for Computer Applications in Science and Engineering, NASA Langley Research Center.

———. 1984. Analysis and design of polygonal resistors by conformal mapping. *ZAMP.* 35: 692-704.

———. 1986. *Numerical Conformal Mapping.* North-Holland, Amsterdam; reprint of *J. Comput. Math.* 14 (1986), no. 1-2.

———. 1989a. Schwarz-Christoffel mapping in the 1980's. Numerical Analysis Rep. 89-1, Department of Mathematics, MIT.

———. 1989b. SCPACK user's guide. Numerical Analysis Rep. 89-2, Department of Mathematics, MIT, Cambridge, MA.

———, and R. J. Williams. 1986. Conformal mapping solution of Laplace's equation on a polygon with oblique derivative boundary conditions. *J. Comput. Appl. Math.* 14: 227-249; also in *Numerical Conformal Mapping* (L. N. Trefethen, ed.), North-Holland, Amsterdam, 227-249.

Trefftz, E. 1913. Graphische Konstruction Joukowskischer Tragflächen. *Zeitschrift für Mathematik und Physik.* 4:130-131.

———. 1921. Prandlsche Tragflächen- und Propeller Theorie. *ZAMM.* 1: 206-218..

Tricomi, F. G. 1957. *Integral Equations.* New York: Wiley Interscience.

Trim, D. W. 1990. *Applied Partial Differential Equations.* Boston: PWS-Kent.

Truckembrodt, E. 1980. *Fluidmechanik*, Vol. 2. Springer-Verlag.

Trummer, M. R. 1986. An efficient implementation of a conformal mapping method based on the Szegö kernel. *SIAM J. Numer. Anal.* 23: 853-872.

———. 1986. A note on integral means of the derivative in conformal mapping. *SIAM Journal on Numerical Analysis.* 23: 853-872.

U. S. Centennial of Flight Commission. 2004. *Airfoil Diagram.* htttp://www.centennialofflight.gov/ essay/Dictionary/angleofattack/D15.htm.

Ueberhuber, C. W. 1997. Monte Carlo techniques, §12.4.4 in *Numerical Computation 2: Methods, Software, and Analysis.* Berlin: Springer-Verlag. 124-125; 132-138.

Ugural, A. C. 1981. *Stresses in Plates and Shells.* New York: McGraw-Hill.

Upton, Roger, Karim Haddad, and J. Sørensen. 2007. Conformal mapping techniques for consumer products. Presented at Noise-Con 2007, Institute of Noise Control Engineering, Reno, NV.

van Dyke, M. 1964. *Perturbation Methods in Fluid Mechanics.* Academic Press, New York.

———. 1975. *Perturbation Methods in Fluid Mechanics.* The Parabolic Press.

Vecheslavov, V. V., and V. I. Kokoulin. 1974. Determination of the parameters of the conformal mapping of simply connected polygonal regions. *Zh. vychisl. Mat. Fiz.* 13

(4), 1973, 865-872 (Russian). Translated in *USSR Comp. Math. and Mat. Phys.* 13: 57-65.

Vekua, I. N. (И. Н. Векуа). 1942. On the linear boundary problem of Riemann. *Trudy Tbilissk. Mat. Inst.* 11: 109-139 (Russian).

———. 1976. On one method of solving the first biharmonic boundary value problem and the Dirichlet problem. *Amer. Math. Soc. Transl. (2).* 104: 104-111; translation of Об одном методе решения основной бигармонической кравой задачи и задачи Дирихле, 120-127 (Russian).

Vertgeim, B. A. (Б. А. Вертгеим). 1958. Approximate construction of some conformal mappings. *Doklady Akad. Nauk SSSR.* 119: 12-14 (Russian).

Veyres, C., and V. Fouad Hanna. 1980. Extension of the application of conformal mapping techniques to coplanar lines with finite dimensions. *International Journal of Electronics.* 48: 47-56.

Visik, M. I., and L. A. Lyusternik. 1957/1962. Regular degeneration and boundary layer for linear differential equations with small parameters. *Uspekhi Mat. Nauk.*, 12: 3-122; Amer. Math. Soc. Transl. 20, 1962 (Russian).

Vladimirov, V. S. (В. С. Владимиров). 1984. *Equations of Mathematical Physics.* Mir Publishers, Moscow.

Vladimirsky, S. 1941. Sur la représentation conforme des domaines limités intérieure ment par des segments rectilignes et arcs circulaires. *Comptes Rendus Acad. Sci. Paris* . 212: 379-382.

Volkovyskiĭ, L. I. (Л. И. Волковыский). 1963. Determination of the type of certain classes of simply connected Riemann surfaces. *Am. Math. Soc. Transl., Series 2,* 32: 83-114; translation of Определние типа некоторых классов односвязных Римановых поверхностей, Мат. Сб. (N. S.), 23(65), 1948, 229-258.

von Karman, T., and E. Trefftz. 1918. Potential-stromung um gegebene Tragflachenquerschnitte. *Zeitschrift für Flugtechnische Moforluftsch.* 9: 111-116.

von Koppenfels, W. 1937. Das hypergeometrische Integral als Periode der Vierecksabbildung. *Sitzungsberichte Akad. Wissen. Wien.* 146: 11-22.

———. 1939. Konforme Abbildung besonderer Kreisbogenvierecke. *J. Reine Angew. Math.* 181: 83-124.

———, and F. Stallmann. 1959. *Praxis der konformen Abbildung.* Springer-Verlag, Berlin.

von Wolferdorf, L. 1984. Zur Unität der Lösung der Theodorsenschen Integralgleichung der konformen Abbildung. *Z. Anal. Anwendungen.* 3: 523-526.

Vvedensky, D. 1993. *Partial Differential Equations with Mathematica.* Addison-Wesley, Workinham.

Wandell, Brian C. 1991. *Transmission Line Design Handbook.* Boston: Artech House.

Walker, M. 1933. *Conjugate Functions for Engineers.* Oxford.

Walker, J. S. 1988. *Fourier Analysis.* Oxford University Press, Oxford.

Walsh, J. L. 1956. Recent advances at Stanford in the application of conformal mapping to hydrodynamics. *Transactions of the American Mathematical Society.* 82: 128-146.

———. 1969. *Interpolation and Approximation by Rational Functions in the Complex Domain*, 5th ed. American Mathematical Society, Providence, RI.

Warner, G., and R. Anderson. 1981. Numerical conformal mapping for undergraduates. *Intern. J. of Electrical Engr. Education.* 18: 359-372.

Warschawski, S. E. 1932. Über das Randverhalten der Ableitung der Abbildungsfunktion bei konformer Abbildung. *Math. Zeit.* 35: 321-456.

———. 1932. Über einen Satz von O. D. Kellogg. *Nachr. Akad. Wiss. Göttingen, Math. Phys. Kl.* 73-86.

———. 1935. On the higher derivatives at the boundary in conformal mapping. *Trans.*

Am. Math. Soc. 38: 326.

———. 1942. Conformality in connection with functions of two complex variables. *Transactions of the American Mathematical Society.* 51: 280-335.

———. 1945. On Theodorsen's method of conformal mapping of nearly circular regions. *Quart. Appl. Math.* 3: 12-28.

———. 1950. On conformal mapping of nearly circular regions. *Proc. Amer. Math. Soc.* 1: 562-574.

———. 1951. Conformal mapping of the interior of an ellipse onto a circle. *Proceedings of the American Mathematical Society.* 2: 254-261.

———. 1955. On the solution of the Lichtenstein-Gershgorin integral equation in conformal mapping, I, Theory. *National Bureau Standards Appl. Math.* Series 42: 7-30.

———. 1956. Recent results in numerical methods of conformal mapping, in *Proc. Symp. Appl. Math.*, Vol. 6, McGraw-Hill, New York, 219-250.

———. 1961. On differentiability at the boundary in conformal mapping. *Proc. Am. Math. Soc.* 12: 614-620.

———. 1966. On conformal mapping of certain classes of Jordan domains. *Arch. Ratnl. Mech. Anal.* 22: 201-209.

———, and G. E. Schober. 1966. On conformal mapping of certain classes of Jordan domains. *Arch. Ratnl. Mech. Anal.* 22: 201-209.

Washizu, K. 1975. *Variational Methods in Elasticity and Plasticity*, 2nd Ed. New York: Pergamon Press.

Watson, G. N. 1944. *A Treatise on the Theory of Bessel Functions*, 2nd ed. Cambridge University Press, Cambridge.

Wayland, H. 1970. *Complex Variables Applied in Science and Engineering.* Van Nostrand Reinhold, New York.

Weber, Ernst. 1950. *Electromagnetic Fields, I: Mapping of Fields.* Wiley, New York.

Wegmann, R. 1978. Ein Iterationsverfahren zur konformen Abbildung. *Numer. Math.* 30: 453-466.

———. 1979. Ein Iterationsverfahren zur konformen Abbildung. *ZAMM* . 59: T85-86.

———. 1984. Convergence proofs and error estimates for an iterative method for conformal mapping. *Numer. Math.* 44: 435-461.

———. 1986. An iterative method for conformal mapping. *J. Comput. Appl. Math.* 14: 7-18; also in *Numerical Conformal Mapping* (L. N. Trefethen, ed.), North-Holland, Amsterdam, 7-18.

———. 1986. An iterative method for the conformal mapping of doubly connected regions. *J. Comput. Appl. Math.* 14: 79-98; also in *Numerical Conformal Mapping* (L. N. Trefethen, ed.), North-Holland, Amsterdam, 79-98.

———. 1986. An Efficient Implementation of a Conformal Mapping Method Based on the Szego Kernel. *SIAM Journal on Numerical Analysis.* 23: 1199-1213.

Weinberger, H. F. 1965. *A First Course in Partial Differential Equations.* Xerox, Lexington, MA.

Weinzierl, S. 2000. *Introduction to Monte Carlo Methods.* 23 June 2000. http://xxx.lanl.gov/abshep-ph/0006269/.

Weisstein, Eric W. 2003. *CRC Concise Encyclopedia of Mathematics.* 2nd ed. Chapman & Hall/ CRC, Boca Raton, FL.

Wen, C. P. 1970. Coplanar-waveguide directional couplers. *IEEE Trans. Microwave Theory and Tech.* 18: 318-322.

Wen, G.-C. 1992. *Conformal Mappings and Boundary Value Problems*, Translations of Mathematical Monographs, Vol. 106, American Mathematical Society, Providence, RI.

Wenig, F. Der Hodograpf der Gittersrömung als Weg zur Ermittlung günstiger Turbinen-

und Propellerprofile. *Proc. 3rd Ontern. Congr. Appl. Mechs.*, 437.

Wendland, W. 1980. On Galerkin collocation methods for integral equations of elliptic boundary value problem, in *Numerical Treatment of Integral Equations* (J. Albrecht and L. Collatz, eds.) Birkhäuser, Basel.

———, and W. Niethammer. 1980. Ein Iterationsverfahren zur Berechnung des Szegö-Kerns mit der integralgleichung von Kerzman und Stein. *Math. Forshungsinstitut Oberwolfach, Konstructive Verfahren in der komplexen Analysis.* 31.

Wermer, J. 1955. Polynomial approximation on an arc in C^3. *Annals Math.* 62: 269-294.

Wheeler, H. A. 1964. Transmission-line properties of parallel wide strips by a conformal mapping approximation. *IEEE Trans. on Microwave Theory and Techniques.* 12: 280-289.

———. 1965. Transmission-line properties of parallel strips separated by a dielectric sheet. *IEEE Trams. Microwave Theory and Techn.* MTT-13: 172-185.

———. 1978. Transmission-line properties of a strip line between parallel plates. *IEEE Trams. Microwave Theory and Techn.* MTT-26: 866-876.

White, G. N. 1962. Difference equations for plane thermal elasticity. *LAMS-2745.* Los Alamos, NM: Los Alamos Scientific Laboratories.

Whiteman, J. R. 1975. *A Bibliography for Finite Elements.* New York: Academic Press.

Whittaker, E. T., and G. N. Watson. 1962. *A Course of Modern Analysis*, 4th ed. Cambridge University Press, Cambridge, UK.

Witoszynski, C. 1924. Modification du principle de circulation. *Proc. 1st Intern. Congr. Appl. Mechs.*, 418.

Wilson, H. B. 1963. A method of conformal mapping and the determination of stresses in solid propellant rocket grains. *Rep. S-38*, Rohm and Haas Co., Huntsville, AL.

Wirtinger, W. 1927. Über die konforme Abbildung der Halbebene auf ein Kreisbogendreieck. *Atti Pontif. Accad. Sci. Nuovi Lincci.* 80: 291-309.

Wittich, H. 1947. Konforme Abbildung einfach zusammenhängender Gebiete. *ZAMM*, 25/27: 131-132.

Wilkinson, W. 1960. *Non-Newtonian Fluids.* New York: Pergamon Press.

Woods, L. C. 1961. *The Theory of Subsonic Planev Flow.* Cambridge University Press.

Wöhner, W. 1965. Konvergenz- und Fehleruntersuchungen zu einem Näherungsverfahren der konformen Abbildung einfach zusammenhängender Gebiete. *Wiss. Z. Hochschule für Verkehrswesen Dresden.* 12: 17-20.

Wolfram, S. 1996. *The Mathematica Book*, 3rd ed. Wolfram Media, Champaigne, IL, and Cambridge University Press.

Wong, S. H., and I. R. Circic. 1985. Method of conformal transformation for the finite-element solution of axisymmetric exterior-field problems. *J. Comut. and Math. in Electrical Engrg.* (COMPEL). 4:123-135.

Wu, J. C., and J. F. Thompson. 1973. Numerical solutions of time dependent incompressible Navier-Stokes equation using integro-differential formulation. *Comput. Fluids.* 1: 197-215.

Yamashita, Shinji. 1978. Applications of conformal mapping to potential theory through computer graphics. *Transactions of the American Mathematical Society.* 245: 119-138.

Yan, Y. 1994. A fast Numerical solution for a second kind integral equations with a logarithmic kernel. *SIAM J. Numer. Math.* 31: 477-498.

Yokota, S. 1927. Discontinuous flow past an aerofoil. *Phil. Mag.* VII 3: 216. Zarankiewicz, K. 1934. Sur la représentation conforme d'un domaine doublement convexe sur un anneau circulaire. *Comptes Rendus Acad. Sci. Paris.* 198: 1347-1349.

Young, E. C. 1972. *Partial Differential Equations.* Allyn and Bacon, Boston.

———. 1934. Über ein numerisches Verfahren zur konformen Abbildung zweifach zusammenhängender Gebiete. *ZAMM.* 14: 97-104.

Zachmanoglou, E. C., and D. W. Thoe. 1976. *Introduction to Partial Differential Equations with Applications.* Williams & Wilkins., Baltimore.

Zarankiewicz, K. 1934. Sur la représentation d'un domaine doublement connexe sur un anneau circulaire. *C. R. Acad. Sci. Paris.* 198: 1347-1349.

Zauderer, E. 1983. *Partial Differential Equations of Applied Mathematics.* Wiley, New York.

Zemach, C. 1986. A conformal map formula for difficult cases. *J. Comput. Appl. Math.* 14: 207-215; also in *Numerical Conformal Mapping* (L. N. Trefethen, ed.), North-Holland, Amsterdam, 207-215.

Zienkiewicz, O. C., and Y. K. Cheung. 1965. Finite elements in the solution of field problems. *The Engineer.* 220: 507-510.

———, and G. S. Holister (Eds.) 1966. *Stress Analysis.* London: Wiley.

———, and R. L. Taylor. 1989. *The Finite Element Method,* vol. 1: Linear Problems. New York: McGraw-Hill.

Zmorovich, V. A. (В. А. Зморович). 1935. Deux problémes du domaine des représentations conformes. *J. de l'Institut Math. de l'Académie des Sciences de l'Ukraine.* 3/4: 215-222.

Zygmund, A. 1935/1959. *Trigonometric Series.* Monografie Matematyczne, Warsaw; also Cambridge University Press, London, 1959.

Zwillinger, D. 2002. *CRC Standard Mathematical Tables and Functions.* 31st ed. Boca Raton: CRC Press.

Index

D

N

O

P

Q

R

S

T

U

V

W

P[illegible]ted in [illegible]ted States
[illegible] Baker & [illegible] [illegible]her Servi[illegible]

Printed in the United States
by Baker & Taylor Publisher Services